Molecular Biology

A SERIES OF BOOKS IN BIOLOGY

CONSULTING EDITOR: Cedric I. Davern

Molecular Biology

DAVID FREIFELDER
University of California, San Diego

Second Edition

Jones and Bartlett Publishers
BOSTON   LONDON

Editorial, Sales, and Customer Service Offices

Jones and Bartlett Publishers
One Exeter Plaza
Boston, MA 02116
1-617-859-3900
1-800-832-0034

Jones and Bartlett Publishers International
P. O. Box 1498
London W6 7RS
England

Library of Congress Cataloging in Publication Data

Freifelder, David Michael, 1935–
    Molecular biology.

    Bibliography: p.
    Includes index.
    1. Molecular biology.   2. Prokaryotes.   I. Title.
QH506.F73 1986   574.8'8   86-15347

ISBN 0-86720-069-3

Book and Cover Design   Hal Lockwood

Illustrator   Donna Salmon, Assisted by Cyndie Clark-Huegel, Evanell Towne,
John and Judy Waller, Dorothy Beebe, Kelly Solis-Navarro, Brenda
Booth, and Jack Vitkus

Manuscript Editor   Kirk Sargent

Production   Bookman Productions

Composition   Graphic Typesetting Service

Printed in the United States of America
10   9   8   7

To my parents,
MORRIS and FLORENCE,
and to my children,
RACHEL and JOSHUA

The Jones and Bartlett Series in Biology

Contents

5   DNA Technology   121

6   The Physical Structure of Protein Molecules   141

7   Macromolecular Interactions and the Structure of Complex Aggregates   173

8 The Genetic Material   207

9 DNA Replication   223

10 Repair   277

14 Translation II: The Machinery and the Chemical Nature of Protein Synthesis   415

15 Regulation of the Activity of Genes and of Gene Products in Prokaryotes   453

16   Regulation in Eukaryotes   501

17   Bacteriophages I: Lytic Phages   551

18   Bacteriophages II: The Lysogenic Life Cycle   595

19   Plasmids   619

20   Genetic Recombination Between Homologous DNA Sequences   641

21 Transposable Elements   679

22 Recombinant DNA and Genetic Engineering   705

23 Eukaryotic Viruses   737

24 Tumor Viruses and Oncogenes   779

Publisher's Note: A separate *Student Study Guide* (ISBN 0-86720-070-7) is available to accompany *Molecular Biology, Second Edition* (ISBN 0-86720-069-3). The *Student Study Guide* contains the following: • Complete worked out solutions to each problem in the text • Complete chapter summaries • Drill questions with answers.

Preface to the Second Edition

For a scientist, it is great fun to work in a field, such as molecular biology, that is experiencing an information explosion. However, for an author, such a field is less desirable, since soon after the long and tiring task of writing, reviewing, and producing a book has been completed, one must think about the next edition. Indeed, the rate of acquisition of significant information, which increased dramatically in the late 1970s with the development of DNA sequencing and recombinant DNA techniques, shows no sign of slowing down, and one year after *Molecular Biology* appeared, it became clear that work on a new edition had to begin.

Several changes from the format and point of view of the first edition have been essential. The two most significant ones are the following. First, since the recombinant DNA technology and DNA sequencing are now part of daily laboratory life, these topics must be presented early in the book. Thus, a new chapter (Chapter 5), which covers the general topic of nucleic acid technology, was written. Second, these and other new techniques have armed molecular biologists with the ability to examine eukaryotic cells in greater detail than ever before; thus, the molecular biology of eukaryotes, which is the focus of most current research, is now found throughout the book. This second change required considerable thought because a textbook on molecular biology must both provide students with the basic information and foundation of the field and give them the latest tools and thoughts for future research. Therefore, one cannot omit the vast body of knowledge of prokaryotic molecular biology. Inclusion of both the fundamentals and the topics of current interest in eukaryotic molecular biology would produce a 2000-page book, so difficult choices about what to include have had to be made. My rationale has been the following: (1) the fundamental material, obtained from studies of prokaryotes, which forms the basis of the field, has been retained; (2) discussions of both the classical systems (such as the *lac* operon) that have laid down the principles of molecular biological thought and those prokaryotic systems that are still actively being studied (for example, the *trp* and *gal* operons, and phages T7 and λ), have been kept, but treatment of less popular systems (e.g., the *hut* operon, *B. subtilis* phages, details of suppression mechanisms) have been discarded. To increase the available space in the book, I have also deleted very elementary material, such as the organic chemicals of biological systems (Chapter 2 of the first edition), and highly specialized topics, such as the structure of the bacterial cell wall and details of the methods for handling and counting phages and viruses (which are more appropriate in texts on microbial genetics and virology). Furthermore, in order to make the book more applicable to current research I have increased the usage of eukaryotic systems to exemplify certain points, whenever possible. Finally, the topic of eukaryotic regulation, which was relegated to the last chapter of the first edition, has been moved to follow the chapter on prokaryotic regulation, both to give equal emphasis and to enable the two topics to be compared while they are fresh in the minds of the reader.

SPECIFIC CHANGES IN THE SECOND EDITION

The following lists present in greater detail what changes have been made.

1. **Topics deleted.** Organic chemistry, structure of the bacterial cell wall, repair of cross links and of x-ray damage, mutation selection and detection, Devoret test, old-style filter hybridization, numerous experiments that elucidated the genetic code, integrative suppression, genetic analysis to study tRNA, streptomycin dependence, gramicidin synthesis, *hut* operon, phage plating techniques, regulation of T4 transcription, phages ϕ6 and Mu and *B. subtilis* phages, RecF pathway, counting and detection of viruses, herpes viruses, parvoviruses, togaviruses, vaccinia virus, yeast mating-type interconversion.

2. **Topics updated.** Z-DNA, structure of the *E. coli* chromosome and of eukaryotic chromosomes, DNA-protein binding, DNA replication initiation and termination, polymerase III subunits, editing function, gyrase mechanism, ColE1 replication, excision repair, SOS repair, promoter structure, transcription initiation, eukaryotic RNA polymerases, pauses

in transcription and termination of transcription in prokaryotes, amino-acyl synthetases, translation initiation signals in prokaryotes and eukaryotes, AUG site selection in eukaryotes, signal hypothesis, *trp* and *gal* operons, repression in attenuated systems, ribosome structure, regulation of synthesis of rRNA and ribosomal proteins, exceptions to the universal code, phages φX174 and λ, RecA protein and synapsis, mechanism of homologous recombination, transposition, genetic engineering, adenovirus, poliovirus, retroviruses, SV40, viral transformation, gene amplification, eukaryotic cellular polyproteins.

3. New topics. DNA cruciforms, restriction polymorphisms, structure of centromeres and telomeres, protein domains, mechanism of binding of various proteins (cAMP-CRP, cII protein, EcoRI, Klenow fragment) to DNA, mechanism of formation of deletions, apurinic sites and mutagenesis, eukaryotic RNA polymerases, footprinting, mechanisms of RNA splicing, ribozymes, RNase P, enhancers, hypersensitive sites, upstream activation sites, transcription factors, transcription initiation in eukaryotes, eukaryotic promoter structure, small nuclear ribonucleoproteins (snurps), Northern and Western blots, SOS regulon, pho (phosphate) regulon, heat-shock regulon, regulation by control of enhancer activity, regulation via processing, regulation in yeast, methylation and its possible role in regulation, recombination in fungi, transformation in yeast, double-strand break gap repair model, eukaryotic transposable elements, retroposons, M13 cloning system, gene therapy, synthetic vaccines, oncogenes.

THE STUDY GUIDE

A significant addition is a short auxiliary text, the **Study Guide for Molecular Biology.** This short book has been designed as a learning aid and follows *Molecular Biology, Second Edition* chapter by chapter. It has the following features: (1) a chapter summary for the purpose of emphasis and review, (2) a listing of all key terms that appear in boldface in the text, (3) simpler and more extensive explanations of certain phenomena presented in the text, (4) a set of drill questions (simple questions that only test whether the student has read and understood the very basic material), (5) hundreds of questions of a more substantive nature, based on the material in the chapter, and (6) numerous advanced and challenging problems that will require thinking beyond the material presented in the chapter. Complete answers are given to all questions. Problems of types 4 and 5 have been taken primarily from the book *Problems for Molecular Biology.* I hope that the *Study Guide* will be as useful as the one that I wrote for another of my books, *General Genetics.*

AN OVERVIEW OF *MOLECULAR BIOLOGY*

In the preface to the first edition, I explained my philosophy of teaching molecular biology and presented a detailed outline of *Molecular Biology.* For instructors and readers who have not used the first edition some of this information is repeated in the following paragraphs.

Molecular Biology is a multipurpose college textbook, based in part on a course that I taught for many years at Brandeis University. Its primary use will be in undergraduate courses emphasizing basic molecular processes (such as the synthesis of DNA, RNA, and protein) and genetic phenomena in both prokaryotic and eukaryotic cells. Students who wish to make full use of the text should have knowledge of basic biology and general chemistry and have completed or be enrolled in a course in organic chemistry.

Molecular Biology differs in organization from other molecular biology texts: it follows my philosophy that molecular biology must emphasize both molecules and biology, and that to be molecular, it must also be chemical and physical—quite a tall order for a single book. With this goal in mind, I chose to present a biological introduction to the subject first (**Chapter 1**) before getting into the complexity of molecular detail, so the student will know the phenomena that molecular biologists would like to explain and will be familiar with the biological systems used in research. The basic life cycles of particular biological systems used by molecular biologists (*E. coli,* phage, viruses, yeasts, and animal cells), and the means for handling these systems are described. Chapter 1 also includes a brief presentation of bioenergetics, which the student lacking a formal course in biochemistry will find essential to understanding later portions of the book. A second introductory discussion, **Chapter 2,** reviews those aspects of genetics needed by the reader (recombination, complementation, mapping) and, even more important, explains how genetic analysis is used in molecular biology; its predictive value is emphasized in particular.

The first major discussion following the introductory material is concerned with the properties of macromolecules; this is a notable feature of the organization of this book. Early presentation of this material is based on my experience that students have an easier time understanding the great variety of phenomena in which macromolecular interactions play a role if they have first acquired a thorough understanding of macromolecular structure. For example, the student should have a clear understanding of the significance of molecular shape and the interactions that give rise to the shape of a particular molecule; how, otherwise, can the concepts of templates, active sites, specificity of binding, and mutation be understood? **Chapter 3** presents the general properties of macromolecules and some of the more common techniques used in their study. **Chapter 4** treats nucleic acids in considerable detail, with respect to both their chemistry and physical properties. A goal in the treatment of nucleic acids is to provide the student with the information needed to understand the mechanisms of replication and transcription, which are examined in later chapters. Some features of modern nucleic acid technology are presented in **Chapter 5.** Here the student learns about how nucleic acids are isolated, base sequencing, restriction enzymes, and some of the more basic aspects of the recombinant DNA technology. This chapter enables the instructor to use these techniques in later analysis of experiments. More sophisticated aspects of the recombinant DNA technology are reserved for Chapter 22 when genetic engineering is discussed; this topic must be delayed until gene expression is understood.

Chapter 6 continues with macromolecules, in particular, proteins. The chapter has three goals—to give the student an appreciation of the variety of possible protein structures, to relate protein structure to biological function, and to indicate how a single amino-acid change in a mutant can have a profound effect on protein function. In recent years knowledge of macromolecular structure has reached the point where examination of complex systems of macromolecules—for example, antibodies, chromatin, membranes, complexes of DNA with sequence-specific proteins, and so forth—is profitable; these and other topics are presented in **Chapter 7.** Some of these topics can be omitted from a short course, but the sections on immunoglobulins, DNA-protein binding, and chromatin should be included, for knowledge of these structures will be required in later chapters of the book.

Following the treatment of macromolecules is **Chapter 8,** a general discussion of the properties that the genetic material should possess. This chapter includes a brief history of the classic experiments.

The treatment of what is traditionally thought of as molecular biology—namely, DNA, RNA, and protein synthesis—begins in **Chapter 9** with DNA replication, an uncommon starting point. I have chosen this organization for four reasons: (1) the essential concepts of a template and of chemical polarity are easily introduced with this approach; (2) DNA replication exemplifies, in a simple way, the kinds of geometric problems faced by many biochemical systems; (3) this topic shows how complicated a "simple" process can be; and (4) the concepts of both variability of genetic material (for example, mutational changes) and protection of existing information (for example, repair processes) are more obvious here than in other topics. In this and subsequent chapters, both prokaryotic and eukaryotic systems are described, with prokaryotic systems given in greater detail, as more is understood about these systems. Similarities and differences between prokaryotes and eukaryotes are pointed out repeatedly.

Chapters 10 and **11,** which are concerned with DNA repair and mutations, respectively, follow logically from a treatment of DNA replication. In these chapters I have discussed the imperfections of biological systems and how cells deal with the problem of damage to their DNA by many environmental agents and reagents.

Having completed the first eleven chapters the reader will be familiar with the structure of the important intracellular molecules and will have examined systems for synthesis of one type of macromolecule. In so doing, the foundation will have been laid for understanding how a cell manages to do what it has to do at the appropriate time and how it responds to environmental factors. Thus, the reader is ready to be presented with the mechanisms of RNA synthesis and protein synthesis in prokaryotes and eukaryotes, which is done in Chapters 12–14. **Chapter 12** deals with transcription; as in Chapter 9, the basic phenomenon is described first and then elaborated in considerable detail. Special attention has been paid to essential features (such as recognition sites) of the macromolecules required for transcription. Chapters 13 and 14 are concerned with protein synthesis. This immense topic has been divided into two parts—information transfer (**Chapter 13**) and chemical synthesis (**Chapter 14**). The structures of the relevant molecules (tRNA and the aminoacyl synthetases) and particles (ribosomes) are described twice—first in a simple way and then in considerable detail. The chemistry of protein synthesis is presented fairly completely and, in some cases, as somewhat advanced material. However, chapter sections have been arranged in such a way that more detailed parts may be omitted, if desired. For example, a student can learn about protein synthesis by reading an overview consisting of a few pages, but if detail is wanted, sections consisting of nearly twenty pages can be used.

Having completed all units on macromolecular synthesis, the reader is then ready to study how such synthesis is regulated. **Chapter 15** is concerned with regulation in prokaryotes and presents the operon concept. The *lac* operon serves as the basis for the chapter because most of the concepts and language of regulation were developed by investigating this system. The reader is then exposed to a variety of operons—the negatively regulated, the positively regulated, those with one promoter or with more than one promoter, degradative systems and biosynthetic systems, and so forth. **Chapter 16** presents regulation in eukaryotes, including both unicellular organisms, such as yeast, and differentiated animal cells. The unusual organization of eukaryotic DNA, gene families, and various modes of DNA deletion, amplification, and recombination are important topics. This chapter will necessarily be frustrating to the reader because more questions are raised than answered. Also, because the field of eukaryotic regulation is advancing so rapidly, the instructor will find it necessary to provide the latest information, for no textbook can be up to date in this field. I have chosen to present systems that exemplify general principles and are reasonably well understood.

Chapters 17 and **18** treat the bacteriophages, for these species exemplify the principles of regulation and of macromolecular synthesis. The study of phage biology is not simple because, though the life cycles of most phages follow the same basic pattern, details of the modes of nucleic acid replication and regulation often differ from one phage species to the next. Many phage species are examined in these chapters—each one selected to demonstrate some feature of phage biology that can be easily observed with that phage.

Chapter 17 deals exclusively with the lytic life cycle; in Chapter 18, lysogeny, primarily of phage λ, is examined.

Features of genetic exchange—natural systems such as plasmids, transposable elements, and homologous recombination, and the artificial process of genetic engineering—form the next unit. Instructors will certainly vary with respect to the time allotted to these topics. Thus, basic phenomena are described briefly early in each chapter (sometimes in an overview section). In many courses the material on plasmids presented in Chapter 5 will be sufficient and **Chapter 19** can be omitted. An extensive overview in the first few pages of **Chapter 20** may be sufficient for learning about transposable elements. **Chapter 21,** which presents the molecular features of genetic recombination (genetic mapping is presented in Chapter 1), contains fairly complex material. The different recombinational systems (transformation, meiotic recombination, etc.) are presented without reference to one another, so particular ones can be selected by the instructor. If the topic is not included in a course, the instructor should be sure that the students read the section on the RecA protein.

Genetic engineering, introduced in Chapter 5 as a tool of molecular biology, is presented in **Chapter 22** in terms of the new biotechnology. Here the student will learn many of the techniques and how genetic engineering can be used in industry, medicine, and agriculture.

The final unit of the book deals with eukaryotic viruses, which exhibit great variety in structure and in life cycles. Here the reader can see that just as the study of phage has given important insight into macromolecular processes and regulation in bacteria, study of viruses gives information about regulation in eukaryotes. Of course, animal viruses possess a great deal of intrinsic interest also, both as highly regulated systems and as agents of disease. Many viruses are examined in **Chapter 23;** each one exemplifies a particular mode of transcription, replication, or production of viral proteins, the mode varying with the particular genetic material possessed by that virus. Two points emphasized in this chapter are that different virus types use rather different strategies for generating mRNA and that viruses use a variety of mechanisms to deal with the problem of making a sufficient number of protein types from a rather limited amount of nucleic acid. **Chapter 24** includes tumor viruses and here the student is introduced to the oncogene. This chapter has been especially difficult to write because "accepted" interpretations of numerous experiments with oncogenes come and go like the wind. With the help of several advisors, I have included only those interpretations that are, at present, believed to be able to stand the test of time.

An introductory textbook must necessarily contain a large number of terms not previously encountered by the reader. To aid in recognizing these terms and in finding the definitions at a later time, new terms are printed in **boldface** type where they first appear in the text.

ACKNOWLEDGMENTS

I would like to thank the following people for help of many kinds: Arthur C. Bartlett and Donald Jones of Jones and Bartlett, Publishers, Inc., for recognizing that the time was right for a second edition; Jack Vitkus, who rendered all of the new art; Hal Lockwood of Bookman Productions, who supervised and carried out efficient production of this and the first edition; my many colleagues who read selected segments of the first edition and suggested how I should update and improve it, and the numerous people who provided me with material, both published and unpublished, for the second edition (they are listed following this preface); and my friend, Peter Geiduschek, the first user of *Molecular Biology,* who spent many hours with me discussing how to improve the first edition and who read several of the completed chapters of the second edition.

David Freifelder
May 1986

Reviewers of *Molecular Biology*

1

Systems and Methods of Molecular Biology

For a century or so, many biologists believed that living cells possessed some vital force that would be lost if a cell was broken. Thus, the vitalists thought that little, possibly nothing, could be learned about life by studying anything other than the intact cell. They were wrong, as events were to prove. When it was first decided to break open a living cell and study its inner workings, molecular biology was born.

A BRIEF OVERVIEW OF MOLECULAR BIOLOGY

The term molecular biology was first used in 1945 by William Astbury, who was referring to the study of the chemical and physical structure of biological macromolecules. By that time, biochemists had discovered many fundamental intracellular chemical reactions, and the importance of specific reactions and of protein structure in defining the numerous properties of cells was appreciated. However, the development of molecular biology had to await the realization that the most productive advances would be made by studying "simple" systems such as bacteria and bacteriophages (bacterial viruses), which yield information about basic biological processes more readily than animal cells. Although bacteria and bacteriophages are themselves quite complicated biologically, they enabled scientists to identify DNA as the molecule that contains most, if not all, of the genetic information of a cell.

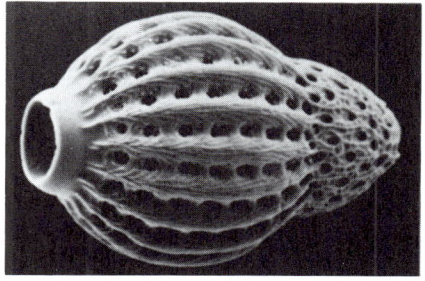

The skeleton of a radiolarian, showing the remarkable symmetry of these marine organisms. Symmetry is a feature of many microorganisms and should be explainable by physical and chemical principles. [From G. Shih and R. Kessel, *Living Images*. Jones and Bartlett (1982).]

Following this discovery, the new field of molecular genetics moved ahead rapidly in the late 1950s and early 1960s and provided new concepts at a rate that can be matched only by the development of quantum mechanics in the 1920s. The initial success and the accumulation of an enormous body of information enabled researchers to apply the techniques and powerful logical methods of molecular genetics to the subjects of muscle and nerve function, membrane structure, the mode of action of antibiotics, cellular differentiation and development, immunology, and so forth. Faith in the basic uniformity of life processes was an important factor in this rapid growth. That is, it was believed that the fundamental biological principles that govern the activity of simple organisms, such as bacteria and viruses, must apply to more complex cells; only the details should vary. This faith has been amply justified by experimental results.

The acquisition of information continued at a more-or-less constant rate through the early 1970s, and knowledge of bacteria and viruses became quite profound. However, understanding of more complex cells, such as animal and plant cells, lagged behind and occasionally seemed to be at an impasse. In the late 1970s a breakthrough occurred—the development of the recombinant DNA technology, or genetic engineering, as it is now called. This provided new tools, and in time information about all cells became available at an extraordinary rate. This second edition of *Molecular Biology,* following three years after the first edition, is a reflection of this information explosion.

HOW MOLECULAR BIOLOGY WILL BE PRESENTED IN THIS BOOK

Molecular biology is an inherently logical discipline, once the ground rules are understood. However, it is enormously complex and crosses traditional boundaries between genetics, biochemistry, cell biology, physics, organic chemistry, and biophysical chemistry, so molecular biology can no longer be thought of as an indivisible subject. Unfortunately, molecular biology also leads the student through a circular process. That is, to understand how a cell replicates, one needs to know about DNA and RNA, yet a clear understanding of DNA and RNA functions requires that one refer to cells. This naturally poses the problem of where to begin.

The approach taken in this book is to explain the systems that must continually be referred to in order to understand basic phenomena. In our initial development of molecular biology we shall look at bacteria and phages; later we shall turn to more complicated systems such as animal viruses and animal cells. These cells and how they are handled will be described in this chapter. Biochemical techniques and knowledge of the chemical conversions that occur in living cells have played an enormous role in all biological research. Thus, we devote the latter part of this chapter to this topic. Genetic techniques provide one of the most powerful experimental tools in molecular biology, especially in the study of microorganisms, and for this reason these techniques will be reviewed in Chapter 2; special attention will be given to the use of mutants to analyze a system, as this is a common approach.

Biological systems contain many types of gigantic molecules (**macromolecules**)—for example, proteins, nucleic acids, and polysaccharides. Discussion of detailed biochemical mechanisms is made easier if the structure of biological macromolecules is understood at the outset. Hence this subject is developed in Chapters 3 through 7.

Armed with the tools of molecular biological research and knowledge of both the structure of macromolecules and the chemical processes occurring in cells, we will be able to proceed to a detailed study of the fundamental activities that determine the biological properties of each cell—namely, DNA replication, RNA synthesis, and protein synthesis—the topics of Chapters 8 through 14. Then we can ask a profound question—since all possible chemical reactions that can occur in a cell do not occur at the same time, what determines the particular intracellular events that transpire at a particular time? Such intracellular regulation is examined in Chapters 15 and 16, in which we shall look at bacteria, bacterial viruses, and eukaryotic viruses.

In Chapters 17 and 18 we will study bacteriophages, both for their intrinsic interest and as a means of seeing how the synthesis of DNA, RNA, and protein are regulated and integrated to yield a complex functioning biological system.

As we approach the end of the book, we will examine the means of creating organisms having new properties. We begin in Chapters 19 through 21 with natural processes for scrambling DNA, such as genetic recombination, and then in Chapter 22 we turn to the new biotechnology, genetic engineering, in which knowledge of fundamental biochemical processes is used to design new organisms having features that are valuable for research, medicine, consumers, and industry.

The vocabulary of molecular biology is very large because it is drawn from so many disciplines. This often makes it difficult for students to familiarize themselves with the subject. Thus, as a study aid, throughout this book, essential terms will be printed in boldface type where each is explained.

PHYSICAL APPROACH TO THE PROBLEMS IN MOLECULAR BIOLOGY

Throughout this book, physical and physicochemical techniques will be applied in order to understand the properties of molecules. There are two reasons to approach molecular biology in this way. One is because structure and function are intimately related; the other is that many biological questions can be answered by data obtained from measurements of physical and physicochemical properties.

The structure-function relation appears repeatedly. For example, collagen, the protein of which tendon is made, is triple-stranded for strength, aggregates side-to-side for additional strength, and has special features that enable long, tough fibers to form. DNA, the single most important molecule in every cell, is double-stranded in order that two copies of its precious genetic information will exist. Cell membranes are rich in nonpolar fatty acids, in order that polar molecules cannot freely pass in and out of the cell; however, other molecules and special proteins provide passageways through the membranes for transit of specific substances. Hundreds of such examples are known and most of the information about them has come from precise physical measurements that enabled molecular structures to be elucidated. Several of these examples will be described in Chapters 3 through 7.

Physical measurements also provide other types of information. For example, by ultracentrifugation one sees that incompletely synthesized protein molecules always sediment together with intracellular particles called ribosomes. This observation was one of many that showed that the ribosome is the structure on which proteins are synthesized. By electron microscopy, DNA molecules are seen to be linear when isolated from a phage but circular when isolated from a phage-infected bacterium. Further studies have shown the reason for this: in an early stage in the phage life cycle, phage DNA becomes circularized. Electrophoretic studies indicate that many presumably pure enzymes have two components, each able to move in an electric field at a distinct rate; these enzymes have multiple forms that are important in metabolic regulation. The electrophoretic behavior of the hemoglobin isolated from a person with sickle-cell anemia differs from that of a normal person; in the sickle-cell hemoglobin, a particular amino acid has been replaced by another amino acid. Infrared absorption spectroscopy of thymine, one of the components of DNA, indicates that both a carbonyl and a hydroxyl group are present; this shows that both keto and enol forms are in equilibrium, which is important in understanding the mechanism of mutation. Hundreds of examples of this type can be taken from the history of numerous subfields of molecular biology. In Chapter 3 several important physical techniques will be described in detail.

THE *IN VITRO* APPROACH TO UNDERSTANDING BIOLOGICAL REACTIONS

At some point in the study of living cells it is almost always necessary to examine individual cellular reactions separately. This approach usually has the advantage of some simplification of a system but ignores the fact that these reactions often interact with other cellular systems. Nonetheless, a great deal of information can be obtained in this way. Thus, one does not usually study directly a phenomenon as complex as overall cell growth but looks separately at the synthesis of DNA, proteins, and other components. In fact, one does not ordinarily even study a process as complex as protein synthesis in its totality, but examines its numerous substages individually.

In many cases, it is not possible to study certain component reactions in a living cell. For example, the polymerization of phosphate-containing

DNA monomers to form DNA molecules cannot be studied by adding these substances to a growth medium, because organic phosphorus–containing compounds are unable to pass through the cell membranes to enter the cell. Likewise, one cannot determine how the activity of an enzyme complex that freely associates and dissociates is regulated when there are 3000 other proteins present. Thus, it has become repeatedly necessary to break open cells and to purify the components of the reactions. Any investigation carried out with broken cells or pure components is called an *in vitro* study; this contrasts with an *in vivo* study, which is done with intact cells.

There are two different types of *in vitro* studies—those that use an unfractionated (crude) extract and those that use a purified or a reconstituted system. A **crude extract** is obtained when cells suspended in a buffer are broken open and nothing is removed other than unbroken cells and, possibly, fragments of the cell wall and cell membrane. A crude extract is often used to assay for the presence of particular enzymes. For instance, the enzyme β-galactosidase, which breaks down the sugar lactose to glucose and galactose, is easily detected in a crude extract and can be measured quantitatively. Crude extracts are often used in early stages of analysis to determine whether a particular enzymatic activity is present. However, their use usually requires great cleverness on the part of the experimenter because so many substances are present in the extract. Conclusions obtained with crude extracts must always be accepted with caution, for often several reactions occur simultaneously that compete with the reaction one thinks is being studied or that even antagonize the reaction. An example of the latter is that, without special care, the synthesis of nucleic acids is unobservable because of the ubiquity of enzymes that depolymerize nucleic acids (**nucleases**).

A more useful and less problematical approach to understanding an intracellular process is to purify the components of the subsystem, determine the chemical and physical properties of each component, and then reconstitute the system. Experimental conditions are then sought in which a process simulates as closely as possible that which occurs, or is believed to occur, within the cell. This is the ultimate *in vitro* experimentation and is often the only way to understand precisely how a system works. The *in vitro* approach requires extraordinary skill both in experimentation and interpretation, because so many misleading possibilities can arise. For instance, one might be misled by choosing conditions that force an enzyme to catalyze a reaction that never occurs within a cell; examples are known of polymerases that have been used to drive depolymerization reactions. Also, essential factors can be lost during purification. For example, in early studies of the synthesis of RNA, excessive purification of the polymerization system caused a protein to be lost that is needed for recognition of specific start sites on a DNA molecule, and this loss prevented essential features of the reaction from being observed. Changes in the physical properties of the molecules are also common. For example, for 96 years DNA molecules were always fragmented during isolation until 1959, when unbroken DNA molecules were first isolated, and it was not until then that it was realized that DNA molecules are huge and that a chromosome contains a single DNA molecule. Also, it is not at all uncommon for protein molecules to be isolated that are either damaged by protein-digesting enzymes or by the physicochemical procedures used in isolation and thus are devoid of biological activity.

The *in vitro* approach is a powerful technique for understanding biological mechanisms. In many cases there is no other way to understand a system fully. We shall see numerous examples of this approach throughout the book.

THE LOGIC OF MOLECULAR BIOLOGY

Three kinds of logical reasoning appear repeatedly in molecular biology—arguments based on efficiency, examination of models, and strong inference. These are described in the following sections.

The Efficiency Argument

Living cells have had hundreds of millions of years to evolve. During this time the rigors of competition and survival have, for free-living cells such as bacteria and yeast, selected for efficiency (though this may have been compromised through accommodation of potential efficiency to other demands on the system). Little energy and material are wasted, inasmuch as wastefulness is disadvantageous in the long run. That is, if among a population of cells a less wasteful mutant arises (i.e., a more efficient cell), this mutant should grow slightly faster than the rest of the population. Ultimately, since population numbers are virtually unlimited, the population will consist entirely of this mutant. Thus, when proposing a mechanism for a biochemical or biophysical process, one usually tries to think of ways in which a system avoids both waste and errors. Experience has shown that this is often a productive approach and it has been an important factor in the early development of new notions of molecular genetics. For instance, the molecular geneticist is continually attuned to the presence of start signals, stop signals, and regulatory sites in DNA and RNA molecules, and similar ways to avoid waste. However, one should not take the efficiency notion unquestionably, because it fails, at the present at least, to explain some very complex aspects of the properties of plant and animal cells. In fact, for these cells, which are present in organized tissue in which there is no competitive pressure, it seems that the only requirement for the existence of a particular mechanism for some process is that it works.

Examination of Models

A model is a tentative explanation of the way a system works. It proposes the components, interactions, and sequences of events. An important function of a model is to suggest additional experiments. Models enable experimenters to make predictions and, to investigate a model, the predictions must be tested experimentally. If predictions do not agree with experimental results, the model must be considered incorrect *as it stands.* Such a contradiction calls for a change in the model. It is important to realize that a model cannot be proved to be correct merely by showing that it makes a correct prediction. However, if it makes *many* correct predictions, it is probably nearly, if not completely, correct. Whereas a rigorous proof of the reality of a model is logically impossible, some models are very persuasive in terms of their elegance and power.

Strong Inference

A great deal of biological reasoning is inductive. That is, something may be observed so many times that one infers that it is a real phenomenon, or a theory may explain so many things that one believes that the theory is correct. Strong inference is a logical process based on exclusion of alternative possibilities. It is one of the most powerful logical approaches in molecular biology. (It is used in other sciences as well, but is especially prevalent in molecular biology.) In the method of strong inference one first states all possible explanations for a particular phenomenon and then experimentally eliminates the possibilities one by one. When only one alternative remains, that one is inferred to be correct. This is true as long as the possibilities are exclusive and all alternatives have been specified. Actually, it is not common in molecular biology (or probably in any branch of science) to consider all possibilities. Usually, the molecular biologist uses intuition and generalizations from other biological systems and considers only those that seem reasonable. For this reason, readers of papers in scientific journals will often encounter the phrase "it is likely that" when experimental data are being interpreted.

A FEW PROBLEMS STUDIED IN MOLECULAR BIOLOGY

It would be foolhardy to try to list all biological phenomena for which a molecular explanation is desired. However, it is useful to list some of the important problems that have been and are being attacked by the methods of molecular biology and to describe the problems of the immediate future.

In the early days of what might be called modern molecular biology, the principal effort went toward understanding genetic phenomena. An account of this effort forms a major part of this book. The two earliest major problems were the identification of the genetic material and the mechanism of protein synthesis. Once DNA was shown to be the genetic substance, two questions immediately arose: how does DNA replicate (the *detailed* answer for which still remains to be discovered), and how is information obtained from a DNA molecule? The latter question became tied to the problem of protein synthesis as soon as it was learned that the amino acid sequence of every protein molecule is determined by the base sequence of its DNA. Basic questions about mechanisms have been answered more or less completely, though there are still many details to be learned.

It also was apparent that the protein-synthesizing mechanisms in bacteria are not identical to those used in animal cells, though they are related. Now, after 30 years of an intensive investigation of phage and bacterial systems, the technology of molecular biology has reached a stage at which more complex systems, such as the animal cell, are profitably being studied. Study of animal cells has been done primarily with cells grown in culture. Progress has been slow for many reasons, not the least of which is the 24-hour doubling time of most cells in culture compared to 25 minutes for many bacteria. In the mid-1970s it was realized that just as the most rapid progress in understanding basic phenomena had been made through study of simpler systems such as phages and bacteria, knowledge of the molecular biology of nucleus-containing cells, such as animal cells, might come faster through concentration of study on a simple, rapidly growing nucleated cell. The baker's yeast, *Saccharomyces cerevisiae,* a unicellular organism, whose doubling time is 70 minutes, has turned out to be a valuable experimental object because its basic mechanisms of macromolecular synthesis and its regulatory mechanisms are similar to those of the more complex cells.

Once basic synthetic mechanisms had become better understood, the question of metabolic regulation arose. Cells rarely make substances they do not need, yet when a particular substance is required, its synthesis usually starts up very soon after the "needed" signal has been received. The nature of the signal and, in general, the identity of all start and stop signals (which are present in many systems) have been and remain important questions. Some complex systems, such as the life cycle of phages, are regulated in an extraordinary fashion; in these systems, the timing, sequence, and amount of synthesis of all components are predetermined by the phage genes and their arrangements on the DNA, and everything is carried out with great efficiency. A few phage systems (particularly *E. coli* phages λ and T7) are very well characterized, and all of the start and stop signals and regulatory elements are known.

The internal concentrations of many cellular substances are remarkably constant. Furthermore, cells are exceedingly selective with respect to the substances allowed to enter the cell. Selectivity is determined by the chemical and physical properties of membranes and by various carrier molecules and pumping systems. The identity of these transport systems, how they work, and the structures of different membranes are exciting questions whose answers are being actively sought.

In the specialized animal cells, such as muscle and nerve cells, there are many mysteries. How does contractility occur? What causes the transmission of an electrical impulse along nerve cells and what maintains the electrical charge across the membranes of resting nerve cells? Why does one type of cell make hemoglobin and another make insulin?

One of the most precise and profitable branches of molecular biology is structural biology, actually the birthplace of modern molecular biology. The driving force for the great activity in this field is the doctrine that no structure exists without a function, and its corollary, that functions are carried out by neatly designed, efficient systems whose activities are usually determined by specific macromolecular structures. For example, the analysis of the structure of DNA led to an understanding of mutation and suggested how replication and genetic recombination might occur. A look at the various forms of RNA opened up the field of protein synthesis, a problem that is still not completely solved. However, the machinery of protein synthesis—that is, the ribosome—has been well characterized, and the sites on the ribosome at which various stages of protein synthesis occur are known. Also, elucidation of the three-dimensional structure of the transfer RNA molecule has explained how the information in DNA is translated into the amino acid sequence of proteins.

At present, structural biology has two subfields—the detailed arrangement of atoms in individual macromolecules and the organization of macromolecules in large aggregates. An example of the former is the structure of DNA, which is already known. Current research focuses on the structure of proteins—especially enzymes. The analysis of enzyme structure shows where on the surface of the enzyme the chemical reaction catalyzed by the enzyme occurs. Biochemists interested in the mechanism of enzyme action anxiously await this information for each enzyme being studied. Unfortunately, definition of structural details at the atomic scale is obtained only by the tedious and difficult technique of x-ray diffraction analysis, and analyzing a structure often takes a very long time. Modern high-speed computers and several other new techniques are shortening the required time. As for the second subfield—the structures of large aggregates—some structures, such as tendon, hair, and wool are well known. However, many questions remain about myosin (the principal muscle protein) and about actin and tubulin (two proteins that are responsible for the shapes of different cell types), and these proteins are being actively studied. Both technique and our understanding of structural rules have advanced to the point that one can attempt to solve the structure of aggregates containing molecules of different types. Thus, a great deal of research effort is directed at present toward the structure of chromosomes, which contain DNA and several proteins, and of membranes, which contain several types of proteins and lipoproteins. Of special interest with respect to membranes is how the structure enables a membrane to be selectively permeable—that is, permeable to some, but not all, molecules. Differences in the structure of the cell membranes of normal and cancer cells are especially tantalizing.

WHAT ISN'T MOLECULAR BIOLOGY?

By now the reader may have the idea that molecular biology includes all biology. This is true in the sense that the ultimate goal of molecular biology is to understand the molecular basis of all biological phenomena. However, certain topics, which are not included in this book, are not usually thought of as molecular biology by those who call themselves molecular biologists. Yet some aspects of even these topics interest molecular biologists. For example, the chemical reactions of metabolism are considered to be biochemistry. If these reactions are regulated simply by the effects of reactant and product concentrations, as in a standard chemical reaction, the study of this regulation remains biochemistry; however, if an enzyme-catalyzed reaction is regulated by alteration of the structure of the enzyme, a molecular biologist would probably lay claim to the topic (though a biochemist might not agree). Similarly, study of the arrangement and structure of the inner components of a cell is usually considered to be cell biology. However, as time passes, this topic is becoming amenable to molecular analysis also. Behavioral biology seems clearly at first not to be molecular biology; however, some people have isolated behavioral mutants of insects and certain protozoa and these too are being subjected to molecular biological analysis. Therefore, the border between molecular biology and other branches of biology is thin and hazy and is apt to be adjusted as molecular biology is extended to fresh territory. Perhaps someday, subdivisions of biological research will no longer be important and the unifying term biology will again be meaningful.

PROKARYOTES AND EUKARYOTES

Living organisms can be classed in two categories—the prokaryotes and the eukaryotes. A **prokaryote** is a cell whose genetic material is present throughout the cell. In a **eukaryote** the genetic material is organized in a well-defined compartment, the **nucleus,** and the DNA is organized transiently into chromosomes, which become compact and visible by light microscopy prior to and in preparation for cell division. Another important distinction between prokaryotes and eukaryotes is in the way that the DNA molecule itself is organized. The DNA of eukaryotes is tightly associated with specific proteins to form nucleoproteins (which are reorganized to form chromosomes) whereas in the prokaryotes the DNA is mostly free of such "structural" proteins. Prokaryotes and eukaryotes are also significantly different in their structural organization. Prokaryotes have a fairly uniform internal structure and may be thought of as a sac containing a solution of several thousand different molecules, plus DNA and ribosomes. In contrast, a eukaryotic cell contains numerous types of particles such as mitochondria, granules, chloroplasts (only in plants), a variety of compartments (nucleus, vacuoles), and numerous membranous structures (Figure 1-1).

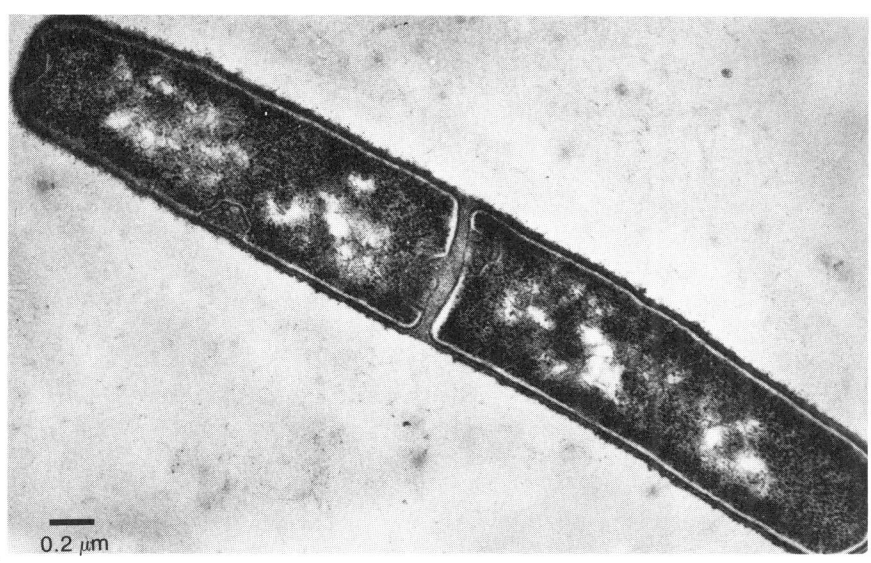

0.2 μm

(a)

Figure 1-1 The organization of a prokaryotic cell and a eukaryotic cell. (a) Prokaryote. An electron micrograph of a dividing bacterium. The layers of the cell wall can be seen. The light areas inside the cell are the DNA; note that the DNA is distributed throughout the cell. (b) Eukaryote. Electron micrograph of a section through a cell of the alga *Tetraspora,* showing the membrane-bound nucleus, the nucleolus (dark central body), and other membrane systems (chloroplasts, mitochondria) that subdivide the cytoplasm into regions of specialized function. This complex structure should be compared to the simpler organization of the bacterium shown in (a). [Courtesy of (a) A. Benichou-Ryter and (b) J. Pickett-Heaps.]

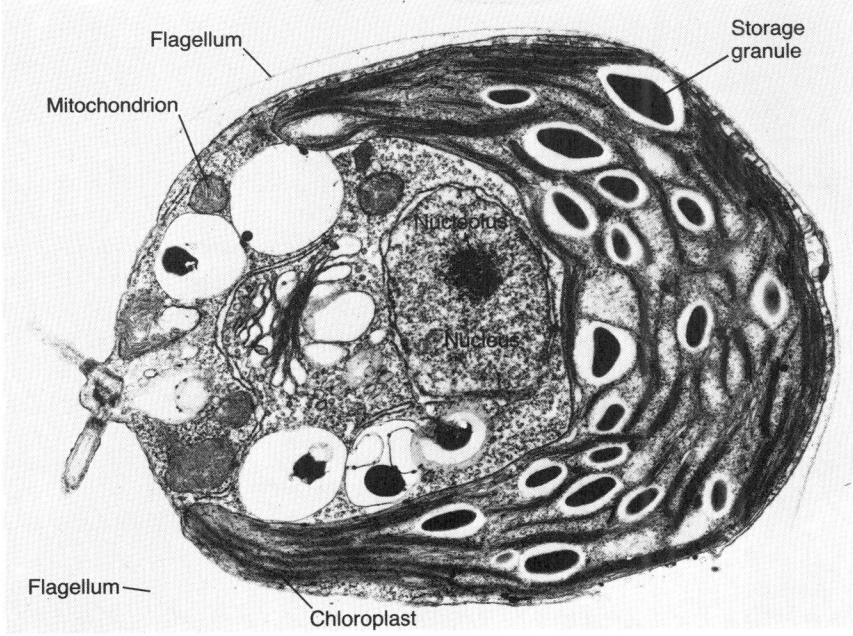

Flagellum

Mitochondrion

Storage granule

Nucleolus

Nucleus

Flagellum

Chloroplast

(b)

The distinction between prokaryotes and eukaryotes is exceedingly important because, for unknown reasons, there are profound biochemical differences between prokaryotes and eukaryotes. For example, the chemistry of assembly of amino acids into proteins occurs by the same mechanism in all organisms; however, the mechanics of protein synthesis in bacteria, which are prokaryotes, differs in many ways from that in eukaryotes, yet the mechanism is the same in all bacteria. Furthermore, the organization of eukaryotic genes, the form of the genetic material in eukaryotes, and the mechanisms for regulation of gene expression in eukaryotes differ from those in prokaryotes.

Some organisms, such as phages and viruses, which are subcellular in organization but dependent on cells as an environment for their reproduction, are neither prokaryotes nor eukaryotes. However, the bacteriophages grow only in bacteria and thus adopt a prokaryotic strategy for reproduction while the animal and plant viruses are obligated to manufacture their component macromolecules by using eukaryotic rules.

Much more is known about prokaryotes than eukaryotes because most of the prokaryotic systems have been studied for a longer time. However, this situation has been changing rapidly, since the recombinant DNA technology has allowed questions about eukaryotes to be addressed by powerful experimental techniques.

In this book prokaryotes and eukaryotes will be discussed separately and compared and contrasted. Usually prokaryotes will be discussed first because they are simpler to understand. In keeping with this, we begin by describing the properties of bacteria that are important in molecular biological research.

BACTERIA

Bacteria are free-living unicellular organisms. They have a single chromosome, which is not enclosed in a nucleus (they are prokaryotes), and compared to eukaryotes they are simple in their physical organization (Figure 1-1). Bacteria have many features that make them suitable objects for the study of fundamental biological processes. For example, they are grown easily and rapidly and, compared to cells in multicellular organisms, they are relatively simple in their needs. The bacterium that has served the field of molecular biology best is *Escherichia coli* (usually referred to as *E. coli*), which divides every 22 minutes at 37°C under laboratory conditions. Thus, a single bacterium becomes 10^9 bacteria in about 20 hours. How one grows bacteria and counts their number is described in the following section. These procedures also apply to the yeasts, which are discussed later in this chapter.

Growth of Bacteria

Bacteria can be grown in a **liquid growth medium** or on a solid surface. A population growing in a liquid medium is called a bacterial **culture.** If the liquid is a complex extract of biological material, it is called a **broth.** An example is tryptone broth, which is the milk protein casein hydrolyzed by the digestive enzyme trypsin to yield a mixture of amino acids and small peptides. If the growth medium is a simple mixture containing no organic compounds other than a carbon source such as a sugar, it is called a **minimal medium.** A typical minimal medium contains each of the ions, Na^+, K^+, Mg^{2+}, Ca^{2+}, NH_4^+, Cl^-, HPO_4^{2-}, and SO_4^{2-}, and a source of carbon such as glucose or glycerol. The best source of carbon is glucose; in minimal media containing glucose, bacteria grow more rapidly than in a minimal medium with any other carbon-containing compound. If a bacterium can grow in a minimal medium—that is, if it can synthesize *all* necessary organic substances, such as amino acids, vitamins, and lipids—the bacterium is said to be a **prototroph.** If any organic substances other than a carbon source must be added for growth to occur, the bacterium is termed an **auxotroph.** For example, if the amino acid leucine is required in the growth medium, the bacterium is a leucine auxotroph; the genetic (phenotypic) symbol for such a bacterium is Leu^-. A prototroph would be indicated Leu^+. Table 1-1 lists standard abbreviations for various nutrients and genetic characteristics encountered in microbiological studies.

Table 1-1   Standard substances used in microbial genetics, which are encountered in this book, and their abbreviations

Substance	Genotype symbol[1]	Substance	Genotype symbol
Amino acids		*DNA and RNA bases*[2]	
Alanine	*ala*	Purine	*pur*
Arginine	*arg*	Pyrimidine	*pyr*
Asparagine	*asn*	Adenine	*ade*
Aspartic acid	*asp*	Cytosine	*cyt*
Cysteine	*cys*	Guanine	*gua*
Glutamic acid	*glu*	Thymine	*thy*
Glutamine	*gln*	Uracil	*ura*
Glycine	*gly*	*Vitamin*	
Histidine	*his*	Biotin	*bio*
Isoleucine	*ile*	*Sugars*	
Leucine	*leu*	Arabinose	*ara*
Lysine	*lys*	Galactose	*gal*
Methionine	*met*	Glucose	*glu*
Phenylalanine	*phe*	Lactose	*lac*
Proline	*pro*	*Antibiotics*	
Serine	*ser*	Ampicillin	*amp*
Threonine	*thr*	Chloramphenicol	*cam*
Tryptophan	*trp*	Kanamycin	*kan*
Tyrosine	*tyr*	Penicillin	*pen*
Valine	*val*	Streptomycin	*str*
		Tetracycline	*tet*

[1]When stating a substance or a phenotype rather than a genotype, the same three-letter symbol is used, but it is capitalized and not italicized—for example, His for histidine.

[2]When stating base sequences, a purine or a pyrimidine is usually abbreviated Pu or Py respectively, if the identity of the base is unimportant. Particular bases will be denoted by A (adenine), G (guanine), T (thymine), U (uracil), and C (cytosine).

Bacteria are frequently grown on solid surfaces. The earliest surface used for growing bacteria was a slice of raw potato. This was later replaced by media solidified by gelatin. Because many bacteria excrete enzymes that digest gelatin, an inert gelling agent was sought. **Agar,** which is a gelling agent obtained from a particular seaweed, is resistant to bacterial enzymes and has been universally used. A solid growth medium is called a nutrient agar, if the liquid medium is a broth, or a minimal agar, if a minimal medium is gelled. Solid media are typically placed in a **petri dish.** In lab jargon a petri dish containing a solid medium is called a **plate** and the act of depositing bacteria on the agar surface is called **plating.**

When bacteria are placed in a liquid medium, they slowly start to grow and divide. After an initial period of slow growth called the **lag phase,** they begin a period of rapid growth in which they divide at a fixed time interval called the **doubling time.** The number of cells per milliliter, the **cell density,** doubles repeatedly, giving rise to a logarithmic increase in cell number; this stage of growth of the bacterial culture is called the **log phase.** This stage of growth continues until a cell density (for *E. coli*) of about 10^9 cells/ ml is reached, at which point nutrients and O_2 become limiting and the growth rate decreases. (This cell density is often five- to tenfold lower for many other bacterial species.) Ultimately, at a cell density of 2 to 3×10^9 cells/ml, no further growth is possible and the cell number becomes constant. This stage is the **stationary phase.** A typical growth curve for a bacterial culture is shown in Figure 1-2.

The terms used in this section are also used in discussing the growth of all microorganisms and frequently of animal cells.

Figure 1-2 A typical growth curve for a bacterial culture. The *y* axis is logarithmic, so the curve is a straight line in log phase.

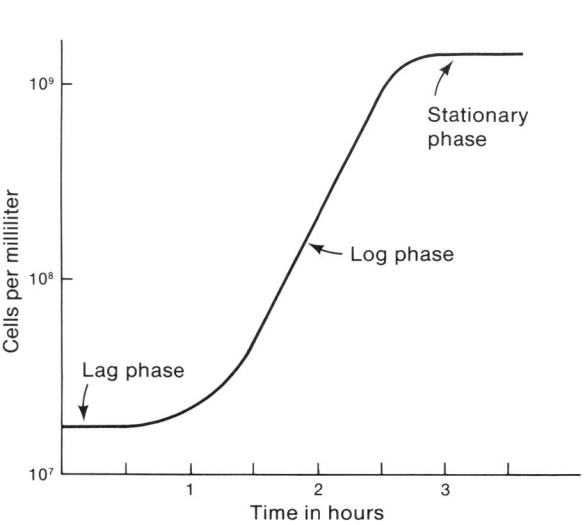

Counting Bacteria

A bacterium growing on an agar surface also divides. Since most bacteria are not very motile on a solid surface, the progeny bacteria remain very near the location of the original bacterium. The number of progeny increases so much that a visible cluster of bacteria appears. This cluster is called a **bacterial colony** (Figure 1-3). Colony formation allows one to determine the number of bacteria in a culture. For instance, if 100 cells are plated, 100 colonies will be visible the next day. If 0.1 ml of a 10^6-fold dilution of a bacterial culture is plated and 200 colonies appear, the cell density in the original culture is $(200/0.1)(10^6) = 2 \times 10^9$ cells/ml.

Identification of Nutritional Requirements of Bacteria

Plating is a convenient way to determine if a bacterium is an auxotroph. This is done in the following way. Minimal agar and nutrient agar plates are prepared. Several hundred bacteria are plated on each plate and the plates are stored overnight in a constant-temperature incubator. Several hundred colonies are subsequently found on the nutrient agar because it contains so many substances that it can satisfy the requirements of nearly any bacterium. If colonies are also found on the minimal agar, the bacterium is a prototroph; if no colonies are found, it is an auxotroph and some required substance is not present in the minimal agar. Minimal plates are then prepared with various supplements. If the bacterium is a leucine auxotroph, the addition of leucine alone will enable a colony to form. If both leucine and histidine must be added, the bacterium is auxotrophic for both of these substances.

Table 1-2 shows the result of a plating experiment in which the nutritional requirements of a bacterium are deduced. The fact that there is no growth with supplement 1 shows that something other than histidine, leucine, and thymine is needed. Growth in the absence of thymine (supplement 2) and of leucine (supplement 3) indicates that neither thymine nor leucine is required. Note, however, that the presence of alanine in supplement 3

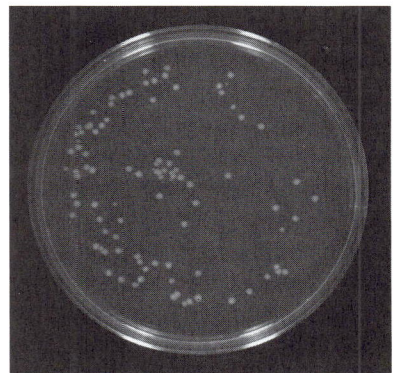

Figure 1-3 A petri dish with bacterial colonies that have formed on agar. [Courtesy of Gordon Edlin.]

Table 1-2 Plating data enabling the determination of the nutritional requirements of a bacterium

Medium supplement	Growth	Conclusion
1. His, Leu, Thy	−	Needs some nutrient
2. Leu, Ala, His	+	Thy not needed
3. Ala, Thy, His	+	Leu not needed
		Ala needed(cf. plate 1)
4. Leu, Ala, Thy	−	His needed (cf. plates 1 and 3)

Note: Abbreviations of names of amino acids are given in Table 1-1, note 1.

overcomes the deficiency of supplement 1, so alanine must be required. Finally, the lack of growth with supplement 4, in which alanine is present, shows that there is still another requirement. Since leucine is not needed, the only significant difference between supplements 3 and 4 is the presence of histidine in supplement 3. Therefore, histidine is also required and the bacterium is auxotrophic for both alanine and histidine.

These techniques apply to all microorganisms.

The Physical Organization of a Bacterium

The intracellular organization of bacteria is poorly understood, but some points are clear (Figure 1-1). The bacterial cell does not have a defined nucleus and hence is a prokaryote. The intracellular material is enclosed in a rigid multilayered cell wall that gives it a defined shape—spherical, rodlike, and so forth (Figure 1-4). If bacteria are treated with the enzyme **lysozyme,** which is isolated from chicken egg white, some of the cell wall components are removed and the rigidity of the wall is lost. All bacteria treated in this way become spherical and ultimately burst. The spherical form, which can be stabilized in suspending media having a high osmotic strength (e.g., 20 percent sucrose), is called a **spheroplast** if some cell wall remains and a **protoplast** if the cell wall is completely removed. Beneath the cell wall is a very thin layer called the **cell membrane** (see Chapter 7). This layer provides a permeability barrier that determines which substances can pass in and out of the cell. Some substances are transported by proteins called **transport proteins** or **permeases;** these will be discussed in Chapter 15.

Metabolic Regulation in Bacteria

Bacteria are well-regulated and highly efficient organisms. For example, they rarely synthesize substances that are not needed. Thus the enzymatic system for synthesizing the amino acid tryptophan is not formed if tryptophan is present in the growth medium, but when the tryptophan in the medium is used up, the enzymatic system will be rapidly formed. The systems responsible for utilization of various energy sources are also efficiently regulated. A well-studied example is the metabolism of the sugar lactose as an alternate carbon source to glucose.

Glucose is metabolized by all living cells by a series of chemical conversions in which the molecule is progressively broken down. Glucose is a fundamental carbon source in the sense that other sugars must either be converted to glucose or to a glucose-degradation product in order to be metabolized. Thus bacteria use glucose more efficiently than lactose. The first step in the metabolism of lactose is its cleavage to form two sugars, glucose and galactose. However, the enzyme needed to catalyze this cleavage

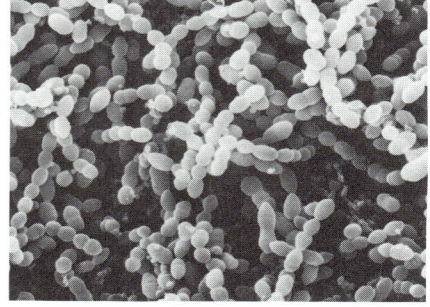

(a)

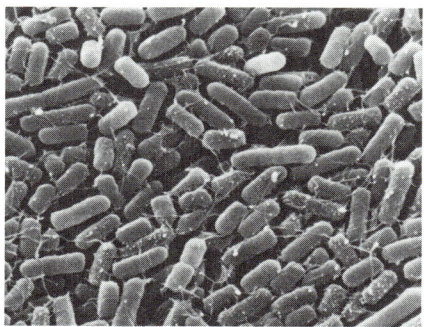

(b)

Figure 1-4 Two forms of bacteria. (a) Cocci (spheres that sometimes form chains). (b) Bacilli (rods). [From G. Shih and R. Kessel, *Living Images: Biological Microstructures Revealed by Scanning Electron Microscopy* (Jones and Bartlett, 1982).]

is not present in cells in significant quantities unless lactose is in the growth medium. If lactose is provided to a cell, through a complex multistep process the enzyme can then be formed, and glucose produced from lactose is broken down and used as a source of energy. The galactose that is also produced is converted to a component that enters the glucose-utilization pathway and is thereby also metabolized. That the system is regulated can be seen by the fact that if a cell is supplied with both glucose and lactose in the growth medium, there is no reason for the cell to synthesize the lactose-cleaving enzyme and, in fact, a simple system prevents the cell from wasting its energy synthesizing this enzyme until all the glucose is utilized. This system is discussed in detail in Chapter 15.

The control of both tryptophan synthesis and lactose degradation are two examples of **metabolic regulation.** This very general phenomenon will be explored extensively throughout the book. Both simple and complex regulatory systems will be seen, all of which act to determine how much of a particular compound is utilized and how much of each intracellular compound is synthesized at different times and in different circumstances. This will demonstrate the length to which the so-called simple cells have gone to utilize limited resources efficiently and to optimize their metabolic pathways for efficient growth.

Metabolic regulation occurs in all living organisms.

BACTERIOPHAGE

Bacteria are subject to attack by smaller organisms called **bacteriophage** or simply **phage.*** These are small particles, of the general class of particles called viruses, and they are capable of growing only inside bacteria. Phage have been the object of choice for a great many types of experiments because they are much simpler than bacteria in their structures (usually having between two and ten components) and life cycles and yet possess the most essential, if minimal, attributes of life. A typical life cycle is outlined briefly in the following sections.

Phage Structures

A typical phage contains only a few different types of molecules—usually several hundred protein molecules of one to ten types (depending on the complexity of the phage) and one nucleic acid molecule. The protein molecules are organized in one of three ways. In the most common mode the

---

*The plural word *phages* refers to different species; the word *phage* can be both singular and plural, and in the plural sense refers to particles of the same species. Thus, T4 and T7 are both phages but a test tube might contain either 1 T7 phage or 100 T7 phage.

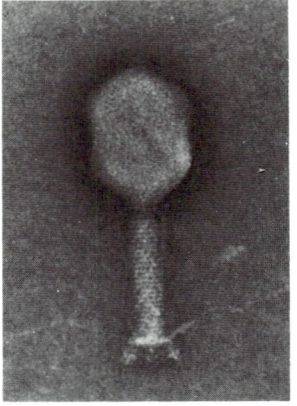

Figure 1-5   An electron micrograph of *E. coli* phage T4. Note the detailed structure of the tail. [Courtesy of M. Wurtz.]

protein molecules form a protein shell called the **coat** or **phage head,** to which a **tail** is generally attached (Figure 1-5); the nucleic acid molecule is contained in the head. Another form of a phage is a tailless head. The least common form is a filament in which the protein molecules form a tubular structure in which the nucleic acid is embedded. Phages are known that contain double-stranded DNA (most common), single-stranded DNA, single-stranded RNA, and double-stranded RNA (least common). Phages containing double-stranded DNA are typically 50 percent DNA by weight, so they are a useful source of DNA for physical studies.

An Outline of the Life Cycle of a Phage

Phage are parasites and cannot multiply except within a host bacterium. Thus, a phage must be able to enter a bacterium, multiply, and then escape. There are many ways by which this can be accomplished. However, a basic life cycle is outlined below and depicted in Figure 1-6.

   The life cycle of a phage begins when a phage particle adsorbs to the surface of a susceptible bacterium. The phage nucleic acid then leaves the phage particle through the phage tail (if the phage has a tail) and enters the bacterium through the bacterial cell wall. In a complicated but understandable way the phage converts the bacterium to a phage-synthesizing factory. Within about an hour, the time varying with the phage species, the infected bacterium bursts (**lyses**) and several hundred progeny phage are released. The suspension of newly synthesized phage is called a **phage lysate.**

   Phage multiply much faster than bacteria. A typical bacterium doubles in about half an hour, while a single phage particle gives rise to more than 100 progeny in the same time period. Each of these phage can then infect

Figure 1-6   A schematic diagram of the life cycle of a typical bacteriophage. The number of phage released usually ranges from 20 to 500.

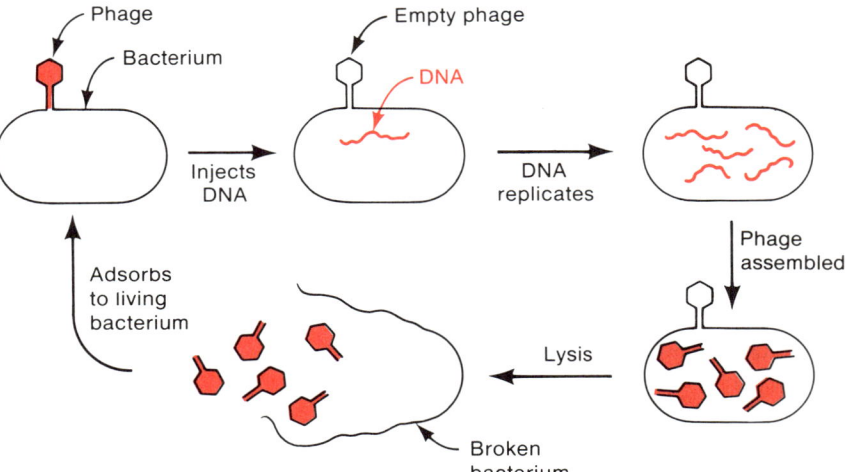

more bacteria and those released in this second cycle of infection can infect even more. Thus, in two hours there are four cycles of infection for both a bacterium and a phage, yet a single bacterium becomes $2^4 = 16$ bacteria and a single phage becomes $100^4 = 10^8$ phage particles.

Counting Phage

Phage are easily counted by a technique known as the **plaque assay**; this technique is performed in the following way.

In a previous section it was explained that 100 bacteria on agar yield 100 colonies. Likewise, if 10^8 bacteria are plated on agar, the 10^8 colonies that result can be made to appear as a confluent, turbid layer of bacteria called a **lawn.** During plating, to achieve maximal uniformity of the turbidity of the lawn that will form, the bacteria are mixed into a small volume of warm liquid agar which is poured onto the surface of the solid medium. The liquid, known as **top agar** or **soft agar,** rapidly hardens, providing a very smooth surface, and the bacteria grow in this thin agar layer with much uniformity (Figure 1-7). If a phage is present in the hardened top agar, it can adsorb to a bacterium in the agar; shortly afterward, the infected bacterium lyses and releases about 100 phage, each of which will adsorb to nearby bacteria; these bacteria in turn will release a burst of phage which then can infect other bacteria in the vicinity. These multiple cycles of infection continue and after several hours, the phage will have destroyed all of the bacteria at a single localized area in the agar, giving rise to a clear, transparent circular region

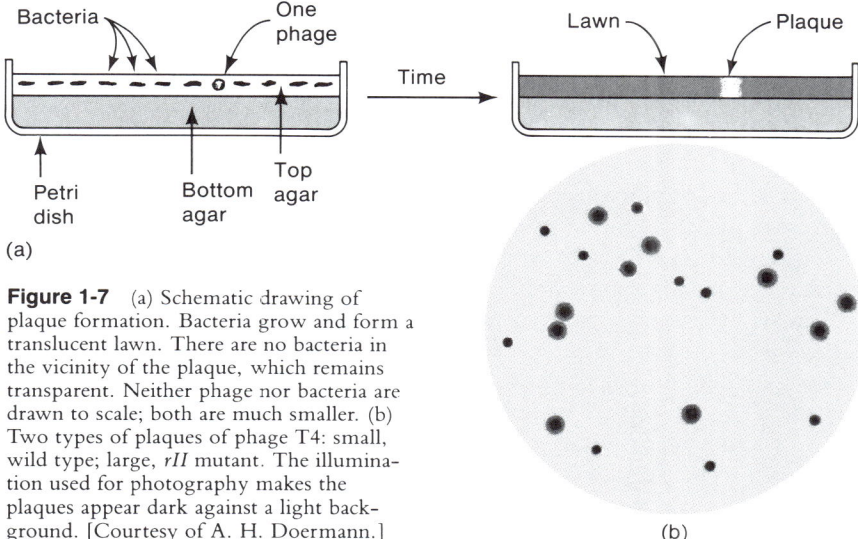

Figure 1-7 (a) Schematic drawing of plaque formation. Bacteria grow and form a translucent lawn. There are no bacteria in the vicinity of the plaque, which remains transparent. Neither phage nor bacteria are drawn to scale; both are much smaller. (b) Two types of plaques of phage T4: small, wild type; large, *rII* mutant. The illumination used for photography makes the plaques appear dark against a light background. [Courtesy of A. H. Doermann.]

in the turbid, confluent layer. This region is called a **plaque.** Since one phage forms one plaque, the individual phage particles put on the plate can be counted.

Regulation of the Phage Life Cycle

The life cycle of a phage is highly regulated, though in a slightly different way from the metabolic regulation of a bacterium. A phage is totally dependent on the metabolism of its host bacterium; thus, usually the regulatory systems of the host controls the basic metabolic processes such as energy generation and synthesis of the precursors of DNA, RNA, and proteins. The job of an infecting phage is to reproduce itself by synthesizing its own nucleic acid and structural proteins, and finally to cause the bacterial cell wall to break in order that progeny phage can escape. This requires a strict temporal regulation. For instance, if the system that causes lysis were to act immediately after infection, the infected bacterium would be destroyed before any new phage had been produced; clearly, the lytic system must act last. A nonproductive infection would also occur if the phage coat were synthesized so early that the incoming nucleic acid molecule was enveloped before phage reproduction could occur. Such failures are avoided by the existence of a phage-specified, finely tuned, temporal regulatory system. The basic program is the same for all phage species, though the particular regulatory mechanisms differ from one species to the next. This temporal regulation will be examined in detail in Chapter 17; in that chapter we shall see how several different phage species manage to determine the amount of each phage protein made at different times after infection.

YEASTS

Yeasts are unicellular organisms that have been used for millenia for producing wine and beer (Figure 1-8). A great deal of early biochemical research was carried out with yeasts rather than bacteria, work stimulated mainly by interest in understanding and improving fermentation. Researchers in beer-producing countries, such as Denmark, have also studied yeast genetics intensively in an effort to produce organisms that can make a better beer.

Yeast cells are propagated in the laboratory and counted in much the same way as bacteria. They grow in liquid suspensions in either chemically defined media (similar to those used for growing bacteria) or in complex broths. They also grow on a solid surface to form colonies. Most yeasts grow at about half the rate of *E. coli* in similar growth media. The multiplication mechanism of all but the fission yeasts differs from the simple splitting of a mature bacterium in that yeast cells do not divide but multiply by budding (Figure 1-8). In contrast with bacteria, which divide to yield

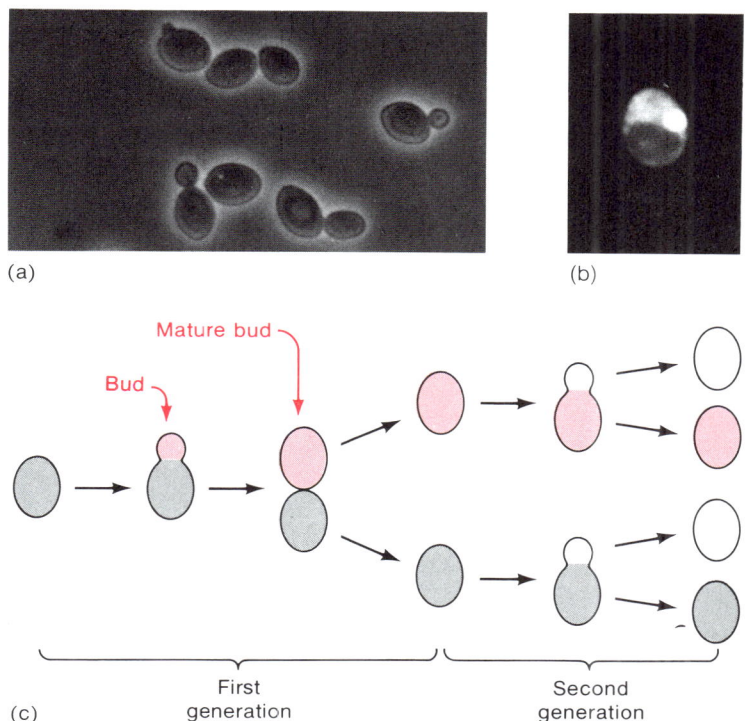

Mature bud

Bud

(a)

(b)

First
generation

Second
generation

(c)

Figure 1-8   (a) A light micrograph of the yeast *Saccharomyces cerevisiae*. Many cells are budding. [Courtesy of Breck Byers.] (b) A fluorescence micrograph of a single cell treated with the fluorescent dye, acridine orange, which binds to DNA (giving green fluorescence) and to RNA (giving orange fluorescence), viewed by ultraviolet light, which excites the fluorescence. The bright spot is the nucleus. The dark region is the vacuole, a liquid-filled sac free of nucleic acids. (c) Budding, showing the mother-daughter relation. The first generation daughter, which is also the second generation mother, is shown in pink.

twin progeny cells, with yeast there is a clear mother–daughter relation in the sense that the mother retains a scar on the cell wall at the site of budding.

Whereas the molecular biology of prokaryotes has undergone enormous advances in the past 25 years, at the present time much of the research activity is directed toward studying eukaryotes, which have a more complex organization. There are many advantages to beginning the study of eukaryotes with a unicellular organism that can be handled as easily as bacteria. Thus, the yeast *Saccharomyces cerevisiae,* which is a eukaryote, has become an important experimental organism in recent years. An initial step in its study has been to see if biochemical mechanisms worked out for *E. coli* and which apply to many other bacteria are also valid for a simple eukaryote. The results have been quite interesting. The basic metabolism of yeast and bacteria is the same. However, the overall process for synthesizing macromolecules (DNA, RNA, protein) are not those of bacteria, but instead those of higher eukaryotes. Thus, yeast has become an outstanding organism for studying eukaryotic macromolecular synthesis, gene organization, and regulation of gene expression.

An additional advantage of yeast as an experimental object is that it has both haploid and diploid (two copies of each chromosome) phases. There are two haploid mating types, *a* and α, which can mate to form a stable *a*α diploid. Under certain conditions a diploid cell can undergo meiosis to form four haploid cells—called **spores**—encased in a sac called an **ascus.** The existence of the mating types means that elegant genetic experiments can be done that complement physicochemical studies and that both the mechanisms of meiosis and the chromosomal interactions responsible for genetic recombination can be studied.

Other unicellular eukaryotes, namely the alga *Chlamydomonas* and the protozoan *Tetrahymena,* also are being used more frequently in eukaryotic molecular biology.

ANIMAL CELLS

In the last decade animal cells have been widely studied. The results have been exciting because we are beginning to understand such complex processes as hormonal regulation and the development of an egg into an adult organism, and to gain some insight into the differences between normal cells and cancer cells. However, research with animal cells has proceeded much more slowly than work with bacteria. There are two reasons for this. First, cells isolated from animals are invariably mixtures of cell types, so experimental results may be composites of effects on different cells. Second, bacteria growing in culture are not significantly different from bacteria in nature but experiments with animal cells require that the cells be removed from the animal and often separated from one another. Cells treated in this way have lost the normal route for receiving nutrients and are definitely in an unnatural state. Many growth media that enable cells to grow in culture have been developed. They have been designed to keep the cells alive as long as possible but, as shown in the next section, they do not always maintain a normal state for the cell.

The Unnatural State of Animal Cells in Culture

In contrast with the behavior of microorganisms, the behavior of most animal cells in culture is far different from that in the animal. There are two main kinds of difference—in culture, animal cells grow as individuals and the cells grow continually.

In nature, most animal cells grow in organized tissue units—that is, they are in contact with one another. Except for special circumstances, in culture animal cells require a substitute surface on which to grow (Figure 1-9). The usual surface that is provided in the laboratory is the bottom of a plastic petri dish that is filled with a liquid growth medium. No agar is used—cells placed in the dish slowly adsorb to the plastic surface. The cells grow and divide until they come in contact with one another and form a confluent monolayer of cells on the bottom of the dish. At this stage growth and cell division stops, as in organized tissue. Continued growth and division

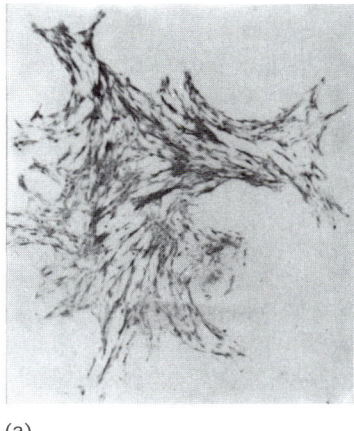

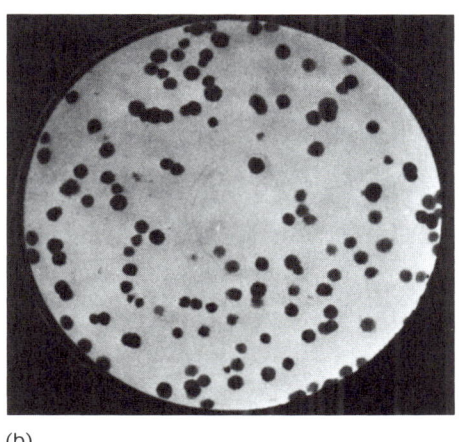

(a)   (b)

Figure 1-9 (a) A microcolony of Chinese hamster fibroblasts which have been growing for a few days on a glass surface. The form of the cluster is determined by the spindle shape of each cell. (b) A petri dish on which are numerous colonies of human HeLa cells. After growth of the cells for about two weeks the plate was filled with a dye solution to stain the individual clones to make them visible. Each clone consists of the progeny of a single cell. [Courtesy of Theodore Puck.]

are obtained only by removing the confluent layer, separating the cells, and placing a small number of dispersed cells in another plastic dish. This handling is certainly an unnatural treatment and, in the course of many generations of growth, results in the accumulation of abnormal (mutant) cells that can tolerate these manipulations and the physiologically alien environment.

A related problem is that indefinite cell division of animal cells is itself exceptional. In an animal, cells grow and divide only until adulthood is reached, a state that is attained in no more than 50 or 60 cell generations in a human. In adulthood, cell division is not a general phenomenon, except in specialized cells such as those in the skin, intestine, mucous membranes, and bone marrow, and those that divide during the healing of wounds. Inasmuch as cells obtained from many tissues other than an embryo have previously received signals that limit growth or terminate growth, their propagation in a culture medium is a novel cellular circumstance.

The result of forcing cells—which in their natural state grow only in contact with other cells and then only for a limited number of generations—to divide for an extended time in cell cultures, is that they become quite different from the cells originally taken from the animal tissue, and are often tumorlike. The number of chromosomes in progeny cells is almost always increased beyond the normal diploid number and is usually not even the same in all cells of one culture. The longer these cells are grown, the more heterogeneous a culture becomes.

Primary Cell Cultures and Established Cell Lines

Two kinds of animal cell cultures are in common use—the primary culture and the established cell line. The distinction between these two types can be understood by examining how a cell culture is established and the characteristics of the culture in further growth. A culture is established by treating fresh tissue with a **proteolytic** (protein-digesting) enzyme, which disperses the cells, and placing a number of individual cells (and sometimes some of the undispersed tissue) in a plastic petri dish containing growth medium. The cells adsorb to the plastic surface, begin to grow and divide after a few days, continue growth for one week to three months (the time depends on the cell type), and then gradually begin to disintegrate and die. During the final few weeks the culture consists mostly of a sick and dying population. A culture showing these characteristics is called a **primary culture** because each cell presumably has the same characteristics of the cells in the original tissue and is following the program of limited cell division set up when the tissue formed.

If a primary culture that has lost its ability to reproduce is maintained for many months by removing some cells and transferring them to a fresh petri dish each time confluent growth is achieved, frequently cells arise that in an unknown way have gained the ability to grow and divide indefinitely (like a bacterium). A population derived from such a cell is called an **established cell line.** Studying primary cultures has a particular advantage even though most of the cells eventually die—namely, that the cells obtained from a particular tissue retain, for at least a few generations, the properties of "normal" cells. There is, of course, a disadvantage too—namely, that the primary population is short-lived; hence, in any experiment that takes more than a few weeks, one does not know whether the results obtained apply to living or dying cells. An established cell line has the advantage of continued growth; however, the cells also are quite variable and often have the capacity to form tumors. Nevertheless, a great many experiments are done with established cell lines because of their immortality. Those cell lines that are used frequently in molecular biological experiments are HeLa (a human tumor cell), 3T3 (a mouse embryo cell), CHO (an apparently normal cell from Chinese hamster ovary), L (a mouse tumor), BHK (Syrian hamster kidney), Balb-c (a mouse tumor), and African green monkey cells.

Certain animal cell types are easy to maintain in culture for short periods of time. For instance, **reticulocytes,** which are hemoglobin-forming cells, are easily obtained from bone marrow as a nearly pure paste of cells, which are easily dispersed. These cells are programmed to make hemoglobin almost exclusively, and have been widely studied as a means of understanding protein synthesis in eukaryotes. Some cell types can also be easily studied while still in the parent tissue. For example, the **oviduct** of the chicken consists almost exclusively of cells whose function is to make egg white proteins. Synthesis of these proteins occurs in response to various sex hormones. Macerated tissue freshly obtained from a chicken oviduct and suspended in standard growth media is one of the most important systems for studying how cells respond to specific hormones.

In one type of animal cell many of the problems just mentioned are not present—namely, in the developing eggs of amphibians and echinoderms, which normally develop in water. In particular, the eggs of the frog, toad, and sea urchin have been especially useful in studies of the molecular biology of differentiation. Eggs have the additional advantage that they are resting cells that can be switched on merely by adding sperm. The egg precursor (the **oocyte**) can also be obtained in large quantities from amphibian ovaries; the oocyte of the South African clawed toad, *Xenopus laevis,* is a favorite object for studying cell differentiation, and will be examined in Chapter 16.

In certain conditions animal cells form colonies on the surface of a plastic petri dish and this method is used occasionally to count the cells (Figure 1-9(b)). However, it is more common to remove the cells from the surface by treatment with trypsin and then either perform a count of a known microvolume with a microscope or use an electronic cell counter. The latter two procedures are always used when cells are grown in liquid suspension.

ANIMAL VIRUSES

Animal viruses, which superficially resemble bacteriophages, consist of a nucleic acid molecule encased in a protein shell; they adsorb to host cells, reproduce within the cells, and often are released in huge numbers by a variety of means. They are usually grown by infecting cultured cells, though occasionally a living animal is used. (Plant viruses are at this time propagated only in living plants.) Counting viruses is quite tedious and inaccurate.

Viruses are very useful tools for studying basic processes in eukaryotes, such as DNA replication and RNA synthesis, for they give one the opportunity to examine eukaryotic systems acting in a way that is controlled by the experimenter—namely, the systems are turned on when the viruses are added. Viruses, whose external shells consist of only one or two types of protein molecules, have also been especially valuable for studying the structure and assembly of macromolecular aggregates. Viruses are, of course, interesting objects of study in their own right.

Most viruses are able to infect only a single cell type (as is the case in nature) and sometimes will grow only in a primary culture. Of special interest are those viruses that reproduce in one cell type yet in other cell types induce a transformation of the host cell to a cancer cell. A well-studied example is polyoma virus, which produces progeny viruses in embryonic mouse cells yet converts hamster cells to tumor cells. We will have more to say about animal viruses and the viral transformation process in Chapter 24.

Animal cells are very fastidious in their growth requirements. They are unable to synthesize many of the amino acids, vitamins, and fatty acids from simple inorganic compounds and sugars. Some amino acids, the so-called essential amino acids, cannot even be obtained by chemical transformation of other amino acids and must be added to the growth medium. (Humans, for example, must obtain at least ten amino acids directly from the diet.) Thus, growth media for animal cells are very complex, usually containing all of the amino acids and vitamins, and several fatty acids. In addition to salts and trace elements, animal cells in culture need fetal blood serum, which is a source of various growth hormones and other complex organic compounds required by individual cell types.

GENERA, SPECIES, AND STRAINS

In the Linnean scheme of classification, genus and species names are assigned to all organisms. For example, the bacterium *Bacillus subtilis* is a member of the genus *Bacillus* and is of the species *subtilis*. Two organisms in the same genus are usually related evolutionarily and exhibit many similar characteristics. Thus, the bacterium *Bacillus brevis* has many characteristics in common with *Bacillus subtilis*. The binary combination of a generic name and a specific epithet describes an organism but sometimes it is necessary to make further divisions. If differences are substantial, the term **variety** is used but if they are small, the term **strain** is used. In molecular biology the term strain is encountered repeatedly. For example, hundreds of times in various parts of the world *E. coli* has been isolated from nature—it grows in the intestines of large animals and has been sampled from individuals and from fecal matter in sewage. By the standard taxonomic criteria these isolates are classified as *E. coli,* yet they may show subtle differences in colony morphology, susceptibility to various phages, motility, and so forth. These differences define strains. Strains are designated by adding an unitalicized capital letter or number after the species name. That is, strains A, B, etc., of *E. coli* are written *E. coli* A, *E. coli* B, etc. Often variation is even found within a strain; a number is then added after the strain letter; for example, K1, K2, etc. The three strains of *E. coli* used most commonly in molecular biology are *E. coli* B (the host for the so-called T phages), *E. coli* C (the host for the phage ϕX174), and *E. coli* K12 (the most widely used strain).

Variants of animal and plant viruses distinguished by immunological criteria are extremely common. The viruses that cause influenza in humans are usually designated by stating a basic immunological type (A, B, C, . . .) and the geographical location where they were first isolated, as with the influenza virus known as type B : Hong Kong that was epidemic in the late 1960s.

METABOLISM AND BIOCHEMICAL PATHWAYS

Throughout this book we will be concerned with cell growth and reproduction. We will be asking how a cell converts simple chemicals in the environment to intracellular small molecules, macromolecules (protein, DNA), and complex structures (membranes, organelles, cell walls), and how genes and complete cells are duplicated. Later we will also ask how these processes are regulated—how certain substances are utilized or synthesized only when needed and how a cell directs a particular molecule down the appropriate pathway. These are chemical questions, but unless great chemical and physicochemical detail is required, only a little chemistry is needed to provide satisfactory answers. In this book it is assumed that the reader has a background in undergraduate chemistry, but not biochemistry. In this section we present those features of cell growth and bioenergetics that are necessary to proceed.

Cell growth requires synthesis of many thousands of compounds, and hence a large number of chemical reactions. These reactions have two common features: (1) the reactions invariably occur in several steps and (2) they are almost always catalyzed by specialized proteins called enzymes. The first point is examined in the following subsections; enzymes will be discussed in detail in Chapter 6.

Degradative and Biosynthetic Reactions

In thinking about biochemical reactions it is worthwhile to separate them into two categories. One of these is the **degradative** or **catabolic reaction** in which molecules are broken down to smaller molecules. An example is the breakdown of glucose. Catabolic reactions serve two functions: (1) they transfer the bond energy released during degradation to a form of energy that can be utilized in other reactions (usually as phosphorus-containing energy-storage compounds), and (2) they provide the small units that are used to build up other essential molecules.

The second class of reactions is the **biosynthetic** or **anabolic reaction,** in which small molecules are built up into compounds that are successively altered to form all essential intracellular constituents. For example, the carbon atoms of glucose are used to form every organic compound if glucose is the sole source of carbon in a growth medium. All of the degradative and biosynthetic reactions of a cell are collectively called the metabolism of the cell.

Degradative and biosynthetic reactions usually consist of a series of reactions. A group of connected reactions is called a **metabolic pathway.** For biosynthesis these pathways are necessary because it is rarely possible for an organism to create a needed compound from a source compound in a single chemical step. In the case of degradation of nutrients, a multistep pathway allows the cell to transfer the bond energy of the initial compound to a large number of energy-storage molecules from which energy can be obtained at a later time when it is needed.

In a series of reactions that constitute a metabolic pathway, each step requires a particular enzyme as a catalyst.

Many metabolic pathways are linear; that is, the series of steps may be written

$$A \xrightarrow{E_1} B \xrightarrow{E_2} C \xrightarrow{E_3} D$$

in which A is a starting material, D is the **product,** B and C are pathway **intermediates** and E_1, E_2, and E_3 are enzymes that catalyze the successive steps. Substances A, B, and C are called **precursors** of D, because providing any one of them to a cell will result in their conversion to D.

Biosynthetic pathways can also branch. This means that an intermediate can be converted into two or more different compounds. An example of a branched pathway is

$$A \longrightarrow B \longrightarrow C \begin{cases} \nearrow D \longrightarrow E \\ \searrow F \longrightarrow G \end{cases}$$

The direction of all chemical reactions can be described in terms of a thermodynamic quantity called the **free energy change,** denoted ΔG. A reaction that proceeds spontaneously in a particular direction is said to have a negative value of ΔG for the reaction as written (from left to right)—that is, the reaction loses free energy as it progresses. A reaction that does not proceed spontaneously but requires an input of energy has a positive value of ΔG. Degradative and biosynthetic reaction sequences can be distinguished by their energetics—that is, by the sum of individual ΔG values. In a degradative pathway, a clear decrease in free energy is measurable ($\Delta G < 0$), so degradation is a spontaneous process (one whose rate is merely accelerated by enzymes). In a biosynthetic process, small molecules are successively built up into larger molecules. If the reaction is considered simply as an assembly of components, it almost always will have a positive value of ΔG and thus cannot occur unless coupled with a "driving" system that loses at least the same amount of free energy that the biosynthetic process requires.

One of the most important features of degradative pathways is that, associated with the degradation, often a series of reactions culminates in the synthesis of compounds containing phosphate esters or pyrophosphate bonds. These important compounds can yield a large negative free energy through hydrolysis; this energy can be utilized by other reactions for which $\Delta G > 0$. This has the effect that *many biosynthetic reactions, which do not occur alone, can be driven in the direction of synthesis when the phosphate bonds are hydrolyzed.* The most important substance of this type is adenosine triphosphate, which will be discussed in the next section.

Energy Metabolism and Storage: ATP

All living cells require a continual supply of free energy for two main purposes: for biosynthesis of both small molecules and macromolecules, and for the active transport of ions and molecules across the cell membrane. Other than chlorophyll-containing cells, which derive energy from light, all cells obtain this free energy ultimately from the oxidation of exogenous

Figure 1-10 Adenosine triphosphate. The large ring at the lower right is the sugar ribose. The adenine-ribose unit is called adenosine.

nutrients. The energy of such an oxidation is not used directly but, by means of a complex and long sequence of reactions, is channeled into a special energy-storing molecule, **adenosine triphosphate (ATP),** whose structure is shown in Figure 1-10. This molecule contains one phosphoester bond (adjacent to the sugar ring) and two pyrophosphate linkages and is able to release energy when needed by hydrolysis of these pyrophosphate bonds. ATP is often said to contain high-energy-phosphate bonds. This is actually a misnomer because the bond energy (i.e., the energy required to break the bond) is not significantly higher than that of chemical bonds in other molecules; rather, what constitutes high energy is the amount of energy released by a reaction with water in which the bond is hydrolyzed and a water molecule is split. Thus, *ATP has a high free energy of hydrolysis.*

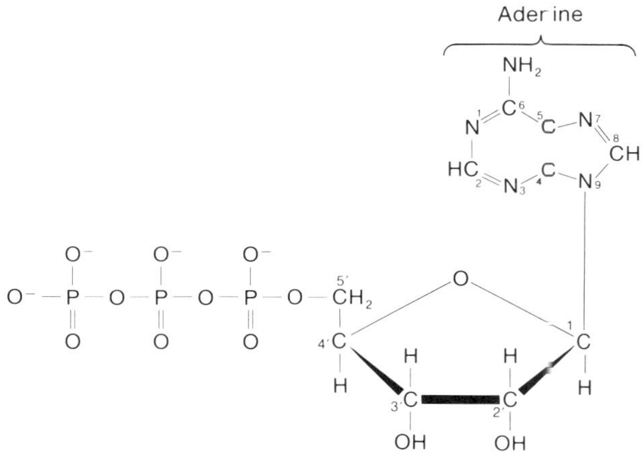

The principal feature of the structure of ATP that gives rise to the large negative free energy of hydrolysis is the physical arrangement of the four negatively charged oxygen atoms, as shown in the figure. These charges repel one another very strongly because they are very close. This electrostatic repulsion is reduced when ATP is hydrolyzed; therefore, hydrolysis releases rather than consumes free energy. This released energy is utilized by means of phosphate-transfer reactions, as will be seen shortly.

ATP can be hydrolyzed in two ways: removal of the terminal phosphate to form adenosine diphosphate (ADP) and inorganic phosphate (P_i) or removal of a terminal diphosphate to yield adenosine monophosphate (AMP) and pyrophosphate (PP_i). The PP_i is usually cleaved further to yield two phosphates. The reactions are

$$\text{I. } ATP + H_2O \rightarrow ADP + P_i + H^+$$
$$\text{II. } ATP + H_2O \rightarrow AMP + PP_i + H^+$$
$$PP_i \rightarrow 2P_i$$

Each of these reactions is enzymatically catalyzed.

Synthesis of ATP

ATP is synthesized primarily from ADP. A cell expends a great deal of work in synthesizing ATP, in order that an adequate supply of free energy will always be available. For example, about 25 distinct enzymatic reactions are coupled to degrade glucose to CO_2 and H_2O, and to utilize as much of the bond energy of glucose as possible. The reaction sequence is a clear example showing why cells have evolved with multistep pathways. That is, the free energy change for complete oxidation of glucose is 688 kilocalories (kcal) per mole (2878 kilojoule/mol), while the free energy required for synthesis of ATP is about 7 kcal/mol (29.3 kJ/mol). Hence, one glucose molecule can provide enough free energy to make 100 ATP molecules. However, it is not possible to design a single chemical reaction in which one molecule of glucose can simultaneously react with oxygen, 100 P_i, and 100 ADP molecules, water, and CO_2. On the other hand, by carrying out the oxidative degradation stepwise, single ATP molecules can be generated in particular steps; the result is the formation of 38 ATP molecules per glucose molecule, or 38 percent efficiency of energy conversion, which thermodynamically is very efficient.

Most ATP molecules are not actually direct products of the glucose degradation pathway. Glucose is degraded in two stages. The first stage, **glycolysis,** is a multistep sequence in which two ATP molecules are generated directly in separate steps (Figure 1-11). The remaining product of glycolysis, pyruvate, is oxidized (although not by molecular O_2) in a pathway known as the **Krebs cycle,** the second stage of glucose degradation (Figure 1-12). At no step in the Krebs cycle is an ATP molecule formed directly. Instead, two molecules of a related compound, guanosine triphosphate (GTP), are formed: each GTP can then generate ATP by the reaction GTP + ADP $\rightleftarrows$ GDP + ATP. The Krebs cycle generates two complex molecules NADH (reduced nicotinamide adenine dinucleotide) and $FADH_2$ (reduced flavin adenine dinucleotide). These molecules contain the unused free energy of glucose remaining after the products of glycolysis have passed through the Krebs cycle. In order that their unused free energy may be extracted, the NADH and $FADH_2$ molecules are oxidized in a complex and incompletely understood series of reactions, collectively called **electron transport** or **oxidative phosphorylation,** yielding 34 additional ATP molecules. Thus, the complete conversion of glucose to CO_2 and H_2O, which is summarized in Figure 1-13, yields 38 ATP molecules.

The student need not memorize the reaction sequences in Figures 1-11 through 1-13; they have been presented mainly to illustrate the stepwise nature of metabolic pathways.

Utilization of the Energy of ATP

The mechanism by which ATP makes its energy available to a cell is the **phosphate-transfer reaction,** which is a special case of a general process

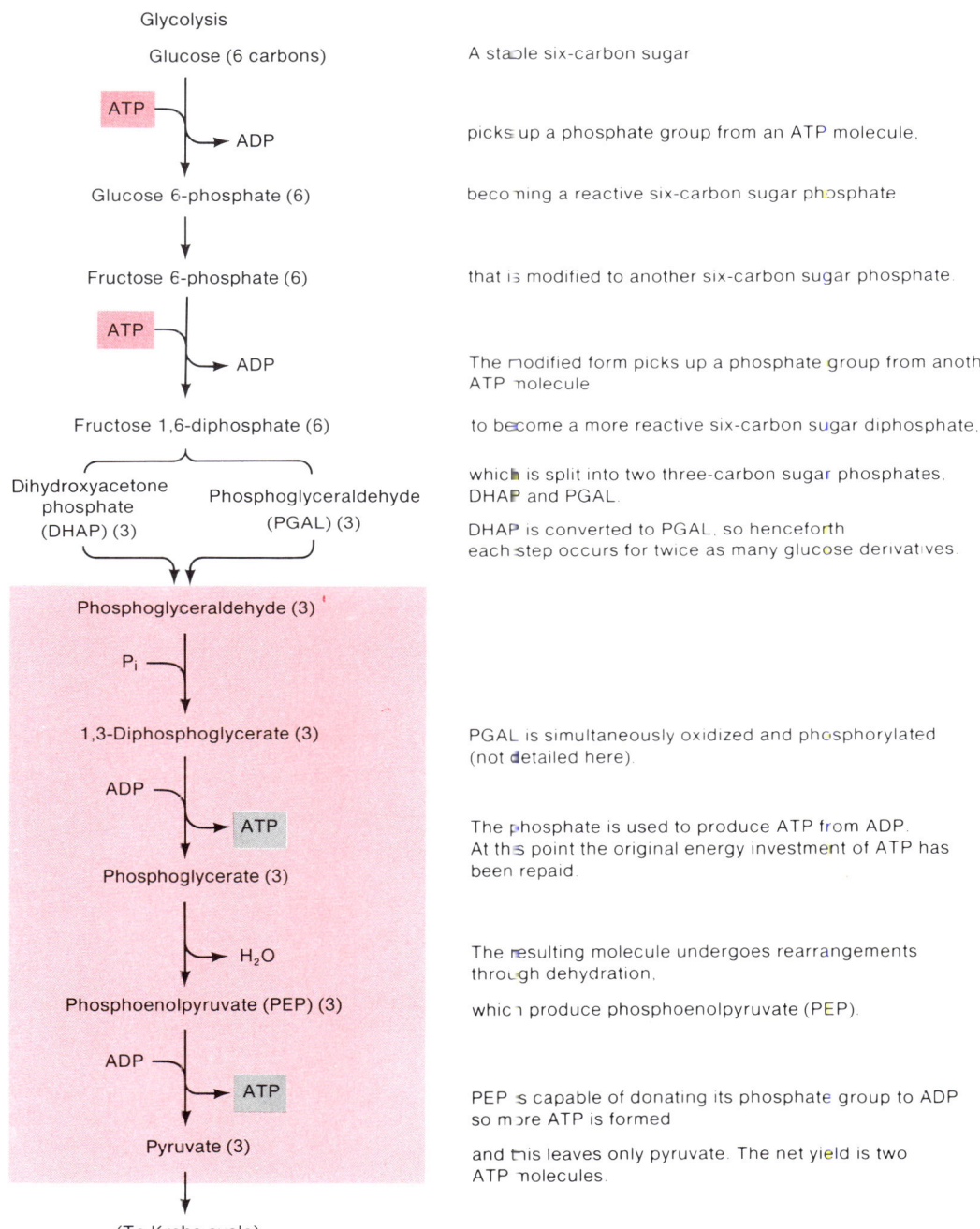

Glycolysis

Glucose (6 carbons)   A stable six-carbon sugar

ATP → ADP   picks up a phosphate group from an ATP molecule,

Glucose 6-phosphate (6)   becoming a reactive six-carbon sugar phosphate

Fructose 6-phosphate (6)   that is modified to another six-carbon sugar phosphate.

ATP → ADP   The modified form picks up a phosphate group from another ATP molecule

Fructose 1,6-diphosphate (6)   to become a more reactive six-carbon sugar diphosphate,

Dihydroxyacetone phosphate (DHAP) (3)   Phosphoglyceraldehyde (PGAL) (3)   which is split into two three-carbon sugar phosphates, DHAP and PGAL.

DHAP is converted to PGAL, so henceforth each step occurs for twice as many glucose derivatives.

Phosphoglyceraldehyde (3)

P_i

1,3-Diphosphoglycerate (3)   PGAL is simultaneously oxidized and phosphorylated (not detailed here).

ADP → ATP   The phosphate is used to produce ATP from ADP. At this point the original energy investment of ATP has been repaid.

Phosphoglycerate (3)

H_2O The resulting molecule undergoes rearrangements through dehydration,

Phosphoenolpyruvate (PEP) (3)   which produce phosphoenolpyruvate (PEP).

ADP → ATP   PEP is capable of donating its phosphate group to ADP so more ATP is formed

Pyruvate (3)   and this leaves only pyruvate. The net yield is two ATP molecules.

(To Krebs cycle)

Figure 1-11 An outline of the glycolysis pathway (left) and an explanation (right). Each step is catalyzed by a separate enzyme. All steps in the shaded area occur twice for each glucose molecule—that is, for each of the three-carbon molecules, DHAP and PGAL. Two ATP molecules shaded in red are used and those shaded in black are produced in the pathway. The number of carbon atoms in each molecule is shown in the parentheses. Not all cofactors for individual reactions are indicated. [After C. Starr and R. Taggart, *Biology* (Wadsworth, 1981).]

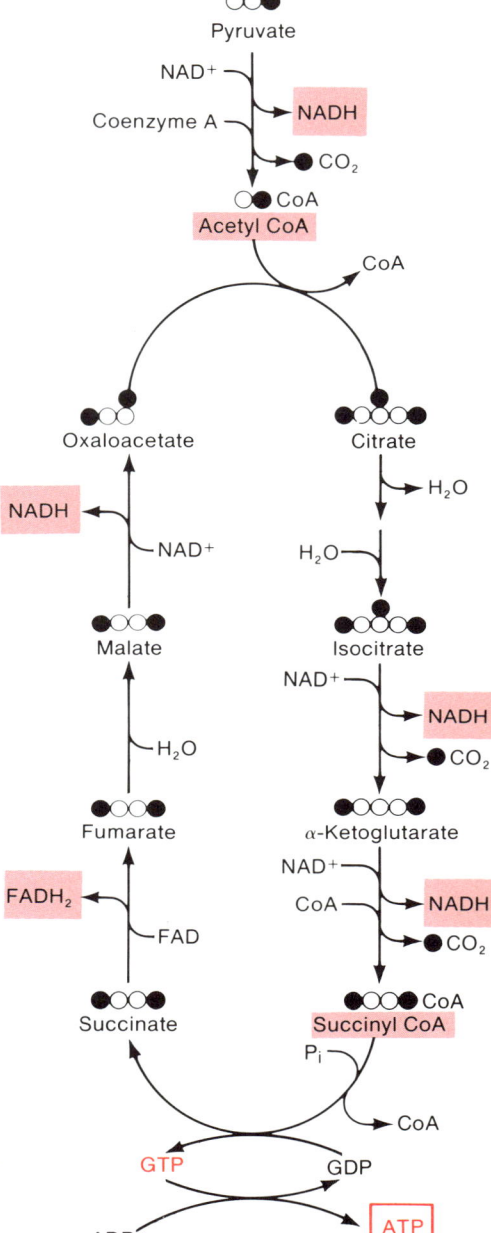

In order for pyruvate to enter the Krebs cycle, a carbon atom must be removed as CO_2; NAD^+ becomes converted to NADH; one of the two remaining carbon atoms is converted to an acid group.

The resulting two-carbon compound is then linked to a molecule of coenzyme A (CoA). Acetyl CoA is then coupled with a four-carbon compound having two acid groups (oxaloacetate). Reaction releases CoA for re-use and produces citrate, a six-carbon compound having three acid groups.

Another NADH molecule is produced by the transfer of two more hydrogen atoms. The product of this last reaction is the same four-carbon compound that the cycle starts with.

After two intermediate reactions, a second molecule of NADH is produced; at the same time an acid group is split off the six-carbon compound and is released as CO_2.

Two more hydrogen atoms are removed, but this time they are passed on to a flavin adenine dinucleotide (FAD), producing $FADH_2$. (There isn't enough energy now to make more NADH.)

The third and last of the original pyruvate carbon atoms is split off as another CO_2 molecule and a third molecule of NADH is produced.

CoA is split off and the released energy is used to form guanosine triphosphate (GTP), which transfers a phosphate group to adenosine diphosphate (ADP) to yield adenosine triphosphate (ATP).

In the process, a new acid group is linked to CoA. The rest of the cycle works to convert this new four-carbon molecule with two acid groups to the related compound oxaloacetate, with which the cycle begins again.

Figure 1-12   The Krebs cycle. Each open circle represents a carbon atom in the molecule and each solid circle represents an acid group (COOH) or a carbon dioxide molecule produced when an acid group is split off. Energy-rich molecules are solid red.

Molecules whose oxidation yields ATP are shaded red. The NADH molecules are ultimately oxidized, producing most of the utilizable energy of the cell. Remember that for each glucose molecule being metabolized, two pyruvate molecules have been formed.

Thus, two turns of the Krebs cycle must occur for each molecule of glucose oxidized. [After C. Starr and R. Taggart, *Biology* (Wadsworth, 1981).]

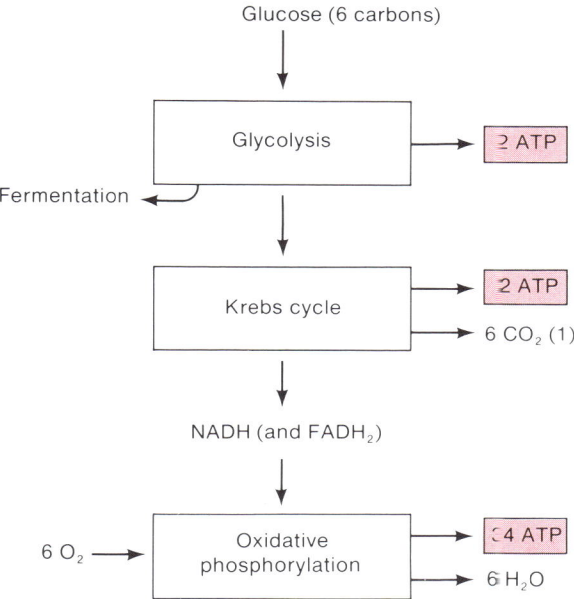

Glucose (6 carbons)

Glycolysis → 2 ATP

Fermentation

Krebs cycle → 2 ATP
→ 6 CO_2 (1)

NADH (and $FADH_2$)

6 O_2 → Oxidative phosphorylation → 34 ATP
→ 6 H_2O

Figure 1-13 Summary of glucose metabolism. The overall oxidative reaction yields 38 ATP molecules. In the absence of oxygen, fermentation occurs, yielding either lactic acid or ethanol, and no ATP molecules form other than the two produced by glycolysis. Note that in the oxidative sequence, 34 of the 38 ATP molecules are produced in the oxidative phosphorylation step; thus, glycolysis and the Krebs cycle may be thought of as a means of converting glucose to NADH and $FADH_2$, rather than a direct source of ATP.

known as the **group-transfer reaction.** A group-transfer reaction is one in which molecules exchange functional groups; for example,

$$AX + BY \rightleftarrows AY + BX$$

The rationale for using a group-transfer reaction can be seen in the following example.

Consider the hypothetical synthetic reaction $M + N \rightarrow Q$, in which M and N are reactants and Q is a product. Let us assume also that at the concentrations of M and N in the cell where the reaction proceeds, the value of ΔG for this reaction is $+3$ kcal/mol. Since $\Delta G > 0$, the reaction will not occur spontaneously. Let us now consider coupling the reaction to the hydrolysis of ATP, for which ΔG is about -13 kcal/mol at the ATP concentration that exists within a cell. If these two reactions could be joined together in some way, the combined value of ΔG would be $+3 - 13 = -10$ kcal/mol. Since this value is less than zero (i.e., there would be excess free energy), the reaction would occur spontaneously. Obviously, simply cleaving of ATP without channeling the free energy into the synthetic reaction is insufficient because the energy released would appear as the kinetic energy of the products (ADP, P_i, and H^+) and would merely be dissipated as heat. However, another reaction

$$M + ATP \rightarrow M\text{—}P + ADP$$

might be possible for which $\Delta G = -5$ kcal/mol. This reaction would happen spontaneously since $\Delta G < 0$. It may be the case then, that M—P can react with N to yield the product Q in the reaction

$$M\text{—}P + N \rightarrow Q + P_i$$

for which $\Delta G = -3$ kcal/mol. Since again $\Delta G < 0$, this reaction would also occur spontaneously. Note that it is the synthesis of M—P, which occurs by transfer of phosphate from ATP to M, that allows the reaction sequence, for which ΔG is $-5 - 3 = -8$ kcal/mol, to proceed. An intermediate substance of this sort (that is, M—P) is called an **activated intermediate.**

This type of mechanism, in which a phosphate or AMP is transferred to a reactant to form an activated intermediate, occurs repeatedly in biosynthetic reactions such as protein synthesis and in processes such as muscle contraction, in which mechanical work is done. We shall see several examples of this in this book. For example, in the first step of catabolism of galactose, an intermediate, galactose 1-phosphate, is synthesized by transfer of a phosphate from ATP. In protein synthesis, AMP rather than a single phosphate group is transferred to the carboxyl end of one amino acid prior to joining it to the amino group of an adjacent amino acid:

Glycine   "Activated" glycine

"Activated" glycine   Alanine   Glycylalanine

In the synthesis of DNA and of RNA, each of the four mononucleotide precursors is converted to a nucleoside triphosphate by two separate phosphate transfers from two ATP molecules. The nucleoside triphosphate, which is the activated intermediate, can then react with the growing end of the nucleic acid chain, as will be seen in Chapter 9.

2

Genetic Analysis in Molecular Biology

Genetics is an abstract and logical system by which the genetic factors determining certain properties of living cells can be located with respect to one another and by which changes in these properties can be expressed in quantitative terms. Genetics alone cannot prove anything about molecular mechanisms because the conclusions it leads to are independent of the molecular basis of biological phenomena. However, genetic analyses have been the major source of intuition about molecular mechanisms and have forced one to consider phenomena that might otherwise be ignored. Genetics-based arguments can be so suggestive of a particular mechanism, and the suggested mechanism has so often turned out to be the correct one, that molecular biologists often view alternative and conflicting ideas as not even worthy of consideration. The following sections are an introduction to genetic analysis of organisms as well as to the genetic phenomena most commonly used in such analyses.

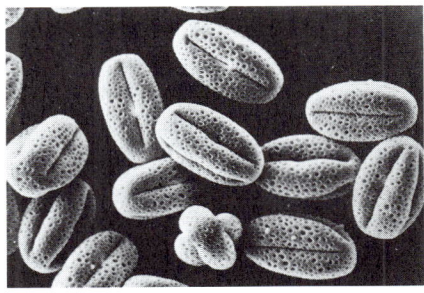

Scanning electron micrograph of the pollen grains of the lemon. The grains carry the genetic information of the male part of the flower. [From G. Shih and R. Kessel, *Living Images*. Jones and Bartlett (1982).]

GENETIC NOTATION, CONVENTIONS, AND TERMINOLOGY

In order to discuss genetics one must first distinguish the phenotype of a cell from its genotype and develop a consistent notation for describing the genetic properties of an organism. The notation we present is used for bacteria and many microorganisms. Unfortunately, a variety of notations are used for eukaryotic organisms that have been studied for a long time, such as maize and *Drosophila*. **Phenotype** refers to an observable property of an organism. Thus, a cell that can synthesize the amino acid leucine is denoted Leu$^+$; if it cannot do so, it is denoted Leu$^-$. Note that the symbol has three letters, is capitalized, and is not italicized. **Genotype** refers to the genetic composition of an organism. The Leu$^-$ cell certainly must have a defective gene that keeps it from synthesizing leucine; this would be denoted *leu$^-$* (or in some books, *leu,* without the minus sign). Note that the genotype is written in lowercase letters and in italics. Several genes may be required to synthesize leucine. These would usually be denoted *leuA, leuB, . . .* , with all letters italicized. A Leu$^+$ (leucine-synthesizing) cell must have one functional copy of every requisite gene; thus, its genotype must be *leuA$^+$ leuB$^+$, . . .* , which would normally be summarized by writing *leu$^+$* unless it is important for some reason to state the genotype of each gene. A Leu$^+$ cell might also be diploid and have a defective gene (e.g., *leuA*) in one chromosomal set. The two haploid sets are separated in their notation by a diagonal line; the genotype would therefore be *leu$^+$/leuA$^-$*.

Some genes are responsible for resistance and sensitivity to certain extracellular products such as antibiotics. The genotypes of a penicillin-resistant cell and a penicillin-sensitive cell are written *pen-r* and *pen-s,* respectively; the phenotypes are correspondingly Pen-r and Pen-s.

In summary, for bacteria the following conventions are used:

1. Abbreviations of phenotypes contain three roman letters (the first capitalized) with a superscript + or − to denote presence or absence of the designated character and "s" and "r" for sensitivity and resistance, respectively.
2. The genotype designation is always lowercase and all components of the symbol are italicized.

This convention has not been adopted for bacteriophages, for which one- and two-letter symbols are widely used, nor for yeast, in which genotypes are usually written with all italicized capital letters.

We shall also have occasion to designate particular mutants of a gene. Mutants are usually numbered in the order in which they have been isolated. Thus the leucine mutants numbered 58 and 79 would be written *leu58* and *leu79*. If one wanted to denote mutants of a particular gene, one would write *leuA58* and *leuB79,* if the mutations were in the *leuA* and *leuB* genes, respectively.

Finally, it is often necessary to denote the protein products of a particular gene. No notation has ever become standard, but in this book the following conventions are used. A gene product, productase, for which the genetic symbol is *prd,* will be written out as productase, or it will be abbreviated and termed the *prd* product or the Prd protein, or it will simply be designated Prd. All three designations are convenient and will be used in keeping with the three common modes seen in scientific literature.

The term **genome** is useful, though it has evolved to have various meanings. It is correctly defined as the genetic complement (the set of all genes and genetic signals) of a cell or virus. With eukaryotes the term is often used to refer to one complete (haploid) set of chromosomes. In laboratory jargon, when discussing bacteria, phages, or most animal and plant viruses, the term has come in use to refer to their single DNA molecule, a classically incorrect but now accepted usage.

Mutants

We have seen that a gene can be either functional or nonfunctional and that these states are denoted by a superscript + or − after the gene abbreviation. The functional form of a gene is sometimes called **wild type** because presumably this is the form found in nature. This term is ambiguous though, because often the − form is the one that is prevalent; for example, many bacterial species isolated from nature carry the genes for metabolism of lactose, yet in many strains the gene is defective—that is, lac^-. In this book we shall minimize the use of the term. The precise genetic term **allele** is used to indicate that there are alternative forms of a gene; sometimes the + and − forms are called the + allele and the − allele, respectively. It is very common to call the defective − form of a gene a mutant form. Strictly speaking, a mutant is an organism whose genotype (or, more precisely, its DNA base sequence) differs from that found in nature. However, it is more convenient (and definitely it is common jargon) to equate the terms mutant gene and defective gene, and we shall use the word mutant in that sense in this book.

Mutants can be classified in several ways. One classification is based on the conditions in which the mutant character is expressed. An **absolute defective mutant** displays the mutant phenotype under all conditions; that is, if a bacterium requires leucine for growth in all culture media and at all temperatures, it is an absolute defective. A **conditional mutant** does not always show the mutant phenotype; its behavior depends on physical conditions and sometimes on the presence of other mutations. An important example of a conditional mutant is a **temperature-sensitive (Ts) mutant,** which behaves normally below 34°C and as a mutant above 39°C (or nearby temperatures); an intermediate state is usually observed between these temperatures. Note that the gene does not mutate above 39°C; rather, the product of the gene is inactive above 39°C. Temperature-sensitive mutants have been of great use in the laboratory because they enable one to turn off the activity of a gene product simply by raising the temperature. In many cases the temperature-sensitive defect is reversible, so the activity of the gene product can be regained by lowering the temperature.

The most widely encountered conditional mutants are the **suppressor-sensitive mutants;** these exhibit the mutant phenotype in some strains but not in others. The difference is a consequence of the presence of particular gene products, called **suppressors.** These suppressor gene products either compensate for the defect in the mutant or, in a variety of ways, enable an altered gene to produce a functional gene product. In the jargon of molecular biology, one says that the phenotype of a suppressor-sensitive mutant "depends on the genetic background." A mutation is sometimes designated as a suppressor-sensitive one by adding the roman letters Am (refers to amber) or Oc (for ochre) in parentheses. (Amber and ochre are terms that arose from a private laboratory joke.) The symbols are roman and capitalized because they represent a phenotype; if the mutant has a number, this is also added. Thus, if mutation 35 in the *leu* gene is a suppressor-sensitive Am type, it would be written *leu35*(Am).

The process of formation of a mutant organism is called **mutagenesis.** In nature and in the laboratory, mutants sometimes arise spontaneously, without any help from the experimenter. This is called **spontaneous mutagenesis.** Mutagenesis can also be induced by the addition of chemicals called mutagens or by exposure to radiation, both of which result in chemical alterations of the genetic material. Mutagenesis and mutations will be discussed in detail in Chapter 11.

The Number of Mutations in a Mutant

Another classification of mutants is based on the number of changes that have occurred in the genetic material (that is, the number of DNA base pairs that have changed). If only one change has occurred, the mutant is called a **single** or **point mutant.** If two changes are present, it is a **double mutant.** Sometimes the mutation occurs by means of removal of all or part of a gene; in that case, the mutation is a **deletion.** Occasionally, other material replaces the deleted gene or gene segment. Such a mutation is called a **deletion-substitution.** These types of mutations are shown schematically in Figure 2-1.

Revertants and Reversion

A mutant organism sometimes regains its original character. This occurs by means of chemical changes in the mutant genetic material, which restore the genetic material to a functional state. The process of regaining the original phenotype is called **reversion,** and an organism that has reverted is called a **revertant.**

Reversion can be spontaneous or be induced by mutagens. The **reversion frequency**—that is, the fraction of cells in a population of mutants that after many generations of growth have regained the functional pheno-

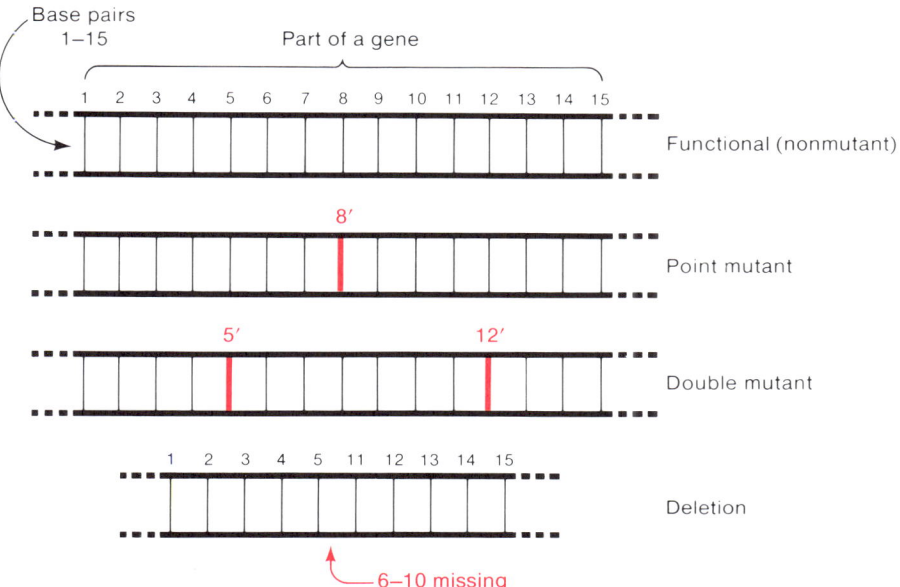

Figure 2-1   Schematic diagram of a normal DNA molecule and various kinds of mutants. In the point mutant, base pair 8′ replaces the normal base pair 8. The double mutant has two base pairs (5′ and 1′) not present in the nonmutant organism. In the deletion, five base pairs (6−10), present in the nonmutant, are absent.

type—is a useful criterion for identifying a point mutant, because a small alteration in the nucleic acid is sufficient to revert a point mutant, and such a change can occur at some measurable, albeit low, frequency. With a deletion, however, the probability that the missing DNA will be replaced with material having an equivalent sequence, thereby restoring a functional gene, is virtually zero. Thus, *reversions are not observed with deletions.*

A revertant bacterium might be detected in the following way. If 100 cells of a *leu*$^-$ bacterium are placed on agar lacking leucine, no colonies will form. If 10^7 *leu*$^-$ bacteria are plated, about 50 colonies arise; these colonies must consist of Leu$^+$ bacteria and are spontaneous revertants. The reversion frequency in this case is $50/10^7 = 5 \times 10^{-6}$. Such a frequency value is characteristic of the reversion of point mutants. The production of a spontaneous revertant is a random process, so the reversion of a double mutant would require two independent events. If each were to occur at a frequency of 5×10^{-6}, then the frequency of reversion of a double mutant would be $(5 \times 10^{-6})^2 = 2.5 \times 10^{-11}$; that is, 25 colonies would be found if 10^{12} double mutants were plated. Usually, no more than 10^9 bacteria can be plated in a single petri dish, so experimentally one would need 40 plates to score one revertant. Reversion will be discussed in detail in Chapters 11 and 13.

Uses of Mutants: Some Examples

Some of the most significant advances in molecular biology have come about by the use of mutants. In the following, the kinds of approaches that have been taken are described.

1. *A mutant defines a function.* For example, the intake of Fe^{3+} ions by bacteria might be by passive diffusion through the cell membrane, or a particular system might be responsible for the process. Wild-type *E. coli* can take in the Fe^{3+} ion from a 10^{-5} *M* solution but mutants have been found that cannot do so unless the ion concentration is very high. This finding indicates that a genetically determined system for Fe^{3+} intake exists, though the observation does not tell what this system is. Temperature-sensitive mutants are especially useful in defining functions. For example, temperature-sensitive mutants of *E. coli* have been isolated that fail to synthesize DNA. These mutants fall into at least ten distinct classes, suggesting that there may be at least ten different proteins required for DNA synthesis.

2. *Mutants can introduce biochemical blocks that aid in the elucidation of metabolic pathways.* The metabolism of the sugar galactose, for example, requires the activity of three distinct genes called *galK, galT,* and *galE.* If radioactive galactose ([^{14}C]Gal) is added to a culture of *gal$^+$* cells, many different radioactive compounds can be found as the galactose is metabolized. At very early times after addition of [^{14}C]Gal, three related compounds are detectable: [^{14}C]galactose-1-phosphate (Gal-1-P), [^{14}C]uridine diphosphogalactose (UDP-Gal), and [^{14}C]uridine diphosphoglucose (UDP-Glu). Different mutant genes will block different steps of the metabolic pathway. If the cell is a *galK$^-$* mutant, the [^{14}C]Gal label is found only in galactose. Thus, the *galK* gene is known to be responsible for the first metabolic step. If the mutant *galT$^-$* is used, Gal-1-P accumulates. Thus, the first step in the reaction sequence is found to be the conversion of galactose to Gal-1-P by the *galK* gene product (namely, the enzyme galactokinase). If a *galE$^-$* mutant is used, some Gal-1-P is found but the principal radiochemical is UDP-Gal. Thus, the biochemical pathway must be

$$\text{Gal} \xrightarrow[\substack{galK \\ \text{product}}]{} \text{Gal-1-P} \xrightarrow[\substack{galT \\ \text{product}}]{} \text{UDP-Gal} \xrightarrow[\substack{galE \\ \text{product}}]{} \text{X}$$

The identity of X cannot be determined from these genetic experiments, but one might guess that it is UDP-Glu (which it is).

3. *Mutants enable one to learn about metabolic regulation.* Many mutants have been isolated that alter the amount of a particular protein that is synthesized or the way the amount synthesized responds to external signals. These mutants define regulatory systems. For example, the enzymes cor-

responding to the *galK, galT,* and *galE* genes are normally not present in bacteria but appear only after galactose is added to the growth medium. However, mutants have been isolated in which these enzymes are always present, whether or not galactose is also present. This indicates that some gene is responsible for turning the system of enzyme production on and off, and this regulatory gene must be responsive to the presence and absence of galactose.

4. *Mutants enable a biochemical entity to be matched with a biological function or an intracellular protein.* For many years an *E. coli* enzyme called DNA polymerase I was studied in great detail. Purified polymerase I is capable of synthesizing DNA *in vitro,* so it was believed that this enzyme was also responsible for *in vivo* bacterial DNA synthesis. However, an *E. coli* mutant (*polA⁻*) was isolated in which the activity of polymerase I was reduced 50-fold, yet the mutant bacterium grew and synthesized DNA normally. This observation suggested strongly that DNA polymerase I could not be the only enzyme that synthesizes intracellular DNA. Indeed, biochemical analysis of cell extracts of the *polA⁻* mutant showed the existence of two other enzymes, polymerase II and polymerase III, which could, when purified, also synthesize DNA. In further study, a temperature-sensitive mutation in a gene called *dnaE* was found to be unable to synthesize DNA at 42°C, though synthesis was normal at 30°C. The three enzymes DNA polymerases I, II, III were isolated from cultures of the *dnaE⁻* (Ts) mutant and each enzyme was assayed. Although polymerases I and II were active at both 30°C and 42°C, polymerase III was active at 30°C but not at 42°C; therefore, polymerase III was determined to be the product of the *dnaE* gene and the enzyme responsible for intracellular DNA synthesis.

5. *Mutants locate the site of action of external agents.* The antibiotic rifampicin prevents synthesis of RNA. When first discovered, it was not known whether rifampicin might act by preventing synthesis of precursor molecules, by binding to DNA and thereby preventing the DNA from being transcribed into RNA, or by binding to RNA polymerase, the enzyme responsible for synthesizing RNA. Mutants were isolated that were resistant to rifampicin. These mutants were of two types—those in which the bacterial cell wall was altered such that rifampicin could not enter the cell (an uninformative type of mutant) and those in which the RNA polymerase was slightly altered. The finding of the latter mutants proved that the antibiotic acts by binding to RNA polymerase.

6. *Mutants can indicate relations between apparently unrelated systems.* Bacteriophage λ, which normally adsorbs to and grows in *E. coli,* fails to adsorb to a bacterial mutant unable to metabolize the sugar maltose. Such failure is not associated with mutants incapable of metabolizing other sugars or with any other phages, and this knowledge implicated some product or agent of maltose metabolism in the adsorption of λ. Similarly, *E. coli*

mutants unable to adsorb the phage $\phi80$ require exceedingly high concentrations of the Fe^{3+} ion in the growth medium if the bacterium is to grow. This is because the adsorption site of $\phi80$ and the protein responsible for transport of the Fe^{3+} ion are structurally related.

7. *Mutants can indicate that two proteins interact.* How this occurs is best shown by a hypothetical example. Suppose that mutants in two genes *a* and *b,* which are responsible for synthesizing the proteins A and B, fail to carry out some process P. Thus, the products of both genes are necessary for this process to occur. These products may act consecutively or interact to form a single functional unit consisting of both products. Interaction as a single unit is often indicated by reversion studies. When revertants of an a^- mutant are sought, it is sometimes found (by additional genetic analysis) that reversion is a result of production of a mutation in gene *b.* When reversion of this type occurs, it is usually observed that reversion of other b^- mutants (those not formed by reversion of an a^- mutant) is often the result of (different) mutations in gene *a.* The interpretation of these results is the following. Proteins A and B interact to form a protein aggregate (Figure 2-2), which is designated an active structure. Some alterations in either A or B prevent aggregation inasmuch as the protein will have been changed at a binding site. A compensating alteration in the other protein can then enable the interaction to occur again. Such an interpretation has frequently been found to be correct.

Figure 2-2 Schematic diagram showing how two separately inactive mutants can combine to make a functioning protein complex. Sites of interaction of proteins A and B are denoted by heavy lines. Components of the active site of the A-B complex are shown in red. Only complexes I and IV have active binding sites.

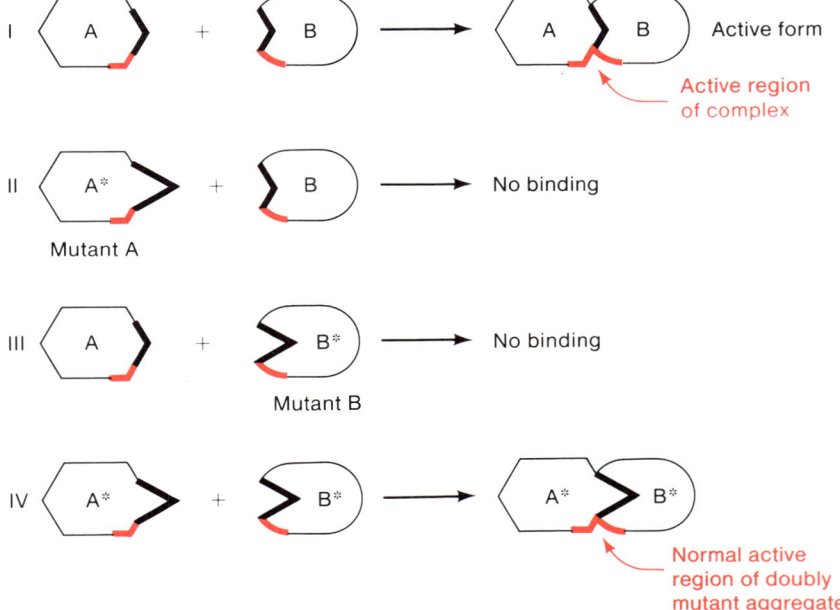

GENETIC ANALYSIS OF MUTANTS*

A great deal of information about molecular mechanisms can be obtained by combining separately isolated mutations in various combinations. This is done by **genetic recombination.** Another important use of genetic recombination is the construction of an array that indicates the positions of genes on a chromosome with respect to one another. When this is done by genetic techniques, the array is called a **genetic map. Physical maps** have also been constructed by using various physical techniques, and some elegant experiments have shown that in some organisms the gene positions in the two maps are identical. Another genetic technique is **complementation.** By this procedure it is possible to determine the number of genes responsible for a particular phenotype and to distinguish genes from regulatory sites. In the following sections both genetic recombination and complementation will be described.

Genetic Recombination

Genetic recombination is the process of combining two genetic loci, initially on two different chromosomes, onto a single chromosome. The molecular mechanism, which is very complex and at present not well understood, is discussed in Chapter 20; for the present purposes it is sufficient to refer to the "scissors-and-tape" mechanism. In this, one assumes that two chromosomes align with one another, that a cut is made in both chromosomes at random but matching points, and that the four fragments are then joined together to form two new combinations of genes. This is a naive and incorrect model; it enables one to account for only the simpler features of genetic exchange, but these features are in fact the only ones that are of concern at this time. The process is depicted in Figure 2-3. Two parental chromosomes having the genotypes a^+b^- and a^-b^+ pair, are cut, and are then joined to form two recombinant chromosomes whose genotypes are a^+b^+ (wild type) and a^-b^- (double mutant).

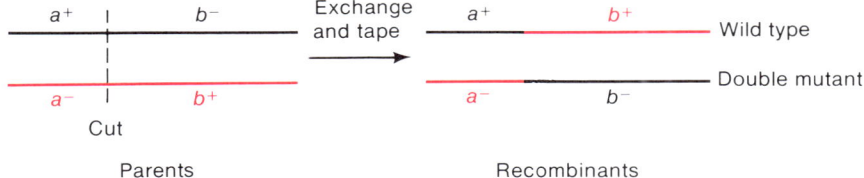

Figure 2-3 A schematic diagram showing genetic exchange.

*This section may be omitted for a course that does not utilize genetic techniques or by readers with experience in genetics.

Genetic recombination frequently occurs when one bacterium is simultaneously infected with two mutant phage. That is, if the parental phage have the genotypes a^+b^- and a^-b^+, as in Figure 2-3, of the hundred-or-so progeny phage released when the infected bacterium lyses, there will be a few a^+b^+ and a^-b^- recombinant phage. The ratio

$$\text{Number of recombinant phage/Number of total phage}$$

is called the **recombination frequency.**

Genetic Mapping

In genetic mapping, the distance along the chromosome between two recombining genetic loci (or mutations) determines the recombination frequency. As long as the two loci are not too near one another, the recombination frequency is proportional to distance, because chromosomal cuts are made at random. Thus, in the following crosses between chromosomes, the genotypes of which are $a^+b^-c^-$ and $a^-b^+c^+$, and the genes of which are in alphabetical order and equally spaced,

$$\frac{a^+ \quad b^- \quad c^-}{\times}$$
$$\overline{a^- \quad b^+ \quad c^+}$$

there will be twice as many a^+c^+ recombinants as a^+b^+ recombinants, because loci a and c are twice as far apart as loci a and b.

Because the recombination frequency is proportional to distance, recombination frequency can be used to determine the arrangement of genes on the chromosome. This can be seen in a simple example (Figure 2-4). Consider three genes a, b, and c, whose arrangement is unknown. Using the notation $p \times q = m$ percent to denote a recombination frequency of m percent between genes p and q, we assume it has been observed that $a \times b = 1$ percent and that $b \times c = 2$ percent. The two arrangements shown in the figure are consistent with these values and can be distinguished by determining the recombination frequency between a and c. Let us assume that this is 1 percent. If that is the case, only arrangement II is possible. The order $c\,a\,b$ for these genes and the relative separation constitute a genetic map.

Any number of genes can be mapped in this way. For instance, consider a fourth gene d, in the preceding example. If $d \times b = 0.5$ percent, d must be located 0.5 unit either to the left or to the right of b. If $a \times d = 1.5$

Figure 2-4   Two arrangements of three genes for which $a \times b = 1$ percent and $b \times c = 2$ percent. See text for discussion.

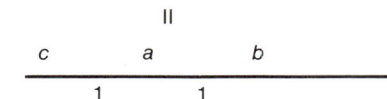

percent, d is clearly to the right of b and the gene order is $c\ a\ b\ d$. If $a \times d$ = 0.5 percent, the gene order would be $c\ a\ d\ b$.

The analysis just given has been oversimplified because the occurrence of multiple exchanges has been ignored. If two cut-and-tape events occur between two genetic markers, recombination will not be observed, because the second event will cancel the effect of the first. Discussion of this important point can be found in any textbook of genetics; however, the effect of multiple exchanges is unimportant for the simple considerations described here.

Linkage and Multifactor Crosses

Mapping can be carried out with a fewer number of crosses if the cross is done with three loci simultaneously (a **three-factor cross**). The procedure utilizes an effect known as **linkage.** Consider the cross shown in Figure 2-5 between two parents whose genotypes are $a^+b^-c^-$ and $a^-b^+c^+$. Instead of measuring recombination frequencies, let us just select all recombinants that are a^+c^+ and then in a second test ask whether they are also b^+ or b^-. If the genes were arranged as in panel I, the recombinants would mostly be b^+, because the distance from a to b is greater than that from b to c. On the contrary, if the genes were located as in panel II, most of the a^+c^+ recombinants would be b^-. This simple analysis locates b with respect to a and c. For arrangements I and II, one says that b and c or a and b, respectively, are linked.

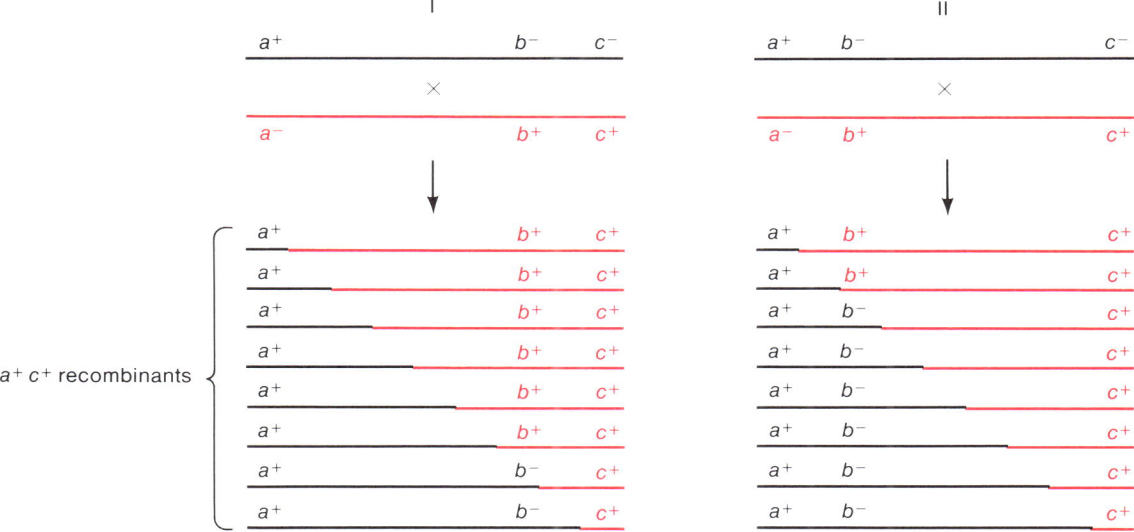

Figure 2-5 A three-factor cross that shows linkage. Eight possible a^+c^+ progeny aris-ing from equally spaced exchange points are shown for each arrangement. In I, 6 of 8 a^+c^+ are $a^+b^+c^+$; in II, 6 of 8 a^+c^+ are $a^+b^-c^+$.

The procedure just described does not give the gene order. That is, if b and c are linked, the gene order could be a b c or a c b and these two orders would not be distinguished by the analysis just given. Gene order can in theory always be determined by the mapping procedures that we have used to distinguish the two arrays shown in Figure 2-4; in the case of two closely linked loci, however, determination of gene order is not feasible from data derived from two-factor crosses, because the data are usually not sufficiently accurate. For example, suppose $a \times b = 7$ percent and $b \times c = 0.2$ percent. If $a \times c = 6.8$ percent, the order is a c $b,$ and if it is 7.2 percent, the order is a b c. However, experimentally observed values might be $a \times b = 7.0 \pm 0.3$ percent and $a \times c = 7.1 \pm 0.3$ percent, so that the order would not be established with certainty in this way. Figure 2-6 shows how a three-factor cross clearly gives the order. A cross is performed between parents having genotypes $a^+b^-c^+$ and $a^-b^+c^-$. Two types of data can be obtained. In the first, the linkage method is used; that is, b^+c^+ recombinants are selected and these are tested to determine whether they are a^+ or a^-. They are usually a^- for order III and a^+ for order IV. In the second and more rapid method, the frequency of $a^+b^+c^+$ recombinants is measured. These recombinants arise by two exchanges in case III but only one exchange in case IV. The frequency for a double exchange is the product of the frequencies of each simple exchange. Thus, for case III, the recombination frequency is $(0.07)(0.002) = 0.00014 = 0.014$ percent but for case IV, in which only a single exchange is needed, the frequency is 0.2 percent (the same as that for forming b^+c^+ recombinants). Hence one three-factor cross yields the gene order unambiguously.

Further discussion of mapping techniques and the analysis of recombination frequencies can be found in the genetics texts listed at the end of the book.

Complementation

As we have seen earlier, a particular phenotype is frequently the result of the activity of many genes. In the study of any genetic system it is always important to know the number of genes and regulatory elements that con-

Figure 2-6 Determination of gene order by a three-factor cross. The frequency of appearance of $a^+b^+c^+$ is much higher for order IV because only one exchange is required.

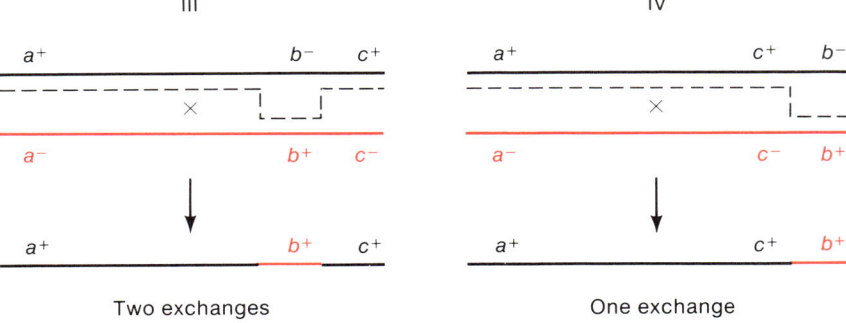

stitute the system. The genetic test used to evaluate this number is called **complementation.**

Complementation is best explained by example. The test requires that two copies of the genetic unit to be tested be present in the same cell. In bacteria this can be done by constructing a **partial diploid**—that is, a cell containing one complete set of genes and duplicates of some of these genes. How cells of this type are constructed is described in a later section. A partial diploid is described by writing the genotype of each set of genes on either side of a diagonal line. As an example of this, $b^+c^+d^+/a^+b^-c^-d^-e^+\ldots z^+$ indicates that a chromosomal segment containing genes $b, c,$ and d is present in a cell whose single chromosome contains all of the genes $a, b, c, \ldots, z$. Usually, only the duplicated genes are indicated, so this partial diploid would be designated $b^+c^+d^+/b^-c^-d^-$. We now consider a hypothetical bacterium that synthesizes a green pigment from the combined action of genes $a, b,$ and c. The genes encode the enzymes A, B, C, which we assume to act sequentially to form the pigment. If there is a mutation in any of these genes, no pigment is made. However, pigment is made by the partial diploid $a^-b^+c^+/a^+b^-c^+$, because the cell contains a set of genes that produce functional proteins A, B, and C; B will be made from the $a^-b^+c^+$ chromosome and A from the $a^+b^-c^+$ chromosome. In a partial diploid $a_1^-b^+c^+/a_2^-b^+c^+$, in which a_1^- and a_2^- are two different mutations in gene a, no pigment can be made because the bacterium will not contain a functional A protein. The two mutations a^- and b^- of the diploid are said to complement one another because the phenotype of the partial diploid containing them is A^+B^+; the mutations a_1^- and a_2^- do not complement one another because the phenotype of the partial diploid containing these mutations is A^-.

Suppose that now a mutation x^- has been isolated but the gene in which the mutation has occurred has not been ascertained. By constructing a set of partial diploids, this gene can be identified by a complementation test. As a start, we might test the genes $a, b,$ and c with the partial diploids a^-/x^- (I), b^-/x^- (II), and c^-/x^- (III). If diploids I and II make pigment, the mutation cannot be in genes a or b; if no pigment is made by diploid III, the mutation must be in gene c. If pigment were made in all three diploids, then the important conclusion that mutation x^- is in none of the genes $a, b,$ or c could be drawn; furthermore, since we have assumed that $a, b,$ and c are each pigment genes, the fact that x is not in any of these genes but prevents pigment formation would be evidence that pigment formation requires at least four genes ("at least" because more genes might still be discovered).

A common approach to the initial characterization of a genetic system is to isolate about 50 mutants and perform complementation tests between them, as will be shown in the example of the next section. The analysis, which may be tedious, is nonetheless a straightforward one. The basic rule is the following:

Rule I. If x_1^- and x_2^- complement, they are in different genes. The converse statement—that is, if two mutations do not complement, they are in the same gene—does not always follow.

There are three explanations for the lack of complementation of two mutations; these are stated in the following rule:

Rule II. If mutations x_1^- and x_2^- fail to complement, then one of the following is true:

(a) They are in the same gene, or
(b) At least one of the mutations is in a regulatory site for the other gene, or
(c) At least one of the mutations yields an inhibitory gene product.

Complementation Analysis: An Example

Use of rules I and II allows every mutation to be placed in a mutually exclusive complementation group. Let us first examine these rules for a set of mutations, 1 through 8, in which rules I and II(a) account for all of the data. After studying this simple example, we will add first one mutation that requires that rule II(b) be invoked and then a second mutation that can be explained by II(c). The data are presented in Table 2-1. To analyze the first example, use rows 1 through 8 of the table. Each + or − entry designates a single pairing of the mutation numbers shown at the top and side of the table. Any − entry denotes that the pair does not complement; any + entry denotes complementation. In particular, notice any − entries that can be aligned in a single direction (i.e., that align down one column or across one

Table 2-1   An example of complementation data

	Mutation number									
	1	2	3	4	5	6	7	8	9	10
Mutation number										
1	−									
2	+	−								
3	+	+	−							
4	+	−	+	−						
5	−	+	+	+	−					
6	+	+	−	+	+	−				
7	+	+	−	+	+	−	−			
8	+	−	+	−	+	+	+	−		
9	−	−	−	−	−	−	−	−	−	
10	±	−	±	−	±	±	±	−	−	−

Note: + = complementation; − = no complementation; ± = weak complementation.
An entry at the intersection of a horizontal row and a vertical column represents the result of one complementation test between two mutations. For example, the + entry at the intersection of row 3 and column 2 indicates that mutations 2 and 3 complement; the corresponding entry for row 2 and column 3 is not given since it is the same as the one just stated—that is, the table is symmetric about the diagonal. The analysis first discussed in the text uses only mutations 1–8, which is the reason for the dashed line between rows 8 and 9.

row). These are noncomplementing mutations within a single gene. For example, in column 1, mutations 1 and 5 fail to complement; other non-complementing (aligning) mutations can be found in columns 2, 3, 4, and 6, and in rows 4, 5, 6, 7, and 8. Three genes can be identified in this way: the first contains mutations 1 and 5; the second contains mutations 2, 4, and 8; and the third contains mutations 3, 6, and 7. (Inasmuch as the table is symmetrical, a row or column need be counted just once.) Single mutations (a row or column) that intersect at a + sign are usually complementing. The three genes (or groups of mutations) found in the table are therefore complementing. We can give each complementing gene an identifying letter for convenience:

Group	Mutation
A	1, 5
B	2, 4, 8
C	3, 6, 7

The simplest explanation for the data is that the phenotype being studied consists of at least three genes.

We now examine the effect of a mutation in a regulatory site rather than in a gene (rule II(b)). Suppose a ninth mutant had been isolated having the complementing property shown below:

	1	2	3	4	5	6	7	8	9
9	–	–	–	–	–	–	–	–	–

Alone these data might suggest that all of the mutations are in the same gene, but the data just analyzed show that this is not the case, for clearly, when mutation 9 is present in a particular chromosome, that chromosome will not yield the products of genes A, B, and C. Mutation 9 is called a **cis-dominant mutation,** because such a mutation prevents expression of related genes residing on the same chromosome in which the mutation is located. Such mutations invariably prove to be regulatory site mutations; that is, they are in some *site* on the chromosome that must be active if gene products in the *same* chromosome and in the particular gene system are to be made. For instance, the mutation might be in a site that signals start of synthesis of the gene products.

This interpretation of mutation 9 has not been done fairly, because mutations 1 through 8 were classified first and then 9 was introduced. Let us consider how all nine mutations would have been classified if they were examined simultaneously. Once again, – entries would be observed and clustered. However, mutation 9 would immediately be anomalous because it would appear in more than one complementation group—in fact, in all groups. Whenever a mutation appears in more than one group, the need for rule II(b) should be suspected and that mutation should temporarily be ignored in making the basic classification. It is, of course, not always the case that a mutation in a regulatory site will fail to complement all other mutations being examined, because the site might be required for activity of only one, or a few, genes in the set. That is, the regulatory mutation would only fail to complement mutations in the genes regulated.

Many examples of *cis*-dominant mutations will be seen later in Chapter 15.

Let us now consider the rare case in which rule II(c) would be applied. Inasmuch as mutation 10 fails to complement mutations 2, 4, and 8, it might appear to be a separate, additional mutation in the group B gene. The + response is weak with mutants 1, 3, 5, 6, and 7, however, and mutant 10 cannot be considered complementary to groups A and C because, tested against the mutants of those groups, 10 would also yield a −. The correct interpretation is that 10 is not only a B-group mutation but that the mutant gene product also inhibits the activity of a good copy of B. Thus, in each cell in which one would expect B to be fully active, the + response is weak because the total activity of the inhibited B is not adequate for a normal + response. In this case, mutation 10 is said to be anticomplementing.

Complementation in Phage-Infected Bacteria and Higher Organisms

So far we have discussed complementation only in bacteria. The concept is also applicable to phages and is a common way to assign phage mutations to particular genes. The equivalent of the partial diploid test is the infection of a bacterium with two mutant phages, neither of which can grow alone; one then notes whether phage are produced by the doubly infected cells. This test is usually done with conditional mutants in the following way. Consider two temperature-sensitive mutations a^-(Ts) and b^-(Ts) in genes a and b of two respective phage; neither gene product is active at 42°C, so the mutants cannot grow at 42°C and phage production depends upon the complementation of separate Ts mutations. The phage are allowed to adsorb at 42°C to a host bacterium and then the infected cell is incubated at that temperature. If only a^-(Ts) or only b^-(Ts) phage have adsorbed, no phage will be produced. However, if both have adsorbed and the Ts$^-$ mutations are in different genes, the infected cell will contain a good copy of product A and a good copy of product B; complementation can then occur and progeny phage will be produced. Note that in addition to the parental types of a^-(Ts) and b^-(Ts) phage that are present in the resulting lysate, there are also some Ts$^+$ and a^-(Ts)b^-(Ts) recombinants.

Complementation tests in higher organisms are carried out in principle like those in bacteria. Higher organisms are diploid, so the test is performed by mating organisms carrying different mutations and selecting progeny in which the two mutations reside on different members of a pair of homologous chromosomes. Construction of these doubly mutant organisms is particularly straightforward when, like yeast, there is a haploid phase, for mutations can be combined simply by mating two haploid strains, each carrying a particular mutation.

GENETIC SYSTEMS IN *E. coli*

Since the turn of the century applications of Mendelian genetics to organisms such as *Drosophila* and various fungi, which are familiar to most biology students, have made considerable contributions to biological thought. However, only quite recently have these genetic systems become amenable to molecular studies. In contrast, study of the genetics of *E. coli* and its phages has provided much of the experimental basis for the concepts of molecular biology that developed in the 1950s through early 1970s. The *E. coli* genetic systems, which are generally not familiar to biology students, will be described in some detail in this section.

The bacterium *E. coli* possesses two mating types: donors or **males,** and recipients or **females.** The determinant of maleness is a small circular piece of extrachromosomal DNA (distinct from the chromosome) denoted by **F** and called the **sex plasmid**; genetically, a male cell is designated F^+. A female cell lacks the F plasmid and is designated F^-. When a culture of males is mixed with a culture of females, male–female pairs form (this is called **conjugation**), each pair being joined by a conjugation bridge. In a way that is not completely understood, pairing signals the replication of F, and one copy of F is transferred to the female in about one minute (Figure 2-7). In contrast with sexual systems in other organisms, *the female is converted to a male* inasmuch as, after the mating, the recipient cell contains the F plasmid, which is the determinant of maleness.

Hfr Cells

Another property possessed by the F plasmid is its ability to become integrated into the bacterial chromosome. (The mechanism for this integration

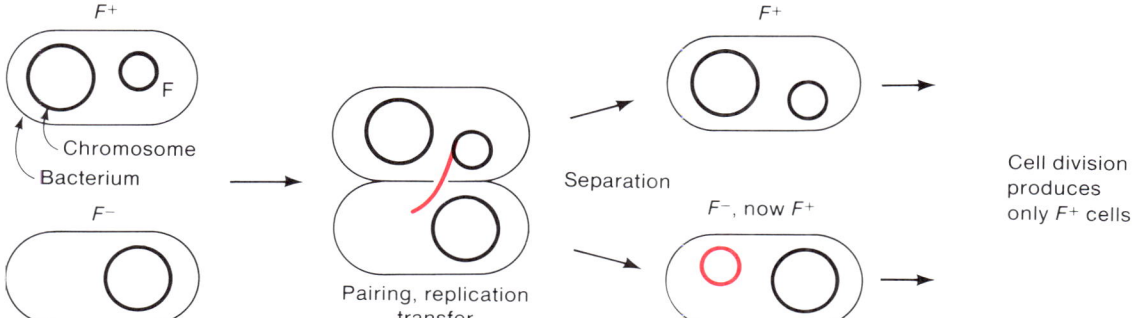

Figure 2-7 Diagram showing the events in an $F^+ \times F^-$ mating.

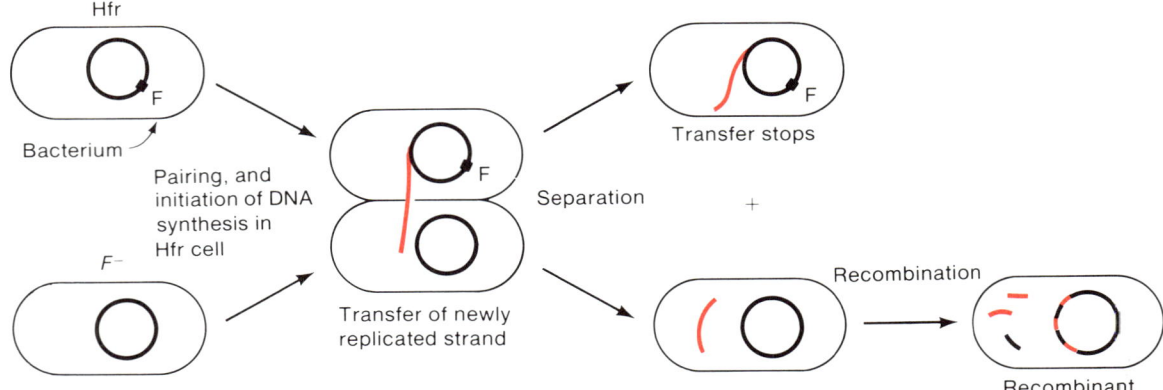

Figure 2-8 A diagram showing mating between an Hfr male bacterium and an F^- female bacterium. The two bacteria form a pair. Then, under the influence of a unit called F, DNA replication begins in the male, adjacent to F, and a replica of F is transferred to the female. By random motion the mating cells break apart. At a later time a portion of the transferred DNA exchanges with the corresponding piece in the female. The intact female chromosome then replicates; the fragments do not replicate and are ultimately lost in the course of cell division.

will be discussed in Chapter 21.) When this occurs, the chromosome remains a single, circular DNA molecule and F behaves as if it were part of this chromosome, increasing the size of the chromosome. The integration process happens rarely but it is possible to purify a culture of cells that are the progeny of the cell in which integration occurred. The cells in such a culture are called **Hfr males.** (Hfr is an acronym for *high frequency of recombination.*) When a culture of Hfr cells is subsequently mixed with an F^- culture, conjugation also occurs as described earlier, though the material transferred is slightly different from that in an $F^+ \times F^-$ mating (Figure 2-8). This time, under the influence of F, DNA replication begins in the Hfr cell and a replica is transferred to the F^- cell; however, the direction of replication is such that a small portion of F is transferred first and the major portion is transferred last. Moreover, because bacteria are very small and in constant motion by being bombarded by solvent molecules (Brownian motion), and because it takes 100 minutes to transfer an entire chromosome, the mating pair usually breaks apart before transfer is completed. Thus, the female receives both a large fragment of the male chromosome, which may contain hundreds of genes, and a small functionless fragment of F. Because of this, the mated female remains a female in an Hfr $\times$ F^- mating.

The presence of the new chromosomal fragment in the female sets in motion a recombination system that causes genetic exchanges to occur, and a recombinant F^- cell often results. Thus, in a mating between an Hfr *leu*$^+$ culture and an $F^- leu^-$ culture, $F^- leu^+$ cells arise. In order to recognize these cells, some means is needed to distinguish a male from a female cell. This is done by using cells that have genetic differences that can be recognized by growth of a colony on agar. A common method is to use antibiotic resistance. For instance, consider an Hfr that is not only Leu$^+$ but also sensitive to streptomycin (Str-s) and a female that is both Leu$^-$ and resistant to streptomycin (Str-r). If the *str* locus is sufficiently far from the origin of transfer that the mating bacterial pairs, which inevitably break apart, have separated before the locus has been transferred, then plating the mated cell mixture onto agar containing streptomycin but lacking leucine (1) will selectively kill the Hfr Str-s cells and (2) will not lead to growth of the F^- cells unless these also possess the *leu*$^+$ allele. Thus, any cell that survives will have the genotype *leu*$^+$ *str-r* and will be a recombinant female. Only these Leu$^+$ Str-r recombinants can form a colony. When a mating is done in this way, the transferred allele that is selected by means of the agar conditions (*leu*$^+$ in this case) is called a **selected marker,** and the allele used to prevent growth of the male (*str-r* in this case) is called the **counterselective marker.**

Mapping by an Hfr × F^- Mating

An important feature of an Hfr × F^- mating is that the transfer of the Hfr chromosome proceeds at a constant rate from a fixed point determined by the site at which F has been inserted in the Hfr chromosome. This means that the times at which particular genetic loci enter a female are proportional to the positions of these loci on the chromosome. Thus, a map can be obtained from the time of entry of each gene. This is done in the following way. An Hfr whose genotype is $a^+ b^+ c^+ d^+ e^+$ *str-s* is mated with an $a^- b^- c^- d^- e^-$ *str-r* female. At various times after mixing the cells, a sample is taken and agitated violently in order to break apart all of the pairs simultaneously. The sample is then plated on five different agars containing streptomycin, each of which also contains a different combination of four of the five substances A through E. Thus colonies that grow on agar lacking A are a^+ *str-r*, those growing without B are b^+ *str-r,* and so forth. All of these data can be plotted on a single graph to give a set of time–of–entry curves, as shown in Figure 2-9(a). Extrapolation of each curve to the time axis yields the time of entry of each locus a^+, b^+, . . . , e^+. These times can be placed on a map, as shown in part (b) of the figure. Use of a second female that is $b^- e^- f^- g^- h^-$ *str-r* can then provide the relative positions of three additional genes. These might give a map such as that in Figure 2-9(c). Since genes *b* and *e* are common to both maps, the two maps can be combined to form a more complete map, as shown in Figure 2-9(d).

The F plasmid can integrate at numerous sites in the chromosome to generate Hfr cells that have different origins of transfer. Each of these Hfr strains can also be used to obtain maps. It has been found that when separate maps are combined, a circle is eventually obtained. For instance, the map obtained from another Hfr might be that shown in Figure 2-9(e), which, when combined with that in panel (d), would yield the circular map shown

Figure 2-9 Construction of the circular map of *E. coli*. (a) Time-of-entry curves for a particular Hfr strain. (b), (c), (d), (e) Four linear maps obtained from four different Hfr types. (f) The circular map derived from the data in panels (a)–(e).

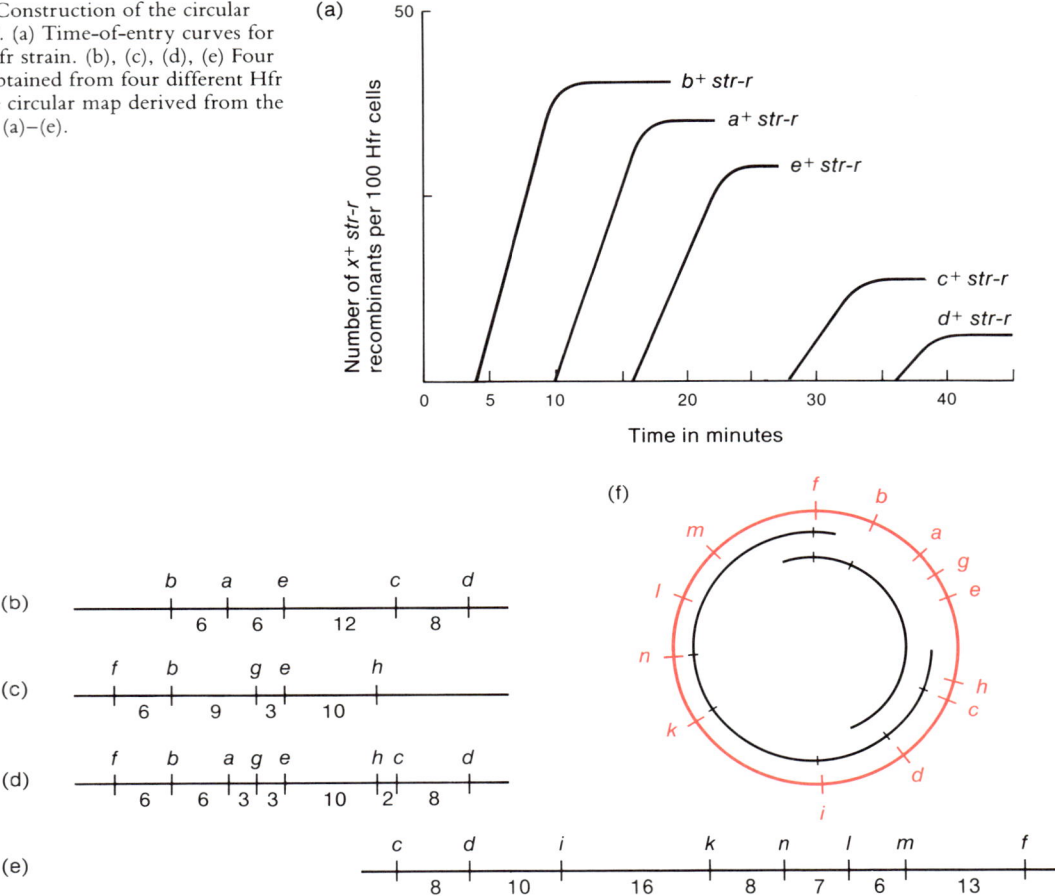

in panel (f). This mapping technique has been used repeatedly with hundreds of *E. coli* genes to generate an extraordinarily useful map—one with great significance in the development of molecular genetics. The discovery in the early 1950s that the genetic map of *E. coli* is circular provided the first suggestion that the *E. coli* chromosome might be a circular DNA molecule, as was later found to be the case.

F′ Plasmids

An Hfr cell is produced when F integrates stably into the chromosome, as we have already stated. At a low frequency, F can also be excised. When this happens, the excised circular DNA is sometimes found to contain genes that

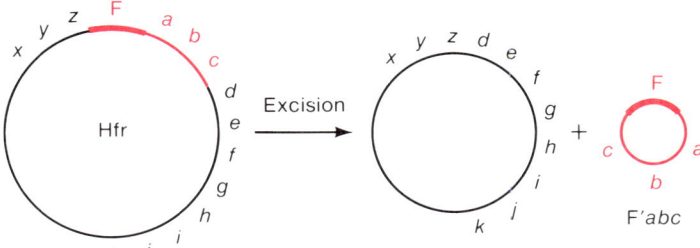

Figure 2-10 Diagram showing excision of an F′ plasmid from an Hfr chromosome.

were adjacent to F in the chromosome (Figure 2-10). A plasmid containing both F genes and chromosomal genes is called an F′ plasmid. It is usual to describe an F′ plasmid by stating the genes it is known to possess—for example, F′*lac pro* contains the genes for lactose utilization and proline synthesis. F′ plasmids can also be transferred from an F′ male to a female. This occurs sufficiently rapidly that the entire F′ is usually transferred before the mating pair breaks apart. Thus, the female recipient is converted to an F′ male in an $F′ \times F^-$ mating.

F′ plasmids have been useful in the production of partial diploid bacteria. For example, if an F′*lac*⁺/*str-s* male is mated with a *lac⁻ str-r* female and a Lac⁺Str-r colony is isolated, the cells in the colony will carry two copies of the *lac* gene—the *lac⁺* allele brought to the female in the F′ and the *lac⁻* allele already present in the female chromosome. To denote this, the genotype of the cell is written F′*lac⁺*/*lac⁻ str-r,* as is usual with partial diploid cells. By convention, genes carried on the F′ plasmid are written at the left of the diagonal line.

In addition to F and F′ plasmids there are numerous other plasmids; their properties and their utility in the laboratory are discussed in detail in Chapters 19 and 22.

3

Macromolecules

A typical cell contains 10^4–10^5 different kinds of molecules. Roughly half of these are small molecules–namely, inorganic ions and organic compounds whose molecular weights usually do not exceed several hundred. The others are polymers that are so massive (molecular weights from 10^4 to 10^{12}) that they are called **macromolecules.** These molecules are of three major classes: proteins, nucleic acids, and polysaccharides, which are polymers of amino acids, nucleotides, and sugars, respectively. There are also subclasses of macromolecules, inasmuch as macromolecules can be modified in various ways—for example, glycoproteins (proteins carrying sugar groups), lipoproteins (proteins carrying lipids or fats), and lipopolysaccharides.

Macromolecules perform a variety of functions. For example, nucleic acids store and carry information, polysaccharides provide an energy reserve and also constitute the external cell wall of plants and many microorganisms, and lipoproteins enclose animal cells and are responsible for determining the substances that can pass in and out of a cell. The most versatile macromolecules—the proteins—catalyze chemical reactions, regulate the flow of information, and are major components of many structural units.

Knowledge of the properties of macromolecules is essential for understanding living processes. This and the next three chapters will emphasize nucleic acids and proteins. Information about other types of macromolecules will be given as needed in other chapters.

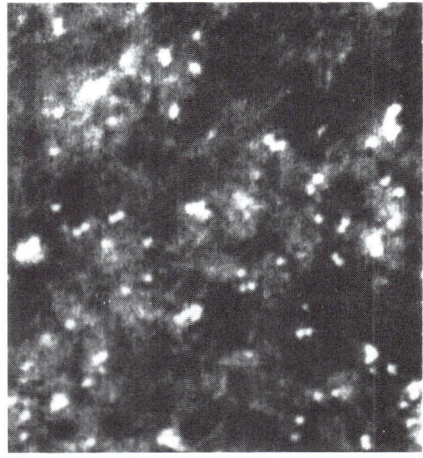

An ultrahigh-resolution scanning transmission electron micrograph showing individual uranium atoms (white dots) and larger aggregates of these atoms. The atoms are supported by a carbon film only a few atoms thick. [Courtesy of John Langmore and Albert Crewe.]

CHEMICAL STRUCTURES OF THE MAJOR CLASSES OF MACROMOLECULES

In this unit we examine the chemical structure of proteins, nucleic acids, and polysaccharides—in particular, the monomers and the chemical linkages by which they are joined. Physical properties of these macromolecules and the interactions that exist between various parts of a single macromolecule are described in later units.

Proteins

A protein is a polymer consisting of many amino acids (a **polypeptide**). Each amino acid can be thought of as a single carbon atom (the α carbon) to which there is attached one carboxyl group, one amino group, one hydrogen atom, and a side chain denoted R (Figure 3-1). The side chains are generally carbon chains or rings to which various functional groups may be attached. The simplest side chains are those of glycine (a hydrogen atom) and alanine (a methyl group). The complete chemical structures of each amino acid are given in Figure 3-2. Their common abbreviations are listed in Table 1-1, Chapter 1. Note that proline differs from the basic structure shown in Figure 3-1 in that the N atom is included in a ring. Properly speaking, proline is an *imino* acid. Because of its structure, proline introduces a kink into a polypeptide chain with a consequent local dramatic effect on the three-dimensional structure of a protein.

To form a protein the amino group of one amino acid reacts with the carboxyl group of another; the resulting chemical bond is called a **peptide bond** (Figure 3-3). Amino acids are joined together in succession to form a linear **polypeptide chain** (Figure 3-4). When the number of peptide bonds exceeds about 15 (the number is arbitrary), the polypeptide is called a **protein.** (A protein may also be an aggregate of several polypeptide chains, as will be seen later.) Thus, a protein is a polymer in which α-carbon atoms alternate with peptide units to form a linear chain having an ordered array of side chains. This linear chain is called the **backbone** of the molecule.

The two ends of every protein molecule are distinct. One end has a free —NH_2 group and is called the **amino terminus;** the other end has a free

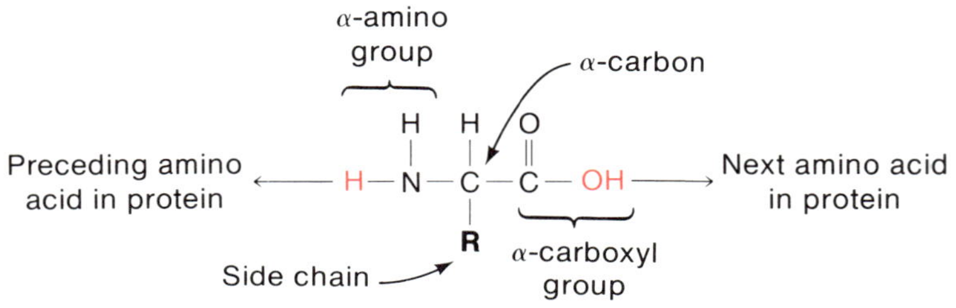

Figure 3-1 Basic structure of an α-amino acid. The NH_2 and COOH groups are used to connect amino acids to one another. The red OH of one amino acid and the red H of the next amino acid are removed when two amino acids are linked together (see Figure 3-3).

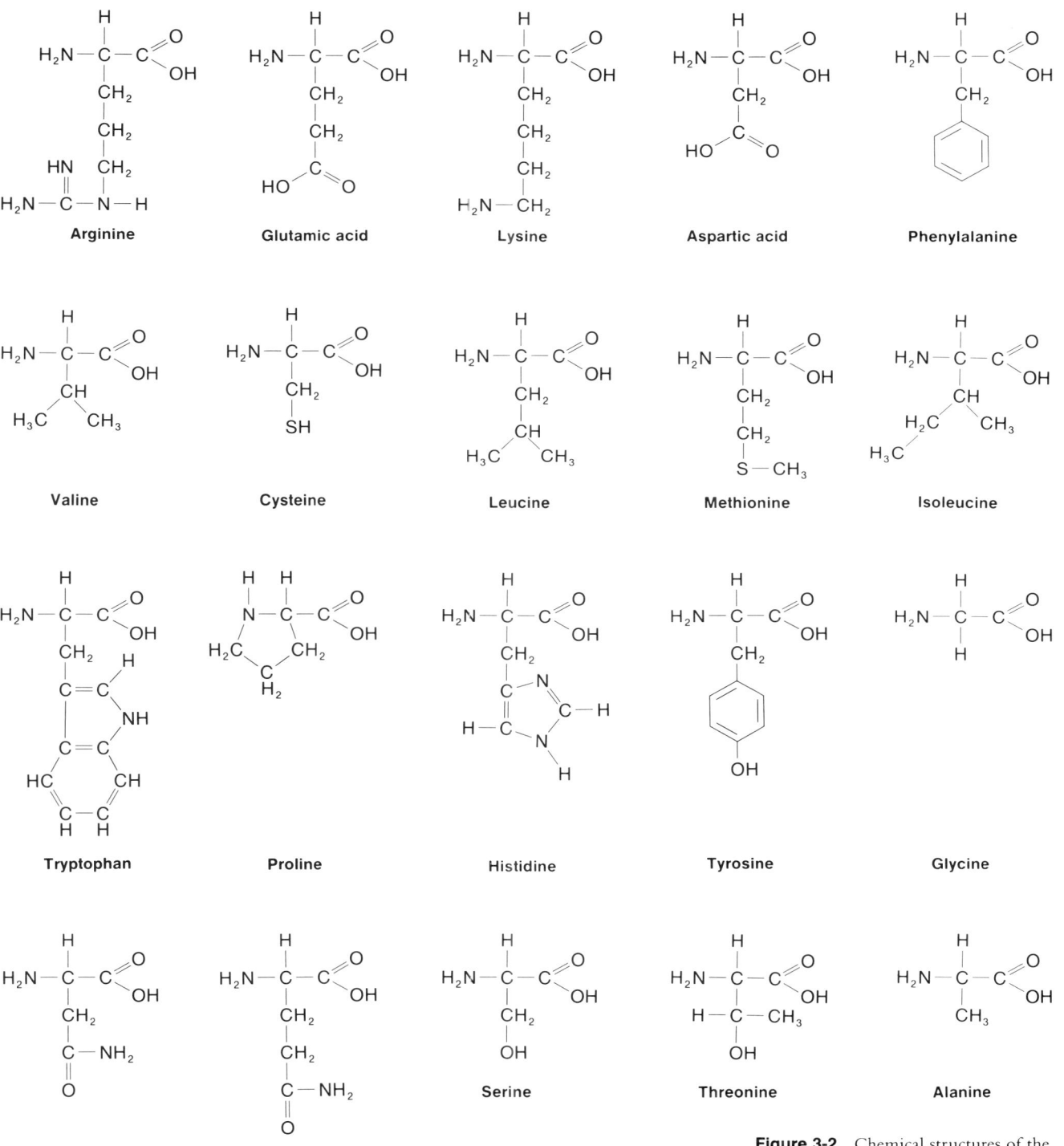

Figure 3-2 Chemical structures of the amino acids.

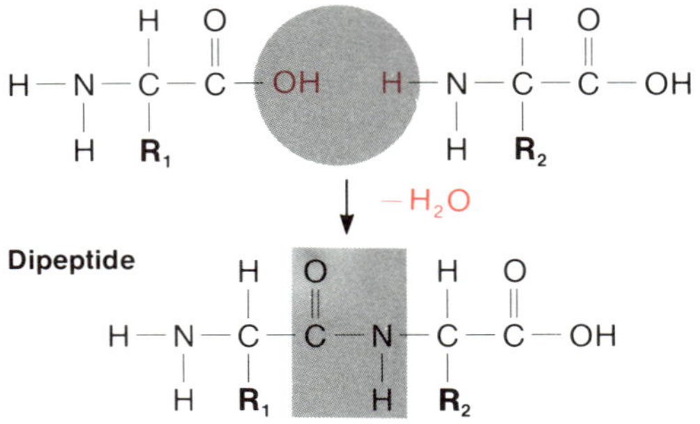

Figure 3-3 Formation of a dipeptide from two amino acids by elimination of water (shaded circle) to form a peptide group (shaded rectangle).

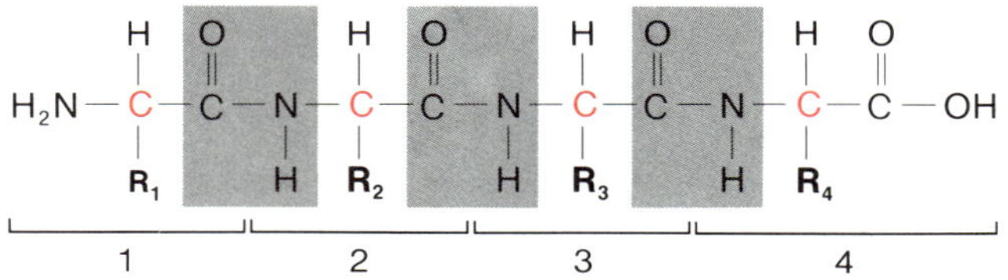

Figure 3-4 A tetrapeptide showing the alternation of α-carbon atoms (red) and peptide groups (shaded). The four amino acids are numbered below.

—COOH group and is the **carboxyl terminus.** The ends are also called the N and C termini, respectively.

The amino acid side chains usually do not engage in covalent bond formation; an exception is the —SH (sulfhydryl) group of cysteine, which often forms a disulfide (S—S) bond by reaction with a second cysteine in the same or different polypeptide chain. This is important in determining the three-dimensional structure of a protein, as will be examined in Chapter 6.

Many proteins contain metal ions engaged in coordination complexes with groups in the side chains. Common metal ions are Zn^{2+}, Cu^{2+}, and Fe^{3+}, which are often bound to histidine and glutamic acid.

Nucleic Acids

A nucleic acid is a polynucleotide—that is, a polymer consisting of nucleotides. Each **nucleotide** has the three following components (Figure 3-5):

1. *A cyclic five-carbon sugar.* This is ribose, in the case of ribonucleic acid (RNA), and deoxyribose, in deoxyribonucleic acid (DNA). The difference in the structure of ribose and 2′-deoxyribose is shown in Figure 3-5. Note that they differ only in the absence of a 2′-OH group in deoxyribose, a difference that makes DNA chemically more stable than RNA, as will be seen later.

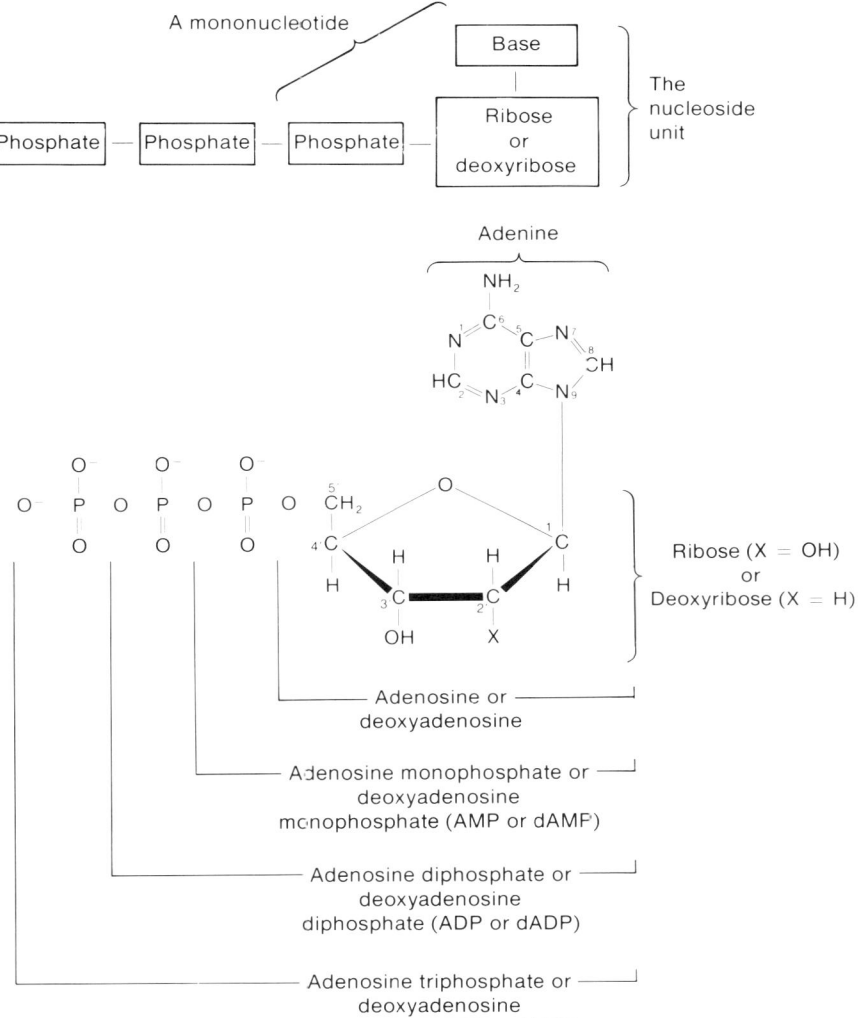

Figure 3-5 Structure of the nucleotides. As an example, the base adenine has been used.

2. *A purine or pyrimidine base* attached to the 1'-carbon atom of the sugar by an *N*-glycosidic bond. The bases, which are shown in Figure 3-6, are the purines, adenine (A) and guanine (G), and the pyrimidines, cytosine (C), thymine (T), and uracil (U). DNA and RNA both contain A, G, and C; however, T is found only in DNA and U is found only in RNA. There are rare exceptions to this rule—namely, T is present in some tRNA molecules and there are a few phages whose DNA exclusively contains U rather than T.

3. *A phosphate* attached to the 5' carbon of the sugar by a phosphoester linkage. This phosphate is responsible for the strong negative charge of both nucleotides and nucleic acids.

Figure 3-6   The bases found in nucleic acids. The weakly charged groups are shown in red. Adenine and guanine are derivatives of purines, the compounds in which all carbon atoms bear H atoms. Cytosine, thymine, and uracil are derivatives of pyrimidine, a compound with three double bonds in the rings and all C atoms bearing H atoms.

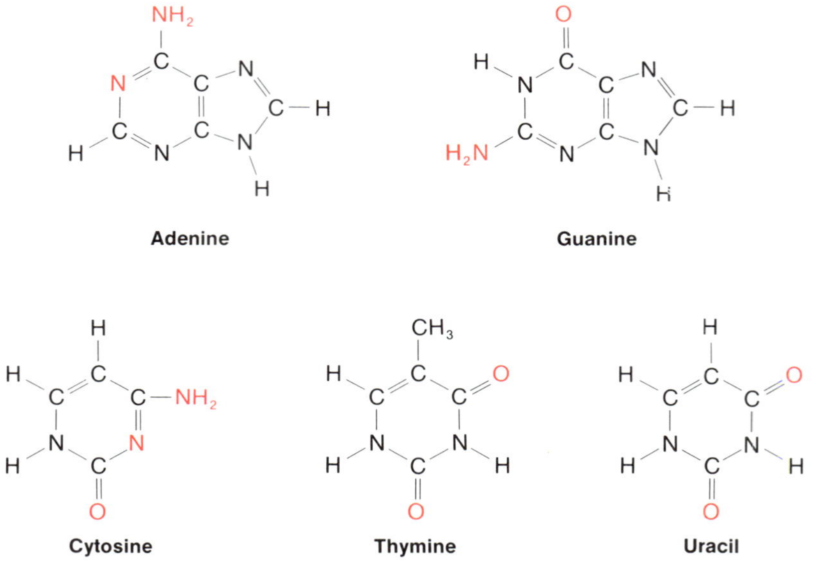

Adenine   Guanine

Cytosine   Thymine   Uracil

A base linked to a sugar is called a **nucleoside;** thus *a nucleotide is a nucleoside phosphate*. The terminology used to describe nucleic acid components is listed in Table 3-1.

The nucleotides in nucleic acids are covalently linked by a second phosphoester bond that joins the 5′-phosphate of one nucleotide and the 3′-OH group of the adjacent nucleotide (Figure 3-7). Thus the phosphate is esterified to both the 3′- and 5′-carbon atoms; this unit is often called a **phosphodiester group.**

The purine and pyrimidine bases are not engaged in any covalent bonds to each other. Thus, a polynucleotide consists of an alternating sugar-phosphate backbone having one 3′-OH terminus and one 5′-phosphate (5′-P) terminus. In the laboratory, polynucleotides can be prepared that have 3′-P and 5′-OH termini but such molecules do not occur naturally.

Figure 3-7   The structure of a dinucleotide. The vertical arrows show the bonds in the phosphodiester group about which there is free rotation. The horizontal arrows indicate the N-glycosidic bond about which the base can freely rotate. A polynucleotide would consist of many nucleotides linked together by phosphodiester bonds.

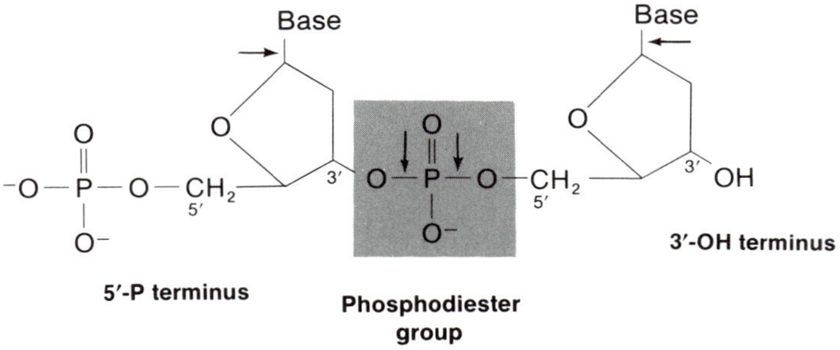

Table 3-1 Nucleic acid nomenclature

Base	Nucleoside[1]	Nucleotide[2]
Purines (Pu)		
Adenine (A)	Adenosine (rA)	Adenylic acid, or adenosine monophosphate (AMP)
	Deoxyadenosine (dA)	Deoxyadenylic acid, or deoxyadenosine monophosphate (dAMP)
Guanine (G)	Guanosine[3] (rG)	Guanylic acid, or guanosine monophosphate (GMP)
	Deoxyguanosine (dG)	Deoxyguanylic acid, or deoxyguanosine monophosphate (dGMP)
Pyrimidines (Py)		
Cytosine (C)	Cytidine (rC)	Cytidylic acid, or cytidine monophosphate (CMP)
	Deoxycytidine (dC)	Deoxycytidylic acid, or deoxycytidine monophosphate (dCMP)
Thymine (T)	Thymidine[4] (dT)	Thymidylic acid, or thymidine monophosphate (TMP)
Uracil (U)	Uridine[5] (rU)	Uridylic acid, or uridine monophosphate (UMP)

[1] Note that the names of purine nucleosides end in -osine and the names of pyrimidine nucleosides end in -idine.

[2] Note that each nucleotide has two names for the same substance.

[3] Guanosine should not be confused with guanidine, which is not a nucleic acid base.

[4] Thymidine is the deoxy- form. The ribo- form, ribosylthymine, is not generally found in nucleic acids.

[5] Uridine is the ribo- form. Deoxyuridine is not commonly found, although deoxyuridylic acid is on the pathway for synthesis of thymidylic acid—i.e., deoxyuridylic acid is methylated to yield thymidylic acid.

A typical RNA molecule is the single-stranded polyribonucleotide that has just been described. However, except in unusual cases, DNA contains two polydeoxynucleotide strands wrapped around one another to form a double-stranded helix. This remarkable structure will be described in the following chapter.

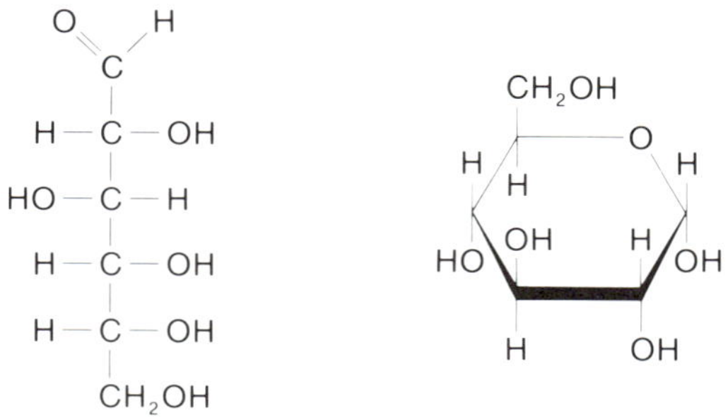

Figure 3-8   Glucose, a simple sugar that is the building block of many polysaccharides. It exists in both ring and linear form. Polysaccharides are also built from sugars that differ from glucose primarily in the relative orientations of H and OH groups.

Glucose (linear)   Glucose (ring)

Figure 3-8   Glucose, a simple sugar that is the building block of many polysaccharides. It exists in both ring and linear form. Polysaccharides are also built from sugars that differ from glucose primarily in the relative orientations of H and OH groups.

Polysaccharides

Polysaccharides are polymers of sugars (most often glucose) or sugar derivatives (Figure 3-8). These are very complex molecules because sometimes covalent bonds occur between many pairs of carbon atoms. This has the effect that one sugar unit can be joined to more than two other sugars, which results in the formation of highly branched macromolecules. These branched structures are sometimes so enormous that they are almost macroscopic. For example, the cell walls of many bacteria and plant cells are single gigantic polysaccharide molecules.

NONCOVALENT INTERACTIONS THAT DETERMINE THE THREE-DIMENSIONAL STRUCTURES OF PROTEINS AND NUCLEIC ACIDS

The biological properties of macromolecules are mainly determined by noncovalent interactions that result in each molecule acquiring a unique three-dimensional structure. In this section the noncovalent interactions that are important and the chemical groups responsible for them are described.

The Random Coil

Linear polypeptide and polynucleotide chains contain several bonds about which there is free rotation. In the absence of any intrastrand interactions* each monomer would be free to rotate with respect to its adjacent monomers;

---

*An intrastrand interaction is an interaction between two regions of the same strand. An interstrand interaction is between two different strands. It is important to remember the distinction between these similar words.

this is limited only by the fact that atoms cannot occupy the same space. The three-dimensional conformation (molecular shape) of such a chain is called a **random coil;** it is a somewhat compact and globular structure that *changes shape continually* owing to constant bombardment by solvent molecules.

Few, if any, nucleic acid or protein molecules exist in nature as random coils because there are many interactions between elements of the chains. These interactions are hydrogen bonds, hydrophobic interactions, ionic bonds, and van der Waals interactions. For example, the bases of nucleic acids attract one another by means of both hydrogen bonds and hydrophobic interactions; and amino acid side chains have attractive interactions of all four types and can repel one another if they have charges of the same sign.

The properties and consequences of these interactions are described in the following sections.

Before discussing these interactions and the conformations they produce, it is essential to realize that the interactions and their consequences take place in an aqueous environment, unless otherwise noted.

Hydrogen-Bonding

The most common hydrogen bonds found in biological systems are shown in Figure 3-9. In nucleic acids, hydrogen-bonding causes intrastrand pairing between nucleotide bases, which, if it were random, would cause a single polynucleotide strand to be more compact than a random coil. In DNA, interstrand hydrogen bonds are responsible for the double-stranded helical structure. In proteins, intrastrand hydrogen-bonding occurs between a hydrogen atom on a nitrogen adjacent to one peptide bond and an oxygen atom adjacent to a different peptide bond. This interaction gives rise to several specific polypeptide chain conformations, which will be discussed shortly.

The Hydrophobic Interaction

A hydrophobic interaction is an interaction between two molecules (or portions of molecules) that are somewhat insoluble in water. The phenomenon is simply that *two molecules (which may be different) that are poorly soluble in water tend to associate*. The explanation is the following. Water molecules form hydrogen bonds with one another and a molecule is soluble if it can hydro-

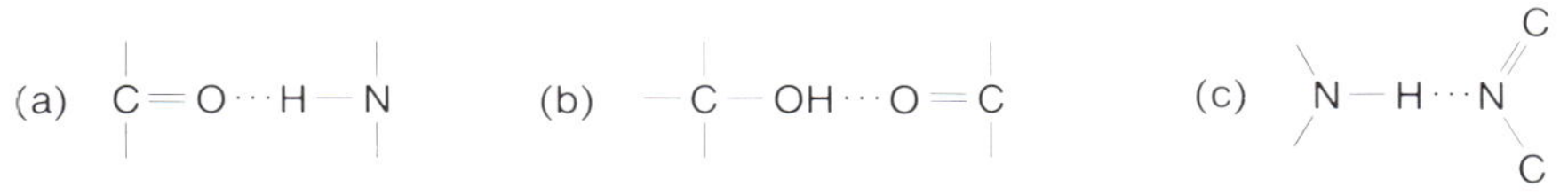

(a) $C = O \cdots H - N$ (b) $- C - OH \cdots O = C$ (c) $N - H \cdots N$

Figure 3-9   Structures of three types of hydrogen bonds (indicated by three dots). (a) A type found in proteins and nucleic acids. (b) A weak bond found in proteins. (c) A type found in DNA and RNA.

Figure 3-10 The meaning of stacking. (a) Three stacked bases. (b) Structures of stacked and (c) unstacked polynucleotides. The stacked polynucleotide is more extended because the stacking tends to decrease the flexibility of the molecule.

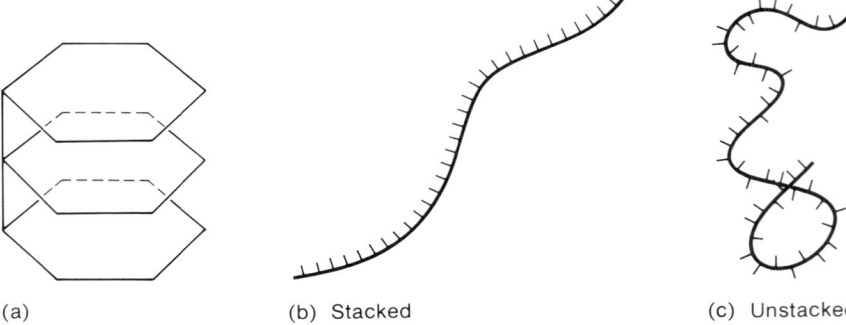

(a)   (b) Stacked   (c) Unstacked

gen-bond with water. An insoluble molecule in water causes (for complicated reasons) a shell of hydrogen-bonded water molecules to form around it. This ordering of water is equivalent thermodynamically to a decrease in entropy,* which is not a probable situation, as is explained by the second law of thermodynamics. Two separate insoluble molecules would form two such shells. If the two molecules were in contact, a single shell would surround the pair and, for strictly geometric reasons, the shell would have fewer than twice the number of ordered water molecules that would surround a single molecule of the solute. In general, the number of ordered water molecules per poorly soluble solute molecule is always smaller if the solute molecules are in contact; hence, *clustering of weakly soluble molecules is thermodynamically favored.* This tendency to cluster is called a **hydrophobic interaction.** Note that no bonds are formed, it is only that a cluster is the more probable arrangement.

Many components in nucleic acids and proteins have the hydrophobic property just described. For example, the bases of nucleic acids are planar organic rings carrying localized weak charges (Figure 3-6). The localized charges are sufficient to maintain solubility, but the large, poorly soluble organic ring portion causes the ordering of water just described, so the bases tend to cluster. The most efficient kind of clustering is one in which the faces of the rings are in contact, an array known as **base-stacking** (Figure 3-10). Note that since the bases are adjacent in the chain, stacking gives some rigidity to single polynucleotide strands and tends to extend them rather than allow a random coil to form. In the following chapter we will see how stacking is of major importance in determining nucleic acid structure. Many amino acid side chains are very poorly soluble, and this causes (1) the benzene ring of phenylalanine to stack and (2) the hydrocarbon chains of alanine, leucine, isoleucine, and valine to form unstacked clusters. Since in a particular polypeptide chain these amino acids are not necessarily adjacent, *hydrophobic interactions tend to bring distant hydrophobic parts of a chain together.*

*The entropy of a system or of a process is a thermodynamic quantity whose value increases as the degree of *disorder* increases. The entropy of the universe *increases* in all processes.

Ionic Bonds

An ionic bond is the result of attraction between unlike charges. Several amino acid side chains have ionized, negatively charged carboxyl groups (aspartic and glutamic acids) and positively charged groups (lysine, histidine, and arginine) that can form such bonds. This tends to bring together distant parts of the chain. Ionic interactions can also be repulsive—namely, between two like charges; thus, it would be unlikely for a polypeptide chain to fold in such a way that two aspartic acids are very near one another. Ionic bonds are the strongest of the noncovalent interactions. However, they are destroyed by extremes of pH, which can change the charge of the groups, and by high concentrations of salt, the ions of which shield the charged groups from one another.

Van der Waals Attraction

Van der Waals forces exist between all molecules and are a result of both permanent dipoles and the circulation of electrons. The dependence of the attractive force between two atoms is proportional to $1/r^6$, in which r is the distance between their nuclei. Thus, the attraction is a very weak force and is significant only if two atoms are very near one another (1 to 2 Å apart). A powerful repulsive force also comes into play if the two outer electron shells overlap. The **van der Waals radius** is defined as the distance at which the attractive and repulsive forces balance precisely. The van der Waals radii of atoms differ from one another; some representative values are shown in Table 3–2.

    The shape of a molecule is in essence the surface formed by the van der Waals spheres of each atom. Figure 3–11 shows the shapes of two molecules when defined in this way. The interaction energy of two atoms separated by the sum of their van der Waals radii is about 1 kcal/mol. Since the average energy of thermal motion at room temperature is about 0.6 kcal/mol and, according to the Boltzmann distribution (a mathematical expression that

Table 3-2 Van der Waals radii of several atoms found in biological molecules

Atom	Radius, Å
H	1.2
O	1.4
N	1.5
S	1.85
P	1.9
C	2.0

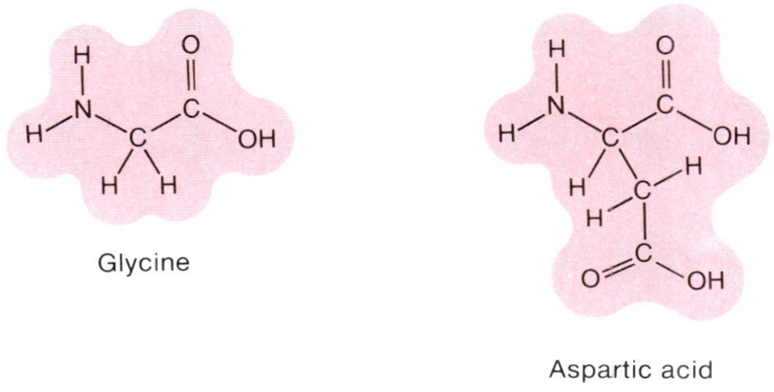

Glycine

Aspartic acid

Figure 3-11 The shapes of two amino acids, glycine and aspartic acid, determined from van der Waals radii. This is only a schematic drawing because the molecules are actually three-dimensional and all bonds are not in the plane of the paper.

relates the mean energy of a system of particles and the fraction of the total number of particles having each energy value), the energy of many molecules will exceed this value (and even exceed 1 kcal/mol), the van der Waals interaction between two atoms is not sufficient to maintain these atoms in proximity. However, if the interactions of *several* pairs of atoms are combined, the cumulative attractive force can be great enough to withstand being disrupted by thermal motion. Thus, two molecules can attract one another if several of their component atoms can mutually interact. However, because of the $1/r^6$-dependence, the intermolecular fit must be nearly perfect. What this means is that *two molecules can bind to one another if their shapes are complementary*. This is true also of two separate regions of a polymer—that is, the regions can hold together if their shapes match. Sometimes the van der Waals attraction between two regions is not large enough to effect this; however, it can significantly strengthen other weak interactions such as the hydrophobic interaction, if the fit is good.

Summary of the Effects of Noncovalent Interactions

The effect of noncovalent interactions is to constrain a linear chain to fold in such a way that different regions of the chain, which may be quite distant in a chain (if the chain were linear), are brought together. An example of a molecule showing the four kinds of noncovalent folding that have been described is shown in Figure 3-12.

In Chapters 4, 6, and 7 we will discuss the specific structures brought about by these interactions and the physical properties of nucleic acids and proteins. In order to appreciate the kinds of experiments that have provided this information, several methods for studying macromolecules must be understood. These are described in the following section.

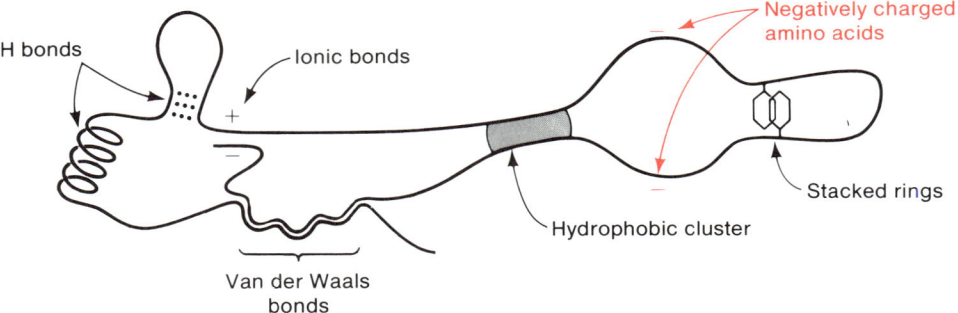

Figure 3-12   A hypothetical polypeptide chain showing attractive (black) and repulsive (red) interactions.

METHODS USED TO STUDY MACROMOLECULES

Several methods of studying macromolecules are used repeatedly in molecular biology. Some of these—e.g., spectrophotometry and chromatography—are usually covered in elementary chemistry courses and will not be discussed here. Three other techniques—ultracentrifugation, electrophoresis, and electron microscopy—will be described briefly in this section, so that the experiments presented in this book can be understood. For additional details, the reader should consult the references at the end of this book.

Velocity Sedimentation

Several important properties of macromolecules can be determined from sedimentation data. These studies cannot be done with an ordinary laboratory centrifuge because this instrument cannot produce a great enough centrifugal force to sediment molecules sufficiently rapidly that their positions are not randomized by diffusion. Modern ultracentrifuges can, however, generate forces as great as 700,000 times gravity, which is more than adequate to cause macromolecules to sediment through a solution.

The velocity with which a molecule moves is mainly a function of two properties of the molecule:

1. *Its molecular weight M.* As M increases, the velocity increases.
2. *Its shape.* The motion of any particle through a fluid is impeded by friction. If a ball and a stick having the same masses are moving through a liquid, the ball, being more compact, will encounter less frictional resistance and, hence, will move faster. This is true of macromolecules also: for a given value of M, the less extended a polymer chain is, the more rapidly it will move.

The ratio of molecular velocity to centrifugal force is called the **sedimentation coefficient, s.** That is,

$$s = \text{velocity/centrifugal force.}$$

The value of s for a particular molecule is often the same in many different solutions, so an s value is frequently considered to be a constant that characterizes a molecule. Furthermore, since the value of s depends on molecular weight and shape, changes in the s value, as experimental conditions are varied, can be used to monitor changes in molecular weight (such as aggregation or degradation) and in shape (such as unfolding an extended molecule to form a random coil.)

For most macromolecules the value of s is between 1×10^{-13} and 100×10^{-13} sec. Thus, in honor of The Svedberg, who invented the ultracentrifuge, 10^{-13} seconds is called one **svedberg** or one S. It is very common

Figure 3-13 Operations in zonal centrifugation. [From D. Freifelder, *Physical Biochemistry*, 2nd ed. (W. H. Freeman and Co., 1982).]

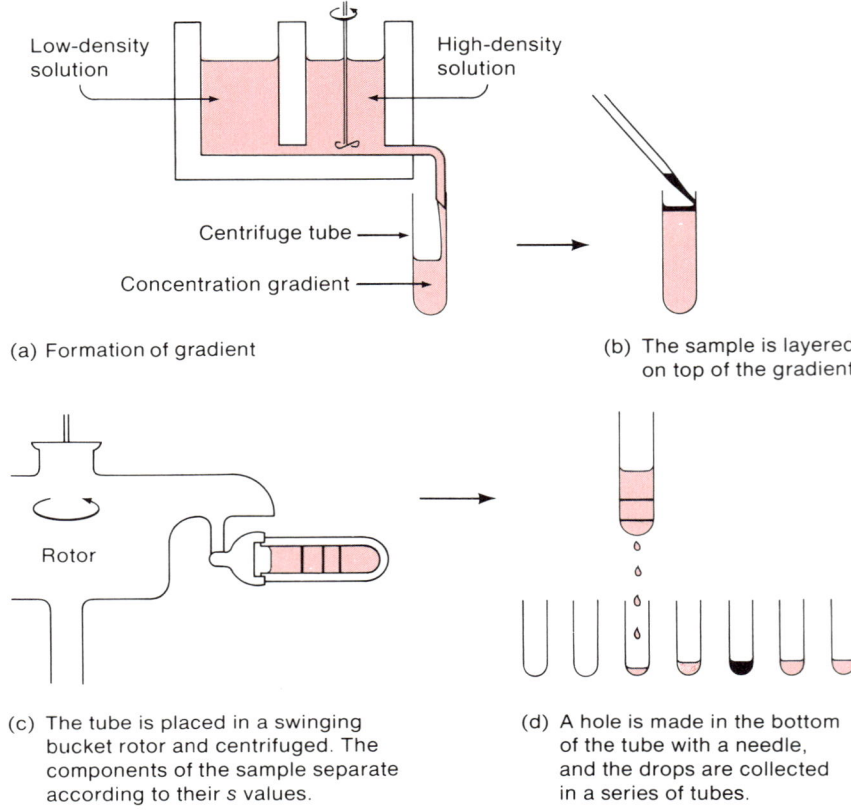

Low-density solution

High-density solution

Centrifuge tube ⟶

Concentration gradient ⟶

(a) Formation of gradient

(b) The sample is layered on top of the gradient

Rotor

(c) The tube is placed in a swinging bucket rotor and centrifuged. The components of the sample separate according to their *s* values.

(d) A hole is made in the bottom of the tube with a needle, and the drops are collected in a series of tubes.

to refer to a molecule whose *s* value is 30 svedbergs as a 30S molecule.

The most common type of sedimentation in use today is called **zonal centrifugation.** In this procedure (Figure 3–13), a centrifuge tube is filled with a sucrose solution (occasionally, glycerol or other solutes are used) whose concentration decreases continuously from the bottom of the centrifuge tube to the top of the tube. Since the density of the solution increases with increasing concentration, the tube contains a solution having a density gradient; the highest density is at the bottom of the tube. The density of the solution of molecules to be sedimented is adjusted to be lower than the density of the sucrose solution at the top of the tube, in order that the sample can be layered on the surface of the sucrose solution, thus forming a band or zone. Because of the density gradient, this procedure is often called **sucrose gradient centrifugation.** After layering the sample, the tube is centrifuged in a swinging bucket rotor for a particular time. After centrifugation is completed, a tiny hole is punched in the bottom of the tube and drops of the solution are collected. These drops represent successive layers of the tube. The drops are assayed for a particular macromolecule to obtain the distribution of the concentration of this macromolecule along the tube.

Equilibrium Centrifugation

Another centrifugation technique, **equilibrium centrifugation in a density gradient,** is also widely used. In this procedure, the macromolecules (usually nucleic acids or virus particles) are suspended in a solution of CsCl whose concentration is chosen so that its density, ρ, is approximately equal to that of the macromolecules. Initially, no sedimentation can occur because the buoyant force of the CsCl solution equals the centrifugal force.*

Under the influence of a powerful centrifugal force, the Cs^+ and the Cl^- ions themselves sediment somewhat, but they do not accumulate on the bottom of the centrifuge tube because the centrifugal force is not great enough to counteract the tendency for diffusion to maintain a uniform distribution of the ions. The result is that after a period of several hours the ions achieve an equilibrium concentration distribution in which there is a nearly linear concentration gradient, and hence a nearly linear density gradient, in the centrifuge tube. The density is maximal at the bottom of the tube. Once a density gradient has formed, macromolecules can begin to sediment. Those in the upper reaches of the tube move toward the bottom, stopping at the position at which their density equals the solution density. Similarly, macromolecules in the lower part of the tube move upward to the same position. In this way the macromolecules form a narrow band in the tube. (For this reason, the technique is often called CsCl "banding.") If the solution contains macromolecules having different densities, each macromolecule forms a band at the position in the gradient that matches its own density, and thus the macromolecules can be separated. The resolution of the technique is extraordinary. For example, DNA molecules with a density of 1.708 g/cm^3 can be separated from other DNA molecules in which the naturally occurring ^{14}N atoms have been substituted by ^{15}N atoms and which therefore have a density of 1.722 g/cm^3, as shown in Figure 3-14.

Electrophoresis

Most biological macromolecules carry an electrical charge and thus can move in an electric field. For example, if the terminals of a battery are connected to the opposite ends of a horizontal tube containing a solution of positively charged protein molecules, the molecules will move from the positive end of the tube to the negative end. The direction of motion obviously depends on the sign of the charge but the rate of movement depends on the magnitude of the charge and, as in sedimentation, on the shape of the molecule (that is, its frictional resistance). The mass of the molecule plays no direct role in

---

*To understand this, one should realize that if a centrifuge tube were to contain particles of wood floating on water, no amount of centrifugal force would make the wood reach the tube bottom, because the value of the force on the water, which is denser than the wood, is greater than the value of the force on the wood.

Figure 3-14 Formation of bands in a CsCl density gradient. The DNA solution contains a mixture of two DNA molecules that differ only in that one contains the normal isotope ^{14}N and the other contains the heavy isotope ^{15}N.

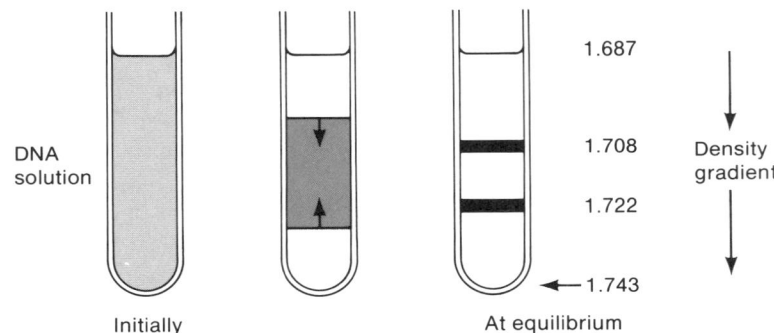

the rate of migration (in contrast with sedimentation) and influences the rate only indirectly when the surface area of the molecule, which affects its frictional coefficient, is a function of its mass.

The most common type of electrophoresis used in molecular biology is zonal electrophoresis through a gel, or **gel electrophoresis.** This procedure can be performed such that the rate of movement depends only on the molecular weight of the molecule, as will be seen below.

An experimental arrangement for gel electrophoresis of DNA is shown in Figure 3-15. A thin slab of an agarose or polyacrylamide gel is prepared containing small slots ("wells") into which samples are placed. An electric field is applied and the negatively charged DNA molecules penetrate and move through the agarose. A gel is a complex network of molecules, and migrating macromolecules must squeeze through narrow, tortuous passages. The result is that smaller molecules pass through more easily and thus the migration rate increases as the molecular weight, M, decreases. For unknown reasons the distance moved, D, depends logarithmically on M,

Figure 3-15 Apparatus for slab-gel electrophoresis capable of running seven samples simultaneously. The liquid gel is allowed to harden in place. An appropriately shaped mold is placed on top of the gel during the hardening in order to make wells for the samples (red). After electrophoresis, the slab is stained by removing the plastic frame and immersing the gel in a solution of the stain. Excess stain is removed by washing. The components of the sample appear as bands, which may be either visibly colored or fluorescent when irradiated with ultraviolet light. The region of the gel in which the components of one sample can move is called a *lane.* Thus, the gel shown has seven lanes. [After D. Freifelder, *Physical Biochemistry,* 2nd ed. (W. H. Freeman and Co., 1982).]

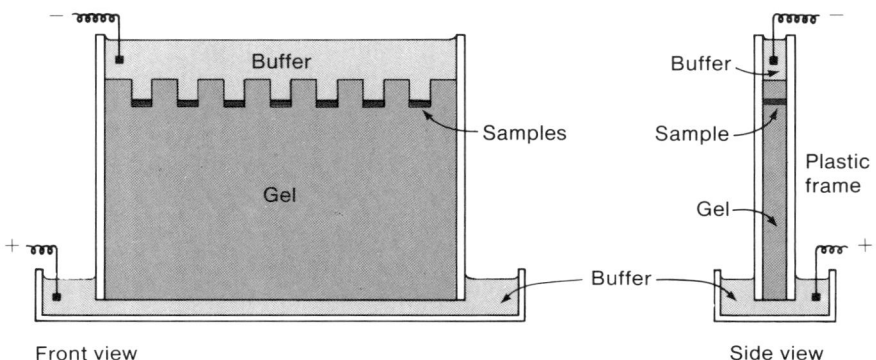

obeying the equation

$$D = a - b \log M,$$

in which a and b are empirically determined constants that depend on the buffer, the gel concentration, and the temperature. Figure 3-16 shows the result of electrophoresis of a collection of DNA molecules.

Gel electrophoresis of proteins is in principle carried out the same way. However, proteins can be either positively or negatively charged, and a sample containing several different proteins must be placed in a centrally located well in order that migration can occur in both directions. The charge per unit mass, which is very small because most amino acids are uncharged, varies from one protein to the next (in contrast with DNA or RNA, which have one negative charge per nucleotide) and proteins come in a variety of shapes, so there is no simple way to predict the migration rate. In the most commonly used technique the detergent sodium dodecyl sulfate (SDS) $H_3C-(CH_2)_{10}-OSO_3^- Na^+$ and the disulfide bond-breaking agent mercaptoethanol

```
        H   H
        |   |
   HO — C — C — SH
        |   |
        H   H
```

are added to the protein solution. One molecule of SDS binds per amino acid, giving each protein molecule the same charge per amino acid residue. Furthermore, *when SDS is bound and there are no disulfide bonds, all proteins have nearly the same shape, namely, the near-random coil.* The net effect is that, as in the case of DNA, the migration rate increases as M decreases and the dependence is logarithmic and described by the equation given above. Because a **poly**acrylamide **g**el is used for protein **e**lectrophoresis, the technique is called the **SDS–PAGE technique.** It is discussed in more detail in a later section.

Electron Microscopy

Macromolecules can be viewed directly by electron microscopy. Two of the many techniques of sample preparation will be described here. Others will be presented, as needed, elsewhere in the book.

The usual method for observing nucleic acids is the **Kleinschmidt spreading technique,** illustrated in Figure 3-17. The nucleic acid molecules, which are considerably larger than any protein molecule, are embedded in a protein film, as shown, and are coated with protein molecules. The protein molecules serve two functions:

1. They cause the nucleic acid molecule to be extended, for the following reason. If a very small volume of a protein solution is layered onto the

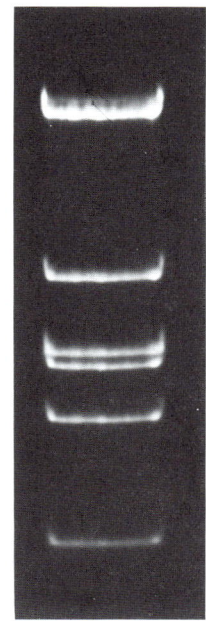

Figure 3-16 A gel electrophoregram of six fragments of *E. coli* phage λ DNA obtained by treating the DNA with the EcoRI restriction endonuclease. The direction of movement is from top to bottom. The DNA is made visible by the fluorescence of bound ethidium bromide. [Courtesy of Arthur Landy and Wilma Ross.]

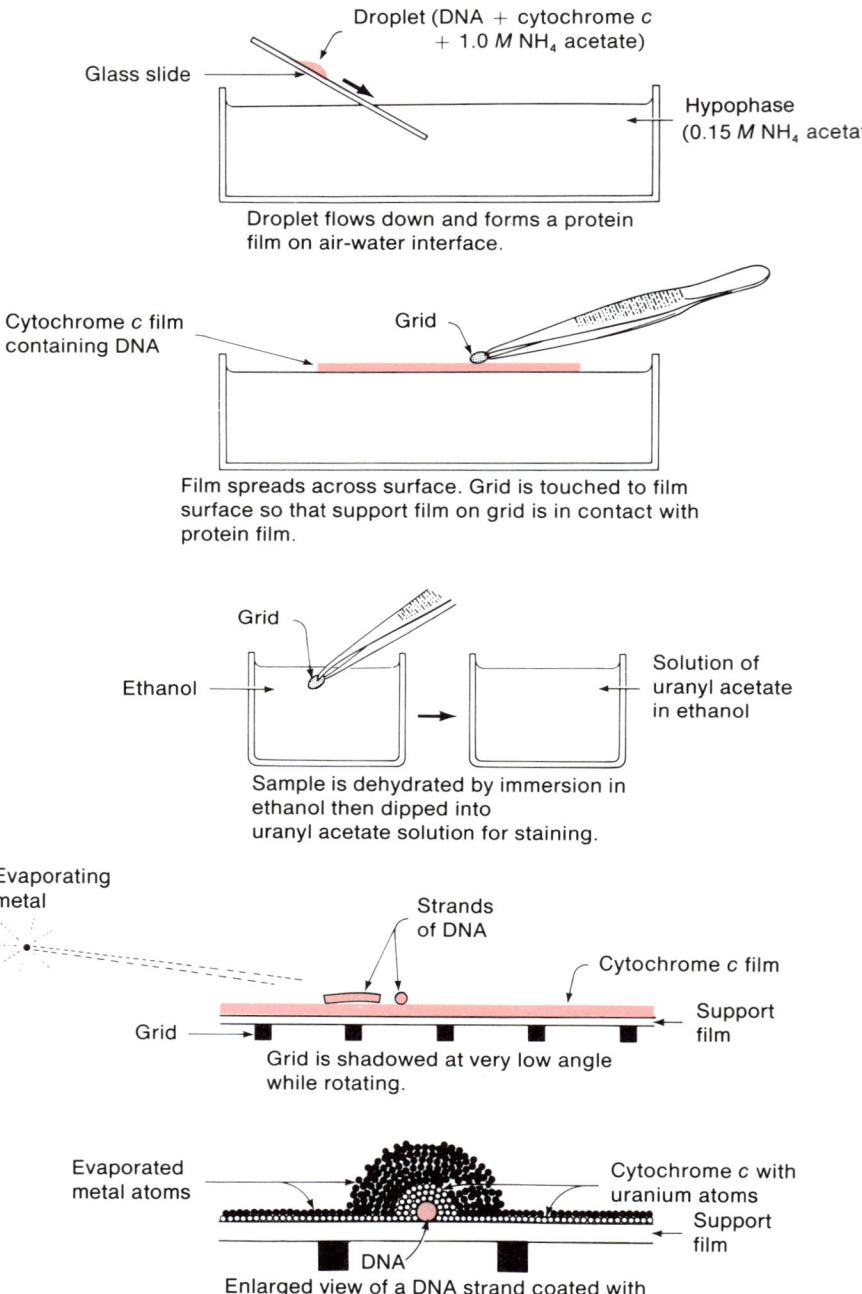

Droplet (DNA + cytochrome *c* + 1.0 *M* NH₄ acetate)

Glass slide

Hypophase (0.15 *M* NH₄ acetate)

Droplet flows down and forms a protein film on air-water interface.

Cytochrome *c* film containing DNA

Grid

Film spreads across surface. Grid is touched to film surface so that support film on grid is in contact with protein film.

Grid

Ethanol

Solution of uranyl acetate in ethanol

Sample is dehydrated by immersion in ethanol then dipped into uranyl acetate solution for staining.

Evaporating metal

Strands of DNA

Cytochrome *c* film

Grid

Support film

Grid is shadowed at very low angle while rotating.

Evaporated metal atoms

Cytochrome *c* with uranium atoms

Support film

DNA

Enlarged view of a DNA strand coated with cytochrome *c*, stained, and shadowed.

Figure 3-17 The Kleinschmidt method for preparing DNA for electron microscopy.

[From D. Freifelder, *Principles of Physical Chemistry,* (Jones and Bartlett, 1982).]

surface of an ionic solution, the protein molecules will spread out to form a monolayer film of protein. If nucleic acid molecules are in the protein solution, they will also spread out as the protein film expands; this spreading enables entire nucleic acid molecules to be seen without the tangling and intramolecular aggregation that is observed if spreading does not occur.

2. Proteins form a thick coating on the nucleic acid molecule and this coating makes the molecule easier to see. Something else is also needed for good observation because molecules containing only small atoms are ordinarily nearly transparent to electrons. Thus, to make a nucleic acid molecule visible, the protein–coated molecule is additionally coated with a thick layer of a heavy metal. This is accomplished either by evaporating platinum or uranium atoms onto the molecule (**shadowing**) or, in the case of uranium, chemically depositing uranium atoms onto the protein coating (**staining**). These heavy atoms are good absorbers of electrons, whereas the background support absorbs relatively few electrons. Thus, the metal ions provide visual contrast, and both shadowing and staining are very useful techniques for observing very long molecules when one is primarily interested in their length and linear topology. An example of an electron micrograph of DNA molecules is shown in Figure 3-18.

The **negative contrast technique** provides an alternative way to visualize molecules that cannot readily be seen by the protein monolayer technique just described and when one is interested in the shape and surface details of the molecule being examined. For example, a protein molecule would not be seen if it were embedded in a protein film. It would also not suffice to coat a single protein molecule with a thick layer of metal atoms because the metal layer would probably be comparable in size to or larger than the protein molecule, and obscure its shape as well. Even with larger particles, such as viruses, whose surface structures are of great interest, a thick metal layer will obscure all surface detail. An alternative is to use a very thin coating of metal atoms. These atoms are sufficiently small that, if the layer is only one or two atoms thick, the atoms will follow the contours of the particle. This can be done by staining with a very small amount of uranium atoms but is more conveniently done by the negative contrast technique. With this technique, the protein molecules or virus particles that are to be viewed are mixed with a solution of phosphotungstic acid (PTA) or uranium salts (usually uranyl acetate or uranyl formate), both of which contain atoms that absorb electrons strongly. The solution is then deposited as microdroplets on a supporting film. The droplets containing the macromolecules or particles of interest dry on the support film. The thickness of metal ions in the droplet is less where the macromolecule is present than elsewhere in the droplet. As shown in Figure 3-19, when the sample is placed in a beam of electrons, more electrons pass through the droplet in the region of a macromolecule; thus a negative image of a protein or a small virus

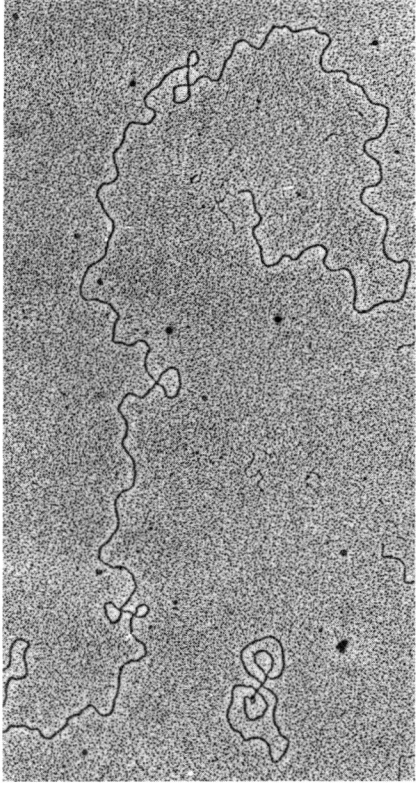

Figure 3-18 Electron micrograph containing a linear *E. coli* phage λ DNA molecule and a smaller circular molecules of *E. coli* phage φX174 DNA. [Courtesy of Manuel Valenzuela.]

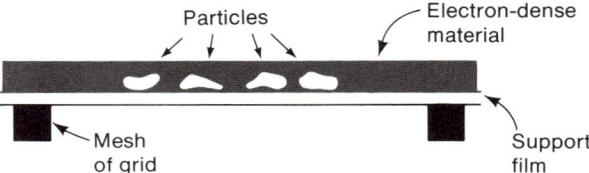

Figure 3-19   The negative-contrast method of visualizing particles by electron microscopy. Four particles are embedded in a substance that absorbs electrons strongly. As the beam passes through the sample, the fraction of the electrons in the beam that is absorbed depends on the total thickness of the substance; therefore, more electrons will pass through the regions containing each particle and the particle will appear bright against a dark background.

particle can be obtained. Electron micrographs of a virus particle and a protein molecule viewed in this way are shown in Figure 3-20.

DETERMINATION OF THE MOLECULAR WEIGHT OF A MACROMOLECULE

The large size of macromolecules eliminates the possibility of determining molecular weight M by procedures used for small molecules. For example, if the freezing-point-depression method, which is a standard procedure for small inorganic and organic molecules, were used for a typical DNA molecule, it would be necessary to detect a depression of about 10^{-6} °C. Thus, a variety of specialized techniques have been developed for nucleic acids and proteins. A few of the more common ones in current use will be described.

Determination of M for DNA Molecules

The means used to determine the molecular weight of a DNA molecule depends on the size of the molecule. That is, there are distinct procedures applicable to small, midsize, and large molecules. These will be discussed separately.

1. *Small DNA molecules ($M < 5 \times 10^6$).* For molecules in this size range, gel electrophoresis is best. Molecules for which M is accurately known are added to the sample and the relative distances migrated are measured. The data are plotted as shown in Figure 3-21.

2. *Midsize DNA molecules ($M = 5$ to 100×10^6).* For molecules in this size range, electron microscopy is most convenient. A measurement of the length of the molecule yields the value of M, since the mass per unit length of double-stranded DNA is 2×10^6 molecular weight units per micrometer.

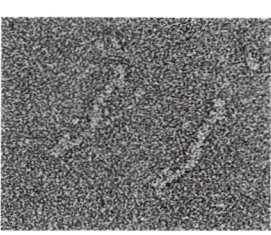

(a)

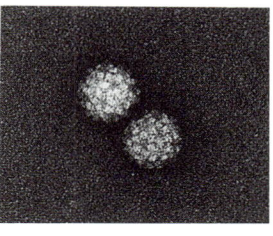

(b)

Figure 3-20   Electron micrographs obtained by the negative contrast procedure. (a) Molecules of the blood-clotting protein, fibrin. (b) Tomato bushy stunt virus particles—note the surface detail, which shows the individual protein molecules of which each particle is composed. [Courtesy of Robley Williams.]

Often errors arise in determining the value of the magnification of the molecules. A common technique that avoids the necessity of knowing this value is to add a molecule of precisely known molecular weight to the sample being studied and to determine the ratio of the lengths of the standard molecule and the molecule of interest. The most commonly used standard molecule is the circular double-stranded replicating form of *E. coli* phage φX174, which contains exactly 5386 nucleotide pairs. The smaller molecule in Figure 3-18 is φX174 DNA.

3. *Large DNA molecules* ($M > 100 \times 10^6$). A variety of specialized techniques are used for very large DNA molecules. These methods are described in the reference by Freifelder (1976, 1982) given at the end of this book.

Measurement of the Molecular Weight of Proteins

For accurate values of M for proteins, various centrifugation procedures (sedimentation equilibrium, Archibald method, Yphantis method), which are not described in this chapter, are used. These methods require several additional precise measurements. Information concerning these procedures can be found in references given at the end of this chapter.

The most convenient procedure for determining the molecular weights of proteins is SDS-PAGE electrophoresis (described earlier). This is a fairly accurate method. It is performed by adding a series of molecules for which M is known to the sample containing molecules of unknown M and plotting the relative distances migrated against log M, exactly as for DNA in Figure 3-21. There is a complication in this procedure—namely, the subunit effect. As mentioned earlier, SDS binds to polypeptide chains in a way that produces a fixed ratio of charge to mass and, furthermore, breaks all noncovalent interactions, so the molecule assumes a random coil conformation. However, many proteins consist of several identical polypeptide chains called subunits, joined together by noncovalent interactions. (This will be discussed in detail in Chapter 6.) Because SDS breaks noncovalent interactions, it dissociates these protein aggregates. Thus, the value of M that is measured is that of a single subunit. To determine M for the entire protein, the number of subunits must be known. This number can be determined in a variety of ways that will not be described here. The problem is slightly more complex if the subunits are not identical, in which case several different components are seen in the gel. Again, the number of each type of subunit can be determined and then the molecular weight of the protein can be calculated from the molecular weight of each subunit.

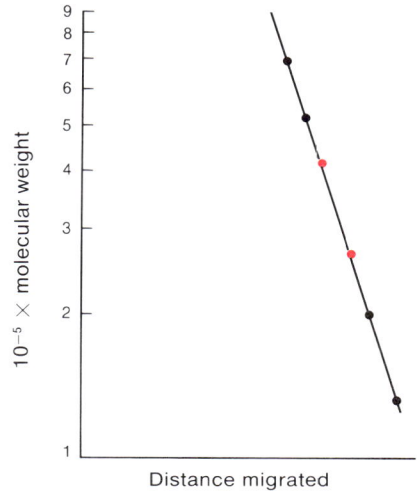

Figure 3-21   Determination of the molecular weights of DNA molecules (red points) by gel electrophoresis with DNA molecules whose molecular weights are known (black points). The line is a plot of the black points; the molecular weights of the molecules being studied are then determined from the positions of the red points.

4

Nucleic Acid Structure

Deoxyribonucleic acid (DNA) is the single most important molecule in living cells and contains all of the information that specifies cellular properties. In this chapter its remarkable structure and many of its physical and chemical properties are described. Some properties of ribonucleic acid (RNA), and how nucleic acids are studied, will also be presented.

PHYSICAL AND CHEMICAL STRUCTURE OF DNA

In the previous chapter the structure of a nucleotide and how nucleotides are joined to form a polynucleotide, or a nucleic acid, were described. In this section we show how nucleotides interact with one another to produce a double helix from two polynucleotides having complementary base sequences, and we present the Watson–Crick model for the DNA double helix.

Base-pairing and the DNA Double Helix

In early physical studies of DNA a variety of experiments indicated that the molecule is an extended chain having a highly ordered structure. The most important technique was x-ray diffraction analysis, by which information was obtained about the arrangement and dimensions of various parts of the

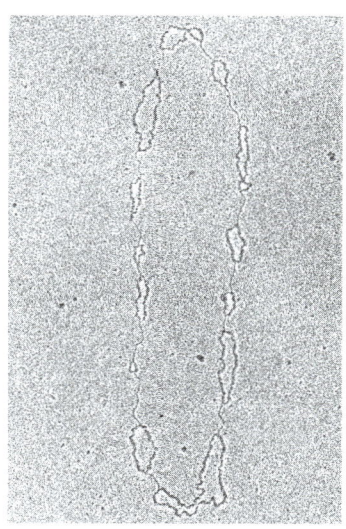

A circular DNA molecule treated with the T4 gene-*32* protein, which binds to single-stranded DNA and has caused unwinding of 13 regions of the DNA. Because of the heavy protein coating, the single-stranded DNA appears thicker than the double-stranded regions [Courtesy of C. Brack.]

molecule. The most significant observations were that the molecule is helical and that the bases of the nucleotides are stacked with their planes separated by a spacing of 3.4 Å.

Chemical analysis of the molar content of the bases (generally called the **base composition**) adenine, thymine, guanine, and cytosine in DNA molecules isolated from many organisms provided the important fact that [A] = [T] and [G] = [C], in which [ ] denotes molar concentration, from which followed the corollary

$$[A+G] = [T+C] \quad \text{or} \quad [\text{purines}] = [\text{pyrimidines}].$$

James Watson and Francis Crick combined chemical and physical data for DNA with a feature of the x-ray diffraction diagram that suggested that two helical strands are present in DNA and showed that the two strands are coiled about one another to form a double-stranded helix (Figure 4-1). The sugar-phosphate backbones follow a helical path at the outer edge of the

Figure 4-1 (a) Diagram of the DNA double helix in the common B form. (b) Space-filling model of DNA, again in the B form. [Courtesy of Sung-Hou Kim.]

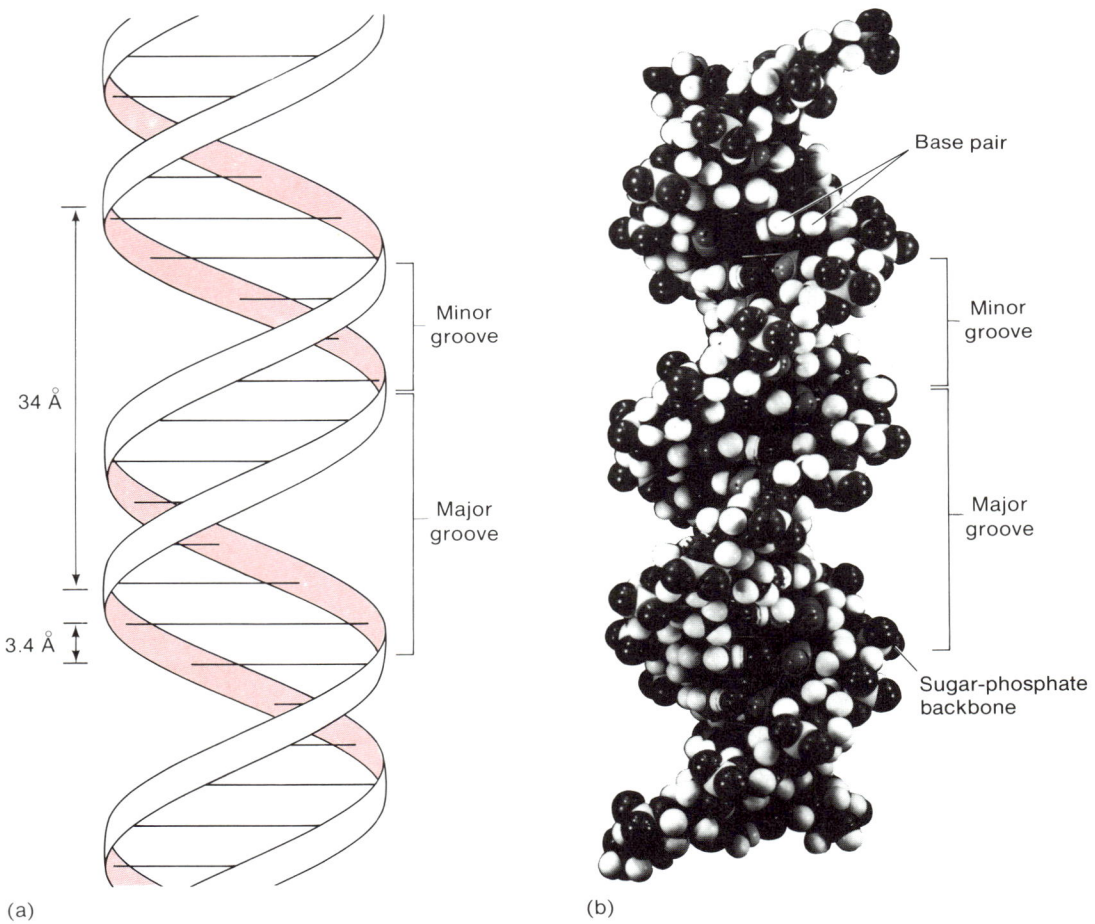

34 Å

3.4 Å

Minor groove

Major groove

(a)

Base pair

Minor groove

Major groove

Sugar-phosphate backbone

(b)

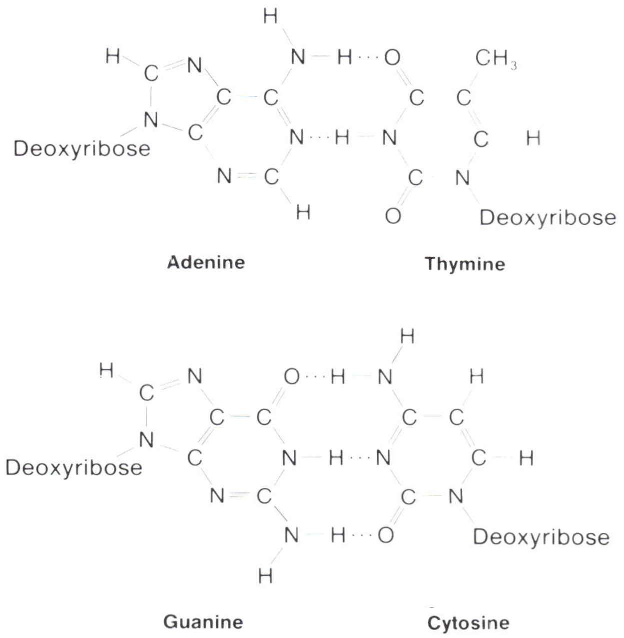

Figure 4-2 The two common base pairs of DNA. Note that hydrogen bonds (dotted lines) are between the weakly charged groups noted in Figure 3-6.

Adenine   **Thymine**

Guanine   **Cytosine**

molecule and the bases are in a helical array in the central core. The bases of one strand are hydrogen-bonded to those of the other strand to form the purine-to-pyrimidine base pairs A·T and G·C (Figure 4-2). Because each pair contains one two-ringed purine (A or G) and one single-ringed pyrimidine (T or C, respectively), the length of each pair (in the sugar-to-sugar direction) is about the same and the helix can fit into a smooth cylinder.

The two bases in each base pair lie in the same plane and the plane of each pair is perpendicular to the helix axis. The base pairs are rotated $36°$ with respect to each adjacent pair, so there are 10 pairs per helical turn. The diameter of the double helix is 20 Å and the molecular weight per unit length of the helix is approximately 2×10^6 per micrometer. Since the molecular weight of a typical bacterial DNA molecule is about 2×10^9, such a molecule is 1 millimeter or 10^7 Å long and is very long and thin indeed.

The helix has two external helical grooves, a deep wide one (the **major groove**) and a shallow narrow one (the **minor groove**); both of these grooves are large enough to allow protein molecules to come in contact with the bases.

Base-pairing is one of the most important features of the DNA structure because it means that the base sequences of the two strands are complementary (Figure 4-3); that is, if one strand has the base sequence AATGCT, the other strand has the sequence TTACGA, reading in the same direction. This has deep implications for the mechanism of DNA replication because, in this way, the replica of each strand is given the base sequence of its complementary strand.

Figure 4-3 A diagram of a portion of a DNA molecule showing complementary base-pairing of the individual strands. The thin lines represent hydrogen bonds.

Antiparallel Orientation of the Two Strands of DNA

The two polynucleotide strands of the DNA double helix are antiparallel—that is, the 3'-OH terminus of one strand is adjacent to the 5'-P (5'-phosphate) terminus of the other (Figure 4-4). The significance of this is twofold. First, in a linear double helix there is one 3'-OH and one 5'-P terminus at each end of the helix. The second and generally more significant point is that the orientations of the two strands are different. That is, if two nucleotides are paired, the sugar portion of one nucleotide lies upward along the chain whereas the sugar of the other nucleotide lies downward. We will see in Chapter 9 that this structural feature poses a constraint on the mechanism of DNA replication.

Because the strands are antiparallel, a convention is needed for stating the sequence of bases of a single chain. The convention is to write a sequence with the 5'-P terminus at the left; for example, ATC denotes the trinucleotide P-5'-ATC-3'-OH. This is also often written as pApTpC, again using the conventions that the left side of each base is the 5' terminus of the nucleotide and that a phosphodiester group is represented by a p between two capital letters.

Variation of Base Composition of the DNA of Different Organisms

Although it is always true that [A] = [T] and [G] = [C], there is no rule governing the ratio of total concentrations of guanine and cytosine to adenine and thymine in a DNA molecule—that is, ([G] + [C])/([A] + [T]). In fact, there is enormous variation in this ratio, ranging from 0.37 to 3.16

Figure 4-4   A stylized drawing of a segment of a DNA duplex showing the antiparallel orientation of the complementary chains. The arrows indicate the 5'→3' direction of each strand. The phosphates (P) join the 3' carbon of one deoxyribose (horizontal line) to the 5' carbon of the adjacent deoxyribose.

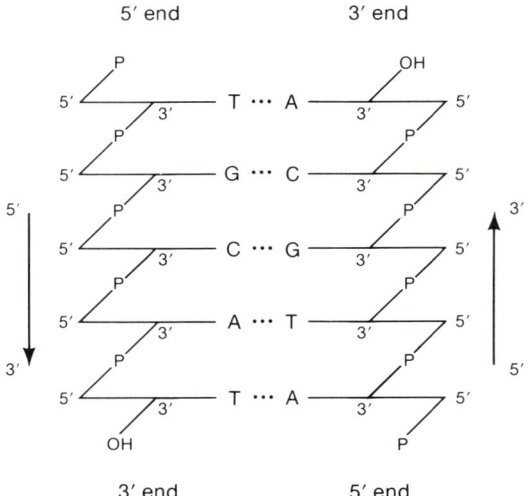

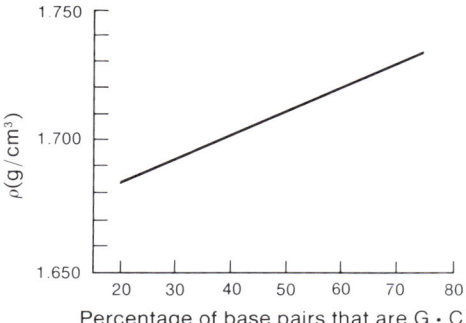

Figure 4-5 Density (ρ) of DNA in CsCl as a function of guanine-plus-cytosine content. The equation that describes the curve is $\rho = $ [percent (G + C) $\times$ 0.00098] + 1.660 g/cm^3.

for different species of bacteria. The usual way that the base composition of the DNA molecule from a particular organism is expressed is not by the ratio just given but by the fraction of all bases that are G·C pairs; that is, ([G] + [C])/[all bases]; this is termed the G + C content or percent G + C. The base composition of hundreds of organisms has been determined. Generally speaking, the value of the G + C content is near 0.50 for the higher organisms and has a very small range from one species to the next (0.49–0.51 for the primates); for the lower organisms the value of the G + C content varies widely from one genus to another. For example, for bacteria the extremes are 0.27 for the genus *Clostridium* and 0.76 for the genus *Sarcina;* E. *coli* DNA has the value 0.50. A convenient way to measure the base composition is by centrifuging the DNA to equilibrium in CsCl (see Chapter 3), because the density of DNA in CsCl solutions is linearly related to the G + C content, as shown in Figure 4-5.

Forms of the DNA Helix

The double helix shown in Figure 4-1 is a right-handed helix. This means that when an observer looks down the axis of the helix in either direction, each strand follows a clockwise path as it moves away from the observer. If the path were counterclockwise, the helix would be left-handed.

Naturally occurring DNA molecules are generally right-handed helices.

In a later section we will examine a synthetic DNA molecule, which in certain conditions forms a left-handed helix, and discuss evidence that some naturally occurring DNA molecules may have regions that are left-handed helices.

The helical structure of DNA was worked out from x-ray diffraction data, as has been stated. This required a sample in which the DNA molecules were oriented; this was accomplished by using a gelatinous fiber drawn from a very viscous solution of DNA. In order to prevent the fiber from drying and becoming disoriented, it was maintained in an atmosphere of high humidity. Two forms of DNA were observed: the A form at 66 percent relative humidity and the B form at 92 percent relative humidity. In the A form there are 11 base pairs per turn of the helix and the planes of the base pairs are tilted 20° away from the perpendicular to the helix axis. In the B form, which is the one shown in Figure 4-1, there are 10 base pairs per turn and the base pairs are perpendicular to the helix axis. The B form is the "classic" Watson-Crick structure. Since DNA is normally studied in solution and since within a cell there is a high water content, it is important to know which of these forms is present in these conditions. A great deal of work has shown that the form in solution is very near that of the B form; the rotation of adjacent base pairs is probably 34.5–35.5° (rather than 36°) with 10.2–10.4 pairs per turn of the helix (instead of 10). The intracellular structure of DNA is considered to be B-like, but, as will be seen, the molecule is often tightly bound to proteins, which certainly causes local alteration of the structural parameters.

The Size of DNA Molecules

In viruses and prokaryotes, the total genome* specifying the virus or cell is usually encompassed in a single DNA molecule (the RNA viruses are an exception to this rule). In eukaryotes, including the unicells such as algae, yeast, and protozoa, the DNA is partitioned into a number of chromosomes, each of which is believed to contain a single gigantic DNA molecule. Table 4-1 shows the molecular weights, M, of the individual DNA molecules of various organisms. The values for many phage, viral, and bacterial DNA molecules are very accurately known (within 2 percent and, in some cases, exactly); the uncertainty in the values for eukaryotic DNA molecules is ±50

Table 4-1   Sizes of various DNA molecules

Organism or particle	Molecular weight, M	Organism or particle	Molecular weight, M
15 plasmid*	1.4×10^6	Vaccinia virus	121×10^6
Polyoma virus	3.2×10^6	Fowlpox virus	178×10^6
Phage 186*	18×10^6	F'450 dimer plasmid*	210×10^6
Phage T7*	25×10^6	*Mycoplasma homina*	5.3×10^8
Phage λ*	32×10^6	Most bacteria	$2.0–2.6 \times 10^9$
F plasmid*	62×10^6	Yeast	6×10^8
F'*lac* plasmid*	95×10^6	*Drosophila* (fruit fly)	7.9×10^{10}
Phage T4*	106×10^6	Human	8×10^{10}

Note: Phages and plasmids marked with an asterisk have *E. coli* as a host. *Mycoplasma homina* is the smallest known bacterium. For yeast, *Drosophila*, and humans the molecular weight of the largest DNA molecule in the organism is given.

*The **genome** is the total DNA content of a haploid organism or, for a diploid organism, the DNA that contains one complete set of genes.

percent in some cases. Note that the range of size is very great among the viruses and phages but much less so for bacteria.

The length of these molecules can be obtained from this relation: 1 micrometer (μm) $= 2 \times 10^6$ molecular weight units. Thus, the molecules listed in the table range from 0.7 μm to 40,000 μm (4 cm!). The width of a DNA molecule is 20 Å $= 0.002$ μm.

In general, the genomes of the more complex organisms require much more DNA than for simpler organisms (though the cells of both the toad and the South American lungfish have considerably more DNA than human cells).

Fragility of DNA Molecules

The great length of DNA molecules makes them extremely susceptible to breakage by the hydrodynamic shear forces resulting from such ordinary operations as pipetting, pouring, and mixing. Unbroken DNA molecules for which $M < 2 \times 10^8$ can usually be isolated with ease from phages and viruses. However, when DNA is isolated from bacteria and higher organisms, unless great care is taken, the DNA molecules are almost always broken. In fact, the mean value of M for a sample is usually about 25×10^6, so bacterial DNA, for instance, is frequently fragmented into about a hundred pieces. For plant and animal cells the yield of unbroken molecules in the least broken samples is usually a small fraction of 1 percent. Later in this chapter and elsewhere in the book we will see several instances in which the fact that the DNA of bacteria and of higher organisms is invariably fragmented by manipulation has important experimental consequences.

THE FACTORS THAT DETERMINE THE STRUCTURE OF DNA

The helical structure of nucleic acids is determined by stacking between adjacent bases in the same strand, and the double-stranded helical structure of DNA is maintained by hydrogen-bonding between the bases in the base pairs. This conclusion, as well as features of the structure that have not yet been described, has come from studies of denaturation. This analysis is presented in this section.

Denaturation and Melting Curves

The free energies of the weaker noncovalent interactions described in Chapter 3 are not much greater than the energy of thermal motion at room temperature; thus, at elevated temperatures the three-dimensional structures of both proteins and nucleic acids are disrupted. A macromolecule in a disrupted state, in which the molecules are in a nearly random coil conformation, is said to be **denatured;** the ordered state, which is presumably that originally present in nature, is called **native.** A transition from the native to the denatured state is called **denaturation.** When double-stranded DNA or native DNA is heated, the bonding forces between the strands are disrupted and the two strands separate; thus, *denatured DNA is single-stranded* (but see later section entitled Structure of Denatured DNA).

A great deal of information about structure and stabilizing interactions has been obtained by studying nucleic acid denaturation. This is done by measuring some property of the molecule that changes as denaturation proceeds—for example, the absorption of ultraviolet light. Originally, denaturation was always accomplished by heating a DNA solution, so a graph of a varying property as a function of temperature is called a **melting curve.** Reagents are known that either break hydrogen bonds or weaken hydrophobic interactions. These are powerful denaturants. Thus, denaturation is also studied by varying the concentration of a denaturant at constant temperature. In the case of DNA, the simplest way to detect denaturation is to monitor the ability of DNA in a solution to absorb ultraviolet light whose wavelength is 260 nanometers (nm). The bases of nucleic acids absorb 260-nm light strongly. A convenient measure of the absorption is the absorbance (A) of a solution—that is, $-\log_{10}$ [(intensity of light transmitted by a solution 1 cm thick)/(intensity of incident light)]. The absorbance at 260 nm, A_{260}, is proportional to concentration, with a value of 0.02 units per microgram DNA per milliliter. More important, the amount of light absorbed by nucleic acids is dependent on the structure of the molecule. The more ordered the structure, the less light is absorbed. Therefore, free nucleotides absorb more light than a single-stranded polymer of DNA or RNA and these in turn absorb more light than a double-stranded DNA molecule. For example, three solutions of double-stranded DNA, single-stranded DNA, and free bases, each at 50 μg/ml, have the following A_{260} values:

Double-stranded DNA	$A_{260} = 1.00$
Single-stranded DNA	$A_{260} = 1.37$
Free bases	$A_{260} = 1.60$

This relation is often described by stating either that double-stranded DNA is **hypochromic** or that the bases are **hyperchromic.**

If a DNA solution is slowly heated and the A_{260} is measured at various temperatures, a melting curve such as that shown in Figure 4-6 is obtained. The following features of this curve should be noted:

1. The A_{260} remains constant up to temperatures well above those encountered by living cells in nature.
2. The rise in A_{260} occurs over a range of 6–8°C.
3. The maximum A_{260} is about 37 percent higher than the starting value.

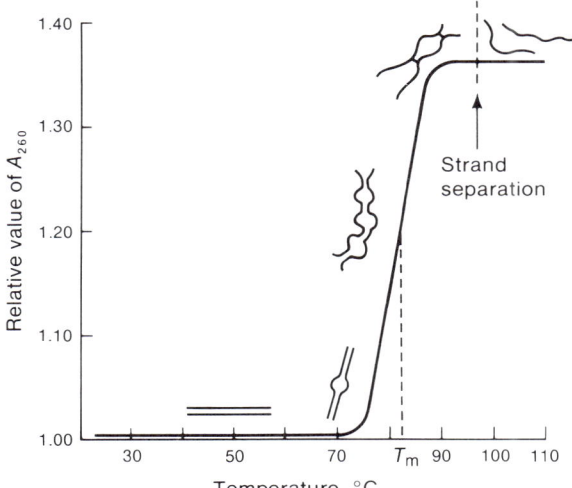

Figure 4-6 A melting curve of DNA showing T_m and possible molecular conformations for various degrees of melting.

The state of a DNA molecule in different regions of the melting curve is also shown in the figure. Before the rise begins, the molecule is fully double-stranded. In the rise region, base pairs in various segments of the molecule are broken; the number of broken base pairs increases with temperature. In the initial part of the upper plateau a few base pairs remain to hold the two strands together until a critical temperature is reached at which the last base pair breaks and the strands separate completely.

A convenient parameter to characterize a melting transition is the temperature at which the rise in A_{260} is half complete. This temperature is called the **melting temperature;** it is designated $\boldsymbol{T_m}$.

In the next section we will see how melting curves are used to understand the interactions that stabilize DNA.

Evidence for Hydrogen Bonds in DNA

A great deal of information is obtained by observing the variation of T_m with base composition and experimental conditions. For example, DNA can be isolated from various bacterial species in which the base compositions vary from 20 percent G + C to 80 percent G + C. The values of T_m from many such DNA molecules are plotted versus percent G + C in Figure 4-7. T_m increases with increasing percent G + C. This fact is interpreted in terms of the relative number of hydrogen bonds in a G·C pair (three) and an A·T pair (two). That is, a higher temperature is required to disrupt a G·C pair than an A·T pair, because the double-stranded structure is stabilized, at least in part, by hydrogen bonds. This conclusion has been substantiated by measuring the value of T_m in the presence of reagents such as urea and formamide (see margin). These two substances are capable of hydrogen–bonding

$$H_2N-\overset{\overset{\displaystyle O}{\|}}{C}-NH_2 \qquad H_2N-\overset{\overset{\displaystyle O}{\|}}{C}-H$$

Urea **Formamide**

Figure 4-7 Plot of T_m versus G + C content of various DNA molecules.

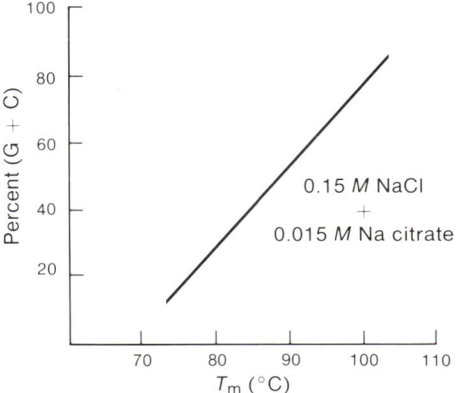

with the DNA bases. Broken and unbroken base pairs are in thermal equilibrium and these reagents compete with one base for pairing to another base; thus, they can maintain the unpaired state at a temperature at which a broken base pair might normally pair again. Hence, permanent melting of a section of paired bases requires less input of thermal energy and T_m is reduced. A reduction of T_m is a characteristic of all known reagents that can form hydrogen bonds with nucleotides.

Evidence for Hydrophobic Interactions in DNA

Any reagent that would either enhance the interaction of weakly soluble substances with water or would disrupt the water shell about the substance should weaken hydrophobic interactions. An example of the former type of substance is methanol, which increases the solubility of the bases. The salt sodium trifluoracetate is an example of the second type. The addition of both of these reagents reduces T_m enormously, which suggests that hydrophobic interactions are also important in stabilizing the DNA structure. (Figure 4-8 shows data for the salt.) This conclusion follows from another experiment. As was discussed in Chapter 3, low solubility of substances favors hydrophobic clustering. The solubility of adenine in solutions containing a wide variety of reagents, such as alcohols, amines, and other soluble organic compounds, was measured. Many of the added substances increased the solubility of adenine and at the same time decreased the T_m of DNA suspended in a solution having the same composition. In fact, the extent to which a reagent lowered T_m correlated well with the increased solubility of the base. To confirm that this is an effect on the bases only, the solubility of sodium pyrophosphate, a highly charged molecule that presumably has the solubility properties of the sugar-phosphate backbone, was also measured. It was found that reagents that increase the solubility of adenine and decrease

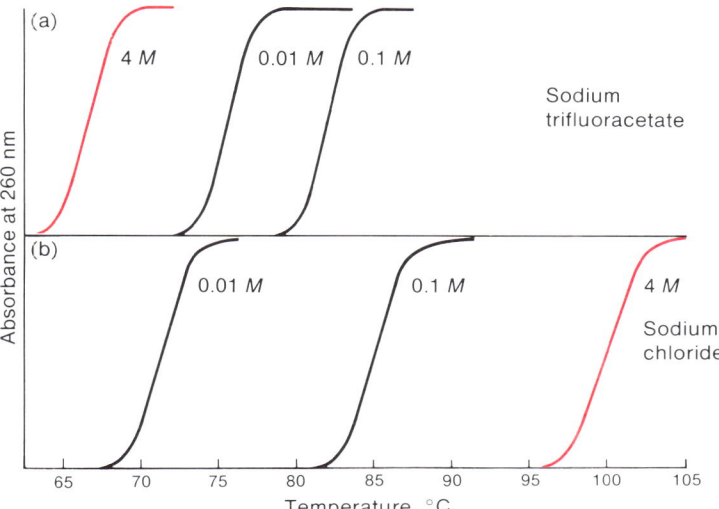

Figure 4-8   A comparison of the effect of sodium trifluoracetate and NaCl on the melting behavior of DNA. With NaCl the melting temperature increases with concentration because the high ionic strength neutralizes the negatively charged phosphates and stabilizes hydrogen bonds. With sodium trifluoracetate the melting temperature rises initially as the concentration increases but then decreases as binding of water by the trifluoracetate ion weakens the hydrophobic interaction.

T_m either have no effect on the solubility of sodium pyrophosphate or decrease it slightly. These results suggest that a hydrophobic interaction is present in DNA. Furthermore, the results imply that a polynucleotide would tend to have a three-dimensional structure that maximizes the contact of the highly soluble phosphate group with water and minimizes contact between bases and water. This explains why, in DNA, the sugar-phosphate chain is on the outside and the bases are on the inside.

Base-stacking

The experiments just described implicate hydrophobic interactions as a stabilizing effect but do not indicate that the bases are stacked. That they do stack became clear when the techniques of optical rotatory dispersion (ORD) and circular dichroism (CD) were applied to single-stranded DNA and to RNA; these experiments showed that a large fraction of the bases in single-stranded polynucleotides are arranged in a helix. A parallel array of bases was even detected in dinucleotides, which suggests that base-stacking is present even in these small molecules. In double-stranded polynucleotides (both DNA and synthetic model compounds) stacking was also found to be present. Furthermore, in all cases it was found that

1. Base-stacking is eliminated by the addition of reagents that weaken hydrophobic interactions.
2. If a DNA sample is heated, base-stacking is reduced during the same transition in which the A_{260} increases.

3. Reagents that break down hydrogen–bonding have no effect on base-stacking of single-stranded polynucleotides but reduce base-stacking of double-stranded DNA to the degree found in denatured DNA.

These results lead to the conclusion that the bases of double-stranded DNA are *more* stacked than those in single-stranded DNA. The increased stacking in double-stranded DNA can be accounted for by considering the effect of hydrogen bonds between the two strands on the structure of each of the individual strands. Both hydrogen bonds and hydrophobic interactions are weak and easily disrupted by thermal motion. Maximum hydrogen-bonding is achieved when all bases are pointing in the right direction. Similarly, stacking is enhanced if the bases are unable to tilt or swing out from a stacked array. Clearly, stacked bases are more easily hydrogen-bonded and correspondingly, hydrogen-bonded bases, which are oriented by the bonding, stack more easily. Thus, the two interactions act cooperatively to form a very stable structure. If one of the interactions is eliminated, the other is weakened; this explains why T_m drops so markedly after the addition of a reagent that destroys either type of interaction.

Cooperativity of Base-stacking

Base stacking itself is also a cooperative interaction. That is, in a sequence of stacked bases, ABCDEFGHIJ, it would be very unlikely for base E to swing out of the stacked array because the plane of the base tends to be parallel to the planes of both D and F. However, the tendency to conform to an orderly stacked array is not so great at the ends of the molecule. That is, the orientation of base A is stabilized by only a single base, namely, B; thus a rapidly moving solvent molecule might crash into A and cause it to rotate out of the stack more easily than collision with E would cause disorientation of E. Since A has a lower probability of being stacked than B, then of course B must also be more easily disoriented than is C. This slight tendency toward instability is also present in a double-stranded molecule. This is most noticeable in the value of T_m for double-stranded polynucleotides containing fewer than twenty base pairs (oligonucleotides). For example, if a molecule having 10^5 base pairs is broken down to fragments having 10^3 base pairs, there is no detectable change in T_m. However, with conditions for which the T_m for a large molecule is 90°C, the value of T_m for a double-stranded hexanucleotide (six nucleotides per strand) can be as low as 30°C; the exact value depends on the base composition and sequence. This effect has several consequences:

1. The ends of a linear double-stranded DNA molecule are usually not hydrogen-bonded, but are frayed (Figure 4-9), with about seven base pairs broken. (Some base sequences stack better than others and are

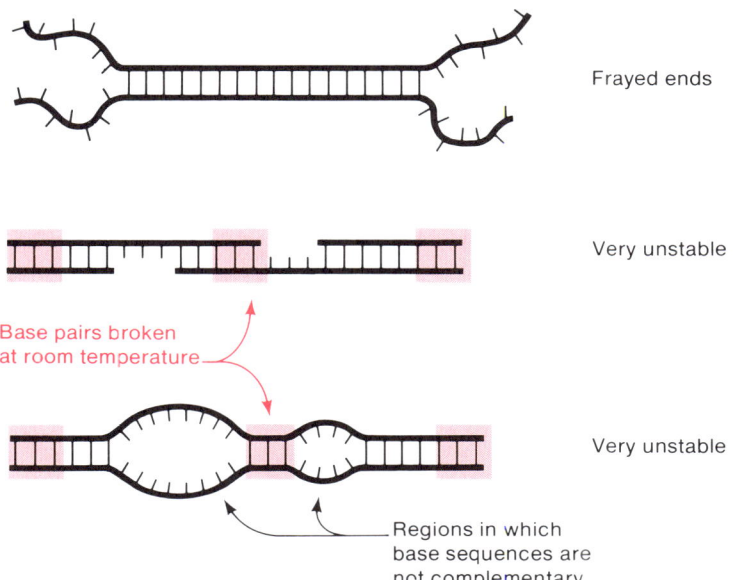

Frayed ends

Base pairs broken
at room temperature

Very unstable

Very unstable

Regions in which
base sequences are
not complementary

Figure 4-9 Several effects of cooperativity
of base-stacking. Each shaded area indicates
base pairs that would be broken at room
temperature; if these tracts contained more
than fifteen base pairs, they would be stable.

even stacked at an end of a molecule. An example is the sequence
CCA, which is found at the 3′ terminus of tRNA molecules.)
2. Short double-stranded oligonucleotides, having fewer than fifteen
   base pairs per molecule, have particularly low values of T_m. A
   double-stranded trinucleotide (three bases per strand) is not stable at
   room temperature.
3. Molecules in which the paired regions are very short and are flanked
   by unpaired regions, such as those shown in the lower part of Figure
   4-9, cannot maintain that conformation at physiological temperatures.

Effect of the Ionic Strength of a Solution on the Structure of DNA

In addition to the cooperative attractive interactions between adjacent DNA
bases and between the two strands, there is an interstrand electrostatic repulsion between the negatively charged phosphates. (There is also an intrastrand
repulsion, which is probably not important.) This strong force would drive
the two strands apart if the charges were not neutralized. By examining the
variation of T_m as a function of the ionic concentration of the buffer solution,
it is found that T_m decreases sharply as salt concentration decreases; indeed,
in distilled water, DNA denatures at room temperature. The interpretation
of the data is the following. In the absence of salt the strands repel one

another. As salt is added, positively charged ions (e.g., Na^+) form "clouds" of charge around the negatively charged phosphates and effectively shield the phosphates from one another. Ultimately all of the phosphates are shielded and repulsion ceases; this occurs near the physiological salt concentration of about 0.2 M. However, T_m continues to rise as the [NaCl] is increased, because the solubility of the bases decreases at high [NaCl], which increases the hydrophobic interaction.

A second effect also occurs—namely, neutralization of the negative charges by *binding* of the Na^+ ion. It might be thought that since ionization of Na^+ salts is always complete, such binding would not occur. However, since the molar concentrations of most DNA solutions are rarely more than 10^{-4} and we are talking about solutions in which [Na^+] is in $10-10^4$-fold excess, it should not be surprising that some Na^+ ions are bound to the DNA. Several lines of evidence indicate that positive ions are bound by DNA. For example, if the molecular weight of a particular DNA molecule is measured first in NaCl and then in CsCl, it is found that the values obtained are different, and that the ratio of these values is 0.75. This is explained by noting the molecular weights of a nucleotide (330 on the average), the Na^+ ion (23), and the Cs^+ ion (137). Thus, a nucleotide of NaDNA has a molecular weight of $23 + 330 = 353$ and CsDNA has a molecular weight of $137 + 330 = 467$. The ratio of these molecular weights, namely, $353/467 = 0.75$, is the measured ratio for the DNA molecules, implying that in NaCl and CsCl solutions, most of the phosphate groups have bound Na^+ and Cs^+ ions, respectively.

Fluctuation of the Structure of the DNA Molecule

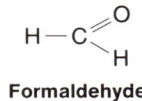

Formaldehyde

An important structural feature of the DNA molecule becomes apparent when the DNA is examined in the presence of formaldehyde (see margin). Formaldehyde can react with the NH_2 group of the bases and thus eliminate their ability to hydrogen-bond. The addition of formaldehyde causes a slow and irreversible denaturation of DNA. Thus, the amino groups must be available to the formaldehyde to allow the reaction, which means that bases are continually being paired and unpaired (that is, hydrogen bonds are breaking and reforming). A related phenomenon is observed when DNA is dissolved in tritiated water (^3H_2O)—there is a rapid exchange between the hydrogen-bonded protons of the bases and the $^3H^+$ ions in the water. These two observations indicate that *DNA is a dynamic structure in which double-stranded regions frequently open to become single-stranded bubbles*. This important phenomenon, called **breathing,** is thought to enable specialized proteins to interact with the DNA molecule and to "read" its encoded information. Note that since a G·C pair has three hydrogen bonds and an A·T pair has only two, breathing occurs more often in regions rich in A·T pairs than in regions rich in G·C pairs.

Denaturation and Strand Separation

Certain changes in the physical properties of DNA solutions that accompany denaturation—for example, a decrease in viscosity and the ability to rotate polarized light—indicate that when hydrogen bonds and hydrophobic interactions are eliminated, the helical structure of DNA is disrupted and the molecule loses its rigidity. This collapse of the ordered structure may be accompanied by complete disentanglement of the two strands. At one time it was thought that DNA strands are so long that complete unwinding of the helix and strand separation would be impossible, but several lines of evidence clearly indicate that strand separation does occur. One of these experiments will be described.

In this experiment DNA was prepared that had one ^{14}N-labeled strand and one ^{15}N-labeled strand. Such a molecule (designated [^{14}N^{15}N]DNA) has a unique density in CsCl equal to the average of the densities of [^{14}N^{14}N] and [^{15}N^{15}N]DNA (i.e., molecules having the same isotope in both strands). The DNA was heated to various temperatures and A_{260} was measured to obtain a melting curve. Each sample was also subjected to equilibrium centrifugation in a CsCl density gradient. It was found that shortly after the maximum A_{260} was reached, two bands appeared in the CsCl solution; these bands had the densities of ^{14}N-labeled and ^{15}N-labeled single-stranded DNA respectively and contained the separated single strands (Figure 4-10).

In the course of studying strand separation, another important fact emerged. If a DNA solution is heated to a temperature at which most but not all hydrogen bonds are broken and then cooled to room temperature, A_{260} drops immediately to the initial undenatured value. Additional experiments show that the native structure is restored. Thus, if strand separation is not complete and denaturing conditions are removed, the helix rewinds. Thus, if two separated strands were to come in contact and form even a single base pair at the correct position in the molecule, the native DNA molecule should reform. We will encounter this phenomenon again when renaturation is described.

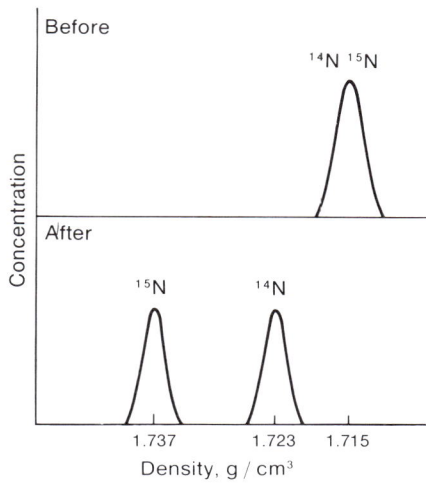

Figure 4-10 Demonstration of strand separation by equilibrium centrifugation in CsCl. Before denaturation, [^{14}N^{15}N]DNA gives a single band. After denaturation two bands result; the density of the [^{14}N]DNA is greater than that of [^{14}N^{15}N]DNA because the density of single-stranded DNA is greater than that of double-stranded DNA by 0.014 g/cm^3, even if the isotopic composition is the same.

Denaturation of DNA by Helix-destabilizing Proteins

There are many proteins that can unwind a DNA helix; these are called **helix-destabilizing** or **melting proteins.** Of these the simplest to understand is the protein made by gene *32* of *E. coli* phage T4, commonly called the *32*-protein. This protein has two properties that enable it to denature DNA—(1) it binds tightly to the bases of single-stranded DNA and (2) the individual molecules of the *32*-protein prefer to line up adjacent to one another along a single strand (Figure 4-11).

Binding of the first molecule of the *32*-protein is made possible by the breathing of the DNA. Once one molecule has invaded the helix, it binds

Figure 4-11 Cooperative binding of gene-*32* protein to single- and double-stranded DNA. At low concentrations the protein molecules are located at random positions but at high concentrations they are adjacent to one another, forming numerous clusters.

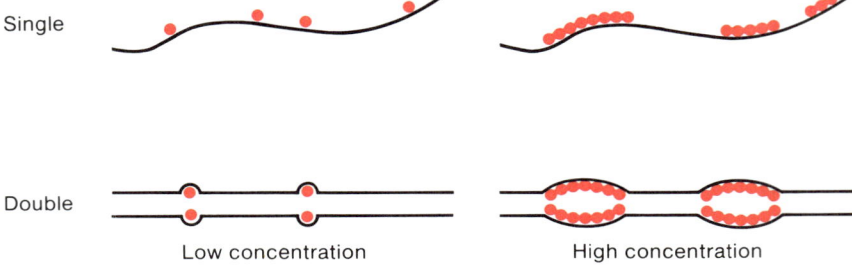

to several adjacent bases. The binding is very tight so the protein stays in place. This not only renders the bases on the opposite strand available for binding, but also destabilizes adjacent base pairs, because paired bases adjacent to unpaired bases cease to be optimally stacked and their hydrogen bonds become less stable. Because of the preference for *32*-proteins to bind in close juxtaposition, another molecule of the *32*-protein will tend to bind next to the first; this breaks still other base pairs, which enables additional protein molecules to bind. The process continues until the DNA is totally denatured.

As will be seen in Chapters 9 and 20, proteins of this type are needed to unwind the helix during replication and to facilitate joining of single strands during genetic recombination.

Denaturation of DNA in Alkali

We have seen by now that the helical structure of polynucleotides is maintained by base-stacking and that two strands are held together by hydrogen-bonding. Most of the studies of denaturation have used heat as a means of disrupting the structure. This can, of course, be used whenever it is necessary to prepare denatured DNA, which is often an essential step in many experimental protocols. However, as will be seen in a later section, phosphodiester bonds may be broken at high temperatures, so the product of heat denaturation often is a collection of broken single strands. At high pH the charge of several of the groups engaged in hydrogen-bonding is changed and a base bearing such a group loses its ability to form these bonds. At a pH greater than 11.3 all hydrogen bonds are eliminated and DNA is completely denatured. Since DNA is quite resistant to alkaline hydrolysis, this procedure is the method of choice for denaturing DNA.

Structure of Denatured DNA

To obtain the data for the melting curves of the sort that have been shown, A_{260} is measured at various temperatures, which are plotted on the *x*-axis.

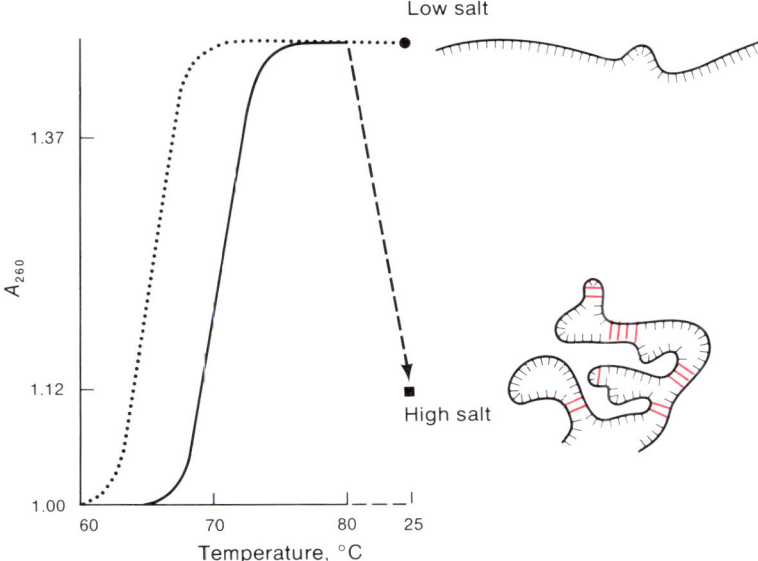

Figure 4-12 The effect of lowering the temperature to 25°C after strand separation has occurred. The types of molecules obtained when the DNA is in a solution having either a low or a high salt concentrations are shown. Base pairs are shown in red.

Denaturation is usually complete at a temperature above 90°C. In most experiments there is a total increase in A_{260} of 37 percent and the solution consists entirely of single strands whose bases are unstacked. However, if the solution is rapidly cooled to room temperature and the salt concentration is above 0.05 $M,$ the value of A_{260} reached at the maximum temperature drops significantly but not totally (Figure 4-12), because in the absence of disrupting thermal motion, random intrastrand hydrogen bonds re-form between distant short tracts of bases whose sequences are sufficiently complementary. Typically, the value of A_{260} drops to 1.12 times the initial value for the native DNA, suggesting that, after cooling, about two-thirds of the bases are either hydrogen-bonded or in such close proximity that stacking is restored. The molecule will be very compact, as shown in Figure 4-12. The situation is quite different if the salt concentration is 0.01 M or less. In this case, the electrostatic repulsion due to unneutralized phosphate groups keeps the single strands sufficiently extended that the bases cannot approach one another. Thus, after cooling no hydrogen bonds are re-formed and base-stacking remains at a minimum.

At a sufficiently high DNA concentration and in a high salt concentration, interstrand hydrogen-bonding should compete with the intrastrand bonding mentioned above. In the following section we will see how this effect can be used to re-form native DNA from denatured DNA.

RENATURATION

A solution of denatured DNA can be treated in such a way that native DNA re-forms. The process is called **renaturation** or **reannealing** and the re-formed DNA is called **renatured DNA.**

Renaturation has proved to be a valuable tool in molecular biology since it can be used to demonstrate genetic relatedness between different organisms, to detect particular species of RNA, to determine whether certain sequences occur more than once in the DNA of a particular organism, and to locate specific base sequences in a DNA molecule.

Requirements for Renaturation

Two requirements must be met for renaturation to occur:

1. The salt concentration must be high enough that electrostatic repulsion between the phosphates in the two strands is eliminated—usually 0.15 to 0.50 M NaCl is used.
2. The temperature must be high enough to disrupt the random, intrastrand hydrogen bonds described in the previous section. However, the temperature cannot be too high, or stable interstrand base-pairing will not occur. The optimal temperature for renaturation is $20-25°$ below the value of T_m.

Renaturation is a slow process compared to denaturation. The rate-limiting step is not the actual rewinding of the helix (which occurs in roughly the same time as unwinding) but the precise collision between complementary strands such that base pairs are formed at the correct positions. Since this is a result only of random motion, it is a concentration-dependent process; at concentrations normally encountered in the laboratory this takes several hours. We shall have more to say about the concentration dependence in the section entitled C_0t Curves.

Mechanism of Renaturation

The molecular details of renaturation can be understood by referring to the hypothetical molecule shown below, having a sequence (shown in red) that is repeated several times.

IA	IB	II	IC
···ATGA···	ATGA···	CCCC···	ATGA···
···TACT···	TACT···	GGGG···	TACT···
IA′	IB′	II′	IC′

Assume that each single strand contains 50,000 bases, and that the base sequences are complementary. However, any short sequence of bases (say, 4–6 bases long) will certainly appear many times in such a molecule and can provide sites for base-pairing. Random collision between noncomplementary sequences such as IA and II' will be ineffective but a collision between IA and IC' will result in base-pairing. However, this will be short-lived, because the bases surrounding these short complementary tracts are not able to pair and stacking stabilization will not occur. Thus, at the elevated temperatures that have brought about strand separation, these paired regions rapidly become disrupted. However, as soon as two sequences such as IB and IB' pair, the adjacent bases will also rapidly pair and the entire double-stranded DNA molecule will "zip up" in a few seconds.

It is important to realize that each renatured native DNA molecule is not formed from its own original single strands: in a solution of denatured DNA, the single strands freely mix so that during renaturation strands join at random. This was shown in an experiment using two DNA samples isolated from *E. coli* grown, in one case, in a medium containing $^{14}NH_4Cl$ and in the other, $^{15}NH_4Cl$. The two DNA samples were mixed, denatured, renatured and centrifuged to equilibrium in a CsCl density gradient. The result was a mixture containing three types of native DNA molecules—25 percent contained ^{14}N in both strands, 50 percent contained ^{14}N in one strand and ^{15}N in the other, and 25 percent contained ^{15}N in both strands, which indicates random mixing of the strands during renaturation. This type of molecular mixing is called **hybridization.** One example of its use is described in the next section.

Filter-binding Assays for Renaturation

Very thin filters (**membrane filters**) made of nitrocellulose are commercially available. These filters bind single-stranded DNA very tightly but fail to bind either double-stranded DNA or RNA. They provide an important method for measuring hybridization by the following technique (Figure 4-13). A sample of denatured DNA is filtered. The single strands bind tightly to the filter along the sugar-phosphate backbone, but the bases remain free. The filter is then placed in a vial with a solution containing a small amount of radioactive denatured DNA and a reagent that prevents additional binding of single-stranded DNA to the filter. After a period of renaturing the filter is washed. Radioactivity is found on the filter only if renaturation has occurred.

The most important use of filter hybridization is the detection of sequence homology between a single strand of DNA and an RNA molecule—this is called **DNA-RNA hybridization** and it is the method of choice to detect an RNA molecule that has been copied from a particular DNA molecule. In this procedure, a filter to which single strands of DNA have been bound, as above, is placed in a solution containing radioactive RNA. After rena-

Figure 4-13 Method of hybridizing DNA to nitrocellulose filters containing bound, single-stranded DNA. In the final step the filter is treated with a single-strand-specific DNase, an enzyme that depolymerizes single-stranded, but not double-stranded DNA.

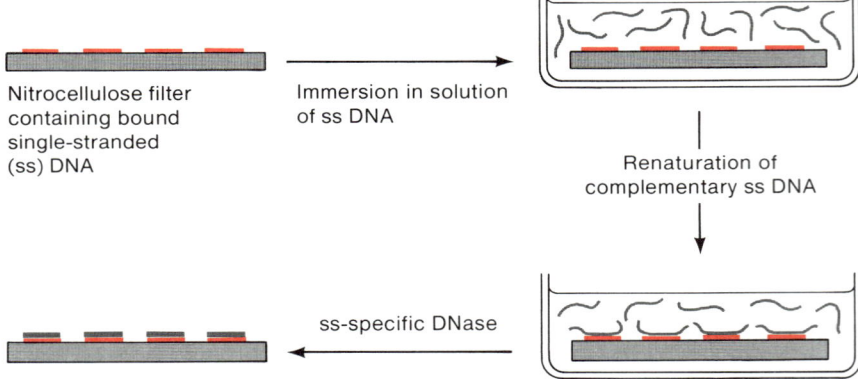

Nitrocellulose filter containing bound single-stranded (ss) DNA

Immersion in solution of ss DNA

Renaturation of complementary ss DNA

ss-specific DNase

turation the filter is washed and hybridization is detected by the presence of radioactive RNA on the filter.

Concentration Dependence of Renaturation: C_0t Curves

The initial event in renaturation is a collision; thus, the renaturation rate obeys the law of mass action and increases with DNA concentration, as shown in Figure 4-14. This concentration dependence has been used to discover some surprising properties of the DNA of eukaryotes and prokaryotes.

Consider two DNA solutions having equal concentrations in terms of g/ml (not molar concentration). The DNA molecules are phage T7 DNA,

Figure 4-14 Dependence of renaturation time on the concentration of phage T7 DNA. The DNA is heated to 90°C to separate the strands and then cooled to 60°C. Renaturation is complete when the relative value of A_{260} reaches 1. The times required for renaturation to be half completed, $t_{1/2}$, are obtained by drawing the red horizontal line, which divides each curve equally in the vertical direction, and then extending the red vertical lines to the time axis.

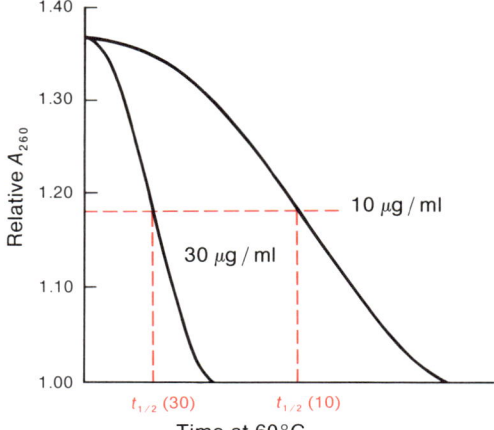

Relative A_{260}

10 μg/ml

30 μg/ml

$t_{1/2}$ (30) $t_{1/2}$ (10)

Time at 60°C

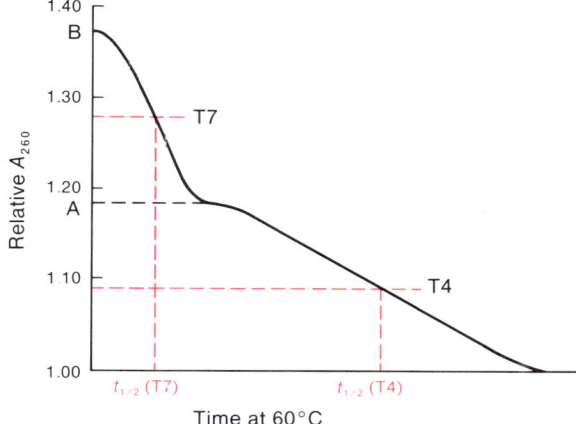

Figure 4-15 Renaturation of a mixture of T7 and T4 DNA molecules, each at 30 µg/ml. Extrapolation (black dashed line) of the inflection region at the junction between the upper and lower components of the curve to the vertical axis yields the early portion of the T4 curve; the ratio of the values of A_{260} at points A and B yields the fraction of the total DNA that is T4 DNA. The times required for renaturation of each type of DNA to be half completed, $t_{1/2}$, are calculated as in Figure 4-14 and shown in red.

whose molecular weight M is 2.5×10^7 and phage T4 DNA, for which $M = 1.1 \times 10^8$. The molar concentration of the T7 DNA is 4.4 times that of the T4 DNA but the A_{260} values of both solutions are the same. We further assume that there is no base sequence homology (that is, no common sequences) between the two DNA molecules. The two solutions are mixed, denatured, and renatured, and A_{260} is measured as a function of time. A curve such as that in Figure 4-15 is obtained. Two features of this curve should be noted. (1) The curve consists of two steps. Each accounts for half of the decrease in A_{260} because the initial DNA concentration (in g/ml) of each was the same. (2) The times at which the renaturation of each component is half-complete differ. Comparison with Figure 4-14 shows that the slower component is the T4 DNA, whose molar concentration is lower.

A very important point about renaturation kinetics is the following. If the two kinds of DNA molecules described in Figure 4-15 were broken into thousands of fragments, the renaturation kinetics would differ from that shown in the figure, but two features—the A_{260} values of each step and the relative values of the time required for 50 percent renaturation—remain unchanged. The first is true because the ratio of the two components and hence the A_{260} values are unchanged. The latter follows from the fact that, whereas the renaturation rates are definitely less for the fragments, the molar concentrations of each base sequence are unchanged; thus, the ratio of the molar concentrations of the T4 and T7 fragments is unaffected by the fragmentation.

Fragmentation of the DNA molecules allows a new feature of the base sequences to be seen. Consider a very long molecule having a repeating base sequence as shown in Figure 4-16(a). This sequence contains 3 percent of the total number of bases of the DNA and there are five copies of the repeated sequence in the DNA. If this large molecule is broken to fragments much

Figure 4-16 (a) A hypothetical DNA molecule having a base sequence that is 3 percent of the total length of the DNA and is repeated five times. The dashed lines represent the nonrepetitive sequences; they account for 85 percent of the total length. (b) A renaturation curve for the DNA in part (a). Time is logarithmic in order to keep the curve on the page. The y-axis—percent unrenatured DNA—is equivalent to the unit, relative value of A_{260}, used in previous figures with $A_{260} = 1.37$ equal to 100 percent.

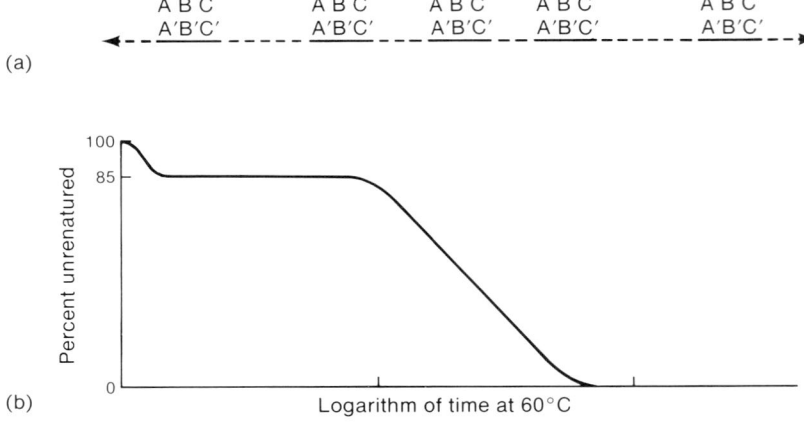

smaller than 1 percent of the total length of the molecule, the molar concentration of this sequence becomes five times larger than that of the remainder but the weight fraction remains $(5 \times 3)/100 = 0.15$. Such a DNA would have a renaturation curve such as that shown in Figure 4-16(b). Thus, this curve, which is obtained with DNA from a single organism, contains two components. The more rapidly renaturing component accounts for 15 percent of A_{260}, from which one can conclude that 15 percent of the DNA (on a weight base) contains a repeating sequence. In the following we show that from the kinetics one can determine the number of bases in the repeating sequence (called its **complexity**), and thus the number of copies of this sequence and the number of bases in a single nonrepeating sequence.

The following equation describes the kinetics of renaturation in terms of the initial DNA concentrations C_0 (expressed in moles of bases per liter), the concentration C of unrenatured DNA at times t (in minutes), and k, a rate constant that depends on the temperature and the size of the DNA fragments:

$$C/C_0 = 1/(1 + kC_0t).$$

Experimentally, one chooses several values of C_0 and then measures C as a function of time; the data obtained are plotted as C/C_0 versus log C_0t. When $C/C_0 = 1/2$, then $C_0t_{1/2} = 1/k$. This is a significant expression because it can be shown that k is inversely proportional to the number N of bases per repeating unit. Thus, the value of $C_0t_{1/2}$ is directly proportional to N. Note that if there are no repeating sequences, so that the DNA molecule itself represents a unique sequence, then N is the number of base pairs in the complete DNA molecule of the organism. Figure 4-17(a) shows the kind of graph obtained. To analyze the sequence complexity of the DNA, one first notes that 53 percent of the sequences have a $C_0t_{1/2}$ of 10^2, 27 percent have a $C_0t_{1/2}$ of 1, and 20 percent have a $C_0t_{1/2}$ of 10^{-3}. These data are then used

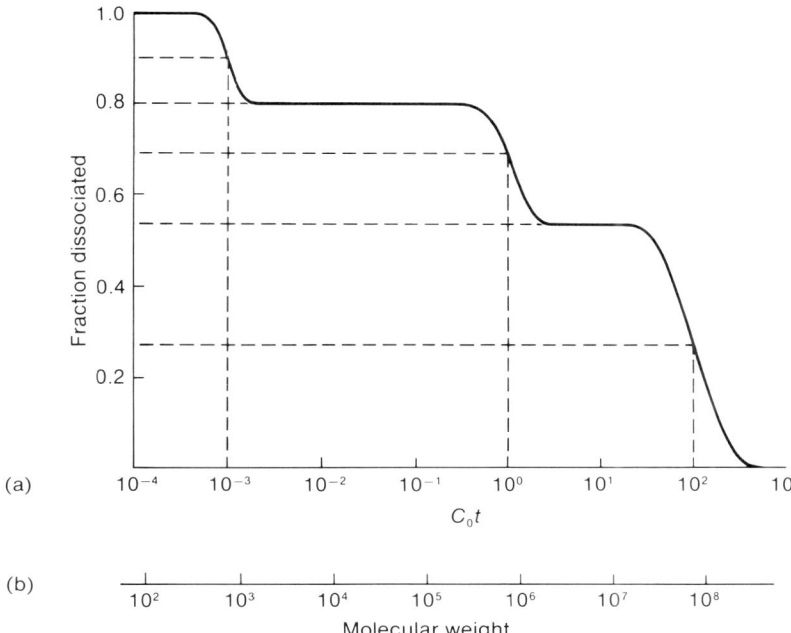

Figure 4-17 C_0t analysis. (a) The C_0t curve analyzed in the text. (b) A standard scale relating the number of nucleotide pairs in a unique sequence and the value of $C_0t_{1/2}$. The horizontal axis in (a) and the scale in (b) are aligned vertically so values can be compared directly.

to determine the number of copies and the sizes of each sequence. From analysis of pure samples of molecules having a unique sequence of known length, a size scale, such as that shown in Figure 4-17(b), can be obtained. However, molecular size cannot be read directly from the observed $C_0t_{1/2}$ values, because the values of C_0 used in plotting the horizontal axis in panel (a) is the *total* DNA concentration in the renaturation mixture, rather than the concentration of each component. To utilize the size scale, one must calculate the concentration of each component; this is done by simply multiplying each observed $C_0t_{1/2}$ value by the fraction of the total DNA that it represents. Thus, the "real" $C_0t_{1/2}$ values are 0.53×10^2, 0.27×1, and 0.20×10^{-3}; the corresponding sizes are 4.2×10^7, 2.2×10^5, and 160 base pairs, respectively. The number of copies of each sequence is inversely proportional to $t_{1/2}$ and hence to the observed (uncorrected) $C_0t_{1/2}$ value. Thus, if one assumes that the genome contains one copy of the longest sequence (possible only in a haploid organism), then the cell from which the DNA has been isolated would contain 1, 10^2, and 10^5 copies of sequences having 4.2×10^7, 2.2×10^5, and 160 bp, respectively. The total number of base pairs per genome would be

$$4.2 \times 10^7 + 100(2.2 \times 10^5) + 10^5(160) = 8 \times 10^7,$$

and the molecular weight of the total cellular DNA would be

$$(8 \times 10^7)(660) = 5.3 \times 10^{10}.$$

Note that if an independent measurement of the total molecular weight of the DNA were to yield the value 1×10^{11}, which is ca. $2(5.3 \times 10^{10})$, then there would be 2, 200, and 2×10^5 copies of the 4.2×10^7-, 2.2×10^5-, and 160-bp sequences, respectively.

Bacterial DNA contains a few repeating units such as the genes for transfer RNA but these account for such a small fraction (0.1 percent) of the DNA that they are not readily detected. However, C_0t analysis has shown that in eukaryotes a significant fraction of the DNA consists of repeated sequences. This will be discussed shortly.

DNA Heteroduplexes

Renaturation has been combined with the technique of electron microscopy in a procedure that allows the localization of common, distinct, and deleted sequences in DNA. This procedure is called **heteroduplexing.** Consider the DNA molecules #1 and #2 shown in Figure 4-18(a). These molecules differ in sequence only in one region. If a mixture of the two molecules is denatured and renatured, hybrid molecules having unpaired single strands are produced, as shown in Figure 4-18(b). Figure 4-19 shows an actual electron micrograph of a heteroduplex. Measurement of the lengths of the single and double-stranded regions yields the endpoints of the regions of nonhomology. Note that the two single strands of the bubble always have the same length.

Figure 4-18 (a) Three DNA molecules to be heteroduplexed. Sequences A and A' of molecules 1 and 2 differ. Neither sequence is present in the deletion molecule 3. The dashed lines indicate reference points. (b) Heteroduplexes resulting from renaturing molecules 1 and 2 or 2 and 3.

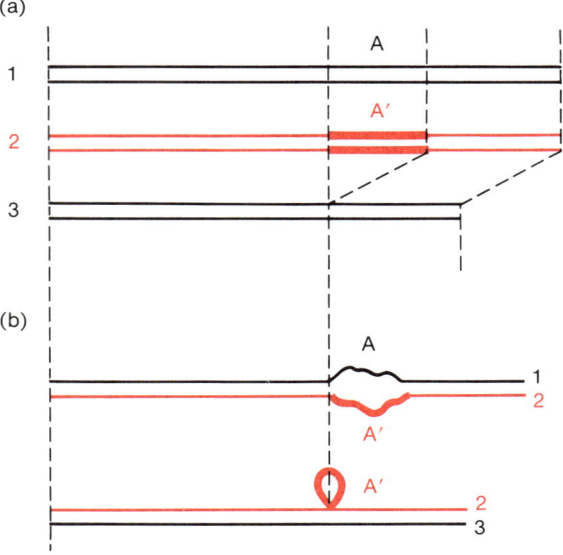

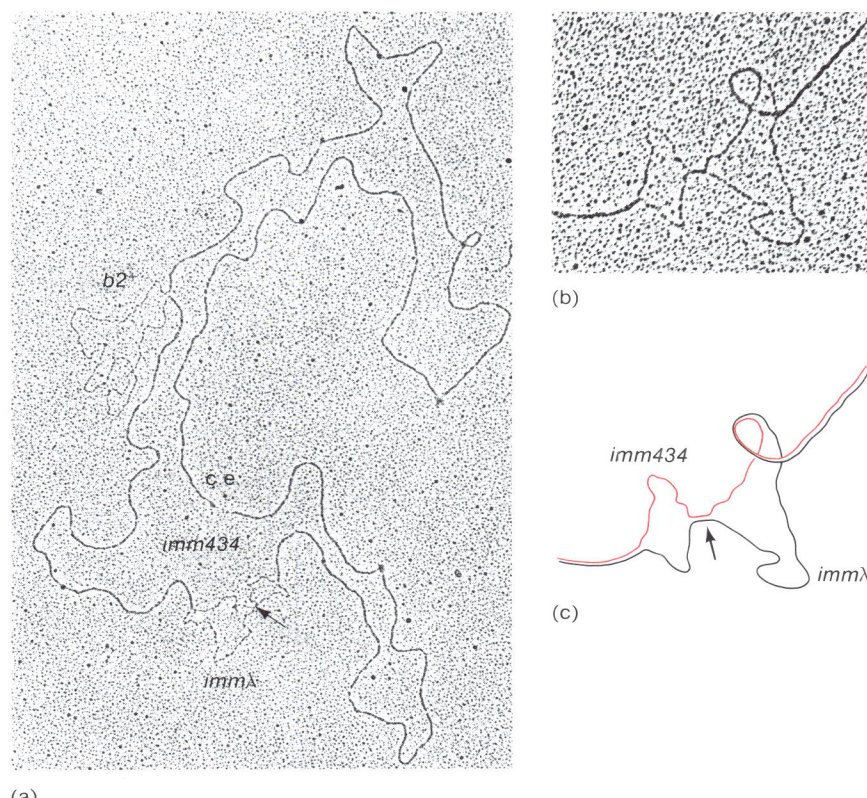

(a)

(b)

(c)

Figure 4-19   An electron micrograph of a heteroduplex between λ*imm*λ *b2* DNA, which carries the *b2* deletion, and λ*imm434 b2*⁺, in which the *imm*λ segment is replaced by the shorter, nonhomologous *imm434* segment. (a) Two bubbles of non-homology are seen. The identity of each single-stranded segment is indicated. The arrow is explained in part (c). (b) An enlargement of the *imm434–imm*λ segment. (c) An interpretive drawing of panel (b). The arrow indicates a region of homology between the *imm434* and *imm*λ segments. The same region is indicated by the arrow in panel (a). [Courtesy of Barbara Westmoreland and W. Szybalski.]

Consider now molecule #3, shown in Figure 4-18(a). In this molecule, region A is deleted. If a hybrid is made between this molecule and molecule #2, the result is a molecule with a single loop, as shown in Figure 4-18(b).

MULTIPLE COPIES OF BASE SEQUENCES IN EUKARYOTIC DNA

C_0t analysis of the total DNA of numerous organisms has shown that there is a profound difference between the sequence organization of prokaryotes and eukaryotes. Figure 4-20 shows C_0t curves for bacterial and mouse DNA. The *E. coli* curve, which is characteristic of all bacteria, is a smooth one-step curve, which indicates that few, if any, sequences exist in multiple copies; that is, each sequence is unique. (Admittedly there could be multiple copies with equal numbers of *all* sequences, but this is unlikely for a haploid organism.) Close examination of C_0t curves from many eukaryotic species shows that there are usually four classes of sequences—**unique** (single copy per

Figure 4-20 C_0t curves for *E. coli* and mouse DNA. The sequences are unique in *E. coli,* so the curve has a single step. In mouse DNA 10 percent consists of about 10^6 copies of a 300-bp sequence, 30 percent consists of 10^3–10^4 copies of several sequences, and 60 percent consists of unique sequences.

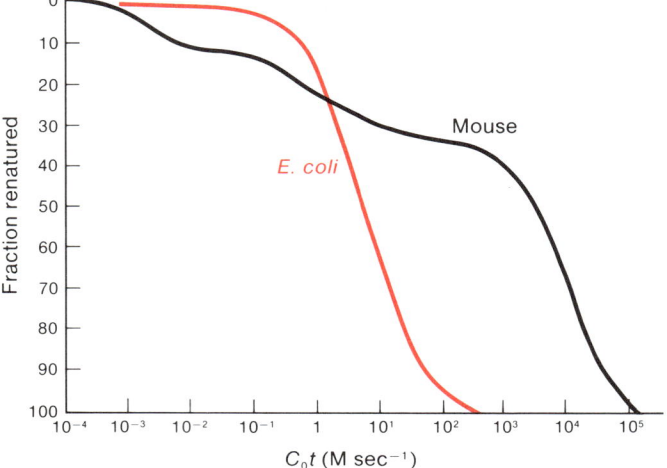

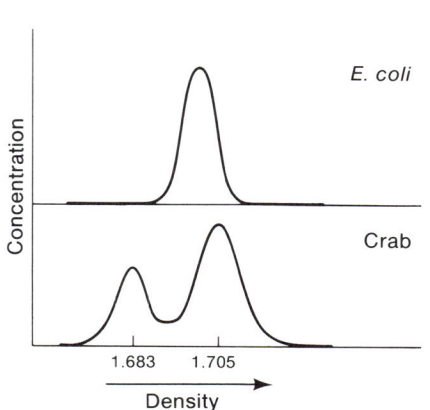

Figure 4-21 The result of equilibrium centrifugation in CsCl of the DNA of the bacterium *E. coli* and the crab *Cancer borealis.* As is common with bacteria, the DNA has a narrow range of densities. The crab DNA consists of two discrete fractions, one of very low density. The minor band is called satellite DNA.

genome), **slightly repetitive** (one to ten copies), **middle repetitive** (ten to several hundred copies), and **highly repetitive** (several hundred to several million copies). The bounds of these classes are arbitrary (and difficult to distinguish between slightly repetitive and middle repetitive sequences), so one must be careful when making generalizations about the classes. The unique sequences account for most of the genes of the cell but frequently for only a few percent of the DNA. Middle repetitive DNA will be encountered in Chapter 21 when eukaryotic transposable elements are described. The highly repetitive sequences are especially intriguing because of their enormous number; the short sequences are notable for their base composition. These short sequences can be detected in another way, as we will now see.

If the DNA of a bacterium is isolated and randomly fragmented into about 500–1000 pieces and then centrifuged in a CsCl density gradient, the DNA forms a single fairly narrow band. This is because the buoyant density of DNA is proportional to the $G+C$ content of the DNA and the average base composition of segments containing several hundred base pairs does not vary much from one section of the DNA to the next. However, if DNA from a crab is analyzed, two bands are found (Figure 4-21). The major band has a base composition of 58 percent $A+T$, whereas the minor band, which is called **satellite DNA,** is 97 percent $A+T$, a most unusual base composition. Satellite bands have been observed in the DNA of many organisms and may comprise 1–30 percent of the total DNA, averaging about 15 percent. An important characteristic of all satellite bands is that they contain **repeated base sequences** having various lengths. For example, in the cow a 1400-bp sequence is repeated again and again and forms a satellite band, whereas in some monkey species a 172-bp sequence is repeated in the satellite band. The crab satellite DNA repeats only two bases—that is, the DNA has the sequence ATATAT . . . , only occasionally interrupted by a G or a C (1 in 30 base pairs). Of particular note is a species of fruit fly, *Drosophila virilis,* which has three satellite DNA bands, each containing a distinct but very closely related repeating heptanucleotide, namely, 5′–A(TC)AAA(TC)T-3′, where the pair of bases in ( ) represents alternate bases at that site. These sequences are repeated 10^7 times in each cell of this species!

The technique of **in situ hybridization** has been used to localize the satellite DNA in the chromosome. In this technique cells are fixed to a glass microscope slide and treated with alkali to denature the DNA in the chromosomes. (The chromosomes remain visually intact.) A droplet containing ^{3}H-labeled copies of purified satellite DNA (made *in vitro*) is placed on the cells on the slide and allowed to renature with the chromosomal DNA. After renaturation is complete, the slide is washed to remove radioactive DNA that has not hybridized, and the sites at which hybridization has occurred is determined by autoradiography (Figure 4-22). The satellite sequences located by this technique have been found to be in the regions of the chromosomes called **heterochromatin.** These are regions that condense earlier in prophase than the rest of the chromosome and are darkly stainable by many standard dyes used to make chromosomes visible (Figure 4-23); sometimes the heterochromatin remains highly condensed throughout the cell cycle. **Euchromatin,** which makes up most of the genome, is visible only during

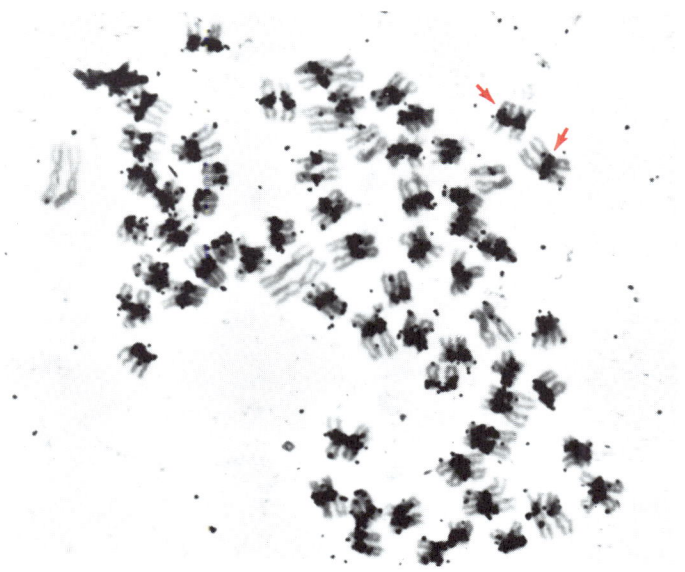

Figure 4-22 Autoradiogram of metaphase chromosomes of the kangaroo rat *Dipodomys ordii;* [^{3}H]RNA copied from purified satellite HS-β sequences were hybridized to the chromosomes to show the location of the satellite DNA. Hybridization occurs mainly in the regions adjacent to the centromeres (arrows). Note that some chromosomes are apparently free of this satellite DNA. They contain a different satellite DNA not tested in this experiment. [Courtesy of David Prescott.]

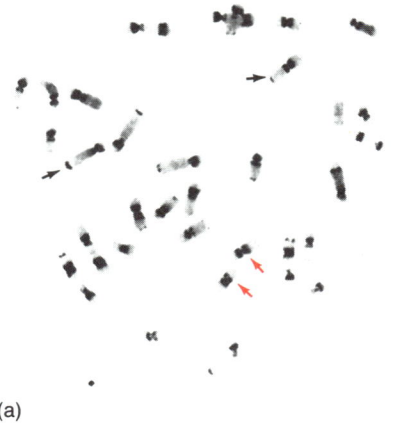

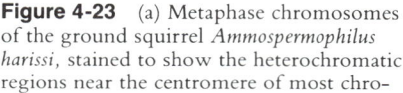

(a)

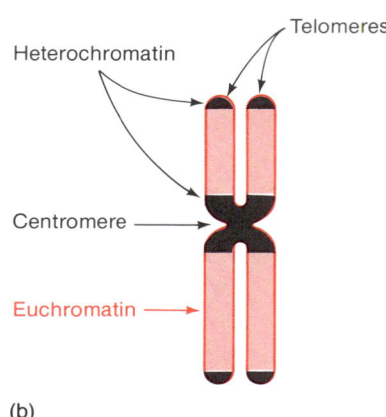

(b)

Figure 4-23   (a) Metaphase chromosomes of the ground squirrel *Ammospermophilus harissi,* stained to show the heterochromatic regions near the centromere of most chromosomes (red arrows) and the telomeres of some chromosomes (black arrows). [Courtesy of T. C. Hsu.] (b) An interpretive drawing.

the mitotic cycle. The major heterochromatic regions are adjacent to the **centromere** (the point of attachment of the chromosome to the mitotic spindle); smaller blocks occur at the ends of the chromosome arms (the **telomeres**) and interspersed with the euchromatin. Different highly repetitive sequences have been purified from *D. melanogaster* and *in situ* hybridization with *Drosophila* cells has shown that each chromosome has its own distinctive types and distribution of these sequences.

Few genes have been located in heterochromatic regions of chromosomes, so heterochromatin has been considered to be genetically inert. However, recent experiments have indicated that it has well defined, though not well understood, functions in the pairing and segregation of homologous chromosomes during meiosis, the structural rearrangement of chromosomes, and the regulation of gene expression.

CIRCULAR AND SUPERHELICAL DNA

The intact DNA molecules of most prokaryotes and viruses are circular, though this was not noticed for many years because the DNA molecules usually broke during isolation, as mentioned in an earlier section. A circular molecule may be a **covalently closed circle,** which consists of two unbroken complementary single strands, or it may be a **nicked circle,** which has one or more interruptions (**nicks**) in one or both strands, as shown in Figure 4-24. With few exceptions, covalently closed circles are twisted, as shown in Figure 4-25. Such a circle is said to be a **superhelix** or a **supercoil.** Let us now examine what is meant by a circular DNA molecule being twisted.

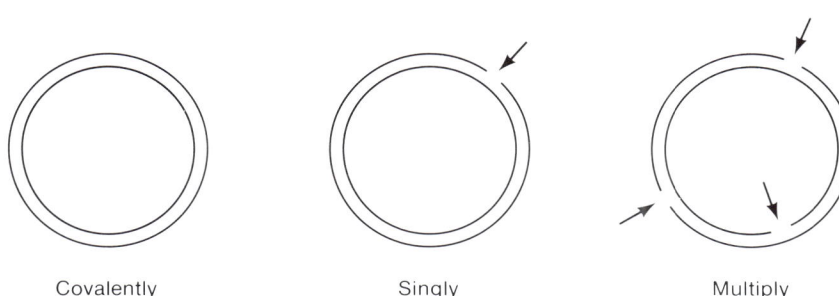

Figure 4-24 A covalently closed circle and two kinds of nicked circles. Arrows indicate the nicks. A nicked circle is also called an open circle.

Covalently
closed circle

Singly
nicked circle

Multiply
nicked circle

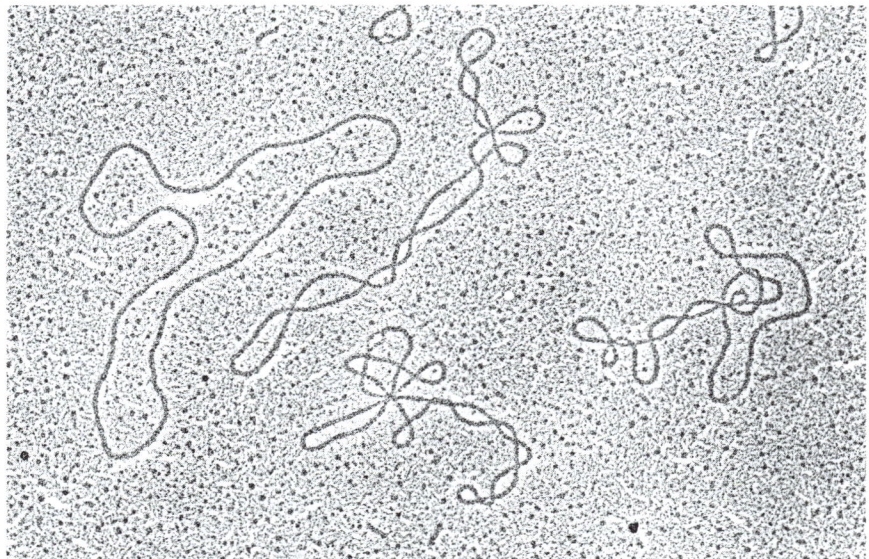

Figure 4-25 Nicked circular and super-coiled DNA of phage PM2. [Courtesy of K. G. Murti.]

Twisted Circles

The two ends of a linear DNA helix can be brought together and joined in such a way that each strand is continuous. If, in so doing, one of the ends is rotated 360° with respect to the other to produce some unwinding of the double helix, and then the ends are joined, the resulting covalent circle will, if the hydrogen bonds re-form, twist in the opposite sense to form a twisted circle, in order to relieve strain. Such a molecule will look like a figure 8 (that is, have one crossover point or **node**). If it is instead twisted 720° prior to joining, the resulting superhelical molecule will have two nodes. The reason for the twisting is the following. In the case of a 720° unwinding of the helix, 20 base pairs must be broken (because the linear molecule has 10 base pairs per turn of the helix). However, a DNA molecule has such a propensity for maintaining a right-handed (positive) helical structure with 10 base pairs per turn that it will deform itself in such a way that the underwinding is rewound and compensated for by negative (left-handed) twisting of the circle. Similarly, the initial rotation might instead be in the sense of overwinding, in which case the joined circle will twist in the opposite sense; this is called a positive superhelix. *All naturally occurring superhelical DNA molecules are initially underwound and hence form negative superhelices.* Furthermore, the degree of twisting—that is, the **superhelix density**—is about the same for all molecules; namely, one negative twist is produced per 200 base pairs, or, 0.05 twists per turn of the helix.

In bacteria, the underwinding of superhelical DNA is not a result of unwinding prior to end-joining but is introduced into preexisting circles by an enzyme called **DNA gyrase,** which will be described in Chapter 9, when DNA replication is examined. In eukaryotes, the underwinding is a result of the structure of chromatin, a DNA-protein complex of which chromosomes are composed. In this complex, DNA is wound about specific protein molecules in a direction that introduces underwinding. Chromatin will be discussed in Chapter 7.

The topological components of twisting has been described in a useful, quantitative way. The **twisting number T,** a property of the double helix itself, is defined as the total number of turns of the double-stranded molecule. For a nicked circle of known size the value of T is calculated as the total number of base pairs divided by the number of base pairs per turn. The **writhe W,** a property of a supercoiled molecule, is the number of turns of the axis of the double-stranded helix in space. It is what one thinks of as the number of supertwists. $W = 0$ for a nicked circle. The **linking number L,** a somewhat more complicated concept, is the total number of times the two strands of the double helix of a closed molecule (a circle, for example) cross each other. If the circle were constrained to lie flat on a surface, L would be the number of times one strand revolves around the other; the number is positive for revolutions in right-handed helical regions and negative for a left-handed helix or a left-handed segment. This number enables one to distinguish positive from negative supercoiling. T, W, and L are related by the expressions

$$L = W + T \quad \text{and} \quad \Delta L = \Delta W + \Delta T$$

in which Δ represents changes in the values. The second equation is important because experimentally one usually measures changes in the linking number (ΔL). The equation shows that a decrease in L corresponds to some combination of negative supercoiling or underwinding, and increase in L reflects a decrease in negative supercoiling and underwinding. Note that $T = L$ for a nicked circle.

An important point about the linking number is that L cannot be changed without breaking a strand, rotating one strand about the other, and rejoining.

Thus, determination of ΔL in a particular process is useful because it tells something about the mechanism of the change; in particular, how enzymes that affect supercoiling do their job. Note that ΔL must be an integer.

Single-Stranded Regions in Superhelices

We have just pointed out that the strain of underwinding can be accommodated by negative supercoiling. Three other arrangements that could counteract the strain of underwinding are possible: (1) The number of base pairs per turn of the helix could change. This does not happen, though, for thermodynamic reasons. (2) All of the underwinding could be taken up by having one or more large single-stranded regions (Figure 4-26). (3) The underwinding could be taken up in part by superhelicity and in part by bubbles. The real situation is alternative 3—that is, a state intermediate between forms (b) and (c) of Figure 4-26—because a DNA molecule is a dynamic structure that breathes. If a circular molecule were made that was not underwound, breathing (the transient unwinding that results from thermal motion) would introduce compensating transient negative twists. If the DNA is initially superhelical, the superhelix density will fluctuate as breathing occurs: the strain produced by the underwinding is relieved in a superhelix both by the superhelicity and by an increase in the number and size of the bubbles and the duration of their existence. Thus, in a supercoil, the fraction of the molecule that is single-stranded at any moment is greater than in a nicked circle. It is likely that a sequence containing more than 90 percent A·T pairs may be permanently unpaired in a superhelical molecule. It is thought that this may play a role in such processes as initiating genetic recombination, the initiation of DNA replication, and the initiation of messenger RNA synthesis.

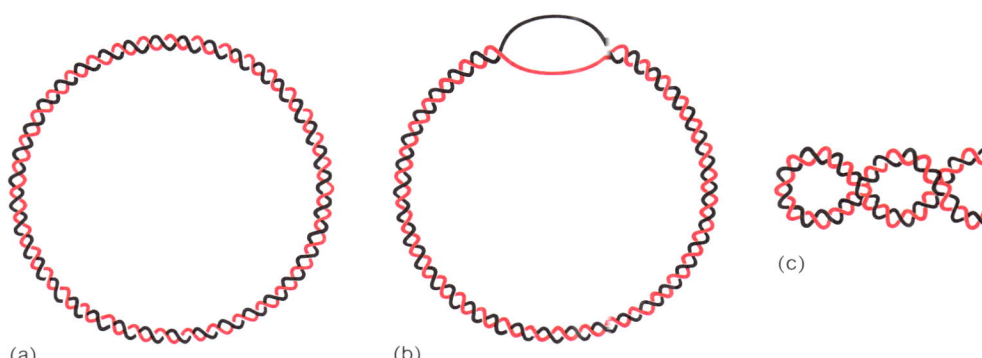

Figure 4-26 Different states of a covalent circle. (a) A nonsupercoiled covalent circle having 36 turns of the helix. (b) An underwound covalent circle having only 32 turns of the helix. (c) The molecule in part (b), but with 4 superhelical turns to eliminate the underwinding. In solution, (b) and (c) would be in equilibrium; the equilibrium would shift toward (b) with increasing temperature.

Experimental Detection of Covalently Closed Circles

In the life cycles of many organisms, the DNA molecules cycle through the various circular forms that have just been described. Several techniques have been developed for distinguishing these forms. Covalently closed (but not necessarily superhelical) circles can be detected in two ways:

1. *Sedimentation at alkaline pH*. Above pH 11.3 all hydrogen bonds are broken and DNA molecules unwind. For a linear DNA molecule two single strands result whose s value is about 30 percent greater than that of the native DNA if the salt concentration is 0.3 M or greater. On the contrary, the two strands of a covalently closed circle cannot separate, so the molecules collapse in a tight tangle whose s value is about three times as great as that of native DNA. If the circle has a single nick, one linear molecule and one single-stranded circle result, the s value of the latter being 14 percent greater than the former (Figure 4-27(a)). Figure 4-27(b) shows a sedimentation pattern for a mixture composed of linear molecules, single circles, and covalently closed circles in an alkaline sucrose gradient.

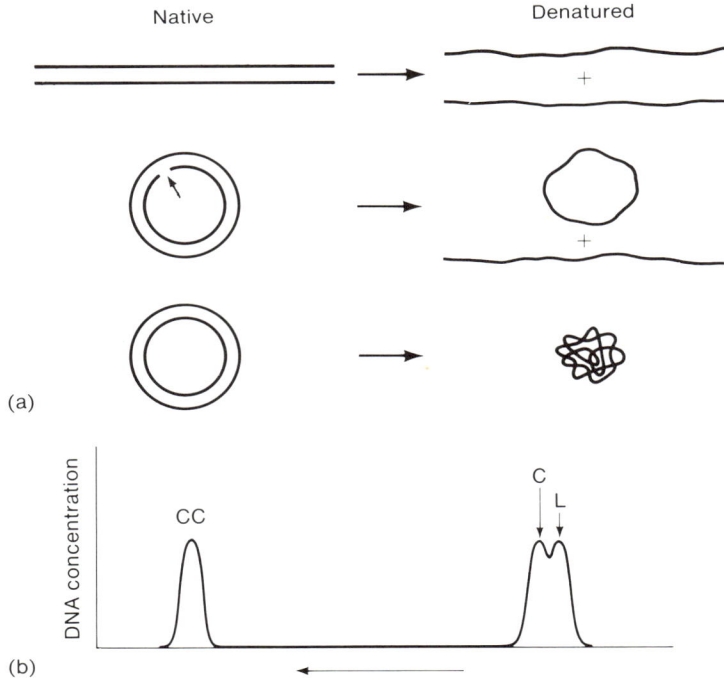

Figure 4-27 (a) Products of the denaturation of different forms of DNA. (b) Separation of covalently closed circles (CC), single circles (C), and linear (L) molecules by sedimentation in alkali. The horizontal axis represents the length of a centrifuge tube. Sedimentation is from right to left.

   2. Equilibrium centrifugation in CsCl containing ethidium bromide. **Ethidium bromide** (Figure 4-28) binds very tightly to DNA (in both dilute and concentrated salt solutions) and, in so doing, decreases the density of the DNA by approximately 0.15 g/cm³. It binds by intercalating between the DNA base pairs, thereby causing the DNA molecule to unwind as more ethidium bromide is bound. Because a covalently closed, circular DNA molecule has no free ends, as it unwinds, the entire molecule twists in the opposite direction. For example, an O-shaped molecule that has bound enough ethidium bromide to produce one clockwise turn will twist in the counterclockwise direction to produce a molecule shaped like the figure 8. As more and more of the molecules intercalate, the 8-shaped molecule will become more twisted. Ultimately, the DNA molecule is unable to twist any more, so no more ethidium bromide molecules can be bound. On the contrary, a linear DNA molecule or a nicked circle does not have the topological constraint of reverse twisting and can therefore bind more of the ethidium bromide molecules. Because the density of the DNA and ethidium bromide complex decreases as more ethidium bromide is bound and because more ethidium bromide can be bound to a linear molecule or an open circle than to a covalent circle, the covalent circle has a higher density at saturating concentrations of ethidium bromide. Therefore, covalent circles can be separated from the other forms in a density gradient, as shown in Figure 4-29.

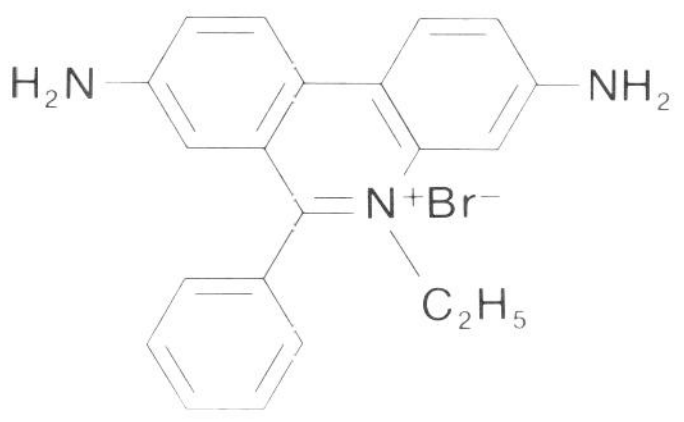

Figure 4-28 Chemical formula for ethidium bromide.

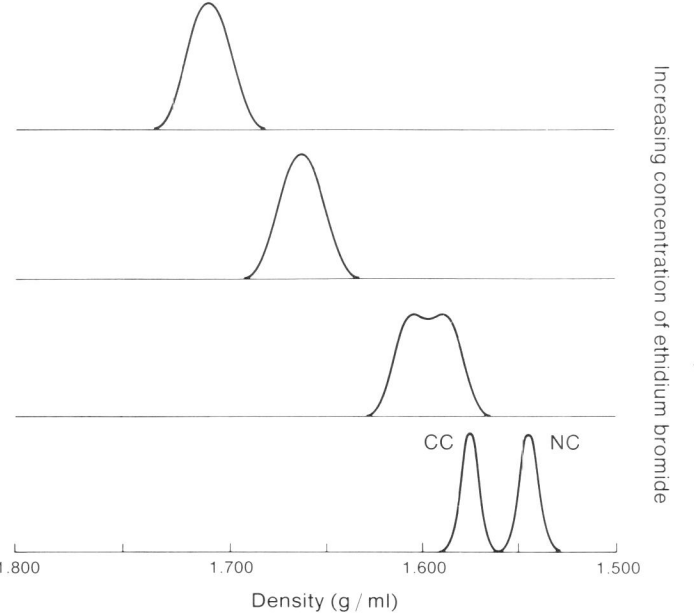

Figure 4-29 Effect of ethidium bromide on the density of DNA in a CsCl solution. A mixture of equal parts of nicked circles (NC) and covalently closed circles (CC) is centrifuged in CsCl containing various concentrations of ethidium bromide. The density of the DNA molecules decreases until, at saturation, the two components separate. The covalent circles bind less ethidium bromide and are therefore at a higher density.

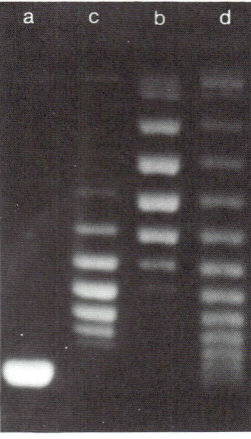

Figure 4-30 A gel electrophoregram showing the conversion of a covalent circle to a supercoil by DNA gyrase. (a) Maximally supercoiled molecules (intense band), used as a position standard. The faint band near the top contains contaminating nicked circles. (b) The species present early in the enzymatic reaction. Each band contains molecules having two more superhelical turns than the molecules in the band just above. (c) A later time—several bands containing molecules with additional twists are present. (d) Still later—by this time some molecules have become maximally supercoiled. Note the order of the lanes: a c b d. The direction of migration is from top to bottom. [Courtesy of James Wang and Tadaatsu Goto.]

Experimental Detection of Superhelicity

Supercoiling is easily detected by electron microscopy, as already shown in Figure 4-25. However, because of stretching and unwinding associated with adhesion of the molecule to the support film, the degree of supercoiling can usually not be quantitated. Gel electrophroresis is useful in this respect. Because of the compactness of a superhelix, the mobility is higher than a nonsuperhelical molecule of the same mass. The difference is sufficiently great that it is possible to separate two DNA molecules whose linking numbers differ by only 1. For example, each of the intermediates present in a reaction mixture in which the enzyme DNA gyrase is introducing twists can be separated, as shown in Figure 4-30. In this reaction, ΔL between adjacent bands is 2.

SPECIAL BASE SEQUENCES AND THEIR STRUCTURAL CONSEQUENCES

An enormous variety of base sequences have been observed in DNA. Most do not have any special features that cause them to be recognized as unusual. Of particular interest are several types of repeated sequences. These sequences are common to regulatory regions and to sites of enzymatic activity and may in some cases impart special properties to either double- or single-stranded nucleic acids. Other sequences in which purines and pyrimidines alternate can cause DNA to form a left-handed helical region.

Repeated Sequences

The **palindrome** is a sequence of the general form

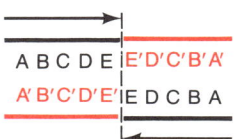

in which A and A′ are complementary bases able to pair. The dashed vertical line is an axis of symmetry: the double-stranded segment to the right of the axis can be superimposed on the one to the left by a 180° rotation in the plane of the page. Other terms used to describe palindromes are **inverted repeat, inverted repetition,** and region of **dyad symmetry,** a term used by crystallographers. Palindromes range in length up to about 50 base pairs. The two inverted sequences may be separated by a spacer: for example,

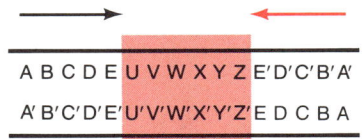

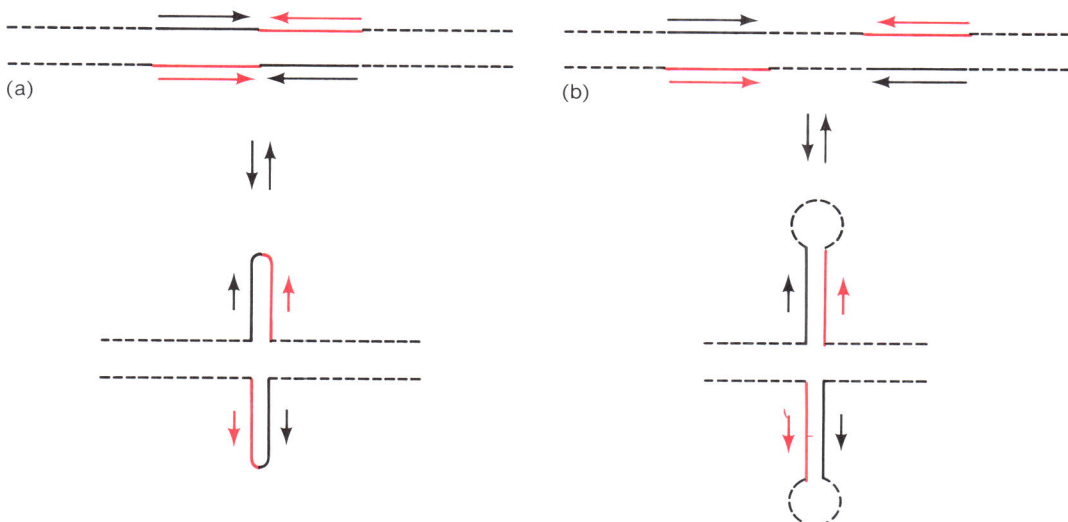

Figure 4-31 Possible alternative forms of a DNA molecule containing two inverted repeats that are (a) adjacent or (b) separated by a spacer. The horizontal arrows denote orientation of the sequences.

in which case only the term inverted repeat is used. Because DNA breathes, molecules containing palindromes and inverted repeats can in theory assume alternate structures (Figure 4-31). Once complementary strands have separated, intramolecular base pairing can cause a double-stranded branch to form between adjacent complementary sequences with, in some cases, higher probability than the original base pairing. Model building and energy calculations show that these **cruciform** structures are somewhat strained compared to standard DNA and that the structure in the left panel of the figure is under greater strain than that in the right. Cruciform structures have been produced in the laboratory (for example, in supercoils in which an inverted repeat has been inserted) but have not yet been detected in DNA isolated from cells.

Both palindromes and interrupted inverted repeats have significant effects on the structure of single-stranded DNA (either obtained by denaturation or the natural single-stranded DNA found in certain phages) and on RNA. There is no simple mechanism with either single-stranded DNA or RNA to prevent intrastrand hydrogen bonding between adjacent or nearby complementary sequences. Thus, a palindrome produces an intrastrand double-stranded segment called a **hairpin** (Figure 4-32(a)) and an interrupted inverted repeat produces a structure consisting of a double-stranded segment with a terminal single-stranded loop, known as a **stem-and-loop** (Figure 4-32(b)).

Figure 4-32 (a) Hairpin and (b) stem-and-loop structures that can form from two types of palindromes in single-stranded nucleic acids.

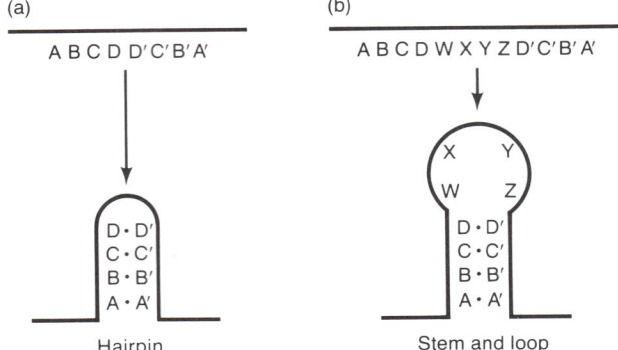

A sequence may also be repeated in the same orientation with or without a spacer.

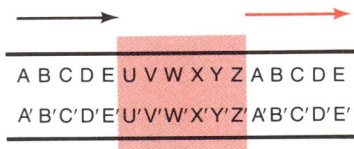

These are called **direct repeats.** They typically consist of tens of base pairs and have large interruptions. They do not provide alternate structures to double-stranded DNA and have no effect on single-stranded molecules.

Left-handed DNA Helices

The Watson-Crick structure of DNA (the B form) and the related A form of DNA are both right-handed helices. Subtle changes in the structure of DNA occur if the salt concentration of the suspending medium is very high and if divalent cations (Ca^{2+}, Mg^{2+}, and Mn^{2+}) are present but the helix remains right-handed. The double-stranded hexanucleotide

$$\overline{\underline{\text{CGCGCG}}}$$
$$\text{GCGCGC}$$

undergoes a drastic and reversible structural change when the NaCl concentration exceeds 2 M or if the $MgCl_2$ concentration is allowed to exceed 0.7 M, namely, the helix becomes left-handed. This left-handed structure is called **Z-DNA** (because of its zig-zag structure). It has been found that this structure is not confined to the poly($dC \cdot dG$) molecule but extends to any polydeoxynucleotide in which purines and pyrimidines alternate—for example,

$$\overline{\underline{\text{ACACACAC}}}$$
$$\text{TGTGTGTG}$$

Furthermore, if an alternating purine-pyrimidine sequence is contained in a long tract of DNA, for example,

···TGATCCGCGCGCGAGTCTT···
···ACTAGGCGCGCGCTCAGAA···

the purine-pyrimidine alternating sequence can still assume the Z configuration at 2 M NaCl while the remainder of the DNA has the B configuration.

Z-DNA is not simply a mirror image of a right-handed B-type helix. First, the sugar-phosphate backbone in Z-DNA follows a zig-zag path, as shown in Figure 4-33. Second, if the Z-DNA region is extensive, the mol-

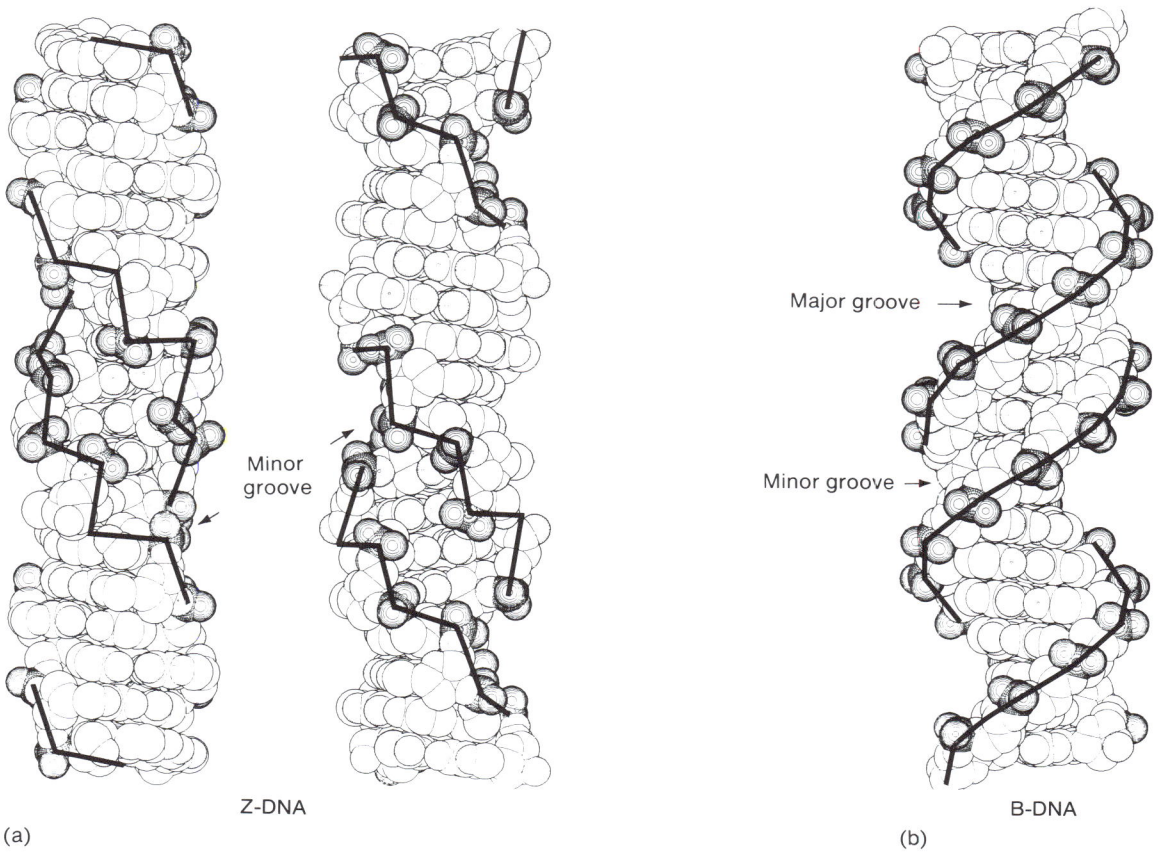

Z-DNA

(a)

B-DNA

(b)

Figure 4-33 Side views of (a) left-handed Z-DNA and (b) right-handed B-DNA. The heavy lines indicate the sugar-phosphate backbone. The zig-zag path of the backbone in Z-DNA is evident. Although it cannot be seen in panel (a), the minor groove of Z-DNA is quite deep, penetrating the helix axis. In contrast, the sugar-phosphate backbone of B-DNA is smooth and the grooves are shallow. [Courtesy of Andrew Wang.]

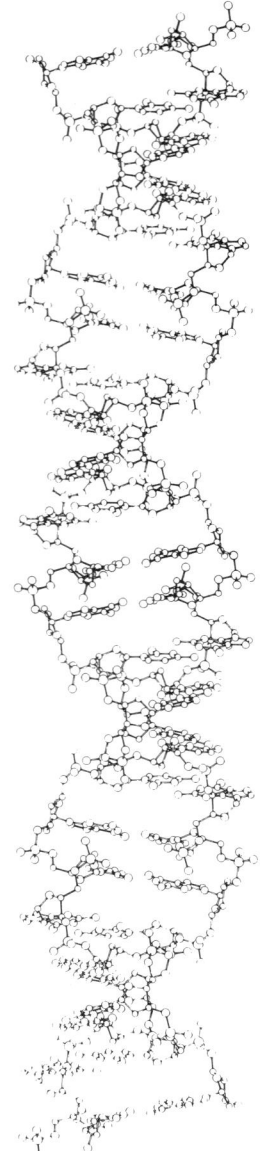

ecule becomes wavy, as shown in Figure 4-34. There are also other differences. Two important differences are the following: (1) In Z-DNA there are 12 base pairs per turn of the helix, in contrast with B-DNA, in which there are 10 pairs. (2) There is only a single groove in the helix, rather than a major and a minor groove, as in B-DNA. The details of the differences between Z-DNA and B-DNA are not important at this point—the significant fact is that the structures are very different.

The immediate question that arises is whether the left-handed structure has biological significance. Certainly the intracellular salt concentration does not ever approach 2 *M,* so *in vivo* salt could not cause the transition. However, it has been found that if potential left-handed sequences are contained in a supercoiled molecule, they form the left-handed structure at salt concentrations roughly equal to that found in cells. Thus, it is believed that left-handed helical regions do exist in intracellular DNA.

A technique called the **fluorescent antibody** technique has been used to show that regions of Z-DNA are present in animal chromosomes. This is the following. A goat is injected with rabbit immunoglobulin; an antibody forms, called goat anti-rabbit-immunoglobulin, which reacts specifically with *any* rabbit immunoglobulin G. A highly fluorescent molecule, fluorescein, is then coupled to the goat antibody. Cells from the salivary gland of the fruit fly *Drosophila,* showing chromosomes, are then prepared for light microscopy. Rabbit antibody to Z-DNA is added and then fluorescent goat anti-rabbit-immunoglobulin. The cells are then washed and observed by fluorescence microscopy. Where Z-DNA is present in the chromosomes,

Figure 4-34 The structure of double-stranded poly(dGC)—an alternating copolymer of dG and dC—showing how the helix axis continuously changes its direction, producing a "wavy" molecule. [Courtesy of S. Arnott and R. Chandrasekaran.]

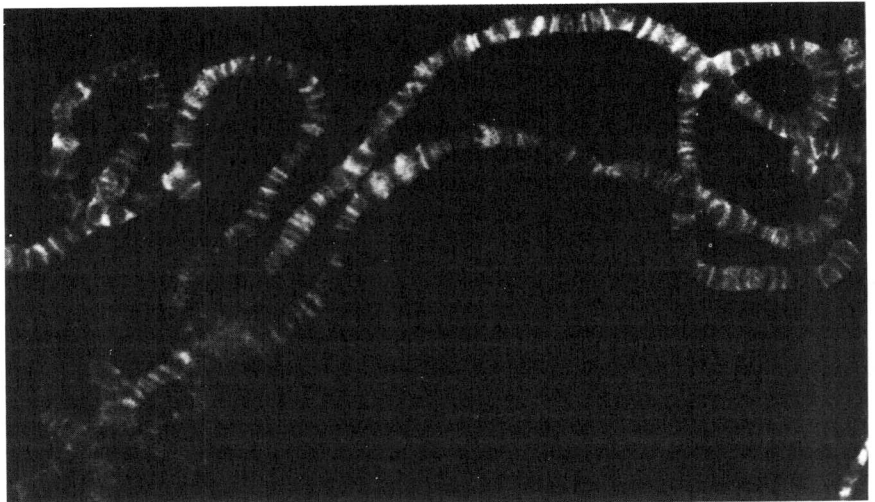

Figure 4-35 Chromosomes of the fruit fly to which a fluorescent antibody to Z-DNA is bound. [From A. Nordheim et al. 1981. *Nature,* 294: 417–422.]

anti-Z-DNA is bound, and fluorescent goat antibody is also bound at the same point. Thus, Z-DNA is made evident by the fluorescence of the goat antibody. Figure 4-35 shows the fluorescence of chromosomes treated by this protocol, indicating that Z-DNA is present in the *Drosophila* chromosomes. Further support comes from very recent experiments in which proteins isolated from *Drosophila* tissue were found to bind to synthetic Z-DNA, but not to ordinary B-DNA.

THE STRUCTURE OF RNA

A typical cell contains about ten times as much RNA as DNA yet we have said little about RNA. With the exception of the RNA of one phage and a few viruses, RNA is a single-stranded polynucleotide. There are four primary types of RNA—ribosomal RNA (of which there are three or four forms), transfer RNA (of which there are about 50 different structures), and messenger RNA (of which there are almost as many different molecules as there are genes), and a class of small eukaryotic RNA molecules found almost exclusively as ribonucleoprotein particles. All of these molecules superficially resemble single-stranded DNA in that single-stranded regions are interspersed with intramolecular double-stranded regions. Between 1/2 and 2/3 of the bases are paired. In single-stranded DNA the pairing is random and the paired regions tend to contain six or fewer pairs. Furthermore, if a sample of identical DNA molecules is denatured and intramolecular hydrogen bonds are allowed to form, the base-pairing pattern may differ from one molecule to the next. On the contrary, in RNA the double-stranded regions may contain up to 50 base pairs and a particular molecule has a definite base-pairing pattern.

The structures of the different classes of RNA molecules will be discussed in detail in Chapters 12 and 13.

DEPOLYMERIZATION OF NUCLEIC ACID

Both DNA and RNA can be hydrolyzed to free nucleotides either chemically or enzymatically.

At very low pH (1 or less), phosphodiester hydrolysis of both DNA and RNA occurs. This is accompanied by breakage of the *N*-glycosidic bond between the base and the sugar such that free bases are produced. This procedure has been used in determining the base composition of nucleic acids. At a pH of about 4 the purine *N*-glycosidic bond is broken. Long-term exposure of DNA to this pH removes all of the purines, yielding a molecule known as **apurinic acid.**

At high pH the behavior of DNA is strikingly different from that of RNA. DNA remains very resistant to pH 13 (about 0.2 phosphodiester bonds broken per million bonds per minute at 37°C) whereas at pH 11 a typical RNA molecule is totally hydrolyzed to ribonucleotides in a few minutes at 37°. However, in one situation DNA is cleaved by alkali: if a base has been removed either by the pH or depurination reaction just described, the phosphodiester bond at the 5′ end of the deoxyribose lacking a base is broken by alkali as rapidly as RNA is cleaved.

Nucleases are enzymes that depolymerize nucleic acids. Most nucleases show chemical specificity and are either deoxyribonucleases (DNase) or ribonucleases (RNase). Many DNases act on only single-stranded or only double-stranded DNA, though some act on either kind. Furthermore, some nucleases act only at the end of a nucleic acid, removing a single nucleotide

Table 4-2   Properties of selected nucleases

Nuclease	Substrate*	Site of cleavage	Product
Pancreatic ribonuclease	RNA	Endonuclease; adjacent to pyrimidines	Mono- or oligonucleotide terminating with pyrimidine nucleoside 3′-P
T1 RNase	RNA	Endonuclease; adjacent to guanosine	Mono- or oligonucleotide terminating with guanosine 3′-P
Pancreatic DNase I	DNA, RNA	Endonuclease	Oligonucleotides
Venom phosphodiesterase	RNA or DNA	Exonuclease at 3′-OH end	5′-P mononucleotides
Spleen phosphodiesterase	RNA or DNA	Exonuclease at 5′-OH end	3′-P mononucleotides
Micrococcal nuclease	ss DNA	Endonuclease	3′-P mononucleotides
S1 nuclease	ss regions of DNA	Endonuclease; any ss site	5′-P mononucleotides
E. coli exonuclease I	ss DNA	Exonuclease at 5′-OH end	5′-P mononucleotides plus a terminal dinucleotide
E. coli exonuclease III	ds DNA	Exonuclease at 3′-OH end	5′-P mononucleotides
E. coli exonuclease VII	ss DNA	Exonuclease at 3′-OH end	5′-P mononucleotides
Phage λ exonuclease	ds DNA	Exonuclease at 5′-P end	5′-P mononucleotides

*ds and ss are abbreviations for double-stranded and single-stranded, respectively.

at a time; these are called **exonucleases** and they may act only at a 3′ or a 5′ terminus. **Endonucleases** act within the strand; some of these are specific in that they cleave only between particular bases.

Nucleases serve a variety of biological functions and have also been useful in the laboratory for removing unwanted nucleic acids and as an early step in determining the base sequence of DNA and RNA molecules. Specific nucleases are described in Table 4-2; others will be described as needed in the text.

The **S1 endonuclease** isolated from *Aspergillus oryzae* has been of special use. This enzyme acts exclusively on single-stranded polynucleotides or on single-stranded regions of double-stranded molecules; it differs from other single-strand-specific enzymes in that the single-stranded region can be as small as one or two bases. We have noted earlier that supercoiled DNA contains somewhat long-lived single-stranded bubbles because of increased breathing. Supercoiled DNA can be broken by S1 nuclease because of these regions. In fact, this nuclease can be used to distinguish supercoiled from both nonsupercoiled covalent circles and nicked circular DNA, both of which are resistant to the enzyme. S1 nuclease makes a double-strand break because it acts on both single-stranded branches of the bubble.

Restriction endonucleases are a class of nucleases, each of which acts on a unique base sequence They are exceedingly important in modern DNA technology and will be described in detail in Chapter 5.

Nucleic Acid Technology

Molecular genetics of the 1980s (and probably the 1990s as well) relies heavily on nucleic acid technology—actually a fairly small number of techniques for isolating, characterizing, and altering segments of DNA. The use of restriction enzymes, cloning, and base-sequencing have become routine and must be understood by researchers and by those who wish to comprehend experimental biology. These topics are the subject of this chapter.

ISOLATION OF NUCLEIC ACIDS

Isolation of DNA is an essential step in many experiments. Although the methods are straightforward, the particular procedure must be tailored to the organism from which the DNA is to be obtained, because the structure and composition of organisms vary. The differences between the techniques follow from how the DNA is enclosed and the fraction of the total dry weight that is DNA (which varies from about 1 percent in complex mammalian cells to about 50 percent in phages). The common feature in all procedures is that a cell or virus is first broken, and then DNA is separated from such other components as protein, RNA, lipid, and carbohydrates. The basic procedures are the following:

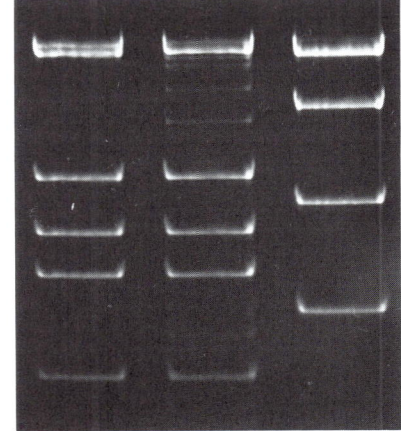

Gel electrophoregram of λ DNA digested with two different restriction enzymes. The center lane, which can be compared to the left lane, contains faint bands representing larger fragments from incomplete digestion of the DNA. [Courtesy of Arthur Landy.]

121

1. *Phages*. The simplest procedure is employed with phages. An aqueous suspension of particles is shaken with phenol, a reagent that does not mix with water. A small amount of phenol enters the aqueous layer, breaking open the protein coat and denaturating the individual protein molecules. Most of the denatured protein either enters the phenol layer or precipitates at the phenol-water interface. The DNA remains in the aqueous layer. Addition of ethanol precipitates the DNA, which can be collected and redissolved in a solution having any desired composition.

2. *Bacteria*. The contents of bacterial cells are enclosed in a multilayered cell wall and an inner cell membrane. These structures cannot simply be removed by exposure to phenol but can be solubilized by successive treatment of a cell suspension with **lysozyme** (an enzyme isolated from chicken egg white), and one of several different detergents, the most common one being sodium dodecyl sulfate (SDS). The result of this treatment is release of the contents of the cells. RNA is then removed by digestion with specific enzymes, protein is removed both enzymatically and with phenol, and the DNA is collected by precipitation with ethanol.

3. *Yeasts and fungi*. These cells are resistant to lysozyme, but other enzymes (one is isolated from snails) will break down their cell walls. Following this step, the procedure follows that used with bacteria.

4. *Higher eukaryotes*. Animal and plant cells have a low ratio of DNA to protein, and for technical reasons substantial loss of DNA occurs if the DNA is directly purified. To minimize this problem, nuclei are usually isolated first; this step increases the ratio of DNA to protein and also avoids contamination of chromosomal DNA by DNA from cytoplasmic organelles (e.g., mitochondria and chloroplasts). The nuclei are broken open in several ways, RNA and protein molecules are enzymatically digested, and the DNA is precipitated.

The special techniques used to isolate supercoiled DNA were described in Chapter 4 in the section entitled Experimental Detection of Covalently Closed Circles (item 2). Isolation of plasmid DNA will be described in a later section.

Isolation of RNA follows the procedure used for DNA with one major change. After the cells are broken open, DNase is added to degrade the DNA. Protein is not usually removed with phenol since a large fraction of the RNA enters the phenol layer and becomes contaminated with protein. Treatment with certain enzymes that degrade proteins (**proteases**) can be used to remove protein. Alternatively, after DNase treatment the solution can be shaken vigorously with chloroform. This solvent is immiscible with water, and shaking creates a large surface area of the chloroform-water interface. Precipitated protein gathers at this interface and is easily removed.

RADIOACTIVE LABELING OF NUCLEIC ACIDS

Many experiments require the use of radioactive nucleic acids. Several techniques are available for labeling biological molecules by growth of cells in radioactive media. The underlying principle is to add to the growth medium a precursor that is incorporated only into the molecule of interest, when this is possible. For DNA this is uncomplicated because thymine and thymidine are utilized only in DNA synthesis. Thus, if [^{3}H]thymidine or [^{14}C]thymidine is put in the growth medium, newly replicated DNA will contain ^{3}H and ^{14}C, respectively. This is by far the most common procedure for preparing radioactive DNA. Sometimes a greater amount of radioactivity per weight of DNA is needed than can be attained with radioactive thymidine; then ^{32}P is added as the ^{32}PO$_4^{3-}$ ion. However, when this is done, both DNA and RNA become labeled, as well as other phosphorus-containing compounds. If only radioactive DNA is needed, labeled cells can be treated with either RNase or dilute alkali to hydrolyze the RNA, as explained in Chapter 4 (Depolymerization of Nucleic Acid). It then only becomes necessary to separate DNA molecules from the free RNA nucleotides, for which there are several simple procedures.

Purified nonradioactive DNA can be rendered exceedingly radioactive by introducing single-strand breaks and using the enzymatic nick-translation procedure described in Chapter 9.

Ribose and uridine are unique to RNA. However, if radioactive ribose (or any other sugar) is added to a growth medium, it would be extensively metabolized and the label would appear in almost all compounds containing carbon. The best precursor for labeling RNA is [^{3}H]- or [^{14}C]uridine, which is converted to uridine triphosphate and then polymerized into RNA. However, some uridine is converted to deoxyuridine monophosphate, which is methylated to give thymidine monophosphate and then incorporated into DNA. Thus the use of radioactive uridine yields both radioactive RNA and radioactive DNA. In order to have radioactive RNA only, it is necessary to treat the cell extract with a DNase to hydrolyze the DNA. RNA can also be labeled with ^{32}P; again, a final step must be removal of DNA with DNase.

The base-sequencing procedure to be described shortly and other analytical techniques requires DNA that is labeled only at a terminus. Such labeling can be obtained enzymatically with **polynucleotide kinase,** an enzyme of unknown biological function. The enzyme commonly used is isolated from *E. coli* that are infected with phage T4. Polynucleotide kinase catalyzes transfer of ^{32}P from [γ-^{32}P]ATP to a 5′-OH group at the terminus of an RNA or DNA molecule. However, naturally occurring nucleic acids possess 5′-P termini, not 5′-OH termini; thus, the 5′-P group must be removed. This is easily done with the enzyme alkaline phosphatase. The reaction sequence that is used for terminal labeling of DNA with ^{32}P is the following:

$$\underline{\hspace{5cm}}\ 5'\text{-P} \xrightarrow[\text{phosphatase}]{\text{Alkaline}} \underline{\hspace{5cm}}\ 5'\text{-OH} + 2\,\text{P}_\text{i}$$
$$\text{P-}5'\ \underline{\hspace{4cm}} \qquad\qquad \text{HO-}5'\ \underline{\hspace{4cm}}$$

$$\underline{\hspace{5cm}}\ 5'\text{-OH} + 2\,^{32}\text{P-P-P-Adenosine} \xrightarrow[\text{kinase}]{\text{Polynucleotide}}$$
$$\text{HO-}5'\ \underline{\hspace{4cm}}$$

$$\underline{\hspace{5cm}}\ 5'\text{-}^{32}\text{P} + 2\,\text{ADP}$$
$$^{32}\text{P-}5'\ \underline{\hspace{4cm}}$$

RESTRICTION ENDONUCLEASES

In Chapter 4 a variety of DNases were described. These have different specificities, such as cutting single-stranded DNA versus double-stranded DNA, but none of the enzymes mentioned cut either adjacent to particular bases or in particular base sequences. However, numerous DNases are known that cut in a particular position in only one base sequence. These enzymes are exceedingly important in modern DNA technology, as will be seen in this section and in Chapter 22.

Properties of Restriction Enzymes

A **restriction endonuclease** is an enzyme that recognizes a specific base sequence in a DNA molecule and makes two cuts, one in each strand, generating 3'-OH and 5'-P termini. Nearly 1000 different restriction enzymes have been purified from hundreds of different microorganisms. All but a few of these enzymes recognize sequences that are palindromes. That is, the recognition site always has a sequence whose general form is

$$
\begin{array}{ccc|ccc}
\text{A} & \text{B} & \text{C} & \text{C}' & \text{B}' & \text{A}' \\
\text{A}' & \text{B}' & \text{C}' & \text{C} & \text{B} & \text{A}
\end{array}
\quad \text{or} \quad
\begin{array}{ccc|ccc}
\text{A} & \text{B} & \text{X} & \text{B}' & \text{A}' \\
\text{A}' & \text{B}' & \text{X}' & \text{B} & \text{A}
\end{array}
\quad \text{or} \quad
\begin{array}{cc|cc}
\text{A} & \text{B} & \text{B}' & \text{A}' \\
\text{A}' & \text{B}' & \text{B} & \text{A}
\end{array}
$$

in which the capital letters represent bases, a prime indicates a complementary base, X is any base, and the dashed line is the axis of symmetry. A few sensitive sequences having more than six bases are known but none have been observed containing fewer than four bases.

There are two major types of restriction enzymes: type I enzymes, which recognize a specific sequence but make cuts elsewhere; and type II enzymes, which make cuts only *within* the recognition site. Type II enzymes are the most important class and we confine our attention to them. All restriction enzymes of this class make two single-strand breaks, one break in each strand. There are two distinct arrangements of these breaks: (1) both breaks at the center of symmetry (generating **flush** or **blunt ends**), or (2) breaks that are symmetrically placed around the line of symmetry (generating **cohesive ends**). These arrangements and their consequences are shown in Figure 5-1. Table 5-1 lists the sequences and cleavage sites for twelve useful restriction enzymes, nine of which generate cohesive sites and three of which yield flush ends.

An important point about these restriction enzymes is this: since a particular enzyme recognizes a unique sequence, *the number of cuts made in the DNA from an organism is limited.* For example, a typical bacterial DNA mol-

Table 5-1 Some restriction endonucleases and their cleavage sites

Microorganism	Name of enzyme	Target sequence and cleavage sites
Generates cohesive ends		
Escherichia coli RY13	EcoRI	G↓A A T T C C T T A A↑G
Bacillus amyloliquefaciens H	BamHI	G↓G A T C C C C T A G↑G
Bacillus globigii	BglII	A↓G A T C T T C T A G↑A
Haemophilus aegyptius	HaeII	Pu G C G C↓Py Py↑C G C G Pu
Haemophilus influenza R$_d$	HindIII	A↓A G C T T T T C G A↑A
Providencia stuartii	PstI	C T G C A↓G G↑A C G T C
Streptomyces albus G	SalI	G↓T C G A C C A G C T↑G
Xanthomonas badrii	XbaI	T↓C T A G A A G A T C↑T
Thermus aquaticus	TaqI	T↓C G A A G C↑T
Generates flush ends		
Brevibacterium albidum	BalI	T G G↓C C A A C C↑G G T
Haemophilus aegyptius	HaeI	$\binom{A}{T}$ G G↓C C $\binom{T}{A}$ C C↑G G
Serratia marcescens	SmaI	C C C↓G G G G G G↑C C C

Note: The vertical dashed line indicates the axis of dyad symmetry in each sequence. Arrows indicate the sites of cutting. The enzyme TaqI yields cohesive ends consisting of two nucleotides, whereas the cohesive ends produced by the other enzymes contain four nucleotides. The enzyme HaeI recognizes the sequence GGCC whether the adjacent base pair is A·T or T·A, as long as dyad symmetry is retained. Pu and Py refer to any purine and pyrimidine, respectively.

(a) Cuts on line of symmetry

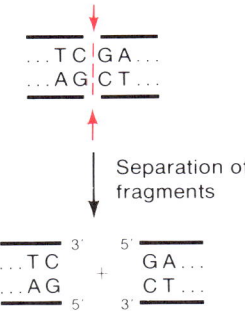

Flush-end molecules

(b) Cuts symmetrically placed around line of symmetry

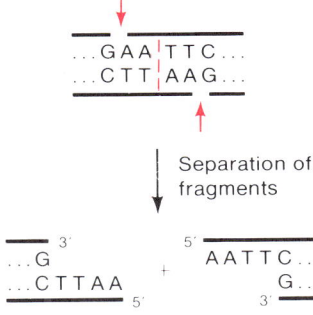

Cohesive-end molecules

Figure 5-1 Two types of cuts made by restriction enzymes. The arrows indicate the cleavage sites. The dashed line is the center of symmetry of the sequence.

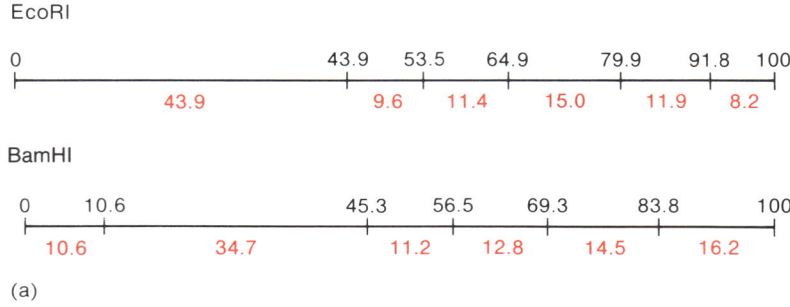

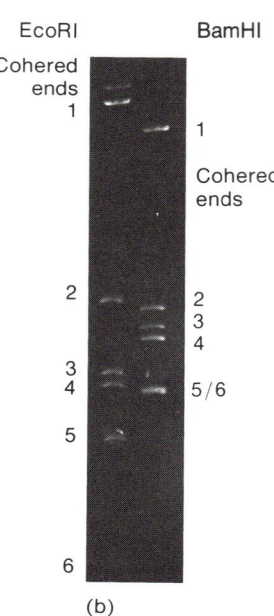

Figure 5-2 (a) Restriction maps of λ DNA for EcoRI and BamHI nucleases. The vertical bars indicate the sites of cutting. The black numbers indicate the percentage of the total length of λ DNA measured from the gene *A* end of the molecule. The red numbers are the lengths of each fragment, again expressed as percentage of total length. (b) A gel electrophoregram of EcoRI and BamHI restriction-enzyme digests of λ DNA. The bands labeled cohered ends contain molecules consisting of the two terminal fragments joined by the normal cohesive ends of λ DNA. Numbers indicate fragments in order from largest (1) to smallest (6). Bands 5 and 6 of the BamHI digest are not resolved. [Courtesy of Dennis Anderson and Lynn Enquist.]

ecule, which contains roughly 3×10^6 base pairs, is cut into a few hundred to a few thousand fragments. Smaller DNA molecules such as phage or plasmid DNA molecules may have fewer than ten sites of cutting (frequently, one or two, and often none). Because of the specificity just mentioned, *a particular restriction enzyme generates a unique family of fragments from a particular DNA molecule.* Another enzyme will generate a different family of fragments from the same DNA molecule. (Some enzymes—isoschizomers—have the same specificity and generate identical families.) Figure 5-2(a) shows the sites of cutting of *E. coli* phage λ DNA by the enzymes EcoRI and BamHI. The family of fragments generated by a single enzyme is usually detected by agarose gel electrophoresis of the digested DNA (Figure 5-2(b)). The fragments migrate at a rate that is a function of molecular weight (Chapter 3) and their molecular weights can be determined by reference to fragments of known molecular weight run concurrently (Figure 3-21).

Determining the Sites of Cuts: Restriction Mapping

It is frequently of value to determine the location of restriction cuts in a particular DNA molecule. A map showing the positions is called a **restriction map.** There are several ways to obtain such a map. One method, applicable to DNA molecules with both linear and circular forms is shown in Figure 5-3. The electrophoretic patterns of enzymatic digests of the linear

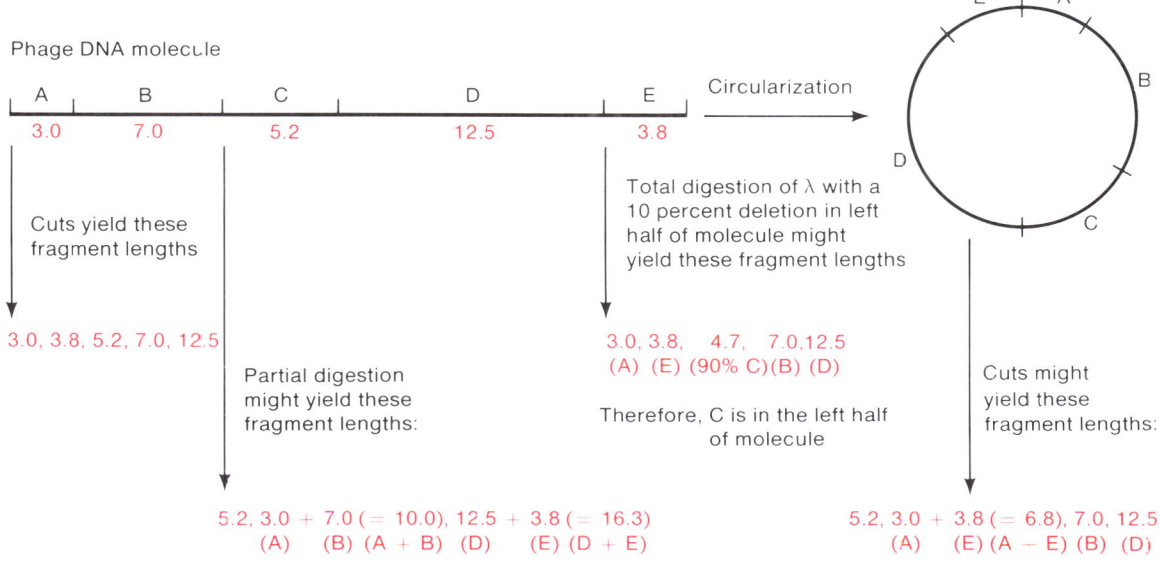

Phage DNA molecule

A	B	C	D	E	Circularization
3.0	7.0	5.2	12.5	3.8	

Cuts yield these
fragment lengths

3.0, 3.8, 5.2, 7.0, 12.5

Partial digestion
might yield these
fragment lengths:

5.2, 3.0 + 7.0 (= 10.0), 12.5 + 3.8 (= 16.3)
(A) (B) (A + B) (D) (E) (D + E)

Therefore, A and B are adjacent;
D and E are adjacent

Total digestion of λ with a
10 percent deletion in left
half of molecule might
yield these fragment lengths

3.0, 3.8, 4.7, 7.0,12.5
(A) (E) (90% C)(B) (D)

Therefore, C is in the left half
of molecule

Cuts might
yield these
fragment lengths:

5.2, 3.0 + 3.8 (= 6.8), 7.0, 12.5
(A) (E) (A − E) (B) (D)

Therefore, A and E are terminal

Figure 5-3 A possible scheme for determining the positions of restriction cuts in a phage λ DNA molecule, which has a linear and a circular form.

and circular forms of the DNA are compared. The pattern for the circles will lack the two bands corresponding to the terminal fragments and will contain a new band that is formed by the joined ends. Thus, the terminal fragments are identified as those that are missing from the circle digest. In a typical digest one or two cuts will not have been made in every molecule; some faint bands formed by the uncut fragments will be seen. The intensity of these bands can be increased by decreasing the reaction time, generating what is called a **partial digest.** In such a digest other bands are weakened compared to the normal digest. The molecular weight of each enhanced band is invariably the sum of the molecular weights of two fragments contained in the weakened bands, which indicates that the two fragments are adjacent in the original uncleaved molecule. If the number of cuts is small (for example, six), this procedure may be sufficient for unambiguous ordering of the fragments. If the molecule does not have a circular form, the terminal fragments can be identified by labeling with ^{32}P by the polynucleotide kinase reaction prior to cleavage and determining which fragments (after cleavage) are radioactive.

The position of some fragments can also be identified by examining genetic deletion mutants. If the deletion does not eliminate a restriction site, a new fragment (some of whose DNA is deleted) will appear in the digest of the deletion mutant and will be smaller, by the size of the deleted DNA, than the corresponding fragment from the wild-type undeleted molecule.

Figure 5-4 Three restriction digests useful in determining restriction maps. Enzyme I yields digest I, enzyme II yields digest II, and enzymes I and II together yield the double digest. To simplify the discussion, the fragments are given in order, though the order would be unknown until the data were analyzed.

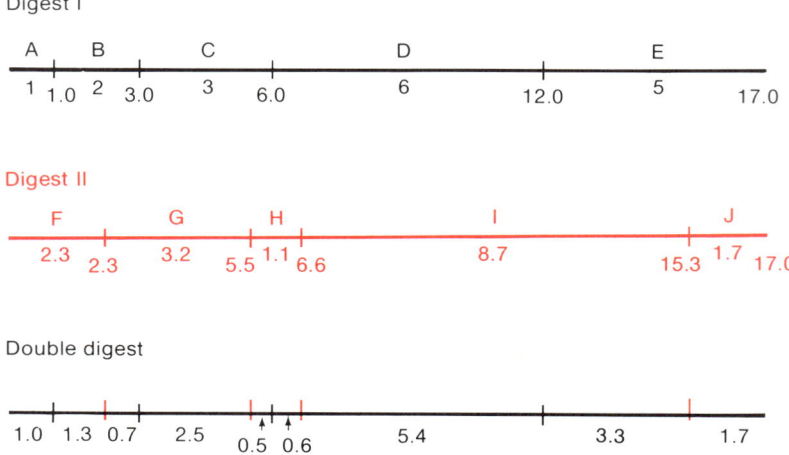

If the deletion instead spans a restriction site, two bands will be absent and a new band will appear; the molecular weight of the new band equals the sum of the molecular weights of the missing fragments minus that of the deletion.

Although the methods just described usually work, the most common way to obtain a restriction map is the **double–digest procedure,** which uses two different restriction enzymes. The procedure is to take three samples of a particular DNA species, treat each of two of these with a separate enzyme and one sample with both enzymes (a **double digest**), and then compare the three sets of fragments. An example of the sets of fragments that might be generated is shown in Figure 5-4. The value of this procedure is that it yields restriction maps for both enzymes simultaneously. Let us assume, as would usually be the case, that the terminal fragments in each digest are identified either by the circularization procedure or by labeling the 5′ termini with ^{32}P. With this information one can then identify the fragments that are adjacent to the terminal fragments. Let us first determine which fragment in digest I is adjacent to fragment A. In the double digest the cut forming fragment F (of digest II) must divide the fragment of digest I adjacent to A, thereby generating a small fragment whose length is $2.3 - 1 = 1.3$ and another fragment of length x; thus, $1.3 + x$ must equal the length of fragment B, C, or D. Excluding the terminal fragments in the double digest, namely, those of length 1 and 1.7, the only possible value of x is 0.7; thus, the fragment adjacent to A must have a length of $1.3 + 0.7 = 2$, which is the length of B. Hence, B is adjacent to A. Now we identify the fragment adjacent to F in digest II. The right end of B must divide the fragment adjacent to F. Since $A + B = 3$ and $F = 2.3$, then $3 - 2.3 + a$ fragment of length p must equal the length of a nonterminal fragment in digest II. The only possible value of p is 2.5, so the fragment adjacent to F has a length of 3.2 and must be G. We now turn our attention to the right end of the molecule. E divides the fragment adjacent to J. Thus, for some fragment of length q in the double digest, $5 - 1.7 + q$ must equal the length of a fragment in digest II. The only possible value of q is 5.4 and the length of the fragment adjacent to J is 8.7—that is, fragment I. This process can be continued to generate the complete restriction maps for both enzymes, as the reader should verify. The process is sometimes more complicated if different fragments having the same size are present.

PLASMIDS

Many bacteria carry, in addition to a chromosome, several copies of a small circular DNA molecule called a **plasmid.** (One example, F, was described in Chapter 1.) They range in size from 0.1 to 5 percent of the chromosome. Plasmids replicate, more-or-less independently of chromosomal replication, and hence are carried from one generation to the next. Plasmid DNA contains genes. Most of the genes are involved in plasmid replication and, for transmissible plasmids such as F (Chapter 1), in DNA transfer. Other genes that confer particular genetic properties may be present. For example, the F′ plasmids, which are derived from Hfr cells, contain chromosomal genes (Chapter 1). The R plasmids possess genes that confer resistance to certain antibiotics; these resistance genes make detection of these plasmids especially straightforward.

Plasmids are important elements in modern DNA technology for two reasons: (1) they can be easily isolated and transferred to other host cells merely by mixing purified plasmid DNA with the desired host cells in appropriate solutions (see below), and (2) by use of restriction enzymes any gene can, in principle, be inserted into the plasmid DNA. In this section we will describe the methods for isolating plasmid DNA and the simplest method for inserting foreign DNA into a plasmid. Much more about plasmids will be given in Chapter 19. A variety of techniques for constructing plasmids with foreign genes, namely, the basic methodology of genetic engineering, will be presented in Chapter 22.

Isolation of Plasmid DNA

Plasmid DNA can be isolated from bacteria in a simple way. Plasmid-containing bacteria are opened by the lysozyme–detergent treatment described earlier. The resulting translucent solution, called a **cell lysate,** is centrifuged. The bacterial chromosome complex, which contains protein and RNA, is very large and compact and moves rapidly to the bottom of the centrifuge tube; the smaller plasmid DNA remains in the clear supernatant, which is

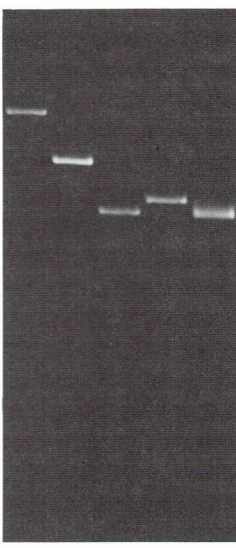

Figure 5-5 A gel electrophoregram showing the migration of five different plasmids. Movement is from top to bottom. Each vertical column ("lane") represents a single plasmid. After migration the gel is soaked in a solution of ethidium bromide, washed, and then illuminated with near-ultraviolet light. The ethidium bromide, which is bound to the DNA, fluoresces. The single intense band in each lane contains supercoiled DNA molecules. The apparatus used for gel electrophoresis is shown in Figure 3-15. [Courtesy of Elaine Cocuzzo and Pieter Wensink.]

called a **cleared lysate.** CsCl is then added to the cleared lysate, and the lysate is centrifuged to equilibrium. Some chromosomal DNA is usually present in the lysate but if the G + C content of the plasmid and the chromosomal DNA differ, the two types of molecules form distinct bands that can be separately removed. In the more common case there is no difference in G + C content and one must make use of the fact that most of the plasmid DNA is supercoiled. Then, the CsCl centrifugation step is performed with ethidium bromide in the lysate in order to introduce a difference in the densities of the supercoil and the linear chromosomal fragments (Chapter 4, Figure 4-29). The supercoiled DNA has a higher density. The only disadvantage of this technique, which is the one most commonly used, is that both nicked circles (which often form accidentally during isolation) and nonsupercoiled replicating molecules are discarded with the chromosomal DNA.

The presence of plasmid DNA in a single bacterial colony can be detected in an elegant way by electrophoresis (Figure 5-5). A single colony is taken and lysed and then subjected to gel electrophoresis. The bacterial chromosome is very large and cannot penetrate the gel (though some fragments do), but the plasmid DNA can do so. The rate of electrophoretic movement of DNA molecules through a gel increases with decreasing molecular weight, so plasmid DNA, if present, will form a narrow band at a position in the gel characteristic of its molecular weight. The band is visualized by staining the gel with ethidium bromide, which binds tightly to the DNA and fluoresces upon irradiation with ultraviolet light. From the distance moved in a particular time interval relative to that for plasmids of known molecular weight, the molecular weight of the plasmid DNA can be calculated, as shown in the figure. This screening technique is very useful in recombinant DNA technology. Plasmid DNA can also be detected in a cleared lysate by this electrophoretic technique.

Cloning a Restriction Fragment in a Plasmid

Earlier we saw that a particular restriction enzyme recognizes only a single base sequence. Furthermore, except for those enzymes producing blunt ends, the cuts generate at each end of a fragment single-stranded termini that are complementary. The first evidence for complementary termini was that fragments produced by many restriction enzymes spontaneously circularize if they are stored in a buffer of high ionic strength and under conditions suitable for renaturation, as shown in Figure 5-6. The complementary termini also enable different fragments to be joined.

The basis of the joining procedure to be described is that the fragments generated by a particular enzyme acting on one DNA molecule have the same set of single-stranded ends as the fragments produced by the same

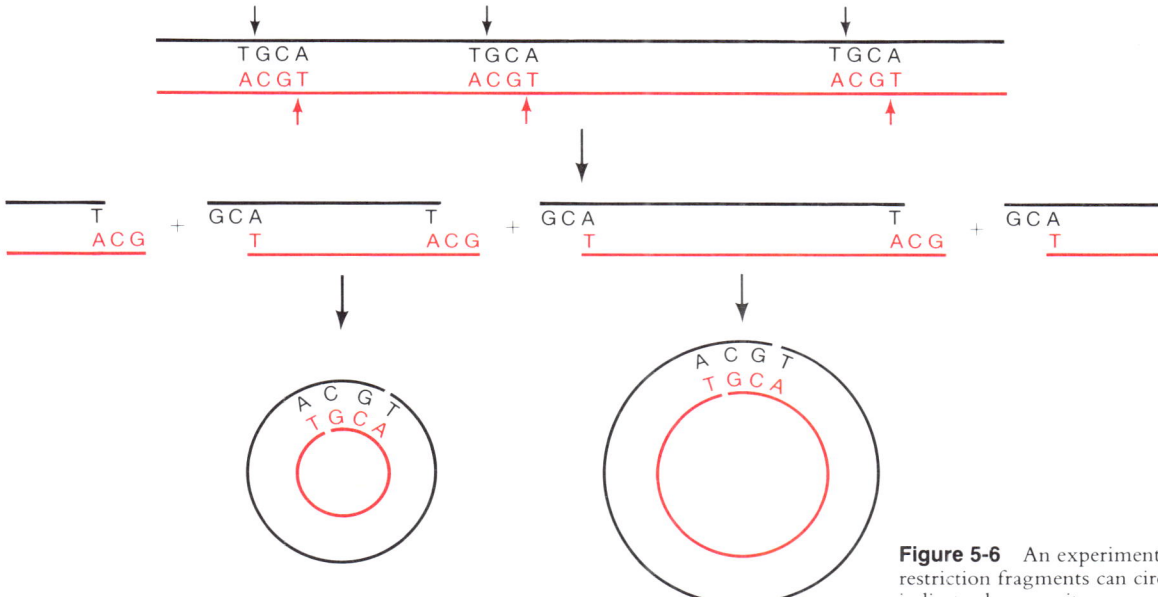

Figure 5-6   An experiment showing that restriction fragments can circularize. Arrows indicate cleavage sites.

enzyme acting on a different DNA molecule (as long as both DNA molecules have sequences recognized by the enzyme). Therefore, fragments from the DNA molecules of two different organisms (for example, a bacterium and a frog) can be joined by renaturation of complementary single-stranded termini. Furthermore, if the joint is ligated after base-pairing, the fragments are joined permanently.

The joining technique takes on special importance if one of the sources of cleaved DNA is a plasmid. Figure 5-7 shows a plasmid DNA molecule that has only one cleavage site for a particular restriction enzyme, which is also used to cleaved frog DNA. If frog fragments are mixed with linearized plasmid DNA and joining is allowed to occur, a circular plasmid DNA molecule containing frog DNA can be generated. The significance of this technique is that the hybrid plasmid can be reestablished in a bacterium; that is, the inserted DNA fragment, which replicates as part of the plasmid, is **cloned.** The technique is simply to mix bacteria and plasmid DNA in a solution of cold $CaCl_2$; in such a solution bacteria can take in DNA molecules. This technique makes it possible to transfer any plasmid that can be isolated to almost any *E. coli* strain. All that is required is that the recipient bacterium possess replication enzymes that are active on the plasmid, which is invariably the case if the plasmid has been isolated from the same species as the recipient. This technique, called **$CaCl_2$ transformation,** is the most common and most important experimental technique for transferring a plasmid from one strain to another.

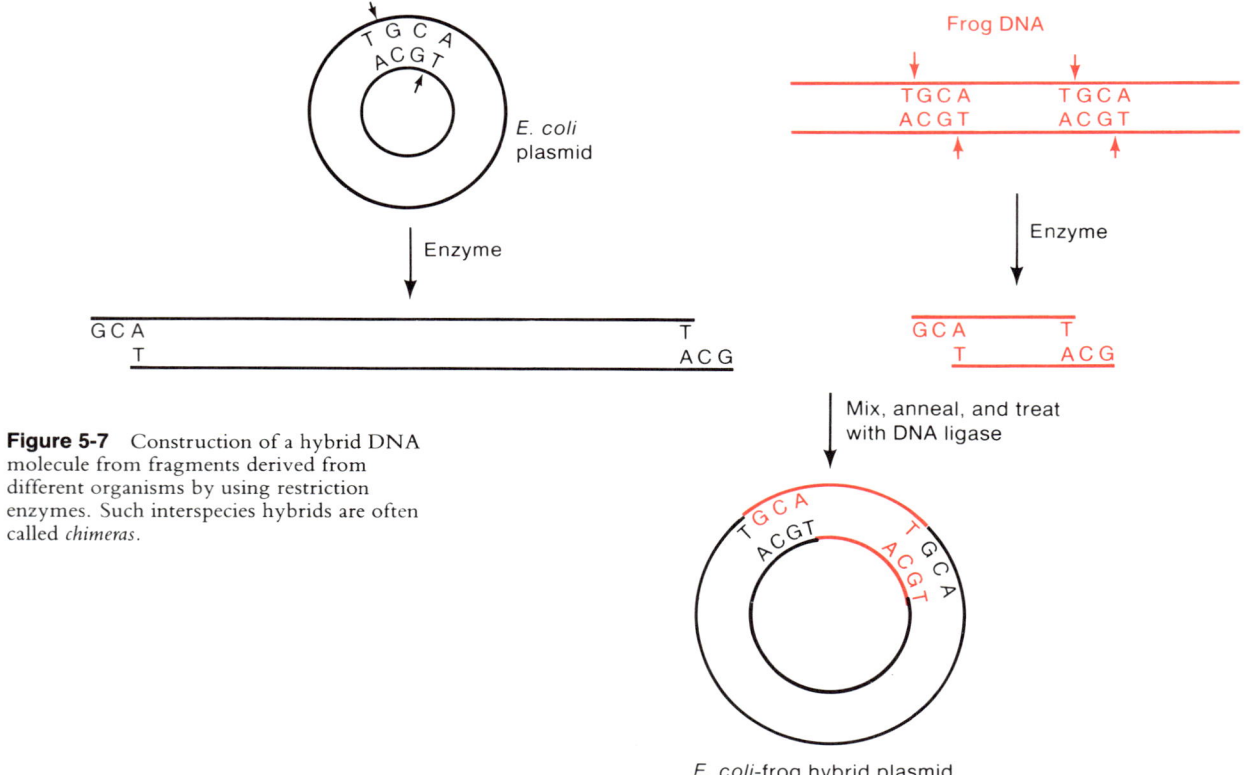

Figure 5-7 Construction of a hybrid DNA molecule from fragments derived from different organisms by using restriction enzymes. Such interspecies hybrids are often called *chimeras*.

PURIFICATION OF COMPLEMENTARY STRANDS OF DNA

A variety of experiments require that the complementary strands of a particular DNA molecule be separately purified. This can be accomplished by two techniques. The first is based on two facts: (1) pyrimidines (C and T) are often present in one strand of DNA in tracts about ten bases long, and (2) the number of such pyrimidine tracts is usually not the same in both strands. If denatured DNA is mixed with a polyribopurine (a mixed polymer of inosine and guanine, poly(I,G), is a favorite), the ribopolymer binds to each strand. The density of RNA in CsCl is much greater than that of DNA, so the DNA–poly(I,G) hybrids have a higher density than the parent DNA strands. In addition, because the number of pyrimidine tracts in the two strands differ, the amount of polymer bound to each strand differs; thus, the two strands have distinct densities and can be separated into two bands by sedimentation to equilibrium in CsCl (Figure 5-8). The material in each band can be collected, and the ribopolymer can be removed by treatment with an

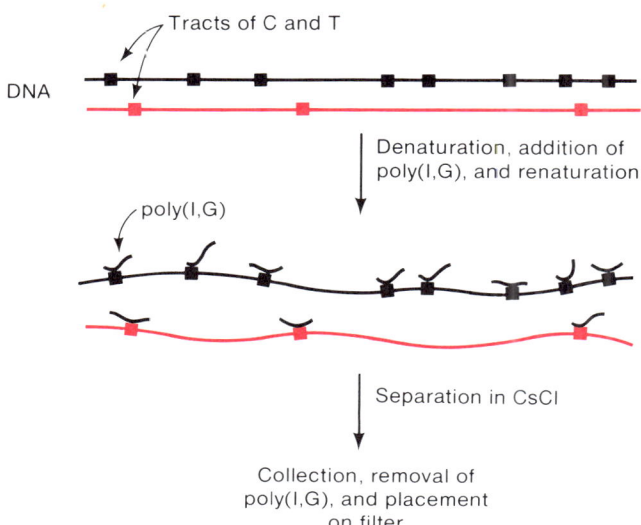

Tracts of C and T

DNA

Denaturation, addition of
poly(I,G), and renaturation

poly(I,G)

Separation in CsCl

Collection, removal of
poly(I,G), and placement
on filter

Figure 5-8 Separation of black and red strands of DNA. The poly(I,G) molecule has a density much greater than that of single-stranded DNA and hence increases the density of the single strands by binding to sequences of cytosine and thymine (pyrimidine tracts). Since the individual single strands usually have different numbers of pyrimidine tracts, the complexes with poly(I,G) will have different densities and therefore come to rest at separate positions when centrifuged to equilibrium in a CsCl density gradient. The complexes are removed from the CsCl and the poly(I,G), which is a polyribonucleotide, is digested with an RNase. The purified single strands of DNA are generally placed on a nitrocellulose filter and used in hybridization experiments. The denser and less dense strands are also frequently called the H (heavy) and L (light) strands.

RNase or with alkali, which hydrolyzes RNA but not DNA (Chapter 4). The second procedure is an electrophoretic one. If DNA is denatured and then electrophoresed through gels containing certain simple buffers, for unknown reasons the complementary strands frequently migrate at different rates through the gel, allowing the two strands to be separated and purified.

HYBRIDIZATION BY BLOTTING

In Chapter 4 we saw that hybridization of RNA and single-stranded DNA could be carried out and measured by filter-binding techniques. With the advent of the use of restriction endonucleases to cleave DNA into unique and fairly small fragments, it became possible to carry out both DNA-DNA and DNA-RNA hybridizations with particular regions of a chromosome. In the earliest experiments individual DNA fragments were painstakingly isolated from electrophoretic gels and hybridization was carried out with these purified fractions. Such tedium can be replaced by the **Southern transfer procedure** (developed by E. Southern and called **blotting**), a method for performing hybridization to a large number of particular DNA segments without the necessity of purifying the individual DNA fragments.

In most analytical hybridization procedures DNA fragments are bound to a nitrocellulose filter. This step is retained in the Southern transfer technique, but a collection of fragments is handled in such a way that all fragments are transferred in a single step from a gel to a sheet of nitrocellulose, significantly simplifying the entire process. The procedure is the following

Figure 5-9 The Southern transfer technique. (a) A stack—consisting of a weight, a gel, a filter, and absorbent material—at the time the weight is applied. (b) A later time—the weight has forced the buffer (shaded area), which carries the DNA, into the nitrocellulose. (c) The lowest layer has absorbed the buffer but not the DNA, which remains bound to the nitrocellulose.

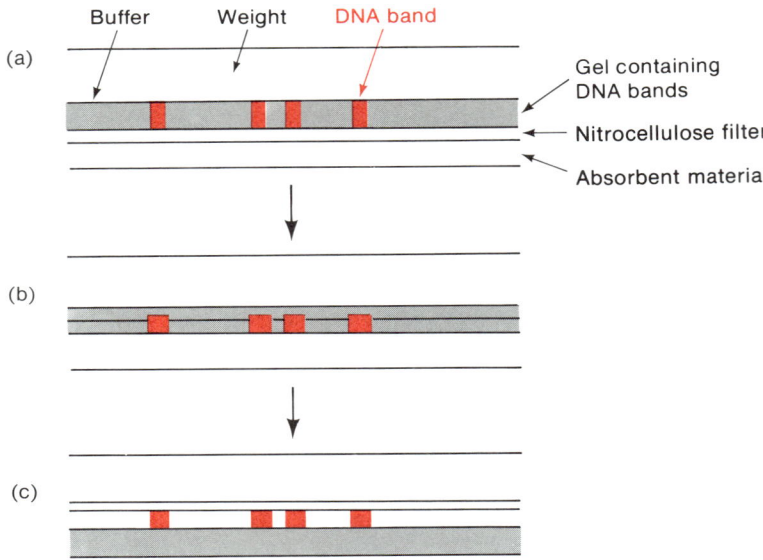

(Figure 5-9). DNA is enzymatically fragmented and then electrophoresed through an agarose gel. Following electrophoresis the gel is soaked in a denaturing solution (usually NaOH), which converts all of the DNA in the gel to the single-stranded form necessary for hybridization. A large sheet of nitrocellulose paper is then placed on top of several sheets of ordinary filter paper; the gel, which is typically in the form of a broad flat slab, is then placed on the nitrocellulose filter and covered with a glass plate to prevent drying. A weight is then placed on the top of the stack, which squeezes the liquid out of the gel. The liquid passes downward through the nitrocellulose filter. Denatured DNA binds tightly to nitrocellulose (Chapter 3); the remaining liquid passes through the nitrocellulose and is absorbed by the filter paper. DNA molecules do not diffuse very much, so if the gel and the nitrocellulose are in firm contact, the positions of the DNA molecules on the filter are identical to their positions in the gel. The nitrocellulose filter is dried and then moistened with a very small volume of a solution of ^{32}P-labeled tester denatured DNA or RNA, placed in a tight-fitting plastic bag to prevent drying, and held at a temperature suitable for renaturation (usually for 16–24 hours). The filter is then removed, washed to remove unbound radioactive molecules, dried, and autoradiographed with x-ray film. The blackened positions of the film indicate the locations of the DNA molecules whose DNA base sequences are complementary to the sequences of the added radioactive molecules. The degree of blackening of the film can be measured quantitatively and under certain conditions is proportional to the amount of DNA or RNA that has hybridized.

The tester radioactive DNA or RNA used in a hybridization experiment is commonly called a **probe.** Often the preparation of a probe is the critical step in a particular hybridization experiment and may be quite difficult. DNA probes are typically prepared by enzymatic techniques (nick translation, Chapter 9) using as starting material a particular DNA fragment obtained by the cloning techniques described in Chapter 22. RNA probes, which are less frequently used, are also enzymatically prepared (Chapter 12) but in bacterial and phage systems they may be isolated from radioactively labeled cells.

A variant of the Southern technique, humorously called **Northern blotting,** is used when it is more convenient to separate RNA molecules electrophoretically. RNA does not bind efficiently to paper so special papers are used to which the RNA can be linked covalently. Bound RNA can be hybridized with probes of radioactive RNA or single-stranded DNA. Still another variant is **Western blotting,** in which proteins are electrophoretically separated and transferred to a paper to which the protein is covalently linked. Western blots are used to detect DNA-binding proteins. Radioactive double-stranded DNA is used as a probe. The association of radioactivity with a protein band indicates that the particular protein is a DNA-binding protein.

DETERMINING THE BASE SEQUENCE OF DNA

Throughout this book, base sequences within DNA molecules will be presented. There are two methods for determining the base sequence of a DNA molecule—the **Maxam–Gilbert procedure** and the **Sanger dideoxy procedure.** Although both methods are equally effective, we shall describe only the Maxam-Gilbert procedure, as it is used more frequently at present. The other procedure is described in a reference listed at the end of the book.

In the Maxam-Gilbert procedure single-stranded DNA (derived from double-stranded DNA by the separation procedures just described) is subjected to several chemical treatments that cleave the DNA molecules and generate a family of short single-stranded fragments. The number of nucleotides in each fragment is determined by gel electrophoresis, which can separate molecules whose lengths differ by only a single nucleotide. Several cleavage protocols are used, each of which is base-specific and provides the base sequence by the following rule:

> If a fragment is found that contains n nucleotides and is generated by a treatment active against a particular base, then that base is present in position $n + 1$ of the DNA strand, the position being counted from the 5′ end.

Thus, if a 36–base fragment results from the protocol that identified guanine, then it is known that guanine is the 37th base in the original molecule.

We will now look more closely at the Maxam-Gilbert procedure, by which a single-stranded fragment containing a maximum of about 300 nucleotides can be sequenced. Such a fragment is typically obtained from a large DNA molecule by digestion with one or more restriction endonucleases. The fragment to be sequenced is processed in the following manner. First, the two 5′-P termini of the double-stranded molecule (which are at opposite ends of any fragment because the two strands are chemically antiparallel) are replaced by ^{32}P by the polynucleotide-kinase procedure described earlier in this chapter. The radioactive DNA is then denatured with alkali and the two single strands are purified, usually by electrophoretic separation. Each of these strands is sequenced separately. (Because the base sequences are complementary, the information is redundant; however, not only do the two determinations serve as a check on one another, but, as will be seen shortly, it is necessary to use the other strand also in order to obtain the complete sequence.)

The sequencing procedure is the following (Figure 5-10). A sample containing the purified single strands is divided into two portions. To one portion (I) is added dimethyl sulfate, which methylates purines; however, G is methylated five times more effectively than A. An essential feature of the protocol is that the reaction is not carried to completion, but only to the extent that about one purine per single strand is methylated. Since methylation occurs at random positions, the particular A or G that is methylated *differs in each strand*. The methylated sample is next divided into two portions, Ia and Ib. Sample Ia is heated, a treatment that removes all methylated bases, leaving the deoxyribose. The sample is then treated with alkali, which cleaves the sugar-phosphate chain at the site of the base that has been removed. This heating-cleavage protocol generates a set of fragments of varying size, and the differing number of nucleotides in each fragment is determined by the different positions of the methylated G or A. Because G is methylated more often than A, sample Ia is said to contain the G-only fragments. Sample Ib is not heated but instead is treated with dilute acid, which removes mainly methylated A (and some G); then it is treated with alkali to cleave the sugar-phosphate chain at the site where an A was removed. Thus, in sample Ib, fragments are produced whose size is determined mainly by the position of the methylated A, but some G—these are called the A + G fragments. Note that every G-only fragment size is also present in the A + G collection.

The two samples Ia (G-only) and Ib (A + G) are now electrophoresed in a 20 percent polyacrylamide gel containing 8 M urea, a denaturant that prevents hydrogen-bonding and hence keeps the fragments single-stranded. After electrophoresis, the bands are located by placing autoradiographic film (a film that responds to radioactivity) on the gel and exposing the film for several days. The single terminal ^{32}P atom, which was enzymatically added before the methylation reaction, is the sole source of the radioactivity. Note that when a 5′-^{32}P-labeled molecule is cleaved, only one of the two fragments produced contains ^{32}P and only that one is detected.

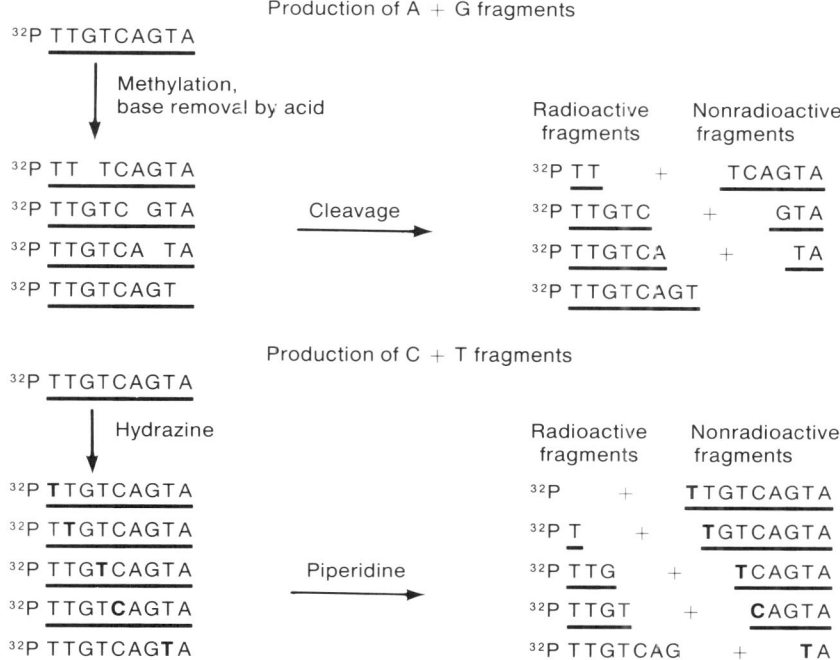

Figure 5-10 A diagram showing the production of the A+G and C+T fragments by the Maxam-Gilbert sequencing procedure. In the hydrazine-treated molecules the bold-faced letters indicate the affected bases. The generation of the G-only and C-only fragments is not shown but is explained in the text.

The positions of A and G in the single strand are determined by the following rules (which are special cases of the rule stated earlier): (1) if a band containing n nucleotides is present in both the (A+G) lane and the G-only lane, then a G exists at position $n+1$ in the original molecule; and (2) if a band containing n nucleotides is present only in the (A+G) lane, then an A exists at position $n+1$ in the original molecule.

So far, the analysis has used only sample I. Sample II is used to identify the positions of cytosine (C) and thymine (T). This sample is also divided into two portions, IIa and IIb, which are reacted with hydrazine in either dilute buffer (sample IIa) or in 2 M NaCl (sample IIb). Hydrazine reacts with C and T (but neither A nor G), but in 2 M NaCl the reaction is with C only. Cleavage at the site of the hydrazine reaction is accomplished by treatment with piperidine, which breaks the sugar-phosphate backbone at the 5′ side of each base that has reacted with hydrazine. The sizes of the fragments are then determined by the positions of both C and T and by C only. Thus, after electrophoresis and autoradiography, the positions of C and T are determined by the following rules: (1) if a fragment containing n nucleotides is

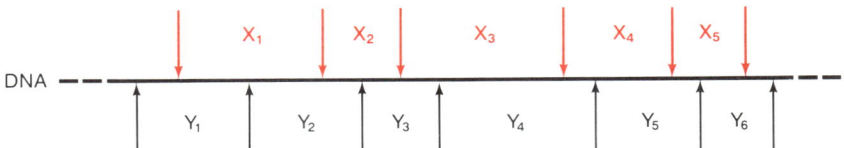

Figure 5-12 Two restriction maps for a segment of DNA. X and Y are simply labels for fragments generated by cuts at sites indicated by arrows. The color indicates a set of sites for a particular restriction enzyme.

Figure 5-11 A portion of a sequencing gel. The sequence is read from the bottom to the top. Each horizontal row represents a single base. Each vertical column represents a sample treated to indicate the position of G, A or G, T or C, and C, respectively. The sequence from a portion of the gel is shown.

present only in the $(C + T)$ lane (sample IIa), there is a T at position $n + 1$ in the original molecule; and (2) if a fragment containing n nucleotides is present in both the $(C + T)$ lane and the C-only lane, there is a C at position $n + 1$ in the original molecule.

All four samples (Ia, Ib, IIa, and IIb) are usually electrophoresed simultaneously, so all bands are seen in a single gel. This enables the sequence to be read directly from the gel. Figure 5-11 is a photograph of a sequencing gel. The shortest fragments are those that move the fastest and farthest. Each fragment contains the original $5'-^{32}P$ group, so the sequence can be read from the bottom to the top of the gel. Note that where a band in the figure is labeled with base X, this means that X was removed from the 3' end of the molecule to generate the fragment.

If only one single strand of the original double-stranded DNA molecule were analyzed in the way just described, the complete base sequence would not be obtained. This is for two reasons. First, if the cleavage is at position $n + 1$ (counting from the ^{32}P-labeled terminus), the number of bases in the fragment is n. This means that no fragment identifies the 5'-terminal nucleotide. Second, the mononucleotide that would identify the penultimate base cannot (for technical reasons) be detected unambiguously. The identity of these bases can be obtained by sequencing the complementary single strand, in which case the two unidentified bases from the first strand are complementary to the 3'-terminal bases of the second strand. Sequencing both strands also has the advantage that the sequence determined from one strand confirms the sequence of the other strand.

Although the gels can resolve two fragments differing in length by one nucleotide, the limit is about 300 nucleotides. To obtain the sequences of molecules containing thousands of nucleotides one uses sets of overlapping fragments. Consider the restriction map shown in Figure 5-12. Each of the X fragments could be sequenced, and, since their order is known, the sequence of the entire DNA molecule would be obtained. There is a danger with this procedure because very small fragments might have been missed when the map was developed. Sequencing the Y fragments, which would provide the sequence across the X-X junctions, enables the total sequence to be unambiguously determined.

PREPARATION OF DNA COMPLEMENTARY TO RNA

Often it is valuable to prepare a double-stranded DNA molecule, one strand of which is complementary to a particular RNA molecule. Whereas this is a fairly unusual biochemical reaction (occurring primarily in the life cycle of certain RNA viruses), it can be carried out simply in the laboratory using the enzyme **reverse transcriptase.** Figure 5-13 shows one of the several ways by which this enzyme can be used. The enzyme is incapable of initiating polymerization; it requires an oligonucleotide that is hydrogen-bonded to the RNA and that bears a 3'-OH group. This oligonucleotide is called a **primer** (Chapter 9). When eukaryotic messenger RNA is used, priming is especially simple, since these RNA molecules are naturally terminated at the 3' end with a long sequence of A's (Chapter 12). Thus, a primer of oligo(T) is used. With other RNA molecules a mixture of random oligonucleotides can be used, since in the mixture there will invariably be an oligonucleotide that can base-pair with the terminus of the RNA. Addition of reverse transcriptase to the primed RNA results in synthesis of a DNA·RNA hybrid. To prepare double-stranded DNA the RNA strand must be removed from the hybrid; this can be done in one of two ways: (1) a ribonuclease activity (RNase H) of reverse transcriptase will under certain conditions remove the RNA; (2) the hybrid can be treated with alkali, which hydrolyzes the RNA and leaves the single-stranded DNA untouched. The DNA strand complementary to the newly synthesized strand is usually made in one of two ways; in both cases the *E. coli* DNA polymerizing enzyme DNA polymerase I, which also needs a primer (Chapter 9), is used. The figure shows a procedure in which priming is again accomplished by addition of a random collection

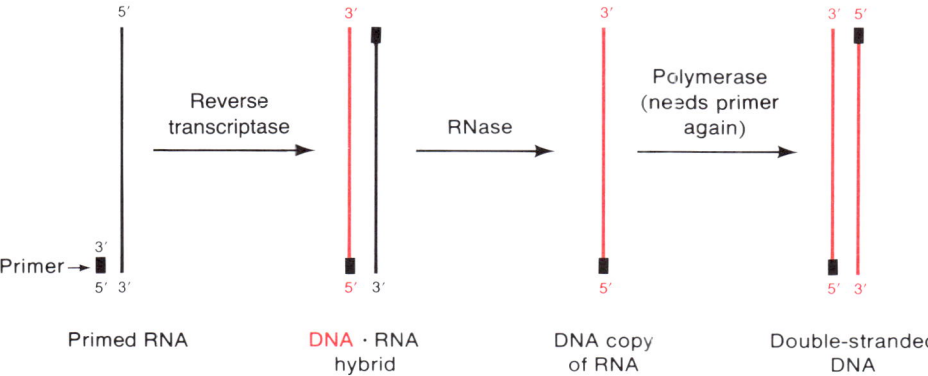

Figure 5-13   Synthesis of double-stranded DNA from an RNA molecule by reverse transcriptase. A primer, which could be either RNA or DNA, is shown.

of oligonucleotides. The second procedure (not shown) makes use of a poorly understood peculiarity of the polymerization reaction catalyzed by reverse transcriptase. If the initial synthesis of the first single strand of DNA is carried out for a longer time than is necessary to complete the strand and if certain solution conditions are used, the DNA strand extends beyond the RNA template strand with a short segment whose sequence is complementary to that of the last few hydrogen-bonded bases. Once the RNA is removed, this extended segment of DNA folds back and forms a hairpinlike structure, thereby eliminating the need for an added primer. However, when polymerase I completes synthesis of the second DNA strand, the result is a full molecular hairpin. **S1 nuclease,** an enzyme specific for single-stranded DNA or single-stranded segments of double-stranded DNA (Chapter 4), can be used to cleave the hairpin; the result is a proper double-stranded DNA molecule. Double-stranded DNA made by copying an RNA molecule is called **complementary DNA** or **c-DNA.** Its principal use will be described in Chapter 22.

The Physical Structure of Protein Molecules

All proteins are polymers of amino acids, yet each species of protein molecule has a unique three-dimensional structure determined principally by the amino acid sequence of that protein, in contrast with DNA, which has a universal structure—the double helix. This makes the study of proteins very complex but, on the other hand, the diversity of protein structures enables these molecules to carry out the thousands of different processes required by a cell.

The study of the detailed structure of proteins is beyond the scope of this book, being heavily dependent on a variety of optical techniques, especially the mathematically complex technique of x-ray diffraction. For this reason the discussion of proteins in this chapter will be a survey rather than a detailed description. For further information the reader should consult the references at the end of this book.

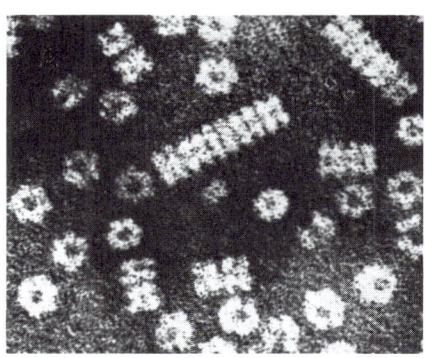

Electron micrograph of the cylindrical multi-subunit respiratory protein hemocyanin, isolated from the snail *Helix promatia*. Two views are seen: the cylinder end-on and a side view (squares).

SIZES OF PROTEIN MOLECULES

In the previous chapter we saw that nucleic acid molecules are very large, having molecular weights often as high as 10^{10}. Protein molecules are much smaller; in fact, the molecular weight of a typical protein molecule is comparable to that of the smallest nucleic acid molecules, the transfer RNA molecules. The molecular weights of hundreds of different proteins have been measured. Typical polypeptide chains have molecular weights ranging from 15,000 to 70,000. The average molecular weight of an amino acid is 110, which means that typical polypeptide chains contain 135 to 635 amino acids. The sizes of some proteins are outside this range. For example, the polypeptide hormones found in higher animals contain 10–50 amino acids, and thus are relatively small. The largest known polypeptide chain has a molecular weight of 165,000 and therefore contains about 1500 amino acids. It will be seen later in this chapter that many proteins consist of several polypeptide chains joined by various noncovalent interactions. The molecular weights of these multisubunit proteins typically range from 75,000 to 200,000; the largest known enzyme, DNA polymerase III holoenzyme, has nine subunits and a molecular weight of 760,000.

The length of a typical polypeptide chain, if it were fully extended, would be 1000 to 5000 Å. A few of the longer fibrous proteins, such as myosin (from muscle) and tropocollagen (from tendon) are in this range—with lengths of 1600 Å and 2800 Å, respectively. However, most proteins are highly folded and their longest dimension usually ranges from 40 to 80 Å. This folding is described in the next section.

STRUCTURE OF A POLYPEPTIDE CHAIN

In Chapter 3 the basic chemical features of protein molecules were described. That is, a protein is a polymer of amino acids in which carbon atoms and peptide groups alternate to form a linear polypeptide chain and specific groups—the amino-acid side chains—project from the α-carbon atom (Chapter 3, Figure 3-1). The term "linear" requires careful consideration, for, as will be seen in the following sections, a polypeptide chain is highly folded and can assume a variety of three-dimensional shapes; each shape in turn consists of several standard elementary three-dimensional conformations and other conformations that may be unique to that molecule.

The Folding of a Polypeptide Chain

A fully extended polypeptide chain, if it were to exist, would have the conformation shown in Figure 6-1. (The chain is not perfectly straight because the C—N and C—C bonds in which the α-carbon atom participates are not colinear.) Such an extended zig-zagged molecule could not exist without stabilizing interactions to maintain the extension. In fact, a single polypeptide is never completely extended but is folded in a complex way. If there were free rotation about every bond in the chain and no interaction between different parts of the chain, the folding would be random. Instead, three rules govern the manner of folding:

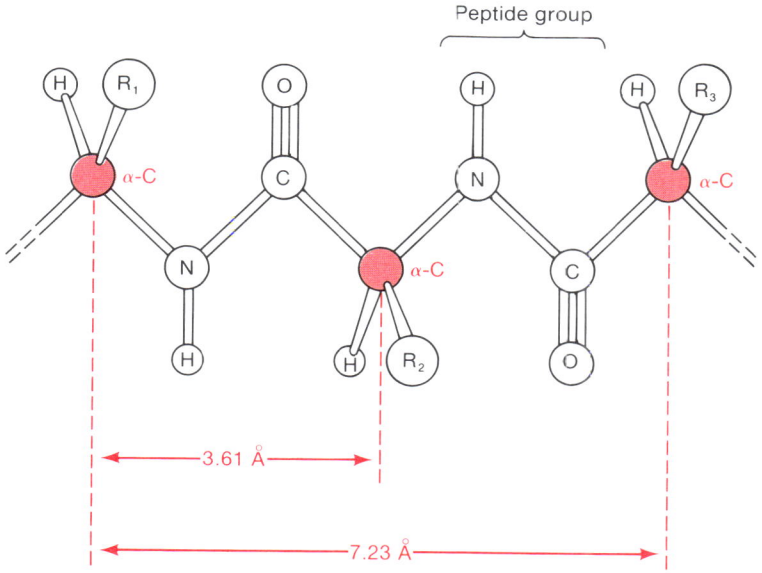

Peptide group

Figure 6-1   The conformation of a hypothetical fully extended polypeptide chain. The length of each amino acid residue is 3.61 Å; the repeat distance is 7.23 Å. The α-carbon atoms are shown in red. Side chains are denoted by R.

1. The peptide bond has a partial double-bond character (Figure 6-2) and hence is constrained to be planar. Free rotation occurs only between the α-carbon atom and the peptide group. Thus, the polypeptide chain is flexible but is not as flexible as would be the case if there were free rotation about all of the bonds.

2. The side chains of the amino acids cannot overlap. Thus, the folding can never be truly random because certain orientations are forbidden.

3. Two charged groups having the same sign will not be very near one another. Thus, like charges tend to cause extension of the chain.

(a)

(b)

Rigid peptide groups

Figure 6-2   The planarity of the peptide group. (a) Two forms of the peptide bond in equilibrium. In the form at the right the double bond creates a rigid unit. Owing to the equilibrium, the peptide group has a partial double-bond character and is fairly rigid.

(b) The peptide groups in a protein molecule. They are planar and rigid, but there is freedom of rotation about the bonds (red arrows) that join peptide groups to α-carbon atoms (red).

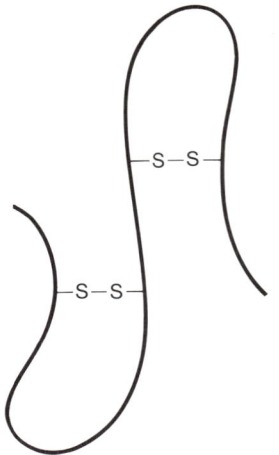

Figure 6-3   A polypeptide chain in which four cysteines are engaged in two disulfide bonds.

In addition to these rules, folding behaves according to general tendencies, a few of which are the following:

1. Amino acids with polar side chains tend to be on the surface of the protein in contact with water.
2. Amino acids with nonpolar side chains tend to be internal. Very hydrophobic side chains tend to cluster. (These two tendencies have been described by likening a protein molecule to an oil droplet with a polar coat.)
3. Hydrogen bonds tend to form between the carbonyl oxygen of one peptide bond and the hydrogen attached to a nitrogen atom in another peptide bond. This hydrogen-bonding gives rise to two fundamental polypeptide structures called the α helix and the β structure, which will be described in the following section.
4. The sulfhydryl group of the amino acid cysteine tends to react with an —SH group of a second cysteine to form a covalent S—S (**disulfide**) bond. Such bonds pose powerful constraints on the structure of a protein (Figure 6-3). Most proteins contain several cysteines and it is not generally possible to predict which cysteines will be paired.

These tendencies should not be considered invariant because there are many exceptions. Nevertheless they indicate that the structure of a protein may be changed markedly by a single amino acid substitution—for example, a polar amino acid for a nonpolar one; similarly, the change might be minimal if one nonpolar amino acid replaces another nonpolar one. This notion will be encountered again in Chapter 11 when mutations are considered.

The three-dimensional shape of a polypeptide chain is a result of a balance between each of the rules and tendencies just described and can be very complex. However, in examining many polypeptide chains, it has become apparent that certain geometrically regular arrays of the chain are found repeatedly in different polypeptide chains and in different regions of the same chain. These are the arrays resulting from hydrogen-bonding between different peptide groups. They are described in the next section.

Hydrogen Bonds and the α Helix

In the absence of any interactions between different parts of a polypeptide chain, a random coil would be the expected conformation. However, hydrogen bonds easily form between the H of the N—H group and the O of the carbonyl group. Figure 6-4 shows several types of hydrogen-bonding possibilities, each of which causes a folding of the polypeptide chain. Such bonding occurs when peptides can hydrogen-bond to one another more strongly than they can form hydrogen bonds with water, which is often the case.

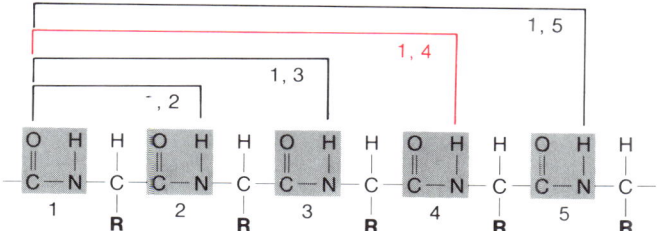

Figure 6-4 Possible hydrogen bonds between the C=O group in peptide unit 1 (lower numbers) and the N—H groups of other peptide units. The red line shows the hydrogen bond present in the α helix. The C=O and N—H bonds have been drawn in the same direction for clarity.

Studies of the effect, on proteins, of urea,

$$H_2N - \overset{\overset{\displaystyle O}{\|}}{C} - NH_2$$

a molecule that forms hydrogen bonds with a variety of substances, indicates that hydrogen bonds are present in polypeptide chains. That is, addition of urea causes most proteins to undergo a structural transition from some definite shape to the random coil conformation. A large number of proteins contain regions having a repeat distance of 5.5 Å. The existence of such repetition implies that some order is present in this region. However, since the distance repeated is less than 7.2 Å (the value that would be present if the chain were fully extended), such a region must contain a polypeptide segment that is foreshortened in some way. Linus Pauling and Robert Corey showed that these facts were consistent with a helical structure, which they named the **α helix.** In the α helix the polypeptide chain follows a helical path that is stabilized by hydrogen-bonding between peptide groups. Each peptide group is hydrogen-bonded to two other peptide groups, one three units ahead and three units behind the chain direction (Figure 6-5). The helix has a pitch of 5.4 Å, which is the repeat distance, a diameter of 2.3 Å, and contains 3.6 amino acids per turn. Thus, it is a much tighter helix than the DNA helix (Figure 6-6). *The side chains, including those that are ionized, do not participate in forming the α helix.*

In the absence of all interactions other than the hydrogen-bonding just described, the α helix is the preferred form of a polypeptide chain because, in this structure, all monomers are in identical orientation and each forms the same hydrogen bonds as any other monomer. Thus, polyglycine, which lacks side chains (the R group of glycine is a hydrogen atom) and hence cannot participate in any interactions other than those just described, is an α helix.

If all monomers are not identical or if there are secondary interactions that are not equivalent, the α helix is not necessarily the most stable structure. Then, not only is it true that the amino acid side chains do not participate in forming the α helix, but also that the side chains are responsible for preventing α-helicity. A striking example of the disruptive effect of a side

Figure 6-5 Properties of an α helix.
(a) The two hydrogen bonds in which peptide group 4 (red) is engaged. The peptide groups are numbered below the chain.
(b) An α helix drawn in three dimensions, showing how the hydrogen bonds stabilize the structure. The red dots represent the hydrogen bonds. The hydrogen atoms that are not in hydrogen bonds are omitted for the sake of clarity.

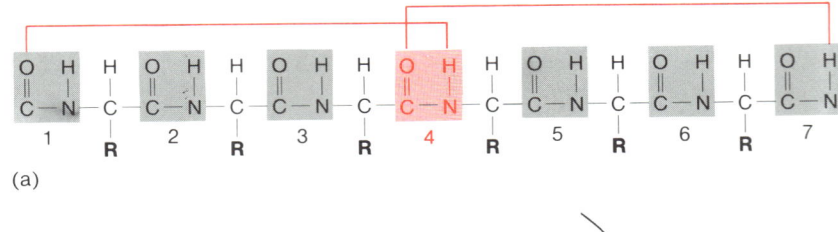

(a)

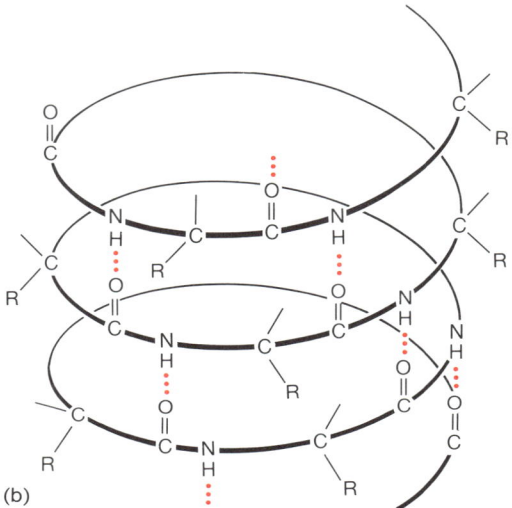

(b)

Figure 6-6 A schematic diagram of a cross-section of one turn of an α helix. The α-carbon atoms are red and the side chains are black. The carboxyl C is pink and the amino N is gray. The polypeptide is rotating clockwise and advancing away from the viewer. The van der Waals radii of the backbone atoms are larger than drawn and almost totally fill the core of the helix.

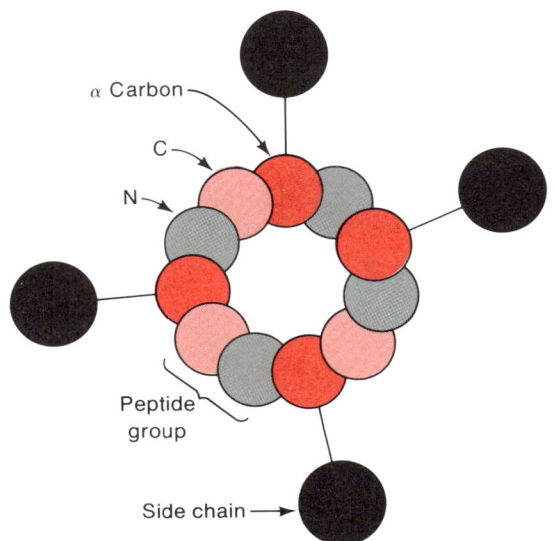

α Carbon

C

N

Peptide group

Side chain →

chain on an α helix is evident with the synthetic polypeptide polyglutamic acid, a polypeptide containing only glutamic acid. At a pH below about 5, the carboxyl group of the side chain is not ionized and the molecule is almost purely an α helix. However, above pH 6, when the side chains are ionized, electrostatic repulsion totally destroys the helical structure. With the synthetic polypeptide polylysine the pH dependence is reversed since the NH_2 group in the side chain of lysine (the ε-NH_2 group) is charged below pH 10.

If the amino acid composition of a real protein is such that the helical structure is extended a great distance along the polypeptide backbone, the protein will be somewhat rigid and fibrous (not all rigid fibrous proteins are α helical, though). This structure is common in many structural proteins, such as the α-keratin in hair.

β Structures

Another common hydrogen–bonded conformation is the **β structure.** In this form, the molecule is almost completely extended (repeat distance = 7 Å) and hydrogen bonds form between peptide groups of polypeptide segments lying adjacent and parallel with one another (Figure 6-7(a)). The side chains lie alternately above and below the main chain.

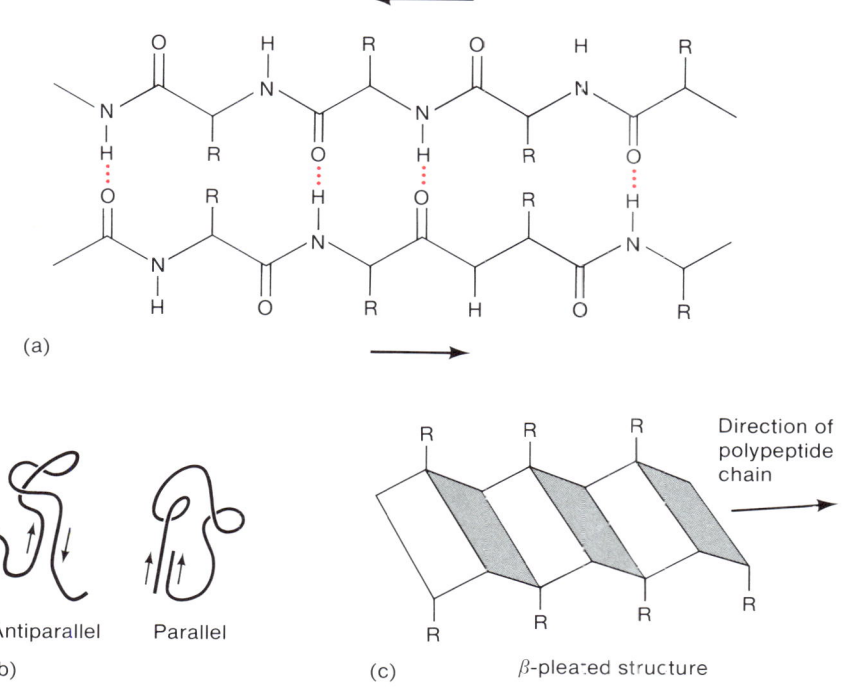

(a)

(b)

Antiparallel   Parallel

(c)   β-pleated structure

Direction of polypeptide chain

Figure 6-7 β structures. (a) Two regions of nearly extended chains are hydrogen-bonded (red dots) in an antiparallel array (arrows). The side chains (R) are alternately up and down. (b) Antiparallel and parallel β structures in a single molecule. (c) A large number of adjacent chains forming a β pleat.

Two segments of a polypeptide chain (or two chains) can form two types of β structure, which depend on the relative orientations of the segments. If both segments are aligned in the N-terminal-to-C-terminal direction or in the C-terminal-to-N-terminal direction, the β structure is **parallel.** If one segment is N-terminal to C-terminal and the other is C-terminal to N-terminal, the β structure is **antiparallel.** Figure 6-7(b) shows how both parallel and antiparallel β structures can occur within a single polypeptide chain.

When many polypeptides interact in the way just described, a pleated structure results called the **β-pleated sheet** (Figure 6-7(c)). These sheets can be stacked and held together in rather large arrays by van der Waals forces and are often found in fibrous structures such as silk.

Fibrous and Globular Proteins

Few proteins are pure α helix or β structure; usually regions having each structure are found within a protein. Since these conformations are rigid, a protein in which most of the chain has one of these forms is usually long and thin and is called a **fibrous protein.** In contrast are the quasi-spherical proteins called the **globular proteins** in which α helices and β structures are short and interspersed with randomly coiled regions and compact structures.

The fibrous proteins are typically responsible for the structure of cells, tissues, and organisms. Some examples of structural proteins are collagen (the protein of tendon, cartilage, and bone), and elastin (a skin protein). Some of the fibrous proteins are not soluble in water—examples are the proteins of hair and silk.

The catalytic and regulatory functions of cells are performed by proteins that have a well-defined but deformable structure. These are the globular proteins, of which the catalytic proteins, or enzymes, are the most widely studied. Globular proteins are compact molecules having a generally spherical or ellipsoidal shape. Large segments of the polypeptide backbone of a typical globular protein are α-helical. However, the molecule is extensively bent and folded. Usually, the stiffer α-helical segments alternate with very flexible randomly coiled regions, which permit bending of the chain without excessive mechanical strain. Numerous segments of the chain, which might be quite distant along the backbone, form short parallel and antiparallel β structures; these also are responsible in part for the folding of the backbone. The α helix and β structures are called the **secondary structure** of the molecule. The extensive folding of the backbone is usually called the **tertiary structure** or **tertiary folding.**

A very important distinction can be made between secondary and tertiary structure; namely, secondary structure results from hydrogen-bonding between peptide groups whereas tertiary structure is formed from α helices and β structures and several different side–chain interactions.

The most prevalent interactions responsible for tertiary structure are the following:

1. Ionic bonds between oppositely charged groups in acidic and basic amino acids.
2. Hydrogen bonds between the hydroxyl group in tyrosine and a carboxyl group of aspartic or glutamic acids.
3. Hydrophobic clustering between the hydrocarbon side chains of phenylalanine, leucine, isoleucine, and valine.
4. Metal–ion coordination complexes between amino, hydroxyl, and carboxyl groups, ring nitrogens, and pairs of SH groups.

Hydrophobic clustering (item 3 above) is the most important stabilizing feature.

Figure 6-8 shows a schematic diagram of a hypothetical protein (in two dimensions) in which several of these interactions determine the structure. This figure should be examined carefully for it indicates the role of different features of a protein molecule in determining the overall conformation of the molecule. One can see the following:

1. Disulfide bonds and hydrophobic interactions bring distant amino acids together.
2. Hydrogen bonds sometimes bring distant amino acids together, but usually a single hydrogen bond makes a more subtle change in position.

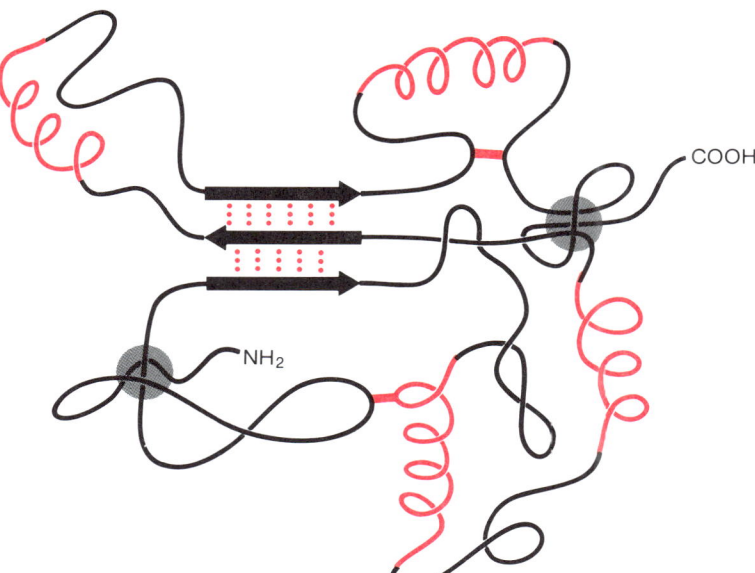

Figure 6-8 A schematic diagram of the path of the backbone of a polypeptide, showing possible ways in which the polypeptide may be folded. Heavy black arrows represent β structure; the red dots joining the arrows represent hydrogen bonds. Note that the interacting regions are not always located in nearby sequences within the polypeptide chain. Helical regions are drawn in red. Two heavy red bars represent disulfide bonds. The two shaded areas are hydrophobic clusters.

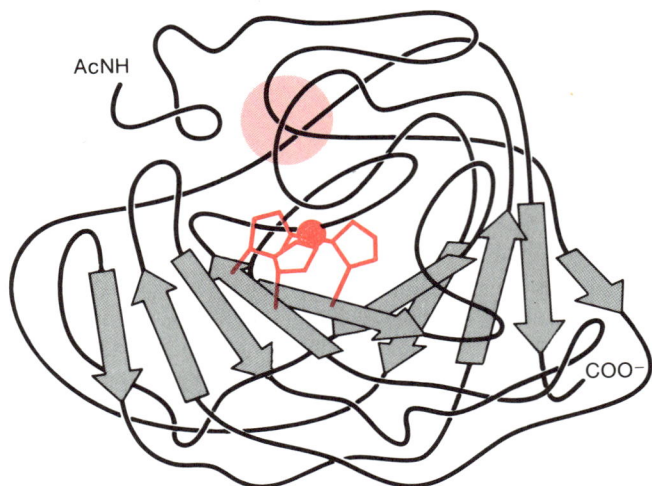

Figure 6-9 An idealized drawing of the tertiary structure of human carbonic anhydrase. Shown are the peptide chain and the three histidines (red) that coordinate to a zinc ion at the active site. Individual β-sheet strands are drawn as arrows from the amino to carboxyl ends. Note the twist of the β sheets. A hydrophobic cluster is shaded in red. [After K. K. Kannan et al. 1971. *Cold Spring Harbor Symp. Quant. Biol.*, 36: 221.]

3. Electrostatic interactions bring amino acids together or keep them apart, depending on the signs of the charges.
4. A β structure brings distant segments of the polypeptide backbone together and creates rigidity.
5. An α helix makes adjacent regions of a polypeptide backbone stiff and linear.
6. Van der Waals forces produce specific interactions between clusters of amino acids which may or may not be nearby in the polypeptide chain. (This is not shown in the figure.)

Figure 6-9 is an idealized drawing of the three-dimensional structure of the enzyme carbonic anhydrase, showing a β structure, a hydrophobic cluster, and three amino acids in a coordination bond with a metal ion.

Domains in Proteins

Many proteins, especially those from higher eukaryotes, consist of independently folded regions called **domains.** What is meant by independently folded is shown in Figure 6-10. Note that the black domain and the red domain could be physically separated by a single cut in the backbone near the black/red junction, for no attractive interactions exist between the red folded portion and the black folded portion.

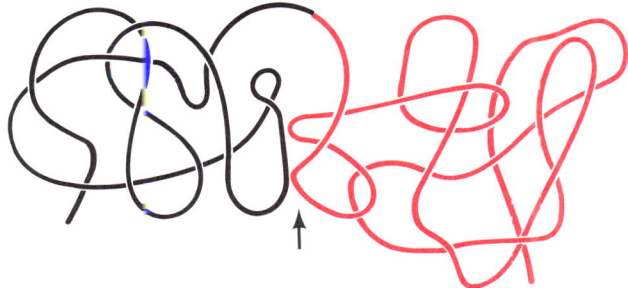

Figure 6-10   A diagram of a hypothetical polypeptide chain folded into two domains, one black and one red. In multifunctional proteins each domain may contain a binding site. In proteins having a single function a binding site frequently occurs at the contact point between the domains (arrow).

BINDING SITES OF PROTEINS

Each protein in a cell carries out a particular function, which may be structural, enzymatic, or regulatory. The structural proteins are usually fibrous, whereas the other two functions are generally carried out by globular proteins utilizing highly specific interactions between the protein molecule and one or more intracellular molecules. These interactions occur on the surface of the protein, primarily with the amino acid side chains.

Because of the folding of the polypeptide chain, the side chains produce a unique surface conformation of charged and hydrophobic groups. Usually a localized portion of the surface, having a size of 6−10 Å, is responsible for an interaction between the protein and other molecules. This is called a **binding site.** In an enzyme, the binding site is the catalytically active region and is called the **active site.** It is frequently the case that a binding site is formed by bringing together amino acids from distant parts of the polypeptide chain rather than from adjacent amino acids. When this occurs, most, if not all, of the intrachain interactions contribute to the creation of the binding site. (This enables one to understand how, in a mutant, substitution of one amino acid for another at some place other than at the binding site can markedly affect or even destroy a binding site by altering the manner of folding of other parts of the chain.)

The binding of a small molecule to a binding site can occur in many ways, a few of which are the following (Figure 6-11):

1. Two charged groups in the protein can be spaced in such a way that they make a precise match with two other charged groups on a small molecule (panel (a)).
2. A hydrophobic cluster of side chains in the protein can provide a nonpolar surface onto which a molecule having a large nonpolar group can adsorb (panel (b)).
3. Hydrogen-bonding groups in the protein can be arranged to facilitate hydrogen bonds to complementary groups on a small molecule (panel (c)).

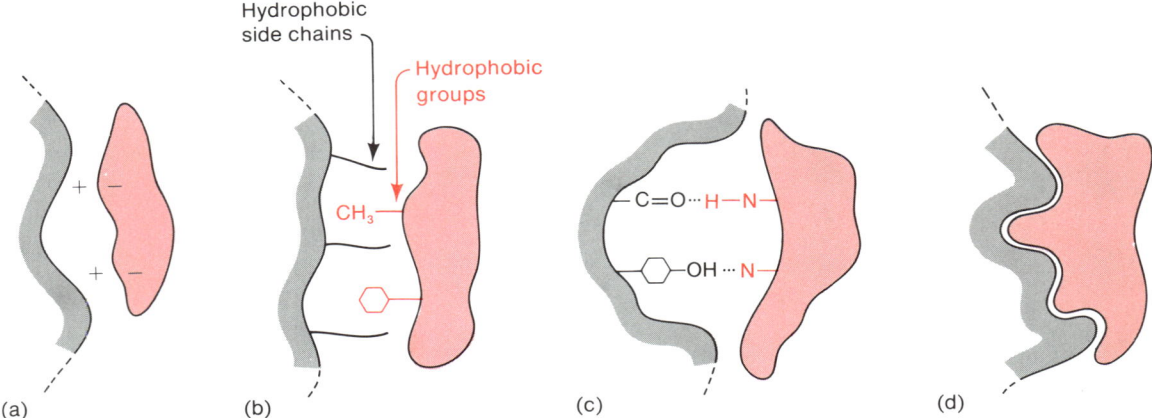

Figure 6-11 Four types of binding of small molecules (red) to the binding site of a protein (shaded). (a) Electrostatic. (b) Hydrophobic. (c) Hydrogen bonds. (d) Van der Waals bonds.

4. An electronic conformation is formed on the protein surface in an arrangement that allows attractive van der Waals interactions between the protein and many portions of the molecule being bound. In this way the very weak van der Waals forces can sum to produce very tight binding if both molecules have precisely the right shapes. This is called **steric fit** (panel (d)).

Usually several of the modes of binding are simultaneously present. For example, it is almost always the case with enzymes that the shape of the active site nearly complements the shape of the molecule on which the enzyme acts. This allows a fairly strong van der Waals attraction to be present and has the effect that molecules that do not fit are excluded from the binding site. The binding is usually stabilized by one or possibly two of the attractive interactions just listed—namely, ionic bonds, hydrophobic interactions, or hydrogen bonds.

Binding a small molecule or a portion of a larger molecule to the binding site of a protein may cause the protein to undergo a small shape change because binding often alters the strength of the weak forces that determine the shape of a protein. Alteration in shape serves two important functions. First, the shape change of an enzyme can improve contact between the active site and the substrate molecule on which an enzyme acts. This property of enzymes is called **induced fit.** Second, through adjustments in shape the biological activity of molecules such as enzymes can be regulated. If an enzyme has two binding sites, a regulatory molecule can bind to one site and thereby affect the active site in a way that either facilitates or prevents binding to the active site. This phenomenon will be discussed in greater detail in Chapter 15.

It should not be thought that the molecules that bind to proteins are necessarily smaller or of different composition than proteins. The surface of one protein molecule can also bind another protein molecule. Some examples are protein-hydrolyzing enzymes (**proteases**) and proteins that aggregate to form complex structures such as membranes and virus shells (see Chapter 7). A particularly common event is self-aggregation to form multisubunit complexes. This phenomenon is described in the following section.

PROTEINS WITH SUBUNITS

A polypeptide chain usually folds such that nonpolar side chains are internal—that is, isolated from water. However, it is rarely possible for a polypeptide chain to fold in such a way that *all* nonpolar groups are internal. Thus, it is often the case that nonpolar amino acids on the surface form clusters in an effort to minimize contact with water. A protein molecule having a large hydrophobic patch can further reduce contact with water by pairing with a hydrophobic patch on another protein molecule. Similarly, if a molecule has several distantly located hydrophobic patches, a structure consisting of several protein molecules in contact effectively minimizes contact with water. The protein is then said to consist of identical **subunits;** this is in fact a very common phenomenon, with two, three, four, and six subunits occurring most frequently. (A multisubunit protein may also contain unlike subunits. This will be discussed shortly.)

One can take the point of view expressed above that multisubunit proteins exist because they dispose of hydrophobic groups in an efficient way. An alternate and probably more correct view is that many proteins have evolved to fold in such a way as to form hydrophobic patches in order that subunit assembly will occur. Such evolution has been valuable because multisubunit systems have the following advantages:

1. *Subunits are an economical way to utilize DNA.* To synthesize a protein containing 6000 amino acids requires an amount of DNA containing 18,000 base pairs (three pairs for each amino acid) or 0.04 percent of the coding capacity of a typical bacterium. If the protein is assembled from six identical subunits, only 3000 DNA base pairs are needed. One might wonder if the resulting proteins are comparable—if, for example, a multisubunit protein is as able to form binding sites as a single larger protein is. It is, because just as a binding site is usually formed by bringing together distant amino acids in a particular polypeptide chain, a binding site can be formed by conjunction of amino acids from separate subunits.

2. *The activity of multisubunit proteins is very efficiently and rapidly switched on and off.* Throughout this book we shall see examples of regulation of activity of enzymes. Often an enzyme must be active for one instant and inactive the next, as, for example, in cycles of muscular contraction and relaxation. In many cases, activity is prevented by dissociation of a multisubunit protein and restored by reassociation. One way to accomplish this

cycle is the following. A signal or effector molecule binds to one subunit whose shape is changed very slightly. As a consequence of the shape change, the subunit falls away from the aggregate and the activity is lost. Once free from the aggregate, the subunit undergoes another shape change that also disrupts its binding to the effector molecule. Having lost the effector, the subunit then regains its original shape and returns to the aggregate, thus re-forming an active protein. This kind of switching is very rapid, achieves a reduction of activity to the zero level, and is accomplished with a very small expenditure of energy. Details of this modulation of activity are described in the section that follows.

The multisubunit proteins that have been described so far consist of several identical subunits. Different polypeptide chains can also aggregate and form proteins made up of nonidentical subunits and in fact this is quite common. For example, hemoglobin, the oxygen carrier of blood, consists of four subunits, two each of two different types; likewise, RNA polymerase, which catalyzes synthesis of RNA, has five subunits of which four are different.

It is quite rare for a very small protein to contain subunits. However, since so many proteins do consist of subunits, there is an implication that these larger proteins are "better" than smaller proteins. This is probably too strong a generalization in that sometimes the multiplicity of subunits may just have been an evolutionary accident. For example, a mutation could occur that allows a particular polypeptide chain to self-aggregate; this could in turn result in the formation of an active site that would catalyze a novel reaction. If this enzyme were of value to the cell in which the mutation occurred, selective pressure would retain the structure even though a simpler enzyme might have sufficed. There are two cases in which a large protein does have an advantage. First, in some processes the molecules to be bound are so large or so complicated that a small protein would find it difficult to produce an effective active site within itself. Second, when a binding reaction is regulated (and many are), the binding protein needs at least two binding sites—one for the molecule being bound and one for the regulator. It may not be possible for a small-molecular-weight molecule to have several binding sites. Furthermore, as will be seen shortly, regulation of multisubunit proteins can be accomplished with elegance and fine attunement.

A FEW MEANS OF STUDYING PROTEIN STRUCTURE

Many of the methods used to study the structure of protein molecules are complex physicochemical techniques such as x-ray diffraction, circular dichroism, nuclear magnetic resonance, and fluorescence spectroscopy. These techniques are described in texts on physical biochemistry listed in the references at the end of the book. Other methods are simple; a few of these are described in this section.

Denaturation of Proteins

As in the case of nucleic acids, protein molecules can be denatured by exposure to reagents that eliminate hydrogen bonds and hydrophobic interactions and by disruptive agents such as high temperature. When a protein contains disulfide bonds, which is often the case, renaturation is often rapid and complete following removal of the denaturant. Thus, reagents such as mercaptoethanol ($HOCH_2CH_2SH$) that break disulfide bonds are usually incorporated into a denaturing mixture.

Denaturation studies are usually carried out to learn something about three-dimensional structure. However, the analyses are beyond the scope of this book and the interested reader should consult the references given at the end of the book. One simple procedure is commonly used in molecular biological research though. If a protein is treated with a denaturant and the molecular weight is observed to decrease, the existence of subunits is indicated. Mild denaturants, unaccompanied by mercaptoethanol, are frequently used in the separation and purification of individual subunits.

Renaturation of Proteins

Renaturation is a process studied in an effort to understand how a polypeptide chain folds. Since a single amino acid substitution often alters the three-dimensional structure of a protein, the idea evolved that the structure of a protein molecule should be entirely determined by its amino acid sequence. What this means is that a polypeptide chain, guided by side-chain interactions, should spontaneously fold and refold until it achieves the conformation having minimum free energy. Furthermore, the appropriate disulfide bonds should form. With this view, it was expected that a denatured protein would re-fold to form the native molecule if appropriate conditions were created. Such results were in fact achieved with the enzyme ribonuclease A (pancreatic RNase). This protein is a single chain containing 124 amino acids; eight of these are cysteines that form four disulfide bonds. If RNase is treated with a denaturant such as urea and then the urea is removed, renaturation occurs immediately, because the disulfide bonds provide reference points for proper folding of the chain. If mercaptoethanol is also added, the disulfide bonds are cleaved and RNase becomes a random coil. If the urea and the mercaptoethanol are slowly removed, perfect renaturation occurs, including formation of the four correct disulfide bonds (disulfides can form spontaneously by oxidation in air). This latter finding is remarkable because there are 105 possible ways to form four disulfide bonds from eight cysteines. The interpretation of this observation is that the folding of this protein is determined exclusively by the amino acid sequence and that the proper disulfide bonds are formed because, during folding, the cysteines are correctly placed for joining. Evidence that disulfide bond formation does not direct the folding comes from an experiment in which the mercaptoethanol was removed first and oxidation was allowed to occur prior to removal of the urea—that is, while the RNase was a random coil. With this protocol, the native molecule was not formed.

From the experiment described we can certainly conclude that, for ribonuclease A at least, the tertiary structure is determined by the amino acid sequence and is the one having the minimum free energy. Similar observations have been made for a few other proteins but, in general, it is not precisely the case. That is, perfect renaturation is not usually attained, though in many cases a structure results that is near that of the native molecule, which means that correct folding must be directed in some way. The following idea, which is not yet proved, is probably the correct explanation.

Protein molecules are synthesized in an N-terminus-to-C-terminus direction (Chapter 14). It seems likely that some folding begins before synthesis is complete, so certain parts of the molecule might form a stable conformation that is not altered. For instance, consider a protein that is arbitrarily divided into ten segments. If the protein is denatured and then allowed to renature, segments 2 and 8 might interact so rapidly and strongly that all subsequent folding is determined by this initial interaction. However, if folding occurs prior to completion of synthesis, segments 2 and 6, whose interaction is less strong than that between 2 and 8, might interact first and this interaction might be stabilized either by a disulfide bond or pairing of segments 1 and 4. When the molecule is complete, segments 3 and 8 might join. Thus, if this were the case, folding would be determined *stepwise* by the amino acid sequence and the native conformation would not be the one having the least free energy. This sequential process may be responsible for the proper folding of the enzyme lysozyme and may apply to other proteins also.

Hydrolysis of Proteins

Proteins can be hydrolyzed to free amino acids by 1 M HCl at high temperature.* There are also numerous hydrolytic enzymes, whose principal role is in digestion (for example, stomach pepsin and intestinal trypsin) or in destruction of bacteria and viruses by phagocytic cells. These enzymes are called **proteolytic enzymes** or **proteases.** Most proteases act only in particular regions of a protein. For example, carboxypeptidase, an exopro-

---

*A disadvantage of this procedure compared to the enzymatic methods is that tryptophan is destroyed by acid.

tease, removes amino acids one by one from the carboxyl terminus of a protein chain, and trypsin, an endopeptidase, cleaves only on the carboxyl side of the basic amino acids arginine and lysine. These enzymes not only have functional biological roles but are useful in the laboratory.

REGULATION OF THE ACTIVITY OF PROTEINS

The activity of many proteins varies with both intracellular and extracellular conditions. In this unit we examine several ways by which the activity of a protein is regulated.

End-Product Inhibition

If a cellular enzyme catalyzes a reaction A→B and the product B is supplied exogenously to the cell, there is no reason for the cell to allow the enzyme to continue to carry out the reaction; in fact, if A has other uses, it would be efficient to turn off the activity of the enzyme. A simple way to accomplish this would be for the enzyme to have two binding sites—one for A and one for B. If B were bound the enzyme could undergo a conformational change which would eliminate the binding site for A. In this way B would prevent the enzyme from carrying out the conversion. This does occur and is an example of **end-product inhibition.** The role of this process in metabolic regulation will be examined further in Chapter 15.

Regulation of Multisubunit Proteins—Allostery

Many enzymes contain several subunits for reason of economy of genetic information, as stated earlier. These enzymes are also regulated. However, a common arrangement is that the binding sites for the molecule that is acted on (the **substrate**) and the inhibitor (which may be the product) are located on different subunits. If binding of the inhibitor prevents binding of the substrate, the information that a site on one subunit is occupied must somehow be transmitted to the other subunit. This can be mediated by the subunit contact regions in the following way. Binding of the inhibitor molecule alters the shape of the subunit to which it is bound and this results in changes in the sites on this subunit with which it interacts with other subunits (Figure 6-12(b)). If the subunits remain in contact, all subunits adjoining the first will undergo a shape change at their respective subunit-interaction sites and this in turn alters the binding site of other subunits. Proteins capable of undergoing such modification are called **allosteric proteins.**

So far only the reduction of binding activity of a protein by an inhibitor has been considered. The reverse can also occur—that is, an inactive protein

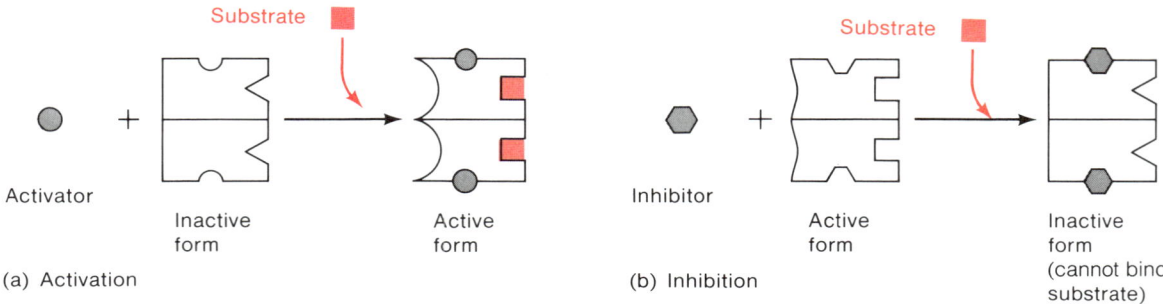

Figure 6-12 An example of allosteric (a) activation and (b) inhibition. The effector and the substrate can bind to the same or different subunits.

can be activated. In other words, a substrate-binding site on one subunit may be not quite right for binding but may acquire the right conformation when an activator molecule is bound to an activation site on another subunit (Figure 6-12(a)). Thus, some proteins undergo allosteric activation, whereas others undergo allosteric inhibition.

The two cases just discussed may be designated on→off (inhibition) and off→on (activation) processes. Many multisubunit proteins are subject to more subtle changes—that is, their binding activity is modulated by a regulator.

In addition to a second molecule affecting the binding of a substrate molecule it is fairly common that the binding activity of a protein is modulated by the substrate itself. For example, suppose an enzyme that carries out the conversion A→B is to be only weakly active when the concentration of A is low but must be very active when it is high. This would be difficult to accomplish with a single-subunit protein. However, consider a multisubunit protein in which *each* subunit can bind A; the affinity of each subunit for A might usually be low but once A has bound to a subunit, the binding could cause a conformational change that increases the affinity of the binding sites on the other subunits. If there is little A, the initial binding will occur at a low rate and in general only one binding site will be occupied, so the conversion of A to B will occur infrequently. If the concentration of A is high and occupation of the first binding site increases the affinity of the other binding sites, then these binding sites will be rapidly filled. When one A is converted to B, the binding site for A will immediately be occupied again, so the conversion rate will be very high.

This kind of modulation of protein activity occurs commonly. An important example is hemoglobin, whose ability to bind O_2 increases with O_2 concentration.

An Allosteric Enzyme: Aspartyl Transcarbamoylase

The best-understood example of an allosteric protein is the *E. coli* enzyme aspartyl transcarbamoylase (ATCase), which has sites for both an activator and an inhibitor. This enzyme catalyzes one step in the synthesis of the RNA

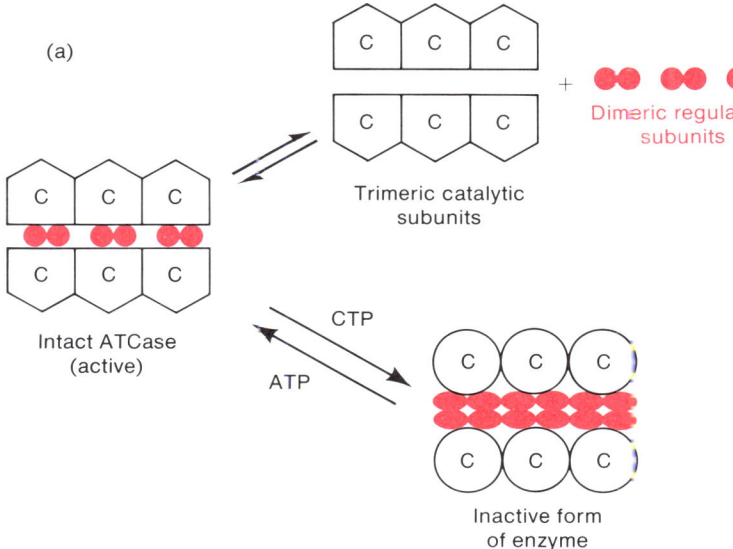

(a)

C C C

Dimeric regulatory
subunits

Trimeric catalytic
subunits

C C C
C C C

Intact ATCase
(active)

CTP

ATP

C C C
C C C

Inactive form
of enzyme

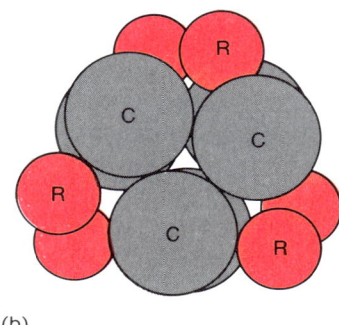

(b)

Figure 6-13 Aspartyl transcarbamoylase. (a) Its subunit structure (two trimeric catalytic subunits—labeled C—and three dimeric regulatory subunits—labeled R) and the conversion to an inactive form by CTP. (b) Three-dimensional arrangement of the subunits in the active molecule.

precursor cytidine triphosphate (CTP) from glutamine. The overall reaction sequence is

$$\text{Glutamine} + CO_2 + \text{ATP} \longrightarrow \text{Carbamoyl phosphate}$$

Aspartate ─┐ Aspartyl transcarbamoylase

$$\text{CTP} \longleftarrow \text{UTP} \longleftarrow \text{UMP} \longleftarrow \text{Carbamoyl aspartate}$$

The enzyme consists of twelve subunits of two types. Six of these are catalytic subunits (C), which form two trimers. The remaining subunits are regulatory subunits (R); these form three dimers. Each R dimer is in contact with two C subunits, but each in a different trimer. This is shown schematically in Figure 6-13(a). The actual arrangement of the subunits has been determined by x-ray crystallography and is shown in Figure 6-13(b).

As an allosteric enzyme, ATCase behaves in the following way. The enzyme is stimulated by the RNA precursor ATP and inhibited by CTP, which is not the immediate product of the reaction but the ultimate product (Figure 6-14). The enzyme can be easily dissociated in the laboratory to yield the two C trimers and the three R dimers and these can be purified. When this is done, it is found that aspartate binds only to a trimer (one aspartate per C subunit); ATP and CTP each bind only to a dimer (one ATP binding site and one CTP per R subunit). Alone, a C trimer is able to synthesize carbamoyl aspartate but the activity is unaffected by the concentration of ATP or CTP. Thus activation and inhibition require that the R dimers be in contact with the C trimers. It has also been found that the R subunits undergo a change in shape when CTP is bound and that there is also a change in shape of the C subunits if the R dimers and the C trimers are associated, in exact accord with the principles of the allosteric effect.

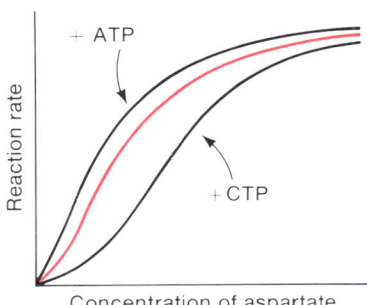

Reaction rate

+ ATP

+ CTP

Concentration of aspartate

Figure 6-14 Allosteric effects in aspartyl transcarbamoylase. ATP is an activator increasing the reaction rate; CTP is an inhibitor and reduces the reaction rate. The red curve results when neither ATP nor CTP is present. [After J. C. Gerhart, *Curr. Top. Cell Regul.*, (1970) 2:275.]

A COMPLEX PROTEIN: IMMUNOGLOBULIN G

As has already been mentioned, many proteins consist of several polypeptide chains, which may or may not be identical. In this section the structure of one such protein, immunoglobulin G, is described. It is not an allosteric protein. Immunoglobulin G has been selected as an example because its structure is well understood and its remarkable mode of synthesis is discussed in detail in Chapter 16.

The immunoglobulins (**antibodies**) are the proteins of the immune system. Their function is to interact with specific foreign molecules (**antigens**) and thereby render them inactive. This interaction is called the **antigen-antibody reaction.** The best understood immunoglobulin is immunoglobulin G (**IgG**); other classes of immunoglobulins are IgA, IgM, IgD, and IgE. In this section, we shall examine only the IgG class, itself comprising several subclasses defined by slight structural differences.

Cleavage of the disulfide bonds of IgG yields two polypeptides chains whose molecular weights are about 25,000 and 50,000. The lighter polypeptide is called an **L chain;** the heavier one is an **H chain.** IgG is a tetramer containing two L chains and two H chains. A schematic structure of IgG is shown in Figure 6-15. Experimentally, IgG can be cleaved with the proteolytic enzyme papain, which causes each of the H chains to break, as shown in the figure, thus producing three separate fragments. The two units that

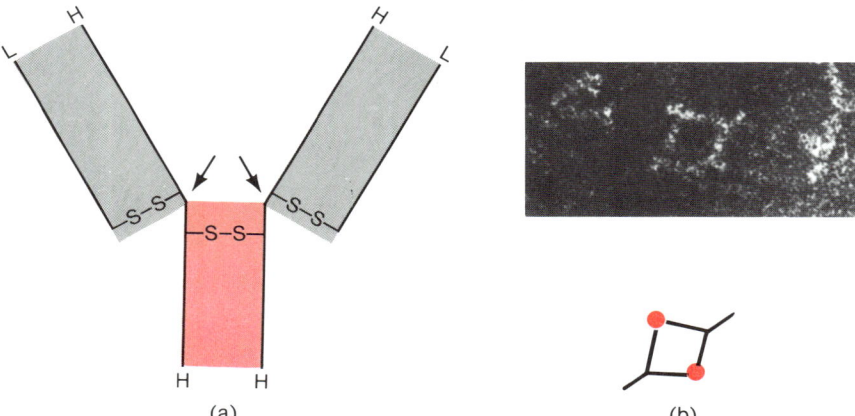

(a)   (b)

Figure 6-15   (a) The subunit structure of immunoglobulin G. There are two L polypeptides and two H polypeptides. Disulfide bonds join each L strand to an H strand and join the two H strands to one another. Treatment with papain cleaves the H strands at the arrows yielding two F_{ab} units (shaded black) and one F_c unit (shaded red). (b) An electron micrograph showing two immunoglobulin G molecules joined at the antigen binding site. The joining is a result of binding two antigen molecules not visible in the micrograph but shown as red dots in the interpretive drawing. [Courtesy of Robin Valentine.]

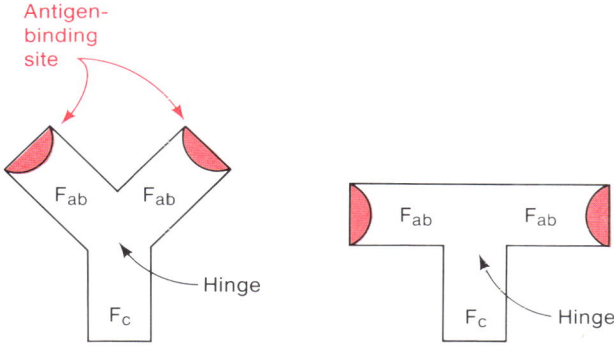

Figure 6-16   Immunoglobulin G is a Y-shaped molecule. It contains a hinge that gives it segmental flexibility.

consist of an L chain and a fragment of the H chain equal in mass to the L chain are called F_{ab} fragments (the subscript stands for "antigen-binding"). The third unit, consisting of two equal segments of the H chain, is called the F_c fragment.

Two sites on an IgG molecule can bind antigen. Each site is at the end of an F_{ab} segment, as shown in Figure 6-16. The F_c segment is not involved in antigen-antibody binding but is used in later processes needed to destroy the antigen.

The amino acid sequences of many subclasses of IgG molecules, each capable of combining with only a single kind of antigen molecule have been determined. All immunoglobulins are structurally similar inasmuch as L and H chains both have what is called a **variable (V)** and a **constant (C)** region.

The V and C regions have separate functions in IgG. The V regions confer antigen-binding specificity, whereas the C regions are responsible for the overall structure of the molecule and for its recognition by other components of the immune system. One might expect, then, that the V regions of both the L and H chains would be closely associated and located in the antigen-binding regions of the IgG molecule. That this is indeed the case is shown in the schematic diagram shown in Figure 6-17.

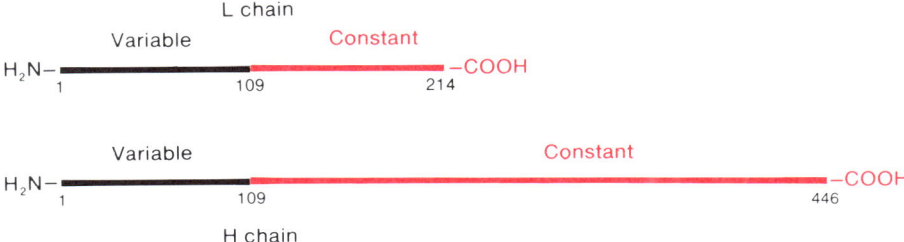

Figure 6-17   The positions and lengths of the variable and constant regions of the IgG L and H chains. The numbers refer to the amino acid positions counted from the NH_2 terminus.

The C regions of the L chains (C_L) of all types of IgG molecules have an identical amino acid sequence. Likewise, the C regions of all H chains (C_H) are identical for all IgG, though different from the C_L sequences. The V regions differ from one type of IgG to the next, however, for reasons that will be seen in Chapter 16. The comparative sizes of the C_L, V_L, C_H, and V_H regions are shown in Figure 6-18.

Some of the similarities between amino acid sequences in different parts of an IgG molecule are quite striking. For example, the C region of the H chain can be arbitrarily divided into three segments C_{H1}, C_{H2}, and C_{H3}, whose amino acid sequences are quite similar though not identical to one another. Furthermore, the sequence of C_L resembles the C_H sequences, though generally they are different. The V_L and V_H sequences also are nearly the same. This means that the IgG molecule, which already has twofold symmetry, consists of four domains, each of which has twofold symmetry, as shown in Figure 6-18, in which the intensity of the shading indicates two homologous regions. The symmetry of the molecules is made especially evident in Figure 6-19, which shows the three-dimensional structure of IgG. Note how in each domain the two components are somewhat wrapped around each other to form what is known as the **immunoglobulin fold.** The point of contact in the fold consists of two antiparallel β sheets between which are buried many hydrophobic amino acids.

Several other features of the IgG molecule should be noted, as they are often also found in multichain and multisubunit proteins whose function is to bind other molecules. The most striking property is that many regions of symmetry exist. Although the kinds of symmetry differ from one molecule to the next, symmetry of some kind is a common element. For example, hemoglobin, which has four subunits, has several planes of symmetry

Figure 6-18 The arrangement and relative sizes of the variable and constant regions of the IgG (a) L chain and (b) H chain. Note that the lengths of the variable regions are the same.

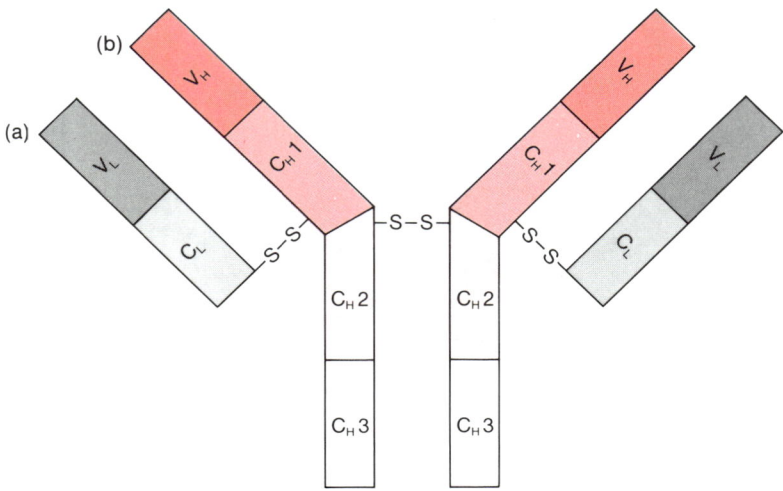

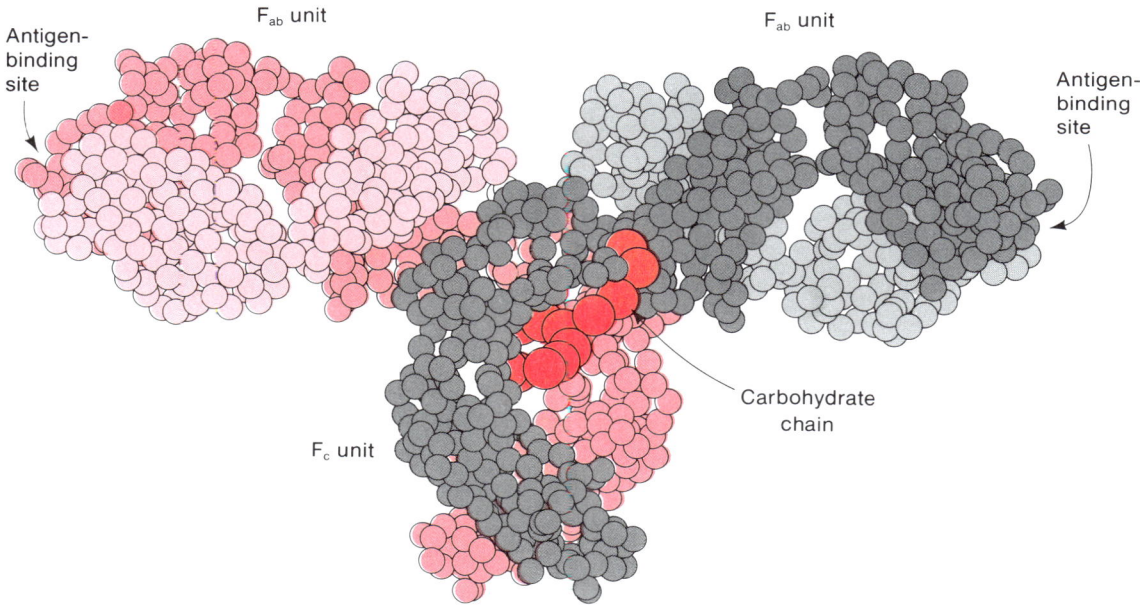

Figure 6-19 Schematic drawing of the three-dimensional structure of an IgG molecule. One of the H chains is shown in dark pink, the other in dark gray. One of the L chains is shown in light pink, the other in light gray. [After E. W. Silverton, M. A. Navia, and D. R. Davies, *Proc. Nat. Acad. Sci.*, (1977) 74:5142.]

that are utilized in arranging particular regions of the molecule for binding oxygen; the major muscle protein myosin has symmetric domains that interact with other muscle proteins; and the Cro protein made by *E. coli* phage λ, which binds to a symmetric base sequence in DNA, binds as a symmetric dimer (described in the following chapter). A second notable feature of IgG, common also to other molecules designed to bind two identical molecules, is that the structure contains multiple subunits capable of providing binding sites. IgG normally binds two antigen molecules—one at each of the two V-region binding sites. This activity is important in the mechanism for ridding the organism of an antigen after it has been bound to IgG. A third element of the structure of IgG is the use of extended hydrophobic segments (as in the immunoglobulin fold) as a means of joining subunits. A fourth feature is the joining of regions in two different subunits to generate a binding site. This is shown clearly in Figure 6-20, in which an antigen that is a small molecule (in this case, vitamin K) is bound to one of the antigen binding sites.* The antigen fits neatly into a cleft formed where the H chain passes the end of the L chain on the IgG molecule, and interacts with amino acids in both chains.

*Vitamin K is not a substance against which an antibody would normally be made. However, by special techniques an animal's immune system can be tricked into making an antibody to such a small molecule. Experimentally, such antibodies are valuable because their antigen-binding sites are small and simple and more amenable to study than large complex sites.

Figure 6-20 Mode of binding of a derivative of vitamin K (shown in black) to an
antibody-combining site. Amino acid residues from the H chain are shown in pink
and those from the L chain in red. Note that
the antigen (vitamin K) is bound by both
the L and H chains. [After I. M. Amzel, R.
Poljak, F. Saul, J. Varga, and F. Richards,
Proc. Nat. Acad. Sci., (1974) 71:1427.]

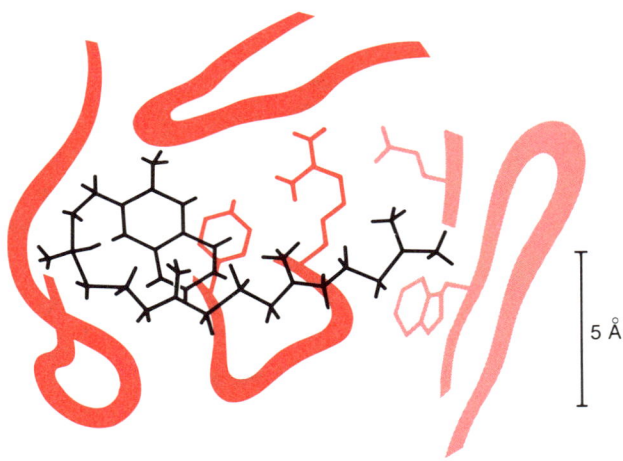

5 Å

ENZYMES

Enzymes are special proteins that catalyze chemical reactions. Their catalytic
power exceeds all man-made catalysts. A typical enzyme accelerates a reaction 10^8- to 10^{10}-fold, though there are enzymes that increase the reaction
rate by a factor of 10^{15}. Enzymes are also highly specific in that each catalyzes
only a single reaction or set of closely related reactions. Furthermore, only
a small number of reactants, often only one, can participate in a single catalyzed reaction. Since nearly every biological reaction is catalyzed by an
enzyme, a very large number of distinct enzyme molecules is required.

Enzyme reactions are studied for a variety of reasons. For example,
biochemists are interested in understanding detailed reaction mechanisms
and therefore study the effect of pH, substrate modification, and inhibitors
on reaction rates. The molecular biologist interested in regulation of enzyme
synthesis or enzyme activity will choose a particular enzyme and measure
the amount of enzyme activity as a function of various biological parameters.
Enzymes are also useful laboratory tools inasmuch as they can be used to
hydrolyze unwanted substances and prepare chemical compounds. Finally,
the mechanisms by which complex processes occur, such as DNA and protein synthesis, include a large number of enzymes and other factors acting
together, and understanding the processes requires detailed knowledge of
the enzymatic reactions.

The detailed mechanism of catalysis by particular enzymes is beyond
the scope of this book. However, all enzymes have certain general features,
the knowledge of which is important for understanding molecular biological
phenomena. These features are described in this unit.

The Enzyme–Substrate Complex

In any reaction that is enzymatically catalyzed, one reactant always forms a tight complex with the enzyme. This reactant is called the **substrate** of the enzyme; in descriptive formulas, it is denoted S. The complex between the enzyme E and the substrate is called the **enzyme–substrate** or **ES complex.**

Initially the enzyme and the substrate are bound by weak bonds, as in Figure 6-11, though in a few cases a covalent bond forms. The site on the enzyme at which the substrate binds is the **active site.** The extraordinary selectivity in enzyme catalysis is almost entirely a result of specificity of enzyme–substrate binding. After the ES complex forms, the substrate is usually altered in some way that facilitates further reaction. When the substrate is in the altered state, the ES-complex is said to be active and is usually denoted (ES)*. The (ES)* complex then engages in one or a series of transformations, which result in conversion of the substrate to the product and dissociation of the product from the enzyme. The extent to which ES forms is determined by the strength of binding between E and S; this is called the **affinity** of E and S. A useful measure of affinity is the **Michaelis constant,** K_M. This is the substrate concentration at which the reaction rate of the enzyme (which increases with substrate concentration) is half-maximal; it is the same as the concentration at which half of the enzyme molecules in the solution have their active sites occupied by a substrate molecule.

For most enzymatic reactions, formation of ES is reversible in the sense that ES can dissociate, yielding free E and free S. Usually, dissociation of the ES-complex is more rapid than conversion of the complex to enzyme and product; when this is the case, the value of K_M is a measure of the strength of the ES binding. That is, a high value of K_M indicates weak binding, and a low value of K_M means strong binding. The strength of binding depends on several conditions, such as the presence of particular ions, the overall ion concentration, and sometimes the presence of inhibitors. Thus, knowing the numerical value of K_M is often of use when one is trying to understand the mechanism of regulation of the activity of a particular enzyme. For most enzymes K_M ranges from 10^{-6} to 10^{-1} M, which shows that the affinity of E and S varies widely for different enzymes.

Mechanisms of Formation of the Enzyme–Substrate Complex

Catalysis by an enzyme occurs in several steps—binding of the substrate, conversion to the product, and release of the product. The initial step, formation of the ES complex, is in principle easy to understand and is often considered in thinking about molecules. The subsequent rearrangements are chemical phenomena and will not be discussed in this book.

Figure 6-21 Two modes of enzyme-substrate binding. (a) Lock-and-key. The active site of the enzyme by itself is complementary in shape to that of the substrate. (b) Induced-fit. The enzyme changes shape upon binding a substrate. The active site has a shape complementary to that of the substrate only after the substrate is bound.

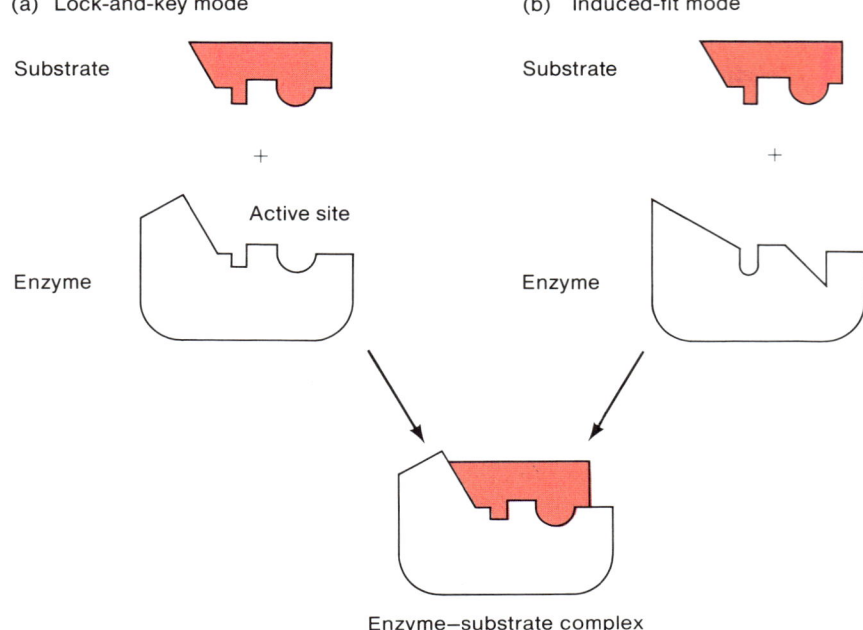

There are two major mechanisms of enzyme binding, **lock-and-key** and **induced fit** (Figure 6-21). In the lock-and-key mode, the shape of the active site of the enzyme is complementary to the shape of the substrate. In the induced fit mode, the enzyme changes shape upon binding the substrate and the active site has a shape that is complementary to that of the substrate only *after* the substrate is bound. For every enzyme-substrate interaction examined to date, one of these two mechanisms applies. It is often the case, though, that the substrate itself undergoes a small change in shape; in fact, the strain to which the substrate is subjected is often the principal mechanism of catalysis—that is, the substrate is held in an enormously reactive conformation.

Molecular Details of an Enzyme-Substrate Complex

The first detailed analysis of enzyme-substrate binding was carried out by using hen egg-white lysozyme. This enzyme cleaves certain bonds between sugar residues in some of the polysaccharide components of bacterial cell walls and is responsible for maintaining sterility within eggs. A schematic diagram of lysozyme is shown in Figure 6-22. The 19 amino acids that are part of the active site are printed in red; it should be noticed that they form

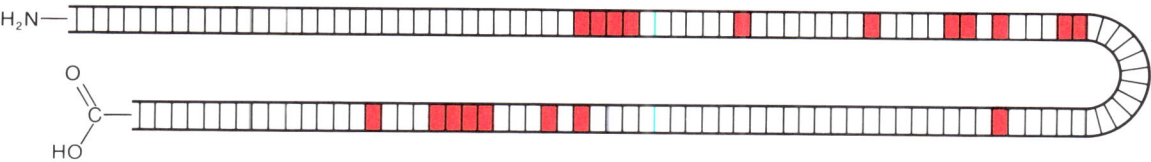

Figure 6-22 Schematic diagram of the amino acid sequence of lysozyme showing that the amino acids (red) in the active site are separated along the chain. Folding of the chain brings these amino acids together.

widely separated clusters along the chain. Only when the chain is folded do they come into proximity and form the active site. The folding of the chain is shown in Figure 6-23. The deep cleft indicated by the arrow is the active site. This is seen more clearly in the space-filling model shown in Figure 6-24. The substrate is a hexasaccharide segment that fits into the cleft and is distorted upon binding. The enzyme itself changes shape when the substrate is bound. A variety of interactions (van der Waals, hydrogen, and ionic bonds) stabilize the binding. Another enzyme, yeast hexokinase A, has been studied in order to examine what structural changes occur on substrate bind-

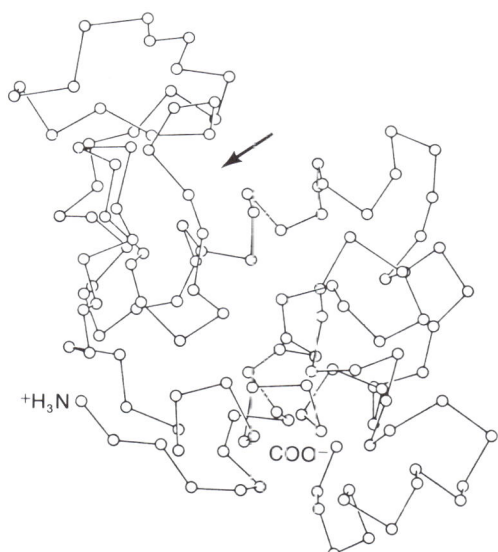

Figure 6-23 Three-dimensional structure of lysozyme. Only the α-carbon atoms are shown. The active site is in the cleft indicated by the arrow. [Courtesy of Dr. David Phillips.]

Figure 6-24 A space-filling molecular model of the enzyme lysozyme. The arrow points to the cleft that accepts the polysaccharide substrate. (C atoms are black; H, white; N, gray; O, gray with slots.) [Courtesy of John A. Rupley.]

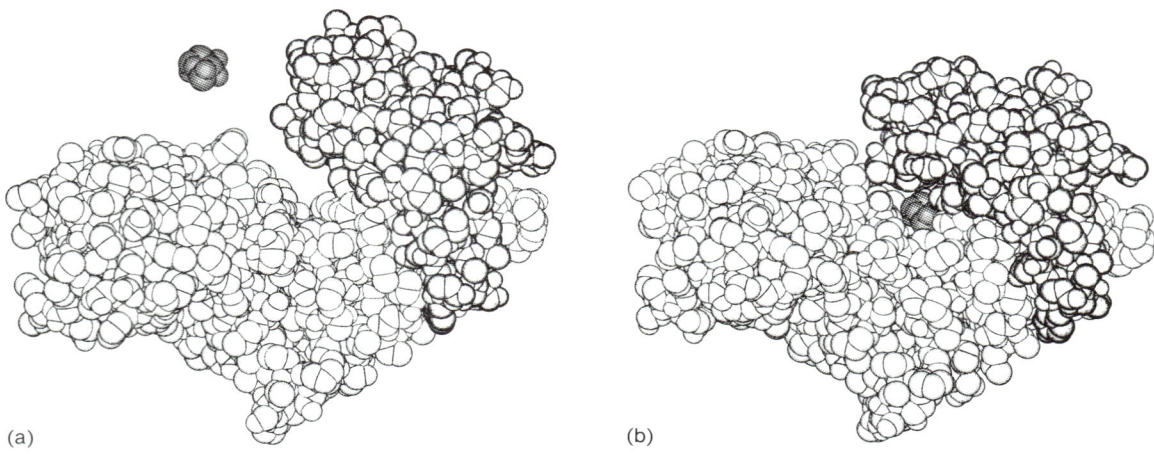

(a)   (b)

Figure 6-25 Structure of yeast hexokinase A. All atoms except hydrogen are shown separately. (a) Free hexokinase and its substrate, glucose. (b) Hexokinase complexed with glucose. Note the marked change in enzyme structure that accompanies glucose binding. [With permission, from J. Bennett and T. A. Steitz, *J. Mol. Biol.,* (1980) 140, 211. Copyright Academic Press, Inc. (London Ltd.)]

ing; these changes are now well documented. This is shown in the pair of space-filling models shown in Figure 6-25.

Dissociation of an Enzyme–Substrate Complex: Turnover

The rate at which a product forms from an ES complex is characterized by the average elapsed time between complex formation and dissociation of the product. This property is commonly expressed as the number of complexed substrate molecules converted to the product per enzyme molecule per second; this is called the **turnover number.** These considerations are summarized by the following statement in which E, S, and P refer to enzyme, substrate, and product, respectively:

$$E + S \rightleftarrows ES \rightleftarrows (ES)^* \rightarrow E + P$$

DETECTION OF ENZYMATIC ACTIVITY

There are numerous ways to study enzyme reactions. Two of these procedures, optical assays and radioactivity assays, are very commonly used and are described next.

Optical Assays of Enzyme Activity

In an optical assay of an enzyme, some component in the reaction mixture is detected by the ability of that component to absorb light of a particular wavelength. Since it is always possible to find a range of concentrations of the substances in which the absorbance A and the concentration are proportional, the concentration of the substances can be measured. The optically absorbing substance can be a reactant, in which case A decreases with time, or a product, in which case A increases.

There are very few substances in cells that absorb visible light and not many more that absorb in the ultraviolet. Some examples of the latter are the nucleic acid bases, the amino acids tryptophan, tyrosine, histidine, and phenylalanine, and a few cofactors and vitamins. (When such substances are present, they may be assayed.) However, if the protein concentration of a mixture is very high, the ultraviolet absorption of the proteins may overwhelm the absorbance of the substance of interest. For this reason, biochemists have attempted to design substrates that generate a product that absorbs in a range of wavelengths for which there are no naturally occurring absorbing compounds. An example of such a substrate is o-nitrophenyl galactoside (ONPG); this is used to assay the enzyme β–galactosidase, which we will encounter frequently in this book.

The enzyme β–galactosidase catalyzes cleavage of lactose to form galactose and glucose:

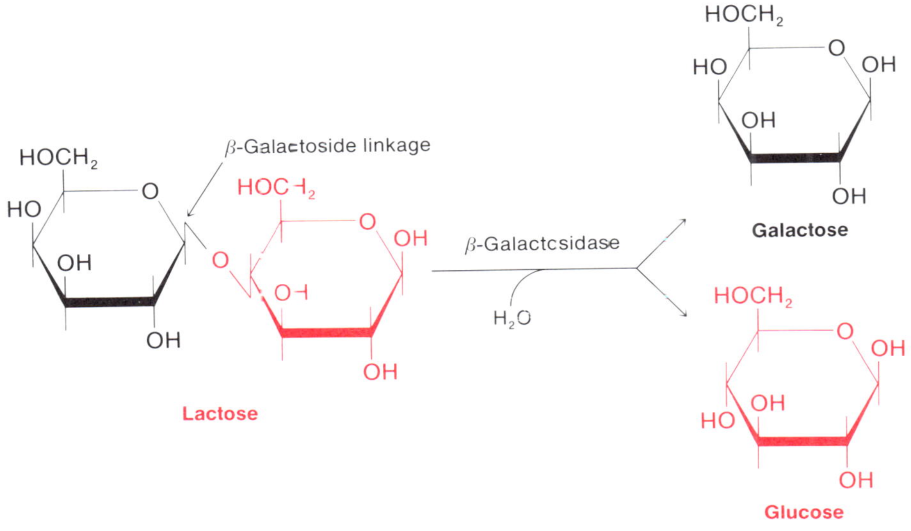

The bond broken is a β–galactoside linkage, as indicated by the arrow. This bond is also present in ONPG. The enzyme can hydrolyze ONPG, which is colorless, yielding galactose and *o*–nitrophenol, which is intensely yellow:

o-Nitrophenylgalactoside
(colorless)   Galactose   *o*-Nitrophenol
                                                                          (yellow)

Thus, the activity of the enzyme can easily be followed by assaying the concentration of *o*-nitrophenol at a wavelength of 420 nm (blue light).

Radioactivity Assay for Enzymes

In a radioactivity assay a reactant that is radioactive is added to a reaction mixture and either the appearance of another radioactive substance or the loss of the radioactive reactant is measured. There is enormous variety in such assays. One of the most common assays is used to measure polymerization. This assay is based upon the following fact—all proteins and nucleic acids are insoluble in 1 M trichloracetic acid (TCA), whereas all amino acids and nucleotides are TCA-soluble. This property allows one to measure protein synthesis by adding a radioactive (e.g., ^{14}C) amino acid to a reaction mixture containing the other 19 nonradioactive amino acids and the appropriate enzymes and factors. After a period of time, TCA is added and the mixture is filtered and washed with TCA. The ^{14}C-labeled amino acid is soluble and passes through the filter but if ^{14}C-protein has been made, it will be precipitated and ^{14}C will be retained on the filter. Counting the filter-bound radioactivity is then a measure of the extent of reaction. Synthesis of DNA can also be measured in this way by using a mixture containing the four deoxynucleotide precursors of DNA, one of which is radioactive—for example, [^{14}C]thymidine triphosphate (TTP)—and other appropriate components of the mixture. After reaction, TCA is added and the mixture is filtered; [^{14}C]TTP passes through the filter but any DNA that has been synthesized is retained on the filter.

Detection of Enzymatic Activity in a Crude Extract

It is often desirable, especially in preliminary studies, to be able to detect an enzymatic activity in a cell extract—that is, a solution obtained by breaking open a concentrated cell suspension. The interpretation of such experiments requires caution, as we shall see.

An enzyme is a catalyst and hence only alters the rate of reaction without affecting reaction equilibrium. This means that it accelerates both the forward and the back reaction by the same factor. This has an important consequence in the laboratory—namely, that an enzyme can be detected by assaying for the product of *either* of these reactions. Usually the forward reaction is the one that is detected because the formation of its product is favored by equilibrium. However, in choosing assay conditions in a cell extract, substances may be added to the assay mixture that can affect the equilibrium. Thus, if an enzymatic reaction in which B is converted to A is detected in a cell extract, one cannot be immediately sure that this is a biologically significant reaction because the assay conditions might be such that the biological reaction is being driven backwards. For example, the biological reaction might be $A + H^+ \rightarrow B + C$ but sufficient C might be present and the pH high enough in the extract to drive the reaction from right to left.

PREPARATION OF RADIOACTIVE PROTEINS

In many types of experiments it is desirable to label proteins with radioactivity. If the protein is of microbial origin, this is usually done by adding radioactive protein precursors to the growth medium. Any radioactive amino acid will suffice, but addition of some amino acids will result in the labeling of other molecules as well. For instance, glycine can be utilized as a general carbon source, so radioactivity will appear in a wide variety of compounds. Similarly, an aspartate label can appear in nucleic acids because aspartate is in the pathway for purine biosynthesis. The amino acids leucine and phenylalanine cannot be converted to any macromolecule other than proteins; therefore, the ^{3}H- and ^{14}C-labeled forms are the precursors of choice for preparing radioactive proteins.

Sulfur-35 in the form of the sulfate ion is also a useful label since it is incorporated into the two sulfur-containing amino acids, methionine and cysteine. The single disadvantage of the use of ^{35}S is that sulfur is also contained in some RNA molecules (there is thiouridine in some tRNA molecules) and that sulfur-containing amino acids are not present in all proteins. The former problem can be eliminated by treating a radioactive sample with RNase to hydrolyze the RNA.

Many proteins that are studied are derived from animal tissue and it is not possible to label these by feeding radioactive amino acids to an animal because the radioactivity will be diluted substantially by the normal food of the animal. The usual procedure is to purify the protein first and then make it radioactive by iodination. In an appropriate reaction mixture containing KI and ^{125}I$_2$, iodination of tyrosine and histidine occurs and it is possible to make highly radioactive proteins. This procedure is generally useful but sometimes has the side effect that iodination eliminates enzymatic activity if tyrosine or histidine are in the active site of an enzyme.

Macromolecular Interactions and the Structure of Complex Aggregates

In the preceding chapter a great deal of information was given about the structure of individual proteins and nucleic acids. We have also explained that many proteins contain two or more subunits, which may or may not be identical. This is true also for nucleic acids, though it is rarer and the number of subunits is usually quite small. As will be seen throughout this book, the existence of interactions between different macromolecules is the rule rather than the exception, and structural components of cells are invariably assemblies of macromolecules. For example, nucleic acids are often associated with proteins, as in chromosomes (which are DNA-protein complexes), extracellular viral nucleic acids are encased in protein shells, and bone and cartilage are complex assemblies of proteins and other macromolecules. Proteins can also interact with lipids to produce membranes, such as those that separate the contents of a cell from the environment and that separate different intracellular components from one another. Finally, polysaccharides form extraordinarily complex structures such as the cell walls of bacteria and of plants.

The study of such complex structures and how they are formed is called structural biology. Some structures are almost completely understood and many more are actively being studied. In this chapter only a few structures will be described. These have been selected by two criteria—they are either generally important or illustrate general principles, and their structures are reasonably well understood. The reader interested in exploring the subject further should consult the references given at the end of the book.

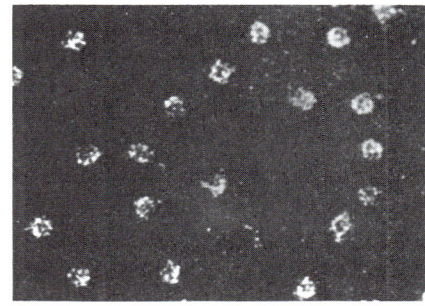

Dark field electron micrograph of chick erythrocyte nucleosomes. [Courtesy of Ada Olins.]

COLLAGEN—A MULTIPROTEIN ASSEMBLY

Collagen is the most abundant protein in mammals and one of the most common proteins in the world. It is the major protein component of tendon, cartilage, bone, skin, and blood vessels. The principal function of collagen is to provide a fiber (Figure 7-1) with great tensile strength, and the molecule is well designed to accomplish this end. Collagen is an especially good example of a protein with a hierarchical organization. It is synthesized by cells called **fibroblasts,** and secreted to the extracellular space (Figure 7-2) as a precursor called **procollagen.** It is then assembled as follows:

Procollagen filament (0–4 nm)→Protofibrils (11–15 nm)→Microfibrils (30–200 nm)→Collagen fibrils (0.2–0.4 μm)→Fibers (1–10 μm)→Tendon, bone, cartilage, etc. (macroscopic).

The amino acid composition and sequence of the polypeptide chains from which collagen is assembled have several striking properties:

Figure 7-1 (a) An electron micrograph of a collagen fiber (large dark object at the right) which has been mechanically teased apart to release a single collagen fibril. The pattern of bands should be compared with Figure 7-6. These bands result from the binding of ions containing heavy metals to charged amino acids. A careful look shows that some bands are clustered to form darker units. (b) Individual isolated fibrils on which metal has been deposited to enhance the major bands. (c) An enlarged view of a stained collagen fibril that has been treated briefly with pepsin. Notice how the fibril is splayed at the bottom and shows the individual tropocollagen molecules. [Courtesy of Peter Davison.]

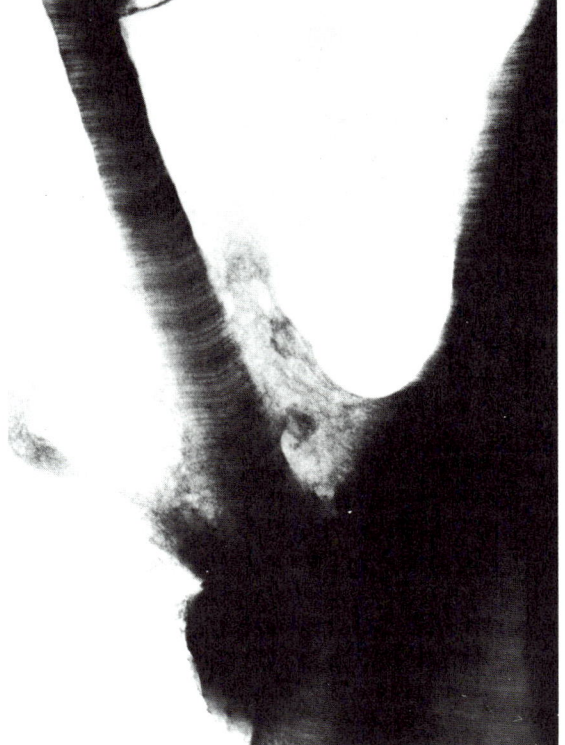

(a)

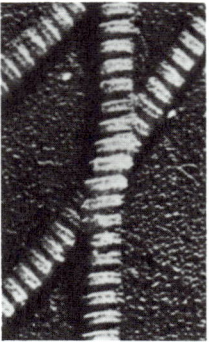

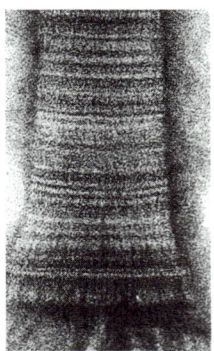

(c)

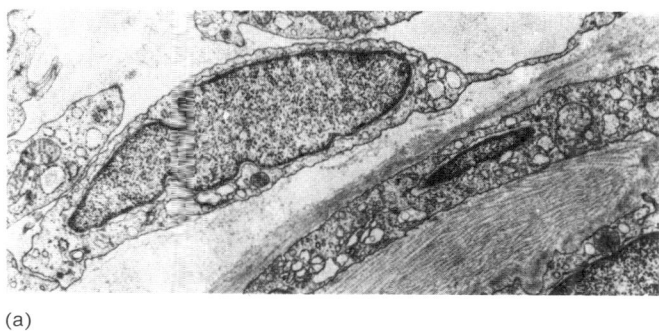

(a)

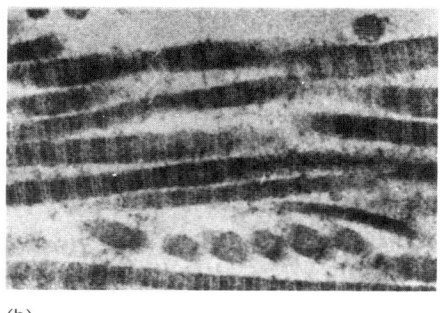

(b)

Figure 7-2   (a) A cross-section of connective tissue showing a fibroblast (large central object) and microfibrils in the intercellular space (fibers at lower right). (b) An enlarged view of the microfibrils. [Courtesy of H. Hecker.]

1. In each chain, which contains nearly 1000 amino acids, one-third of the amino acids are glycine. This value is much higher than that for a typical protein, in which about 5 percent of the amino acids is glycine; that is, the amino acid sequence can be represented as

   . . . Gly–A–B–Gly–C–D–Gly–E–F–Gly . . .

   in which A, B, C, . . . are amino acids other than glycine.
2. One-fourth of the amino acids are proline or a derivative, hydroxyproline (Figure 7-3). Proline is an amino acid somewhat infrequently encountered and hydroxyproline is not found in other proteins.
3. Many of the lysines are also hydroxylated (Figure 7-3).
4. The sequence glycine-proline-hydroxyproline occurs frequently in collagen, whereas in most proteins, recurring amino acid sequences are unusual.

These amino acids and their arrangement are responsible for the more important properties of collagen.

The single-stranded monomer of collagen does not exist in nature except during one stage of collagen synthesis. Instead, three strands of this monomer are joined to form **tropocollagen.** In part, the structure of tropocol-

Figure 7-3   Formulas for hydroxyproline and hydroxylysine.

**Hydroxyproline
(Hyp)**

**Hydroxylysine
(Hyl)**

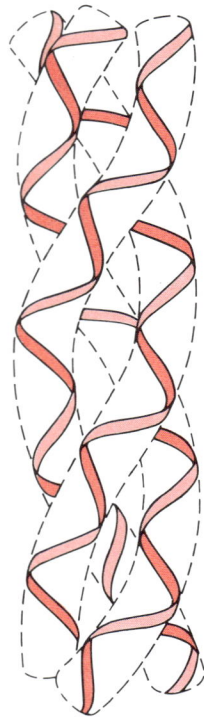

Figure 7-4 Diagram of the triple-stranded collagen helix. Each polypeptide chain is a helix and the helices interact to form the triple-stranded structure.

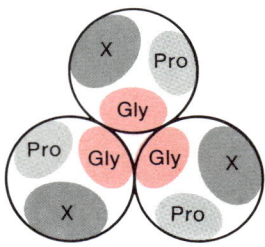

Figure 7-5 A schematic diagram of a cross section of triple-stranded collagen. Each heavy circle represents the outline of each α-helical strand. The crowding in the inner part of the triple helix is evident—glycine is the only amino acid whose side chain will fit. The shape labeled X can be any amino acid.

lagen is the result of the large amount of proline present in the individual polypeptide chains. In Chapter 3 it was pointed out that in proline the N atom that engages in linkage to adjacent amino acids is in a ring, so proline fails to form a typical peptide bond. The bond it forms is much less flexible and thereby causes each polypeptide strand to be rather extended. This extension exposes most of the potentially interactive groups, enabling the individual strands to interact by means of interstrand hydrogen bonds; the structure that forms is a **triple-stranded helix** (Figure 7-4), which is the first stage in forming a fiber with great tensile strength. These hydrogen bonds and also van der Waals interactions pull the strands together to form a very tight structure in which the proline and most of the other side chains are external and the glycines are internal. Examination of the triple helix explains why collagen has evolved to have glycine in every third position. The inner region of a triple helix is very crowded, as shown in Figure 7-5. The α-carbon atoms of glycine are shown in the cross-section of the triple helix; clearly no space is available for the bulky side chains of other amino acids. Thus glycine enables the structure to have maximum tightness and increase its strength.

In order to form a long fiber some means is needed to align tropocollagen molecules end to end. However, there is no direct way to form a strong end-to-end aggregate. Tropocollagen contains numerous clusters of positively and negatively charged amino acids on its surface, which enables tropocollagen molecules to aggregate side to side. Having so many charges of both signs on a *single* chain would cause the molecule to fold back on itself were it not for the rigidity of the triple helix. These charges instead cause side-to-side aggregation. If the tropocollagen molecules were aligned precisely side to side, a broad unit would result whose length would equal the length of one tropocollagen molecule and whose width would be determined by the number of tropocollagen molecules in the unit. Such aggregates have been prepared in the laboratory and are shown in Figure 7-6. Patterns of dark bands are seen if the units are reacted with uranyl acetate, which binds to negative charges. If phosphotungstic acid, which binds to positive charges, is used as a stain, the band patterns are identical, which shows that the positively and negatively charged amino acids occupy approximately the same positions.

Another type of aggregate can be prepared in the laboratory. This structure is a long fiber that is easily disrupted. Its band pattern differs from that shown in the figure. An analysis of the band pattern shows that in the form in Figure 7-6 the tropocollagen molecules are in a parallel array; in long fibers they are in an antiparallel array, as shown in Figure 7-7. Band patterns have also been observed with native collagen—that is, collagen that has not been formed in the laboratory from tropocollagen (Figure 7-1). These band patterns are different from those observed in synthetic material, which indicates that in natural collagen the tropocollagen molecules form neither parallel nor antiparallel arrays. An analysis of the band patterns of natural collagen shows that the tropocollagen molecules are aggregated side to side,

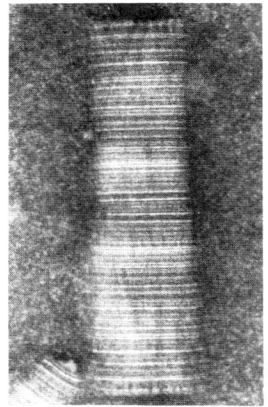

Figure 7-6 Electron micrograph of disaggregated and reconstituted collagen in the so-called "segment-long-spacing" form, in which the monomers are parallel and joined side by side. The band pattern differs from that of native collagen shown in Figure 7-1(a). [Courtesy of Peter Davison.]

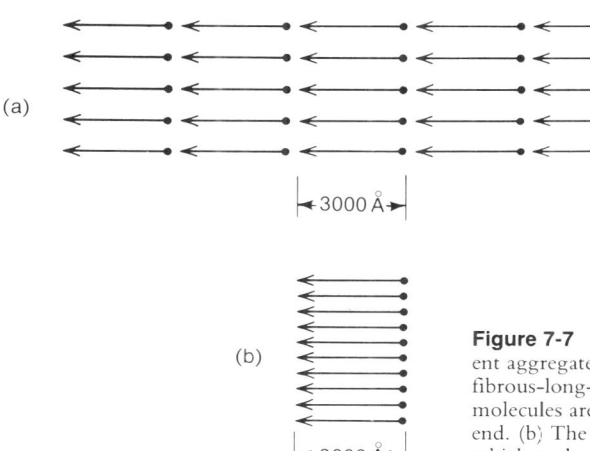

(a)

(b)

Figure 7-7 Diagrams showing two different aggregates of tropocollagen. (a) The fibrous-long-spacing form, in which the molecules are antiparallel and joined end to end. (b) The segment-long-spacing form, in which molecules are parallel.

but they are in a **quarter-staggered array** (Figure 7-8). This arrangement allows the generation of long fibers without the necessity for end-to-end aggregation. This staggering is actually a better way to generate strength. If collagen were formed by end-to-end joining of either of the arrays shown in Figure 7-6, one would expect the end-to-end joining of the aggregates to be cooperative. Thus, if a single end-to-end interaction were broken down (perhaps by accidental chemical attack), this would weaken adjacent pairs and possibly cause a cascading breakdown that could lead to cleavage of the fiber. However, in the quarter-staggered array, breakdown of the interaction between any pair of neighboring molecules does not weaken the fiber significantly.

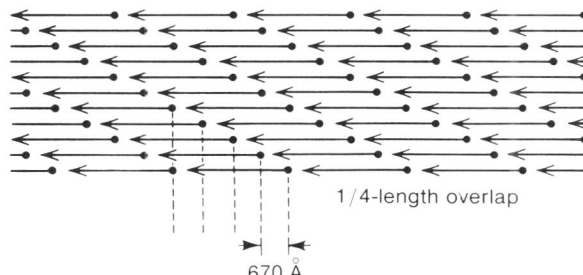

1/4-length overlap

670 Å

Figure 7-8 Schematic diagram showing the quarter-staggered array of tropocollagen in native collagen. The gap between each arrowhead and tail is about 400 Å. In reality, the structure is helical, as illustrated in Figure 7-4.

In addition to the charge attraction just described, a second interaction strengthens the quarter-staggered array. This interaction utilizes the collagen-specific amino acids, hydroxyproline and hydroxylysine, some of which engage in covalent cross-links between adjacent polypeptide chains in tropocollagen and between adjacent tropocollagen chains.

Collagen serves not only to form fibers of great tensile strength but also as a matrix for formation of bone. This is a result of another structural feature of the molecule. Between the head and tail of each successive tropocollagen molecule in the quarter-staggered array is a gap (Figure 7-8), which plays a role in bone formation. Bone is formed by the deposition of crystals of hydroxyapatite $(Ca_{10}(PO_4)_6(OH)_2)$ on a collagen fiber. The crystallization process can be carried out *in vitro*. It is found that initially microcrystals form on a collagen fiber at intervals of 670 Å, which is exactly the spacing of the gaps. Crystallization does not occur on the synthetic structures shown in Figure 7-7, neither of which possess gaps.

A COMPLEX DNA STRUCTURE: THE *E. coli* CHROMOSOME

The chromosome of *E. coli,* and presumably of most bacteria, is a single supercoiled double-stranded circular DNA molecule. For *E. coli* the total length of the circle is about 1300 μm, whereas the cylindrical bacterium has a diameter and a length of about 1 and 3 μm, respectively (Figure 7-9). Clearly the bacterial DNA must be highly folded when it is in a cell.

When *E. coli* DNA is isolated by a technique that avoids both DNA breakage and protein denaturation, a highly compact structure called a **nucleoid** is found. This structure contains a single DNA molecule, a fixed amount of protein, and variable amounts of RNA; the RNA is probably not part of the structure, but associates with it during isolation. An electron micrograph of the *E. coli* chromosome is shown in Figure 7-10. Two features of the structure should be noted: (1) The DNA is arranged in a series of loops and (2) each loop is supercoiled. A dense region containing membranous material is commonly seen in the central part; it is thought to be an artifact of the isolation procedure. The degree of condensation of the nucleoid is affected by a variety of factors and questions exist about the intracellular state of the nucleoid.

Treatment of the nucleoid with very tiny amounts of a DNase that produces single-strand breaks, followed by sedimentation of the treated DNA, gives some insight into the physical structure of the chromosome. If a supercoiled DNA molecule receives one single-strand break, the strain of underwinding is immediately removed by free rotation about the opposing sugarphosphate bond, and all supercoiling is lost (Chapter 4). Since a nicked circle is much less compact than a supercoil of the same molecular weight, the nicked circle sediments much more slowly. Thus, one single-strand break causes an abrupt decrease (by about 30 percent) in the *s* value. However, if

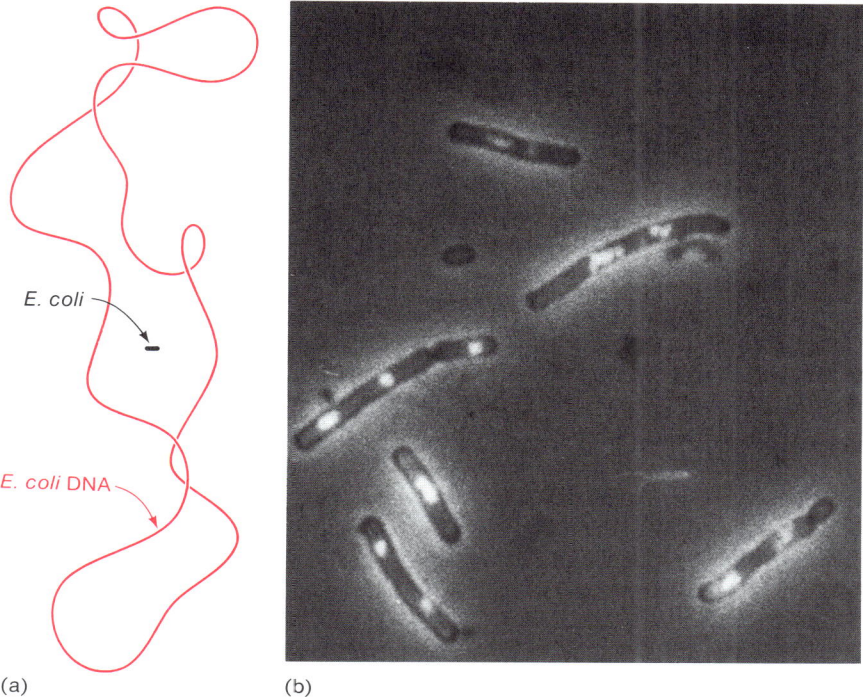

(a)

(b)

Figure 7-9 (a) Schematic diagram showing the relative sizes of *E. coli* and its DNA molecule, drawn to the same scale except for the width of the DNA molecule, which is enlarged approximately 10^6 times. (b) The localization of DNA in *E. coli*. Bacteria were exposed to a fluorescent dye that binds to DNA and then observed by fluorescence microscopy. The mode of sample preparation causes the DNA to condense slightly; in a living cell, the DNA occupies about twice as much space. [Courtesy of Todd Steck and Karl Drlica.]

E. coli

E. coli DNA

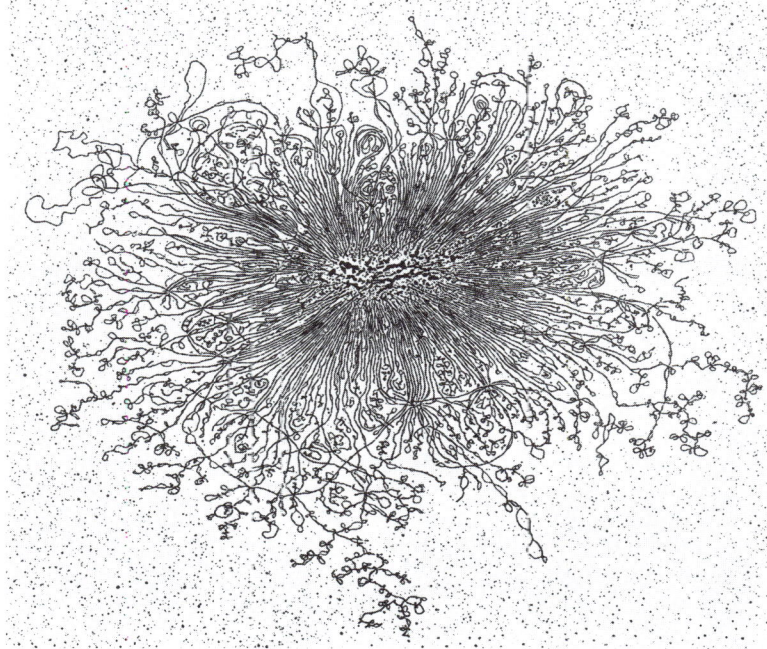

Figure 7-10 An electron micrograph of an *E. coli* chromosome showing the multiple loops emerging from a central region. [Bluegenes #1. © 1983. All rights reserved by Designergenes Posters Ltd., P.O. Box 100, Del Mar, CA, 92014–100, from which posters, postcards, and shirts are available.]

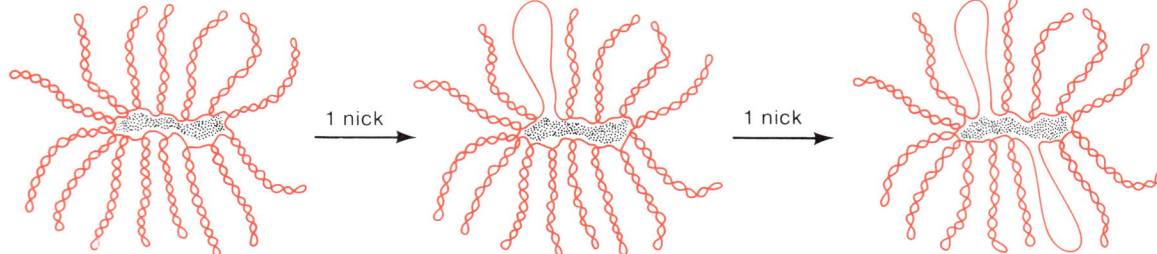

E. coli chromosome

Figure 7-11 A schematic drawing of the highly folded supercoiled _E. coli_ chromosome, showing only 15 of the 40–50 loops attached to proteins (stippled region) of unknown organization, and the opening of a loop by a single-strand break (nick).

single-strand breaks are introduced one by one by a nuclease into the _E. coli_ chromosome, the _s_ value decreases continuously; that is, the structure does not change in an all-or-none fashion but proceeds through a large number of intermediate states. After about 45 breaks the form of the chromosome remains constant. This clearly indicates that free rotation of the entire DNA molecule does not occur when one single-strand break is introduced. The structure of the _E. coli_ chromosome that has been deduced from these data is shown in Figure 7-11. The DNA is assumed to be fixed to proteins in a way (not yet understood) that prevents free rotation and produces supercoiled loops of DNA. One single-strand break would then cause one supercoiled loop to become open, and each subsequent break would, on the average, open another loop. This notion has been confirmed by electron microscopy: as nuclease treatment continues, the number of nonsupercoiled loops increases.

The enzyme DNA gyrase, which plays an important role in DNA replication (Chapter 9), is responsible for the supercoiling. If coumermycin, an inhibitor of _E. coli_ DNA gyrase, is added to a culture of _E. coli_ cells, the chromosome quickly loses its supercoiling. Indirect measurements of the number of binding sites for gyrase on the chromosome indicate that there are roughly 45 sites. The spatial distribution of these sites is not known but it is tantalizing to think that there may be one binding site in each supercoiled loop.

For reasons that are unclear _E. coli_ has a system that controls the degree of supercoiling. In addition to DNA gyrase, which introduces negative superhelical twists, _E. coli_ (and other bacteria) contains an enzyme, **topoisomerase I** (formerly called the ω protein), which removes superhelicity; _in vitro,_ purified topoisomerase I converts a supercoil to a nonsupercoiled covalent circle. Mutants (_topA_) lacking topoisomerase I activity have been found. Nucleoids isolated from these mutants have increased supercoiling, about 32 percent greater than normal. Interestingly, although in the laboratory the mutants show no evidence for growth defects, they often acquire secondary mutations in or near the gyrase gene; these secondary mutations reduce gyrase activity slightly and cause a restoration of near-normal superhelix density to the nucleoid. The significance of these observations is unknown.

CHROMOSOMES AND CHROMATIN

The DNA of all eukaryotes is organized into morphologically distinct units called **chromosomes** (Figure 7-12(a)). Each chromosome contains a single enormous DNA molecule. For example, the DNA molecule in a single chromosome of the fruit fly *Drosophila* has a molecular weight greater than 10^{10} and a length of 1.2 cm. (Since the width of all DNA molecules is 20 Å or 2×10^{-7} cm, the ratio of length to width of *Drosophila* DNA has the extraordinary value of 6×10^6.) These molecules are much too long to be seen in their entirety by electron microscopy because, at the minimum magnification needed to see a DNA molecule, the field of view is only about 0.01 cm across. However, the DNA can be visualized by autoradiography (Figure 7-12(b)). The pattern of grains in autoradiographic film above a DNA molecule is 100 times wider than the DNA; thus, a much lower magnification can be used and the entire molecule will be included in the field of view.

A chromosome is much more compact than a DNA molecule and in fact a DNA molecule cannot spontaneously fold to form such a compact structure because the molecule would be strained enormously. Instead, DNA is made compact by a hierarchy of different types of folding, each of which is mediated by one or more protein molecules.

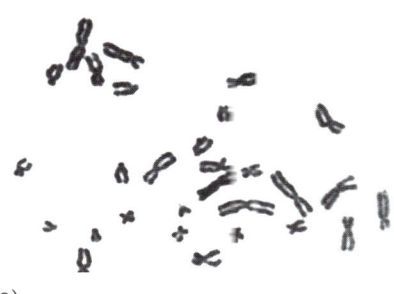

(a)

Figure 7-12 (a) Human chromosomes from a cell at metaphase. Each chromosome is partially separated along its long axis prior to separation of the two daughter chromosomes. The constriction is the site of attachment to the mitotic spindle. [Courtesy of Theodore Puck.] (b) Autoradiogram of a DNA molecule from *Drosophila melanogaster*. The contour length of this DNA is 1.2 cm [From R. Kavenoff, L. C. Klotz, and B. H. Zimm, *Cold Spring Harbor Symp. Quant. Biol.*, (1974) 38: 4.]

(b)

Histones and Chromatin

The DNA molecule in a eukaryotic chromosome is bound to very basic proteins called **histones.** The complex comprising DNA and histones is called **chromatin.** There are five major classes of histones—H1, H2A, H2B, H3, and H4. Histones have an unusual amino acid composition in that they are extremely rich in the positively charged amino acids lysine and arginine. The lysine-to-arginine ratio differs in each type of histone. (In the older literature H1 was called the lysine-rich histone and H3 and H4 were called arginine-rich.) The positive charge of the histones is one of the major features of the molecules, enabling them to bind to the negatively charged phosphates of the DNA. This electrostatic attraction is apparently the most important stabilizing force in chromatin since if chromatin is placed in solutions of high salt concentration (e.g., $0.5\ M$ NaCl), which breaks down electrostatic interactions, it dissociates to yield free histones and free DNA. Chromatin can also be reconstituted by mixing purified histones and DNA in a concentrated salt solution and gradually lowering the salt concentration by dialysis. The result shows that no other components are needed to form chromatin.

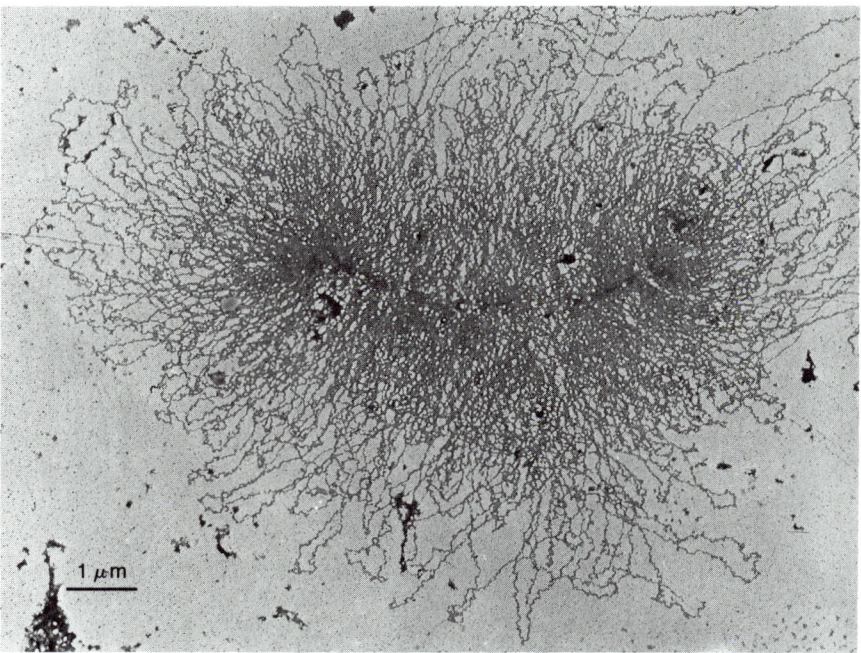

Figure 7-13   Electron micrograph of a partially disrupted anaphase chromosome of the milkweed bug *Oncopeltus fasciatus,* showing multiple loops of 30-nm chromatin at the periphery. [From V. Foe, H. Forrest, L. Wilkinson, and C. Laird, *Insect Ultrastructure,* (1982) 1: 222.]

Reconstitution experiments have been carried out in which histones from different organisms are mixed. Usually, almost any combination of histones work, because, except for H1, the histones from different organisms are very much alike. In fact, the amino acid sequences of both H3 and H4 are nearly identical (sometimes one or two (of 102) of the amino acids differ) from one organism to the next. Histone H4 from the cow differs by only two amino acids from H4 from peas—arginine for lysine and isoleucine for valine—which shows that the structure of histones has not changed in the 10^9 years since plants and animals diverged. Clearly histones are special proteins.

The Structural Hierarchy of Chromosomes

As a cell passes through its growth cycle, the structure of its chromatin changes. In a resting cell the chromatin is dispersed and fills the entire nucleus. Later, after DNA replication has occurred, the chromatin condenses about 100-fold and chromosomes form. Chromosomes have been isolated and gradually dissociated, and chromosomes at various degrees of dissociation have been observed by electron microscopy (Figure 7-13). The chromosome is first broken down into thick fibers of varying width. These are composed of fibers 25–30 nm wide, formed from smaller fibrils 10 nm wide, which appear like a string of beads. A schematic diagram of the hierarchy of DNA folding is shown in Figure 7-14. The beadlike structure is also seen when chromatin isolated from resting nuclei is examined (Figure 7-15). This string of beads is chromatin. The beads, which have a diameter of 100 Å, are connected by DNA strands 20 Å wide. The beadlike particles are an orderly aggregate of histones and DNA.

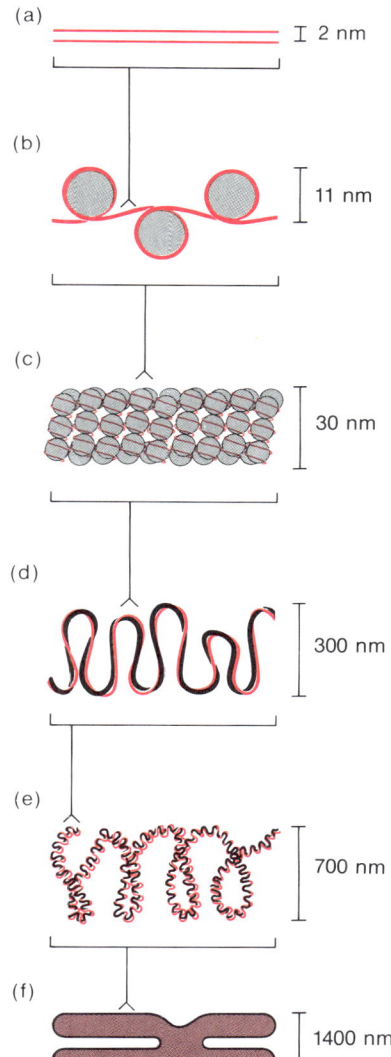

Figure 7-14 Various stages in the condensation of (a) DNA and (b–e) chromatin forming (f) a metaphase chromosome. The dimensions indicate known sizes of intermediates, but the detailed structures are hypothetical.

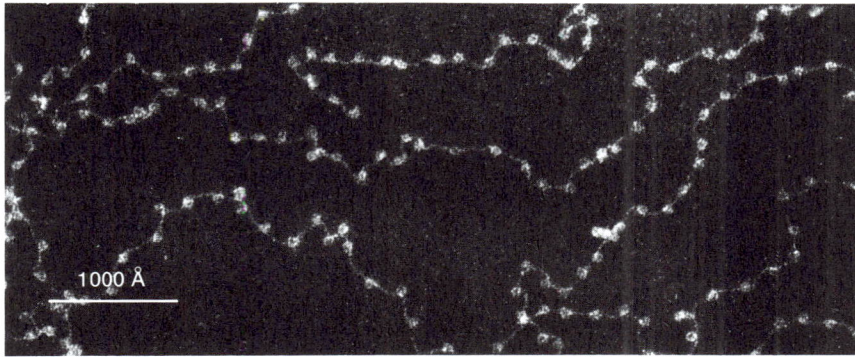

1000 Å

Figure 7-15 Electron micrograph of chromatin. The beadlike particles have diameters of nearly 100 Å. [Courtesy of Ada Olins.]

Nucleosomes

Prior to electron microscopic studies, the effect of various DNA endonucleases on chromatin also suggested that chromatin contains repeating units. Treatment of chromatin with micrococcal nuclease (which cannot attack DNA that is in contact with protein) yields a collection of small particles containing DNA and histones (Figure 7-16). If digestion is incomplete, fractionation by centrifugation yields individual particles and aggregates of two, three, and four particles (Figure 7-16). If, after enzymatic digestion, the histones are removed, DNA fragments having roughly 200 base pairs, or a multiple of 200, are found. After a long period of digestion the multiple-size units are not found and all of the DNA has the 200-bp-unit size.

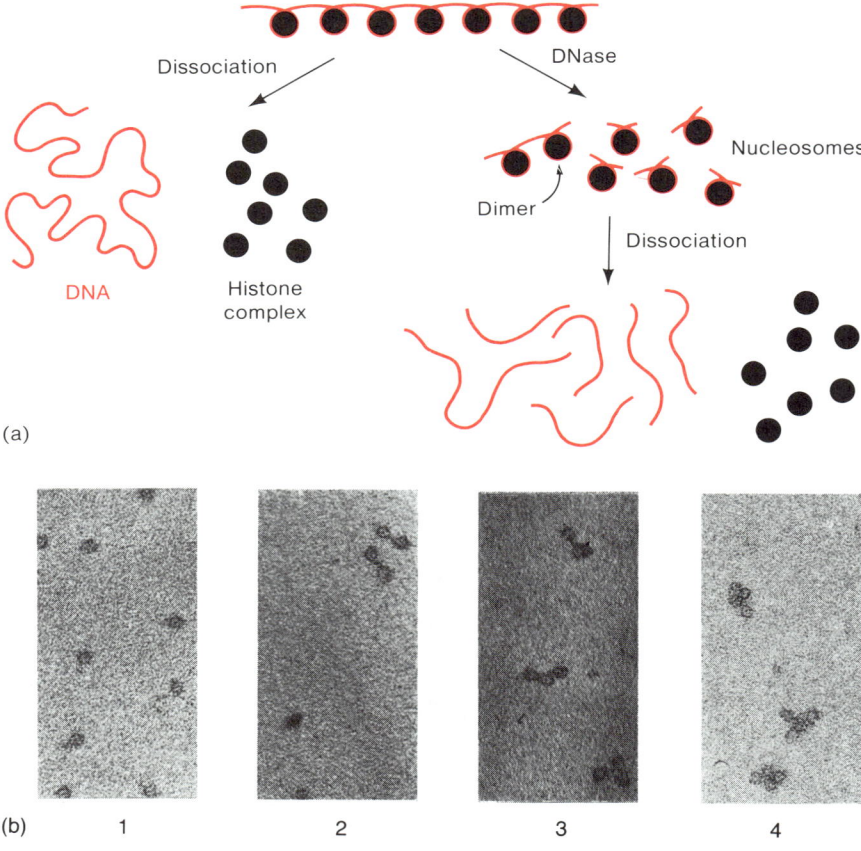

(a)

(b)   1   2   3   4

Figure 7-16   (a) The DNase-digestion method for production of individual nucleosomes and 200-bp fragments. (b) Electron micrograph of incompletely digested chromatin fractionated by centrifugation through a sucrose concentration gradient, showing (1) nucleosome monomers, (2) dimers, (3) trimers, and (4) tetramers. [From J. T. Finch, M. Noll, and R. D. Kornberg, *Proc. Nat. Acad. Sci.,* (1975) 72: 3321.]

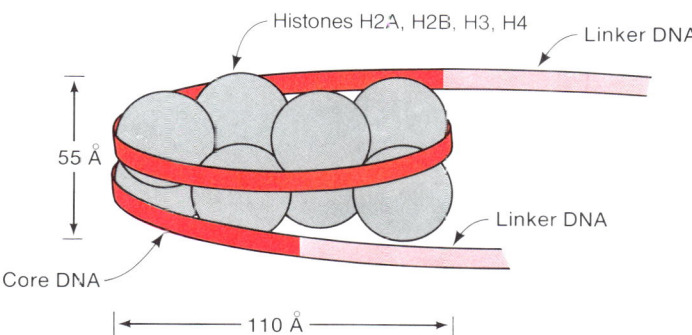

Histones H2A, H2B, H3, H4

Linker DNA

55 Å

Linker DNA

Core DNA

|← 110 Å →|

Figure 7-17   Schematic diagram of nucleosome core particle. The DNA molecule is wound $1\frac{3}{4}$ turns around a histone octamer (2 molecules each of histones H2A, H2B, H3, and H4). Histone H1 (not shown) is bound to the linker DNA. Note that the two linker units point in the same direction.

The beadlike particles are called **nucleosomes.** Each nucleosome has the following composition: one molecule of histone H1; two molecules each of histones H2A, H2B, H3, and H4; and a DNA fragment. Prolonged treatment of the nucleosomes with the nuclease gradually removes additional amounts of DNA. All histones remain associated with the DNA until the number of base pairs is <160, at which point H1 is lost. More bases can be removed but the number cannot be reduced to less than 140 base pairs. The structure that remains is called the **core particle.** It contains an **octameric protein disc** consisting of two copies each of H2A, H2B, H3, and H4, around which the 140-bp segment is wrapped like a ribbon (Figure 7-17). Thus a nucleosome consists of a core particle and **linker DNA** to which H1 is attached. The overall structure of the chromatin fibril is that shown in Figure 7-18. Possibly, H1 also binds to adjacent core particles to make a more compact structure.

The binding of DNA and histone has been examined. About 80 percent of the amino acids in the histones are in α-helical regions. Preliminary x-ray-diffraction data suggest that many of the extended α-helical regions lie in the larger groove of the DNA helix and that the complex is stabilized by an electrostatic attraction between the positively charged lysines and arginines of the histones and the negatively charged phosphates of the DNA.

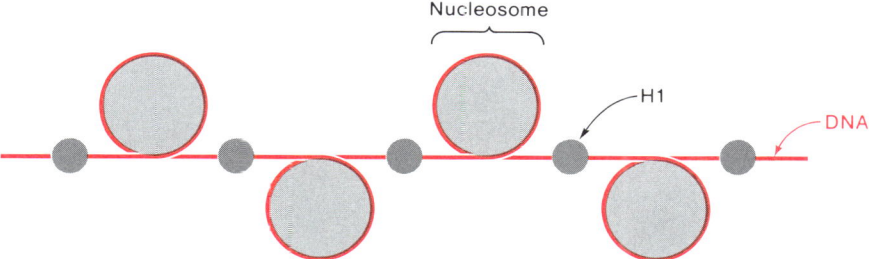

Nucleosome

H1

DNA

Figure 7-18   Schematic diagram of chromatin showing histone H1 binding to the linker DNA. Somehow H1 increases the DNA winding to two full turns, and the nucleosomes assume a zigzag, possibly stacked array.

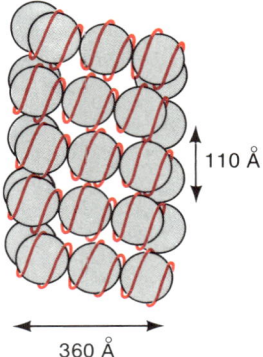

Figure 7-19 A proposed solenoidal model of chromatin. Six nucleosomes (shaded) form one turn of the helix. The DNA double helix (shown in red) is wound around each nucleosome. [After J. T. Finch and A. Klug, *Proc. Nat. Acad. Sci.*, (1976) 73: 1900.]

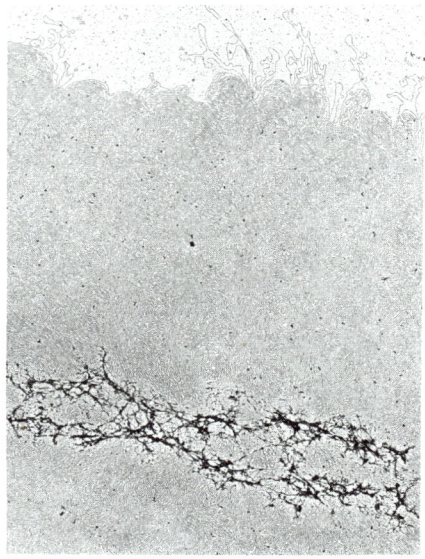

Figure 7-20 Electron micrograph of a segment of a human metaphase chromosome from which the histones have been chemically removed. The dense network near the bottom of the figure is the protein scaffold on which the chromosome is assembled. [Courtesy of Ulrich Laemmli.]

The DNA content of nucleosomes varies from one organism to the next, ranging from 150 to 240 bp per unit. The core particles of all organisms have the same DNA content (140 bp); thus, the observed variation results from different sizes of the linker DNA between the nucleosomes (namely, 10–100 bp). Surprisingly, in the same organism the lengths of the linkers may also vary from one tissue to another (for instance, brain versus liver). The significance of this variation is unknown.

Assembly of Nucleosomes—The 30-nm Fiber and the Solenoidal Model of Chromosome Structure

Chromosomes have several orders of organization (Figure 7-14). The nucleosome is the first; in this structure a segment of DNA containing 150–240 bp and whose length is 510–726 Å when extended in solution is compacted into a cylinder about 110 Å across and 55 Å high. The next level of organization is the 30-nm fiber. This fiber has been isolated and characterized by several techniques.

The currently accepted model for the first stages of chromatin folding is the solenoidal model shown in Figure 7-19. In this structure a linear beaded chain is arranged as a hollow helix that is 360 Å in diameter and that has a repeating unit of 110 Å. Other experiments suggest that these helices are arranged in a series of loops. The looped structure has the degree of compactness of DNA in a resting nucleus. Chromosomes are about 50–100 times as compact as the looped structure allows. Further compacting is probably provided by some type of **scaffold** consisting of nonhistone proteins and possibly RNA. Electron micrographs have been taken of metaphase chromosomes stripped of histones. The DNA is unfolded somewhat but, as shown in Figure 7-20, it remains tightly associated with a fairly dense region (defined to be the scaffold) in a form having an enormous number of loops. Condensation and organization around the scaffold probably serve only to create a discrete unit that can be moved along the mitotic spindle prior to cell division.

Centromeres and Telomeres

The **centromere** is a specific region of the eukaryotic chromosome that becomes visible as a distinct morphological entity (appearing as a thin region) along the chromosome during condensation. It is responsible for chromosome movement during both mitosis and meiosis, functioning, at least in part, by providing an attachment site for one or more spindle fibers. Electron microscopic analysis has shown that in some organisms—for example, the yeast *Saccharomyces cerevisiae*—a single spindle-protein fiber is attached to centromeric chromatin. The chromatin segment of the centromeres of yeast

(a)

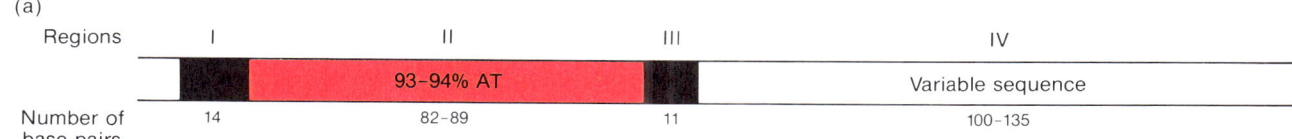

(b)

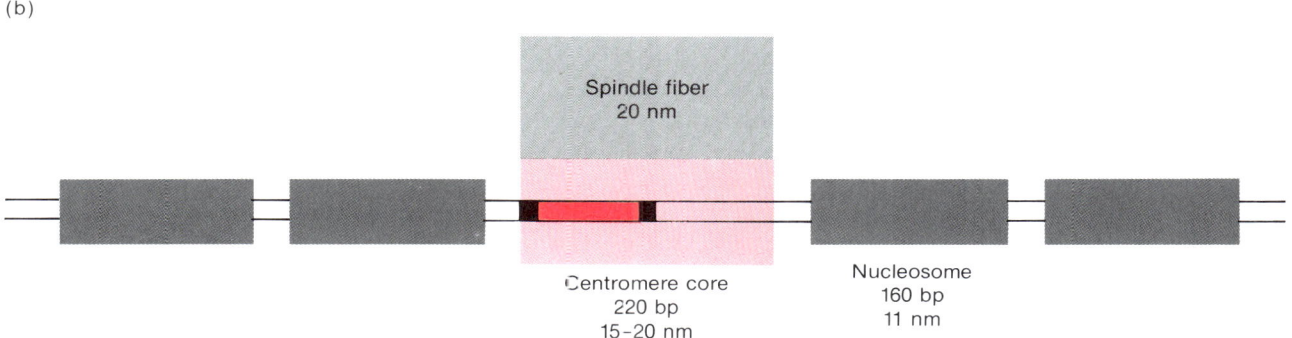

Figure 7-21 A yeast centromere. (a) Diagram of the DNA showing the major regions. The base sequences of regions I–III are nearly the same; those of region IV vary from one centromere to the next. (b) Positions of the centromere core and the nucleosomes on the DNA. The DNA is wrapped around histones in the nucleosomes; the organization and composition of the centromere core are unknown. [After K. S. Bloom, M. Fitzgerald-Hayes, and J. Carbon, *Cold Spring Harb. Symp. Quant. Biol.,* (1982) 47: 1175.]

has been cloned in an *E. coli* plasmid. It has a unique structure in that it is exceedingly resistant to the action of various DNases and has been isolated as a protein-DNA complex containing 220–250 base pairs. The nucleosomal constitution and DNA base sequences of the centromeres of four different yeast chromosomes have been determined; several common features of the base sequences are shown in Figure 7-21(a). There are four regions—labeled I–IV; the sequences in regions I, II, and III are nearly identical, but that of region IV varies from one centromere to another. Region II is noteworthy in that 93 percent of the 82–89 base pairs are A·T pairs. The centromeric DNA forms a structure (the centromeric core particle) that contains more DNA than a typical yeast nucleosome core particle (160 base pairs) and is larger. This structure is responsible for the resistance of centromeric DNA to DNase. It is not known whether histones or other proteins form the centromeric particle. The spindle fiber is believed to be attached directly to this particle (Figure 7-21(b)).

Whether the base-sequence arrangement of the yeast centromeres is typical of eukaryotic centromeres remains to be determined. In higher eukaryotes the chromosomes are about 100 times larger than yeast chromosomes and several spindle fibers are usually attached to each chromosome; thus, it

Figure 7-22   A telomere sequence of the protozoan *Tetrahymena*. The vertical red lines separate the repeated 6-bp sequence. The arrows point to sites of missing nucleotides.

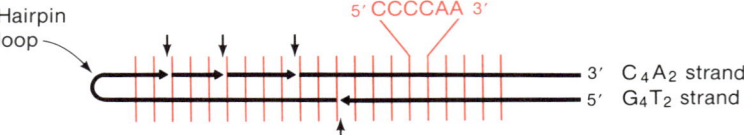

is possible that the centromeres are larger and more complex than those of yeast. Furthermore, the yeast genome is free of highly repetitive sequences, whereas the centromeric regions of the chromosomes of many higher eukaryotes contain large amounts of heterochromatin, consisting of repetitive satellite DNA, as described in Chapter 4.

Telomeres, the complex structures at the ends of eukaryotic chromosomes are essential for chromosome stability, based on genetic and microscopic observation. For example, in both maize and *Drosophila,* chromosomes that lack telomeres (broken chromosomes) often fuse end to end. If they do not, they usually fail to replicate. A telomere of the protozoan *Tetrahymena* has been isolated. It has an unusual structure, with 20–70 repeats of the sequence 5'-CCCCAA-3', a hairpin terminus, and several gaps in which the first cytosine of the adjacent repeated sequence is absent (Figure 7-22). In Chapter 9 the problem of replication of a linear DNA molecule will be described. Telomeres are thought to be responsible for the completion of the replication of the linear DNA molecule contained in a eukaryotic chromosome.

INTERACTION OF DNA AND PROTEINS THAT RECOGNIZE A SPECIFIC BASE SEQUENCE

In the formation of chromatin the histones do not bind to specific base sequences but recognize only the general DNA structure. There are, however, numerous proteins that bind only to particular sequences of bases, and we will encounter many such proteins in this book. These proteins are of three types: (1) regulatory proteins, which turn on and off the activity of particular genes; (2) restriction endonucleases; and (3) polymerases, which initiate synthesis of DNA and RNA from particular base sequences.

For many DNA-binding proteins, the specific sequence of bases to which the protein binds is known. In a few cases, one of which we now describe, the particular features of the DNA that are recognized and the mode of binding are understood.

Binding of the Cro Protein to Phage λ DNA

The Cro protein made by *E. coli* phage λ is a regulatory protein that binds to a specific base sequence in λ DNA. It is a small basic protein that contains 66 amino acids and binds to a sequence of DNA containing 17 base pairs

(Figure 7-23). Both the amino acid sequence of the protein and the base sequence of the interaction region in DNA are known.

The three-dimensional structure of the Cro protein has been determined by standard techniques of x-ray diffraction and is shown in Figure 7-24. It is a simple structure (compared to larger proteins), consisting of three strands of antiparallel β sheet (residues 2–6, 39–45, and 48–55) and three α helices (residues 7–14, 15–23, and 27–36). Residues 63–66 are not shown in the figure as they are not ordered in the crystal.

Three procedures have been used to determine some of the points of contact between the Cro protein and the 17-bp site in the DNA. In the first procedure, purified Cro protein and λ DNA were mixed using conditions that cause formation of the Cro-DNA complex. The Cro-DNA complex was then treated with dimethyl sulfate, which adds a methyl group to guanine in the N-7 position and to adenine in the N-3 position of the purine ring. The complex was then dissociated and the Cro protein removed. By direct analysis of the base sequence of the DNA, the positions of the methylated guanines and adenines were determined. Certain guanines and adenines were resistant to methylation; presumably, in the complex, Cro contacts these bases and prevents the dimethyl sulfate from approaching the bases. In the second procedure, free DNA was treated with dimethyl sulfate but so briefly that only a few guanines and adenine were methylated in a given

Figure 7-23 The base sequence of one of the λ Cro protein binding sites. This sequence is called *oR3*. The adenines and guanines shown in Figure 7-25 are printed in red. [After W. F. Anderson et al, *Nature*, (1981) 290: 754.]

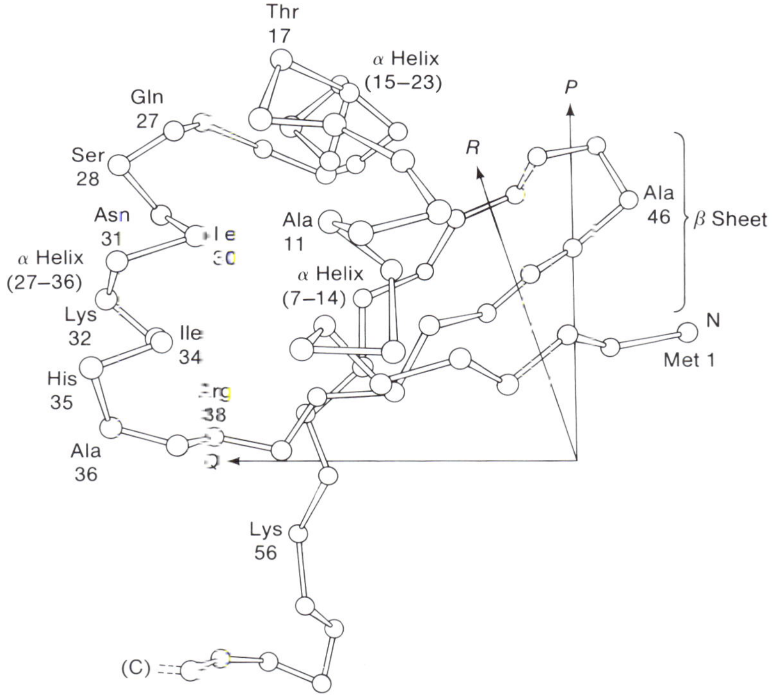

Figure 7-24 Three-dimensional structure of the λ Cro protein. Segments consisting of α helices and β structure are indicated. The arrows P, Q, and R indicate the three geometric axes referred to in other figures. The complete C-terminus is not shown.

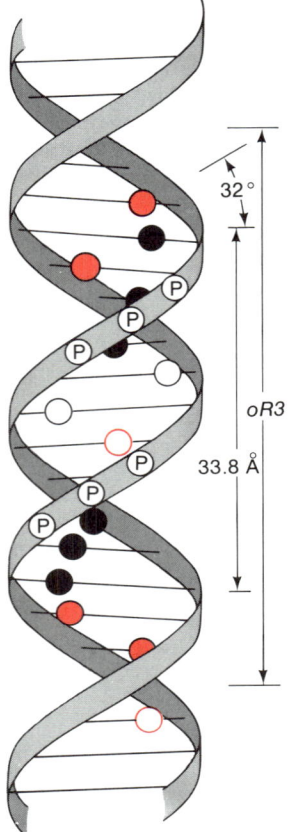

Figure 7-25 Points at which the *oR3* region of λ DNA makes contact with Cro. Phosphates are labeled P; guanines and adenines in contact are solid black and red circles; guanines and adenines not in contact are open red and black circles, respectively. The solid circles are the bases shown in red in Figure 7-23. The bases in contact are only in the wider (major) groove of the DNA. Contact with adenine has not been established by the dimethyl sulfate procedure since N-3 of adenine is in the narrow groove, in which there is very little contact. The lengths and angles should be compared with Figure 7-26. [After W. F. Anderson et al., *Nature,* (1981) 290: 754.]

molecule. Then, Cro was added in excess. Free DNA and Cro-DNA complexes were separated and the methylated sites in the free DNA were identified; these represent contact sites whose methylation prevented binding of Cro. The third procedure identified the phosphates that contact Cro. In this procedure phosphates were ethylated and free DNA is added. As in the second procedure, free DNA was separated from Cro-DNA complexes and the positions of ethylation of the free DNA were noted, identifying the contact points. Some information about binding to thymine has also been obtained by using DNA in which one or two thymines have been replaced by a thymine analogue, 5-bromouracil (Chapter 11). Substitution at critical sites inhibits binding. At present (1985) no information is available about cytosine contacts.

Figure 7-25 shows the Cro-binding segments of λ DNA drawn as a double helix. Guanines, adenines, and relevant phosphates are indicated. The guanine N-7 atoms (the potential methylation sites) face the reader in the wide groove. The adenine N-3 atoms face the reader in the narrow groove. Note that the guanines (solid circles) and the phosphates that contact the protein are all located on a single side of the helix, which, from a geometrical point of view, seems reasonable. Furthermore, the contact points are only in the major (wider) groove of the DNA. Two additional points should be noted: (1) bases to which Cro is bound are on both strands (see also Figure 7-24); and (2) the contact points on the DNA are nearly symmetrically arranged.

Analysis of the arrangement of Cro protein molecules in a water-containing crystal has shown that the proteins are in an array having twofold symmetry, shown in Figure 7-26. In solution Cro also dimerizes, and it is likely that together the two monomers have the same symmetry that the molecules have in a crystal. This observation is particularly significant because the Cro-binding DNA sequence is also somewhat symmetric, and such a symmetric array is a reasonable basis for binding to a nearly symmetric base sequence. A significant feature of the Cro dimer is the 34 Å spacing between the α_3 helices, precisely the spacing required for the two α helices to fit into successive major grooves on a single face of a DNA molecule. Figure 7-27 shows the presumed structure of the Cro-DNA complex. In panel (a) the symmetry of the structure is readily apparent, as inferred from model building.

Examination of a detailed molecular model (not shown) indicates the following: (1) the α_3 helix fits into the major groove and several amino acids on its surface can make hydrogen bonds and van der Waals bonds with groups on the DNA bases; (2) amino acid side chains elsewhere in the molecule can form both hydrogen bonds and stronger ionic bonds with the phosphates indicated in Figure 7-25. Other analyses show that binding to the phosphates provides the major part of the binding strength, and binding of α_3 provides the base-sequence specificity. It has been hypothesized that when Cro first comes into contact with DNA, it binds at the initial contact site only to the phosphates and then slides along the DNA helix until the two α_3 helices come into contact with the particular bases in O_R.

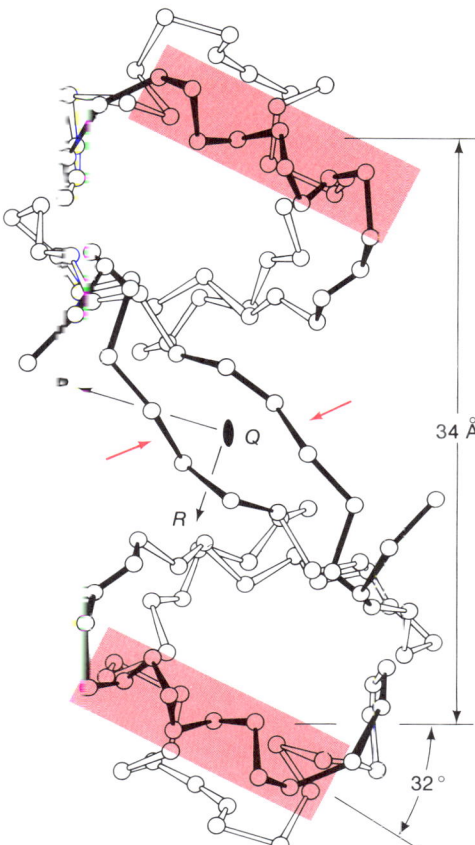

Figure 7-26   Arrangement of two Cro molecules in a protein crystal. Regions of the protein backbone closest to the viewer and presumably in contact with DNA are drawn solid. These regions include two α helices 34 Å apart, inclined at 32°, and also a pair of extended strands (arrows) close to the Q axis. The pink screens cover the α_3 helices referred to in the text. [After W. F. Anderson et al., *Nature*, (1981) 290: 754.]

Since the elucidation of the structure of the Cro dimer, the structures of several other regulatory DNA–binding proteins have also been worked out—for example, the *E. coli* CRP protein (Chapter 15), and the repressor and Cro proteins of *E. coli* phages λ and 434 and of *Salmonella typhimurium* phage P22. The relevant DNA sequences have also been examined by the ethylation and dimethyl sulfate procedures. The results of these experiments have led to several important conclusions:

1. Each protein forms a symmetric dimer.
2. Each monomer has two α helices corresponding in position and orientation to the α_2 and α_3 helices of Cro, with a 34 Å separation between the α_3 helices, allowing binding in the major groove.

Figure 7-27 (a) Structure of the Cro-*oR3* complex. The DNA is rotated 90° clockwise relative to that in Figure 7-25, so the contact points are on the right side of the molecule. A pair of α helices, related by twofold symmetry, occupy successive major grooves of the DNA, and the two extended chains, indicated in Figure 7-26, run parallel to the axis of the DNA. (b) A view of the Cro-DNA complex at an angle that the orientation of the Cro dimer is the same as that shown in Figure 7-26. The open circles on the DNA backbone indicate the phosphates; the solid circles indicate those phosphates whose ethylation hinders binding. The α₃ helices are shaded in pink. [After W. F. Anderson et al., *Nature,* (1981) 290: 754.]

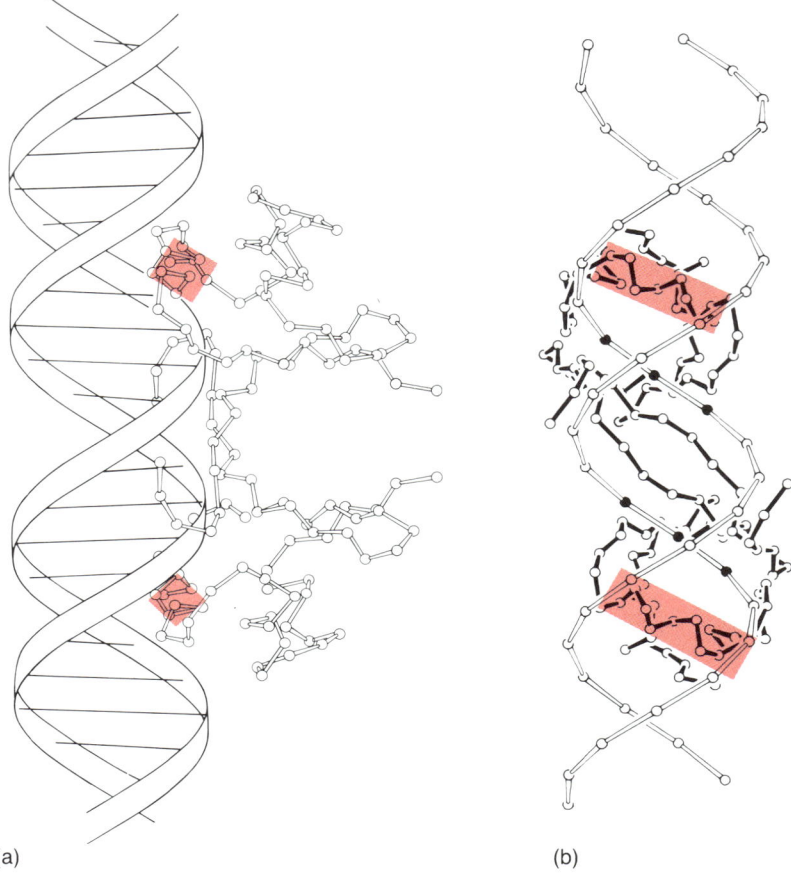

(a)   (b)

3. The total number of α helices and β sheets per monomer are not common features of different DNA-binding proteins.
4. Amino acid side chains on the surface of an α_3 helix are, for each protein, able to hydrogen-bond to particular groups on the bases.

The fourth point has led to a renaming of α_3 as the **recognition helix.** A complete computer search of all known protein structures (stored in the Brookhaven Data Bank) has shown that the symmetric α_2-α_3 unit with 34 Å spacing is not found in any protein other than certain DNA-binding proteins. Thus, it has been suggested that this unit and binding in the major groove of DNA may be essential features of many DNA-binding proteins, with binding specificity residing in α_3. Recall that it was suggested that the α helices in histones also bind to the major groove of DNA, even though there is no sequence specificity.

The amino acid sequences of the α_3 helices of these proteins have been examined and they are not the same, which is to be expected when the base sequences of the target DNA differ. However, both λ Cro and λ cI proteins bind to the same base sequence, and the amino acid sequences of the α_3 helices of each differ; the same can be said of the 434 Cro and 434 cI proteins, which also share target sequences. Thus, a one-to-one recognition code between base sequences and amino acid sequences does not exist and the rules for recognition remained to be discovered.

The conclusions presented so far about the structure of the DNA-protein complexes have been derived from the structures of the DNA-binding proteins, rather than the structures of the complexes, by fitting each structure to the DNA molecule in the B form. To make this construction, an assumption has been made, namely, that neither the DNA segment nor the proteins change structure when binding occurs. In 1985 several important experiments were reported that placed these conclusions on a firm basis. In the first experiment crystals of the complex of the cI protein of phage 434 (434R) to the 434 O_R region were prepared, and the structure of the complex was determined by x-ray diffraction. The x-ray analysis confirmed the conclusions we have stated, though the DNA segment was found to be bent slightly in the complex. An important observation was that each amino acid thought to hydrogen-bond to a particular base was located such that the predicted hydrogen-bonding was possible. However, the resolution of the analysis was insufficient to prove that the hydrogen bonds actually exist in the complex. The second experiment proved this point.

Phage P22 has a DNA-binding system similar to that of the Cro-cI-O_R system of λ and 434. However, the amino acid sequences and the base sequences of the 434 and P22 systems differ. The x-ray diffraction analysis of the 434R-O_R(434) system described in the preceding paragraph showed which amino acid in α_3 could hydrogen-bond with each base. An x-ray analysis of the corresponding P22 protein (P22R) indicates the amino acids in α_3 of P22R that probably hydrogen-bond with O_R(P22). By using site-directed mutagenesis (Chapter 11) and genetic engineering (Chapter 22), a hybrid protein was created, consisting mostly of 434R, but in the α_3 helix the presumed hydrogen-bonding amino acids were replaced by the corresponding amino acids of P22R. This hybrid protein was denoted 434R[α_3(P22R)]. When 434R[α_3(P22)] was mixed with either purified O_R(434) or O_R(P22), binding occurred only to O_R(P22), proving conclusively that specific base recognition resides in α_3. *In vivo* experiments with 434 phages carrying 434R[α_3(P22R)] instead of 434R also showed that the hybrid phage failed to recognize O_R(434). In a control experiment the α_3 regions of 434R and 434 Cro were exchanged; since both proteins recognized O_R(434), it came as no surprise that the hybrid protein retained the ability to bind to O_R(434).

Recently the structures of two other DNA-binding proteins, the EcoRI restriction enzyme and the Klenow fragment of DNA polymerase I, which do not have regulatory functions, have been elucidated. EcoRI (Chapter 5) binds to a specific palindrome, and x-ray diffraction analysis has been carried out on a DNA-protein complex with a 13-bp fragment containing the EcoRI

(a) Target Sequence

TCGCGAATTCGCGA
AGCGCTTAAGCGCT

(b) Kinks

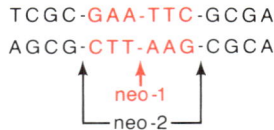

TCGC-GAA-TTC-GCGA
AGCG-CTT-AAG-CGCA

Figure 7-28 Binding of EcoRI to DNA. (a) The target sequence. (b) Alterations of the sequence produced by the protein binding. The dashes indicate the positions of the kinks.

target sequence (Figure 7-28(a)). The protein contains α helices that bind in the major groove of DNA. However, a hitherto unobserved feature of the DNA in the complex was observed. Three kinks in the DNA backbone occur, as shown in Figure 7-28(b). The central kink, termed neo-1, is a symmetric change in which the DNA unwinds by 25°, thereby enlarging the major groove slightly, which is necessary for entry of the protein. The other kinks (neo-2) are symmetric, do not include unwinding, and involve a bend of the DNA helix. Thus, as we saw with the 434R-O_R(434) complex, some deformation of DNA accompanies the binding of protein. However, the structural deformation accompanying kinking is greater than the bending seen with 434R. It has been suggested that neo-1 kinking may be a common feature of many DNA-protein interactions in which binding is to a specific symmetric sequence.

DNA polymerase I is an important enzyme involved in both DNA repair and DNA synthesis. It has several catalytic activities. A particular fragment of the enzyme containing the polymerizing activity can be prepared; it is called the Klenow fragment. The structure of the fragment is much more complicated than the small proteins just described because it is a large protein with nearly 600 amino acids. The Klenow fragment is also a different type of DNA-binding protein in that it does not bind to a unique sequence; as polymerization proceeds, it moves along the DNA. The polypeptide chain folds to form 19 α helices and 13 β sheets. The structure resembles a clenched right hand. The deep cleft where the fingertips touch the palm is 20–24 Å wide and 25–35 Å deep (roughly the dimensions of a DNA molecule) and is the DNA binding site. The cleft contains two α helices that in space-filling molecular models fit in the major groove of the DNA; these α helices are analogous to α_2 and α_3 of the various Cro and repressor proteins. Thus, in this system in which the protein is not a dimer and the binding site is not symmetric, the basic rule for DNA binding—insertion of an α helix into the major groove of DNA—seems to be followed.

BIOLOGICAL MEMBRANES

Biological membranes are organized assemblies consisting mainly of proteins and lipids. The structures of all biological membranes have many common features, but small differences exist to accommodate the varied functions of different membranes.

There are many types of membranes, each having its own function. A fundamental role of membranes is to separate the contents of a cell from the environment. However, cells must take up nutrients from the surroundings, so the enclosing membrane of a cell must be permeable. However, the permeability of the membrane must be selective, in order that the concentrations of compounds within a cell can be controlled; otherwise, intracellular concentrations would be the same as extracellular concentrations. Selective permeability is attained by means of restricting free passage of most intra- and extracellular substances and allowing transport of specific substances by molecular pumps, channels, and gates. External membranes also have a signal function. For instance, they often contain receptors for signal molecules such as hormones. The external membrane is also responsible for transmitting electrical impulses, as in nerve cells.

There are also numerous types of intracellular membranes. These membranes serve to compartmentalize various intracellular components and to provide surfaces on which certain molecules can be adsorbed prior to chemical reaction with other adsorbed molecules. Some internal membranes even contain enzyme systems. For example, photosynthesis takes place within the inner membrane of chloroplasts in plant cells.

Membrane Structure: The Basic Bilayer

Figure 7-29 shows an electron micrograph in which the membrane that encloses a red blood cell is seen in cross-section. The most notable feature of this membrane is that it consists of two layers—it is a **bilayer.** This structure is a consequence of the chemical nature of the lipids that form the membrane. There are three major classes of membrane lipids: the phospholipids (most prevalent), cholesterol, and the glycolipids (least prevalent). The chemical structure of one phospholipid is shown in Figure 7-30. This molecule, like all phospholipids, has a central, somewhat polar, carbohydrate moiety (glycerol in this case) linked to a highly polar group, phosphorylcholine (shown in red) and to two nonpolar long hydrocarbon chains. The two tails usually contain different numbers of carbon atoms (range, 14–24) and hence have different lengths; one tail typically has one or more *cis* double bonds, each of which causes a kink in the chain. Figure 7-30 shows two schematic drawings of this molecule. The essential feature is a polar "head" group to which hydrocarbon "tails" are attached. When placed in water, such molecules (with distinct polar and nonpolar segments) tend to aggregate, because only the polar head is capable of interaction with the polar water molecules. The hydrocarbon tails are brought together by hydrophobic interactions—that is, they cluster because they are unable to interact with water. If the length of the hydrocarbon tail satisfies certain geometric constraints, a collection of such molecules forms a **micelle,** a spherical array of molecules in which the nonpolar tails form a hydrocarbon microdroplet enclosed in a shell composed of the polar heads (Figure 7-31(a)). The geometric constraints limit the size of a stable micelle and, as the size of the hydrocarbon tail increases, the array becomes ellipsoidal (Figure 7-31(b)), with the ratio of the lengths of the major axis to the minor axis of the ellipsoid increasing with length of the tail. Above a certain tail length, the only stable configuration is an ellipsoid whose major axis is "infinitely"

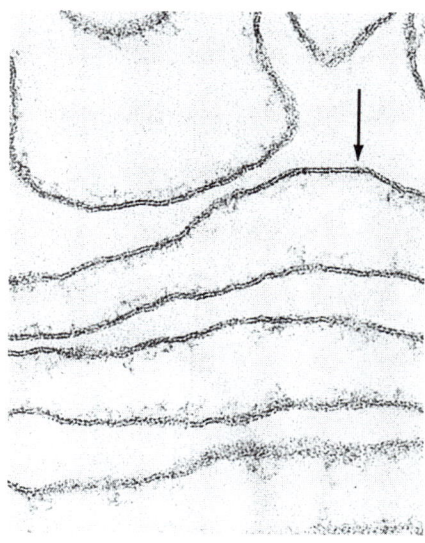

Figure 7-29 Electron micrograph of a preparation of plasma membranes from red blood cells. A single membrane is denoted by an arrow. [Courtesy of Vincent Marchesi.]

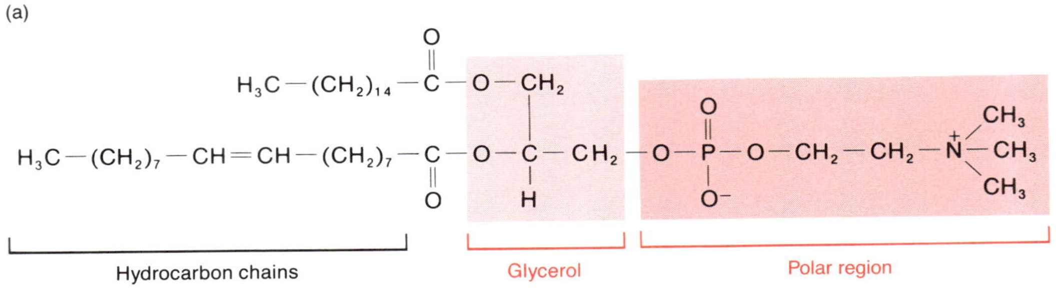

(a)

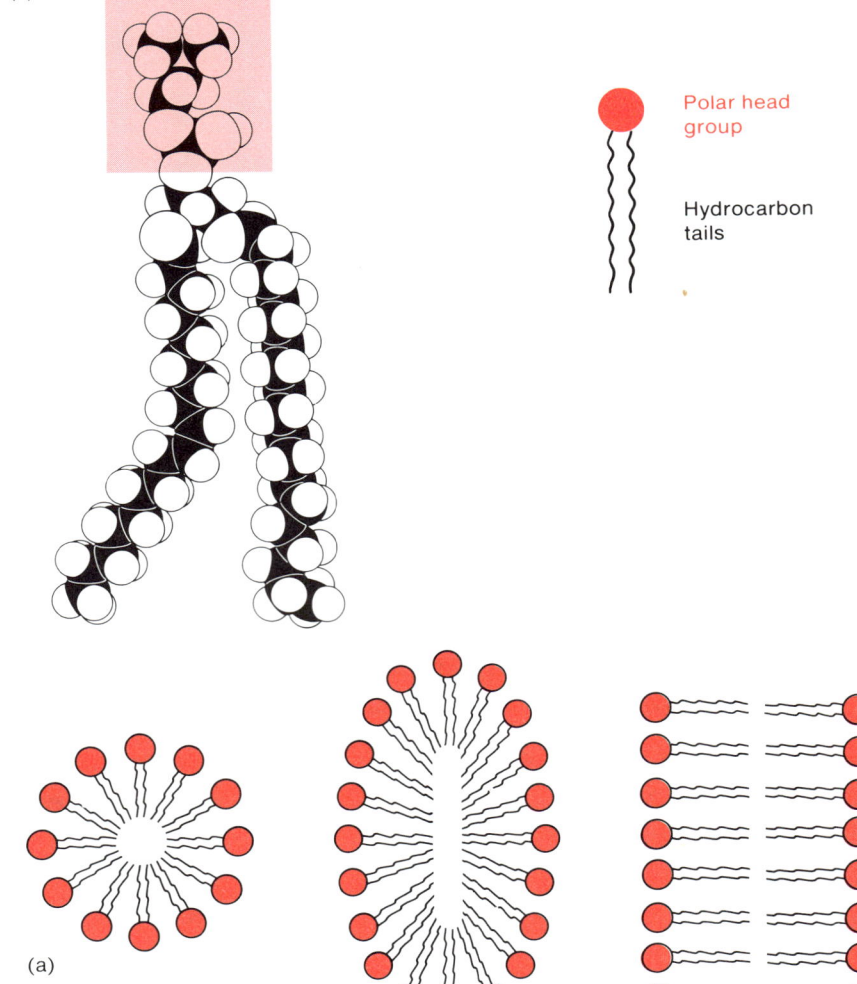

Hydrocarbon chains Glycerol Polar region

Figure 7-30 (a) Chemical structure of a phosphatidyl choline (1-palmitoyl-2-oleoyl-phosphatidyl choline). (b) Space-filling model of a phosphatidyl choline molecule. (c) The essential features of a phospholipid or glycolipid molecule.

(b) (c)

Polar head group

Hydrocarbon tails

Figure 7-31 (a) Diagram of a section of a micelle formed from phospholipid molecules. (b) An ellipsoidal micelle. (c) Diagram of a section of a membrane bilayer formed from phospholipid molecules.

(a) (b) (c)

long—that is, an extended lipid bilayer sheet, as shown in part (c) of the figure. The lipid bilayer can also be stabilized by van der Waals attractive forces between the hydrocarbon tails.

A variety of physicochemical studies of artificial membranes show that the phospholipids in a membrane drift laterally, indicating that the membrane is fluid. The fluidity is affected by temperature and in fact at temperatures that might be encountered by living cells phase changes can occur with the membrane becoming a semisolid gel. Furthermore, when membranes consisting of several different lipids are formed in the laboratory, the individual lipids separate from one another at critical temperatures. Living cells could not tolerate these effects of temperature and in fact certain features of natural membranes are responsible for maintaining constant fluidity and permeability of these membranes. Recall that the two hydrocarbon chains of phospholipids are different and that one usually contains a double bond. These chains have quite different temperatures for phase transitions, but because they are both covalently linked to glycerol, physical separation cannot occur. This has the effect of broadening any phase transition; that is, the transition would occur gradually over a range of temperature rather than sharply at a critical temperature. Furthermore, the membranes contain several different phospholipids, which widens the transition even more. Earlier we mentioned that cholesterol is also a membrane component (Figure 7-32(a)). Cholesterol also broadens phase transitions but has another effect. Its ring system is quite rigid and very hydrophobic. The rigid ring system interacts significantly with the hydrocarbon chains (Figure 7–32(b)) and strengthens the membrane structure. The function of the glycolipids is unknown; they may participate in intercellular adhesion in some tissues.

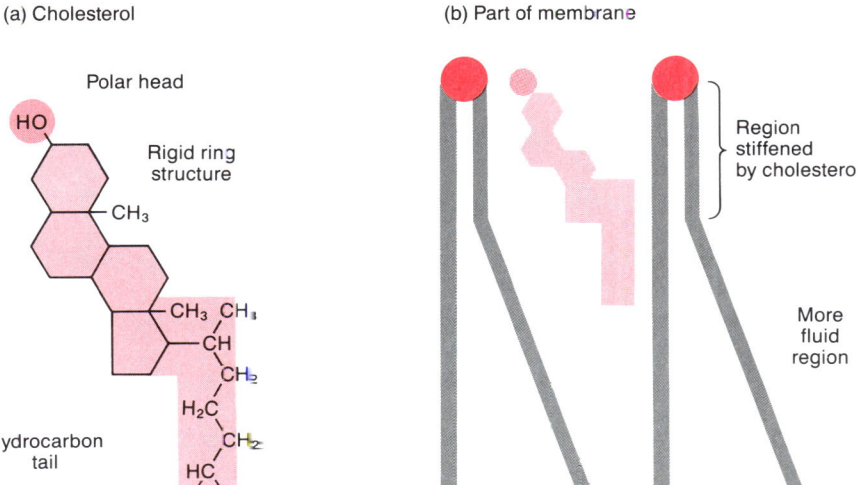

(a) Cholesterol

Polar head

HO

Rigid ring
structure

CH₃

Hydrocarbon
tail

CH₃ CH₃
CH
CH₂
H₂C
CH₂
HC
H₃C CH₃

(b) Part of membrane

Region
stiffened
by cholesterol

More
fluid
region

Figure 7-32 (a) Structure of cholesterol. (b) Drawing showing the region of a lipid membrane stiffened by cholesterol.

Vesicles

The ends of a lipid bilayer are unstable because the hydrocarbon chains are exposed to water. Thus lipid bilayers tend to close upon themselves to form hollow bilayer spheres known as **vesicles** (Figure 7-33). A lipid bilayer is a fluid structure that can be mechanically disrupted. Once disrupted, however, they spontaneously re-form to avoid contact between the exposed hydrocarbon tails and water. Often a vesicle is broken in more than one position, so several vesicles form on the reclosing of fragments. An extended structure such as a complete cell membrane will usually form a large number of small vesicles, if it is extensively fragmented and allowed to re-form. This protocol allows one to study the permeability of a membrane (Figure 7-34). Either extended membranes or vesicles are suspended in a solution of a substance to be tested—for example, an amino acid. The suspension is then violently agitated, which breaks the membranes into small fragments or breaks the vesicles. When the agitation ceases, the vesicles re-form, but some amino acid molecules are trapped inside. The vesicles can then be collected and suspended in a solution lacking the amino acid. One can then determine whether in time the amino acid appears in the external fluid. Such passage will not occur if the membrane is impermeable to the amino acid. Otherwise, the rate of appearance of the amino acid is a measure of the permeability of the membrane to that molecule. Other molecules can be added either to the original (amino acid) solution or to the external fluid to determine their effect on membrane permeability. By appropriate choice of ionic conditions during disruption of natural membranes and reconstitution of vesicles it is possible to prepare vesicles with the outer side of the membrane on either the inside or outside of the vesicles. In this way one can study the two membrane surfaces and both inward and outward passage of molecules.

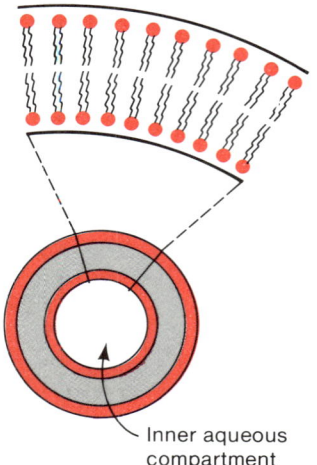

Inner aqueous compartment

Figure 7-33 Schematic diagram of a lipid vesicle. The membrane is enlarged to show the double layer.

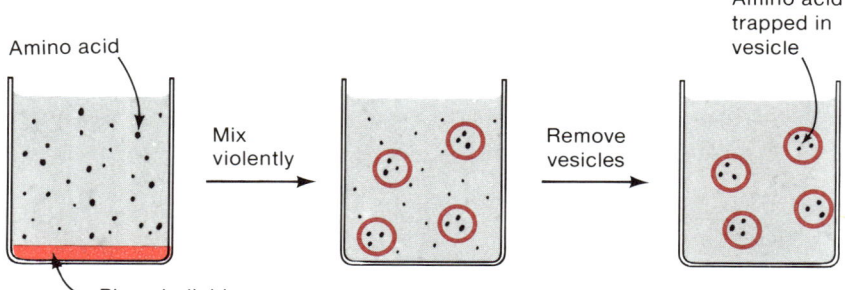

Figure 7-34 Method of preparation of a suspension of lipid vesicles containing amino acid molecules.

Modulation of the Basic Bilayer Structure: Permeability and Transport

Experiments of the sort just described, using synthetic vesicles made of a single phospholipid, indicate that the lipid bilayer is impermeable to all ions and highly polar molecules. How, then, do these molecules get in and out of cells? Naturally occurring biological membranes contain many protein molecules (about 50 percent by weight), and membranes having different functions contain different proteins. These proteins, which act in a variety of ways, are responsible for the transport of all molecules having polar regions and for most nonpolar molecules as well.

There are two classes of membrane proteins—the **integral membrane proteins,** which are contained wholly or in part within the membrane, and the **peripheral proteins,** which lie on the membrane surface and are bound to the integral proteins. A technique known as **freeze–fracture electron microscopy** shows that many protein molecules are present in membranes. This technique is shown schematically in Figure 7-35. A small volume of a suspension containing cells is frozen and then fractured. The line of fracture lies within some of the lipid bilayers in the suspension so the surface exposed by the fracture is the inner surface of the bilayer. An electron micrograph of the plasma membrane of a red blood cell prepared by the freeze-fracture method is shown in Figure 7-36. The outer surface of the membrane is smooth but the inner surface is covered with many globular protein molecules.

Numerous physical measurements have shown that the integral membrane proteins can freely diffuse laterally throughout the bilayer. The degree of motion is influenced by the fluidity of the lipid layer, which is determined by the features described a few pages back. This phenomenon plus several

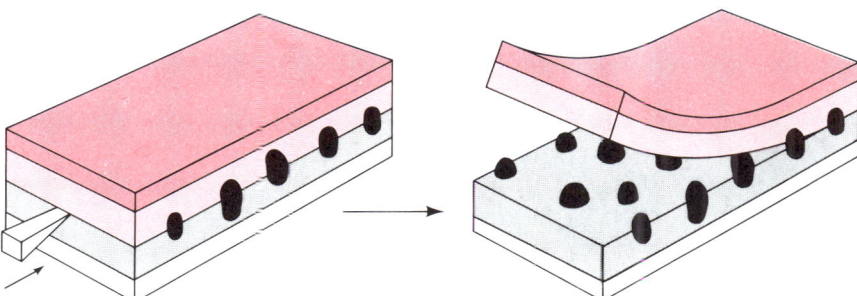

Figure 7-35 Technique of freeze-fracture electron microscopy applied to membranes. The membrane sample is frozen and struck with a sharp point in the direction of the cleavage plane. The membrane can then be split to show whether internal particles (black ellipses) are present.

Figure 7-36 An electron micrograph of the plasma membrane of a red blood cell, prepared by the freeze-etch technique. The interior of the membrane, which has been exposed by fracture of the membrane, contains numerous globular particles having a diameter of about 75 Å. These particles are termed integral membrane proteins. [Courtesy of Vincent Marchesi.]

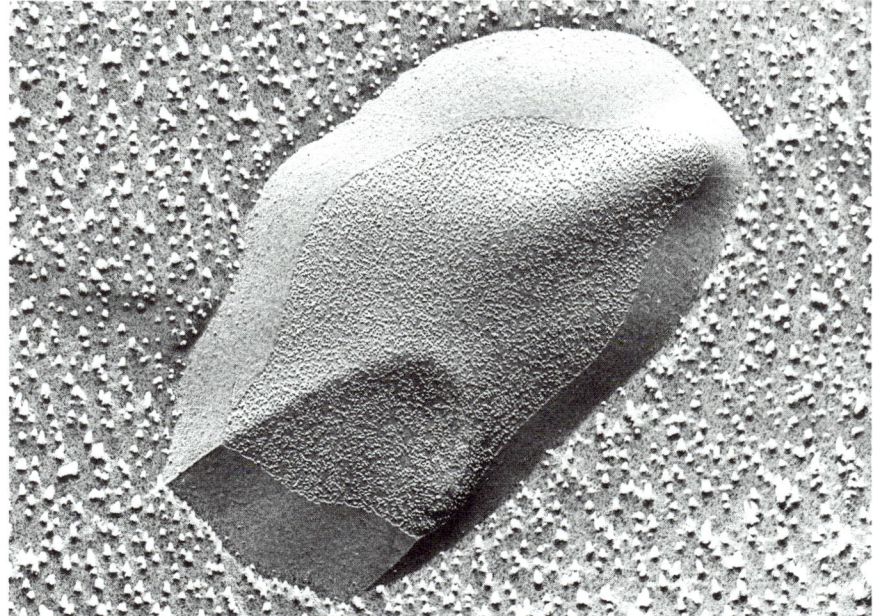

related observations led to the **fluid mosaic model** of membrane structure (Figure 7-37). It has also been shown that the proteins do not spontaneously rotate (flip-flop) from one side of the membrane to another. This is because the integral proteins also have polar and nonpolar regions, as shown in the figure; and inasmuch as the external region is polar and the internal region is nonpolar, rotation would require passage of the polar region through the nonpolar center of the bilayer.

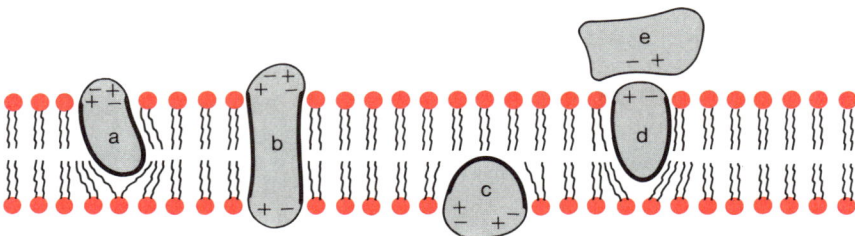

Figure 7-37 The structure of a membrane according to the fluid mosaic model. Four integral proteins (a)–(d) are embedded to various degrees into a lipid bilayer such that the hydrophobic surface of each protein (heavy lines) is in the membrane and the polar region (indicated by + and −) are external. Peripheral proteins (e.g., (e)) are on the surface and bound to a polar region of an integral protein. Integral proteins can drift laterally but cannot flip-flop.

An interesting experiment also demonstrated that the integral membrane proteins drift randomly and freely and, in addition, enabled the rate of movement to be measured. This experiment made use of a cell fusion technique (Chapter 24) in which animal cells from different species are stimulated to fuse by infection with Sendai virus inactivated by ultraviolet light. The experiment is shown in Figure 7-38. Antibodies were prepared against the integral membrane proteins of human cells and mouse cells. The antibody to mouse protein was covalently linked to rhodamine, a molecule that fluoresces red when activated with the appropriate wavelength of light. The human antibody was linked to fluorescein, which fluoresces green. Immediately after fusion the two fluorescent antibodies were added and the joined cells (heterokaryons) were observed with a fluorescence microscope. Each heterokaryon had a red half and a green half. Observation at various times showed a gradual mixing of the colors. After 40 minutes at 37°C both red and green fluorescence covered the cell surface, indicating that the human and mouse integral membrane proteins were completely and randomly mixed.

Since different proteins protrude from the two sides of the membrane, the membrane is an asymmetric structure having (taking the cell as reference) an inside and an outside. It is this asymmetry that determines the direction of movement of molecules entering and leaving a cell.

There are many modes of transport of molecules across the bilayer, and only a few will be mentioned. One mode makes use of channels through

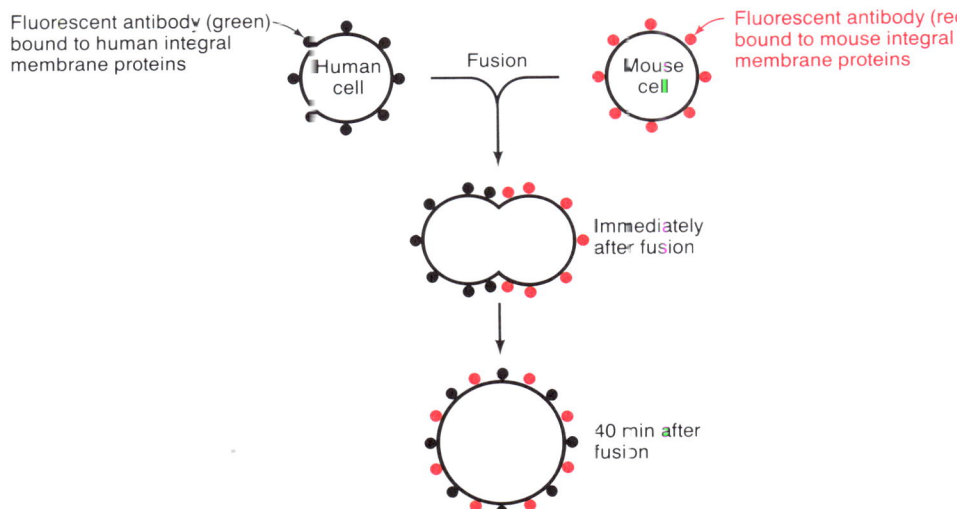

Figure 7-38   An experiment demonstrating drifting of integral membrane proteins. The fluorescent antibodies are *not* present throughout the experiment but are added at various times (e.g., before fusion, after fusion, and later) to visualize the degree of mixing.

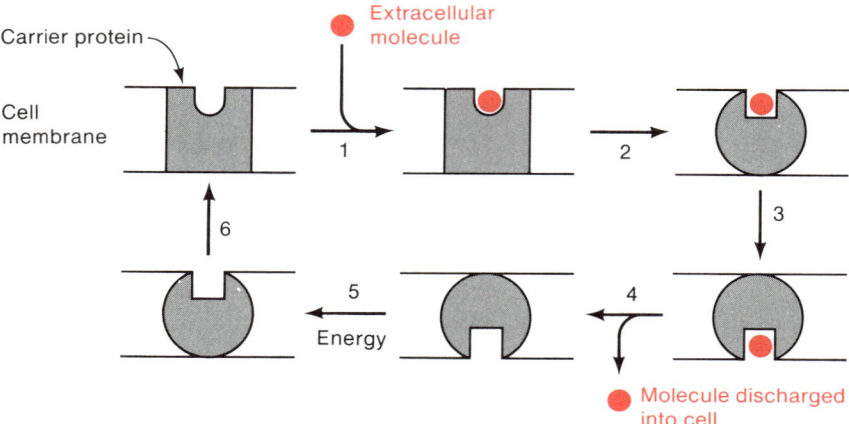

Figure 7-39 A schematic model for active transport. (1) An extracellular molecule binds to a carrier protein. (2) As a result of binding, the carrier protein changes shape. (3) The carrier protein rotates. (4) The bound molecule is discharged to the intracellular space. (5) By means of an energy-requiring process the carrier rotates again. (6) The original shape of the carrier is restored.

the membrane. These channels usually are passages through the integral membrane proteins; the passages will have an abundance of polar amino acids if their function is to allow transit of polar substances. Often the channels can be opened and closed by means of conformational changes of the membrane proteins. Other transport systems utilize chemical reactions that convert the substance to be transported to a molecule that can enter the membrane and then, after transit of the modified molecule, restore the original molecule at the other side of the membrane; these chemical mechanisms are usually very complex, consume a great deal of energy, and are poorly understood. The interesting model shown in Figure 7–39 uses a carrier protein that can act like a revolving door. An integral carrier protein binds the molecule to be transported and undergoes a change in shape and charge distribution that allows the protein to rotate. During one of the rotations the molecule to be transported is released at the other side of the membrane. Upon release, the carrier protein undergoes a second conformational change that causes it either to lose the ability to rotate or to bind the transported molecule. This step is a necessary one because otherwise the carrier molecule would bring as many molecules into the cell as it would take out of the cell.

Transport of this sort is an energy–requiring process; a certain amount of evidence suggests that the energy is needed to restore the molecule to the "ready" state—that is, facing the right direction and having the right binding site.

SELF-ASSEMBLY OF A COMPLEX AGGREGATE: TOBACCO MOSAIC VIRUS

Several aggregates, consisting of two or more macromolecules and having well-defined structures, have been described. However, nothing has been said yet about how such aggregates form. Basically there are two possibilities:

1. The structures form spontaneously from the subunits without an additional source of energy; this is called **self-assembly.**
2. Aggregation is directed, possibly by enzymes or some template, and requires a source of energy.

Examples are known of aggregates formed in both ways. In this section one of the best understood examples of self-assembly is described—namely, the formation of a particle of tobacco mosaic virus.

Assembly of Tobacco Mosaic Virus

Tobacco mosaic virus (TMV) is a rodshaped virus 3000 Å long and 180 Å in diameter (Figure 7-40(a)). There are 2130 identical subunits in the protein coat; these are closely packed in a helical array around a single-stranded RNA molecule that consists of 6390 nucleotides (Figure 7-40(b)). The RNA is deeply buried in the protein and each protein subunit interacts with three nucleotides.

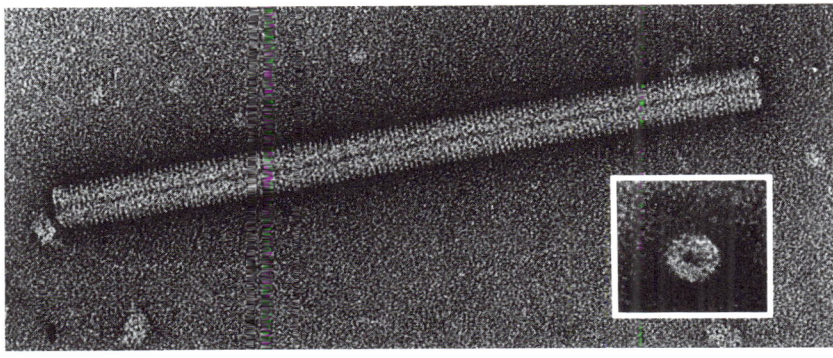

(a)

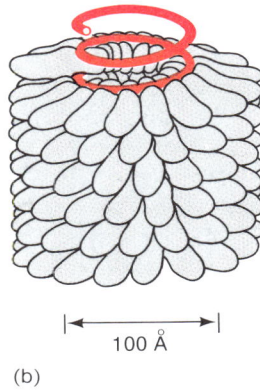

100 Å

(b)

Figure 7-40   (a) Electron micrograph of tobacco mosaic virus. The insert shows a cross-sectional view of a broken particle, in which the hollow central core can be seen. [Courtesy of Robley Williams.] (b) Model of a part of a tobacco mosaic virus particle, showing the helical array of protein subunits around a single-stranded RNA molecule. [After A. Klug and D. L. D. Caspar, *Advan. Virus Res.*, (1960) 7: 274.]

Figure 7-41 Schematic diagram showing the conversion of a TMV protein disc into the helical "lock-washer" form. [After A. Klug, *Fed. Proc.*, (1972) 31: 40.]

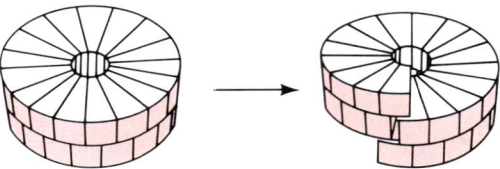

The simplest mechanism for the assembly of TMV would be stepwise addition of single protein subunits to the RNA but such a process would be too slow. About 17 coat subunits would have to be added to a flexible RNA molecule before the complex could close on itself, by forming a turn of the helix, and thereby acquire stability. Instead, a stable complex of many subunits is first formed and then the complex is added to the RNA. The coat-protein subunits spontaneously aggregate to form a two-layered disc consisting of 34 subunits. Each layer of a disc is a ring of 17 subunits, which is nearly the same as the number of subunits ($16\frac{1}{3}$) in one turn of the TMV helix. The two-layered disc is a key intermediate in the assembly of TMV.

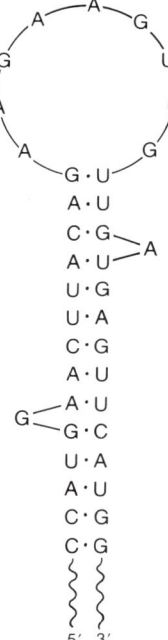

Figure 7-42 A portion of the initiation region in TMV RNA for the assembly of the virus particle.

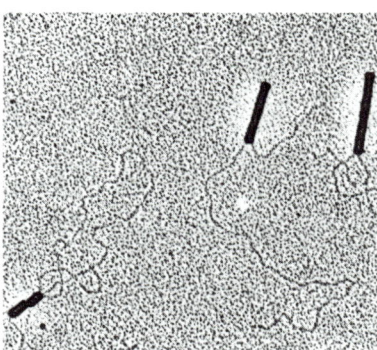

Figure 7-43 Electron micrograph of partially reconstituted tobacco mosaic virus particles. Note that two branches of the RNA emerge from each growing particle indicating that assembly does not begin at a terminus of the RNA molecule. From the relative lengths of the branches the position on the RNA at which assembly begins can be determined. The length of the longest rod is about 2000 Å. [Courtesy of G. Lebeurier and A. Nicolaieff.]

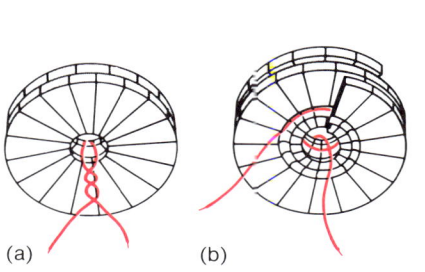
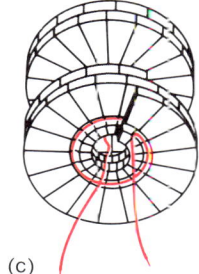
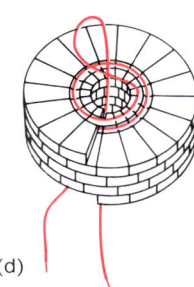

(a)　　　(b)　　　(c)　　　(d)

Figure 7-44 Model for the assembly of tobacco mosaic virus. (a) The initiation region of the RNA is looped into the central hole of the protein disc yielding (b) the helical "lock-washer" form. (c) Additional discs add to the looped end of the RNA. (d) One of the RNA tails is continually pulled through the central hole to interact with incoming discs.

An important property of the disc is that the subunits can slide over each other to form a two-turn helix; the disc resembles and is called a lock-washer (Figure 7-41).

The two-layer disc interacts rapidly with TMV RNA. A particular base sequence is specifically recognized by the disc and serves to initiate assembly of the virus. This initiation region was first isolated by adding a few discs to TMV RNA to coat the initiation region and then digesting the rest of the RNA with an RNase. After such treatment the protected fragment always contained a common core of about 65 nucleotides bound very tightly and specifically to discs. The base sequence of this initiation region strongly suggests that it forms a hairpin structure with a base-paired stem and a loop (Figure 7-42). It is thought that the loop binds the first disc to start the assembly of the virus. The initiation loop is far from either end of the RNA—roughly 5300 nucleotides from the 5′ end and 1000 nucleotides from the 3′ end of the RNA. Two "tails" of RNA emerge from one end of a growing TMV particle (Figure 7-43). The length of the 3′ tail is rather constant throughout most of the assembly process, whereas the 5′ tail becomes shorter as the virus particle becomes longer.

The currently accepted model for the formation of TMV particles is shown in Figure 7-44. Assembly starts with the insertion of the initiation loop into the central hole of a two-layered protein disc. The loop binds to the first turn of the disc and the adjacent base-paired stem opens. This interaction transforms the disc into the helical lock-washer form and traps the viral RNA. The viral helix is now started. Another disc then adds to the newly formed loop of RNA that protrudes from the central hole. A new loop is formed after the addition of each disc by the drawing up of the 5′ tail through the central hole of the growing virus particle. The virus is completed when the 3′ tail becomes coated, which occurs by an unknown mechanism.

The Genetic Material

The genetic material of any organism is the substance that carries the information determining the properties of that organism. Furthermore, it is responsible for transferring genetic information from parent to progeny. In almost all organisms the genetic material is DNA; the exceptions are a few bacteriophages and numerous plant and animal viruses in which the genetic material is RNA.

Before the birth of modern molecular biology it was widely believed that the genetic material consisted wholly of protein molecules. This idea was based on the concept that of all of the known intracellular molecules, only proteins were sufficiently complex and variable in structure and in chemical properties to contain the great variety of information required of the genetic substance. DNA was not considered a possible genetic material, though it was known to be present only in the nuclei of cells; for reasons now only of historical interest, it was thought to be too simple to be a gene and was assumed instead to be a structural component of chromosomes. Two experiments, described later in the chapter, reversed this view and, in so doing, helped to create the science of molecular genetics.

Once the significance of DNA and its remarkable structure (as described in Chapter 4) was appreciated, a variety of experiments were performed that led to what is known as the Central Dogma—that is, a statement of the way in which the information contained in DNA is translated into protein structure. This concept is described in the following section.

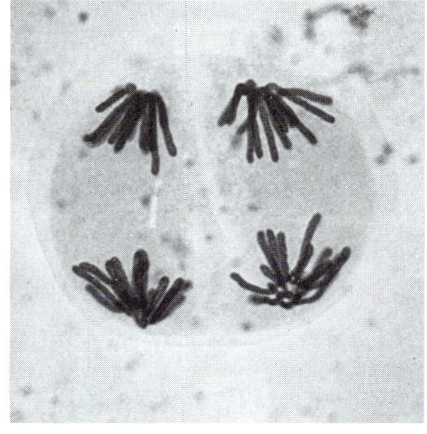

Chromosomes segregating in the second meiotic division during formation of pollen in *Lilium longiflorum*. Such observations gave the earliest indication that genes reside on chromosomes. [Courtesy of Herb Stern.]

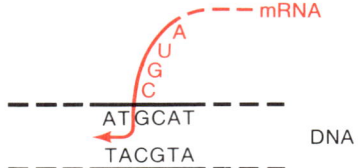

Figure 8-1 Schematic diagram showing how mRNA is copied from DNA. The upper strand of the DNA molecule is being copied and the mRNA molecule is elongating by addition of nucleotides at the end of the molecule denoted by the arrow. As the head of the arrow progresses, the mRNA strand is released at the other end, so at any instant only a few RNA bases are hydrogen-bonded to DNA.

THE CENTRAL DOGMA

What is commonly called the Central Dogma is the statement that *DNA is copied to make RNA and RNA is the template for protein synthesis.* This statement is based on the recognition that (1) amino acids do not bind to DNA molecules, so a cell cannot produce a protein molecule with a particular amino acid sequence simply by aligning amino acids along a DNA molecule; and (2) in the long run, there is much less risk of damage to a cell's single DNA molecule if it is not used directly as a template for the synthesis of each protein molecule. Considering these two points, the Central Dogma states that an intermediate molecule, now known to be a **messenger RNA (mRNA) molecule,** is copied from the DNA (Figure 8-1) and then used repeatedly for protein synthesis. This statement is usually written as *DNA→mRNA→protein.*

THE GENETIC CODE AND THE ADAPTOR HYPOTHESIS

There are only four bases in DNA and twenty amino acids in proteins, so some combination of the bases is needed to specify a particular amino acid. The set of such combinations is called the **genetic code;** this is discussed in detail in Chapter 13. A base sequence corresponding to a particular amino acid is a **codon.**

The surface distinctions among the twenty amino acids are neither specific enough nor sufficiently large for patterns of bases to be recognized by means of van der Waals forces or hydrogen-bonding; thus, recognition requires another intermediate—an adaptor RNA molecule. This adaptor, which we know as **transfer RNA (tRNA),** contains a site to which a particular amino acid is attached, and a base sequence called the **anticodon,** which hydrogen-bonds via complementary base pairing to a given codon (Figure 8-2). Thus the process of protein synthesis may be stated briefly as follows:

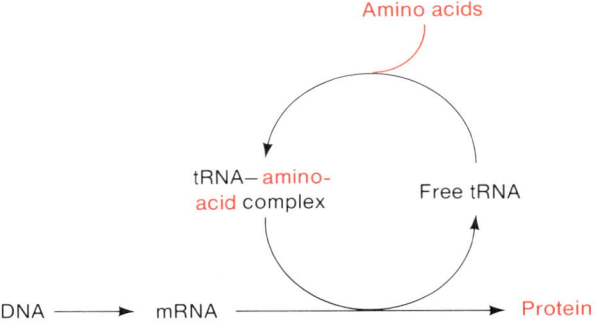

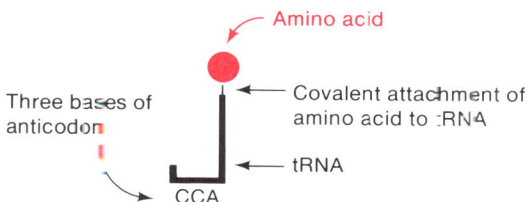

Figure 8-2   A schematic drawing of a tRNA molecule to which an amino acid (red) is covalently linked and which has an anticodon CCA corresponding to a codon UGG.

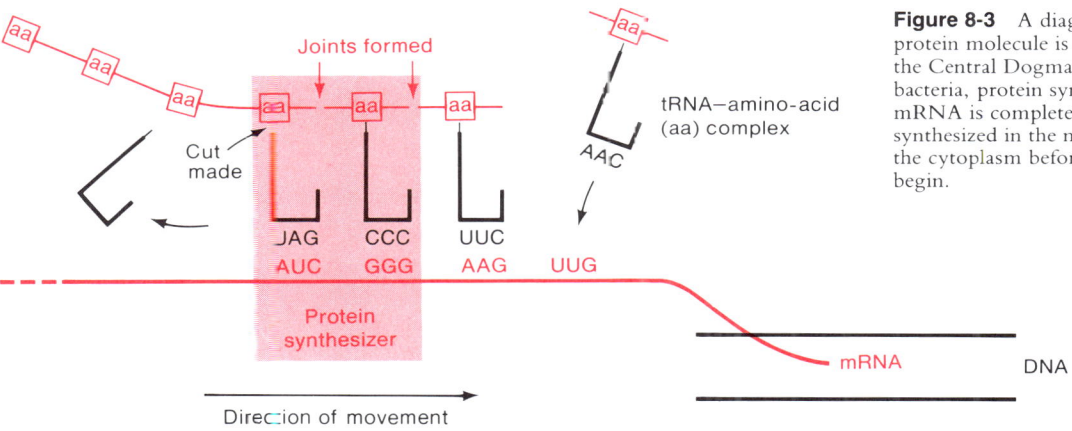

Figure 8-3   A diagram showing how a protein molecule is synthesized according to the Central Dogma. See text for details. In bacteria, protein synthesis begins before the mRNA is completed. In eukaryotes, RNA is synthesized in the nucleus and released into the cytoplasm before protein synthesis can begin.

In outline, the mechanism of protein synthesis can be depicted as in Figure 8-3. An mRNA molecule is copied from a DNA molecule and then is laid onto the surface of a protein-synthesizing unit. Elsewhere appropriate tRNA–amino-acid complexes have formed and these are then aligned along the mRNA molecule. Peptide bonds are made between successive amino acids; finally, the tRNA–amino-acid bond is broken and the completed protein is removed. Details of these processes, which are oversimplified in the figure, are presented in Chapters 12 through 14.

IDENTIFICATION OF DNA AS THE GENETIC MATERIAL

Three experiments, simple but done with great care, identified DNA as genetic material. They are presented in this section both for historical interest and to enable the reader to appreciate the revolutionary effect they had on biological thinking.

The Transformation Experiments

The development of the current idea that DNA is the genetic material began with an observation in 1928 by Fred Griffith, who was studying the bacterium responsible for human pneumonia—i.e., *Streptococcus pneumoniae* or *Pneumococcus*. The virulence of this bacterium was known to be dependent on a surrounding polysaccharide capsule that protects it from the defense systems of the body. This capsule also causes the bacterium to produce smooth-edged (S) colonies on an agar surface. It was known that mice were normally killed by S bacteria. Griffith isolated a rough-edged (R) colony mutant, which proved to be both nonencapsulated and nonlethal. He subsequently made a significant observation—namely, whereas both R and heat-killed S were nonlethal, a mixture of live R and heat-killed S was lethal. Furthermore, the bacteria isolated from a mouse that had died from such a mixed infection were only S—i.e., the live R had somehow been replaced by or **transformed** to S bacteria. Several years later it was shown that the mouse itself was not needed to mediate this transformation because when a mixture of R and heat-killed S was grown in a culture fluid, living S cells were produced. A possible explanation for this surprising phenomenon was that the R cells restored the viability of the dead S cells; but this idea was eliminated by the observation that living S cells grew even when the heat-killed S culture in the mixture was replaced by a cell *extract* prepared from broken S cells, which had been freed from both intact cells and the capsular polysaccharide by centrifugation (Figure 8-4). Hence, it was concluded that the cell extract contained a **transforming principle,** the nature of which was unknown.

Figure 8-4 The transformation experiment.

Preparation of transforming principle from S strain

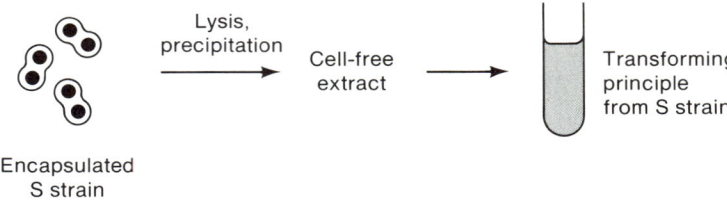

Encapsulated
S strain

Addition of transforming principle to R strain

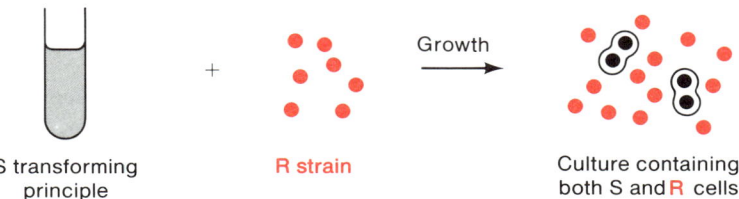

S transforming
principle

R strain

Culture containing
both S and R cells

The next development occurred some 15 years later when Oswald Avery, Colin MacLeod, and Maclyn McCarty partially purified the transforming principle from the cell extract and demonstrated that it was DNA. These workers modified known schemes for isolating DNA and prepared samples of DNA from S bacteria. They added this DNA to a live R bacterial culture; after a period of time they placed a sample of the S-containing R bacterial culture on an agar surface and allowed it to grow to form colonies. Some of the colonies (about 1 in 10^4) that grew were S type (Figure 8-4). To show that this was a permanent genetic change, they dispersed many of the newly formed S colonies and placed them on a second agar surface. The resulting colonies were again S type. If an R colony arising from the original mixture was dispersed, only R bacteria grew in subsequent generations. Hence the R colonies retained the R character, whereas the transformed S colonies bred true as S. Because S and R colonies differed by a polysaccharide coat around each S bacterium, the ability of purified polysaccharide to transform was also tested, but no transformation was observed. Since the procedures for isolating DNA then in use produced DNA containing many impurities, it was necessary to provide evidence that the transformation was actually caused by the DNA alone.

This evidence was provided by the following four procedures.

1. Chemical analysis of samples containing the transforming principle showed that the major component was a deoxyribose-containing nucleic acid.
2. Physical measurements showed that the sample contained a highly viscous substance having the properties of DNA.
3. Experiments demonstrated that transforming activity is not lost by reaction with either (a) purified proteolytic (protein-hydrolyzing) enzymes—trypsin, chymotrypsin, or a mixture of both—or (b) ribonuclease (an enzyme that depolymerizes RNA). Thus, the active principle was neither protein nor RNA.
4. Treatment with materials known to contain DNA-depolymerizing activity (DNase) inactivated the transforming principle.

In drawing conclusions from their experiments Avery, MacLeod, and McCarty avoided stating explicitly that DNA was *the* genetic material and concluded at the end of their work only that nucleic acids have "biological specificity, the chemical basis of which is as yet undetermined." The problem they faced in persuading the scientific community to accept their conclusion was that, whatever the genetic material was, it had to be a substance capable of enormous variation in order to contain the information carried by the huge number of genes. At the time, DNA was considered to be a tetranucleotide and this notion certainly seemed incompatible with any strong contention that DNA should be considered the sole component of the genetic material. Furthermore, the consensus was that genes were made of chromosomal protein—an idea that, in the course of a 40-year period, had logically evolved from the recognition that protein composition and structure varied greatly between organisms. For this reason the transformation experiments had little initial impact and those people who supported the genes-as-protein theory posed the following two alternative explanations for the results: (1) the transforming principle might not be DNA but rather one of the proteins invariably contaminating the DNA sample; (2) DNA somehow affected capsule formation directly by acting in the metabolic pathway for biosynthesis of the polysaccharide and permanently altering this pathway.

The first point should have already been discounted by the original work because the experiments showed insensitivity to proteolytic enzymes and sensitivity to DNase. However, since the DNase was not a pure enzyme, the possibility could not be eliminated conclusively. The transformation experiment was repeated five years later by Rollin Hotchkiss with a DNA sample whose protein content was only 0.02 percent, and it was found that this extensive purification did not reduce the transforming activity. This result supported the view of Avery, MacLeod, and McCarty but still did not prove it. The second alternative, however, was cleanly eliminated, also by Hotchkiss, with an experiment in which he transformed a penicillin-sensitive bacterial strain to penicillin resistance. Since penicillin resistance is totally distinct from the rough-smooth character of the bacterial capsule, this showed that the transforming ability of DNA was not limited to capsule synthesis. Interestingly enough, most biologists still remained unconvinced that DNA was the genetic material and it was not until Erwin Chargaff showed in 1950 that a wide variety of chemical structures in DNA were possible—thus allowing biological specificity—that this idea was thoroughly accepted.

The Chemical Experiments

The hypothesis of a tetranucleotide structure for DNA arose from the belief that DNA contained equimolar quantities of adenine, thymine, guanine, and cytosine. This incorrect conclusion arose for two reasons. First, in the chemical analysis of DNA, the technique used to separate the bases before identification did not resolve them very well, so the quantitative analysis was poor. Second, the DNA analyzed was usually isolated from eukaryotes (and in eukaryotes the four bases are nearly equimolar) or from bacteria whose DNA happened to have nearly equal amounts. Using the DNA of a wide variety of organisms, Chargaff applied new separation and analytical techniques and showed that the molar concentrations of the bases could vary widely. Thus it was demonstrated that DNA could have variable composition, a primary requirement for a genetic substance. Upon publication of Chargaff's results, the tetranucleotide hypothesis quietly died and the DNA-gene idea began to catch on. Shortly afterward, workers in several laboratories found that, for a wide variety of organisms, somatic cells have twice the DNA content of germ cells, a characteristic to be expected of the genetic material, given the tenets of classical chromosome genetics. Although it could apply just as well to any component of chromosomes, once this result was revealed, objections to the work of Avery, MacLeod, and McCarty were no longer heard, and the hereditary nature of DNA rapidly became the fashionable idea.

The Blendor Experiment

An elegant confirmation of the genetic nature of DNA came from an experiment with *E. coli* phage T2. This experiment, known as the **blendor experiment** because a kitchen blendor was used as a major piece of apparatus, was performed by Alfred Hershey and Martha Chase, who demonstrated that the DNA injected by a phage particle into a bacterium contains all of the information required to synthesize progeny phage particles.

A single particle of phage T2 consists of DNA (now known to be a single molecule) encased in a protein shell (Figure 8-5(a)). The DNA is the only phosphorus-containing substance in the phage particle; the proteins of the shell, which contain the amino acids methionine and cysteine, have the only sulfur atoms. Thus, by growing phages in a nutrient medium in which radioactive phosphate ($^{32}PO_4^{3-}$) is the sole source of phosphorus, phage containing radioactive DNA can be prepared. If instead the growth medium

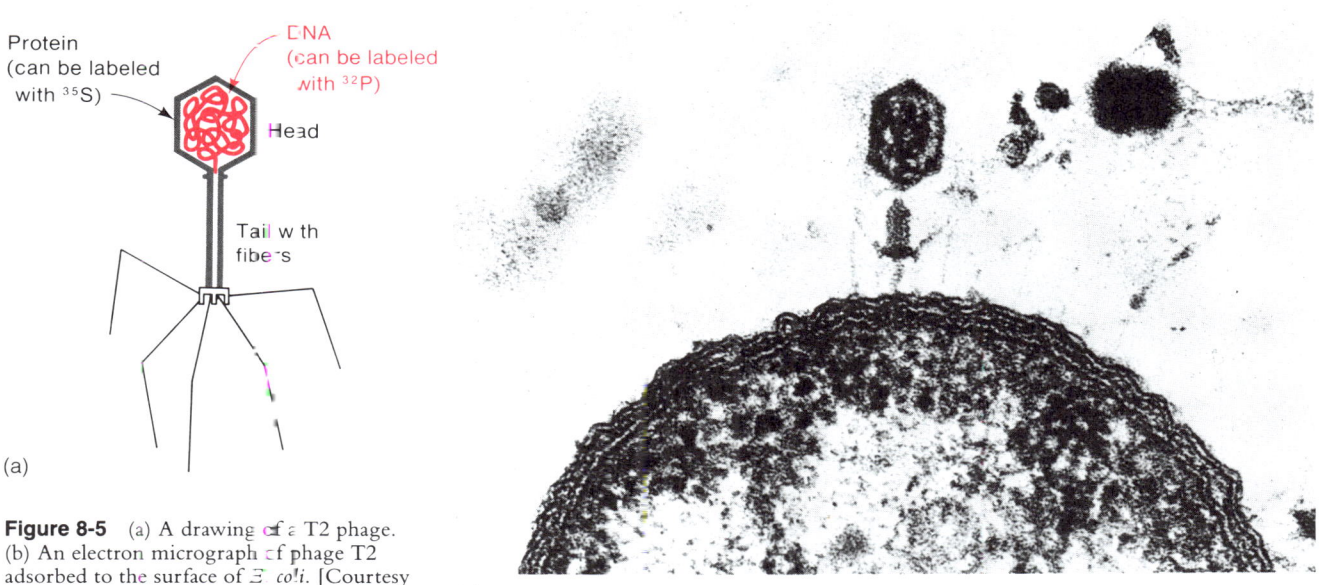

(a)

Protein
(can be labeled
with ^{35}S)

DNA
(can be labeled
with ^{32}P)

Head

Tail with
fibers

Figure 8-5 (a) A drawing of a T2 phage. (b) An electron micrograph of phage T2 adsorbed to the surface of *E. coli*. [Courtesy of Lee Simon and Thomas Anderson.]

(b)

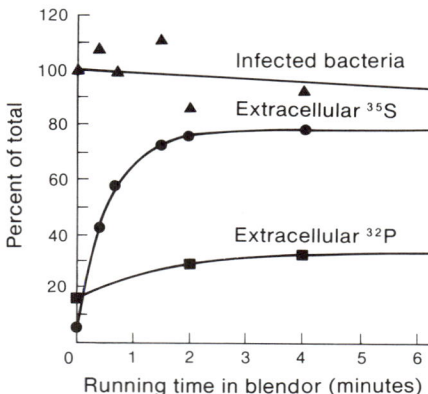

Figure 8-6 Data from the blendor experiment. The ordinate presents the percent of ^{35}S and ^{32}P removed from bacteria infected with radioactively labeled T2 phage and the percent of infected bacteria surviving as infective centers, after the times of agitation in a blendor shown on the abscissa. [After A. D. Hershey and M. Chase, *J. Gen. Physiol.*, (1952) 36: 39.]

contains radioactive sulfur as $^{35}SO_4^{2-}$, phage containing radioactive proteins are obtained. If these two kinds of labeled phage are used in an infection of a bacterial host, the phage DNA and the protein molecules can always be located by their radioactivity. Hershey and Chase used these phage to show that ^{32}P but not ^{35}S is injected into the bacterium.

Each phage T2 particle has a long tail by which it attaches to sensitive bacteria (Figure 8-5(b)). Hershey and Chase showed that an attached phage can be torn from a bacterial cell wall by violent agitation of the infected cells in a kitchen blendor. Thus, it was possible to separate an adsorbed phage from a bacterium and determine the component(s) of the phage that could not be shaken free by agitation—presumably, those components had been injected into the bacterium.

In the first experiment ^{35}S–labeled phage particles were adsorbed to bacteria for a few minutes. The bacteria were separated from unadsorbed phage and phage fragments by centrifuging the mixture and collecting the sediment (the pellet), which consisted of the phage-bacterium complexes. These complexes were resuspended in liquid and blended. The suspension was again centrifuged; and the pellet, which now consisted almost entirely of bacteria, and the supernatant were collected. It was found that 80 percent of the ^{35}S was in the supernatant and 20 percent was in the pellet (Figure 8-6). The 20 percent of the ^{35}S that remains associated with the bacteria was shown many years later to consist mostly of phage tail fragments that adhered too tightly to the bacterial surface to be removed by the blending. A very different result was observed when the phage population was labeled with ^{32}P. In this case 70 percent of the ^{32}P remained associated with the bacteria in the pellet after blending and only 30 percent was in the supernatant. Of the radioactivity in the supernatant roughly one-third could be accounted for by breakage of the bacteria during the blending. (The remainder was shown some years later to be a result of defective phage particles that could not inject their DNA.) When the pellet material was resuspended in growth medium and reincubated, it was found to be capable of phage production. Thus, the ability of a bacterium to synthesize progeny phage is associated with transfer of ^{32}P, and hence of DNA, from parental phage to the bacteria (Figure 8-7).

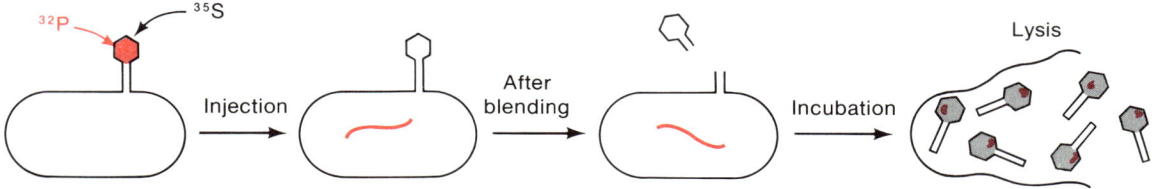

Figure 8-7 The blendor experiment. The parental DNA molecule is drawn in red. During incubation the DNA replicates and progeny phage are produced. Some, but not all, of the parental DNA appears in these progeny. Although not known to Hershey and Chase, progeny DNA molecules engage in genetic recombination resulting in dispersal of parental DNA among the progeny phage, as shown.

Another series of experiments, known as **transfer experiments,** supported the interpretation that genetic material contains ^{32}P but not ^{35}S. In these experiments progeny phage were isolated, after blending, from cells that had been infected with either ^{35}S- or ^{32}P-containing phage and the progeny were then assayed for radioactivity, the idea being that some parental genetic material should be found in the progeny. It was found that no ^{35}S but about half of the injected ^{32}P was transferred to the progeny. This result indicated that though ^{35}S might be residually associated with the phage-infected bacteria, it was not part of the phage genetic material. The interpretation (now known to be correct) of the transfer of only half of the ^{32}P was that progeny DNA is selected at random for packaging into protein coats and that all progeny DNA is not successfully packaged.

The Question of RNA as a Genetic Substance

The transformation and blendor experiments settled once and for all the question of the chemical identity of the genetic material. The absolute generality of the conclusion remained in question, though, because several plant and animal viruses were known to contain single-stranded RNA and no DNA. These particles became understandable shortly afterwards as the role of RNA in the flow of information from gene to protein became clear—that is, that DNA stores genetic information for protein synthesis and the pathway from DNA to protein always requires the synthesis of an RNA intermediate which is copied from a DNA template. Thus, an organism that lacks DNA utilizes the base sequence of RNA both for storage of information and as a template from which the amino acid sequence of proteins can be obtained. RNA does not serve these two functions as efficiently as the DNA→RNA system, but it does work.

In the following section we discuss the properties that should be possessed by the genetic material and indicate how DNA satisfies the criteria.

PROPERTIES OF THE GENETIC MATERIAL

The genetic material must have the following properties:

1. Ability to store genetic information and to transmit it to the cell as needed.
2. Ability to transfer its information to daughter cells with minimal error.
3. Physical and chemical stability, in order that information is not lost.
4. Capability for genetic change, though without major loss of parental information.

We will see in the following sections that DNA is particularly suited to be the genetic material.

Storage and Transmission of Genetic Information by DNA

The information possessed and conveyed by the DNA of a cell is of several types:

1. The sequence of amino acids in every protein synthesized by the cell.
2. A start and a stop signal for the synthesis of each protein.
3. A set of signals that determine whether a particular protein is to be made and how many molecules are to be made per unit time.

This information is contained in the sequence of DNA bases. The amino acid sequence and the start and stop signals are not obtained directly from the DNA base sequence but via RNA intermediates. That is, DNA serves as a template for synthesis of specific RNA molecules (mRNA). The base sequence of the mRNA is complementary to that of one of the DNA strands and it is from the mRNA sequence that the amino acid sequence is translated. In particular, the base sequence is "read" in order in groups of three bases called **triplets** or **codons,** and each group corresponds to a particular amino acid or to a start or stop codon. This two-stage process has the advantage that the DNA molecule neither has to be used very often nor has to enter the protein-synthetic mechanism. Since protein synthesis occurs continually, one can appreciate why DNA evolved with bases having groups capable of forming hydrogen bonds. First, by specific patterns of hydrogen-bonding— that is, base-pairing—the base sequence of DNA is easily transcribed into an RNA molecule by means of an enzymatic system that adds a base to the polymerizing end of an RNA molecule only if that base can hydrogen-bond with the base being copied. This system of transcription is discussed in detail in Chapter 12. The use of hydrogen-bonding enables the cell to use less genetic material to hold information than if van der Waals forces had been selected in the course of evolution; this is because van der Waals forces are so weak that the error frequency in transmitting information would be much greater than with hydrogen bonds unless more bases (or other molecules) were used for complementary binding.

The process of base recognition is facilitated by the breathing (or localized denaturation) of DNA. That is, while the base pairs are temporarily broken, a polymerizing enzyme (RNA polymerase) can penetrate the helix and read the base sequence, as must be necessary for RNA synthesis.

Specific base sequences are also recognized by particular protein molecules, which are regulatory elements, that determine whether a protein is to be made. These proteins bind directly to sequences of six or more base pairs but do not require that the base pairs separate, as they bind in external grooves of the DNA molecule (Chapter 7). These regulatory proteins recognize specific configurations of electrons and bind by means of electrostatic and van der Waals forces, hydrophobic interactions, and hydrogen bonds.

Transmission of Information from Parent to Progeny

When a cell divides, each daughter cell must contain identical genetic information; that is, each DNA molecule must become two identical molecules, each carrying the information that was contained in the parent molecule. This duplication process is called **replication.** Once again, the value of bases capable of hydrogen-bonding is apparent. That is, to make a replica, the replication system need only require that the base being added to the growing end of a new chain be capable of hydrogen-bonding to the base being copied.

Chemical Stability of DNA and of Its Information Content

In long-lived organisms a single DNA molecule may have to last one hundred years or more. Furthermore, the information contained in the molecule is passed on to successive generations for millions of years with only small changes. Thus, DNA molecules must have great stability.

The sugar-phosphate backbone of DNA is extremely stable. The C—C bonds in the sugar are resistant to chemical attack under all conditions other than strong acid at very high temperatures. The phosphodiester bond is a little less stable and can be hydrolyzed at room temperature at pH 2, but this is not a physiological condition. In considering the stability of the phosphodiester bond, one can see immediately why 2'-deoxyribose rather than ribose has evolved as the constituent sugar in DNA. At alkaline pH, RNA hydrolyzes to free nucleotides exceedingly rapidly (see Chapter 4). Even at pH 8, RNA molecules are gradually fragmented by phosphodiester-bond breakage. The reaction mechanism, which is called a β-elimination reaction, requires that a hydroxyl group be present on the sugar. Except for the 3'-OH terminal nucleotide, there is only a single OH group in the RNA backbone, namely, the 2'-OH group. Thus, in DNA, which uses 2'-deoxyribose, there is no free OH group and hydrolysis by means of this pathway is not possible. This has the effect that at room temperature and pH range 8–9 there is no detectable hydrolysis of DNA (less than one bond broken per million bonds per month).

An alteration in the chemical structure of a base definitely means loss of genetic information. In a cell there are certainly a large number of chemical compounds that can attack a free base. In considering this problem one can see immediately the value of the double helix. One aspect of this is that the molecule is redundant in the sense that identical information is contained in both strands; that is, the base sequence of one strand is complementary to that of the other strand. In fact, there exist in cells elegant repair systems that can remove an altered base and then by reading the sequence on the complementary strand, can replace the correct base. (Repair will be discussed in detail in Chapter 10.) The more important aspect of this double-helical structure is that the nature of the bases and the duplex structure of DNA provides extreme protection against chemical attack. The bases are hydrophobic rings having charged groups that contain the genetic information. It is therefore the charged groups that need protection. Hydrogen-bonding between each of these groups provides the first line of defense. The hydrophobicity of the bases causes stacking and this vastly reduces the area of the base that is vulnerable to attack. Hydrogen-bonding of the two strands places all of the stacked bases into one gigantic cylindrical cluster in the central region of which are the bases. DNA stacks so extensively that water is almost completely excluded from a stacked array of bases. Since any potentially harmful compound must usually be water-soluble, these molecules will find it difficult to come into close contact with the "dry" stack of bases. Certainly the likelihood of reaching the buried charged groups is small.

The N-glycosidic bond, which joins a DNA base to the 1'-carbon of a sugar, is stable except at extremes of pH. The bond would be less stable if the bases were not rings. Methylation of the purine ring on the 7-nitrogen decreases chemical stability enormously; this is the mode of action of the highly mutagenic and destructive nitrogen and sulfur mustards used in chemical warfare in World War I—namely, any of these various mustards provides an alkyl group that attaches to the 7-nitrogen of guanine. The guanine then falls off the sugar, leaving a hydroxyl group on the 1'-carbon, allowing phosphodiester hydrolysis to occur.

The bases themselves, with the exception of cytosine, are very stable. At a very low rate, however, cytosine is deaminated to form uracil.

Deamination is a disastrous change because the deamination product, uracil, pairs with adenine rather than with guanine. This has two effects: (1) an incorrect base will appear in mRNA and (2) an adenine instead of a guanine will occur in newly replicated DNA strands (Figure 8-8). There exists an intracellular system, which will be described in Chapter 9, for removing uracil from DNA and replacing it with thymine. The necessity for eliminating a uracil formed by deamination explains why DNA utilizes thymine and not uracil as the base complementary to adenine. If uracil were the normal DNA base, there would be no way to distinguish a correct uracil from an incorrect uracil produced by deamination of cytosine. By using thymine, the cell can follow the rule: *Always remove a uracil from DNA because it is unwanted.*

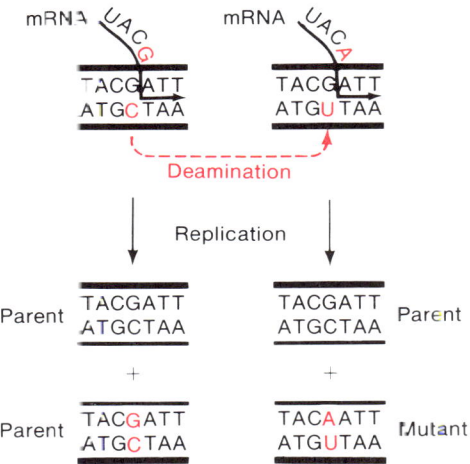

Figure 8-8 The effect of deamination of cytosine to form uracil in the base sequences of mRNA and of two daughter DNA molecules. The C→U transition is shown in red in the uppermost panel. "Parent" and "mutant" refer to base sequences before and after deamination. Newly replicated DNA is indicated by a thin line.

It is not obvious why RNA uses uracil and not thymine nor why DNA evolved with cytosine rather than with some base that would not deaminate. This may have been an evolutionary accident. It is possible, though, that the original DNA and RNA molecules both contained cytosine and uracil because these were the only pyrimidines available in the primordial sea. Cells then gained the ability to methylate uracil to form thymine because this provides a cell with a criterion for eliminating the result of a cytosine→uracil conversion. It is significant that the final step in the synthesis of thymidylic acid is the methylation of deoxyuridylic acid. The thymidylic acid is then converted to the triphosphate needed for incorporation into DNA (see Chapter 9).

The Ability of DNA to Change: Mutation

All of the information contained in a cell resides in its DNA. Thus, if a cell is to be able to improve through time—that is, to evolve—then the base sequence of its DNA must be capable of change. Furthermore, the new sequence must persist so that progeny cells will have the new property. The process by which a base sequence changes is called **mutation.** There are two main mechanisms of mutation:

1. A **chemical alteration** of the base that gives it new hydrogen-bonding properties and causes a new base to be present in a newly replicated daughter molecule.

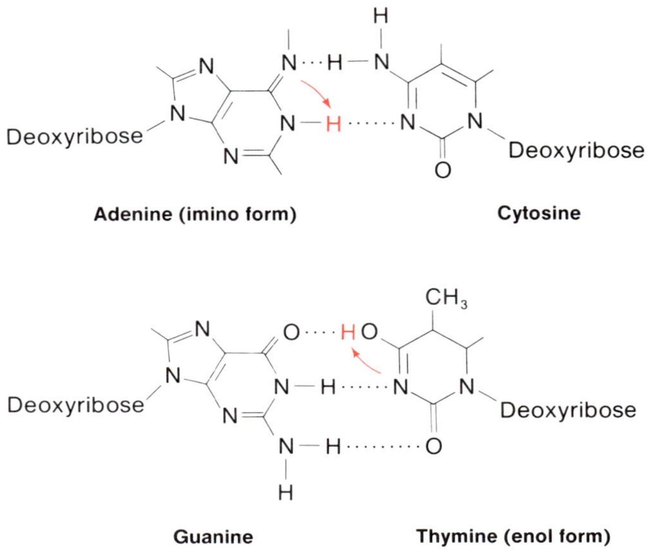

Figure 8-9 Base-pairing between the rare imino form of adenine and cytosine and the enol form of thymine and guanine. The red H is the one that has moved from the more common position. Compare with Figure 4-2, which shows the standard base pairs.

2. A **replication error** by which an incorrect base or an extra base is accidentally inserted in a daughter molecule.

On the average, mutational changes are deleterious and lead to cell death. Therefore, it is important that not too many mutations occur in a single DNA molecule; otherwise the rare advantageous alteration would always occur in a cell destined to die by virtue of lethal mutations. Thus, the mutation rate must be low or controlled. This is accomplished in two ways. First, the hydrophobic, water-free core of the DNA molecule reduces the accessibility of the DNA to attacking molecules, as we discussed earlier. Second, the cell has evolved several repair mechanisms (discussed in Chapter 10) for correcting alterations and replication errors. These repair systems are not entirely efficient though and allow mutations to occur at a rate that is very low but useful in the long run. These mutations are, as we have said, usually deleterious, so it is important that the parental information is not lost. Such loss is prevented in two ways: (1) The other members of the species retain the parental base sequence, and (2) a double-stranded molecule is redundant. Normally only one strand is altered, and DNA replicates in such a way that after cell division the DNA molecule in each daughter cell receives only one of the parental single strands. Thus, it is possible for one of the daughter DNA molecules to be normal and the other to be mutant; the mutant may not be able to survive but the cell with the parental base sequence will. If the mutant is better equipped to survive and multiply than the parent organisms or any other member of the species, then after a great many generations, Darwin's principle of survival of the fittest will lead to ultimate replacement of the parental genotype by the mutant phenotype in nature.

The second mechanism of mutation—replication errors—follows from the tautomeric properties of the bases adenine and thymine. The rare imino form of adenine can form a stable hydrogen bond with cytosine and the enol form of thymine can pair with guanine (Figure 8-9). It may be the case that in the primordial seas nucleic acid molecules having other bases also arose spontaneously but only those having tautomeric bases survived because their mutability allowed adaptation to changing circumstances and evolution.

9

DNA Replication

Genetic information is transferred from parent to progeny organisms by a faithful replication of the parental DNA molecules. Usually the information resides in one or more double-stranded DNA molecules. Some bacteriophage species contain single-stranded instead of double-stranded DNA. In these systems replication consists of several stages in which single-stranded DNA is first converted to a double-stranded molecule, which then serves as a template for synthesis of single strands identical to the parent molecule. Viruses containing single- and double-stranded RNA molecules are also known; these organisms use several different modes of replication, some of which include double-stranded DNA as an intermediate. The modes of replication of each of these types of molecules differ in detail, though certain fundamental features are common to each mode.

Replication of double-stranded DNA is a complicated process that is not completely understood. This complexity results in part from the following facts: (1) a supply of energy is required to unwind the helix; (2) the single strands resulting from the unwinding tend to form intrastrand base pairs; (3) a single enzyme can catalyze only a limited number of physical and chemical reactions and many reactions are needed in replication; (4) several safeguards have evolved that are designed both to prevent replication errors and to eliminate the rare errors that do occur; and (5) both circularity and the enormous size of DNA molecules impose geometric constraints on the replicative system, and how these fit into the system has to be understood.

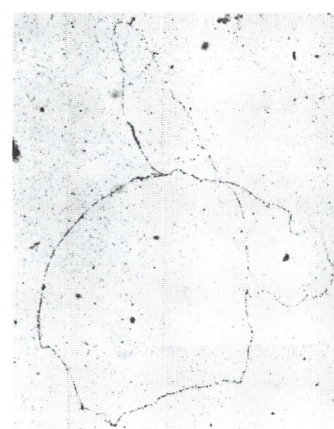

Autoradiogram of the intact replicating chromosome of an *E. coli* bacterium that has been allowed to incorporate [^{3}H] thymine into its DNA for slightly less than two generation periods. The continuous lines of dark grains were produced by electrons emitted during a two-month storage period by decaying ^{3}H atoms in the DNA molecule [From J. Cairns, *Cold Spring Harbor Symp. Quant. Biol.* (1963) 28: 44.]

To add to the difficulty for the researcher, there is not a unique mode of replication common to all organisms having double-stranded DNA.

This chapter begins by examining a few general features of the replication process but then, out of necessity, we will plunge into complicated topics. An attempt has been made to explain the sequence of events in a logical order; however, the reader will probably find that a full understanding of some stages of replication can only be had after understanding certain phenomena associated with other stages and described elsewhere in the chapter.

THE BASIC RULE FOR REPLICATION OF ALL NUCLEIC ACIDS

All genetically relevant information contained in any nucleic acid molecule resides in its base sequence, so the prime role of any mode of replication is to duplicate the base sequence of the parent molecule. The specificity of base pairing—adenine with thymine and guanine with cytosine—provides the mechanism used by all replication systems. Furthermore,

1. Nucleotide monomers are added one by one to the end of a growing strand by an enzyme called a **DNA polymerase.**
2. The sequence of bases in each new or **daughter strand** is complementary to the base sequence in the old or **parent strand** being copied—that is, if there is an adenine in the parent strand, a thymine will be added to the end of the growing daughter strand when the adenine is being copied.

In the following section we consider how the two strands of a daughter molecule are physically related to the two strands of the parent molecule.

THE GEOMETRY OF DNA REPLICATION

The production of daughter DNA molecules from a single parental molecule gives rise to several topological problems, which result from the helical structure and great size of typical DNA molecules and the circularity of many DNA molecules. These problems and their solutions are described in this section.

Semiconservative Replication of Double-Stranded DNA

The purpose of DNA replication is to create daughter DNA molecules that are identical to the parental molecule. Originally, two modes of replication were suggested which differ in whether or not the two single strands of the parent molecule become rearranged after one round of replication.* These were called the **semiconservative** and **conservative** modes.

In the semiconservative mode, first proposed by Watson and Crick, each parental DNA strand serves as a template for one new or daughter strand and as each new strand is formed, it is hydrogen-bonded to its parental template (Figure 9-1). Thus, as replication proceeds, the parental double helix unwinds and then rewinds again into two new double helices, each of which contains one originally parental strand and one newly formed daughter strand. At the time this model was proposed, DNA denaturation was not understood and strand separation was, for a variety of reasons, thought to be impossible. This provided the impetus for proposing the conservative mode. In the conservative mode the two strands of the parental molecule remain entwined before and after replication and unwind at the replication site only as much as is sufficient to allow the base sequence there to be read by the polymerizing enzyme. From this reading a single daughter molecule is produced; the parental molecule remains intact. Thus, in conservative replication, complete unwinding of the parental helix is not necessary. The correctness of the semiconservative model was demonstrated by the following experiment.

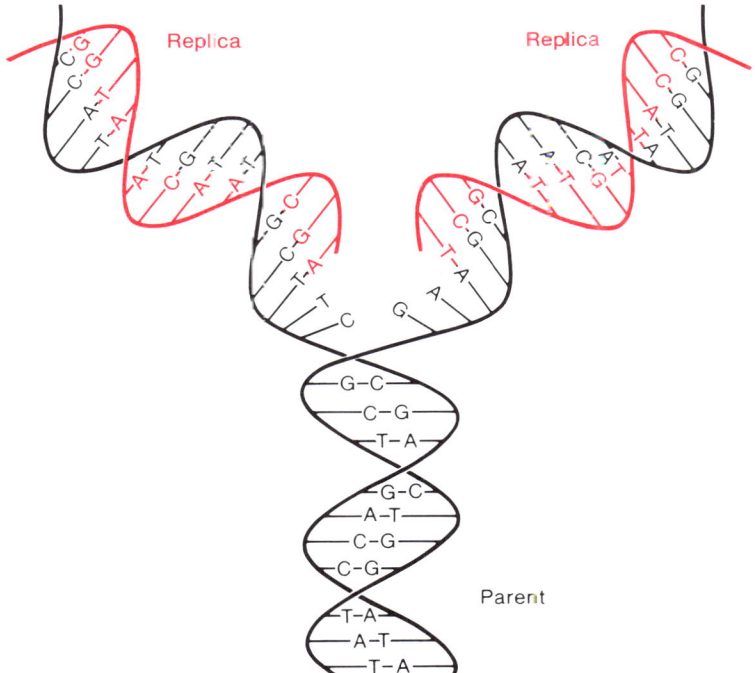

Figure 9-1 The replication of DNA according to the mechanism proposed by Watson and Crick. The two replicas consist of one parental strand (black) plus one daughter strand (red). Each base in a daughter strand is selected by the requirement that it form a base pair with the parental base.

*The term "round of replication" refers to a single transit of the replication system along a DNA molecule.

The conservative and semiconservative modes differ in the composition of daughter molecules. In the conservative mode, of any two "daughter" DNA molecules one is in fact also the original "parent" molecule, and in the semiconservative mode, the parental strands are distributed equally between the two daughter molecules, one parental strand forming one of the two strands in each daughter. Matthew Meselson and Franklin Stahl developed a simple method by which the parental and daughter strands could be distinguished. A culture of bacteria was grown for many generations in a growth medium containing ^{15}N-labeled NH_4Cl as the sole source of nitrogen (called a "heavy medium"). In this way the parental DNA molecules were labeled with the heavy isotope, ^{15}N, thereby increasing the density of the DNA. The cells were then transferred to a medium containing the common isotope of nitrogen, ^{14}N (light medium). At various times after the transfer, samples of cells were collected and the DNA was isolated. The DNA molecules were fragmented during isolation; Figure 9-2 represents the composition of the fragments at one and two generations after the density shift. In the semiconservative mode after one generation all daughter molecules would have one ^{15}N strand and one ^{14}N strand (this is called a **hybrid** molecule); hence all daughter molecules would have the same density (hybrid density)—namely

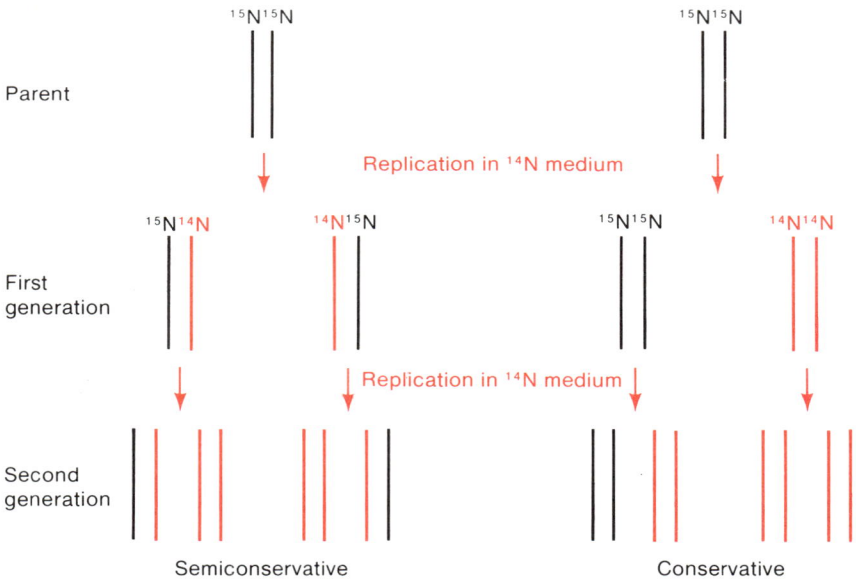

Figure 9-2 Distribution of parental ^{15}N-labeled strands (black) during two rounds of replication in ^{14}N (red) medium, according to the semiconservative and conservative modes of replication.

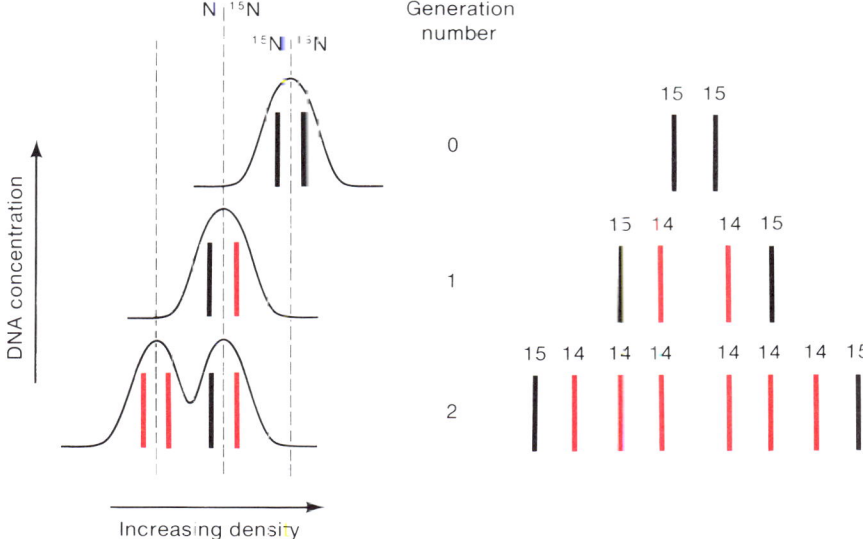

Figure 9-3 The type of data obtained in the Meselson-Stahl experiment. *E. coli* were grown for many generations in ^{15}N medium (black) and then at zero time transferred to ^{14}N medium (red). After one generation, all DNA had hybrid density. DNA of less than hybrid density is not seen before the first generation, because the chromosome is fragmented into about 200 pieces during isolation. At the right, the state of the isolated DNA at various times is shown.

midway between that of $[^{15}N^{15}N]$ and $[^{14}N^{14}N]$DNA molecules.* In the conservative mode both $[^{15}N^{15}N]$DNA and $[^{14}N^{14}N]$DNA molecules would result; hence, the two first-generation daughter molecules (one of which is the parent molecule) would differ in density. After two generations the predicted distributions in density also differ, one from the other, as shown in the figure. Meselson and Stahl used density gradient centrifugation in CsCl (Chapter 3) to measure the densities of the molecules present as a function of time after the change from heavy to light medium. Their results, shown in Figure 9-3, that all DNA had a hybrid density after one round of replication, indicated that the semiconservative mode is correct.

A second experiment confirmed the structure of the $[^{14}N^{15}N]$DNA found after one generation. In this experiment the hybrid DNA was denatured by heating to 100°C and then centrifuged in CsCl. The heated DNA yielded two bands having the densities of denatured (single-stranded) $[^{14}N]$- and $[^{15}N]$DNA (Figure 4-10), proving that the DNA of hybrid density did in fact consist of one ^{14}N strand and one ^{15}N strand.

*Double-stranded DNA molecules containing either ^{14}N or ^{15}N in *both* strands are denoted $[^{14}N^{14}N]$DNA and $[^{15}N^{15}N]$DNA, respectively. A molecule containing ^{14}N in one strand and ^{15}N in the other is denoted $[^{14}N^{15}N]$DNA.

The Geometry of Semiconservative Replication

Unwinding a double helix in semiconservative replication presents a mechanical problem. Either the two daughter branches at the Y-fork shown in Figure 9-1 must revolve around one another or the unreplicated portion must rotate. If the molecule were fully extended in solution, there would be no problem and rotation of the unreplicated portion would be the simpler motion, as there would be less friction with the solvent. However, since in *E. coli* the molecule is 600 times longer than the cell that contains it (so that it must be repeatedly folded, as was shown in Chapter 7), such rotation is unlikely. A simple solution would be to make a single-strand break in one parental strand ahead of the growing fork; this would enable a small segment of the unreplicated region to rotate, thereby eliminating the geometric problem (Figure 9-4). The only requirement would then be to re-form the broken bond. However, this repair would have to occur before the replication fork had passed the nick; otherwise a daughter strand would be lost, as shown in the figure.

In bacteria most DNA molecules replicate as circular structures. This introduces a geometric problem that is more severe than that just described. This problem is solved by an enzyme, DNA gyrase, which will be discussed in the next section.

The Geometry of the Replication of Circular DNA Molecules

Most bacteriophage species contain linear DNA molecules which, by a variety of mechanisms, circularize after infection. The DNA molecules of the bacteria examined to date are also circular. The first demonstration that *E.*

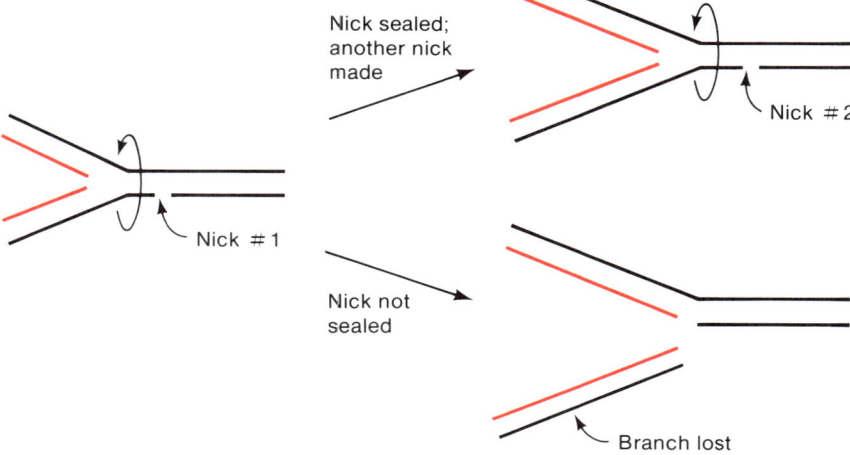

Figure 9-4 Mechanism by which a nick ahead of a replication fork allows rotation. If the nick is not sealed, a newly formed branch is lost.

Nick sealed; another nick made

Nick #2

Nick #1

Nick not sealed

Branch lost

coli DNA replicates as a circle came from an autoradiographic experiment. Cells were grown in a medium containing [^{3}H]thymine so that all DNA synthesized would be radioactive. The DNA was isolated without fragmentation and placed on photographic film. Each ^{3}H-decay exposed one grain in the film and after several months there were enough grains to visualize the DNA with a microscope; the pattern of black grains on the film located the molecule. One of the now-famous autoradiograms from this experiment is shown at the opening of this chapter. Electron micrographs of replicating circular molecules of plasmids, phages, and viruses have also been obtained (Figure 9-5). A replicating circle is schematically like the Greek letter θ (theta), so this mode of replication is usually called **θ replication.**

The unwinding problem in θ replication is formidable because lack of a free end makes rotation of the unreplicated portion impossible. An advancing nick, as in Figure 9-4, or a swivel at the replication origin (the point of initiation of replication) would solve the problem. These suggestions, though not quite correct, anticipate the correct mechanism.

As replication of the two daughter strands proceeds along the helix, in the absence of some kind of swiveling the nongrowing ends of the daughter strands would cause the entire unreplicated portion of the molecule to become overwound (Figure 9-6). This in turn would cause *positive* supercoiling (see Chapter 4) of the unreplicated portion. This supercoiling obviously cannot increase indefinitely because, if it were to do so, the unreplicated portion would become coiled so tightly that no further advance of the replication fork would be possible. This topological constraint could be avoided by the simple nicking-sealing cycle just described but the twists are removed in another way. As discussed in Chapter 4, most naturally occurring circular DNA molecules are *negatively* supercoiled. Thus, initially the overwinding motion is no problem because it can be taken up by the underwinding already present in the negative supercoil. However, after about 5 percent of the circle

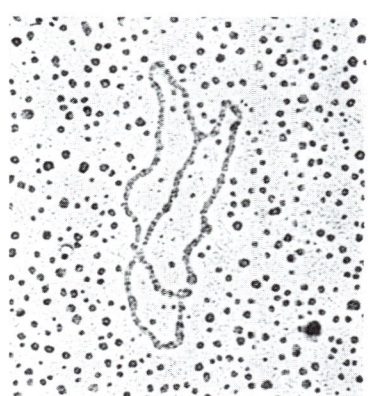

(a)

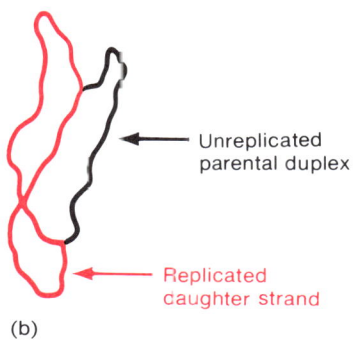

(b)

Unreplicated
parental duplex

Replicated
daughter strand

Figure 9-5 θ replication. (a) Electron micrographs of a ColE1 DNA molecule (molecular weight = 4.2×10^6) replicating by the θ mode. (b) Interpretive drawing showing parental and daughter segments. [Courtesy of Donald Helinski.]

Figure 9-6 Drawing showing that the unwinding motion (curved arrows) of the daughter branches of a replicating circle lacking positions at which free rotation can occur causes overwinding of the unreplicated portion.

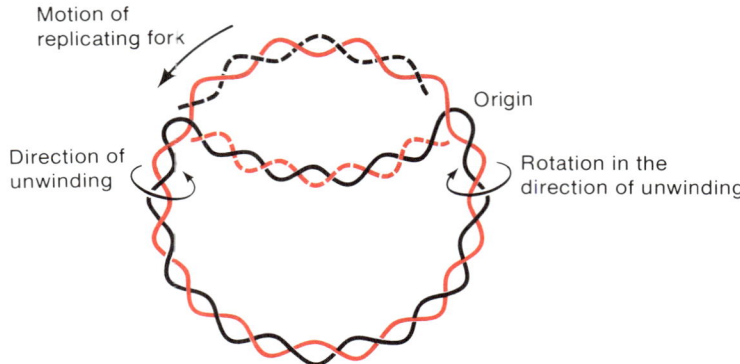

is replicated, the negative superhelicity is used up and the topological problem arises.

Most organisms contain one or more enzymes called **topoisomerases.** These enzymes can produce a variety of topological changes in DNA; the most common are production of negative superhelicity and the removal of superhelicity. In *E. coli* there is an enzyme called **DNA gyrase** (Eco topoisomerase II) which is able to produce negative superhelicity, and it is DNA gyrase that is responsible for removing the positive superhelicity generated during replication. That is, positive superhelicity is removed by gyrase introducing negative twists by binding ahead of the advancing replication fork. (Topoisomerase I, an enzyme that eliminates negative superhelicity, was described briefly in Chapter 7, in the description of the *E. coli* nucleoid. It does not seem to play a role in DNA replication.) The evidence for this point comes from experiments using drugs that inhibit DNA gyrase (e.g., coumermycin, nalidixic acid, oxolinic acid, and novobiocin) and from the study of gyrase mutants. Addition of any of these drugs to a growing bacterial culture inhibits DNA synthesis; in a phage-infected cell they prevent supercoiling of injected phage DNA molecules (that is, the molecules circularize but do not supercoil). Proof that these effects are due to inhibition of DNA gyrase comes from isolating strains of *E. coli* that can grow normally in the presence of one or more of these antibiotics. With these strains the antibiotics have no effect on either DNA replication or supercoiling. If DNA gyrase is isolated from both wild-type *E. coli* and the mutant *E. coli* strain and each form is tested for binding of the antibiotics, it is found that only the wild-type enzyme binds the antibiotic. (This approach of probing for an activity with an inhibitor and comparing *in vitro* and *in vivo* results is used in many problems in molecular biology.)

DNA gyrase does have a DNA-breaking–sealing activity but this activity cannot by itself introduce negative superhelical turns. How twisting is done is explained in the following section.

Mechanism of Action of DNA Gyrase*

If a nonsupercoiled covalent circular DNA molecule is exposed to DNA gyrase and ATP, the DNA molecule undergoes a series of structural transitions, each of which introduces *two* superhelical twists into the molecule. Thus *the total number of superhelical twists introduced into a particular molecule is always an even number.* This fact gave the first clue to understanding how gyrase works.

The mechanism for gyrase activity, called the **sign inversion mechanism,** is shown in Figure 9-7. Let us first examine molecules I and II of the figure. If a single gyrase molecule were to bring together the two sites in molecule I indicated by the arrows and, in so doing, twist the molecule in the direction shown by the circular arrow, a cross-over region or a **node** would be produced. By definition this node would be a positive node. Because the molecule is circular and the base pairs are most stable when maintaining their 36° rotation with respect to one another in the double helix, the lower part of molecule II apparently twists in the opposite sense, generating a negative node. Molecule II would be stable as long as gyrase remained bound to the positive node. At this point the net twisting of the circular DNA molecule would still be zero. In molecule III the portion of the upper node that is farthest from the viewer would then be cleaved (a double-strand break) and moved *in front of* the other portion, which would convert the positive node to a negative node. Sealing of this double-strand break would yield molecule IV, which would have two negative twists.

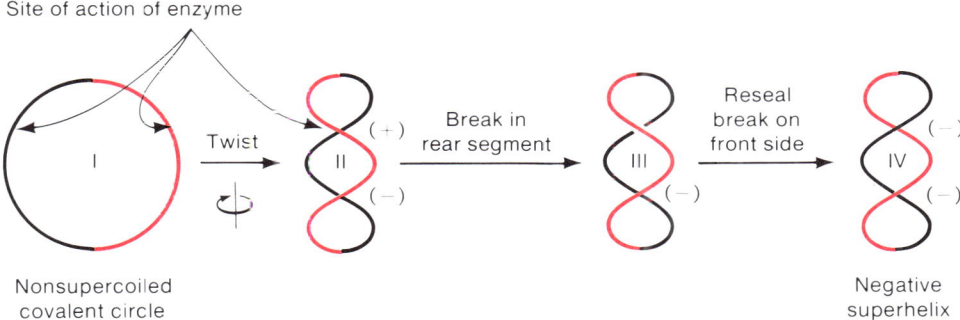

Site of action of enzyme

I — Twist → II (+) (−) — Break in rear segment → III (−) — Reseal break on front side → IV (−) (−)

Nonsupercoiled covalent circle

Negative superhelix

Figure 9-7 The sign inversion model for superhelix formation. The left and right halves of the circle are in different colors only to facilitate seeing whether one segment passes in front of or behind another segment. (+) and (−) refer to positive and negative nodes respectively. Note that the enzyme acts only on the upper node.

*This section can be omitted in a less comprehensive course.

Figure 9-8 Model for negative supercoiling of DNA by gyrase. Step I represents a gyrase molecule consisting of two α and two β (shaded) monomers with a section of a circular duplex DNA wrapped in a right-handed superhelix. The solid red DNA molecule contains the site of double-strand breakage at its middle. The DNA, beginning about 85 bp from the cleavage site, is to be passed through the transient gap. The enzyme contains a hole (concealed in step I) that is exposed by opening the molecule about a hinge between the α subunits (as in step II). To effect supercoiling, a right-handed DNA loop in step I is converted into a left-handed one (step IV) by the following steps: II, DNA breakage and opening of the enzyme; III, passing of DNA through the gap by exchange of the stippled DNA between β protomers; and IV, closing the enzyme and resealing the broken DNA. This sequence of steps introduces two negative supercoils. Step V represents a way of resetting the DNA to complete the cycle; the resealed DNA section is released, followed by reopening of gyrase to allow escape of the passed DNA section. [From A. Morrison and N. R. Cozzarelli, *Proc. Nat. Acad. Sci.*, (1981) 78: 1416.]

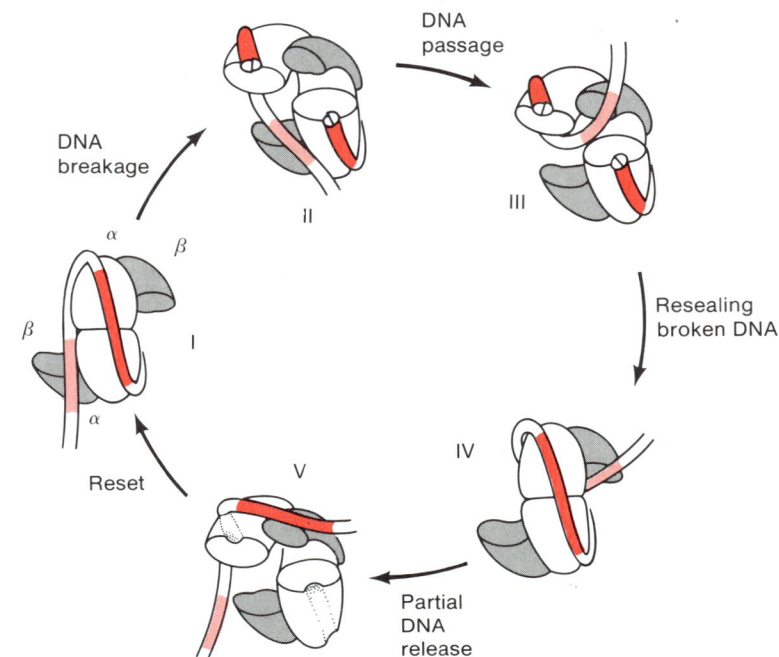

This model, which in principle is the one that occurs, has the following requirements: (1) Gyrase must be capable of binding to two DNA segments, which may be quite distant, and it must have at least two binding sites. (2) The two free ends in molecule III must be held together; otherwise the circle would completely open up and become a linear molecule. (3) A mechanism is needed to pass the two free ends from one side of the unbroken strand to the other.

The first requirement is met by the mode of binding of DNA to gyrase. The DNA molecule is wrapped around the gyrase in a manner resembling (but not identical to) the coiling of DNA around a histone octamer in a nucleosome (Chapter 7). Wrapping brings distant regions into apposition. The second requirement is fulfilled by the subunit structure of the enzyme. Gyrase consists of four subunits, two α and two β subunits. In the cleavage shown in molecule III the 5′-P terminus of each end is covalently joined to one α subunit. This linkage has two important effects. First, the two ends are attached to a single gyrase molecule and therefore cannot fly apart. Second, the energy gained in breaking the phosphodiester bonds is stored in the DNA—5′-P–gyrase linkage, so no activation (such as triphosphate formation) is needed for rejoining the ends after sign inversion. The third requirement is met by the ability of the subunits to move with respect to one another.

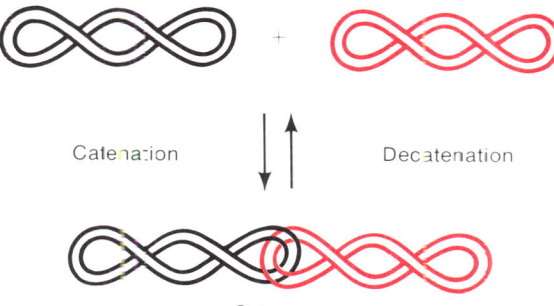

Catenation   Decatenation

Catenane

The details of the supercoiling reaction are not yet resolved, though a model has been proposed that includes the known properties of DNA gyrase. This model, which is shown in Figure 9-8, may not be correct in detail but is probably near enough to the correct mechanism to be worthy of careful study. It is presented for the student interested in how an enzyme can move molecules or parts of molecules. Details of the model are explained in the figure legend. Not shown in the drawing is the nature of the strand break; it is not a clean double-strand break but consists of two single-strand breaks staggered by four base pairs.

Catenation and Decatenation by DNA Gyrase

The essence of the activity of DNA gyrase is its ability to make a double-stranded break in a DNA molecule, pass another segment of DNA through the broken strand, and then reseal the break. The enzyme does not require that the broken segment and the segment passing through are part of the same molecule. This is made clear by the phenomenon of catenation and decatenation (Figure 9-9). **Catenation** is the linking of two circular DNA molecules to form a chain and decatenation is the reverse process. One catenane will produce two DNA circles. These two reactions can be carried out if the two units both bind to a single gyrase molecule in such a way that a segment of each circle can pass through a segment of the other. Decatenation is probably important in DNA synthesis because, when a circle replicates, the products under certain conditions are two catenated circles; this will be discussed later.

ENZYMOLOGY OF DNA REPLICATION

The enzymatic synthesis of DNA is a complex process, primarily because of the need for high fidelity in copying the base sequence and for physical separation of the parental strands. The number of steps that must be completed is far too great to be accomplished by a single enzyme and, in fact, about twenty proteins are known at present to be necessary. Thus, in an effort to provide some understanding with a minimum of confusion, each step in the process will be treated separately. We will consider the basic chemistry of polymerization, the source of the precursors, the problems raised by the chemistry of polymerization, the means of initiating and terminating synthesis, and the mechanisms for eliminating replication errors.

The Polymerization Reaction and the Polymerases

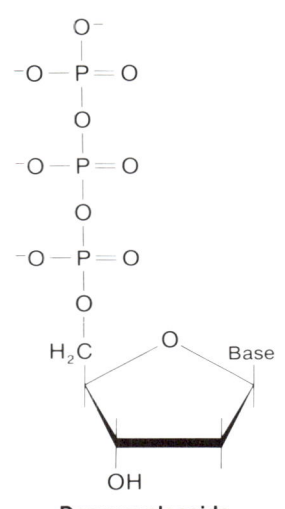

**Deoxynucleoside
triphosphate**

In 1957, Arthur Kornberg showed that extracts of *E. coli* contain a DNA polymerase (now called **polymerase I** or **pol I**). This enzyme is able to synthesize DNA from four precursor molecules, namely, the four deoxy-nucleoside 5′-triphosphates (dNTP), dATP, dGTP, dCTP, and dTTP, as long as a DNA molecule to be copied (a **template** DNA) is provided. Neither 5′-monophosphates nor 5′-diphosphates, nor 3′-(mono-, di-, or tri-) phosphates can be polymerized—only the 5′-triphosphates are substrates for the polymerization reaction; soon we will see why this is the case. Some years later, it was found that pol I, though playing an essential role in the replication process, is not the major polymerase in *E. coli;* instead, the enzyme responsible for advance of the replication fork is polymerase III or pol III.* Pol III also exclusively uses 5′-triphosphates as precursors and requires a DNA template before polymerization can occur. Pol I and pol III have many features in common, and in fact a few types of DNA molecules replicate by using only pol I. The overall chemical reaction catalyzed by both DNA polymerases is:

$$\text{Poly(nucleotide)}_n\text{-3}'\text{-OH} + \text{dNTP} \rightleftarrows \text{Poly(nucleotide)}_{n+1}\text{-3}'\text{-OH} + \text{PP}_i$$

in which PP_i represents pyrophosphate cleaved from the dNTP.

We have already mentioned (Chapter 1) that nucleoside monophosphates cannot be added to the end of a growing DNA strand because the free energy of hydrolysis of a phosphodiester bond is large and negative; hydrolysis and not polymerization is the preferred direction of reaction. The rate of spontaneous hydrolysis of these bonds is very low but it does occur; as a consequence, most cells contain an auxiliary enzyme called a ligase, which seals nicks that arise spontaneously. (We will see shortly that this is not the main function of ligase.) In a nucleoside triphosphate, the strong negative charge on adjacent phosphate groups mutually repel one another and hence weaken the P—O bonds; this repulsion is reflected in the rather

---

*The numbers I and III refer only to the order in which the enzymes were first isolated, not to their relative importance. Another polymerase, pol II, has also been isolated from *E. coli*. It plays no role in DNA replication and its biological function is unknown; mutant cells lacking pol II grow normally in the laboratory. For convenience, the abbreviation "pol" will be used in this chapter to refer to specifically named polymerases.

large negative value of the free energy of hydrolysis for these substances, which tends to compensate for the large positive free energy of formation of the phosphodiester bond in DNA. However, the value of the total free energy for the polymerization reaction written above is still slightly greater than zero ($+0.5$ kcal/mol = about 2400 joule/mol), so equilibrium still lies to the left, as written. Indeed, if excess pyrophosphate (one of the products of cleavage of the triphosphate) and a polymerase are added to a solution containing a partially replicating DNA molecule, the polymerase catalyzes depolymerization. In order to drive the reaction to the right, pyrophosphate must be removed, and this is done by a potent **pyrophosphatase,** a widely distributed enzyme that breaks down pyrophosphate to inorganic phosphate. Hydrolysis of pyrophosphate has a large negative free energy, so essentially all of the pyrophosphate is rapidly removed.

The energetics of the polymerization reaction explains why polymerization has not evolved to make use of diphosphate precursors, which also have a large, negative free energy of hydrolysis. If the diphosphates were used as precursors, the free energy change would still be positive for the polymerization reaction and the product would be inorganic phosphate, which cannot be broken down further, but this is not easily removed by known biochemical reactions. Hence if the polymerization depended on diphosphate precursors, no straightforward way would exist to drive the reaction in the direction of polymerization.

In the following section, the synthesis of nucleoside triphosphates is described. This section may be omitted but, if used, the reader should pay special attention to the synthesis of thymidine triphosphate (TTP), because it introduces a complication that will be described later.

Source of Precursors

There are two distinct pathways for the synthesis of the DNA precursors—the **de novo pathway** and the **salvage pathway.** In the latter, free bases and nucleosides obtained either by degradation of nucleic acids or from the growth medium are built up to nucleoside monophosphates by a variety of reactions. For further information, see the references at the end of the book.

In the *de novo* pathway, ribonucleoside monophosphates are synthesized from phosphoribosylpyrophosphate, amino acids, CO_2, and NH_3 (rather than from free bases). The nucleoside monophosphates (NMP) are then converted to nucleoside diphosphates (NDP) by enzymes called **kinases.** The kinases are specific for each base but nonspecific with respect to ribose or deoxyribose. The reaction involves cleavage of ATP and is:

$$NMP + ATP \rightarrow NDP + ADP$$

The NDP are then converted to their deoxy form by an enzyme, **ribonu-cleotide reductase.** This enzyme is not base-specific: it acts on all ribo-NDP. The enzyme **nucleoside diphosphate kinase,** which is neither base-nor sugar-specific, then forms the triphosphate. Because of the lack of base specificity of both this enzyme and ribonucleotide reductase, the nucleotide dUTP, which is not an immediate precursor for either DNA or RNA, is synthesized. As will be seen later, bacteria devote considerable effort toward preventing incorporation of uracil (U) into DNA and removing the few molecules of uracil that do become part of the DNA.

The synthesis of TTP requires special consideration since thymidine is unique to DNA. Thymidine monophosphate (TMP) is formed by methylation of dUMP; the reaction is catalyzed by the enzyme **thymidylate synthetase** and the methyl group by which uracil and thymine differs is obtained from 5,10-methylene tetrahydrofolate.

$$\text{dUTP} \longrightarrow \text{dUMP} + \text{tetrahydrofolate} \xrightarrow{\text{Thymidylate synthetase}} \text{TMP} + \text{dihydrofolate}$$

$$\searrow \text{PPi}$$

Recognizing this mode of synthesis has been important for at least two reasons:

1. Mutants lacking thymidylate synthetase (*thy*⁻ mutants) have been isolated. Exogenous thymine or thymidine, which can be built into TMP by the salvage pathway, must be provided to these mutants in order for them to grow. Since there are no salvage enzymes that convert thymine to any compound other than thymidine and TMP, addition of radioactive thymine or thymidine to growth medium allows radiolabeling of DNA without labeling any other compound. This technique has been important in the study of replication *in vivo*. In eukaryotes it is often not possible to obtain *thy*⁻ mutants. However, addition of fluorouracil deoxyriboside (FUDR), a thymidine analogue in which fluorine replaces the methyl group, inhibits thymidylate synthetase. FUDR is phosphorylated and then forms a ternary complex with the enzyme and tetrahydrofolate, thereby inactivating the enzyme.

2. Inhibitors of tetrahydrofolate synthesis also introduce a requirement for exogenous thymine or thymidine, without which the cells die. A drug that inhibits the conversion of dihydrofolate to tetrahydrofolate—methotrexate—is commonly used in cancer chemotherapy.

Properties of Polymerases I and III

Pol I and pol III have many features in common. Both enzymes can only polymerize deoxynucleoside-5'-triphosphates and can do so only while copying a template DNA. Furthermore, polymerization can only occur by

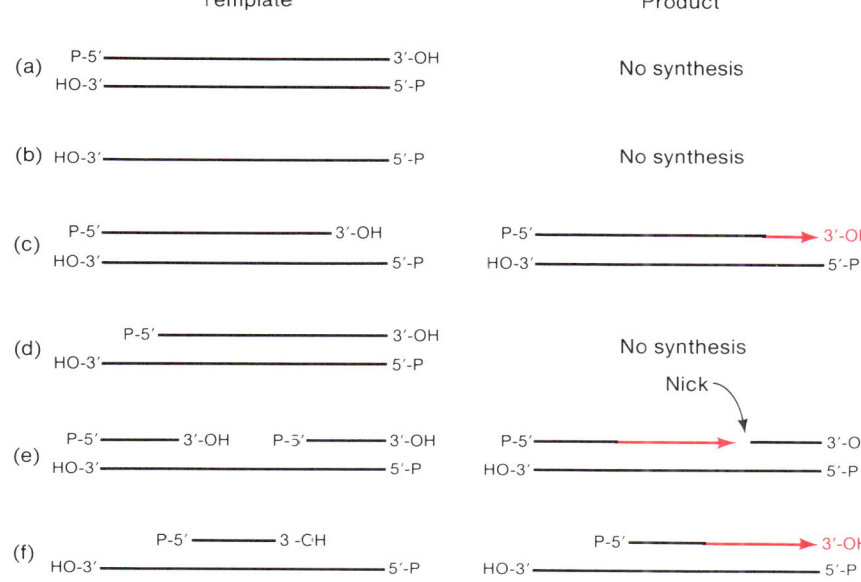

Figure 9-10 The effect of various templates used in DNA polymerization reactions. A free 3'-OH on a hydrogen-bonded nucleotide at the strand terminus and a non-hydrogen-bonded nucleotide at the adjacent position on the template strand is needed for strand growth. Newly synthesized DNA is red.

addition to a **primer**—that is, an oligonucleotide hydrogen-bonded to the template strand and whose terminal 3'-OH group is available for reaction (that is, a "free" 3'-OH group). The meaning of a primer is made clear in Figure 9-10, which depicts six potential template molecules; of these, only three can be said to be active—(c), (e), and (f)—each of which has a free 3'-OH group. The lack of activity with (d) and the direction of synthesis with (e) and (f) indicate that nucleotides do not add to a free 5'-P group. The lack of any synthesis with (a) or (b) indicates that addition to a 3'-OH group cannot occur if there is nothing to copy. Thus, we draw two conclusions:

1. Both a primer with a free 3'-OH group and a template are needed.
2. Polymerization consists of a reaction between a 3'-OH group at the end of the growing strand and an incoming nucleoside 5'-triphosphate. When the nucleotide is added, it supplies another free 3'-OH group. Since each DNA strand has a 5'-P terminus and a 3'-OH terminus, strand growth is said to proceed in the 5'→3' (5'-to-3') direction.

These two points are summarized in Figure 9-11

All known polymerases (both DNA and RNA polymerases) are capable of chain growth in only the 5'→3' direction; that is, the growing end of the

Figure 9-11 Schematic diagram of a replicating DNA molecule showing the distinction between template and primer and the meaning of 5'→3' synthesis.

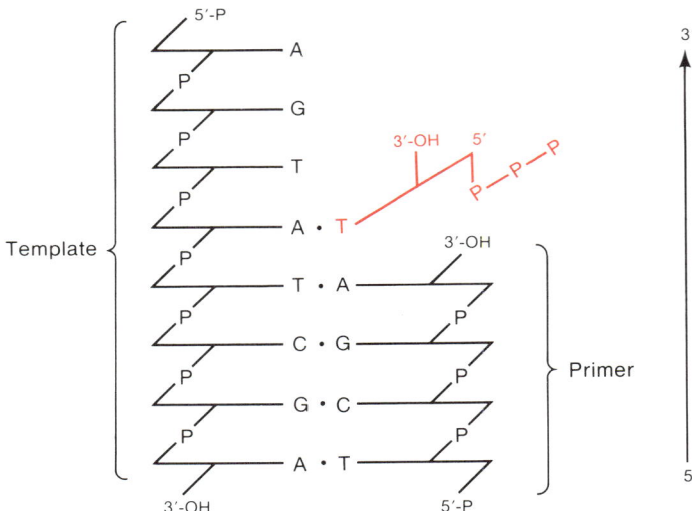

polymer must have a free 3'-OH group. The observation that chain growth proceeds in only one direction introduces what is probably the greatest complication in the entire replication process; this will be described shortly.

Pol I has several activities other than its ability to polymerize; some of these are nevertheless important during the polymerization process. Notably, polymerase I functions as a 3'→5' exonuclease, a 5'→3' exonuclease, and an endonuclease, and can perform nick translation and strand displacement. These functions are described in the following sections.

The 3'→5' Exonuclease Activity

Occasionally polymerases add a nucleotide to the 3'-OH terminus that cannot hydrogen-bond to the corresponding base in the template strand. This may be purely a mistake or may result from the tautomerization of adenine and thymine discussed in Chapters 10 and 11. Inasmuch as such a nucleotide alters the information content of the daughter DNA molecule, it is reasonable that cells would have evolved mechanisms for correcting such errors. All cells do, in fact, possess numerous repair systems that can excise a damaged (chemically altered) or an unpaired base (see Chapter 10). In the case of a damaged base, the distinction between the correct base (i.e., the undamaged member of the pair) and the altered base is clear. However, if a standard base (i.e., A, T, G, or C) incapable of pairing is inserted, these repair systems have no simple way of distinguishing the correct from the incorrect. Thus it is essential that the unpaired base be removable while its incorrectness is recognizable—namely, as an unpaired base at the 3'-OH terminus of a growing strand.

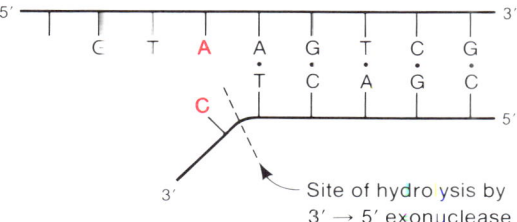

Figure 9-12 The 3′→5′ exonuclease activity of DNA polymerase I showing the site of hydrolysis. The C that is removed (red) does not base-pair with the A being copied (red).

Pol I responds to an unpaired terminal base by terminating polymerizing activity because the enzyme requires a primer that is correctly hydrogen-bonded. When such an impasse is encountered, a 3′→5′ exonuclease activity, which may be thought of simply as pol I running backwards or in the 3′→5′ direction, is stimulated, and the unpaired base is removed (Figure 9-12). After removal of this base, the exonuclease activity stops, polymerizing activity is restored, and chain growth begins again. This exonuclease activity is called the **proofreading** or **editing function** of pol I.

The 5′→3′ Exonuclease Activity

Figure 9-13 shows several substrates and the products of the 5′→3′ exonuclease activity of pol I. From these results five features of the 5′→3′ exonuclease can be deduced.

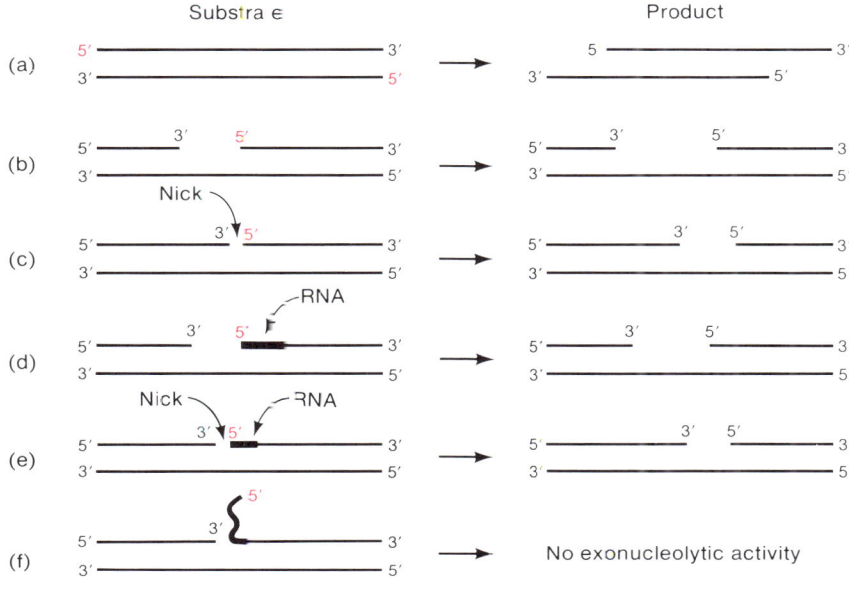

Figure 9-13 Several substrates and products after activity of the polymerase I 5′→3′ exonuclease. The 5′ terminus attacked by the enzyme is shown in red. (a) 5′-terminal nucleotides are removed from each end of the molecule. (b) A gap is enlarged by cleavage at a 5′ terminus. (c) A gap is formed by removal of a 5′-terminal nucleotide at a nick. The gap is then enlarged as in (b). (d) RNA (heavy line) at the boundary of a gap is removed by cleavage at the 5′ terminus of the RNA. (e) A gap is formed from a nick at a DNA-RNA boundary by removal of 5′-P-terminal ribonucleotides. (f) A non-hydrogen-bonded 5′ terminus is resistant to the exonuclease; there is an endonucleolytic activity, as described in Figure 9-15.

1. Nucleotides are removed from the 5′-P terminus only; although not shown in the figure, they are removed one by one.
2. More than one nucleotide can be removed by successive cutting.
3. The nucleotide removed must have been base-paired.
4. The nucleotide removed can be either of the deoxy or the ribo type.
5. Activity can be at a nick as long as there is a 5′-P group.

We will see shortly that the main function of the 5′→3′ exonuclease activity is to remove ribonucleotide primers.

Nick Translation

The 5′→3′ exonuclease activity at a single-strand break (a nick) can occur simultaneously with polymerization. That is, as a 5′-P nucleotide is removed, a replacement can be made by the polymerizing activity (Figure 9-14). Since pol I cannot form a bond between a 3′-OH group and a 5′-monophosphate, the nick moves along the DNA molecule in the direction of synthesis. This movement is called **nick translation.**

Strand Displacement

Experimental conditions can be chosen such that polymerization will occur at a nick (see nick translation, above) without concomitant 5′→3′ exonuclease activity. The growing strand then displaces the parental strand (Figure 9-14). This is thought to be an important step in the mechanism of genetic recombination (Chapter 20). Of all *E. coli* polymerases known to date, pol I is the only one capable of carrying out an unaided displacement reaction. In other strand displacement reactions, auxiliary proteins are required and ATP is cleaved to fuel the unwinding of the helix; this will be discussed when the events at a replication fork are described.

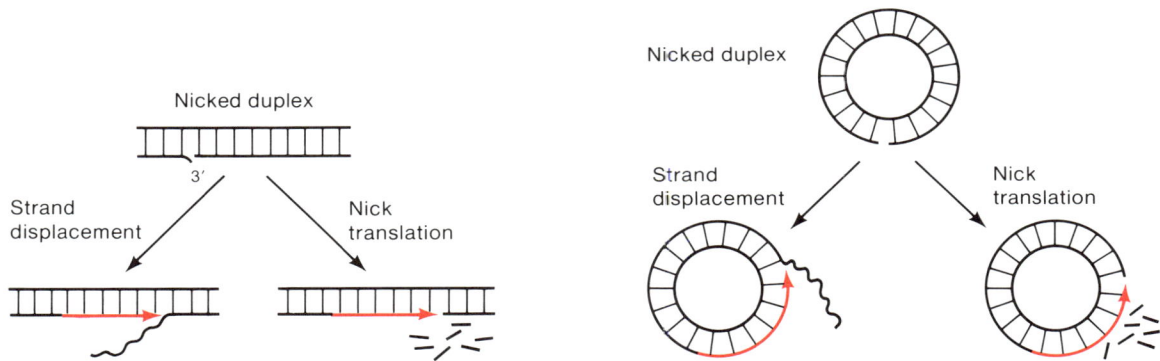

Figure 9-14   Strand displacement and nick translation on linear and circular molecules. In nick translation a nucleotide is exonucleo-lytically removed for each nucleotide added. The growing strand is shown in red.

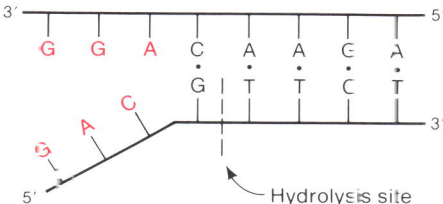

Figure 9-15   The endonuclease activity of polymerase I. Base pairing does not occur between the bases shown in red.

An Endonuclease Activity

The $5' \rightarrow 3'$ exonuclease activity is also capable of an endonuclease activity, as shown in Figure 9-15. An endonucleolytic cut is made between two base pairs that follow a 5'-P-terminated segment of unpaired bases. This is unimportant in normal replication but may be a stage in a major excision-repair system (Chapter 10).

Polymerase III

E. coli polymerase III is a very complex enzyme. In its most active form it is associated with nine (or more) other proteins to form the **pol III holoenzyme,** occasionally termed pol III. The term holoenzyme refers to an enzyme that contains several different subunits and retains some activity even when one or more subunits is missing. The smallest aggregate having enzymatic activity is called the **core enzyme.** The activities of the core enzyme and the holoenzyme are usually very different. Genes encoding five of the subunits have been identified; these are called *dnaE, anaN, dnaQ* (also called *mutD*), *dnaX,* and *dnaZ.* The DnaE protein (the α subunit) possesses the polymerizing activity and the DnaQ protein (the ε subunit) has the editing ($3' \rightarrow 5'$ exonuclease) activity. The other subunits confer catalytic efficiency and high processivity (continual movement of the enzyme along the DNA without dissociation). Pol III shares with pol I a requirement for a template and a primer but its substrate specificity is much more limited. For instance, pol III cannot act at a nick nor is it active with single-stranded DNA primed by either a DNA or RNA nucleotide fragment without the presence of additional protein factors. Pol III cannot carry out strand displacement either, and another system is needed to unwind the helix in order that a replication fork will be able to proceed (this will be discussed in detail in a later section). Pol III also possesses a $5' \rightarrow 3'$ exonuclease activity; however, the enzyme acts only on single-stranded DNA so it cannot carry out nick translation. The biological role of the $5' \rightarrow 3'$ exonuclease activity of pol III is unknown.

Although pol III holoenzyme is the major replicating enzyme in *E. coli,* less is known about it than about pol I, because it is a more complex enzyme and until recently has been difficult to isolate in sufficient quantity for detailed examination. Study of pol III is currently an active field of research.

Later in this chapter, when the events at the growing fork catalyzed by the *E. coli* replication system are examined, it will be seen that pol I and pol

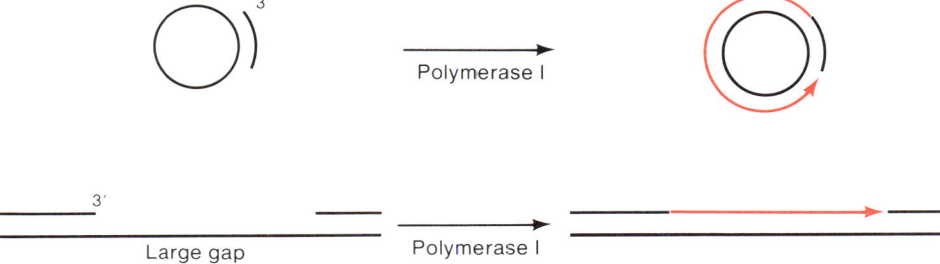

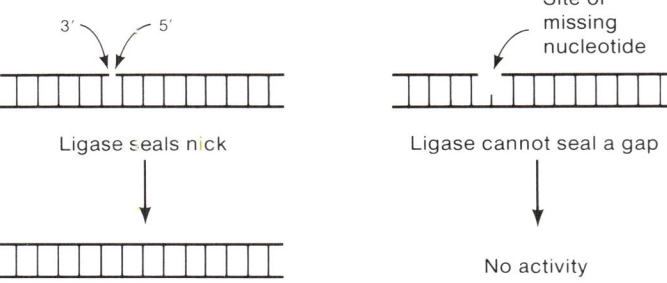

Figure 9-16   The activities of polymerase I on two substrates. Newly synthesized DNA is shown in red. Note that, in the product, the growing daughter strand is never covalently joined to the single-stranded DNA ahead of the red arrowhead.

Figure 9-17   The action of DNA ligase. A nick having a 3′-OH and a 5′-P terminus is sealed (left panel). If one or more nucleotides are absent, the gap cannot be sealed.

III holoenzyme are both essential for *E. coli* replication.* However, a requirement for two polymerases is not common to all organisms; for instance, *E. coli* phage T4 synthesizes its own polymerase, and this enzyme is capable of carrying out all necessary polymerization functions.

DNA Ligase

Figure 9-16 illustrates the results of the activity of polymerase I on two different primed single strands. In the upper molecule a single-stranded circle is copied; in the lower molecule a gap is filled. Similar results would be obtained with polymerase III if appropriate accessory factors were present. The figure shows that neither replication from a primed circular single strand nor gap filling results in a continuous daughter strand. Discontinuity results because no known polymerase can join a 3′-OH and a 5′-*mono*phosphate

---

*When mutants of *E. col* lacking pol I were first isolated and found to be viable, it was concluded that pol I is a nonessential enzyme. However, the explanation for the viability of such mutants is that all known pol I mutants have a residual activity and make some functional enzyme.

group. The joining of these groups is accomplished by the enzyme **DNA ligase,** which functions in replication and other important processes.

E. coli DNA ligase can join a 3′-OH group to a 5′-P group as long as both groups are termini of adjacent base-paired deoxynucleotides—the enzyme cannot bridge a gap (Figure 9-17).

In the usual polymerization reaction, the activation energy for phosphodiester bond formation comes from cleaving the triphosphate. Since DNA ligase has only a monophosphate to work with, it needs another source of energy. It obtains this energy by coupling ligation with hydrolysis of either ATP or nicotinamide adenine dinucleotide (NAD); the energy source depends upon the organism from which the DNA ligase is obtained. The *E. coli* DNA ligase uses NAD.

DISCONTINUOUS REPLICATION

In the model of replication shown in Figure 9-1 both daughter strands are drawn as if replicating continuously. However, no known DNA molecule replicates in this way—instead, *one of the daughter strands is made in short fragments, which are then joined together.* The reason for this mechanism and the properties of these fragments are described in the following sections.

Fragments in the Replication Fork

As we have just seen, pol I and pol III can add nucleotides only to a 3′-OH group. Examination of the growing fork indicates that if both daughter strands grew in the same overall direction, only one of these strands would have a free 3′-OH group; the other strand would have a free 5′-P group because the two strands of DNA are antiparallel (Figure 9-18). Thus one of the following must be true:

1. There is another polymerase that can add a nucleotide to the 5′ end; that is, it would catalyze strand growth in the 3′→5′ direction. However, no such polymerase exists.

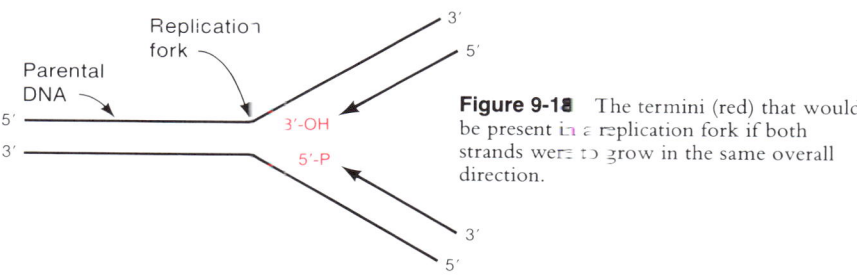

Figure 9-18   The termini (red) that would be present in a replication fork if both strands were to grow in the same overall direction.

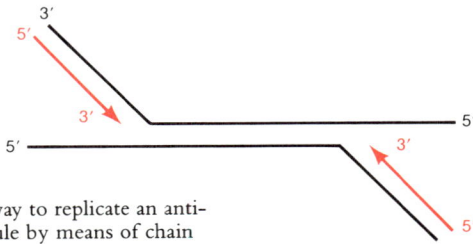

Figure 9-19 One way to replicate an anti-parallel DNA molecule by means of chain growth. Daughter strands are shown in red.

2. The two strands both grow in the $5' \rightarrow 3'$ direction but from opposite ends of the parental molecule, as shown in Figure 9-19. If this were correct, a significant fraction of the unreplicated molecule would have to be single-stranded.

3. The two strands both grow in the $5' \rightarrow 3'$ direction at a single growing fork and hence do not grow in the same direction along the parental molecule. This predicts that some newly made DNA consists of fragments.

Many experiments have shown that alternative 3, the mechanism shown in Figure 9-20, is the correct one.

Detection of Fragments

In 1968, Reiji Okazaki demonstrated that in *E. coli* newly synthesized DNA consists of fragments that later attach to one another to generate continuous strands. The presence of fragments supported the discontinuous replication model of Figure 9-20. Okazaki did two experiments to demonstrate this. In the first experiment, [^{3}H]dT was added to a growing bacterial culture in order to label the new DNA strands with radioactivity, and 30 seconds later the cells were collected and all of their DNA was isolated. This is called a **pulse-labeling experiment.** The DNA was then sedimented in alkali, which causes strand separation. The type of data obtained (Figure 9-21(a)) showed that the most recently made ("pulse-labeled") DNA sediments very slowly in comparison with the single strands obtained from the parental DNA (even though these strands are usually broken in the course of isolation); from the s value it was estimated that the fragments of pulse-labeled DNA range in size from 1000 to 2000 nucleotides, whereas the isolated parental DNA is usually 20 to 50 times as large. In the second experiment, the bacteria were pulse-labeled for 30 seconds; then, the [^{3}H]dT was replaced with nonradioactive dT, and the bacteria were allowed to grow for several minutes. This is called a **pulse–chase experiment,** and it allows one to examine the current state of molecules synthesized at an earlier time. Okazaki observed

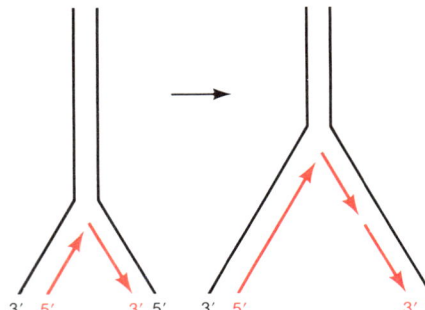

Figure 9-20 The discontinuous replication mode—one way to accommodate $5' \rightarrow 3'$ polymerization with a single overall direction of movement of the fork.

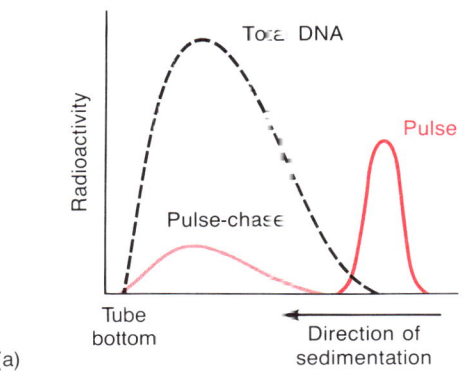

(a)

Figure 9-21 (a) The type of data obtained by alkaline sedimentation of pulse-labeled DNA (red) and pulse-chased DNA (black). Total DNA is the sum of the nonradioactive and radioactive DNA, as might be indicated by the optical absorbance. The s value of the sedimenting material increases from right to left. (b) The location of radioactive DNA (red) at the time of pulse-labeling and after a chase. The radioactive molecules present in alkali are shown. The fragmentation resulting from removal of uracil (see later in text) accounts for the fact that all pulse-labeled DNA has a low s before the chase.

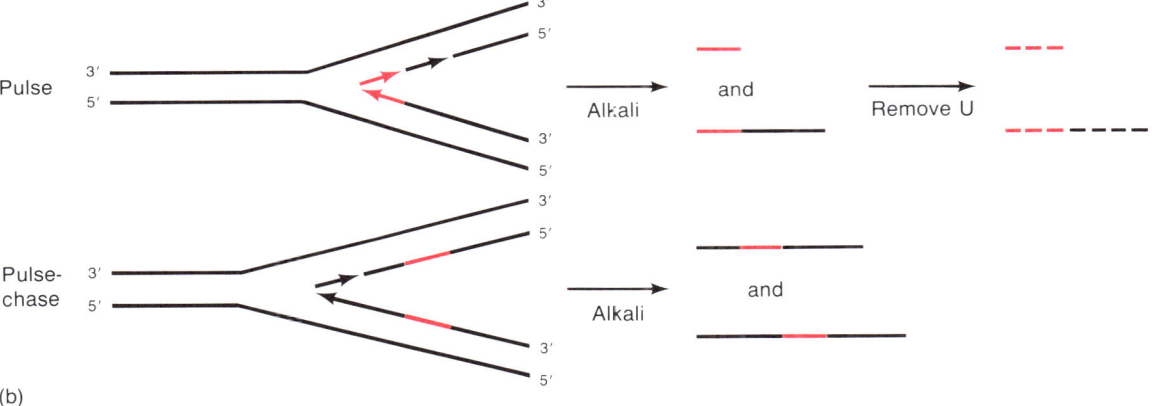

(b)

that the s value of the radioactive material increased with time of growth in the nonradioactive medium. These experiments are presented and interpreted in Figure 9-21 in terms of the discontinuous replication mode shown in Figure 9-20. Apparently the joining of the pulse-labeled fragments to the growing daughter strands caused the label to sediment with the bulk of the DNA.

These fragments, which are widely known as **Okazaki fragments,** have the properties predicted by the discontinuous model; that is, they are initially small and then become large as they are attached to previously made DNA. However, the model of Figure 9-20 predicts that only half of the radioactivity should be found in small fragments whereas the data shown in Figure 9-21(a) indicate that all of the newly synthesized DNA consists of fragments. This result should be surprising because there is no reason why the DNA of the 3'-OH-terminated strand should be synthesized discontinuously. The answer is that this strand is made continuously and is fragmented *after* synthesis. Why this occurs is explained in the following section.

Uracil Fragments

Earlier it was explained that both TTP and dUTP are present in cells and that pol III (and pol I, as well) cannot distinguish dUTP from TTP because both are equally able to form hydrogen bonds with adenine. Thus, some dUTP is incorporated into DNA to make A·U pairs. In order to minimize this incorporation, *E. coli* possesses an enzyme, **dUTPase,** that converts dUTP to dUMP, which can no longer be incorporated into DNA. However, the action of dUTPase is not perfectly efficient; some dUTP survives and is incorporated.

In Chapter 8 (Figure 8-8) the deamination of cytosine to uracil was described. After one round of replication such a deamination would lead to replacement of a G·C pair by an A·U pair, which would become an A·T pair after another round of replication. Since this would be mutagenic, cells have evolved a mechanism for replacing the unwanted U by a C. The first step of this repair cycle (Figure 9-22) is removal of the uracil by the enzyme **uracil *N*-glycosylase.** This enzyme cleaves the *N*-glycosidic bond and leaves the deoxyribose in the backbone. A second enzyme, **AP endonuclease,** makes a single cut, freeing one end of the deoxyribose (AP stands for apurinic acid, a polynucleotide from which purines have been removed by hydrolysis of the *N*-glycosidic bonds). This is followed by removal of the deoxyribose and several adjacent nucleotides (probably by a second enzyme that acts at the other side of the apurinic site), after which pol I fills the gap with the correct nucleotides. This sequence, endonuclease–enlargement–polymerase, is an example of a general repair mechanism called excision-repair, which will be described in the next chapter.

Uracil *N*-glycosylase cannot distinguish between a uracil arising by deamination and a uracil resulting from incorporation of dUTP into DNA. Thus, whenever a dUTP molecule is incorporated, uracil *N*-glycosylase goes to work. The steps of repair following the activity of this enzyme are fairly slow; hence, after removal of uracil, the phosphodiester bond can be

Figure 9-22   Scheme for repair of cytosine deamination. The same mechanism could remove a uracil that is accidentally incorporated.

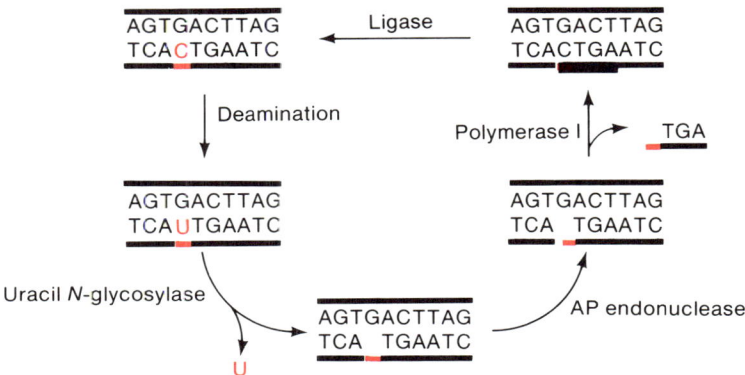

hydrolyzed in alkali (see Chapter 4). Thus, all newly synthesized DNA will appear to be fragmented. The question then is whether any of the fragments observed in Figure 9-21(a) are true precursor fragments arising as suggested in Figure 9-20. The answer comes from studies with two bacterial mutants, one lacking dUTPase (dut^-) and the other lacking uracil-N-glycosylase (ung^-). The relevant experiments show the following:

1. In a dut^- mutant, the Okazaki fragments are smaller than in a dut^+ mutant because dUTP is not hydrolyzed; there is more uracil in the DNA, and fragments result when uracils are excised.
2. In an ung^- mutant, roughly half (instead of all) of the newly made DNA consists of fragments, because this mutant lacks the ability to excise uracil, so no fragments are generated. The size of these ung-independent fragments is about the same as that of the ung-dependent ones; thus, there is about 1 incorporated dUTP per 1000 nucleotides or about 1 uracil per 250 thymines in newly made DNA.
3. An $ung^- dut^-$ double mutant behaves like an ung^- mutant.

These results argue that about half of the Okazaki fragments are true precursor fragments and that the discontinuous-synthesis model in Figure

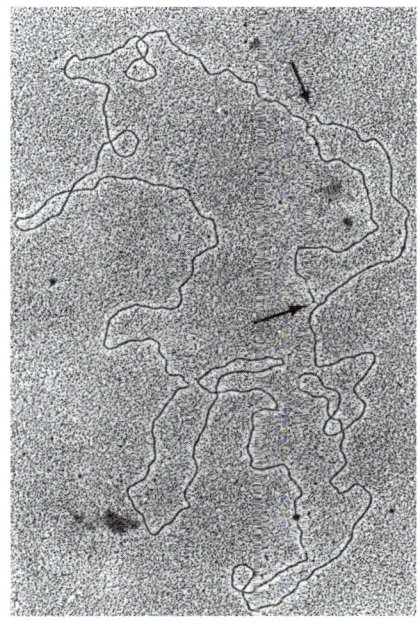

(a)

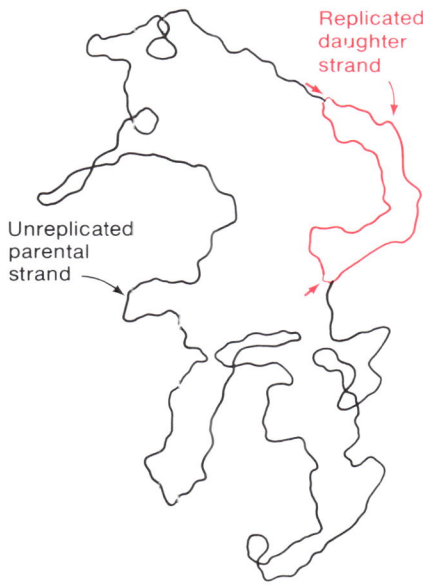

(b)

Figure 9-23 (a) A replicating θ molecule of phage λ DNA. The arrows show the two replicating forks. The segment between each pair of thick lines at the arrows is single-stranded; note that it appears thinner and lighter. (b) An interpretive drawing. [Courtesy of Manuel Valenzuela.]

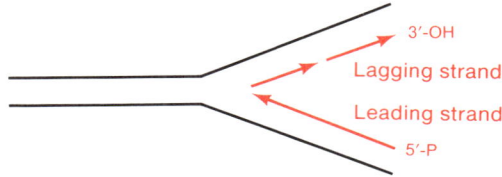

Figure 9-24   A growing fork showing the direction of growth of the leading and lagging strands (both in red).

9-20 is correct. Further support for this model comes from high-resolution electron micrographs of replicating DNA molecules, showing a short single-stranded region on one side of the replication fork (Figure 9-23). This region results from the fact that synthesis of the discontinuous strand is initiated only periodically; perhaps a particular base sequence or some other signal is required for initiation. In fact, the 3′-OH terminus of the continuously replicating strand is always ahead of the discontinuous strand. This had led to the use of the convenient terms, **leading strand** and **lagging strand,** for the continuously and discontinuously replicating strands, respectively (Figure 9-24).

The experiments just described have been carried out with *E. coli.* (The electron micrograph is of phage λ DNA.) Discontinuous replication is apparently a universal phenomenon, as it has been observed with numerous bacterial species, bacteriophages, animal viruses, and eukaryotes. In eukaryotes, the fragments are only about 100–200 bases long (1000–2000 for bacteria); however, whether these smaller fragments represent the actual size of the precursor fragment or simply a higher level of incorporation of dUTP is not known since *ung⁻* mutants of eukaryotes have not been isolated.

In the next sections we examine how synthesis of a precursor fragment is initiated. In order to do so, we must first consider the general question of initiation and priming of replication. This will also put us in a position to understand (1) how fragments are attached to one another, and (2) the role of polymerase I in replication of *E. coli* DNA.

The RNA Terminus of Precursor Fragments

Pol III cannot lay down the first nucleotide to initiate chain growth but requires a primer. There are two ways that might be envisioned to produce a primer: (1) A pool of oligonucleotide fragments might exist that, by binding to parental strands, can provide a free 3′-OH end, which pol III can extend. (2) A polymerizing enzyme, distinct from pol III, synthesizes a primer oligonucleotide, which is then extended by pol III; this is the correct explanation.

In *E. coli* initiation of synthesis of the leading strand and of the precursor fragments of the lagging strand occurs by distinct mechanisms, possibly because initiation of leading strand synthesis requires priming of double-stranded DNA, whereas in initiation of synthesis of a precursor fragment, single-stranded DNA is primed (that is, the strand to be copied is already unwound). Analysis of the mechanism of priming will be examined shortly. At this point, the most important feature of priming is that in every case examined so far, the primer for both leading and lagging strand synthesis is a short RNA oligonucleotide that consists of 1 to 60 bases; the exact number depends on the particular organism. This RNA primer is synthesized by copying a particular base sequence from one DNA strand and differs from a typical RNA molecule in that after its synthesis *the primer remains hydrogen-bonded to the DNA template.* In bacteria two different enzymes are known that synthesize primer RNA molecules—**RNA polymerase,** which is the same enzyme that is used for synthesis of most RNA molecules, and **primase** (the product of the *dnaG* gene). Experimentally, these enzymes can be distinguished *in vivo* by their differential sensitivities to the antibiotic **rifampicin**—RNA polymerase, but not primase, is inhibited by the antibiotic. In *E. coli,* initiation of leading strand synthesis is rifampicin-sensitive, presumably because RNA polymerase is involved in initiation; however, initiation of precursor fragment synthesis is resistant to the drug, as it uses primase. In all cases, the growing end of the RNA primer is a 3'-OH group to which pol III can easily add the first deoxynucleotide; the 5'-P end of the RNA chain, which remains free, has a 5'-triphosphate group. Thus, a precursor fragment has the following structure, while it is being synthesized:

PPP-5'   RNA   DNA   3'-OH

The Joining of Precursor Fragments

The precursor fragments are ultimately joined to yield a continuous strand. This strand contains no ribonucleotides, so assembly of the lagging strand must require removal of the primer ribonucleotides, replacement with deoxynucleotides, and then joining. In *E. coli* the first two processes are accomplished by DNA polymerase I and joining is catalyzed by DNA ligase. How this is done is shown in Figure 9-25. Pol III extends the growing strand until the RNA nucleotide of the primer of the previously synthesized precursor fragment is reached. Pol III can go no further since its 5'→3' exonuclease activity is very weak and is inactive on base-paired DNA; it cannot join a 5'-triphosphate at the terminus of a polymer (i.e., on the primer) to a 3'-OH group on the growing strand and it cannot carry out strand displacement. Thus pol III moves away from the 3'-OH terminus, leaving a nick. *E. coli* DNA ligase cannot seal the nick because a triphosphate is present; even if an additional enzyme could cleave the triphosphate to a monophos-

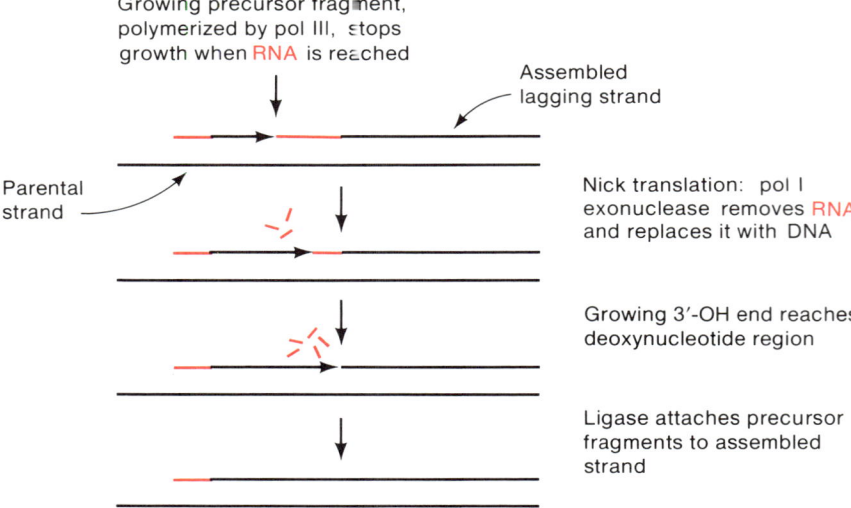

Growing precursor fragment, polymerized by pol III, stops growth when RNA is reached

Assembled lagging strand

Parental strand

Nick translation: pol I exonuclease removes RNA and replaces it with DNA

Growing 3′-OH end reaches deoxynucleotide region

Ligase attaches precursor fragments to assembled strand

Figure 9-25 Sequence of events in assembly of precursor fragments. RNA is indicated in red. The replication fork (not shown) is at the left.

phate, DNA ligase would be inactive when one of the nucleotides is in the ribo form. However, recall that pol I works efficiently at a nick as long as there is a 3′-OH terminus. In this case the enzyme carries out nick translation, probably proceeding into the deoxy section but, there DNA ligase can compete with pol I and seal the nick. Thus, the precursor fragment is assimilated into the lagging strand. By this time, the next precursor fragment has reached the RNA primer of the fragment just joined and the sequence begins anew.

It seems as if everything might be simpler if the primer were made by a special DNA polymerase that could initiate strand growth. However, this is not the course that evolution has taken.

Synthesis of precursor fragments follows synthesis of the leading strand. In the next section, how the leading strand advances into the parental double helix is described. Once this is understood, we can return to the question of how RNA primer synthesis begins—a surprisingly complicated process.

EVENTS IN THE REPLICATION FORK

DNA replication requires not only an enzymatic mechanism for adding nucleotides to the growing chains but also a means of unwinding the parental double helix. In this section we will see that these are distinct processes and that the unwinding mechanism is closely related to the initiation of synthesis of precursor fragments.

Advance of the Replication Fork and the Unwinding of the Helix

The pol III holoenzyme cannot carry out strand displacement, as has already been discussed, because it is unable to unwind the helix. In order for a helix to be unwound, hydrogen bonds and hydrophobic interactions must be eliminated, and this requires energy. Pol I is able to utilize both the free energy of hydrolysis of the triphosphate group and the binding energy of forming a new hydrogen bond to unwind the parental molecule as it synthesizes the leading strand. No other known polymerase can do that; usually a helix must first be unwound in order for a DNA polymerase to advance. Helix-unwinding is accomplished by enzymes called **helicases.** The helicase active in *E. coli* DNA replication is an enzyme called the **Rep protein.** This protein hydrolyzes ATP and utilizes the free energy of hydrolysis in an unknown way to unwind the helix, hydrolyzing two ATP molecules per base pair broken.

In *E. coli* the pol III holoenzyme synthesizing the leading strand is not immediately behind the advancing Rep protein (Figure 9–26). Thus, the Rep protein leaves in its wake two single-stranded regions, a large one on the strand to be copied by precursor fragments and a smaller one just ahead of the leading strand. In order to prevent the single-stranded regions from annealing or from forming intrastrand hydrogen bonds, the single-stranded DNA is coated with *E. coli* single-strand binding protein (**SSB protein**). This is one of a class of binding proteins that bind tightly to both single-stranded DNA and to one another. As pol III holoenzyme advances, it must displace the SSB protein in order that base-pairing of the nucleotide being added can occur. It is not known whether pol III holoenzyme carries out this displacement by itself or whether it is aided by the protein produced by the *dnaB* gene. DnaB protein also moves along a DNA molecule fueled by ATP hydrolysis; it will be described in more detail when initiation of synthesis of precursor fragments is discussed.

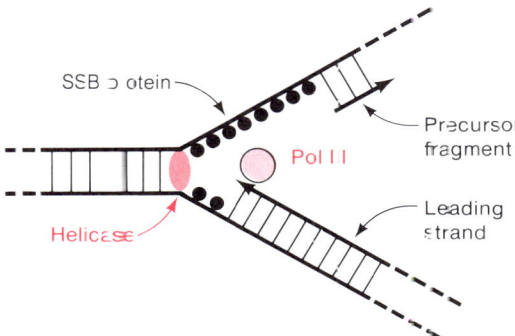

Figure 9-26   The unwinding events in a replication fork.

The mechanism of initiation of replication of precursor fragments is closely connected to both the unwinding by the helicase and the action of the SSB protein. This is described in the following section.

Initiation of Replication from a Single-stranded Template: Initiation of Precursor Fragments

A direct study of the mechanism of initiation of the synthesis of precursor fragments in bacteria has not yet been carried out because experimentally the problem is very complex. Instead, attention has been directed to several *E. coli* phages having single-stranded DNA, in which the first step in replication is the conversion of single-stranded circular DNA to a double-stranded covalent circle. Since some of the phages studied use *E. coli* enzymes exclusively, it is thought that they can serve as a model for understanding initiation of precursor fragment synthesis.

The simplest mechanism of initiation seen so far occurs with the single-stranded circular DNA obtained from phage G4. This DNA molecule has a small palindromic segment which forms a double-stranded hairpin. This structure and part of the replication sequence are shown in Figure 9-27. The initial step, which is common to the replication of all single-stranded molecules, is the coating of all single-stranded regions with SSB protein. The double-stranded region remains uncoated and is the site of binding of primase, which synthesizes a small segment of RNA. Pol III holoenzyme then extends this RNA primer and, displacing the SSB protein, proceeds to replicate around the circle. When the 5′-P terminus is reached, pol I replaces the pol III holoenzyme and removes the RNA, replacing it with DNA.

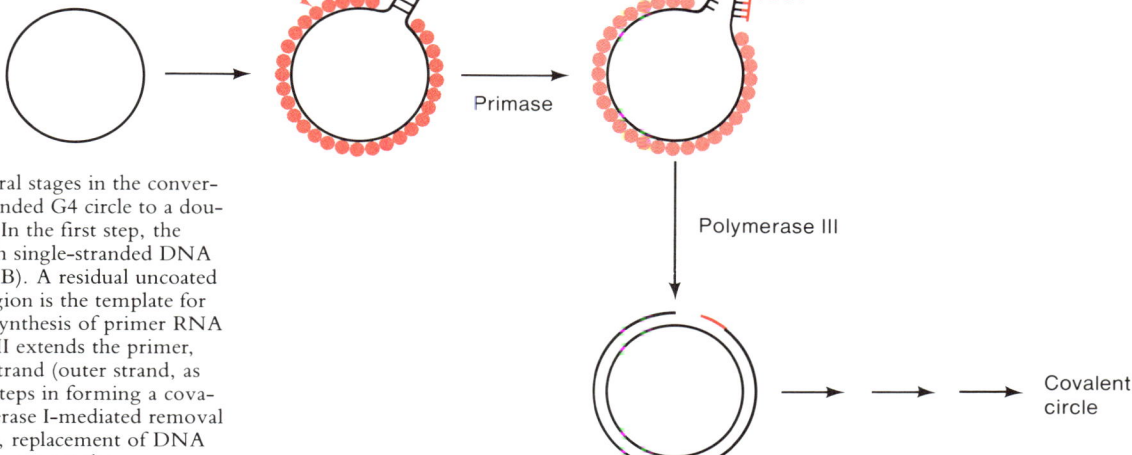

Figure 9-27 Several stages in the conversion of a single-stranded G4 circle to a double-stranded circle. In the first step, the DNA is coated with single-stranded DNA binding protein (SSB). A residual uncoated double-stranded region is the template for primase-catalyzed synthesis of primer RNA (red). Polymerase III extends the primer, forming daughter strand (outer strand, as shown). The final steps in forming a covalent circle—polymerase I-mediated removal of the RNA primer, replacement of DNA bases, and ligation—are not shown.

Ligase seals the nick and replication is complete. Note that with G4 DNA, initiation begins at a unique sequence and that only SSB protein and DnaG primase are needed to synthesize the primer; thus, primer synthesis is rifampicin-resistant.

Initiation of copying the single-stranded circular DNA of phage M13 is somewhat more complicated. This DNA also has a hairpin loop that is not coated by SSB protein, and this loop is the priming region. However, initiation of primer synthesis requires several other factors, and the RNA primer is synthesized by *E. coli* RNA polymerase rather than primase. Thus, primer synthesis is sensitive to rifampicin. Following initiation the overall mechanism for production of a double-stranded circle is the same for G4 and M13.

The most complicated system studied so far and the one that is most closely related to *E. coli* DNA synthesis is that of the circular single-stranded DNA of phage φX174. This DNA molecule does not have a hairpin loop. The primer is synthesized by primase (thus, synthesis is rifampicin-resistant). The first step in primer synthesis is the formation of a complex called a **preprimosome,** which contains six proteins: n, n′, n″, i, DnaB, and DnaC proteins. The protein n′ binds to single-stranded DNA and acquires a bound ATP molecule, and then primase joins the preprimosome, forming a unit called the **primosome.** The protein n′ uses the energy of ATP hydrolysis to move the primosome along the DNA until a priming site, which is chosen at random, is found (Figure 9-28). Although the details are unclear, the DnaB protein alters the structure of the DNA at the site that has been selected, and this alteration enables primase to initiate synthesis of an RNA primer. Replication then proceeds as with G4 DNA—namely, by pol III holoenzyme, pol I, and ligase, in sequence. The important characteristics of the prepriming reaction are the lack of a specific sequence at which initiation occurs, the use of a complex of six proteins, the requirement for ATP, and synthesis of RNA by primase.

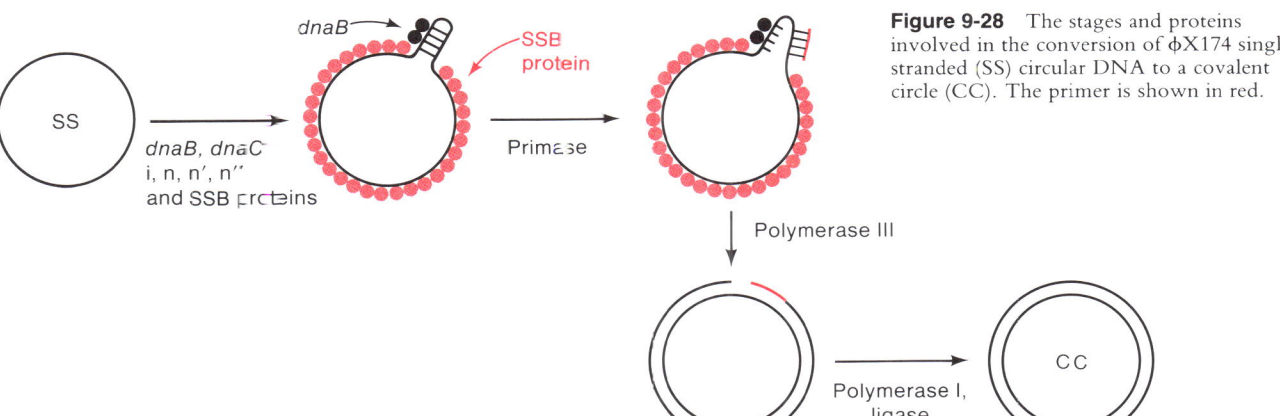

Figure 9-28 The stages and proteins involved in the conversion of φX174 single-stranded (SS) circular DNA to a covalent circle (CC). The primer is shown in red.

Many single-stranded DNA molecules have now been studied and all are replicated by one of the three modes just described. It is thought that these may be the only modes, at least in *E. coli* systems. The initiation of synthesis of precursor fragments in *E. coli* is known from *in vivo* studies to require primase, DnaB and DnaC proteins, and SSB. Little information is available about the other proteins, though it is hypothesized that the initiation of synthesis of precursor fragments uses the same mechanism as described for φX174. Since each precursor fragment must be initiated, it is assumed that the primosome moves along the DNA (the motion is thought to be a property of DnaB) following the motion of the replication fork, allowing primase to synthesize primers repeatedly at various intervals. In the following section this hypothesis and other information are combined to generate a reasonable picture of the events at a replication fork.

A Summary of Events at the Replication Fork

Figure 9-29 summarizes the proposed events that occur in or near a replication fork in *E. coli*. A helicase, driven by ATP hydrolysis and probably aided by binding of SSB, unwinds the helix. Temperature-sensitive mutants of the DnaB protein fail to synthesize any DNA at all at restrictive temperatures, so the DnaB protein may also participate in some way in the unwind-

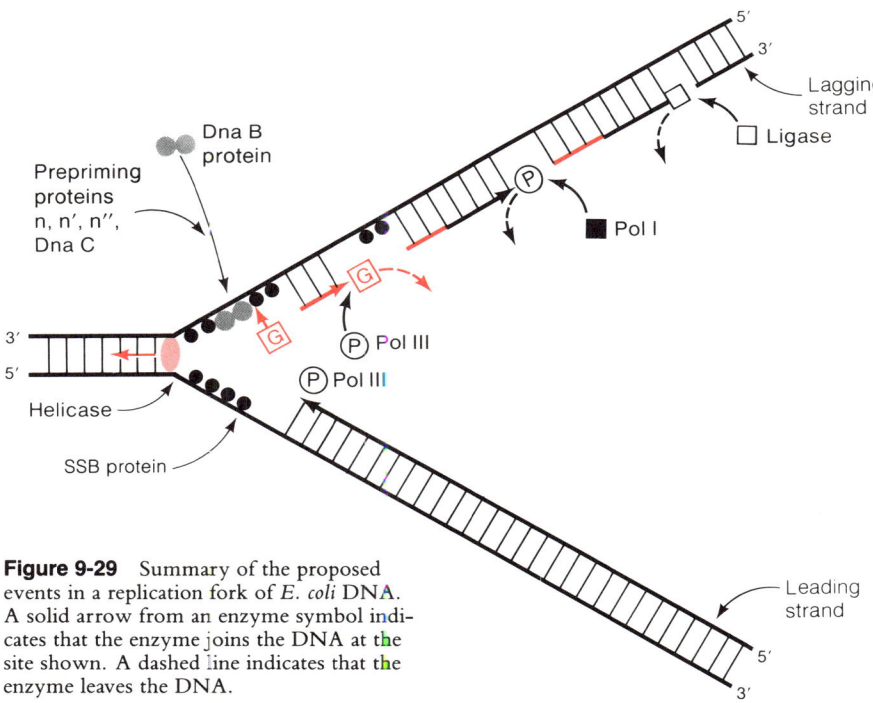

Figure 9-29 Summary of the proposed events in a replication fork of *E. coli* DNA. A solid arrow from an enzyme symbol indicates that the enzyme joins the DNA at the site shown. A dashed line indicates that the enzyme leaves the DNA.

ing process. The unpaired bases are coated with SSB. The leading strand advances along one parental strand by nucleotide addition catalyzed by the pol III holoenzyme. The DnaB protein complex moves along the other parental strand, prepriming it so that primase will synthesize a primer RNA. Pol III holoenzyme adds nucleotides to the primer, thereby synthesizing a precursor fragment. This synthesis continues up to the primer of the preceding precursor fragment; at this point pol I replaces pol III holoenzyme; by nick-translation the RNA is removed and replaced by DNA. Once the RNA is gone, DNA ligase seals the nick, thereby joining the precursor fragment to the lagging strand. When pol III is displaced by pol I, the pol III does not dissociate from the DNA but moves back along the DNA (in the direction of advancement of the fork) until it encounters the next primer.

The advance of pol III holoenzyme is continually delayed by $3' \rightarrow 5'$ exonuclease editing. Furthermore, uracil residues that are accidentally incorporated into DNA are removed by uracil N-glycosylase. The deoxyribose is removed by AP endonuclease, the gap is (often) enlarged and then filled by pol I, and then ligase makes the final seal.

The advance of the replication fork continues until replication is completed. An unresolved question is how the rates of growth of the leading and lagging strands are coordinated. How termination is accomplished is described in a later section.

The reader must surely be impressed with the complexity of this process. However, in some systems replication is somewhat simpler. For instance, with *E. coli* phage T7, which is discussed in detail in Chapter 17, priming is accomplished by a phage RNA polymerase and all replication is performed by a single DNA polymerase encoded in the phage DNA; none of the known *E. coli* enzymes are needed. Phages T4 and λ replicate in a somewhat more complicated manner than T7 but still the mechanism is far simpler than *E. coli* replication. Why should the process in *E. coli* be so complex? This may be a consequence of the requirement for fidelity. The larger the DNA molecule, the greater is the probability that a replication error will be made during a round of replication. Furthermore, a bacterial DNA is replicated only once per generation; a phage replicates many times in an infected cell, so one defective replica would not significantly affect the successful outcome of the infection. Thus, in order to reduce the error frequency to a tolerable level, organisms with a larger DNA molecule and a single replication event need more replication proteins, since each protein can be designed to minimize or correct a particular type of error.

So far, a large number of *E. coli* replication genes have been mentioned. These genes and their properties, when known, are listed in Table 9-1.

We now leave our analysis of the replication fork and examine how the fork itself is created at the outset; what is the process by which synthesis of double-stranded DNA is initiated? This is one of the least understood subjects of replication, especially because it may occur in several ways; it is described in the following section.

Table 9-1 Replication genes and proteins either isolated from *E. coli* or characterized by the existence of mutants

Gene	Function or gene product
dnaA	Initiation of a new round of replication in *E. coli*.
dnaB	Chain elongation. Component of primosome.
dnaC	Initiation of a new round of replication and of precursor fragments; binds to *dnaB* protein. Component of primosome.
dnaD	Same as *dnaC* gene; designation no longer in use.
dnaE	Polymerizing and editing activity of pol III.
dnaF	Ribonucleotide reductase.
dnaG	Primase for precursor fragments.
dnaH	Similar to *dnaA*; poorly characterized.
dnaI	Chain elongation; details unknown.
dnaJ	Initiation; details unknown.
dnaK	Initiation; details unknown.
dnaN	Chain elongation; subunit of pol III holoenzyme.
dnaP	Initiation; details unknown; subunit of pol III holoenzyme.
dnaQ	Fidelity; subunit of pol III holoenzyme.
dnaX	Chain elongation; subunit of pol III holoenzyme.
dnaZ	Chain elongation; subunit of pol III holoenzyme.
polA	Polymerase I. Chain elongation. Its $5' \rightarrow 3'$ exonuclease removes primer RNA. Required to join a precursor fragment to the lagging strand.
gyrA	α subunit of DNA gyrase, which introduces negative superhelical twists into nonsupercoiled DNA and removes positive twists in the unreplicated portion of replicating DNA. Inactivated by nalidixic acid.
gyrB	β subunit of DNA gyrase (see above). Inactivated by coumermycin.
lig	DNA ligase. Seals nicks; required to join a precursor fragment to the lagging strand.
dut	dUTPase.
rep	Helicase. Presumably unwinds DNA at the replicating fork.
ung	Uracil *N*-glycosylase; removes uracil from DNA.
rpoA,B,C,D	Subunits of RNA polymerase; synthesizes primer in some systems.
ssb	Single-stranded DNA-binding protein. Required for advance of the replication fork, priming, and all chain growth. Prevents formation of inter- and intrastrand hydrogen bonds.
Not defined	Proteins i, n, n', n''. Prepriming of φX174 DNA synthesis and presumably of precursor fragments. Components of the primosome.

INITIATION OF SYNTHESIS OF THE LEADING STRAND

All known double-stranded DNA molecules initiate a round of replication at a unique base sequence, called the **replication origin** or ***ori.*** The sequence is specific to each organism, though there is one example in which two organisms have the same sequence.

Initiation can occur in two ways—***de novo* initiation,** in which the leading strand is started afresh, and **covalent extension,** in which the leading strand is covalently attached to a parental strand. In this section we discuss *de novo* initiation; covalent extension is discussed in a later section in which rolling circle replication is described. A particular topological conformation associated with *de novo* initiation is also described in this section.

De Novo Initiation in *E. coli*

De novo initiation remains poorly understood. The single feature common to all bacterial and phage systems examined to date is the requirement for an RNA primer synthesized by an RNA-polymerizing enzyme. Additional proteins are also needed. For instance, initiation of *E. coli* DNA synthesis requires the products of the genes *dnaA, dnaH, dnaJ, dnaK,* and *dnaP* (Table 9-1). However, the biochemical activity of these gene products is not known.

A feature that has been observed in several phage systems is the requirement for a phage-encoded protein that makes a nick in one strand at the origin. Two examples are the ϕX174 gene-*A* product and the gene-*O* and gene-*P* proteins of phage λ. How the nicking is connected to the initiation event is not known. In one system a ribonuclease, RNase H, active on RNA that is hydrogen-bonded to DNA, is required. It is possible that this nuclease activity terminates primer synthesis and allows a switch from RNA synthesis to chain elongation by the polymerase.

A novel technique has been used to study initiation of bacterial synthesis in *E. coli.* By genetic engineering, *oriC,* the 245-bp sequence of bases at which replication of the *E. coli* chromosome begins, has been cloned in a plasmid. Initiation of replication of isolated *oriC* plasmid DNA has been studied in an *in vitro* system consisting of highly purified components. The result, which are preliminary (1985), have been difficult to interpret, but one point is clear: one class of proteins is required for initiation and a second class serves to prevent initiation at secondary sites outside of *oriC.* The mechanism of initiation is not clear because initiation in this system occurs several ways, depending on the proteins introduced into the reaction mixture. For example, the primer RNA can be made by RNA polymerase alone, by primase alone, or by a sequence of events utilizing both RNA polymerase and primase. Since *in vivo* initiation of replication of the *E. coli* chromosome is inhibited by rifampicin, the reaction using primase alone is not considered to occur to any significant extent *in vivo*. The system using both RNA polymerase and primase utilizes a prepriming and a priming step, so by analogy with the initiation of precursor fragments, this mechanism is considered by some workers to be the more likely one. The tentative sequence of events may be the following: (1) RNA polymerase transcribes *oriC*, somehow "activating" the sequence; (2) DnaA, DnaB, DnaC, DNA gyrase, SSB, ATP (and possibly a histone-like protein called HU) form a prepriming complex; (3) primase then joins the complex and synthesizes the RNA primer. It should be noted that the *in vitro* system with the *oriC* plasmid fails to require the products of the *dnaH, dnaJ, dnaK,* or *dnaP* genes, which are required for initiation of *E. coli* chromosome synthesis *in vivo*. Thus, the *in vitro* system probably does not yet reflect the entire process that occurs in the cell.

All known prokaryotic DNA molecules, with only few exceptions, replicate as circles and hence initiate within the helix. Even those molecules that replicate as linear molecules, as well as eukaryotic chromosomes, which contain linear molecules, usually initiate within the helix rather than at one end. Thus, because of the requirement for $5' \rightarrow 3'$ synthesis, initiation of all DNA molecules creates a replication "bubble" consisting of one single-stranded and one double-stranded branch. This is called a D loop and its properties will be examined in the next section.

The DNA molecules of human adenovirus and of *Bacillus subtilis* phage φ29 are linear and initiate at a terminus. Remarkably, a deoxynucleotide covalently linked to a protein is used as a primer, rather than RNA. The initiation mechanism for adenovirus DNA, which is quite similar to that for φ29, is described in Chapter 23.

D Loops

In *de novo* initiation, synthesis of the leading strand precedes that of the lagging strand. Thus, before synthesis of the first precursor fragment begins, a replication bubble exists that consists of one double-stranded branch, made up of one parental strand paired with the leading strand, and one single-stranded branch, which is the unreplicated, second parental strand (Figure 9-30). Since the leading strand displaces the unreplicated parental strand, the bubble is called a **displacement loop** or **D loop.** Such a conformation is ordinarily a transient one, existing only until synthesis of precursor fragments begins—which may require that the leading strand release a specific sequence (in single-stranded form) that can be used for prepriming. However, in certain circumstances, namely in a replication system that does not employ DNA gyrase to relieve topological constraints, a D loop may be long-lived.

In the initial stages of replication of a naturally occurring circular DNA molecule, advance of the replication fork does not require the presence of DNA gyrase, because such a circular DNA molecule initially is negatively

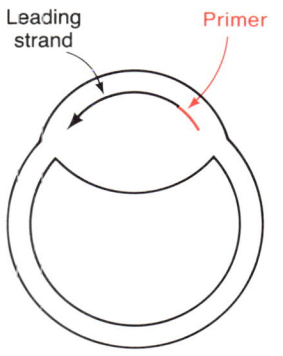

Figure 9-30 A circular DNA molecule with a D loop.

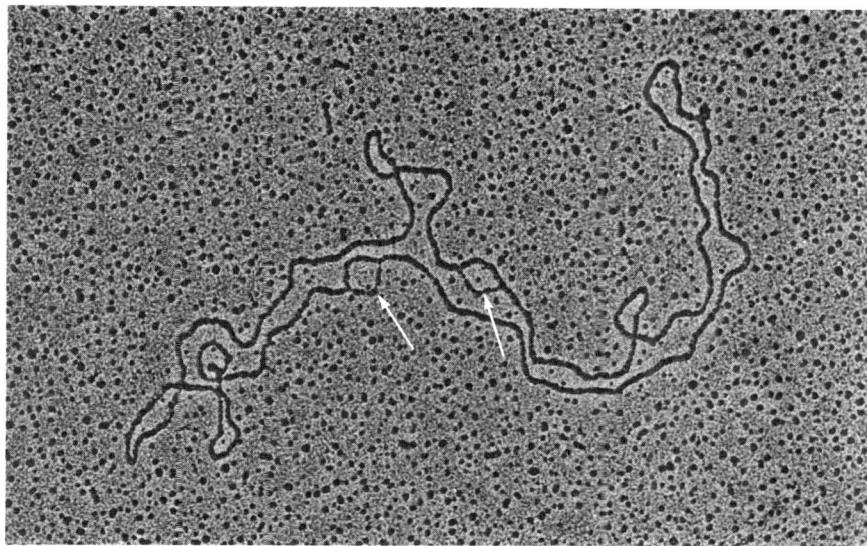

Figure 9-31   An electron micrograph of a dimer of mouse mitochondrial DNA showing diametrically opposing D loops (arrows). The single strand in each D loop appears thinner than the double-stranded DNA. The total length of the molecule is 10 μm. [Courtesy of David Clayton.]

supercoiled and the negative twists compensate for the positive turns introduced by movement of the fork. However, once the fork has advanced sufficiently that the negative twists are used up, a topological constraint-relieving system such as gyrase is needed. The superhelix density of all naturally occurring DNA molecules is 0.05 twists per turn of the helix. This means that, if positive superhelicity is forbidden, the leading strand could move roughly 5 percent of the distance along a circular, negatively super-helical molecule before gyrase would be needed. A replication complex cannot in fact move along a double helix if the movement would produce more than one or two turns of overwinding. Thus, if neither gyrase nor any other unwinding system were present, replication would cease when the unreplicated portion had lost its supercoiling. The molecular conformation of the bubble would depend upon whether the first precursor fragment had been initiated; if it had not, the structure in Figure 9-30 would result.

Such a structure is seen in animal cell mitochondrial DNA (Figure 9-31). The molecular weights of these molecules in most species is about 10^7 and the molecular weight of a precursor fragment is typically about 10^6, so the lack of double-stranded DNA in both branches is consistent with the fact that only 4 percent of the DNA (or 2×10^5 molecular weight units of the leading strand) is replicated. The fact that 70 percent of all replicating mitochondrial DNA molecules are in the D-loop conformation indicates that indeed replication has ceased at this point. Exactly what happens to allow replication to proceed is not known precisely. Mitochondrial DNA replication takes about one hour, which is about 100 times slower than bacterial DNA. After synthesis is complete, another 40 minutes elapses before the daughter molecules are supercoiled. These results are interpreted as follows. It is assumed that either a gyraselike enzyme, which is very inefficient, is present, or a simple nicking-sealing system is present. The system does not work continuously, so there are repeated pauses in replication to remove topological constraints; the system may only be active in response to slightly overwound DNA.

Initiation by Covalent Extension: Rolling Circle Replication

There are numerous instances in which, in the course of replication, a circular phage DNA molecule gives rise to linear daughter molecules in which the base sequence of the DNA present in the phage is repeated numerous times, forming a concatemer (Figure 9–32). These concatemers are usually an essential intermediate in phage production. Likewise, in bacterial mating, a linear DNA molecule is transferred by a replicative process from a donor cell to a recipient cell. Both phenomena are consequences of initiation by covalent extension, an event that gives rise to a replication mode known as **rolling circle replication.**

Consider a duplex circle in which, by some initiation event, a nick is made having 3′-OH and 5′-P termini (Figure 9–33). Under the influence of a helicase and SSB protein a replication fork can be generated. Synthesis of a primer is unnecessary because of the 3′-OH group, so leading strand synthesis can proceed by elongation from this terminus. At the same time, the parental template for lagging strand synthesis is displaced. The polymerase used for this synthesis is apparently pol III holoenzyme. (In this circumstance we can ignore the fact that the enzyme cannot ordinarily carry out a displacement reaction, because in this case, displacement is a result of a coupling of helicase, SSB protein, and pol III holoenzyme.) The displaced parental strand is replicated in the usual way by means of precursor fragments. The result of this mode of replication is a circle with a linear branch; it resembles the Greek letter *sigma* and is called **σ replication** or **rolling circle replication.**

There are four significant features of rolling circle replication:

1. The leading strand is covalently linked to the parental template for the lagging strand.

Figure 9-32 A concatemer consisting of the repeating unit ABC . . . XYZ. Note that the definition of concatemer does not make any requirements about the terminal sequences.

XYZABC   XYZABC   XYZABC   XYZABC

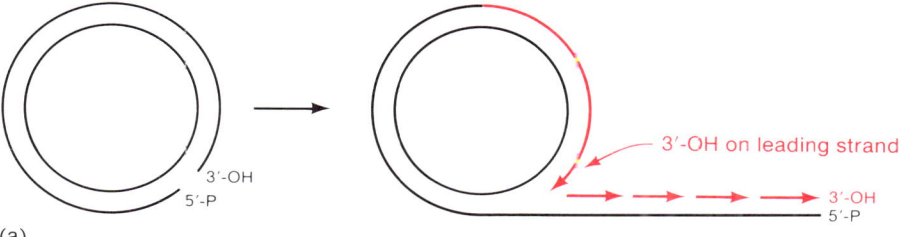

(a)

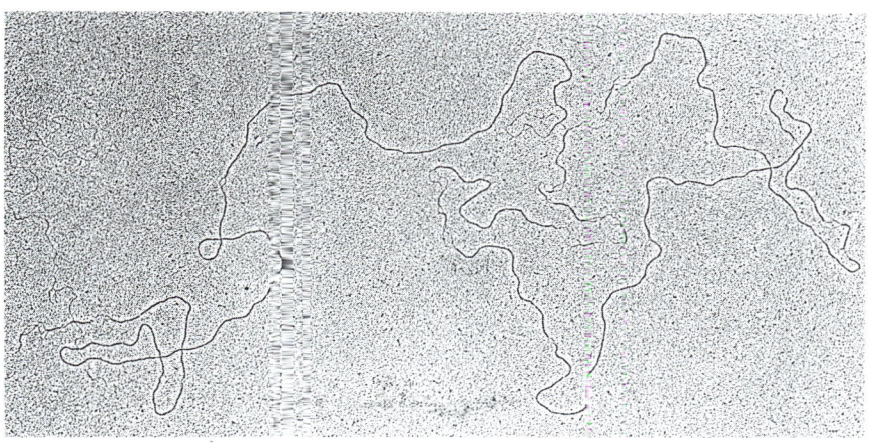

(b)

Figure 9-33 (a) Rolling circle or σ replication. Newly synthesized DNA is shown in red. (b) An electron micrograph of a rolling circle of phage λ DNA isolated from infected *E. coli*. The thin regions are partially denatured, as described in Figures 9-36 and -37, and served as references in the original experiment (*Proc. Nat. Acad. Sci.*, (1973) 70: 1768). [Courtesy of Ross Inman.]

2. Before precursor fragment synthesis begins, the linear branch has a free 5′-P terminus.
3. Rolling circle replication continues unabated, generating a concatemeric branch.
4. The circular template for leading strand synthesis never leaves the circular part of the molecule.

A variant of the rolling circle mode, called **looped rolling circle replication,** is used to generate a progeny single-stranded circle from a double-stranded circular template. For phage φX174 this is accomplished in the following way (Figure 9-34). A phage protein called the gene-*A* protein makes a nick at the origin and in the process becomes covalently linked to the newly formed 5′-P terminus. Using the Rep and SSB proteins and pol III holoenzyme, chain growth occurs from the 3′-OH group, displacing the broken parental strand (which we call the (+) strand). This strand becomes coated with SSB protein and is not replicated. Leading strand synthesis con-

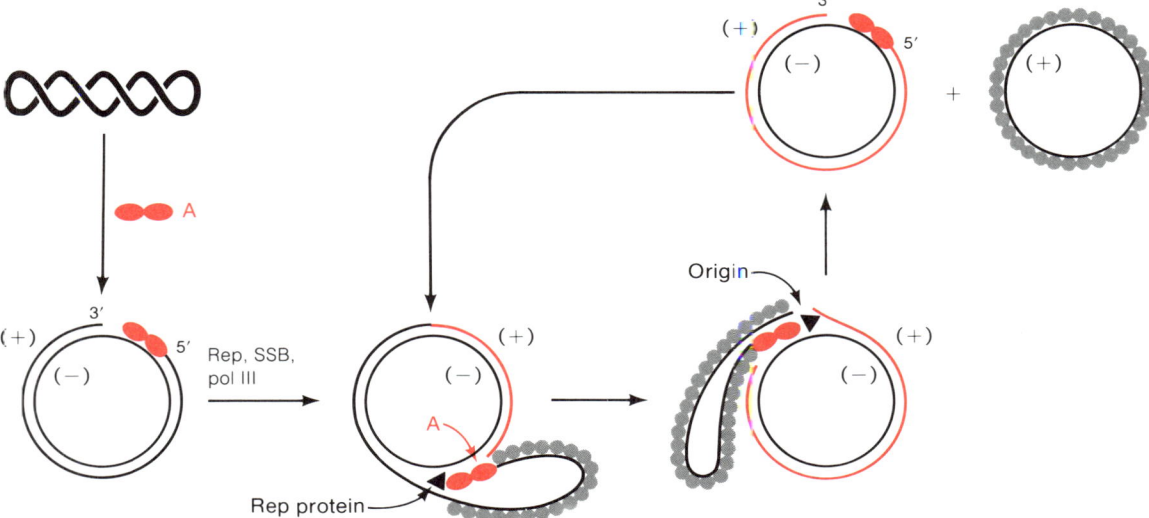

Figure 9-34 A diagram of looped rolling-circle replication of phage φX174. The gene-A protein nicks a supercoil and binds to the 5′ terminus of a strand (known as the + strand) whose base sequence is the same as that of the DNA in the phage particle. Rolling circle replication ensues to generate a daughter strand (red) and a displaced + single strand that is coated with SSB protein and still covalently linked to the A protein. When the entire + strand is displaced, it is cleaved from the daughter + strand and circularized by the joining activity of the A protein. The cycle is ready to begin anew. Note that the − strand is never cleaved.

tinues until the origin is reached. At this point, the *A*-gene product binds to the 3′-OH group of the (+) strand and, using the energy obtained from the original nicking event, it joins the 3′-OH and 5′-P groups of the (+) strand, dissociates, and attaches to the newly synthesized (+) strand. This process can continue indefinitely, generating numerous circular (+) strands. Note that in looped rolling circle replication, the displaced strand never exceeds the length of the circle, in contrast with ordinary σ replication.

BIDIRECTIONAL REPLICATION

In this section we examine the events following a D-looplike initiation step in which chain growth occurs continuously. It will be seen that unless special rules are imposed, replication will be bidirectional—that is, there will be two replication forks.

Somewhat after initiation of synthesis of the leading strand at the origin, the first precursor fragment is synthesized. This is shown in part I of Figure 9-35, in which the overall direction in which the replicating fork moves is counterclockwise. In our earlier analysis of lagging strand replication, it was noted that synthesis of each precursor fragment is terminated when the growing end reaches the primer of the previously synthesized fragment.

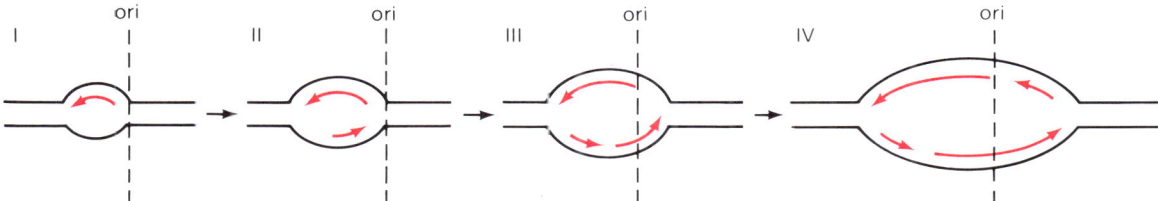

Figure 9-35 The formation of a bidirectionally enlarging replication bubble. I. The leftward-leading strand starts at *ori*. II. The leading strand has progressed far enough that the first rightward precursor fragment begins. III. The leftward-leading strand has progressed far enough that the second rightward precursor fragment has begun. The first rightward precursor fragment has passed *ori* and has become the rightward-leading strand. IV. The rightward-leading strand has moved far enough that the first leftward precursor fragment has begun. There are now two complete replication forks.

However, in the case of the first precursor fragment, there can be no previously made fragment. We may consider two possible events: (1) there is a termination signal for synthesis of the first fragment, or (2) there is no such signal and the precursor fragment continues to grow. In the second case, the precursor fragment is equivalent to a leading strand for a second replication fork, moving clockwise, as shown in the figure. Clockwise replication requires the synthesis of precursor fragments in the second replication fork but this can be achieved by the standard mechanism. The result of these events is that the DNA molecule will have two replication forks moving in opposite directions around the circle. This is called **bidirectional replication.** If alternative (1) occurs, there is a single fork and replication is unidirectional. There is no particular reason to have such a stop signal and, in fact, bidirectional replication has the advantage of halving the time required to complete the process. Bidirectional replication was first detected with *E. coli* phage λ and it has since been observed with many other phages, in bacteria and plasmids, and in animal and plant cells. Unidirectional replication has also been observed but less frequently.

The method used to study the direction of replication is called **denaturation mapping.** It is done as follows. If a DNA molecule is heated to a temperature at which melting is just detected, single-stranded bubbles form in the regions having very high A + T content. If formaldehyde (which reacts with the amino groups on the DNA bases) is added and the DNA is cooled, the bubbles persist. DNA treated in this way can be observed by electron microscopy and the position of the bubbles can be noted (Figure 9-36). These positions are fairly constant from one molecule to the next (Figure 9-37). When the technique is applied to a replicating molecule, the bubbles serve as fixed reference points against which the positions of a branch point can be plotted. Figure 9-38 shows the kinds of molecules that would be expected for unidirectional and bidirectional replication of a hypothetical molecule. Note that (1) the relation between the positions of the bubbles, and the branch points differ for the two modes and (2) a replication fork is identified by its changing distance from a bubble.

Figure 9-36 An electron micrograph of a partially denatured circular phage λ DNA molecule. With this method of sample preparation the single strands are very thin and faint compared to the double-stranded DNA. The arrows indicate the replication forks. [Courtesy of Manuel Valenzuela.]

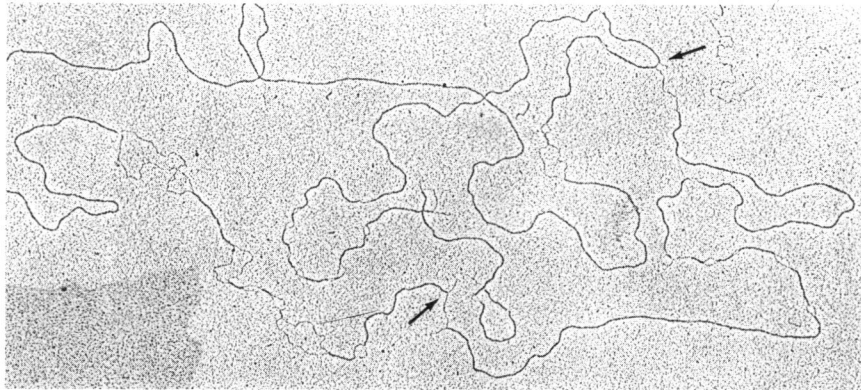

(a)   Map

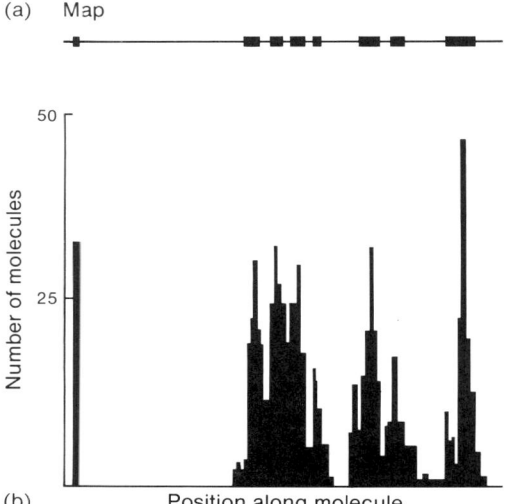

(b)

Figure 9-37 A denaturation map of *E. coli* phage λ DNA showing (a) the size and positions of most denaturation bubbles and (b) the number of molecules whose strands are separated at the positions shown, observed in a single experiment.

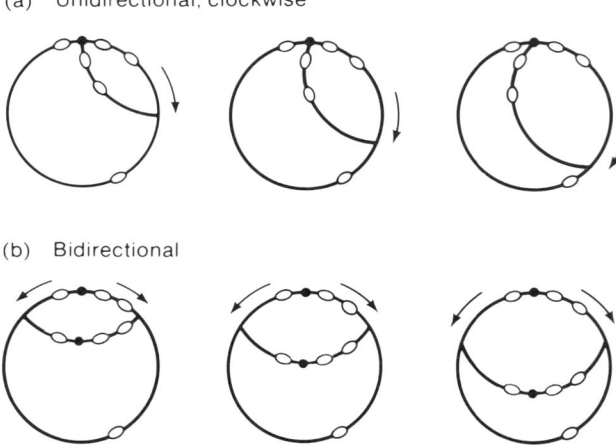

Figure 9-38 A diagram showing the relative positions of branch points and a replication bubble for a DNA molecule replicating (a) unidirectionally and clockwise or (b) bidirectionally. The arrows indicate the direction of movement of the replication fork and the dots indicate the replication origin.

In unidirectional replication, one branch point remains at a fixed position with respect to the bubbles; this position defines the replication origin. In bidirectional replication both branch points move with respect to the bubbles, so each branch point is a replication fork. If both replication forks move at the same rate, the origin is always at the midpoint of each branch of the replication loop. With phage λ it is found that the midpoints of all replication loops (that is, loops of all sizes) are at the same position with respect to the denaturation bubbles; this indicates that the two growing forks move at the same rate and also locates the origin. It has also been observed that there is one single-stranded region in each replication fork at opposite ends of each double-stranded branch of the loop; this is expected of a replication fork since each fork has a lagging strand (Figures 9-23 and 9-24).

Bidirectional replication has been widely observed with phage, bacterial, and plasmid DNA. A small number of phages and plasmids use the unidirectional mode exclusively, so a stop signal must exist that prevents chain growth of the first precursor fragment past the origin. One plasmid uses a surprising variant of the bidirectional mode in that initiation of movement of one fork occurs at a much later time than the primary initiation event.

Bidirectional replication is the major mode of chain growth with eukaryotes. However, since eukaryotic DNA is almost always fragmented in the course of isolation, a different technique was required for observing the replication forks of eukaryotic DNA. The method used which employed autoradiography, is shown in Figure 9-39(a). A culture of living cells was pulse-labeled with [³H]thymidine, the DNA was isolated, and then autoradiograms were prepared. Except in the rare case of labeling at the time of

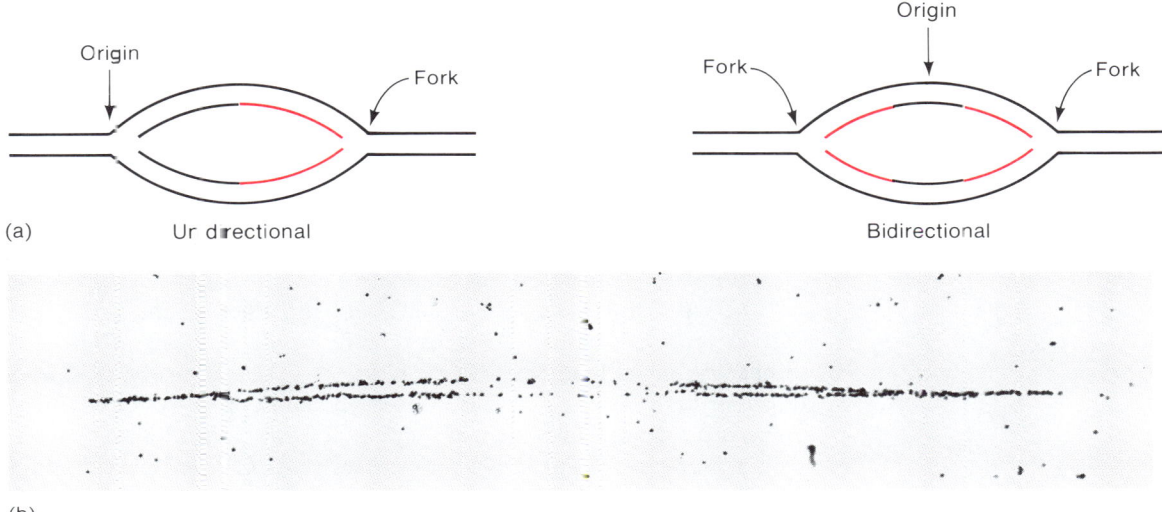

Figure 9-39 (a) Expected patterns of labeling for unidirectional and bidirectional replication. Radioactive DNA is shown in red. (b) Actual pattern of labeling. as seen in an autoradiogram. [From D. M. Prescott and P. Kuempel, *Meth. Cel Biol.*, (1973) 7: 147.] Note the very faint pattern of grains between the labeled V sections. This occurs because the cells had been grown for one generation in a medium having a very small amount of [³H]thymidine prior to pulse-labeling; this was done to allow the strands connecting the Vs to be visible.

initiation, a replication bubble will become radioactive only adjacent to the fork. Remembering that only radioactive DNA is visualized in an autoradiogram, we can expect that a unidirectionally replicating bubble will appear as a labeled V, whereas in bidirectional replication both forks are labeled and this will appear as two Vs whose arms point at one another. Figure 9–39(b) shows an actual autoradiogram that demonstrated bidirectional replication in animal cells.

TERMINATION OF REPLICATION

We have discussed initiation of a round of replication by formation of a replication fork and the events that occur while the fork moves along a parental DNA molecule. A round of replication is completed when a molecule is totally replicated. The termination event(s) is not well understood and must differ in circular and linear molecules; these two cases are considered in the following sections.

Termination of Replication of a Circle

In a unidirectionally replicating molecule, replication terminates at the origin. In a bidirectionally replicating molecule this clearly is not the case. There are two possible modes of termination: (1) there is a defined termination sequence, or (2) two growing points collide and termination occurs wherever the collision point happens to be. In both cases, termination might occur exactly halfway around the circle (at the antipode of the origin). Both termination modes have been observed. In several plasmids it has been observed by denaturation mapping that one growing fork stops at a fixed position before the antipode is reached; the other fork advances until this termination site is reached. Why this first fork stops moving is not known. Possibly, a termination protein sits at this point or some base sequence signals dissociation of the replication machinery.

    Termination has a topological problem. When double-stranded circular DNA replicates semiconservatively, the result is a pair of circles that are linked as in a chain. Such a structure is called a **catenane** (Figure 9-9). Catenated molecules have been observed in numerous systems, and evidence is accumulating to indicate that they result from replication. Apparently, they are a precursor to the separated circles that ultimately result. Figure 9-9 shows that DNA gyrase is capable of decatenating two circles, and it has been believed for some time that gyrase is the enzyme responsible for separation of daughter molecules. Support for this hypothesis comes from a study of replication of the *E. coli* nucleoid (Chapter 7) in a bacterial mutant that makes a temperature-sensitive DNA gyrase. Nucleoids isolated from these cells grown at temperatures at which gyrase was active (permissive temperature) appeared in a microscope as a spherical object. If the cells were instead grown at a nonpermissive temperature (gyrase inactive), paired spheres accumulated, whose size and appearance was consistent with the presence of two completely synthesized *E. coli* chromosomes. Treatment of isolated nucleoid pairs with gyrase caused them to separate into individual units.

Decatenation of replicating SV40 viral DNA (a bidirectionally replicating circle) seems to require a particular termination site. In normal replication the products of one round of replication include both catenanes and separated circles with short gaps in one strand. If a piece of DNA is inserted into SV40 such that the replication forks collide at a point other than the antipode, only catenanes result.

Termination of the replication of a linear DNA molecule lacks the topological problem just described. However, a more serious difficulty occurs, as will be seen in the next section.

Termination of Linear DNA Molecules

E. coli phage T7 DNA replicates as a linear molecule. The origin is located 17 percent of the total distance from the left end of the molecule and replication is bidirectional (Figure 9-40). Initially there is a single replication bubble (molecule I), and when the leftward fork reaches the terminus, the molecules assume a Y form (molecule III). A problem in termination can be seen by examining the synthesis of the final precursor fragment in each fork (Figure 9-41). Assuming that priming could occur at each terminus, we can see that two linear molecules would result, each having a single terminal RNA segment (only one is shown in the figure). Ribonucleases exist that can remove this RNA but once it has been removed, either a 5'-OH or a 5'-P group (depending on the particular RNase) would remain at the ends of the molecule, and no known polymerase can act at these termini. The molecules are nevertheless completed and no RNA remains. How this is accomplished is not known, but the following hypothesis, which accounts for all of the facts, has been made and is widely believed

T7 DNA has a stretch of several hundred nucleotides that is repeated at each end of the DNA molecule (Figure 9-42). This is called **terminal redundancy** and it means that unreplicated bases at the right and left ends of the daughter molecule are complementary to free single-stranded sequences at the opposite end of the molecule. It is hypothesized that these sequences on two separate molecules pair, and the excess single-stranded material is removed by a DNase, as shown in the figure. The result is a dimer called a **concatemer.** Each unit of the dimer contains a replication origin so the dimer can also replicate. The result of this round of replication is a pair of daughter dimers, each of which has complementary single-stranded termini. Once again, reannealing and trimming can occur to yield a tetramer. This process can continue indefinitely to form a gigantic linear concatemer. At any time

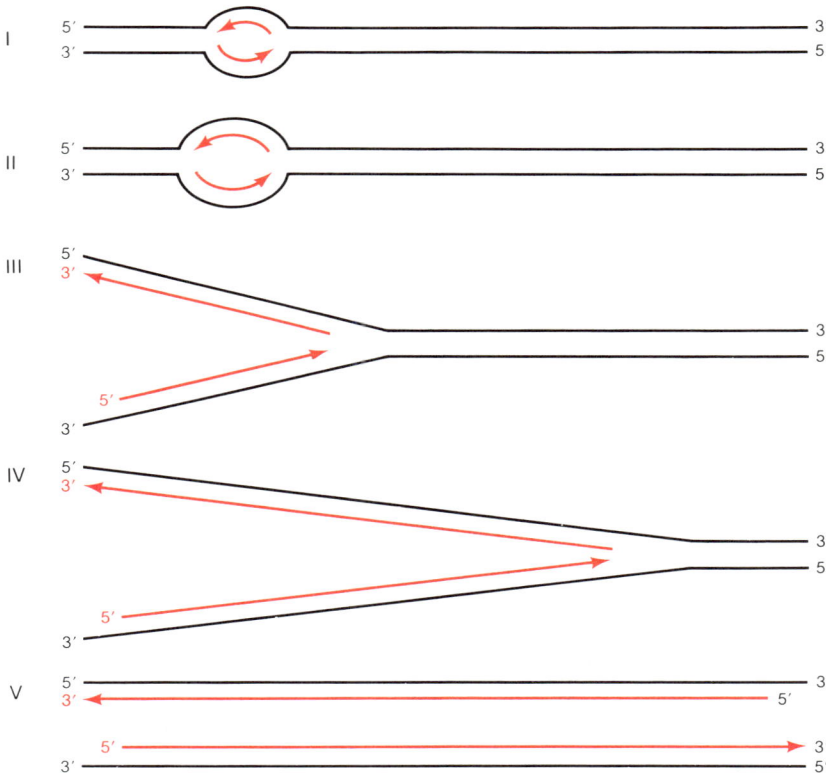

Figure 9-40   Origin and direction of replication of T7 DNA. Bidirectional replication starts at a point about 17 percent from the left end. Because the two forks move at equal rates, the left end is finished first. Note that the region at the 3' end of each template strand remains unreplicated. The daughter strands are shown in red. The arrows show the direction of chain growth but the precursor fragments are drawn as if already joined to the lagging strand.

Figure 9-41   Events at the terminus of a replicating linear molecule. This is an expanded view of the upper molecule of row V in Figure 9-40. I. Synthesis of the final precursor fragment begins by extension from an RNA primer. II. The precursor fragment in I is joined to the lagging strand. III. RNA is removed, leaving a 5'-P-terminated gap that cannot be filled by any known polymerase.

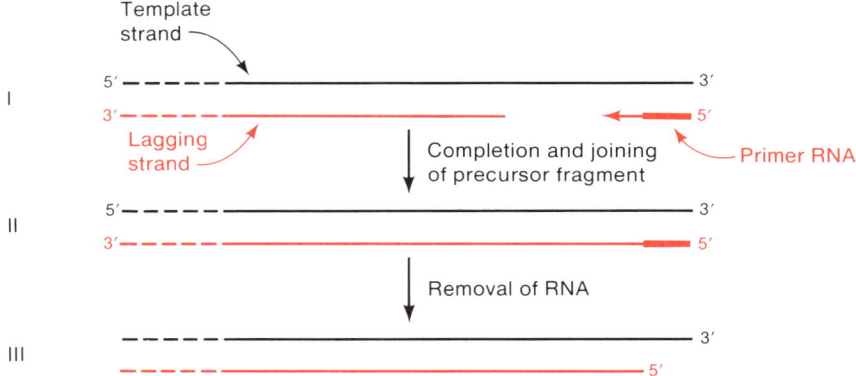

Two daughter molecules

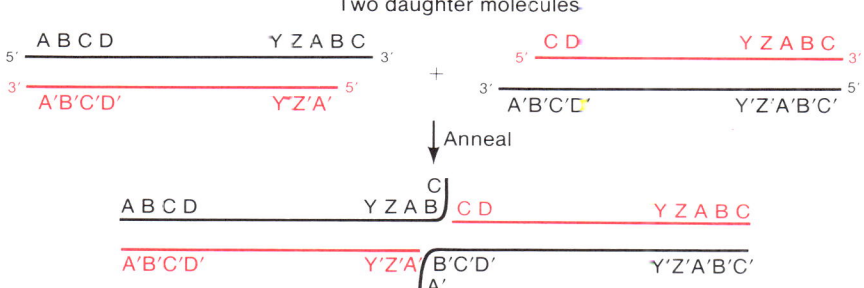

Figure 9-42   A hypothesis for the generation of concatemers from two incomplete daughter molecules shown in the lowermost section of Figure 9-41. Note that the color coding of the single strands is the same as in Figure 9-41. Annealing of the daughter molecules is made possible by the terminally redundant sequence ABC. The excess unmatched single strands bearing A' and C are presumably removed by a DNase. When the dimer replicates, two daughter molecules result that are terminally incomplete, as in Figure 9-41. These can anneal to form tetramers.

linear T7 DNA molecules can be cut out of this structure. Huge concatemers of T7 DNA have been isolated from phage-infected cells, which lends support to this hypothesis.

A eukaryotic chromosome contains a single linear DNA molecule and hence has a termination problem. Replication of these molecules does not include a concatemeric stage, so the mechanism proposed for T7 cannot apply. Little is known about how replication of eukaryotic chromosomes is terminated except that telomeres (Chapter 7) are essential. A variety of models have been proposed but none have been proved. Several are based on the observation that telomeres have a hairpin structure. Termination of replication of a hairpin is straightforward, though separation of daughter molecules is a significant problem (Figure 9-43).

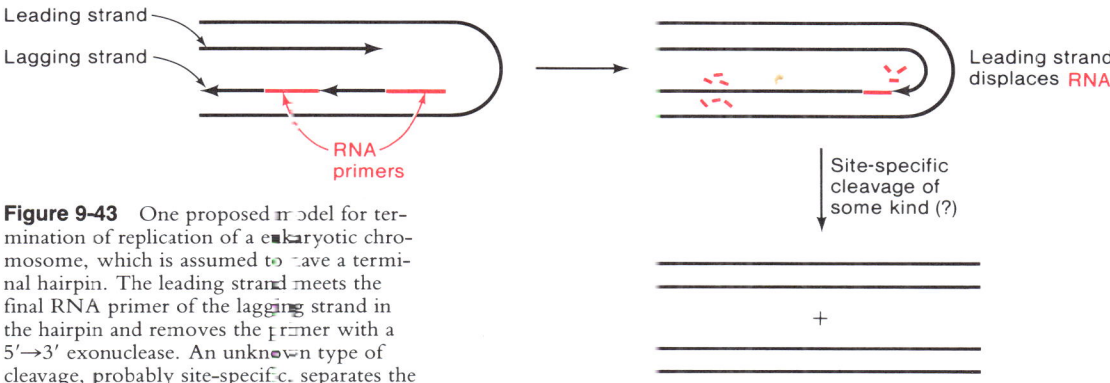

Figure 9-43   One proposed model for termination of replication of a eukaryotic chromosome, which is assumed to have a terminal hairpin. The leading strand meets the final RNA primer of the lagging strand in the hairpin and removes the primer with a $5' \rightarrow 3'$ exonuclease. An unknown type of cleavage, probably site-specific, separates the daughter strands.

METHYLATION OF DNA
AND MISMATCH REPAIR

In Chapter 4 the existence of methylated bases, particularly cytosine and adenine, was mentioned. The methylation of adenine, which in *E. coli* is accomplished by the product of the *dam* gene, has been studied carefully. Two important observations are that the methylated adenine is normally contained in the sequence G-A-T-C, indicating that there is sequence specificity, and that there is a gradient of methylation along a newly synthesized daughter strand, the least methylation occurring at the replication fork. The parental strand is always uniformly methylated. Different methylation of parental and daughter strands during replication is important for fidelity of replication.

As already described, both pol I and pol III occasionally catalyze incorporation of an "incorrect" base, which cannot form a hydrogen bond with the template base in the parental strand; such errors are usually corrected by the editing function of these enzymes. However, the integrity of the base sequence of DNA is so important that a second system exists for correcting the occasional error missed by the editing function. This correction system is called **mismatch repair.** In mismatch repair a pair of non-hydrogen-bonded bases is recognized as incorrect and a polynucleotide segment is excised from one strand, thereby removing one member of the unmatched pair. The resulting gap is filled in by pol I, which presumably uses this "second chance to get it right" to form only correct base pairs; then the final seal is made by DNA ligase.

If it is to correct but not create errors, the mismatch repair system must be able to distinguish the correct base in the parental strand from the incorrect base in the daughter strand. If it were unable to do this, the correct base might sometimes be replaced by the complement to the incorrect base, thereby producing a mutation. Studies with *dam*⁻ (methylation-defective) mutants have suggested how this distinction is made. For any genetic locus the mutation frequency in a *dam*⁻ mutant is much higher than in a *dam*⁺ bacterium. This indicates that incorrectly incorporated bases are less frequently corrected in a *dam*⁻ mutant than in the wild type. The reason for this is that the mismatch repair system recognizes the degree of methylation of a strand and *preferentially excises nucleotides from the undermethylated strand* (Figure 9-44). The daughter strand is always the undermethylated strand, as its methylation lags somewhat behind the moving replication fork; the parental strand is fully methylated, having been methylated in the previous round of replication.

The mismatch repair system also counteracts the mutagenic effects of tautomeric bases. Such molecules, in their rare forms, can occasionally be misincorporated into DNA. At the time of incorporation, the base will be correctly hydrogen-bonded to the template strand, so it is not recognized by the editing function as incorrect. However, when the base later assumes

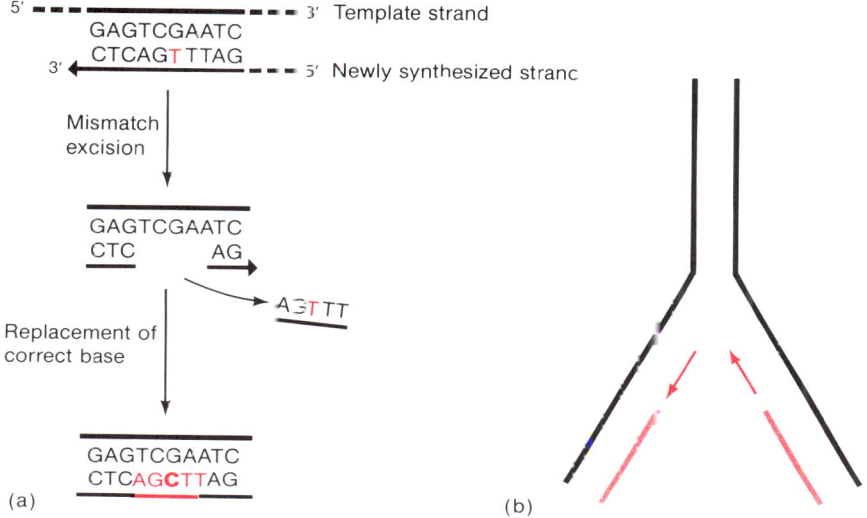

Figure 9-44 Mismatch repair. (a) Excision of a short segment of a newly synthesized strand, and repair synthesis. (b) Methylated bases in the template strand direct the excision mechanism to the newly synthesized stand containing the incorrect nucleotide. The regions in which methylation is complete are pink; the regions in which methylation may not be complete are solid red.

its normal structure, a mismatched base pair will be present. The mismatch repair system can correct such errors, but if the elapsed time is so great that the region of the daughter strand containing the incorrect base has become methylated, the mismatch repair system will be unable to distinguish parental and daughter strands, and mutations can occur.

REPLICATION OF EUKARYOTIC CHROMOSOMES

The replication of eukaryotic chromosomes presents many problems not found in the prokaryotes because of the enormous size of eukaryotic chromosomes and the geometric complexity imposed by the organization of the DNA into nucleosomes. How these problems are handled is described in this section. Termination has already been discussed briefly.

Eukaryotic DNA Polymerases

The enzymology of replication in eukaryotes is poorly understood, primarily because *in vitro* systems consisting of purified components have yet to be developed. In bacteria, mutants can be isolated that lack a particular replication factor; extracts of these mutants serve as assay systems for purifying the missing factor from an extract of wild-type cells. Such mutants are not available for the higher eukaryotes, but are being isolated in eukaryotic unicells such as yeast. A yeast system will soon be available.

Animal cells have three DNA polymerases: pol α, pol β, and pol γ. Pol α is located in the nucleus and is responsible for replication of the chromosome. Pol β is also nuclear, but its function is unknown. Pol γ is found both in the nucleus and in mitochondria and is believed to be used solely to replicate mitochondria DNA. Pol α lacks the $3' \to 5'$ and $5' \to 3'$ exonuclease activities of bacterial polymerases; these activities reside in distinct proteins. Little more of significance is known about animal DNA polymerases.

Multiple Forks in Eukaryotic DNA

The rate of movement of a replication fork in *E. coli* is about 10^5 base pairs per minute. In eukaryotes the polymerases are much less active and the rate ranges from 500 to 5000 base pairs per minute. Since a typical animal cell contains about 50 times as much DNA as a bacterium, the replication time of an animal cell should be about 1000 times as great as that of *E. coli,* or about 30 days. However, the duration of the replication cycle is usually several hours, and this is accomplished by having multiple initiation sites. For instance, the DNA of the fruit fly *Drosophila* has about 5000 initiation sites, each separated by about 30,000 bp, and each site replicates bidirectionally. The number of sites is regulated in a way that is not understood. For example, in the round of replication following fertilization of *Drosophila* eggs, the number of initiation points reaches 50,000, and it takes only three minutes to replicate all of the DNA. An example of a fragment of this rapidly replicating *Drosophila* DNA is shown in Figure 9-45.

The enormous number of growing forks in eukaryotic cells is reflected in the number of polymerase molecules. In *E. coli* there are between 10 and 20 molecules of pol III holoenzyme. However, a typical animal cell has 20,000 to 60,000 molecules of polymerase α.

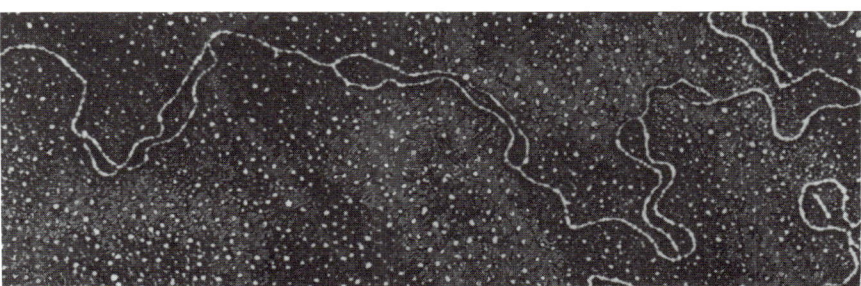

Figure 9-45 Replicating DNA of *Drosophila melanogaster* showing many replicating eyes. The molecular weight of the segment shown is roughly 20×10^6. [Courtesy of David Hogness.]

Replication of Chromatin

Replication of double-stranded DNA proceeds through both a polymerization step (nucleotide addition) and a dissociation step (strand separation). Replication of chromatin, which is the form that DNA has in eukaryotes, proceeds through an additional dissociation step—namely, dissociation of DNA and histone octamers—and a histone-DNA reassociation step (see Chapter 7 for a discussion of the structure of the octamers contained in nucleosomes). In chromatin, DNA is wrapped around a histone octamer to form a nucleosome, and if the DNA were never unwrapped from the histone spool, severe geometric problems would arise at the growing fork. Moreover, after DNA dissociates from the histones, newly formed DNA must rejoin with the nucleosomal octamers in order that each daughter molecule will be organized into nucleosomes, just as the parent was. Clearly some mechanism is needed for regulating the histone-DNA dissociation and reassociation reactions.

Examination of replication forks in DNA that has not been deproteinized during isolation indicates that nucleosomes form very rapidly after replication. For example, Figure 9-46 shows that all portions of a replication eye have the beadlike appearance characteristic of nucleosomes.

The synthesis of histones occurs simultaneously with DNA replication—that is, histones are made in the cell as they are needed, so the cell does not contain an appreciable amount of unassociated histone molecules. In light of this, we would like to know whether newly synthesized histones mix with parental histones in the octamers associated with daughter DNA molecules. Three models have been considered (Figure 9-47): (1) Parental histone octamers are conserved; i.e., the octamers do not dissociate. (2) The parental histone octamers totally dissociate to monomers. Parental monomers mix at random with new histone monomers, re-form octamers, and then rejoin the DNA molecule. (3) The parental histone octamers dissociate not to monomers but to tetramers (or possibly dimers), which then mix with new histones prior to reassembly of nucleosomes. Several critical experiments have demonstrated unambiguously that model 1 is correct.

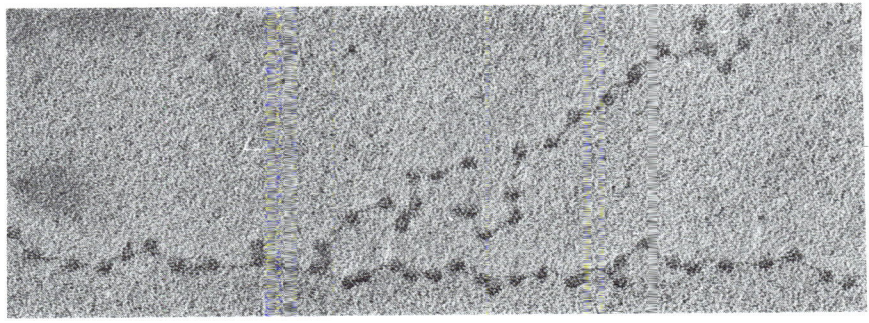

Figure 9-46   A replicating fork showing nucleosomes on both branches. The diameter of each particle is about 110 Å. [Courtesy of Harold Weintraub.]

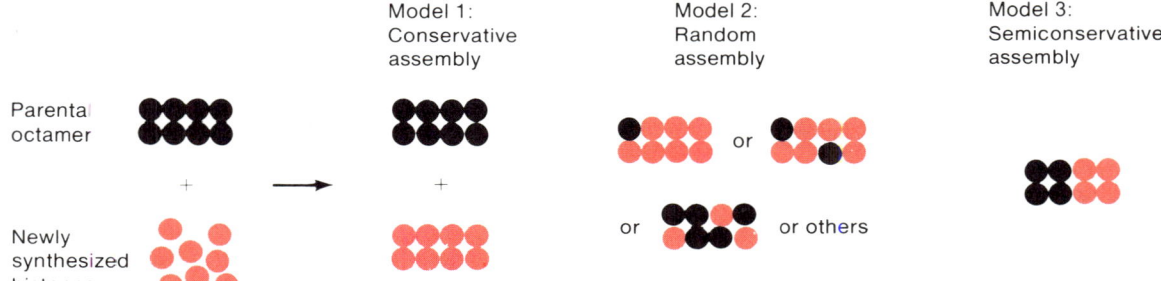

Parental octamer

Newly synthesized histones

Model 1: Conservative assembly

Model 2: Random assembly

or   or

or   or others

Model 3: Semiconservative assembly

Figure 9-47 Three models for the assembly of newly made histone octamers.

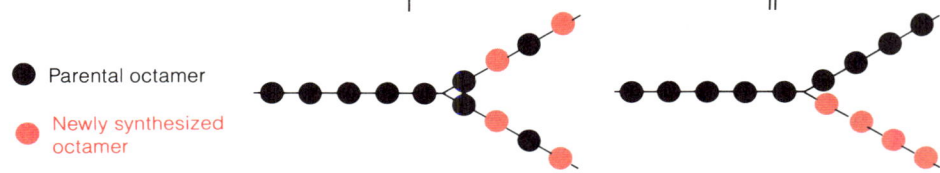

● Parental octamer

● Newly synthesized octamer

I   II

Figure 9-48 Schematic diagram showing two arrangements of parental and newly synthesized octamers in the replicating region. Arrangement II is the correct one.

Other experiments show how the conserved histone octamers are arranged on the daughter strands. Two possibilities are shown in Figure 9-48. In the first panel the parental octamers are distributed between both daughter strands, which would imply that the octamers dissociate completely from the DNA during replication. In the second arrangement all parental octamers are located on one daughter strand. The results indicated that arrangement II is the correct one. This conclusion leads to a clear prediction about the appearance of chromatin replicated in the absence of protein synthesis. Since histones are made as they are needed, replication forks should be seen in which one branch is free DNA. The reason and confirmation of this point is shown in Figure 9-49.

The final question is whether the parental octamers are associated with a particular strand (leading or lagging strand). So far, it has been possible to answer this question for only one system—an animal virus, SV40, whose DNA is also organized in nucleosomes (which may not be true of all viral DNA). In this system the parental octamers are associated exclusively with the leading strand.

Since DNA replication in each eye is bidirectional and the parental octamers are always associated with leading strand synthesis, then each branch of an eye is covered with a long tract of parental octamers and a long tract of new octamers. The dividing point between these tracts is the replication origin of each eye, as shown in Figure 9-50.

DNA made during a period of
inhibition by cycloheximide

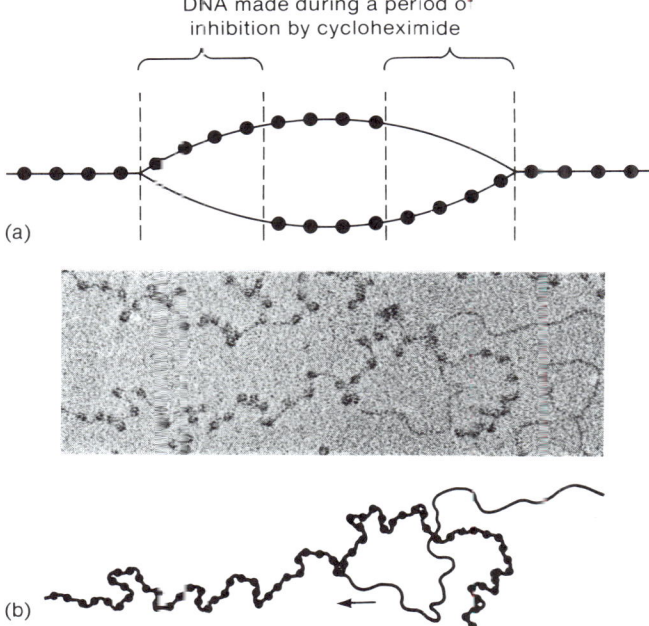

(a)

(b)

Figure 9-49 Replication of chromatin in the presence of cycloheximide, an inhibitor of protein syhthesis. (a) A schematic diagram of an eye. (b) Replicating chromatin from cycloheximide-treated cells. Top panel: a replication fork. Lower panel: an interpretive drawing. Particles have a cross-section of 110 Å [Courtesy of H. Weintraub.]

Origin

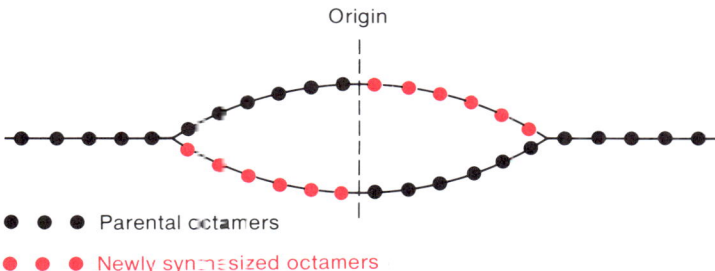

● ● ● Parental octamers

● ● ● Newly synthesized octamers

Figure 9-50 The arrangement of parental and newly synthesized octamers in an eye, when parental octamers are associated with the lagging strand.

Even though assembly of nucleosomes normally accompanies replication, we may ask whether replication is a prerequisite. The question has been answered in one system. SV40 viral DNA was injected into oocytes of *Xenopus laevis*. The DNA was unable to replicate in these cells. Because of the low concentration of histones, a small fraction of the DNA molecules bound histone octamers. Then, histones were injected into the cells. Very quickly the SV40 DNA assembled into nucleosomes, indicating that assembly does not require DNA replication.

We may also inquire whether assembly is a spontaneous process or requires additional factors. Recall from Chapter 7 that chromatin can be formed *in vitro* by incubating DNA and histones in solutions of high salt concentration. The required concentrations are much higher than that in cells. If physiological (low-salt) concentrations are used, histone DNA aggregates form, but they lack the regular structure of native chromatin. However, nucleosome assembly of SV40 DNA occurs in *Xenopus* oocyte cell-free extracts at low salt concentration. Fractionation of the extracts has yielded a nuclear protein essential for assembly. The protein is called **nucleoplasmin;** it has since been found in many eukaryotic species. The mechanism of action of nucleoplasmin has not been elucidated.

10

Repair

There is no single molecule whose integrity is as vital to the cell as DNA. Thus, in the course of hundreds of millions of years there have evolved efficient systems for correcting the occasional mistakes that occur during replication. Two of these have already been discussed in the previous chapter—namely, the editing function of both DNA polymerase I and polymerase III and the uracil *N*-glycosylase system. DNA is also subject to damage in many ways, by environmental agents such as radiation as well as by intracellular chemicals; thus, it is understandable that repair systems exist for eliminating a variety of altered structural features of DNA. However, repair must not be perfect; if so, mutations would never persist and a species acquiring the ability to repair DNA alterations perfectly would no longer evolve.

In this chapter we examine what is recognized by the repair systems and how the correct structure is restored. We begin by describing the principal kinds of damage that can occur in a DNA molecule.

ALTERATIONS OF DNA MOLECULES

There are three distinct mechanisms for altering the structure of DNA: (1) base substitutions during replication, (2) base changes resulting from the inherent chemical instability of the bases or of the *N*-glycosidic bond, and (3) alterations resulting from the action of other chemicals and environmen-

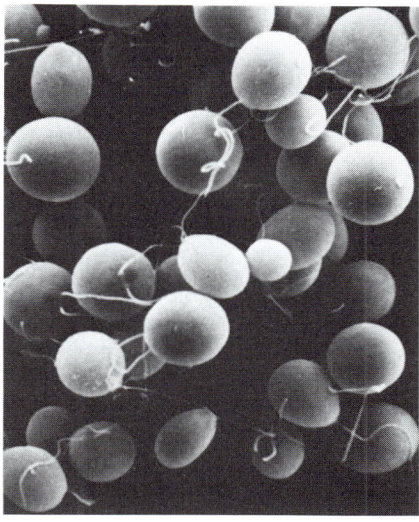

Scanning electron micrograph of the alga *Chlamydomonas*. Microorganisms are continually exposed to solar radiation and to hazardous chemicals and hence have efficient repair systems. [From G. Shih and R. Kessel. 1982. *Living Images.* Jones and Bartlett.]

277

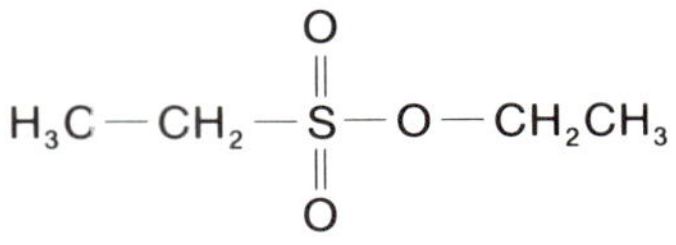

Ethyl ethane sulfonate

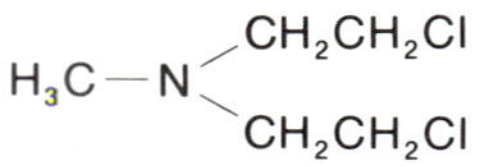

**Methyl-bis-(β-chloroethyl)amine,
a nitrogen mustard**

Figure 10-1 Alkylating agents.

tal agents. These mechanisms are responsible for the occurrence of the following defects:

1. *An incorrect base in one strand that cannot form hydrogen bonds with the corresponding base in the other strand*. This defect can result from a replication error that by chance is not corrected by the editing function; presumably this failure is rare. A more common event is spontaneous deamination of cytosine to uracil, followed by replacement of the uracil by thymine in subsequent rounds of replication. Less frequent than the cytosine → uracil → thymine transition is occasional deamination of adenine to form hypoxanthine, which forms base pairs with cytosine instead of thymine. Hypoxanthine N-glycosylase corrects this alteration. Deamination will not be discussed further in this chapter as the mechanism for repair of this type of alteration has already been explained in Chapter 9 (see Figure 9-22).

2. *Missing bases*. The N-glycosidic bond of a purine nucleotide is spontaneously broken at physiological temperatures, though at a very low rate. This process is called **depurination** because the purine is lost from the DNA. The rate of spontaneous depurination is about one purine removed per 300 purines per day at pH 7 and 37°C, which amounts to about 10^4 purines per day in a mammalian cell and 0.25 purines per day per generation time for a bacterium. The rate is increased as the pH is lowered or as the temperature is elevated. Noxious chemicals called **alkylating agents** (Figure 10-1), once used in chemical warfare and now used in cancer treatment, react primarily with guanine, adding an alkyl group to N-7 of the purine ring. This alkylation weakens the N-glycosidic bond and greatly increases the rate of depurination.

Since breakage of the N-glycosidic bond is the first step in the N-glycosylase repair systems, restoration of the correct structure uses the apurinic acid (AP) nuclease pathway already shown in Figure 9-21.

3. *Altered bases*. Bases can be changed into strikingly different compounds by a variety of chemical and physical agents. For instance, ionizing radiation (such as the β particles emitted by naturally occurring radioisotopes or x rays) can break purine and pyrimidine rings and can cause several types of chemical alterations—the most frequent substitutions are made in thymine. Free radicals produced in many metabolic reactions can also cause a variety of significant changes. The best-studied altered base is the intrastrand dimer formed by two adjacent pyrimidines as a result of ultraviolet radiation. The most prominent of these dimers is the thymine dimer shown in Figure 10-2. The significant effects of the presence of thymine dimers are the following: (1) the DNA helix becomes distorted as the thymines, which are in the same strand, are pulled toward one another (Figure 10-3) and (2) as a result of the distortion, hydrogen-bonding to adenines in the opposing strand, though possible (because the hydrogen-bonding groups are still pres-

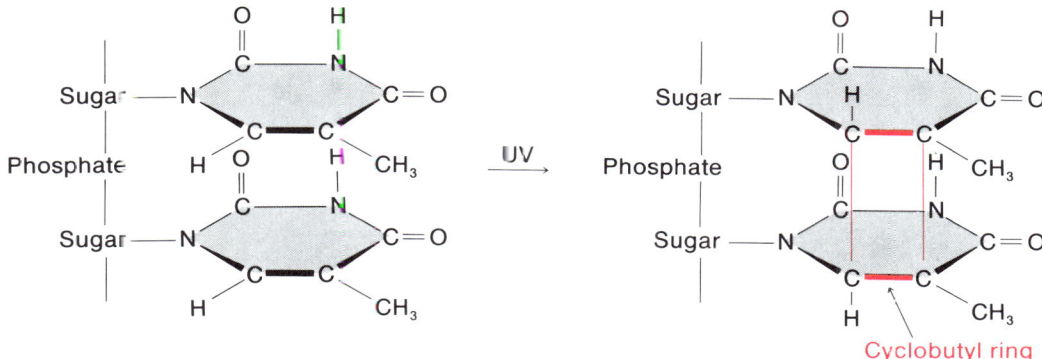

Figure 10-2 Structure of a cyclobutylthymine dimer. Following ultraviolet (UV) irradiation, adjacent thymine residues in a DNA strand are joined by formation of the bond, shown in red. Although not drawn to scale, these bonds are considerably shorter than the spacing between the planes of adjacent thymines, so the double-stranded structure becomes distorted. The shape of the thymine ring also changes as the C=C double bond (heavy horizontal line in left panel) of each thymine is converted to a C—C single bond (horizontal red lines in right panel) in each cyclobutyl ring.

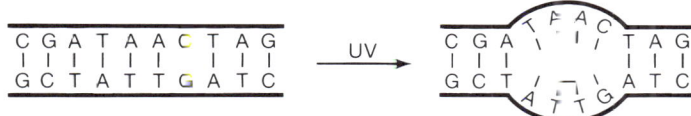

Figure 10-3 Distortion of the DNA helix caused by two thymines moving closer together when joined in a dimer. The dimer is shown as two joined lines.

ent), is significantly weakened; this structural distortion causes inhibition of advance of the replication fork, as will be discussed in a later section.

4. *Single-strand breaks.* A variety of agents can break phosphodiester bonds. Among the more common chemicals are peroxides, sulfhydryl-containing compounds (for example, cysteine) and metal ions such as Fe^{2+} and Cu^{2+}. Ionizing radiation produces strand breaks both by the action of secondary electrons produced by a β particle or an x-ray photon passing nearby and by production of free radicals in water (e.g., ·OH) that can attack the bond. In the case of energetic radiation the bases adjacent to the broken bond are usually damaged also. DNases that are present in cells probably also make frequent phosphodiester scissions.

5. *Double-strand breaks.* If a DNA molecule receives a sufficiently large number of randomly located single-strand breaks, two breaks may be situated opposite one another, resulting in breakage of the double helix. Actually, the breaks need not be directly opposite one another to constitute a double-

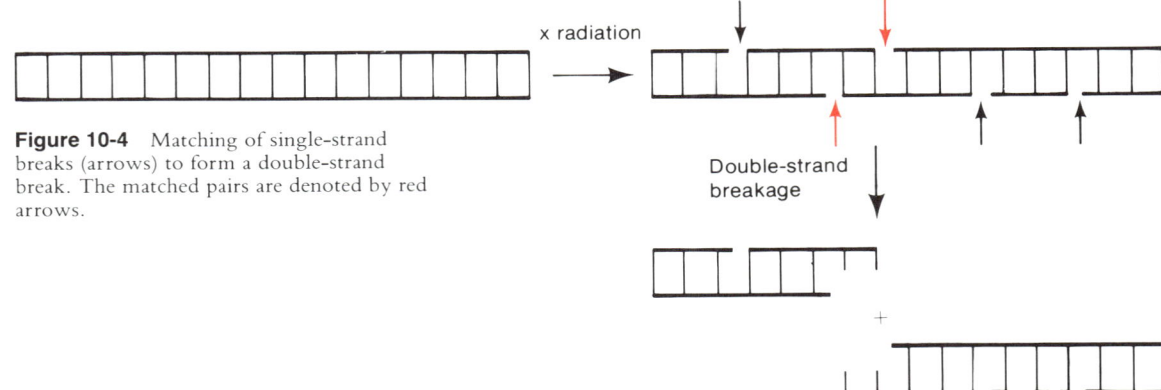

Figure 10-4 Matching of single-strand breaks (arrows) to form a double-strand break. The matched pairs are denoted by red arrows.

strand break but can be separated by a few base pairs (Figure 10-4). Double-strand breaks also form in a single event. This can result from exposure to highly ionizing radiation (for example, x rays, naturally occurring cosmic rays, electron beams, γ rays, and radiation beams of various kinds used in radiotherapy). When highly ionizing radiation passes through water, large clusters of electron pairs and free radicals form in volumes of about 5 Å in diameter. These clusters cause local, multiple alterations. Double-strand breaks arising in this way are invariably accompanied by adjacent base damage.

6. *Cross-linking.* Some antibiotics (for example, mitomycin *C*) and some reagents (the nitrite ion) can form covalent linkages between a base in one strand and an opposite base in the complementary DNA strand. This prevents strand separation during DNA replication and also causes a local distortion of the helix.

Each of the defects just described can be corrected by one of several repair systems.

Repair was first observed and is best understood in bacteria. However, it is a widespread and probably universal phenomenon and is now well documented in eukaryotic microorganisms and in mammalian cells. In the following section we show how repair is recognized both biologically and chemically.

BIOLOGICAL INDICATION OF REPAIR

The existence of repair was first suggested by several observations of the effect of ultraviolet light on the ability of bacteria to form colonies. Once it was realized that DNA is the genetic material, it was expected that ultraviolet light in a wavelength range absorbable by DNA would be able to cause chemical damage and hence either death or mutagenesis in an irradiated cell. The means of investigating this possibility is to obtain a **survival curve.** This is done in the following way.

Samples are drawn at intervals from a population of bacteria irradiated by ultraviolet light. The samples are plated on nutrient agar and the colonies that form are counted. The proportion of cells able to produce colonies is plotted as a function of ultraviolet dose; it is found that colony-forming ability decreases as the dose increases. An example of a survival curve is shown in Figure 10-5. If different wavelengths of ultraviolet light are used, it is found that the steepness of the survival curve varies. A graph of steepness (actually the slope) vs. wavelength matches the absorption spectrum of DNA; this provides strong evidence that DNA is the target molecule for the effect.

There are many types of repair. The first type that was recognized is called **photoreactivation**—that is, the survival of various ultraviolet-irradiated bacteria is higher if the bacteria are exposed to intense visible light (for example, sunlight) after ultraviolet irradiation but before plating (Figure 10-6(a,b)). Thus, visible light eliminates some of the damage introduced by the ultraviolet light.

Another response to damage that shows that repair occurs is **liquid-holding recovery.** If ultraviolet-irradiated cells are held in a nonnutrient buffer for several hours before plating, the surviving fraction for a particular dose is increased (Figure 10-6(c)). When first observed, it was suggested that delaying cell growth or perhaps merely DNA replication created additional

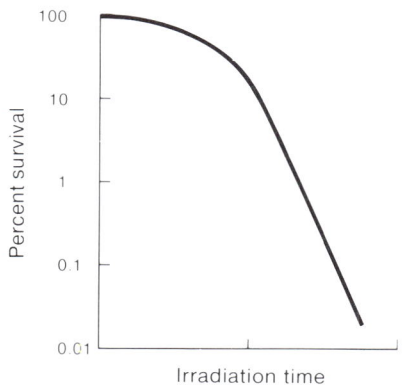

Figure 10-5 A typical ultraviolet-light survival curve for a bacterium. Initially, the curve is fairly flat because initial damage does not cause killing; this could either mean that several lethal hits are needed to kill a cell or that, at low doses, a great deal of the damage is repaired. Note that the y-axis is logarithmic.

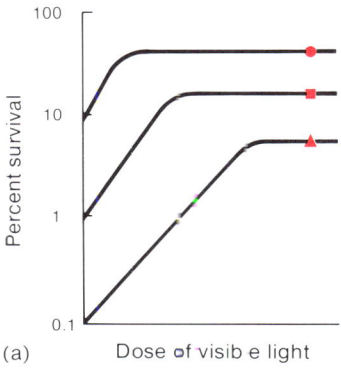

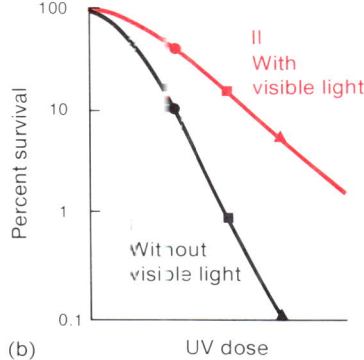

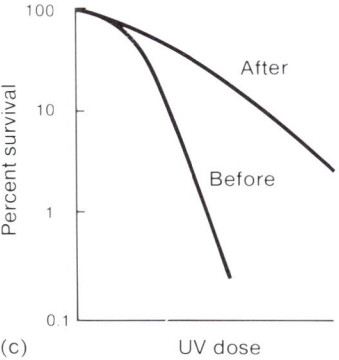

Figure 10-6 Types of repair. (a) Increase in survival of three different samples of ultraviolet-light-irradiated bacteria as a function of the dose of visible light. This is called photoreactivation. (b) A pair of survival curves showing the effect of postirradiation with visible light. Curve I consists of the points taken from the y-axis of part (a) and curve II is a plot of the red points taken from the plateaus in part (a). (c) Survival curves before and after incubation in buffer following ultraviolet-light irradiation. This is called liquid-holding recovery.

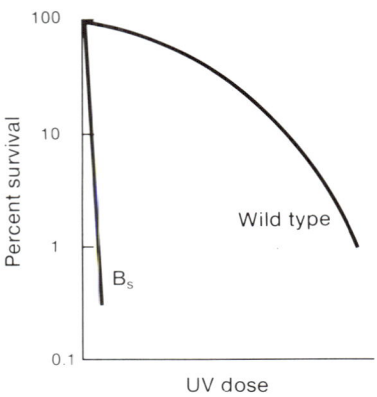

Figure 10-7 Survival curves showing the great sensitivity to ultraviolet light of the mutant *E. coli* B$_s$.

time for some repair process to occur. Furthermore, since liquid-holding recovery takes place without light from any source, it was hypothesized that *E. coli* possesses two distinct repair processes—a light-dependent one (photoreactivation) and a light-independent one. Proof of the existence of two repair systems came from the isolation of an extraordinarily ultraviolet-sensitive *E. coli* mutant called B$_s$ (Figure 10-7). The mutant will not recover its growth potential after ultraviolet irradiation when held in a buffer in the dark but it does photoreactivate normally. The isolation of another mutant called *pho⁻*, which cannot photoreactivate but can undergo liquid-holding recovery, confirmed the conclusion that there are two repair systems. Liquid-holding recovery is now known to be a manifestation of a general phenomenon called **dark repair.** The chemical mechanisms of both photoreactivation and dark repair are described in a later section.

Other mutants have been found that indicate the existence of other repair systems. For example, *E. coli rorA⁻* is excessively sensitive to x rays but not to ultraviolet light, indicating that there is a system that specifically repairs x-ray damage. It should be noted that the existence of these mutants exemplifies how the isolation of a mutant indicates that a function exists.

It might seem surprising that cells should have evolved systems for repairing damage caused by ultraviolet light, which is not an agent commonly encountered in nature by most cells. To understand this, one must remember that the properties of cells reflect ancient history. During the period when life was evolving in the primordial seas, the earth was not enveloped by a stratospheric ozone layer. Without this layer a very high flux of ultraviolet light would have reached the surface of the earth and would have been very damaging to primitive organisms. Thus, acquiring the ability to repair ultraviolet damage had great survival value and this ability has persisted for eons because, as will be seen shortly, the dark repair system is also able to repair other types of damage. How photoreactivation has persisted is less clear since it seems to be active only against the pyrimidine dimers formed by ultraviolet irradiation.

IDENTIFICATION OF THE THYMINE DIMER AS A PRIMARY LESION THAT IS REPAIRABLE

If purified DNA whose thymines are ³H-labeled is ultraviolet-irradiated and then acid-hydrolyzed to obtain free bases, [³H]thymine dimers can be isolated and the mole fraction of thymine recoverable as dimers will increase with ultraviolet dose. Thymine dimers can also be isolated from purified [³H]DNA obtained from ultraviolet-irradiated cells. In an ultraviolet-irradiated population of bacteria, in which only half of the bacteria are killed, thousands of dimers are present in each cell immediately after irradiation. This number is very large because dimers are repairable, as shown by the following experimental results:

1. A population of bacteria is given a single dose of ultraviolet light and then irradiated with various doses of high-intensity visible light. The DNA is isolated, the number of thymine dimers per cell is measured, and it is found that the number of thymine dimers per cell decreases continually with increasing doses of visible light. No radioactive molecules other than thymine are detected. Thus, photoreactivation, which increases survival, converts thymine dimers back to two thymine monomers.

2. A population of bacteria is ultraviolet-irradiated and then incubated for various periods of time in a nonnutrient buffer (that is, it is subjected to liquid-holding recovery). During this period the number of thymine dimers present in the DNA continually decreases and at the same time, thymine dimers appear both in the buffer and in the intracellular fluid. Thus, during the dark repair process, thymine dimers are excised from the DNA.

In the following sections the biochemical mechanisms for photoreactivation, dark repair, and two other processes will be described.

BIOCHEMICAL MECHANISMS FOR REPAIR OF THYMINE DIMERS

There are four major pathways for dealing with thymine dimers in DNA, which can be subdivided into two classes—light-induced repair and light-independent pathways. The latter can be accomplished by three distinct mechanisms: (1) excision of the damaged bases (**excision repair**), (2) reconstruction of a functional DNA molecule from undamaged fragments (**recombination repair**), and (3) tolerance of the damage (**SOS repair**). The chemical mechanisms for each of these repair processes are described in this section.

Photoreactivation

Photoreactivation is an enzymatic cleavage of thymine dimers activated by visible light (300–600 nm). An enzyme, called the **PR enzyme** or **photolyase,** has been isolated from almost all cells, from bacteria to animals. In a way that is not known the enzyme-DNA complex absorbs light (the light-absorbing portion of the complex has never been identified) and uses the light energy to cleave the C—C bonds of the cyclobutyl rings shown in Figure 10-2. The activity of photolyase is shown in Figure 10-8. First, the enzyme binds specifically to a thymine dimer (cytosine dimers and cytosine-thymine dimers also are formed by ultraviolet irradiation but much less frequently than thymine dimers; photolyase is also active against these dimers). If light is absorbed, the thymine dimer is monomerized; finally, photolyase dissociates from the now-normal pair of thymines.

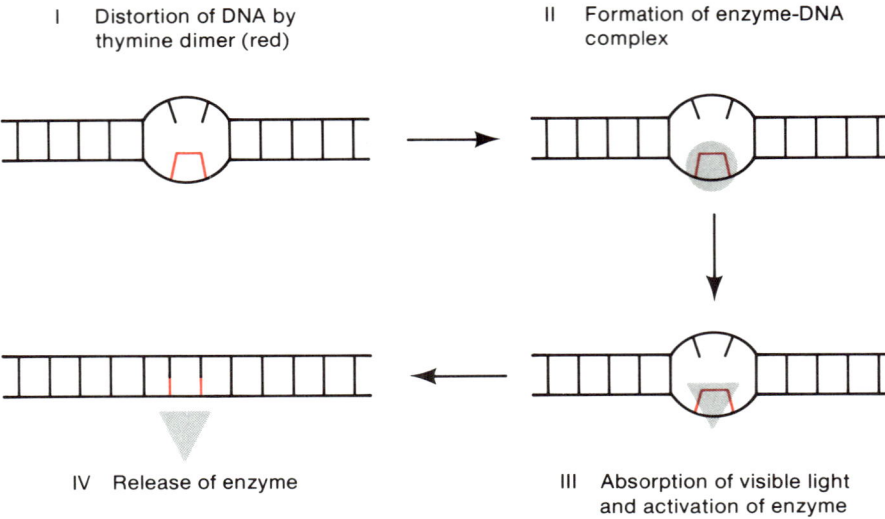

I   Distortion of DNA by
    thymine dimer (red)

II   Formation of enzyme-DNA
     complex

IV   Release of enzyme

III   Absorption of visible light
      and activation of enzyme

Figure 10-8   Scheme for enzymatic photoreactivation of a thymine dimer.

Excision Repair

Excision repair is a multistep enzymatic process (Figure 10-9). Two distinct mechanisms have been observed for the first step, an **incision** step. In *E. coli* a repair endonuclease recognizes the distortion produced by a thymine dimer and makes two cuts in the sugar–phosphate backbone: one is 8 nucleotides to the 5′ side of the dimer and another is 4−5 nucleotides to the 3′ side (panel(a)). At the 5′ incision site a 3′-OH group is produced, which polymerase I uses as a primer and synthesizes a new strand while displacing the DNA segment that carries the thymine dimer. The final step of the repair process is joining of the newly synthesized segment to the original strand by DNA ligase. The excised fragment is ultimately degraded to single nucleotides plus a thymine dimer dinucleotide by the combined activity of numerous scavenging exo- and endonucleases.

In several other systems (e.g., the bacterium *Micrococcus luteus* and *E. coli* phage T4) the incision step occurs in two distinct stages (panel (b)). The first is an enzymatic cleavage of the *N*-glycosidic bond in the 5′ thymine nucleotide of the dimer; the responsible enzyme is a pyrimidine dimer glycosylase, an enzyme of the type that removes uracil from DNA (Chapter 9). Incision of the strand is completed by a second activity of the glycosylase, namely, an endonuclease activity that recognizes a deoxyribose lacking a base (like apurinic acid nuclease); the enzyme makes a single cut at the 5′ side of the remaining thymine in the dimer site. This cleavage produces a 3′-OH group, but since it is on a base-free deoxyribose, it cannot be used to prime DNA synthesis. In a way that is not understood, the deoxyribose

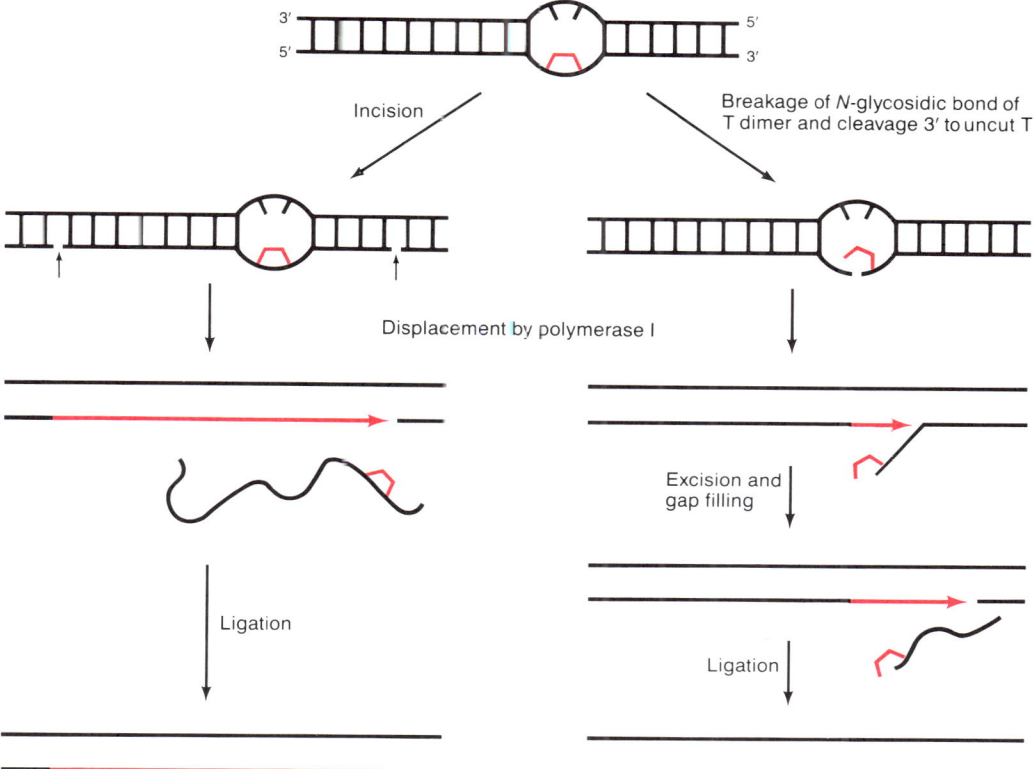

Figure 10-9 Two modes of excision repair. (a) The *E. coli* mechanism. Two incision steps are followed by gap-filling and displacement by polymerase I. (b) The *M. luteu* mechanism. A pyrimidine dimer glycosylase breaks an *N*-glycosidic bond and makes a single incision. Pol I displaces the strand, which is removed by an exonucleo-lytic event. In both mechanisms the final step is ligation.

is removed, and pol I acts at the new 3′-OH group, displaces the strand, and fills the gap. Removal of the displaced strand requires an excision step, which is carried out in part by the $5′ \rightarrow 3′$ exonuclease activity of polymerase I and partly by the associated endonuclease activity. Other nucleases can also carry out the excision step; it has been found that excision still occurs in a mutant lacking the $5′ \rightarrow 3′$ exonuclease of polymerase I and its associated endonuclease activity.

The incision-excision activities of mammalian cells are more complicated, require a larger collection of enzymes, and are poorly understood.

The incision activity of *E. coli* is determined by a complex of the products of three genes called *uvrA, uvrB,* and *uvrC.* The products of the *uvrA* and *uvrB* genes are two subunits of a protein complex (called ultraviolet-endo I) that has endonuclease activity. The UvrC product is necessary for maximum endonuclease activity *in vivo,* but its precise role is not known.

Proof that the Uvr system is responsible for excision repair *in vivo* comes from studies with mutants. Mutation in any of these genes renders *E. coli* exceedingly sensitive to killing by ultraviolet light (Figure 10-10) and when *uvrA⁻* and *uvrB⁻* mutants are irradiated and then incubated, thymine dimers are not found in the intracellular or extracellular fluid. That these phenomena are due to a loss of endonucleotic activity was shown by the following results. An analysis of the formation of single-strand breaks in ultraviolet-irradiated DNA (performed by measurement of the size of the DNA obtained when the isolated DNA is sedimented in alkali, which denatures the DNA) shows that breaks occur in the DNA of *uvr⁺* bacteria but not in that of *uvrA⁻* and *uvrB⁻* mutants.

The Uvr system is able to repair lesions other than thymine dimers. These lesions have in common either the displacement of bases, as in thymine dimer formation, or the addition of bulky substituents on the bases. It is assumed that the incision enzyme recognizes a helix distortion.

The bacterial excision-repair system is also active on phage DNA. For most phages the survival of irradiated phage is greater on wild-type bacteria than on bacterial mutants lacking the excision repair system. Furthermore, pyrimidine dimers are removed from irradiated phage DNA by the bacterial enzymes. A few phages (e.g., T4) encode their own repair systems.

One inherited human disease, xeroderma pigmentosum, results from inability to carry out excision repair. It is a result of mutations in genes that encode the ultraviolet-light incision system. Cells cultured from patients with the disease are killed by much smaller ultraviolet doses than are cells from a normal person. Furthermore, their ability to remove thymine dimers from their DNA is very much reduced. Patients with this disease develop skin lesions when exposed to sunlight and commonly develop one of several kinds of skin cancer.

Recombination Repair

Figure 10-10 shows that an *E. coli uvrA⁻* mutant is much more sensitive to ultraviolet light than is the wild type. However, it can be shown that an ultraviolet dose yielding an average of one lethal event per *uvrA⁻* cell produces about 300 thymine dimers per cell, none of which is excised. This number seems rather large and suggests that the cells possess another repair system. Evidence for such a system comes from examination of cells mutant in the gene *recA*, a gene that is essential for genetic recombination in *E. coli* (hence the symbol *rec*); as mentioned in Chapter 1, in an Hfr × F⁻ bacterial mating, no recombinants are found if the F⁻ cell is *recA⁻*. An ultraviolet irradiation survival curve for a *recA⁻* mutant shows that it is very ultraviolet-sensitive (Figure 10-10), which suggests that the *recA* gene must be a component of a repair system. Excision of thymine dimers occurs normally in a *recA⁻* mutant so *recA*-mediated repair clearly differs from excision repair.

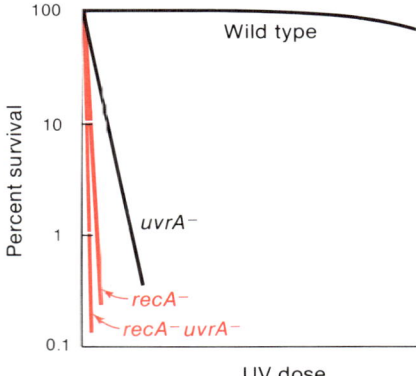

Figure 10-10 Survival curves of *E. coli* showing the sensitization to ultraviolet light resulting from the *uvrA⁻*, *recA⁻*, and *recA⁻uvrA⁻* mutations.

The existence of two repair systems is confirmed by a quantitative analysis of the survival curve of a $uvrA^- recA^-$ double mutant, which is more ultraviolet-sensitive than either of the single mutants (Figure 10-10) and for which the number of thymine dimers per lethal event is roughly one. It has been established that the Uvr system is responsible for removal of many of the thymine dimers, whereas the Rec system eliminates the dire effects of most of those that are not removed.

In order to discuss the mechanism of recombination repair, it is necessary to know the effect of a thymine dimer on DNA replication. When polymerase III reaches a thymine dimer, the replication fork is temporarily stalled. A thymine dimer is still capable of forming hydrogen bonds with two adenines because the chemical change in dimerization does not alter the groups that engage in hydrogen bonding. However, the dimer introduces a distortion into the helix, and when an adenine is added to the growing chain, polymerase III reacts to the distorted region as if a mispaired base had been added; the editing function then removes the adenine. The cycle begins again—an adenine is added and then it is removed; the net result is that the polymerase is stalled at the site of the dimer. (The same effect would occur if, instead of a dimer, radiation or chemical damage resulted in formation of a base to which no nucleoside triphosphate could base-pair.) Evidence that such a phenomenon occurs after ultraviolet irradiation is the existence of an ultraviolet-light-induced *idling* process—that is, rapid cleavage of deoxynucleoside triphosphates to monophosphates without any net DNA synthesis (i.e., without advance of the replication fork). A cell in which DNA synthesis is permanently stalled cannot complete a round of replication. However, stalling is only temporary, for there are two different ways by which DNA synthesis can restart—**postdimer initiation** and **transdimer synthesis.** These are responsible for **recombination repair** and **SOS repair,** respectively. In this section recombination repair is described; SOS repair is discussed in the succeeding section.

One way to deal with a thymine dimer block is to pass it by and initiate chain growth beyond the block (Figure 10-11). Such postdimer initiation does occur after a pause of about five seconds per thymine dimer, but the mechanism, which appears to involve unprimed reinitiation, is unknown. The result of this process is that the daughter strands have large gaps, one for each unexcised thymine dimer. There is no way to produce viable daughter cells by continued replication alone, because the strands having the thymine dimer will continue to turn out gapped daughter strands, and the first

Figure 10-11   Blockage of replication by thymine dimers (represented by joined lines) followed by re-starts several bases beyond the dimer. The black region is a segment of ultraviolet-light-irradiated parental DNA. The red region represents synthesis of a daughter molecule from right to left. The daughter strand contains gaps.

set of gapped daughter strands would be fragmented when the growing fork enters a gap. However, by a recombination mechanism called **sister–strand exchange** proper double-stranded molecules can be made.

The essential idea in sister-strand exchange is that a single-stranded segment free of any defects is excised from a "good" strand on the homologous DNA segment at the replication fork and somehow inserted into the gap created by excision of a thymine dimer (Figure 10-12). The combined action of polymerase I and DNA ligase joins this inserted piece to adjacent regions, thus filling in the gap. The gap formed in the donor molecule by excision is also filled in completely by polymerase I and ligase. If this exchange and gap filling are done for each thymine dimer, two complete daughter single strands can be formed and each can serve in the *next* round of replication as a template for synthesis of normal DNA molecules. Note that the system fails if two dimers in opposite strands are very near one another because then no undamaged sister strand segments are available to be excised. The molecular details of recombinational repair are not known, so the model shown in Figure 10-12 must be considered to be a working hypothesis that is consistent with all known facts.

Recombination repair is an important mechanism because it eliminates the necessity for delaying replication for the many hours that would be needed for excision repair to remove all thymine dimers. It may also be the case that some kinds of damage cannot be eliminated by excision repair— for example, alterations that do not cause helix distortion but do stop DNA

Figure 10-12 Recombination repair. I. A molecule containing two thymine dimers (red boxes) in strands a and d is being replicated. II. By postdimer initiation, a molecule is formed whose daughter strands b and c have gaps. If repair does not occur, in the next round of replication, strands a and d would yield gapped daughter strands, and strands b and c would again be fragmented. III. A segment of parental strand is excised and inserted into strand c. IV. The gap in strand a is next filled in by repair synthesis. Such a DNA molecule would probably engage in a second exchange in which a segment of c would fill the gap in b. DNA synthesized after irradiation is shown in red. Heavy and thin lines are used for purposes of identification only.

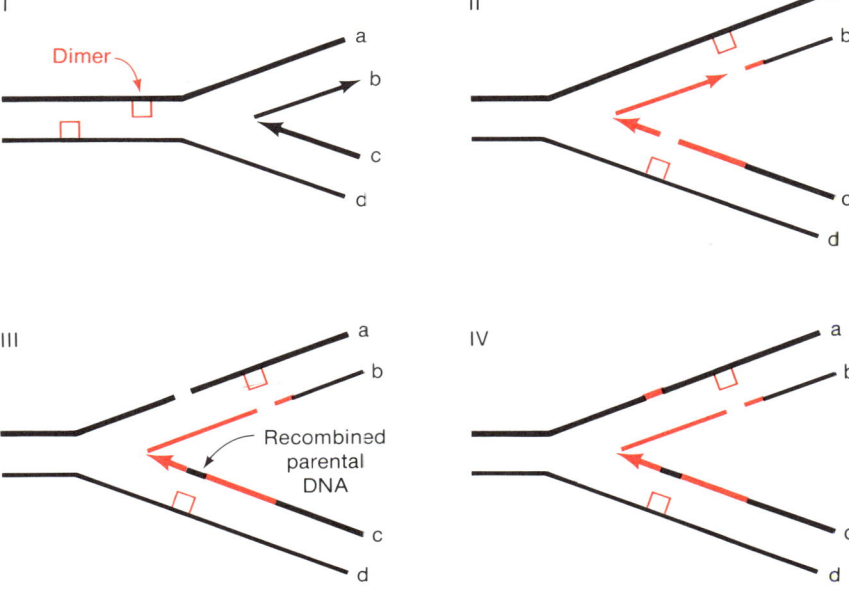

synthesis. Recombination repair has been demonstrated in several bacterial species, but has not yet been proved to occur in animal cells.

Recombination repair also occurs with ultraviolet-irradiated phages of some types; this is indicated by the fact that, if a population of a phage that lacks its own repair system is ultraviolet-light irradiated, fewer phage will plate on a $recA^-$ host than on a $recA^+$ host.

Since recombination repair occurs after DNA replication, in contrast with excision repair, it has been called postreplicational repair. Recently, the term **daughter-strand gap repair** has become adopted since only the gaps formed opposite dimers, rather than the dimers themselves, are repaired.

SOS Repair

SOS repair includes a **bypass system** that allows DNA chain growth across damaged segments at the cost of fidelity of replication. It is an error-prone process; even though intact DNA strands are formed, the strands contain incorrect bases. (The principle involved is that survival with mutations is better than no survival at all.) SOS repair is not yet thoroughly understood but is thought to invoke a relaxation of the editing system in order to allow polymerization to proceed across a dimer (transdimer synthesis) despite the distortion of the helix. A great deal of evidence now indicates that SOS repair is the major cause of ultraviolet-induced mutagenesis.

An important feature of SOS repair is an absolute requirement for a functional *recA* gene (that is, mutagenesis caused by ultraviolet irradiation does not occur in a $recA^-$ bacterium). The role of the RecA product is twofold. One of these, a regulatory function, will be described in Chapter 15; the other is a direct effect on editing. The RecA protein binds tightly to single-stranded DNA but only very weakly to double-stranded DNA. The distortion resulting from a pyrimidine dimer produces a short stable single-stranded region to which RecA binds. (Whether RecA binds to the dimer strand or to the opposite strand is not known.) When DNA polymerase III encounters a dimer site to which RecA is bound, RecA interacts with the ϵ subunit of the polymerase (the subunit responsible for editing) and inhibits the editing function, thereby avoiding the idling response and allowing the replication fork to advance. Most of the time pol III can utilize the residual base-pairing ability of the thymine dimer and put two adenines in the daughter strand. However, mispairing is enhanced by the distortion, which normally would invoke the editing response. The presence of RecA at the dimer site inhibits editing and causes the mispaired base to remain in the daughter strand as a mutation.

SOS repair also requires two other genes, *umuC* and *umuD*. The role of the gene products is unknown, but three hypotheses have been suggested: (1) they facilitate tight binding of RecA at the small distortion, (2) they facilitate binding of pol III to the distorted region, or (3) they enable pol III

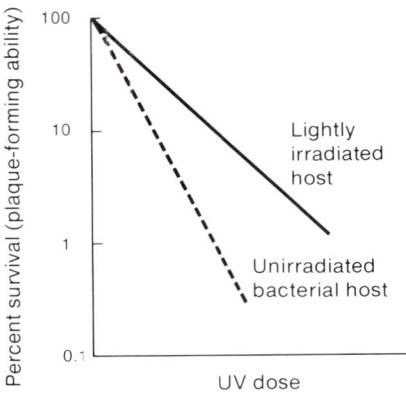

Figure 10-13 W reactivation of ultraviolet-light-irradiated phage λ. The dashed line shows the survival curve (for plaque-forming ability) obtained when λ phage irradiated with various doses of ultraviolet light are plated on unirradiated bacteria. The solid line represents survival of plaque-forming ability, when ultraviolet-light-irradiated λ are plated on lightly irradiated bacteria.

to leave the damaged site either by releasing pol III from RecA or RecA from the DNA.

Another important feature of SOS repair is that the system is *induced* as a result of damage to the DNA. In particular, sufficient amounts of RecA protein and UmuC and UmuD proteins are not available until some time after damage has occurred. The best evidence for this point comes from an analysis of the repair of ultraviolet-light-inactivated phage by irradiated bacterial host cells.

Most phages do not need a fully functional (or even viable) host to produce progeny phage particles. For instance, a study of the result of infecting ultraviolet-irradiated *E. coli* with phage λ showed that if the host is irradiated with a dose of ultraviolet light yielding 10 percent survival of the ability to form colonies, there is no loss of the ability to support phage infection. However, if the λ particles are also irradiated, the irradiated *E. coli* cells are better able to support growth of the irradiated λ than are unirradiated cells. That is, the survival of an ultraviolet-irradiated λ is higher on an irradiated host than on an unirradiated host (Figure 10-13). This phenomenon, which is called **UV-reactivation** or **W reactivation** (for Jean Weigle, who discovered it), clearly involves a repair process and occurs only if the host is Rec$^+$; the phenomenon does not require the *uvr* genes. Although more phage survive, the surviving population contains a higher proportion of phage mutants when the irradiated host is used; that the mutation frequency increases suggests that the repair mechanism makes use of an error-prone replication system. It is now known that this phenomenon is the SOS system in action. The SOS system has been activated or turned on by ultraviolet irradiation of the host. How the system is turned on will be explained in Chapter 15.

Up to this point, we have discussed mutations that arise at the dimer site, indeed the major effect of SOS repair. A less frequent event is the production of mutations elsewhere in the DNA (this has been termed untargeted mutagenesis); a related phenomenon is the increased mutation frequency observed in cells in which the SOS system has been turned on by altering control elements rather than by irradiation. The following explanation has been suggested. It is known that when the SOS system is turned on, the intracellular concentration of RecA protein becomes very high. RecA binds weakly and transiently to double-stranded DNA owing to breathing. With a large amount of RecA protein in the cell the number of sites in a DNA molecule at which, at any instant, RecA is bound may be great enough that pol III encounters a RecA molecule and temporarily loses editing ability. A localized lack of editing would increase the mutation frequency somewhat.

Summary of the Repair of Thymine Dimers

Four repair systems have been described—excision repair, photoreactivation, recombination, and SOS repair. These systems differ both in the source of energy—light in the case of photoreactivation and ATP hydrolysis for the other three—and in the mechanisms used to allow functional daughter DNA molecules to be formed. Excision repair and photoreactivation both repair the template, recombination repair forms a new template, and SOS repair

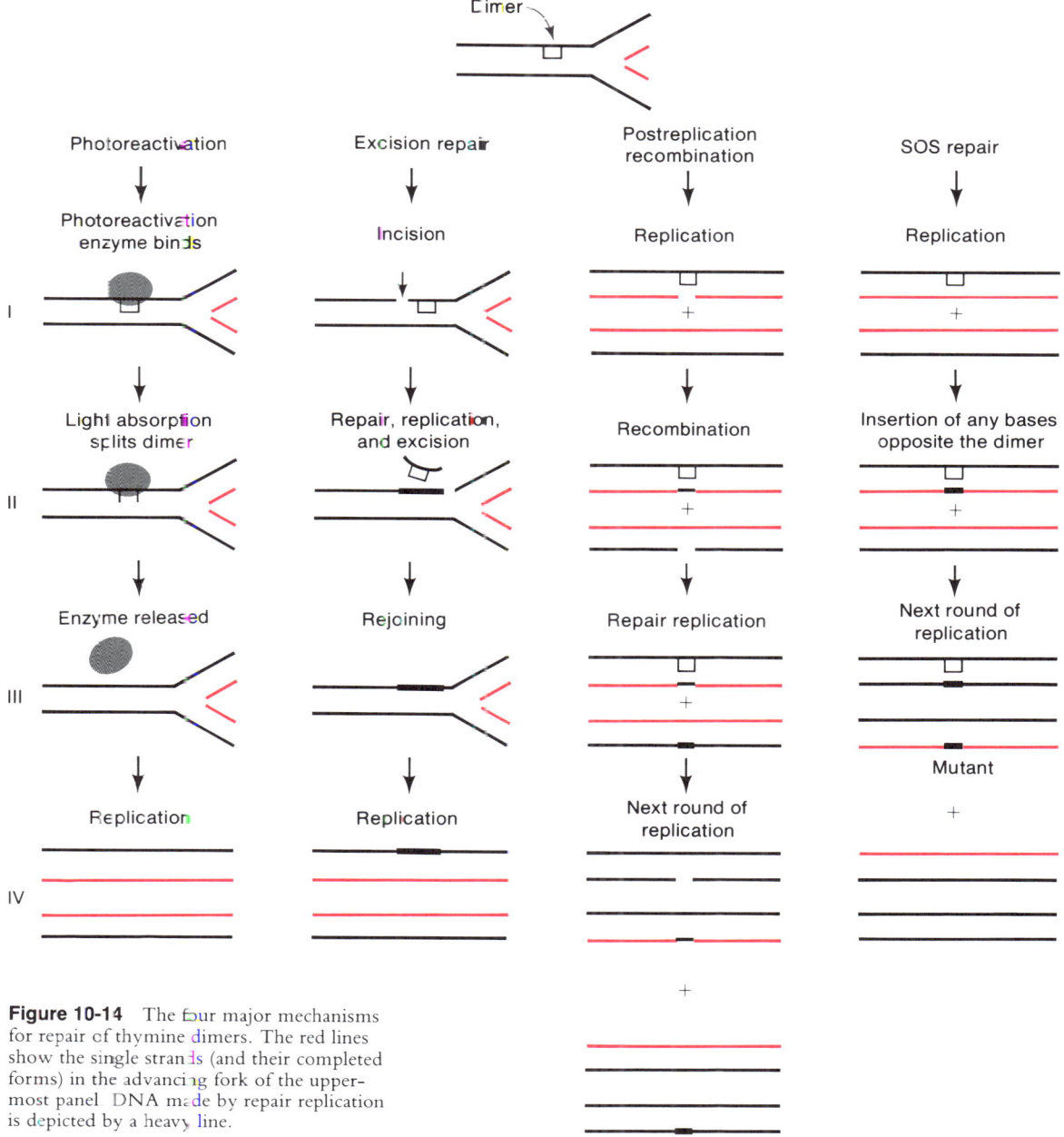

Figure 10-14 The four major mechanisms for repair of thymine dimers. The red lines show the single strands (and their completed forms) in the advancing fork of the upper-most panel. DNA made by repair replication is depicted by a heavy line.

ignores the damage and forms uninterrupted daughter strands, despite the presence of damaged segments of the template. Figure 10-14 summarizes these pathways. It should be noted that photoreactivation is the only mechanism in which there is no interruption of the sugar-phosphate backbone and that SOS repair is the only process that leads to mutation.

WHY DO DAMAGED CELLS DIE?

We have discussed a variety of repair systems that seem to be able to fix everything, yet cells still are killed by various forms of radiation and toxic chemicals. Why? There are many answers to this question. One trivial reason is that there are probably some alterations that are not repairable—for instance, a base might be altered to yield a replication block that cannot be overcome. Also, damaged regions may be opposite one another or so clustered that there is nothing to use as a template for repair. A second possibility is that the repair systems themselves can be damaged; this is not unlikely since it takes a very large amount of damage before lethality ensues in a wild-type organism. The repair systems might also be saturated by such large amounts of damage that some lethal imbalance in the protein/DNA ratio occurs. Finally, some of the repair systems may be sufficiently error-prone that mutations sometimes occur in essential genes.

11

Mutagenesis, Mutations, and Mutants

In previous chapters the use of mutants has been encountered repeatedly. In this chapter we explain what a mutant actually is, how a mutant is created, and how it is detected. We begin with the terminology.

TERMINOLOGY

The terminology used in discussing mutants is somewhat confusing because of the similarity of four words—**mutant, mutation, mutagen,** and **mutagenesis.** These terms have the following meanings

Mutant refers to the genetic state of the organism. That is, if one of the ensemble of characteristics (the **phenotype**) that comprises a so-called normal organism is different from the "wild-type" character, then that organism is in that respect said to be a mutant. Thus, for *E. coli,* which in nature is able to utilize galactose as a carbon source, its wild-type state is Gal$^+$, and a Gal$^-$ cell, unable to utilize galactose, is a mutant. The terminology "wild type" is a misnomer, however, and often does not refer to the state of the organism in nature because many organisms are defective (mutant) in some particular respect, even in their most common natural state. For example, *E. coli* isolated from nature is usually unable to utilize lactose (it is Lac$^-$), yet the Lac$^+$ phenotype is invariably called wild type and Lac$^-$ is called mutant.

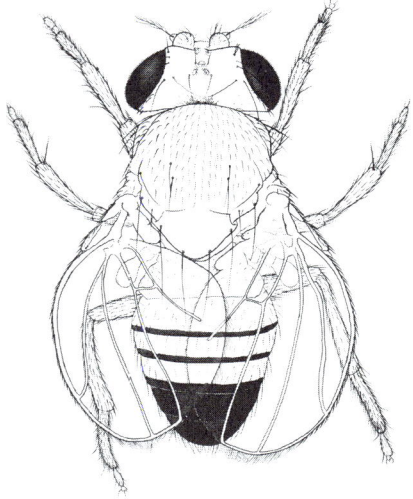

Drawing of a *Drosophila melanogaster* carrying the mutation *rotund,* which produces roundish, rather than the normally long thin wings. This historic drawing was made by Edith Wallace, artist for Thomas Hunt Morgan, in the early 1900s. [Courtesy of Dan Lindsley.]

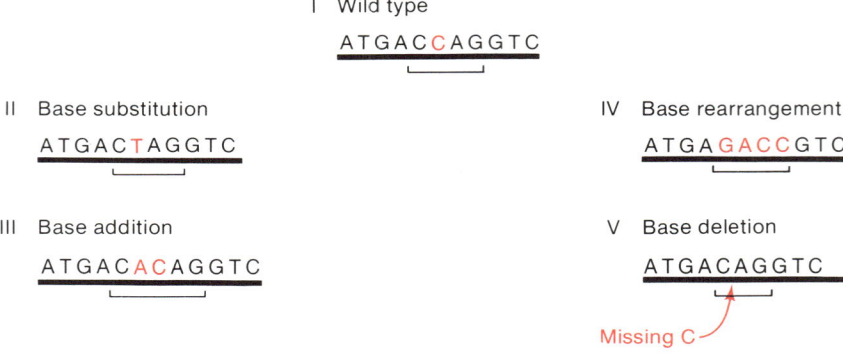

I   Wild type

ATGAC**C**AGGTC

II   Base substitution

ATGAC**T**AGGTC

III   Base addition

ATGAC**AC**AGGTC

IV   Base rearrangement

ATGA**GACC**GTC

V   Base deletion

ATGACAGGTC

Missing C

Figure 11-1   Four types of mutations. Only the base sequence in one DNA strand is shown. Changes are shown in red. The horizontal brackets indicate the affected segment.

Mutation refers to any change in the base sequence of DNA. The most common change is a substitution, addition, rearrangement, or a deletion of one of more bases (Figure 11-1). A mutation need not give rise to a mutant phenotype, as we will see shortly.

A **mutagen** is a physical agent or a chemical reagent that causes mutations. For example, nitrous acid reacts with some DNA bases, changing their identity and hydrogen-bonding properties; thus, it is a mutagen.

Mutagenesis is the process of producing a mutation. If it occurs in nature without the addition of a known mutagen, it is called **spontaneous mutagenesis** and the resulting mutations are **spontaneous mutations.** If a mutagen is used, the process is **induced mutagenesis.** Unfortunately, the term mutation is sometimes used when mutagenesis is meant.

TYPES OF MUTATIONS AND THEIR NOTATION

Mutations can be categorized in several ways. One distinction is based on the nature of the change—specifically, on the number of bases changed. Thus, we may distinguish a **point mutation,** in which there is only a single changed base pair, from a multiple mutation, in which two or more base pairs differ from the wild-type sequence. A point mutation may be a **base substitution,** a **base insertion,** or a **base deletion,** but the term most frequently refers to a base substitution.

A second distinction is based on the consequence of the change in terms of the amino acid sequence that is affected. For example, if there is an amino acid substitution, the mutation is a **missense mutation.** If the substitution produces a protein that is active at one temperature (typically 30°C) and inactive at a higher temperature (usually 40–42°C), the mutation is called a **temperature-sensitive** or **Ts mutation.** According to the most recent convention, a *gal*⁻ temperature-sensitive mutation is written *gal*⁻ (Ts). In Chapter 13 it will be seen that sometimes there is no amino acid that corresponds to a new base sequence; in that case, termination of synthesis of the protein occurs at that point and the mutation is called a **chain termination mutation** or a **nonsense mutation.** There are three kinds of nonsense mutations, each representing one of three particular base sequences that do not correspond to any amino acid. They are called **amber** (Am), **ochre** (Oc), and **opal** (no symbol agreed on). A gal^- amber mutant and a gal^- ochre mutant would be designated gal^-(Am) and gal^-(Oc), respectively.

When mutants are isolated in the laboratory, they are usually collected in large numbers. Normally each mutation is given a number that states the chronological order in which it was "isolated." Thus, the fourth gal^- mutation is written $gal4$. If $gal82$ and $gal223$ were Am and Ts mutations, respectively, the designations would be $gal82$(Am) and $gal223$(Ts). Even though there is a "standard" notation, some variation is still seen. For example, $gal3$(Ts) and $gal5$(Am) are frequently seen written as $gal\,ts3$ and $gal\,am5$, respectively.

Often a phenotype (which for bacteria is always capitalized), such as the ability to synthesize tryptophan (Trp^+), is determined by several genes encoding all of the enzymes in the biosynthetic pathway; such genes are usually distinguished by capital letter suffixes, as $trpA$, $trpE$, and so forth. Mutation 103 in the $trpA$ gene is written $trpA103$. The notation for bacterial mutations is summarized in Table 11-1. Other notations are used for viruses and various eukaryotes.

Table 11-1   Summary of notation used to designate bacterial mutants and mutations

Phenotype or genotype	Designation*
Phenotype	
Lacking in ability to make a substance; the substance must be supplied	Sub^-
Possessing ability to make a substance	Sub^+
Resistance to an antibiotic	Ant-r
Sensitivity to an antibiotic	Ant-s
Genotype	
Wild-type gene for making a substance	sub^+
Mutant gene for making a substance	sub^-†
Mutant $subA$ gene	$subA^-$
Mutation in $subA$ gene having a temperature-sensitive phenotype	$subA^-$(Ts)
Mutation in $subA$ gene having an amber phenotype	$subA^-$ (Am)
Mutation number 63 in $subA$ gene	$subA63$
Gene for resistance or sensitivity to a particular antibiotic	ant
Genotype conferring resistance to an antibiotic	$ant\text{-}r$
Genotype conferring sensitivity to an antibiotic	$ant\text{-}s$

*The arbitrary abbreviations "sub" and "ant" mean substance and antibiotic in this table.

†The notation sub for sub^- is widespread. Because of possible ambiguity, the superscript minus is used throughout this book.

BIOCHEMICAL BASIS OF MUTANTS

In common usage the term mutant refers to an organism in which either the base sequence of DNA or the phenotype has been changed. These definitions are the same (except for the case of a silent mutation, which will be discussed shortly) since the base sequence of DNA determines the amino acid sequence of a protein. The chemical and physical properties of each protein are determined by its amino acid sequence, so a single amino acid change is capable of inactivating a protein. This was first demonstrated by Vernon Ingram, who found that the mutant hemoglobin molecule obtained from patients with sickle-cell anemia differed from normal hemoglobin only in that a glutamic acid in the normal protein is replaced by valine in the mutant.

From the discussion of protein structure in Chapter 6 it is easy to understand how an amino acid substitution can change the structure and, hence, the biological activity of a protein. For instance, consider a hypothetical protein whose three-dimensional structure is determined entirely by an interaction between one positively charged amino acid (for example, lysine) and one negatively charged amino acid (aspartic acid). A substitution of methionine, which is uncharged, for the lysine would clearly destroy the three-dimensional structure, as would substitution of histidine, which is positively charged, for asparagine, which is uncharged. Similarly, a protein might be stabilized by a hydrophobic cluster, in which case substitution of a polar amino acid for a nonpolar one would also be disruptive.

An amino acid substitution does not always lead to a mutant phenotype. For instance, a hydrophobic cluster might be virtually unaffected by a replacement of one leucine by another nonpolar amino acid such as isoleucine. Similarly, a negatively charged amino acid might successfully substitute for another negatively charged amino acid. When an amino acid substitution has no detectable effect on the phenotype of a cell, it is called a **silent mutation.** There is also another type of silent mutation, namely, a base change without an amino acid alteration; this is a result of the degeneracy of the genetic code (Chapter 13).

The shapes of proteins are determined by such a variety of interactions that sometimes an amino acid substitution is only partially disruptive. For instance, an isoleucine might substitute successfully for leucine and be silent, but replacement with a more bulky amino acid such as phenylalanine might cause subtle stereochemical changes, though a hydrophobic cluster is preserved. This could be manifested as a reduction, rather than a loss, of activity of an enzyme. For example, a bacterium carrying such a mutation in the enzyme that synthesizes an essential substance might grow very slowly (but it would grow) unless the substance is provided in the growth medium. Such a mutation is called a **leaky mutation;** these mutations are not particularly useful for most genetic studies.

Generally speaking, the following types of amino acid substitutions are expressed as nonleaky mutations: polar to nonpolar, nonpolar to polar, change of sign of a charge, small side chain to bulky side chain, sulfhydryl to any other side chain, hydrogen-bonding to non-hydrogen-bonding, any change to or from proline (which changes the shape of the polypeptide backbone), and any change in a substrate-binding site.

So far, we have discussed mutants that arise from amino acid substitutions. Three other types of alterations occur that invariably eliminate activity of the protein totally:

1. Deletion of bases, which causes one or more amino acids to be absent in the completed protein.
2. A frameshift, which will be discussed shortly. In a frameshift all amino acids starting from a particular point (the site of the mutation) are different, so the mutant contains a protein that can be extensively different from the wild-type protein.
3. A chain termination mutation, in which a base change results in the production of a signal for stopping protein synthesis. Such mutations cause premature termination of the protein chain at the site of mutation, making the mutant protein a fragment of the wild-type protein.

In the following section we will see how an altered base sequence can arise.

MUTAGENESIS

The production of a mutation requires that a change occur in the base sequence. This can occur spontaneously by replication errors or can be stimulated to occur in five main ways: (1) removal of an incorrectly inserted base is prevented; (2) a base is inserted that tautomerizes and allows a substitution to occur in subsequent replication; (3) a previously inserted base is chemically altered to a base having different base-pairing specificity; (4) one or more bases are skipped during replication; or (5) one or more extra bases are inserted during replication. In the following sections we describe mutagens that act by one or more of these mechanisms and address the question of spontaneous mutagenesis. The information that follows is also presented in Table 11-2.

Base-Analogue Mutagens

A base analogue is a compound sufficiently similar to one of the four DNA bases that it can be built into a DNA molecule during normal replication. Such a substance must be able to pair with a base in the template strand or the editing function will remove it. However, if the base has two modes of hydrogen-bonding, it will be mutagenic.

Table 11-2   Types of mutagens

Mutagen	Mode of action	Example	Consequence
Base analogue	Substitutes for a standard base during replication and causes a new base pair to appear in daughter cells in a later generation	5-Bromouracil	$A \cdot T \rightarrow G \cdot C$, and $G \cdot C \rightarrow A \cdot T$
		2-Aminopurine	$A \cdot T \rightarrow G \cdot C$
Chemical mutagen	Chemically alters a base so that a new base pair appears in daughter cells in a later generation	Nitrous acid	$G \cdot C \rightarrow A \cdot T$, and $A \cdot T \rightarrow G \cdot C$
		Hydroxylamine	$G \cdot C \rightarrow A \cdot T$
		Ethyl methane sulfonate (EMS)	$G \cdot C \rightarrow A \cdot T$ $G \cdot C \rightarrow C \cdot G$, and $G \cdot C \rightarrow T \cdot A$
		NNG	Same as EMS
		Ultraviolet light	All single base-pair changes are possible.
Intercalating agents	Addition or deletion of one or more base pairs	Acridines	Frameshifts
Mutator genes	Excessive insertion of incorrect bases or lack of repair of incorrectly inserted bases	——	All single base-pair changes are possible.
None	Spontaneous deamination of 5-methylcytosine (MeC)	——	$G \cdot MeC \rightarrow A \cdot T$

Note: Italicized changes in base pairs are transversions; those that are not italicized are transitions.

The substituted base 5-bromouracil (BU) is an analogue of thymine inasmuch as the bromine has about the same van der Waals radius as the methyl group of thymine (Figure 11-2). In subsequent rounds of replication BU functions like thymine and primarily pairs with adenine. Thymine can sometimes (but rarely) assume an enol form that is capable of pairing with guanine, and this conversion occasionally gives rise to mutations in the course of replication. The mutagenic activity of 5-bromouracil stems in part from a shift in the keto-enol equilibrium caused by the bromine atom; that is, the enol form exists for a greater fraction of time for BU than for thymine. Thus, if BU replaces a thymine, in subsequent rounds of replication, it occasionally generates a guanine, which in turn specifies cytosine, resulting in formation of a G·C pair (Figure 11-3). BU can also induce a change from G·C to A·T. The enol form is actually sufficiently prevalent that BU is sometimes (but infrequently) incorporated into DNA in that form. When that occurs, BU is acting as an analogue of cytosine rather than thymine. However, even though it may become part of the DNA by temporarily having the base-pairing properties of cytosine, the keto form is the predom-

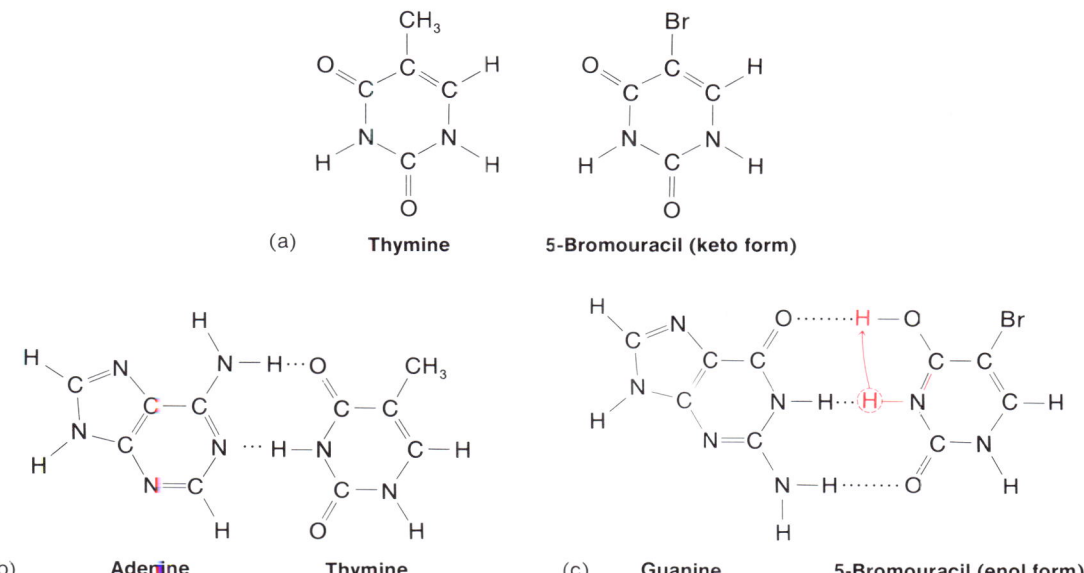

Figure 11-2 Mutagenesis by 5-bromouracil. (a) Structural formulas of thymine and 5-bromouracil. (b) A standard adenine- thymine base pair. (c) A base pair between between guanine and the enol form of 5-bromouracil. The red H in the dashed cir- cle shows the position of the H in the keto form. When tautomerization occurs, the red double bond forms.

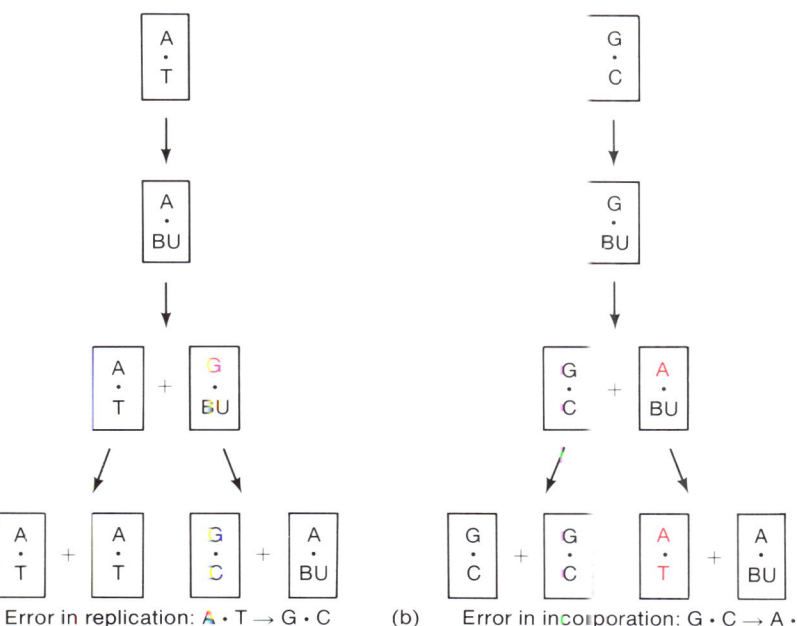

(a)   Error in replication: A · T → G · C

(b)   Error in incorporation: G · C → A · T

Figure 11-3 Two mechanisms of 5-bro- mouracil- (or BU)-induced mutagenesis. (a) During replication, BU, in its usual keto form, substitutes for T and the replica of an initial A·T pair becomes an A·BU pair. In the first mutagenic round of replication the BU, in its rare enol form, pairs with G. In the next round of replication, the G pairs with a C, completing the transition from an A·T pair to a G·C pair. (b) During replica- tion of a G·C pair a BU, in its rare enol form, pairs with a G. In the next round of replication the BU is again in the common keto form and it pairs with A, so the initial G·C pair becomes an A·T pair. The replica of the A·BU pair produced in the next round of replication is another A·T pair.

inant form; hence, in subsequent rounds of replication BU will usually pair like thymine. Thus, a G·C pair, which, as a result of an incorporation error, is converted to a G·BU pair, ultimately becomes an A·T pair, as shown in Figure 11-3(b). Thus, BU can stimulate a transition from A·T to G·C as well as from G·C to A·T.

Recent experiments suggest that BU is also mutagenic in another way. The concentrations of the nucleoside triphosphates in most cells is regulated by the concentration of thymidine triphosphate (TTP). This regulation results in appropriate relative amounts of the four triphosphates for DNA synthesis. One part of this complex regulatory process is the inhibition of synthesis of deoxycytidine triphosphate (dCTP) by excess TTP. The BU nucleoside triphosphate also inhibits production of dCTP. When BU is added to the growth medium, TTP continues to be synthesized by cells at the normal rate while the synthesis of dCTP is significantly reduced. The ratio of TTP to dCTP then becomes quite high and the frequency of misincorporation of T opposite G increases. The editing function and mismatch repair systems are capable of removing incorrectly incorporated thymine, but in the presence of BU the rate of misincorporation can exceed the rate of correction. An incorrectly incorporated thymine that persists will pair with adenine in the next round of DNA replication, yielding a GC→AT change in one of the daughter molecules.

Both base-pair changes induced by BU maintain the original purine (Pu)-pyrimidine (Py) orientation. That is, the original and the altered base pairs both have the orientation Pu·Py—namely, A·T and G·C. If the original pair was T·A, the altered pair would be C·G—that is, Py·Pu for both the original and the altered pairs. A base change that does not change the Py·Pu orientation is called a **transition.** Base analogue mutations are always transitions. Later we will see changes from Pu·Py to Py·Pu and Py·Pu to Pu·Py; when such a change of orientation occurs, the mutation is called a **transversion.** Note that BU induces transitions in both directions—AT→GC by the tautomerization route, and GC→AT by the misincorporation route. Thus, mutations that are induced by BU can also be reversed by it.

Chemical Mutagens

A chemical mutagen is a substance that can alter a base that is already incorporated in DNA and thereby change its hydrogen-bonding specificity. Three commonly used chemical mutagens are nitrous acid (HNO_2), hydroxylamine (HA), and ethylmethane sulfonate (EMS). The chemical structures of these are shown in Figure 11-4.

Nitrous acid primarily converts amino groups to keto groups by oxidative deamination. Thus cytosine, adenine, and guanine are converted to uracil (U), hypoxanthine (H), and xanthine (X), respectively. These bases can form the base pairs U·A, H·C, and X·C. Therefore, the changes are

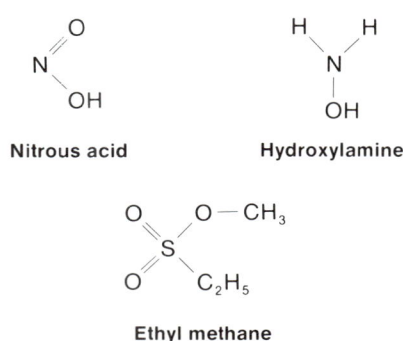

Nitrous acid   **Hydroxylamine**

Ethyl methane sulfonate

Figure 11-4   Structures of three chemical mutagens.

G·C→A·T and A·T→G·C as cytosine and adenine respectively are deaminated. Guanine is converted to xanthine (X), a change that should not be directly mutagenic since both guanine and xanthine pair with cytosine. However, GC→AT transitions have been observed with certain single-stranded DNA phages at sites of a guanine, which suggests that xanthine probably has an undiscovered tautomeric form able to pair with T.

Hydroxylamine reacts specifically with cytosine and converts it to a modified base that pairs only with adenine, so a G·C pair ultimately becomes an A·T pair.

Ethylmethane sulfonate (EMS), an **alkylating agent,** is a potent mutagen used extensively with eukaryotes. Nitrous acid and hydroxylamine, which are exceedingly useful in prokaryotic systems, are not particularly effective as mutagens in higher eukaryotes (because the chemical conditions necessary for reaction are not easily obtained), whereas the alkylating agents are highly effective. Many sites in DNA are alkylated by these agents; of prime importance for the induction of mutations is the addition of an alkyl group to the hydrogen-bonding oxygen of guanine and thymine. These alkylations impair the normal hydrogen-bonding of the bases and cause mispairing of G with T, leading to the transitions AT→GC and GC→AT. EMS also reacts with adenine and cytosine. In both phage T4 and *D. melanogaster* bases that have been altered by EMS such that their normal hydrogen-bonding is impaired mispair about half the time. A second important mechanism for mutagenesis by alkylating agents in many organisms is error-prone repair, such as SOS repair (Chapter 10). Evidence that an error-prone system leads to mutagenesis by alkylating agents in the yeast *Saccharomyces cerevisiae* comes from studies of strains in which the error-prone repair system is inactive (owing to a mutation in one of the genes responsible for activity of the system). In these strains the frequency of mutations induced by many alkylating agents is greatly reduced. The reduction is so great that it is clear that in yeast mutagenesis by alkylating agents is almost entirely a result of error-prone repair.

Another phenomenon resulting from alkylation of guanine is **depurination,** or loss of the alkylated base from the DNA molecule by breakage of the *N*-glycosidic bond joining the purine nitrogen and deoxyribose. Depurination is not always mutagenic, since the gap left by loss of the purine can be repaired by the apurinic acid endonuclease system described in Chapter 10. However, sometimes the replication fork may reach the apurinic site before repair has occurred. When this happens, replication stops just before the apurinic site. If the cell has a functional SOS repair system (to avoid the idling response described in Chapter 10), replication re-starts after a brief pause. However, with high probability an adenine nucleotide is put in the daughter strand opposite the apurinic site. Since the original parental base (which was removed) was a purine, the base pair at that site will be a mismatch (Pu·A) and after replication the base pair at that site will change orientation from Pu·Py to Py·Pu (Figure 11-5), the first example we have

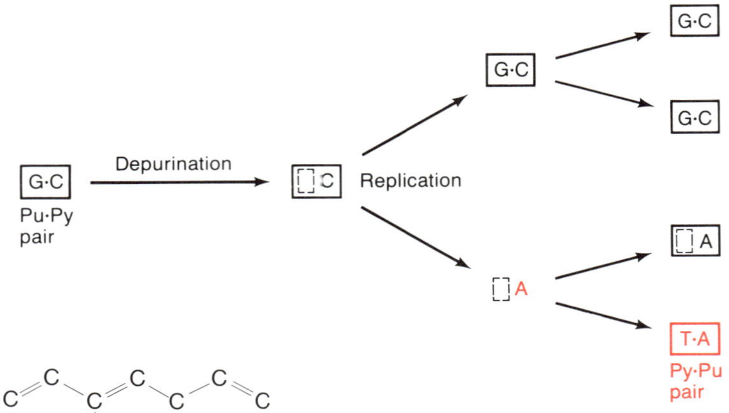

Figure 11-5 Production of a transversion by insertion of an adenine when DNA polymerase encounters an apurinic site.

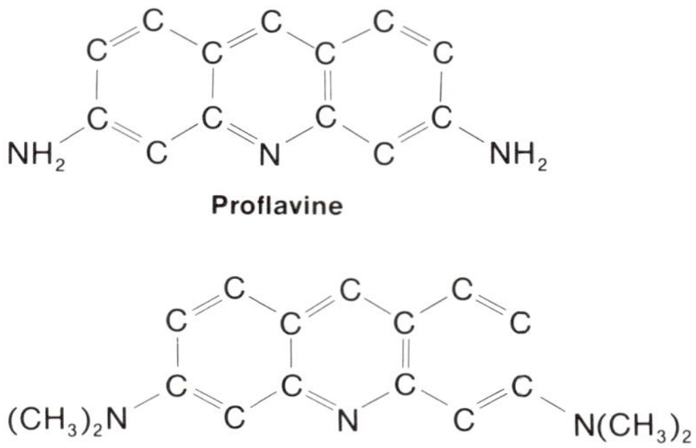

Figure 11-6 Structures of two mutagenic acridine derivatives.

Proflavine

Acridine orange

seen of a transversion. Treatment of phages with low pH buffers or at high temperature also produces depurinations (see Chapter 4). On replication of phages treated in this way, numerous transversions occur, in agreement with the adenine–insertion mechanism just suggested.

Mutagenesis by Intercalating Substances

Acridine orange, proflavine, and acriflavine (Figure 11-6), which are substituted acridines, are planar, three-ringed molecules whose dimensions are roughly the same as those of a purine-pyrimidine pair. In aqueous solution, these substances form stacked arrays and are also able to stack with a base pair; this insertion occurs between bases in adjacent pairs, a process called **intercalation.** Since the thickness of the acridine molecule is approximately that of a base pair and because the two bases of a pair are normally in contact, the intercalation of one acridine molecule causes adjacent base pairs to move apart by a distance equal to that of the thickness of one base pair (Figure 11-7). This has bizarre effects on the outcome of DNA replication, though the mechanism of action of the mutagen is not known. When DNA containing intercalated acridines is replicated, additional bases appear in the

Figure 11-7 Separation of two base pairs (shown in red) by an intercalating agent.

sequence (Figure 11-8). The usual addition is a single base, though occasionally two bases are added. Deletion of a single base also occurs but this is far less common than base addition. Mutations of this sort are called **frameshift mutations.** This is because the base sequence is read in groups of three bases when it is being translated into an amino acid sequence and the addition of a base changes the reading frame (Figure 11-8).

Mutator Genes

In *E. coli* there are genes that in a mutant state cause mutations to appear very frequently in other genes throughout the genetic map. These genes are called **mutator genes.** This is a misnomer, because the function of each gene is probably to keep the mutation frequency low; that is, it is only when the product of a mutator gene is itself defective that there is widespread production of mutations. In a few cases the mechanisms by which these genes act are known.

DNA polymerases make occasional errors and correct these errors by means of 3′→5′-exonuclease editing functions. An alteration in the region of the polymerase molecule that reduces or eliminates the editing function constitutes a mutator. Of course, mutations are not being created—it is only that random replication errors are not being corrected.

A second mutator gene is the *dam* gene described in Chapter 9. This gene is responsible for the methylation of DNA. If a misincorporated base is missed by the 3′→5′ editing function, a mismatch repair system comes into play and removes one member of the pair. The greater methylation of the parental strand compared to the methylation of the newly synthesized daughter strand instructs the mismatch repair system to remove the incorrect base from the daughter strand. In a *dam⁻* mutant there is no methylation of the parent (or any other) strand and the mismatch repair system frequently excises the parental (correct) base and inserts a base that can pair with the daughter (incorrect) base. Note that this gives rise to a mutation that does not require additional replication for expression.

Site-Specific Mutagenesis

Mutations are introduced into organisms in the laboratory for a variety of reasons. When a mutant is needed solely as a genetic marker, methods that introduce mutations at random sites are generally sufficient. However, in modern molecular genetics many experimental approaches to understanding molecular mechanisms of such phenomena as gene expression and gene regulation require mutations in particular base sequences (for example, binding sites). This need has led to the development of a variety of procedures for introducing mutations at specific sites. A few of these procedures will be outlined in this section. In each procedure the gene to be altered is carried

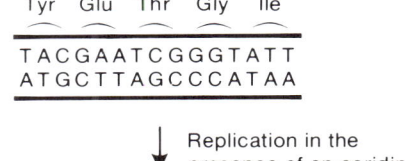

Figure 11-8   A base addition (red) resulting from replication in the presence of an acridine. The change in amino acid sequence read from the upper strand in groups of three bases is also shown in red.

on a plasmid. The plasmid is isolated, manipulated, and then returned to a bacterium for replication and gene expression. (If the gene is not naturally carried on a plasmid, recombinant DNA techniques can be used to produce such a plasmid.)

1. *Production of deletions.* The plasmid is treated with a restriction endonuclease to make two cuts surrounding a site to be deleted. The small fragment is isolated and discarded, and the ends of the larger piece are rejoined. The gene of interest now contains a deletion. With detailed restriction maps for many restriction enzymes it is also possible to delete a small amount of bases in a particular region.

2. *Point mutations at random sites in a particular region of DNA.* As in (1), a small fragment is cleaved out and purified. This fragment is then exposed to a chemical mutagen that induces base changes of the desired type. It is then rejoined to the larger portion of the plasmid, which has not been treated

Figure 11-9 A method for site-specific mutagenesis.

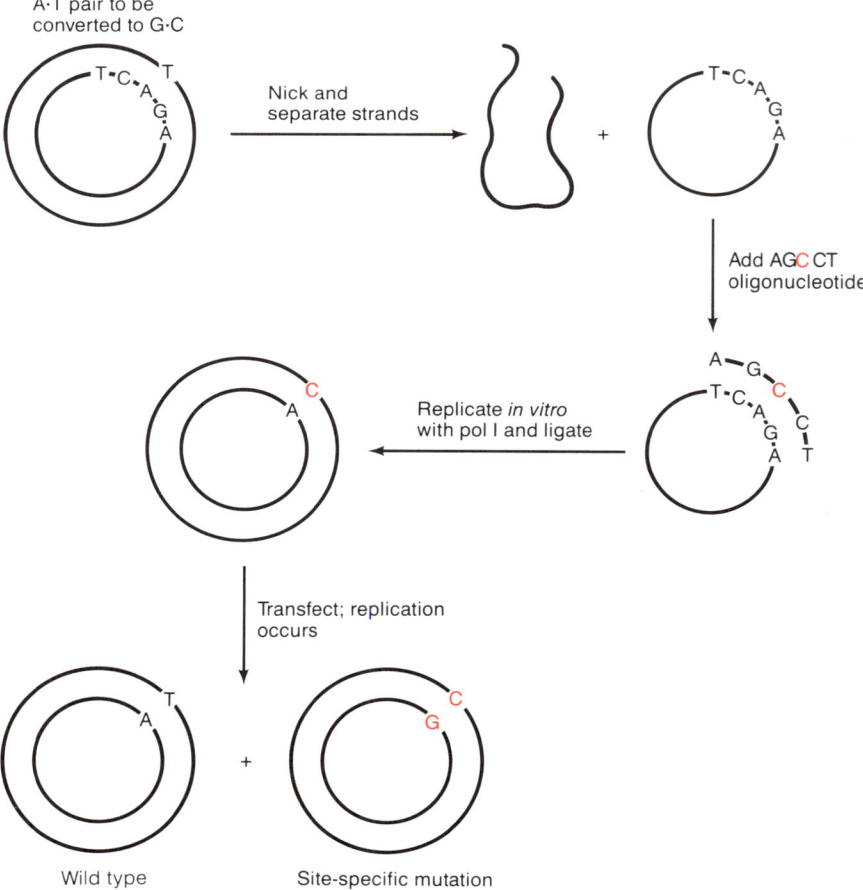

with the mutagen. Cells are transfected and, after two rounds of DNA replication, the gene contains one or more point mutations in the desired region.

3. *Point mutations at a particular base pair* (Figure 11-9). The plasmid is nicked with an endonuclease, using conditions that yield on the average one single-strand break per plasmid. The DNA is then denatured and intact circles are isolated. Short single-stranded oligonucleotides with defined sequences are easily synthesized, and one such oligonucleotide is made that has a mutant base at the desired site. This oligonucleotide is renatured with the intact single-stranded circle and a fragment of DNA polymerase lacking $5' \rightarrow 3'$ exonuclease activity (the Klenow fragment, Chapter 7) is added with the four nucleoside triphosphates. The oligonucleotide primes the circles and the circles are copied, yielding a double-stranded circle. (The circles that are not complementary to the oligonucleotide are not primed and hence are not replicated.) The replicated circles are sealed with DNA ligase and covalent circles are isolated. These circles are then transfected into a bacterium and DNA replication occurs. Each circle contains a base-pair mismatch at the mutant site. Mismatch repair occurs at random, generating some wildtype sequences and some with a mutation at a particular site. Genetic tests are used to detect the cells with the mutant sequence.

MUTATIONAL HOT SPOTS

In any protein some amino acids can be replaced by others without changing the phenotype. Furthermore, in Chapter 13 it will be seen that all base substitutions do not lead to amino acid changes. Thus, it should be expected that within a particular gene, mutations will not be distributed randomly but should be confined to certain sites. An average gene contains 1000–2000 base pairs and changes in about half of these will produce amino acid substitutions. Of these substitutions perhaps half again will produce mutant proteins and one might expect about 200–500 distinct mutational sites in a typical gene. Furthermore, if several thousand mutations are independently isolated, each of the distinct sites might be represented by more or less the same number of isolates; however, this is not the case.

Detailed study of the distribution of mutations in the *rII* gene of *E. coli* phage T4 and the *E. coli lacZ* and *trpE* genes indicated that some sites, termed **hot spots,** are altered 20–500 times more frequently than the average mutant site. There are probably many explanations for hot spots in induced mutagenesis. For example, with chemical mutagenesis certain bases might be in regions of the DNA that breathe more often and hence are more susceptible to attack. However, the explanation for spontaneous mutagenesis is quite different.

About one percent of the cytosines in a typical bacterial DNA molecule (5 percent for eukaryotes) are in a methylated form—5-methylcytosine (MeC).

Like cytosine, MeC spontaneously deaminates. When cytosine is deaminated, uracil is formed and this is removed by uracil N-glycosylase. However, when MeC is deaminated, the result is thymine; therefore the G·MeC pair becomes a G·T pair, which in subsequent replication yields an A·T pair. The mismatch repair system (Chapter 10) can convert the G·T pair back to a correct G·C pair, if the system is correctly instructed by the methylation pattern. However, if spontaneous deamination occurs in a nonreplicating DNA molecule (e.g., in a resting cell or in a phage), both strands will be equally methylated. In this case the mismatch repair system receives no signal indicating that the G·C pair is the correct one and could just as well convert the G·T pair to an A·T pair. Thus the mutation frequency can be very high at a MeC site. A given G·MeC→A·T transition will of course only produce a mutant if the change causes an amino acid substitution that affects the activity of the gene product. Such MeC sites do not occur very often, so hot spots are not particularly frequent. Nonetheless, sequencing of several genes and of hot-spot mutants has shown that, indeed, MeC accounts for most of the hot spots seen in spontaneous mutagenesis.

Some hot spots yield large deletions rather than point mutations. The base sequences of these regions indicate that the deletion is often bounded by a sequence that is repeated in the genome. Two possible mechanisms for the production of a deletion in molecules having repeated nucleotide sequences are illustrated in Figure 11-10. These mechanisms are recombinational excision and a particular type of replication error.

REVERSION

So far we have discussed changes from the wild-type to the mutant state. The reverse process, in which the wild-type phenotype is regained, also occurs; this process is called **back mutation, reverse mutation,** or, most commonly, **reversion.** A reverse mutant is called a **revertant.**

An appreciation of what a revertant is can be had by examining how a bacterial revertant is detected. If 10^4 Leu$^-$ bacteria are plated on agar lacking leucine, no colonies will form, but if 10^7 bacteria are plated, a few colonies will appear. If one of these colonies is collected and suspended in a buffer, and a few hundred cells of the suspension are plated, each cell will form a new colony, because the colony consists of bacteria that make their own leucine; these bacteria are called Leu$^+$ revertants.

Reversion is a result of spontaneous mutagenesis and thus it is a nearly random process. It is important to realize that revertants are not formed as a result of the absence of leucine but rather that they preexisted in the population and were merely selected by growth without leucine. However, it is possible in the laboratory to stimulate reversion by mutagenesis.

One way that the wild-type phenotype may be restored is to regain the wild-type genotype (that is, the wild-type base sequence). However, this is not always what happens because reversion can occur in several ways.

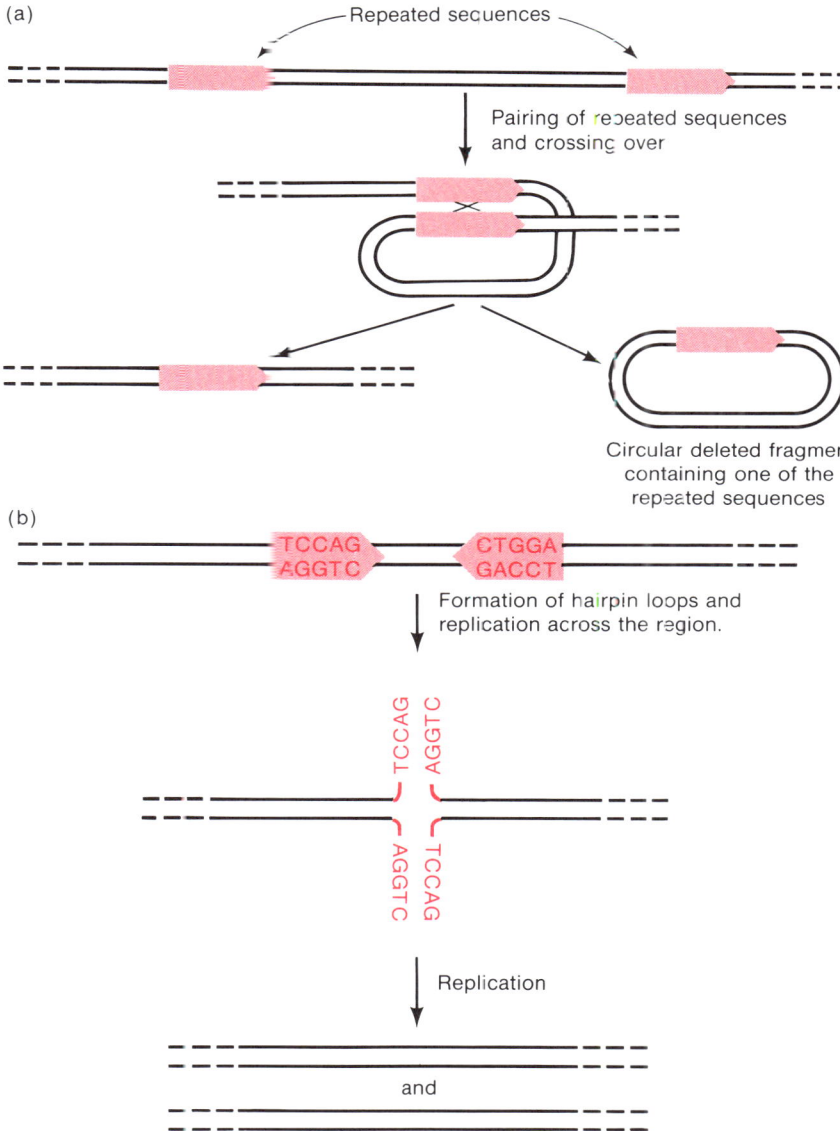

(a)

Repeated sequences

Pairing of repeated sequences and crossing over

Circular deleted fragment containing one of the repeated sequences

(b)

TCCAG
AGGTC

CTGGA
GACCT

Formation of hairpin loops and replication across the region.

TCCAG
AGGTC

AGGTC
TCCAG

Replication

and

Daughter molecules with the red region deleted.

Figure 11-10 Two mechanisms for the production of deletions in molecules with repeated nucleotide sequences. (a) Crossing over (homologous recombination) between paired direct repeats. (b) Replication across an aberrant form of a molecule with an inverted repeat.

Second–Site Revertants

The most useful type of revertant in molecular biological studies has been a kind that does not faithfully recreate the wild–type base sequence. In these revertants the reversion does not occur at the site of the mutation but instead entails a mutation at a second site. Such mutations are often called **second-site** or **suppressor mutations.** In some cases the reverse mutation does not even occur within the mutated gene, and we may distinguish between intragenic and intergenic reversion. One example of intergenic reversion will be discussed shortly, but the major treatment of this phenomenon is in Chapter 13.

The somewhat large value of the frequency of reversion makes one suspect that spontaneous reversion rarely results in a restoration of the wild-type base sequence. In a population of about 10^9 leu^+ cells, only about 100 leu^- mutants are typically present, and the mutations creating these mutants will in general be distributed over roughly 100 different sites in the DNA. In a second population there may also be 100 mutants; some may be mutant at sites present in the first population but most will be located at other sites. In many populations there will be, roughly speaking, mutants representing about 500 different sites. Thus, if we were to insist that a reversion were to occur at one particular site (in order to revert a particular mutant to the wild-type base sequence), then on the average only $100/(500 \times 10^9)$ or one cell in 5×10^9 would be mutated at that site. If base changes occur at random and a change from any one base pair to one of the other three is equally probable, then the original base sequence would be restored in about one cell in 1.5×10^{10} cells. However, it is usually found that in a population derived from a single Leu$^-$ mutant, about one in 10^8 cells is Leu$^+$. For any particular mutant, the number just stated can vary widely, but it is frequently observed that the fraction of the mutant population with the revertant phenotype is much too high to be explained by a return to the wild-type base sequence. The explanation is simply that reversion events occurring at many different sites can produce the same phenotype. This has been confirmed by biochemical data in which it has been shown that the amino acid sequence in a revertant is rarely the wild-type sequence and that the original mutational amino acid substitution is still present. The following discussion shows that this result is not unexpected.

Consider a hypothetical protein containing 97 amino acids whose structure is determined entirely by an ionic interaction between a positively charged $(+)$ amino acid at position 18 and a negative one $(-)$ at position 64 (Figure 11-11). If the $(+)$ amino acid were replaced by a $(-)$ amino acid, the protein would clearly be inactive. Three kinds of reversion events would restore activity (Figure 11-11(a)): (1) The original $(+)$ amino acid could be put back. (2) A different $(+)$ amino acid could be put at position 18. (3) The $(-)$ amino acid at position 64 could be replaced by a $(+)$ amino acid; this second-site mutation would restore the activity of the protein. A possibility which would not generally work but which might be effective in a specific case is to insert a $(+)$ amino acid at position 17 or 19.

Figure 11-11(b) shows another, but more complicated, example of intragenic reversion. In this case the structure of a protein is maintained by a hydrophobic interaction. The replacement of an amino acid with a small

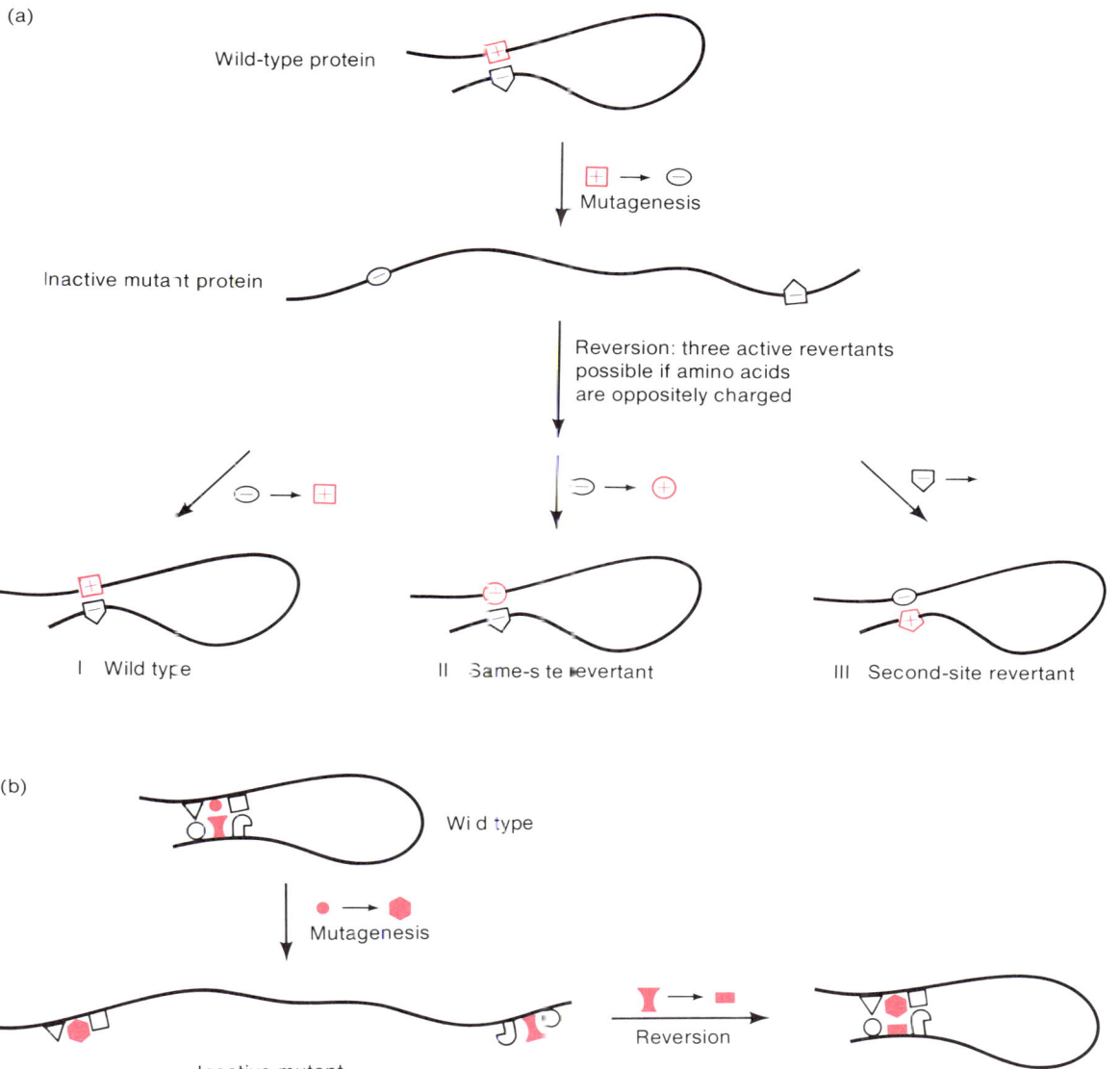

Figure 11-11 Several mechanisms of reversion. In panel (a) the charge of one amino acid is changed and the protein loses activity. The activity is returned by (I) restoring the original amino acid, or (II) by replacing the (−) amino acid by another (+) amino acid, or (II) by reversing the charge of the original (−) amino acid. In each case the attraction of opposite charges is restored. In panel (b) the structure of the protein is determined by interactions between six hydrophobic amino acids. Activity is lost when the small circular amino acid is replaced by the bulky hexagonal one and is restored when space is made by replacing the concave amino acid by the small rectangle.

side chain by a bulky phenylalanine changes the shape of that region of the protein. A second amino acid substitution providing space for the phenylalanine could restore the protein structure.

The analysis of second-site amino acid substitutions has been an important aid in determining the three-dimensional structure of proteins because the following rule is often obeyed:

> If a substitution of amino acid A by amino acid X, which creates a mutant, is compensated for by a substitution of amino acid B by amino acid Y, then A and B are either three-dimensional neighbors or are both contained in two interacting regions.

Second–Site Revertants of Frameshift Mutations

Revertants of frameshift mutations almost always occur at a second site. It is of course possible that a particular added base could be removed or a particular deleted base could be replaced by a spontaneous event, but this would not occur very often. Second-site reversion of a frameshift mutation has two requirements illustrated in Figure 11-12: (1) the reverting event must be very near the original site of mutation, so very few amino acids are altered between the two sites; and (2) the segment of the polypeptide chain in which both changes occur must be able to withstand substantial alterations. Frameshift mutations will be considered in greater detail in Chapter 13, when the genetic code is examined.

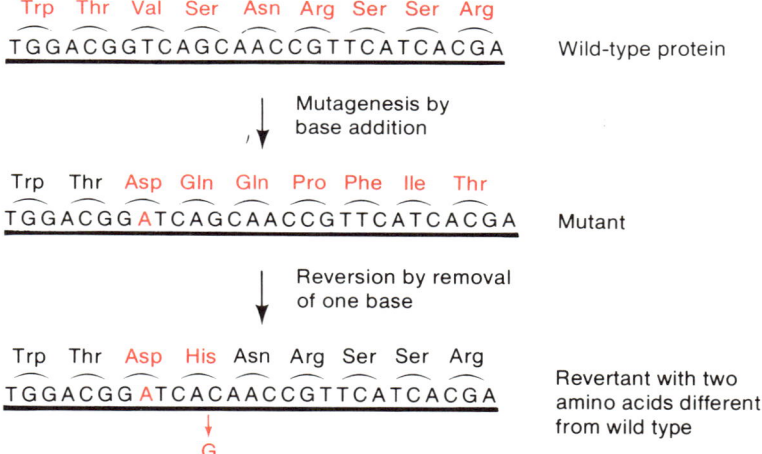

Figure 11-12   Reversion by base deletion from an acridine-induced base-addition mutant.

Intergenic Reversion

Intergenic reversion refers to a mutational change in a second gene that eliminates or suppresses the mutant phenotype. One type of suppression, which occurs when two proteins interact, has already been examined (Figure 1-11) and occurs in the following way. A mutation in the binding site of protein A prevents the protein from interacting with protein B. Another mutation in the binding site of protein B alters this binding site such that the mutant B protein can bind to the mutant A protein; thus, the interaction between the two proteins is restored. The occurrence of intergenic reversion of this kind is an important indicator of the interaction between two proteins and is a splendid example of a genetic result with profound implications for molecular structure.

A second type of intergenic reversion has the remarkable property that the second-site mutation not only eliminates the effect of the original mutation but also suppresses mutations in many other genes as well. This type will be examined in detail in Chapter 13.

Reversion as a Means of Detecting Mutagens and Carcinogens

In view of the increased number of chemicals used and present as environmental contaminants, tests for the mutagenicity of these substances have become important. Furthermore, most carcinogens are also mutagens, so mutagenicity provides an initial screening for these hazardous agents. One simple method for screening large numbers of substances for mutagenicity is a reversion test using nutritional mutants of bacteria. In the simplest type of reversion test a compound that is a potential mutagen is added to solid growth media, known numbers of a mutant bacterium are plated, and the number of revertant colonies that arise is counted. A significant increase in the reversion frequency above that obtained in the absence of the compound tested would identify the substance as a mutagen. However, simple tests of this type fail to demonstrate the mutagenicity of a number of potent carcinogens. The explanation for this failure is that some substances are not directly mutagenic (or carcinogenic), but are converted to active compounds by enzymatic reactions that occur in the liver of animals and have no counterpart in bacteria. The normal function of these enzymes is to protect the organism from various noxious substances that occur naturally by chemically converting them to nontoxic substances. However, when the enzymes encounter certain manmade and natural compounds, they convert these substances, which may not be themselves directly harmful, to mutagens or carcinogens. The enzymes are contained in a component of liver cells called the **microsomal fraction.** Addition of the microsomal fraction of the rat liver to the growth medium as an activation system has been used to extend the sensitivity and usefulness of the reversion test system. The use of the microsomal fraction is the basis of the **Ames test** for carcinogens.

In the Ames test histidine-requiring (His⁻) mutants of the bacterium *Salmonella typhimurium,* containing either a base substitution or a frameshift mutation, are used to test for reversion to His⁺. In addition, the bacterial strains have been made more sensitive to mutagenesis by the incorporation of several mutant alleles that inactivate the excision-repair system and make the cells more permeable to foreign molecules. Since some mutagens act only on replicating DNA, the solid medium used contains enough histidine to support a few rounds of replication but not enough to permit formation of a colony. The medium also contains a potential mutagen to be tested. Rat-liver microsomal fraction is spread on the surface of the medium and bacteria are plated. If the test substance is a mutagen or is converted to a mutagen, colonies form. A quantitative analysis of reversion frequency can also be carried out by incorporating known amounts of the potential mutagen in the medium. The reversion frequency depends on the concentration of the substance being tested and, for a known carcinogen or mutagen, correlates roughly with its known effectiveness.

The Ames test has now been used with thousands of substances and mixtures (such as industrial chemicals, food additives, pesticides, hair dyes, and cosmetics) and numerous unsuspected substances have been found to stimulate reversion in this test. A high frequency of reversion does not mean that the substance is definitely a carcinogen but only that it has a high probability of being so. As a result of these tests, many industries have reformulated their products: for example, the cosmetic industry has changed the formulation of many hair dyes and cosmetics to render them nonmutagenic. Ultimate proof of carcinogenicity is determined from testing for tumor formation in laboratory animals. The Ames test and several other microbiological tests are used to reduce the number of substances that have to be tested in animals since to date only a few percent of more than 500 substances known from animal experiments to be carcinogens failed to increase the reversion frequency in the Ames test.

Restriction Polymorphisms to Detect Mutant Sites in Humans

Every DNA molecule has a unique restriction map for a particular restriction enzyme. If a mutation occurred in a target sequence for a restriction enzyme and the DNA was exposed to the enzyme, one cut normally made in the wild-type DNA would not be made, and the pattern of bands formed after gel electrophoresis of the enzyme digest would differ: two bands present in the digest of wild-type DNA would be missing, and a new band would be present whose molecular weight equals the sum of the molecular weights of the missing bands. If a radioactive probe were available for a particular region of the DNA, e.g., a radioactive RNA copy of a particular gene, band changes could be detected by Southern blotting.

Such changes are called **restriction polymorphisms,** and they have been used to detect base changes in particular human genes. For example, there is a wide variety of mutant hemoglobins, caused by numerous mutations in the β-globin gene. By using [^{32}P]-labeled β-globin RNA as a probe (prepared enzymatically from a cloned β-globin gene template), DNA from humans can be cleaved with various restriction enzymes and screened for particular β-globin mutations. In the case of hemoglobin diseases this technique is not particularly valuable since the mutant hemoglobins themselves can be detected (though the technique is more sensitive and faster than analysis of hemoglobins). However, the technique is becoming exceedingly important in screening for genetic diseases for which the biochemical basis is unknown—in particular, diseases that do not become evident until after breeding age. Huntington's disease is an example. This disease, caused by a dominant allele, produces complete mental and physical debilitation, but does not show up in affected individuals until they are past the age of about 45; unfortunately, by this time the individual has usually mated and the allele has, with a certain probability, been passed on to progeny. Knowledge that one is a carrier could be used in family planning and conceivably to eliminate the allele from the human population. Furthermore, the allele could be detected in a fetus, and, if desired, pregnancy could be terminated.

In attempting to detect a mutation in an unknown gene one is faced with the problem of finding a probe. For Huntington's disease this has been done by brute-force preparation of radioactive copies of fragments covering nearly the entire human genome (some technical tricks are used to reduce the amount of work). In addition, a probe can only be identified by performing Southern blots of a DNA obtained from numerous individuals in a family carrying the disease and digesting with a variety of restriction enzymes in the hope that one will show a restriction polymorphism. The appearance of a polymorphism in a known carrier of the disease and the lack thereof in unaffected elderly adults both identifies the probe and the polymorphism. This has recently been accomplished for Huntington's disease in several families and is being attempted for other genetic diseases. The use of restriction polymorphisms is a splendid example of the application of molecular biological techniques to the elimination of human disease.

12

Transcription

Gene expression is accomplished by the transfer of genetic information from DNA to RNA molecules and then from RNA to protein molecules. RNA molecules are synthesized by using a portion of one strand of DNA as a template in a polymerization reaction that is catalyzed by enzymes called **RNA polymerases.** The process by which RNA molecules are initiated, elongated, and terminated is called **transcription**.

Two aspects of transcription must be considered—(1) the enzymology, and (2) the signals that determine where on a DNA molecule transcription begins and stops. The enzymatic mechanism for forming chemical linkages between nucleotides in transcription has been fairly well understood for many years. The recognition sites were extremely difficult to study until the development, more than a decade ago, of gene cloning techniques (Chapter 5), which allow the isolation of particular segments of a larger DNA molecule, and, a few years later, of a procedure for determining the base sequence of DNA (Chapter 4). In every case cited in this chapter, information about recognition sites has been obtained by a combination of these techniques.

Throughout this book base sequences of both DNA and RNA molecules will be presented. For DNA there is no reason to give the sequences in both strands (which would be redundant), so a convention has been adopted by which only the sequence of the **noncoding** DNA strand is stated. This strand has been chosen since, for a transcribed region, the base sequence on this strand is that of the RNA synthesized from that region. This convention

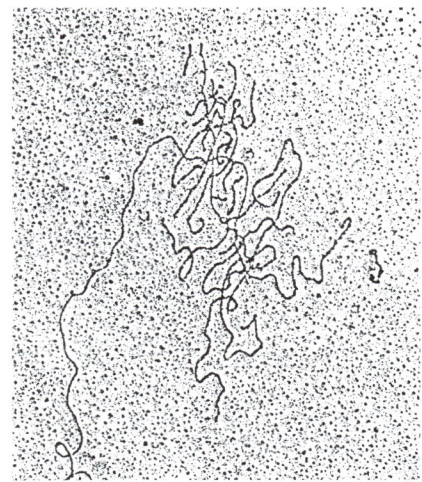

A prokaryotic transcription complex. The extended molecule is DNA. The short branches and the tangled mass at the end are RNA molecules still bound to the DNA. The RNA is coated with a binding protein that extends the molecule and improves contrast. [Courtesy of J. Meyer.]

has the advantage that the base sequences corresponding to amino acids (the codons) and to start and stop signals for protein synthesis are in their standard forms and easily recognizable.

We begin our examination of transcription by describing the characteristics of the enzymatic mechanism of polymerization of RNA.

ENZYMATIC SYNTHESIS OF RNA

In this section we describe the basic features of the polymerization of RNA, the identity of the precursors, the nature of the template, the properties of the polymerizing enzyme, and the mechanisms of initiation, elongation, and termination of synthesis of an RNA chain.

Basic Features of RNA Synthesis

The essential chemical characteristics of the synthesis of RNA are the following:

1. The precursors in the synthesis of RNA are the four ribonucleoside 5′-triphosphates (NTP) ATP, GTP, CTP, and UTP. On the ribose portion of each NTP there are two OH groups—one each on the 2′- and 3′-carbon atoms (Figure 12-1).
2. In the polymerization reaction a 3′-OH group of one nucleotide reacts with the 5′-triphosphate of a second nucleotide; pyrophosphate is removed and a phosphodiester bond results (Figure 12-2). This is the same reaction that occurs in the synthesis of DNA.
3. The sequence of bases in an RNA molecule is determined by the base sequence of the DNA. Each base added to the growing end of the RNA chain is chosen by its ability to base-pair with the DNA strand used as a template; thus, the bases C, T, G, and A in a DNA strand cause G, A, C, and U, respectively, to appear in the newly synthesized RNA molecule.
4. The DNA molecule being transcribed is double-stranded, yet in any particular region only one strand serves as a template. The meaning of this statement is shown in Figure 12-3.

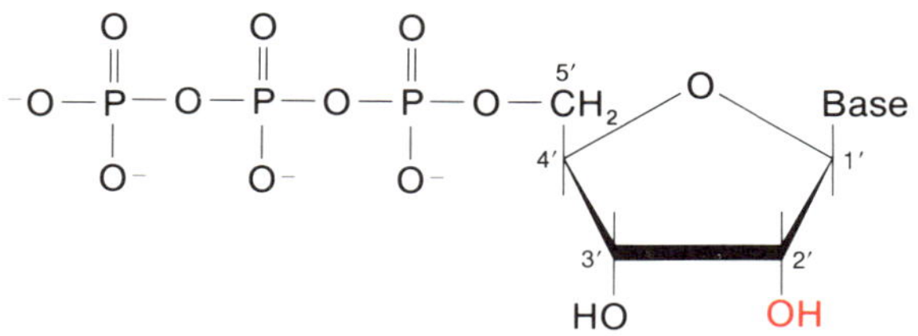

Figure 12-1 A ribonucleoside 5′-triphosphate. The 2′-OH group shown in red is replaced by an H in a *deoxy*nucleotide.

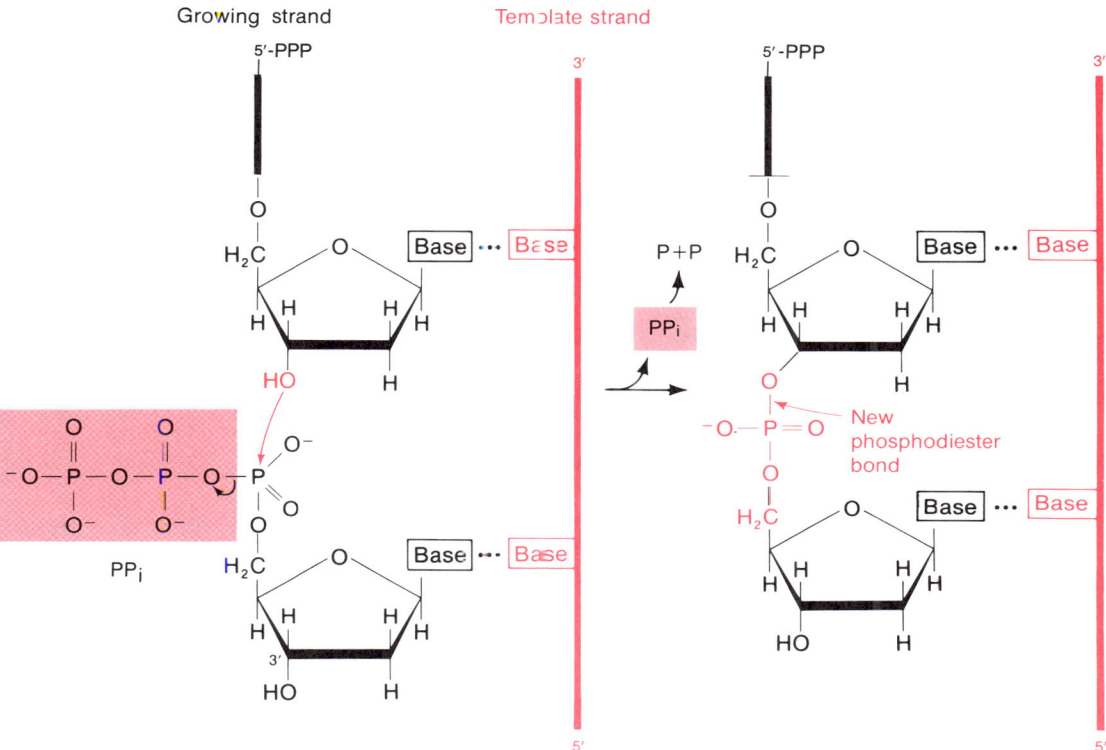

Figure 12-2 Mechanism of the chain-elongation reaction catalyzed by RNA polymerase. The red arrow joins the reacting groups. The pyrophosphate group (shaded in red) and the red hydrogen do not appear in the RNA strand. The DNA template strand and the newly synthesized RNA strand are antiparallel and complementary, as in double-stranded DNA.

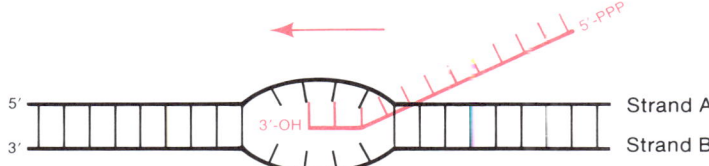

Figure 12-3 An RNA strand (shown in red) is copied only from strand A of a segment of a DNA molecule. No RNA is copied from strand B in that region of the DNA molecule. However, elsewhere, for example, in a different gene, strand B might be copied (see Figure 12-18); in that case, strand A would not be copied in that region of the DNA. The RNA molecule is antiparallel to the DNA strand being copied and is terminated by a 5'-triphosphate at the nongrowing end. The red arrow shows the direction of RNA chain growth. The drawing is quite schematic; in reality, the bubble and the DNA·RNA paired regions are considerably larger, as shown in Figure 12-13.

5. The RNA chain grows in the $5' \rightarrow 3'$ direction: that is, nucleotides are added only to the 3'-OH end of the growing chain—this direction of chain growth is the same as that in DNA synthesis.

6. RNA polymerases, in contrast with DNA polymerases, are able to initiate chain growth; that is, *no primer is needed.*

7. Only ribonucleoside 5'-triphosphates participate in RNA synthesis and the first base to be laid down in the initiation event is a triphosphate. Its 3'-OH group is the point of attachment of the subsequent nucleotide. Thus, the 5' end of a growing RNA molecule terminates with a triphosphate (Figure 12-3).

The overall polymerization reaction may be written as

$$n\text{NTP} + \text{XTP} \xrightarrow[\text{Mg}^{2+}]{\text{DNA, RNA-P}} (\text{NMP})_n - \text{XTP} + n\text{PP}_i$$

in which XTP represents the first nucleotide at the 5' terminus of the RNA chain, NMP is a mononucleotide in the RNA chain, RNA-P is RNA polymerase, and PP_i is the pyrophosphate released each time a nucleotide is added to the growing chain. The Mg^{2+} ion is required for all nucleic acid polymerization reactions.

Most of the features just described and other properties of RNA synthesis have been determined by experiments using the following simple assay for polymerization. RNA is insoluble in trichloracetic acid and forms a precipitate, but the nucleotides are soluble in the acid. Thus, a reaction mixture is prepared containing DNA, Mg^{2+}, RNA polymerase (which may be in a crude extract), three nonradioactive NTP, and one radioactive NTP labeled either in the base with ^3H or ^{14}C or in the phosphate attached to the ribose with ^{32}P. Following an incubation period, acid is added to precipitate the nucleic acids and the mixture is filtered. If RNA has been synthesized, radioactive material will be retained on the filter; the amount of this radioactivity is proportional to the amount of RNA synthesized.

The synthesis of RNA consists of four discrete stages: (1) binding of RNA polymerase to a template at a specific site, (2) initiation, (3) chain elongation, and (4) chain termination and release. A discussion of these stages is facilitated if the structure of RNA polymerase is understood. In the following section we will describe the properties of the RNA polymerase of *E. coli,* which is the best understood RNA polymerase and a prototype for bacterial RNA polymerases. The RNA polymerases of mammalian cells will be described in a later section.

E. coli RNA Polymerase

E. coli RNA polymerase consists of five subunits—two identical α subunits and one each of types β, β', and σ. It is one of the largest enzymes known with a total molecular weight of 465,000 (Figure 12-4). *E. coli* mutants containing altered subunit structures have been isolated.

The σ subunit dissociates from the enzyme during the elongation stage of RNA polymerization (to be discussed later). The term **core enzyme** is used to describe the σ-free unit—namely, $\alpha_2\beta\beta'$. The complete enzyme, $\alpha_2\beta\beta'\sigma$ is called the **holoenzyme.** We will use the name RNA polymerase when the holoenzyme is meant, or, when synthesis is in progress, for the core enzyme.

RNA polymerase is sufficiently large that it can come into contact with many DNA bases simultaneously. An estimate of the size of the region of the DNA where contact is made is obtained by selectively degrading adjacent DNA bases with DNase, a procedure known as the **DNase protection method** (Figure 12-5) RNA polymerase is bound to DNA; then a DNA endonuclease is added to the mixture. The nuclease degrades most of the DNA to mono- and dinucleotides but leaves untouched DNA segments in close contact with RNA polymerase; these segments vary from 41 to 44 base pairs. If RNA polymerase holoenzyme were to be added to the total DNA complement of *E. coli* and then DNase was added, the protected fragments would consist of about a thousand different DNA segments, each representing a particular gene or set of adjacent genes. To study details of the DNA-enzyme binding process, it is obviously desirable to examine a single binding sequence. This can be obtained by using a cloned gene in a protection experiment.

Another test, called **footprinting,** detects regions of both strong and weak binding and shows that the complete contact region is about 70 bp. The footprinting technique (which can be used with any DNA-binding protein) is shown in Figure 12-6. A particular piece of double-stranded DNA having a known number and sequence of base pairs and binding RNA polymerase is labeled in one strand at its 5′ terminus with ^{32}P and then is allowed to interact with the polymerase. A DNA endonuclease is then added, but so briefly that no DNA molecule receives more than one single-strand break. Nicking occurs at all positions except those protected by RNA polymerase. The DNA is then isolated and denatured. The radioactively labeled DNA consists of a set of molecules whose sizes are determined by the distance

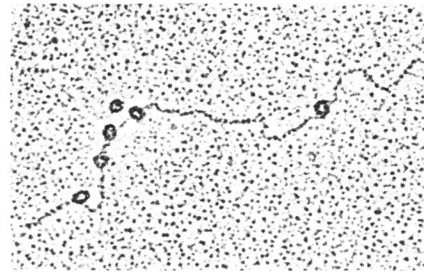

(a)

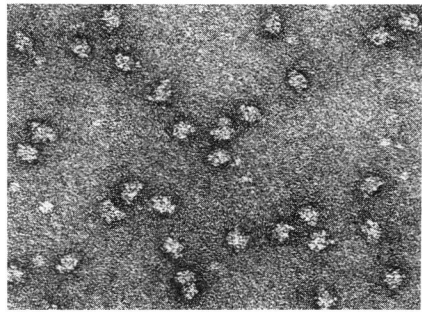

(b)

Figure 12-4 Electron micrographs of *E. coli* RNA polymerase. (a) Molecules bound to DNA. (b) The holoenzyme viewed by negative-contrast electron microscopy (× 270,000). [Courtesy of Robley Williams.]

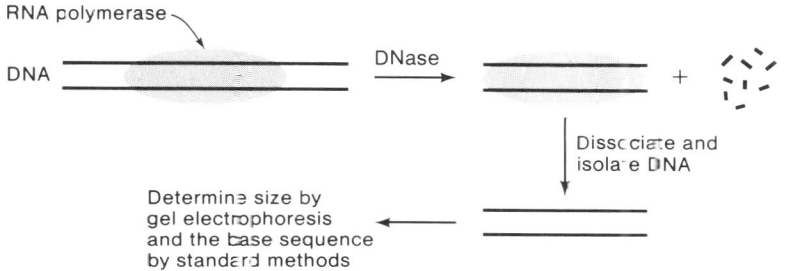

Figure 12-5 The DNase-protection method for isolating a region of the DNA in contact with RNA polymerase.

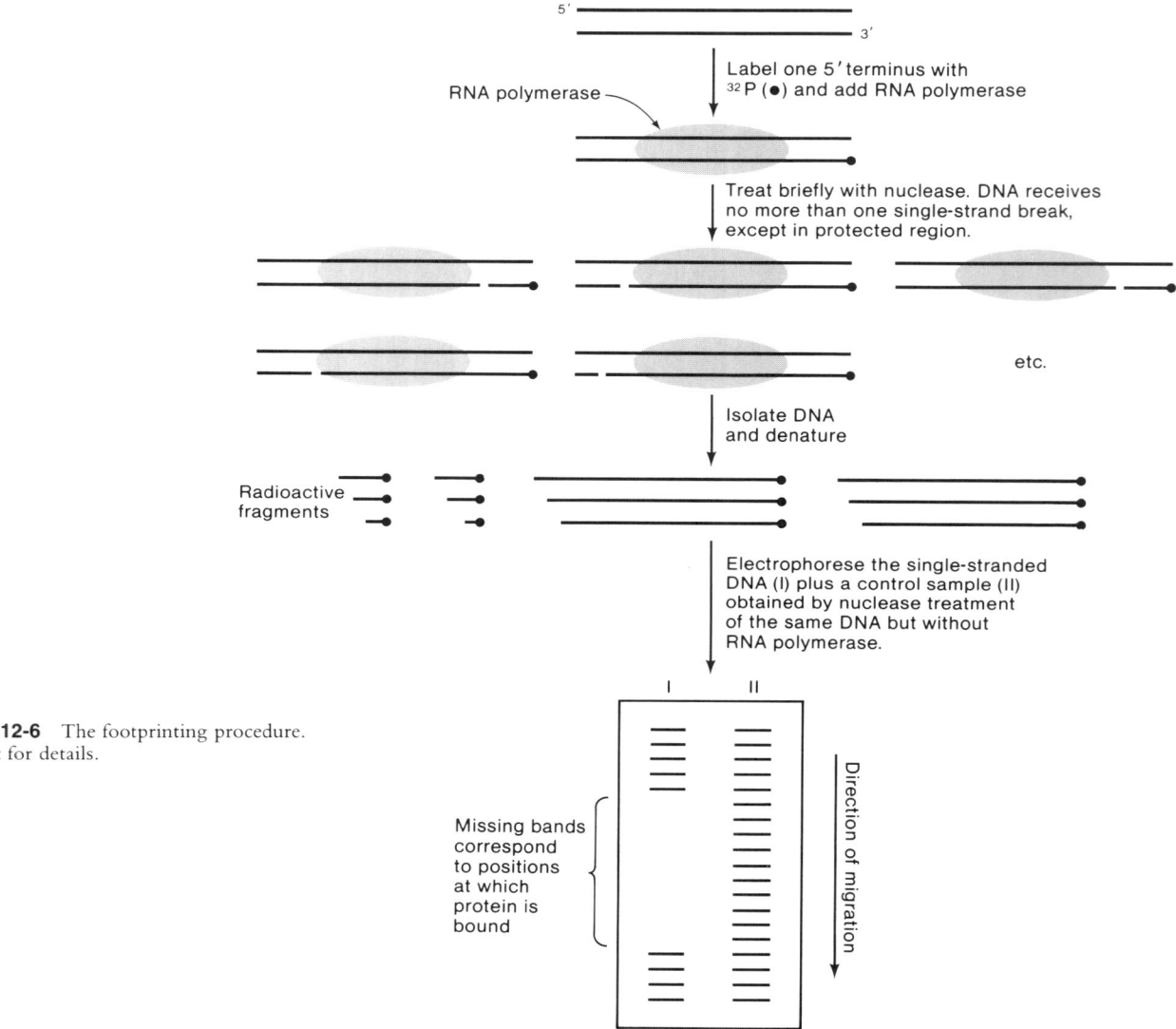

5′

3′

Label one 5′ terminus with
^{32}P (●) and add RNA polymerase

RNA polymerase

Treat briefly with nuclease. DNA receives
no more than one single-strand break,
except in protected region.

etc.

Isolate DNA
and denature

Radioactive
fragments

Electrophorese the single-stranded
DNA (I) plus a control sample (II)
obtained by nuclease treatment
of the same DNA but without
RNA polymerase.

I   II

Missing bands
correspond
to positions
at which
protein is
bound

Direction of migration

Figure 12-6   The footprinting procedure.
See text for details.

from the labeled 5′ end to the nicks. If the DNA contains n base pairs and RNA polymerase is *not* added, n sizes of DNA fragments will be present. However, if RNA polymerase binds to x base pairs and thereby prevents access of the DNA to the nuclease, only $n - x$ different sizes of DNA fragments will be represented. These fragments are separated by gel electrophoresis, as in the DNA base-sequencing procedure (Chapter 5). As shown in the figure, two DNA samples are compared, one without RNA polymerase (to obtain the positions of the n bands), and one with RNA polymerase (to determine the positions of the missing bands). If a third sample is subjected to base-sequencing analysis, the sequence of the protected region is also obtained.

Specific points of contact within the contact region can be identified with another protective method—this one, the **dimethyl sulfate protection method** (used to study the DNA–Cro-protein interaction examined in Chapter 7). Dimethyl sulfate methylates A and G (not C or T) but cannot react with these bases if they are in close contact with a protein. When a protein molecule binds to a long sequence of DNA bases, many bases are not in close contact with the protein because the surfaces of the protein and DNA molecules are not completely complementary in shape. Thus, the A's and G's in the actual contact region will be unavailable for methylation; the sites of contact can be identified by determining the positions of nonmethylated A and G in the endonuclease-protected region with and without bound RNA polymerase. One might wonder why the entire contact region is resistant to DNase yet only specific contact points are resistant to methylation. This is explained in part by the different relative sizes of the DNase and dimethyl sulfate molecules. The separation of a DNA base that is in the contact region (but not in direct contact from RNA polymerase is usually only a few angstrom units; this space is sufficient for a small molecule such as dimethyl sulfate to approach the base but not great enough for a bulky enzyme molecule to reach the adjacent phosphodiester bond.

The large size of the *E. coli* holoenzyme compared to typical enzymes raises the question of whether some complex feature of the polymerization reaction necessitates such a large enzyme. In fact, the RNA polymerases synthesized by certain *E. coli* phages are considerably smaller proteins (mol. wt., ca. 110,000), yet elongate RNA chains about five times more rapidly than the *E. coli* enzyme. Clearly, the polymerization reaction itself does not require the huge multisubunit *E. coli* enzyme. A clue to understanding the difference in the sizes of the *E. coli* and phage enzymes comes from (1) attempting to use the phage enzymes to transcribe *E. coli* DNA and (2) studying the gene organization of the phages. First, the phage polymerases can transcribe at best a small fraction of the *E. coli* genes. Second, the phage genes are arranged in only a few transcription units, so there are only a few RNA polymerase binding sites. In *E. coli*, RNA polymerase must be able not only to recognize about 1000 transcription units, which are signalled by hundreds (if not a thousand) of different binding sites but also to respond to a large number of regulatory proteins that alter the ability to recognize a binding site. It seems likely that these multiple requirements necessitate the large multisubunit *E. coli* enzyme. Eukaryotic RNA polymerases are even more complex.

Site Selection: I. The Promoter

The first step in transcription is the binding of RNA polymerase to a DNA molecule. Binding occurs in particular regions called **promoters,** which are

```
            mRNA
            start
              ↓
C G T A T A A T G T G T G G A
G G T A C G A T G T A C C A C A
A G T A A G A T A C A A A T C G
G T G A T A A T G G T T G C A
C T T A T A A T G G T T A C A
C G T A T G T T G T G T G G A
G C T A T G G T T A T T T C A
G T T T T C A T G C C T C C A
A G G A T A C T T A C A G C C A
T G T A T A A T A G A T T C A
G G C A T G A T A G C G C C C G
G C T T T A A T G C G G T A G
```

Figure 12-7   Segments of the noncoding strand of protected regions from various genes showing the common sequence of six bases (pink) known as the Pribnow box. The start point for mRNA synthesis is indicated by the red letters. The "conserved T" is underlined.

sequences in which several interactions occur. The existence of promoters was first demonstrated by the isolation of a particular class of Lac⁻ mutations in *E. coli*. These mutations not only eliminate gene activity but also are noncomplementable (because they are *cis*-dominant) and prevent synthesis of the RNA transcript of the *lac* gene. These mutations are called **promoter mutations.**

Several events must occur at a promoter: RNA polymerase must recognize a specific DNA sequence, attach in a proper conformation, locally open the DNA strands in order to gain access to the bases to be copied, and then initiate synthesis. These events are guided by the base sequence of the DNA, the polymerase σ subunit (without which the promoter is not recognized) and, for some promoters, also by auxiliary proteins. The details of these events are not known, but the process can be broken down into three parts—(i) template binding at a polymerase recognition site, (ii) binding to an initiation site, and (iii) establishment of what is termed an open-promoter complex (shown schematically later in Figure 12-10). The approach to elucidating these steps for many genes has been to isolate the DNA segment (the promoter) that is protected by RNA polymerase from DNase digestion, then isolate a larger segment (2–3 times as large) that includes the protected region, determine the base sequence in the larger segment, and look for common features in the sequences (Figure 12-7). The specific sites of contact are also determined by the dimethyl sulfate protection method. This is important because one might expect that the specific contact sites would be in the regions common to all promoters.

The RNA molecules synthesized *in vitro* from each of these promoter regions must also be sequenced if one wishes to identify the initiation sequence, which is the sequence of the first few bases that are transcribed; this sequence is the complement of the bases at the 5′ terminus of the newly synthesized RNA molecule. Additional information is obtained by determining the sequence of bases in promoters having mutations that either eliminate initiation *in vivo* or change the requirements for initiation (Figure 12-8). The rationale is that if a base change affects promoter activity, that base must be contained in the promoter.

Figure 12-7 shows portions of several promoter sequences in *E. coli* and its phages (each promoter sequence is recognized by *E. coli* RNA polymer-

Figure 12-8   A region of the noncoding strand of the promoter for the *lac* gene showing six mutations (red arrows) that affect promoter activity; Δ means a base deletion. The Pribnow box is shaded in red. Many base changes are known; all are either in or near the Pribnow box or are clustered around base −35 and thus define an important site.

ase) and their important features. In a region 5–10 bases to the left of the first base copied into mRNA is the right end of a sequence called the **Prib-now box.*** A basic sequence derived from a large set of observed similar sequences is called a **consensus sequence;** it is obtained by comparing a large number of sequences from a particular region. If the sequences were totally unrelated, one would expect each base to occur at each position 25 percent of the time. Examination of more than 100 *E. coli* promoters has shown that the frequency of occurrence of the bases is the following, in which the subscript is the frequency:

$$T_{80}A_{95}T_{45}A_{60}A_{50}T_{96}$$

Furthermore, there are few Pribnow boxes that differ from the consensus by more than two bases, and most that differ do so by only one base.

The Pribnow box is thought to orient RNA polymerase, such that synthesis proceeds from left to right (as the sequence is drawn), and to be the region at which the double helix opens to form the open–promoter complex (see below).

Before enough sequences were known that the Pribnow box was recognizable, the first base transcribed was chosen as a reference point and numbered +1. The direction of transcription was called **downstream;** all **upstream** bases, which are not transcribed, were given negative numbers starting from the reference. The Pribnow box is enclosed between −21 and −4, depending on the particular promoter.

Mutations in the Pribnow box that prevent initiation of transcription have been observed for several genes. Two of these are shown in Figure 12-8. These mutations clearly indicate the importance of this sequence. Other bases outside of the Pribnow box are important too, as indicated by the other mutations shown in the figure.

Examination of the complete sequence of the region protected by RNA polymerase indicates that for many (but not all) promoters, there is a second important region, to the left of the Pribnow box, whose sequences in different promoters have common features (Figure 12-9). This six–base sequence, which is called the **−35 sequence** and has the consensus **TTGACA,** may be the initial site of binding of the enzyme, when the sequence is present.

The earliest events that take place when RNA polymerase binds to a promoter and the precise roles of the −35 sequence and the Pribnow box are not completely clear. The haziness derives from two facts: (1) whereas the core enzyme binds to DNA, it binds at random positions rather than only to promoters; and (2) the holoenzyme (core enzyme plus the σ subunit) binds specifically to promoters. Thus, one would expect the σ subunit to be the promoter-recognition element and make direct contact with the DNA.

---

*In recent years it has become common to use the word "box" to indicate a sequence that occurs repeatedly as a transcription or regulatory signal. All sequences found in Pribnow boxes are considered to be variants of a basic sequence TATAAT.

```
C C A G G C  T T T A C A  C T T T A T G C T T C C G G C T C G  T A T G T T  G T G T G G A  A T T G
C T T T T T  G A T G C A  A T T C G C T T T G C T T C T G A C  T A T A A T  A G A C A G G  G T A A
G G C G G T  G T T G A C  A T A A A T A C C A C T G G C G G T  G A T A C T  G A G C A C A  T C A G
G T G C G T  G T T G A C  T A T T T T A C C T C T G G C G G T  G A T A A T  G G T T G C A  T G T A
A T T G T T  G T T G T T  A A C T T G T T T A T T G C A G C T  T A T A A T  G G T T A C A  A A T A
C G T A A C A C T T T A  C A G C G G C G C G T C A T T T G A  T A T G A T  G C G C C C G  C T T
```

−35 Sequence   Pribnow box   mRNA start

Figure 12-9 Base sequences in the non-coding strand of six different RNA polymerase-protected regions showing the similarity between the −35 sequences. In each case, mutations that eliminate promoter activity have been found in the −35 sequence. The Pribnow boxes rather than the mRNA start points are aligned. In these examples, 18 nucleotides separate the −35 sequence and the Pribnow box; many promoters have other numbers of separating nucleotides.

However, although several experiments indicate that the holoenzyme binds first to the −35 region in a highly specific interaction, paradoxically no evidence has been obtained for direct binding of σ to the −35 sequence, or to any other part of the promoter (the required experiments are surprisingly difficult to carry out). In the absence of solid evidence, direct binding of σ is considered to be an open question and the most likely picture of the early stages in binding is thought to be the following: σ, which is quite a large protein, probably makes initial contact with either or both the −35 sequence and the Pribnow box and then mediates a conformational change of the core enzyme that enables various regions of the giant enzyme to contact both sequences. In this model, direct contact of σ with the DNA could be short-lived, which would account for the difficulty in detecting the contact. An alternative model is that σ never contacts the DNA but, in forming the holoenzyme, causes a conformational change of the core enzyme that enables it to recognize the promoter; however, there are enough experiments that give hints about direct binding that this explanation is considered less likely.

After binding of the holoenzyme to the Pribnow box (or possibly concomitant with binding) the conformation called the **open–promoter complex** forms; this highly stable complex is the active intermediate in chain initiation. The DNA helix in an open-promoter complex is locally unwound, starting about 10 bp from the left end of the Pribnow box and extending about 20 bp past the position of the first transcribed base (Figure 12-10). It is thought that RNA polymerase itself induces this unwinding and undergoes a conformational change itself in so doing. This melting is necessary for pairing of incoming ribonucleotides.

The promoters discussed in this section are classified as **high-level** or **strong promoters.** There are also **weak promoters** in which recognition and/or binding by RNA polymerase is less strong. In a given time period the number of RNA molecules synthesized from genes with weak promoters is much less than from a strong promoter with the result that fewer mRNA

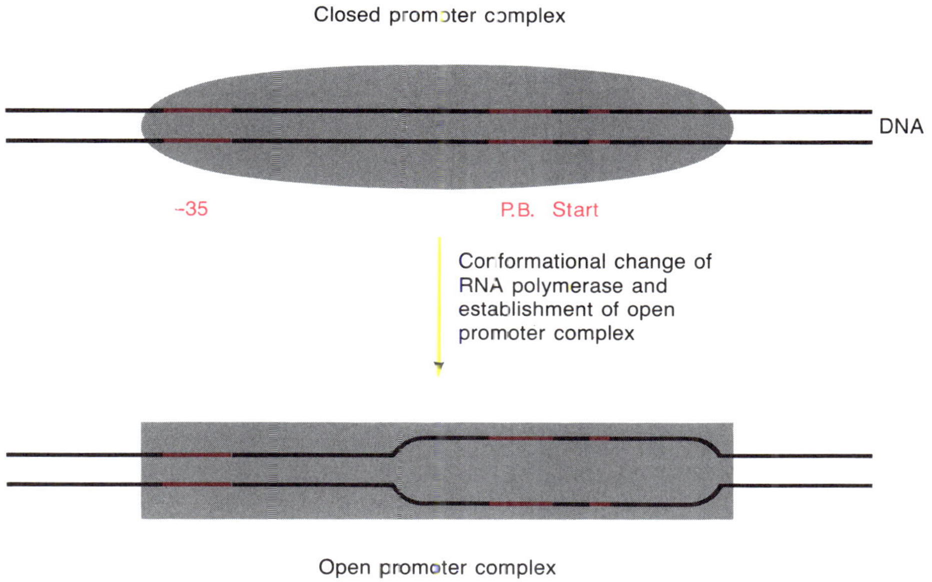

Closed promoter complex

DNA

−35

P.B. Start

Conformational change of
RNA polymerase and
establishment of open
promoter complex

Open promoter complex

Figure 12-10 Stages of binding of RNA polymerase holoenzyme to a prokaryotic promoter. First, the huge enzyme covers the −35 sequence, the Pribnow box (P.B.), and the transcriptional start site; this binding is somehow mediated by the σ subunit. This early stage is called the closed promoter complex. Then, the enzyme undergoes a conformational change, yielding the open promoter complex, in which DNA strands are separated in a large region.

molecules are made per unit time by genes with weak promoters. Promoter strength is one factor that determines the number of copies of each protein molecule present in the cell. The difference between weak and strong promoters lies in the sequences of the −35 and −10 regions.

Site Selection: II. Auxiliary Proteins

Some bacterial promoters require an activator protein for effective initiation. In general, the mode of action of these proteins is not well understood. However, some information is available about the two phage λ promoters, *pI* and *pre,* both of which are inactive unless an auxiliary protein, the λ-encoded cII protein, is present. The site of action of this protein is a region upstream from the Pribnow box; there is no −35 sequence in these promoters but the site of action is where this sequence would be. This region consists of two identical tetranucleotides flanking a hexanucleotide. Because there are 10 bp per helical turn, the two tetranucleotides are on the same side of the helix and the central hexanucleotide is on the other side. The cII protein binds to the DNA in the major groove containing each tetranucleotide, as we saw for the DNA-binding proteins described in Chapter 7. The

```
AATGTGAGTTAGCTCACTCA
TTACACTCAATCGAGTGAGT
```

Figure 12-11 Base sequence of both strands of the principal CRP protein binding site (site 1) of the *E. coli lac* promoter showing the axis of the inverted repeat (vertical line) and the symmetric elements (shown in red).

hexanucleotide sequence, which is quite different from the −35 consensus sequence, is a binding site for RNA polymerase, but *only when cII protein is already bound to the tetranucleotides*. Without bound cII protein, RNA polymerase makes sequence-specific contacts with the hexanucleotide but the interaction is too weak to be stable. However, even though RNA polymerase and cII bind to opposite sides of the DNA, because of the great size of the enzyme (which wraps partly around the helix), specific contact is made between RNA polymerase and cII protein and these contacts provide the necessary stability to allow transcription to be initiated.

Another well-studied system is the *E. coli lac* promoter. This promoter includes a −35 sequence but does not bind RNA polymerase significantly unless the **cyclic AMP (cAMP) receptor protein (CRP)** is also bound. There are two CRP-cAMP binding sites in the *lac* promoter, one in the −70 to −50 segment (site I) and another in the −50 to −40 segment (site II). Site I contains an inverted repeat sequence, which seems to be characteristic of most strong binding sites (Figure 12-11). Site II is a very weak binding site but when a CRP-cAMP complex is bound to site I, the ability of CRP-cAMP to bind to site II is very much enhanced. Once site II is occupied, RNA polymerase binds tightly to the promoter region, presumably by binding to CRP-cAMP, as shown in Figure 12-12 (see the discussion of the Cro protein in Chapter 7). In the absence of CRP, RNA polymerase holoenzyme binds to the Pribnow box, but a **closed-promoter complex** results—dissociation of the DNA strands in the initiation region does not occur. Many questions remain about this system. (1) Little information is available about how the CRP-cAMP complex facilitates the formation of the open-promoter complex. At one time it was thought that the complex is a DNA-melting agent, but that is now known to be incorrect. Possibly, CRP-cAMP causes a conformational change in bound RNA polymerase that induces strain in the region of the DNA to which it is bound, thereby weakening the helix. (2) Although it was pointed out in Chapter 7 that specific contact is made between CRP-cAMP and RNA polymerase when bound to the promoter

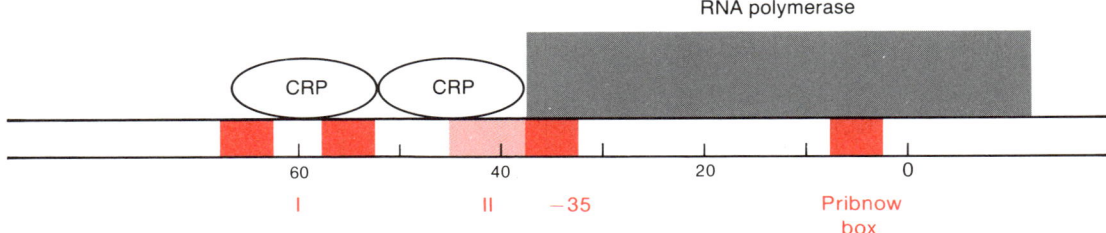

Figure 12-12 A hypothetical arrangement of two CRP protein molecules and RNA polymerase on the *lac* promoter. I and II are the CRP sites mentioned in the text.

of the *E. coli lac* system the situation must be more complicated. In phage λ, the separation of the cI binding site and the RNA polymerase binding sites of the λ *pI* and *pre* promoters is the same, which is expected when a specific binding interaction is involved. However, CRP-cAMP is required for initiation of transcription of many *E. coli* genes (Chapter 15) yet the separation of the CRP-cAMP binding site and the RNA polymerase binding site varies considerably. How one protein bound to DNA can specifically influence the binding of RNA polymerase at varying distances from the protein is completely unclear. Many suggestions have been given but at present none have experimental support. When eukaryotic transcription is examined, we will see that such variable distance is the rule. In eukaryotes RNA polymerase may never recognize a promoter by itself but needs a diversity of auxiliary proteins that bind to sites whose distance from the transcription start site vary by hundreds of nucleotides.

RNA Chain Initiation

Once the open-promoter complex has formed, RNA polymerase is ready to initiate synthesis. RNA polymerase contains two nucleotide binding sites called the **initiation site** and the **elongation site.** The initiation site primarily binds purine triphosphates, ATP and GTP, and ATP is usually the first nucleotide in the chain. Thus, the first DNA base that is transcribed is usually thymine. The initiating nucleoside triphosphate binds to the enzyme in the open-promoter complex and forms a hydrogen bond with the complementary DNA base. The elongation site (also called the catalytic site) is then filled with a nucleoside triphosphate that is selected by its ability to hydrogen-bond with the next base in the DNA strand. The two nucleotides are then joined together, the first base is released from the initiation site, and initiation is complete. In some way, the details of which are not understood, the RNA polymerase and the template strand move relative to each other, so the two binding sites and the catalytic site are shifted by exactly one nucleotide. The dinucleotide remains hydrogen-bonded to the DNA. The initiation phase is not yet complete since polymerization of the first 6–10 nucleotides is different from the process that occurs henceforth. The main difference is that the strict one-by-one-base template reading (called **processivity**) is not yet established. Evidence is that during this initial period, short oligonucleotides are often released from the DNA, indicating that RNA polymerase stops and restarts at the initiation site. It is thought that this is a result of RNA polymerase remaining anchored to some upstream site. In an unknown way, RNA polymerase finally becomes locked in to forward motion and the elongation phase begins.

The drug **rifampicin** is useful in studying initiation. It binds to the β subunit of RNA polymerase blocking the transition from the chain initiation phase to the elongation phase; it is an inhibitor of chain initiation but not elongation. Rifampicin and related drugs have been used extensively in investigations of transcription, as they can be added to a growing culture for *in vivo* studies or to a reaction mixture for *in vitro* studies and will rapidly inhibit the initiation of new chains. Since it takes only a few seconds to complete a growing RNA chain, RNA synthesis is quickly terminated.

How selection of the first base copied occurs is not known. It is always within six to nine bases of the right (3') end of the Pribnow box, but neither the precise position nor the identity of adjacent bases shows any obvious pattern.

Chain Elongation

After several nucleotides (eight is the best estimate) are added to the growing chain, RNA polymerase undergoes a conformational change and loses the σ subunit. This marks the transition from the stuttering of the initiation phase, just described, to the stable forward movement of the elongation phase. Thus, *most elongation is carried out by the core enzyme* (Figure 12-13).

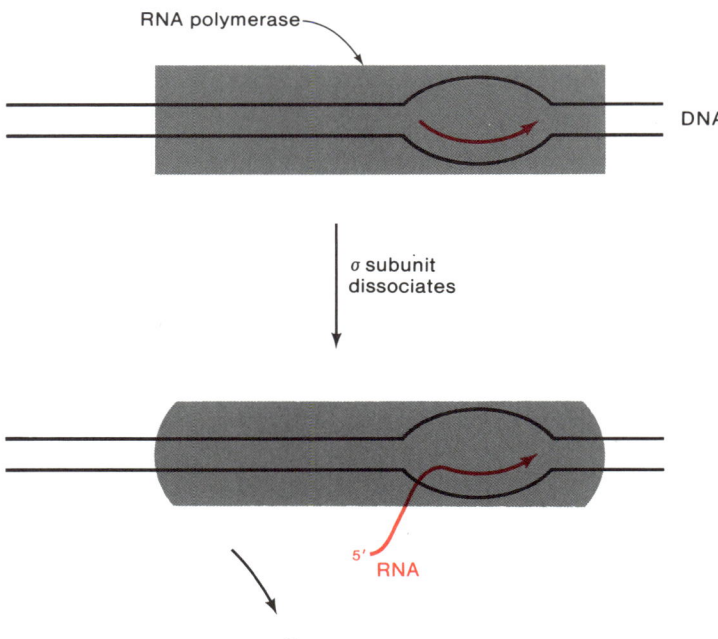

Figure 12-13 Dissociation of the σ subunit and change in conformation of RNA polymerase core enzyme after about eight bases are transcribed. The σ subunit is not shown as a separate part of the holoenzyme since, as mentioned earlier in the text, its location with respect to the holoenzyme and the DNA is not known.

The core enzyme moves along the DNA, binding a nucleoside triphosphate that can pair with the next DNA base and opening the DNA helix as it moves. The DNA helix recloses as synthesis proceeds. The newly synthesized RNA is released from its hydrogen bonds with the DNA as the helix re-forms. Roughly 12 RNA bases are paired to the DNA in the open region.

A peculiarity of the chain elongation reaction is that it does not occur at a constant rate; that is, synthesis markedly slows down when particular regions of DNA are passed, then continues at the normal rate, slows down again, accelerates again, and so forth. This reduction in rate is called a **pause.** Analysis of pausing along stretches of DNA of known sequence shows that pausing frequently follows sequences that form hairpins in the RNA, but at least half of the pause sites have no recognizable features. One very recent report suggests that pausing in nonhairpin regions may be associated with an increase in rate of phosphoester cleavage by RNA polymerase. Since enzymes only alter the rate and not the equilibrium of a reaction, RNA polymerase catalyzes both polymerization and degradation, with polymerization predominating vastly Detailed study of the elongation process shows that occasionally RNA polymerase actually runs in reverse (like the editing function of DNA polymerase, Chapter 9), either remaining in place (polymerization and degrading repeatedly) or moving backward. This may be the cause of pausing at certain sites, with unrecognized features of the base sequence upsetting the normal equilibrium. This explanation must be considered quite preliminary, as pausing is a poorly understood phenomenon. Pausing seems to participate in termination, as will be seen in the following section.

Termination and Release of Newly Synthesized RNA

Termination of RNA synthesis occurs at specific base sequences within the DNA molecule. These sequences are of two types, simple terminators and those that require auxiliary **termination factors.** About 100 factor-independent termination sequences have been determined in bacteria and each has the characteristics shown in Figure 12-14. There are three important regions:

1. There is an inverted repeat containing a central nonrepeating segment; that is, the sequence in one DNA strand would read like ABCDE-XYZ-E'D C'B'A'. Thus, this sequence is capable of intrastrand base pairing, forming a stem-and-loop in the RNA transcript and possibly a cruciform structure in the DNA strands. It is possible though not yet proved that the stem-and-loop serves a purpose independent of its role in termination—namely, to render the newly synthesized RNA resistant to degradation by RNase II, an intracellular ribonuclease that is inactive against double-stranded RNA.

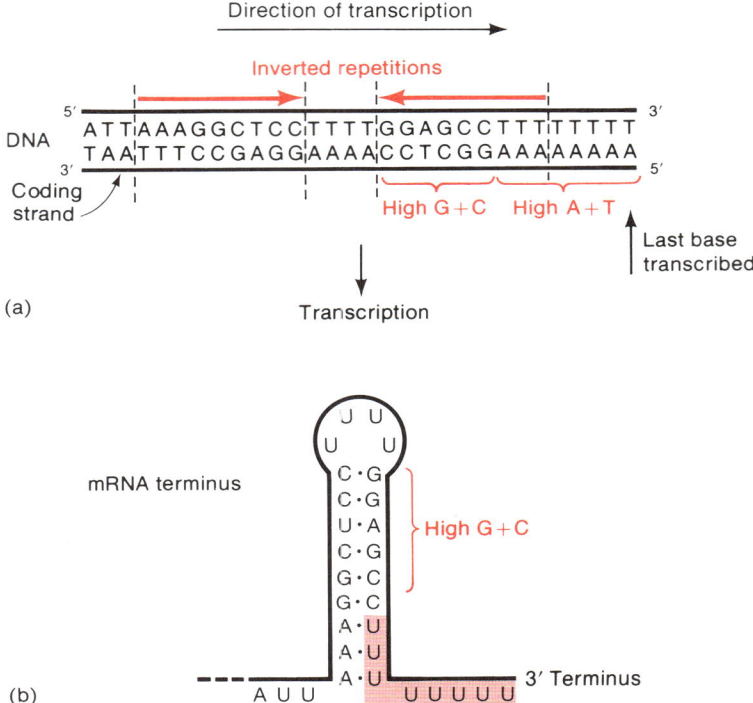

Figure 12-14 Base sequence of (a) the DNA of the *E. coli trp* operon at which transcription termination occurs and of (b) the 3′ terminus of the mRNA molecule. The inverted-repeat sequence is indicated by reversed red arrows. The mRNA molecule is folded to form a stem-and-loop structure. The relevant regions are labeled in red; the terminal sequence of U's in the mRNA is shaded in red.

2. Near the loop end of the putative stem (sometimes totally within the stem) there is a high-G+C sequence. RNA polymerase usually slows down when synthesizing the corresponding RNA segment.

3. This is followed by a sequence of A·T pairs (which may begin in the putative stem) with A in the template strand, yielding, in the RNA, a sequence of 6–8 U's. The importance of this sequence is made clear by introducing deletions in the DNA that remove some of the A·T pairs in this segment. RNA polymerase still pauses in the high G+C region, but termination does not occur.

Transcription does not terminate at a unique site: in a collection of RNA molecules terminating at one site, some end with five U's and some with six U's.

How these features lead to termination is not known. It is currently believed that termination occurs when an RNA hairpin that causes a pause

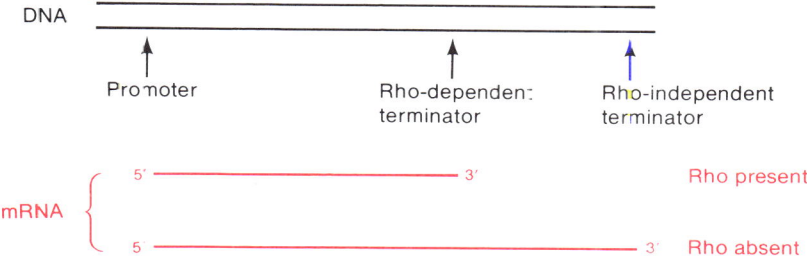

Figure 12-15 The effect of Rho on termination of RNA synthesis. This effect is easily demonstrated *in vitro*. If Rho is added to a reaction mixture *after* RNA polymerase has passed the Rho-dependent terminator, RNA synthesis continues until the Rho-independent terminator is reached. The red lines represent the completed mRNA molecules.

is formed. Evidence supporting this hypothesis is that mutations in termination sequences that remove a pause also eliminate termination.

We mentioned above that there are two types of termination sequences. The type just described is self-terminating in that it depends only on the DNA base sequence. The second class requires a termination protein called Rho (Figure 12-15). Rho-dependent termination sequences have a different form from the Rho-independent sequences: the region in which termination occurs is preceded by a large stem-and-loop that is followed by a long tract (70−80 nucleotides) free of double-stranded segments. It is believed that the large stem-and-loop may serve to ensure that the region that follows is free of any secondary structure. The unpaired segment seems to require one or more short G + C-rich regions upstream from the termination site. A poly(U) segment is not present. Rho is a protein that binds tightly to RNA. When bound to segments rich in C (especially repeating C's), it acquires a powerful ATP-cleaving activity that is essential to its action in termination. It does not bind to either DNA or to RNA polymerase *in vitro;* however, since certain mutations in the RNA polymerase β subunit eliminate Rho-dependent termination and other mutations in the β subunit compensate for mutations in the *rho* gene that eliminate Rho activity, it seems likely that Rho and RNA polymerase interact *in vivo* in some way. The mechanism of Rho-mediated termination is not known with any certainty. The current hypothesis is that as polymerization proceeds and C-rich segments of RNA are made, the ATPase activity of Rho increases until no further polymerization is possible, because nucleoside triphosphates cannot reach RNA polymerase without being degraded. However, this explanation must be incomplete, because certain features of Rho-dependent termination vary with the particular terminator. For example, with some sequences synthesis stops without Rho but Rho is required to prevent chain growth from continuing after a delay. At one terminator, synthesis stops completely but Rho is needed to release the RNA. Clearly, Rho has a multifaceted activity.

The final step in termination is dissociation of both the core enzyme and the RNA from the DNA, a poorly understood step. Possibly, and this is strictly speculative, when RNA polymerase core enzyme no longer advances, the reverse reaction (degradation of phosphoester bonds) predominates, degrading the RNA right through the DNA-RNA hybrid region. Since without the σ subunit, the core enzyme cannot re-start transcription, the core enzyme also leaves the DNA.

Following dissociation of the core enzyme and the DNA, the core enzyme interacts with a free σ subunit to re-form the holoenzyme. Thus, transcription includes a σ cycle—σ falls off the holoenzyme when the elongation phase begins and rejoins the core enzyme after the core enzyme falls off the DNA (Figure 12-16).

Mutations sometimes occur that generate a new termination site well before the natural termination site; a shorter mRNA is then produced. Genetic mapping techniques indicate that these mutations have the following properties. They are localized at a single site, but if the site is in one gene of a polygenic mRNA molecule, they reduce the activity of all genes downstream from the mutation. Such mutations are called **polar mutations.** (Other types of polar mutations are described in Chapters 13 and 15.) The expla-

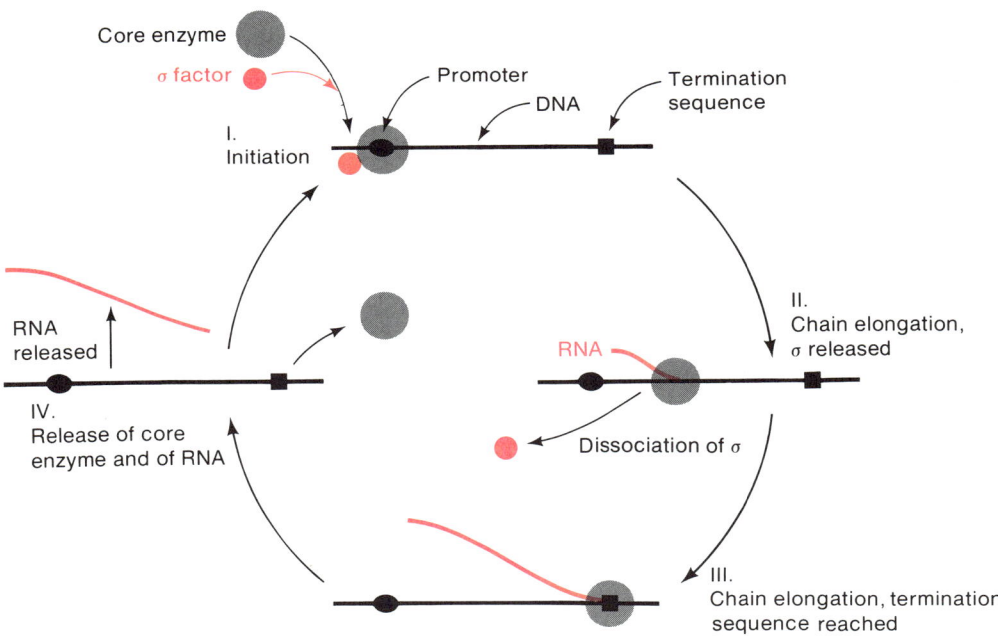

Figure 12-16 The transcription cycle of *E. coli* RNA polymerase showing dissociation of the σ subunit shortly after chain elongation begins, dissociation of the core enzyme during termination, and re-formation of the holoenzyme from the core enzyme and the σ subunit. A previously joined core enzyme and σ subunit will rarely become rejoined; instead, reassociation occurs at random.

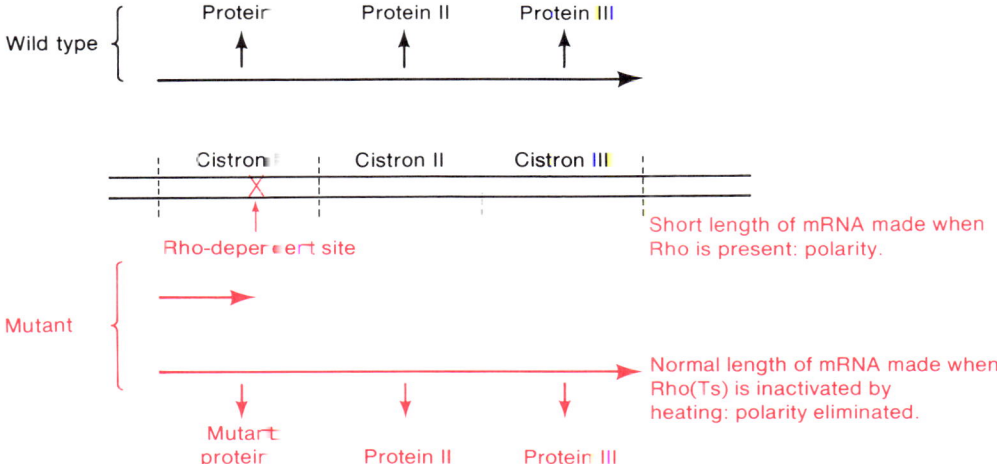

Wild type — Protein I Protein II Protein III

Cistron I Cistron II Cistron III

Rho-dependent site

Short length of mRNA made when Rho is present: polarity.

Mutant

Normal length of mRNA made when Rho(Ts) is inactivated by heating: polarity eliminated.

Mutant protein Protein II Protein III

Figure 12-17 The effect of a polar mutation on mRNA synthesis in a polycistronic system. When premature termination is prevented by eliminating Rho activity, the polar effect is gone, though a mutant protein is still made. Usually, the mutation is slightly upstream from the Rho-dependent site.

nation for this type of polarity is that the downstream genes are not transcribable because the mutation has created or enhanced a termination site upstream from the normal one. An example of a polar mutation is one that generated a Rho-dependent termination site in the first gene of a cluster of three genes responsible for metabolizing galactose (Figure 12-17). The mutant was Gal$^-$ and the products of genes II and III were not made. The connection between this mutation and Rho was made evident when a temperature-sensitive mutation in the *rho* gene was introduced into the bacterial strain carrying the *gal*$^-$ polar mutation. When the temperature was raised from 34°C to 42°C, Rho was inactivated and the activities of genes II and III were restored.

CLASSES OF RNA MOLECULES

There are three major classes of RNA molecules—messenger RNA (an informational molecule), ribosomal RNA (a structural molecule), and transfer RNA (both informational and structural). In eukaryotes a fourth class of RNA molecules located in small ribonucleoprotein particles also exists; these are discussed later in the chapter. All are transcribed from DNA base sequences. Significant differences exist between the structures and modes of synthesis of the RNA molecules of prokaryotes and eukaryotes, though the basic mechanisms of their functions are nearly the same. The greatest amount of information about these molecules has been obtained from studies with bacteria and bacterial cell extracts, so it is here that we begin. Transcription in eukaryotes and the structure and synthesis of eukaryotic RNA molecules are discussed in a later section.

Structure of Messenger RNA

The base sequence of a DNA molecule determines the amino acid sequence of every polypeptide chain in a cell, though amino acids have no affinity for DNA. Thus, instead of a direct pairing between amino acids and DNA, a multistep process is used in which the information contained in the DNA is converted to a form in which amino acids can be arranged in an order determined by the DNA base sequence. This process begins with the transcription of the base sequence of one of the DNA strands (the **coding strand**) into the base sequence of an RNA molecule and it is from this molecule—messenger RNA (mRNA)—that the amino acid sequence is obtained by the protein-synthesizing machinery of the cell. As we will see in Chapter 13, the nucleotide sequence of the mRNA is then read in groups of three bases (a group of three is called a **codon**) from a start codon to a stop codon, with each codon corresponding either to one amino acid or a stop signal.

A DNA segment corresponding to one polypeptide chain plus the translational start and stop signals for protein synthesis is called a **cistron** and an mRNA encoding a single polypeptide is called monocistronic mRNA. In prokaryotes it is very common for an mRNA molecule to encode several different polypeptide chains; in this case it is called a **polycistronic mRNA** molecule. The segment of RNA corresponding to a DNA cistron is often called a **reading frame,** since it is "read" by the protein-synthesizing system, as will be seen in Chapters 13 and 14. The cistrons contained in polycistronic mRNA often correspond to the proteins of a single metabolic pathway. For example, in *E. coli* the three proteins required to metabolize galactose are synthesized from a single mRNA molecule and the ten enzymes needed to synthesize histidine are encoded in another mRNA molecule. The use of polycistronic mRNA is a way for a cell to regulate synthesis of related proteins coordinately. For example, the usual way to regulate synthesis of a particular protein is to control the synthesis of the mRNA molecule that encodes it. With a polycistronic mRNA molecule the synthesis of several related proteins—in similar quantities and at the same time—can be regulated by a single signal.

The sizes of mRNA molecules vary within a broad range. The smallest proteins contain about 50 amino acids; correspondingly, 150 nucleotides (3 per amino acid) would be needed in a monocistronic mRNA molecule. A protein of more typical size would have 300–600 amino acids, corresponding to 900–1800 mRNA nucleotides. In prokaryotes polycistronic mRNA is actually more common than monocistronic mRNA, and a typical mRNA size contains 3000–8000 nucleotides.

In addition to reading frames and start and stop sequences for transla-

DNA

Figure 12-18 Schematic drawing showing that complementary DNA strands can be transcribed, but not from the same region of DNA. Promoters are indicated by black arrowheads, and termination sites by black bars. Promoters are present in both strands. Termination sites are usually located such that transcribed regions do not overlap, but this is not always the case.

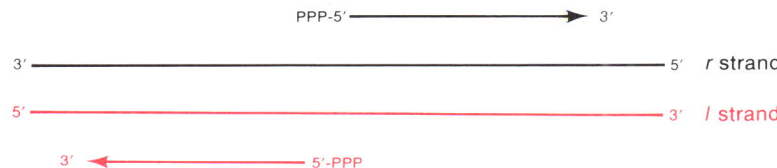

Figure 12-19 A drawing showing the definitions of the *l* strand and the *r* strand of a DNA molecule. The arrows represent transcripts.

tion, other regions in mRNA are significant. For example, translation of an mRNA molecule (that is, protein synthesis) seldom starts exactly at one end of the RNA and proceeds to the other end; instead, initiation of synthesis of the first polypeptide chain of a polycistronic mRNA may begin hundreds of nucleotides from the 5′-P terminus of the RNA. The section of nontranslated RNA before the coding regions is called a **leader;** in some cases, the leader contains a regulatory region. Untranslated sequences are usually found at both the 5′-P and the 5′-OH termini and a polycistronic mRNA molecule typically contain intercistronic sequences (**spacers**) usually tens of bases long.

We have previously stated that mRNA is transcribed from only one DNA strand at a given place along a gene (we will see later that exceptions to this more-or-less general rule occur). However, not all mRNA molecules are synthesized from the same DNA strand. That is, in an extended segment of a DNA molecule, mRNA would be seen to be growing in both directions (Figure 12-18).

When discussing RNA synthesis, the following terminology is often found in the older literature and remains common for certain phages that have been studied for a long time. The two strands of DNA are called *l* and *r* strands; *l* and *r*, which stand for left and right, refer to directions of synthesis of the mRNA when the DNA strands are depicted in a standard but arbitrary way (horizontally). That is, rightward-moving mRNA is transcribed from the *r* strand. According to this convention, since the growth of mRNA proceeds from the 5′ end to the 3′ end, and since a DNA-RNA hybrid is an antiparallel structure, the *r* strand is drawn with the 3′-OH terminus at the left, as shown in Figure 12-19.

The Lifetime of Prokaryotic mRNA

An important characteristic of prokaryotic mRNA is that its lifetime is short compared to other types of prokaryotic RNA molecules. For bacteria the half-life of a typical mRNA molecule is a few minutes. This feature, which may seem terribly wasteful, has an important regulatory function. If a protein is no longer required, a cell need only turn off synthesis of the mRNA that encodes the protein; soon afterwards, none of that particular mRNA will remain, and synthesis of the protein will no longer occur. Of course, this also means that in order to maintain synthesis of a particular protein, the mRNA molecules encoding these proteins must be synthesized continuously. Continuous synthesis is small payment by the cell for the ability to regulate the synthesis of specific proteins; that is, in the overall metabolism of a cell much less ATP is consumed than would be used if synthesis of proteins encoded in the mRNA continued long after the proteins were no longer needed.

The short lifetime of bacterial mRNA is one criterion used to identify mRNA in prokaryotes. A common experimental technique to determine whether a particular molecule or class of molecules is mRNA is the pulse-chase experiment (Chapter 9). RNA is labeled briefly by growing bacteria in the presence of [^{3}H]uridine; bacteria are then grown in a medium containing no [^{3}H]uridine and a high concentration of nonradioactive uridine. Different species of RNA are then separated by gel electrophoresis or centrifugation and detected by their radioactivity. A radioactive stable RNA molecule will be present through many generations, whereas radioactive mRNA molecules will decrease with a half-life of 2−3 minutes. A difficulty with this technique is that bacteria contain some long-lived mRNA molecules and these would be misclassified. The best criterion for identifying a molecule as mRNA is to isolate it and determine in an *in vitro* protein-synthesizing system whether the particular RNA can cause synthesis of protein.

Degradation of mRNA proceeds enzymatically primarily from the 5′-P terminus. The enzymology is not completely understood.

Ribosomal RNA and Transfer RNA

During the synthesis of proteins, genetic information is supplied by mRNA. RNA also plays other roles in protein synthesis. For example, proteins are synthesized on the surface of an RNA-containing particle called a **ribosome;** these particles consist of several classes of ribosomal RNA (**rRNA**), which are stable molecules, and a large number of proteins that have various functions. Amino acids do not line up against the mRNA template independently during protein synthesis but are aligned by means of a set of about 50 adaptor RNA molecules called transfer RNA (**tRNA**), also a stable species. Each tRNA molecule is capable of "reading" three adjacent mRNA bases and placing the corresponding amino acid at a site on the ribosome at which a peptide bond is formed with an adjacent amino acid (Chapter 8, Figure 8-3). Neither rRNA nor tRNA is translated into polypeptide chains. The roles of ribosomes and of tRNA molecules will be explained in detail in the following two chapters. In this chapter we are concerned only with their synthesis, which, as we will see, involves posttranscriptional modification of a remarkable sort.

The synthesis of both rRNA and tRNA molecules is initiated at a promoter and completed at terminators; in this respect, their synthesis is no different from that of mRNA. However, the following three properties of these molecules indicate that neither rRNA nor tRNA molecules are the immediate products of transcription (**primary transcripts**):

1. The molecules are terminated by a 5'-monophosphate rather than the expected triphosphate found at the ends of all primary transcripts.
2. Both rRNA and tRNA molecules are much smaller than the primary transcripts.
3. All tRNA molecules contain bases other than A, G, C, and U and these "unusual" bases (as they are called) are not present in the original transcript.

All of these molecular changes are made after transcription by processes collectively called **posttranscriptional modification** or, more commonly, **processing.** In the following sections we will consider processing of prokaryotic rRNA and tRNA. Eukaryotic processing is discussed later.

Processing of tRNA Molecules

A well-understood tRNA molecule from the point of view of its synthesis is the *E. coli* $tRNA_I^{Tyr}$ molecule, a molecule containing 85 nucleotides of known sequence (Chapter 13). In *E. coli* there are two copies of the $tRNA_I^{Tyr}$ gene, that is, two identical adjacent copies of the DNA from which this tRNA is transcribed. Each gene consists of about 350 (not 85) nucleotide pairs separated by a "spacer" of 200 nucleotide pairs (Figure 12-20). The two genes are transcribed as a single RNA molecule that is cut up after transcription is complete. In order to simplify the study of the synthesis of $tRNA_I^{Tyr}$, genetic techniques have been used to create a transcription unit containing only a single 350-bp gene. The transcription start site is 41 bp before (upstream from) the 5' end of the tRNA base sequence and a stop site is 224 bp downstream from the 3' terminus of the tRNA. This transcript is processed by a series of steps shown in Figure 12-21, which may be grouped into the following three stages:

1. *Formation of the 3'-OH terminus*. This process involves the action of an endonuclease that recognizes a hairpin loop (I in the figure) and an exo-

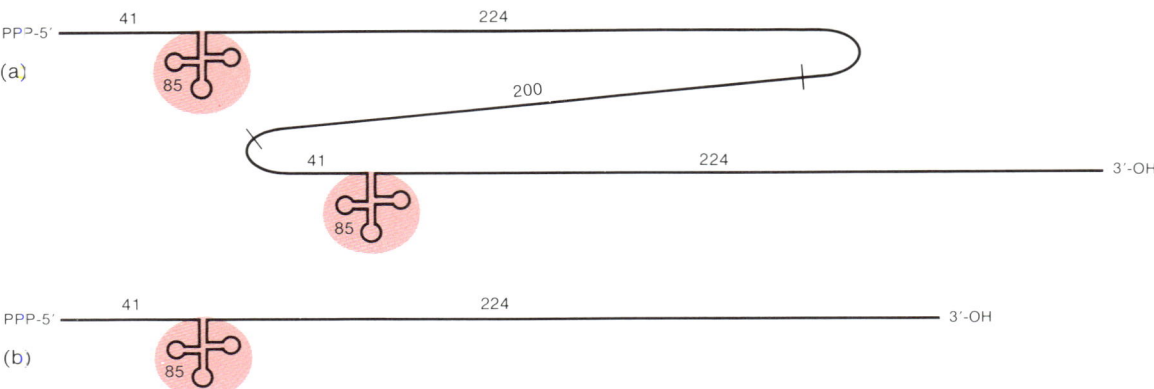

Figure 12-20 The *E. coli* tRNA$_I^{Tyr}$ gene showing (a) the complete transcript with two adjacent identical tRNA segments and the spacer region and (b) the single genetic unit used to study processing. The numbers indicate the number of bases in each segment of the transcript. The tRNA sequences are shaded in red.

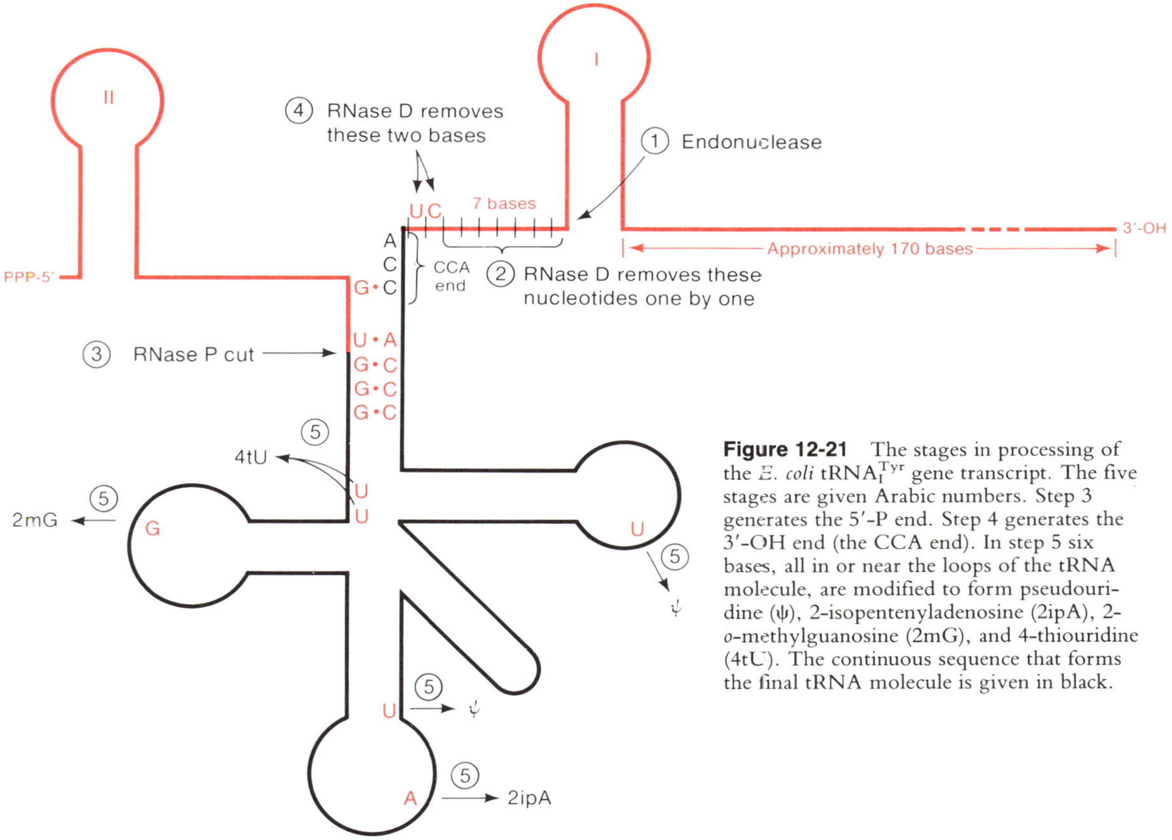

Figure 12-21 The stages in processing of the *E. coli* tRNA$_I^{Tyr}$ gene transcript. The five stages are given Arabic numbers. Step 3 generates the 5′-P end. Step 4 generates the 3′-OH end (the CCA end). In step 5 six bases, all in or near the loops of the tRNA molecule, are modified to form pseudouridine (ψ), 2-isopentenyladenosine (2ipA), 2-*o*-methylguanosine (2mG), and 4-thiouridine (4tU). The continuous sequence that forms the final tRNA molecule is given in black.

nuclease that recognizes the three-base sequence CCA. After endonuclease digestion at site 1, the seven bases upstream are removed by an exonuclease called RNase D (step 2). This enzyme initially stops two bases short of the CCA terminus, though it later removes these two bases after the 5′ end is processed. This leaves a molecule called **pre-tRNA** that is easily isolated from E. coli and that has the structure

$$\text{P-5}'\text{—(41 bases)—tRNA—(2 bases)—3}'\text{-OH}$$

2. *Formation of the 5′-P terminus.* The 5′-P terminus is formed by an enzyme called RNase P, which is probably responsible for generation of the 5′-P terminus of all E. coli tRNA molecules. Evidence for this comes from studies with an E. coli mutant in which RNase P is inactive at 42°C. When this mutant is grown at 42°C, large RNA molecules accumulate that contain tRNA sequences and the hairpin loop II (Figure 12-21) at the 5′ terminus. RNase P removes the excess RNA from the 5′ end of a precursor molecule by an endonucleolytic cleavage (step 3) that generates the correct 5′ end and a single fragment. RNase P does not recognize a specific base sequence at the cleavage site or anywhere else, but instead responds to the overall three-dimensional conformation of the tRNA molecule with its several hairpin loops and then makes a cut at just the right place. (The evidence for this statement is that base changes in the segment to be removed do not affect the cleavage unless the alteration causes extensive disruption of the stem-and-loop arrangement.) Once the 5′-P terminus has been formed, RNase D removes the two 3′-terminal nucleotides (step 4), leaving a tRNA molecule having the correct length.

RNase P is an unusual enzyme in that it contains 86 percent RNA and 14 percent protein by weight. Furthermore, *the RNA possesses the catalytic activity* and the protein serves instead to ensure the correct folding of the RNA, in order to maximize the catalytic activity. Other RNA molecules with enzymatic activity will be encountered when processing of eukaryotic rRNA is discussed. The recent discovery of these catalytic RNAs (termed **ribozymes**) indicate that the long-standing dogma that all enzymes must be proteins is invalid.

3. *Production of the modified bases.* The final modification is to produce the altered bases in the tRNA (step 5). Enzymes that act only on bases at specific sites in tRNA produce the necessary changes. In the case of tRNA_I^{Tyr}, two uridines are converted to pseudouridine (ψ), two uridines to two 4-thiouridines (4tU), one guanosine to 2′-o-methylguanosine (2mG), and one adenosine to isopentenyladenosine (2ipA), as shown in Figure 12-21. These and other modified bases are formed in most tRNA molecules.

All tRNA molecules are terminated by CCA-3′-OH. The precursor shown in Figure 12-21 contains this sequence, so the terminus is generated by the appropriate cut. However, the precursors of some tRNA molecules lack a terminal CCA. With these molecules the CCA is added by the enzyme **tRNA nucleotidyl transferase.**

Multiple copies of a particular tRNA molecule are commonly found in a single transcription unit.* For example, there are four copies of one of the tRNA^{Leu} molecules in its precursor molecule. The occurrence of different tRNA molecules in a single transcript is also frequent. For example, one tRNA^{Ser} and one tRNA^{Thr} are present in a single unit in *E. coli*. In the following section in which processing of ribosomal RNA is described, we will see that some tRNA molecules are even obtained from the transcript that contains rRNA.

Processing of Ribosomal RNA in *E. coli*

Bacterial ribosomes contain three kinds of rRNA (Chapter 13). They are called 5S rRNA, 16S rRNA, and 23S rRNA and contain 120, 1541, and 2904 nucleotides, respectively. These molecules, plus several tRNA molecules, are cleaved from a continuous transcript having more than 5000 nucleotides of known sequence. At present seven distinct primary transcripts containing rRNA have been isolated from *E. coli;* they differ by the identity of the tRNA molecules and the location of the tRNA sequences with respect to the rRNA sequences. A diagram of one of these transcripts is shown in Figure 12-22. This transcript contains four different tRNA molecules and the segments are, from the 5′ end to the 3′ end,

16S rRNA–tRNA^{Ile}–tRNA^{Ala}–23S rRNA–5S RNA–tRNA^{Asp}–tRNA^{Trp}

The general pattern, 16S–spacer–23S–5S–spacer, in which tRNA is in the spacer regions, is retained in numerous primary transcripts, but the number of tRNAs is variable.

A rRNA transcript is usually cut while it is being synthesized, by several enzymes acting in sequence. The first cuts are usually made by RNase III, which cleaves double-stranded RNA in the double-stranded stem regions by making two single-strand breaks in complementary sequences but not opposite one another. A 5′-P and a 3′-OH group are generated but these are not the termini of the rRNA. Several enzymes are required to complete the processing.

The sequence of processing events is not the same in all rRNA transcripts or in all organisms, but the basic pattern of excision of all rRNA components from a single precursor seems to be a general phenomenon.

---

*This phenomenon occurs for several reasons. For example, multiple copies of a gene allow a greater rate of synthesis of a particular tRNA molecule than if only a single copy is made of a gene; they also permit cells to generate suppressors (Chapter 13).

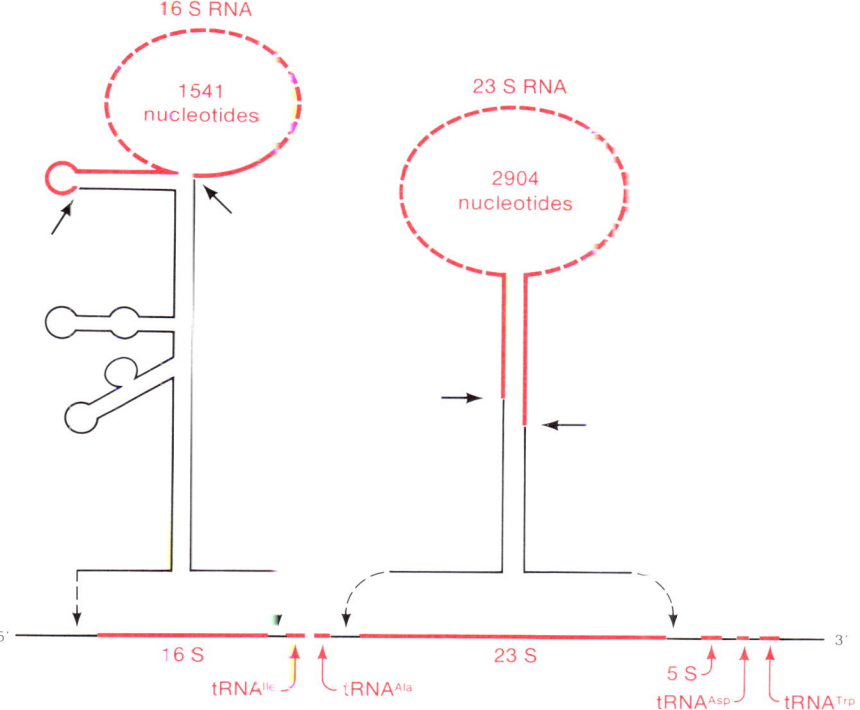

Figure 12-22 A schematic diagram of one of the *E. coli* rRNA transcripts from which 5S, 16S, and 23S rRNA molecules are excised. The regions containing the 16S rRNA and 23S rRNA molecules are shown in expanded form above the line designating the transcript. The arrows indicate the termini of the 16S rRNA and 23S rRNA molecules.

Although RNA sequences are discarded, which appears somewhat wasteful, this mechanism provides a constant ratio of the 16S, 23S, and 5S RNA molecules. Since one molecule of each of these three is present in a ribosome and since the molecules are used nowhere else in the cell, efficiency would demand that if each were transcribed separately, some means would be needed to maintain a 1:1:1 ratio. We will see in Chapters 15 and 16 that such efficiency is common in prokaryotes, but less often in eukaryotes.

Processing of bacterial mRNA has not been observed but a few cases are known in which phage mRNA is processed: *E. coli* phages T4, T7, and φ80. However, this processing is not essential for phage multiplication, since when mutant bacteria unable to carry out RNA processing (e.g., bacteria lacking RNase III) are infected, a normal number of phage progeny is produced. Eukaryotic mRNA is extensively processed; this is described later in this chapter.

Summary of Processing in Prokaryotes

Although a great deal remains to be learned about the processing of pro-karyotic RNA, the following facts seem to be well established.

1. All stable RNA molecules are processed; mRNA molecules are not processed, except for a few phage RNAs.
2. About ten nucleases account for all of the cuts. The endonucleases always generate a 5′-P and a 3′-OH group.
3. The 3′-OH ends are generally formed by exonucleases; 5S rRNA may be an exception.
4. Many processing enzymes are not sequence-specific but recognize large structural features.
5. Bases other than A, U, G, and C are formed by enzymatic modi-fication of bases already present in otherwise completed molecules.

TRANSCRIPTION IN EUKARYOTES

The basic features of the transcription and the structure of mRNA in eukary-otes are similar to those in bacteria. However, there are five notable differences:

1. Eukaryotic cells contain three classes of nuclear RNA polymerase and these are responsible for the synthesis of all RNA.
2. Both the 5′ and 3′ termini are modified; a complex structure called the **cap** is found at the 5′ end of all mRNAs, and a long (up to 200 nucleotides) sequence of polyadenylic acid—poly(A)—is found at the 3′ end of most (but not all) mRNA molecules.
3. The mRNA molecule that is used as a template for protein synthesis is usually a small fraction of the size of the primary transcript. During mRNA processing **intervening sequences,** or **introns,** are excised and the surrounding fragments are rejoined.
4. With almost no exceptions, eukaryotic mRNA molecules are monocistronic.
5. Many mRNA molecules are very long-lived.

These points are illustrated in Figure 12-23, which shows a schematic dia-gram of a typical eukaryotic mRNA molecule and how it is produced. Clearly the production of mRNA in eukaryotes is not a simple matter of copying the DNA.

The Transcription Unit Concept

In prokaryotes a mRNA molecule can encode multiple proteins and all encoded proteins can be translated. This is not the case for eukaryotes for, as will be discussed in detail in Chapter 14, only one kind of protein molecule can be

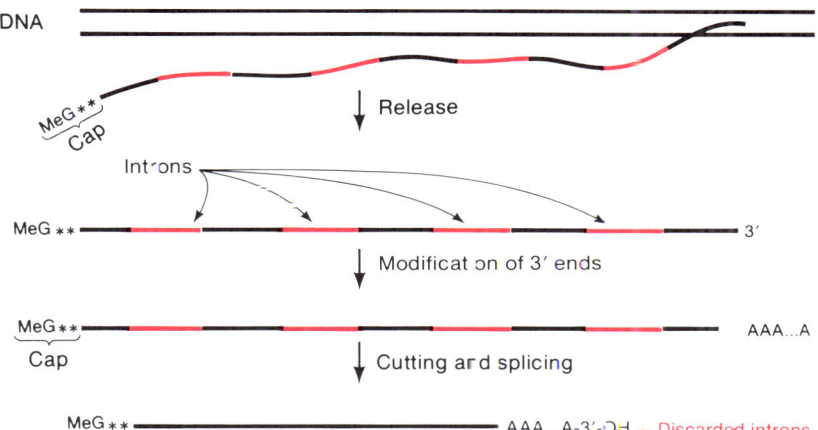

Figure 12-23 Schematic drawing showing production of eukaryotic mRNA. The primary transcript is capped before it is released. Then, its 3'-OH end is modified, and finally the intervening regions are excised. MeG denotes 7-methylguanosine, and the two asterisks indicate the two nucleotides whose riboses are methylated.

translated from a particular mRNA molecule (in other words, eukaryotic translation is essentially monocistronic). Nonetheless, in eukaryotes many of the primary transcripts contain information for synthesizing more than one protein. It has been useful in eukaryotic molecular biology to define a segment of DNA that can be transcribed as a **transcription unit** and to distinguish a simple transcription unit—one that carries information for only one protein—from a complex transcription unit—one that carries information for more than one protein molecule. If only one protein molecule can be translated from a single mRNA molecule, how can one say that a complex transcription unit encodes two or more proteins? As we will see shortly, a primary transcript can be altered such that it terminates at one of several poly(A)-addition sites, generating distinct mRNA molecules; also, a particular transcript with a poly(A) tail can be processed in different ways (that is, different introns can be removed), producing mRNA molecules having totally different coding sequences. This variability has important consequences for regulation in eukaryotes.

Eukaryotic rRNA Genes

Eukaryotic ribosomes contain four RNA molecules. They are referred to by their sedimentation coefficients as 5S, 5.8S, 18S, and 28S rRNA molecules (see Chapter 14 for details). The organization of the genes for these RNA species in higher eukaryotes differs from that observed in prokaryotes. The main differences are (1) the 5S species resides in a single RNA molecule and (2) the gene order in the other RNA, which contains the sequences for the other three rRNAs, is not the same as that for prokaryotes, that is, the smallest rRNA is between the larger two rather than at one end (Figure 12-24).

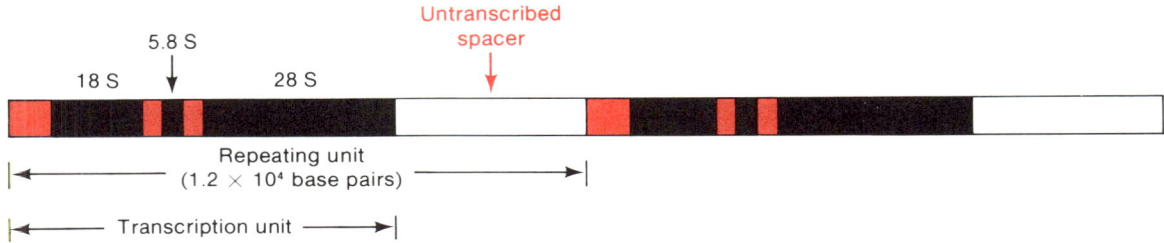

Figure 12-24 Organization of the genes in the 40S precursor of *Xenopus* rRNA. The solid colors represent the transcribed region. The primary transcript encompasses the red and black segments. The red regions are removed from the primary transcript by the processing enzymes. The black regions represent the three desired rRNA molecules.

In the nucleolus of a typical animal cell several hundred copies of a DNA sequence (ribosomal DNA or rDNA) encode the 18S, 5.8S, and 28S rRNA molecules (in that order). The 5S rRNA genes are located outside the nucleolus. In several species of *Xenopus* (toads), which are the best-understood systems, the three ribosomal genes comprise a 40S transcription unit (Figure 12-24) from which three RNA segments are excised by processing enzymes giving rise to 18S, 28S, and 5.8S rRNA molecules. Each transcription unit is separated from its neighbors by a large nontranscribed spacer segment, as shown in the figure. The repeating unit contains about 12,000 nucleotide pairs.

The gene for the 5S rRNA, which contains 120 bp, is contained in a separate transcription unit. Neighbor transcription units are separated by an untranscribed segment of about 600 bp (Figure 12-25). The 5S rRNA genes are also arranged in tandem—in *Xenopus* there are several enormous units containing several thousand copies of the gene. Altogether, there are 24,000 copies of the 5S RNA gene!

All eukaryotic rRNA transcripts require processing, which occurs by events similar to those observed in prokaryotes—namely, trimming of the 5′ and 3′ termini and excision of the unwanted regions of the transcript. Processing of eukaryotic rRNA has been most widely studied in cultured mammalian cells. The sequence of events is shown in Figure 12-26. Several endonucleases are involved, one of which closely resembles *E. coli* RNase III. Note that the rRNA sequences do not contain introns. This is a constant

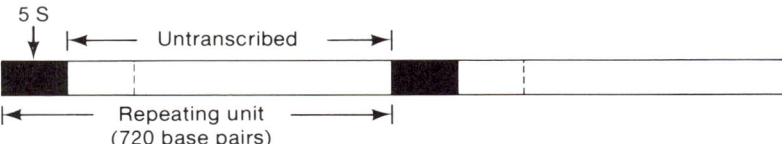

Figure 12-25 Organization of the genes for the 5S rRNA of *Xenopus*. Only the black regions, which are 120 base pairs long, are transcribed. Between each black region and the dashed line is a segment whose base sequence is very near that of the 5S rRNA; however, oddly, it is not transcribed.

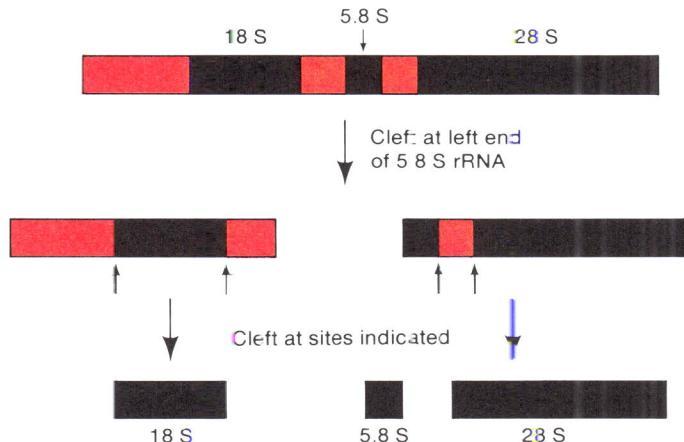

18 S   5.8 S   28 S

Cleft at left end
of 5.8 S rRNA

Cleft at sites indicated

18 S   5.8 S   28 S

Figure 12-26 Representative stages in the formation of mammalian rRNA from a primary transcript.

feature of the rRNAs of the higher eukaryotes. Some of the rRNA of the lower eukaryotes, in particular, the protozoa, contain introns; the rRNA of one of these, *Tetrahymena*, will be examined when splicing mechanisms are described.

Formation of Eukaryotic tRNA Molecules

Eukaryotic tRNA molecules are also excised from large transcripts (called **pre-tRNA**), which may contain one or more tRNA sequences. The enzymology of processing is not yet fully understood, though an enzyme similar to *E. coli* RNase P (which is utilized for processing of tRNA) has been purified from mouse cells. The best information at present comes from studies with yeast.

Many, but not all, yeast tRNA genes contain introns of 10–30 bases; thus, not only external processing (as in prokaryotes) but also splicing is necessary to produce mature yeast tRNA (Figure 12-27). The processing mechanism in yeast also differs from that in *E. coli* in that the 3′-terminal CCA sequence is not present in the primary transcript but is added as part of the processing operation.

Small RNA Molecules

Eukaryotic cells contain a class of small RNA molecules (100–300 nucleotides). Those localized in the nucleus are called **small nuclear RNA (snRNA),** and the cytoplasmic species are **small cytoplasmic RNA (scRNA).** Their distinguishing feature is their association with specific protein molecules,

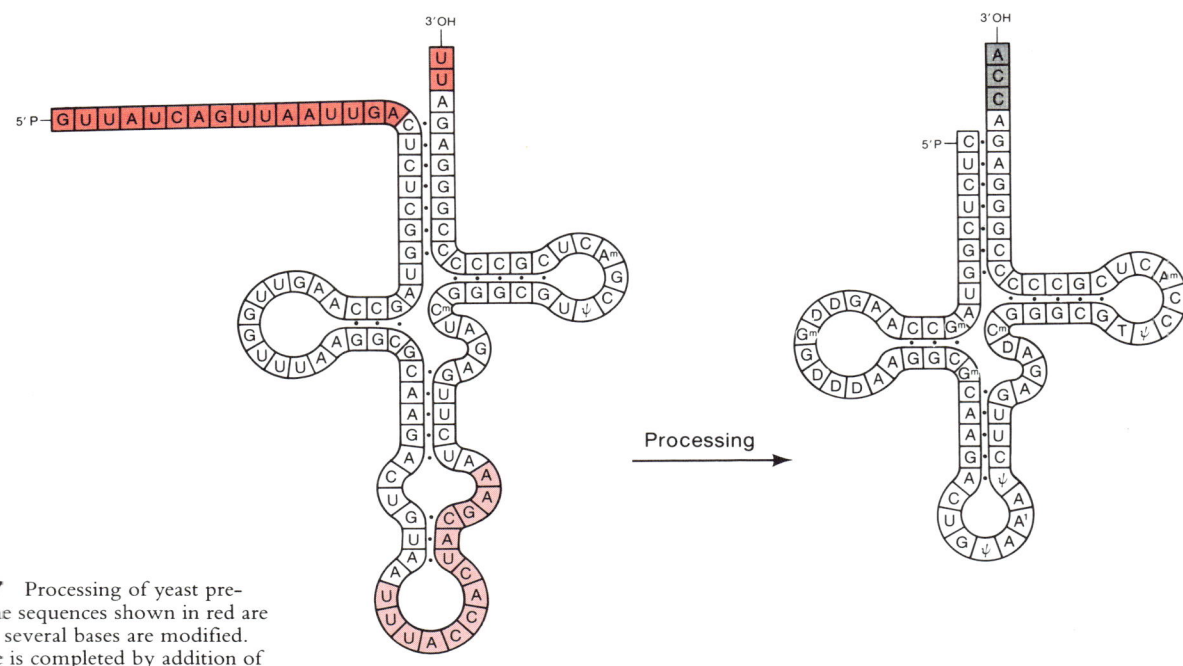

Figure 12-27 Processing of yeast pre-tRNA^{Tyr}. The sequences shown in red are removed and several bases are modified. The molecule is completed by addition of CCA to the 3′ end.

Early transcript

Processing

Mature tRNA

forming small *ribo*nucleoproteins—the *small nuclear RNP (**snRNP**) and the small cytoplasmic RNP (**scRNP**).* The different snRNPs are designated U1, U2, . . . , and each contains a unique RNA molecule (U1 RNA, U2 RNA, etc). Some of the proteins are common to different snRNPs. We will see that some snRNPs participate in RNA processing. Little is known about the function of the scRNPs at this time.

RNA Polymerases of Eukaryotes

The first RNA polymerase that was isolated was obtained from mammalian cells. However, difficulties in purifying what proved years later to be several enzymes, and the successes in the molecular biology of *E. coli,* hampered further study of eukaryotic polymerases. Even at this time, knowledge of the eukaryotic polymerases is scant compared to that of *E. coli* RNA polymerase.

Three different eukaryotic RNA polymerases, denoted I, II, and III, are known. Each is responsible for the synthesis of particular classes of RNA:

*In laboratory speech and in some recent texts, they are called **snurps** and **scyrps,** respectively. However, we shall only use the correct terms given in the text.

RNA polymerase I makes only rRNA, RNA polymerase II synthesizes all mRNA, and RNA polymerase III makes tRNA and the 5S RNA of ribosomes. All are found in the nucleus. Each is a high-molecular-weight (over 500,000) protein consisting of two large subunits and up to ten small subunits (the number depends on the organism). The functions of the individual subunits are not known. The class II enzymes are highly sensitive to inhibition by α-amanitin (the toxic product of toadstool mushrooms), the class III are somewhat sensitive, and the class I enzymes are resistant to this inhibitor. (In yeast, RNA polymerase III is sensitive and RNA polymerase I is resistant.) Each of the enzymes can be isolated in a variety of chromatographically and electrophoretically distinct forms, the significance of which is unknown. The biochemical reaction catalyzed by the eukaryotic RNA polymerases is the same as that catalyzed by *E. coli* RNA polymerase.

RNA Polymerase II Promoters

Identification and characterization of promoters in eukaryotes is considerably more difficult than with prokaryotes. The structure of eukaryotic promoters is generally more complex than the prokaryotic ones. The most noticeable difference is the existence of lengthy sequences, hundreds of base pairs upstream from the transcription start site, that control the rate of initiation; these are called **upstream sites.** Furthermore, initiation is a more complicated process and, for most genes, requires a diversity of specific proteins (**transcription factors**) that bind to particular DNA sequences. The complexity may derive in part from the fact that eukaryotic DNA is in the form of chromatin, which is inaccessible to RNA polymerases. Upstream sites and transcription factors will be described shortly.

Several approaches, which have been applied to about 100 genes from various organisms and different tissues, have been quite successful in dissecting promoter structure. Each is based on the ability to isolate a single gene in large quantities by cloning techniques. In one procedure, individual gene samples are treated with restriction enzymes that cleave the cloned gene at either one end of the gene or the other. Each time a particular segment is removed, transcription of the remaining DNA is attempted *in vitro,* using mixtures of partially purified enzymes, or by microinjection into the nucleus of an oocyte of *Xenopus laevis* (the African clawed toad), where transcription occurs *in vivo.* Once the promoter itself is either cut or removed, transcription will no longer occur. In this way, the promoter region can be identified, isolated, and sequenced. In another procedure individual bases are altered *in vitro* (thereby creating potential promoter mutations), and the DNA molecules are transcribed by one of the two procedures just mentioned.

An analysis of nucleotide sequences and studies of transcription with *in vitro* systems have led to the following picture of RNA polymerase II promoters.

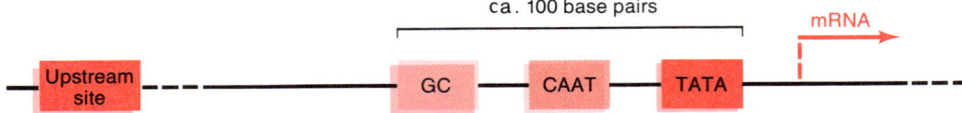

Figure 12-28   Sequences found in and near some RNA polymerase II promoters. Only the TATA box is represented in almost all promoters. The CAAT box occurs much less frequently, and the GC box has only occasionally been observed. An upstream site is very common but is not considered to be part of the promoter.

1. There is a nearly universal sequence, about 25 bp upstream from the transcription start site, whose consensus is

    TATAAAT

    It is known as the **TATA** or **Hogness box** and is similar in sequence to the Pribnow box. The TATA box is usually flanked by high-G + C sequences.

2. Many, but certainly not all, promoters have another common sequence in the −75 region with the consensus

    GG(T/C)CAATCT

    in which T and C are equally frequent at the third position from the left. This sequence is occasionally called the **CAAT box.** A third element recognized in a few promoters is the GC box, which has the consensus GGGCGG (Figure 12-28).

3. The TATA box, and probably the other sequences (when they are present), are sites of binding of transcription factors, but not of RNA polymerase.

4. RNA polymerase II probably does not interact directly with any part of the promoter by itself and is strictly dependent on the presence of the bound proteins mentioned in item 3. This is probably the major difference between prokaryotic and eukaryotic promoters.

5. The TATA box probably determines the base that is first transcribed. Evidence is that in some systems experimental deletion or alteration of the TATA box does not prevent initiation; instead, initiation occurs at a variety of nearby places.

Deletion studies have also shown that a few genes require a fourth sequence in the −85 to −110 region.

Eukaryotic Promoters for RNA Polymerase III

RNA polymerase III is responsible for synthesis of a 120-base rRNA molecule called 5S RNA and for tRNA molecules. Its promoters differ signifi-

Gene

tRNA $_{\text{I}}^{\text{Met}}$

5S RNA

VAI

Figure 12-29 Three genes transcribed by RNA polymerase III, showing the internal promoters (red) determined by deletion mapping. The first two genes are in *X. laevis;* the third is in adenovirus.

cantly from RNA polymerase II promoters, as they are *downstream* from the transcription start site and *within the transcribed DNA* (Figure 12-29). For example, in the 5S RNA gene of *X. laevis* the promoter is between 45 and 95 nucleotides downstream from the startpoint. The mechanism for this seemingly anomalous relative position of the promoter and the start point is not known but can be explained merely by assuming that the binding sites on RNA polymerase III are reversed with respect to the transcription direction, compared to RNA polymerase II. That is, simply speaking, whereas RNA polymerase II reaches forward to find the start point, RNA polymerase III reaches backward.

In *Xenopus* transcription of the 5S-RNA gene has been studied in some detail. Initiation of transcription requires three transcription factors. One of these, called **TFIIIA** or the **40-kd protein,** binds to the DNA, protecting bases 45–96. It has at least three binding sites: two for the internal promoter (also called the internal control region) and one for RNA polymerase III. The 40-kd factor is responsible for specific transcription of the 5S-rRNA genes by RNA polymerase III and is important in developmental regulation. There are so many contact points for this protein in the major groove of the DNA that it is not clear whether in a preinitiation state RNA polymerase III contacts the DNA or only the 40-kd protein. Ultimately, RNA polymerase must contact the DNA, at least sufficiently to read the base sequence. Once the 40-kd protein is bound to the DNA, many cycles of transcription occur without dissociation of the complex; in fact, the complex even survives cell division. One explanation given for this phenomenon is that the 40-kd protein binds tightly only to the noncoding DNA strand, so it can be passed in each transcription cycle by RNA polymerase III, which is presumably bound only to the coding strand during chain elongation.

Hypersensitive Sites, Upstream Activation Sites, and Enhancers

If free DNA molecules are exposed to most DNases, the DNA is rapidly degraded to oligo- and mononucleotides. However, with chromatin there is little breakage because the histones reduce access of the DNA to the enzymes.

When chromatin isolated from a single cell type is treated with certain DNases, specific regions of the DNA are cleaved at a higher rate than the bulk of the DNA; and the positions of these regions vary from one cell type to another. These regions are called **hypersensitive sites;** their importance is that their existence and positions often correlate with transcriptional activity of particular genes. For example, in chromatin isolated from chicken reticulocytes (which synthesize globin), sites in the globin cluster are preferentially broken, whereas in the oviduct (which synthesizes ovalbumin) the chromatin contains hypersensitive sites in the ovalbumin region. In contrast, there are no such sites in the ovalbumin region of reticulocyte chromatin nor in the globin region of oviduct chromatin. A more suggestive relation between hypersensitive sites and gene activity can also be seen in a study of the *Sgs4* gene of *Drosophila melanogaster*. A mutant has been found in which the hypersensitive region is deleted but in which the coding region of the Sgs4 protein is intact. This mutant fails to make Sgs4 protein.

The structural features of chromatin that make a region of DNA hypersensitive are not known. In two cases, namely, SV40 virus and the β-globin gene of young chickens, the DNA of the hypersensitive region lacks the typical nucleosome structure and may even be protein-free. The location of hypersensitive sites indicates that they have some connection with transcription. For example, they are not present within coding sequences or in introns but are invariably several hundred base pairs upstream from the start point for transcription, near or in regions that can bind transcription factors (Figure 12-30). Currently, the most important question about hypersensitive regions is the cause-effect relation between the sites and transcription initiation; that is, are genes active because the region is hypersensitive or does initiation cause hypersensitivity? At present, little data are available that have bearing on this question but, recently (1985), a system has been developed that should yield information. That is, a particular protein has been isolated from

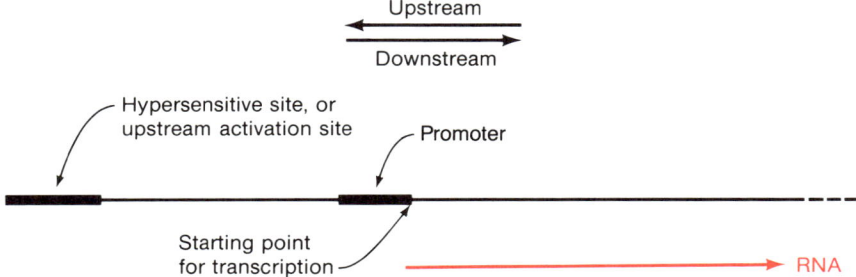

Figure 12-30   The relative sizes and locations (drawn approximately to scale) of a hypersensitive site or upstream activation site (at roughly the same position) and the promoter in a typical eukaryotic gene. The transcribed region would be about ten times longer than drawn. The meaning of the terms "upstream" and "downstream" is also shown.

the nuclei of chick erythrocytes in which the β–globin gene is expressed; this protein renders chromatin from other cells (cells that do not normally produce β-globin) sensitive to DNase-I cleavage adjacent to that gene. If the chromatin is fragmented into small pieces, the protein is found to be associated with a nucleotide sequence 100–200 bp long, extending from a region 5′ to the hypersensitive site, past the site and stopping short of the TATA box. This region is so large that more than one protein molecule must interact with the region. It is not yet known whether binding of these proteins must precede transcription or if transcription enables these proteins to be bound.

Upstream elements of importance have also been observed in other experiments. For example, in several yeast genes mutations have been isolated that eliminate or reduce transcription. These mutations define regions called **upstream activation sites,** whose properties are quite different from the transcription-regulating promoter-operator regions in prokaryotes. In a bacterial operon the operator and promoter are adjacent (within fewer than 10 bp), and binding of a regulatory protein to an operator either blocks (a repressor) or stimulates (a molecule of the cAMP-CRP type) the binding of RNA polymerase to the promoter. If a DNA fragment is inserted between a bacterial promoter and its operator, the ability to regulate transcription is lost; that is, a repressor cannot prevent initiation of transcription nor can a positive regulator stimulate transcription. In contrast, upstream activation sites are usually 50–300 nucleotide pairs from the promoter, and their locations can be experimentally changed by as many as a few hundred nucleotide pairs without substantially diminishing their effect on gene expression. Upstream sites are believed to be binding sites for transcription factors.

Another type of region whose location can also be changed without detracting from its effectiveness is the **enhancer.** The prototype enhancer was discovered in the animal virus SV40. It is a sequence of 72 nucleotide pairs located more than 100 nucleotide pairs upstream from the gene T and is necessary for optimal expression of that gene. Enhancer sequences of comparable size (though often 20–30 bp larger) are associated with cellular genes. Seven properties of these sequences are general features of enhancers.

1. They have no promoter activity themselves.
2. Experimental deletion of an enhancer (by restriction-enzyme cutting and splicing) reduces transcription of the associated gene roughly 100-fold.
3. By experimental manipulation an enhancer can be moved up to a thousand base pairs upstream, downstream, and even to the other side of the promoter without significant loss of activity. As the distance between an enhancer and a promoter increases, the enhancing effect is gradually reduced.
4. By genetic engineering an enhancer can be cut from its normal site and reinserted in reverse orientation with respect to the promoter

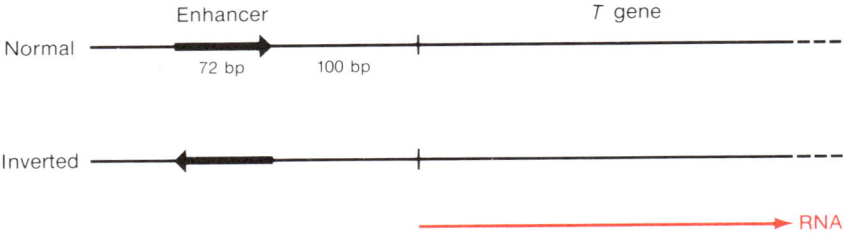

Figure 12-31   Position of the enhancer sequence near the virus SV40 *T* gene with respect to the transcription start site. If the normal sequence is experimentally inverted, RNA is still made at a high rate. Deletion of the enhancer reduces transcription more than a hundredfold.

without loss of its enhancing activity (Figure 12-31). This is the distinguishing feature of enhancers.

5. When two or more promoters are downstream from an enhancer, the magnitude of enhancement is greater for the nearest promoter.
6. The efficiency of enhancement varies with the promoter.
7. Enhancers can be isolated and spliced into regions in other cells. Study of a variety of enhancers and cells shows that different enhancers act preferentially or exclusively within cells from particular species, tissues, or cell types. This specificity is believed to derive from the presence of species- and cell-specific transcription factors that bind to the enhancer.

In a few cases enhancers are located in hypersensitive sites, but the generality of this observation is unknown.

Little is known with certainty about the molecular basis of enhancement, and theories abound. The two currently (1985) most popular hypotheses are that (1) enhancers either direct the formation of a particular DNA conformation or alter chromatin structure and thereby favor promoter function, or (2) they are bidirectional entry sites for transcription factors or RNA polymerase. In one well-studied case, a gene regulated by the hormone glucocorticoid, the hormone binds to a receptor protein, forming a complex that is a transcription factor that binds to an enhancer sequence; after binding, high-level transcription begins.

Structure of 5′ and 3′ Termini of Eukaryotic mRNA Molecules: "Caps" and "Tails"

The 5′ terminus of a eukaryotic mRNA molecule is not a 5′ triphosphate group; the group is altered by the formation of the 5′-5′ linkage of a methylated guanosine derivative, 7-methylguanosine (7-MeG). The first step in forming this structure is a reaction between GTP and the terminal triphosphate of the RNA,

$$\text{Gppp} + \text{pppApNpNp} \rightarrow \text{GpppApNpNp} + \text{PP} + \text{P}$$

by the enzyme guanylyl transferase. This is followed by the methylations of the initial terminal G or A to form either 2'-*o*-methylguanosine (2'-oMeG) or 2'-*o*-methyladenosine (2'-oMeA). The resulting structural unit

(7-MeG)—5'-PPP-5'—(2'-oMe(G/A))—3'-P-5'—nucleoside-3'-P,

in which P and PPP refer to mono- and triphosphate groups, respectively, is called a **cap.** The structure of a generalized 5' terminus of a mRNA molecule is shown in Figure 12-32. In yeast and in the slime molds, the most abundant cap structure is that designated cap 0, in which 7-MeG is added to the 5' terminus and no further methylation occurs. This type of structure is not found in animal cells. In viral mRNA and most animal cell mRNA, methylation of ribose also occurs on the second nucleotide, giving the structure known as cap 1. In some cases, methylation also occurs in the third nucleotide, giving the cap 2 structure. It seems likely at present that each mRNA molecule has a specific cap structure.

Capping occurs shortly after initiation of synthesis of the mRNA and precedes all other modification. The biological significance of capping has not yet been unambiguously established but it is believed that it is required for efficient protein synthesis. The best evidence comes from studies in which mRNA is injected into the cytoplasm of a cell. Cap-binding proteins in the cytoplasm attach to the cap and stimulate formation of the translation complex between mRNA and the ribosome and certain proteins needed to initiate translation (eIF, see Chapter 14). Capping may function to protect the mRNA from degradation by nucleases; if uncapped mRNA is injected, it is usually rapidly broken down to nucleotides.

Most animal mRNA molecules are terminated at the 3' end with a poly(A) tract 20–200 nucleotides long (depending on the particular RNA). The poly(A) is synthesized by a nuclear enzyme, poly(A) polymerase. The adenylate residues are not added to the 3' terminus of the primary transcript. Instead, RNA polymerase passes the site of addition of poly(A), endonucleolytic cleavage occurs, and poly(A) is added. A sequence, AAUAAA (this is in the RNA, not the DNA), 10–25 bases upstream from the poly(A)-addition site is a component of the system for recognizing the site of polyadenylation. The role of this sequence has been elucidated by studying transcription of cloned genes in which the sequence has been altered, for instance, to AAGAAA. In each case, the alteration markedly reduces the efficiency of cleavage, but if cleavage occurs, poly(A) is added. Hence, AAUAAA must signal the cleavage site only, rather than the polyadenylation reaction.

Not all mRNA molecules are polyadenylated. For example, in animal cells histone mRNA lacks poly(A) and the AAUAAA sequence. Termination of synthesis of this mRNA occurs at a unique position that remains the 3'

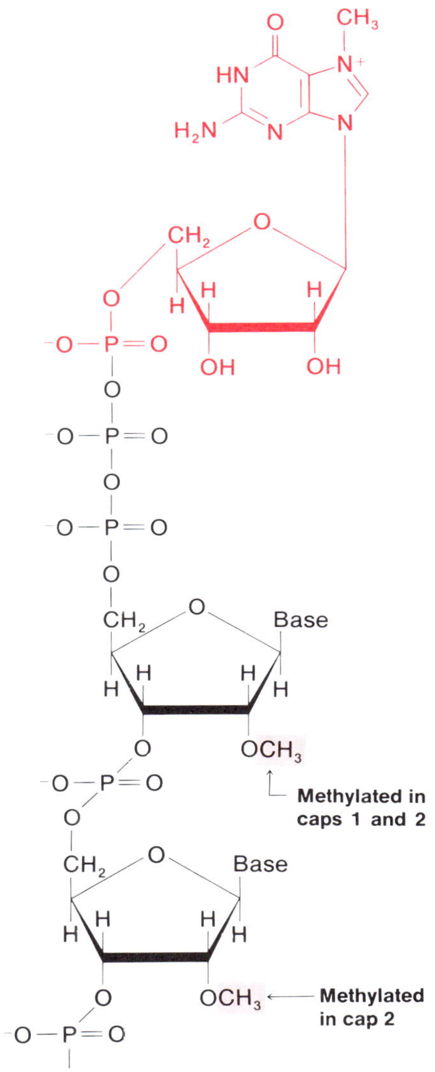

Figure 12-32 Structure of caps at the 5' ends of eukaryotic mRNA molecules. There are three types of caps (0, 1, and 2). All contain 7-methylguanosine (shown in red) attached by a diphosphate linkage to the 5' end of the mRNA molecule. In cap 0, none of the riboses are methylated. In cap 1, one is methylated, and in cap 2, two are methylated.

Figure 12-33 The role of U7 RNA in processing of the sea urchin histone primary transcript. Base-pairing between histone RNA and U7 RNA (shaded) results in the indicated cut and removal of the 3′-terminal segment shown in red.

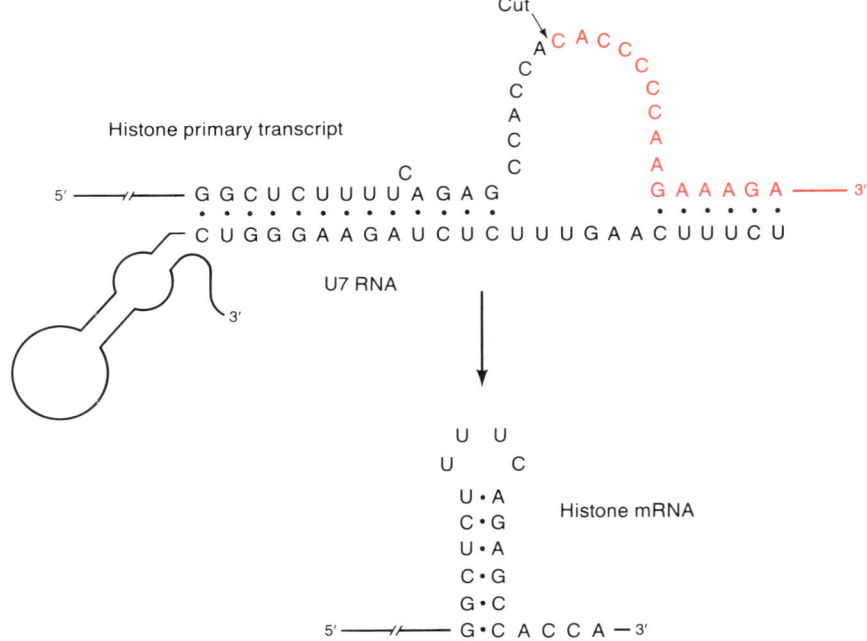

end. In yeast, termination and polyadenylation are coupled processes; however, some transcripts lack the AAUAAA sequence and polyadenylation occurs. The function of the poly(A) terminus is to increase the stability of mRNA. However, since some mRNA molecules lack poly(A), its presence cannot be obligatory for successful translation. The splicing events described in the next section always follow addition of poly(A). However, if polyadenylation is inhibited, splicing occurs normally. Thus, poly(A) has no recognizable role in the further processing of mRNA.

Some snRNPs have been implicated in the formation of the 3′ ends of mRNA. The best understood system is the formation of sea urchin histone mRNA, which, as just stated, lacks a poly(A) tail. The 5′-terminal sequence of the RNA component of the U7 snRNP is complementary to two sequences near the 3′ terminus of the histone primary transcript (Figure 12-33). It has been suggested that U7 RNA base-pairs with the histone primary transcript, producing a single-stranded loop that somehow marks the correct place to be cut. After cutting, a stem-and-loop forms just upstream from the 3′ terminus. This same stem-and-loop exists in the primary transcript and is linearized by U7 RNA. Possibly, the stem-and-loop prevents cutting. Pure U7 RNA does not stimulate cutting *in vitro;* evidence that the U7 RNA must be part of a snRNP comes from studies in which antibodies directed against snRNP proteins prevent cutting.

Preliminary evidence suggests that another snRNP (U4) may be required for polyadenylation: addition of certain antisera to snRNP proteins inhibits polyadenylation of some mRNA species *in vitro* without inhibiting production of the correct 3' terminus.

RNA Splicing

A characteristic of most of the primary transcripts of higher eukaryotes and some transcripts in all eukaryotes is the presence of untranslated intervening sequences (**introns**) that interrupt the coding sequence and are excised from the primary RNA transcript (Figure 12-23).

In the processing of RNA in higher eukaryotes the amount of discarded RNA ranges from 50 to nearly 90 percent of the primary transcript. The remaining segments (**exons**) are joined together to form the finished mRNA molecules. The excision of the introns and the formation of the final mRNA molecule by joining of the exons is called **RNA splicing.** The 5' segment (the cap) of a primary transcript is never discarded and hence is always present in the completely processed (finished) mRNA molecule; the 3' segment is also usually retained. Thus the number of exons is usually one more than the number of introns. The number of introns per primary transcript varies considerably (Table 12-1). Furthermore, within different primary transcripts the introns are variously distributed and have many different sizes (Figure 12-34), but introns are usually longer than exons.

One must be sure to appreciate the difference between introns in mRNA primary transcripts and the regions excised from the prokaryotic primary transcripts containing tRNA and rRNA sequences. In the latter, the excised sequences always *surround*, but do not interrupt, each tRNA and rRNA molecule: ligation of RNA termini does not occur. However, introns are

Table 12-1 Translated eukaryotic genes in which introns have been demonstrated

Gene	Number of introns
α-Globin	2
Immunoglobulin L chain	2
Immunoglobulin H chain	4
Yeast mitochondria cytochrome *b*	5
Ovomucoid	5
Ovalbumin	7
Ovotransferrin	16
Conalbumin	17
α-Collagen	52

Note: At present the genes for histones and for interferon are the only known translated genes in the higher organisms that do not contain introns.

Primary transcript

Excise introns

mRNA

Figure 12-34 A diagram of the conalbumin primary transcript and the processed mRNA. The 16 introns, which are excised from the primary transcript, are shown in black.

Figure 12-35 RNA splicing with adenovirus. (a) An electron micrograph of a DNA-RNA hybrid obtained by annealing a single-stranded segment of adenovirus DNA with fully processed capped and polyadenylated mRNA encoding the major virion protein. The loops are single-stranded DNA. [Courtesy of Tom Broker and Louise Chow.] (b) An interpretive drawing. RNA and DNA strands are shown in red and black, respectively. Four regions do not anneal—three single-stranded DNA segments corresponding to the introns, and the poly(A) tail of the mRNA molecule.

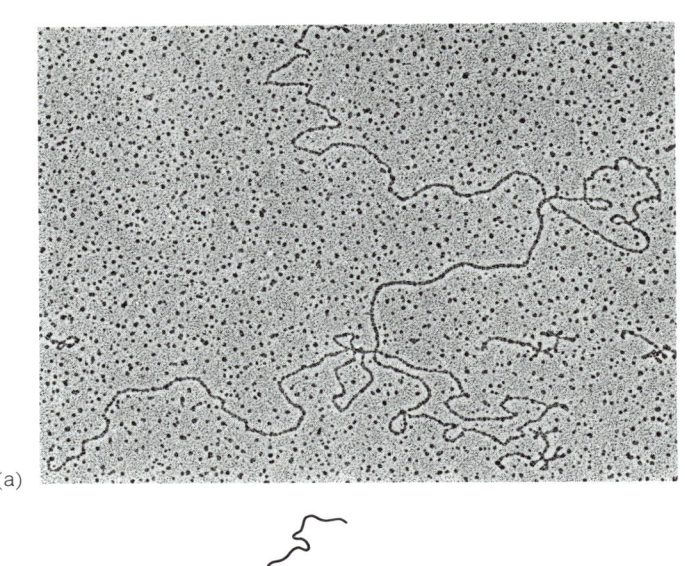

(a)

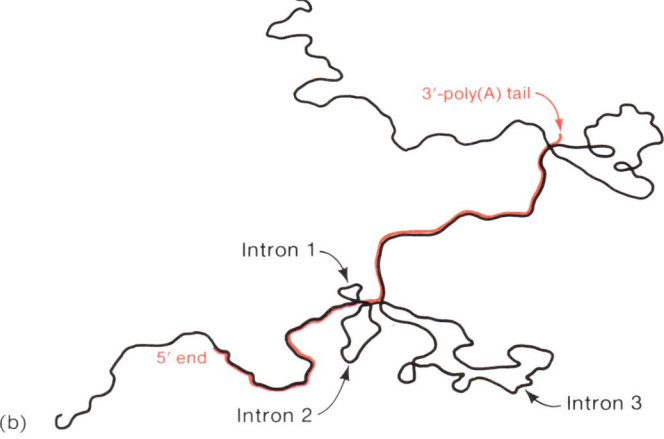

3'-poly(A) tail

Intron 1

5' end

Intron 2

Intron 3

(b)

removed from *within* the coding sequence and *must* be removed in order to obtain the correct amino acid sequence of the encoded protein.

The existence and the positions of introns in a particular mRNA molecule are demonstrable by electron microscopic examination of a DNA–RNA hybrid formed from the coding DNA strand of a gene and the corresponding fully processed mRNA molecule. For example, a hybrid consisting of an adenovirus mRNA and the corresponding DNA is shown in Figure 12-35. A variety of other techniques are also used to measure intron and exon length (these are described in references given at end of this book).

The splicing reaction is remarkably precise: cuts are made at unique positions in transcripts that contain thousands of bases. The demand in fidelity of the excision and splicing reaction is extraordinary, for if an error of one base were made, the correct reading frame would be destroyed. Such fidelity is invariably achieved by recognition of particular base sequences; in this case, they are expected to be at or near the intron–exon junctions.

Base-sequence studies of the regions surrounding several hundred different introns indicate that common sequences can be found at each end of an intron (Figure 12-36). The sites at which cutting occurs are always 5′ to GU and 3′ to AG, as shown in the figure. This so-called **GU–AG rule** may be stated as

The base sequence of an intron begins with GU and ends with AG.

The sites of cutting adjacent to GU and AG are called the **5′** and **3′ splice sites,** respectively. In a later section we will see how these sites may be selected for splicing. A few complete introns have been sequenced and only occasionally have special features have been noted.

Splicing occurs in the nucleus after capping and addition of poly(A). This location explains the enormous size distribution of nuclear RNA. Most animal mRNA molecules are monocistronic (Chapter 13), so one might expect the mRNA to have a small range of sizes. However, the amount of RNA that is discarded as a result of excision and splicing is so variable that the primary transcripts range in molecular weight from about 6×10^4 to 3×10^6. Furthermore, introns are excised one by one and ligation occurs before the next intron is excised; thus, the number of different nuclear RNA molecules present at any instant is huge. Translation does not occur until processing is complete. Hence, before splicing was discovered, the nucleus appeared to contain a significant amount of possibly untranslated RNA. The collection of RNA molecules of widely varied sizes was given the name **heterogeneous nuclear RNA** (hnRNA), a term that is still sometimes used for nuclear RNA. After processing is completed, the mature mRNA is transported to the cytoplasm to be translated; it is not known what prevents the primary transcript and partially processed molecules from being transported. An attractive idea is that splicing occurs on the nuclear membrane and is linked to transport through the nucleopores to the cytoplasm.

Exon	Intron	Exon

...AG GUAAGU...6PyNC AGG N...

Figure 12-36 Common features of the base sequence in and surrounding an intron (red). Py = pyrimidine; N = any nucleotide. The 5′-GU and AG-3′ sequences are underlined.

The biological significance of splicing and its evolutionary origin are unclear, for it would seem easier to synthesize mRNA molecules directly without splicing. However, whatever the origins of splicing, one would expect that in some organisms splicing would evolve until it had a purpose. In fact, in a few organisms splicing has definite value. For instance, the virus SV40, whose DNA is so small that it can only encode a few genes, uses alternative splicing patterns, which enables multiple proteins to be encoded in a single DNA segment. This type of economy has been observed frequently in the small animal viruses, which, without splicing, would have an inadequate amount of DNA for their necessary functions. In other viruses and in some higher eukaryotes alternative splicings of a single transcript, occurring at different times, enable different proteins encoded in the RNA to be synthesized at different times or stages of development. Some examples will be given in Chapters 16 and 23. However, these uses do not provide an adequate explanation for the ubiquity of splicing, since most genes do not exhibit alternative splicing patterns.

Since the left and right ends of all introns are GU and AG, respectively, and since introns are very long, it seems possible that a unit consisting of two introns and the exons between them could be excised. Such a splicing event would be totally destructive to the mRNA, as an exon would be lost. It seems unlikely that such events occur since the frequency of splicing errors in proteins such as collagen, whose primary transcript contains 52 introns (Table 12-1) would be so high that such proteins would not be made. Thus, some rule, as yet unknown, determines how 5′ and 3′ splice sites are matched. A scanning mechanism, such as "find a GU and move on to the first AG," would eliminate the excision of two exons but cannot account for the alternative splicing patterns that will be described in Chapters 16 and 23, and the phenomenon of preferred splicing order that follows. The identity of intron termini raises the question whether introns are excised in a definite order or randomly. If excision occurs in a specific order, then only particular RNA intermediates should arise; if random, all intermediates should exist. This question has been investigated for only a few genes, where it has been observed that intron excision does not occur in a unique order but in a preferred order. For example, the chick ovomucoid gene has seven introns (A–G). Random excision would yield more than 300 intermediates, but only a few are seen. Intermediates lacking one intron have usually lost E or F. When two are lacking, they are always E and F. Introns G and D tend to be removed next, followed by B, and finally by A and C. What features of an RNA molecule might affect the order of excision? There is no definite answer but an educated guess is that intramolecular base pairing determines which intron(s) can be excised first. Once excision occurs, the RNA conformation would probably be altered, making another intron excisable. Presumably conformational changes would occur after each splicing event until all introns are removed.

SPLICING MECHANISMS*

Several mechanisms for RNA splicing are used for different RNA species. Three are discussed in this section.

Splicing of tRNA Precursors (Pre-tRNA)

A recurrent problem in understanding splicing is how the two splice sites are brought together. In one case, removal of the intron in the anticodon loop of some pre-tRNAs may be straightforward (Figure 12-27). This intron

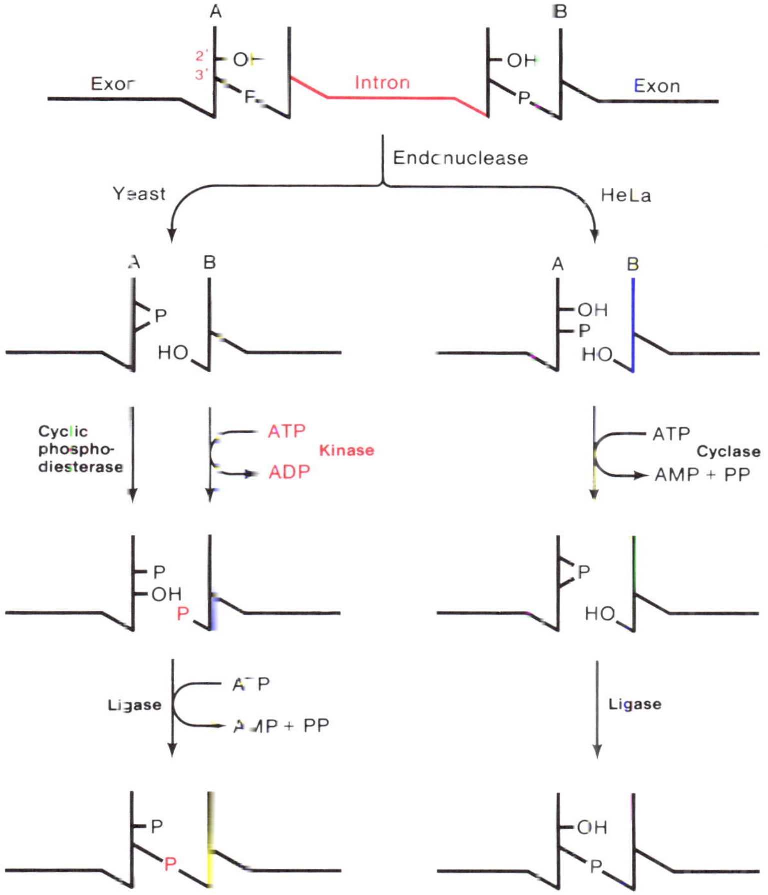

Figure 12-37 Steps that occur in removing introns from yeast and human (HeLa) cell pre-tRNA. See text for details.

*This section may be omitted in courses in which chemistry is deemphasized.

does not interfere with the remainder of the molecule assuming the standard tertiary structure of mature tRNA (Figures 13-8 and 13-9, Chapter 13), so cutting can occur without separating the splice joints from one another. Two different pre-tRNA splicing reactions are known, one in yeast and wheat, and the other in human (HeLa) cells (Figure 12-37). Both begin with an endonuclease cleaving the pre-tRNA at both splice sites, thereby excising the intron. In yeast and wheat (left panel) the nuclease leaves a cyclic phosphate on the nucleotide of the splice site at the left. This is cleaved by a cyclic phosphodiesterase, which yields a 3′-OH group. The nucleotide at the right carries a 5′-OH group, which is converted to a 5′-P group by a kinase, using ATP as the phosphate source. Splicing of the 3′-OH and 5′-P group is accomplished by an RNA ligase. The 2′-P group remains at the junction. In contrast, in HeLa cells the endonuclease yields a 3′-P and a 5′-OH group. A cyclase uses ATP cleavage to produce a high-energy 2′-3′ cyclic phosphate, which reacts with the 5′-OH group to join the two nucleotides. This is a ligase reaction quite different from ligations commonly seen in other organisms. The lack of this unusual ligase in yeast and wheat is what necessitates the cleavage of the cyclic phosphate and the phosphorylation of the 5′-OH group. The overall process of splicing is the same in both pathways; only the chemistry of ligation differs.

Splicing of rRNA Precursors (Pre-rRNA) in *Tetrahymena*: Splicing Without a Protein Enzyme

The splicing processes of nuclear pre-rRNA in the protozoan *Tetrahymena* and of several mitochondrial pre-mRNAs in the fungi *Neurospora* and yeast show two common features: (1) the introns all contain a set of conserved sequences that cause the introns to fold into a common secondary and tertiary sequence that brings the exons into proximity; and (2) remarkably, splicing occurs *in vitro* in the absence of any protein.

The best-understood reaction is that of *Tetrahymena* (Figure 12-38). The process begins with a reaction between a 3′-OH group of a free guanosine and the phosphodiester bond at the 5′ end of the intron. The result of this reaction (which is called a transesterification reaction) is that the guanosine becomes covalently linked to the 5′ end of the intron RNA. Then the newly formed 3′-OH group of the lefthand exon attacks the P at the 3′ end of the intron in a second transesterification that yields the spliced exons and the excised intron. After its excision the intron is converted to an RNA circle by another transesterification in which the 3′-OH group of the intron attacks an internal P, 15 nucleotides from the 5′ end of the molecule. The terminal guanosine is then released as the 5′ terminus of a 16-unit oligonucleotide.

The most striking feature of this splicing reaction is that it is nonenzymatic in the sense that no protein is involved. Detailed study of the process has shown that the intramolecular folding of the intron lowers the activation

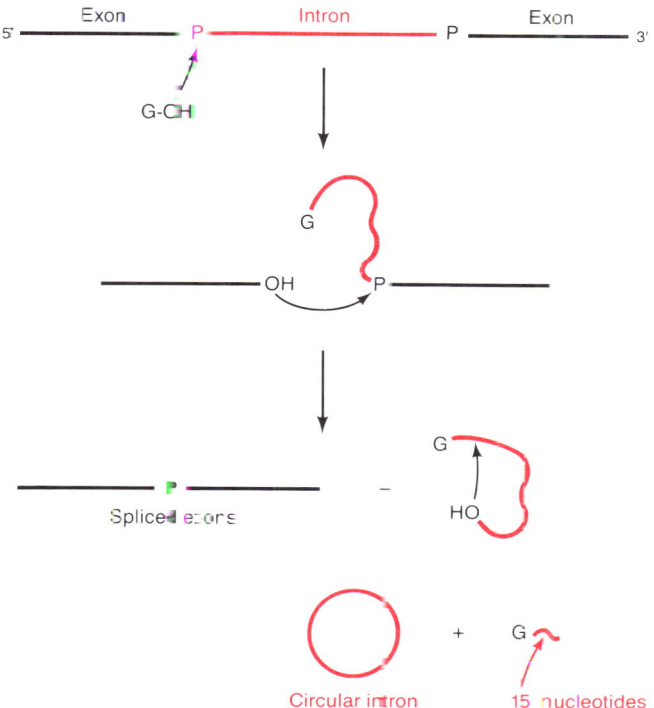

Figure 12-38   The self-splicing reaction that occurs in *Tetrahymena* pre-rRNA. See text for details.

energy of the reaction so in a sense the RNA can be considered catalytic, as we saw earlier for RNase P. However, since the intron does not emerge from the reaction unchanged (it has lost 15 nucleotides and gained a guanosine), it cannot be considered an enzyme in the strictest sense (also, enzymes work on external substrates and the RNA catalyzes an intramolecular process). A second significant feature of the reaction is that only particular bonds in the RNA are reactive; what specifies these bonds is not yet known.

Splicing of Precursor mRNA (Pre-mRNA) of Higher Eukaryotes

A notable feature of the splice joint in the pre-mRNA is the presence of consensus sequences surrounding the 5′ and 3′ termini of each intron, as discussed earlier. The consensus sequences at the intron termini presumably identify the sites of cutting and aid in bringing the two sites together for correct joining. In a few observed cases (for example, yeast) one additional sequence is required for splicing: TACTAACA must occur within the intron about 20–60 nucleotides from the 3′ splice site. However, the generality of this observation has not yet been ascertained.

Figure 12-39 Mechanism of formation of mRNA from pre-mRNA, via the lariat intermediate. See text for details.

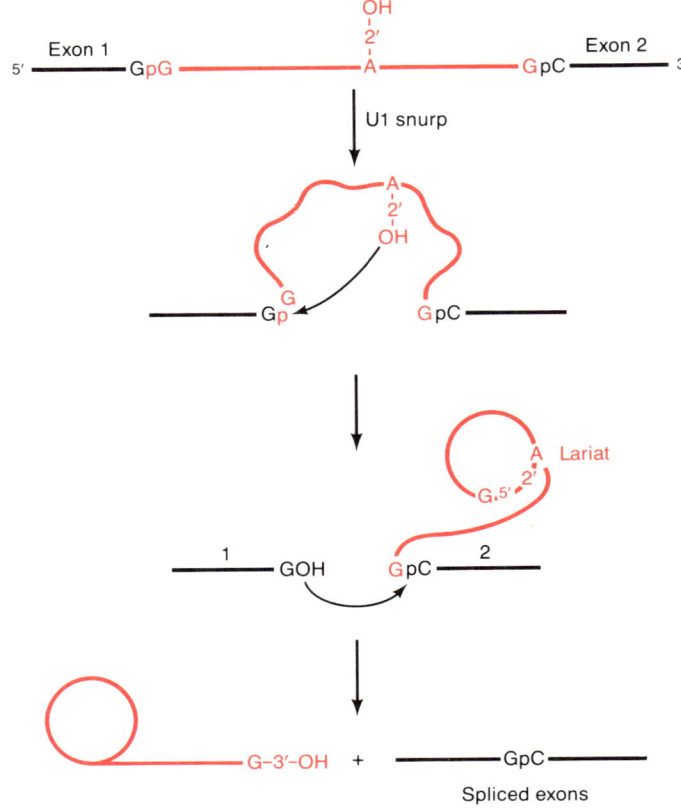

Pre-mRNA splicing requires the snRNP called **U1,** which contains a 165-nucleotide RNA molecule and several polypeptides. Part of the sequence of the U1 RNA is complementary to and can bind to the 5′ splice site, forming RNA-RNA base pairs with this site. The hypothesis that base-pairing of the 5′ consensus sequence and U1 RNA is essential for splicing is supported by three observations made with an *in vitro* splicing system: (1) purified U1 snRNP can bind to and protect the 5′ splice site of several pre-mRNAs from digestion with an RNase; (2) addition of antibodies directed against various proteins in the U1 snRNP inhibits the splicing reaction, and (3) removal of the 5′ terminus of U1 RNA by an RNase prevents splicing. There is no evidence for binding of U1 snRNP to the 3′ splice site, so how this site is recognized and brought to the 5′ site is not known.

Splicing of nuclear pre-mRNA proceeds by a mechanism quite different from the other splicing reactions we have seen. It includes an unusual branched intermediate called a **lariat** (tailed circle), shown in Figure 12-39. An adenosine residue *within* the intron reacts with the conserved G residue at the 5′ end of the intron. In yeast this adenosine (italicized) is part of the TAC-

TAACA sequence. In mammalian pre-mRNA molecules an adenosine residue (possibly in another consensus sequence) in a similar position in the intron serves the same function; it is not clear how this A is selected from the many adenosines in the intron, but it is known to be important. If the lariat A is altered by mutation to C, the adjacent A is not used and splicing does not occur either *in vitro* or *in vivo*. This reaction produces a left exon with a 3'-OH group and a right exon joined to the 5' terminus of the intron, which is in lariat form. Note that the circular part of the lariat contains the unusual 2'-5' linkage between the participating A and G nucleotides. The 3'-OH group of the left exon then undergoes a transesterification reaction with the 3' end of the intron, resulting in joining of the exons and release of the lariat. These reactions require proteins, some of which are likely to be those of the U1 snRNP.

Little is known about how the 5' and 3' splice sites are joined and how the 3' site participates in the overall process. One experiment shows that some feature of the 5' site is required for 3' cutting. In this experiment mutations of the highly conserved GUAUGU 5' terminus of each intron to AUAUGU blocks the reaction. Cutting at the 5' splice site still occurs but the lariat is a dead end—no 3' cut is made. A second experiment utilizing β-globin pre-mRNA and a mammalian cell extract suggest that the 3' site influences what occurs at the 5' site. Cleavage and lariat formation did not occur with RNA substrates whose 3' splice site had been altered either by deletion or substitution of a few bases. In contrast, with yeast pre-mRNA and a yeast extract, deletion of the 3' splice site does not interfere with lariat formation. These observations suggest that substantial differences may exist between the splicing mechanisms of the lower and higher eukaryotes.

MEANS OF STUDYING INTRACELLULAR RNA*

Several procedures are used to study RNA metabolism *in vivo*. In most techniques, radioactive RNA is prepared by adding ³H-labeled or ¹⁴C-labeled uridine to growth medium. However, this also produces radioactive DNA, because uridine can be metabolized to cytosine and thymine; this is significant when studying RNA metabolism because, when RNA is isolated from cells, it is usually contaminated with DNA. The contaminating DNA can be removed by treatment of the sample with pancreatic DNase, for this enzyme degrades the DNA to mononucleotides and small oligonucleotides, which can be easily separated from the RNA.

Because of the almost continuous range in size of intracellular RNA molecules, it is rarely possible to separate a particular mRNA from other RNA species quantitatively just by centrifugation or electrophoresis. Thus,

*This section may be omitted in courses that do not include examination of experimental techniques.

DNA-RNA hybridization is the method of choice. One important procedure for studying prokaryotic and phage mRNA synthesis is filter hybridization, described in Chapter 4 (Figure 4-15). Briefly, single-stranded DNA containing a particular gene is bound to a filter, which is subsequently incubated with an extract containing radioactive RNA. If a DNA-RNA hybrid forms, radioactivity is retained on the filter even after extensive washing.

In performing filter hybridization one needs a pure sample of a DNA molecule that contains a sequence complementary to that of the RNA molecule of interest. Furthermore, in a typical hybridization experiment 0.1–2 μg of DNA is needed. When studying phage or viral mRNA, the necessary DNA is readily available because it can be obtained from purified phage and viruses; and since it is straightforward to prepare several milligrams of phage or viral DNA, the required amounts present no problem. In studying bacterial mRNA it is usually not feasible to use the total bacterial DNA because any particular gene would consist of only about 0.03 percent of the total DNA. This small percentage gives rise to two problems—(1) regions of the DNA other than the gene of interest may have some degree of homology with the mRNA, and (2) in order to have the necessary quantity of the DNA of the gene, 0.3–6 *milli*grams of DNA would be needed, which vastly exceeds the capacity of the usual nitrocellulose filters to hold DNA. Furthermore, with total bacterial DNA a particular mRNA would not be detected if it were in a mixture containing many bacterial mRNA molecules. For studies of bacteria the solution to this problem is to "clone" the gene—in other words, to create a new DNA molecule, which can be easily isolated, that consists mostly of sequences that are nonhomologous to the mRNA of interest, and that contains the particular gene as a major component. A standard procedure is to use the recombinant DNA technology (Chapters 5 and 22) to insert the gene into the DNA of an *E. coli* phage or plasmid. Thus, simply by isolating phage or plasmid DNA one will have purified DNA, roughly 3 percent of which is the gene sequence. If necessary, by use of restriction enzymes this sequence can be excised and purified. Following is an example of filter hybridization to study phage mRNA metabolism.

Figure 12-40 shows the use of filter hybridization to compare the amounts of phage X mRNA synthesized in *E. coli* in two time intervals, 3–4 minutes and 8–9 minutes after infection, and to study the effect of phage infection on *E. coli* RNA synthesis. Phage X mRNA and *E. coli* RNA of all classes become labeled after incubation of the infected cells in the presence of [^{14}C]uridine; however, only phage X mRNA will be retained on a filter to which phage X DNA has been bound. During the interval 3–4 minutes, [^{14}C]uridine is incorporated into RNA to the extent of 6000 cpm;* hybridization indicates that the synthesis of phage X mRNA occurs to the extent of 2000 cpm. In the 8–9 minute interval the incorporation increases to 8500

---

*The notation "cpm" symbolizes counts of radioactive decays per minute in which "minute" refers to the counting time.

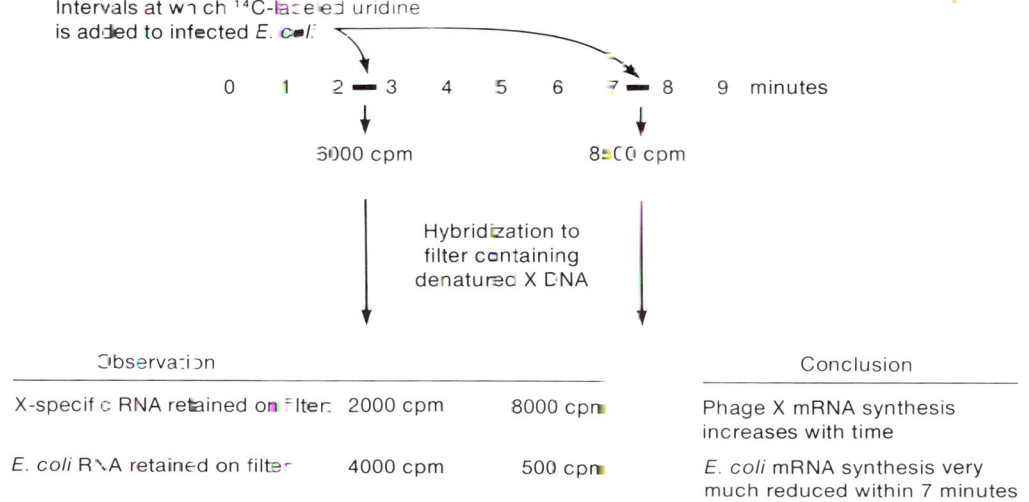

Intervals at which ^{14}C-labeled uridine is added to infected E. coli

0 1 2 — 3 4 5 6 7 — 8 9 minutes

6000 cpm   8500 cpm

Hybridization to
filter containing
denatured X DNA

Observation			Conclusion
X-specific RNA retained on filter:	2000 cpm	8000 cpm	Phage X mRNA synthesis increases with time
E. coli RNA retained on filter	4000 cpm	500 cpm	E. coli mRNA synthesis very much reduced within 7 minutes

Figure 12-40 Use of hybridization to monitor phage mRNA synthesis and shutdown of E. coli RNA synthesis during infection with phage X. See text for a description of the experiment.

cpm; now, 8000 cpm is hybridizable and is phage X mRNA. Thus, the rate of phage X mRNA synthesis increases fourfold (from 2000 to 8000 cpm) and E. coli RNA synthesis is shut down after phage infection, from 6000 − 2000 = 4000 cpm to 8500 − 8000 = 500 cpm in the two time intervals.

By using restriction enzymes DNA molecules can be broken into defined fragments containing particular genes or clusters of genes, and these fragments can easily be separated by gel electrophoresis, purified, and bound to filters. Thus, labeled mRNA can be hybridized to DNA bound to a collection of filters and in this way shown to be transcribed from particular DNA segments and hence particular genes. Additional information could be obtained if the complementary strands of either the cloned genes or of particular restriction fragments were available. With such DNA on the filter one could determine which strand is the template and hence the direction of transcription.

In studying most eukaryotic mRNA the concentration problem is more formidable than with prokaryotes since a particular gene may be only 0.0001 percent (or less) of the DNA. In these cases, cloning techniques to provide a DNA probe are essential. For highly differentiated cells—for example, reticulocytes, whose major gene product is globin (the protein component of hemoglobin)—more than half of the mRNA present in the cell can be a single species. In these cases the mRNA and its precursors can be radioactively labeled sufficiently for detection in a filter hybridization experiment. However, for other experiments blotting techniques are essential. For example, unlabeled RNA is isolated from unlabeled cells, separated by electro-

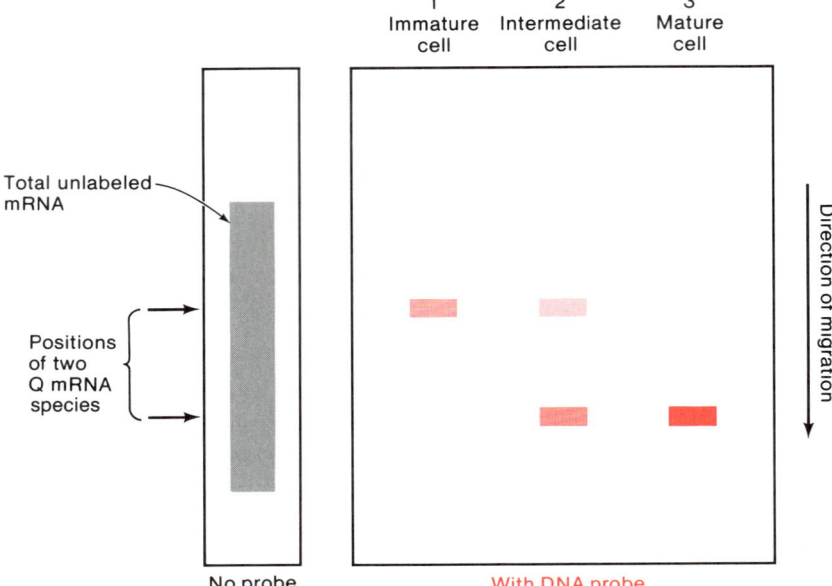

Figure 12-41   The use of Northern blotting to study RNA synthesis. The narrow panel at the left shows that the total mRNA is widely and uniformly spread through the electrophoretic gel; this RNA is not readily detectable, because it is not radioactive. The larger panel shows the locations of a radioactive single-stranded DNA probe that is complementary to the two Q mRNA species.

phoresis, and transferred to specially treated paper (Northern blotting). Hybridization with a highly labeled DNA probe complementary to the RNA followed by quantitative autoradiography enables one to identify and measure the amount of a particular RNA in a eukaryotic cell. Figure 12-41 shows hypothetical data for a cell that goes through two stages of differentiation until it makes a large amount of a protein Q. The undifferentiated cell makes a small amount of Q mRNA. As differentiation begins and a new cell type arises (second lane), a new mRNA becomes evident. In the fully differentiated cell type the original mRNA is absent and a new mRNA is present. Thus, one can conclude that differentiation involves suppression of very weak synthesis of one Q mRNA species and activation of synthesis of another Q mRNA in large amounts.

Translation

I: THE INFORMATION PROBLEM

The synthesis of every protein molecule in a cell is directed by intracellular DNA. There are two aspects to understanding how this is accomplished—the **information** or **coding problem** and the **chemical problem.** By the information problem is meant the mechanism by which a base sequence in a DNA molecule is translated into an amino acid sequence of a polypeptide chain. The chemical problem refers to the actual process of synthesis of the protein: the means of initiating synthesis; linking together the amino acids in the correct order; terminating the chain; releasing the finished chain from the synthetic apparatus; folding the chain; and, often, postsynthetic modification of the newly synthesized chain. The overall process is called **translation.**

Translation is a complex process requiring the participation of about 200 distinct macromolecules. Our approach in Chapters 13 and 14 is to outline the general features of the process and then to discuss the various stages in some detail. The structure of one major component, the ribosome, is actively being studied; this topic will be discussed in detail in sections that may be omitted by the beginning student.

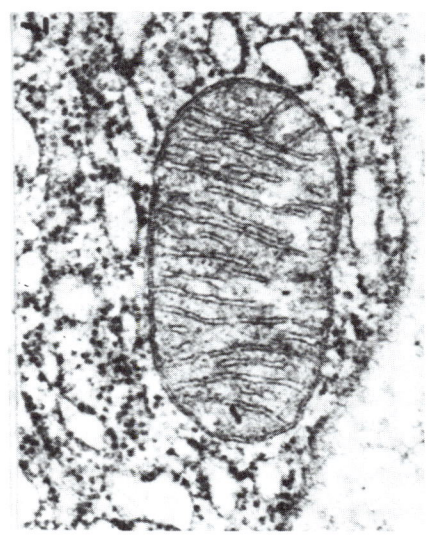

An electron micrograph of a mitochondrion of a mouse cell. The genetic code of mitochondrial DNA differs slightly from the "universal" code of nuclear DNA.

OUTLINE OF TRANSLATION

Protein synthesis occurs on intracellular particles called **ribosomes.** These particles, which in prokaryotes consist of three RNA molecules and about

52 different protein molecules, contain the enzymatic system needed to form a peptide bond between amino acids, a site for binding the mRNA, and sites for bringing in and aligning the amino acids in preparation for assembly into the finished polypeptide chain. Amino acids themselves are unable to interact with the ribosome and cannot recognize bases in the mRNA molecule. Thus there exists the collection of carrier molecules described in the previous chapter, **transfer RNA (tRNA).** These molecules contain a site for amino acid attachment and a region called the **anticodon** that recognizes the appropriate base sequence (the **codon**) in the mRNA. Proper selection of the amino acids for assembly is determined by the positioning of the tRNA molecules, which in turn is determined by hydrogen–bonding between the anticodon of each tRNA molecule and the corresponding codon of the mRNA. There are two sites on a ribosome for binding tRNA molecules, the **P site** and the **A site.**

In outline, protein synthesis consists of several stages (Figure 13-1):

1. An initiation complex forms in which the P site is filled with an initiator tRNA molecule (bearing methionine) and in which the mRNA is correctly positioned such that a start codon matches up with the initiator tRNA.
2. The A site is occupied by an amino acid–tRNA complex (**aminoacyl–tRNA**); the particular aminoacyl–tRNA that is chosen is determined by the codon on the mRNA that is adjacent to the start codon.
3. The first amino acid (the one on the initiator tRNA) is transferred to and covalently joined to a second amino acid (the one attached to the tRNA in the A site) while both are still attached to tRNA.
4. The first tRNA molecule is then ejected from the P site.
5. The aminoacyl–tRNA molecule in the A site advances to the P site.

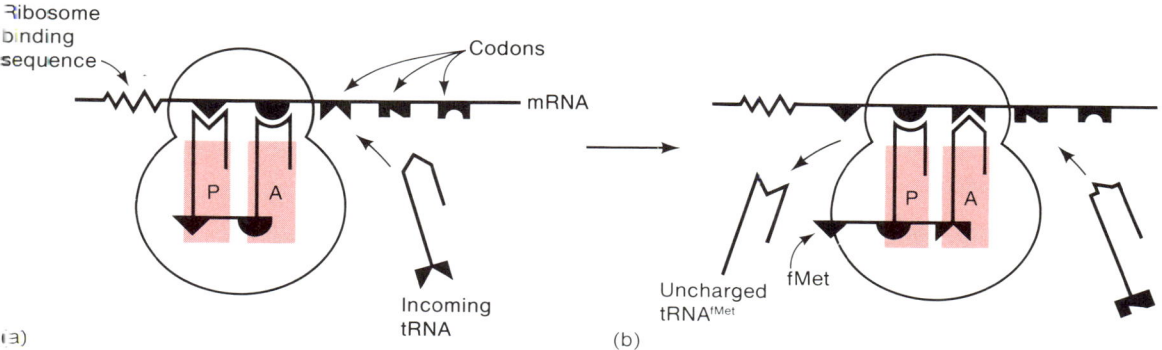

Figure 13-1   Schematic diagram of the beginning of polypeptide synthesis. In (a) the P site is occupied by fMet-tRNA^fMet and the A site is occupied by the tRNA that matches the second codon. In (b) the mRNA has advanced by one codon. The fMet is separated from its tRNA and leads the new polypeptide chain. The ribosome binding sequence is a base sequence in prokaryotic mRNA that binds to the ribosome; eukaryotic mRNA lacks such a sequence, and ribosome binding occurs in a different way.

6. The aminoacyl-tRNA corresponding to the third codon occupies the A site.

7. The process continues until a stop codon in the mRNA occupies the A site. Then, the completed polypeptide chain is released from the ribosome.

Clearly there must be many different tRNA molecules, because each amino acid must be brought to the ribosome in a way that ensures that it corresponds to the base sequence of the mRNA. Thus, there are specific tRNA molecules that correspond to each amino acid.

The formation of aminoacyl-tRNA is catalyzed by a set of enzymes called **aminoacyl-tRNA synthetases.** Each enzyme recognizes specific sites (a recognition site and an amino acid attachment site) on each tRNA molecule as well as one of the 20 amino acids, in order that the appropriate amino acid will be attached to the correct tRNA molecule.

In this chapter we shall be examining the various stages of the information problem in detail.

THE GENETIC CODE

The **genetic code** is the collection of base sequences (codons) that correspond to each amino acid and to stop signals for translation.

Since there are 20 amino acids, there must be more than 20 codons to include signals for starting and stopping the synthesis of particular protein molecules. If one assumes that all codons have the same number of bases, then each codon must contain at least three bases. The argument for this conclusion is the following. A single base cannot be a codon because there are 20 amino acids and only 4 bases. Pairs of bases also cannot serve as codons because there are only $4^2 = 16$ possible pairs of four bases. Triplets of bases are possible because there are $4^3 = 64$ triplets, which is more than adequate. We will see shortly that the code is indeed a triplet code and that all 64 codons carry information of some sort. In many cases, several codons designate the same amino acid—that is, the code is **redundant** or **degenerate.**

Two Types of Codes

There are several ways in which a codon could be read from a mRNA molecule. The two most important alternatives originally considered are the **overlapping** and **nonoverlapping codes** (Figure 13-2). In an overlapping code, each base serves as the first base of some codon; in a nonoverlapping code, each base is used in only one codon. The overlapping code is certainly more economical in that a particular region of the DNA can carry the information for synthesis of three times as many proteins as a nonoverlapping code. However, an overlapping code has a particular disadvantage in that a

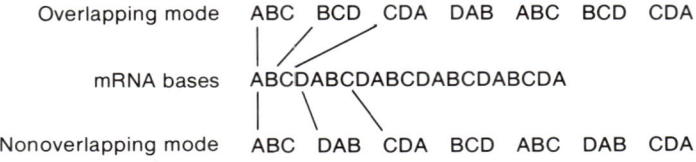

Overlapping mode   ABC BCD CDA DAB ABC BCD CDA

mRNA bases ABCDABCDABCDABCDABCDA

Nonoverlapping mode ABC DAB CDA BCD ABC DAB CDA

Figure 13-2 Two ways of reading the bases of a mRNA molecule. In the overlapping mode the seven codon triplets are read from nine bases; in the nonoverlapping mode 21 bases are needed.

single mutagenic base change could alter as many as three amino acids, whereas only one would be altered if the code were an overlapping one. Since most mutations are deleterious, the rare advantageous mutations that are the raw material of evolution would generally be lost because they would usually be accompanied by the deleterious ones. The two kinds of codes were distinguished experimentally by examining the amino sequence of wild-type and mutant proteins made from particular viral genes. In each case observed it was found that the mutant protein was altered by one and not three amino acids. Thus, the code was shown to be of the nonoverlapping type.

We have argued that the genetic code must be at least a three-letter code but have not ruled out codes having more than three letters. A great deal of chemical evidence clearly identified the sequences of three bases specifying particular amino acids but, prior to this work, a brilliant genetic experiment was performed that clearly indicated that the code is a **triplet code.** This experiment, characteristic of the classic period of molecular biology, is described because it shows the great power of genetic arguments in molecular biology.

A Genetic Argument for a Triplet Code

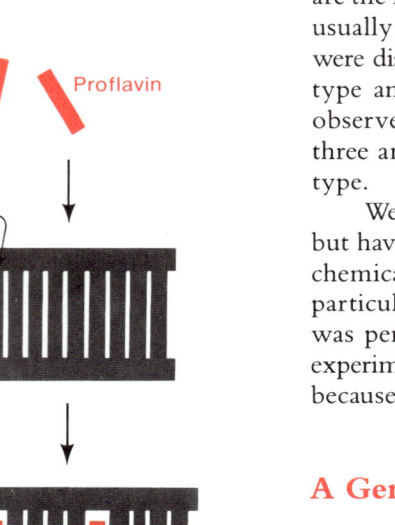

Figure 13-3 A segment of a DNA molecule showing the intercalation of proflavin. In normal DNA there is virtually no space between the base pairs. Proflavin, whose thickness equals that of a base pair, pushes the base pairs apart and thereby lengthens the DNA molecule. Proflavin binds tightly to DNA because it stacks between the bases. The proflavin-induced distortion of the parent DNA strand interferes with replication and can lead to two types of mutations in the daughter strands. In one instance an extra base is added, leading to an insertion mutation; in the second a base is omitted, leading to a deletion mutation.

The molecule proflavin (a derivative of the molecule acridine) binds tightly to a DNA molecule. In 1961, before the triplet code had been determined, it was believed that proflavin binds by intercalating between the DNA bases (Figure 13-3), though there was no proof of this idea. Proflavin is also highly mutagenic, but only if it is applied during a replication cycle. That is, treatment of a phage suspension with proflavin is not mutagenic, but if proflavin is added to a phage-infected culture of bacteria, mutant phage progeny are produced.

Acridine-induced mutants (Chapter 11) have one property that distinguishes them from other kinds of mutants—namely, they do not revert to wild type by treatment with base-substitution mutagens; such a property is a characteristic of a deletion, but this possibility is eliminated by the fact that reversion of an acridine-induced mutant can be achieved by further culture of the mutant in a growth medium containing mutagenic acridines. Since it is far too improbable that the wild-type base sequence could be restored by

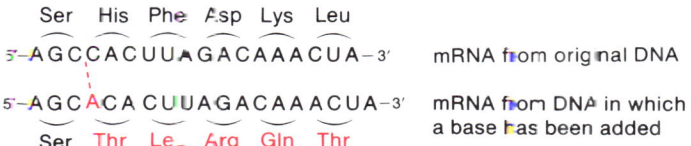

Figure 13-4　Change in the amino acid sequence of a protein caused by addition of an extra base.

further mutagenesis, it was concluded that acridine mutants were of a new type, and the key to understanding these mutants was that they arose only if the phage replicated in the presence of a substance thought to intercalate. As an explanation of how acridine mutations are caused, it was proposed that intercalation of an acridine molecule moves two adjacent bases apart by the thickness of a single base, and that during replication an additional base is occasionally inserted when DNA polymerase meets an already intercalated molecule (Figure 13-4) Thus, acridine-induced mutants were in fact base-addition mutants. (We will see shortly that base deletion also occurs.)

Clearly, a base addition must be mutagenic, because it upsets the phase of reading of the code for a specific protein: every amino acid downstream from the added base will be different, as shown in the figure. Reversion of an acridine-induced mutant could be explainable as the removal of an extra base; however, the reversion was sufficiently high that removal of a particular added base seemed unlikely. Ultimately, an understanding of this phenomenon, as well as proof of the three-letter nature of the code, came from a study of a collection of acridine-induced mutants in the gene called *rIIB* of *E. coli* phage T4.

The T4 *rIIB* gene can be subdivided, genetically at least, into two regions that we term *tolerant* and *intolerant* (Figure 13-5). The tolerant region is defined by the fact that it is very difficult to obtain base substitution mutants in this

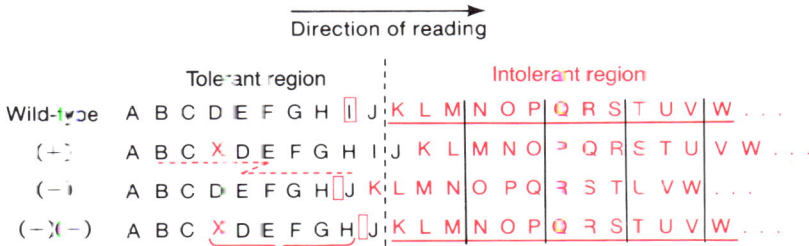

Figure 13-5　A schematic diagram of the T4*rIIB* gene showing the tolerant (black) and intolerant (red) regions. A, B, C, etc. represent consecutive bases. In the (+) mutant, base X is inserted between C and D. In the (−) mutant base I is deleted. The dashed red line shows the crossover that generates the (+)(−) recombinant. In (+)(−) the horizontal red brace denotes the sequence in the tolerant region that differs from the wild-type sequence. The intolerant regions are the same in wild type and (+)(−) and in each case are in the same reading frame—namely, beginning at the eleventh letter. The reading frame in the intolerant region is indicated by the vertical black line.

region and, in fact, the region can be deleted without full loss of function of the gene. (This means that the tolerant region of the *rIIB* protein can have many amino acid changes without loss of function.) However, by proflavin mutagenesis, many mutations in the tolerant region can be found, and these mutations are reverted with high frequency by growth in the presence of an acridine. Normally, if two phage strains, each carrying a mutation in the same gene, are crossed by mixed infection of a bacterium, some wild-type phage progeny arise by genetic recombination between the two genetic loci. However, in experiments to investigate acridine-induced mutagenesis, it was found that when these mutants were crossed, wild-type phage progeny did not always arise. In fact, acridine-induced mutants could be placed in two distinct classes, which were arbitrarily termed (+) and (−). These classes were distinguished by the following properties:

1. A cross between two (+) mutants did not yield wild-type phage.
2. A cross between two (−) mutants did not yield wild-type phage.
3. A cross between one (+) and one (−) mutant did yield wild-type phage.

These results were interpreted in the following way. The (+) mutants were thought to have one added base and the (−) mutants to lack one base. A double mutant of the type (+)(+) or (−)(−) was thought to have two additional bases or to lack two bases, respectively. In both cases reading of the code would be shifted two bases out of phase and a functional protein would not be made. However, in a (+)(−) double mutant, the advanced reading frame *following* the (+) site would be incorrect, but the correct reading frame would be restored at the (−) site. Between the two mutant sites, the amino acid sequence would not be that of the wild-type phage but nonetheless the (+)(−) and (−)(+) phage strains would be functional, because both mutant sites were in the tolerant region. These interpretations proved correct. The point is that *as long as the reading frame is restored before the intolerant region is reached, a functional (though not always perfect) protein can be produced.* By this reasoning a reversion event of a (+) mutant must often be the production of a (−) mutation at a second site.

The fact that a double (+)(+) mutant never had the wild-type phenotype meant that the code could not be a two-letter code; if it were a two-letter code, the reading frame would be restored in this combination. The critical experiment to test for a triplet code was the construction of a triple mutant. As expected, the triple mutants (+)(+)(+) and (−)(−)(−) had the wild-type phenotype, whereas the mixed triples, (+)(+)(−) and (+)(−)(−), were still mutant. How the combination of three mutants of the same type can yield a wild-type phage is shown in Figure 13-6.

This elegant experiment provided strong evidence for the triplet code.* In later experiments using the lysozyme gene of phage T4, the amino acid

*Note that it did not prove it, since a six-letter code was not ruled out; if a (+) and a (−) were always additions or deletions of two bases, the code would be a sextuplet code.

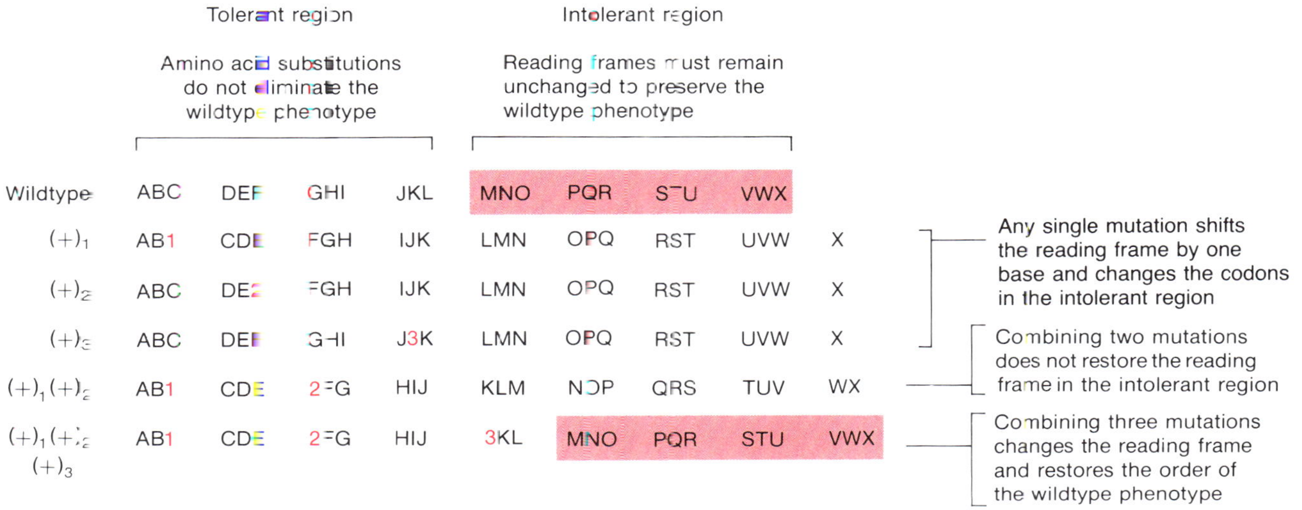

	Tolerant region				Intolerant region				
	Amino acid substitutions do not eliminate the wildtype phenotype				Reading frames must remain unchanged to preserve the wildtype phenotype				
Wildtype	ABC	DEF	GHI	JKL	MNO	PQR	STU	VWX	
$(+)_1$	AB1	CDE	FGH	IJK	LMN	OPQ	RST	UVW	X
$(+)_2$	ABC	DE2	FGH	IJK	LMN	OPQ	RST	UVW	X
$(+)_3$	ABC	DEF	G3I	J3K	LMN	OPQ	RST	UVW	X
$(+)_1(+)_2$	AB1	CDE	2FG	HIJ	KLM	NOP	QRS	TUV	WX
$(+)_1(+)_2$ $(+)_3$	AB1	CDE	2FG	HIJ	3KL	MNO	PQR	STU	VWX

Any single mutation shifts the reading frame by one base and changes the codons in the intolerant region

Combining two mutations does not restore the reading frame in the intolerant region

Combining three mutations changes the reading frame and restores the order of the wildtype phenotype

Figure 13-6 A diagram showing that to combine three base additions (1, 2 3, shown in red), but not two, restores the reading frame in the intolerant region (shaded red) of the T4 *rIIB* protein, if the code is a triplet code. An extra amino acid is present in the tolerant region of $(+)_1(+)_2(+)_3$ recombinants, but this does not lead to a mutant phenotype.

sequence of the wild-type protein and acridine-induced single mutants were compared and it was found that indeed there was a site after which all amino acids are changed in the mutant. The amino-acid sequence of double mutants of the type $(+)(-)$ also showed that a short sequence of amino acids in the double mutant differed from that of wild-type, as expected, if $(+)$ and $(-)$ mutants were not in adjacent codons. Once the genetic code was worked out, it was shown also that many $(-)$ mutants were actually $(+)(+)$ mutants—that is two bases had been added rather than a single base deleted. Interestingly, when this occurs, the two added bases are adjacent to one another.

Elucidation of the Base Composition of the Codons

The first step in the elucidation of the genetic code was taken in 1961 when a mixture of ribosomes, tRNA molecules, radioactive amino acids, a cell fraction (of unknown composition) known to be required for protein synthesis, and the synthetic polynucleotide polyuridylic acid (poly(U)) was incubated, and polyphenylalanine was synthesized. From this simple result and previous knowledge that the code is a triplet code, it was immediately proved that the codon UUU corresponds to the amino acid phenylalanine. Variations on this basic experiment made it possible to identify all U-containing codons. For instance, if a single guanine is added to the terminus of a poly(U) chain, it is found that the polyphenylalanine is often terminated by leucine. Thus, UUG must be the leucine codon. If many guanines are

Figure 13-7 Polypeptide synthesis from UUUU . . . UUGGGGGGG in three different reading frames showing the origin of the incorporation of glycine, leucine, and tryptophan.

added at the terminus, the polyphenylalanine is sometimes terminated by glycine, indicating that GGG corresponds to glycine (Figure 13-7). At low frequency leucine and tryptophan are also found at the carboxyl terminus of the polyphenylalanine. These must come from the codons UUG and UGG at the transition point between U and G. Since it had already been shown that UUG is the leucine codon, this observation also identified UGG as the codon for tryptophan.

Similar experiments were carried out with mixed polymers of uridine, adenosine, and cytidine—poly(U,A) and poly(U,C)—to identify the remaining U–containing codons. Other homopolymers were also studied. Poly(A) results in the formation of polylysine, which indicates that AAA is a lysine codon and allows experiments with mixed polymers of the type just described to be performed to identify the A–containing codons. Similarly, poly(C) was found to produce polyproline. Poly(G) fails to act as a synthetic mRNA because it forms a triple-stranded helix and cannot be translated.

Other experiments were used to identify the codons that could not be identified by the procedure just described and to confirm the assignments already known. For instance, an alternating polymer, poly(AC)—that is, . . . ACACAC . . .* This polymer will be read in groups of threes as . . . ACA CAC ACA CAC ACA CAC . . . ; thus, the protein made will consist of two alternating amino acids. This polymer yields

. . . Thr-His-Thr-His-Thr-His . . .

From other experiments it was known that threonine has a (2A,1C) codon and histidine has a (2C,1A) codon, so the codon assignments, Thr = ACA and His = CAC, could be made.

In another type of experiment synthetic trinucleotides were observed to stimulate the binding of tRNA molecules to ribosomes. For example, if a mixture is prepared consisting of ribosomes, all of the tRNA molecules (to which amino acids, *one* of which is radioactive, are already bound), and the trinucleotide UUU, and then the ribosomes are isolated, it is found that only phenylalanyl-tRNA^{Phe} is associated with the ribosomes in any appre-

*Remember the convention for abbreviating copolymers. Poly(X,Y) is a polynucleotide in which bases X and Y are randomly arranged; poly(XY) is the alternating copolymer XYXYXY . . . . Also, in any polynucleotide ABC . . . XYZ, the direction of the chain is P-5′-ABC . . . XYZ-3′-OH, or, using the p notation, pApBpC . . . pZ.

ciable amount. This type of experiment was repeated with all 64 trinucleotides and in most cases unambiguous results were obtained; 50 codons were confirmed in this way.

Identification of the Stop Codons

The stop codons were also identified by using polynucleotides having a defined sequence. For example, the polymer poly(GUAA) can be read as:

GUA AGU AAG UAA GUA AGU AAG . . .

G UAA GUA AGU AAG UAA GUA AGU . . .

GU AAG UAA GUA AGU AAG UAA GUA . . .

AGU AAG UAA GUA AGU AAG UAA . . .

However, large polypeptides are not synthesized. Only two peptides are found—the tripeptide Val-Ser-Lys and the dipeptide Ser-Lys. From the known codon assignments—namely, Val = GUA, Ser = AGU, and Lys = AAG— these peptides can be lined up in four reading frames:

1   GUA AGU AAG UAA GUA AGU AAG . . .

      Val   Ser   Lys   ?

2   G UAA GUA AGU AAG UAA GUA AGU . . .

        ?   Val   Ser   Lys   ?

3   GU AAG UAA GUA AGU AAG UAA GUA . . .

     Lys   ?

4   AGU AAG UAA GUA AGU AAG UAA . . .

      Ser   Lys   ?

Note that synthesis stops in each case just before a UAA codon, which indicates that UAA must be a stop codon. Similar experiments identified the codons UAG and UGA as stop codons. Later it was found that *there is no tRNA that can bind to UAA,* which is the reason that it is a stop codon.

As a laboratory joke, the three stop codons were given the following names—UAG, **amber** codon; UAA, **ochre** codon; UGA, **opal** codon— which are now in common use. We shall occasionally use these terms.

Each of the codon assignments, including those for the stop codons, has been amply confirmed by direct sequencing of DNA molecules that encode proteins whose amino acid sequences are known. For example, there are instances in which a wild-type protein of known sequence, containing lysine—let us say, ABC-Lys . . . KLM—is converted to the mutant fragment ABC; this clearly indicates a mutation from the lysine codon AAG to the stop codon UAG.

As will be seen in Chapter 14, in protein synthesis each codon is recognized by a tRNA molecule having a sequence that can form hydrogen bonds with the codon. The convention used to describe the anticodon base sequence requires special mention. By convention, base sequences are always stated in the $5' \rightarrow 3'$ direction. That is, the codon AUG is accurately written 5'-AUG-3'. Thus the anticodon corresponding to AUG is CAU (5'-CAU-3') and not UAC, because of the antiparallel nature of all double-stranded DNA and RNA segments. Actually, many base pairs other than the standard A·U and G·C pairs are used in codon-anticodon pairing, as will be seen in a later section.

Identification of the Start Codons

In the *in vitro* system, protein synthesis can start at any base in the polymer used to direct the synthesis. However, *in vivo* protein synthesis cannot begin at any base in an RNA molecule. Instead, a start codon is needed. The codon **AUG** *is the most commonly used start codon;* in a few instances, the codon GUG is used. In all DNA molecules whose base sequences have been compared to amino acid sequences, the AUG codon appears in the same reading frame as the base sequence corresponding to a particular protein. In prokaryotes special signals, which we will describe shortly, designate a particular AUG codon as a start codon and thereby define the reading frame. In eukaryotes, position identifies the start codon. If an AUG is not designated as a start codon, it is simply read as an internal codon corresponding to methionine. When used to initiate polypeptide chain growth in prokaryotes, the AUG codon causes initiation of a polypeptide chain with a modified amino acid, N-formylmethionine (fMet); in eukaryotes, ordinary unmodified methionine is used.

Universality of the Code

The experiments with the *in vitro* systems that have been described used components isolated from the bacterium *E. coli*. These experiments have been repeated with ribosomes and tRNA molecules obtained from many species of bacteria, yeast, plants, and animals, including mammals. Furthermore, for numerous genes and their protein products, the base sequences and the amino acid sequences have been compared. With the exception of the mitochondria of numerous eukaryotes and *Mycoplasma capricolum,* a small bacterium (discussed later in this chapter), the same codon assignments can be made for all organisms that have been examined. Thus, the genetic code is said to be universal.

A summary of the codon assignments is shown in Table 13-1.

Table 13-1   The "universal" genetic code

First position (5' end)	Second position				Third position (3' end)
	U	C	A	G	
U	Phe	Ser	Tyr	Cys	U
	Phe	Ser	Tyr	Cys	C
	Leu	Ser	Stop	Stop	A
	Leu	Ser	Stop	Trp	G
C	Leu	Pro	His	Arg	U
	Leu	Pro	His	Arg	C
	Leu	Pro	Gln	Arg	A
	Leu	Pro	Gln	Arg	G
A	Ile	Thr	Asn	Ser	U
	Ile	Thr	Asn	Ser	C
	Ile	Thr	Lys	Arg	A
	Met	Thr	Lys	Arg	G
G	Val	Ala	Asp	Gly	U
	Val	Ala	Asp	Gly	C
	Val	Ala	Glu	Gly	A
	Val	Ala	Glu	Gly	G

Note: The boxed codons are used for initiation. GUG is very rare.

Redundancy of the Code

We have now accounted for four codons as signals—three stop codons and one start codon. The remaining 60 codons, as well as internal and start AUG, all correspond to amino acids. Since there are only 20 amino acids, there are many cases in which several codons direct the insertion of the same amino acid into a protein chain. Thus, the genetic code is highly **redundant,** as shown in Table 13-1.

Examination of the codon assignments shows that the redundancy is not random, in that multiple codons corresponding to a single amino acid are not distributed haphazardly throughout the table. In fact, with the exception of serine, leucine and arginine, all **synonyms** (codons corresponding to the same amino acid) are in the same box—that is, *the codons differ by only the third base.* For example GGU, GGC, GGA, and GGG all code for glycine. Furthermore, redundancy is the rule: there are multiple codons except for tryptophan and methionine.

Certain features of the redundancy are quite regular—for example:

1   Pairs of codons of the general form XYC and XYU always code for the same amino acid.

2. The pairs XYG and XYA usually code for the same amino acid.

The structural basis for this phenomenon will become evident later when we examine codon–anticodon binding.

What is the biological significance of the redundancy? The simplest explanation is that it minimizes the deleterious effects of mutations. If redundancy were not the case, then 44 of the 64 codons would not code for an amino acid—that is, they would be stop codons. Thus, two-thirds of the base changes could not possibly lead to improvement of an organism and the evolutionary process would be seriously limited.

Genetic Confirmation of Codon Assignments

The codon assignments shown in Table 13-1 are completely consistent with all chemical observations and have been confirmed independently by a comparison of the amino acid sequences of wild-type proteins against base-substitution mutant proteins. In every case the amino acid substitution can be accounted for by observing the codons corresponding to the two differentiating amino acids. For example, in tobacco mosaic virus coat protein a Pro→Ser substitution has been observed. This can be accounted for by the changes CCC→UCC, CCU→UCU, CCA→UCA, and CCG→UCG. Similarly, in mutant hemoglobins the substitutions of a particular glutamic acid by both valine and lysine have been observed. This is consistent with GAA as the glutamic acid codon and changes to GUA and AAA for valine and lysine, respectively.

Protein fragments have also been isolated from cells carrying chain termination mutations. These are also easily explained as a mutation from a sense codon to a stop codon—for example, a change from the tyrosine codon UAC to the stop codon UAG.

THE DECODING SYSTEM

The genetic code provides the relation between the base sequence of the coding region of a mRNA molecule and the amino acid sequence of the protein translated from that molecule. However, amino acids do not by themselves line up along a mRNA molecule. This decoding operation—that is, the conversion of the base sequence within a mRNA molecule to an amino acid sequence of a protein—is accomplished by a system consisting of two different types of molecules. One type is a set of small adaptor RNA molecules called **transfer RNA (tRNA)** and the other is a set of enzymes, the **aminoacyl synthetases.** The structures and the modes of action of these molecules are described in the sections that follow.

Transfer RNA and the Aminoacyl Synthetases

There are many different tRNA molecules in a particular cell. Each tRNA molecule has several important regions, three of which are of concern to us at this point. One of these regions is a contiguous sequence of three bases that can hydrogen-bond by base-pairing to each codon; this sequence is the **anticodon.** A second site is the **amino acid attachment site;** the amino acid that binds to this site corresponds to the particular codon in mRNA that forms base pairs with the anticodon of the tRNA. In order that the aminoacyl synthetase match the amino acid and the anticodon correctly, the enzyme must be able to distinguish one tRNA molecule from another. Thus each tRNA molecule must have a **recognition region.** Each of these regions will be discussed in detail shortly.

The name of the amino acid that is linked to a particular tRNA molecule by a specific aminoacyl synthetase is also appended when naming both types of molecules. Thus, one says that leucyl-tRNA synthetase attaches leucine to tRNA^{Leu}. Because of the redundancy of the code, there is sometimes (but not always) more than one tRNA molecule for a particular amino acid; these may be denoted tRNA$_\text{I}^{\text{Leu}}$ and tRNA$_\text{II}^{\text{Leu}}$ or sometimes tRNA$_\text{UUA}^{\text{Leu}}$ and tRNA$_\text{CUA}^{\text{Leu}}$. A tRNA molecule and its corresponding aminoacyl synthetase are called **cognates;** the terms cognate tRNA or cognate synthetase are also used when referring to one member of a pair.

The Cloverleaf Structure of tRNA

All tRNA molecules are small, single-stranded nucleic acids ranging in size from 73 to 93 nucleotides. The base sequence of a tRNA, namely, yeast tRNA$^{\text{A.a}}$ was first determined in 1965. It was noticed that several segments of this molecule could form double-stranded regions that would give the molecule a cloverleaf structure in which open loops are connected to one another by double-stranded stems (Figure 13-8). More than 200 different tRNA molecules from bacteria, yeast, plants, and animals have been sequenced, and for each sequence base pairing can be arranged to produce the cloverleaf conformation; study of many of these molecules by physical methods has shown in each case that the cloverleaf is the correct conformation. By careful comparison of these sequences certain features have been found to be common to almost all of the molecules. This has led to the idea of a "consensus" tRNA molecule consisting of 76 nucleotides arranged in a cloverleaf form (Figure 13-8). By convention, the nucleotides are numbered 1 through 76 starting from the 5'-P terminus. This 76-base sequence is sufficiently fundamental that in tRNA molecules having more than 76 bases the additional ones can usually be recognized as additions to the standard sequence. The most common additions follow standard bases 17, 20, and 47; the "extra" bases are numbered 17:1, 17:2, 20:1, 20:2, 47:1, 47:2, and so forth.

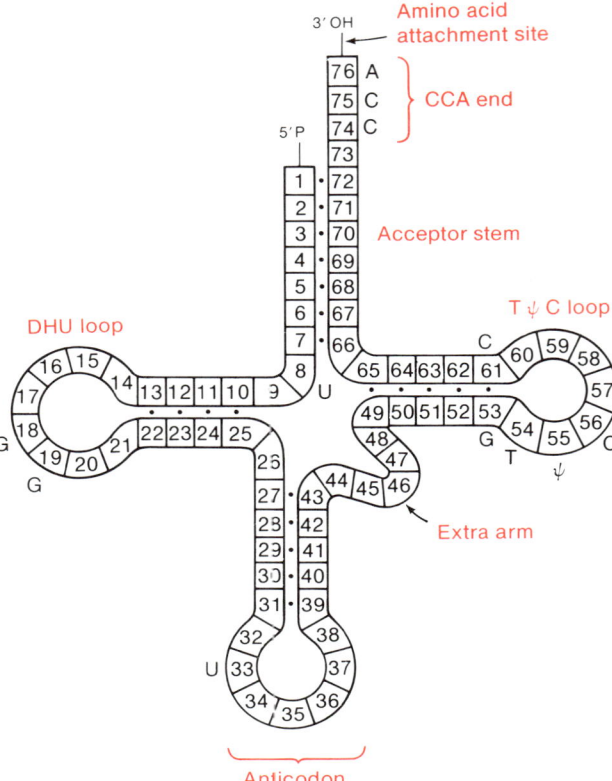

Figure 13-8   The currently accepted "standard" tRNA cloverleaf with its bases numbered. A few bases present in almost all tRNA molecules are indicated.

The standard tRNA molecule has the following features:

1. Bases in positions 8, 11, 14, 15, 18, 19, 21, 24, 32, 33, 37, 48, 53, 54, 55, 57, 58, 60, 61, 74, 75, and 76 are invariant in that they are the same in nearly all tRNA molecules whose sequences are known.

2. The 5′-P terminus is always base-paired. It is thought that this contributes to the stability of tRNA.

3. The 3′-OH terminus is always a four-base single-stranded region having the base sequence XCCA-3′-OH, in which X can be any base. This is called the **CCA** or **acceptor end.** The adenine in the CCA sequence is the site of attachment of the amino acid by the cognate synthetase.

4. There are many so-called "modified" bases in tRNA. A few of these, dihydrouridine (DHU), ribosylthymine (rT), pseudouridine (ψ), and inosine (I) occur frequently and in particular regions. Others are found in a variety

of positions. In some cases the substituents of the purine and pyrimidine rings are so numerous and complex that the base is called "hypermodified." In only a few cases is the significance of these unusual bases known.

5. There are three large single-stranded loops. The lowermost or **anticodon loop** contains seven bases. The anticodon occupies positions 34 through 36. It is almost always preceded by two pyrimidines (usually base 33 is uridine) and followed by a modified purine. Thus the anticodon loop has the general sequence

$$5'-Py-U-XYZ-Pu \text{ (modified)}-Variable base$$

Anticodon

The loop containing bases 14 through 21 is called the **DHU loop.** It is not constant in size in different tRNA molecules but may contain up to three extra bases between bases 17 and 18 and one to three extra bases following base 19. The loop containing bases 54 through 60 almost always contains the sequence TψC and is called the **TψC loop.**

6. There are four double-stranded regions called stems, or arms. The stems have no invariant bases and often contain base pairs other than G·C and A·T, such as G·U. The three stems attached to each loop are given the name of the corresponding loop, as for example, the anticodon stem.

7. An additional loop, containing bases 44 through 48, is also present. In the smallest tRNA molecules it contains four bases, lacking base 47, whereas in the largest tRNA molecule it contains 21 bases with 16 bases present between bases 47 and 48. This highly variable loop is called the **extra arm.**

Three-Dimensional Structure of tRNA

Although the tRNA molecule is conveniently depicted as a planar cloverleaf, x-ray crystallographic analysis shows that its three-dimensional structure is actually more complicated. Figure 13-9(a) shows the skeletal model of yeast tRNA^{Phe}; an interpretive drawing is in panel (b) of the figure. The major features of the three-dimensional structure are the following:

1. All base pairs proposed in the cloverleaf model exist.
2. The molecule is folded and has two helical double-stranded branches, each about 60 Å long and mutually perpendicular. One branch consists of the acceptor stem and the TψC stem; the other consists of the stems of the DHU and anticodon loops. Thus, the molecule is L-shaped.

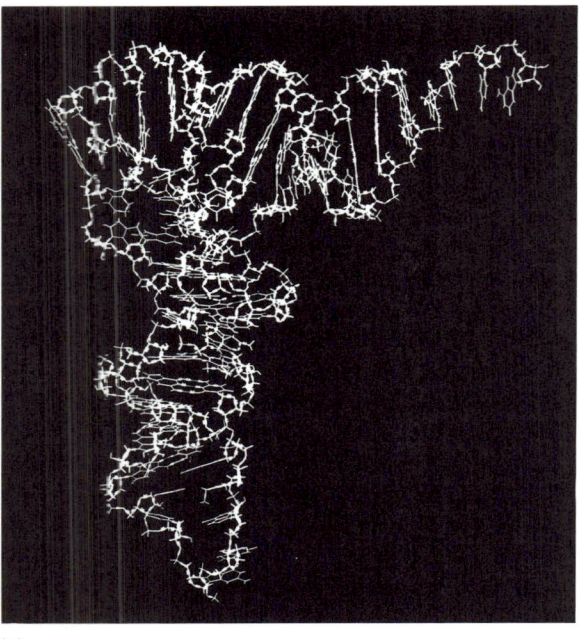

(a)

(b)

Figure 13-9 (a) Photograph of a skeletal model of yeast tRNA^{Phe}. (b) Schematic diagram of the three-dimensional structure of yeast tRNA^{Phe}. [Courtesy of Dr. Sung-Hou Kim.]

3. The CCA acceptor terminus is at one end of the L and the anticodon loop is at the other end, about 80 Å away.
4. The TψC loop, which is thought to interact with ribosomal RNA, is at the corner of the L.
5. All bases except five (16, 17, 20, 47, and 76) are stacked, even though half of the bases are not in double-stranded helical regions. Even the bases of the single-stranded 3'-terminal segment (except for the terminal adenine) are stacked.
6. There are nine base pairs that are not shown in the cloverleaf model. These are called **tertiary base pairs** because they are responsible for the three-dimensional folding. Only one of these, a G·C pair, is a standard Watson-Crick pair.
7. In addition to the tertiary base pairs described in item 6, the tertiary structure is stabilized by hydrogen bonds between bases and different ribose units and between ribose units in separate nucleotides.

Two observations make it likely that all tRNA molecules are structurally similar: (1) Most of the tertiary base pairs are formed from bases that are conserved in all tRNA molecules. (2) Different tRNA molecules can co-crystallize, forming a uniform crystal; a low-resolution x-ray crystallographic analysis of these mixed crystals does not indicate that two types of molecules are present and yields molecular dimensions that are the same as those of a single kind of molecule. This suggests that the two tRNA molecules in a mixed crystal have common structural features. Mixed crystals containing many different tRNA molecules have been tested with similar results.

Attachment of an Amino Acid to a tRNA Molecule

When an amino acid is attached to a tRNA molecule, the tRNA is said to be **aminoacylated** or **charged.** This is designated in several ways. For example, if the amino acid is serine, the charged tRNA would be written seryl-tRNA or seryl-tRNA^{Ser}. There are cases in which a tRNA molecule is mischarged. If tRNA^{Leu} were mischarged with serine, this would be written seryl-tRNA^{Leu}. The term "uncharged tRNA" refers to a tRNA molecule lacking an amino acid.

There are probably two reasons that cells have evolved with acylated tRNA as an intermediate in protein synthesis. The first and obvious reason is to align the amino acids according to the order of the codons. The second reason is a thermodynamic one—namely, that peptide-bond cleavage, rather than peptide-bond formation, is thermodynamically favored. This situation is reversed by the simple expedient of first forming an amino acid ester with a 2′-OH or 3′-OH group of the 3′-terminal nucleotide of tRNA (Figure 13-10). Thus, the amino acid ester is the activated intermediate of protein synthesis.

Acylation is accomplished in two steps, both of which are catalyzed by an aminoacyl synthetase. In the first, or activation, step an aminoacyl AMP is synthesized in a reaction between an amino acid and ATP:

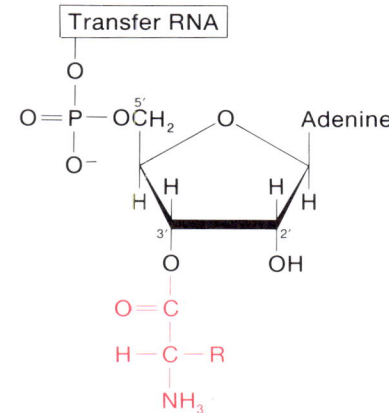

Figure 13-10 Mode of attachment of an amino acid (shown in red) to a tRNA molecule. The amino acid is covalently linked to the 2′-OH or 3′-OH group of the terminal adenosine of tRNA.

In the second, or transfer, step the aminoacyl-AMP, which is bound by the enzyme, reacts with tRNA to form acylated tRNA and AMP:

$$\text{Aminoacyl-AMP} + \text{tRNA} \rightleftarrows \text{Aminoacyl-tRNA} + \text{AMP}$$

The transfer reaction is an equilibrium reaction in which neither direction predominates. As in DNA synthesis, the reaction is driven to the right by hydrolysis of pyrophosphate; the overall reaction is

$$\text{Amino acid} + \text{ATP} + \text{tRNA} + \text{H}_2\text{O} \rightleftarrows \text{Aminoacyl-tRNA} + \text{AMP} + 2\text{P}_i$$

The Aminoacyl Synthetases

There is at least one and usually only one aminoacyl synthetase for each amino acid. For a few amino acids specified by more than one codon, more than one synthetase exists. The enzymes are clearly very specific and must be able to recognize both a particular tRNA molecule and a particular amino acid. Because of the great structural similarities between all tRNA molecules, one would expect the synthetases to be similar, but this is not the case. There are four classes of synthetases: (1) those consisting of a single polypeptide chain, (2) those having two identical chains, (3) those containing four chains of two types, with two copies of each type, and (4) one enzyme with four identical subunits. Furthermore, the molecular weights of the subunit range from 33,000 to 113,000 and the molecular weights of the holoenzyme range from 54,000 to 380,000. The synthetases are definitely a heterogeneous group of proteins.

Recognition of a tRNA Molecule by Its Cognate Synthetase

Each synthetase must be able to recognize its cognate tRNA molecule. The recognition region on the tRNA molecule spans the entire molecule rather than being a localized site like the codon binding and amino acid attachment. Two experimental approaches—**photoactivated cross-linking** and **RNase-protection**—have been used to determine the regions of a tRNA molecule that are in contact with a synthetase. In both methods the synthetase and the tRNA molecule are first allowed to bind to one another. In the cross-linking technique (Figure 13-11), the synthetase-tRNA complex is irradiated with

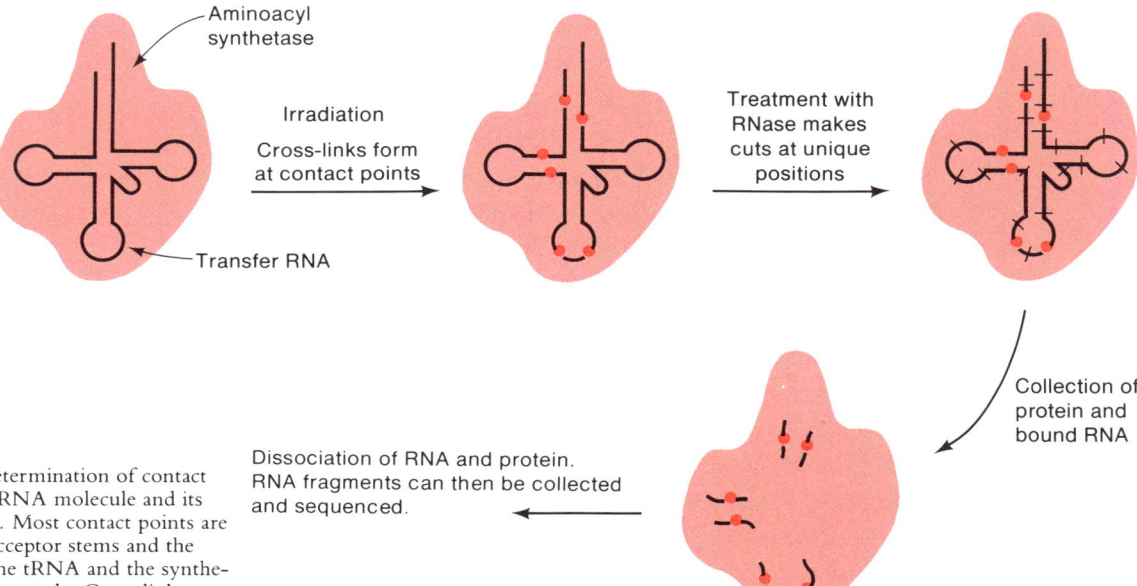

Figure 13-11 Determination of contact points between a tRNA molecule and its cognate synthetase. Most contact points are in the DHU and acceptor stems and the anticodon loop. The tRNA and the synthetase are not drawn to scale. Cross links are shown as red dots.

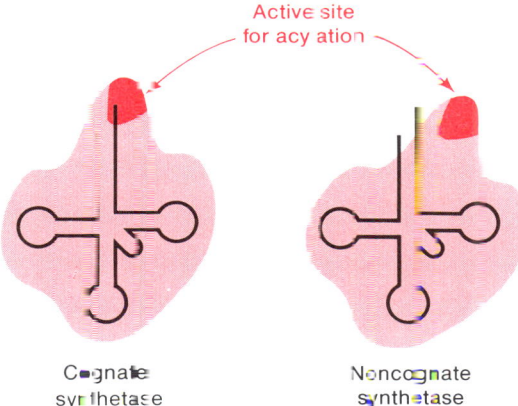

Active site
for acylation

Cognate
synthetase

Noncognate
synthetase

Figure 13-12   Diagram showing how an aminoacyl synthetase can bind to a noncognate tRNA yet be unable to charge the tRNA molecule. The tRNA and the aminoacyl synthetase molecules are not drawn to scale.

ultraviolet light, which causes chemical bonds (cross-links) to form between bases and amino acids in the protein. The cross-links can occur only if the bases are within 2 or 3 Å of an amino acid. After irradiation, the tRNA molecule is fragmented and oligonucleotides bound to proteins are identified by electrophoresis. Each oligonucleotide fragment comes from a particular (and known) segment of the tRNA molecule. In the RNase–protection method the complex is treated with a collection of RNases that digest all of the RNA that can be reached by the enzymes. Points of contact between the tRNA molecule and the synthetase are protected from the nucleases. The results of such studies with several synthetase-tRNA complexes indicate that there are many points of contact, primarily in the DHU and acceptor stems and the anticodon loop. All contact points are on the same side of the three-dimensional structure. The odd nucleosides are rarely engaged in contact; instead, they are probably involved in attachment to the ribosome. There are several examples, in which a synthetase can bind to a tRNA molecule that is not its cognate, and again, there are many points of contact. However, aminoacylation does not occur in these pairs because the CCA terminus is not positioned at the active site of the enzyme (Figure 13–12). Thus, it seems that the specificity of aminoacylation is not derived exclusively from binding ability but also from the three-dimensional structure of each tRNA molecule, which serves to place the 3′-terminal adenosine at the aminoacylation site of the synthetase

Error Correction in Aminoacylation

The location of each amino acid in a growing polypeptide is determined only by the codon-anticodon interaction. This is known from the following experiment. A sample of tRNA was acylated in a reaction mixture that contained all of the synthetases, 19 nonradioactive amino acids, and

[^{14}C]cysteine. The only radioactive charged tRNA was [^{14}C]cysteinyl-tRNA^{Cys}. The charged tRNA molecules were then hydrogenated with Raney nickel, which converted the [^{14}C]cysteine to [^{14}C]alanine; then, the only radioactive tRNA molecules were [^{14}C]alanyl-tRNA^{Cys}. These molecules were then used in an *in vitro* protein-synthesizing system containing hemoglobin mRNA as the only mRNA molecule. Hemoglobin was synthesized and isolated, and its amino acid sequence was determined. It was found that [^{14}C]alanine is present at sites normally occupied by cysteine, which showed that the protein-synthesizing apparatus does not examine the identity of each amino acid but looks only at the tRNA molecule. Thus, the burden of producing the correct amino acid sequence is on the synthetase.

Several amino acids are structurally similar, so the synthetases might make occasional mistakes. If the error frequency is high, it seems reasonable to expect that an editing mechanism, as we saw with DNA synthesis, has evolved. Valine and isoleucine constitute such a possibly ambiguous pair of amino acids and, in fact, isoleucyl-tRNA synthetase forms valyl-AMP, which remains bound to the synthetase, at a frequency of about one per 225 activation events. This would mean that 1/225 of all isoleucine positions in proteins could contain a valine. For a typical protein containing 500 amino acids of which 25 were isoleucines, about one copy in nine could be altered. Since there are at least ten known examples of misacylation, there could be an error in almost every protein molecule. However, this does not occur.

The editing mechanism that corrects the valine–isoleucine error is a multistep process in which the isoleucyl tRNA synthetase first transfers the valine from valyl AMP to tRNA^{Ile}, and then cleaves the valine from the valyl-tRNA^{Ile}. The effectiveness of this and other synthetases that cleave misacylated tRNA is such that only about one protein per thousand contains one incorrect amino acid. The hydrolysis is carried out by the isoleucyl-tRNA synthetase itself.

When an incorrect complex is formed, it has been noticed that the side chain of the incorrect amino acid is never larger than that of the correct amino acid. This observation has led to a scheme that predicts what mis-acylations will occur and when hydrolytic editing will occur. This scheme is called the **double-sieve mechanism.** According to this mechanism, amino acids larger than the correct amino acid are never activated because they are too large to fit into the active site of the synthetase. Thus, the size of the active site forms a sort of sieve. Those amino acids that are smaller than the correct one can be activated, though at a lower rate since they fit less well into the active site. It is further proposed that the hydrolytic site is too small for the correct amino acid; only amino acids that are smaller than the correct one can be removed by hydrolysis. Thus, the hydrolytic site forms the second sieve. There are a few cases in which two amino acids have the same size and nearly the same shape—for example, valine and threonine; indeed, threonine is activated by valine-tRNA synthetase. However, threonine is removed from threonyl-tRNA^{Val}. This is explained in terms of obvious chemical differences between the two side chains—that the threonine hydroxyl group forms a hydrogen bond that draws the amino acid into the hydrophilic hydrolytic site.

Since error correction occurs by hydrolysis of an incorrect amino acid, one might ask how alanyl-tRNA^{Cys} persisted in the Earley nickel experiment presented at the beginning of this section. The explanation is that hydrolysis of misacylated tRNA occurs almost immediately after misacylation, while the tRNA is still bound to the synthetase. In fact, there are no examples of proofreading of free charged tRNA.

THE CODON-ANTICODON INTERACTION

The fidelity of protein synthesis is determined by two features of the system—correct charging of a particular tRNA molecule by the cognate aminoacyl synthetase and the codon-anticodon pairing. The latter is the subject of this section.

Presence of Inosine in the Anticodon Loop

Once it was established that 61 of the 64 possible codons represent amino acids, it was expected that there would be 61 distinct tRNA molecules, each with an anticodon that is the complement of one of the 61 codons. The one-codon–one-tRNA idea was demolished when the first tRNA, yeast tRNA^{Ala}, was sequenced; this molecule responds to three (not one) of the four GCX alanine codons, namely, GCU, GCC, and GCA. Study of the sequence of the non-hydrogen-bonded regions in the tRNA molecule indicated that there is only one place in the molecule at which the base sequence is 5′-GC-3′. Thus, assuming that the anticodon consists of three adjacent bases, these two bases had to be part of the anticodon.

However, this assignment requires an anticodon sequence of 5′-IGC-3′ (Figure 13-13) in which I is the unusual nucleoside inosine. (Inosine is not a base but the nucleoside ribosylhypoxanthine.) Hence, since the tRNA molecule responds to the three codons GCU, GCC, and GCA, then inosine either does not form hydrogen bonds or it can pair with each of the bases U, C, and A. If inosine fails to form hydrogen bonds, then tRNA^{Ala} ought to respond to GCG, which is not the case. This suggests that inosine does form hydrogen bonds but not of the strict Watson–Crick type, as shown in the figure.

The chemical origin of inosine in the anticodon is rather interesting. Sequence analysis of a large number of tRNA molecules has shown that *no anticodon starts with 5′-A*. Whenever A appears in the first position in the unmodified tRNA transcript, it is deaminated by an enzyme called **anticodon deaminase** to hypoxanthine (the base of inosine).

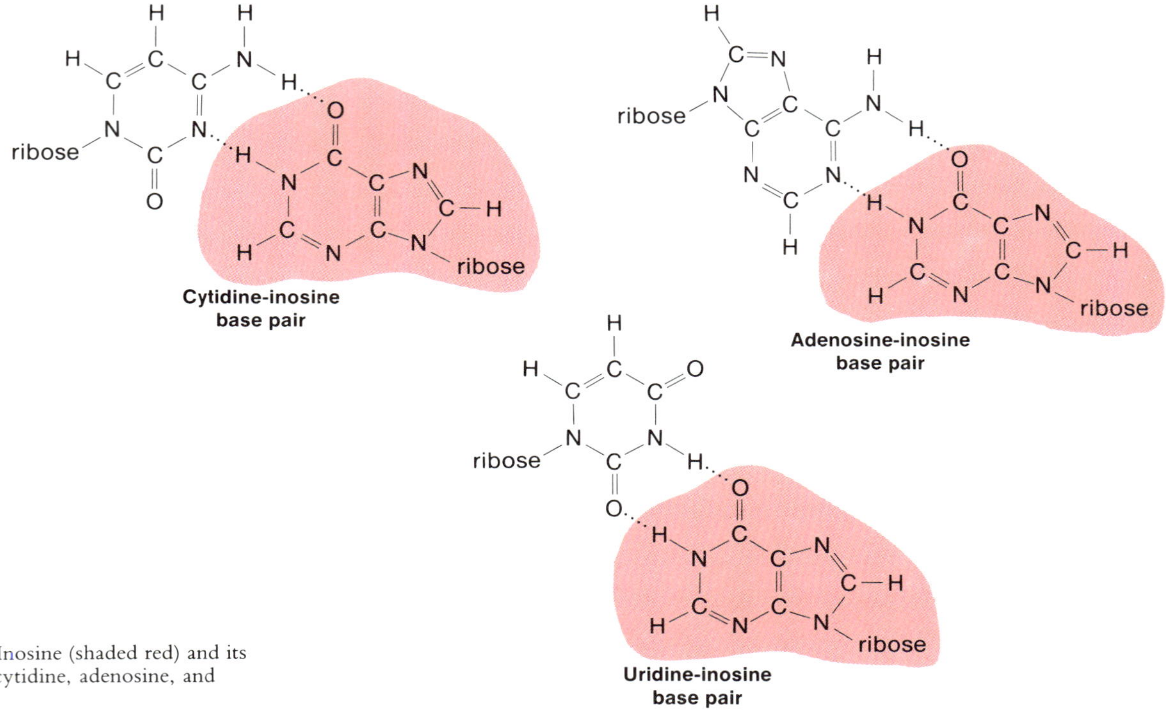

Figure 13-13   Inosine (shaded red) and its base pairs with cytidine, adenosine, and uridine.

Redundancy as a Response of one Anticodon to Several Codons: The Wobble Hypothesis

The pattern of the redundancy of the code suggests that something is missing in the explanation of codon-anticodon binding; the most striking aspect of the redundancy is that in many, but not all, codons, the identity of the third codon base appears to be unimportant. This can be seen in Table 13-1.

In 1965, Francis Crick made a proposal, known as the **wobble hypothesis,** that explains both the response of some tRNA molecules to several codons and the pattern of redundancy of the code. Up to that time it was generally assumed that no base pair other than G·C, A·T, or A·U would be found in a nucleic acid. This is true of DNA because the regular helical structure of double-stranded DNA imposes two steric constraints: (1) two purines cannot pair with one another because there is not enough space for a planar purine–purine pair, and (2) two pyrimidines cannot pair because they cannot reach one another. Crick proposed that since the codon-anticodon interaction involves only six bases, the paired structure might not be the standard double helix. By model-building he showed that at the third position of the codon, play in the structure (this play he called wobble), allowed nonstandard base-pairing, namely, the base pairs listed in Table 13-2. The possibility of forming the four base pairs shown in the table—namely A·I, U·I, C·I, and G·U—explains how a single tRNA molecule can pair with more than one codon for the same amino acid.

The wobble hypothesis explains the arrangement of synonyms in the code. The following points can be made (compare Figure 13-13 and Table 13-2):

Table 13-2	The wobble pairings
Third position codon base	First position anticodon base
A	U, I
G	C, U
U	G, I
C	G, I

1. The codons XYC and XYU are always synonyms.
   a. If the anticodon to XYC is GY'X', this anticodon can also pair with XYU, because G can pair with U in the third position of a codon.
   b. If the anticodon is IY'X', then pairing also occurs with XYU, XYA, and XYC. Thus, no anticodon can pair only with XYC and not with XYU, or only with XYU and not with XYC.
   c. If the anticodon to XYU were AY'X', then XYU would pair with two tRNA molecules (having anticodons GY'X' and IY'X'). However, A is not found in the first position of an anticodon (except in the mitochondrial anticodon for glycine), because the enzyme anticodon deaminase acts as this position to convert A to I.

2. The codon XYA has the anticodons IY'X' or UY'X'.
   a. If the anticodon is IY'X', then XYA is synonymous with both XYU and XYC.
   b. If the anticodon is UY'X', then XYA is synonymous with XYG.

Thus, *every codon ending with A is redundant.* This point also explains the number of different tRNA molecules that correspond to a particular amino acid. For example, in boxes in Table 13-1 in which there are four codons for one amino acid, there is one tRNA molecule that recognizes codons ending only in G and a second tRNA that recognizes XYA, XYC, and XYU.

3. The codons AUG (Met) and UGG (Trp) are nonredundant (they are both single entries in the universal code table). Figure 13-14 and Table 13-2 indicate that a codon XYG can have an anticodon CY'X' that responds to no other codon.

4. There is no tRNA having an anticodon complementary to UGA. The anticodon for the Cys codons UGU and UGC is GCA and the only anticodon for tryptophan is CCA, which pairs with UGG. If UCA were a codon for tryptophan, it would pair with both UGA and UGG.

5. Finally, UAU and UAC for Tyr, and UAA and UAG for Stop, are explained by having GUA as the Tyr anticodon and supposing that no tRNA has an anticodon UUA.

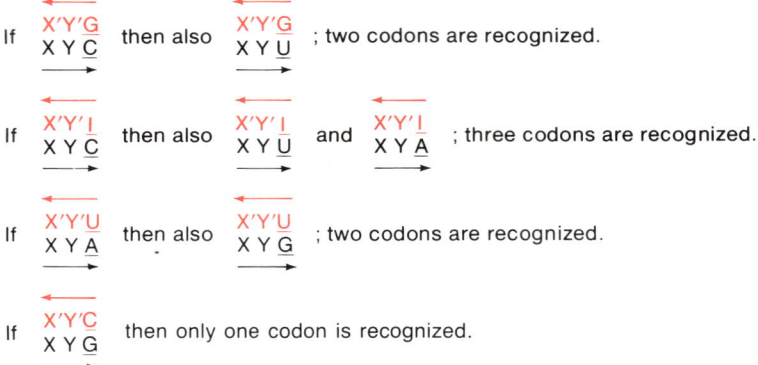

Figure 13-14 Synonyms resulting from wobble pairing. In each case the codon is black and the anticodon is red. Codons in the same horizontal row are synonyms. The arrows show the 5'→3' direction for both codon and anticodon. The bases of the "wobble pair"—that is, the third base of the codon and the first base of the anticodon—are underlined.

In the examples considered so far, most of the redundancy has been explained by the presence of the modified base inosine in the anticodon. This arrangement is true for yeast but not for all other organisms. In *E. coli* another modified base has recently been discovered in the wobble position of tRNA^{Val} (it probably exists in other organisms, too). This base, uridine-5-oxyacetic acid, can pair with U, A, and G. Another modified base in this position in some organisms is 2-thiouridine, which pairs only with A. Five other modified bases have also been observed in the first position of the anticodon; their pairing characteristics vary. Apparently, the first base in the anticodon can be modified to produce either less or greater base-pairing specificity.

The reader must not be led to the conclusion at this point that, because of the varying kinds of redundancy, the codon assignments vary from one organism to the next. This is certainly not the case (except, as will be described later, for mitochondria and *Mycoplasma*). The differences in the first anti-codon positions merely determine the number of different tRNA molecules corresponding to a particular amino acid.

Polypeptide synthesis is occasionally initiated with GUG, rather than the usual AUG. This use of GUG may indicate a slight wobble in the first, rather than the third, position of the codon in this special case. First-position wobble has also been observed in certain yeast strains that translate, to a limited extent, UAA and UAG as glutamine by wobble pairing with the normal glutamine anticodons UUG and CUG, respectively.

The reader is urged to refer constantly to Tables 13-1 and 13-2 while reviewing these paragraphs.

These findings can be generalized as follows:

Every tRNA molecule reads either one, two, or three codons, the number depending on whether the first base of the anticodon, which is the wobble base, is C (1 codon), U or G (2 codons), or a modified base (3 codons). The redundancy of the code is caused in part by wobble in the pairing of the third codon base.

The modified bases have probably evolved in the wobble position because they maximize the number of codons that can be read by a single tRNA molecule.

THE SPECIAL PROPERTIES OF THE PROKARYOTIC INITIATOR tRNA^{fMet}

The initiator tRNA molecule in prokaryotic organisms—tRNA^{fMet}—has several biological properties that distinguish it from all other types of tRNA molecules:

1. The tRNA is activated at the outset with an amino acid—methionine—that is subsequently further modified prior to binding to the ribosome. That is, tRNA^{fMet} is charged with methionine by methionyl-tRNA synthetase. (This same enzyme also charges tRNA^{Met} with methionine.) The methionine of charged tRNA^{fMet} is immediately recognized by another enzyme, tRNA methionyl transformylase, which transfers a formyl group from N-10-formyltetrahydrofolate (fTHF) to the NH_3^+ group of the methionine to form N-formylmethionine (fMet; Figure 13-15):

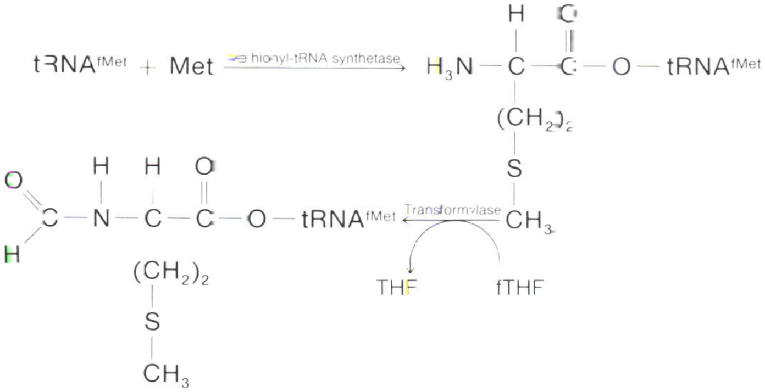

Transformylase does not catalyze formylation of methionyl-tRNA^{Met}; thus, there must be structural differences between tRNA^{Met} and tRNA^{fMet}

2. Ribosomes contain two tRNA binding sites, the A site and the P site. After several preliminary steps, a formylated, charged tRNA^{fMet} molecule enters the P site without prior occupation of the A site, whereas all

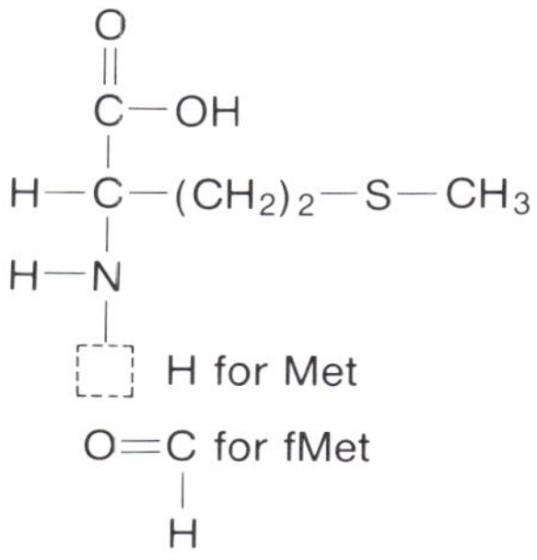

Figure 13-15 Comparison of the chemical structures of methionine and N-formyl-methionine.

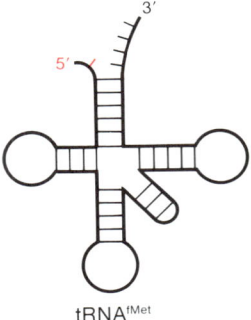

tRNA^{fMet}

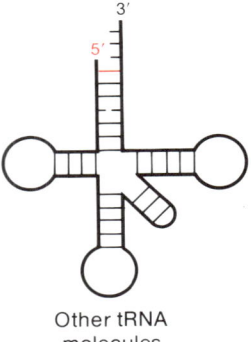

Other tRNA
molecules

Figure 13-16 A notable difference between *E. coli* tRNA^{fMet} and other *E. coli* tRNA molecules—that is, the 5′-terminal base is not hydrogen-bonded.

other aminoacyl-tRNA molecules bind initially to the A site and move to the P site *only after a peptide bond has been formed*. Thus, some structural feature (or more than one) must be present in tRNA^{fMet} that is absent in all other tRNA molecules. The most striking difference is evident from the base sequence—that is, in the tRNA^{fMet} molecules of all prokaryotes examined so far the 5′-terminal base is not hydrogen-bonded to the opposite base in the acceptor stem (Figure 13-16). An x-ray crystallographic analysis of the structure of *E. coli* tRNA^{fMet} has revealed the following. The overall structure of tRNA^{fMet} is very similar to other tRNA molecules. Nevertheless, a hydrogen bond, which in most tRNA molecules occurs between the uracil adjacent to the anticodon and a ribose in the anticodon loop, is absent in tRNA^{fMet}. This gives the anticodon loop a more open structure. There are also small differences in the acceptor stem and the DHU loop.

The codon-anticodon interaction of tRNA^{fMet} differs from that of other tRNA molecules that there may be some wobble in the *first* codon base. The major codon recognized by tRNA^{fMet} is AUG, but tRNA^{fMet} also interacts significantly with the valine codon GUG and weakly with the leucine codons UUG and CUG. Possibly this wobble results from the lack of the hydrogen bond in the anticodon loop, as just mentioned. The three codons—GUG, UUG, and CUG—are occasionally used as start codons and each is recognized by tRNA^{fMet}.

Eukaryotic initiator-tRNA molecules differ from the prokaryotic initiator molecule in several ways. For example,

1. Eukaryotic organisms produce both a normal tRNA^{Met} and an initiator tRNA, which is also charged with methionine, but the methionine is not formylated. In eukaryotes, *the first amino acid in a growing polypeptide chain is Met and not fMet*. The codon for both kinds of tRNA molecules in eukaryotes is AUG just as for prokaryotes.
2. The sequence AUC or AψC instead of TψC is present in the TψC loop.

The eukaryotic initiator tRNA molecules are like the prokaryotic tRNA^{fMet} in lacking hydrogen-bonding of the 5′-terminal base.

At this point, the reader might be perplexed by two questions: (1) How is an AUG start codon distinguished from an internal AUG codon for Met? This question will be answered when initiation of protein synthesis is discussed in Chapter 14. (2) How are polypeptides successfully synthesized on templates such as poly(U) and poly(U,C), which clearly do not possess an AUG codon? The answer in this case concerns the Mg^{2+} concentration in the reaction mixture that is used. As will be seen in Chapter 14, in the usual sequence of events in protein synthesis a tRNA^{fMet} molecule interacts with a mRNA molecule and a free 30S ribosome, and then a 50S ribosome joins the complex. When the Mg^{2+} concentration is 0.005 to 0.01 M, the joining of the 30S and 50S subunits occurs only if a start codon is present. However, the poly(U) system uses 0.02 M MgCl$_2$; at this concentration of Mg^{2+} free

Table 13-3 Number of tRNA genes in various organisms or particles

Organism	Number of tRNA genes[*]
Mitochondria	22
E. coli	62
Yeast	300
Drosophila (fruit fly)	500
Xenopus (toad)	8000

[*]These are numbers per haploid set of chromosomes. The E. coli number is a minimum value. The others are approximate.

30S and 50S ribosomal subunits join together spontaneously, and the resulting particle is able to bind mRNA and any tRNA molecule whose anticodon matches a codon in the mRNA. Thus, in 0.02 M $MgCl_2$ the specificity is lost and initiation can occur at any codon.

TRANSFER RNA GENES

Because of wobble, only 32 tRNA molecules are needed for the 61 different codons; however, the actual number of tRNA genes is much greater, as shown in Table 13-3. There are two reasons for this great multiplicity of tRNA genes: the existence of (1) minor tRNA molecules (an infrequently occurring form of a major tRNA species having the same anticodon but difference base sequences elsewhere) and (2) duplicate copies of a particular gene. For instance, in E. coli there are two identical copies of the $tRNA_I^{Tyr}$ gene and both are in a single transcription unit. Even more striking is $tRNA_I^{Leu}$ of E. coli, for which there are four (or possibly five) copies in one transcription unit and at least one more copy in another transcription unit. However, duplication is not always the case, because $tRNA^{Trp}$ is transcribed from only one gene. In the next section, we will see that having multiple tRNA genes provides a means for elimination of deleterious effects of certain mutations, thereby providing another tool for evolution.

SUPPRESSORS

In Chapter 11 conditional mutations were described. These are mutations that yield the wild-type phenotype in certain circumstances (genetic background, temperature, etc.) and a mutant phenotype in other conditions. Of particular interest in this section are the **suppressor-sensitive mutations,** which behave like the wild type when a suppressor molecule is present. An important example of this phenomenon is a phage mutant that can grow in one strain of bacteria (denoted Sup⁺) but fails to grow in other strains. We will see that suppression is based on changes in the decoding system.

Suppressor–sensitive mutations are of two main types: **nonsense** or **chain termination** mutations and **missense** or **amino acid substitution** mutations (Chapter 11). The nonsense type will be considered first.

Genetic Detection of a Nonsense Suppressor

Chain termination mutations are very common, for they can arise in many ways. For example, a single base change in any of the codons AAG, CAG, GAG, UCG, UUG, UGG, UAC, and UAU (not UAA) can give rise to the amber chain termination codon UAG. If such a mutation occurs within a gene, a mutant protein with little or no function will result because there is no tRNA molecule whose anticodon is complementary to UAG. Thus, only a fragment of the wild-type protein is produced and this will usually fail to function unless the mutation is very near the carboxyl terminus of the protein.

In certain bacterial mutants a chain termination mutation does not cause termination; this can be seen in the following simple system. Consider a bacterium B that has mutated such that chain growth terminates at a Tyr site in the *lac* gene, making the bacterium genotypically and phenotypically Lac⁻ (Figure 13-17). We now seek a Lac⁺ revertant of B by mutagenizing

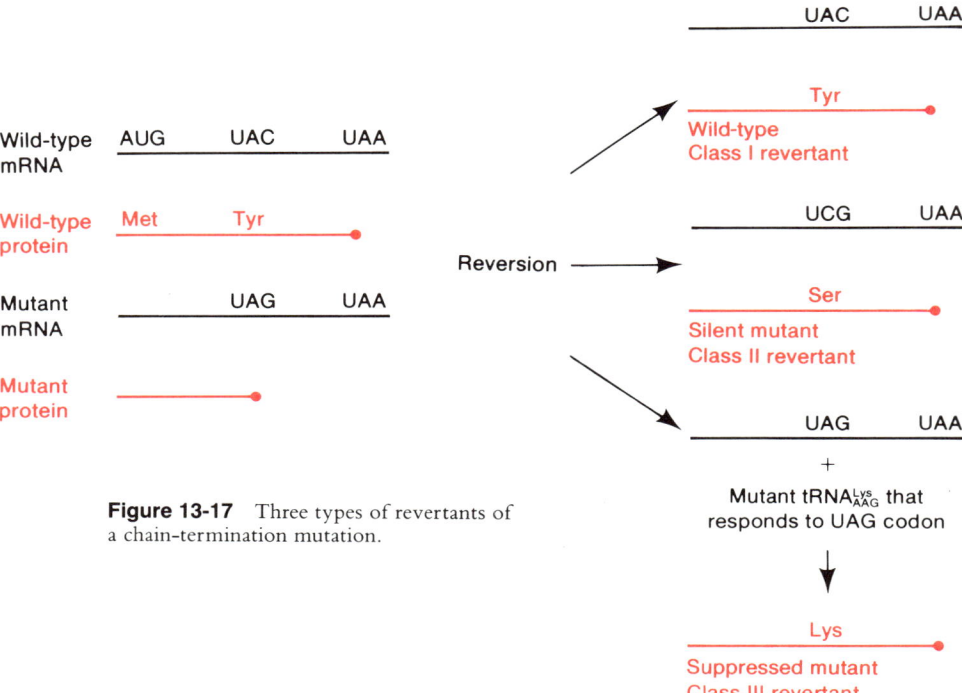

Figure 13-17   Three types of revertants of a chain-termination mutation.

a population of B, allowing several generations of growth, and plating the culture on agar selective for Lac$^+$ colonies. Three classes of revertants will be found, as shown in the figure. In class I, a rare class, the chain termination mutation is reversed by a base substitution mutation that converts UAG back to UAC. In this class the complete protein chain will be present, the tyrosine will be restored at the correct position, and the protein will have the wild-type amino acid sequence. In class II a new amino acid, serine, is present in the now complete protein chain, and the base sequence is likewise altered from the original wild-type sequence. In this case the chain termination mutation has been converted to a silent mutation that has a wild-type or near wild-type phenotype. In the figure this has happened as a result of a base substitution mutation converting UAG to UCG, the serine codon, and the substitution of serine for tyrosine does not markedly alter function; that is, serine is an acceptable amino acid at this position in the protein. The class III revertants are those of interest for this discussion. These revertants produce a complete polypeptide chain usually by supplying some other amino acid at the mutant site without in any way altering the original mutant base sequence—that is, the codon remains UAG. Class III revertants also typically have another property: a mutant phage with a chain termination mutation in an essential protein can often form a plaque on a class III revertant. The revertant cell has gained the ability to ignore (or **suppress,** to use the genetic term) UAG-type chain termination mutations. Thus, somehow the mutant UAG codon is being translated into an amino acid.

Nonsense Suppressor tRNA

The molecular explanation for the suppression phenomenon described in the foregoing paragraphs is that the class III revertant, which is called a suppressor-containing bacterium or a **suppressor mutant,** contains an altered tRNA molecule—one that has the anticodon CUA, which can pair with UAG. This mutant tRNA is called a **suppressor tRNA.**

How can such a mutant tRNA molecule arise? Since a tRNA molecule is simply the product of a tRNA gene, these mutants occur in exactly the same way as any other mutant—by an error in replication or repair. It is important to realize that a suppressor tRNA does not arise in response to a chain termination mutation; rather, it forms spontaneously or as a result of mutagenesis, and it has been selected (often in the laboratory) from a population of bacteria containing both it and a chain termination mutation by growth in a medium in which only cultures of cells having a wild-type phenotype can form a colony.

What has been mutated in the production of a suppressor tRNA? Clearly it must be a normal tRNA gene. Thus, in the example given previously, a tRNA^{Lys} molecule whose anticodon is CUU, has been altered to have the anticodon CUA, which can hydrogen-bond to the codon UAG.

Inasmuch as a single base change is sufficient to alter the complementarity of an anticodon and a codon, there are (at most) eight tRNA molecules having a complementary anticodon that, with a single changed base, will also suppress a UAG codon. Thus, the following amino acids (whose codons are also indicated) can be put at the site of a chain termination codon: Lys (AAG), Gln (CAG), Glu (GAG), Ser (UCG), Trp (UGG), Leu (UUG), and Tyr (UAC and UAU). Note that these are the same amino acid codons that can be altered by mutation to form a UAG site.

Suppressors also exist for chain termination mutants of the ochre (UAA) and opal (UGA) type. These too are mutant tRNA molecules whose anticodons are altered by a single base change.

In conventional notation, suppressors are given the genetic symbol *sup* followed by a number (or occasionally a letter) that distinguishes one suppressor from another. A cell lacking a suppressor is designated *sup0* or *sup*⁻.

Biochemical Features of Nonsense Suppression

Several features of nonsense suppression should be recognized:

1. Not every UAG suppressor can restore a functional protein by suppressing each UAG chain termination mutation. Thus, a UAG codon produced by mutating the leucine UUG codon might be suppressed by a suppressor tRNA that inserts tyrosine or serine, but might not be able to tolerate a substitution by the electrically charged amino acids lysine or glutamic acid.

2. Suppression may be incomplete, for two reasons. First, replacement of leucine by tyrosine or serine may restore activity sufficient for survival but the activity may be subnormal because the amino acids are not hydrophobic. Second, the nonsense codon is usually translated as sense by a suppressor tRNA only part of the time (that is, chain termination is still fairly frequent), so the concentration of the protein produced is lower than that of the wild-type protein.

3. An ochre suppressor also suppresses the amber codon, but an amber suppressor does not always suppress an ochre codon. This is a result of wobble. An ochre (UAA) suppressor must have the anticodon UUA. Because G·U is a wobble pair, interaction of UUA with UAG (amber) is possible with G·U in the wobble (third) position. An amber suppressor may have the anticodons CUA or UUA. The UUA anticodon can pair with the ochre UAA but the CUA cannot (see Table 13-2). Thus, ochre mutations are not suppressed by amber suppressors.

4. A cell can survive the presence of a suppressor only if the cell also contains two or more copies of the tRNA gene. Clearly if a tRNA^Ser mol-

ecule that reads the UCG codon is mutated, then the UCG can no longer be read as a sense codon. This will lead to chain termination wherever UCG occurs, and a cell harboring such a mutant tRNA molecule will fail to terminate virtually every protein made by the cell. However, as mentioned earlier, there are multiple copies of most tRNA molecules and minor tRNA molecules having the same anticodons as the major molecules. Thus, in any living cell containing a suppressor tRNA, there must always be an additional copy of a wild-type tRNA that can function in normal translation. The exception to this generalization is the tRNA^{Trp} molecule of *E. coli*. There is only a single copy of this tRNA gene; thus an amber suppressor that inserts tryptophan is not possible in an ordinary bacterium.

5. Some tRNA molecules can mutate to yield either an amber or an ochre suppressor. For example, tRNA^{Tyr} which has the anticodon QUA (Q is a modified base called queosine), can mutate to CUA to be an amber (UAG) suppressor and to UUA to be an ochre (UAA) suppressor.

6. Although suppression of chain termination is typically a result of an anticodon mutation, it can also result from an alteration elsewhere in the molecule. One example is the opal *sup9* suppressor, in which a base change *in the DHU stem* somehow alters the specificity of codon-anticodon binding. This suppressor is an especially interesting one since it is an alteration of the tRNA^{Trp} molecule of which there is only a single copy. Clearly, some mechanism is needed to allow translation of the tryptophan codon if a cell containing *sup9* is to survive. In fact, the *sup9* suppressor retains the ability to respond to the tryptophan codon in addition to the UGA codon. Furthermore, it does not recognize the UGA codon with high efficiency (it is a so-called **weak suppressor**) but at a sufficient rate to allow completion of the prematurely terminated protein at a level adequate for survival. Why the *sup9* suppressor is able to respond to both UGC and UGA with the anticodon CCA is not known. Presumably, the alteration in the DHU stem affects the orientation of the tRNA molecule on the ribosome and produces ambiguous pairing between the codon and the anticodon.

Normal Termination in the Presence of a Suppressor tRNA

If a cell contains an amber suppressor, then proteins terminated by a single UAG codon will not be completed and the existence of a suppressor tRNA should be lethal. There are two ways that this problem can be avoided:

1. Protein factors active in termination (Chapter 14) respond to chain termination codons even though a tRNA molecule that recognizes the codon is present, that is, suppression is weak. For example, if the probability of

recognition is only x percent, then only x percent of the prematurely terminated mutant proteins would be completed and only x percent of normally terminated proteins would be ruined by lack of termination.

2. Normal chain termination may utilize pairs of distinct termination codons such as the sequence UAG–UAA. Thus, the existence of a UAG suppressor would not prevent termination of a doubly terminated protein.

It is likely that both of these mechanisms are responsible for the existence of viable suppressor-containing organisms. Suppression of potentially lethal UAG and UGA mutations is very efficient; termination is prevented more than 50 percent of the time. (Note that the *activity* of the suppressed mutant may not always be as high as 50 percent; this is because the particular amino acid insertion is not always well tolerated by the protein rather than because of inefficient suppression of chain termination.) A partial explanation for the viability of organisms containing UAG and UGA suppressors is that when either UAG or UGA is used as a natural stop signal, it usually occurs adjacent to or 3–6 base pairs away from a second and different stop codon in the same reading frame.

Chain termination is accomplished most commonly by a single UAA codon. This is consistent with the observation that UAA suppressors are typically very inefficient, preventing termination of mutationally induced UAA codons between 1 and 5 percent of the time. Furthermore, cells containing UAA suppressors are generally unhealthy, growing rather slowly compared to cells lacking any suppressor or containing either a UAG or UGA suppressor. Presumably, several percent of the normal proteins, perhaps those needed critically, are damaged by not being terminated at the right time. The reader can see that an understanding of how cells survive having suppressors is incomplete. Presumably this reflects a lack of understanding of how normal chain termination occurs. It is likely that unknown factors are active in normal chain termination; some of these factors may be the secondary structure of the mRNA or special ribosomal features that reduce the recognition of natural termination signals by suppressors. Some support for the existence of such factors is the following: there are numerous instances of leaky chain termination mutations—that is, a UAG codon within a gene and in the correct reading frame, at which termination does not always occur. In lab jargon, it is said that **read-through** occurs past the mutant site. Furthermore, mutations in ribosomal proteins sometimes increase the efficiency of termination of leaky mutations.

Suppression of Missense Mutations

Suppression can also occur for missense mutations. For example, a protein in which valine (nonpolar) has been mutated to aspartic acid (polar), resulting in loss of activity, can be restored to the wild-type phenotype by a missense suppressor that substitutes alanine (nonpolar) for aspartic acid. Such a substitution can occur in three ways: (1) a mutant tRNA molecule may recognize two codons, possibly by a change in the anticodon loop; (2) a mutant tRNA molecule can be recognized by a noncognate aminoacyl synthetase and be misacylated; and (3) a mutant synthetase can charge a noncognate tRNA molecule. Numerous examples of the first two classes of suppressors are known, but only one altered synthetase has been observed. Suppression of missense mutations is necessarily inefficient. If a suppressor that substitutes alanine for aspartic acid worked with, say, 20 percent efficiency, then in virtually every protein molecule synthesized by the cell at least one aspartic acid would be replaced, which is a situation that a cell could not possibly survive. The usual frequency of missense suppression is about 1 percent. The missense suppressor that inserts glycine at an AGA codon is 25 percent efficient. However, the AGA codon does not occur very frequently, so damage to normal proteins is minimal. In this way a small amount of a functional, essential protein is made and thereby a mutant cell is able to survive. However, missense suppression still introduces a significant number of defective proteins of all other types and, as a result, a cell carrying a missense suppressor usually grows slowly and is generally unhealthy. The principle that applies to these cells is that it is better to be sick than dead.

Frameshift Suppressors

When a frameshift mutation occurs—for example, a base addition—all downstream codons are read incorrectly and the entire amino acid sequence following the mutation is incorrect. There is no way that either a missense or a nonsense suppressor could allow synthesis of an active protein. Restoration of activity is possible only by recreating the original reading frame. However, in bacteria three types of spontaneous revertants of frameshift mutations occur (Figure 13-18). The most straightforward type is that in which the added base is simply removed. A second type is an intragenic second-site suppressor in which either a nearby base is removed or two extra bases are added. In both cases, the reading frame is restored and as long as the additional amino acids between the mutant and revertant sites can be tolerated, protein activity returns. Sometimes this type of reversion requires the presence of a chain termination suppressor (Figure 13-18(a), revertant II). The third type of reversion is produced by a true intergenic suppressor, some of which are mutant tRNA molecules having a four-letter anticodon. An example of such a suppressor is a mutant tRNA^{Gly} in which the anticodon loop possesses an extra C—the anticodon contains the sequence CCCC. It is possible that this allows pairing with a sequence of four G's, though other explanations have also been given. For instance, it has been proposed that the additional base is not engaged in pairing but is merely a spacer that forces the next codon read to be displaced by one base.

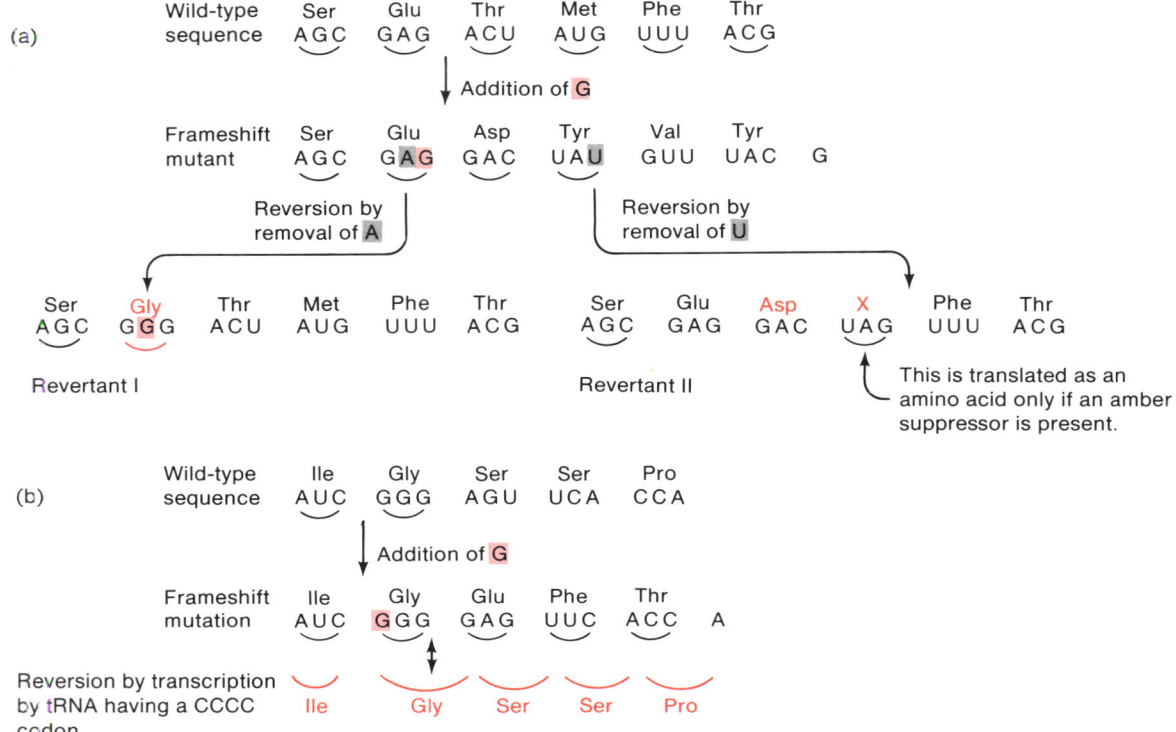

Figure 13-18 Reversion of frameshift mutants. (a) Reversion by eliminating a base, to restore the reading frame. In revertant I there is a Gly (red) where there is a Glu in the wild-type protein. Revertant II is functional only if the UAG codon is read by an amber suppressor; amino acids Asp and X replace Thr and Met, respectively. (b) Reversion by means of a suppressor that can read a four-base unit.

SELECTION OF THE CORRECT AUG CODON FOR INITIATION

The sequence AUG serves both as an initiation codon and the codon for methionine. Since methionine occurs within protein chains, there must be some signal in the base sequence of the mRNA that identifies particular AUG triplets as start codons. Base-sequencing of numerous genes has shown that in prokaryotes a particular sequence provides the signal and that in eukaryotes location of an AUG with respect to the start of the mRNA is usually the critical factor.

The Initiation Signal of a Prokaryotic mRNA

We will see in Chapter 14 that in the presence of charged tRNA^{fMet} and several proteins required for initiation, a ribosome and a mRNA molecule

5′ AAUCUU**GGAGG**CUUUUUUAUGGUUCGUUG■ 3′ Phage φX174 gene-*A* protein

5′ UAACU**AAGGA**UCAAAUGCAUGUCUAAGAGA 3′ Phage Qβ replicase

5′ UCCU**AGGAGGU**UUGACCUAUGCGAGCUUUU 3′ Phage R17 gene-*A* protein

5′ AUGUAC**UAAGGAGGU**UGUAUGGAACAACGC 3′ Phage λ gene-*cro* protein

Pairs with   **Pairs with**
16S RNA   **tRNA^fMet**

Figure 13-19 Sequences of initiation regions for protein synthesis in four phage mRNA molecules.

form a stable complex. If no other tRNA molecules are present, protein synthesis is stalled at this initiation stage. If the complex is treated with an RNase that hydrolyzes RNA only if it is free and single-stranded, all of the RNA is hydrolyzed except for that which is bound to the ribosome. This protected fragment can be isolated and its base sequence determined. This type of experiment has been performed with hundreds of prokaryotic mRNA molecules encoding proteins of known amino acid sequences (Figure 13-19). The protected RNA fragment is usually 30–40 nucleotides long and always contains an AUG (or, infrequently, a GUG) codon. This codon is known in each case to be a start codon because the adjacent bases (in the 3′ direction) constitute the codons for the initial amino acids of the proteins. The AUG is usually 20–30 bases from the 5′ end, so at least some of the preceding sequence (the **leader**) is recovered. The most notable feature of the leader region is the presence of a sequence whose consensus is

5′-AGGAGGU-3′

4–7 bases ahead of the initiator triplet. This sequence defines the start codon by providing a binding site for the ribosome and is called the **Shine-Dalgarno sequence,** after the workers who first recognized its significance. The importance of the Shine-Dalgarno sequence becomes evident when the 3′-terminal region of the 16S rRNA of several bacteria, a sequence that is virtually the same in all bacterial species, is examined. This region contains the sequence

3′-AUUCCUCCA-5′

which can base-pair with the Shine-Dalgarno sequence (Figure 13-20). This

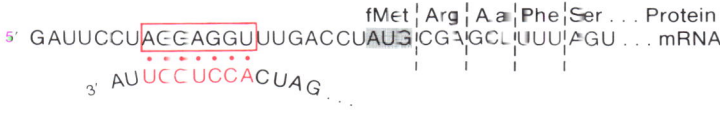

Figure 13-20 Base-pairing between the purine-rich region (in box) in the initiator region of mRNA and the complementary region (red) near the 3′ end of rRNA. The AUG start codon is shaded.

pairing brings the AUG codon to a start site on the ribosome. The evidence, other than the sequence homology just described, that supports this conclusion is the following:

1. An intact 3′-OH terminus of 16S RNA is essential for initiation of protein synthesis.

2. Certain ribosomal proteins (soon to be described) involved in initiation are located in the ribosome near the 3′ terminus of the 16S RNA molecule; antibiotics that inhibit initiation bind to protein components right next to 3′-terminal sequences of 16S RNA.

3. Mutations that reduce the efficiency of translation (the number of protein molecules made per unit time) have been isolated and many of these are base changes in the Shine–Dalgarno sequence.

4. Direct binding of the initiator region of an mRNA molecule and the 3′-terminal region of the 16S RNA molecule has been detected. This was done in the following experiment. A 30-nucleotide fragment of the mRNA molecule encoding the initiator region for the synthesis of the A protein of *E. coli* phage R17 was isolated and allowed to form an initiation complex with *E. coli* ribosomes. A substance called colicin E3, which makes a single endonucleolytic cut 59 nucleotides from the 3′ terminus of 16S RNA, was then added to the complex. The mixture was treated with a detergent that dissociates ribosomes and releases RNA. The RNA that was released is a mRNA–rRNA hybrid molecule (Figure 13-21). In this hybrid the mRNA and the rRNA fragments were held together by hydrogen bonds. The paired

Figure 13-21　Binding of mRNA to a complementary sequence in 16S rRNA. The AUG start codon is underlined.

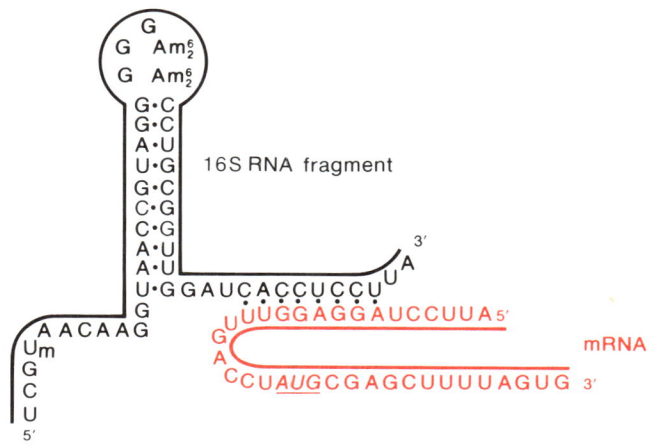

sequences are those shown in Figure 13-20. Although the Shine-Dalgarno sequence is essential for recognition of an AUG start codon, it is not sufficient: several examples have been observed of Shine-Dalgarno sequences adjacent to AUG triplets that are not used as codons. The additional features of the mRNA base sequence needed for initiation are not yet known.

A major factor in the frequency of initiation may be the strength of mRNA-16S RNA base-pairing. This is suggested by a study of the efficiency by which the *E. coli* phage R17 *A* protein is synthesized with ribosomes obtained from other bacterial species. The 3′-terminal sequences of the 16S RNA molecules of six bacterial species were determined and it was observed that the length and strength (number of G C versus A U pairs) of the base-paired regions are not the same. The amount of *A* protein synthesized with those different ribosomes correlates with the thermal stability of the base-paired region.

Repeated Initiation in Polycistronic mRNA

Many prokaryotic mRNA molecules are polycistronic: they contain sequences specifying the synthesis of several proteins. A polycistronic mRNA molecule must possess a series of start and stop codons. If t encodes three proteins the minimal requirement would be the sequence

Start, protein 1, stop–start, protein 2, stop–start, protein 3, stop.

Actually, such an mRNA molecule is probably never so simple in that the leader sequence preceding the first start signal may be several hundred bases long and there is usually a sequence called a **spacer** 5–20 bases long between one stop codon and the next start codon. Thus the structure of a tricistronic mRNA is more typically that shown in Figure 13-22. We have already mentioned that stop codons sometimes occur in pairs. However, in addition to this pairing that occurs in the same reading frame, there may be one or more out-of-phase stop codons in the spacer. For instance,

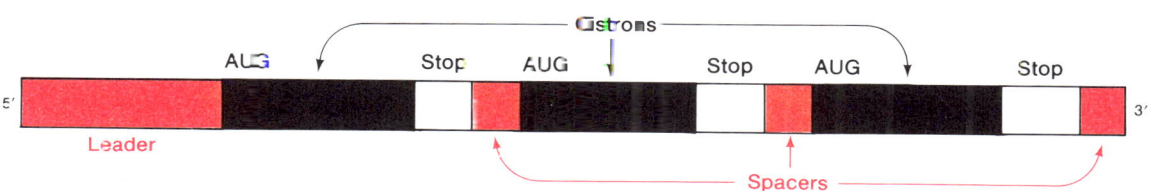

Figure 13-22 Arrangement of cistrons and untranslated regions (red) in a typical polycistronic mRNA molecule.

Figure 13-23   A portion of a typical spacer sequence. Usually, the sequence is longer than that shown.

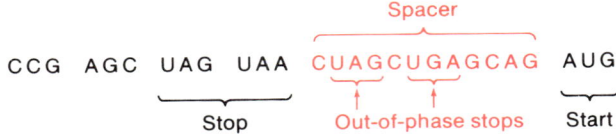

the type of sequence shown in Figure 13-23 might occur. It is thought that the out-of-phase stops provide insurance against two possibilities—(1) on occasion the reading frame may "slip," or (2) a frameshift mutation might alter the reading frame. Having stop codons in all reading frames in the spacer region allows the translational system to have a fresh start at the following AUG.

The mechanism for initiating synthesis of the first protein molecule in a polycistronic mRNA is no different from that in a monocistronic mRNA. However, in a polycistronic mRNA, if the second protein is to be made, protein synthesis must start again after the stop codon of the first protein is reached. Usually, either the start codon of the second protein is so near the preceding stop codon that synthesis reinitiates before the ribosome and the mRNA dissociate or a second initiation signal locks the protein-synthesizing system onto the second cistron. In general, the number of copies of each protein synthesized from a polycistronic mRNA is not the same. The main reasons are the variability of the strength of binding of the ribosome to the initiation sites for each cistron, and the distance between a stop codon and the next start codon.

Initiator tRNA and Ribosome Binding Sites in Eukaryotes

The basic pattern of protein synthesis in eukaryotes is quite similar to that for prokaryotes and in Chapter 14 it will be seen that the differences are mostly in detail. However, the mechanism by which the correct reading frame in eukaryotic mRNA is selected is quite different from that which has been described for prokaryotes. Interestingly, the sequence of the 3′ terminus of mammalian 18S rRNA (which corresponds functionally to the 16S rRNA of prokaryotes) has virtually the same base sequence as the 16S rRNA, except that the sequence CCUCC (part of the sequence complementary to the Shine-Dalgarno sequence) is absent (Figure 13-24). The following list describes the essential features of initiation in eukaryotes:

1. In eukaryotes the initiating tRNA molecule carries Met and not fMet. As in prokaryotes a true tRNA^{Met} molecule and an initiator tRNA^{Met} molecule exist; in eukaryotes the latter is denoted tRNA$_{\text{f}}^{\text{Met}}$. This tRNA responds only to the AUG codon and never to GUG or other rare start codons.

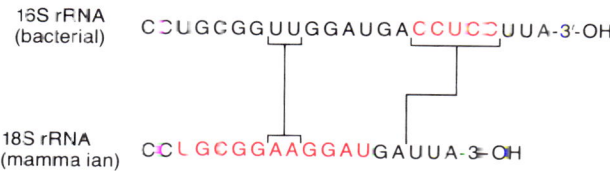

16S rRNA
(bacterial)

CCUGCGGUUGGAUGACCUCCUUA-3'-OH

18S rRNA
(mammalian)

CCUGCGGAAGGAUGAUUA-3-OH

Figure 13-24 The nearly common 3'-terminal sequences of bacterial 16S rRNA and 18S mammalian rRNA. Red bases are involved in binding mRNA. Differences are indicated by the vertical lines.

2. In eukaryotes most mRNA molecules are monocistronic. Furthermore, eukaryotic ribosomes cannot translate all of the cistrons of a prokaryotic polycistronic mRNA. If bacterial polycistronic mRNA is added to an *in vitro* protein-synthesizing system that uses eukaryotic ribosomes, only the cistron nearest to the 5' terminus is translated (Figure 13-25). The lack of polycistronic mRNA in eukaryotes agrees with genetic mapping results, which indicate that related gene products are usually in different parts of the same chromosome or on different chromosomes. This is quite different from prokaryotes, in which related genes are often adjacent.

3. In eukaryotic mRNA molecules the initiator AUG codon is, with only few exceptions (see item 4 below), the AUG triplet nearest to the 5' terminus of the mRNA molecules. The AUG is typically 50–100 bases from the 5' end and part of the consensus sequence **PuXXAUGG,** in which X can be any base. If a new AUG triplet is introduced (either by mutation or by genetic engineering) nearer to the 5' terminus than the wild-type initiator is, initiation begins at the new AUG codon (Figure 13-26). Furthermore, if the normal AUG codon is eliminated, the next AUG sequence downstream (in the 3' direction from the natural AUG start codon) serves as an initiator codon.

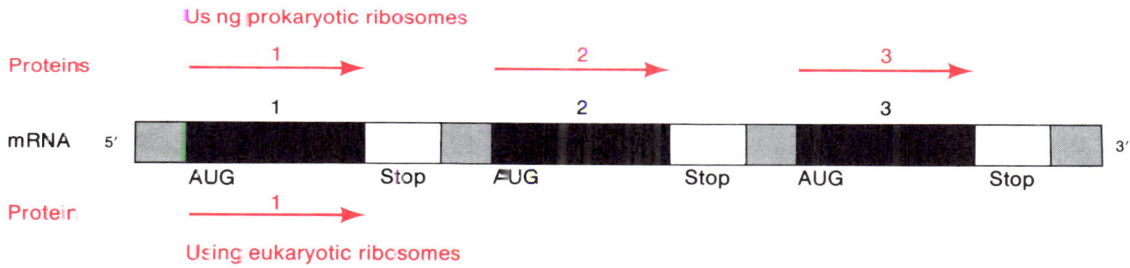

Figure 13-25 Difference in the products translated from a tricistronic mRNA molecule by the ribosomes of prokaryotes and eukaryotes. The prokaryotic ribosome translates all of the cistrons but the eukaryotic ribosome translates only one cistron—the one nearest the 5' terminus of the mRNA. Translated sequences are in black, stop codons are white, and the leader and spacers are shaded.

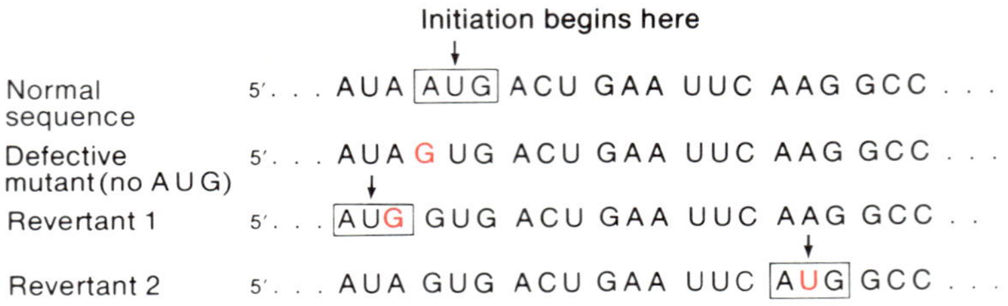

Initiation begins here
↓

Normal sequence	5'... AUA $\boxed{\text{AUG}}$ ACU GAA UUC AAG GCC ...
Defective mutant (no AUG)	5'... AUA G UG ACU GAA UUC AAG GCC ...
Revertant 1	5'... $\boxed{\text{AUG}}$ GUG ACU GAA UUC AAG GCC ..
Revertant 2	5'... AUA GUG ACU GAA UUC $\boxed{\text{AUG}}$ GCC ...

Figure 13-26 A segment of the yeast iso-1-cytochrome c mRNA. Initiation normally begins at the arrow. The defective mutant has lost the normal AUG start codon and cannot initiate. Revertants 1 and 2 of the defective mutant have each gained a new AUG codon and can initiate from either of these codons. Base changes are shown in red; start codons are boxed.

4. The rate of initiation (the number of initiation events per unit time) varies from one gene to the next. This variation depends in part on the sequence surrounding the AUG. For example, PuXXAUGG is the most efficient one and XXXAUGPy is inactive. The inactivity or weak activity of certain sequences explains the fact that in some mRNAs the 5'-proximal AUG is not used for initiation; in these cases the second AUG is used. In Chapter 23 it will be seen that in poliovirus the seventh or eighth AUG is used.

5. If a new AUG is introduced ahead of the natural AUG and if a stop codon is present fairly near the natural AUG, initiation at the natural AUG occurs, indicating that the ribosome remains attached to the mRNA for some time (not yet determined). If the distance following the stop codon is large, the natural AUG will not be used. If a natural AUG is eliminated by mutation, initiation does not usually occur if the second AUG is far from the 5' end.

6. A few viral RNAs have two AUG codons and both are recognized to yield alternative proteins. The 5'-proximal AUG is still preferred but is typically not embedded in the consensus sequence given in item 3. This is a second example (see item 4) of ribosomes occasionally missing the first AUG if the optimal adjacent bases are lacking.

These and many related results have led to the **scanning hypothesis** for initiation in eukaryotes (Figure 13-27). The idea is that a ribosomal subunit attaches to the mRNA at or near the 5' end of a mRNA molecule (most likely at the 5' cap) and then drifts along the molecule in the 5'→3' direction until it encounters an AUG sequence. At this point a translation-initiation complex (to be described in Chapter 14) forms and the reading frame is established. When a stop codon is reached, the ribosome leaves the mRNA

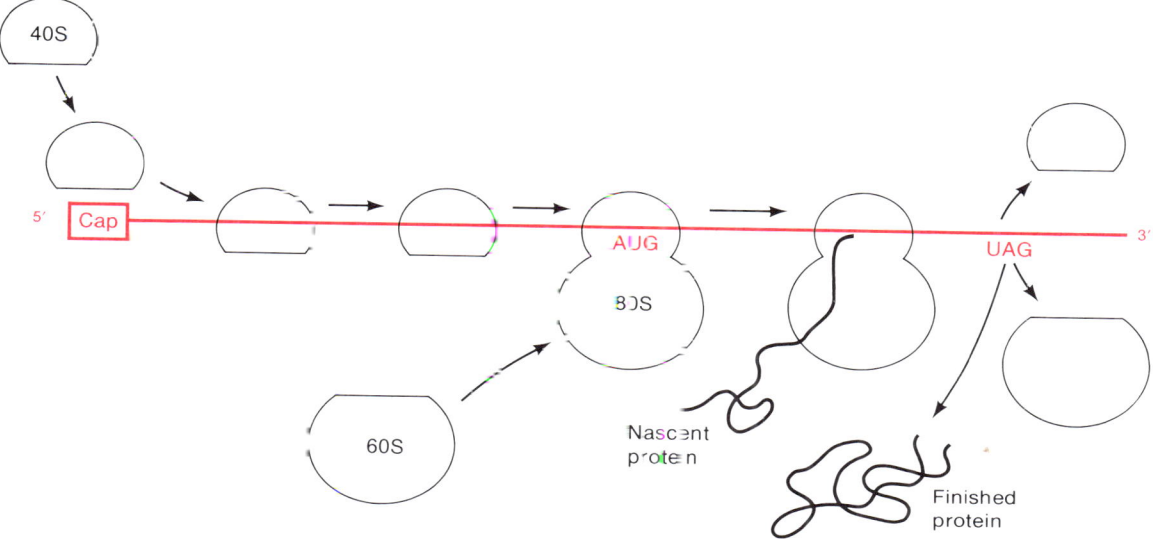

Figure 13-27 The scanning hypothesis. A 40S ribosome subunit attaches at or near the cap and drifts in the 3' direction. It stops at an AUG codon and is joined by a 60S ribo-somal subunit to form an active 80S ribo-some. The 80S ribosome dissociates when reaching a stop codon.

shortly afterwards, which explains why a natural AUG following a natural stop codon does not serve for initiation.

The scanning hypothesis is supported by the fact that certain circular eukaryotic mRNA molecules (circularized in the laboratory), which can be translated *in vitro* with *E. coli* ribosomes, are not translated if eukaryotic ribosomes are used; if, however, the RNA is nicked (to linearize the molecule), *in vitro* translation with eukaryotic ribosomes occurs normally. Other experiments also support the hypothesis.

OVERLAPPING GENES

In all that has been said so far about coding and signal recognition, an implicit assumption has been that the mRNA molecule is scanned for start signals to establish the reading frame and that reading then proceeds in a single direction within the reading frame. The idea that several reading frames might exist in a single segment was supposedly eliminated in the early 1940s when it was observed that in *Neurospora crassa* mutations never affect more than one gene. This view, called the one-gene–one-enzyme hypothesis, has through the years been supported by an enormous number of observations and justified by noting that, if there were overlapping reading frames, several constraints would be placed on the amino acid sequences of two protein translated from the same portion of mRNA. However, because the code is highly redundant, the constraints are actually not so rigid. For instance, each of the base sequences shown below

I.   UCU CCU GCA AUU CGU A

II.   UCC CCA GCG AUC CGC A

III.   UCA CCG GCC AUA CGA A

specifies the amino acid sequence Ser-Pro-Ala-Ile-Arg, yet in the reading frame in which the shaded C is the first codon position, the amino acid sequences are

I.   Leu-Leu-Gln-Phe-Val
II.   Pro-Gln-Arg-Ser-Ala
III.   His-Arg-Pro-Tyr-Glu

Further inspection of the codon assignments shows that there are several hundred sequences of five amino acids that can be obtained by a one-base displacement of a sequence that otherwise translate as Ser-Pro-Ala-Ile-Arg. Thus, the constraint resulting from multiple reading frames is not as great as was initially thought. Clearly, if there are three reading frames, a single DNA segment would be utilized with maximal efficiency. The disadvantage of such a system is that evolution may be slowed, because single-base-change mutations would be deleterious more often than if there were a unique reading frame. Nonetheless, some organisms have evolved that use overlapping reading frames, namely, small animal viruses and the smallest phages.

E. coli phage φX174 contains a single strand of DNA consisting of 5386 nucleotides of known base sequence. If a single reading frame were used, at most 1795 amino acids could be encoded in the sequence and, if 110 is taken as the molecular weight of an "average" amino acid, at most 197,000 molecular weight units of protein could be made. However, the phage makes eleven proteins with a combined molecular weight of 262,000. This paradox is explained by translation occurring in several reading frames from three mRNA molecules (Figure 13-28). For example, the sequence for protein B is contained totally in the sequence for protein A but translated in a different reading frame. Similarly, the protein E sequence is totally within the sequence for protein D. Protein K is initiated near the end of gene A, includes the base sequence of B, and terminates in gene C; synthesis is not in phase with either gene A or gene C. Of note is protein A′ (also called A*), which is formed by reinitiation within gene A and in the same reading frame; it terminates at the stop codon of gene A, so the amino acid sequence of A′ is identical to a segment of protein A. In total, five different proteins obtain some or all of their primary structure from shared base sequences in φX174. This phenomenon, known as **overlapping genes,** has been observed in the related phage G4 and in the small animal virus SV40. It should be realized that in prokaryotes the single structural feature responsible for gene overlap is the location of each AUG initiation sequence.

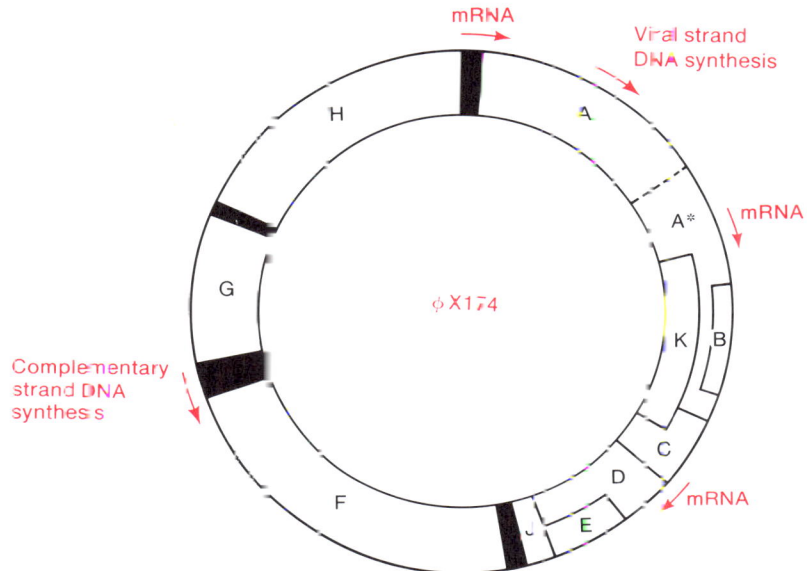

Figure 13-28 Genetic map of phage φX174 showing the start points for mRNA synthesis and the boundaries of the individual protein molecules. The solid regions are spacers.

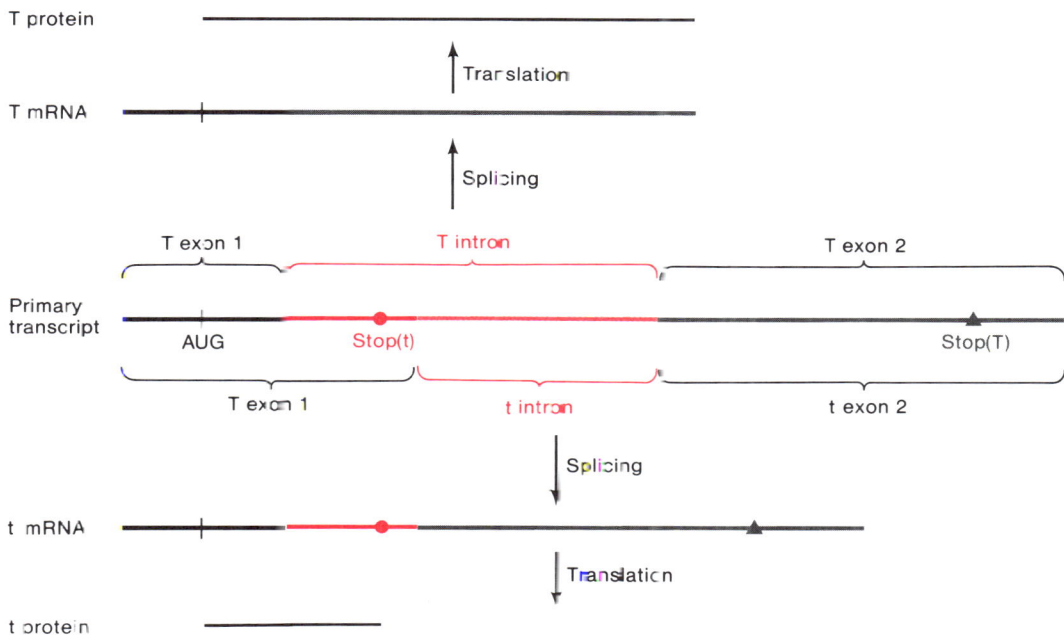

Figure 13-29 Alternate splicing patterns of the T-t primary transcript to yield either t protein or T protein. The T intron consists of the entire t intron and part of t exon 1.

Gene overlap occurs in certain animal viruses, but because of the mechanism of AUG selection in eukaryotic system, how it occurs is different from that in prokaryotes: in eukaryotes, alternative splicing patterns are used. A well-studied example is a segment of the DNA of the virus SV40 that encodes two proteins—T (called "big" T) and t ("little" t) antigens. The two proteins have the same N-terminal amino acid sequences but different C-terminal sequences (Figure 13-29). The T and t introns have the same right terminus but different left termini, so the solid red segment, which contains a stop codon, is part of the t exon 1 but excised when T mRNA is formed. Thus, stop codon 1, terminator t, and stop codon 2 terminate T.

THE GENETIC CODE OF MITOCHONDRIA

Mitochondria are small particles contained in the cells of all multicellular and many unicellular eukaryotes. Their principal function is to generate ATP and other energy-storage compounds. Their structure and composition are somewhat unusual; a single mitochondrion is a highly convoluted membrane system containing a variety of enzymes used in generating energy, a set of tRNA molecules that are not found elsewhere in the cell, and a circular DNA molecule containing $1-2 \times 10^4$ base pairs. This DNA molecule encodes many of the mitochondrial enzymes and is the template for synthesis of all of the mitochondrial tRNA molecules.

In 1979, after 16 years of believing in the universality of the genetic code, the world of molecular biology was jolted by the discovery that the genetic code of mitochondria is not the same as the "universal" code. In 1980 it was found that several mitochondrial codes exist for different organisms. The high point of this study came in April, 1981, when the complete base sequence of human mitochondrial DNA was reported. In this sequence were found the genes for 12S and 16S ribosomal RNA, 22 different tRNA molecules, three subunits of the enzyme cytochrome oxidase (whose amino acid sequence is known), cytochrome *b,* and several other enzymes. A few mitochondrial mRNA molecules were also isolated and sequenced, so their sites of transcription initiation could be located on the DNA. Comparing the amino acid sequences of several proteins, and the base sequences of the mRNA molecules and the DNA confirmed the fact that the mitochondrial genetic code is not the same as the universal code, a point that has profound implications for the evolution of mitochondria.

The human mitochondrial code is shown in Table 13-4; entries given in red are those that differ from the universal code (compare to Table 13-1). The differences are striking in that most are in the initiation and termination codons. That is, in mammalian mitochondria

1. UGA codes for tryptophan and not for termination.
2. AGA and AGG are termination codons rather than codons for arginine.

Table 13-4 The genetic code of human mitochondria

First position (5' end)	Second position				Third position (3' end)
	U	C	A	G	
U	Phe	Ser	Tyr	Cys	U
	Phe	Ser	Tyr	Cys	C
	Leu	Ser	Stop	Trp	A
	Leu	Ser	Stop	Trp	G
C	Leu	Pro	His	Arg	U
	Leu	Pro	His	Arg	C
	Leu	Pro	Gln	Arg	A
	Leu	Pro	Gln	Arg	G
A	Ile	Thr	Asn	Ser	U
	Ile	Thr	Asn	Ser	C
	Met	Thr	Lys	Stop	A
	Met	Thr	Lys	Stop	G
G	Val	Ala	Asp	Gly	U
	Val	Ala	Asp	Gly	C
	Val	Ala	Glu	Gly	A
	Val	Ala	Glu	Gly	G

Note: The red entries are found in mitochondria but not in the universal code. Boxed codons are used as start codons. The mitochondrial codes of other organisms exhibit further differences.

3. AUA and AUU are initiation codons, as well as AUG. Both AUA and AUG also code for methionine. AUU also codes for isoleucine, as in the universal code.
4. AUA codes for methionine (and initiation, as shown in point 3) instead of isoleucine.

Maize mitochondria use CGG for tryptophan rather than for arginine, and CGU, CGC, and CGA for arginine. Yeast mitochondria use CUX, where X is any base, for threonine rather than leucine. Both maize and yeast use AGA and AGG for arginine. Evidently, various mitochondrial codes can differ from each other as well as from the universal code.

The number of mammalian mitochondrial tRNA molecules is 22, which is less than the minimum number (32) needed to translate the universal code. This is possible because in each of the fourfold redundant sets—for example, the four alanine codons, GCU, GCC, GCA, and GCG—only one tRNA molecule (rather than two, as explained earlier) is used. In each set of four tRNA molecules the base in the wobble position of the anticodon is U or a modified U (not I). It is not yet known whether this U is base-paired in the codon–anticodon interaction or the U manages to pair weakly with each of the four possible bases. For those codon sets that are doubly redundant— for example, the two histidine codons, CAU and CAC—the wobble base always forms a G·U pair, as in the universal code.

The structure of the human mitochondrial tRNA molecule is also different from that of the standard tRNA molecule (except for mitochrondrial tRNA$_{UU\_\_}^{Leu}$). The most notable differences are the following:

1. The universal sequence CTψC__A is lacking in mitochondrial tRNA.
2. The "constant" 7-bp sequence of the TψC loop varies from three to nine bases.
3. The invariant bases U8, A14, G15, G18, G19, and U48 of the standard tRNA molecule are not invariant in mitochondrial tRNA.

In standard tRNA molecules, each of these bases participates in bonds that produce the folded L-shaped molecule. Thus, the mitochondrial tRNA molecule seems to be stabilized by fewer interactions. The three-dimensional configurations of these molecules are not known with certainty; possibly they differ from the standard L-shape and mitochondrial tRNA engages in a different type of interaction with the ribosome than standard tRNA molecules do.

Most DNA molecules contain long noncoding segments (spacers) between genes. In mitochondrial DNA there are either a few bases or none. In each case there is a start codon (AUG, AUA, or AUU) at the 5′ end of the mRNA molecule or within a few bases of it. This differs from the usual arrangement for eukaryotes in that ordinarily eukaryotic mRNA molecules commence with a short leader sequence thought to be responsible for binding to the ribosome. Such a leader is not present in the mitochondrial mRNA molecules, which suggests that the 5′ terminus itself of a mitochondrial mRNA molecule binds to the ribosome.

There is one final surprising point about mitochondrial mRNA. Whereas initiation resembles the eukaryotic mode and mitochondria are found only in eukaryotes, proteins are initiated with fMet, the prokaryotic initiator amino acid.

OTHER EXCEPTIONS TO THE UNIVERSAL CODE

Mycoplasma capricolum, a small bacterium, uses UGA as its codon for tryptophan instead of as a stop codon. DNA of this organism is usually high in A·T pairs (75 percent), and perhaps this tendency toward high A + T content has led to the replacement of UGG by UGA for tryptophan, and replacement of UGA stop codons by UAA. In contrast to mitochondria, *M. capricolum* uses AUA for isoleucine, but not methionine. Some ciliated protozoa, including

Tetrahymena and *Stylonichia,* use UAA for glutamine rather than as a stop codon. A glutamine tRNA with anticodon UUA pairing with UAA has been isolated from *Tetrahymena thermophila* and sequenced. Perhaps this difference helps to protect these protozoa against invasion by viruses that use the universal code.

Translation

II THE MACHINERY AND THE CHEMICAL NATURE OF PROTEIN SYNTHESIS

In the preceding chapter we described how the information in a mRNA molecule is converted to an amino acid sequence having specific start and stop points. We called this the informational problem. In this chapter we examine the chemical problem of attachment of the amino acids to one another. We begin with the structure of the ribosome—a complex nucleo-protein particle on which assembly occurs—and the physical role played by the ribosome in protein synthesis.

RIBOSOMES

A ribosome is a multicomponent particle that contains several of the enzymatic activities needed for protein synthesis and serves to bring together a single mRNA molecule and charged tRNA molecules in the proper position and orientation that the base sequence of the mRNA molecule is translated into an amino acid sequence. The chemical composition of the ribosome of prokaryotes is well known, though there are still questions about the composition of the ribosomes of eukaryotes. However, both share the same general structure and organization. Studies of the physical structure of the subunits and the sites pertinent to protein synthesis are still incomplete, but are progressing. The properties of bacterial ribosomes are constant over a wide range of species. The E. coli ribosome is best understood and, although

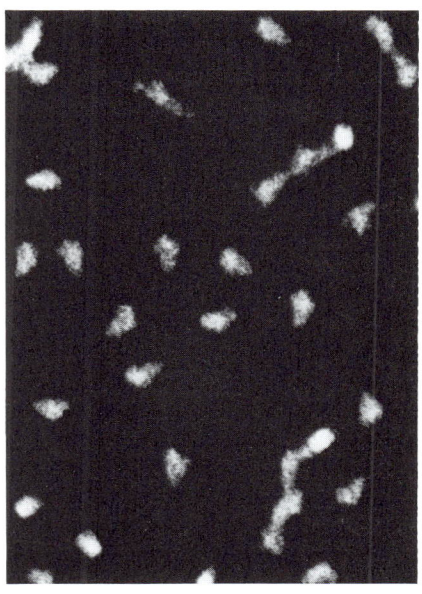

A dark-field electron micrograph showing a field of 30S subunits of the *E. coli* ribosome. [Courtesy of Alex Korn.]

its structure differs slightly from that of eukaryotic ribosomes, it serves as a useful model for discussion of all ribosomes; thus, we begin by describing its chemical composition.

Chemical Composition of Prokaryotic Ribosomes

A prokaryotic ribosome is a nucleoprotein particle consisting of two subunits. The intact ribosome and the subunits are, for historical reasons, named by stating their sedimentation coefficients. Thus the intact particle is called a **70S ribosome** (because its s value is 70 svedbergs) and the subunits, which are unequal in size and composition, are termed **30S** and **50S subunits.**

A 70S ribosome consists of one 30S subunit and one 50S subunit. In the laboratory the stability of the 70S ribosome is determined by the Mg^{2+} concentration in its environment. In $0.0005 \, M \, MgCl_2$ the ribosome is almost completely dissociated into its subunits, whereas in $0.005 \, M \, MgCl_2$ (roughly the concentration within bacteria) there is very little dissociation. Thus, the following equation describes the equilibrium:

$$30S \text{ subunit} + 50S \text{ subunit} \xrightleftharpoons[\text{Low}[Mg^{2+}]]{\text{High}[Mg^{2+}]} 70S \text{ ribosome}$$

At physiological concentrations of the Mg^{2+} ion the 70S ribosome is the preferred form and

the 70S ribosome is the form active in protein synthesis.

The molecular weights of the *E. coli* ribosome and its subunits are

70S ribosome	2.5×10^6
50S subunit	1.6×10^6
30S subunit	0.9×10^6

These values are characteristic of the values of all prokaryotic ribosomes.

Both the 30S and the 50S subunits can be further dissociated into RNA and protein molecules under appropriate conditions (Figure 14-1). The composition of each subunit is the following:

30S subunit: one 16S rRNA molecule + 21 different proteins

50S subunit: one 5S rRNA molecule + one 23S rRNA molecule + 31 different proteins

The proteins of the 30S subunit are termed $S1, S2, \ldots, S21$ (S for "small subunit"); there is one copy of each of these protein molecules in the 30S subunit. The proteins of the 50S subunit are denoted by a capital L (for large subunit).

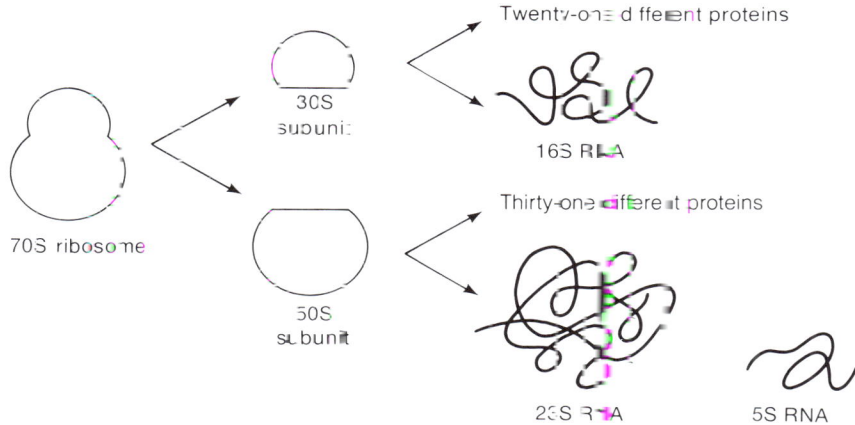

Twenty-one different proteins

30S
subunit

16S RNA

Thirty-one different proteins

70S ribosome

50S
subunit

23S RNA   5S RNA

Figure 14-1   Dissociation of a prokaryotic ribosome. The configuration of two overlapping circles will be used throughout this chapter, for the sake of simplicity. The correct configuration is shown in Figure 14-3.

Each 50S subunit contains one copy of each protein molecule except for proteins $L7$ and $L12$, of which there are four copies in total in a 50S subunit, and $L26$, which is not a true component of the 50S subunit. The number of copies of $L7$ relative to those of $L12$ depends on the bacterial growth rate. These protein units are distinct molecules with two exceptions. Protein $L7$ is a terminal-N-acetylated form of $L12$ and the protein designated $L8$ is a complex consisting of one molecule of $L10$ to which $L7$ and $L12$ are bound. This complex results from a functional association that probably exists in the ribosome; sometimes the complex does not dissociate during isolation. Furthermore, a protein designated $L26$ (of which only one molecule per five 50S subunits is observed) is actually protein $S20$ of the 30S subunit; occasionally, when a 70S ribosome is dissociated, $S20$ leaves the 30S subunit and becomes associated with a 50S subunit. The total number of chemically distinct protein molecules in a 70S ribosome is $21 + 31 + 2$ (the extra $L7$ and $L12$) = 54. Most of these are very basic proteins containing up to 34 percent basic amino acids (arginine + lysine). This basicity probably accounts in part for their strong association with the RNA, which is acidic. The amino acid sequences of most of the ribosomal proteins are known.

The sizes of the RNA components are the following:

5S rRNA:	120 nucleotides
16S rRNA:	1541 nucleotides
23S rRNA:	2904 nucleotides

The base sequence of each of these molecules is known. Roughly 70 percent of the bases in each RNA molecule are internally paired to form a highly complex structure with numerous stems and loops.

The structure of the *E. coli* 70S ribosome is summarized in Figure 14-1.

Chemical Composition of Eukaryotic Ribosomes

Although the basic features of eukaryotic ribosomes are similar to those of bacterial ribosomes, all eukaryotic ribosomes are somewhat larger than those of prokaryotes. They contain a greater number of proteins (about 80) and an additional RNA molecule. The biological significance of the differences between prokaryotic and eukaryotic ribosomes is unknown.

A typical eukaryotic ribosome has an *s* value of about 80 and consists of two subunits, a **40S** and a **60S subunit.** These sizes may vary over a 20 percent range from one organism to the next, in contrast with bacterial ribosomes, whose sizes are nearly the same for all bacterial species examined. The best-studied eukaryotic ribosome is that of the rat liver. The components of the subunits of eukaryotic ribosomes are:

40S subunit: one 18S rRNA molecule + about 30 proteins

60S subunit: one 5S, one 5.8S, and one 28S rRNA molecule + about 50 proteins

The 5.8S, 18S, and 28S rRNA molecules of eukaryotes correspond functionally to the 5S, 16S, and 23S molecules of bacterial ribosomes. The bacterial counterpart to the eukaryotic 5S rRNA is very likely present as part of the 23S rRNA.

The molecular weights of the 80S ribosome of the rat liver and its 40S and 60S subunits are:

80S ribosome:	4.3×10^6
60S subunit:	2.8×10^6
40S subunit:	1.4×10^6

The precise number of proteins found in the eukaryotic ribosome is uncertain because several of the proteins are not always found in purified subunits; thus, it is unclear whether these are ribosomal proteins that have been accidentally removed during purification or nonribosomal proteins that sometimes contaminate the samples.*

Eukaryotic ribosomal proteins are highly basic, like the *E. coli* ribosomal proteins. Study of eukaryotic ribosomal proteins has been hampered by their insolubility in ordinary buffers.

A comparison of the properties of *E. coli* and eukaryotic ribosomes is given in Table 14-1. We have already stated that there are more components

---

*The problem of ascertaining that components associated after isolation are also associated *in vivo* is ubiquitous in all biology. Usually many additional experiments are needed before the question can be answered.

Table 14-1 Characteristics of the ribosomes of prokaryotes and eukaryotes (e.g., rat liver)

Property	Prokaryote*	Eukaryote (rat liver)[†]
Small subunit		
s value (total)	30S	36.9S (termed the 40S subunit)
Molecular weight	0.9×10^6	1.44×10^6
s value (RNA)	16S	18S
Molecular weight (RNA)	0.51×10^6	0.7×10^5
Proteins	21	≈ 30
Range of molecular weight	8300–25,800	11,200–41,500
Total molecular weight	0.39×10^5	0.74×10^6
Large subunit		
s value	50S	56.3S (termed the 60S subunit)
Molecular weight (total)	1.6×10^6	2.8×10^6
s value and molecular weights of corresponding RNA molecules	5S: 40,000 23S: 0.98×10^6	5S: 39,000 5.8S: 51,000 28S: 1.7×10^6
Total molecular weight of RNA	1.02×10^6	1.79×10^6
Proteins	31	49
Range of molecular weight	5300–24,600	11,500–41,800
Total molecular weight	0.78×10^6	1.05×10^6

*The values given are for *E. coli*; they vary little from one prokaryote to the next.

[†]The values given may vary from one eukaryote to the next and may be as much as 15 percent greater.

in eukaryotic ribosomes (both RNA and proteins) than in prokaryotic ribosomes; it should be noted also that the eukaryotic components usually also have a higher molecular weight.

Reconstitution of *E. coli* Ribosomes

A major breakthrough in understanding bacterial ribosomal structure came about when the *E. coli* ribosome was dissociated and reconstituted from a mixture of purified ribosomal RNA and protein molecules. By using this technique it is possible (1) to determine whether a component is necessary in order to have a functional ribosome and (2) to identify the function of each component. An important observation was that assembly of a 30S subunit requires nothing other than the 16S RNA molecule and the 21 type-S proteins, and that a reconstituted particle has all the biological functions of the natural subunits. Thus, no nonribosomal components (for example, enzymes or special factors) are needed to form an intact particle *in vitro*. The same observation was made for the 50S subunit. Thus, the formation of a ribosome *in vivo* is a self-assembly process.

The rRNA-protein interaction is a highly specific one. For instance, mRNA plus ribosomal proteins will not interact to form a compact particle.

The relatedness of the ribosomes of prokaryotes is shown by the fact that the 16S rRNA of *E. coli* can be substituted by 16S rRNA of other bacteria in a reconstitution experiment. However, reconstitution will not occur if the 18S rRNA of a eukaryote (e.g., yeast) is mixed with the proteins of a bacterial 30S subunit, or if bacterial 16S rRNA and eukaryotic proteins are mixed. This is one of many known examples of a molecular difference between prokaryotes and eukaryotes.

Reconstitution experiments have been carried out with various mixtures, each lacking one ribosomal protein, in order to determine whether all proteins are essential and to elucidate the sequence of events during assembly. In these experiments it was found that sometimes an apparently intact particle forms, but the particle is unable either to combine with the complementary particle to form a 70S ribosome or to perform certain steps in protein synthesis. When mixtures of various combinations of proteins were studied, it was observed that there are stable intermediates. For example, there is a stable structure containing 16S rRNA and five proteins (S4, S8, S15, S18, and S20). Additional proteins added singly do not bind to this structure. However, a mixture of S5, S7, S12, S16, and S18 causes a transition to a second stable form containing all eleven proteins.

Reconstitution experiments can also give information about the functions of individual proteins. However, the analysis of such experiments is complicated by the fact that often there is not a one-to-one correspondence between a particular protein and a particular function—that is, several proteins may be required for a particular function, or a protein may participate in two different functions. A useful procedure is the reconstitution of mixtures obtained from wild-type and mutant proteins. For instance, the antibiotic streptomycin is an inhibitor of protein synthesis. Bacterial mutants that are streptomycin-resistant (Str-r) have been isolated and with these mutants protein synthesis is not inhibited by streptomycin *in vitro* or *in vivo*. If a mixture is prepared consisting of wild-type 16S rRNA and all of the wild-type S-proteins with the substitution of S12 from a streptomycin-resistant (Str-r) bacterium for the S12 from a Str-s bacterium, the reconstituted ribosomes are not inhibited by streptomycin (Figure 14-2). Thus, protein S12 determines the response to the drug. One might be tempted to conclude that S12 is the binding site for streptomycin. However, there are other possible interpretations of the data. For example, S12 might be responsible for arranging two other proteins to form a streptomycin binding site and a mutant S12 might just prevent formation of the site. Another possibility is that S12 is adjacent to the binding site and that a mutant S12 sterically hinders binding.

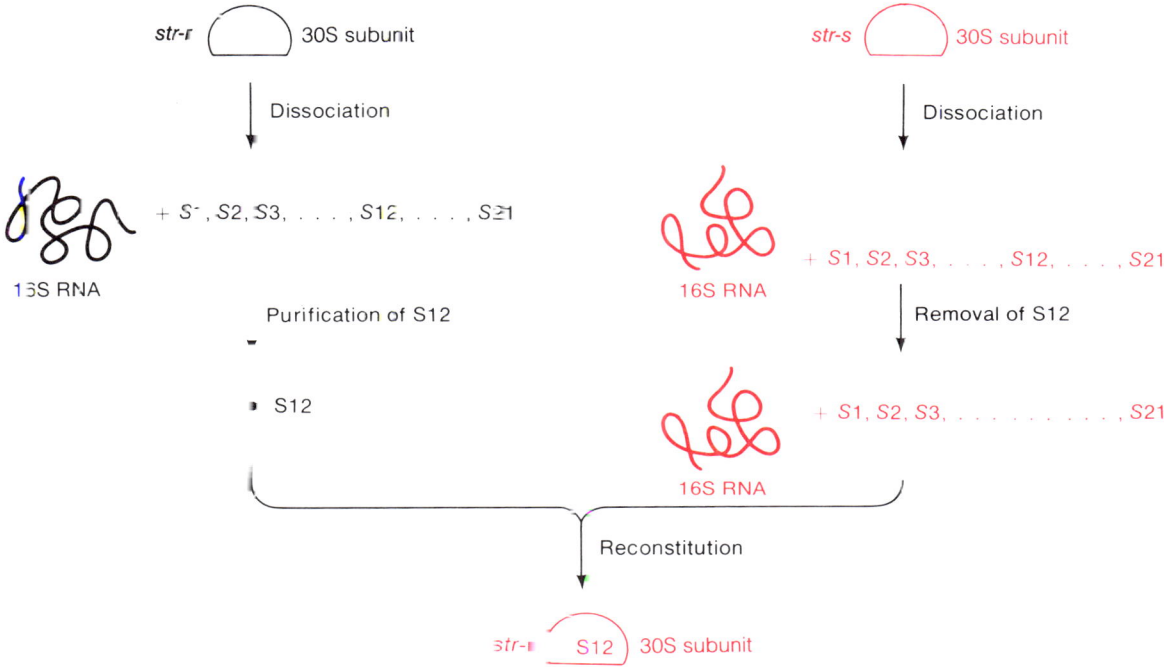

Figure 14-2 A reconstitution experiment showing that protein S12 is the determinant of resistance to streptomycin. Only the 30S subunit is shown to form a functional ribo- some, the 50S subunit is added after reconstitution.

The technique of reconstitution with mutant proteins has been used to examine a variety of drug-resistance mutations. It has been quite useful because if the particular protein molecule can be shown to bind the anti- biotic, and if the step in protein synthesis inhibited by the antibiotic can be determined, this information can be used to identify one of the functions of that protein molecule in protein synthesis.

Physical Structure of Prokaryotic Ribosomes*

Figure 14-3 shows an electron micrograph of a field of *E. coli* 70S ribosomes, and its two kinds of subunits. In any electron micrograph of particles, indi- vidual subunits are viewed from a variety of angles depending on how each particle has been deposited on the surface viewed. By studying the features of many 30S, 50S, and 70S particles in many orientations, it is possible to reconstruct a three-dimensional model of the ribosome (Figure 14-4). Sev- eral models have been proposed, but the asymmetric subunit model shown

*If methodology is not included in the course, the student should read the first paragraph, examine Figures 14-3(a) and 14-4, and omit the remainder of the section.

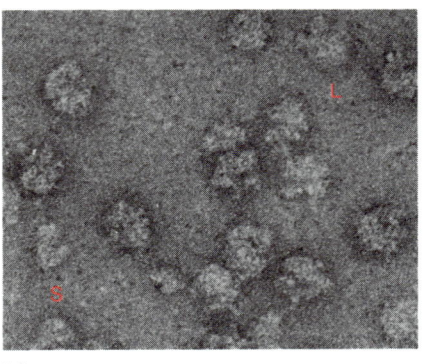

(a)

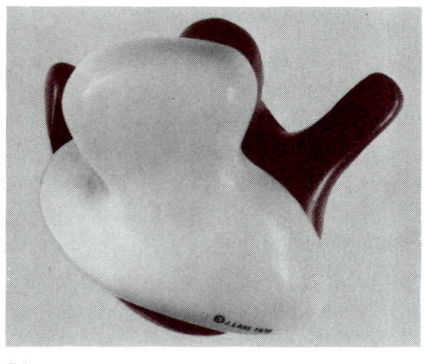

(b)

Figure 14-3   Ribosomes. (a) An electron micrograph of 70S ribosomes from *E. coli*. Some ribosomes are oriented as in the model shown in panel (b). A few ribosomal subunits that also lie in the field are identified by the letters *S* and *L*, which stand for small and large. To interpret the variety of observed shapes, refer to Figure 14-4. (b) A three-dimensional model of the *E. coli* ribosome. The 30S subunit is white and the 50S subunit is red. [Courtesy of James Lake.]

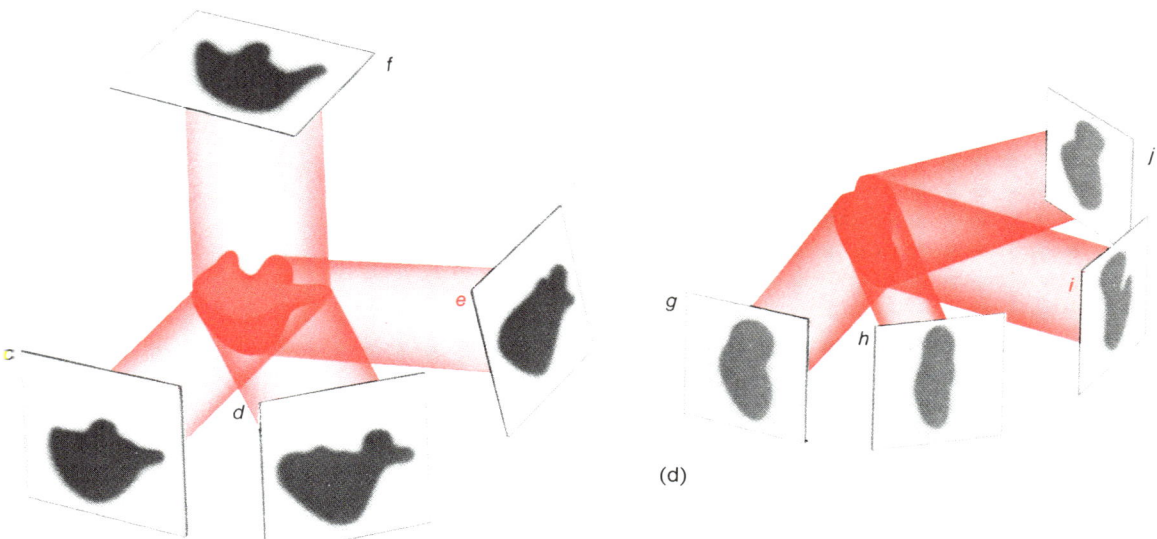

(d)

Figure 14-4   Method of deducing molecular shape from electron micrographs. The drawings show how the observed shapes (black) depend on the orientation of the particle. (a) Large subunit. (b) Small subunit. The letters *c, d, e,* and *f* identify images of the large subunit; *g, h, i,* and *j* identify images of the small subunit. Each image is a two-dimensional projection of a subunit that results from directing a beam of electrons through a three-dimensional structure and the uranium salt that surrounds it. Electron micrographs do not show the surface of a ribosome; thus, the three-dimensional shape of the ribosome is inferred from the multiple projections. [Reprinted with permission of the author from James A. Lake, "The Ribosome" in *Scientific American*, August 1981, pp. 84–97.]

in Figure 14-3(b) is generally agreed to fit the data on protein distributions best.

A variety of methods have been used to locate the various protein molecules with respect to one another and with respect to the RNA molecules. Most of these are quite detailed and beyond the scope of this book. Some, which are of general interest, are the following:

1. *Protection of particular regions of RNA by specific proteins.* A particular type of rRNA molecule is mixed with a single purified ribosomal protein and then the mixture is treated with RNase. (This is the same procedure that was encountered in determining the sequences of promoters in DNA and ribosome binding sites in mRNA.) If the protein fails to bind, the rRNA is totally digested by the enzyme. If there is binding, a region will be protected and its sequence can be compared to the known sequence of the ribosomal rRNAs. By using this procedure the following results have been obtained:

RNA	Bound proteins
5S	*L*5, *L*18, *L*25
16S	*S*4, *S*7, *S*8, *S*15, *S*17, *S*20
23S	*L*1, *L*2, *L*3, *L*4, *L*6, *L*7, *L*10, *L*11, *L*13, *L*14, *L*16, *L*20, *L*23, *L*24

2. *Linkage of proteins by cross-linking reagents.* Certain chemicals cause a covalently linked bridge (**cross-link**) to form between two ϵ-amino groups of lysine. If a 30S or 50S subunit or an intact ribosome is treated with these reagents and then the particle is dissociated, pairs of covalently linked neighboring proteins form, which can be detected by gel electrophoresis. Each pair is then exposed to a reagent that cleaves the cross-link and the component proteins are identified by gel electrophoresis—that is, by comparison of the electrophoretic diagram with that obtained with purified proteins (Figure 14-5). It is important to realize that the production of a cross-link between two proteins does not mean that the protein molecules are in physical contact, because the cross-link itself has length and need only span the distance between the two molecules. Information about relative distances can be obtained by using cross-linking agents of different lengths.

3. *Localization by neutron scattering* Neutron scattering is a complex technique whose methodology and interpretation is beyond the scope of this book. Like x-ray diffraction, it can be used to localize individual atoms. Its application to the study of ribosomes is based on the fact that ^1H and ^2H nuclei scatter neutrons in very different ways. In practice, deuterated ribosomal proteins are prepared either by growth of bacteria in deuterated growth medium or by prolonged incubation of purified ribosomes in ^2H_2O (in which ^1H-^2H exchange occurs). Individual deuterated proteins are then iso-

Figure 14-5 The principle of the cross-linking technique to identify proteins of the 30S subunit that are in contact. Panel (a) shows the result of gel electrophoresis of the 21 proteins of a 30S subunit. Each band represents a particular protein. Panel (b) shows a hypothetical distribution obtained with proteins isolated from a 30S subunit that had been exposed to a cross-linking agent. Two new bands, I and II, are seen. If the material in band I is isolated and the cross-link is cleaved, gel electrophoresis shows two bands (panel (c)) whose positions match those of band i and ii of panel (a); thus, I = i + ii. A similar experiment with band II is shown in panel (d); here, II = iii + iv.

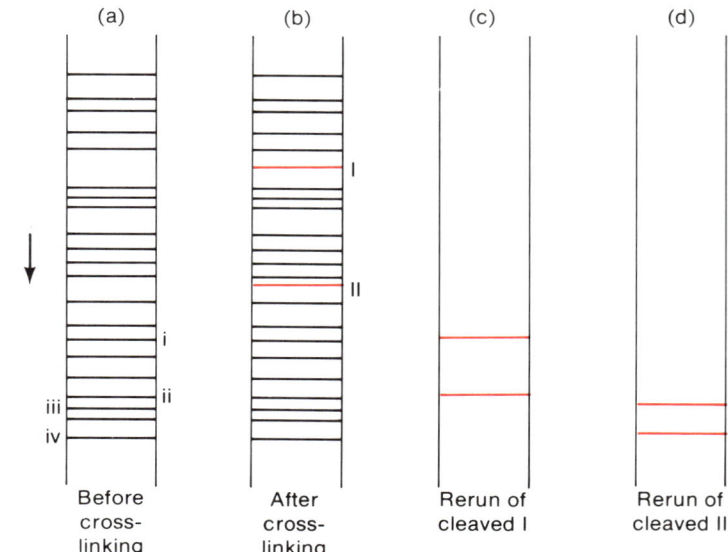

lated and purified. A ribosomal subunit is then reconstituted by using two different deuterated proteins—for example, $S2$ and $S6$—all the other being nondeuterated S proteins. The reconstituted particle is then subjected to neutron scattering. Analysis of the scattering pattern yields the distance between the deuterated protein molecules. By using many different deuterated pairs, a three-dimensional protein map can be obtained.

The interface between the 30S and 50S subunits is of special interest because, as we will see shortly, the subunits associate at one step of protein synthesis and then dissociate at a later step. This interface has been studied by several procedures. If the cross-linking technique is applied to a sample of 70S ribosomes, some S-L protein pairs form, which identifies the proteins in the two subunits that contact one another. Interparticle contact between proteins and RNA molecules has also been examined. For example, proteins such as $S11$ and $S12$ (from the 30S subunit) protect 23S RNA (from the 50S subunit) from digestion by RNase. Observing the interference of aggregation by antiprotein antibodies is also a useful method. Thus, if a particular S protein is in the interface region, an antibody directed against it will prevent formation of the 70S complex. This method has also been used with 50S subunits and antibodies directed against L proteins.

Physical Structure of Eukaryotic Ribosomes

The techniques just described have also been used to study the structure of eukaryotic ribosomes. There are no major differences between the structures of prokaryotic and eukaryotic ribosomes apart from the larger number of proteins in eukaryotic ribosomes, of which some contain phosphorylated serine residues. In fact the proteins in eukaryotic ribosomes resemble those of prokaryotes quite closely in several cases there is such close correspondence in the amino acid sequences that an antibody directed against a particular prokaryotic protein can interact with the analogous eukaryotic protein. The interpretation of these and other similarities is that the properties of the ribosomal proteins that are most important in ribosome function have been conserved throughout evolution. The similarities also suggest that prokaryotic and eukaryotic ribosomes did not arise independently, but that either one evolved from the other or both evolved from a common ancestor.

Chloroplasts and **mitochondria** (eukaryotic intracellular organelles) also possess ribosomes. These subunits are similar to but distinct from, the major eukaryotic type of ribosome and more closely resemble the bacterial type. Both the size and chemical composition of chloroplast ribosomes are very much like those of bacteria, a fact which is thought to have some evolutionary significance. The s values of the subunits of mitochondrial ribosomes varies enormously (30–40S for the small subunit and 40–60 for the large subunit) from one organism to the next.

PROTEIN SYNTHESIS

Protein synthesis can be divided into three stages—(1) **polypeptide chain initiation,** (2) **chain elongation,** and (3) **chain termination.** The main features of the initiation step are binding of mRNA to the ribosome, selection of the initiation codon, and binding of the charged tRNA bearing the first amino acid (fMet or Met). In the elongation stage there are two processes— joining together two amino acids by peptide bond formation and moving the mRNA and the ribosome with respect to one another in order that each codon can be translated successively. This movement is called **translocation.** In the termination stage the completed protein is dissociated from the synthetic machinery and the ribosomes are released to begin another cycle of synthesis. These three stages will be described separately, first in an overview and then in some detail. However, before entering this complex subject two simpler features of protein synthesis will be described—namely, the direction of reading of mRNA and the direction of polypeptide chain growth.

Directions of Polypeptide Chain Growth and of Translation of mRNA

Every protein molecule has an amino terminus and a carboxyl terminus. *Synthesis begins at the amino terminus;* in a protein having the sequence NH_2-

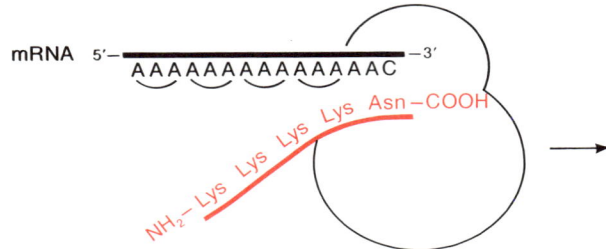

Figure 14-6 A ribosome translating a synthetic mRNA molecule, showing the direction of movement of the ribosome and the direction of polypeptide chain growth.

Met-Ser-Ala-Glu-Ala-COOH, the Met is the starting amino acid and Ala is the last amino acid added to the chain.

Translation of mRNA molecules occurs in the 5′→3′ direction. The meaning of this statement is made clear by examining the original experiment that proved the point. In this experiment (Figure 14-6) the synthetic polynucleotide

$$\text{P-5′-AAAAAA . . . AAC-3′-OH}$$

was used as an mRNA in an *in vitro* protein-synthesizing system not requiring an AUG start codon. The codons AAA and AAC correspond to lysine and asparagine, respectively. The protein that is translated from this mRNA is

$$\text{H}_2\text{N-Lys-Lys- . . . -Asn-COOH}$$

Since the protein is synthesized from the amino to the carboxyl terminus, the direction of mRNA translation is 5′→3′. If the direction had been 3′→5′, the protein would have been

$$\text{H}_2\text{N-Gln-Lys- . . . -Lys-COOH}$$

An Overview of Protein Synthesis*

In bacteria protein synthesis begins by the association of one 30S subunit (not the 70S ribosome), one mRNA molecule, a charged tRNA^{fMet}, three proteins known as **initiation factors,** and guanosine 5′-triphosphate (GTP). These molecules comprise the **30S preinitiation complex** (Figure 14-7). Following formation of the 30S preinitiation complex, a 50S subunit joins to the 30S subunit to form a 70S initiation complex (Figure 14-7). This

*For a course in which fine details of initiation, translocation, and termination are not included, this section should be followed by the section entitled "The Role of GTP."

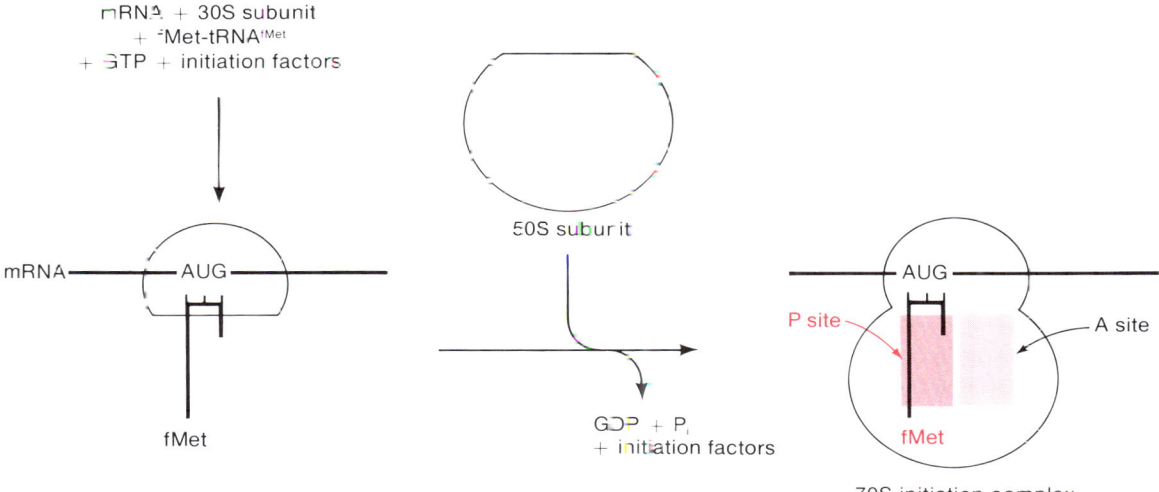

Figure 14-7 Early steps in protein synthesis: in prokaryotes: formation of the 30S preinitiation complex and of the 70S initiation complex. Details are shown in Figure 14-12.

joining process requires hydrolysis of the GTP contained in the 30S preinitiation complex. There are two tRNA binding sites, which overlap the 30S and 50S subunits. These sites are called the **aminoacyl** or **A site** and the **peptidyl** or **P site;** each site consists of a collection of segments of S and L proteins and 23S rRNA. The 50S subunit is positioned in the 70S initiation complex such that the tRNA^{fMet}, which was previously bound to the 30S preinitiation complex, occupies the P site of the 50S subunit. Positioning tRNA^{fMet} in the P site fixes the position of the anticodon of tRNA^{fMet} such that it can pair with the initiator codon in the mRNA. Thus, *the reading frame is unambiguously defined upon completion of the 70S initiation complex.**

The A site of the 70S initiation complex is available to any tRNA molecule whose anticodon can pair with the codon adjacent to the initiation codon. However, entry to the A site by the tRNA requires a helper protein called an **elongation factor, EF** (specifically **EF-Tu**) (Figure 14-8). After occupation of the A site a peptide bond between fMet and the adjacent amino acid can be formed. Once it was thought that the blockage of the NH$_2$ group of fMet by the formyl group was responsible for peptide bond formation between the COOH group of fMet and the NH$_2$ group of the adjacent amino

*If the 30S preinitiation complex and the 70S initiation complex formed only with tRNA^{fMet}, the polyribonucleotide technique for the determination of codon assignments would not have worked. However, at Mg^{2+} concentrations above the physiological intracellular value the system becomes less fastidious. Under these conditions the ribosome will bind to any base sequence in the mRNA and, for initiation, the P site can be occupied by any tRNA whose anticodon can pair with the mRNA base sequence that is present at the P site.

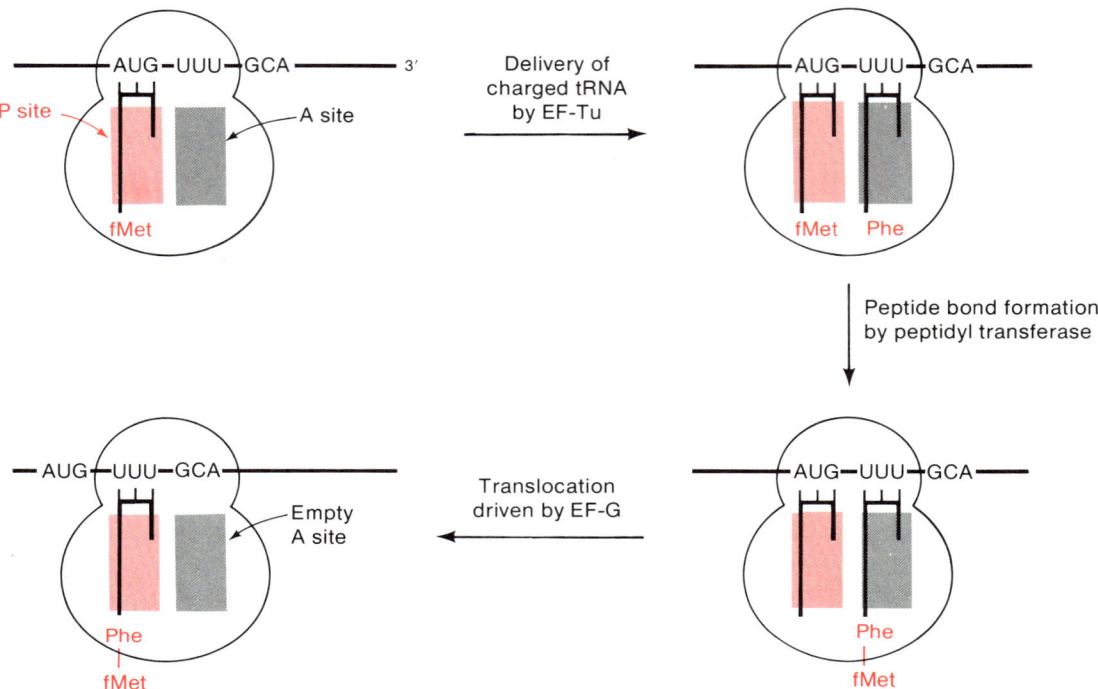

Figure 14-8 Elongation phase of protein synthesis: binding of charged tRNA, peptide bond formation, and translocation.

acid. However, in eukaryotic systems the starting amino acid is Met and not fMet, and protein synthesis proceeds in the correct direction. Presumably, the relative orientation of the two amino acids in the A and P sites determines the linkage that is made.

The peptide bond is formed by an unusual enzyme complex called **peptidyl transferase.** The active site of peptidyl transferase consists of portions of several proteins of the 50S subunit. As the peptide bond is formed, the fMet is cleaved from the $tRNA^{fMet}$ in the P site. The chemistry of peptide bond formation is shown in Figure 14-9.

After the peptide bond forms, an uncharged tRNA occupies the P site and a dipeptidyl-tRNA is in the A site. At this point three movements, which together comprise the translocation step, occur: (1) the deacylated $tRNA^{fMet}$ leaves the P site, (2) the peptidyl-tRNA moves from the A site to the P site, and (3) the mRNA moves a distance of three bases in order to position the next codon at the A site (Figure 14-8). The translocation step requires the presence of another elongation protein **EF-G** and hydrolysis of GTP. The movement of the mRNA by three bases is probably dependent on the movement of the tRNA from the A site to the P site and, in fact, it is likely that mRNA translocation is a consequence of the tRNA motion. Supporting this view is the fact that the distance that mRNA moves is not

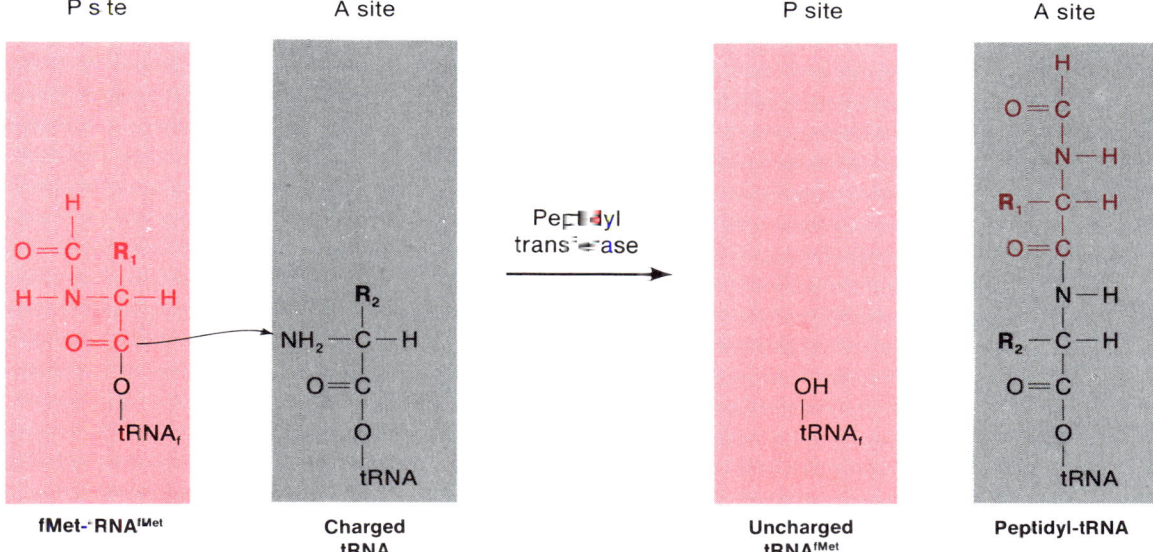

Figure 14-9 Peptide bond formation in prokaryotes. In eukaryotes the chemistry is the same but the initiating tRNA is not $tRNA^{fMet}$, but $tRNA_{init}^{Met}$.

fixed, whereas the translocation distance is normally three bases; if a glycyl frameshift suppressor having an extra base in the anticodon loop is present, the displacement in the translocation step is four bases instead of three.

After translocation has occurred, the A site is again available to accept a charged tRNA molecule having a correct anticodon. If a $tRNA^{fMet}$ molecule, whose anticodon is the same as that of $tRNA^{Met}$ molecule, were to enter the A site (because an internal AUG site were present), protein synthesis would stop because a peptide bond cannot form with the blocked NH_2 group of fMet. However, inasmuch as the factor EF-Tu is needed to facilitate tRNA entry into the A site, this misadventure is prevented, since EF-Tu cannot bind to $tRNA^{fMet}$.

When a chain termination codon is reached there is no aminoacyl-tRNA that can fill the A site and chain elongation stops. However, the polypeptide chain is still attached to the tRNA occupying the P site. Release of the protein is accomplished by release factors (RF), proteins that in part respond to chain termination codons. There are two such release factors in E. coli—RF1, which recognizes the UAA and UAG codons, and RF2, which recognizes UAA and UGA. Why the number of release factors is not one (i.e., useful for all codons) or three (one for each stop codon) is not known. (In eukaryotes there is only one release factor.) Each release factor forms an activated complex with GTP; this complex binds to a termination codon and alters the specificity of peptidyl transferase. In the presence of release factors peptidyl transferase catalyzes the reaction of the bound peptidyl moiety

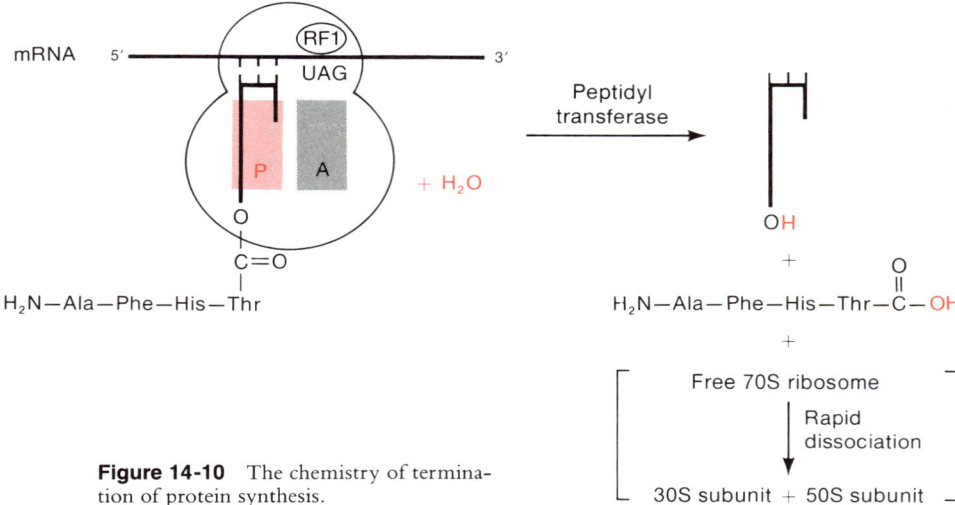

Figure 14-10　The chemistry of termination of protein synthesis.

with water rather than with the free aminoacyl-tRNA (Figure 14-10). Thus, the polypeptide chain, which has been held in the ribosome solely by the interaction with the tRNA in the P site, is released from the ribosome. The 70S ribosome dissociates into 30S and 50S subunits and the system is ready to start synthesis of a second chain.

In the following three sections details of initiation, translocation, and termination are described. The reader for whom the overview is sufficient should proceed to the section entitled "The Role of GTP."

Initiation in Prokaryotes

Initiation of protein synthesis consists of several stages. The first stage, formation of a preinitiation complex, requires the cooperative interaction of three nonribosomal proteins called **initiation factors.** In bacteria these factors are designated IF1, IF2, and IF3. They are not required when using synthetic polynucleotides as mRNA at a high Mg^{2+} concentration, but are essential with natural mRNA molecules and at physiological Mg^{2+} concentrations.

At the intracellular concentration of Mg^{2+}, the 70S ribosome is rarely dissociated, yet formation of the preinitiation complex requires a free 30S subunit. This subunit is supplied by IF1 and IF3, whose primary function is to shift the equilibrium toward dissociation. IF1 does not itself cause dissociation, but increases the rate of dissociation and association (Figure 14-11), which increases the period of exposure of an IF3 binding site on the 30S ribosome and in the region that contacts the 50S subunit. Thus, when-

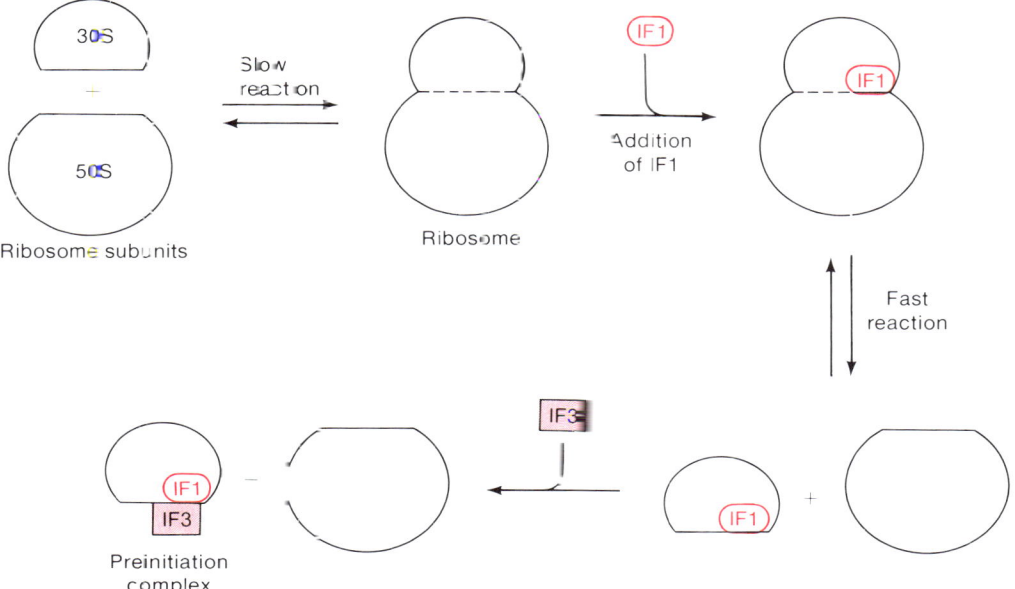

30S

Slow
reaction

Ribosome subunits

50S

Ribosome

IF1

IF1

Addition
of IF1

Fast
reaction

IF3

IF1

+

IF1

IF3

Preinitiation
complex

Figure 14-11 The role of IF1 and IF3 in
forming the preinitiation complex.

ever a 70S ribosome becomes dissociated. an IF3 may bind to this site tightly,
thereby preventing reassociation. IF2 can then join this complex to form a
unit now consisting of one 30S subunit, IF1, IF2, and IF3, and called the
preinitiation complex (Figure 14-11). Cross-linking studies indicate that IF3
binds very close to the 3'-OH terminus of 16S rRNA, and there is also
evidence that IF3 may have the ability to denature base-paired regions—in
particular, the 3'-OH terminus of 16S rRNA. The denatured 3'-OH terminus of the 16S rRNA also includes the Shine-Dalgarno sequence needed
to bind mRNA (Chapter 13). At this point the exact sequence of events is
unclear. However, IF2 mediates binding of one molecule of GTP and charged
tRNA^{fMet}. This addition only occurs if a formyl group is present, which is
shown by two experimental results: (1) binding does not occur if tRNA^{fMet}
is acylated with Met, but does occur once the Met is formylated, and (2)
misacylated tRNA^{fMet} binds if the amino acid is formylated (for example,
formylvalyl-tRNA^{fMet} will bind). IF3 then leaves the complex when the
tRNA^{fMet} binds. It is not known whether mRNA joins the complex before
or after binding of the tRNA^{fMet} and it is thought that binding can occur in
either order. Binding between the mRNA and the Shine-Dalgarno sequence
of the 16S rRNA is aided by the ribosomal proteins S1 and S12.

The loss of IF3 during binding of charged tRNA^{fMet} frees the site on
the 30S subunit that binds to the 50S subunit. Thus, the 50S subunit binds

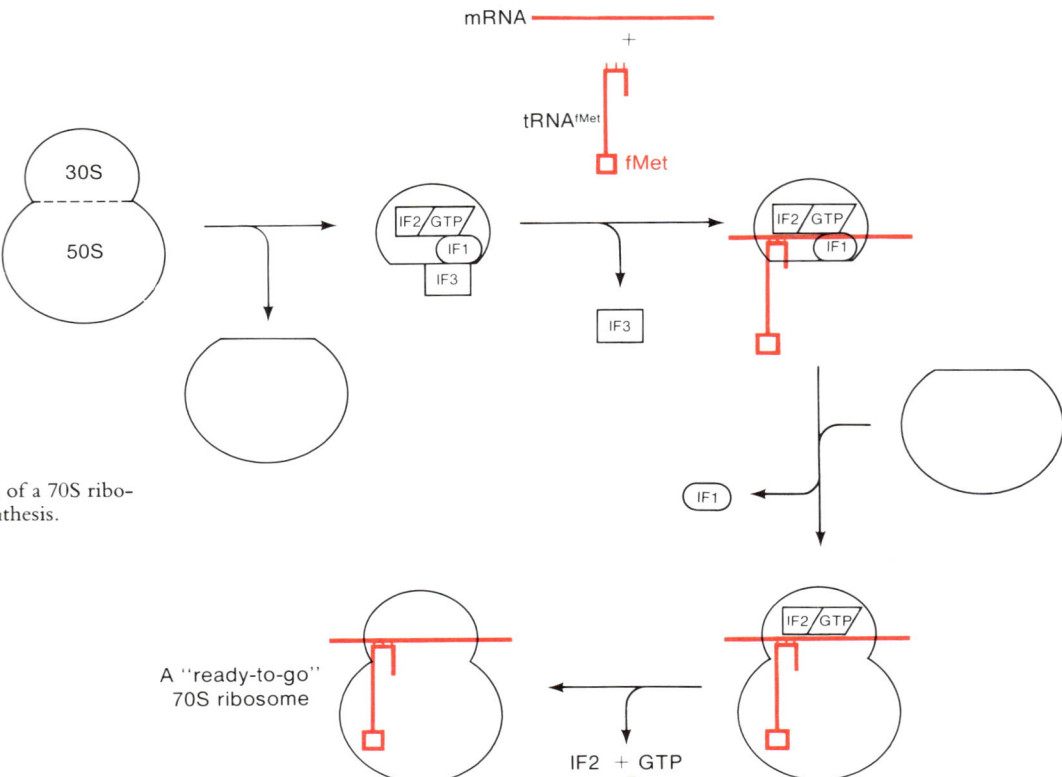

Figure 14-12 Formation of a 70S ribosome ready for protein synthesis.

to the complex and, in so doing, causes IF1 to fall off of the 30S subunit (Figure 14-12). The P and A sites then form by a conjunction of S and L proteins. The P site is occupied by tRNA^{fMet}, but IF2 occupies the A site. IF2 must vacate the A site if a second charged tRNA molecule is to enter that site. In a way that is not understood, a GTPase is activated in the A site and the GTP bound to IF2 is hydrolyzed to yield GDP and P$_i$: IF2 cannot remain bound without the GTP, so the IF2 leaves the A site; this departure is aided by IF1 in an unknown way. The positioning of the fMet in the P site is determined by interactions of fMet and tRNA^{fMet} mediated by specific ribosomal proteins. For instance, certain cross-linking agents produce linkages between fMet and the proteins $L2$ and $L27$.

The ribosome is now ready for the elongation stage.

Chain Elongation in Prokaryotes

Three nonribosomal proteins—the elongation factors, EF-Tu, EF-Ts, and EF-G—are required for chain elongation. These important proteins together comprise about 5–10 percent of the soluble protein of the cell.

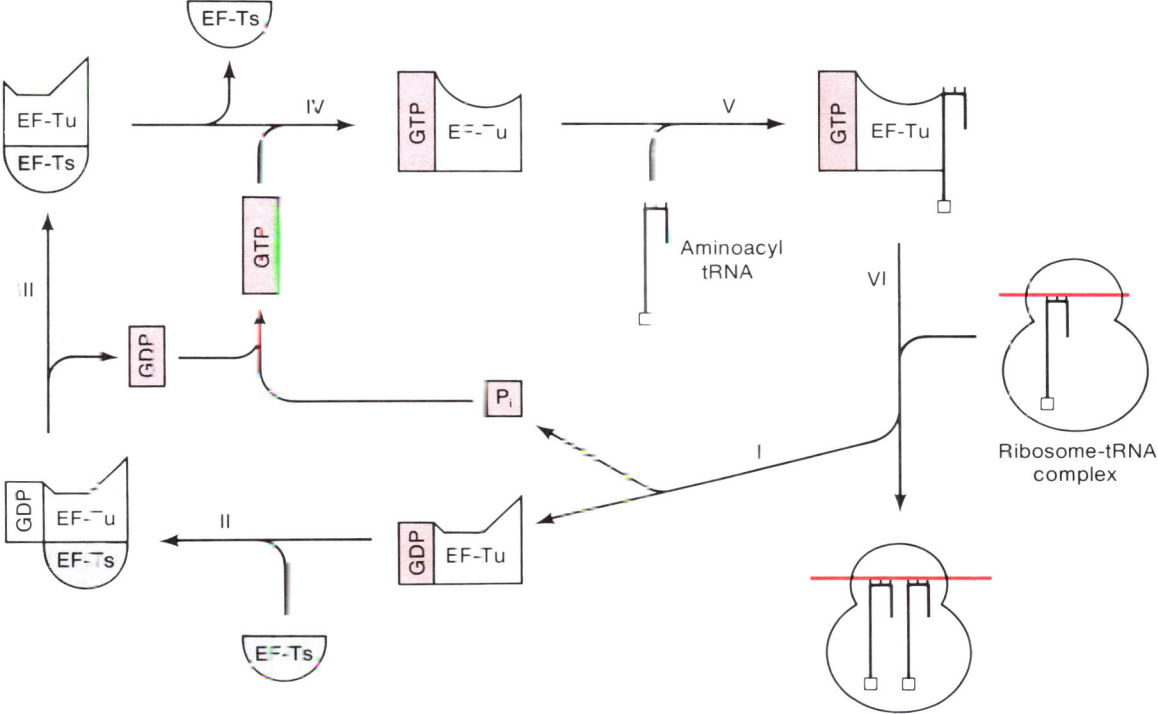

Figure 14-13 The cycle of the elongation factor EF-Tu. The steps shown are the following: I. Formation of GDP–EF-Tu. II. Binding of EF-Ts. III. Formation of EF- Tu–EF-Ts. IV. Formation of active EF-Tu. V. Binding of tRNA by active EF-Tu. VI. Transfer of tRNA to a ribosome.

EF-Tu is a carrier protein needed for the binding of charged tRNA to the A site of an active 70S ribosome. It acts in a multistep process that begins with the interaction of GTP with EF-Tu to form a binary complex (Figure 14-13). The binding of GTP causes a conformational change in EF-Tu, which allows a charged tRNA molecule to bind to the binary complex to form a ternary complex. In the ternary complex EF-Tu is situated near the CCA terminus of the acceptor stem of the tRNA molecule. The anticodon loop is, however, completely free for pairing with the codon.

The ternary complex is able to bind to the A site. There is some evidence that, as a result of codon–anticodon binding, the aminoacyl-tRNA undergoes a small conformational change that exposes the TψCG sequence in the TψC loop. This is thought to allow pairing of the tRNA to a sequence AAGC in the 5S RNA, thereby stabilizing binding to the A site. Once the ternary complex is in the A site, a ribosomal protein having GTPase activity causes hydrolysis of the GTP bound to EF-Tu to yield bound GDP and free P

The GDP-(EF-Tu) complex lacks the ability of the GTP-(EF-Tu) complex to bind to acylated tRNA, and its dissociation from the tRNA leaves the acylated tRNA in the A site. Removal of EF-Tu likewise is necessary for the formation of a peptide bond; the evidence for this requirement is that the addition of substances that either prevent GTPase activity or prevent removal of EF-Tu also inhibit peptide bond formation. Presumably, until EF-Tu is removed, the charged tRNA is not positioned correctly for bond formation to occur.

GDP binds more tightly to EF-Tu than does GTP. Hence, some means is required either to regenerate GTP in the binary complex or to dissociate the complex. The latter is done by the factor EF-Ts, which displaces the GDP, forming an (EF-Ts)-(EF-Tu) intermediate. That is,

$$\text{GDP-(EF-Tu)} + \text{EF-Ts} \rightleftharpoons \text{(EF-Tu)-(EF-Ts)} + \text{GDP}$$

The released GDP joins the internal pool of metabolic intermediates and is reconverted to GTP. The EF-Ts is displaced from the (EF-Ts)-(EF-Tu) complex when GTP-(EF-Tu) re-forms. That is,

$$\text{(EF-Ts)-(EF-Tu)} + \text{GTP} \rightleftharpoons \text{GTP-(EF-Tu)} + \text{EF-Ts}$$

EF-Ts is unable to bind to this complex, presumably because it does not recognize EF-Tu after the GTP-induced conformational change. In summary, the sole function of EF-Ts is to regenerate GTP-(EF-Tu) from GDP-(EF-Tu) by intermediate formation of (EF-Ts)-(EF-Tu); that is, the net reaction is

$$\text{GDP-(EF-Tu)} + \text{GTP} \xrightleftharpoons{\text{EF-Ts}} \text{GTP-(EF-Tu)} + \text{GDP}$$

The system has now reached the stage at which both the P and A sites are occupied and a peptide bond can form. This simple reaction is unusually complex and poorly understood. The problem in understanding this reaction is that although there is a peptidyl transferase enzymatic activity, no such enzyme has been isolated. The reaction occurs in a region of the 50S ribosomal subunit known as the **peptidyl transferase center.** Presumably, portions of the polypeptide chain of at least ten different 50S ribosomal proteins contribute to formation of the active site for this enzymatic activity. The overall chemical reaction in forming a peptide bond is shown in Figure 14-9.

Formation of the peptide bond is accompanied by cleavage of the bond connecting fMet and tRNA^{fMet}. (The cleavage enzyme, **tRNA deacylase,** is a ribosomal component.) Deacylated tRNA binds very poorly to the P site, so this tRNA leaves the ribosome immediately after peptide bond formation. In addition, the binding of the peptidyl-tRNA molecule that is now situated in the A site is weakened. The binding of peptidyl-tRNA to the P site is always strong, so there should be a tendency for the peptidyl-tRNA to move from the A site to the P site. This movement, which is called **translocation,** does occur, but is not dependent on random dissociation and reassociation (a random dissociation-reassociation mechanism could lead to premature termination). A third elongation factor, EF-G, controls this process. As in several previous binding events, the process in which EF-G participates begins by the formation of a complex with GTP. This is then followed by binding to the ribosome, hydrolysis of GTP and release of GDP. There are numerous theories that attempt to explain translocation and the role of EF-G, but the mechanism of translocation remains obscure. What is certain is that a ribosome cannot bind both EF-Tu and EF-G, so these alternate in binding to the ribosome during elongation.

Chain Termination

The presence of chain termination codons in the A site stimulates, in an unknown way, the appearance of release factors RF1 and RF2, each of which responds to a particular codon. A third factor, RF3, is also involved in termination; however, all that is known about RF3 is that it binds GTP and somehow enhances the binding of RF1 and RF2 to the ribosome. Chain termination is not a well-understood process. RF1 and RF2 have been purified and their amino acid sequences determined. They consist of 323 and 339 amino acids, respectively, and are about 35 percent homologous. This homology presumably reflects the common functional features of each protein; namely, both bind to a ribosome, recognize a termination codon, and interact with the peptidyl transferase complex. The homology also suggests that at one time only one ancestral RF gene existed, and the present genes resulted from an ancient gene duplication followed by evolutionary divergence. The existence of a single RF in eukaryotes, having the function of both RF1 and RF2, is consistent with this notion.

The Role of GTP

GTP, like ATP, is an energy-rich molecule. Generally, when such molecules are hydrolyzed, the free energy of hydrolysis is used to drive reactions that otherwise are energetically unfavorable. This does not seem to be the case in protein synthesis. A review of the reaction sequence indicates that *the role of GTP is to facilitate binding of protein factors* either to tRNA or to the ribosome. Furthermore, hydrolysis of GTP to GDP and P_i always precedes dissociation of the bound factor. Comparison of the structure of the free factor and the factor-GTP complex indicates that the factor undergoes a slight change in shape when GTP is bound. Thus, GTP is an allosteric effector—namely, a small molecule that can induce a shape change in a macromolecule by binding to it. Since it is easily hydrolyzed by various GTPases, the use of GTP as a controlling element allows a cyclic variation in macromolecular shape. That is, when GTP is bound, the macromolecule has an active conformation and when the GTP is hydrolyzed or removed, the molecule resumes its inactive form.

The differing roles of GTP and ATP may be summarized as follows: *GTP is used in binding reactions and in translocation, whereas ATP is used for peptide bond formation.*

Posttranslational Modification of Proteins

The protein molecule ultimately needed by a cell often differs from the polypeptide chain that is synthesized. There are several ways in which the modification of the synthesized chain occurs:

1. In prokaryotes, fMet is never retained as the NH_2-terminal amino acid. In roughly half of all proteins the formyl group is removed by the enzyme **deformylase,** which leaves methionine as the NH_2-terminal amino acid. In both prokaryotes and eukaryotes the fMet or Met, and possibly a few more amino acids, are often removed; this removal is catalyzed by a hydrolytic enzyme called **aminopeptidase.** This hydrolysis may sometimes occur while the chain is being synthesized and sometimes after the chain is released from the ribosome. The choice of deformylation versus removal of fMet usually depends on the identity of the adjacent amino acids. That is, deformylation predominates if the second amino acid is arginine, asparagine, aspartic acid, glutamic acid, isoleucine, or lysine, whereas fMet is usually removed if the adjacent amino acid is alanine, glycine, proline, threonine, or valine.

2. Newly created NH_2-terminal amino acids are sometimes acetylated.

3. Some amino-acid side chains may also be modified. For example, in collagen a large fraction of the proline and lysine are hydroxylated (Chapter 7, Figure 7-3). Phosphorylation of serine, tyrosine, and threonine (on the OH group) occurs in many organisms. Also, various sugars may be attached to the free hydroxyl group of serine or threonine to form glycoproteins. Finally a variety of prosthetic groups such as heme and biotin are covalently attached to some enzymes.

4. Two distant sulfhydryl groups in two cysteines may be oxidized to form a disulfide bond. This is very common.

5. Polypeptide chains may be cleaved at specific sites. For instance, chymotrypsinogen is converted to the digestive enzyme chymotrypsin by removal of four amino acids from two different sites (Figure 14-14). In some cases the uncleaved chain represents a storage form of the protein that can be cleaved to generate the active protein when needed. This is true of many

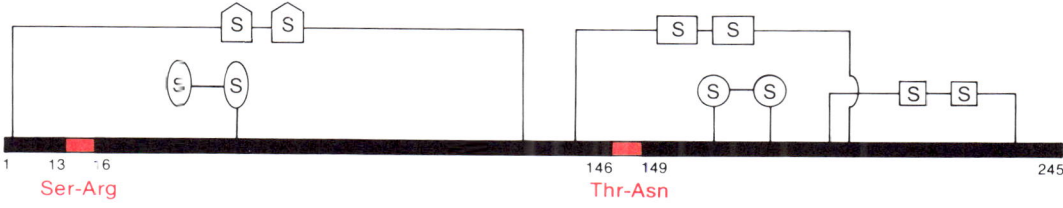

Excision of two dipeptides

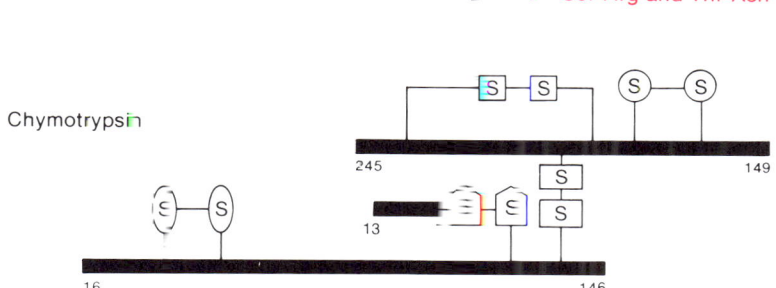

Ser-Arg and Thr-Asn

Figure 14-14 Conversion of chymotrypsinogen to chymotrypsin by removal of two pairs of amino acids (shown in red). The various disulfide bonds are given different shapes for reference only. The numbers refer to amino acid positions.

mammalian digestive enzymes—for example, pepsin is formed by cleavage of pepsinogen. An interesting precursor is a huge protein synthesized in cells infected with poliovirus; this molecule is cleaved at several sites to yield different active proteins and hence is called a **polyprotein.** We will see several examples of polyproteins in Chapters 16 and 23.

COMPLEX TRANSLATION UNITS

The unit of translation is almost never simply a ribosome traversing an mRNA molecule, but is a more complex structure, of which there are several forms. Some of these structures are described in this section.

After about 25 amino acids have been joined together in a polypeptide chain, the AUG initiation site of the encoding mRNA molecule is completely free of the ribosome. A second initiation complex then forms. The overall con-

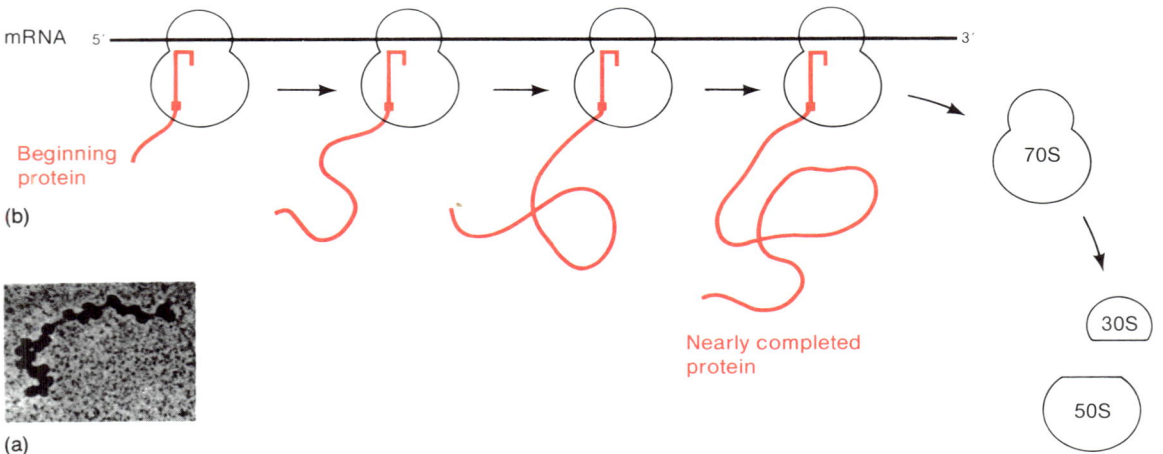

(b)

(a)

Figure 14-15   Polysomes. (a) Electron micrograph of an *E. coli* polysome. [Courtesy of Barbara Hamkalo.] (b) Diagram showing relative movement of the 70S ribosome and the mRNA, and growth of the protein chain.

formation is of two 70S ribosomes moving along the mRNA at the same speed. When the second ribosome has moved along a distance similar to that traversed by the first, a third ribosome is able to attach. This process—movement and reinitiation—continues until the mRNA is covered with ribosomes at a density of about one 70S subunit per 80 nucleotides. This large translation unit is called a **polyribosome** or simply a **polysome.** This is the usual form of the translation unit in all cells.

An electron micrograph of a polysome and an interpretive drawing are shown in Figure 14-15. The following features of a polysome should be noted:

1. The ribosomes move along the mRNA in the 5′→3′ direction.
2. There is a gradient in the length of the nascent polypeptide chain increasing in the 5′→3′ direction. This is because a ribosome nearer the 3′ end has been synthesizing protein for a greater time than one nearer the 5′ end.
3. In prokaryotes, the spacing between adjacent ribosomes is approximately constant. The codons being translated are about 80 nucleotides apart.
4. The total number of ribosomes per polysome is proportional to the length of the polypeptide chain and inversely proportional to the spacing between adjacent ribosomes. A rough value is about 1 ribosome per 25 codons. Thus, since typical proteins have a molecular weight ranging from 3 to 5×10^4, the segment of mRNA encoding them would be bound to about 11–18 ribosomes. In prokaryotes

most mRNA molecules are polycistronic, so polysomes with 40–50 ribosomes would be expected. Polysomes of this size have been detected, though not often because they are very fragile and easily broken.

The use of polysomes has a particular advantage to a cell—namely, the overall rate of protein synthesis is increased compared to the rate that would occur if there were no polysomes. Clearly, ten ribosomes traversing a single segment of mRNA can make ten times as many protein molecules per unit time as a single ribosome can.

Coupled Transcription and Translation in Prokaryotes

An mRNA molecule being synthesized has a free 5' terminus; hence, since translation occurs in the 5'→3' direction, each cistron contained in the mRNA is synthesized in a direction appropriate for immediate translation. That is, the ribosome binding site is transcribed first, followed in order by the AUG codon, the region encoding the amino acid sequence, and finally the stop codon. Thus, in bacteria, in which no nuclear membrane separates the DNA and the ribosome, there is no obvious reason why the 70S initiation complex should not form before the mRNA is released from the DNA. With pro-karyotes this process does indeed occur and is called **coupled transcrip-tion-translation.** This coupled activity does not occur in eukaryotes, because the mRNA is synthesized and processed in the nucleus and later transported through the nuclear membrane to the cytoplasm where the ribosomes are located. In prokaryotes, coupling of transcription and translation is the rule, though a few phage mRNA molecules are first processed after release from the DNA and then translated.

Coupled transcription-translation speeds up protein synthesis in the sense that translation does not have to await release of the mRNA from the DNA. Translation can also be started before the mRNA is degraded by nucleases.

Coupled transcription-translation was first detected by sedimentation analysis in which it was found that DNA, RNA polymerase, mRNA, ribo-somes, and nascent protein chains sedimented as a unit. Definitive evidence came from the remarkable electron micrograph shown in Figure 14-16(a). This micrograph shows a DNA molecule to which are attached a number of mRNA molecules, each associated with ribosomes. The micrograph is interpreted in Figure 14-16(b). Transcription of DNA begins in the upper left part of the micrograph. The lengths of the polysomes increase with distance from the transcription initiation site, because the mRNA is farther from that site and hence longer. Note that the longest polysome is shorter than the total distance from the initiation site; this is probably because some degradation of mRNA has occurred from its 5' end.

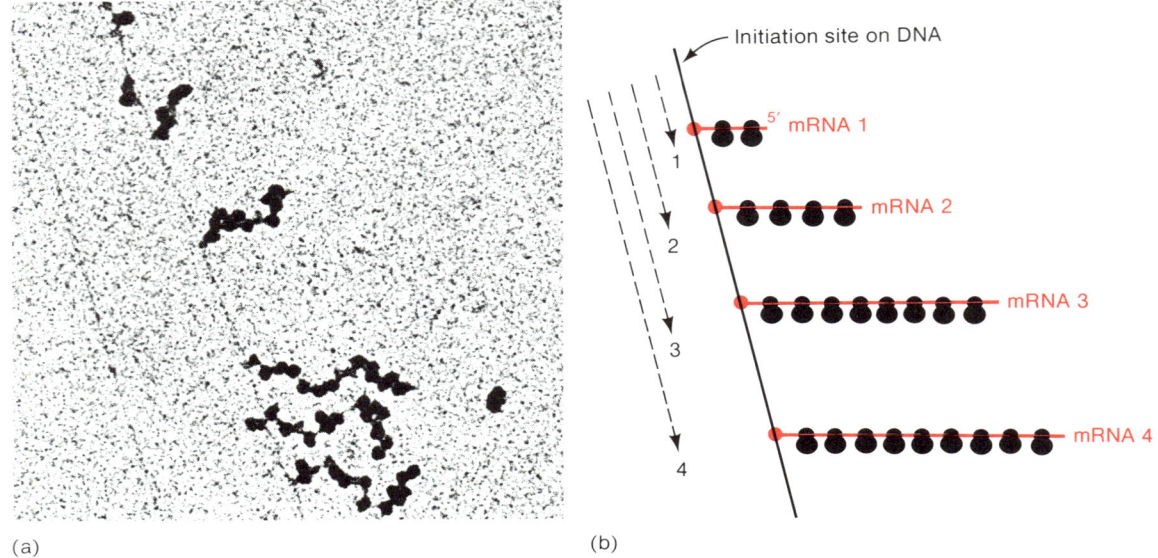

(a)   (b)

Figure 14-16 (a) Transcription of a section of the DNA of *E. coli* and translation of the nascent mRNA. Only part of the chromosome is being transcribed. The dark spots are ribosomes, which coat the mRNA. [From O. L. Miller, Barbara A. Hamkalo, and C. A. Thomas, *Science*, 169, 392 (1977)] (b) An interpretation of the electron micrograph of part (a). The mRNA is in red and is coated with black ribosomes. The large red spots are the RNA polymerase molecules; they are actually too small to be seen in the photo. The dashed arrows show the distances of each RNA polymerase from the transcription initiation site. Arrows 1, 2, and 3 have the same length as mRNA 1, 2, and 3; mRNA 4 is shorter than arrow 4, presumably because its 5′ end has been partially digested by an RNase.

SOME NUMERICAL PARAMETERS OF PROTEIN SYNTHESIS

In the analysis of growth rates and of metabolic regulation the numerical values of various rates of synthesis are needed. These values for growth at 37°C are independent of the doubling time of bacteria, as influenced by nutrition, and are listed in Table 14-2.

A typical protein molecule contains about 500 amino acids, so a protein molecule is synthesized in about one-half minute. Note that the polypeptide elongation rate is one-third that of the mRNA elongation rate, which is consistent with coupled transcription-translation and a coding ratio of three nucleotides per amino acid. From the half-life of mRNA and the peptide elongation rate, the average number of amino acids polymerized during the half-life of an mRNA molecule is known to be 1836. If a typical polycistronic mRNA molecule contains information for synthesizing three proteins, then roughly one copy of each protein molecule is made per half-life. This clearly indicates the need for continual synthesis of mRNA if a bacterium is to make hundreds of copies of a particular protein molecule in the course of one generation.

Table 14-2 Values of some parameters of protein and RNA synthesis in *E. coli*

Parameter	Value at 37°C*
Polypeptide elongation rate	17 amino acids/sec
mRNA elongation rate	55 nucleotides/sec
mRNA half-life	1.3–1.8 min
Amount of mRNA bound per ribosome	80 nucleotides

*37°C is the temperature for optimal growth of *E. coli*. At 30°C, a temperature commonly used in experiments, these values should be multiplied by 0.66.

INHIBITORS AND MODIFIERS OF PROTEIN SYNTHESIS

Many inhibitors of protein synthesis are known. They have been extremely useful in the study of protein synthesis because a particular substance typically blocks a single step in the complex protein synthetic process. The use of an inhibitor as an experimental reagent yields three types of information. First, if a particular step is blocked and other steps are not, then the uninhibited steps must precede the blocked step. By using this technique, the sequence of some of the stages in protein synthesis have been worked out. Second, blocking a step sometimes causes the accumulation of a previously unrecognized molecule. Such a molecule is likely to be a short-lived intermediate obtained from the preceding step. Third, if a step is blocked, the ribosomal protein responsible for that step can often be identified; this can be done either by showing that a particular ribosomal protein binds the antibiotic or, in the case of a microorganism, by first isolating a mutant that is resistant to the antibiotic and then determining which protein is altered.

Antibiotics that Affect Protein Synthesis

Many antibacterial agents (**antibiotics**) have been isolated from fungi. They have been widely used both clinically and as reagents for unraveling the details of protein synthesis and RNA and DNA synthesis. Table 14-3 lists a few of the hundreds of antibiotics that inhibit protein synthesis in bacteria. Some of them have only limited clinical usefulness because they also inhibit the growth of animal cells and hence are toxic to both bacterium and host.

A particularly well-studied antibiotic is **puromycin** (Figure 14-17). Puromycin is chemically similar to the aminoacyl part of a charged tRNA molecule, as shown in the figure, and it competes effectively with charged tRNA molecules for the A site. In addition, puromycin has an α-amino group that can form a peptide bond with the carboxyl group of a growing peptide chain in a reaction catalyzed by peptidyl transferase. When this bond forms, the polypeptide is cleaved from the tRNA in the P site and the pur-

Table 14-3 Antibiotic inhibitors of protein synthesis in prokaryotes

Antibiotic	Action
Streptomycin	Binds to the S12 protein of the 30S ribosomal subunit and thereby inhibits binding of tRNA^{fMet} to the P site. Also causes misreading in a system that is in the act of synthesis
Neomycin, kanamycin	Same as streptomycin
*Chloramphenicol (also called chloromycetin)	Inhibits peptidyl transferase of 70S ribosome
Tetracycline	Inhibits binding of charged tRNA to 30S particle
Erythromycin	Binds to free 50S particle and prevents formation of the 70S ribosome. Has no effect on an active 70S ribosome
*Puromycin	Causes premature chain termination by acting as an analogue of charged tRNA
*Fusidic acid	Binds to EF-G. GTP is still hydrolyzed and translocation occurs. However, EF-G and GDP are not released from ribosome, so the ribosome cannot bind another aminoacyl-tRNA
Kasugamycin	Inhibits binding of tRNA^{fMet}
Lincomycin	Inhibits peptidyl transferase complex
Kirromycin	Binds to EF-Tu; stimulates formation of (EF-Tu)-GTP and binding of ternary complex to ribosome; and inhibits release of EF-Tu
Thiostrepton	Prevents translocation by inhibiting EF-G

* Also active in eukaryotes. See Table 14-4.

omycin leaves the A site. Puromycin does not bind to the P site, so the polypeptide, which has a terminal puromycin, falls away from the ribosome. Thus, in the presence of puromycin, polypeptide chain elongation is blocked.

Another well-studied antibiotic is streptomycin (Figure 14-18). It interferes with the binding of tRNA^{fMet} to the P site and thereby inhibits the initiation of synthesis of a polypeptide chain. Streptomycin also has another effect that is evident when its concentration is sufficiently low that initiation can still occur. In this case extensive misincorporation of amino acids is observed. This happens because streptomycin alters codon-anticodon recognition—that is, it induces misreading of the code. For example, in an *in vitro* system using poly(U) as mRNA it is found that in addition to phenylalanine (UUU codon), four other amino acids are sometimes incorporated—isoleucine (AUU), leucine (CUU), tyrosine (UAU), and serine (UCU).

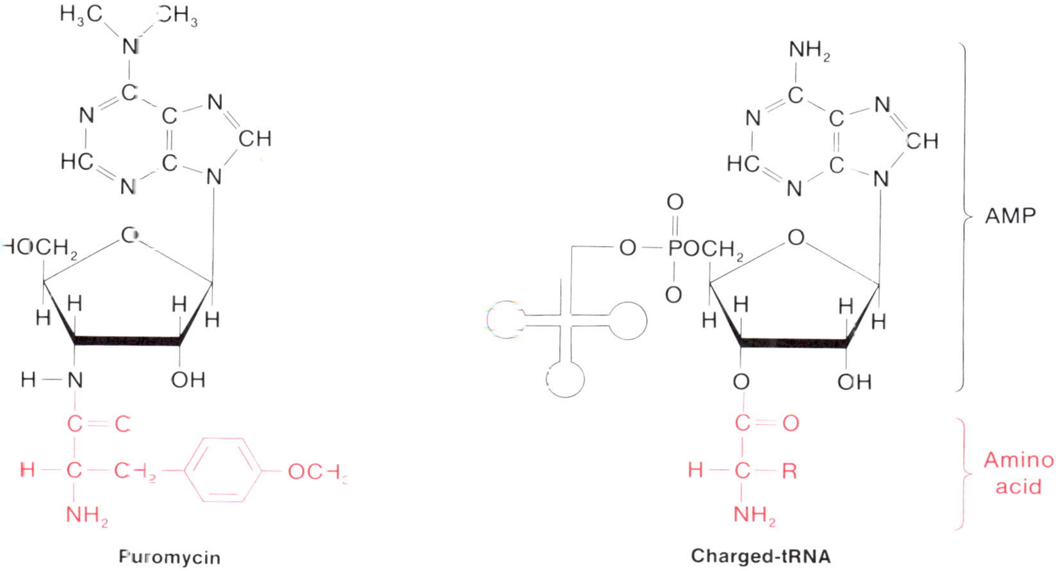

Figure 14-17 Puromycin resembles the aminoacyl terminus of an aminoacyl-tRNA. The cloverleaf represents all of the tRNA except the terminal AMP, whose structure is drawn out. The amino acid in each structure is shown in red.

Isoleucine and leucine are substituted much more often than tyrosine and serine, which suggests that the first base of a codon is misread more often than the second base. Mutants have been isolated that are resistant to the drug (they are Str-r). These mutants have an altered S12 ribosomal protein that enables them to initiate polypeptide synthesis in the presence of streptomycin.

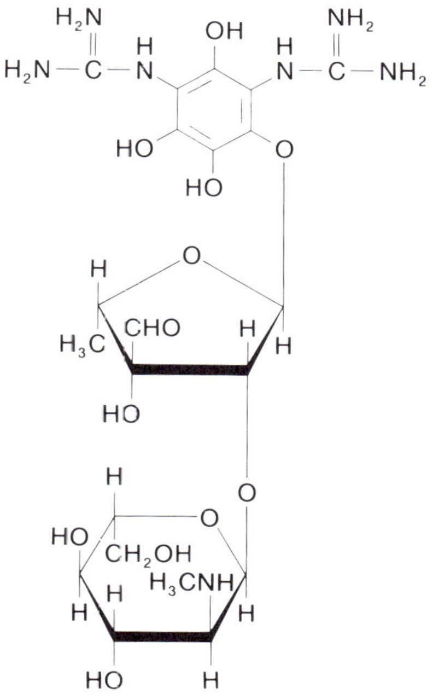

Inhibitors of Protein Synthesis in Eukaryotes

The nontoxic antibiotics that have no effect on eukaryotes are nontoxic to them either because they fail to penetrate the eukaryotic cell membrane (which is quite common) or because they do not bind to eukaryotic ribosomes. The former type are effective inhibitors of protein synthesis *in vitro*, in systems obtained either from prokaryotes or eukaryotes. The differences in effectiveness of antibiotics on the two classes of cells *in vivo* is the basis for the use of these drugs. That is, they kill bacterial cells but not animal cells.

Some antibiotics are active against both bacterial and mammalian cells. One example is **chloramphenicol** (also called chloromycetin), which inhibits peptidyl transferase in both bacterial and mitochondrial ribosomes, though normal cytoplasmic ribosomes are unaffected. Such a drug can still be clin-

Figure 14-18 Structure of streptomycin.

Table 14-4 Inhibitors of protein synthesis in eukaryotes

Inhibitor	Action
Abrin, ricin	Inhibits binding of aminoacyl tRNA
Diphtheria toxin	Enzymatic catalysis of a reaction between NAD^+ and eEF2 to yield an inactive factor. Inhibits translocation
*Chloramphenicol (also called chloromycetin)	Inhibits peptidyl transferase of mitochondrial ribosomes. Inactive against cytoplasmic ribosomes
*Puromycin	Causes premature chain termination by acting as an analogue of charged tRNA
*Fusidic acid	Inhibits translocation by altering an elongation factor
Cycloheximide (also called actidione)	Inhibits peptidyl transferase
Pactamycin	Inhibits positioning of $tRNA_f^{Met}$ on the 40S ribosome
Showdomycin	Inhibits formation of the $eIF2$-$tRNA_f^{Met}$-GTP complex
Sparsomycin	Inhibits translocation

*Also active on prokaryotic ribosomes. See Table 14-3.

ically useful if there is a concentration range in which the antibacterial effect is substantial and the toxic effect to the host is weak. However, because of their potential for host toxicity, these antibiotics tend to be used only in serious infections when all else fails.

There are also many drugs that act either mainly or significantly on eukaryotes. Some of these are toxins of bacterial origin while others are synthetic. An example of the former is the toxin produced by *Corynebacterium diphtheriae,* the bacterium that causes diphtheria. This toxin is an enzyme that causes covalent modification of the eukaryotic elongation factor needed for translocation and thereby inhibits that step. **Cycloheximide** is a chemical inhibitor of the peptidyl transferase complex of the 60S subunit and hence inhibits formation of the peptide bond. Substances like cycloheximide are commonly used in cancer chemotherapy.

Table 14-4 lists a few of the better known inhibitors of protein synthesis in eukaryotic cells.

PROTEIN SYNTHESIS IN EUKARYOTES

The basic pattern of protein synthesis in eukaryotes is similar to that in bacteria. (This discussion does not apply to mitochondria and chloroplasts, which will be discussed later.) That is, a ribosomal subunit known as $43S_n$, which already contains some initiation factors, first forms a complex with the initiator tRNA and with several factors to form a $43S_n$ **preinitiation complex.** This is followed sequentially by binding of mRNA and the 60S ribosomal subunit to form an active 80S initiation complex. Chain elongation then occurs until a termination codon is reached. Each step requires several protein factors. To distinguish these factors from those used by prokaryotes, a lower case "e" precedes the usual symbol. Thus, the first initiation, elongation, and release factors are written eIF1, eEF1, and eR1, respectively.

The process of protein synthesis in eukaryotes is not known in as much detail as in prokaryotes because there are more factors and not all of the factors have yet been purified. There are many differences in detail between protein synthesis in prokaryotes and eukaryotes. Rather than describe the complex (and incompletely understood) sequence of events in eukaryotes, it is more informative to examine the differences between the systems.

Differences Between Protein Synthesis in Eukaryotic and Prokaryotic Cells

The following differences have been noted between protein synthesis in eukaryotic and prokaryotic cells:

1. In eukaryotes the initiating amino acid is methionine and not fMet. The initiating tRNA, which responds only to AUG, is designated $tRNA_{init}^{Met}$ or $tRNA_f^{Met}$ to distinguish it from the $tRNA^{Met}$ used in translating internal AUG codons. This has already been described in Chapter 13.

2. At least nine initiation factors plus GTP are required for binding of $tRNA_f^{Met}$ to the preinitiation complex. Two of these are similar to the factors in *E. coli* that prevent binding of the large subunit to the small subunit. One of these factors, eIF3, is a large complex containing nine protein subunits and having a molecular weight of 7×10^5. An enzyme called **initiating tRNA hydrolase,** which cleaves the bond between methionine and $tRNA_f^{Met}$ removes the initial tRNA molecule after the first peptide bond forms, is also present in the complex. In prokaryotes the corresponding enzyme, tRNA deacylase, is a ribosomal component.

3. Binding of $tRNA_f^{Met}$ must occur before mRNA can bind, whereas for prokaryotes the mRNA can bind either before or after binding of the initiator tRNA. For binding of mRNA two other initiation factors are needed and ATP must be cleaved to form ADP and P_i. The reason for the cleavage of ATP is unknown. Binding occurs initially at or near the 5' cap and is mediated by a cap-binding factor (which is unnecessary for uncapped viral

mRNA); the mRNA moves with respect to the preinitiation complex until the anticodon of the prior bound $tRNA_f^{Met}$ pairs with the first AUG encountered. No Shine-Dalgarno sequence is needed. The fact that the AUG codon nearest the 5′ terminus is almost always the initiating codon is a significant difference between eukaryotes and prokaryotes and plays an important role in metabolic regulation, as will be seen in Chapters 16 and 23.

4. More factors are needed for binding of the 60S subunit than for binding of the bacterial 50S subunit.

5. At least four elongation factors are needed by eukaryotes. These factors probably differ in structure and size in different tissues, often forming aggregates containing as many as 50 monomers. The factors, which in terms of function roughly correspond to the *E. coli* factors, are listed in the following chart:

Elongation factor	*E. coli* factor
$eEF1_\alpha$	EF–Tu
$eEF1_\beta$	EF–Ts
$eEF1_\gamma$	None
eEF2	EF–G

The reactions that are catalyzed in eukaryotic protein synthesis do not exactly correspond to those in prokaryotic protein synthesis though.

6. Little is known about termination in eukaryotes, though release factors have been purified from several systems. Surprisingly, release in *in vitro* systems requires the presence of one of four tetranucleotides—UAAA, UAGA, UGAA, or UAGG.

Two other differences between eukaryotes and prokaryotes in the overall process of protein production should be noted. (1) The most striking is that transcription and translation are not coupled in eukaryotes. The mRNA is synthesized in the nucleus, where it is processed, and then transported to the cytoplasm, where the ribosomes are located. (2) The second point concerns the lifetime of mRNA. In prokaryotes degradation of mRNA occurs continuously and while translation is in process. Thus the mRNA is continually being shortened and translation can no longer be initiated once the Shine-Dalgarno sequence has been degraded. The half-life of a typical bacterial mRNA is about 1.8 minutes. Eukaryotic mRNA is very stable—possibly because of the 5′ cap; degradation (by unknown enzymes) occurs very slowly and a typical half-life is several hours. Some of the more stable mRNA molecules have a half-life of a few days. Hemoglobin mRNA, whose half-life is several hours, is translated about 17,000 times on the average before it is degraded.

Bound Ribosomes in Eukaryotic Cells: The Endoplasmic Reticulum

The ribosomes in most prokaryotes are distributed throughout the cell without any apparent spatial organization (except for a minor fraction that is bound to the inner membrane just under the cell wall). However, in most types of eukaryotic cells there are two major classes of ribosomes—attached ribosomes and free ribosomes. The attached ribosomes are bound to an extensive internal network of lipoprotein membranes called the **endoplasmic reticulum** (Figure 14-19). The membranes of the endoplasmic reticulum to which ribosomes are bound are called the rough endoplasmic reticulum; those devoid of ribosomes comprise the smooth endoplasmic reticulum. There is no structural difference between a free and an attached ribosome, and attachment occurs after protein synthesis begins, in part by means of the nascent polypeptide (explained shortly); whether a ribosome does or does not attach apparently depends upon the particular protein that is synthesized.

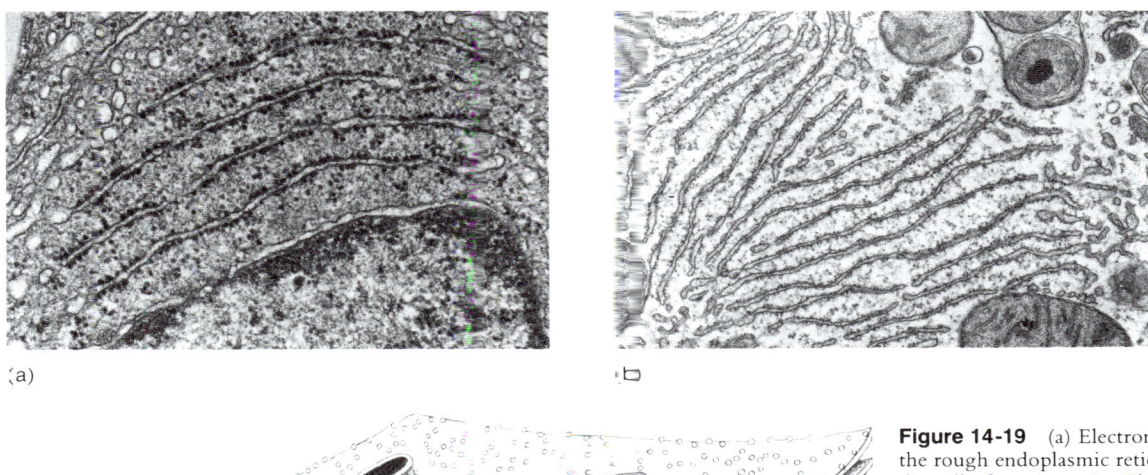

(a)

(b)

(c)

Figure 14-19 (a) Electron micrograph of the rough endoplasmic reticulum in the Leidig cell of a guinea pig. The dark circles are ribosomes. (b) The more complex endoplasmic reticulum of the rat liver. [Courtesy of Daniel Friend.] (c) A three-dimensional representation of the rough endoplasmic reticulum. [From S. Wolfe, *Biology of the Cell* (Wadsworth, 1981).]

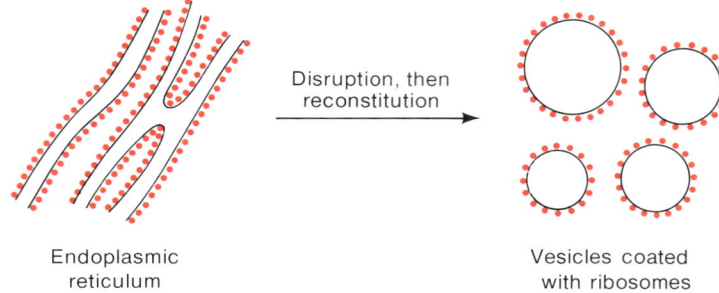

Figure 14-20   An experiment that demon-
strated that ribosomes are on the "outer" surface of the membranes of the endoplasmic reticulum.

The endoplasmic reticulum is a highly convoluted membrane system. Sometimes the membranes enclose a discrete region of the cell called a **cisterna.** A large fraction of the system probably forms one or more large, irregularly shaped cisternae. In this sense, the membrane system has an inside and an outside, and ribosomes are only bound to the outside. Proof of this point comes from an experiment explained in Figure 14-20. In this experiment rough endoplasmic reticulum is broken into small fragments. If these fragments are allowed to reunite, small spherical vesicles form and the ribosomes are found only on the outer surface of the vesicles. These vesicles can be used as a source of ribosomes in an *in vitro* protein-synthesizing system. When this is done, the newly synthesized protein molecules are found inside the vesicles (Figure 14-21)—this important finding provides a key to understanding the function of the endoplasmic reticulum.

Cells responsible for secreting large amounts of a particular protein (for instance, hormone-secreting cells) have a very extensive endoplasmic reticulum. Furthermore, most (but not all) proteins destined either to be secreted

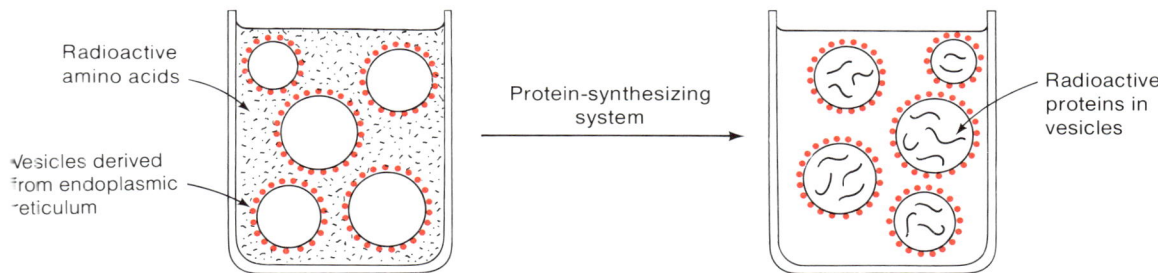

Figure 14-21   An experiment showing that protein synthesized on ribosomes bound to the endoplasmic reticulum passes through the membrane. Vesicles derived from the endoplasmic reticulum and coated with ribosomes are incubated in a solution containing a protein-synthesizing system and radioactive amino acids. Radioactive protein is synthesized; all of the protein is contained in the cavity of the vesicles.

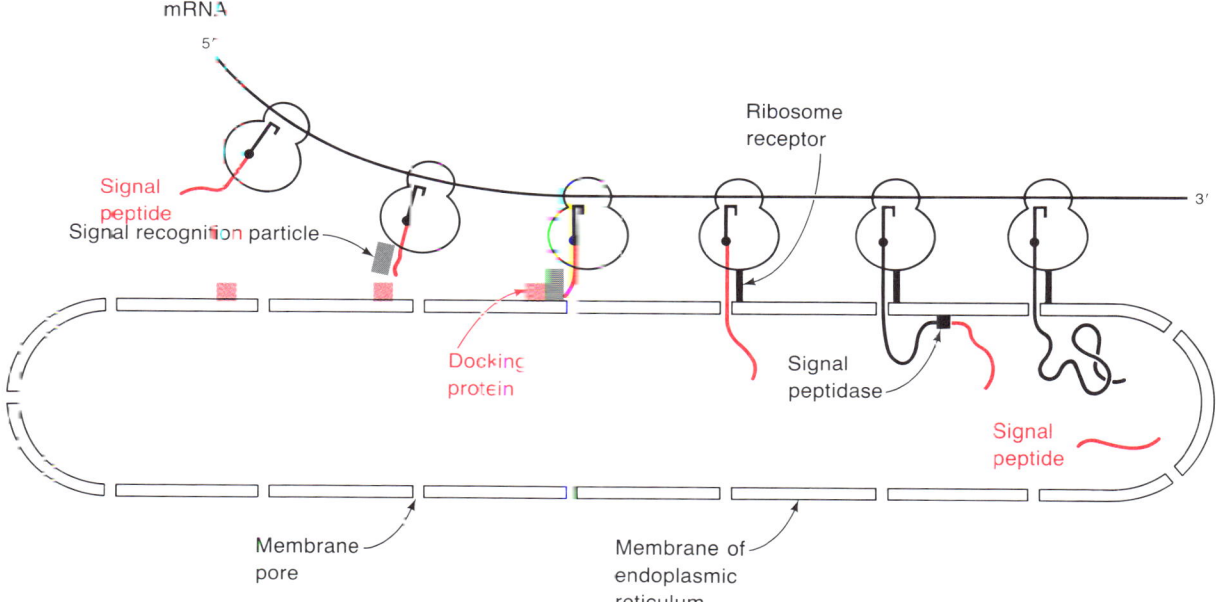

Figure 14-22 Signal hypothesis for the synthesis of secretory and membrane proteins. Shortly after initiation of protein synthesis the amino-terminal sequence of the polypeptide chain binds a signal recognition protein, which then binds to a docking protein. The signal peptide is released from the signal recognition protein as the ribosome binds to a ribosome receptor, which is adjacent to a pore. Translation continues with the signal peptide passing through the pore. Once through the endoplasmic reticulum, the signal sequence is excised by the signal peptidase within the vesicle. When protein synthesis is completed, the protein remains within the vesicle and the ribosome is released.

by the cell or to be stored in natural intracellular vesicles such as lysosomes (which contain degradative enzymes) are synthesized by attached ribosomes. These proteins are initially found in the cisternae of the endoplasmic reticulum. In contrast, most (but not all) proteins destined to be free-floating in the cytoplasm are made on free ribosomes.

The current explanation for the formation of the rough endoplasmic reticulum and the mechanism by which newly synthesized proteins pass through the membrane is called the **signal hypothesis,** which is shown in Figure 14-22. The basic idea of the signal hypothesis is that the signal for attachment of the ribosome to the membrane is a sequence of very hydrophobic amino acids near the amino terminus of the growing polypeptide chain. When protein synthesis begins, the ribosome is free. The hydrophobic amino-terminal sequence interacts with a cytoplasmic ribonucleoprotein particle called the **signal recognition particle (SRP).** The SRP then binds to a protein that is in or on the endoplasmic reticulum; this protein is called both the **SRP receptor** and the **docking protein.** In an unknown way the ribosome and the signal sequence detach from the SRP and the ribosome binds to the **ribosome receptor,** a protein in the endoplasmic reticulum

NH₂–Met–Arg–Ser–Leu–Leu–Ile–Leu–Val–Leu–Cys–Phe–Leu–Pro–Leu–Ala–Ala–Leu–Gly┃Gly–Lys . . .

NH₂–Met–Lys–Trp–Val–Thr–Phe–Leu–Leu–Leu–Leu–Phe–Ile–Ser–Gly–Ser–Ala–Phe–Ser┃Arg . . .

Figure 14-23 Signal sequences of two secretory proteins. The hydrophobic amino acids are printed in red. The vertical bars denote the points of cleavage by the signal peptidase.

and probably adjacent to a pore in the membrane. There is some evidence that translocation may stop while the signal sequence is bound to the SRP and then resumes when the ribosome binds to the ribosome receptor. At any rate, as protein synthesis proceeds, the protein moves through the membrane to the cisternal side of the endoplasmic reticulum. A specific protease termed the **signal peptidase** then cleaves off the amino-terminal signal sequence.

Two important experimental results support the signal hypothesis: (1) If *in vitro* synthesis of several secretory proteins is carried out with free ribosomes, the resulting proteins contain NH₂-terminal sequences, about 20 amino acid residues long, that are not present in the proteins isolated from intact cells. (2) The amino acids in these NH₂-terminal extensions are very rich in hydrophobic amino acids (Figure 14-23).

Some of the integral proteins of the membrane of the endoplasmic reticulum are made by the bound ribosomes. These proteins do not pass through, but remain within, the membrane. Presumably these proteins possess amino acid sequences that prevent transfer through the membrane before the carboxyl terminus is reached.

WHY IS PROTEIN SYNTHESIS SO COMPLICATED?

The reader cannot fail to notice the extraordinary complexity of protein synthesis, and it is natural to ask why it is so complicated. At least in part, the answer is surely to guarantee accuracy while allowing protein synthesis to occur rapidly. Most of the ribosomal proteins seem to be designed to form a surface on which two tRNA molecules are precisely positioned both with respect to one another and with respect to the codons. The various protein factors probably also serve this function, though why they are not ribosomal components is not obvious. The need for the dissociation cycle of the 70S ribosome is also unclear. It is difficult to conceive of a system in primordial time being as complicated as the contemporary one. More likely, a primitive highly error-prone system initially existed and in time it was improved as natural selection acted on spontaneous mutations. Whereas the resultant complexity indeed seems horrendous, if a new capability in a system either reduced the fraction of defective molecules that were synthesized and/or increased the rate of polypeptide chain growth, a cell having the new capability would grow slightly faster than other less efficient cells. Thus, in time the new cell type would become the predominant species even if its synthetic mechanisms were very complicated. The critical factors are always the survival value or the growth rate, not simplicity.

Compared to prokaryotes, eukaryotes have 27 more ribosomal proteins and one more RNA molecule, and the ribosomes, as well as all the components of the ribosomes, are larger than those of prokaryotes. Furthermore, eukaryotic protein synthesis requires more accessory factors. Presumably, this larger number of components reflects the greater needs of a cell that is more complex than a bacterium. The greater stability of eukaryotic mRNA compared to prokaryotic mRNA, which is a valuable attribute in slowly growing cells often committed to production of large quantities of a few proteins is probably in part responsible for the large number of factors in eukaryotic protein synthesis. For example, rapid short-term changes in the overall rate of protein synthesis necessitated by environmental fluctuations are accomplished in bacteria by changing the rate of production of particular mRNA molecules (Chapter 15); when synthesis is turned off, the concentration of the mRNA rapidly decreases. This mechanism cannot be the only one in eukaryotes, for which some environmental responses must be mediated by regulation of translation of stable mRNA molecules. That is, changes in the concentration of a particular protein must often be effected by changing the rate of initiation of protein synthesis. One would expect this to necessitate having a multifactor system that can respond to various regulatory signals. In Chapter 16 some examples of regulated systems in which regulation depends on inactivation of accessory factors will be seen.

Regulation of the Activity of Genes and of Gene Products in Prokaryotes

The number of molecules of a protein produced per unit time from a particular gene differs from gene to gene. We have seen in the previous chapters that this can result from varying efficiencies of promoter recognition by RNA polymerase and of initiation of translation. However, this is only one way that the flow of genetic information (**gene expression**) is regulated. For example, the products of many genes are needed only occasionally, and for such a gene the product is usually not present in significant amounts except when circumstances demand it. Thus, not only are there quantitative differences in the amounts of the different gene products that are present but gene activity can apparently be regulated.

In this chapter we consider why gene activity is regulated and the basic mechanisms of metabolic regulation, and we describe numerous examples of well-understood regulated prokaryotic systems. Regulation of eukaryotic systems is described in Chapter 16.

GENERAL ASPECTS OF REGULATION

In this section some of the general features of regulated systems are discussed.

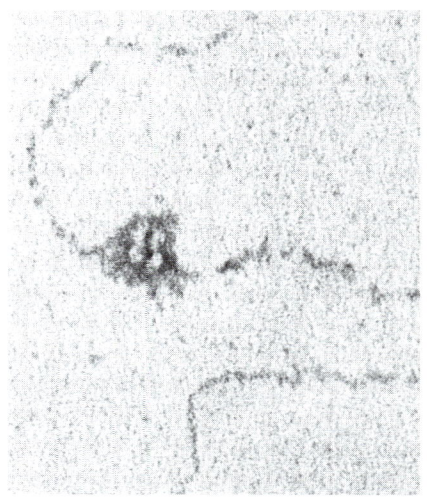

An electron micrograph showing the *E. coli* phage λ repressor bound to its operator in λ DNA. [Courtesy of C. Brack.]

Why Gene Activity is Regulated

Natural selection improves efficiency. Among the unicellular organisms, any mutation that increases the overall efficiency of cellular metabolism should enable a mutant cell to grow slightly faster than a wild-type organism. Thus, if enough time is allowed, a mutant cell line will outgrow a wild-type one. For example, in a population of 10^9 bacteria with a 30-minute doubling time if one bacterium is altered such that it has a 29.5-minute doubling time, in about 80 days of continued growth, 99.9 percent of the population will have a 29.5-minute doubling time. (In this calculation it is assumed that the growing culture is repeatedly diluted; otherwise, unless the volume were greater than that of the earth, the culture would stop growth in a day or so.) This time may seem very long on the laboratory time scale but it is infinitesimal on the evolutionary scale. Thus it is reasonable on this basis alone that regulated systems, in which efficiency has been improved, should have evolved.

There are two principal ways for a cell to become more efficient: (1) it can develop a new process that requires less free energy or that simply proceeds at a greater rate, or (2) it can eliminate waste. The second mode prevails in bacteria and is the concern of this chapter.

There are several more-or-less-general rules of intracellular regulation in bacteria. These are the following:

1. Molecules that are only occasionally needed are synthesized only when the need arises.
2. An enzymatic activity that uselessly consumes energy or consumes a molecule that is the substrate of a second enzyme is usually inhibited.
3. If a cell has a choice of utilizing more than one catabolic pathway for energy utilization, it will choose the pathway that yields the greater amount of energy per unit time.
4. An alteration of a biosynthetic pathway that reduces the production of defective molecules is efficient and hence valuable.

General Mechanisms of Metabolic Regulation in Bacteria

The most elementary regulatory mechanisms used in bacteria obey the following rule:

A system is turned on when it is needed and off when it is not needed.

This type of on-off activity is accomplished by regulating transcription—that is, mRNA synthesis is either allowed or prevented—or by regulating enzyme activity. Actually, there are no known examples of switching a system completely off. When transcription is in the "off" state, there always remains a basal level of gene expression. This often amounts to only one or two transcription events per cell generation and thus very little protein synthesis. For convenience, when discussing transcription, we will use the term "off," but it should be kept in mind that what is meant is "very low." We will also see examples in this chapter of systems whose activity is switched from fully on to partly on (or even slightly on) rather than to off.

In bacterial systems in which several enzymes act in sequence in a single metabolic pathway, it is often the case that either all of these enzymes are present or all are absent. This phenomenon, which is called **coordinate regulation,** results from control of the synthesis of a single polycistronic mRNA that encodes all of the gene products. There are several mechanisms for this type of regulation, as will be seen in the next major unit of this chapter.

Enzyme activity is commonly regulated by the concentration of the reaction product or, in the case of a biosynthetic pathway, by the product of the reaction sequence. This mode of regulation is called **feedback inhibition** or **end-product inhibition.** Small effector molecules are also frequently used either to activate or inhibit a particular enzyme. Regulation of enzyme activity will be discussed in a later unit of this chapter.

Types of Regulation of Transcription

There are several common patterns of regulation of transcription. These depend on the type of metabolic activity of the system being regulated. For example, in a catabolic (degradative) system the concentration of the substrate of an enzyme early in the pathway (often the enzyme catalyzing the first step) often determines whether the enzymes in the pathway are synthesized. In contrast, in an anabolic (biosynthetic) pathway it is often the case that the final product is the regulatory substance. Even in a system in which a single type of protein molecule is translated from a monocistronic mRNA, the protein may "autoregulate" in the sense that the transcriptional activity of the promoter is determined directly by the concentration of the protein. The molecular mechanisms for each of the regulatory patterns vary quite widely but usually fall in one of two major categories—**negative regulation** or **positive regulation.**

In negative regulation, an inhibitor is present in the cell and the inhibitor keeps transcription turned off. An anti-inhibitor, generally called an **inducer,** is needed to turn the system on. In positive regulation, an effector molecule (which may be a protein, a small molecule or a molecular complex) activates a promoter; an inhibitor is not overridden in positive regulation. Negative and positive regulation are not mutually exclusive, and some systems are both positively and negatively regulated and need two regulators. Properties of negative and positive control are summarized in Table 15-1.

Table 15-1   Types of regulation

Binding of regulator to DNA	Positive	Negative
Yes	On	Off
No	Off	On

Note: If a system is both positively and negatively regulated, it is "on" when the positive regulator is bound to the DNA and the negative regulator is not bound to the DNA.

A degradative system may be regulated either positively or negatively. In a biosynthetic pathway, the final product of the pathway often negatively regulates its own synthesis; in the simplest mode, absence of the product encourages synthesis and presence of the product inhibits its synthesis.

In this chapter we will examine a variety of regulated bacterial systems. We begin with the best-understood system, the genes responsible for lactose utilization, as study of the lactose system has provided both the language and the principles required to understand regulation.

THE LACTOSE SYSTEM AND THE OPERON MODEL

Metabolic regulation was first studied in detail in the system responsible for degradation of the sugar lactose, and most of the terminology used to describe regulation came from genetic analysis of this system. We begin with a description of the original observations.

Genetic and Biochemical Evidence for the Existence of Two Proteins Required to Degrade Lactose

In *E. coli* two proteins are necessary for the metabolism of lactose. These proteins are the enzyme **β-galactosidase,** which cleaves lactose to yield galactose and glucose (Figure 15-1) and a carrier molecule, **galactoside permease,** which is required for the entry of lactose (and other galactosides) into the cell. The existence of two proteins was first shown by a combination of genetic experiments and biochemical analysis.

First, hundreds of Lac$^-$ mutants (unable to use lactose as a carbon source) were isolated. By genetic manipulation some of these mutations were moved from the *E. coli* chromosome to an F'*lac* plasmid (a plasmid carrying the genes for lactose utilization) and then partial diploids having the genotypes F'*lac$^-$*/*lac$^+$* or F'*lac$^+$*/*lac$^-$* were constructed. (The genotype of the plasmid is given to the left of the diagonal line and that of the chromosome to the right.) It was observed that these diploids always produced a Lac$^+$

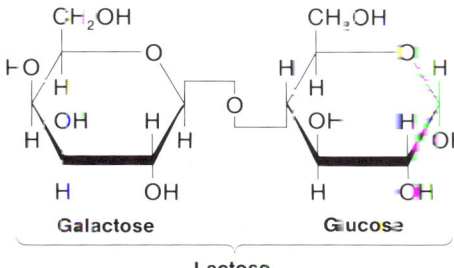

Figure 15-1 Structure of lactose.

phenotype (that is, they made β-galactosidase) which showed that none of the lac^- mutants make an inhibitor that prevents functioning of the lac gene. Partial diploids were also constructed in which both the chromosome and the F'lac plasmid were lac^-. Using different pairs of lac^- mutants, it was found that some pairs were phenotypically Lac$^+$ and some were Lac$^-$. This complementation test (Chapter 1) showed that all of the mutants initially isolated fell into one of two groups, which were called $lacZ$ and $lacY$. (For the sake of simplicity of notation, in the discussion of the lac system, one-letter designations for genetic elements will be used. Thus, the genes $lacA$, $lacI$, $lacY$, $lacZ$ and the noncoding regions ("sites") $lacO$ and $lacP$ will simply be written a, i, y, z, o, and p, respectively.) Mutants in the two groups z and y have the property that the partial diploids F'$y^- z^+ / y^+ z^-$ and F'$y^+ z^- / y^- z^+$ have a Lac$^+$ phenotype and the genotypes F'$y^- z^+ / y^- z^+$ and F'$y^+ z^- / y^+ z^-$ have the Lac$^-$ phenotype. The existence of two complementation groups was good evidence that there are at least two genes in the lac system.

Experiments in which cells were exposed to $[^{14}C]$lactose showed that radioactive material cannot enter a y^- cell, whereas a z^- mutant readily takes it up. Treatment of a y^- cell in various ways that disrupt the cell membrane and make it very permeable enables a y^- cell to take up $[^{14}C]$lactose. This experiment shows that the y gene is probably concerned with lactose transport; other experiments, which will not be described, showed that the z gene is the structural gene for β-galactosidase. A final important result was obtained by genetic mapping: the y and z genes are adjacent.

Evidence that the *lac* System is Regulated: Inducible and Constitutive Synthesis and Repression

The on-off nature of the lactose-utilization system is shown by the following observations:

1. If a culture of *E. coli*, whose genotype is lac^-, is growing in a medium lacking lactose or any other β-galactoside, the intracellular concentrations of β-galactosidase and permease are exceedingly low—roughly one or two

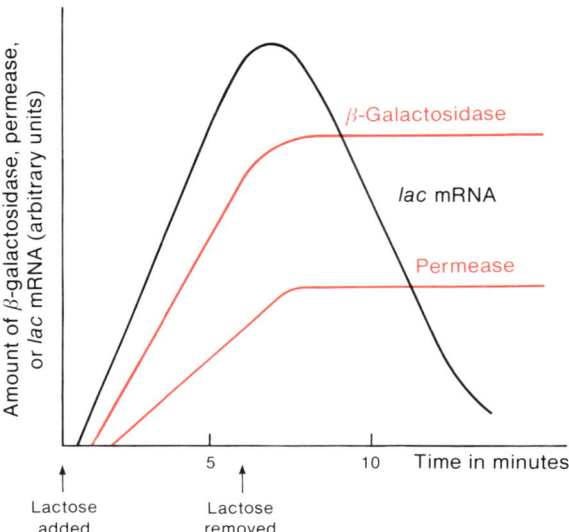

Figure 15-2 The "on-off" nature of the *lac* system. *Lac* mRNA appears very soon after lactose is added; β-galactosidase and permease appear at the same time but are delayed with respect to mRNA synthesis because of the time required for translation. When lactose is removed, no more *lac* mRNA is made and the amount of *lac* mRNA decreases owing to the usual degradation of mRNA. Both β-galactosidase and permease are stable. Their concentration remains constant even though no more can be synthesized. A third protein of the *lac* system, thiogalactoside transacetylase, is synthesized coordinately with β-galactosidase and permease. This protein, the product of the *a* gene, will be discussed later.

molecules per bacterium. However, if lactose is present in the growth medium, the concentration of these proteins is about 1 percent of the total cellular protein (about 10^5 molecules per cell).

2. If lactose is added to a *lac*$^+$ culture growing in a lactose-free medium (also lacking glucose, a point that will be discussed shortly), both β-galactosidase and permease are synthesized simultaneously, as shown in Figure 15-2. Furthermore, hybridization of mRNA labeled with ^{32}P at various times after addition of lactose, with DNA obtained from a hybrid phage λ*lac,* which carries the *lac* genes, shows that the addition of lactose triggers synthesis of *lac* mRNA.

These facts led to the view that the lactose system is **inducible** and that lactose is an **inducer.** Studies of the mechanism of induction have provided the foundation of our understanding of genetic regulation.

Lactose is rarely used in experiments to study induction because the β-galactosidase that is synthesized catalyzes the cleavage of lactose; thus, the lactose concentration continually decreases and this complicates the analysis of many types of experiments (for example, kinetic experiments). Instead, two sulfur-containing analogues of lactose in particular isopropylthioga-lactoside, IPTG, and thiomethylgalactoside TMG,

Isopropylthiogalactoside
(IPTG)

Thiomethylgalactoside
(TMG)

which are effective inducers without being substrate of β-galactosidase, are used. Inducers having this property are called **gratuitous inducers.**

E. *coli* mutants have been isolated that make *lac* mRNA (and hence β-galactosidase and permease) in the presence as well as absence of an inducer. These unregulated mutants are called **constitutive.** A variety of partial diploid cells containing constitutive mutants were constructed and it was found that these mutants are of two types, termed i and o. The characteristics of the mutants are shown in Table 15-2. The i^- mutants behave like typical minus mutations in most genes and are recessive (entries 3, 4). Since *lac* mRNA synthesis is off in an i^+ cell and on in an i^- mutant, the i gene is apparently a regulatory gene whose product is an inhibitor that keeps the system turned off. An i^- mutant lacks the inhibitor and thus is constitutive. One good copy of the i-gene product is present in an i^+/i^- partial diploid, so the system is inhibited. The i-gene product is called the **lac repressor.** When these observations were first made, it was unclear whether the *lac* repressor was a protein or an RNA molecule that was transcribed from the i gene and does not encode a protein. However, the discovery of amber (chain termination) mutants in the i gene indicated strongly that the repressor is a protein and in 1967 the *lac* repressor was purified. (Properties of this protein will be discussed shortly.) Genetic mapping experiments placed the i gene adjacent to the z gene and established the gene order i z y.

Table 15-2 Characteristics of partial diploids carrying various combinations of i and o mutants

Genotype	Constitutive (C) or inducible (I)
1. $F'o^cz^+ / o^+z^+$	C
2. $F'o^+z^+ / o^cz^+$	C
3. $F'i^-z^+ / i^+z^+$	I
4. $F'i^+z^+ / i^-z^+$	I
5. $F'o^cz^+ / i^-z^+$	C
6. $F'o^cz^- / o^+z^+$	I
7. $F'o^cz^+ / o^+z^-$	C

Dominance of o^c Mutations: The Operator

A striking property of the o^c mutations is that in certain cases, they are dominant (entries 1, 2, and 5, Table 15-2). The significance of the dominance of the o^c mutations becomes clear from the properties of the partial diploids shown in entries 6 and 7. Both combinations are Lac$^-$, because there is a functional z gene. However, entry 6 shows that β-galactosidase synthesis is inducible even though an o^c mutation is present. The difference between the two combinations in entries 6 and 7 is that, in entry 6, the o^c mutation is carried on a DNA molecule that also has a z^- mutation, whereas in entry

7, o^c and z^+ are carried on the same DNA molecule. Thus, o^c causes constitutive synthesis of β–galactosidase only when o^c and z^+ are on the same DNA molecule. Confirmation of this conclusion came from an important biochemical observation: an immunological test capable of detecting a mutant β–galactosidase showed that the mutant enzyme is synthesized constitutively in a $o^c z^- / o^+ z^+$ partial diploid (entry 6), whereas the wild–type enzyme is synthesized only if an inducer is added.* Genetic mapping experiments showed that all o^c mutations are located between genes i and z, so the gene order of the four elements of the *lac* system is i o z y. Together these observations lead to the conclusion that o^c mutations define a *site* or a noncoding region of the DNA rather than a gene (because mutations in coding genes should be complementable) and that the o region determines whether synthesis of the product of the adjacent z gene is inducible or constitutive. The o region is called the **operator.**

The Operon Model

Regulation of the *lac* system is explained by the **operon model,** which has the following features (Figure 15-3):

1. The products of the z and y genes are encoded in a single polycistronic mRNA molecule.†
2. The promoter for this mRNA molecule is immediately adjacent to the o region. Promoter mutations (p^-) completely incapable of making both β–galactosidase and permease have been isolated and are located between i and o.
3. The operator is a sequence of bases (in the DNA) to which the repressor protein binds.
4. When the repressor protein is bound to the operator, transcription of *lac* mRNA cannot be initiated.
5. Inducers stimulate mRNA synthesis by binding to the repressor. This binding alters the three–dimensional structure of the repressor so it cannot bind to the operator. Thus, in the presence of an inducer the operator is unoccupied and the promoter is available for initiation of mRNA synthesis. This is often called **derepression.**

---

*The test is carried out as follows. Antibody to purified β-galactosidase is prepared. This antibody also reacts with a mutant β-galactosidase as long as the structural differences between wild type and mutant are not too great. This is called **cross-reaction** and the mutant protein is called cross-reacting material, or **CRM**. Thus, the presence of CRM, which can be detected by a variety of standard immunological procedures, is indicative of the presence of a mutant protein.

†This mRNA molecule contains a third gene, denoted a, which encodes an enzyme **thiogalactoside transacetylase** that is not directly involved in lactose metabolism. Therefore, this gene will be discussed only briefly in this chapter.

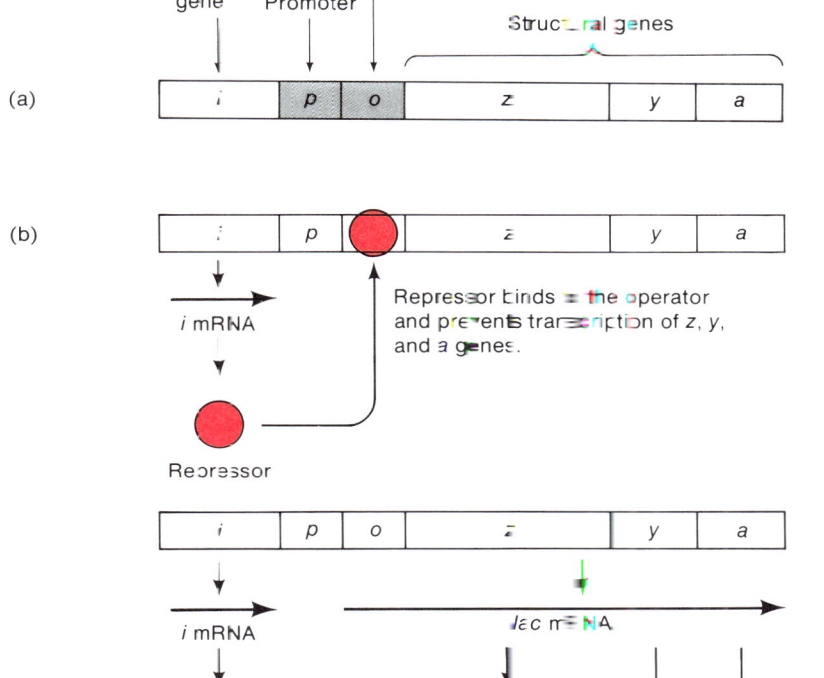

(a)

(b)

Figure 15-3 (a) Genetic map of the *lac* operon, not drawn to scale: the *p* and *o* sites are actually much smaller than the genes. (b) Diagram of the *lac* operon in (I) repressed and (II) induced states. The inducer alters the shape of the repressor, so the repressor can no longer bind to the operator.

This simple model explains many of the features of the *lac* system and of other negatively regulated genetic systems. However, we will see in a later section that this explanation is incomplete as the *lac* operon is also subject to positive regulation.

Background Constitutive Synthesis

As it is stated, the operon model fails to explain two anomalies. First, inducers must penetrate a cell in order to bind to repressor molecules, yet transport of inducers requires permease, and permease synthesis requires induction.

Thus, we must explain how the inducer gets into a cell in the first place. There are only two possible explanations—either some inducer can enter a cell without permease or some permease is made without inducer. We will see in a moment that the latter is correct.

A second apparent paradox is that in recent years it has been shown that lactose (galactose-1,4-glucose) does not bind to the repressor and that the true inducer is a lactose isomer called allo-lactose (glucose-1,6-galactose). However, allo-lactose is formed from lactose by the catalytic action of β-galactosidase, so induction of the synthesis of β-galactosidase by lactose requires that β-galactosidase be present. Both of these seeming anomalies have the same explanation—in the uninduced state there is a small amount of *lac* mRNA synthesized (roughly one mRNA molecule per cell per generation). This synthesis, which is called **background constitutive synthesis,** or **basal synthesis,** occurs because binding is never infinitely strong. Thus, even though the repressor binds very strongly to the operator, it occasionally comes off and, for the instant that the promoter is free, an RNA polymerase molecule may be able to initiate transcription.

Response of a Lac⁺ Bacterial Culture to Lactose

We are now able to describe in molecular terms the sequence of events following addition of a small amount of lactose to a growing Lac⁺ culture. Consider bacteria growing in a medium in which the carbon source is glycerol. Each bacterium contains one or two molecules of β-galactosidase and of lactose permease. Lactose is then added. A few lactose molecules are transported into the cell by a few permease molecules and a small amount of allo-lactose is then made by the few β-galactosidase molecules. An allo-lactose molecule then binds to a repressor molecule that is sitting on the operator, and the repressor is inactivated and falls off the operator. Synthesis of mRNA then begins and, from these RNA molecules, hundreds of copies of β-galactosidase and permease are made. This allows lactose molecules to pour into the cell. Most of the lactose molecules are cleaved to yield glucose and galactose, but many molecules are converted to allo-lactose molecules, which bind to and inactivate all of the intracellular repressor molecules. (Repressor is made continuously, though at a very low rate, so there is usually sufficient allo-lactose to maintain the cell in the derepressed state.) Thus, mRNA is synthesized at a high rate and the concentration of permease and β-galactosidase becomes quite high. The glucose produced by the cleavage reaction is used as a source of carbon and energy. (The galactose formed by the cleavage is converted to glucose by a set of enzymes, the synthesis of which is also inducible. This inducible system, in which galactose is the inducer, is called the *gal* operon. It is discussed in another section of this chapter.)

Ultimately all of the lactose in the growth medium and within the cells is consumed. As a result of the falling concentration of allo-lactose, newly

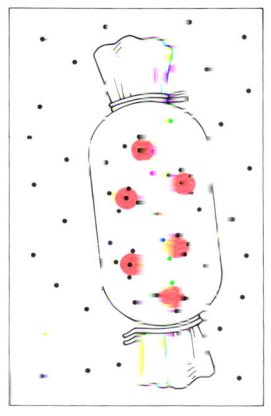

Zero time   At equilibrium

Figure 15-4 Equilibrium dialysis. A dialysis bag filled with a cell extract containing macromolecules (red circles) is placed in a solution containing IPTG (black circles), which can bind to the repressor. At equilibrium, the concentration of free IPTG is the same inside and outside the bag. Because the repressor binds some of the small molecules, the total concentration of small molecules is greater inside the bag than outside.

made repressor, whose synthesis has been proceeding unabated, unaffected by the presence or absence of lactose, is no longer inactivated by allo-lactose; thus, repression is reestablished, thereby eliminating further synthesis of mRNA. In bacteria most mRNA molecules have a half-life of only a few minutes (Chapter 14); in less than one generation there is little remaining *lac* mRNA and synthesis of β-galactosidase and permease ceases. These proteins are quite stable but are gradually diluted out as the cells divide. Note that if lactose were added again to the growth medium one generation after the original lactose had been depleted, cleavage of lactose would begin immediately because the cells would already have adequate permease and β-galactosidase.

Purification of the *lac* Repressor

An important step in proving the principal hypothesis of the operon model was the isolation of the *lac* repressor and the demonstration of its expected properties. This was accomplished by using the ability to bind radioactive IPTG (one of the gratuitous inducers) as an assay for the repressor. This binding was detected by equilibrium dialysis, as shown in Figure 15-4; by monitoring the IPTG-binding capacity of various protein fractions obtained from a cell extract, the protein was partially purified. The *lac* repressor consists of four identical protein subunits, each containing 360 amino acids and each capable of binding one molecule of IPTG. The crude unfractionated

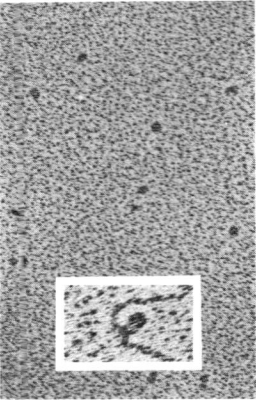

Figure 15-5 Electron micrograph of *E. coli lac* repressor (× 107,000). Inset shows one repressor molecule bound to *lac* DNA (× 215,000). [Courtesy of Robley Williams.]

cell extract binds about 20–40 molecules of IPTG per cell, so there are roughly 5–10 repressor molecules per cell. Proof that the IPTG-binding molecule is indeed the repressor came from the observation that IPTG-binding protein is absent in extracts of i^- mutants.

Since the number of repressor molecules is extremely small, these molecules must be translated from no more than one or two repressor mRNA molecules transcribed per generation time. The number of mRNA molecules is so small that either repressor synthesis itself is regulated or the mRNA is transcribed from a weak promoter. Both mechanisms have been observed for regulation of repressor synthesis in other operons, but for the *lac* repressor the second explanation is correct—that is, repressor mRNA is transcribed constitutively from a weak promoter. The reason for the small number of repressor molecules is made clear from the properties of several mutants in which the weak promoter is converted to a strong promoter. These mutants are noninducible because it is not possible to fill a cell with enough inducer to overcome repression. These repressor-overproducers have been extremely valuable experimentally because high concentrations of repressor (about 1 percent of the cellular protein) have in turn meant that very large amounts of repressor could be purified—amounts sufficient for physical study and the determination of the amino acid sequence.

With purified repressor the specific binding of repressor to the operator sequence and the inhibition of this binding by an inducer have been demonstrated. The source of *lac* DNA in these studies was a λ phage variant (a transducing phage, see Chapter 18) that, as a result of genetic manipulations, carried the *E. coli lac* genes. Figure 15-5 shows an electron micrograph of pure *lac* repressor bound to a DNA molecule.

An important procedure for studying repressor-operator binding is the nitrocellulose filter assay. Proteins are retained by these filters but DNA molecules are not; if a mixture of repressor and [^{14}C]DNA is passed through such a filter, ^{14}C will be retained on the filter if the protein and the DNA form a complex. By means of this test the data shown in Table 15-3 have been obtained. These results indicate that the repressor binds to the operator (i.e., it fails to bind to an o^c mutant) and that IPTG prevents this binding, which confirms a major prediction of the operon model.

Table 15-3 Demonstration of repressor-operator binding by the filter-binding assay

Mixture applied to filter	^{14}C bound to filter
[^{14}C]*lac* DNA	No
[^{14}C]*lac* DNA + repressor	Yes
[^{14}C]*lac* DNA + repressor + IPTG	No
[^{14}C]*lacO^c* DNA + repressor	No

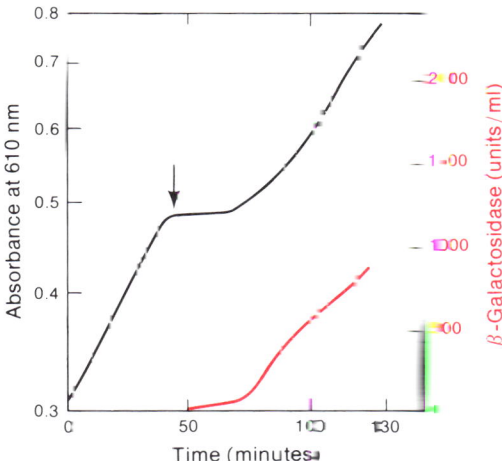

Figure 15-6 An experiment showing the switch from glucose metabolism to lactose metabolism when a culture containing both glucose and lactose exhausts the supply of glucose. At the time indicated by the arrow, glucose is in such short supply that the cell mass stops increasing. No β-galactosidase is present because synthesis of *lac* mRNA did not occur while glucose was present. As the glucose concentration drops, the concentration of cAMP increases until *lac* mRNA and β-galactosidase synthesis begins (red curve). The absorbance at 610 nm is a measure of the total mass of the cells; the enzyme activity is measured in an arbitrarily defined unit.

The Effect of Glucose on the Activity of the *lac* Operon

The function of β-galactosidase in lactose metabolism is to form glucose by cleaving lactose. (The other cleavage product, galactose, is also ultimately converted to glucose by the enzymes of the *gal* operon, as already mentioned.) Thus, if both glucose and lactose are present in the growth medium, then in the interest of efficiency, there is no reason for a cell to turn on the *lac* operon. Indeed, cells behave according to this logic; that is, no β-galactosidase is formed until all of the exogenous glucose is consumed (Figure 15-6). The reason for the lack of β-galactosidase synthesis when glucose is present is that no *lac* mRNA is made. There are two ways that this inhibition of *lac* mRNA synthesis might occur: (1) glucose prevents inactivation of the repressor by the inducer, or (2) something else is needed for initiating *lac* mRNA synthesis other than removal of the repressor. We will see shortly that the second alternative is the correct one.

The inhibitory effect of glucose on expression of the *lac* operon is indirect and involves a complicated and poorly understood process by which some glucose metabolite (possibly in the glycolytic pathway) regulates the synthesis of the substance **cyclic AMP (cAMP)**, whose structure is shown in Figure 15-7. This substance is universally distributed in animal tissues

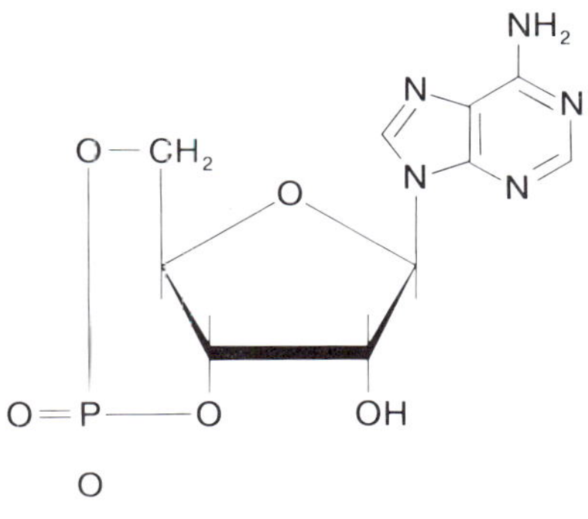

Figure 15-7 Structure of cyclic AMP.

Table 15-4   Concentration of cyclic AMP in cells growing in media having the indicated carbon sources

Carbon source	cAMP concentration
Glucose	Low
Glycerol	High
Lactose	High
Lactose + glucose	Low
Lactose + glycerol	High

where it plays an important role in the action of many hormones. Cyclic AMP is synthesized from ATP by the enzyme **adenylate cyclase,** which is encoded in the *cya* gene. In a bacterial culture that is starved of an energy source (for example, by incubation in a medium lacking a carbon source), the intracellular concentration of cAMP is high. If a culture is growing in a medium containing glucose, the cAMP concentration is very low. In a medium containing glycerol or any carbon source that cannot enter the glycolytic pathway, the cAMP concentration is high (Table 15-4). The significance of these results is that *cAMP is a mediator of activity of the* lac *operon.*

E. coli (and presumably other bacteria as well) contain a protein formerly called the catabolite activator protein (CAP) and now called the **cyclic AMP receptor protein (CRP),** which is encoded in a gene called *crp* and which forms a complex with cAMP. Mutants of either the *crp* or *cya* genes are unable to synthesize *lac* mRNA, indicating that both CRP and cAMP are required for *lac* mRNA synthesis. Further study has shown that CRP and cAMP form a complex, denoted **cAMP-CRP,** which is needed to activate the *lac* system. This requirement is independent of the repression system

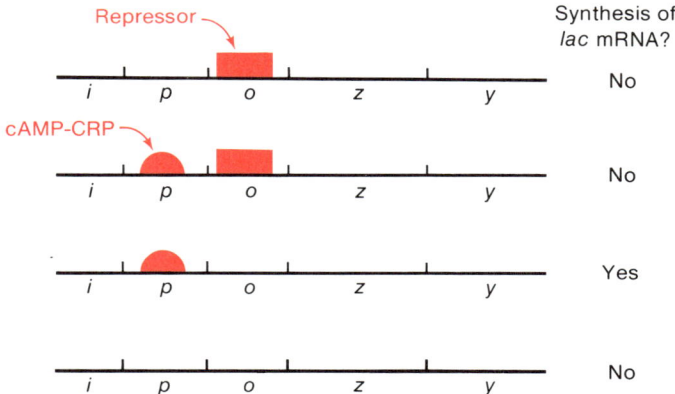

Figure 15-8   Three states of the *lac* operon showing that *lac* mRNA is made only if cAMP-CRP is present and repressor is absent.

since *crp* and cyclase mutants are unable to make *lac* mRNA even if an i^- or o^c mutation is present. We now know that the complex must be bound to a base sequence in the DNA in the promoter region in order for transcription to occur. Thus, *cAMP-CRP is a positive regulator*, in contrast with the repressor, and the *lac* operon is independently regulated both positively and negatively (Figure 15-8).

Other Systems Responding to cAMP

There are many other sugars (e.g., galactose, maltose, arabinose, sorbitol) that are converted to glucose or some intermediate in glycolysis during their degradation. The enzymes responsible for metabolism of each of these sugars are synthesized by inducible operons and, as might be expected, each operon cannot be induced if glucose is present. These are called **catabolite-sensitive operons.** A simple genetic experiment shows that each of these operons is regulated by cAMP-CRP. A single mutation in a sugar operon (e.g., lac^- or mal^-) arises spontaneously at a frequency of roughly 10^{-6}. A double mutation, $lac^- mal^-$, would arise at a frequency of 10^{-12}, which for all practical purposes is unmeasurable. However, double mutants that are phenotypically $Lac^- Mal^-$ or $Gal^- Ara^-$ do arise at a measurable frequency. These apparent double mutants are not the result of mutations in the two sugar operons but always turn out to be crp^- or ya^-. Furthermore, if a $Lac^- Mal^-$ mutant is found, it is always also $Gal^- Ara^-$, $Lac^- Ara^-$, etc. Biochemical experiments with a few of these catabolite-sensitive operons indicate that binding of cAMP-CRP occurs in the promoter region in each of these systems.

The reason for the requirement for cAMP in the degradation of arabinose (and certain other pentoses) is not obvious since these sugars are not metabolized to glucose. This indicates the more general nature of the role of cAMP. Remember that the concentration of glucose, which is perhaps the optimal carbon source, determines the concentration of cAMP—when the glucose supply is inadequate, cAMP is made in high amounts. Thus, presence of much cAMP is a signal that not enough glucose is available for growth. When the glucose supply is exhausted, the increased cAMP concentration puts a great many sugar operons in a state of readiness; thus, if any of these sugars is presented to the cell, the response can be rapid. If glucose is present, the signal is absent because such a state of readiness is unnecessary.

Regulatory Region of the DNA of the *lac* Operon

The segment of DNA comprising the *lac* operator has been isolated by the nuclease-protection procedure that we have seen in the study of nucleosomes

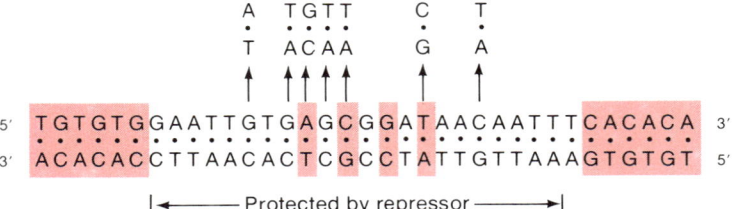

Figure 15-9   Base sequence of the *lac* operator. The symmetrically related regions are shown in red. The arrows point to observed mutational changes that make the operon constitutive.

(Chapter 7) and of RNA polymerase binding sites (Chapter 12). In this procedure purified repressor was adsorbed to DNA that had been isolated from a transducing phage variant (called λ*lac*) carrying the *E. coli lac* operon. Then DNase was added and a short 24-bp fragment, which survived enzymatic degradation, was isolated. The sequence of bases in this fragment and in a slightly larger fragment isolated in a different way is shown in Figure 15-9. This sequence has a characteristic that will be seen repeatedly in base sequences recognized by regulatory elements—namely, the sequence is an inverted repeat. Proof that this palindromic sequence is itself the operator came from studies of operator mutants. A mutant operator sequence could not be isolated by the DNase-protection procedure because the repressor does not bind to a mutant operator. However, by using recombinant DNA techniques (Chapter 22), longer fragments containing the operator were cloned and isolated. The operator was then identified in the fragment by locating the operator sequence shown in Figure 15-9; this procedure also defined the base sequence flanking the operator. A mutant operator, which may differ by one or two bases from the wild-type sequence, can be identified by looking for a sequence related to the wild-type sequence between the flanking sequences already identified. This experiment was done by obtaining fragments from several λ*lac* phage particles carrying different *o^c* mutations and observing the base changes shown in Figure 15-9. Note that a single base change either in the inverted repeat sequence or in other base pairs eliminated the ability of the operator to bind the repressor.

The cAMP-CRP binding site has also been isolated by the DNase-protection method. Its sequence is shown below.

Note that this is also an inverted repeat, symmetric about the arrow, except for the two base pairs enclosed in the dotted lines. Two mutations known

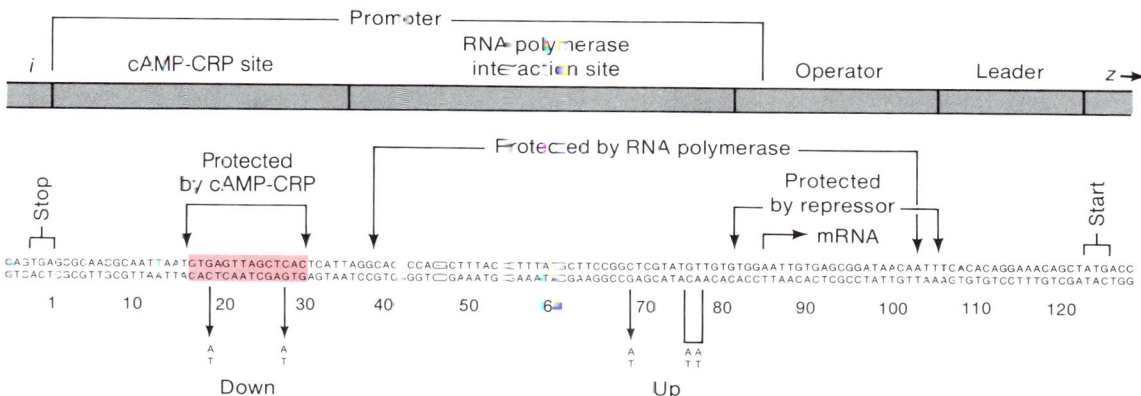

Figure 15-10 DNA base sequence of the regulatory region of the *E. coli lac* operon. The shaded red region indicates the inverted repeat in the cAMP-CRP binding site. Regions protected against DNase digestion by various proteins are shown. Base-pair changes associated with promoter mutations yielding an inactive promoter (down) or an especially active or (cAMP-CRP)-independent promoter (up) are indicated. The start and stop codons in the *z* and *i* gene products respectively are also indicated.

by genetic and physiological tests to be "promoter-down" mutations (that is, they prevent mRNA synthesis) are base changes in this sequence and these mutations prevent binding of cAMP-CRP; this confirms that binding of cAMP-CRP is a prerequisite for transcription from the *lac* promoter.

A 130-bp sequence containing the cAMP-CRP binding site, the RNA polymerase interaction site, and the operator, and flanking sequences has been identified. This sequence is shown in Figure 15-10. The start point for synthesizing mRNA has been determined by sequencing the 5′-P end of purified *lac* mRNA and finding the complementary sequence in the DNA. The start codon of β-galactosidase and the termination sequences for the *i* gene have also been identified from the known amino acid sequences of β-galactosidase and the repressor. Several features of this entire region should be noted:

1. The initiation site for mRNA synthesis is in the operator and, when the repressor is present, the site is covered.
2. The operator is protected from DNase digestion by bound RNA polymerase.
3. The cAMP-CRP binding site is very near to, but does not overlap, the RNA polymerase interaction site.
4. The regulatory region is 115 bp long or roughly 12 turns of the DNA helix.
5. The order of the loci, namely *i p* → *z*, corresponds to that determined by genetic analysis.

The Mechanism of Action of the cAMP–CRP Complex and the Repressor

In order for *lac* mRNA synthesis to occur, cAMP–CRP must be bound to the promoter and the operator must not be occupied by a repressor (Figure 15-8). The precise mechanisms by which cAMP–CRP and the repressor respectively stimulate and inhibit transcription are not yet known in detail. However, a reasonable model has been obtained both from a determination of the promoter bases contacted by RNA polymerase and by the repressor, as well as by the following facts:

1. RNA polymerase can bind to *lac* DNA *in vitro* in the absence of any other proteins, but binding is stronger with cAMP–CRP present.
2. RNA polymerase bound to *lac* DNA cannot be displaced by addition of the repressor.
3. If the repressor is first bound to the operator, RNA polymerase cannot form a stable complex, but a weak and very transient binding to the RNA polymerase interaction site occurs.

The currently accepted hypothesis for the mechanism of initiation, derived from the information just given, is the following:

1. RNA polymerase forms a loose complex with the *lac* promoter region, probably the closed promoter complex described in Chapter 12.
2. When cAMP–CRP is bound to the DNA, formation of the stable open promoter complex is facilitated. This is probably a result of certain α helices in CRP (those not bound to the DNA, see Chapter 7) providing a protein-protein binding site for RNA polymerase, which stabilizes the binding of RNA polymerase to the DNA and orients the enzyme such that an open-promoter complex is formed.
3. The repressor may work in one of two ways: it may be an antimelting protein, which would prevent formation of an open promoter complex, or, when bound, it could prevent access of RNA polymerase to the bases needed to form an open complex.
4. Once an open-promoter complex forms, RNA polymerase then initiates synthesis of *lac* mRNA.

A recent observation, which has not yet been confirmed or examined, further suggests another possible explanation for the role of cAMP–CRP. In one experiment, a second binding site for RNA polymerase was found in the region of the *lac* promoter protected by cAMP–CRP. This site, which does not lead to the production of functional *lac* mRNA, competes with the *lac* mRNA binding site for RNA polymerase. When cAMP–CRP is bound, the competing secondary RNA-polymerase binding site is blocked. If this observation and its interpretation should prove to be valid, cAMP–CRP would then be acting as a repressor, and point 2 above will have to be reexamined.

Physiological Properties of an Incomplete System

If regulation is valuable to a cell, a mutant in which an operon is unregulated should be disadvantaged in some way. So far the only handicap that has been observed with i^- and o^c mutants is a very slight decrease in growth rate. Whereas in the laboratory this difference is barely observable, on an evolutionary time scale a mutation that introduces regulation to a system and increases the growth rate slightly would confer a significant selective advantage to a cell.

In certain conditions (which conceivably might arise more frequently in nature), constitutive *lac* mutants grow significantly slower than the wild-type strain. An analysis of these conditions suggests a possible reason for the presence of the fourth gene of the *lac* operon the *a* gene, mentioned earlier. This gene encodes the enzyme thiogalactoside transacetylase, which can transfer an acetyl group from an acetyl donor to many galactosides, though lactose itself cannot be acetylated; this, we will see, is a significant point.

There exist in nature many substituted galactosides that can be cleaved by β-galactosidase. However, often the substituted galactose moiety produced by the cleavage reaction cannot be metabolized further and accumulates to very high levels. This is often detrimental because several galactose derivatives are inhibitory to normal growth when they are present at high concentrations. When such galactosides are present in the growth medium, i^- and o^c mutants that are also a^- grow more slowly than the a^+ counterparts. The reason is that the acetylated forms can freely diffuse through the cell membrane and leave the cell.

It is reasonable to wonder why a cell should expend the energy needed to synthesize transacetylase even when lactose is the only galactoside present, for alternatively this enzyme could be independently regulated. However, independent regulation, uncoupled with β-galactosidase synthesis, is clearly not optimal, since the enzyme is needed only when β-galactosidase is made. It is possible that a secondary regulatory system could be imposed on the *lac* operon but this might be even more costly in energy. It may be that in nature, the substituted galactosides are prevalent in environments in which lactose is the primary carbon source, so having the *a* gene directly coupled to the *z* gene is advantageous.

Difference in the Amount of β-Galactosidase, Permease, and Transacetylase Translated from a Single *lac* mRNA Molecule

The ratios of the number of copies of β-galactosidase, permease, and transacetylase, are 1 : 0 5 : 0.2. These ratios, which are the results of millions of years of evolution, probably reflect the needs of a cell when exposed to a β-galactoside as the sole carbon source. These differences, which are examples of *translational* regulation, are achieved in two ways:

1. Translating ribosomes may detach from *lac* mRNA following chain termination. The frequency with which this occurs is a function of the probability of reinitiation at each subsequent AUG codon. Thus, there is a gradient of synthesis from the 5′ terminus to the 3′ terminus of the mRNA. This is true with most polycistronic mRNA molecules.

2. At any given instant, there are more complete copies of the *z* gene than the *a* gene message. Two factors appear to contribute to this: (1) some of the *lac* transcripts terminate between the *z* and *y* genes, and (2) the degradation of *lac* mRNA is initiated more frequently by endonucleolytic scissions in the *a* gene than in the *z* gene.

Throughout this chapter, it will be seen that this mode of regulation occurs repeatedly in prokaryotes—that is,

The *overall* expression of activity of an operator is regulated by controlling transcription of a polycistronic mRNA, and the *relative* concentrations of the proteins encoded in the mRNA are determined by controlling the frequency of translation of each cistron.

Fusions of the *lac* Operon with Other Operons

Gene fusions, which may or may not occur in nature, are of great value in research. One well-studied example is the coupling of the *lac* operon to the *pur* operon, a system responsible for the synthesis of purines. This operon is regulated in a rather different way from the *lac* operon, though how it is regulated is not relevant to this discussion. The *pur* operon is located "downstream" (in the direction of transcription) from the *lac* operon in the *E. coli* chromosome (Figure 15-11). Between these two operons is a gene *tsx*, which governs sensitivity to the phage T6. At one time a z^+ *tsx-s* culture was taken

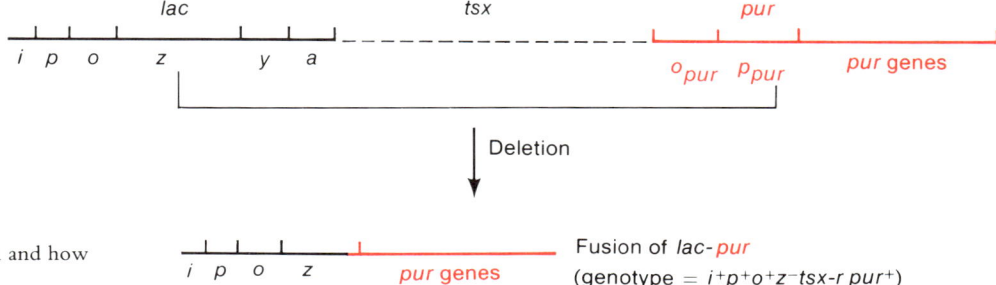

Figure 15-11   The *lac-pur* fusion and how it arose.

Fusion of *lac-pur*
(genotype = $i^+p^+o^+z^-$*tsx-r pur$^+$*)

and a *tsx-r* mutant (which is easy to obtain simply by adding T6 to the culture) that was also z^- was isolated. This mutation was found to be a deletion beginning within the *z* gene, including all of the *tsx* gene, and extending into the *pur* operon past the *pur* operator and promoter. This deletion removes the RNA polymerase termination sequence in the *lac* operon; thus, when initiation begins from the *lac* promoter, an mRNA molecule is made that contains the proximal region of the *z* gene and all of the *pur* genes. Thus, in the fused system transcription of the *pur* operon is induced by β–galactosides.

With genetic and recombinant DNA techniques that allow relocation of the *lac* operon to other parts of the *E. coli* chromosome it has been possible to fuse many genes to the *lac* promoter. This has practical value for biochemists because the *lac* promoter is a very strong promoter and increased synthesis of proteins transcribed from weak promoters can often be accomplished. In some cases the amount of a particular protein recovered from a cell suspension can be increased a thousandfold.

Other fusions in which the *z* gene is fused (downstream) to the promoter-operator region of other operons have proved valuable in studies of regulation. Some operons have been difficult to study since the gene products are not easily measured quantitatively. However, β–galactosidase can be assayed simply and accurately. In the fusion strains β–galactosidase activity has been used to determine the conditions needed to turn on an operon of interest.

THE GALACTOSE OPERON

Galactose is another sugar that *E. coli* can utilize as a carbon source. It is metabolized by three enzymes, **galactokinase, galactose transferase,** and **galactose epimerase,** which act in a sequence of steps to yield the overall reaction

$$\text{Galactose} - \text{ATP} \rightarrow \text{Glucose-1-phosphate} + \text{ADP} + \text{H}^+$$

The galactose (*gal*) operon is regulated in principle like the *lac* operon in that there is a repressor and a role of cAMP; however, there are several differences that are quite interesting and possibly reflect the dual role of galactose in cellular metabolism. For example, optimally the two states of the *gal* operon are high-level and low-level on, rather than off–on.

Genetic Elements of the *gal* Operon

The genes corresponding to the three *gal* enzymes (I, II, III) are *galK, galT,* and *galE,* respectively. Together with an operator (*galO*), a promoter region (*galP*) and a repressor gene (*galR*), these genes form an operon. However, the arrangement of the elements is strikingly different from that of the *lac*

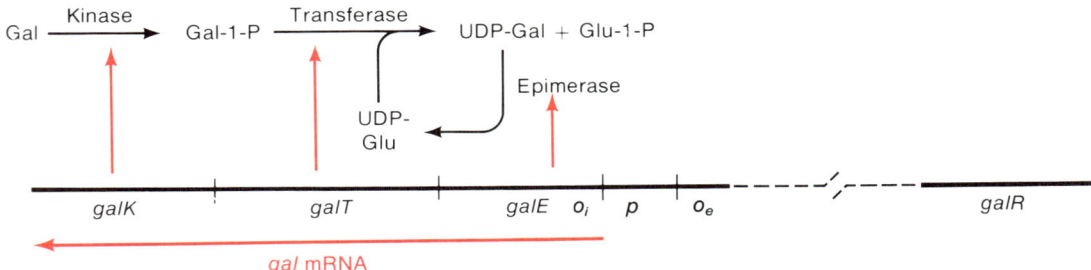

Figure 15-12   The enzymes and elements of the *E. coli gal* operon. Abbreviations: Gal, galactose; Gal-1-P, galactose 1-phosphate; UDP-Gal, uridinediphosphogalactose; Glu-1-P, glucose 1-phosphate; UDP-Glu, uridinediphosphoglucose. The *galR* repressor gene is very far from the remainder of the operon. Vertical red arrows connect genes and corresponding enzymes.

operon. One difference is that *galR,* the structural gene for the repressor, is very far from the cluster consisting of the operator, promoter, and the structural genes (Figure 15-12). Actually, there is no reason for a repressor gene to be adjacent to an operator, since the repressor is a diffusible substance, and in fact, there are as many examples of operons with adjacent repressor and structural genes (e.g., *lac*) as of those with distant repressor genes. A second difference is that there are two operators: one termed *galO_e* is upstream and adjacent to *galP* (as in the *lac* operon), but the other, termed *galO_i*, is 95 bp downstream from *galO_e* and within the *galE* gene. The action of the *gal* repressor (which has been purified) is like that of the *lac* repressor in that the *galR^-* and *galO^c* mutations (in both operators) confer a constitutive phenotype and purified repressor binds to both of the operator sequences. The repressor–operator system, with its dual operators, is not well understood and will not be discussed further. Instead, we will be concerned with the fact that the *gal* operon has two promoters.

Two *gal* Promoters and the Effect of cAMP-CRP on the Activity of Each One

Galactose is utilized less efficiently than glucose, and one might expect the *gal* operon to be noninducible in the presence of glucose. However, it is inducible when glucose is present. Several promoter mutants have been isolated and, surprisingly, all fall into only two classes. One class of mutant has the property that high-level synthesis of the *gal* enzymes fails to occur only when glucose is present. The other class of mutant is, instead, defective in enzyme synthesis when glucose is absent. Both observations are explained by the fact that the *gal* operon has two promoters. The mRNA molecules synthesized from these two promoters differ in length by only five nucleo-

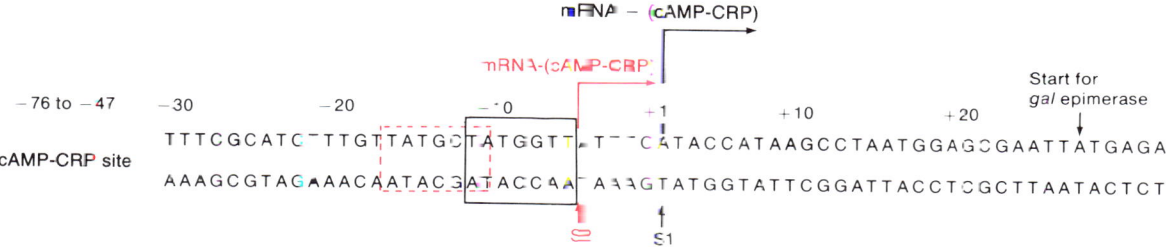

Figure 15-13 Base sequence of the operator-promoter region of the *gal* operon. The +1 refers to the S1 start site for the (cAMP-CRP)-dependent mRNA. The two Pribnow sequences preceding each start site are boxed. S2 is the start site for the (cAMP-CRP)-independent mRNA.

tides, so both transcripts contain the *galK, galT,* and *galE* genes; in fact, the information content of these two mRNA molecules is identical. The properties of these two promoters are best understood by examining the base sequence of the promoter–operator region.

A DNA fragment consisting of 139 base pairs and containing the promoter–operator region has been isolated and sequenced. The sequence of the relevant portion of this fragment is given in Figure 15-13, which shows the start points S1 and S2 for the two *gal* mRNA molecules; these sites have been located by isolating these mRNA molecules and determining the base sequences of their 5′ termini. S1 and S2 differ functionally in the following way:

1. Transcription from S1 occurs only when glucose is absent and fails to occur in *crp⁻* and *cya⁻* mutants. In vitro this transcript is made only if cAMP-CRP is present.
2. Transcription from S2 occurs primarily when glucose is present; cAMP-CRP is not required for initiation at S2 and in fact inhibits the transcription from S2.

Apparently, glucose fails to inhibit induction of the *gal* operon because one of the start sites does not require cAMP-CRP and thus remains active even though glucose is present. Hence, the promoter mutation that prevents synthesis of the *gal* enzymes only when glucose is present must be a mutation in S2, and the other mutation, which prevents synthesis when glucose is absent, must be a mutation in S1.

The mechanisms by which cAMP-CRP stimulates S1 and inhibits S2 are not clearly known. Two cAMP-CRP binding sites have been identified by DNase protection experiments., one from −25 to −50 and the other from −50 to −68 (Figure 15-13). Binding to the second site requires binding of cAMP-CRP and RNA polymerase to the first site. The significance of this complex arrangement is unknown.

Possible Value of Two Promoters in the *gal* Operon

Why does the *gal* operon have two transcription initiation sites? The answer is uncertain but may have to do with the fact that galactose has two roles in cellular metabolism. Galactose not only serves as a carbon source when it is all that is available but the related compound uridinediphosphogalactose (UDPGal) is a precursor in the synthesis of the *E. coli* cell wall. In the absence of exogenous galactose, UDPGal is formed from UDP-glucose in a reaction catalyzed by galactose epimerase, the product of the *galE* gene. Thus, if the cell is to grow, it must be capable of synthesizing the epimerase at all times. This could be accomplished by having a second *galE* gene that is not part of the *gal* operon. Alternatively, since the epimerase is required for cell wall synthesis in very small quantities, the background constitutive synthesis might be sufficient. In fact, the constitutive level of *gal* mRNA synthesis is higher than that of the *lac* operon, and this mRNA is apparently the source of the enzyme when galactose is absent.

Let us suppose that S1 were the only promoter. Since the activity of this promoter depends on cAMP-CRP, the epimerase could not be made constitutively when glucose is present. Suppose instead that S2 were the only promoter. In this case, galactose would fully derepress (induce) the operon even when glucose is present and this would be wasteful. Thus, for the sake of both necessity and economy a cAMP-CRP-independent promoter (S2) is needed for background constitutive synthesis and a cAMP-CRP-dependent promoter (S1) is needed to regulate high-level synthesis; furthermore, the regulation is efficient only if S2 is inhibited by cAMP-CRP.*

THE ARABINOSE OPERON

Arabinose is another sugar (a pentose) that can serve as a carbon source for metabolism. It is found in the cell walls of many plants and is released in the human intestine after vegetables are eaten. It is not absorbed by the intestine and hence provides a source of carbon for bacterial inhabitants of the intestine, such as *E. coli*. Since the availability of arabinose is sporadic, one would expect the ability to utilize this sugar to be inducible, as indeed it is.

Three genes, *araB*, *araA*, and *araD*, are needed for arabinose degradation in *E. coli* and form a cluster abbreviated *araBAD* (Figure 15-14). Two other genes, *araE* and *araF*, which are necessary for transport of arabinose across the cell membrane, are quite far from the linked cluster. *AraE* and *araF* encode

---

*The jargon term "high-level" is in common usage to mean both high concentration or high rate (the correct meaning is usually obtained from context) in contrast with "low-level," referring to low concentration or low rate.

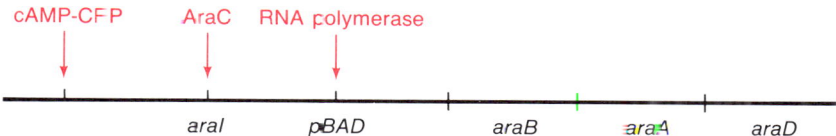

Figure 15-14   A portion of the genetic map of the *ara* operon, showing the three genes (*araB*, *araA*, *araD*) for the metabolic enzymes, and three binding sites in the regulatory region.

a membrane protein and an arabinose-binding protein located between the cell membrane and the cell wall; these two genes are not understood and will not be discussed further. Adjacent to the *araBAD* cluster is a complex promoter region and a regulatory gene *araC*, whose properties differ markedly from what we have seen in the *lac* and *gal* operons. That is, in regulating the activity of the promoter *pBAD* (the promoter for the *araB, araA, araD* gene cluster) the AraC protein functions primarily as a positive control element.* This contrasts with the *lac* and *gal* repressors, which are solely negative regulators, and the cAMP-CRP protein, which is usually a positive regulator.

The *ara* operon is inducible, as we have said, and the inducer is arabinose itself. That is, in a wild-type operon *araBAD* mRNA is made only when arabinose (or certain analogues) is present. Also, *araBAD* mRNA is not made if glucose is present or by *cya⁻* and *crp⁻* mutants, which indicates that cAMP-CRP is needed for transcription from *pBAD*.

Positive Regulation by the AraC Protein

The positive regulatory activity of the AraC protein is shown by the following results:

1. All point and deletion mutations in the *araC* gene, which are denoted *araC⁻*, are unable to synthesize *araBAD* mRNA.

Knowing nothing else, one might also conclude either that an *araC⁻* cell contains a mutant repressor that cannot bind inducer or that the *araC* locus is the promoter. These possibilities are eliminated by the next two results.

*AraC also acts as a negative regulator, both in controlling its own synthesis and in a poorly understood effect on the *pBAD* promoter, but this will not be discussed in this book.

2. Partial diploids having the genotype F′araC⁺/araC⁻ are fully inducible. This means that the araC⁻ allele is recessive. If araC⁻ mutants
were noninducible repressors, then araC⁻ should be dominant to
araC⁺.

3. Partial diploids of the genotype F′araC⁺araA⁻/araC⁻araA⁺ are fully
inducible for synthesis of the araA product. If araC were a promoter,
the araA⁺ gene, being on the same DNA molecule as the araC⁻
mutation, could not be transcribed. Since the araC product complements in a *trans* configuration, it must be a diffusible molecule.

Other genetic evidence—namely, that there are amber mutants in the
araC gene—indicates that the araC product is a protein. These genetic results
have been confirmed by purification and sequencing of the AraC protein.
Details of the experimental results just given need not be remembered; the
important point is that synthesis of araBAD mRNA requires a functional
AraC protein.

The conclusion that both cAMP-CRP and the AraC protein (when
acting as a positive regulator) are required to initiate transcription has been
confirmed by *in vitro* experiments of two sorts. The first experiments show
that no araBAD mRNA is synthesized in a reaction mixture containing either
(a) only ara DNA and RNA polymerase or (b) DNA, RNA polymerase,
and only one of the proteins (AraC protein or cAMP-CRP)—that is, both
of the proteins must be added. In the second series of experiments, ara DNA
was examined by electron microscopy (which is capable of visualizing bound
RNA polymerase molecules); one result of these experiments was that no
RNA polymerase is bound unless the other two proteins are present.

How the AraC protein acts to turn on transcription from *pBAD* is not
clearly understood. A binding site, *araI*, (Figure 15-14) is just upstream from
the RNA polymerase binding site in *pBAD*. AraC protein itself does not
bind to *araI* if arabinose is absent. Physical experiments with purified AraC
protein have shown that arabinose does bind to the AraC protein and induces
a conformational change. However, *in vitro* the binding of this form of AraC
to *araI* is fairly weak. Since cAMP-CRP is required for transcription from
pBAD in vivo, one might expect that addition of cAMP-CRP would stimulate or strengthen this binding; however, such stimulation is not observed.
The true state of affairs is quite complex and not well understood. Suffice it
to say that there are two effects: (1) cAMP-CRP, probably rather indirectly,
facilitates binding of a complex containing arabinose and AraC protein to
araI, and (2) bound AraC protein significantly stabilizes the interaction of
RNA polymerase with *pBAD*. Together these effects allow an open–promoter complex to form. For additional information and current speculation,
see articles by Schleif given in the references at the end of this book.

The *ara* operon has been described in considerably less detail than other
operons in this chapter. It has been presented to give an example of a system
regulated primarily by a positive control element. The reader might ask why
the *ara* operon is regulated positively rather than by a repressor. Twenty

years ago, when the *lac* operon was considered to be the model for all operons, such a question would have seemed very important. However, in time it has been realized that a variety of regulatory strategies have evolved and the criterion for survival of a system over millions of years is merely that the mechanism works.

THE TRYPTOPHAN OPERON, A BIOSYNTHETIC SYSTEM

The tryptophan *(trp)* operon is responsible for the synthesis of tryptophan. Regulation of this operon is based on the simple principle that when tryptophan is present in the growth medium, there is no need to activate the *trp* operon. Thus, we should expect there to be a regulatory system that turns *trp* transcription off when adequate tryptophan is present and turns it on when tryptophan is absent—in other words, tryptophan, or a related compound, should be active in repression rather than induction. Furthermore, since the *trp* system forms a biosynthetic rather than a degradative pathway, there should be no inhibition by glucose; indeed, there is not, and cAMP-CRP plays no role in the activity of the *trp* operon.

In nature a situation may arise in which there is a small exogenous supply of tryptophan, but not enough to allow normal growth if synthesis of tryptophan were to be totally shut down. There are two ways to avoid tryptophan starvation when there is a suboptimal supply of the amino acid. (1) Repression can be prevented until the external concentration of tryptophan exceeds a critical level. This mechanism is not particularly efficient because more of the biosynthetic enzymes may be made than are needed. (2) A modulating system could be utilized in which the amount of transcription in the derepressed state is determined by the concentration of tryptophan. This mechanism is more efficient than the first mechanism and is the one that is used in the *trp* operon (and in other amino-acid biosynthetic operons as well).

The Elements of the *trp* Operon

Tryptophan is synthesized in five steps, each requiring a particular enzyme. The genes encoding these enzymes in *E. coli* are adjacent to one another in the same order as their use in the biosynthetic pathway; they are translated from a single polycistronic mRNA and are called *trpE, trpD, trpC, trpB,* and *trpA.* The *trpE* gene is the first one translated. Adjacent to the *trpE* gene are the promoter, the operator, and two regions called the **leader** and the **attenuator,** which are designated *trpL* and *trpa* (not *trpA*), respectively (Figure 15-15). The repressor gene *trpR* is located very far from this gene cluster, like the repressor in the *gal* operon.

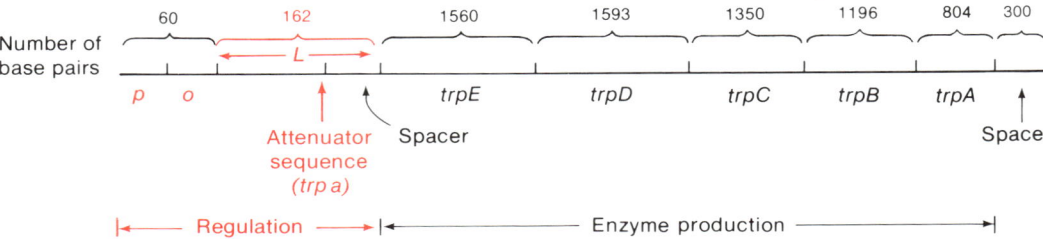

Figure 15-15 The *E. coli trp* operon. For clarity, the regulatory region is enlarged with respect to the coding region. The proper size of each region is indicated by the number of base pairs. *L* is the leader. The regulatory elements are shown in red.

The *trp* Repressor-Operator System

Mutations in the *trpR* gene and in the operator cause constitutive initiation of *trp* mRNA synthesis, as in the *lac* operon. The protein product of the *trpR* gene, which is often called the *trp* **aporepressor,** does not bind to the operator unless tryptophan is present. The aporepressor protein and the tryptophan molecule join together to form an active repressor that binds to the operator. The reaction scheme is

$$\text{Aporepressor alone} \rightarrow \text{No repressor (Transcription occurs)}$$

$$\text{Aporepressor} + \text{Tryptophan} \rightarrow \text{Active repressor}$$
$$+$$
$$\text{Operator}$$
$$\downarrow$$
$$\text{Inactive promoter} \quad \text{(Transcription does not occur)}$$

Thus, only when tryptophan is present does an active repressor molecule inhibit transcription. When the external supply of tryptophan is depleted (or reduced substantially), the equilibrium in the equation above shifts to the left, the operator is unoccupied, and transcription begins. This is the basic "on-off" regulatory mechanism.

The promoter and operator regions overlap significantly and binding of the active repressor and RNA polymerase are competitive. *In vitro,* if repressor is added first, RNA polymerase cannot bind, and vice versa.

The base sequence of the promoter-operator region and of the adjacent region through the start point for *trp* mRNA synthesis is shown in Figure 15-16. The operator is an inverted repetitive sequence with 18 of 20 base pairs participating in the symmetry.

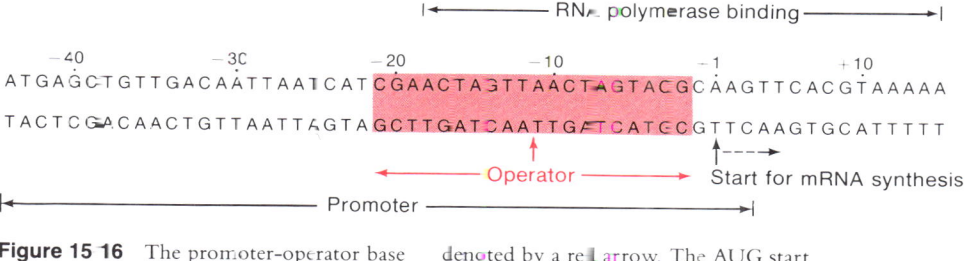

Figure 15-16 The promoter-operator base sequence of the *trp* operon. The operator is shaded in red and the center of symmetry is denoted by a red arrow. The AUG start codon for the first gene in the *trp* operon is at position +162.

The repressor-operator mechanism should be a sufficient on-off switch for the *trp* operon. However, in operons responsible for the biosynthesis of amino acids an additional mechanism allows a finer control in which the enzyme concentration is varied according to the amino acid concentration. This control is effected by (1) premature termination of transcription before the first structural gene is reached and (2) regulation of the frequency of this termination by the concentration of the amino acid. How this is accomplished is described in the following section.

The Attenuator and the Leader Polypeptide

Beginning at the 5′ end of the *trp* mRNA molecule, there are 162 bases before the start codon of the *trpE* gene. This segment of the mRNA is called the leader (a general term used for such regions; see Chapter 12). Within the leader is a sequence of bases (bases 123 through 150) which, if deleted, causes a sixfold increase in the synthesis of the *trp* enzymes in either a derepressed cell or a constitutive mutant cell. Thus, bases 123–150 must have regulatory activity. Furthermore, after initiation of mRNA synthesis, unless there is no

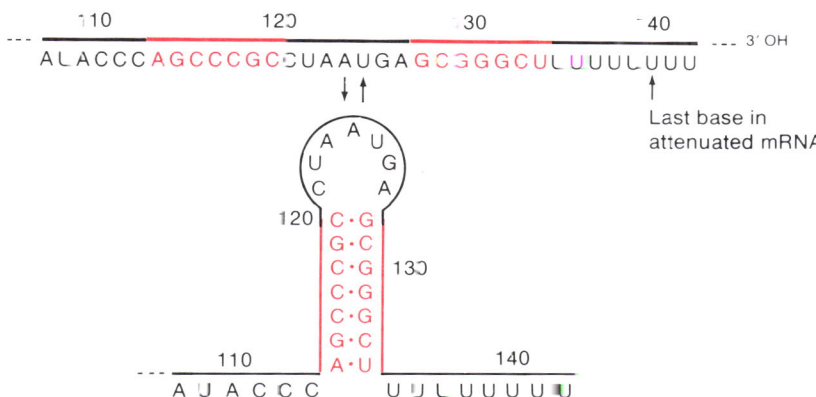

Figure 15-17 The terminal region of the *trp* leader mRNA (right end of *L* in Figure 15-15). The base sequence given is extended past the termination site at position 140 to show the long stretch of U's. The red bases form an inverted repeat sequence that could lead to the stem-and-loop configuration shown (segment 3–4, Figure 15-20).

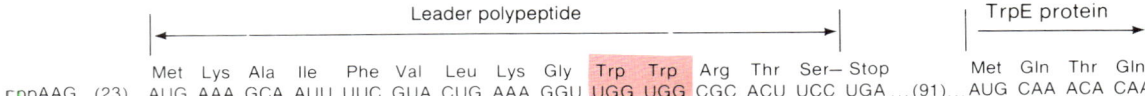

									Leader polypeptide							TrpE protein

Met Lys Ala Ile Phe Val Leu Lys Gly **Trp Trp** Arg Thr Ser– Stop   Met Gln Thr Gln
pppAAG…(23)…AUG AAA GCA AUU UUC GUA CUG AAA GGU **UGG UGG** CGC ACU UCC UGA …(91)…AUG CAA ACA CAA

Figure 15-18 The sequence of the *trp* leader mRNA showing the leader polypep-tide, the two Trp codons (shaded red), and the beginning of the TrpE protein. The numbers (23 and 91) refer to the number of bases whose sequences are omitted for clarity.

(a)   Met Lys His Ile Pro **Phe Phe Phe** Ala **Phe Phe Phe** Thr **Phe** Pro Stop
   5′ AUG AAA CAC AUA CCG UUU UUC UUC GCA UUC UUU UUU ACC UCC CCC UGA 3′

(b)   Met Thr Arg Val Gln Phe Lys **His His His His His His His** Pro Asp
   5′ AUG ACA CGC GUU CAA UUU AAA CAC CAC CAU CAU CAC CAU CAU CCU GAC 3′

Figure 15-19 Amino acid sequence of the leader peptide and base sequence of the cor-responding portion of mRNA from (a) the phenylalanine operon and (b) the histidine operon. The repeating amino acid is shaded in red.

tryptophan at all, most of the mRNA molecules are terminated in this region, yielding an RNA molecule consisting of only 140 nucleotides and stopping short of the genes encoding the *trp* enzymes; this explains why deletion of this region results in increased gene expression. This region, in which ter-mination occurs and is regulated, is called the **attenuator.** Careful exami-nation of the base sequence (Figure 15-17) around which termination occurs shows that it contains the usual features of a termination site—namely, a possible stem-and-loop configuration in the mRNA followed by a sequence of eight A·T pairs, as explained in Chapter 12.

A variety of experiments indicate that attenuation requires the presence of charged tRNA^{Trp}, which suggests that some part of the leader is trans-lated. Examination of the leader sequence shows an AUG codon and a later in-phase UGA stop codon; when translation begins at the AUG codon, a leader polypeptide consisting of fourteen amino acids is synthesized (Figure 15-18).

The leader polypeptide has an interesting feature—namely, at positions 10 and 11 there are adjacent tryptophan codons. This is significant because in the histidine operon, which also has an attenuator system (i.e., prema-turely terminated mRNA), there is a similar base sequence that could encode a leader polypeptide having *seven* adjacent histidine residues (Figure 15-19). We will see the significance of these repeated codons shortly.

Base-pairing in the mRNA Leader

The *trp* leader has been completely sequenced. The most notable features are four segments denoted 1, 2, 3, and 4, which are capable of base-pairing

(Figure 15–20) in two different ways—namely, forming either the base-paired regions 1–2 and 3–4 or just the region 2–3. By use of the enzyme RNase T1, which cannot digest base-paired RNA, it has been shown that the two base-paired regions 1–2 and 3–4 are present in purified *trp* leader mRNA. The base-paired region 3–4, which, by its location, should be in the terminator recognition region, has been shown to be necessary for termination—that is, several mutations in this region both prevent transcription termination *in vivo* and eliminate resistance of that region to RNase T1 *in vitro*. It should be noted that because sequences 2 and 3 are paired in the duplex segments 1–2 and 3–4, then the region 2–3 cannot be present simultaneously with 1–2 and 3–4. This also means—and this is the essential point—that if conditions were somehow right for formation of region 2–3, then neither of the regions 1–2 or 3–4 could be present. In fact, if regions 1 and 4 are deleted, 2–3 pairing is demonstrable. Together these facts have led to a theory of the mechanism of premature termination in the *trp* leader; the theory, which is presented in the following section, is widely accepted though not yet proved rigorously.

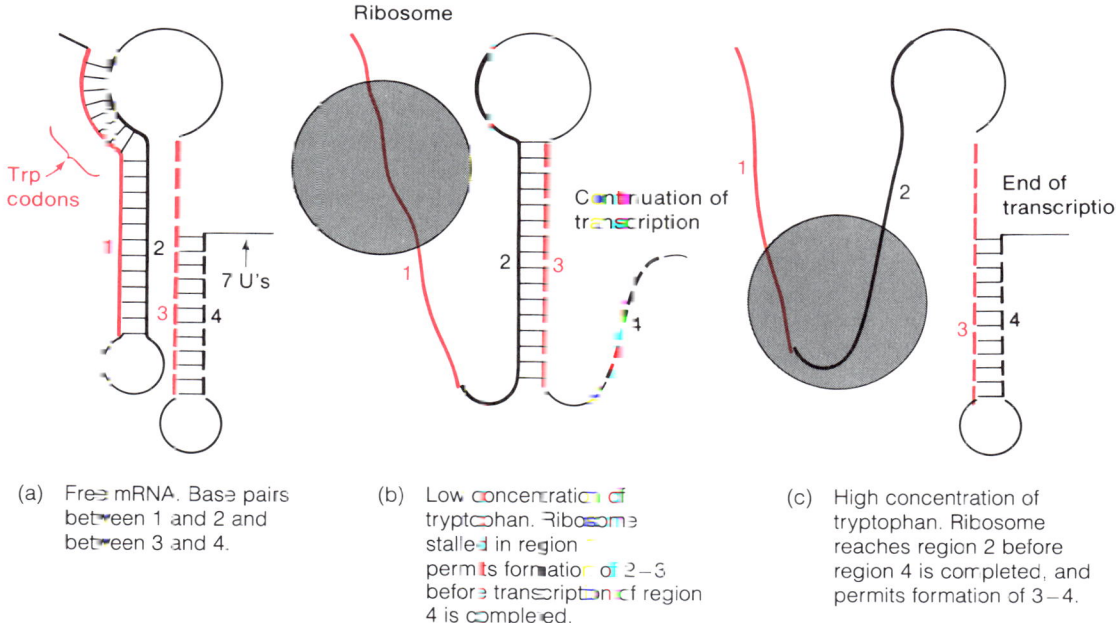

(a) Free mRNA. Base pairs between 1 and 2 and between 3 and 4.

(b) Low concentration of tryptophan. Ribosome stalled in region 1 permits formation of 2–3 before transcription of region 4 is completed.

(c) High concentration of tryptophan. Ribosome reaches region 2 before region 4 is completed, and permits formation of 3–4.

Figure 15-20   The accepted model for the mechanism of attenuation in the *E. coli trp* operon.

The Mechanism of Attenuation

This theory (Figure 15-20) proposes that mRNA termination is mediated through translation of the leader peptide "gene." Because there are two tryptophan codons in this gene, the ability to translate the sequence should be sensitive to the concentration of charged $tRNA^{Trp}$—that is, if the amount of tryptophan is limiting, there will be insufficient charged $tRNA^{Trp}$ and hence translation across the tryptophan codons should be very slow. It is also assumed that transcription and translation are coupled, as is usually true in bacteria, and all base pairing is eliminated in the segment of the mRNA that is in contact with the ribosome. Figure 15-20 shows that the end of the *trp* leader peptide is in segment 1. Usually a translating ribosome is in contact with about ten bases in the mRNA past the codons being translated; hence, when the final codons are being translated, segments 1 and 2 are not paired. In a coupled transcription-translation system, the leading ribosome is not far behind the RNA polymerase. Thus, if the ribosome is in contact with segment 2 when synthesis of segment 4 is being completed, then segments 3 and 4 are free to form the duplex region 3−4 without segment 2 competing for segment 3. The presence of the 3−4 stem-and-loop configuration (the terminator) allows termination to occur when the terminating sequence of seven U's is reached. If there is no added tryptophan, the concentration of charged $tRNA^{Trp}$ becomes limiting and occasionally a translating ribosome is stalled for an instant at the tryptophan codons. These codons are located sixteen bases before the beginning of segment 2. Thus, segment 2 is free before segment 4 has been synthesized and the duplex 2−3 region (the antiterminator) can form. In the absence of the 3−4 stem and loop, termination does not occur and the complete mRNA molecule is made, including the coding sequences for the *trp* genes. Thus, if tryptophan is present in excess, termination occurs and little enzyme is synthesized; if tryptophan is absent, there is no termination and the enzymes are made.

An implicit assumption in the attenuation model is that the movements of the ribosome and of RNA polymerase are coupled. This assumption is necessary because if RNA polymerase moved far ahead of the ribosome, a structure with an intact termination sequence—for example, the structure shown in panel (a) of Figure 15-20—could form well before the ribosome reaches the adjacent Trp codons. Recently, such coupling has been observed in an *in vitro* transcription-translation system and results from a transcription pause at position 92 in the attenuator (between segments 2 and 3). The signal for the pause is the formation of structure 1−2. The pause is relieved by entry of the ribosome into the leader-peptide region. Since the mechanism of pausing is unknown (Chapter 12), little more can be said about the phenomenon other than that it does provide the molecular coupling between transcription and translation that is needed for the attenuation mechanism to function.

Repression versus Attenuation

The *trp* repressor-operator system does not operate as a simple on-off switch but can yield intermediate levels of operon expression. It is known that the synthesis of *trp* mRNA is partially repressed at all times in a cell growing in the absence of added tryptophan because the concentration of the *trp* enzymes is tenfold greater in a cell having a mutant (inactive) repressor than in a wild-type cell. This observation implies that in a wild-type cell, if the endogenous concentration of tryptophan were to fluctuate for any reason, the equilibrium between active and inactive repressor would shift to maintain a usable supply of tryptophan. Thus it is not at all clear why the attenuation system is needed.

Repression has been studied independently of attenuation in a cell containing a gene fusion linking the *lacZ* gene to the *trp* promoter-operator region, lacking the attenuator. The activity of β-galactosidase as a function of concentration of tryptophan in the growth medium is a measure of the response of the repressor-operator system. Comparison of the behavior to that of an intact *trp* operon yields the contribution of the attenuation system. It was found that repression and attenuation are responsible for an 80-fold and 6–8-fold variation, respectively, of expression of the operon, for a total variation of 500–600-fold. Furthermore, repression is the dominant regulatory mechanism at higher concentrations of tryptophan; attenuation is not relaxed until starvation for tryptophan becomes severe (implying that charging of tRNA^{Trp} occurs when the concentration of tryptophan is quite low).

Many operons responsible for amino acid biosynthesis are regulated by attenuators equipped with the base-pairing mechanism for competition described for the *trp* operon. So far this has been described for the histidine, threonine, leucine, isoleucine-valine, and phenylalanine operons of the bacteria *E. coli, Salmonella typhimurium,* and *Serratia marcescens,* though in less detail than for the *trp* operon. Except for the phenylalanine operon, each lacks a repressor-operator system and is regulated solely by attenuation. Since these operons are regulated adequately (though the range of expression is not as great as that of the *trp* operon), one might ask why the *trp* operon (and the *phe* operon) has a dual regulatory system. This question has given rise to considerable speculation. The most obvious explanation is that the separate effects expand the range of tryptophan concentration in which regulation occurs. Whereas this is certainly true, there is more to it, because the *trp* repressor also regulates the activity of the *aroH* gene.

A common intermediate in the synthesis of the aromatic amino acids is made by three distinct enzymes, each having the same enzymatic activity. (Enzymes of this sort, called **isozymes,** are discussed in greater detail in a later section of this chapter, entitled Feedback Inhibition.) All three enzymes are needed only when all three aromatic amino acids must be synthesized. Thus, each enzyme is independently regulated by an amino acid-specific repressor. It has been suggested that in the distant past, the *trp* operon was regulated only by attenuation, and *aroH* was regulated as it is now, by the tryptophan–activated *aroH* repressor, which for historical reasons we have called the *trp* repressor. The existence of a tryptophan–sensitive repressor allowed the possibility of evolution of a sequence in the *trp* promoter into an operator that can bind the tryptophan–sensitive *aroH* repressor.

RELATIVE POSITIONS OF PROMOTERS AND OPERATORS

We have seen that repressor genes need not be near the structural genes they regulate—for example, the *trpR* gene is quite far from the *trp* structural genes. However, the nature of operator function is such that a promoter and its corresponding operator must be nearby. Here we ask whether the orientation of the promoter and operator with respect to the transcription start site is important. Examination of many operons (Figure 15-21) shows that it is not. Note that in some cases, the operator is embedded in the promoter and in others it a distinct element.

REGULONS

So far, we have examined systems in which synthesis of one polycistronic mRNA containing structural genes is regulated. Other, more complex sys-

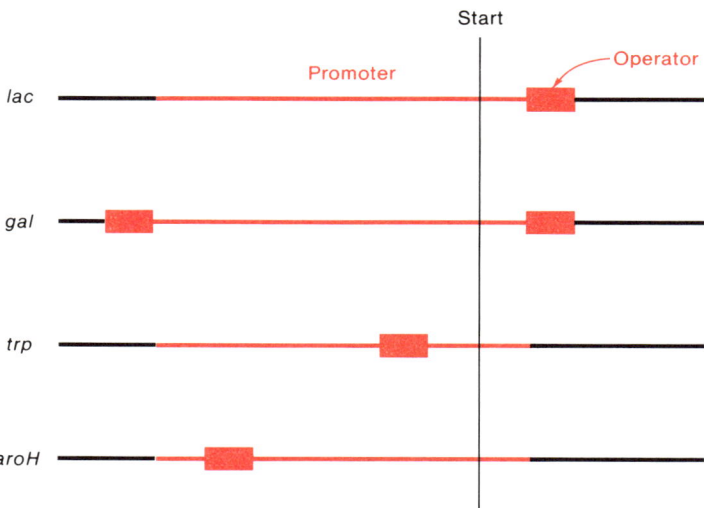

Figure 15-21 Four different arrangements of operators (red rectangles) and promoters for *E. coli* genes. Other arrangements have also been observed. Transcription begins at the start site indicated by the vertical black line and proceeds to the right.

tems have also been studied in which a regulatory molecule controls synthesis of many mRNA molecules transcribed from regions scattered throughout the chromosome. The term **regulon,** originally proposed in 1964 but rarely used, has recently been adopted for these multi-mRNA systems.

Most regulons, because of their complexity, are poorly understood. Two of these, the heat-shock and pho (phosphate) regulons will be presented descriptively; a third, the SOS system will be given in slightly greater detail.

The HTP Regulon

If a culture of *E. coli* growing at 30°C is heated to 40°C, synthesis of a variety of proteins is abruptly increased. Some proteins are not synthesized at all at the lower temperature; others are made, but the rate of synthesis increases greatly. The magnitude of the increase in production of these proteins depends on the size of the temperature shift. Seventeen of these proteins are called **heat-shock proteins;** their genes constitute the **high-temperature protection** or **HTP** regulon. Once growth of the culture becomes adjusted to the higher temperature, production of these proteins is reduced unless the temperature is so high that growth of the cells becomes irregular. In these conditions the 17 proteins amount to 15 percent of the cellular protein. At 50°C growth is totally inhibited.

In *E. coli* a single protein, the product of the *atpR* gene is a positive effector of all mRNAs of the regulon. The protein has been isolated and sequenced and its C-terminal end has been found to be homologous to the RNA polymerase σ subunit. Several experimental results indicate that the heat-shock response involves a reprogramming of RNA polymerase by this σ-like element. Six of the heat-shock proteins have been identified. These are the DnaJ and DnaK proteins (involved in DNA synthesis), the normal σ subunit the GroES and GroEL proteins (used in protein processing), the Lon protein (used in protein degradation), and lysyl-tRNA synthetase—proteins connected to all major macromolecular processes in the cell.

The biological function of the heat-shock response is unknown. However, the fact that the response has been observed in numerous bacterial genera and in a wide variety of eukaryotes (e.g., *Drosophila*), attests to its generality and importance.

The Pho Regulon

Inorganic phosphate is an essential nutrient in all organisms; it is required for the synthesis of fundamental small molecules, such as ATP and GTP, and of precursors for DNA, RNA, and membrane. When bacteria sense a deficiency of inorganic phosphate, a number of enzymes and structural proteins are rapidly synthesized; these enable the cell to increase the efficiency of uptake of inorganic phosphate from the surrounding medium (by creating new transport systems) and to utilize other forms of phosphate. The genes whose products are formed in response to the phosphate deprivation are scattered throughout the *E. coli* chromosome, forming several clusters, some of which are operons. Other genes, whose function is either to sense the concentration of phosphate or to positively regulate synthesis of some of the structural genes, are also part of the pho regulon. This regulon includes at least 24 genes whose transcription is positively regulated by the *phoB* gene, and this gene is in turn positively regulated by the genes *phoM* and *phoR*. The *phoR* product is also a negative regulator. How this occurs, as well as many other features of the regulon, remain obscure. Interestingly, 40–60 other genes are also regulated by the concentration of inorganic phosphate, but do not depend on the *phoB, phoM,* or *phoR* genes; thus, these genes are not considered to be part of the pho regulon.

The SOS Regulon

In Chapter 10 the inducible SOS repair system was described. It was pointed out that this system, which allows the frequency of replication errors to increase when necessary, must be regulated in order to keep the normal error frequency low. Since the repair system is only needed following certain types of DNA damage, it seems reasonable that some feature of the damage would be the inducer. The same might be said of other ultraviolet repair systems, such as RecA-mediated recombination repair and Uvr excision repair. Indeed, these systems are inducible and the same regulatory elements control these systems that regulate SOS repair. Several (>11) other genes representing functions not directly related to repair are also in the SOS regulon.

Critical to the regulation of the regulon are the genes *lexA* and *recA; lexA* encodes the common repressor of the SOS operons and *recA* is a component of the induction system. The LexA repressor has been purified and found to bind to an operator sequence (called the SOS box) adjacent to each gene or operon. The SOS boxes at different loci are not identical but are related through a consensus sequence. Of importance, as will be seen shortly, is the fact that an SOS box is also adjacent to the *lexA* gene. Thus, LexA regulates its own synthesis by self-repression. The RecA product is a protein with several activities: it binds to single- and double-stranded DNA, it has an ATPase activity, and it is a protease. The proteolytic activity is very weak against most proteins, but it is very strong with certain repressors in phage systems and cleaves each repressor at a unique site. We will return to this important point shortly. Following DNA damage of the sort inflicted by ultraviolet irradiation, there is a burst of synthesis of *recA* mRNA. However, this does not occur in *recA*⁻ mutants, which is an observation quite different from what has been seen in the operons described so far. For example, if a cell has a *galK⁻* mutation, addition of galactose still turns on the *gal* operon even though the *galK* product that is synthesized is defective and has no enzymatic activity. With the *recA* operon it appears that functional RecA protein is required to turn the operon on. The reason is that the RecA protein has inducing activity.

Let us consider how the SOS regulon should be regulated. In the absence of DNA damage its gene products are unnecessary. Following the occurrence of DNA damage the system must be active; once repair is complete, the regulon should optimally be shut down rapidly. The three features of the system that enable these regulatory events to occur are (1) a damage-induced activation of the protease function of the RecA protein, (2) a vulnerable proteolytic site in the LexA protein, and (3) the self-repression of transcription of the *lexA* gene.

The RecA protein in an undamaged cell lacks protease activity. However, if both single-stranded DNA and ATP are present, purified RecA protein is a protease. It has been shown that RecA binds to thymine dimer sites (produced by ultraviolet irradiation), possibly to the short segment of single-stranded DNA generated by the distortion (Chapter 10). Binding of RecA protein to DNA is thought to cause a conformational change in the protein, which thereby acquires protease activity. However, some experiments suggest that an unidentified small molecule produced by the damage may also be necessary for activation of RecA, but this action has been debated for more than a decade. At any rate, RecA is activated in some way by DNA damage and its protease activity cleaves the LexA molecule. In the absence of functional LexA protein, transcription of all operons of the SOS regulon rapidly increases about 50-fold, yielding an adequate supply of the Uvr enzymes, RecA, and SOS repair proteins. Since LexA represses its own synthesis, it is also made in abundance. However, the large amount of RecA protein, much of which has been activated to a protease, continues to cleave LexA; thus, all proteins of the SOS regulon continue to be made. Once all repair is completed, the inducing signals (whatever they may be) are no longer present and RecA loses its proteolytic activity. Thus, LexA is no longer cleaved. During the period in which RecA was active, LexA was made at a rapid rate (because it is self-repressed); thus, without RecA protease activity, it rapidly accumulates, binds to the SOS boxes, turns off the SOS regulon and the state of the cell existing before DNA damage occurred is reestablished.

REGULATION OF TRANSLATION

In various places in this chapter translational regulation has been mentioned. Typically this refers to the fact that the number of copies of each protein translated from a polycistronic mRNA varies from gene to gene. Usually there is a gradient of translation decreasing from the 5′ terminus to the 3′

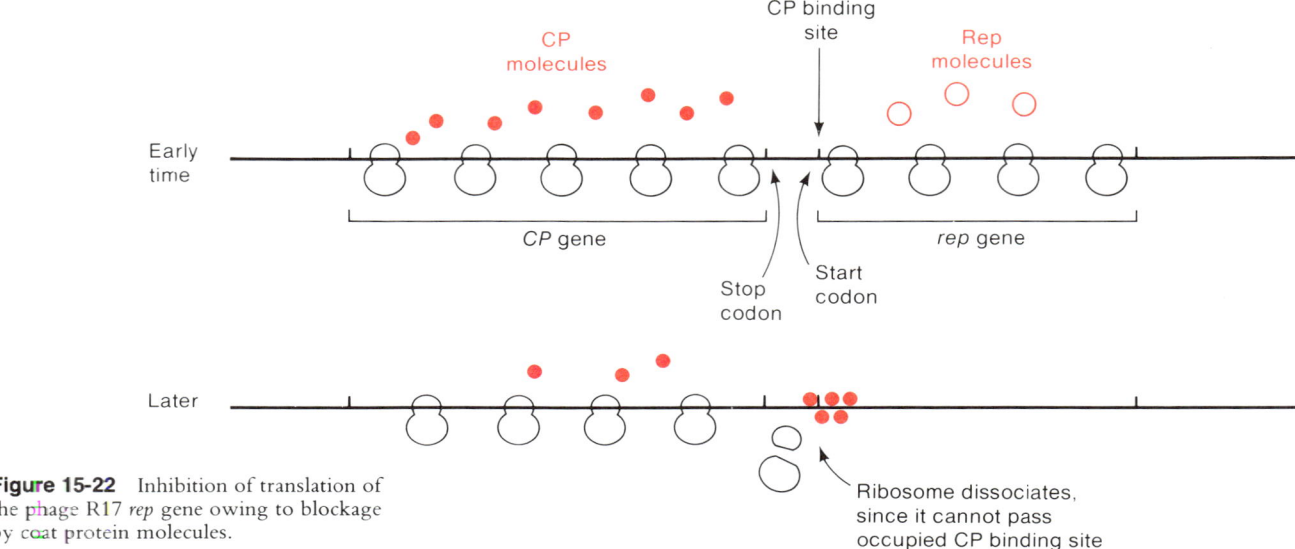

Figure 15-22 Inhibition of translation of the phage R17 *rep* gene owing to blockage by coat protein molecules.

terminus of the mRNA, and several processes account for this phenomenon—namely, varying efficiencies of initiation of translation, different spacing between chain-termination codons and a subsequent AUG codon (which affects the probability that the ribosome and mRNA will dissociate), and differential sensitivity of various regions of the mRNA to degradation. However, these are not regulatory in that translation efficiency does not vary in response to a signal. Inhibition of translation of a particular gene by its gene product has been observed. The first system elucidated was the regulation of production of an enzyme in *E. coli* infected with phage R17. This phage encodes only three gene products—two structural proteins (the A protein and the phage coat protein) and an enzyme (replicase). Far more coat protein is needed than replicase, which is used only in catalytic amounts. The mRNA molecule has a binding site for the coat protein between the termination codon of the coat protein gene and the AUG codon of the replicase gene. As the coat protein is synthesized, this binding site gradually is filled with protein molecules, blocking the ribosome from translating the replicase region (Figure 15-22). A more comprehensive example of translational regulation will be seen when we describe how synthesis of ribosomes is regulated.

REGULATION OF THE SYNTHESIS OF RIBOSOMES

The growth rates of all bacterial species vary with the composition of the growth medium. In minimal media having an efficiently utilized carbon source such as glucose, E. coli cells divide roughly every 45 minutes at 37°C; with a poorer carbon source such as proline the doubling time is about 500 minutes. In rich media containing glucose, amino acids, purines, pyrimidines, vitamins, and fatty acids, a cell does not have to synthesize these substances and hence it can grow very rapidly; typically, the generation time is about 25 minutes in rich media.

A ribosome has a limited capacity for protein synthesis (15 amino acids per sec at 37°C, independent of carbon source); thus, at different overall growth rates, at which different rates of protein synthesis occur, the number of ribosomes per cell varies (Table 15-5).

If a bacterial culture is transferred from a growth medium in which growth is rapid to one in which growth is slow (this is called a "downshift"), the ribosome content of each cell decreases from the higher value for the rapid medium to the lower value characteristic of the slow medium. This is reasonable since otherwise each cell would possess more ribosomes than it would need. The decrease is accomplished by allowing DNA synthesis to proceed without synthesis of rRNA. Some signal must be sensed by each slowly growing cell and this signal must inhibit transcription of the rRNA genes. This observation, plus the fact that in any growth medium the rate of synthesis of rRNA is proportional to the rate of production of ribosomes makes it clear that synthesis of rRNA is regulated. Furthermore, the 52 ribosomal proteins (r-proteins) are all made at a rate that maintains a constant ratio to rRNA (at all growth rates); and the amounts of tRNA, initiation factors, elongation factors, and aminoacyl-tRNA synthetases are related to the number of ribosomes per DNA molecule. Thus, synthesis of all major components of the protein synthesis apparatus must be regulated. In this section we will only be concerned with rRNA and r-proteins.

Genetic mapping experiments indicate that all of the genes encoding the 52 different ribosomal proteins of E. coli occupy a small region of the chromosome, but it is known that these genes are transcribed as about 20 polycistronic mRNA molecules. The rRNA genes occur in 7 nearly identical but separated regions scattered throughout the chromosome. These regions include genes for certain tRNA molecules, as was indicated in Chapter 12; the remaining tRNA genes are found in several other regions.

Table 15-5 Some characteristics of E. coli growing at different growth rates

Doubling time, minutes	Ribosomes per cell	Ribosomes per DNA molecule*
25	69,750	15,500 (4.5)
50	16,500	6,800 (2.4)
100	7,250	4,200 (1.7)
300	2,000	1,450 (1.7)

*The number in parentheses is the number of DNA molecules per cell.

A constant ratio of rRNA to ribosomes and r-proteins to ribosomes, at all growth rates, is regulated by two feedback systems. In the first, ribosomes are made in slight excess of that which is needed to attain the appropriate rate of protein synthesis, and free nontranslating ribosomes inhibit (directly or indirectly) the synthesis of rRNA. This is called ribosome feedback regulation. In the second, certain r-proteins inhibit translation of an mRNA that encodes the one or more r-proteins. This is called translational repression. Some of the experiments that support the existence of these regulatory systems illustrate general methods of molecular biology and hence they will be described.

We first examine regulation of synthesis of rRNA. The principle underlying the experiment to be described is that if the amount of RNA transcribed from a set of operons, each having the same promoter, is not proportional to the number of operons and is less than demanded by proportionality, then it is likely that the RNA itself or some product of the RNA inhibits transcription. The experiment used three cultures of E. coli, each growing on the same medium and hence containing the same number of ribosomes per cell. The three cultures are described in Table 15-6. Cells in culture I contain the usual seven rRNA operons. Each operon made rRNA in roughly equal amounts, and the rRNA was assembled into ribosomes. Cells in culture II also contained the seven chromosomal rRNA operons plus seven copies of a plasmid with a single copy of one rRNA operon (marked to make it distinguishable from the chromosomal operons. Thus, 14 rRNA operons, rather than 7, were present in each cell. However, culture II made the *same* amount of rRNA per cell (not twice as much). Furthermore, half of the rRNA was transcribed from the chromosomal operons and half from the plasmids. Thus, each of the 14 operons was being transcribed at half the normal rate, indicating that repression of some kind was occurring. Culture III was like culture II except that the rRNA made by the 7 plasmids had a deletion that prevented it from being assembled into ribosomes. These

Table 15-6 Amount of rRNA made by bacterial cultures containing different numbers of rRNA operons

| Culture | Number of operons per cell | | | Amount of rRNA made per operon* | Total rRNA made* |
	Chromosomal	Plasmid (normal rRNA)	Plasmid (defective rRNA)		
I	7	0	0	1	7
II	7	7	0	$\frac{1}{2}$	7
III	7	0	7	1	14

*Arbitrary units.

cells made twice as much rRNA as culture I or culture II, with half from plasmid operons. The plasmid-encoded (defective rRNA) did not contribute to regulating transcription of rRNA operons, which implies that only rRNA that can be assembled into ribosomes can lead to repression. The interpretation of this and other experiments is the following: (1) all functional rRNA is assembled into ribosomes, (2) ribosomes are made in slight excess, and (3) the excess (nontranslating) ribosomes repress transcription of the rRNA operons. How the repression occurs (e.g., direct interaction of a free ribosome with a promoter versus some indirect effect) is not known.

A variant of the experiment just described shows that the repressive effect acts on the promoter regions of the rRNA operons. In this experiment a plasmid was constructed containing an rRNA operon but in which the rRNA promoter was replaced by the promoter-operator region of *E. coli* phage λ. A λ repressor that binds to the λ operator is also included. However, the λ repressor contains a temperature-sensitive mutation that renders the repressor inactive above 39°C. When a culture containing the plasmid was warmed above 39°C, rRNA was transcribed from the plasmid rRNA operon. The total amount of rRNA made per cell increased markedly compared to that made by plasmid-free cells and transcription of the chromosomal rRNA operons was depressed considerably, accounting for only 20 percent of the total rRNA made. The interpretation of the experiment is that the excess rRNA results in production of an excess of nontranslating ribosomes, which repress transcription of the rRNA operons having a normal promoter but not those having a λ promoter-operator region. Thus, the target of the nontranslating ribosomes must be in the promoter. Note how the model accounts for the change in ribosome content with growth rate. If a culture is transferred to a medium that allows a high rate of protein synthesis, all ribosomes will suddenly be engaged in protein synthesis and there will be no nontranslating ribosomes. Thus, the rate of rRNA synthesis will increase and the number of ribosomes will increase until a small number of nontranslating ribosomes are produced, which turns off transcription of the rRNA genes.

A similar type of experiment shows that synthesis of r-proteins is not regulated at the transcriptional level. The genes encoding the 52 r-proteins are organized into 20 operons. Addition of an appropriate plasmid can increase the number of copies of one of these operons in a cell, say, tenfold; this would increase the rate of synthesis of the corresponding mRNA tenfold, but *the rate of production of r-proteins encoded in the mRNA remains that observed with one copy of the operon*. Thus, the amount of the r-proteins that is synthesized is not proportional to the amount of mRNA encoding them, so clearly translation and not transcription is being limited. An understanding of the mode of translational regulation came from *in vitro* translation experiments. In these experiments a single species of r-protein mRNA was translated and the inhibitory effect of each r-protein encoded in the mRNA was tested. It was observed that translation of each mRNA could be inhibited by the addition of *one* (a particular one) of the encoded r-proteins. Further analysis showed that the translational repression is a result of binding of that r-protein to a base sequence near the ribosome binding site. A variant of the *in vitro* experiment completes the story—namely, addition of the particular rRNA (5S, 18S, or 23S) to which the r-protein binds in the ribosome prevents translational repression. A base sequence analysis of the mRNAs and the rRNA shows a similar sequence and a common structure (a stem-and-loop) in the binding sites of each RNA for a particular r-protein. Thus, competition exists between rRNA and r-protein mRNA for a particular r-protein. Binding studies show that each repressing r-protein species binds preferentially to the rRNA. Thus, *as long as rRNA is available for ribosome production, r-proteins will bind to rRNA and synthesis of r-proteins will continue.* Studies with each r-protein mRNA show that when translational repression is not occurring, each mRNA is completely translated, yielding equal numbers of each r-protein encoded in a particular mRNA species. Furthermore, translational repression ensures that all r-proteins are synthesized at the same rate and coupled to the synthesis of ribosomes. This means that the synthesis of all r-proteins is ultimately regulated by the rate-limiting component in ribosome synthesis, that is, the rRNA.

Since ribosomes function only in protein synthesis, production of tRNA should also be coupled to that of ribosome assembly. Recall from Chapter 12 that many tRNA genes are part of the rRNA operons, so synthesis of these tRNA molecules is necessarily connected to that of rRNA. Experiments similar to those described above that showed that nontranslating ribosomes control transcription of the rRNA operons have also been carried out for the tRNA species that are not made by the rRNA operons; transcription of these tRNA genes is also apparently regulated by the number of non-translating ribosomes.

If a growing culture is depleted of an amino acid (for example, by removing a required amino acid from a medium in which an auxotroph is growing), protein synthesis rapidly stops. This is associated with complete absence of transcription of the rRNA genes. Also, accompanying amino acid starvation two unusual ribonucleotides are produced—guanosine-5'-diphosphate-3'-diphosphate (ppGpp) and guanosine-5'-triphosphate-3'-diphosphate (pppGpp).

The inhibition of rRNA synthesis associated with amino acid depletion is called the **stringent response.** The factors responsible for the stringent response are gradually being revealed by studying mutants in which rRNA synthesis continues unabated during amino acid starvation. The phenotype of these mutants is described as **relaxed.** The mutations map in several genes designated *rel*. The only *rel* gene that is understood at all is *relA*.

The product of the *relA* gene is a protein known as the **stringent factor;** it is an enzyme that is responsible for the synthesis of ppGpp. The stringent factor is located exclusively in the 50S ribosome (but only in about 1 in 200 ribosomes); during normal protein synthesis the stringent factor is inactive and virtually no ppGpp is synthesized. Activation of the stringent factor occurs only when two conditions are met: (1) the 50S particle must be in an intact 70S ribosome that is bound to mRNA and engaged in translation, and (2) the A site of the ribosome must be occupied by an uncharged tRNA molecule rather than charged tRNA. In normal growth conditions the concentration of uncharged tRNA is much lower than that of charged tRNA, which means that the A site is rarely occupied by uncharged tRNA. However, in the case of starvation for an amino acid, the tRNA species that normally carries that amino acid will be uncharged. In this way amino acid deprivation leads to production of ppGpp. The ppGpp signals that something has stopped protein synthesis and that additional rRNA should not be made.

A great deal of effort has gone into attempting to understand how ppGpp regulates the synthesis of rRNA; most evidence has been conflicting and little can be said positively at present. Whether it plays a role in the regulation of tRNA synthesis discussed earlier in this section is unknown.

UNREGULATED CHANGES IN GENE EXPRESSION

In the systems described so far, a bacterial operon responds to a signal of some kind. Here we see an example of a change in expression that may occur at random.

Many bacteria possess flagella (Figure 15-23)—that is, long protein-containing hairlike appendages that propel the bacteria through liquid media or across solid surfaces. Flagella consist primarily of a single protein called **flagellin.**

Strains of the bacterium *Salmonella typhimurium* have two genes called *H1* and *H2* that are responsible for the synthesis of two distinct types of flagellin. In any particular bacterial cell only one of the two *H* genes is transcribed; thus, an individual bacterium makes either H1-type flagella or H2-type flagella. If a bacterium making the one type of flagella is allowed to grow for many generations, in time a bacterium having the other type of flagella will arise. The frequency of appearance of the other type is about 10^{-5}–10^{-3} switches per cell per generation (the exact frequency depending on the individual strain). This phenomenon is called **phase variation.**

The *H1* and *H2* genes have been isolated and partially sequenced. There is nothing unusual about the *H1* gene, but the *H2* gene exists in two forms. At one end of *H2* is a 970-bp segment that exists in one orientation in some of the isolates and is inverted in others. Furthermore, only one orientation can be isolated from a population that makes H1-type flagella, whereas the other is isolated exclusively from population in which H2-type flagellin is made. Thus, it appears that inversion of this 970-bp element is related to the activity of the *H1* and *H2* genes. The key to understanding what is happening

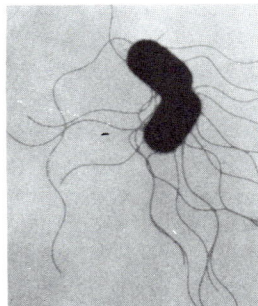

Figure 15-23 Electron micrograph of two cells of *Salmonella typhimurium* with flagella. [Courtesy of Mel Simon.]

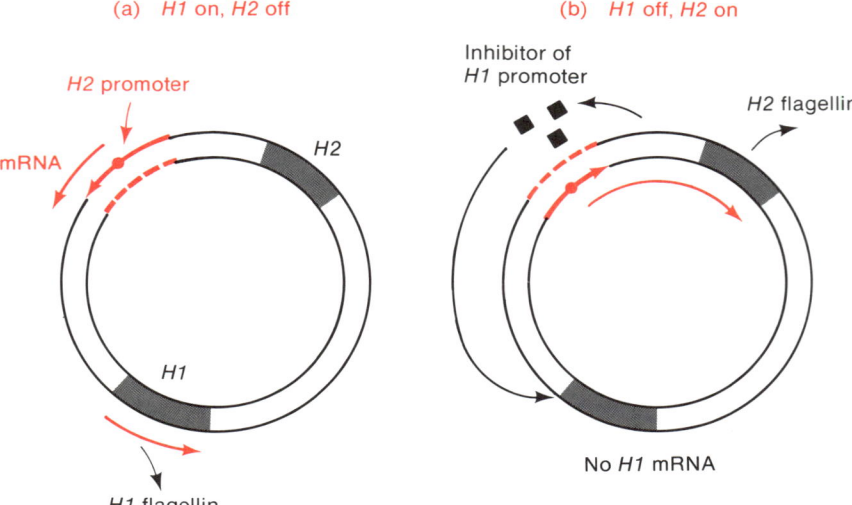

(a) *H1* on, *H2* off

H2 promoter

mRNA

H2

H1

H1 flagellin

(b) *H1* off, *H2* on

Inhibitor of
H1 promoter

H2 flagellin

No *H1* mRNA

Figure 15-24   Regulation of the two flagellin genes in *Salmonella typhimurium*. The heavy red region is an invertible segment that contains the *H2* promoter. The red arrows represent mRNA molecules.

is that when this element is in one orientation, *H2* is transcribed, and when it is in the other orientation, *H2* is inactive. The explanation for the two states of activity is shown in Figure 15-24. The invertible element of 970 base pairs carries the promoter for *H2*. Thus, in one orientation initiation of mRNA synthesis at this promoter leads to transcription of the *H2* gene, whereas in the off orientation, transcription is away from *H2*. This invertibility accounts for the activation and deactivation of the *H2* gene but not the regulation of *H1*. When *H2* is transcribed, a polycistronic mRNA is made that encodes not only flagellin but also a protein that inhibits transcription of the *H1* gene. Thus, in the *H2*-on orientation *H2* flagellin is made and *H1* flagellin is not, whereas in the *H2*-off orientation, *H1* flagellin but not *H2* flagellin is made.

This system is an example of a **site–specific inversion.** That is, somehow the sequence of 970 base pairs is excised from the chromosome and then reinserted in the opposite orientation. Further analysis of the products encoded in the *H2* polycistronic mRNA indicates that in addition to the *H2* promoter and the inhibitor of *H1* transcription, the invertible element also contains a gene whose product is needed for the inversion. To date, no signal that stimulates inversion has been observed and the inversion may occur at random. Such random events have been observed in several prokaryotes and eukaryotes. The selective advantage of these events is rarely known.

In Chapter 16 it will be seen that site-specific exchanges are used in programming eukaryotic cells—for example, to produce antibodies.

FEEDBACK INHIBITION

In this section a mode of regulation that does not alter transcription or translation is described. If a culture of bacteria in which the *trp* operon has been derepressed is suddenly exposed to tryptophan repression is rapidly established, and there is little further synthesis of the *trp* enzymes. However, the *trp* enzymes persist and were it not for a second type of regulatory mechanism, wasteful synthesis of tryptophan and needless consumption of precursors and energy would occur. This state of affairs is avoided by **feedback** or **end-product inhibition,** a mechanism by which the activity of the enzymes is turned off by the product of the pathway. In bacteria, feedback inhibition is common in biosynthetic pathways.

In the pathway for tryptophan synthesis the first two reactions are catalyzed by the products of the *trpD* and *rpE* genes (Figure 15-25). These proteins form a tetramolecular aggregate consisting of two subunits of each protein. The *trpE* subunits produce anthranilate (from chorismate and glutamine) and the *trpD* subunits convert the anthranilate to phosphoribosyl anthranilate (PRA). The *trpE* subunit also contains a binding site for tryptophan that is occupied only when there is an excess of tryptophan. Occupation of this site causes a conformational change in the *trpE* subunit and this change destroys the enzymatic activity (Figure 15-26). The tetramer is an allosteric protein (see Chapter 6) and the conformational change in the *trpE* subunit induces a similar change in the *trpD* subunit, whose activity is thereby inhibited.

The most common type of feedback inhibition is effected by an inhibition of the first enzymatic step of a biosynthetic pathway by the product of the pathway. For example, in the pathway

$$A \xrightarrow{\ 1\ } B \xrightarrow{\ 2\ } C \xrightarrow{\ 3\ } D$$

catalyzed by enzymes 1, 2, and 3, the product D would act on enzyme 1, which is clearly the most economical mode of inhibition. Note that this means that enzyme 1 must have two binding sites—one for the substrate A

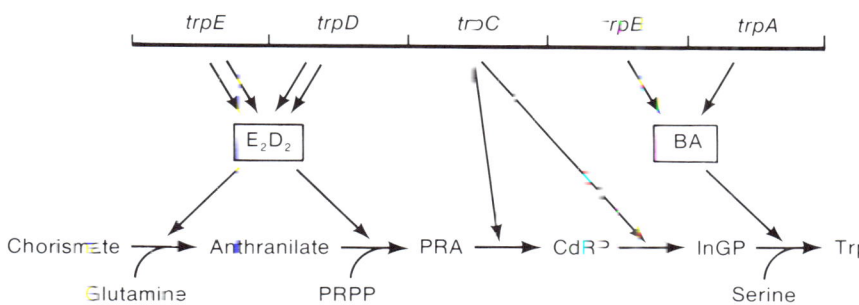

Figure 15-25 Sites of action of the *trp* gene products. The products of genes *E* and *D* and of *B* and *A* form active tetramers, E_2D_2 and A_2B_2 respectively. The site of feedback inhibition is E_2D_2. The components of the reaction sequences have been abbreviated.

Figure 15-26 Inactivation of the anthranilate complex E_2D_2 by tryptophan.

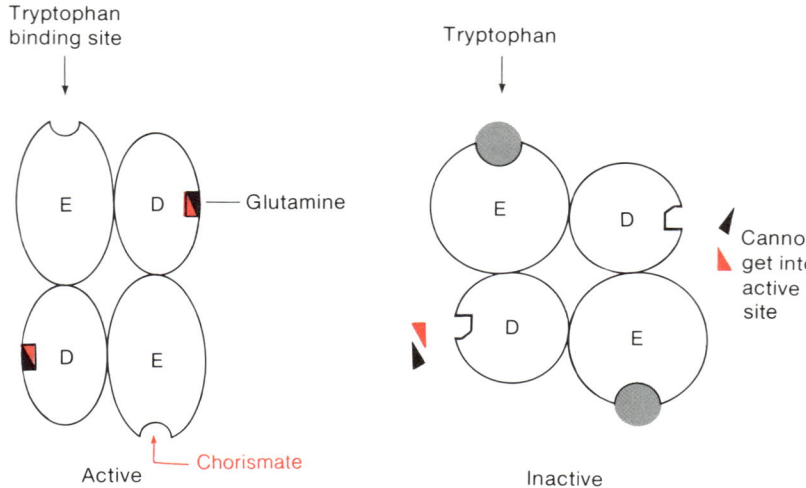

and one for the product D. Often binding of the product inhibits binding
of the substrate. This can be accomplished by an overlap of the binding sites
or by a conformational change that weakens or eliminates the substrate-
binding site. An example of the second mechanism was described in Chapter
6—namely, the inhibition of aspartyl transcarbamoylase by CTP.

Some biosynthetic pathways are branched—that is, they are responsible
for the synthesis of two products from a common precursor. A hypothetical
example of such a pathway is the following:

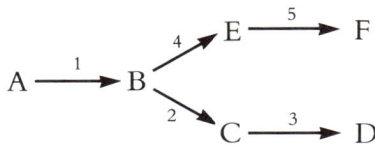

In this type of pathway it would be undesirable for a single product to inhibit
enzyme 1 because both pathways would be inhibited. In general, the most
economical kind of inhibition prevails—namely, D inhibits enzyme 2 and F
inhibits enzyme 4. In this way, neither D nor F inhibits the synthesis of the
other. Conversion of A to B is wasteful if both D and F are present; therefore,
in many pathways it is found that D and F together inhibit enzyme 1. There
are four major ways in which an enzymatic step can be inhibited by two
different molecules: (1) enzyme 1 is inhibited by its product B, which accumulates when enzymes 2 and 4 are blocked; (2) enzyme 1 has binding sites
for both D and F, and when both sites are filled the enzyme is inactivated;
(3) D and F both inhibit enzyme 1 slightly—for example, d-fold and f-fold,
respectively, and together they inhibit the enzyme $(d \times f)$-fold; (4) in some

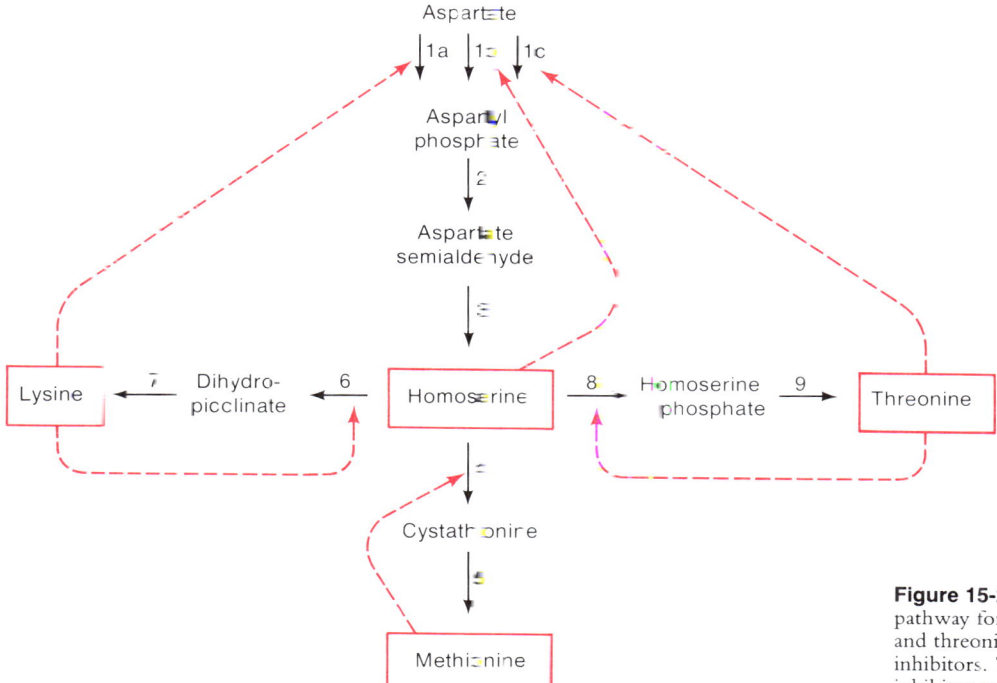

Figure 15-27 Feedback inhibition of the pathway for synthesizing lysine, methionine, and threonine. The boxed molecules are inhibitors. The dashed arrows lead from an inhibitor to the enzyme that is inhibited.

pathways, A is converted to B by two different enzymes (**isoenzymes**), one inhibited by D and one by F.

The elegance of feedback inhibition of a branched pathway is shown in Figure 15-27, which shows a well-studied multibranched pathway in which lysine, methionine, and threonine are synthesized from aspartic acid. Note how each amino acid inhibits an enzyme immediately after a branch, namely, enzymes 4, 6, and 8. Three isoenzymes, 1a, 1b, and 1c, are separately inhibited by lysine, homoserine, and threonine, respectively. The three isoenzymes have different enzymatic activity; each synthesizes an amount of aspartyl phosphate that is needed to form the appropriate amounts of the product that inhibits it. Thus, when an isoenzyme is inhibited, the correct amount of aspartyl phosphate is still made.

16

Regulation in Eukaryotes

The regulatory systems of prokaryotes and eukaryotes are quite different from each other. Prokaryotes are generally free-living unicellular organisms that grow and divide indefinitely as long as there are appropriate environmental conditions and an adequate supply of nutrients. Thus, their regulatory systems are geared to provide the maximum growth rate in a particular environment—except when growth would be detrimental. This optimization is accomplished by limiting cell synthesis and enzymatic activity to that necessary to achieve the maximum growth rate—that is, by eliminating waste—and with mechanisms that enable a cell to respond rapidly to changing conditions. In a prokaryote, since the DNA is not encased in a membrane, it is always available to receive signals present in the cytoplasm. Thus, the on-off aspect of protein synthesis is often regulated by controlling initiation of transcription, as was seen in Chapter 15.

The eukaryotes, except for the unicells such as yeast, algae, and protozoa, have different requirements. In a developing organism—for example, in an embryo—a cell must not only grow and produce many progeny cells but also the cells must undergo considerable change in morphology and biochemistry and maintain the changed state—that is, they must differentiate. However while growing and dividing, these cells have a somewhat easier time than bacteria in that the composition and concentration of their growth media do not change in time—examples of such media are blood, lymph, or other body fluids, or, in the case of marine animals, sea water.

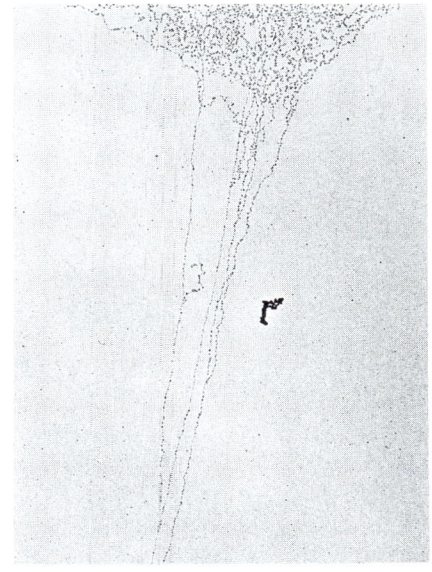

An electron micrograph of the amplified chorion-gene region showing several DNA branches. See also Figure 16-8 for a diagram. [Courtesy of Oscar Miller.]

Finally, in an adult organism, growth and cell division are stopped in most cell types and each cell needs only to maintain itself. Many other examples could be given; however, the main point is that because the needs of a typical eukaryotic cell differ from those of a bacterium, the regulatory mechanisms of eukaryotes and prokaryotes are basically different. From this reasoning one might expect the regulatory mechanisms of freely growing unicellular eukaryotes to resemble those of bacteria, but this is not usually the case. Somehow, in ancient times enclosure of the nucleus in a membrane was a great step forward in evolution and was accomplished by, or perhaps preceded, significant changes in gene organization and regulation.

This chapter is concerned primarily with eukaryotic cells that form organized tissue. Most of the information comes from detailed studies of the DNA of mammals, amphibians (toads of the genus *Xenopus*), insects (the fruit fly *Drosophila*), birds (usually the chicken), and echinoderms (the sea urchin). Some features of gene expression in the yeast *Saccharomyces* will also be presented.

Knowledge of regulation in eukaryotes is much less detailed than for prokaryotes. Studies with multicellular eukaryotes have been hampered by two problems—the inability to obtain regulatory mutants and to manipulate genes with ease, and lack of an *in vitro* RNA polymerase II transcription system. Nonetheless, a great deal of information is available, though the framework for understanding the facts is not always clear; thus, this chapter is presented as a compendium of observations illustrating the variety of regulatory mechanisms used by different cells.

SOME IMPORTANT DIFFERENCES IN THE GENETIC ORGANIZATION OF PROKARYOTES AND EUKARYOTES

There are numerous differences between prokaryotes and eukaryotes with respect to transcription, translation, and the spatial organization of DNA. Some of these differences were mentioned in earlier chapters and others will be discussed in this chapter. Seven of these differences will come up repeatedly in discussing regulation of eukaryotic gene expression; these are the following:

1. In a eukaryote only a single polypeptide chain can be translated from a processed mRNA molecule (Chapter 13); thus, operons of the type seen in prokaryotes are not found in eukaryotes.
2. The DNA of eukaryotes is bound to histones (to form chromatin) and to numerous nonhistone proteins. Only a small fraction of the DNA is bare.
3. A large fraction of the base sequences in the DNA of higher eukaryotes is untranslated.

4. Eukaryotes possess mechanisms for rearranging DNA segments in a controlled way and for increasing the number of genes when needed. (This also occurs in prokaryotes but seems to be rare.)

5. Introns are present in most eukaryotic genes.

6. In prokaryotes, transcriptional regulatory sites are small, near to, and usually upstream from the promoter, and binding of proteins to such sites *directly* stimulates or inhibits binding of RNA polymerase. In eukaryotes, the regulatory regions are much larger and may be hundreds of bases away from promoters. They bind proteins but the proteins are usually too far away to interact with the promoter at the same time.

7. In eukaryotes RNA is synthesized in the nucleus and must be transported through the nuclear membrane to the cytoplasm where it is translated. Such extreme compartmentalization probably does not occur in prokaryotes.

We shall see throughout this chapter how these features are incorporated into particular modes of regulation.

GENE FAMILIES

In prokaryotes closely related genes are often organized in operons and are transcribed as part of a polycistronic mRNA. Thus, the entire system is under control of a single promoter and the system can be turned on and off by controlling the availability of the promoter to RNA polymerase. This method is of no use in eukaryotes without modification inasmuch as eukaryotic RNA is usually monocistronic. However, many related eukaryotic genes can be functionally grouped as a set of genes called a **gene family.** Activity of members of the set is usually, but not always, correlated. Recombinant DNA techniques have been used to clone members of gene families, and on occasion it has been found that all genes in a family can be cloned on a single plasmid, which clearly indicates that such a family comprises a gene cluster. More often, a cluster of related genes is sufficiently large that cloning of the entire cluster on one plasmid is unlikely and it is instead found that two or three genes of a family are cloned as a unit. This partial recovery is useful in determining the order of the genes. For example, if one clone contains genes *A* and *B,* another *B* and *C,* and a third *C* and *D,* the gene order in the chromosome must be *ABCD.* Cloning experiments have also shown that many genes whose products are related but not coordinately regulated are also clustered; the term gene family has therefore been expanded to include any cluster of functionally related genes (such as the tRNA genes).

Gene families are currently classified as simple multigene families, complex multigene families, and developmentally controlled complex multigene families. An example of each type is shown in Figure 16-1.

(a) Simple multigene family

Single-stranded
5S rRNA

(b) Complex multigene families

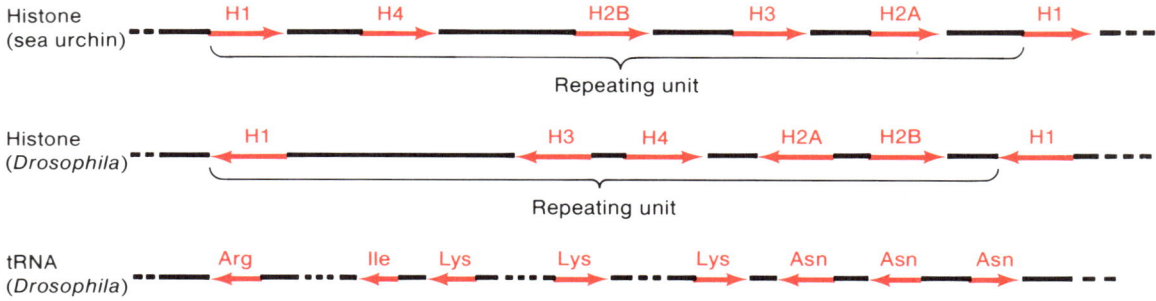

Histone
(sea urchin)

Histone
(*Drosophila*)

tRNA
(*Drosophila*)

(c) Developmentally controlled complex multigene family

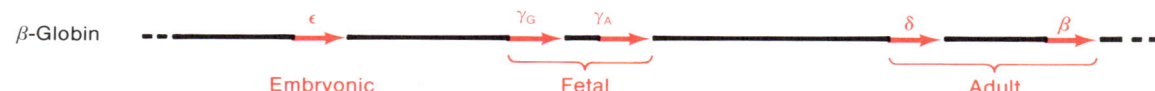

β-Globin

Figure 16-1 Five examples of gene families. The arrows indicate the direction of transcription. Genes are shown in red; spacers are black.

Simple Multigene Families

A simple multigene family is one in which one or a few genes are repeated in a tandem array. The simplest example is the set of genes for 5S rRNA, which has been studied most carefully in the toad *Xenopus laevis*. This set of rRNA genes forms a gigantic array in which each 5S rRNA gene sequence is separated from an adjacent gene sequence by a spacer to form a gene cluster. The sizes of the spacers vary from two to six times the length of the 5S rRNA gene. Each 5S rRNA gene is transcribed as a separate RNA molecule that is processed to generate the finished 5S rRNA molecule (Chapter 12, Figure 12-27). In addition to clustering of the 5S rRNA genes the clusters themselves exist in several copies in *Xenopus,* each copy containing hundreds to thousands of 5S rRNA genes.

The 5.8S, 18S, and 28S rRNA genes of *Xenopus* comprise another simple multigene family, in which a single transcription unit containing all three rRNA sequences is separated by spacers. The 5.8S, 18S, and 28S rRNA molecules are cleaved from the transcript by processing enzymes (Chapter 12, Figure 12-26). The number of copies of this three-gene rRNA sequence ranges from a few hundred to several thousand per diploid cell of various types. The number tends to correlate with the amount of protein that must be synthesized.

The rRNA genes provide an interesting contrast between prokaryotes, unicellular eukaryotes, and multicellular eukaryotes. For example, in *E. coli*, 5S, 16S, and 23S rRNA molecules comprise a single transcription unit from which each molecule is cleaved. In yeast, on the other hand, the rRNA molecules form a single *repeated* unit separated by spacers; however, the 5S RNA is transcribed separately from the 5.8S–18S–28S unit. (This grouping probably has evolutionary significance since the 5.8S rRNA of eukaryotes corresponds functionally to the 5S rRNA of prokaryotes.) Finally, in the multicellular eukaryotes the 5S RNA genes comprise a distinct family, as has just been seen.

Complex Multigene Families

The complex multigene families generally consist of a cluster of several related genes, each transcribed independently and separated by a spacer. Three examples are shown in Figure 16-15); each example has features that indicate that there are several types of complex multigene families.

First, let us examine the histone-gene family of the sea urchin. In this animal there are five histone proteins synthesized by five genes, contained in a small DNA segment and separated by spacers. This five-gene unit is tandemly repeated about 1000 times. Each of the five genes is separately transcribed as a single intron-free monocistronic RNA. Note that each gene is transcribed in the same direction—that is, from the same DNA strand. In a way to be discussed shortly, the rate of transcription and translation of each gene is regulated, first, such that histones are made at the right time for chromatin replication and, second, so that equimolar amounts of histones H2A, H2B, H3, and H4, and half as much H1 (the molar proportion in chromatin), are produced. The purpose of having multiple copies of the five-gene unit is undoubtedly to enable the cell to make large amounts of histone rapidly when DNA replication is occurring. Recently, it has been shown that in a particular cell all of the copies of the five-gene unit are not transcribed; different units are copied in different tissues and at different specific stages of embryonic development, suggesting that there may be subclasses of specific histones formed at different times. More will be said about regulation of the histone genes in a later section.

Examination of the histone-gene family of *Drosophila* shows that, in contrast with the sea urchin, the five genes are transcribed in both directions, as shown in the Figure 16-1. In yeast the organization of the histone genes differs even more; there are two quite distant gene clusters, each containing one H2A gene and one H2B gene, whereas the other histone genes are separate and scattered throughout the chromosome.

The tRNA gene cluster of *Drosophila* shown in the figure has two characteristics not seen in the histone clusters—namely, there are several copies of the same gene (for example, three tRNA^{Lys} genes) in the cluster, and not every tRNA gene is in the cluster. Each of the genes in this cluster is separately transcribed and both directions of transcription are used. Other tRNA genes are in clusters found elsewhere in the DNA, even in other chromosomes. The clustering of tRNA genes is not a general rule for the unicellular eukaryotes; for example, 175 of the 360 known yeast tRNA genes have been cloned in plasmids, and in only a few cases are two tRNA genes contained in one recombinant plasmid. Furthermore, the eight tRNA^{Tyr} genes in yeast are located on six different chromosomes.

Developmentally Controlled Complex Gene Families

Hemoglobin is a tetramer containing two α subunits and two β subunits. However, there are several different forms of both α and β subunits, differing by only one or a few amino acids, and the forms that are present depend on the stage of development of an organism. For example, the following temporal sequence shows the subunit types present in humans at various times after conception (Figure 16-2):

	Embryonic (<8 weeks)	Fetal (8–41 weeks)	Adult (birth → henceforth)
α-like	$\zeta_2 \rightarrow \zeta_1$	α	α
β-like	ϵ	γ^G and γ^A	β and δ

During the embryonic period the ζ_2 α-like chain, which appears first, is gradually replaced by the ζ_1 form. In the fetal and adult stages, the α type is present; the two β-like types, γ^G and γ^A (called so because one has a *g*lycine at a site at which the other has an *a*lanine), are roughly equimolar. In the adult stage, there is 50 times as much of the β type as the δ form. The result of these changes is that embryonic hemoglobin contains two ζ_2 chains and two ϵ chains ($2\zeta_2, 2\epsilon$), whereas after birth 98 percent contains two α and two β subunits (customarily called hemoglobin A) and 2 percent contains two α and two δ units (hemoglobin A$_2$).

Both the α- and β-like genes form separate clusters. The β cluster is shown in Figure 16-1(c). A remarkable property of each cluster is that the order of the genes is the order in which they are expressed in development. (This is not true of all developmentally controlled gene families.) Neither

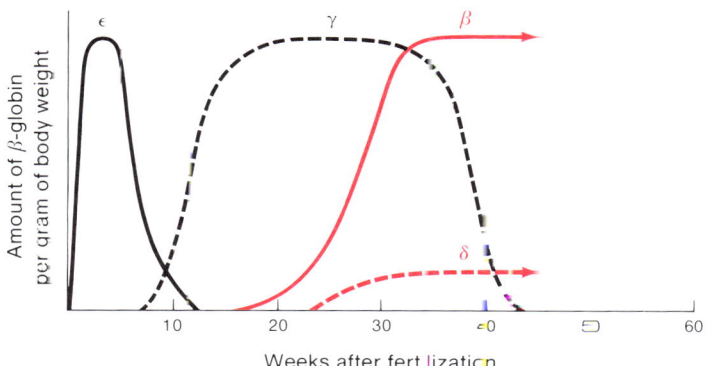

Figure 16-2 The variation of the concentrations of the different β-like globins in the course of development of a human.

the significance of the different forms nor the way in which synthesis is programmed is known in any detail.

The human α and β globin clusters have been cloned and analyzed for repetitive sequences (Chapter 4, Figure 4-20). The base sequences of some regions have also been determined. The most important observation that has emerged from this analysis is that *repetitive sequences are interspersed between the individual genes.* This fact suggests that the repetitive sequences may have a regulatory function.

Examination of the globin clusters in many different organisms gives some indication of how clusters arose and of the evolutionary factors that may have provided for their maintenance. Primitive fish, marine worms, and insects have a single globin gene, whereas in amphibians α and β genes are closely linked on a single chromosome. The lower mammals and birds have different forms of both α and β genes, and the clusters are on different chromosomes. Thus, it has been postulated that ancestral globin (which may have appeared about 800 million years ago) was the product of a single gene and that the globin families evident today arose by a series of gene duplications, mutations, and transpositions from an ancestral gene. The following hypothetical (but likely) sequence of events has been suggested. The small animals were able to function with the limited O_2-carrying capacity of a single hemoglobin molecule. Spontaneous mutation gave rise to an altered globin that in a heterozygote was able to form a tetrameric molecule capable of carrying more O_2. For reasons beyond the scope of this book, the tetrameric molecule can deliver a larger fraction of its bound O_2 to tissue than the monomers can, which enabled animals of larger size to evolve. Later, during the evolution of mammals one of the two β-chain genes underwent mutation and duplication, again giving rise to the γ form of globin found in the fetus. Fetal hemoglobin has an even higher affinity for O_2 than adult hemoglobin and thus is advantageous for the rapidly developing fetus, perhaps enabling a more complex organism to arise. Further mutation and duplication occurred during primate evolution, giving rise to additional forms of hemoglobin.

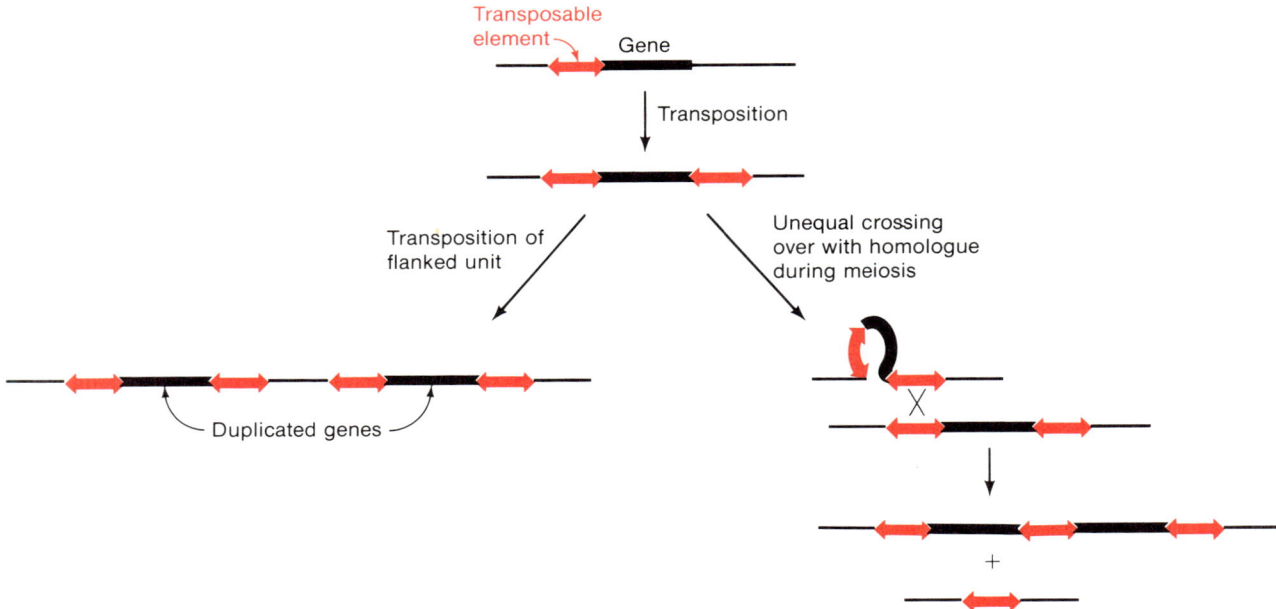

Figure 16-3   Two of several possible mechanisms for gene duplication. Both begin with a transposition event, that is, movement of a mobile element (Chapter 21) from one site in the chromosome to another. Left pathway: a second transposition of the unit consisting of the gene flanked by two mobile elements leads to gene duplication. Right pathway: unequal but reciprocal crossing over between two transposable elements forms one chromosome with a duplication and a second chromosome with a deletion. The right pathway might occur without a transposable element if the original gene was flanked by two sequences that were sufficiently homologous for unequal crossing over to occur. A third mechanism might also occur in the right pathway if rare non-homologous recombination occurred between flanking nonhomologous sequences.

Of particular genetic interest is the mechanism for gene duplication and for subsequent spread of the duplicated genes throughout a population. Several mechanisms are theoretically available for gene duplication (Figure 16-3). For example, a simple transposable element (a mobile DNA sequence, discussed in Chapter 21) located on one side of a gene may transpose to the other side of the gene, leaving behind a copy of the element in the original position. The complex element might then transpose, as a unit, to a new location, leaving a copy at the original location and producing a gene duplication. Alternatively, nonhomologous recombination (see Chapter 20, Figure 20-1) might lead to production of a duplication in one member of a pair of homologous chromosomes and chromosome loss in the other. Subsequent crossover events would generate triplications and more copies, as shown in Figure 16-4. How an advantageous mutant might be maintained along with the original form of the gene is illustrated in Figure 16-5.

The scheme just suggested is supported by the existence of remnants of past events of the type described. Study of various segments of the globin gene clusters in several organisms has indicated the presence of nonfunctional

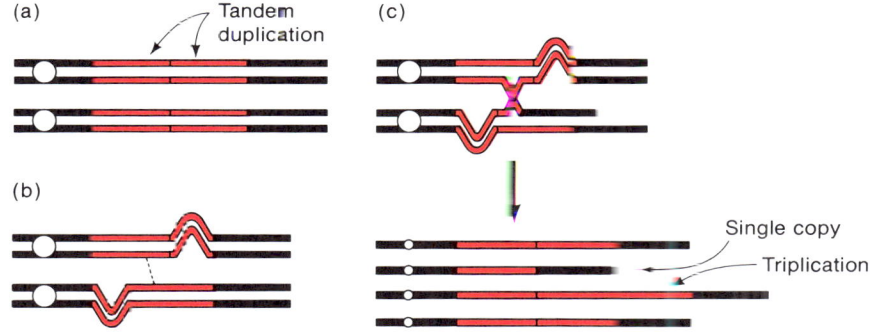

(a) Tandem duplication

(b)

(c)

Single copy

Triplication

Figure 16-4 An increase in the number of copies of a chromosome segment resulting from unequal crossing over of tandem duplications (red). (a) Normal pairing of chromosomes with a tandem duplication. (b) Mispairing. The right element of the lower chromosome is paired with the left element of the upper chromosome. The dashed line indicates a potential site of crossing over. (c) Crossing over within the mispaired duplication yields the four-chromosome configuration shown. One chromosome carries a single copy of the duplicated region, and another chromosome carries a triplication.

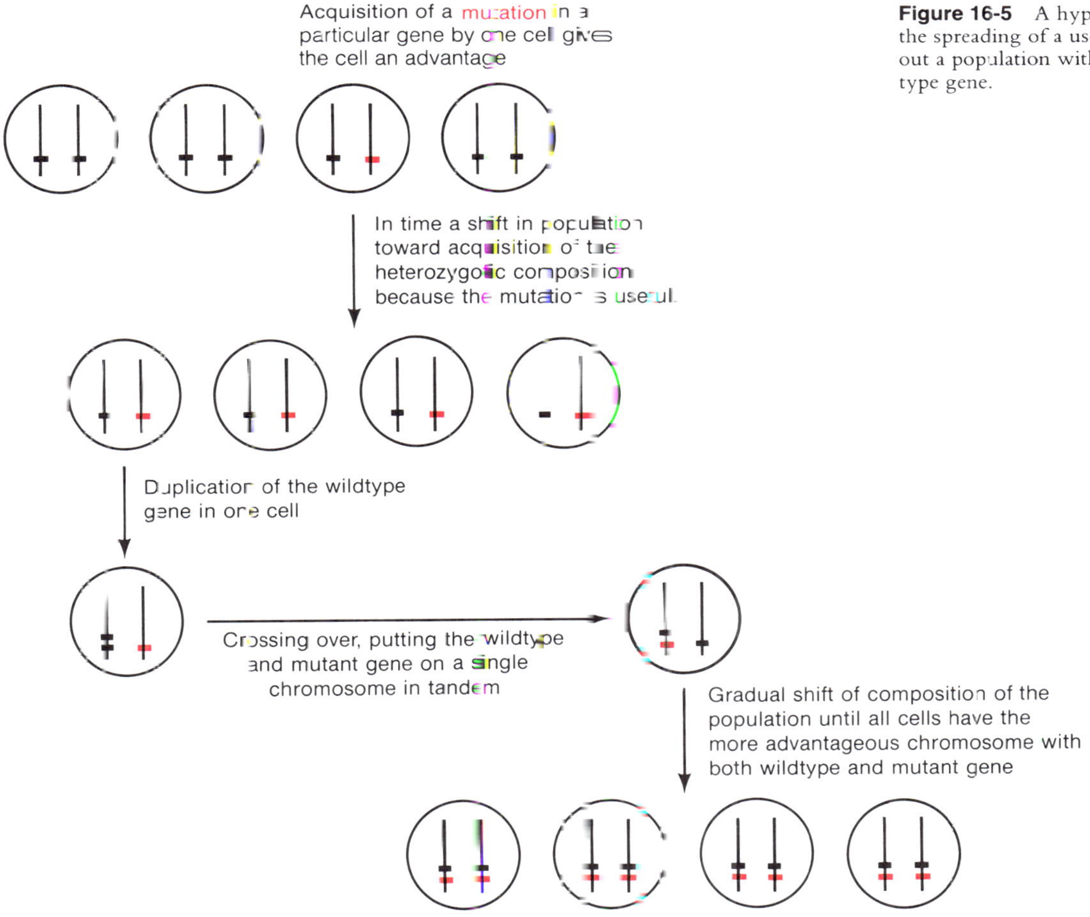

Acquisition of a mutation in a particular gene by one cell gives the cell an advantage

Figure 16-5 A hypothetical mechanism for the spreading of a useful mutation throughout a population without loss of the wild-type gene.

In time a shift in population toward acquisition of the heterozygotic composition because the mutation is useful.

Duplication of the wildtype gene in one cell

Crossing over, putting the wildtype and mutant gene on a single chromosome in tandem

Gradual shift of composition of the population until all cells have the more advantageous chromosome with both wildtype and mutant gene sequences that are nearly homologous to normal globin genes. The sequences are called **pseudogenes,** and are considered to be relics of evolution, derivatives of once-functional genes that were duplicated. Presumably a mutation occurred at one time that inactivated the gene product. Since at least one functional copy of the gene remained, there would have been no strong selective pressure to eliminate the mutant gene, and in time additional mutations would have accumulated. Pseudogenes are widespread in gene clusters and are frequently defective in transcription, intron excision, and translation.

ABUNDANCE CLASSES OF EUKARYOTIC mRNA MOLECULES

We have seen in discussing prokaryotes that the concentrations of individual species of mRNA molecules vary from one operon to another. In an "on" state the range of concentration from the most abundant species to the least abundant species is at most a factor of a few hundred. In eukaryotes the variation is much greater. Furthermore, it is likely that the synthesis of the most abundant mRNA molecules is regulated by mechanisms that differ from those for the other classes of mRNA. Thus, it has been useful to define (arbitrarily) three classes of eukaryotic mRNA molecules: **low–copy–number** mRNA molecules, which appear at a concentration of one to several copies per cell; **moderately prevalent** mRNA molecules, for which there may be several hundred copies per cell; and **superprevalent** mRNA molecules, for which there may be 10^4 copies per cell. Molecules of this latter class are synthesized in developing eggs and in cells committed to making an enormous amount of a single protein such as hemoglobin.

Regulation of the low–copy–number and the moderately prevalent mRNA molecules is not understood in as much detail as regulation of the synthesis of the superprevalent mRNA molecules.

REGULATORY STRATEGIES IN EUKARYOTES

In Chapter 15 it was seen that there are many mechanisms of regulating gene expression in prokaryotes. However, almost all are ways to control transcription initiation or termination. Translational regulation occurs also but is not particularly common. In Chapter 12 we saw that production of mRNA in eukaryotes is much more complicated than in prokaryotes and hence many points of control are possible—for example, in synthesis of the primary transcript, production of mRNA, and transport to the cytoplasm. These points may also have several levels of regulation; for example, promoter availability could be determined by the state of the chromatin, the activity or concentration of transcription factors, or enhancer availability. Production of mRNA from primary transcripts could be regulated by control of splicing and polyadenylation. Regulation of the lifetime of mRNA is rarely encountered in prokaryotes because of the usual need to switch the synthesis of regulated gene products on and off fairly promptly. However, in eukaryotes extension of the lifetime of mRNA is a reasonable strategy when a large amount of a particular protein is needed and a long time is available to achieve the required amount.

As pointed out earlier, prokaryotes must be able to respond to a changing environment and hence must be able to switch genes both on→off and off→on. However, in the differentiation of eukaryotic cells a gene or gene family is usually turned on or off permanently. This less demanding type of switching allows other regulatory strategies to be used in such cell types—for example, gene loss, gene inactivation, gene amplification, and gene rearrangement.

It is often necessary to maintain equal numbers or a constant ratio of particular proteins. In prokaryotes this is accomplished by use of polycistronic mRNA. However, the monocistronic nature of eukaryotic mRNA requires another mechanism. One of these is the production of proteins that are cleaved to produce other proteins.

In the following sections of this chapter we will describe each of these regulatory strategies, plus a few more.

GENE ALTERATION

The production of superprevalent mRNA molecules is usually associated with a change in the stage of differentiation—for example, an adult erythroblast produces a huge amount of mRNA from which adult globins may be translated, whereas little or no globin is produced by precursor cells that have not yet become erythroblasts. One way that such a change can occur is a permanent change in the gene(s) or the number of copies of the gene(s). Such changes are of three major types—**loss, amplification,** and **rearrangement** of genes. The latter category consists of transcriptional, post-transcriptional, and translational control. Each of these modes will be discussed in the following sections.

Gene Loss

One way to eliminate the activity of certain genes as a cell differentiates is to remove the genes from the cells. Some protozoa, insects, and crustaceans do this, but in a gross way—that is, entire chromosomes are lost. In these organisms only those cells destined to produce germ cells maintain the complete DNA complement. Chromosome loss has not been observed in the higher organisms.

Many protozoa have two types of nuclei—a small body called the **micronucleus** and a large body called the **macronucleus.** Usually, the germ cells contain only a micronucleus, but as the cell differentiates (as many protozoa do), the micronucleus undergoes a variety of changes, the result being a cell possessing both a micronucleus and a macronucleus. The micronucleus contains inactive DNA and is simply a storehouse of genetic information for the production of future germ cells. The DNA molecules of the macronucleus are the templates for all transcription.

The formation of the macronucleus of the protozoan *Oxytrichia* has been studied carefully. The DNA of the micronucleus in the germ cell is cleaved during differentiation into linear fragments ranging in size from 0.5 to 20 kilobase pairs. Most of the DNA of the micronucleus is completely degraded during the fragmentation. These fragments replicate repeatedly until the cell contains roughly 1000 copies of 17,000 different fragments that constitute the macronucleus. At this point, the macronucleus has several hundred times as much DNA as the micronucleus yet more than half of the DNA sequences are absent. The mechanisms for this selective gene loss is not known. A particularly interesting observation is that most of the fragments have the same base sequence at each end. This terminal repetition might represent a site of cleavage by a site-specific enzyme like a restriction endonuclease.

Gene loss has not yet been detected in the higher organisms; this mode of regulation may be confined to organisms fairly low in the evolutionary scale. It is of course impossible to determine with certainty that gene loss does not occur at all in the higher organisms since loss of one gene in 50,000 might be difficult to detect. However, the following experiment, performed with frog eggs, suggests that no *essential* gene is lost during development. In this experiment the nucleus of a fertilized frog egg was removed by microsurgery in such a way that the egg remained intact. A second nucleus was also taken from a cell in the intestine of a tadpole. The intestinal nucleus was injected into the nucleus-free egg. Many of these repaired eggs grew and divided and, remarkably, developed into living frogs. If genes essential for growth and development had been lost during differentiation, the egg would have lacked information to form a living frog. Although some non-essential DNA might be lost, it seems reasonable to conclude from this experiment that the nucleus of the differentiated cell had not lost any genetic material; it is said that the nucleus of the differentiated cell is **totipotent.**

Gene Amplification

Gene amplification is a means by which a cell can produce vast quantities of a specific gene product. The best-understood example of gene amplification occurs in the development of the oocytes (eggs) of the toad *Xenopus laevis,* in which the genes for rRNA increase in number about 4000-fold. The precursor to the oocyte, like all somatic (non-germ) cells of the toad, con-

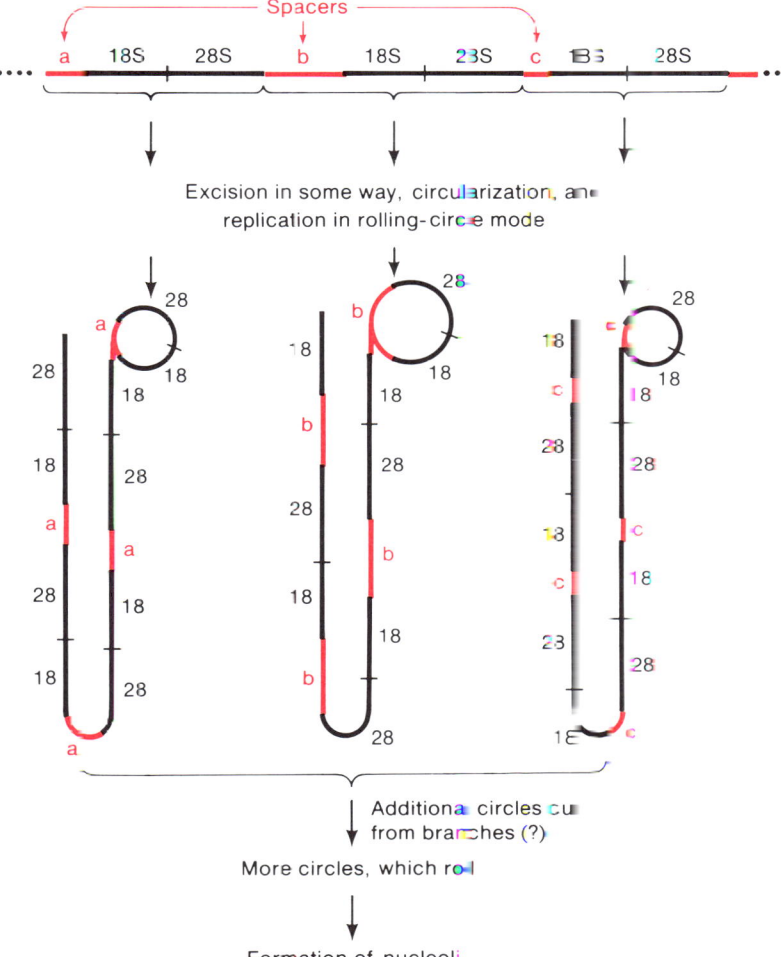

Figure 16-6 A hypothetical mechanism for amplification of rDNA genes in *Xenopus*. The spacers, a, b, and c, have different lengths.

tains about 500 rRNA-gene (rDNA) units; after amplification there are about 2×10^6 copies of each unit, which amounts to about 75 percent of the nuclear DNA of the oocyte. This enables the oocyte to synthesize 10^{12} ribosomes, which are required for the large amount of protein synthesis that occurs during the various cell cleavage stages following fertilization.

Prior to amplification the rDNA region consists of 500 rDNA units arranged in tandem. Each unit contains one gene each for 18S and 28S RNA and the genes are separated by a spacer (Figure 16-6, upper portion). The size of the spacer is not constant but varies from one unit to the next. The entire rDNA segment is contained in a single nucleolus (a nuclear body containing rDNA, found in most animal cells). During amplification, over a three-week period during which the oocyte develops from a precursor cell,

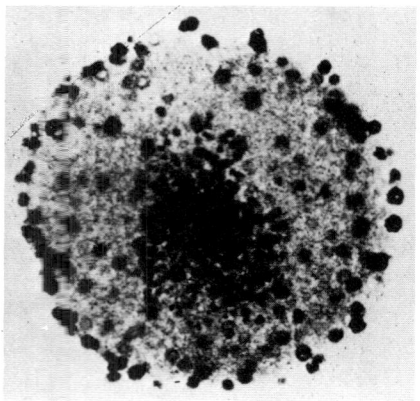

Figure 16-7 Micrograph of an isolated nucleus from a *Xenopus* oocyte stained to show the hundreds of nucleoli that are formed in the amplification of the ribosomal RNA genes. [From D. D. Brown and I. B. Dawid. *Science* (1968), 160: 272.]

the amount of rDNA increases 4000-fold and the number of nucleoli increases to several hundred (Figure 16-7). Both during and after amplification the total amount of rDNA no longer consists of a single contiguous DNA segment but instead is present as a large number of small circles and replicating rolling circles. The precise mechanism of formation of these structures is not known but some information about it is based on the size of the spacer between the 18S and 28S components contained in the nonreplicating and rolling circles. Figure 16-6 shows a schematic diagram of the rDNA contained in three rolling circles. The essential feature of these molecules is that each molecule contains a spacer having only one size. If a large segment of chromosomal rDNA containing adjacent rDNA units were excised and replicated, the rolling circles would have spacers of different sizes. This and several other experimental results suggest that a single rDNA unit is excised from the chromosome, as shown in the figure. The excised molecule circularizes (or perhaps is excised as a circle) and begins to replicate by the rolling circle mode, generating long branches.

Curiously, in most of the rolling circles the circular portion does not contain one rDNA unit but often two to four and as many as sixteen. It is unclear how a monomeric excised circle can be converted to a multimeric circle. This question may be related to the following point. An rDNA unit has a length of 4.3 μm and requires roughly one-half hour to form a unit-sized branch by rolling circle replication. Gene amplification proceeds for about 72 hours, and in this time each monomeric segment can generate 144 copies. Since all monomers are not excised from the chromosome, the amplification must be less than a factor of 144. However, the observed value is roughly a factor of 4000, so at least some of the excised monomers must be replicated to yield a template for further DNA synthesis. There are several possible ways that this might occur. For example, a circle might replicate several times via the θ mode prior to initiation of rolling circle replication; however, θ molecules are rarely seen. Another possibility is that DNA segments, which may often contain several rDNA units, could be excised from the linear branch of a rolling circle and then circularize. This could account for the existence of the circle mentioned above containing tandem repeats of single rDNA units. Clearly, more work is required to elucidate this phenomenon.

Ultimately, each replicating unit forms one of the many nucleoli of the mature oocyte (Figure 16-7). Amplification stops at this stage. It is of interest that not all rDNA genes are amplified in a particular oocyte and that the particular genes that are amplified vary from one oocyte to the next.

Once the oocyte is mature, the excess rDNA serves no purpose and it is slowly degraded. The signal for this degradation is unknown. Following fertilization the chromosomal DNA replicates and mitosis ensues. This occurs repeatedly as the embryo develops. During this period the extrachromosomal rDNA does not replicate; degradation continues and by the time several hundred cells have formed, none of this DNA remains.

Amplification of rRNA genes during oogenesis has been observed in a large number of organisms including insects, amphibians and fish, and in protozoa during formation of the macronucleus.

Recently, amplification of a gene that produces protein has also been observed in *Drosophila;* the genes that produce chorion proteins (components of the sac that encloses an egg) are amplified in ovarian follicle cells just before maturation of an egg. In *Drosophila* a precursor cell called an oogonium undergoes four rounds of cell division to produce 16 cells. One of these cells is the oocyte, which is destined to become an egg, and the other 15 are nurse cells, whose function is to provide the oocyte with a large number of proteins and other macromolecules necessary for egg formation. The nurse cells are able to produce large amounts of material because in their formation many rounds of DNA replication have occurred without mitotic separation of the daughter DNA molecules (chromatids). That is, each chromosome consists of many identical DNA molecules; the chromosomes are **polytene**, like the giant chromosomes of the salivary gland (used extensively in cytogenetic studies). Thus, the first step of amplification is chromatid amplification. The nurse cells are attached to the oocyte by cytoplasmic bridges, which allow gene products to enter the enlarging oocyte. At a later stage of oogenesis the nurse cells are responsible for synthesis of large amounts of chorion proteins, which are needed to enclose the oocyte in an eggshell. To synthesize the required amount, amplification of the chorion genes occurs by means of local DNA replication from a replication origin in the chorion gene cluster. Repeated replication initiation occurs, producing nested replication bubbles, as shown in Figure 16-8(a). The extra

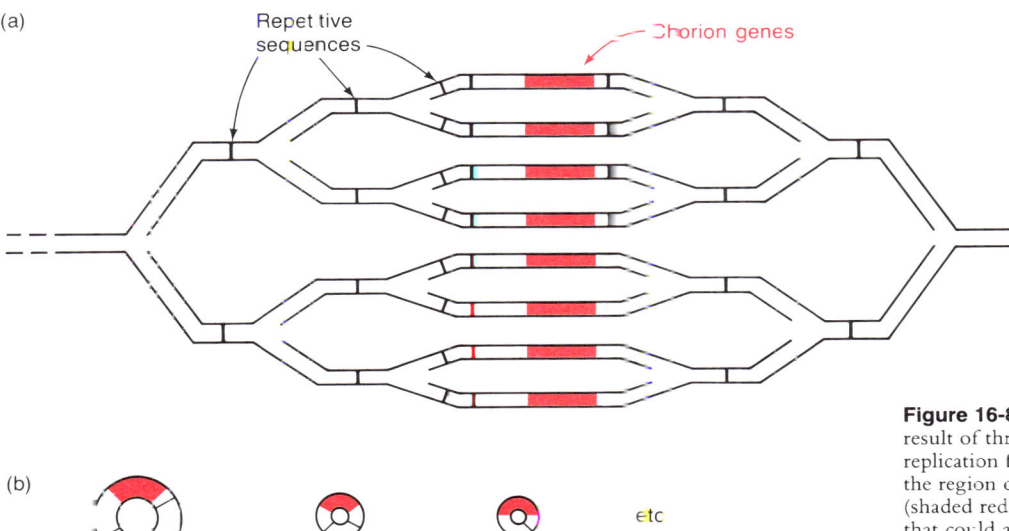

Figure 16-8   Gene amplification. (a) The result of three rounds of initiation of DNA replication from a single replication origin in the region containing the chorion genes (shaded red) (b) Three circular molecules that could arise by recombination between repetitive sequences.

copies of the chorion genes in the bubble are not excised and circularized but remain part of the chromosome. In this case, as with the rRNA genes, amplification enables the cells to produce a large amount of protein in a short time. We will see later that if a large amount of protein is to be synthesized but a long time is available for the synthesis, gene amplification is unnecessary, for this can be accomplished by increasing the lifetime of the mRNA.

It should be noted that amplification of this sort could not survive mitosis, since completion of a replication cycle and segregation of daughter molecules would ultimately yield molecules with only one gene cluster (lacking a bubble). However, mitotic segregation does not occur in this system, for the nurse cells are a "dead end" and disintegrate after the egg forms. In other organisms and in other genes, differentiation of somatic cells sometimes involves gene amplification, and after amplification many rounds of cell division occur, forming a clone of specialized cells. Such systems also utilize repeated DNA replication to increase the number of gene copies, but the copies are excised, forming circles and linear fragments. The excision results from homologous recombination between repetitive sequences (Figure 16-8(b)). These circles and fragments lack centromeres (Chapter 7), so they cannot segregate in an ordered fashion during mitosis. However, the number of circles and linear fragments is so great that *random* segregation into daughter cells is sufficient to ensure that each daughter cell receives multiple copies of the amplified gene.

Gene Rearrangement: The Joining of Coding Sequences in the Immune System

A gene could be turned on by moving it from a location far from its promoter to a site near its promoter. One example of such relocation is the phase variation of *Salmonella* described in Chapter 15. A few examples are known in eukaryotes (e.g., mating type interconversion in yeast), but they are too complex to be included in this book. Furthermore, it is not clear that the relocation is either regulated or responsive to an external or internal signal. In this section we examine instead the activation of gene expression by a programmed juxtaposition of different segments of structural genes, namely, in the production of **immunoglobulins (antibodies).**

Humans can synthesize more than 10^6 different antibody molecules—"different" in the sense of being an antibody to a different target molecule or **antigen.** If each antibody molecule were encoded in a distinct gene, a significant fraction of the DNA of a single cell would have to be used for antibody synthesis. The amount of DNA required could be reduced substantially if an antibody contained three subunits A, B, and C and there were 100 different *A* genes, 100 different *B* genes, and so on. In this way, $100 \times 100 \times 100 = 10^6$ different proteins would be generated by 300 genes. If each of the genes were active in every cell—that is, if 100 different proteins were synthesized in every antibody-producing cell—then each cell would produce 10^6 different antibodies. However, it is known that a mature antibody-producing cell makes only one type of antibody, so there must be some mechanism that is responsible for programming each cell.

In this section it will be seen that such programming does occur. In outline, the germ-line cells contain the set of all genes whose coding sequences can be combined to generate all possible antibody molecules. In the course of cell division and development of the immune system, most of these genes are removed from the daughter cells and discarded. However, the removal is programmed in such a manner that in the adult a cell will exist that is capable of producing each (but only one) type of antibody. This population consists of roughly 10^6 cells, a small fraction of the 10^{12} somatic cells in an adult human. The ability to produce large quantities of a particular antibody comes about by a complex response of the immune system to a particular antigen; in this response, the antigen triggers multiplication of the cell capable of producing the particular antibody that can neutralize that antigen. As a result of this multiplication, an antibody-producing clone forms, in which each cell can synthesize the particular antibody.

The immune system is very complex in that it consists of many cell types and can respond to a large number of signals. In this section we will be concerned primarily with how the base sequence of an antibody-producing gene is formed. This will be explained in five parts by describing (1) the properties of the amino acid sequence of an antibody molecule, (2) the genes whose coding sequences are joined, (3) the base sequence of a spliced gene, (4) possible signals for gene splicing, and (5) the final splicing event. Following this we discuss briefly how a clone of antibody-producing cells may be instructed to produce a specific antibody.

Properties of the Amino Acid Sequence of an Antibody Molecule

Immunoglobulin G (**IgG**) is one of several classes of immunoglobulins; it is a tetramer containing two **L chains** and two **H chains** (Chapter 6). Determination of the amino acid sequences of many different IgG molecules (different in that they respond to different antigens) has shown that both the L and H chains comprise two distinct segments. One of these, the **constant** or **C region,** has a nearly invariant amino acid sequence (there is one amino acid that can vary) for a particular class of immunoglobulins. The other segment is called the **variable** or **V region** because its amino acid sequence differs between any two IgG molecules that recognize different antigens. The locations and the sizes of the V and C regions of the L and H chains are shown in Figures 6-17, 6-18, and 6-19 (Chapter 6). In most classes of immunoglobulins, the C regions are used in the transport of the molecules across the membrane of the antibody-producing cell to the blood stream and the lymphatic system and in the final disposal of the neutralized antigen. The V regions are responsible for recognition and binding of a specific antigen.

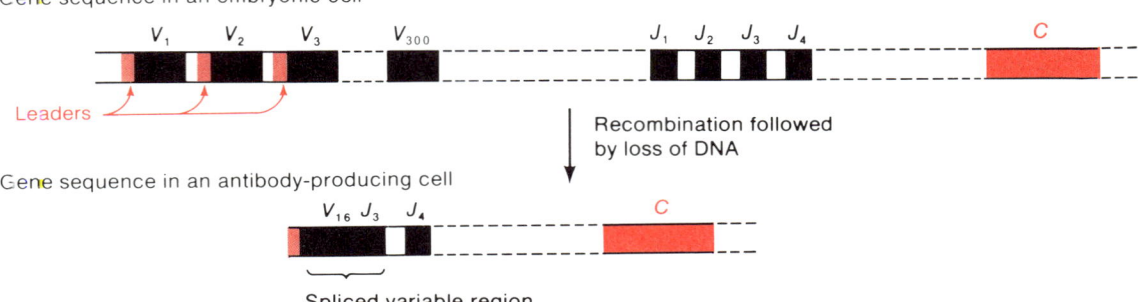

Gene sequence in an embryonic cell

Leaders

Recombination followed by loss of DNA

Gene sequence in an antibody-producing cell

Spliced variable region

Figure 16-9 DNA splicing in the formation of the variable region of an L chain. The V_{16}–J_3 joint has been removed, presumably by a site-specific recombinational event. The V genes to the left of V_{16} have been removed either by homologous recombination or by site-specific recombination within the leader sequences.

Note in Figure 6-17 that the V regions of the L and H chains have the same number of amino acids, although their amino acid sequences are quite different.

There are actually two distinct types of L chains. These are called κ and λ and have significantly different amino acid sequences. Functionally these types have many similarities and their differences are unimportant for the present discussion. We will only describe how the amino acid sequence of the κ type is determined; the basic mechanism for the λ type is similar, differing only in detail.

Genes Whose Coding Sequences Are Joined

Many genes can be used to form a κ-type L chain. These genes are of three types—*V*, *J*, and *C*. There are roughly 300 different *V* genes (which are responsible for the synthesis of the first 95 amino acids of the variable region), 4 different *J* genes (which encode the final 12 amino acids of the variable region and join the V and C regions), and 1 copy of the *C* gene (which encodes the constant region). In an embryonic cell the *V* genes form a tight cluster, the *J* genes form a second tight cluster quite far from the *V*-gene cluster, and the *C* gene follows not far after the *J*-gene cluster. All of the genes are on the same chromosome, as shown in Figure 16-9. Note that each *V* gene is preceded by leader regions where transcription can be initiated; the *J* and *C* genes are not preceded by leaders.

The Base Sequence of a Spliced Gene

Genetic regions encoding particular IgG molecules have been cloned from various mouse cell lines, each producing a particular IgG molecule.* The

*These cell lines were derived from cancerous tumors called **myelomas.** A myeloma is a tumor in which a single cell programmed to make one type of antibody has multiplied to yield a large clone of cells, each of which produces only that antibody. A single tumor can be isolated and individual cells can be grown indefinitely in cell culture; the cells secrete a large quantity of a pure antibody that can be recovered from the culture medium.

V-J-C region has also been cloned from mouse embryo cells, which have not yet been committed to antibody synthesis and contain an unaltered master set of genes for antibody synthesis. For each clone obtained from an antibody-producing cell line it has been found that a large segment of the embryonic DNA sequence is not present and that the missing segment is always a sequence between the particular *V* gene that encodes the first 95 amino acids of the V region and the *J* gene that encodes the last 12 amino acids of the V region. This is explained by a gene rearrangement in which DNA between the particular *V* and *J* genes is deleted. An example of one such rearrangement is shown in Figure 16-9. Many different gene sequences encoding particular IgG proteins have been cloned and it has been found that in each clone a different segment of DNA is not present. For example, in the figure the DNA between V_{16} and J_3 is absent; in another clone there might be a V_{210}-J_1 junction instead. In both cases a *V* gene and a *J* gene have become *spliced* together to form a complete gene for the variable region. Note that this is DNA splicing and not the RNA splicing discussed in Chapter 12.

Studies of the base sequence of cloned IgG DNA sequences show that the junction between a particular *V* gene and a particular *J* gene is not always the same. That is, the two terminal triplets of the juxtaposing *V* and *J* genes can exchange at any one of four sites that yields a triplet. The meaning of this statement is shown in Figure 16-10. In the example shown there are three possible amino acids at the joint, which adds diversity to the number of possible variable regions.

Since there are 300 different *V* genes, 4 different *J* genes, and (on the average) 2.5 different amino acids at the junction, there are, then, $300 \times 4 \times 2.5 = 3000$ different variable regions. The H chain genes are organized in a similar but not identical way (there are four different types of genes) and there are about 5000 variable regions in the H chains. Thus, since each IgG molecule contains two identical H chains and two identical L chains, about $3000 \times 5000 = 1.5 \times 10^7$ different IgG molecules can be formed from the V_L, J_L, V_H, and J_H genes. This is ample to account for the diversity of antibody molecules. The value of 1.5×10^7 different antibody specificities is actually an underestimate because there is an additional source of antibody variability. The V regions are susceptible to a high rate of somatic mutation, which occurs during development of the antibody-producing cell. These mutations allow different clones to produce different polypeptide sequences, even if they have undergone exactly the same V-J joining.

Possible Signals for Gene Splicing

The splicing of the *V* and *J* genes just described is thought to occur by means of a site-specific recombination mechanism. The signal for the *V-J* joining is probably a pair of specific base sequences to the right and left of the *V* and *J* genes, respectively. A base sequence

$$V\text{-gene}–CACAGTG–11\text{ bases}–ACAAAAAC$$

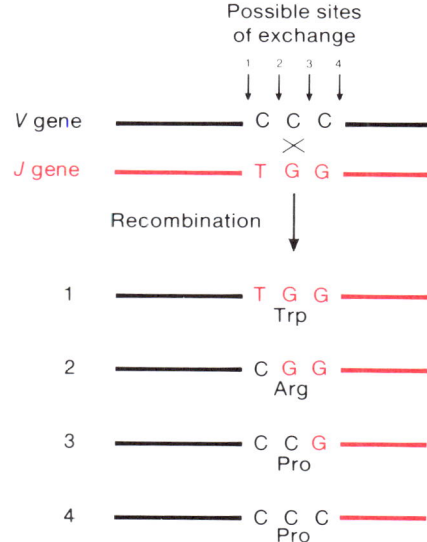

Possible sites
of exchange

Figure 16-10 Four possible junctions at a *V-J* joint giving rise to three different amino acids.

consisting of a 7-bp palindrome, an 11-bp spacer, and an 8-bp (A + T)-rich sequence is immediately to the right of each *V* gene. To the left of each *J* gene is the sequence

GTTTTTGT–22 bases–CACTGTG–*J* gene

consisting of (reading right to left) of a 7-bp palindrome identical (except for the fourth base) to the one that is adjacent to the *V* gene, a 22-bp spacer, and an 8-bp (A + T)-rich sequence almost complementary to the one near the *V* gene. Presumably, those sequences are recognized by the splicing system. Neither the enzymes required for splicing, the nature of the chemical exchange, nor the chemical events that trigger splicing are known.

The Final Splicing Event: RNA Splicing

Figure 16-11 shows that the DNA splicing event does not fully generate an L chain sequence: (1) the spacer between the *J* and *C* gene remains, and (2) the actual L chain has amino acids derived from only one *V* gene and one *J* gene and the spliced DNA usually contains many *V* and *J* genes. The correct amino acid sequence is obtained by a final RNA-splicing event, as shown in the figure. Note how the particular *V-J* joint determines the RNA splicing pattern. For example, the RNA removed is always a segment between the leader and the *V* gene and between the *C* gene and the right end of the *J* gene in the *V-J* joint.

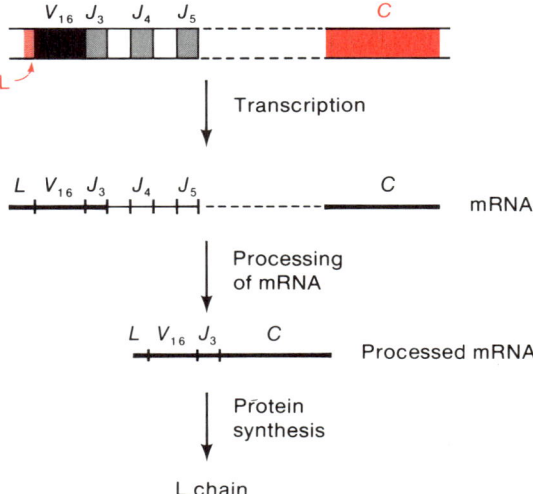

Figure 16-11 Production of the L chain of a particular IgG molecule. Solid regions indicate coding sequences used to generate the final L chain.

Diversity of H Chains

The mechanism of generating diversity of amino acid sequence in the H chains is similar to that for the L chains. The H chain system also has many *V* and *J* genes but it is more complicated than the L chain system in two ways: (1) there is an additional large set of genes, called *D*, between the *V* and *J* genes, so many *V-D* and *D-J* joints are possible and (2) there are eight different *C* regions. Choice of the *C* region determines the particular class of immunoglobulin (e.g., IgD, IgG, IgM etc.). It is interesting that the 7-bp palindrome described above, which signals *V-J* joining in the L chain, is also present in corresponding positions in the *C*-chain system.

Diploidy in Antibody-Producing Cells

An interesting point arises because antibody-producing cells are diploid: since only one type of IgG molecule is produced by a particular cell, then either the same rearrangement occurs in both chromosomes or one of the *V-J-C* regions is silent. Existing evidence suggests that during development of an antibody-producing cell from an embryonic cell, gene rearrangement stops as soon as one complete functional gene is assembled. Genes still in the embryonic configuration and incompletely spliced variable region genes are either not expressed or their products cannot function.

Programming

A major question is whether DNA recombination occurs *in response to* a particular antigen. The best evidence suggests that the events occur at random in the course of development and that when differentiation is complete, there are about 1.5×10^7 different antibody-producing cell types, each able to make a single type of antibody. How, then, does an organism know when to make a huge amount of a particular antibody in response to a particular antigen? The answer to this question is not known with certainty but the following mechanism, called **clonal selection,** is probably correct—at least in outline. Each antibody-producing cell type makes a small amount of specific antibody. Some of this antibody becomes bound to the cell surface. When such a cell is exposed to the specific antigen that can react with the specific antibody, a complex antigen–antibody network forms, because each antibody molecule can interact with two antigen molecules and each antigen can interact with two antibody molecules (Chapter 6). These interactions, coupled with an energy-requiring movement of the surface antibody molecules, cause the formation of a single antigen-antibody network called a **cell cap** (Figure 16-12). The cell membrane then folds in around the cap and the cap is taken into the cell (this process is called **endocytosis**). Then, in a way that is totally obscure, cell division is stimulated and extensive production of antibody ensues. Thus, a clone of cells is generated which synthesizes a large amount of the specific antibody.

Figure 16-12 The capping phenomenon. Rabbit lymphocytes that have made immunoglobulin G (IgG) (spread uniformly over the cell surface) are exposed to an antigen. The antigen and IgG molecules combine and, by an unknown energy-requiring process, the IgG molecules move toward one another. If the antigen is multivalent (has multiple binding sites), a complex called a cap forms. By a process called endocytosis the cap is taken into the cell for processing, after which the cell becomes programmed to make large quantities of the antibody. If the antigen is coupled with fluorescein, a fluorescent cap can be seen with a fluorescence microscope; if the antigen is radioactive, an autoradiogram can be prepared and the film will be blackened over one region of each cell.

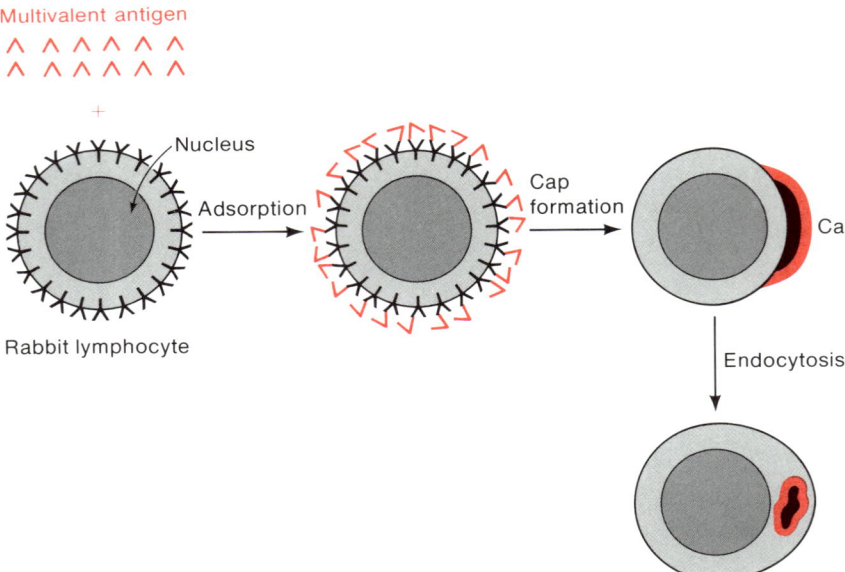

REGULATION OF SYNTHESIS OF PRIMARY TRANSCRIPTS

Control of synthesis of a significant number of gene products occurs by means of regulation of transcription. This is probably the major mechanism for proteins encoded in low-copy-number and moderately prevalent mRNAs; the mechanism is also used for superprevalent mRNAs. In this section examples of several different ways that transcription can be switched on and off are described. We begin by considering some features of eukaryotic gene organization that make it clear that regulation in eukaryotes cannot occur by the direct mechanisms used in prokaryotic operons.

Features of Gene Organization that Affect Regulation

A major feature of the prokaryotic operon model and its many variants is that, adjacent to each promoter, a base sequence is acted upon by some regulatory molecule, and this action determines whether transcription occurs. The high frequency with which regulatory sites and promoters are adjacent and the simplicity of this regulatory mechanism have made it hard to envision any system in which the regulatory site and the gene promoter are separated. In fact, our programming to think in these terms, resulting from more than 20 years of studying bacterial operons, has contributed significantly to our difficulty in understanding eukaryotic regulation, for in eukaryotic systems *regulatory sites are rarely adjacent to promoters*.

Another problem in understanding regulation in eukaryotes is to explain how sets of genes that are very far apart in the DNA (even in different chromosomes) can be turned on and off simultaneously. One suggestion is that each gene in a set is preceded by the same regulatory sequence, so a single regulatory element can activate or deactivate all of the genes in the set. This mechanism is similar to that proposed for bacterial regulons (Chapter 15). However, in eukaryotic systems there often seems to be intermediate regulatory elements that are themselves turned on by some signal. An oversimplified way by which this might occur is shown in Figure 16-13.

Another feature of regulation in eukaryotes is that it is common for certain gene products to be synthesized in response to several different signals. For instance, two different hormones might induce synthesis of the same protein. Moreover there are sets of genes that are activated in different combinations by different signals. For example, one signal might activate genes P and Q, another might activate Q and R, and a third might activate P and R. Such phenomena can be accounted for by assuming that some

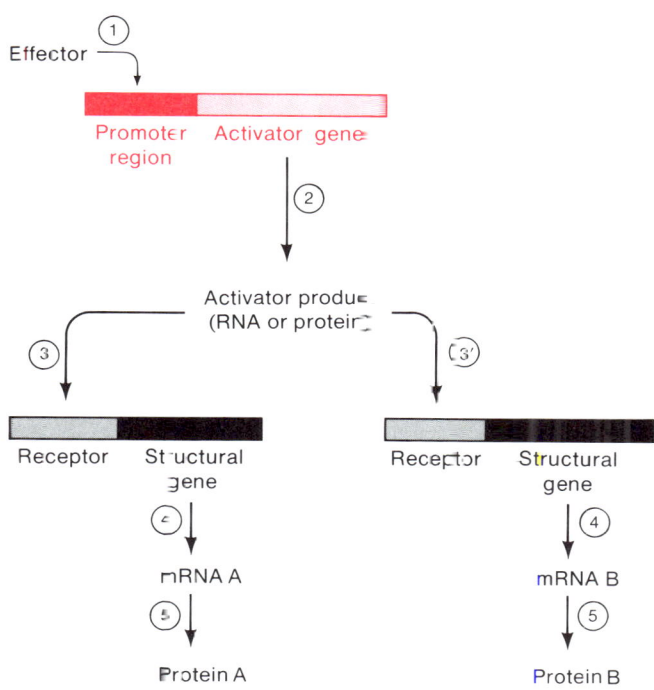

Figure 16-13 A simple way by which a eukaryotic gene might be regulated. Each step is numbered in order of their occurrence. Receptors might be activated either by the activator RNA in step 3 or an activator protein in step 3'.

Figure 16-14 A possible organization of multiply regulated gene families. In the upper panel. units having several activator genes can activate several structural genes. In the lower panel, a single activator turns on several structural genes, each having the same receptor.

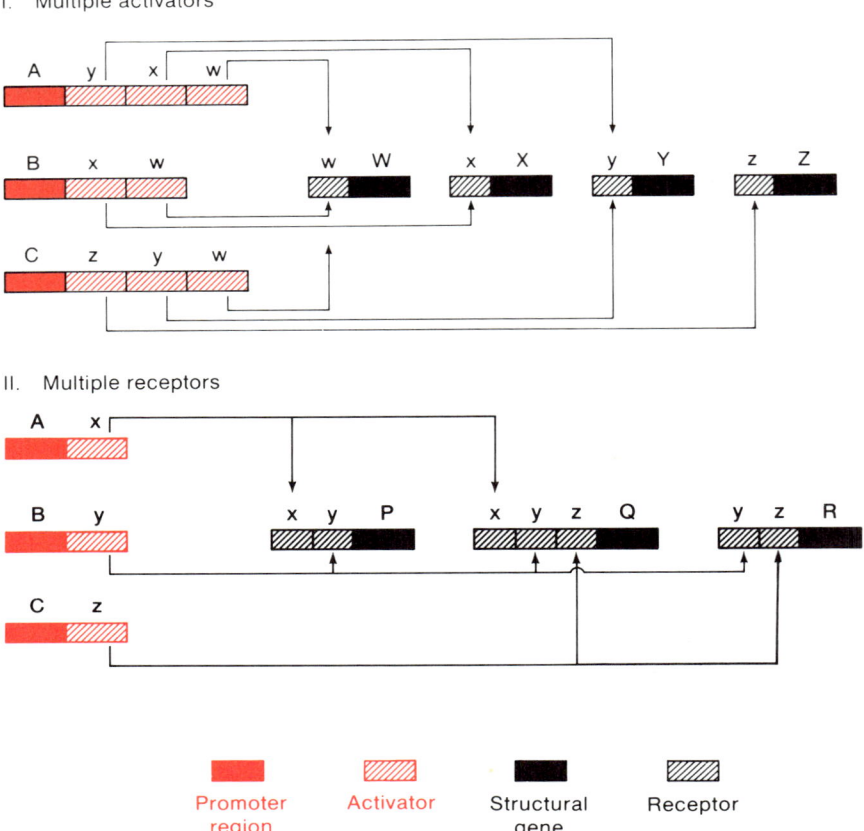

genes are adjacent to two (or more) receptors, only one of which needs be occupied to cause activation (a multicomponent receptor) or by having single receptors and multicomponent signal molecules. How such systems might function is shown in Figure 16–14.

Models of the type shown in Figures 16–13 and 16–14 are surely over–simplified. They have been presented mainly to provide a simple framework for thinking about certain features of eukaryotic regulation.

Regulatory Units in Yeast

Yeast, being free–living unicellular organisms, must be able to respond quickly to environmental fluctuations, as bacteria do. One set of yeast genes that behaves in this way is the *GAL* system, which is responsible for metabolism of galactose. This system consists of numerous genes located on several chromosomes. Here we will consider only the regulation of the *GAL1* gene, whose regulatory region is shown in Figure 16-15. It consists of an upstream activation site (UAS$_G$), a 250-bp spacer, the TATA box, and the *GAL1* gene, in that order. The product of the gene, *GAL4*, located on another chromosome, is a positive control element that is activated by the binding of galactose. Activated GAL4 protein binds to a 15-bp sequence in UAS$_G$ and thereby causes initiation of transcription from a site a few bases downstream from the TATA box. The distance between UAS$_G$ and the *GAL1* TATA box is not critical; for example, by genetic engineering it can be moved 500 bp upstream without affecting the inducibility of the *GAL1* gene. The UAS$_G$ is apparently a general activator of transcription. Evidence is that if it is placed several hundred base pairs upstream from the TATA box of another yeast gene, *CYC1*, transcription of the *CYC1* gene then becomes inducible by galactose.

Manipulation of the spacer between the UAS$_G$ and the *GAL1* TATA box has given some insight into the relation between the two sequences. Since binding of the galactose-activated GAL4 protein to UAS$_G$ stimulates transcription, it seems reasonable that RNA polymerase should bind sufficiently near the activated UAS$_G$ to contact the GAL4 protein (at least, this is the prokaryotic way of thinking about it). However, the TATA box is the site of formation of the open-promoter complex, and RNA polymerase II cannot possibly span the distance between UAS$_G$ and TATA, in contrast with prokaryotic systems in which RNA polymerase can contact both a −35 sequence and the Pribnow box. There are three simple ways by which RNA polymerase could reach TATA from UAS$_G$:

1. The DNA could bend and bring UAS$_G$ and TATA next to one another. Since bending probably occurs at a unique sequence and UAS$_G$ can be relocated, either with respect to TATA or to another structural gene (*CYC1*), this idea seems unlikely.
2. Binding of activated GAL4 protein to UAS$_G$ could change the structure of chromatin, and the structural alteration could be propagated downstream to TATA, which in this model is assumed to be the only RNA polymerase binding site.
3. RNA polymerase could slide along the DNA from UAS$_G$ to TATA.

The third hypothesis has been investigated by inserting a transcription-termination site between UAS$_G$ and TATA. This insertion eliminates induci-

Figure 16-15 The yeast *GAL1* region. The *GAL4* gene product, with bound galactose, acts on UAS$_G$, resulting in synthesis of *GAL1* RNA.

bility of *GAL1* transcription by galactose. Since RNA polymerase presumably falls off DNA at termination sites, this result supports the sliding hypothesis. However, explanation 2 remains a possibility since a termination protein could bind to the termination site and thereby prevent the hypothetical change in chromatin structure from reaching TATA. A counterargument is that termination proteins have not been detected in eukaryotes.

A second experiment also shows that meddling with the spacer between UAS_G and TATA can interfere with gene expression. In this experiment the *E. coli lexA* operator (Chapter 15) was inserted between UAS_G and TATA. This alone had no effect on *GAL1* inducibility; however, if the *lexA* gene was also introduced, such that LexA protein was synthesized, *GAL1* transcription was inhibited. There was no inhibitory effect of an occupied *lexA* operator if it was located upstream from UAS_G. These results show that a bound protein between UAS_G and TATA prevents transcription initiation, which is consistent with both hypotheses 2 and 3 above.

An observation in another yeast system (mating-type interconversion) suggests that in some cases a chromatin change may be the key event. Several thousand base pairs upstream from the UAS for this system is a site called the silencer S. Products of several genes called *SIR* interact with this site. When S is occupied, several promoters within a few thousand base pairs of S are inactive. This result has been interpreted in terms of a *SIR*-induced change in chromatin structure that blocks transcription.

The *CYC1* gene, which synthesizes isocytochrome *c*, has two upstream sites UAS_c1 and UAS_c2, located at positions -275 and -225 from the transcription start site. The activity of both sites is mediated by heme, which is a cofactor of the cytochrome. Specific positive regulatory elements are encoded by the genes *HAP1* and *HAP2*. Both proteins bind heme and when heme is bound, they interact with their respective UAS. Each UAS is separately functional and their effects are additive. Having tandem UAS enables yeast to respond to distinct regulatory stimuli. In this case, the different signal molecules determine the amount of isocytochrome c that is made—that is, amount 1, amount 2, or amount $1+2$.

The examples just given are of positively regulated systems. Yeast also has a negatively regulated system. The enzyme alcohol dehydrogenase is encoded in two nearly identical genes, *ADRI* and *ADRII*. Transcription of *ADRI* is constitutive but transcription of *ADRII* is inhibited by glucose. A regulatory region upstream from *ADRII* is responsible for glucose repression: if it is deleted, glucose-insensitive constitutive transcription occurs. This site differs from the UAS discussed so far in that its distance to TATA is critical; for example, if it is displaced upstream by an insertion, glucose repression is lost. The region is not gene-specific in that if it is moved (by genetic engineering) upstream of *ADRI*, transcription of *ADRI* becomes glucose-repressible. This region behaves like a prokaryotic operator; however, no corresponding repressor has been detected.

Transcriptional Control by Hormones

Of the known regulators of eukaryotic gene activity those studied most extensively are the **hormones**—extracellular substances, either small molecules, polypeptides, or small proteins, that are carried from hormone-producing cells to target cells. It has been found that many of the steroid hormones (e.g., estrogen, progesterone, aldosterone, glucocorticoids, and androgens) and general metabolic hormones (e.g. insulin) frequently act by turning on transcription. If a hormone regulates transcription, it must somehow signal the DNA. The penetration of a target cell by a hormone and its transport to the nucleus is a complex process that is understood only in outline, as shown in Figure 16-16. Steroid hormones (H) are hydrophobic molecules and pass freely through the cell membrane. A target cell contains a specific cytoplasmic receptor (R) that forms a complex (H-R) with the hormone. The receptor R usually undergoes some modification (conformational or chemical) after the H-R complex has formed; the modified form of the receptor is denoted R'. The H-R' complex then passes through the nuclear membrane and enters the nucleus. In the nucleus either the H-R' complex or possibly the hormone alone binds to chromatin and transcription begins. The binding is very likely to be mediated by specific DNA-binding proteins, though it is possible that in some systems direct binding to the DNA occurs.

A well-studied example of induction of transcription by a hormone is the stimulation by **estrogen** of the synthesis of ovalbumin in the chicken oviduct. When chickens are injected with estrogen, oviduct tissue responds

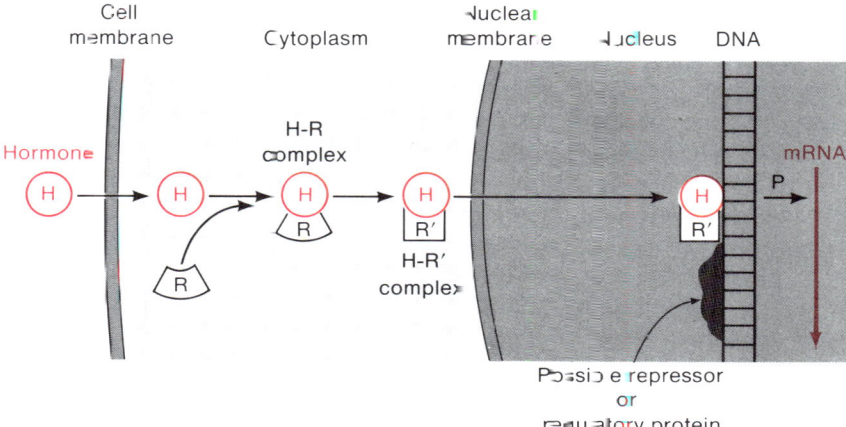

Figure 16-16 Schematic diagram showing how a hormone H reaches the DNA and triggers transcription by binding to a cytoplasmic receptor. A regulatory protein may prevent the H-R' complex from reaching a promoter (P) or may stimulate binding to the promoter.

by synthesizing ovalbumin mRNA. This synthesis continues as long as estrogen or an estrogenlike hormone (e.g., diethylstilbesterol) is administered. Once the hormone is withdrawn, the rate of synthesis decreases. Both before giving the hormone and 60 hours after withdrawal, no ovalbumin mRNA is detectable. Experiments of this sort must be interpreted with caution to avoid a potential artifact arising from methods used to measure transcription and the existence of introns. Specific mRNA, such as ovalbumin mRNA, is detected by hybridization of labeled mRNA to a plasmid containing c-DNA (a DNA copy of RNA, Chapter 5). Suppose a primary transcript had a great many large introns. This might interfere with hybridization with c-DNA. If, compared to processed mRNA, the primary transcript binds poorly to c-DNA, the ability to detect mRNA might only indicate that processing rather than *de novo* synthesis has occurred. There are two straightforward ways to demonstrate that a hormone stimulates transcription rather than processing. One way is to show that the newly synthesized RNA contains introns, which can be done if cloned intron-containing c-DNA of the natural gene is used in the hybridization test. The other method is to measure the size of the RNA that appears. If transcription is being turned on by the hormone, the primary transcript, which is larger than the processed mRNA, should be detected; furthermore, it should appear sometime before the processed mRNA is observed. If only processing were stimulated, there would be no change in the amount of the large RNA that is detected; only the smaller processed mRNA would appear to increase in concentration. It was just such experiments that indicated that estrogen stimulates transcription of the ovalbumin primary transcript.

When estrogens are given to chickens, only the oviduct synthesizes mRNA; other tissues fail to be stimulated because they lack the cytoplasmic hormone receptor. This type of deficiency is invariably the cause of insensitivity to a particular hormone. How the receptor is synthesized in some but not all cells is not known; this is another problem of differentiation.

We have said nothing so far about how a hormone that turns on transcription might do so. Without invoking complicated details we may examine at least five possible basic mechanisms:

1. A hormone binds directly to the DNA and facilitates binding of RNA polymerase or of a protein (transcription factor) that is needed for the polymerase interaction with the DNA.
2. A hormone binds to an effector protein (such as the CRP protein in the *lac* operon) that must bind to a site on the DNA prior to RNA polymerase binding.
3. A protein already bound to the DNA is activated by the hormone and the activated form is needed for RNA polymerase binding.
4. A hormone is an inducer that inactivates a repressor.
5. A hormone causes a structural change in chromatin that makes the DNA available to RNA polymerase.

In the first three mechanisms, transcription is positively regulated; in the fourth and fifth, it is negatively regulated. At present only a few hormone-induced systems have been studied and there is no clear picture yet.

Regulation Mediated Through Transcription Factors

In Chapter 12 we saw that in a great many eukaryotic systems initiation of transcription requires the formation of transcription complexes between DNA, RNA polymerase, and one or more proteins called transcription factors. As more systems are being examined, it is becoming clear that regulation frequently involves transcription factors. Two mechanisms have been observed: (1) regulation of formation of the transcription complex, and (2) regulation of binding of RNA polymerases to a preexisting transcription complex by a small molecule. In this section, we examine several examples of such effects.

The *Xenopus* genome contains two multigene families encoding 5S rRNA (Figure 16-1). These are called oocyte-type and somatic-type families. The somatic type, which comprises two percent of the 5S-rRNA genes, accounts for more than 95 percent of the synthesis of 5S rRNA in somatic cells, whereas in oocytes the oocyte-type genes are primarily transcribed. This differential gene expression results from control of formation of the transcription complex at the internal control region (Chapter 12, Figure 12-31). If chromatin is isolated from somatic cells, the somatic-type 5S-rRNA genes are found to be part of an active transcription complex, which includes a protein called the **40-kd transcription factor.** The oocyte-type genes do not have this transcription factor bound to them and in fact are prevented from binding the factor by a structure (not yet understood) that contains histone H1. This difference can be reproduced *in vitro*. If somatic-cell chromatin is stripped of 40-kd protein by washing in a concentrated salt solution, transcription of 5S-rRNA genes will not occur. Addition of 40-kd factor restores the ability. Addition of H1 prior to the 40-kd factor prevents formation of the transcription complex, but it is not inhibitory if the 40-kd factor is added first. Note that H1 has the features of a prokaryotic repressor.

The role of the 40-kd factor in embryological development has not been completely elucidated, but the following sequence of events is believed to be important. Each *Xenopus* cell contains about 800 somatic-type 5S-rRNA genes and about 40,000 oocyte-type 5S-rRNA genes. The 50-bp internal control regions are not the same but differ by 2 bp, a difference believed to reduce substantially the affinity of the 40-kd protein for the oocyte genes. Young oocytes contain an excess of 40-kd protein, so all internal control regions are activated. For unknown reasons, the amount of factor decreases considerably during late oogenesis and meiosis. After fertilization and during development of the embryo, the total amount remains constant but the amount per cell drops because of cell division. During this period no 5S rRNA of either class is made. The reasons are unknown, but a variety of possibilities may be suggested: for example, insufficient RNA polymerase III, insufficient amount of other factors needed in the transcription complex, and complexing of the internal control region in nucleosome form. At any rate, at the 4000-cell stage synthesis of 5S rRNA begins. At this time there are about ten 40-kd factor molecules per 5S-rRNA gene, and both oocyte-type and somatic-type 5S-rRNA genes are transcribed. After 3–4 more cell divisions the ratio of factor to gene is about 1, and the oocyte-type 5S-rRNA genes are fully repressed. Presumably, during chromatin replication the DNA-protein complex dissociates, and at the lower concentration of the 40-kd factor repressive binding of H1 predominates. From this point on, 40-kd factor is synthesized as cells divide, maintaining a ratio of about 0.25 molecules per 5S-rRNA gene. Thus, the oocyte-type 5S-rRNA genes are not transcribed in a somatic cell simply because there are fewer 40-kd factor molecules than 5S-rRNA genes. Presumably, the higher affinity of the factor for the internal control region of the somatic genes causes these genes to be transcribed preferentially.

The expression of the histone genes is also mediated by transcription factors. Newly synthesized histones are needed only when DNA is being replicated, in order to form chromatin (Chapter 8), and indeed in the mouse, transcription of the various histone genes increases 20-fold during the DNA-replication phase of the cell cycle. The histone H4 gene has an upstream promoter 70–110 bp from its transcription start site. Transcription occurs only after a particular transcription factor is bound to the promoter. Thus, regulation of transcription of the H4 gene could result from either activation or availability of the transcription factor. In this case, the latter is probably the case, for preliminary evidence suggests that the concentration of the transcription factor increases early in the replication cycle. Thus, regulation of the H4 gene is indirect (Figure 16-13), depending on activation of the production of the transcription factor. How this occurs is not known. Since all histones (H2A, H2B, etc.) are either needed at the same time or not needed at all, one might expect that each histone gene would utilize the same transcription factor. However, evolution has taken another course; the H2A and H3 genes of the mouse each have their own transcription factor. It seems likely that production of each transcription factor would be controlled by the same signal, but no information is available concerning this point.

In yeast (but not in the mouse), expression of the histone genes has another requirement. By genetic engineering these genes have been moved with respect to a natural DNA replication origin and it has been found that gene expression requires that the gene be very near the origin. Possibly, changes in chromatin structure associated with replication initiation are required for transcription factors to bind, but this idea is speculative.

Another system regulated by transcription factors is the heat-shock family of *Drosophila*. As explained in Chapter 15 (the heat shock regulon of *E. coli*), when many organisms are stressed by heating above their optimal temperature range, synthesis of a set of proteins called the heat shock proteins is induced. In bacteria and in higher eukaryotes the heat-shock genes are widely dispersed throughout the genome. The *Drosophila* protein Hsp70 has been examined fairly carefully. Heating the flies causes a 1000-fold increase in the number of Hsp70 mRNA molecules. The TATA box for the *hsp70* gene is 10 bp upstream from the start site. About 60 bp further upstream is a regulatory sequence called HSE (heat shock regulatory element) to which a transcription factor (HSTF) binds and thereby turns on transcription. A second HSE is located 800 bp further upstream

Understanding induction of *hsp70* mRNA requires knowledge of the effect of heating. Two possibilities have been considered: (1) HSTF is normally bound to the nearby HSE but the binding is too weak to initiate transcription; when heated, HSTF undergoes a conformational change to a tight-binding form and transcription begins. (2) HSTF is normally bound weakly to the distant HSE; heating causes it to dissociate from this site and either by random collisions between free HSTF molecules and the nearby HSE or drifting along the DNA, HSTF reaches the nearby HSE and initiates transcription. Experimental data are not yet available to distinguish these possibilities.

An interesting phenomenon has been discovered in the β-globin region. A hypersensitive site (Chapter 12) can be found in β-globin DNA isolated from erythrocytes of nine-day chicks, when the gene is being expressed. The site is absent from younger chicks, in which the gene is not expressed. A protein has been isolated from nine-day cells that induces the hypersensitive site; if this protein is added to chromatin from younger chick cells, the β-globin hypersensitive site forms. This protein binds to a specific base sequence in the region that is cleaved. One may hypothesize that binding of this protein displaces the histones or at least weakens its interaction with the DNA, thereby altering the structure of chromatin at that site and rendering the DNA sensitive to nuclease degradation.

The systems just described are examples of regulation via control of binding of a transcription factor. Preliminary evidence provides one example of *activation* of a bound transcription factor by a small molecule. In this case, a hormone-receptor complex must join to the bound transcription factor before RNA polymerase can bind to the promoter

Regulation of Enhancer Activity

The presence of an enhancer can markedly increase the rate of transcription for promoters located several hundred base pairs upstream or downstream (Chapter 12). Although the mechanism by which enhancers work is unknown, one would expect that control of enhancer activity would be another regulatory strategy. Enhancers can be experimentally relocated throughout the genome and will exert their enhancing effect on any nearby gene; thus, they are not gene-specific. On the other hand, they often show species and cell-type specificity. For example, the enhancer for mouse immunoglobulin H chains is active in myeloma cells (which normally make immunoglobulin), but not in fibroblasts (which do not make antibody); this suggests that specific proteins, which may be transcription factors, are needed for the enhancing activity. One such factor, needed for the SV40 and polyoma virus enhancer, has been isolated from human HeLa cells. A recent (1985) experiment shows that the activity of the polyoma enhancer can be inhibited. Two plasmids were prepared: (I) one contained the polyoma enhancer and the human β-globin gene and (II) the other contained the *EIA* regulatory gene of adenovirus (Chapter 23). When plasmid I was put into HeLa cells, strong transcription of the β-globin gene occurred. However, when both plasmids were put in the cells, very little β-globin RNA was made. Several other experiments provided evidence that the EIA protein is the inhibitor of the enhancer. Whether the EIA protein binds directly to the enhancer or if inhibition requires some product whose synthesis is stimulated by EIA protein is unknown. The latter might be more likely inasmuch as in a cell infected with adenovirus the EIA protein normally stimulates transcription of both cellular and viral genes.

Normal cells do not contain either the *EIA* gene or the polyoma enhancer, so this is quite an artificial system. Nonetheless, the results just described show that an enhancer can be inactivated and hence raises the possibility that there may be enhancer repressors. However, no direct evidence is available to support this idea.

Methylation

Most cells contain several DNA-methylating enzymes. Some are simply the methylating activity of restriction enzymes, which methylate restriction sites and thereby prevent self-destruction. Others methylate cytosine (Figure 16-17(a)) in particular CpG sequences. In a few cases, particularly in vertebrates, methylation of particular CpG sequences prevents transcription of some genes. It has been suggested that such methylation may be generally important in eukaryotic regulation and development, though this notion is quite controversial. The principal difficulty with the idea is that methylation is rare in invertebrates and nonexistent in insects, and in mammals there are numerous examples in which methylation has no detectable effect on transcription. Evidence for a regulatory role of methylation is the following:

1. Certain genes are heavily methylated in cells in which the gene is not expressed and unmethylated in cells in which the gene is expressed.
2. *In vitro* methylation of the upstream site of a cloned γ-globin gene, followed by $CaCl_2$ transformation of the affected DNA into susceptible cells, prevents transcription. Methylation outside of the upstream sequence does not inhibit transcription.

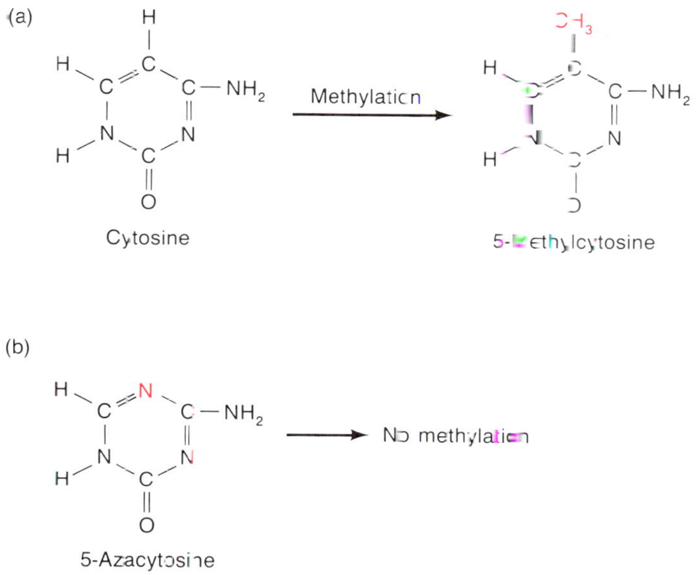

(a)

(b)

Figure 16-17 (a) Methylation of cytosine from 5-methylcytosine and (b) structure of 5-azacytosine, showing that the methylatable C atom is replaced by an N atom. 5-Azacytidine has a ribose attached to the lower right N atom.

3 If the base analogue 5-azacytidine (Figure 16-17(b)) is added to cultures of growing cells, newly made DNA contains the analogue, and methylation of CpG sites containing the analogue fails to take place (5-azacytidine is also a general inhibitor of many methylases). Cells in such cultures gain the ability to make proteins whose synthesis is normally turned off (i.e., in cells growing in medium lacking the base analogue).

4. The so-called **housekeeping genes,** which provide for general cell function and which are continuously transcribed, are rarely methylated in or near their initiation regions.

The fact that methylation is prevalent in vertebrates and its frequency decreases in animals lower on the evolutionary scale suggests that methylation may have arisen relatively late in evolution and hence might be expected to play a regulatory role only with the more recently evolved genes. This could account for the variability of the effects of methylation on mammalian genes. Clearly, the role of methylation will remain controversial until more information is obtained.

Undifferentiated and precursor cells often replicate. If methylation actually prevents gene expression in some of the cell types, an inhibitory methylated site must be inherited as a methylated site in a daughter strand during DNA replication. One possible mechanism is that certain sequences are always methylated unless methylation is somehow prevented. However, this seems not be the case; furthermore, the methylatable sequence in a daughter strand is not identical but complementary to that of the methylated sequence in the parental strand. A property of certain DNA methylases shows how a methylated parental strand can direct methylation of the appropriate daughter sequence: namely, these enzymes only methylate CpG (embedded in certain surrounding sequences) and only when the CpG in the opposite strand is already methylated (Figure 16-18). In this way, the methylation pattern of parental DNA strands is inherited by the daughter strands.

Heavy methylation has been offered as an explanation of X-chromosome inactivation. The sex-chromosome composition of human males and females is XY and XX, respectively. Males with more than one X chromosome (for example, XXY) have physiological and developmental problems and are usually mentally retarded, which suggests that gene dosage of the X chromosome is important. Indeed, early in the development of females, one of the X chromosomes in each cell is made transcriptionally silent, presumably to maintain equal gene dosage in both males and females. The X that is inactivated is selected at random. Evidence for X-chromosome inactivation is the following: (1) One of the X chromosomes is highly condensed and heterochromatic. (2) Females are mosaic for X-linked genes for which they are heterozygous. For example, a female that is heterozygous for a gene that controls production of sweat glands has patches of skin that

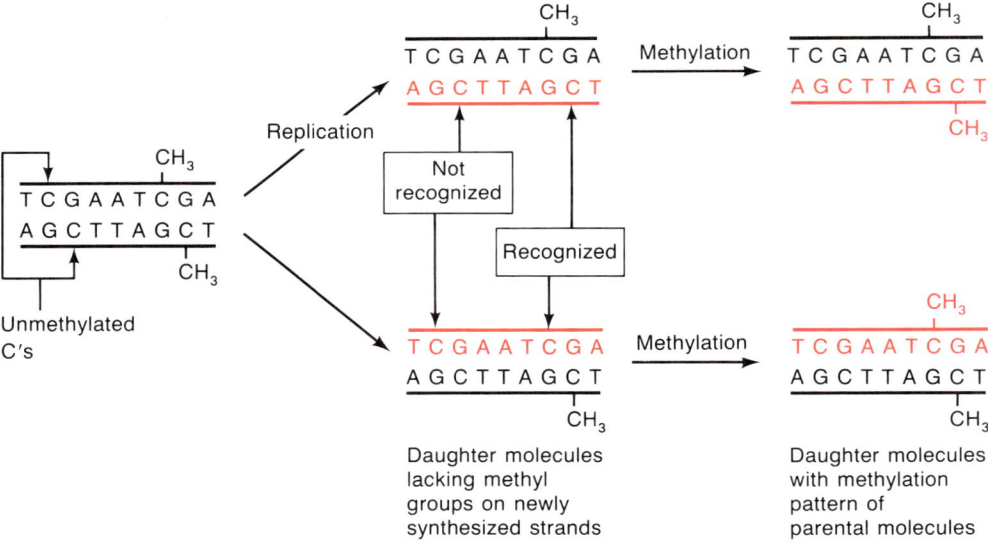

Figure 16-18   The mechanism by which the pattern of methylation in parental DNA is inherited in daughter molecules. The rule is that a C in a CpG sequence can be methylated only if the C in the complementary CpG sequence is already methylated. For clarity only, methyl groups have been drawn outside the sugar-phosphate strands.

perspire (the X containing a mutant sweat allele is inactive) and patches that do not (the X containing the wild-type gene is inactive). It has been suggested that the DNA of the inactive X chromosomes is methylated. There has been no direct measurement of methylation, and the main evidence is the following. If DNA isolated from a cell line of female cells heterozygous for various X-linked genes is used in a transformation experiment, genes from one X chromosome are expressed and those from the other are not expressed. The result clearly shows that one of the X chromosomes is inactive. However, if the cells are grown for several generations in medium containing 5-azacytidine and then DNA is isolated and used for transformation, genes from both X chromosomes are functional. Thus, the inactive X chromosome has been activated by replication in the presence of 5-azacytidine, presumably by preventing methylation. This interpretation of this experiment has been quite controversial.

Opposing Transcription Units

Usually only a single strand of each DNA segment is transcribed, as we have seen repeatedly. However, several regions have been observed in which both strands are transcribed. Such a dually transcribed region does not include a gene in its entirety in each strand; rather, the initial or terminal portions of the two genes (which are necessarily transcribed in opposite directions) overlap. The two opposing transcription events often interfere with one another to various degrees, and such interference can be regulatory. For example, transcription of gene A could inhibit initiation of transcription of gene B in the complementary strand, so turning off transcription of A would turn on transcription of B. In adenovirus, an effect of transcription termination and subsequent processing has been observed. In this case the two transcription units overlap in the terminal regions. Transcription of one of the units (X) causes premature termination of the primary transcript of the other (Y); this premature termination (and subsequent polyadenylation) determines the pattern of splicing and hence the sequence in the mRNA. Late in the life cycle of the virus, transcript X is no longer made, so Y is not terminated; this more extended primary transcript is processed in another way, which allows the expression of other genes contained in the transcription unit.

REGULATION OF PROCESSING

Initiation of transcription ultimately leads to production of a primary transcript, which in higher eukaryotes is processed to form an mRNA. Alternative processing patterns can yield different mRNAs. One example comes from chicken skeletal muscle in which two forms of the muscle protein myosin, called alkali light chains LC1 and LC3, are produced. The myosin gene has two different TATA sequences, which yield two different primary transcripts. These two transcripts are processed differently to form mRNA molecules encoding distinct forms of the protein (Figure 16-19). In *Drosophila* the mRNA is processed in four different ways, the precise mode depending on the stage of development of the fly. One class of myosin is found in pupae and another in the later embryo and larval stages. How the mode of processing is varied is not known.

Some of the best examples of regulation of processing are found in adenovirus (Chapter 23). Adenovirus has only a few promoters but makes numerous primary transcripts from which are generated a large number of mRNAs (see Figure 23-35, Chapter 23). This allows a fairly small amount of DNA to be used efficiently in that proteins are translated in all reading frames and different regions of the DNA are, by virtue of distinct splicing patterns, used to form several proteins. The amount of each mRNA is regulated with respect to the time after infection. For example, the primary transcript for the EIB region is differentially spliced to yield several different mRNAs. Two of these are shown schematically in Figure 16-20. RNAs 1 and 2 are both formed shortly after the primary transcript is made, though more of mRNA-2 is made. Later in the viral life cycle, mRNA-1 is not made and mRNA-2 is abundant. If cycloheximide, an inhibitor of protein synthesis, is added to the infected cells before the shift in splicing pattern takes place, the shift does not occur. This inhibition implies that a newly synthe-

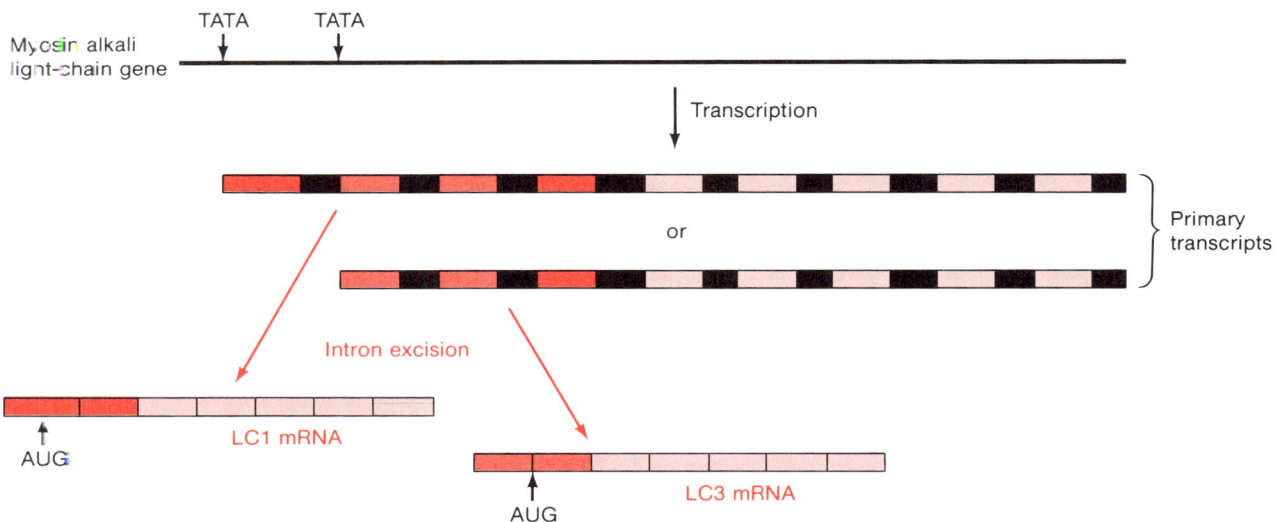

Figure 16-19   The chicken *LC1/LC3* gene. Two distinct TATA sequences lead to production of different primary transcripts, which contain the same coding sequences. Two modes of intron excision lead to the formation of distinct mRNA molecules, which encode proteins having different amino-terminal regions and the same carboxyl-terminal regions.

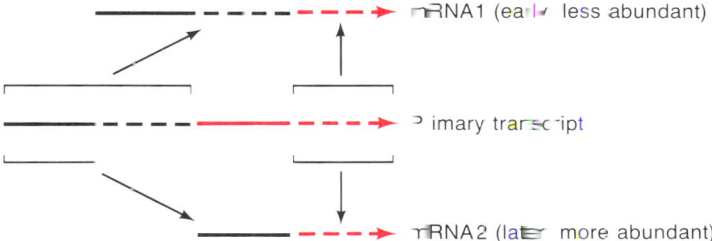

Figure 16-20   Two alternative splicing patterns of the adenovirus EIB primary transcript. Colors and dashes are for identification purposes only.

sized protein is a positive effector of the shift. Cycloheximide has similar effects on other mRNA species derived from other primary transcripts at late times.

Often two cells make the same protein but in different amounts, even though in both cell types the same gene is transcribed. This phenomenon is frequently associated with the presence of different mRNA molecules, which are not translated with the same efficiency. In the synthesis of α-amylase in the rat different mRNAs from the same gene result from the use of different processing sequences. The rat salivary gland produces more of the enzyme than the liver, though the same coding sequence is transcribed. In each cell type the same primary transcript is synthesized, but two different splicing mechanisms are used. The initial part of the primary transcript is shown in Figure 16-21. The coding sequence begins 50 base pairs within exon 2 and

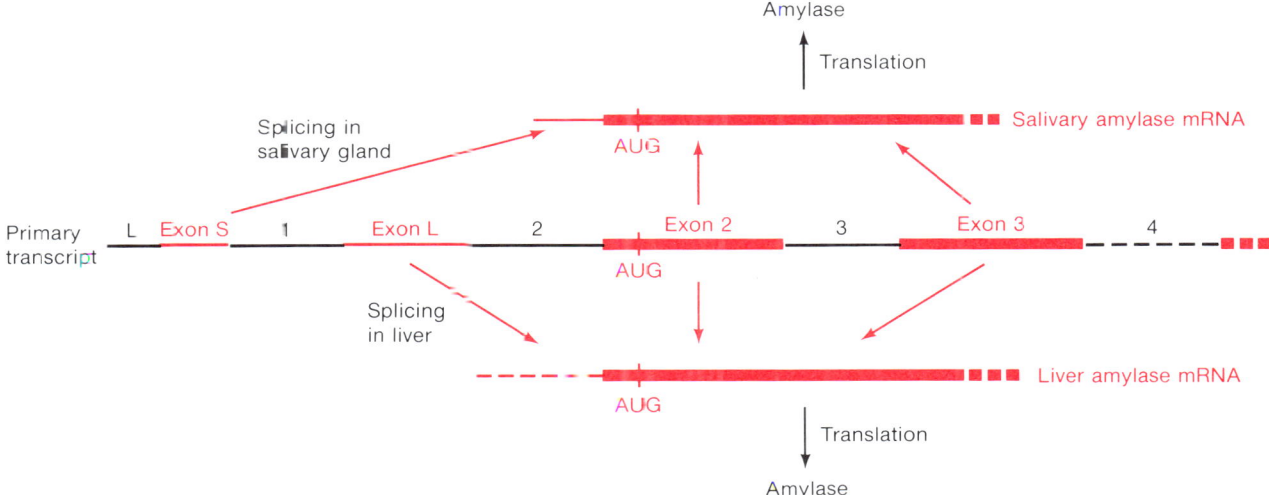

Figure 16-21   Production of distinct amylase mRNA molecules by different splicing events in cells of the salivary gland and liver of the mouse. The leader L and the introns are in black. The exons are red. The coding sequence begins at the AUG codon in exon 2.

is formed by joining exon 3 and subsequent exons. In the salivary gland the primary transcript is processed such that exon S is joined to exon 2 (that is, exon L is removed as part of introns 1 and 2). In the liver, exon L is joined to exon 2, because exon S is removed along with intron 1 and with the leader L. The exons S and L become alternate leaders of amylase mRNA, which somehow results in translation at different rates.

Another type of regulation of processing involves choice of different sites of polyadenylation. One example is the differential synthesis of the hormone calcitonin in different tissues; another is the synthesis of two forms of the μ chain of immunoglobulins. In both cases, the differential processing includes distinct patterns of intron excision (i.e., splicing) but these are necessitated by an earlier event in which differential poly(A) sites are selected from the primary transcript. That is, when the poly(A) site nearer the promoter is selected, a splice site used in the larger primary transcript is not present, so a different splice pattern results. Thus, slightly different proteins are translated.

Adenovirus has a single promoter for all RNA made late in the cycle of infection. The primary transcripts terminate at five major polyadenylation sites (Figure 23-35). Each termination site influences the splicing pattern by allowing particular introns or intron termini to be present or not. The five sets are not used with equal frequency, with the result that the amounts of mRNAs encoding various genes are not the same. This is the major mechanism for determining the relative amounts of the different viral structural proteins synthesized late in the life cycle.

POLYPROTEINS: MULTIPLE PROTEINS FROM ONE mRNA

In prokaryotes coordinate regulation of the synthesis of several gene products is accomplished by regulating the synthesis of a single polycistronic mRNA molecule encoding all of the products. The analogue to this arrangement in eukaryotes is the synthesis of a **polyprotein,** a large polypeptide that is cleaved after translation to yield individual proteins. Each protein can be thought of as the product of a single gene. In such a system the coding sequences of each gene in the polyprotein unit are not separated by stop and start codons but instead by specific *amino acid sequences* that are recognized as cleavage sites by particular protein-cutting enzymes. Polyproteins have been observed with up to eight cleavage sites (poliovirus, Chapter 23, Figure 23-11); the cleavage sites are not cut simultaneously, but are cut in a specific order. Use of a polyprotein serves to maintain an equal molar ratio of the constituent proteins, as was seen earlier in the production of *Xenopus* rRNA; moreover, delay in cutting at certain sites introduces a temporal sequence of production of individual proteins, a mechanism frequently used by animal viruses (Chapter 23, Figures 23-8 and 23-11).

Many examples of polyproteins are known. The synthesis of the RNA precursors uridine triphosphate and cytidine triphosphate proceeds by a biosynthetic pathway that at one stage has the reaction sequence

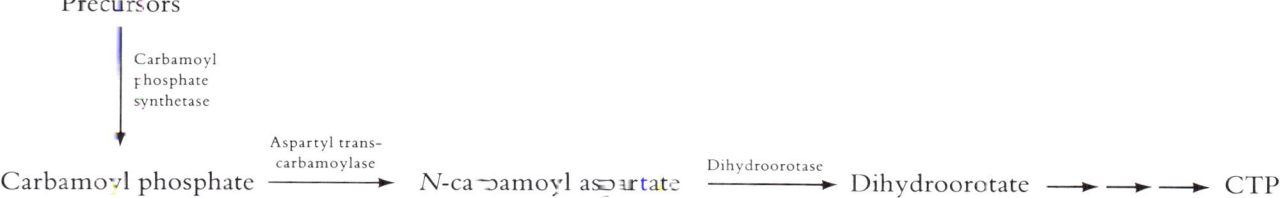

Precursors

Carbamoyl phosphate synthetase

Carbamoyl phosphate $\xrightarrow{\text{Aspartyl trans-carbamoylase}}$ N-carbamoyl aspartate $\xrightarrow{\text{Dihydroorotase}}$ Dihydroorotate $\longrightarrow$ $\longrightarrow$ $\longrightarrow$ CTP

In bacteria each of the three enzymes are made separately. In yeast and *Neurospora* only two proteins are made; one is dihydroorotase and the other is a large protein that is cleaved to form carbamoyl phosphate synthetase and aspartyl transcarbamoylase. Mammals synthesize a single tripartite protein, which is cleaved to form the three enzymes.

Some polyproteins are cleaved in different ways in different tissues. An example is proopiomelanocortin, a polyprotein that is the source of several hormones synthesized in the pituitary gland. In the anterior lobe of the pituitary the polyprotein is cleaved first to release β-lipotropin from the C-terminal end (Figure 16-22). Then the N-terminal fragment is cleaved, releasing adrenocorticotropic hormone (ACTH). No more cleavage occurs in this tissue. However, in the intermediate lobe β-lipotropin is cleaved further, releasing the C-terminal polypeptide, β-endorphin. The ACTH is also cut, forming α-melanotropin. In summary, a single round of cleavage generates two hormones in the anterior lobe; in the intermediate lobe a second round of cleavage converts these hormones to two new hormones.

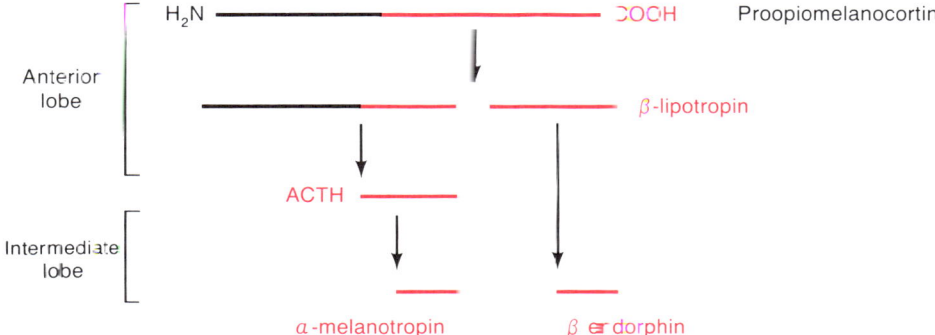

Figure 16-22 Cleavage of the pituitary polyprotein proopiomelanocortin to form four hormones. β-lipotropin and ACTH are formed in the anterior lobe, and α-melanotropin and β-endorphin are produced in the intermediate lobe. The vertical arrows indicate the points of cleavage. Colors are for reference only.

Each of the polyproteins just described is the source of equal numbers of distinct proteins (one copy of each). This is not always the case. For example, the precursor to enkephalin contains six copies of Met-enkephalin and one copy of Leu-enkephalin.

TRANSLATIONAL CONTROL

By translational control is meant the regulation of the number of times a finished mRNA molecule is translated. There are three ways in which translation of a particular mRNA may be regulated: (1) by the lifetime of the mRNA; (2) by the probability of initiation of translation; and (3) by regulation of the rate of overall protein synthesis. In this section we will see examples of each of these modes of regulation.

Extension of the Lifetime of mRNA

The silk gland of the silkworm *Bombyx mori* predominantly synthesizes a single type of protein, silk fibroin. Since the worm takes several days to construct its cocoon, it is the *amount* and not the rate of fibroin synthesis that must be great; hence, the silkworm can manage with a fibroin mRNA molecule that is very long-lived.

The fibroin gene is thought to have a strong promoter, and when a signal (of which the exact nature is unknown) is received to initiate transcription, about 10^4 fibroin mRNA molecules are made in a period of several days. An "average" eukaryotic mRNA molecule has a lifetime of about three hours before it is degraded. However, fibroin mRNA survives for several days during which each mRNA molecule is translated repeatedly to yield 10^5 fibroin molecules. Thus, each gene succeeds in synthesizing 10^9 protein molecules in four days. If a fibroin-producing cell were diploid, 2×10^9 protein molecules would be synthesized per diploid set of chromosomes. The whole silk gland makes 300 μg or 10^{15} molecules of fibroin in four days; therefore, about 5×10^5 diploid cells would be needed. However, the silk gland has an unusual feature in that it develops from a single diploid cell. When the chromosomes of the original cell double and separate by mitosis, the cell does not divide; hence, a tetraploid cell results. Chromosome doubling and mitotic separation without cell division continue until the silk gland is mature; at this time the gland consists of a single giant cell with a ploidy of approximately 10^6. Each of the 10^6 sets of chromosomes contains a single active fibroin gene, which is responsible for the synthesis of 10^9 fibroin molecules; in this way, the $10^6 \times 10^9 = 10^{15}$ protein molecules are made by a single exceptional cell. How cell division is inhibited during formation of the silk gland is unknown.

Production of a large amount of a single type of protein by means of prolonged mRNA lifetime is common in highly differentiated cells. For example, cells of the chicken oviduct, which makes ovalbumin (for egg white), contain only a single copy of the ovalbumin gene per haploid set of chromosomes, but the cellular mRNA is long-lived.

It is instructive to compare this mode of protein production, which is sometimes called **translational amplification,** with the gene amplification utilized to generate the 10^{12} rRNA molecules needed in the *Xenopus* oocyte. The silkworm is able to produce a huge number of protein molecules because there are two stages of amplification—that is, the DNA template for protein synthesis is amplified 10^4-fold by synthesis of 10^4 mRNA molecules and then each RNA template is amplified 10^5-fold by being translated repeatedly. However, in the synthesis of rRNA transcription is the final step—there is no translation. Thus, in order to get a large overall multiplicative factor an amplification step must occur prior to transcription—that is, the genes themselves must be amplified, as is the case.

In the silkworm, fibroin synthesis need not be rapid, as we have stated. However, in other organisms and in other cellular systems accelerated protein synthesis is needed and, in these cases, translational amplification is inadequate. For example, in the cleavage of fertilized eggs, the histone needed to form new chromatin must be synthesized at a rate of about 3×10^5 molecules per minute to keep pace with DNA synthesis. Using ordinary translation rates, about 1000 molecules can be synthesized per minute. The necessary high rate of protein synthesis is achieved here by having several hundred initial copies of each histone gene. Note that this is not an example of the type of gene amplification discussed earlier in this chapter since this copy number is present in all cells.

Differential half-lives of specific mRNA molecules play a role in the life cycle of adenovirus (Chapter 23). Transcriptional unit 1B produces two mRNA molecules—a 14S species and a 20S species—which differ because of distinct modes of processing of the primary transcript. The ratio of the amount of 14S mRNA to 20S mRNA is considerably greater late in the growth cycle of the virus than at early times. This is because of an increased half-life of the 14S mRNA.

Translational and transcriptional control are sometimes combined. For example, insulin (which regulates the synthesis of a large number of substances) and prolactin (another hormone) are required together for production of casein (milk protein) in mammary tissue. Both hormones are needed to initiate transcription, but prolactin, in addition, increases the lifetime of casein mRNA. None of the mechanisms are known.

Control of the Initiation of Translation

Another example of translational regulation is that of **masked mRNA.** Unfertilized sea urchin eggs maintain a state of readiness by storing large quantities of mRNA for many months in the form of ribonucleoprotein particles. This RNA is inactive, but within minutes after fertilization, trans-

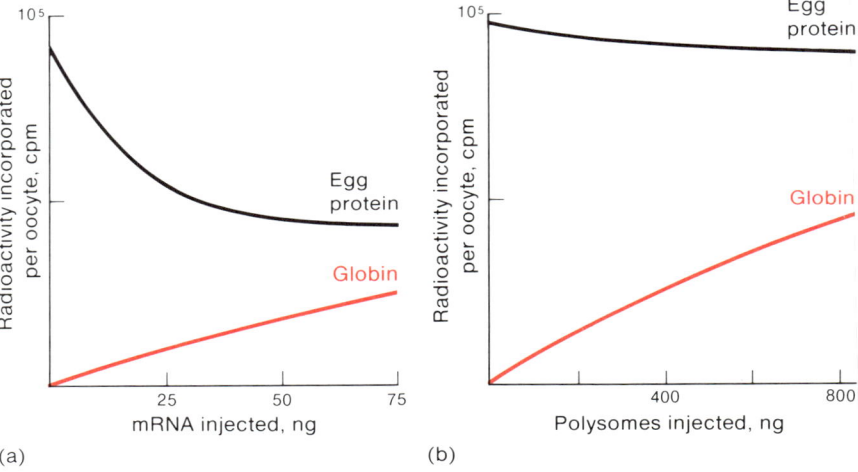

Figure 16-23   Effect of injected reticulocyte globin mRNA (panel (a)) or reticulocyte polysomes (panel (b)) on incorporation
of radioactive amino acids into oocyte
proteins.

lation of these mRNA molecules begins. The mechanisms for stabilization
of the mRNA protection against nucleases and for activation are unknown.

Since mature unfertilized eggs need to maintain themselves but do not
have to grow or change their state, the rate of protein synthesis in eggs is
generally low. This is not due to an inadequate supply of mRNA but to a
limitation of some element required to initiate translation. This conclusion
is derived from an unusual type of experiment, the data for which is shown
in Figure 16-23. Toad eggs are sufficiently large that small volumes of liquid
can easily be injected into them with a hypodermic syringe. If globin mRNA
(isolated from a reticulocyte) plus a radioactive amino acid are injected into
a toad egg, radioactive globin is synthesized. Figure 16-23(a) shows the result
of injecting increasing amounts of globin mRNA. As more mRNA is added,
more radioactive globin is made, but the amount of normal egg proteins
synthesized decreases. This result suggests that in the egg the ability to utilize
mRNA is saturated—as more exogenous mRNA is added, it competes with
endogenous mRNA for the protein synthesis apparatus. However, if reticulocyte polysomes, which contain globin mRNA, are injected instead of
purified globin mRNA, synthesis of egg proteins is only slightly reduced,
as shown in Figure 16-23(b). Thus, there must be an ample supply of amino
acids, elongation factors, and termination factors. Together these experiments show that the limiting factor is either the number of ribosomes or
some molecule needed to form a polysome. There are more than enough
ribosomes, so some other element of the system is implicated. The element
is called the **recruitment factor.**

The synthesis of some proteins is regulated by direct action of the protein on the mRNA. For instance, the concentration of one type of immu-

noglobulin is kept constant by self-inhibition of translation. This protein, like all immunoglobulins, consists of two H chains and two L chains. The tetramer binds specifically to H-chain mRNA and thereby inhibits initiation of translation. How L-chain synthesis is regulated is unknown.

Autoregulation

In Chapter 15 we saw that many genes, especially those that are made nearly continuously, are autoregulated. A few examples of autoregulation have also been seen in eukaryotes. Most cells of eukaryotic tissue contain a cytoplasmic network of filaments called microtubules, which are made of the protein tubulin. This protein is also the main constituent of the spindle, the organelle used to segregate chromosomes during mitosis. The drugs colchicine and vinblastine inhibit polymerization of tubulin and hence cause an increase in the intracellular concentration of free tubulin. Addition of either of these drugs to cultured cells causes a loss of tubulin mRNA molecules. Furthermore, if tubulin is injected into mammalian cells with a microneedle, new synthesis of tubulin rapidly decreases. These experiments indicate that tubulin inhibits its own synthesis. However, in both experiments synthesis of the tubulin primary transcript is not prevented; thus, the inhibition occurs either at the level of RNA processing or of translation. No further information is available.

Regulation of the Rate of Overall Protein Synthesis

Cells that synthesize only a single protein sometimes regulate production of that protein by modulating overall protein synthesis. The production of hemoglobin by reticulocytes is regulated in this way. Reticulocytes are the penultimate cell type in the line of differentiation leading to a red blood cell. In most organisms reticulocytes have no nuclei (the nucleus is extruded when an erythroblast becomes a reticulocyte) and utilize the globin mRNA molecules transcribed from the nucleus of the erythroblast to synthesize the α- and β-globin subunits. Other enzymatic systems are responsible for making hemin, the prosthetic group of hemoglobin. There is one hemin molecule per globin subunit, and for optimal efficiency each type of globin (α and β) should be synthesized at twice the rate of synthesis of hemin. Maintaining the correct ratio of the rates is accomplished by a two-phase regulatory system in which (1) hemin represses its own synthesis, and (2) hemin inactivates an inhibitor of globin synthesis. With this system, if there is more hemin than globin, hemin synthesis is shut off and globin synthesis is stimulated, but, on the other hand, if there is not enough hemin, globin synthesis is repressed and hemin synthesis is turned on. In this way, the amounts of hemin and globin are kept in balance. The mechanism by which the synthesis of hemin is regulated is unknown, but how hemin controls globin synthesis is well understood, as we will see.

Rabbit reticulocytes do not have a nucleus and hence have no DNA. Therefore, globin synthesis in these cells cannot be transcriptionally regulated; regulation must be translational. The mode of regulation is depicted schematically in Figure 16-24. Initiation of translation requires the formation of a ternary complex between $tRNA_{init}$, the initiation factor eIF-2, and a third protein called eIF-2-stimulating protein, or ESP. Reticulocytes contain a protein kinase (a protein-phosphorylating enzyme) called the **hemin-controlled repressor,** or HCR. This inhibitor phosphorylates the small subunit of eIF-2 and thereby prevents eIF-2 from complexing with ESP. HCR is also activated by a cAMP-dependent protein kinase; this kinase contains two pairs of identical subunits called R and C and is denoted R_2C_2. The R and C subunits have regulatory and catalytic roles respectively. The two R subunits are each able to bind cAMP. When cAMP is present, R_2C_2 dissociates and the C subunits, which now have the ability to phosphorylate HCR, are released. The active HCR then phosphorylates eIF-2, thereby preventing initiation of protein synthesis. Hemin binds to the R subunit, causing a structural change in R that prevents R from binding cAMP. Thus, if there is excess hemin, HCR is not activated and protein synthesis can begin.

It should be noted that hemin binds to both globin and HCR, though binding is stronger to globin. Thus, when globin is synthesized, it binds

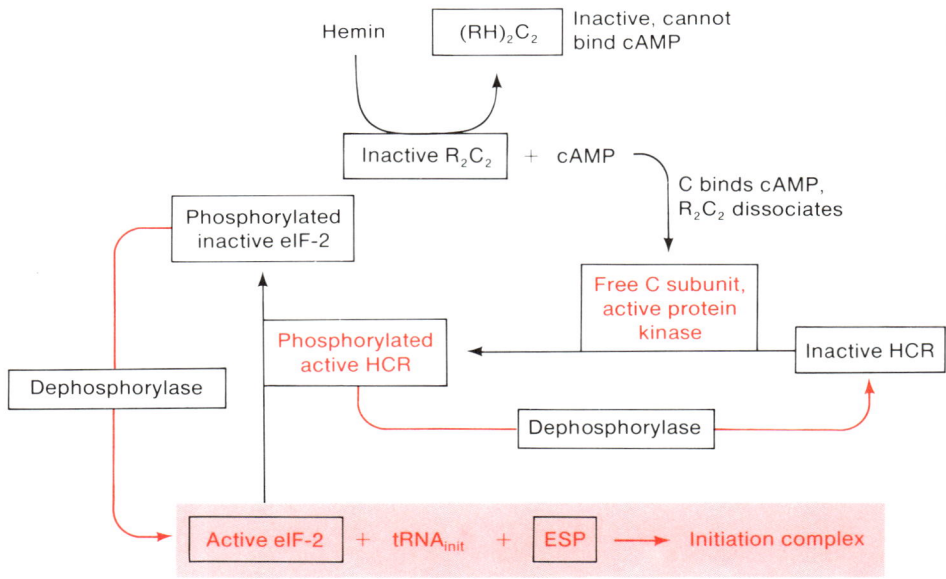

Figure 16-24   Regulation by hemin of protein synthesis in reticulocytes. Proteins are boxed; active forms are in red; inactive forms in black. The reaction shaded in red leads to protein synthesis. Red arrows (dephosphorylations) promote protein synthesis; black arrows inhibit protein synthesis.

hemin, thereby reducing the concentration of free hemin, and this allows HCR to be activated. Globin synthesis is then inhibited until more hemin is made. Without the binding of hemin to F_2C_2, the dephosphorylases deactivate HCR and thereby activate eIF-2, thus allowing globin synthesis to catch up to the hemin. Note that globin synthesis is not specifically turned on and off but that overall protein synthesis is modulated. This is possible only in a cell such as this one in which only a single type of protein is made.

REGULATION OF GENE EXPRESSION IN PLANT CELLS BY LIGHT

The cells of many higher plants lose their chlorophyll and hence their green color, when held in darkness for several days. This loss (called etiolation) is caused by lack of production and loss of the enzymes that catalyze chlorophyll synthesis. When etiolated plants are reexposed to light, within a few hours synthesis of more than 60 photosynthetic enzymes, chloroplast rRNA, and chlorophyll occurs. The resynthesis of one of the subunits of a photosynthetic enzyme, ribulose diphosphate carboxylase (RDC), has been studied in detail. Synthesis of this subunit is mediated by the molecule **phytochrome,** a protein bearing a covalently bound light-absorbing pigment. In the dark phytochrome is inactive; in the light it is converted to an active form that participates in the induction of transcription of the gene encoding the RDC small subunit (and of several other genes). Its mechanism of action is not known, but the protein is thought to be a light-activated transcription factor. Synthesis of the large subunit is also regulated by phytochrome, but several other proteins are also involved, and the phenomenon is poorly understood.

Phytochrome also regulates its own synthesis (autoregulation). In the dark the phytochrome gene is transcribed, but when light activates the protein, transcription is inhibited. Note that this role of light is the reverse of the light-induced activation of synthesis of photosynthetic enzymes.

Gene expression in the colonial alga *Volvox* is also regulated by light. The population of mRNA molecules present in the dark is essentially the same as those present during the light period. However, a shift from dark to light is accompanied by a dramatic increase in translation. The mechanism of this and similar phenomenon in other algae is not known.

REGULATION OF THE SYNTHESIS OF VITELLOGENIN: AN EXAMPLE

Vitellogenin is the protein precursor to the egg yolk proteins phosvitin and lipovitellin, found in amphibians and birds. Vitellogenin is an especially interesting protein because its synthesis is subject to regulation at several points, and many of the features of regulation that have been discussed can be seen to act in a single pathway. Vitellogenin synthesis has been studied most carefully in the toad *Xenopus laevis*.

There are at least four genes in the vitellogenin gene family. These are called *A1, A2, B1,* and *B2.* The translation products of the class-A genes differ from the products of the class-B genes by about 20 percent of the amino acids. The product of a subgroup-1 gene differs from the product of a subgroup-2 gene by about 5 percent of the amino acids. Synthesis of each of the four gene products is regulated by a common system. However, the cleavage pattern of the polyprotein gene products differ.

Synthesis of Yolk Proteins

The overall pathway for the production of phosvitin and lipovitellin is shown in Figure 16-25. The three main stages are the following:

1. Vitellogenin is synthesized in the liver of a mature female; synthesis is stimulated by female sex hormones.
2. It is then secreted into the bloodstream and thereby transported to the ovaries.
3. In the ovary it is selectively taken up by developing eggs (oocytes) in which the vitellogenin is cleaved into the yolk proteins.

Details of the synthesis of vitellogenin are the following.

The synthesis of the yolk proteins is regulated by hormones that control transcription. The transcript is processed, a polyprotein is synthesized and modified, and finally the individual polypeptide chains are cleaved from the modified polyprotein. The individual steps are the following (Figure 16-24):

1. The initial stimulus for vitellogenin production comes from the hypothalamus (part of the brain), which triggers the pituitary gland to secret the hormone gonadotrophin.

2. Gonadotrophin is taken up by cells of the ovarian follicle, which encloses the developing egg. The follicle cells respond by synthesizing estrogen, another hormone, which is secreted into the bloodstream.

3. The estrogen binds to receptors in the plasma membrane of liver cells, which respond by synthesizing vitellogenin. This effect of estrogen makes the synthesis of vitellogenin an especially useful experimental system because the liver cells of both hens and roosters can synthesize vitellogenin but only hens synthesize estrogen. Thus, if roosters are used, it is possible to turn the system on at will by injection of estrogen into the bloodstream, and this makes it possible to look carefully and in a controlled way at early events.

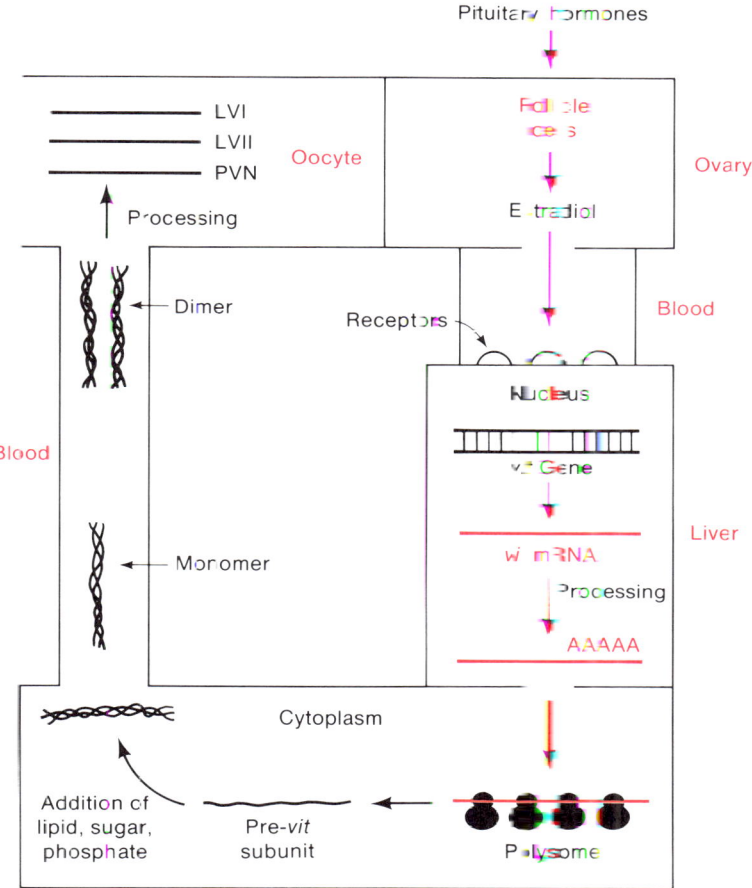

Figure 16-25 Scheme summarizing the stimulation by estradiol of the synthesis of vitellogenin and its processing to form phosvitin (PVN), lipovitellin I and II (LV), and phosvettes I and II (PVT). The results of processing of a type-A vitellogenin in the oocyte are shown. For type B there would be four proteins—LVI, LVII, PVTI, and PVTII.

4. The estrogen is carried to the nucleus of the liver cell, where it turns on transcription of the vitellogenin gene. The mode of stimulation of vitellogenin synthesis by estrogen differs slightly from the basic mechanism of hormone stimulation already described. At the time of first exposure to estrogen the nucleus of each liver cell has about 200 estrogen receptors. During the first 12 hours the number of receptors increases to 1000 as a result of *de novo* synthesis; this number is retained by the cells for eight days. This increase is thought to explain the following phenomenon. If an animal is given a small dose of estrogen on day 1 and then it is given a second injection

several days later when vitellogenin synthesis is no longer occurring, it is found that the rate of synthesis of vitellogenin mRNA is much greater after the second dose than after the first one.

5. The transcript is capped, poly(A) is added, introns are removed by processing enzymes, the mRNA is transported to the cytoplasm, and polysomes form.

Interestingly, the lifetime of vitellogenin mRNA is increased by the presence of estrogen. As long as estrogen is present, the half-life of the mRNA is about 24 hours; if, after continuous administration of estrogen for several days, the hormone is withdrawn, the half-life changes to 3 hours.

6. Translation of the processed mRNA occurs. However, translation does not begin immediately, although overall protein synthesis continues normally. It is thought that one or more translation factors are needed to initiate translation of vitellogenin mRNA and that these factors are also produced in response to estrogen. The evidence for this is the following: if the hormone is added to a male animal and then, after vitellogenin synthesis has begun, it is withdrawn, synthesis of vitellogenin stops because no more vitellogenin mRNA is being made and the preexisting mRNA has been degraded. However, if estrogen is given again, vitellogenin is synthesized as soon as vitellogenin mRNA is made. That is, there is no delay in translation, because the protein factors are already present.

7. The capacity for translation increases. Once vitellogenin synthesis begins, the liver cells gain the capacity to synthesize all proteins more rapidly than before the administration of estrogen. This is shown by an experiment in which the ability of two extracts of liver cells, obtained from animals before and after estrogen is administered, to support poly(U)-mediated synthesis of polyphenylalanine was compared. It was found that much more polyphenylalanine was made by an extract of cells from an estrogen-treated male than from an untreated male. Thus, there are two effects on translation that enable the liver cells to produce a huge amount of vitellogenin: (i) overall protein synthesis is accelerated and (ii) certain cytoplasmic factors enable vitellogenin mRNA to compete very effectively with all other mRNA molecules, in order that vitellogenin is the major protein made.

8. The final step occurring in liver cells is the modification of the protein by the addition of phosphate groups, lipids, and sugars.

9. The completed vitellogenin monomer is secreted into the bloodstream where the modified protein molecules aggregate to form a dimer, which is the usual molecule thought of as vitellogenin.

10. The dimer is taken from the blood by the oocytes. The dimer is a polyprotein and in the oocyte it is cleaved. The cleavage pattern of class-A

and class-B proteins differ. Class-A polyproteins are cleaved to form phosvitin (molecular weight $M = 35,000$), lipovitellin I ($M = 115,000$), and lipovitellin II ($M = 31,000$); class-B polyproteins form lipovitellin I and II, phosvette I ($M = 14,000$), and phosvette II ($M = 19,000$). The significance of the multiple forms of vitellogenin is not known but in both cases the proteins are, after cleavage, assembled in the yolk platelet to yield a mature egg.

Coordinated Synthesis of Eggwhite Proteins

The eggs of amphibians and birds contain both yolk and white. It seems reasonable that the synthesis of both yolk proteins and white proteins should be coordinately regulated, since one is not needed without the other. This does in fact occur and the regulator is again estrogen, which not only turns on synthesis of vitellogenin but also turns on the synthesis of ovalbumin (the major white protein, as described earlier in this chapter).

Bacteriophages

I. LYTIC PHAGES

Bacteriophages, or phages, have played an important role in the development of molecular biology. At the present time a few phages are the most completely understood of any organisms. Because of their lesser complexity than bacteria and higher cells and the availability of an enormous number of mutants, phages have been extraordinarily useful in the study of replication, transcription, and regulation.

GENERAL PROPERTIES OF PHAGES

A bacteriophage is, first of all, a bacterial parasite. By itself, a phage can persist, but it can neither grow nor replicate except within a bacterial cell. Most phages possess genes encoding a variety of proteins. However, all known phages use the ribosomes, protein-synthesizing factors, amino acids, and energy-generating systems of the host cell, and hence a phage can grow only in a metabolizing bacterium.

Each phage must perform some minimal functions for continued survival. These are the following:

1. Protection of its nucleic acid from environmental chemicals that could alter the molecule (for example, break the molecule or cause a mutation).

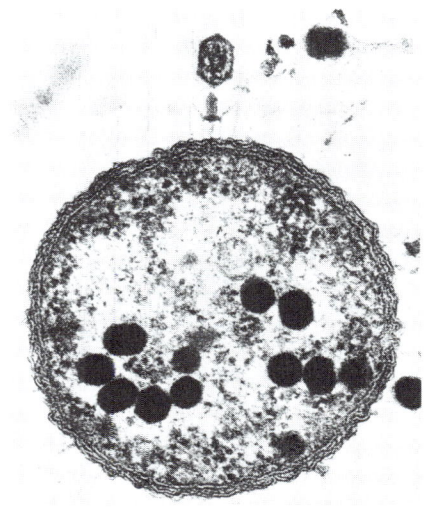

An electron micrograph of a cross section of *E. coli* in which T4 phage are developing. A phage, whose tail is contracted and in the DNA-injection position, is on the cell surface; possibly, its DNA was the parent of the developing phage particles. [Courtesy of Lee Simon and Tom Anderson.]

2. Delivery of its nucleic acid to the inside of a bacterium.
3. Conversion of an infected bacterium to a phage-producing system which yields a large number of progeny phage.
4. Release of progeny phage from an infected bacterium.

These functions are carried out in a variety of ways by different phage species, as will be seen in this chapter. All of these species have certain features in common but differences in detail show the many ways in which specific biological functions can be accomplished.

An important observation to be made in this chapter is of the degree to which an individual phage particle uses parts of the machinery of the cell. Some phage species have fewer than 10 genes and use almost all of the cellular functions, whereas others have 30–100 genes and are less dependent on the host. A few of the largest phage particles have so many of their own genes that, for certain functions such as DNA replication, they need no host genes. Surprisingly, in a few cases phage genes duplicate host genes; this seeming redundancy in function is discussed in a later section concerned with these so-called nonessential genes.

Phage particles also differ in their physical structures from species to species and often certain features of their life cycles are correlated with their structure. These structural differences are described in the following section. However, before beginning that section the reader is urged to review some of the material about phage biology and techniques for detecting and counting phage presented in Chapter 1.

Structures of Phages

There are three basic phage structures: icosahedral tailless, icosahedral head with tail, and filamentous. (An icosahedron is a quasispherical polyhedron having 20 triangular faces and 12 corners.) Usually the phage particle consists of a single nucleic acid molecule—which may be single- or double-stranded, linear or circular DNA, or single-stranded, linear RNA—and one or more proteins. (The one known exception is phage φ6, which contains three linear double-stranded RNA molecules, whose base sequences differ from one another.) The proteins form a shell, called either the **coat** or the **capsid,** around the nucleic acid; the nucleic acid is thereby protected from nucleases and harmful substances. Figure 17-1 shows electron micrographs of the three basic structures (a drawing of a tailed phage is given in Figure 8-5). The following points are general:

1. In both icosahedral tailless and tailed phages the nucleic acid is contained in a hollow region formed by the capsid and is highly compact. In a filamentous phage the nucleic acid is embedded in the capsid and is present in an extended helical form.

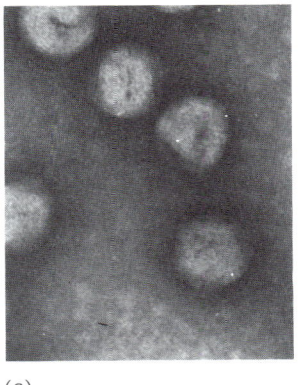
(a)

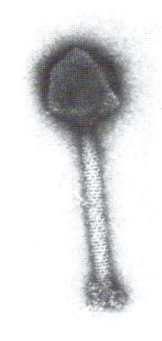
(b)

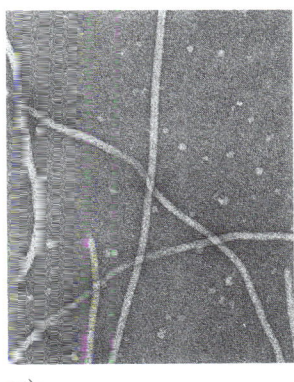
(c)

Figure 17-1 The three major morphological classes of phages (a) Icosahedral, tailless: PM-2; (b) Icosahedral, tailed: SP01; (c) Filamentous: M13. A fourth class, in which double-stranded DNA is encased in a flexible lipid-containing envelope rather than a rigid protein coat, has one representative and is not shown. Courtesy of (a) Christine Brack, (b) Beatrice ten Heggeler, and (c) Robley Williams.]

2. The tail is a complex multicomponent structure often terminated by tail fibers.
3. In icosahedral phages the length of the DNA molecule is very much greater than any dimension of the head.

There are many variations on the basic structure of the tailed phages. For example, the length and width of the head may either be the same or the length may be greater than the width; however short fat heads are not seen. The tail may be very short (barely visible in electron micrographs) or up to four times the length of the head and it may be flexible or rigid. A complex baseplate may also be present on the tail; when present, it typically has from one to six tail fibers.

Stages in the Lytic Life Cycle of a Typical Phage

Phage life cycles fit into two distinct categories—the **lytic** and the **lysogenic** cycles. A phage in the lytic cycle converts an infected cell to a phage factory, and many phage progeny are produced. A phage capable *only* of lytic growth is called **virulent.** The lysogenic cycle, which has been observed only with phages containing double-stranded DNA, is one in which no progeny particles are produced; the phage DNA usually becomes part of the bacterial chromosome. A phage capable of such a life cycle is called **temperate.** * In this section only the lytic cycle is outlined. The lysogenic cycle will be described in the next chapter.

*Most temperate phages also undergo lytic growth under certain circumstances.

There are many variations in the details of the life cycles of different virulent phages. There is, however, what may be called a basic lytic cycle, which is the following (Figure 17-2):

1. *Adsorption of the phage to specific receptors on the bacterial surface* (Figure 17-3). These receptors are very varied and serve the bacteria for purposes other than phage adsorption.

2. *Passage of the DNA from the phage through the bacterial cell wall.* Some types of tailed phages use an injection sequence shown schematically in Figure 17-4. In this process the nucleic acid is transferred into the cell and is never exposed to the medium surrounding the recipient cell. Little is known about the transfer mechanisms for the other phage types. With tailless phages

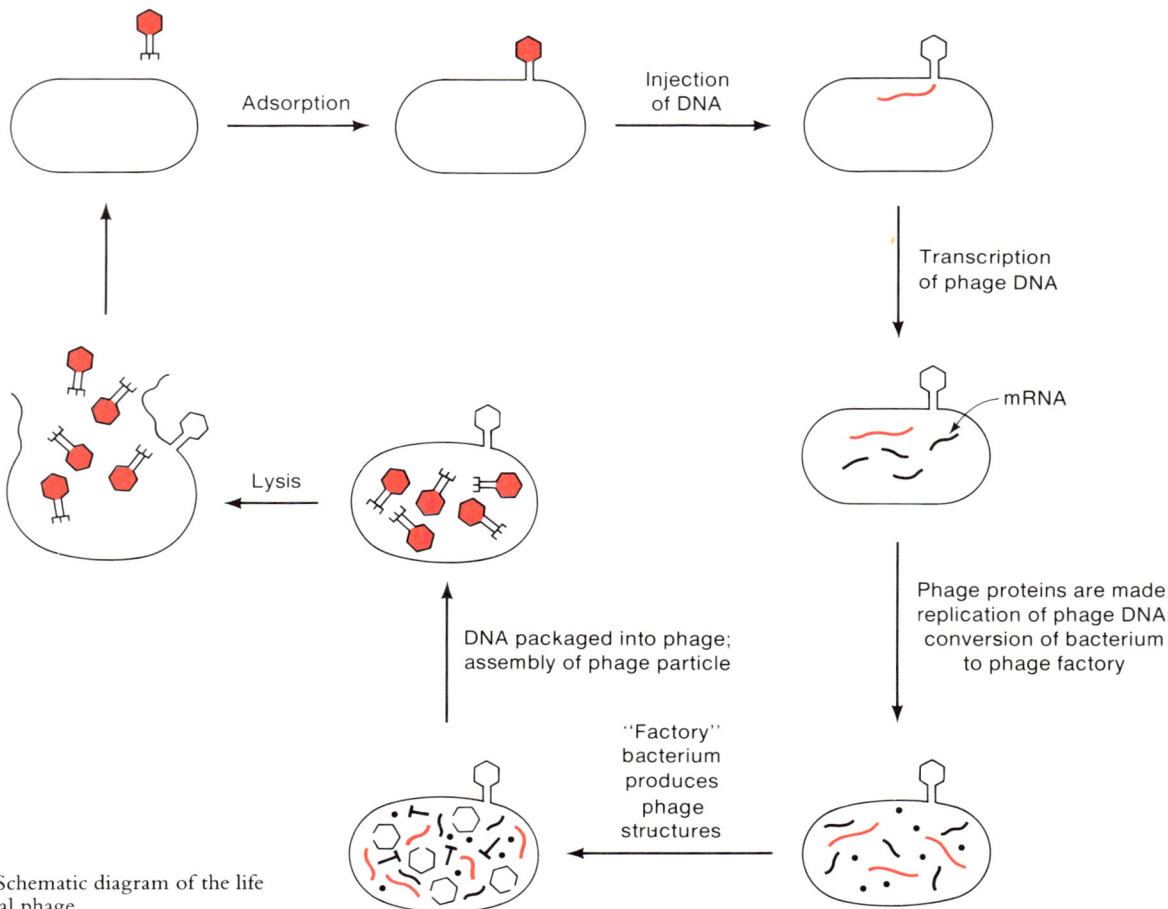

Figure 17-2 Schematic diagram of the life cycle of a typical phage.

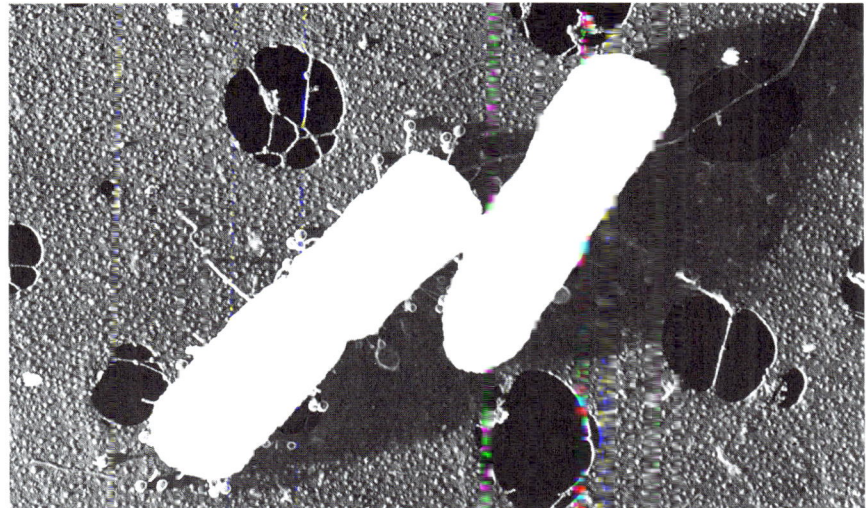

Figure 17-3 An electron micrograph of an *E. coli* cell to which numerous λ phage particles are adsorbed by their long tails. [Courtesy of T. F. Anderson.]

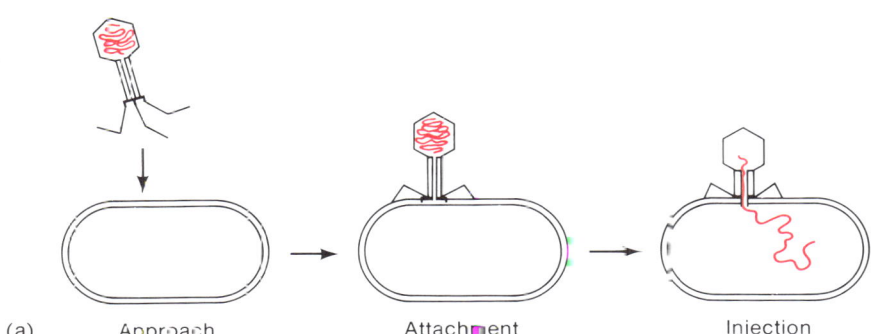

(a)　　Approach　　　　Attachment　　　　Injection

Figure 17-4 (a) Injection sequence of a tailed phage. In the injection stage the tail sheath contracts and drives a core protein tube through the cell wall like a hypodermic syringe. (b) Electron micrograph of T4 phage adsorbed to the cell wall of *E. coli*, observed in thin section. The tail sheath is contracted and the core is fixed firmly against the cell wall. The arrow shows a portion of the core projecting through the cell wall. DNA can be seen entering the cell from the two phage at the right. [Courtesy of Lee Simon.]

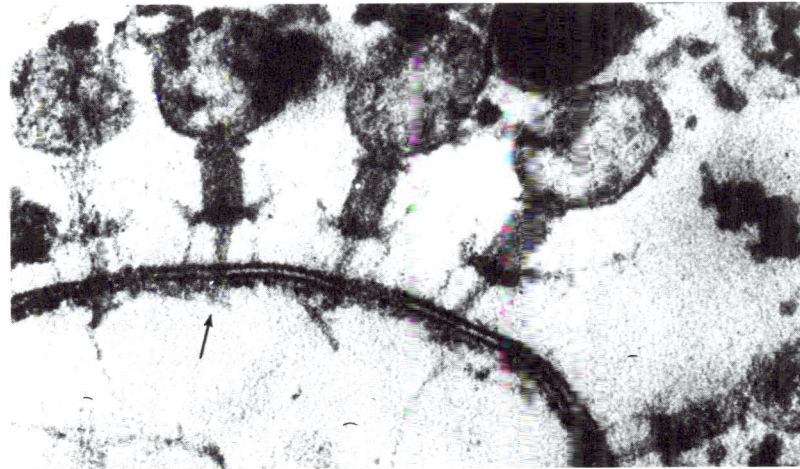

(b)

the nucleic acid is transiently susceptible to nuclease attack, so it is thought that the phage coat may break open and release its nucleic acid first onto the cell wall prior to entering the cell.

3. *Conversion of the infected bacterium to a phage-producing cell.* Following infection by most phages a bacterium loses the ability either to replicate or to transcribe its DNA; sometimes it loses both. This shutdown of host DNA or RNA synthesis is accomplished in many different ways, depending on the phage species.

4. *Production of phage nucleic acid and proteins.* By several mechanisms the phage directs the synthesis of a replicative system that specifically makes copies of phage nucleic acid. This programming is accomplished either by synthesis of phage-specific polymerases or by addition of specificity elements to bacterial enzymes. In both cases many bacterial replication proteins are used. RNA-containing phages are quite constrained in this respect—they must encode their own replication enzymes because bacteria do not contain enzymes that replicate RNA. Transcription is almost always initiated by the bacterial RNA polymerase but after the first transcription event either the bacterial polymerase is modified to recognize phage promoters or a phage-specific RNA polymerase is synthesized. Transcription is regulated and phage proteins are synthesized sequentially in time as they are needed. Usually there is a fairly distinct difference in the time of synthesis of phage-specified enzymes (made early in the life cycle and called **early proteins**) and the structural proteins of the phage particle (made late in the life cycle). The temporal difference is accomplished by the timing of mRNA synthesis. The early proteins are encoded in **early mRNA** and the structural proteins are in **late mRNA.** The RNA molecules of phages containing single-stranded RNA serve as their own mRNA, and thus have no need for transcription; these phages are regulated by controlling the efficiency of translation.

5. *Assembly of phage particles* (**morphogenesis**). Two types of proteins are needed for the assembly process: **structural proteins,** which are present in the phage particle, and **catalytic proteins,** which participate in the assembly process but do not become part of the phage particle. A subset of the latter class consists of the **maturation proteins,** which convert intracellular phage DNA to a form appropriate for packaging in the phage particle. For the icosahedral phages assembly occurs in several stages: (1) Aggregation of phage structural proteins to form a phage head and, when needed, to form a phage tail; at this point, the tail is not attached to the head. (2) Condensation of the nucleic acid and entry into a preformed head. (3) Attachment of the tail to a filled head. With filamentous phages the nucleic acid and the protein form a phage particle in a single step. The mechanism of condensation is unknown. Usually 50–1000 phage particles are produced, the number depending on the particular phage species.

6. *Release of newly synthesized phage.* With most phages an enzyme called a **lysozyme** or an **endolysin** is synthesized late in the cycle of infection. The enzyme causes disruption of the cell wall and breakdown of the cell (**lysis**), so phage are released to the surrounding medium. A few filamentous phages release progeny continuously by outfolding of the cell wall; this process is called **budding** and it does not cause major damage to the cell. Cells infected with such filamentous phages can continue to produce virus particles for very long periods of time.

Properties of a Phage-Infected Bacterial Culture

In the preceding section the events following infection of one cell by a phage were described. In the laboratory one usually infects a bacterial culture with a large number of phage particles. In this and the following three sections the parameters needed to describe such an infection are described.

The adsorption of phage to a bacterial culture is a random process and the variation in the distribution of phage among the cells is described by the Poisson distribution,

$$P(n) = \frac{m^n e^{-m}}{n!}$$

in which $P(n)$ is the fraction of the bacteria to which n phage have adsorbed when m is the average number of adsorbed phage per bacterium (the multiplicity of infection or MOI). That is, the fraction of bacteria infected with 0, 1, 2, 3 . . ., i phage is $P(0)$, $P(1)$, $P(2)$, $P(3)$, . . . , $P(i)$. Thus, if 3×10^8 phage adsorb to 10^8 bacteria ($m = 3$), the values of $P(0)$, $P(1)$, $P(2)$, $P(3)$, . . . are 0.05, 0.15, 0.22, 0.22, ... Since $P(0) = 0.05$, the sum of $P(1) + P(2) + . . . + P(i)$ must equal $1 - 0.05 = 0.95$. In other words, 95 percent of the bacteria are infected by at least one phage.

Note also that the value of $P(0)$ tells the fraction of the phage particles that have adsorbed. For example, in the infection just described, $P(0)$ should equal 0.05 for $m = 3$, if all of the added phage had adsorbed to the bacteria. However, in a particular experiment using 3×10^8 phage and 10^8 bacteria, if one observed that 12 percent of the bacteria remained uninfected, then the value of $P(0)$ would be 0.12; using this value one can calculate from the Poisson term $P(0) = e^{-m}$ that $m = 2.12$, which is the true value of the MOI. Thus, $2.12/3 = 0.71$ or 71 percent of the added phage particles actually adsorbed.

It is possible to measure $P(0)$ in a simple way. A known number of cells is used in an infection. After an adsorption period the bacteria are placed on an agar surface and the fraction of the bacteria able to form a colony is measured. Only uninfected cells can do so. Thus, $P(0)$ is the number of colonies formed divided by the total number of cells. As a check, the number of infected cells can be measured in the following way. First, antibodies that inactivate unadsorbed phage are added to the infected culture.* Then, the phage suspension is plated on a lawn of phage-sensitive cells, where each infected cell, providing it is plated before lysis, will produce a *single* plaque. A cell that can form a plaque in this way is called an **infective center.**

The number of phage produced by an infected cell is called the **burst size.** This is an important parameter in many experiments because it is a measure of the efficiency of phage production. It is measured by determining (by plaque counting) the number of phage produced after lysis of the culture and dividing by the number of infective centers.

Essential and Nonessential Phage Genes

Most of the remainder of this chapter is devoted to a detailed look at particular phages. A distinction will be made between essential and nonessential genes. A gene is usually considered to be nonessential if a mutation in that gene does not prevent plaque formation. If the mutation prevents plaque formation, the gene is essential. These definitions must be interpreted carefully because, whereas a typical burst size for an infected bacterium is 50–100, a burst size of four is usually sufficient for plaque production. Indeed, mutations in many so-called nonessential genes reduce the burst size markedly, though not enough to prevent plaque formation.

A gene may be nonessential for three reasons: (1) An identical gene may be present in the bacterium, or a phage gene may have the same function as a gene present in a bacterium. This duplication may be of some value to the phage, because it might increase the burst size by providing a higher concentration of an essential enzyme; alternatively, in nature there may be hosts that lack the gene and the gene will then be essential for growth. (2) The gene does not duplicate a bacterial gene but in some way increases either the rate of phage production or the burst size. In both cases (1) and (2) there is an evolutionary advantage. (3) The gene is not needed in the laboratory but it enables the phage to cope with special situations met in nature. A fourth possibility—that the gene is always unnecessary—is less likely because a truly useless gene would in general lack the survival advantage needed for the phage to retain it.

Specificity in Phage Infection

More than a thousand phages have been isolated. The ability of a particular phage to infect a bacterium is almost always limited to a single bacterial

---

*Antibodies are obtained from the blood of a rabbit that has been injected with a purified suspension of the phage.

species and often to a few strains of that species. For example, no phage species that infects the genus *Pseudomonas* can also infect *E. coli;* furthermore, phages that grow in *Ps. fuorescens* generally fail to grow in *Ps. aeruginosa*. Among the *E. coli* phages, which are the most extensively studied phages, extraordinary specificity has been observed; for example, the phage φX174 grows well on *E. coli* strain C but fails to grow on most other laboratory strains. There are exceptions, though; for instance, *E. coli* phage T4 is capable of growing on many strains of *E. coli* and on certain species of the genus *Shigella*.

Several factors contribute to this specificity. One of these is the ability to adsorb. For example, *E. coli* phages will not adsorb to any known species of *Pseudomonas*. This is not unexpected inasmuch as there are specific receptors on the cell wall for phage adsorption. It is of course a distinct *disadvantage* to a bacterium to be able to adsorb phages, since this invariably kills the bacterial cell; thus, in the course of evolution cell wall components have evolved to be unrecognizable to most phage species. Phages have also evolved in order to be able to recognize some bacterium; if a particular phage had not been able to do so, it would not exist at the present time.

Another specificity factor is the inability of a "foreign" bacterial RNA polymerase to recognize phage promoters. However, even when adsorption and transcription are possible, with most bacteria there is usually another barrier called **host restriction** and **host modification.** This is a phenomenon in which a bacterium of type X is able to distinguish a phage that has been grown in type X bacterium from one grown in a different type such as Y and is able to prevent the phage grown in Y from carrying out a successful infection. The notation used to discuss this phenomenon is the following: a phage P grown in a bacterium X is denoted P·X. Host modification and restriction are illustrated by the data in Table 17-1. Note that λ·K, which has been grown in *E. coli* strain K, forms plaques at a low efficiency in strain B. Thus, λ·K is **restricted** by strain B. The phage population in these rare plaques (λ·B) has been modified by strain B, so the phage grow efficiently in strain B; however, λ·B now fails to grow in strain K—that is, it is restricted. The molecular explanation for this is the following: *E. coli* B contains a restriction endonuclease (Chapter 5)—specifically, it is the EcoB nuclease—a site-specific nuclease that cuts DNA strands only within a specific base sequence. Phage λ·K contains this sequence; when its DNA is injected into *E. coli* B, the phage DNA is broken. *E. coli* B also contains this sequence and would destroy its own DNA were the sequence not modified. A site-specific methylating enzyme (EcoB methylase) methylates an adenine in the sequence, thereby rendering the sequence resistant to EcoB. When λ·K infects strain B, a few parental phage-DNA molecules in the large population of infected cells are methylated before they are restricted. All progeny DNA molecules are already methylated on one strand and the newly synthesized strands are also methylated rapidly (see Chapter 9), and restriction is avoided. Thus, a small population of phage having the B mod-

Table 17-1 The restriction and modification pattern of *E. coli* phage λ

Bacterial strain	Phage		
	λ·K	λ·B	λ·C
K	1	10^{-4}	10^{-4}
B	10^{-4}	1	10^{-4}
C	1	1	1

Note: Numbers indicate relative plating efficiency.

ification (λ·B) are produced. *E. coli* K also contains a restriction enzyme (EcoK). It attacks a base sequence that is different from the sequence recognized by EcoB. An EcoK methylase also protects *E. coli* K from self-destruction, producing the K modification. A λ phage that has always been grown in strain K—namely, λ·K—is methylated in the EcoK-specific sequence and is resistant to EcoK nuclease. However, λ·B has an unmethylated EcoK sequence, so λ·B DNA is usually broken when a strain K cell is infected. Occasionally, a λ·B DNA molecule escapes restriction and replicates, and its replicas have a methylated K-specific sequence. Thus, the rare progeny phage that result when λ·B successfully infect *E. coli* K are λ·K; they now lack the B modification and are restricted when infecting strain B.

Note in Table 17-1 that λ grown on strain C—that is, λ·C—also fails to grow well in strains B and K, but neither λ·B nor λ·K is restricted by strain C. The reason for the lack of restriction is that strain C has no restriction nuclease active against any base sequence in λ DNA. The λ·C phage are restricted by both strains B and K because, of course, strain C does not have the B and K methylases.

Host restriction and modification is a widespread process, probably serving to destroy foreign DNA.

Table 17-2   Properties of the nucleic acid of several phages

Phage	Host	DNA or RNA	Form	Molecular weight, $\times 10^6$	Unusual bases
ϕX174	E	DNA	ss, circ	1.8	None
M13, fd, f1	E	DNA	ss, lin	2.1	None
PM2	PB	DNA	ds, circ	9	None
186	E	DNA	ds, lin	18	None
B3	PA	DNA	ds, lin	20	None
T7	E	DNA	ds, lin	26	None
λ	E	DNA	ds, lin	31	None
P1	E	DNA	ds, lin	59	None
T5	E	DNA	ds, lin	75	None
SPO1	B	DNA	ds, lin	100	HMU for thymine
T2, T4, T6	E	DNA	ds, lin	108	Glucosylated HMC for cytosine
PBS1	B	DNA	ds, lin	200	Uracil for thymine
MS2, Qβ, f2	E	RNA	ss, lin	1.0	None
ϕ6	PP	RNA	ds, lin	2.3, 3.1, 5.0*	None

Note: Abbreviations used in this table are: E, *E. coli;* B, *Bacillus subtilis;* PA, *Pseudomonas aeruginosa;* PB, *Ps. aeruginosa BAL-31;* PP, *Ps. phaseolica;* ss, single-stranded; ds, double-stranded; circ, circular; lin, linear; HMU, hydroxymethyluracil; HMC, hydroxymethylcytosine.

*ϕ6 contains three molecules.

SPECIFIC PHAGES

On a biological scale of complexity, a phage is a relatively simple form of life. However, phages are sufficiently complex that there is not a single phage type for which all of the molecular details of its life cycle are completely understood. Major progress has been with a small number of phage species. Special features of each phage have resulted in particular phages being more suited to the study of certain processes. For example, regulation of transcription is best understood in phages λ and T7, and the study of morphogenesis has been more successful with phages T4 and λ than with other phages.

In the following sections several of these phages will be described. We will emphasize those features that are well understood. As much as possible, an attempt will be made to give an overview of the life cycle of each phage. Table 17-2 lists some of the important features of several types of phages.

E. coli PHAGE T4

During the period 1939 to 1941 Milislav Demerec isolated a set of phages, each of which grow in a particular strain of *E. coli* (strain B); these phages were named T1, T2, . . . , and T7. Their study revolutionized genetics because, with these phages, physical and biochemical experiments could be coupled to genetic analysis. The genetic material of these phages was later shown to be double-stranded DNA (Chapter 8). The structures and life cycles of some of these phages differ in significant ways, so the phages are grouped in four classes: T1; T2, T4, T6 (called T-even phages); T3, T7; and T5.* The T-even phages have been especially attractive subjects for study, because they are easily grown in large quantities, many mutants can easily be obtained, and their DNA molecules contain an unusual base that enables the phage DNA to be distinguished from host DNA. In recent years the principal effort with this class has been with phage T4, an electron micrograph of which is shown in Figure 17-5. A great deal has been learned about this phage, though because of its large size and complex life cycle certain features of the life cycle are still poorly understood.

Properties of T4 DNA

The DNA of phage T4 is rather large for a phage (Table 17-2); its molecular weight is about 110 million (about 165,000 base pairs). Its length is 55 μm

*T1, T3, T5, and T7 have been called T-odd phages, but because the four phages fall into three classes, the term has no value.

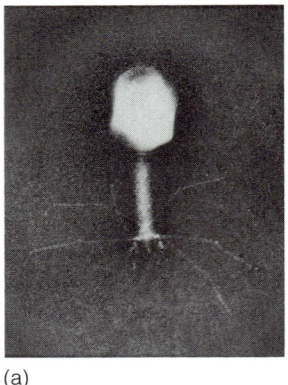

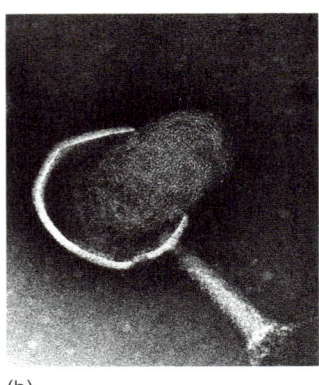

(a)

(b)

Figure 17-5 Two electron micrographs of phage T4. (a) An intact phage particle, showing the tail plate and the six tail fibers.

(b) A phage with a broken head, releasing its DNA. [Courtesy of Robley Williams.]

and the phage head is roughly 0.06 by 0.09 μm; thus, the DNA must be folded tightly. The DNA contains adenine, thymine, and guanine but no cytosine; instead, there is a modified form of cytosine called **5-hydroxy-methylcytosine (HMC),** which base-pairs with guanine (Figure 17-6). This base is further modified by glucosylation—that is, a sugar is coupled to the OH group of HMC. The three T-even phages, which we have said are quite similar, differ chemically in the identity of the sugar. There is a sufficient amount of the sugar attached to the HMC that T4 DNA is virtually cloaked within these molecules; this has the consequence that purified T4 DNA is somewhat resistant to a variety of DNases.

The DNA of T4 is **terminally redundant** (Figure 17-7(a))—that is, a sequence of bases is repeated at both ends of the molecule. The terminal redundancy contains about 1600 base pairs.

5-Hydroxymethylcytosine (HMC)

Glucosylated HMC

Figure 17-6 Nonglucosylated and glucosylated 5-hydroxymethylcytosine. If the

CH_2OH (red) in HMC were replaced by hydrogen, the molecule would be cytosine.

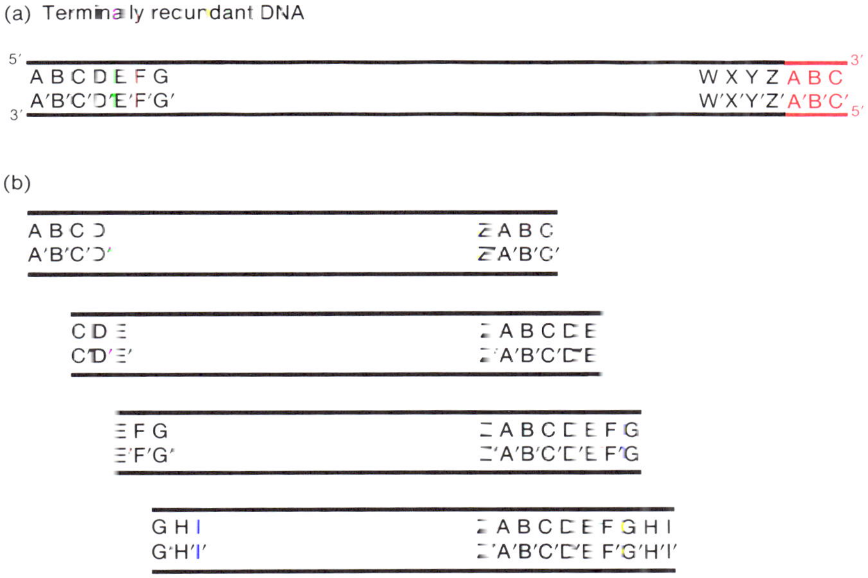

(a) Terminally redundant DNA

5'
A B C D E F G   W X Y Z A B C 3'
A'B'C'D'E'F'G'   W'X'Y'Z'A'B'C'
3'   5'

(b)

A B C D   Z A B C
A'B'C'D'   Z'A'B'C'

C D E   Z A B C D E
C'D'E'   Z'A'B'C'D'E

E F G   Z A B C D E F G
E'F'G'   Z'A'B'C'D'E F'G

G H I   Z A B C D E F G H I
G'H'I'   Z'A'B'C'D'E F'G'H'I'

Figure 17-7   (a) A terminally redundant molecule and its identification by means of exonucleolytic digestion and circularization. A nonredundant DNA cannot be circular-ized in this way. (b) A cyclically permuted collection of terminally redundant DNA molecules.

A remarkable feature of T4 DNA is that although each phage contains one DNA molecule, the molecules differ from phage to phage, even if the population was produced by replication of a single phage particle and its progeny. A sample of T4 DNA molecules is **cyclically permuted,** the meaning of which is shown in Figure 17-7(b). The figure indicates schematically that the termini of the DNA molecules in the population can be found at many different bases within an overall sequence—in reality, probably at any point in the base sequence. Note that cyclic permutation is a property of a phage population, whereas terminal redundancy is a property of an individual phage DNA molecule.

Many phage species produce terminally redundant and cyclically permuted populations of DNA molecules. Terminal redundancy can also be present without cyclic permutation but the converse has not been observed. How these properties arise in phage T4 will be seen when packaging of phage DNA is described.

Genetic Organization of the T4 Phage

To date, 135 genes have been identified in the T4 phage. These genes account for about 90 percent of the DNA; thus, 15 to 20 genes remain to be found.

T4 genes can be divided into two classes—82 metabolic genes and 53 particle-assembly genes. Of the 82 metabolic genes only 22 genes—namely, those involved in DNA synthesis, transcription, and lysis—are essential. The remaining 60 metabolic genes duplicate bacterial genes; particles in which these genes are mutated will grow, though occasionally they will have a smaller burst size.

The T4 Growth Cycle in Brief

The timing of the life cycle of T4 is outlined below. The times are in minutes at 37°C.

$t = 0$ Phage adsorbs to bacterial cell wall. Injection of phage DNA probably occurs within seconds of adsorption.

$t = 1$ Synthesis of host DNA, RNA, and protein is totally turned off.

$t = 2$ Synthesis of first mRNA begins.

$t = 3$ Degradation of bacterial DNA begins.

$t = 5$ Phage DNA synthesis is initiated.

$t = 9$ Synthesis of "late" mRNA begins.

$t = 12$ Completed heads and tails appear.

$t = 15$ First complete phage particle appears.

$t = 22$ Lysis of bacteria; release of about 300 progeny phage.

The main feature to be noticed at this point is the orderly sequence of events. This sequence is also illustrated in part in the electron micrographs shown in Figure 17-8. In the sections that follow a few of these stages are described in detail.

Taking Over the Cell

Shortly after infection, several events occur that enable the phage to turn off many bacterial functions necessary for continued bacterial growth. For example, the host RNA polymerase is modified in such a way that host promoters are recognized poorly. Furthermore, the first phage mRNA encodes DNases that rapidly degrade host DNA. Thus, in time the bacterial DNA is almost totally degraded to nucleotides and is not available as a template for replication or transcription. However, turning off the synthesis of host DNA, RNA, and protein precedes synthesis of these nucleases and also occurs when *E. coli* is infected by phage mutants unable to synthesize these nucleases. Thus another explanation (as yet unknown) is needed for inhibition of host macromolecular synthesis.

Part of the takeover process is a series of events that allow transcription of T4 to occur in an orderly way. The complete pattern of early transcription

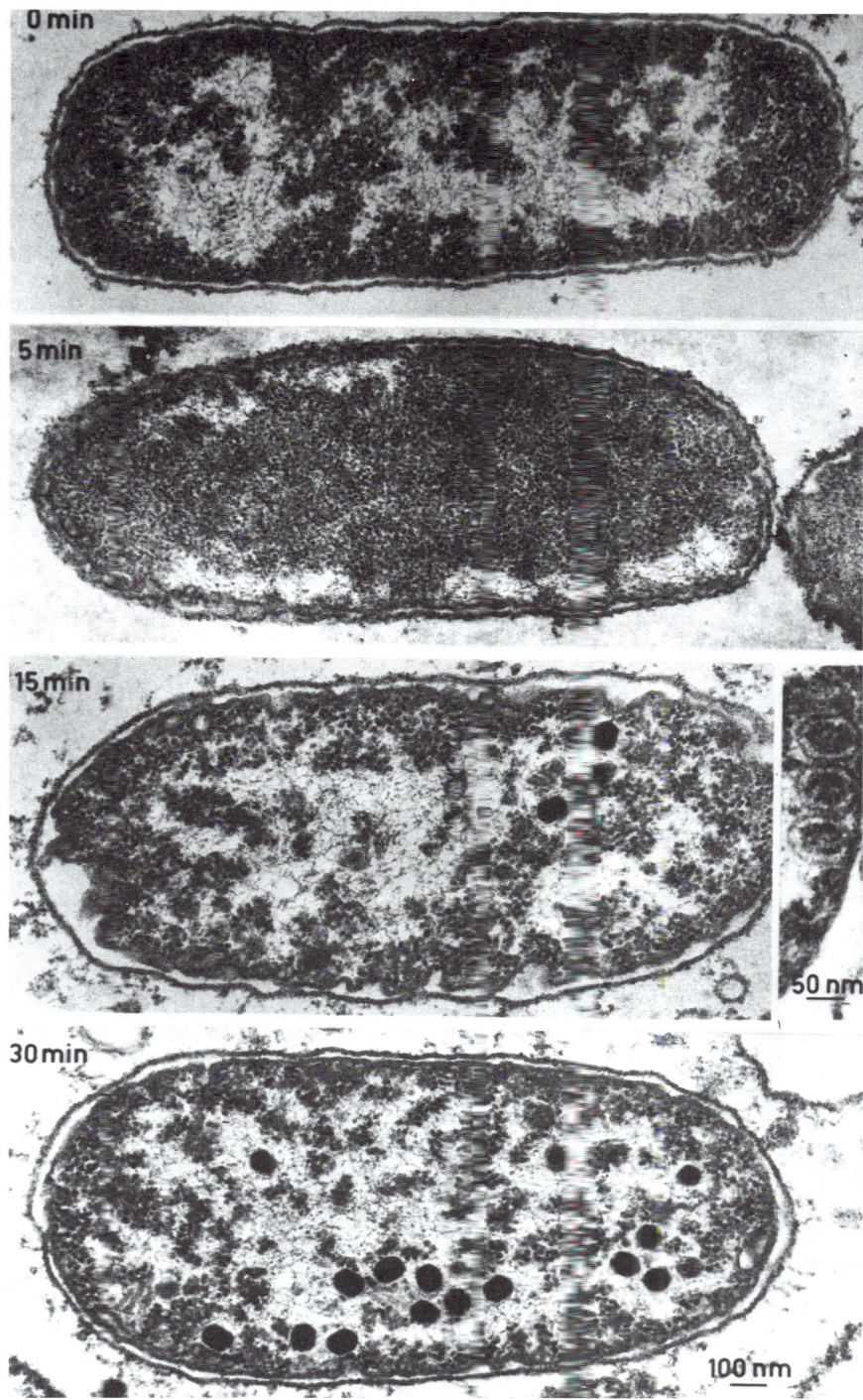

Figure 17-8 Electron microscopic observation of development of bacteriophage T4. The times are indicated in minutes after infection at 30°C. (0 min) Immediately after infection the cell has the morphology of an uninfected cell. The DNA-containing nucleoid is the light region. (5 min) This micrograph shows the nuclear disruption induced by this phage. At this stage the host DNA and its breakdown products are found mainly near the cell wall. The nucleotides of the host DNA will be used for synthesizing phage DNA. (15 min) At about eight minutes after infection the structural proteins used for assembling and maturing the phage start to be made. A few minutes later, the first phage particles appear. From then on, phage appear at a rate of about five particles per minute. It takes about seven minutes to assemble a particle. A prohead (insert) containing little or no DNA is made first. The fine fibrillar material (light region) resembling the bacterial nucleus is the large mass of phage DNA. (30 min) Many finished heads are present. Since this micrograph is of a thin section of the cell the actual number of particles per cell is about twenty times the number seen. [Courtesy of B. Menge, J. v.d. Broek, H. Wunderli, K. Lickfield, M. Wurtz, and E. Kellenberger.]

of T4 DNA is very complex and includes numerous transcription units, many of which have overlapping sequences, and there are several attenuators (similar to those in the *E. coli trp* and *his* operons). These features of the T4 life cycle are beyond the scope of this book. However, the basic pattern, which is true of several phages, is the following:

> Transcription of early mRNA starts at a single class of promoter by means of *E. coli* RNA polymerase (because that is the only polymerase in the cell). Then the polymerase is modified by the activity or addition of phage-specified proteins, with the result that it no longer recognizes host promoters but goes on instead to initiate transcription of later species of mRNA at phage promoters. Successive modifications then provide one means of temporal control of the synthesis of many species of T4 mRNA.

We will see later that other phages ultimately inactivate *E. coli* RNA polymerase and make a phage-encoded polymerase.

Replication of T4 DNA

Five aspects of T4 DNA replication are especially interesting: (1) the source of nucleotides; (2) the synthesis of 5-hydroxymethylcytosine (HMC), which substitutes for cytosine (C); (3) the prevention of incorporation of cytosine; (4) glucosylation of T4 DNA; and (5) the enzymology of replication. The first four are discussed below. The enzymology is quite complex.

1. *Source of T4 DNA nucleotides—degradation of host DNA.* An early event in the T4 life cycle is the degradation of host DNA to deoxynucleoside monophosphates (dNMP). The degradation is initiated by two endonucleases, which are the products of the genes *denA* and *denB*. These two enzymes, which are active *only on cytosine-containing DNA,* cleave the host DNA to double-stranded fragments, which are then degraded to dNMP by a phage-encoded exonuclease. These mononucleotides are then built up to dATP, dTTP, dGTP, and dCTP by the usual *E. coli* enzymes and this provides sufficient dNTP to synthesize 30 T4 DNA molecules. DNA precursors are also formed by *de novo* synthesis but, to ensure an abundant supply of dNTP, five phage-encoded enzymes, which are virtually identical in activity to the *E. coli* enzymes, are also synthesized.

2. *Synthesis of 5-hydroxymethylcytosine (HMC).* T4 DNA contains HMC instead of cytosine. However, *E. coli* does not possess enzymes for forming HMC; therefore, this is accomplished by two phage enzymes, which convert dCMP to dHDP:

$$dCMP \xrightarrow{\text{T4 hydroxymethylase}} dHMP$$

$$dHMP \xrightarrow{\text{HM kinase}} dHDP$$

The *E. coli* enzyme, nucleoside phosphate kinase, which forms all triphosphates in *E. coli,* then converts dHDP to dHTP, the immediate precursor of the HMC in the DNA. These pathways are shown in Figure 17-9.

3. *Prevention of incorporation of cytosine into T4 DNA.* In Chapter 9 we saw that the *E. coli* DNA polymerases cannot readily distinguish dUTP from dTTP because the two molecules have the same hydrogen-bonding properties and that an additional enzymatic mechanism is required to prevent permanent incorporation of uracil into DNA. A similar problem arises with T4 DNA polymerase, for it cannot distinguish dCTP from dHTP, both of which can hydrogen-bond to guanine. It is essential that no cytosine be incorporated into daughter T4 DNA strands, because such cytosine-containing DNA would be a substrate for the T4 endonucleases that degrade host DNA. Also, for unknown reasons, cytosine-containing DNA is not a template for the transcription that occurs late in the life cycle. *E. coli* itself has no use for enzymes that prevent incorporation of cytosine into DNA; thus, the phage must encode these enzymes. For cytosine to become part of daughter DNA molecules, dCMP must be converted to dCDP and then to dCTP. A phage enzyme, called dCTPase, degrades both dCDP and dCTP to dCMP. This process seems somewhat wasteful in that ATP is consumed in forming dCTP; however, the more economical process of preventing formation of dCTP would be difficult to carry out in *E. coli,* since nucleoside phosphate kinase is responsible for the production of all triphosphates. Inhibiting the dCMP → dCDP reaction would also be ineffective because in

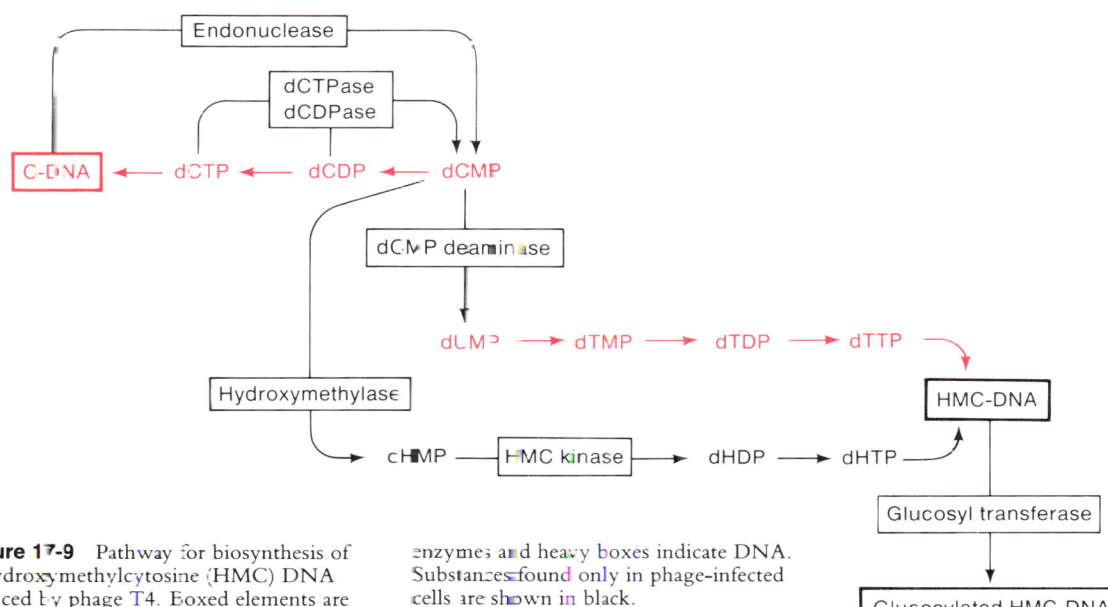

Figure 17-9 Pathway for biosynthesis of 5-hydroxymethylcytosine (HMC) DNA induced by phage T4. Boxed elements are macromolecules; thin boxes indicate enzymes and heavy boxes indicate DNA. Substances found only in phage-infected cells are shown in black.

E. coli most of the deoxynucleoside diphosphates are formed by enzymatic reduction of ribonucleoside diphosphates by a single enzyme (Chapter 9).

Another phage enzyme, dCMP deaminase, converts dCMP to dUMP (in preparation for synthesis of dTMP) (Figure 17-9). This enzyme duplicates the activity of a similar *E. coli* enzyme (and hence is a product of one of the nonessential genes) but has an interesting economic function. The base compositions of *E. coli* DNA and T4 DNA are 50 percent (A + T) and 66 percent (A + T), respectively. In *E. coli,* the ratio of dTTP and dCTP is about 1:1 in keeping with the T:C ratio in DNA. The bacterial and phage dCMP deaminases, acting together, increase the amount of dTMP with respect to dCMP, so the ratio of dTTP to dHTP is 2:1, as is the T:HMC ratio in T4 DNA.

Occasionally some cytosine appears in progeny phage DNA. Presumably, the T4 endonucleases degrade this DNA shortly after synthesis; since the cytosine is in only one strand of each daughter double helix, the DNA can be repaired by normal repair synthesis. Also, as is true of *E. coli* DNA synthesis, occasionally uracil is inserted into progeny DNA (by incorporation of dUTP). The uracil residues are removed by the usual uracil-*N*-glycosylase (see Chapter 9).

4. *Glucosylation of T4 DNA.* The presence of HMC in T4 DNA creates another problem for the phage because *E. coli* possesses an endonuclease that attacks certain sequences of nucleotides containing HMC. To avoid this damage the HMC residues in T4 DNA are glucosylated. This is accomplished by two phage enzymes, α-glycosyl transferase (αgt) and β-glucosyl transferase (βgt), each of which transfers a glucose from uridine diphosphoglucose (UDPG) to HMC that is already in DNA. Thus glucosylation is a postreplicative modification. The *E. coli* endonuclease is inactive against glucosylated DNA; therefore, glucosylation is a protective device. A simple genetic experiment shows that this is the only essential function of glucosylation. A T4 α*gt*⁻ mutant cannot carry out glucosylation, so its newly synthesized DNA is destroyed by the *E. coli* HMC endonuclease. However, if an *E. coli* mutant (*rglB*⁻), which lacks this nuclease, is used as a host bacterium, both T4 α*gt*⁻ and T4 β*gt*⁻ mutants grow normally even though nonglucosylated DNA is produced.

Production of T4 Phage Particles

Production of complete phage particles can be separated into two parts—assembly of heads, tails, and other structures, and packaging of DNA of a sufficient length to provide a little more than one set of genes in the phage head.

Assembly of T4 (and many other phages) has been studied by two techniques, both of which require a large collection of phage mutants unable to produce finished particles. In one method, different cultures of cells, each infected with a particular mutant, are lysed and examined by electron microscopy. This procedure shows that heads are made in the absence of tail synthesis and that tails are made by a mutant unable to synthesize heads; head and tail assembly are independent processes. The second technique is a complementation assay of a type widely used to purify proteins. If two extracts of infected cells, one lacking heads and the other lacking tails, are mixed, phage particles form *in vitro* (Figure 17-10). The "headless" extract can also be fractionated and a component can be isolated that allows a "tailless" extract to make tails. In this way, a protein in the tail assembly pathway can be isolated and identified.

These studies have shown that there are two types of components—**structural proteins** and **morphogenetic enzymes**. Some of the structural components assemble spontaneously to form phage structures, whereas others need the help of enzymes. A few host-encoded factors are also needed for head assembly. For T4 and for phage λ the assembly pathway has been almost completely worked out and can be found in a reference given at the end of the book.

The mechanism of packaging of T4 DNA in a phage head is not yet elucidated, though the process can be carried out *in vitro*. The basic problem is how the long strand of DNA is tightly folded so that it will fit in the phage head. It is thought that packaging begins by the attachment of one end of a DNA molecule to a protein contained in the phage head. Then either a condensation protein or a small, very basic molecule (such as a polyamine) induces folding (Figure 17-11). The best-understood feature of the process is that the molecule is cut from long concatemers. The cuts are not made in unique base sequences in the DNA because if they were, T4 DNA could not be cyclically permuted. Instead, the cuts are made at positions that are determined by the amount of DNA that can fit in a head. Presumably a free end of the DNA molecule enters the head and this continues until there is no more room; then the concatemer is cut. This is known as the **headful mechanism** and it explains how both terminal redundancy and cyclic permutation arise (Figure 17-12). The essential point is that the DNA content of a T4 particle is greater than the length of DNA required to encode the T4 proteins. Thus, when cutting a headful from a concatemeric molecule, the final segment of DNA that is packaged is a duplicate of the DNA that is packaged first—that is, the packaged DNA is terminally redundant. The first segment of the second DNA molecule that is packaged is not the same as the first segment of the first phage. Furthermore, since the second phage must also be terminally redundant, a third phage-DNA molecule must begin with still another segment. Thus, the collection of DNA molecules in the phage produced by a single infected bacterium is a cyclically permuted set. An important bit of evidence that indeed prove the headful mechanism was that if a mutant phage contains a deletion for nonessential genes, the terminally redundant region increases in length by the amount of DNA that is deleted.

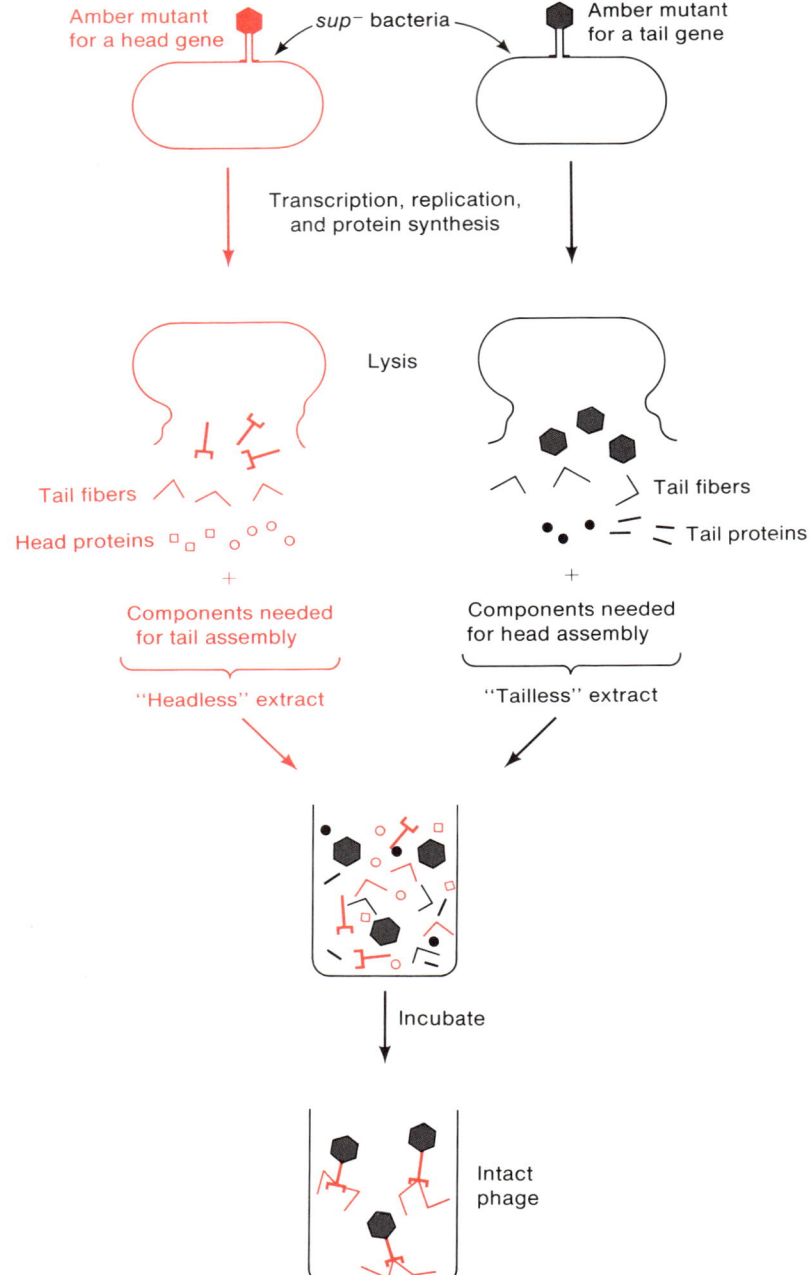

Figure 17-10 Production of intact T4 phage by *in vitro* complementation.

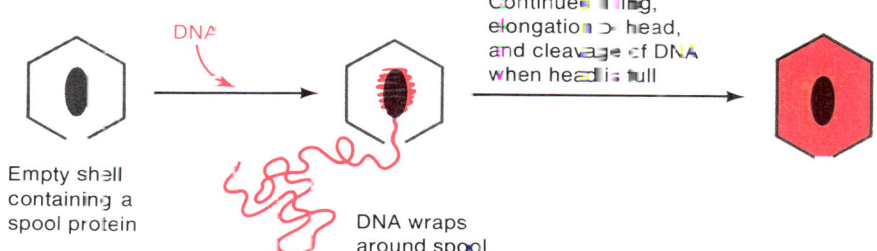

Empty shell
containing a
spool protein

DNA wraps
around spool

Continued filling,
elongation of head,
and cleavage of DNA
when head is full

Figure 17-11 Proposed model for filling a
T4 head. Cleavage and rearrangement of
head proteins is known to occur at several
stages of the process.

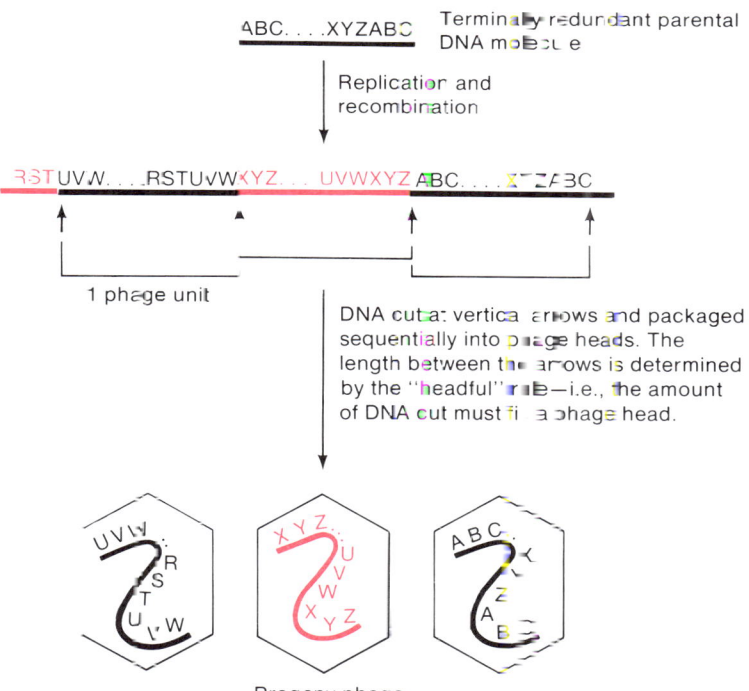

ABC....XYZABC

Terminally redundant parental
DNA molecule

Replication and
recombination

RSTUVW....RSTUVWXYZ...UVWXYZABC....XYZABC

1 phage unit

DNA cut at vertical arrows and packaged
sequentially into phage heads. The
length between the arrows is determined
by the "headful" rule—i.e., the amount
of DNA cut must fill a phage head.

Progeny phage

Figure 17-12 Origin of cyclically permuted T4 DNA molecules. Alternate units
are shown in different colors for clarity only.

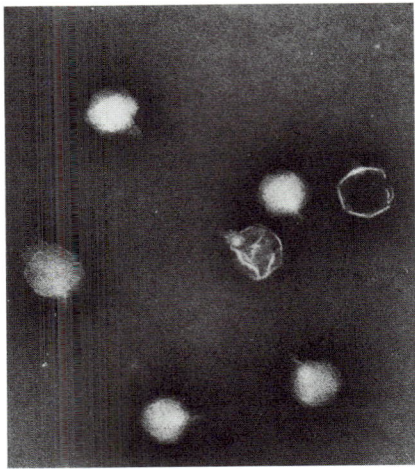

E. coli PHAGE T7

Phage T7 is a midsized phage (Table 17-2). Its DNA has a molecular weight of about 26×10^6 and contains equal amounts of adenine, thymine, guanine, and cytosine and has no unusual bases. It has a terminal redundancy of 160 base pairs but it is not cyclically permuted. The DNA molecule is encased in a head that is attached to a very short tail (Figure 17-13).

The complete sequence of 39,936 base pairs of T7 DNA is known; this information allows one to locate, by direct inspection, all of the promoters, termination sites, regulatory sites, spacers, leaders, initiation codons, termination codons, and genes, and to deduce the amino acid sequence of each protein. There are 50 closely packed genes and 5 potentially overlapping genes, of which 2 are known to be expressed.

Organization of the T7 Genes

An incomplete map of the T7 genes is shown in Figure 17-14 (for clarity some genes have been omitted). Of the 55 known genes, 34 gene products have been identified and the functions of 27 genes are known. As will be seen in a later section, the gene products are synthesized in numerical order starting with gene *0.3*.

Examination of the base sequence indicates economical utilization of DNA. For example, leaders and spacers between genes are generally only a

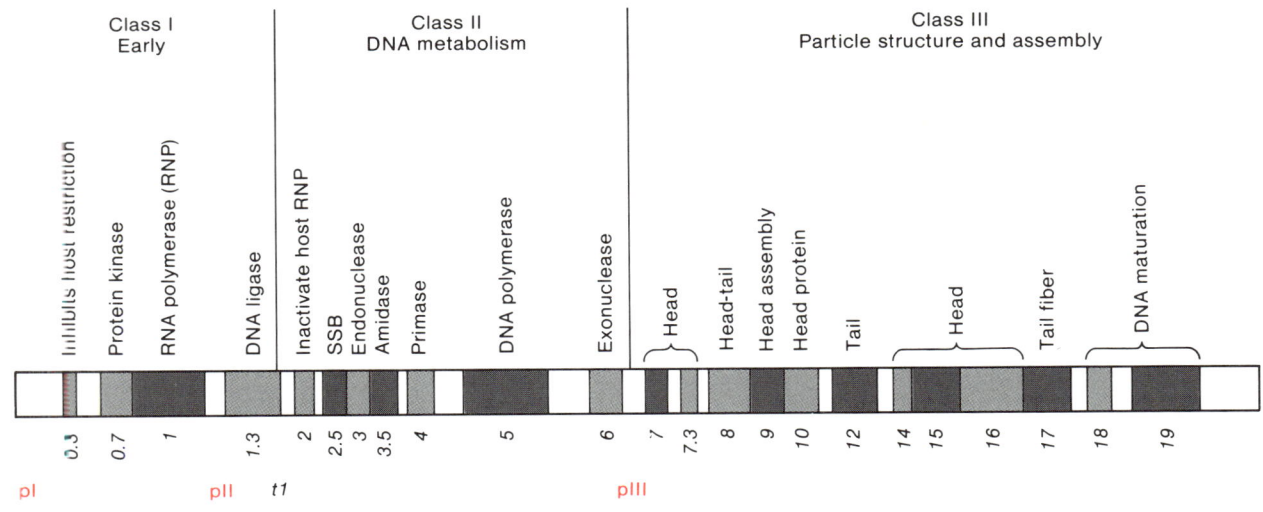

Figure 17-14   Map of phage T7. Genes are numbered, but not all genes are shown. The sizes of the blocks are approximately proportional to protein size. Cross-hatching is used only to distinguish adjacent genes and has no other significance. The promoters, *pI, pII,* and *pIII,* and the terminator *tI* are in red.

few nucleotides long, when present at all whereas in bacterial genetic systems hundreds of nucleotides are used for these. For roughly half of the genes the Shine-Dalgarno sequence used for binding mRNA to a ribosome (Chapter 13) is in the coding sequence of the preceding gene. Furthermore, the termination codon of one cistron often overlap the initiation codon of the next cistron—for example, the sequence UGAUG terminates gene *1.7* and initiates gene *1.8*. There are also several cases in which coding sequences of adjacent genes overlap; the maximum overlap is 36 bases so this overlap is much smaller than that present in phage φX174 (explained later in this chapter) and many animal viruses (Chapter 23). Finally, the catalytic proteins made early in the life cycle tend to be much smaller than the average bacterial protein, ranging from 29 to 883 amino acids.

A notable feature of the organization of the T7 genes is that they are clustered according to function and are arranged to provide a continuous order according to the time of their function. The first gene (Figure 17-14) is essential for survival of the phage DNA after infection; *E. coli* strain B, a common host for T7, contains a restriction endonuclease that cuts T7 DNA, and the product of the first gene inhibits the activity of this nuclease by directly binding to it. The next few genes control transcription; these are followed by genes needed for DNA replication and then by genes encoding the structural proteins of the phage particle. The final genes are used for packaging of newly synthesized DNA in the phage head.

The T7 Growth Cycle

The life cycle of T7 is outlined below (time in minutes at 30°C):

$t = 0$ Adsorption of the phage.

$t = 0-1$ Initiation of slow injection of phage DNA

$t = 2$ Initiation of synthesis of phage mRNA.

$t = 4$ Turning off of host transcription begins.

$t = 8-9$ Initiation of phage DNA synthesis

$t = 8-10$ Initiation of synthesis of structural proteins completion of injection of phage DNA.

$t = 15$ First phage appears.

$t = 25$ Lysis and release of progeny phage.

Most of these stages will be discussed in the following sections.

Injection of T7 DNA into a Host Cell

With T4 and for many other phages, the phage DNA is injected into a host bacterium very soon after adsorption, and injection is complete in less than one minute. In contrast, with T7 it takes about ten minutes for injection of all of the DNA. Why T7 injects slowly is unknown (but no more so than the mechanism of injection of any other phage).

Slow injection is a significant feature of the T7 life cycle because a gene cannot be transcribed until it has been injected. In fact, the timing of transcription, which we discuss in the next section, depends in part on the kinetics of injection.

Regulation of Transcription of T7 DNA

The life cycle of T7 is normally divided into two transcriptional stages called early and late. The early stage uses *E. coli* RNA polymerase to transcribe genes *0.3* through *1.3;* the late stage uses a newly synthesized T7 RNA polymerase. Transcription of T7 DNA is temporally regulated. There are three major classes of transcripts, labeled I, II, and III in Figure 17-15. Each of these is transcribed from the same DNA strand, which is termed the *r* strand (because transcription occurs in the *r*ightward direction when the DNA molecule is drawn in a standard orientation). The class I transcripts are synthesized by *E. coli* RNA polymerase and comprise the early mRNA. There are three distinct promoters for the class I transcripts; they are very near one another, so each mRNA includes the same genes. It is convenient when discussing regulation to consider each of these mRNA molecules as a single molecule called transcript I and to refer to a single promoter termed *pI*. The existence of three promoters probably serves only to increase the net rate of mRNA synthesis threefold.

Synthesis of transcript I begins 2 minutes after infection, the delay probably resulting from the slow injection of the DNA. Transcription is mediated by the bacterial RNA polymerase and stops at a termination site just after gene *1.3*. Transcript I is cleaved by *E. coli* RNase III into five mRNA molecules. However, in laboratory conditions the processing is not essential;

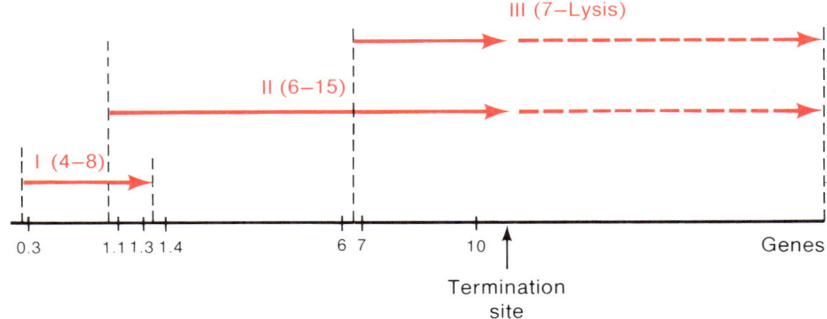

Figure 17-15 A simplified transcription map of T7. The DNA and gene numbers are in black; the three classes of transcripts are in red. The red numbers indicate the time interval (in minutes at 37°C) during which the transcript is made, the termination site indicated by the black arrow is about 90 percent efficient; thus, 10 percent of the class II and III molecules are extended further, as indicated by the dashed lines.

that is, if an *E. coli* mutant lacking RNase III is used as host, the transcript remains intact and phage production seems to be normal. Possibly in nature or in other bacterial hosts processing is required.

Several proteins are translated from transcript I. One is a protein kinase (gene *0.7*) that phosphorylates *E. coli* RNA polymerase, inactivating the enzyme. The inactivation reduces transcription of *E. coli* DNA and is the first step in takeover of the bacterium by the phage. This kinase is not necessary in infection by T7 phage of most laboratory strains of *E. coli;* as we will see shortly, another inhibitor of *E. coli* RNA polymerase is made later. A second protein translated from transcript I is T7 RNA polymerase (gene *1*), which is essential because this enzyme is responsible for all further transcription.

Transcripts II and III also represent classes of mRNA that are initiated at several promoters. These promoters are not recognized by *E. coli* RNA polymerase but require the translation of T7 RNA polymerase from transcript I. However, transcription from all of these promoters does not begin as soon as T7 RNA polymerase is made. Instead, class II transcripts, whose promoters are much weaker than class III ones, are made first. This delay and the difference in the time of initiation of the two classes both have the same explanation—namely, that T7 DNA is injected slowly and that the promoters for class II enter the cell several minutes before entry of the later promoter.

Transcripts II are initiated upstream of the termination site of class I; therefore, gene *1.3* is contained in both transcripts I and II. The reason for this is unclear. The T7 RNA polymerase ignores the site that causes termination of transcript I and each transcript runs rightward to a second termination site (Figure 17-15). This termination site is 90 percent efficient, meaning that 10 percent of the transcripts proceed on to the end of the DNA molecule. Since the termination site is just after gene *10*, which encodes the major head protein, it is believed that its location is a means of increasing the concentration of this gene product with respect to the tail proteins and minor head proteins, whose genes follow the termination site and which are needed in lesser amounts.

A class II transcript gene (gene *2*) makes a second inhibitor of *E. coli* RNA polymerase. Combined with the earlier reduction of activity of the polymerase by the gene-*0.7* protein kinase, the *E. coli* enzyme is completely inhibited. Thus, by eight minutes after infection there is no longer any synthesis of *E. coli* RNA and takeover of the bacterium is complete. Transcript I, which is made by the *E. coli* RNA polymerase, is also no longer made; this is economical because at this point sufficient T7 RNA polymerase has been produced to complete the life cycle and no other products of this transcript are needed.

Transcript II contains genes for DNA replication and also for synthesis of structural proteins. In fact, the structural genes are transcribed very shortly after the replication genes. At first glance this seems inefficient because premature assembly of phage particles and packaging of phage DNA might result in termination of replication or at least a vast reduction of the number of templates for replication. This arrangement contrasts with that found in T4, in which synthesis of the structural proteins is delayed with respect to the replication enzymes. However, the T7 DNA molecule is small and can be replicated rapidly (in nearly 20 seconds), while the maturation proteins (genes *18* and *19*) are translated late and head assembly is slow; there is in fact no packaging of DNA before replication forms many DNA molecules.

Once the DNA replication enzymes have been synthesized, there should be no further need for transcription of this region since the proteins are, for the most part, catalytic. However, termination of synthesis of transcript II would also stop transcription of structural genes, whose products are needed in stoichiometric amounts. Efficiency as well as continued synthesis of structural proteins are both attained with two late mRNA molecules (II and III) and by turning off the synthesis of transcript II. Admittedly, the same result could be achieved if transcript II terminated before encountering the promoter for transcript III, but evolution has taken a different course.

Synthesis of transcript II stops 15 minutes after infection. This is a result of a general shutting down of transcription and the relative strengths of the promoters *pII* and *pIII* for transcripts II and III, respectively. That is, overall transcription is reduced and only the stronger promoter *pIII* remains active. The product of the gene *3.5* (amidase) binds to T7 RNA polymerase and is responsible for the overall inhibition of transcription. This protein is a lysozymelike enzyme and is also responsible for cell lysis.

DNA Replication and Maturation of T7 Phage

The mode of replication of T7 DNA is described in detail in Chapter 9 (see Figures 9-40 through 9-42). It is recommended that the reader review that section. An essential point is that the original parent molecule generates a long, linear concatemer from which unit molecules are cut by the following mechanism.

Packaging of T7 DNA proceeds by a mechanism quite different from that in T4 DNA. For example, (1) the DNA is terminally redundant but not cyclically permuted, and (2) a mutant phage containing a genetic deletion contains less DNA than a wild-type phage. The mechanism of maturation of T7 is not known but a proposal that is likely to be valid is shown in Figure 17-16. The essential point is that by generating a nick with 3′-OH and 5′-P termini, the concatemeric DNA is made susceptible to the action of enzymes that can form the blunt-ended terminally redundant DNA molecule found in T7 phage particles.

A great deal is known about the formation of T7 heads and tails and *in vitro* assembly. This information can be found in the references listed at the end of the book.

I. Nicking

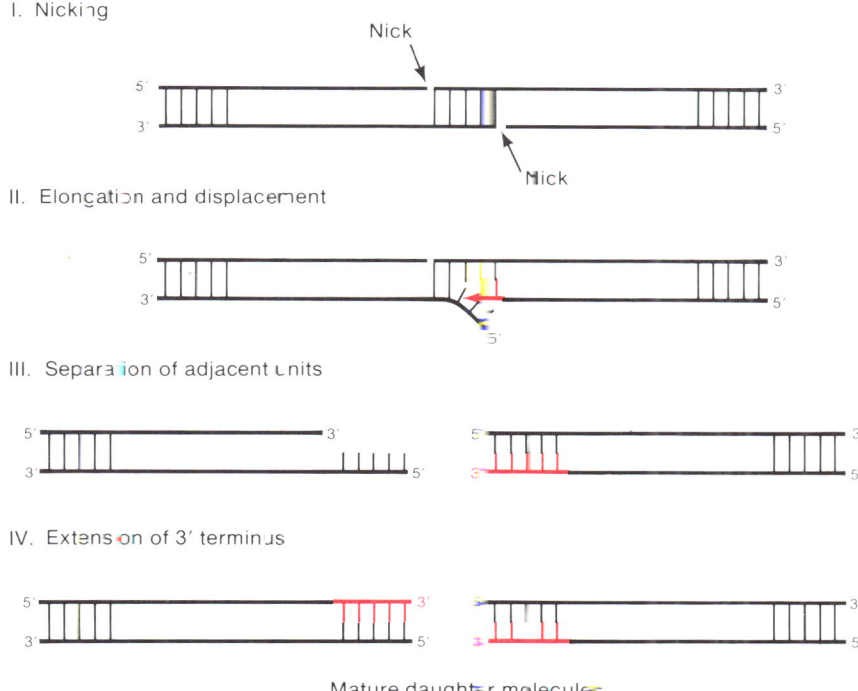

II. Elongation and displacement

III. Separation of adjacent units

IV. Extension of 3′ terminus

Mature daughter molecules

Figure 17-16 Proposed scheme for production of mature T7 DNA from a concatemer. Three redundant regions are designated by showing the base pairs. The steps are the following: I. Nicks are made at each end of a redundant region, generating two 3′-OH termini. II. A DNA polymerase adds nucleotides to one of the 3′ termini produced by nicking, displacing the 5′-terminated parental strand. III. When the displacement reaction reaches the second nick, the two double-stranded fragments separate to yield one completed daughter molecule. IV. Extension of the 3′ end of the incomplete daughter molecule by DNA polymerase forms a second completed daughter molecule.

E. coli PHAGE λ

E. coli phage λ (Figure 17-17) has two alternate life cycles—the lytic and the lysogenic. In this chapter we consider only the former; the lysogenic cycle is described in detail in Chapter 18.

Phage λ DNA and Its Conversion Following Injection

λ is a mic-sized phage (Table 17-2). It contains a linear double-stranded DNA molecule whose molecular weight is 31×10^6. The complete base sequence (48,514 bp) is known. The DNA molecule has an unusual structure—at each end of the molecule the 5′ terminus extends 12 bases beyond the 3′-terminal nucleotide. The base sequences of these single-stranded terminal regions, which are known as **cohesive ends,** are complementary to one another (Figure 17-18). Thus, by forming base pairs between the single-stranded ends, the linear DNA molecule can circularize, yielding an open circle containing two single-strand breaks. No bases are missing in the newly formed

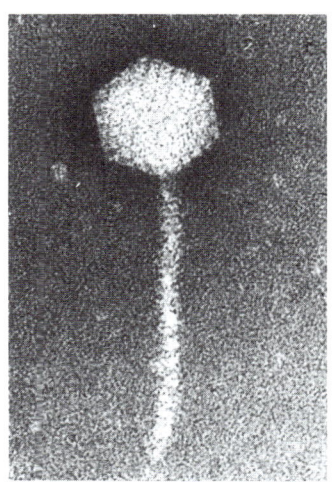

Figure 17-17 Electron micrograph of phage λ. [Courtesy of Robley Williams.]

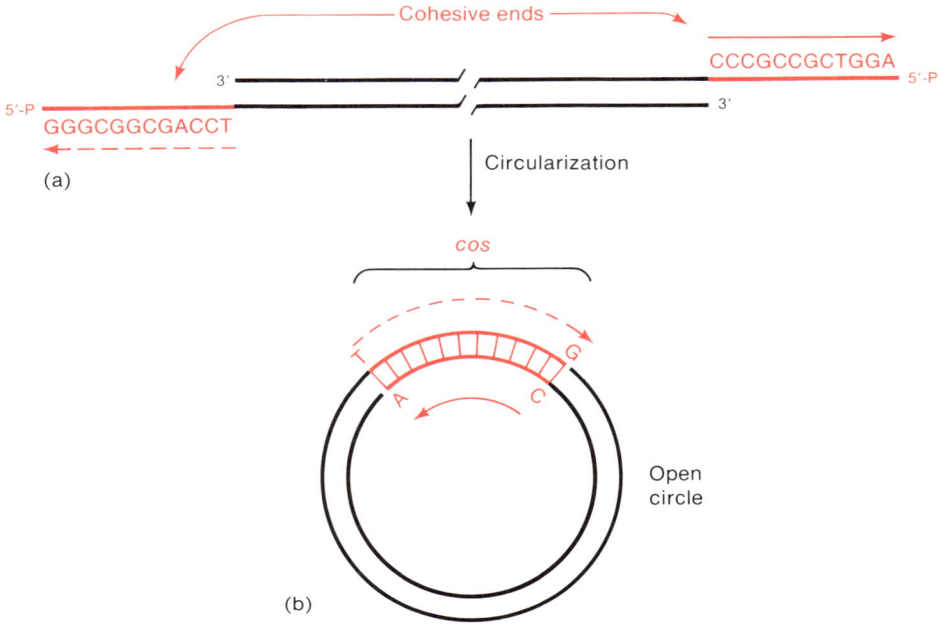

Figure 17-18   (a) A diagram of a λ DNA molecule showing the complementary single-stranded ends (cohesive ends). Note that 10 of the 12 bases are G or C. (b) Circularization by means of base-pairing between the cohesive ends. The double-stranded region that is formed is designated *cos*.

double-stranded region of this circle, so DNA ligase can convert the molecule to a covalent circle. This circularization is easily performed in the laboratory and also occurs in an infected cell within a few minutes following infection.

Organization of λ Genes and Regulatory Sites

Forty-six λ genes have been identified; of these, 14 are nonessential to the lytic cycle but only 7 are nonessential to both the lytic and lysogenic cycles. Most λ proteins have been either identified by gel electrophoresis or purified. All regulatory sites, promoters, and termination sites are known.

The genetic map of λ is shown in Figure 17-19. For reasons that will become clear when lysogeny is discussed (Chapter 18), the map is usually drawn as a circle. As with phage T7, genes of phage λ are clustered according to function. For example, the head, tail, replication, and recombination genes form four distinct clusters. Many λ proteins—for example, regulatory proteins and those responsible for DNA synthesis—act at particular sites in the DNA. In general these proteins are situated adjacent to their sites of action

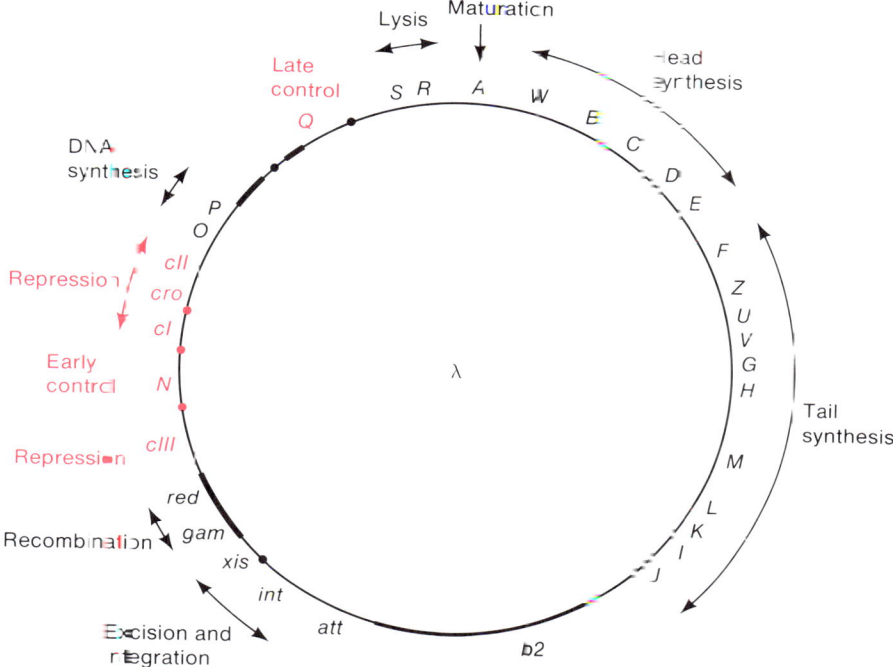

Figure 17-19 Genetic map of phage λ. Regulatory genes and functions are given in red. All genes are not shown. Major regulatory sites are indicated by black solid circles. Regions nonessential for both the lytic and lysogenic cycles are denoted by a heavy line.

(when there is a single site)—for instance, the origin of DNA replication lies within the coding sequence for gene *O*, which encodes a DNA replication-initiation protein.

The Transcription Sequence of λ

With phages T4 and T7 timing of synthesis of the various mRNA molecules is accomplished primarily by mechanisms that determine the availability of promoters—namely, the synthesis of a new RNA polymerase, the modification of the host polymerase, or the time of entry of the promoter in the host cell. In λ the host RNA polymerase is also modified but not for the purpose of recognition of phage promoters. Instead, *the modification enables the polymerase to ignore certain termination sites*. Figure 17-20 shows a genetic map of λ, which includes the three regulatory genes *cro*, *N*, and *Q*; three promoters *pL*, *pR*, and *pR2*; and five termination sites *tL1*, *tR1*, *tR2*, *tR3*, and *tR4*. Seven mRNA molecules are also shown; the L and R series are transcribed leftward and rightward, respectively, from complementary DNA

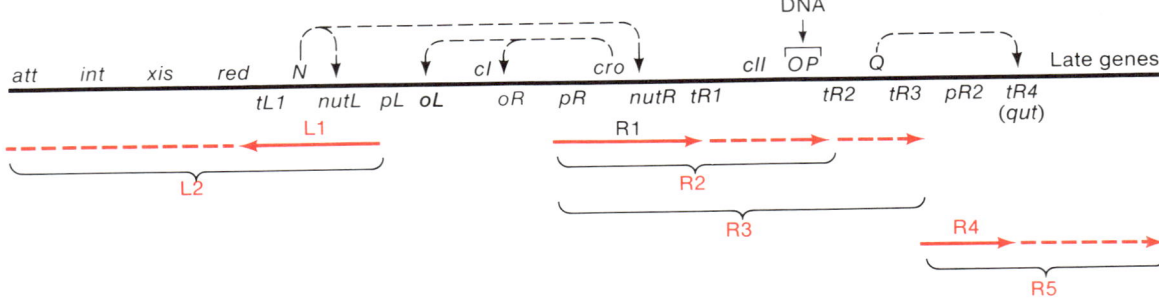

Figure 17-20   A genetic map of the regulatory genes of phage λ. Genes are listed above the line; sites are below the line. The mRNA molecules are red. The dashed black arrows indicate the sites of action of the N, Cro, and Q proteins.

strands. It is recommended that the reader refer to Figure 17–20 continually during the following discussion.

An essential feature of the life cycles of phages T4 and T7 is rapid killing of the host and degradation of the host DNA. Phage λ differs in this respect because in the lysogenic cycle the host must survive. Instead λ multiplies while the host cell continues its normal function. Since inactivation of host functions is not required, the transcription pattern that might be expected is the following. First, the *O* and *P* genes, whose products are necessary for DNA synthesis, would be transcribed. This would be followed by transcription of the genes encoding the structural proteins, and finally the packaging system and the lytic proteins would be made. This is basically what occurs, with the modification that other small transcripts are formed that encode the regulatory proteins responsible for turning transcription on and off at the appropriate times.

The sequence of transcriptional events is described in the list that follows. The genes, sites, and mRNA molecules referred to are depicted in Figure 17–20 and listed in Table 17–3. Two events having the same number, for example, 1A and 1B, begin at the same or nearly the same time.

1A. *Synthesis of transcript L1*. This begins at the promoter *pL* and terminates at the site *tL1*. This transcript encodes only the gene-*N* protein, which is a positive regulatory element needed to allow certain regions of the DNA to be transcribed.

1B. *Synthesis of transcript R1*. This begins at the promoter *pR* and terminates at the site *tR1*. This transcript encodes the gene-*cro* protein, which is a negative regulatory element serving ultimately to turn off transcription from the promoters *pL* and *pR*.

2. *Synthesis of transcript R2*. The terminator *tR1* is a weak one and about half of the R1 mRNA molecules are extended to *tR2*, a strong terminator. The Cro protein, whose function is discussed

Table 17-3   Some λ gene products

Gene	Function
cro	Inhibits transcription from *pL* and *pR*
cII	Delays late transcription
N	Antiterminates at *tL1, tR1,* and *tR2*
O, P	DNA replication
Q	Antiterminates at *tR4*

in item 4, is translated from both of these mRNA molecules. The portion of R2 not present in R1 contains the genes O and P, whose products are needed for DNA replication; the O and P products are made at this time from R2 but their concentrations are sufficient for only one round of replication.

3A. *Antitermination at* tL1 *and synthesis of transcript L2.* Once synthesized, the N protein plus a host protein (called NusA) modify the *E. coli* RNA polymerase; it is then able to ignore the terminator *tL1*. In this way the transcript L1 is no longer made but is extended, forming L2. (The L2 mRNA is translated to yield several products that are not essential but make phage production more efficient.) The mechanism of N-mediated antitermination remains a subject of active study. Sites in the λ DNA called *nut*, located downstream from *pL* and *pR*, are required. The N protein probably binds at these sites and is picked up by an RNA polymerase molecule while moving along the DNA molecule.

3B. *Antitermination at* tR1 *and* tR2 *and synthesis of transcript R3 and production of DNA replication proteins.* Once the N protein has been synthesized, the modified RNA polymerase is able to ignore *tR1*. R1 is no longer made; *tR2* is also bypassed, thereby extending R2 to form R3. This molecule terminates at *tR3*, which is a termination site for RNA polymerase even when it is modified. That is, *tR3* is not sensitive to the antitermination effect of N. The synthesis of R3 provides enough mRNA that the concentrations of the O and P proteins reach values sufficient for DNA replication. The gene-*cII* protein is also made from R2 and R3; it delays late mRNA synthesis.

4. *Turning off of synthesis of transcript L2.* While the preceding events have been occurring, the concentration of the Cro protein has been increasing. This protein is a repressor that binds to the left operator *oL* and blocks initiation of transcription at *pL*. Thus, L1 and L2, which at this point are not needed because sufficient N protein has been made, are no longer synthesized.

5A. *Turning off of synthesis of transcript R3.* Somewhat later than the time when the Cro protein inhibits transcription from *pL*, the concentration of the Cro protein increases to the point that it acts at the operator *oR* to block synthesis of transcript R3 from *pR*.

5B. *Translation of the gene-Q protein from transcript R3 and synthesis of late mRNA.* Since the time of infection, a tiny transcript R4 has been synthesized continually from *pR2*, terminating at *tR4*. This transcript does not encode any known genes but is a leader for the late mRNA. Once R3 has been made (item 3), the product of gene Q is made. This protein is responsible for turning on synthesis of the late mRNA molecule that encodes the structural and assembly proteins, the maturation system, and the lysis enzymes. The Q protein is also an antiterminator that enables *E. coli* RNA polymerase to ignore *tR4*. Thus R4 is then extended to form transcript R5, the late mRNA.

Let us now review the essential features of this highly efficient regulatory system.

1. A λ-specific RNA polymerase is not made; the *E. coli* RNA polymerase is used throughout the life cycle (as is true of T4) and is modified by accessory proteins to alter its specificity toward various DNA base sequences. However, at no time is its activity with respect to promoters modified; instead, its ability to terminate transcription at certain termination sites is impaired.

2. Inhibition of transcription occurs, but only as a result of the repressor activity of Cro on the promoters *pL* and *pR*. Wasted synthesis is avoided by this repressing activity.

3. All structural components are encoded in a single giant mRNA molecule, which is translated sequentially. Thus, synthesis of the complete set of components takes many minutes, thereby delaying synthesis of intact heads and of a functional phage maturation system until the DNA replication system has provided many copies of λ DNA. The result of these delays is that about 30 copies of λ DNA form before the maturation system is established and about 100 completed phage particles form before the onset of lysis.

DNA Replication and Maturation: Coupled Processes

Following synthesis of the *O* and *P* products, replication of circular λ DNA begins. There are two modes of λ DNA replication—θ and rolling circle replication (Figure 17-21; also Chapter 9). The θ mode increases the number of templates for transcription and further replication; the rolling circle mode provides the DNA for phage progeny. As the life cycle proceeds, θ replication stops and the rolling circle becomes the predominant form.

Figure 17-21 Three species of λ DNA present at the time maturation begins. The region containing the joined complementary single-stranded termini (of the linear DNA molecule present in the phage head) is called *cos* (for cohesive site).

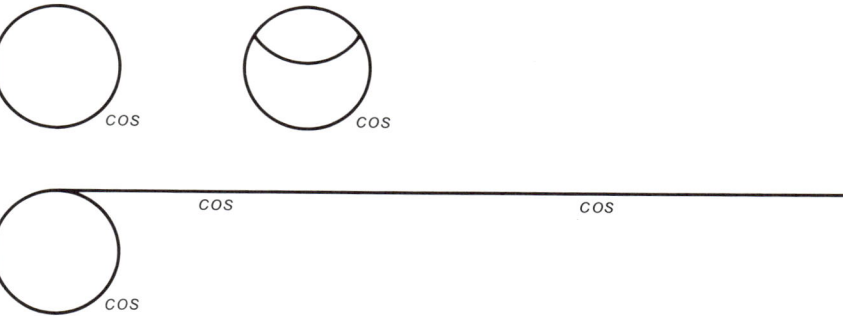

Little is known about the control of initiation of rolling circle replication. However, for our purposes the most important feature is the relation between replication and maturation of the DNA, since the DNA-cutting mechanism of λ differs from those used by T4 and T7.

The DNA found in a λ phage particle is linear and has single-stranded termini. These ends are joined when a circle forms, and the double-stranded region so formed is called a *cos* site (for *cohesive site*). Thus, every monomeric λ circle contains one *cos* site; however, a multimeric branch of a rolling circle contains many *cos* sites.

Since the ends of the DNA molecule in the phage particle are always the single-stranded termini, there must be a mechanism for cleaving a *cos* site to generate these termini. This is accomplished by a sequence-specific cutting system called the **terminase** or **Ter system**, and the DNA-cutting is often called Ter-cutting.

Figure 17-21 shows the three major species of intracellular λ DNA—circles, θ molecules, and rolling circles. By the time heads and tails have been synthesized and the Ter system is active, rolling circles predominate. Note that since a rolling circle has two classes of *cos* sites—the one in the circle and those in the linear branch, some mechanism must exist for preventing cleavage of the one in the circle because if it were broken, replication would cease. This difficulty is avoided by a site requirement in the Ter system: *efficient cleavage of a single* cos *site does not occur if there are two* cos *sites and both are present on a single segment of DNA, a tie can occur.* A λ DNA molecule can be cut from a linear branch by cleavage of two neighboring *cos* sites, but a pair of cuts in which one *cos* site is in the branch and the other is in the circle cannot be made. However Ter cutting does not require that the DNA molecule be linear, inasmuch as a single λ unit can be cut from a dimeric circle that has been formed by genetic recombination. A single cut in a monomeric circle does not occur efficiently. The "two-*cos*-sites" rule explains how the first λ DNA unit is cut from a concatemeric branch of a rolling circle. However, this rule would not allow excision of the second (adjacent) λ unit, because this unit would be flanked by only one *cos* site and a free cohesive end. Hence, only half of the DNA would be usable because only alternate segments of DNA would be packageable (Figure 17-22). The solution to this lack of economy is that a free cohesive end and an adjacent *cos* site are also sufficient for DNA cutting to occur, and allow sequential packaging. Thus, the Ter-cutting rule may be restated as follows: *Ter-cutting requires two* cos *sites or one* cos *site and a free cohesive end on a single DNA molecule.*

Cutting at the *cos* sites and packaging of λ DNA are somehow coupled. In fact, the Ter system is virtually inactive unless the Ter proteins are components of an empty λ head. Thus, when a bacterium is infected with a λ mutant unable to cause formation of an intact phage head (for example, an E^- mutant, which fails to make the major head protein), the linear branch of a rolling circle of DNA is not cleaved. With such a mutant, dimers, which contain two *cos* sites, are also not cleaved.

Figure 17-22 Two rules of packaging. In one mode (black) each λ unit is packaged. In the more limited mode (red) alternate units are packaged. The more economical black mode is used by λ.

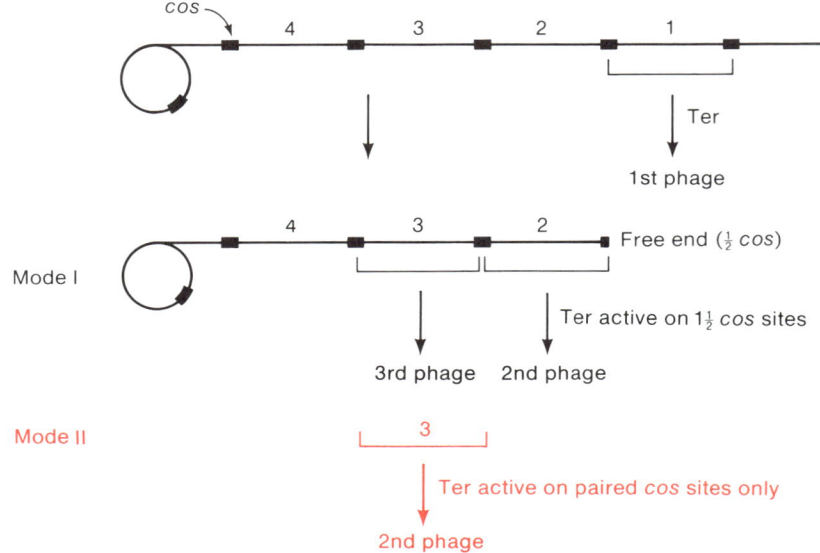

E. coli PHAGE φX174

E. coli phage φX174 contains one circular, single-stranded DNA molecule (Figure 17-23) consisting of 5386 nucleotides, whose base sequence is known. The DNA is enclosed in an icosahedral head composed of three coat proteins (Figure 17-24) and one internal protein.

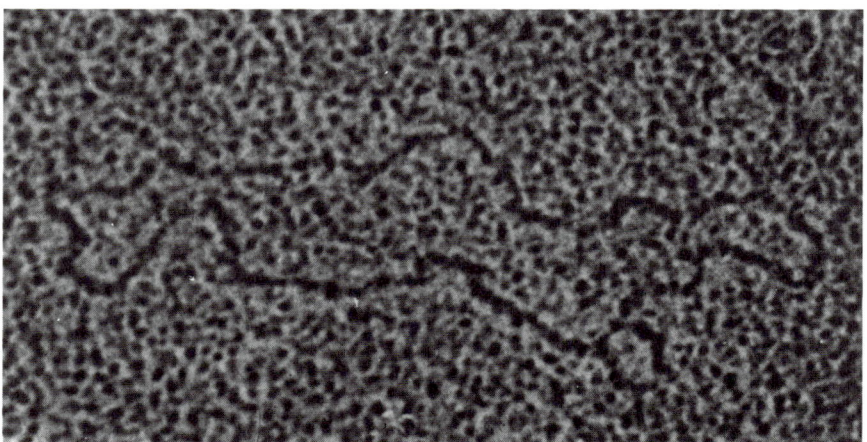

Figure 17-23 A circular single-stranded φX174 DNA molecule. This photo shows the first single-stranded DNA molecule ever seen by electron microscopy. The circumference of the circle is 1.7 μm. [Courtesy of Albrecht Kleinschmidt.]

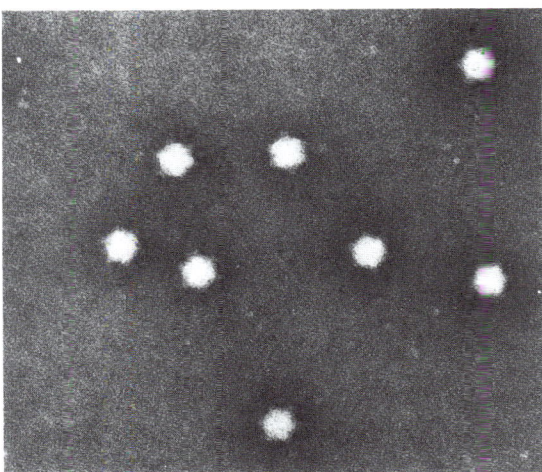

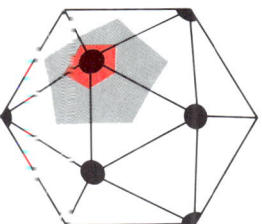

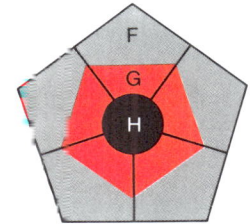

Figure 17-24   Schematic representation of the φX174 subunits in the phage particle (left) and an enlargement of a spike region (right), suggesting orientations of H, G, and the major coat F) proteins.

The phage φX174 has only eleven genes (Table 17-4) and each gene product has been isolated. The number of amino acids contained in these proteins exceeds the coding capacity of this very small DNA molecule. However, φX174 has efficiently evolved to have overlapping genes that are translated in different reading frames, already shown in Figure 13-28.

Phage φX174 adsorbs to *E. coli* by means of a spike protein (Figure 17-24). In contrast with the T phages, which do not inject proteins, the terminal spike protein (of gene *H*) enters the bacterium with the DNA. The

Table 17-4   Phage φX174 genes and functions

Gene	Function
A	RF replication; viral strand synthesis
A*	Turning off of host DNA synthesis
B	Formation of capsid
C	Formation of phage-unit-sized DNA
D	Formation of capsid
E	Lysis of bacterium
F	Major coat protein
G	Major spike protein
H	Minor spike protein; adsorption to host cell
J	Core protein; entry of progeny DNA into phage particle
K	Unknown

H protein is required in an unknown way for converting the bacterium to a phage factory.

The DNA of φX174 is not injected into the bacterium. Instead, release of the phage DNA occurs on the cell surface; this is known because, shortly after adsorption of a phage particle to a bacterium, the phage DNA can be digested if DNase is added to the suspending medium. (Such DNase sensitivity is not true of any of the tailed phages.) Removal of the coat protein is coupled to replication of the parental DNA. The mechanisms of uncoating the DNA and transfer of the molecule to the cell interior, and the role of replication in the transfer process are unknown.

A great deal is known about φX174, but in this section we will only examine replication, transcription, and viral assembly.

Replication of φX174 DNA

The mode of replication of φX174 is especially interesting since the phage has the problem of making an identical copy of a single strand. Clearly, the template cannot yield progeny molecules in a single step.

In order to discuss replication of single-stranded DNA molecules, the following terminology is used. The strand contained in the virus particle or any strand that has the same base sequence is called a (+) strand; a strand having the complementary base sequence is called a (−) strand.

Replication of φX174 DNA occurs in several steps:

1. *Conversion of the parental single-stranded DNA molecule (the viral or (+) strand) to a covalently closed, double-stranded molecule called replicative form I (RFI).* This occurs before transcription begins and hence depends on host enzymes exclusively. The newly synthesized strand (the (−) strand) is the coding strand—the only strand that is transcribed. The enzymatic mechanism for converting the (+) strand to RFI is described in detail in Chapter 9 (Figure 9-28).

2. *Synthesis of many copies of RFI.* The (−) strand is transcribed and a multifunctional protein, the gene-*A* protein, is made. The *A* product makes a single-strand break in the (+) strand between bases 4305 and 4306 (this is called the replication origin, *ori*) and remains covalently linked to the 5′-P terminus, as shown in Figure 17-25. This nicked molecule is called RFII. *E. coli* proteins (synthesized prior to the infection) then cause the parental (+) strand to be displaced from RFII by the looped rolling-circle replication mode discussed in greater detail in Chapter 9. When one round of replication is completed, the displaced (+) strand is cleaved from the looped rolling circle, recircularized, and used as a template for the synthesis of another (−) strand; the result is a new RFI.

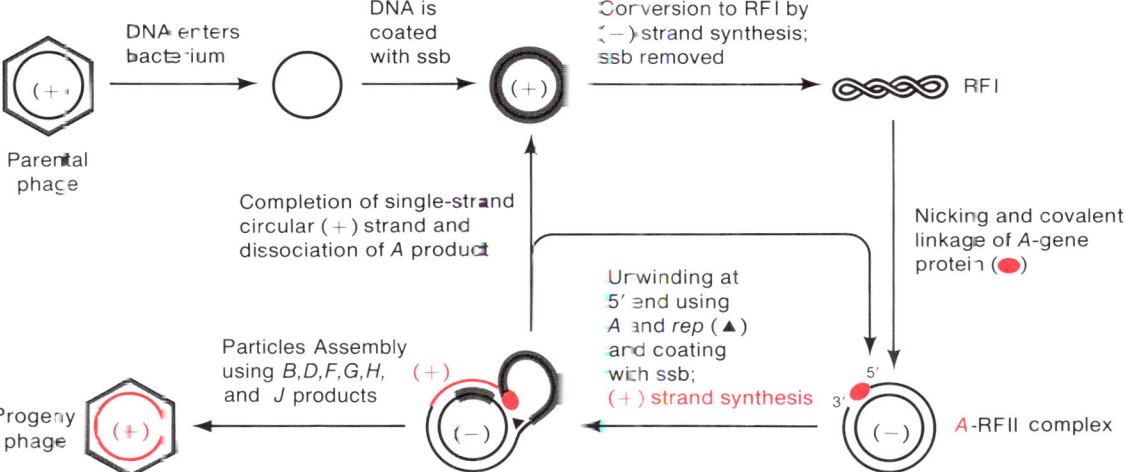

Figure 17-25 The two stages of replication of φX174 DNA. Parental phage DNA is first converted to the double-stranded supercoiled RFI. Then, the gene-*A* protein makes a nick and the looped rolling circle mecha- nism begins, using displacement of the progeny (+) strand made during formation of RFI. In the presence of phage-encoded maturation and coat proteins, progeny (+)- strand circles are encapsidated rather than being converted to RFI. With the exception of the phage-encoded A protein, all proteins used in replication are derived from bacterial genes.

3. *Synthesis of (+) strands for encapsidation.* There is no switch from synthesis of RFI to synthesis of free (+) strand. Instead, the packaging system captures progeny (+) strands before they can serve as templates for further synthesis of (−) strands. The capture is delayed until nearly com- pleted phage heads have been synthesized and begins when there are about 30–40 copies of RFI. At this point, most of the DNA is engaged in looped rolling-circle replication, producing new RFI and some (+) strands for pack- aging. Host DNA synthesis finally stops at this time, partly as a result of an unknown function of the A protein and partly because all of the host DNA polymerase is engaged in synthesis of φX174 DNA. As more heads are made, all of the displaced (+) strands are packaged and synthesis of RFI stops for lack of a template. The mechanism of packaging is described in a later section.

Transcription of φX174 DNA

Phage φX174 has three promoters, all of which are activated simultaneously. Except for lysozyme, no temporal regulation is required. All transcription is from the (−) strand, so transcription cannot occur until the first (−) strand is made; from then on, RFI synthesis continues unabated until a sufficient number of heads are formed to start packaging. Because a φX174 DNA molecule is very small, many rounds of replication are completed before the head proteins are synthesized and assembled (The first completed particle appears about 10 minutes after infection.) Lysozyme activity is delayed because translation is slow and the enzyme is not particularly active; lysis does not occur until 30 minutes after infection, by which time about 500 phage particles have been synthesized.

Packaging of φX174 DNA

Packaging of φX174 DNA requires seven phage proteins, four of which are present in the finished particle. The steps of assembly, which are shown in Figure 17-26, are the following:

1. The head protein (gene *F*) and the spike protein (gene *G*) form pentamers that in the presence of the gene-*B* protein, form an aggregate called a 12S particle.
2. The gene-*H* proteins are added to complete the spike.
3. The gene-*D* protein forms a frame on which a prohead is built from the 12S units. Thus, the prohead contains the proteins B, D, F, G, and H.
4. The gene-*C* protein shown schematically in the figure as binding to the 5′ end of the displaced (+) strand somehow directs the 5′ end

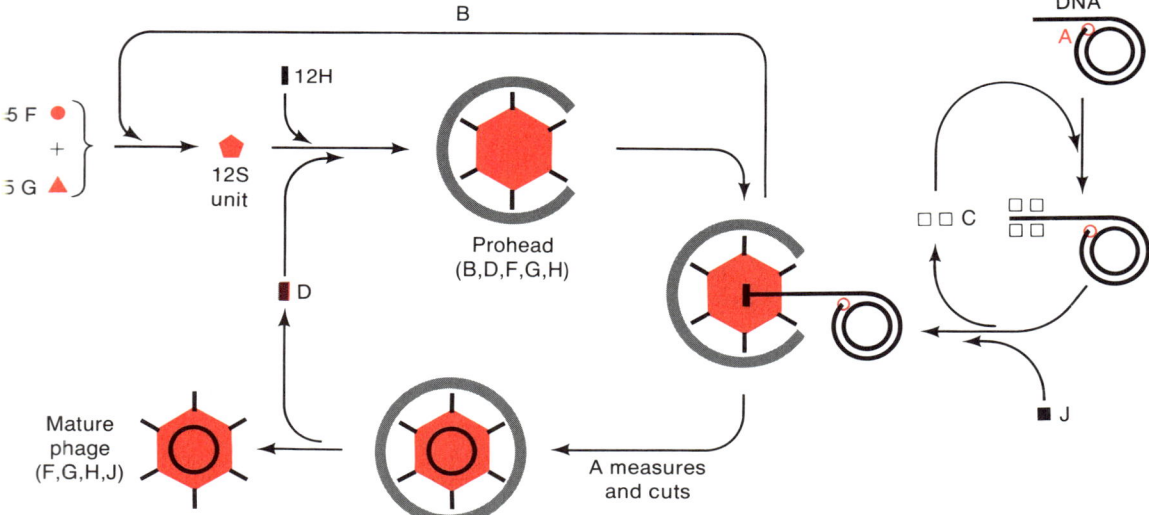

Figure 17-26 Model for assembly of φX174 phage. Letters represent φX174 proteins. The B protein aggregates five molecules of F protein (the major head protein) with five molecules of the G protein (spike protein) to form a 12S unit. The D protein may provide a scaffolding function to form the prohead. The C protein facilitates encapsidation of the DNA. The A protein measures a unit length of DNA and forms the circle. The J protein is needed for stable packaging of single-stranded DNA.

into the prohead. The gene-*J* protein, which probably initially binds to the DNA outside the prohead organizes the DNA inside the head and consolidates the structure.

5. B protein is released from the prohead, probably during synthesis of single-stranded DNA.

6. The A protein cuts the displaced (+) strand from the looped rolling circle and circularizes the displaced strand

7. The D protein is then removed (probably spontaneously following completion of and closing of the head) leading to formation of the finished phage particle.

At the present time it is possible to synthesize *in vitro* both the first RFI and the daughter RFI molecules and to carry out rolling circle replication. Also, if the seven morphogenetic proteins are added, the packaging reaction can be coupled to rolling circle replication. Since *in vitro* coupled transcription-translation systems are also available, it should be possible to synthesize all of the phage proteins from RFI. When these systems are combined, φX174 will probably be the first complete biological organism synthesized totally from molecularly defined components.

FILAMENTOUS DNA PHAGES

Three filamentous *E. coli* phages, M13, fd, and f1 (Figure 17-27), each of which contains a circular single-stranded DNA molecule, have been carefully studied. They are very similar and a picture of the life cycle of each has

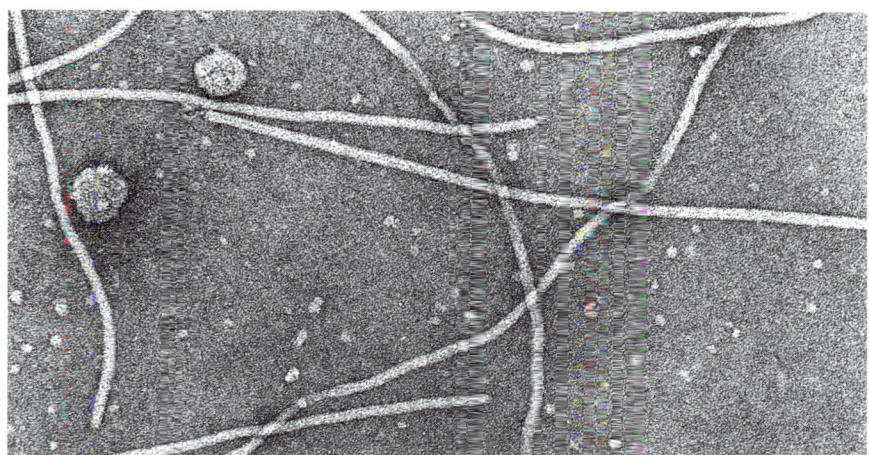

Figure 17-27 Electron micrograph of the filamentous *E. coli* phage M13. The large and small spherical particles are cell debris. [Courtesy of Robley Williams.]

been developed by combining information obtained from each of them. The life cycle, which shows many similarities to that of φX174, will not be given in any detail; only the mode of penetration of the phage DNA and release of completed particles will be described.

Of all known virulent phages, only the filamentous *E. coli* phages neither kill the host cell nor cause lysis. The filamentous phages adsorb to the tip of the F pilus, a protein-containing hairlike structure present on the surface of *E. coli* strain containing the F plasmid. In an unknown way (for which there are many theories), the particle is brought to the bacterial cell wall at the base of the pilus; one idea is that the phage moves down a groove on one side of the pilus and another is that the pilus retracts by absorption of pilus proteins into the bacterium (retraction is an observed phenomenon with pili). What is certain is that the *entire* particle, in intact form, penetrates the cell wall and comes to rest on the inner cell membrane. There, the particle is uncoated and simultaneously the single-stranded DNA is converted to an RFI. The coat is degraded and the protein subunits are re-used in progeny particles. The entry of coat proteins into the cell is unique to these phages.

The details of the replication cycle of filamentous phages are not known. There is a stage in which RFI is formed by a rolling circle mechanism, as with φX174. The switch to (+)-strand synthesis requires the gene-*5* protein, which is a single-stranded DNA-binding protein that coats the tail of the rolling circle. This switch is not coupled to encapsidation as with φX174; instead, linear (+) strands coated with gene-*5* protein are continually cut from the rolling circle and then circularized. Thus, a large pool of complexes of (+) strand and gene-*5* protein are present in the cell.

The major coat protein (gene-*8* protein) is synthesized in large quantities and immediately deposited in the inner cell membrane; no gene-*8* protein is found free in the cytoplasm. This protein is an α helix having three distinct regions—a basic end, a hydrophobic center, and an acidic end (Figure 17-28). It is hypothesized that the hydrophobic center enables the proteins to be situated in the hydrophobic membrane, the acidic end faces the cell wall, and the basic end faces the cell interior, perhaps displacing the gene-*5* protein and binding to the (+) strand. The phage particle, which also contains two other proteins, is assembled in the inner membrane. For each completed particle a bud forms on the cell surface and a phage is extruded (Figure 17-29). This process can continue indefinitely without damage to the cell.

SINGLE-STRANDED RNA PHAGES

Several single-stranded RNA phages have been isolated; the best known are the *E. coli* phages f2, R17, MS-2, and Qβ. Each of these is a tailless icosahedron like φX174 (Figure 17-30). The genetic map of MS-2 is shown in Figure 17-31. Their life cycles, which are very simple, have been elucidated

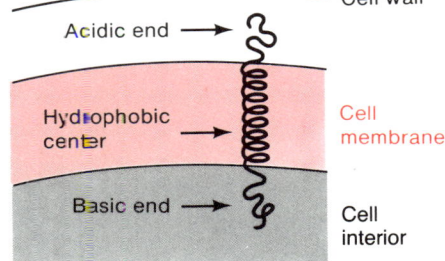

Figure 17-28 Schematic diagram of orientation of the M13 coat protein in the cell membrane.

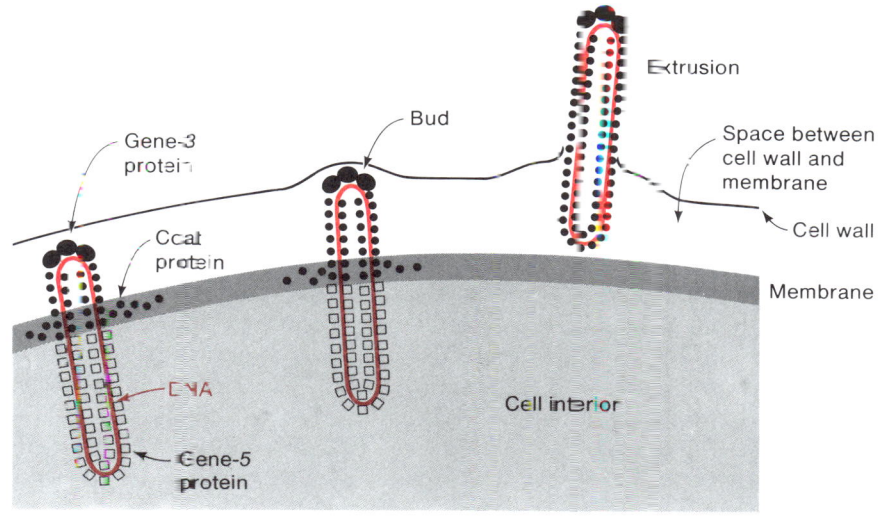

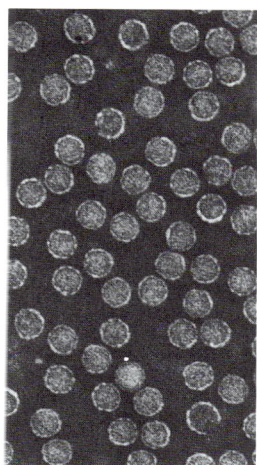

Figure 17-29 Assembly and extrusion of M13 phage. Gene-5 protein covers the DNA and somehow enables the DNA to penetrate the cell membrane. Coat protein molecules in the membrane bind to the DNA, displac- ing the gene-5 protein molecules. The gene-3 protein is attached to one end of the parti- cle and is needed for both adsorption and extrusion.

Figure 17-30 Electron micrograph of phage MS-2. A few of the phage particles are empty. [Courtesy of Robley Williams.]

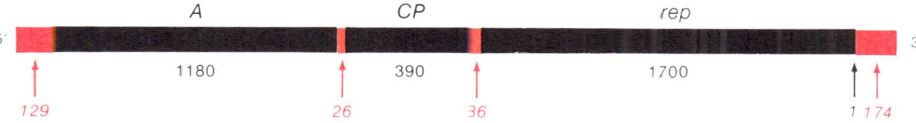

Figure 17-31 The genetic map of phage MS-2 showing the positions and number of bases in each gene. The italicized numbers are the lengths of leaders and spacers.

by comparing different features discovered from each phage. The RNA is a single-stranded linear molecule having a great deal of intramolecular hydro- gen-bonding, and it consists of about 3600 nucleotides (the sequence is known for R17, MS-2, and Qβ) and contains three genes encoding a coat protein (CP), an attachment protein (A), and an RNA replicase (Rep). The RNA molecule is both a replication template and mRNA, so 1) neither DNA nor RNA polymerase is needed in the life cycle and 2) regulation must occur at the translational level. A typical burst size is 5000 to 10,000, which is very large compared to the burst of one to a few hundred for DNA phages. The particles form huge crystalline arrays (Figure 17-32) within each bacterium; the crystals somehow damage the cell membranes, causing cell lysis in the absence of a lytic enzyme. Thus, the timing of lysis is not precise, as with

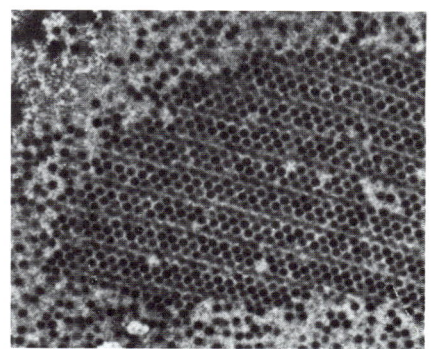

Figure 17-32 A crystalline array of mature MS-2 particles in *E. coli*.

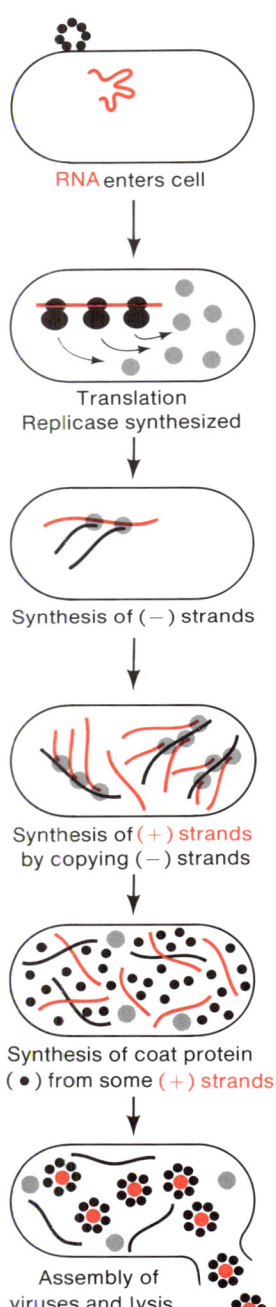

RNA enters cell

↓

Translation
Replicase synthesized

↓

Synthesis of (−) strands

↓

Synthesis of (+) strands
by copying (−) strands

↓

Synthesis of coat protein
(•) from some (+) strands

↓

Assembly of
viruses and lysis

Figure 17-33 Schematic diagram of the life cycle of an RNA phage.

other phages; instead, lysis of a bacterial population occurs gradually from 30 to 60 minutes after infection.

A schematic diagram for the life cycle of a RNA phage is shown in Figure 17-33. The individual stages are the following:

1. *Entry to the cell and binding to ribosomes.* Immediately after entry, a ribosome attaches to a binding site at the beginning of the *CP* gene, which is the centermost gene in the RNA molecule.

2. *Translation and regulation of the amount of each of the three proteins.* The ribosomal binding sites for the A and Rep proteins of R17 are blocked by the presence of stem-and-loop structures in the RNA. However, translation of the *CP* gene opens up the binding site for the Rep protein. Thus, both proteins are made initially but increasing amounts of the CP protein bind to the *rep* site and block translation of the *rep* gene. This inhibition is efficient: about 2×10^6 copies of the CP protein are needed as structural components for 10,000 phage, whereas the replicase is needed only in catalytic amounts. The synthesis of the A protein will be described shortly.

3. *Replication of the phage RNA* (Figure 17-34). Only the Qβ replicase has been studied in detail. It is a tetramer consisting of one Rep molecule and three host proteins—the EF-Ts and EF-Tu translation factors needed for placement of charged tRNA molecules on the ribosome during protein synthesis, and the ribosomal protein S1. The function of this surprising set of host proteins is unknown. The replicase copies the viral (+) strand to generate a (−) strand. While synthesis is proceeding, the (−) strand is in contact

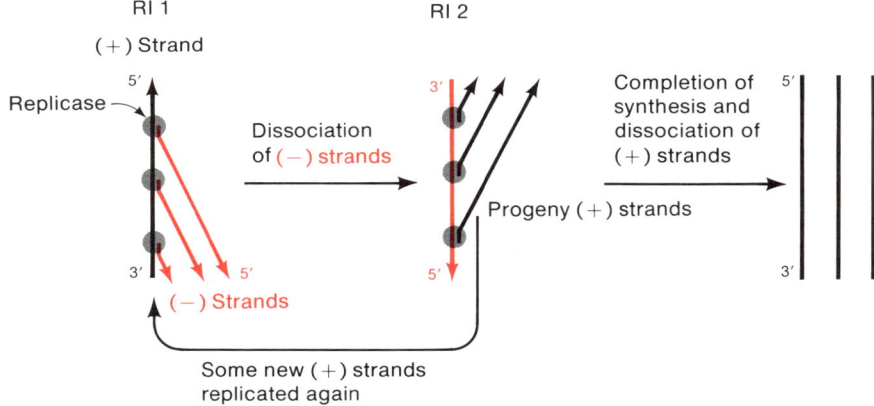

Figure 17-34 Replication of RNA by Qβ replicase. RI stands for replication intermediate.

with the (+) strand only at the polymerization site; for the most part the replicative form is therefore single-stranded. Initiation of several (−) strands occurs before the first (−) strand is complete, and the replicative form is branched. The (−) strands are released and immediately used by the replicase to form (+) strands. Some of the (+) strands return to the ribosomes for synthesis of more CP proteins and others are packaged. All progeny contain (+) strands exclusively.

4. *Synthesis of the A protein*. The ribosomal binding site for the A protein is never available on a free (+)-strand because of base pairing. However, replication of a (−) strand begins at the 3′ terminus adjacent to the A gene. Just after (+)-strand synthesis begins there is a brief period during which the ribosomal binding site for the A protein is free, the complementary segment of the (+) strand having not yet been replicated; during this time the A gene is translated. Only one or two passages are possible before the binding site becomes closed. Therefore, the number of A proteins is maintained roughly equal to the number of (+) strands; this arrangement is economical since each virus particle contains one A protein molecule. Each A protein later becomes bound to one RNA molecule at an unidentified binding site. It is thought that the A protein facilitates the interaction of the RNA with the CP molecules. The A protein remains bound to the RNA and enters the cell in a subsequent infection. Possibly its function is similar to that of the spike protein of the phage ϕX174.

5. *Particle assembly*. Molecules of the CP protein spontaneously aggregate around the newly synthesized (+) strand and form an icosahedral shell.

6. *Cell lysis*. Lysis occurs after about 10,000 phage particles have formed. However, no phage-encoded lytic enzyme is produced and the lytic mechanism is not known.

Bacteriophages

II. THE LYSOGENIC LIFE CYCLE

Each of the phages discussed in the preceding chapter has a life cycle during which the host cell lyses and releases many progeny phage (the most common outcome) or in which phage are continually released without killing the host (the filamentous phages). In this chapter an alternate life cycle, the lysogenic cycle, is described. In this cycle phage are not produced and the host survives and divides indefinitely. However, many bacterial generations later, if environmental conditions are right, the lysogenic cycle can be terminated and a lytic cycle started anew. When this occurs, the host cell is killed and progeny phage are released as in a standard lytic cycle that has been initiated by infection with a phage particle.

One of the fascinating aspects of phage species that can enter a lysogenic cycle is that they can also initiate a lytic cycle of the type described in the preceding chapter. Growth conditions of the bacterium prior to infection and the multiplicity of infection determine the pathway entered by the phage. These alternatives are a result of regulatory systems that differ from those we have seen in the preceding chapter in that they not only have temporal regulation but are sensitive to environmental conditions, such as the supply of nutrients.

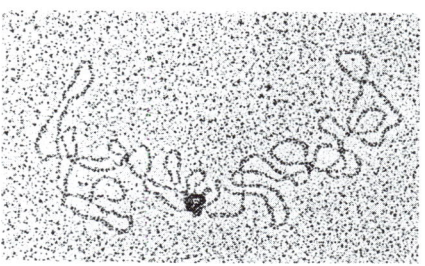

Two genetically engineered plasmid DNA molecules carrying the phage λ *att* site, joined by integrase. [Courtesy of Marc Better.]

THE LYSOGENIC CYCLE

We begin by describing the general features of lysogeny; then we will examine *E. coli* phage λ, the best-understood temperate phage.

Outline of the Lysogenic Life Cycle

There are two types of lysogenic cycles. The most common one, for which the *E. coli* phage λ pathway is the prototype, follows (Figure 18-1):

1. A DNA molecule is injected into a bacterium.
2. After a brief period of transcription, which is needed to synthesize a repressor and an integration enzyme, transcription is turned off by the repressor.
3. A phage DNA molecule, typically a replica of the injected molecule, is inserted into the DNA of the bacterium.
4. The bacterium continues to grow and multiply, and the phage genes replicate as part of the bacterial chromosome.

The second and less common type, for which *E. coli* phage P1 is the prototype, differs from the preceding one in that there is no integration system and the phage DNA becomes a plasmid (an independently replicating circular DNA molecule) rather than a segment of the host chromosome. In this chapter we will mainly consider the first type.

The following terms describe various aspects of the lysogenic life cycle.

1. A phage capable of entering either a lytic or a lysogenic life cycle is a **temperate** phage.
2. A bacterium containing a complete set of phage genes is called a **lysogen.**
3. The process of forming a lysogen by infecting a bacterial culture with temperate phage is **lysogenization.**
4. If the phage DNA is contained within the bacterial DNA, the phage DNA is said to be **integrated.** The process by which this state of

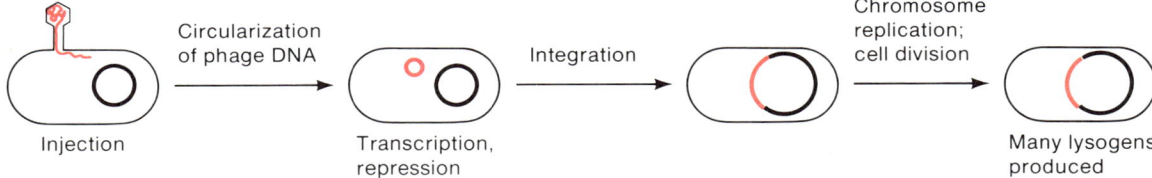

Figure 18-1   The general mode of lysogenization by insertion of phage DNA into a bacterial chromosome.

the DNA is achieved is called **integration or insertion.** Phage DNA in plasmid form is nonintegrated. Both integrated and plasmid phage DNA are called a **prophage.**

General Properties of Lysogens and Lysogenization

Two important properties of lysogens are the following:

1. A lysogen cannot be reinfected by a phage of the type that first lysogenized the cell; this resistance to "superinfection" is called **immunity.**

2. Even after many cell generations, a lysogen can initiate a lytic cycle; in this process, which is called **induction,** the phage genes are excised as a single segment of DNA (Figure 18-2).

The molecular mechanism for immunity and the circumstances that give rise to induction will be described shortly.

More than 90 percent of the thousands of known phages are temperate. These phages are often unable to produce bursts as large as the highly virulent phages such as T4 and T7, but compensate by their ability to multiply in environmental conditions that are not suitable for rapid production of progeny. The meaning of this statement will become clear when we examine the possible outcomes of a single phage encountering a population of bacteria in nature.

Let us first consider a bacterial population that is actively dividing. If a phage can infect one cell and multiply (in a lytic cycle), the number of progeny phage will increase rapidly. However, if the bacteria were growing very slowly owing to exhaustion of nutrients in the surrounding medium, the phage infection might abort if the infected cell were to stop growing during the phage life cycle. (Remember that a phage can only grow in a bacterium that is actively metabolizing.)

When bacteria are starved of nutrients, they degrade their own mRNA and protein before they become dormant. Restoration of nutrients enables

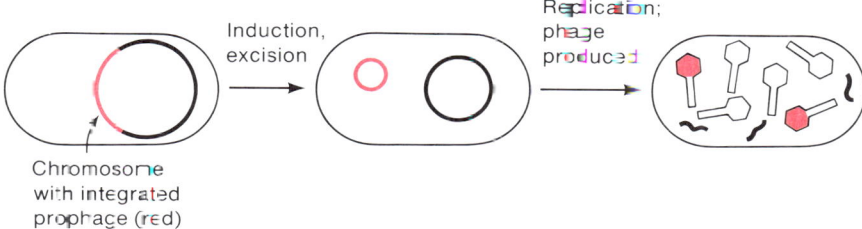

Chromosome
with integrated
prophage (red)

Figure 13-2 An outline of the events in prophage induction. The prophage DNA is in red. The bacterial DNA is omitted from the third panel for clarity.

the bacteria to grow again. This is not true of a phage-infected cell in which the phage life cycle has been interrupted; usually the ability to produce phage is permanently lost, probably because the delicate balance of phage functions is destroyed by the protein and mRNA degradation. Furthermore, the cell dies. This deleterious outcome (permanent loss of both phage and bacteria can be avoided if a lysogenic cycle is possible—that is, if the phage DNA can become dormant. When growth of the bacterium resumes, the phage genes replicate as part of the chromosome. Even though phage production has been suspended for the time being, because of the induction phenomenon, the potential for phage production remains.

Now let us return to an infection of an actively growing bacterial population in which phage are multiplying rapidly. When the number of phage exceeds the number of bacteria, the phage enter their final cycle of multiplication; after lysis occurs, no further multiplication is possible because there are no more sensitive bacteria. It is possible that years could pass before these phage particles might encounter another sensitive host bacterium and during this time various deleterious agents might damage the particles. Until a host cell appears, the phage particles have no chance to increase in number. However, if lysogenization could occur at a high multiplicity of infection (MOI), the phage genes could be maintained indefinitely, since the lysogen would grow whenever nutrients are available. Indeed, the two conditions that stimulate a lysogenic response of a temperate phage are (1) depletion of nutrients in the growth medium and (2) a high multiplicity of infection.

E. coli PHAGE λ

The best-understood temperate phage is *E. coli* phage λ and its lysogenic cycle is the main topic of this chapter. Before proceeding, the reader is advised to review the description of its lytic cycle presented in the preceding chapter.

Immunity to Infection

When a virulent phage forms a plaque on a lawn of growing bacteria, the plaque is clear because all bacteria in the center of the plaque are killed and lysed. However, phage λ forms a plaque with a *turbid center,* as do all temperate phages (Figure 18-3). The turbidity is caused by the growth of phage-immune lysogens in the plaque. Phage and bacteria are usually placed on an agar surface in a ratio of about 1 phage per 10^7 bacteria. The bacteria grow rapidly and the MOI is low, so the lytic cycle ensues. After several lytic cycles, the MOI becomes high and a few cells are lysogenized; since availability of nutrients does not yet limit development of phage or cells, most cells are lysed. When the nutrients in the agar are depleted, the uninfected

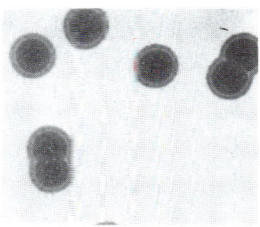

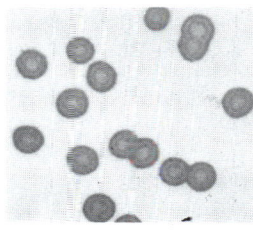

(a)

Early
bacterial
lawn

Mature
bacterial
lawn

1   2   3   4

(b) Development of a turbid plaque

Figure 18-3   (a) Clear (cI^-) and turbid (c^+) plaque of phage λ. [Courtesy of A. D. Kaiser.] (b) Diagram showing the development of a turbid plaque of a temperate phage. (1) Phage is in a bacterial lawn; (2) a small clear plaque (usually invisible) contains a few lysogens (shown as rods); (3) the clear region enlarges but lysogens grow within the plaque. (4) clear region reaches maximum size and lysogens stop growing as nutrient is exhausted.

cells stop growing, and the plaque stops increasing in size. However, since there has been less bacterial growth within the plaque, nutrients are still present there. Therefore, the lysogenic cells, which are immune to subsequent infection by λ, continue to grow, forming a turbid center in the plaque.

The resistance of a λ lysogen to infection by λ is called **immunity.** The cause of the phenomenon is the following. Phage λ contains a repressor-operator system (Figure 18-4). The repressor gene is called *cI;* the repressor protein binds to two operators, *oL* and *oR,* which are adjacent to two promoters *pL* and *pR.* The letters L and R mean leftward and rightward and refer to the direction of synthesis of two early mRNA molecules, when the genetic map is drawn in a standard orientation (see Chapter 17, Figure 17-19).

In a lysogen the cI repressor is synthesized continuously and in slight excess with respect to the operators. Both *oL* and *oR* sites contain bound repressor molecules, so *pL* and *pR* are unavailable to RNA polymerase. Thus, in a lysogen, transcription from the two early promoters is prevented; this is sufficient to keep the prophage in an "off" state, and the lysogen grows indefinitely. If a normal (nonlysogenic) cell is infected by λ, the two operators of the incoming λ DNA molecule will be unoccupied, since phage repressor has not yet been made; thus, transcription from *pL* and *pR* will occur. However, if a phage tries to infect a lysogen, the excess repressor molecules already present in the lysogen bind to the two operators on the infecting DNA molecule before an RNA polymerase can bind to *pL* and *pR.* This operator-binding prevents the phage from proceeding into lytic development. This inhibition is referred to as **resistance to homoimmune superinfection.**

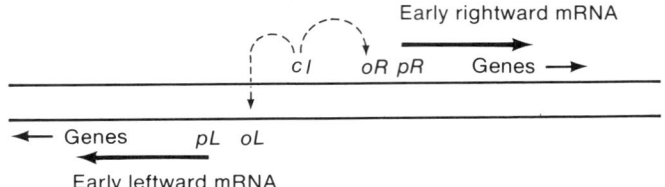

Figure 18-4 The repressor-operator system of λ showing the two early mRNA molecules. Symbols: *cI*, repressor gene; *p*, promoter; *o*, operator; *L*, left; *R*, right. See also Figure 18-13.

What happens to superinfecting DNA? It is able to form a supercoil—phage gene products are not needed for supercoiling—but it cannot replicate. The bacterium is unaffected by the presence of this DNA molecule and grows and divides normally, so the superinfecting DNA is progressively diluted out by bacterial multiplication.

Phage with mutations in the immunity system have been isolated. The two most important types are cI^- and *vir* mutants:

1. A cI^- mutant does not make a functional repressor and hence can engage only in a lytic cycle. Thus, a cI^- mutant makes a clear plaque (Figure 18-3(a)).
2. A *vir* (**virulent**) mutant carries mutations in both *oL* and *oR* and also makes a clear plaque because it cannot establish repression; furthermore, it can superinfect and grow in a lysogen because it is insensitive to the repressor already present in the cell.

There are many temperate phages of *E. coli* other than λ. Two of these, which are closely related to λ (in that they have large regions of homology), are phages 21 and 434. Each of these phages has its own immune system—that is, its own repressor and repressor-specific operators. Thus, the 434 repressor cannot bind to a λ operator and a λ repressor cannot bind to a 434 operator. Such a pair of phages is said to be **heteroimmune** with respect

Table 18-1 Ability of different phages to form plaques on homoimmune and heteroimmune lysogens

Superinfecting phage	Lysogen		
	B(λ)	B(434)	B(21)
λ	−	+	+
434	+	−	+
21	+	+	−

Note: The notation B(P) denotes a bacterium B lysogenic for phage P. + = forms a plaque; − = does not form a plaque.

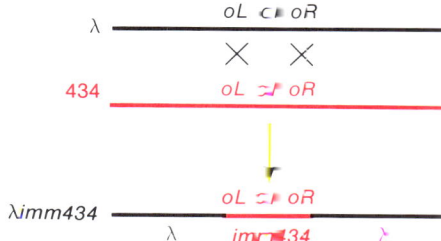

Figure 18-5   Formation of λ*imm434*.

to one another. A temperate phage can form a plaque on a heteroimmune lysogen because the repressor made in the lysogen does not bind to the operator of the superinfecting phage. This is summarized in Table 18-1. The immunity region of the DNA, which includes the *cI* gene, operators, and promoters (and other elements to be described shortly), is given the genotypic symbol *imm*—specifically, *imm*λ, *imm21,* and *imm434*. Interesting hybrid phages, which have been very useful in research, have been created by crossing two heteroimmune phages and selecting a recombinant containing the immunity region of one phage and the remaining genes of the other heteroimmune phage. A prominent example is a λ-434 hybrid, which is genotypically designated λ*imm434* (also, occasionally, λ*434hy*) as shown in Figure 18-5.

Prophage Integration

In the lysogenic cycle of λ and other similar phages, a phage DNA molecule is inserted (integrated) into the bacterial chromosome to form a prophage and repression is established in order to prevent synthesis of deleterious phage proteins.

When λ DNA integrates, it is inserted at a preferred position in the *E. coli* chromosome; this site is between the *gal* operon and the *bio* (biotin) operon and is called the λ attachment site and designated *att*. For most temperate phages there is only one integration site. An important feature of the insertion mechanism was derived from the genetic observation that the gene order in the λ prophage is a permutation of the gene order of the DNA in the phage (Figure 18-6). This permutation was explained by a mechanism of integration shown in Figure 18-7. The essence of the mechanism is circularization of λ DNA followed by physical breakage and rejoining of phage and host DNA—precisely, between the bacterial DNA attachment site and another attachment site in the phage DNA that is located near the center of the phage DNA molecule. A phage protein, **integrase** (its gene designation is *int*), recognizes the phage DNA and bacterial DNA attachment sites and

Figure 18-6   The order of the genes on the DNA molecule in the phage head (vegetative order) and in the prophage (prophage order). The genes have been selected arbitrarily to provide reference points.

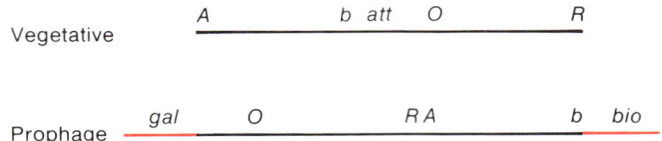

catalyzes the physical exchange. The result is integration of the λ DNA molecule into the bacterial DNA. As a consequence of the circularization and integration, the linear order of the genes in the infecting λ DNA molecule is permuted, as shown in the figure. A question that arose after this mechanism was discovered was: why doesn't the integrase excise the prophage shortly after integration occurs, since the prophage is flanked by two attachment sites? Furthermore, excision would seem to be the preferred reaction since, kinetically, two attachment sites in the same DNA molecule (that is, the host chromosome) ought to interact with each other more rapidly than two attachment sites in different DNA molecules (phage and bacterial DNA). Why this does not happen became clear when it was discovered that the DNA attachment sites in the phage and bacteria are not the same. Thus, they recombine to form **prophage attachment sites** that also differ from the sites joining the bacterial and phage DNA.

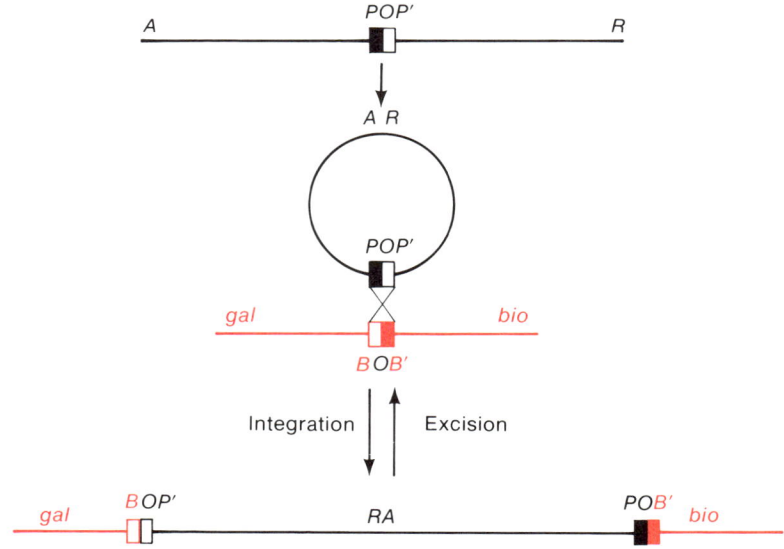

Figure 18-7   The mechanism of prophage integration and excision of phage λ. The phage attachment site has been denoted *POP'* in accord with subsequent findings. The bacterial attachment site is *BOB'*. The prophage is flanked by two new attachment sites denoted *BOP'* and *POB'*.

All of the λ attachment sites have three different components. One of these is common to all sites and is denoted by the letter *O*.* The phage attachment site is written *POP'* (P for phage) and the bacterial attachment site is written *BOB'* (B for bacteria). Thus, in the integration reaction two new attachment sites, *BOP'* and *POB'*, are generated (Figure 18-7). These are often written *attL* and *attR* to designate the left and right prophage-attachment sites, respectively. Integrase cannot by itself catalyze a reaction between *BOP'* and *POB'*, so the reaction

$$BOB' + POP' \xrightarrow{\text{Integrase}} BOP' + POB'$$

Bacterium Phage   Prophage

is irreversible if integrase is the only enzyme present. The result of the integration reaction is that the λ DNA is linearly inserted between the *gal* and *bio* loci and henceforth is replicated simply as a segment of *E. coli* DNA.

Integrase has been purified, the base sequences of the *BOB'* and *POP'* sites are known, and the integration reaction can be carried out *in vitro*. Integrase has DNA-binding activity, binding strongly to *POP'*, and is a type I topoisomerase (Chapter 9); that is, it can cause a strand break in one strand of a double helix, rotate one branch of the broken strand about the continuous strand, and then rejoin the ends. In addition to integrase, the reaction requires a host protein called IHF (integration host factor); this protein also binds to the attachment site, but its biochemical function is not known. Sequencing data show that the sequences of *B*, *B'*, *P* and *P'* are quite different and the sequence of the *O* (as in *BOB'*), often called the **core sequence,** is very rich in A·T pairs (Figure 18-8).

The mechanics of the integration reaction are fairly well known. First, two single-strand breaks are made in each attachment region in the complementary strands seven base pairs apart, as shown in Figure 18-9. Second, an exchange occurs in which two overlapping joints containing strands from two core sequences are formed, as also shown in the figure. The geometry of the integrase reaction is shown in Figure 18-10. The easily melted, high-A + T core sequence in each attachment site opens and then complementary strands from different attachment sites pair to form a four-stranded segment

```
GCTTTTTTATACTAA
CGAAAAAATATGATT
```

Figure 18-8   Base sequence of the core of the λ *att* region.

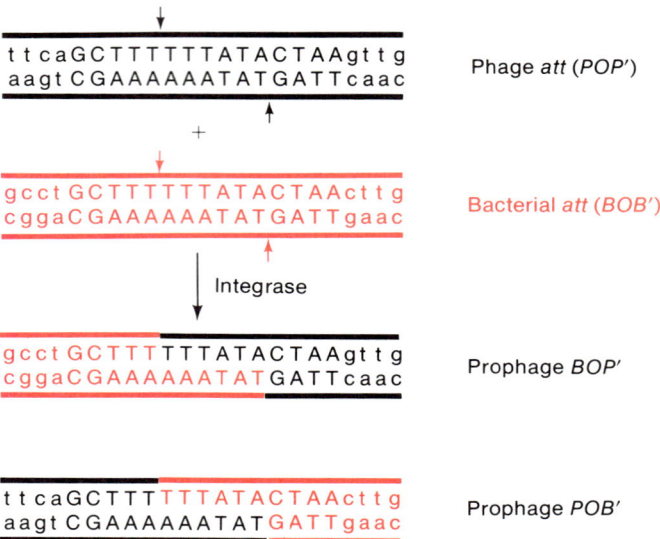

Figure 18-9 The exchange that occurs in the integrase reaction showing the approximate positions of the strand breaks (arrows). The lowercase letters are bases in the flanking sequences B, B', P, or P'; the capital letters are in the O region.

(panel (b)). Pairs of single-strand breaks in homologous regions are made (panel (c)). Then the topoisomerase activity of integrase causes each double-stranded region to rotate (panel (c)) after which corresponding strands from the two core sequences are joined (panel (c)). This process (nick–rotate–join) is repeated on the other side of the base-paired region (panel (d)), and the two newly formed core sequences with overlapping joints separate (panel (e)).

The synthesis of integrase is coupled to the synthesis of the cI repressor. This is efficient because integrase and the repressor are both needed in the lysogenic cycle and neither is needed in the lytic cycle. An outline of the regulatory pathway responsible for this coupling is shown in Figure 18-11. In the absence of a positive regulatory element (the product of the λ gene *cII*) the promoters for both the *int* and *cI* genes are unavailable to RNA polymerase. Shortly after infection, the cII protein is synthesized (see Chapter 17). If the concentration of the protein is high enough, the cII protein binds to sites near the promoters for the *cI* and *int* genes (designated *pre* and *pI*, respectively) and thereby renders them accessible to RNA polymerase. (In this respect the cII protein is similar to the cAMP-CRP complex in the *lac* operon.) RNA polymerase then transcribes both genes and the gene products are made. We will see later that the immediate cause that determines whether a lysogenic or lytic cycle occurs in an infection is mediated through the concentration of the cII protein.

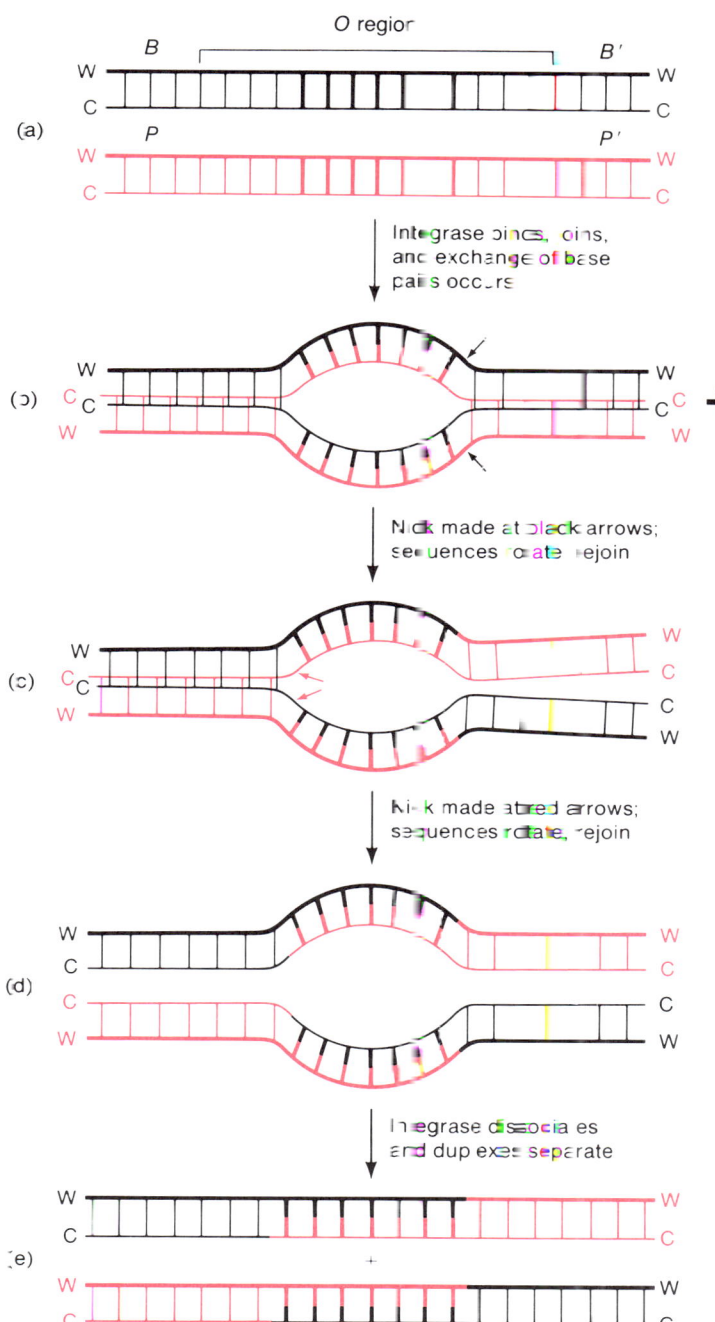

(a)

Integrase binds, joins, and exchange of base pairs occurs

(b)

Nick made at black arrows; sequences rotate, rejoin

(c)

Nick made at red arrows; sequences rotate, rejoin

(d)

Integrase dissociates and duplexes separate

(e)

Figure 18-10 A model for the integrase reaction. The segments shown represent the 15-bp O region and small portions of the flanking *B*, *B'*, *P*, and *P'* sequences. The heavy bars represent the only base pairs that are broken. The letters W and C are only for distinguishing the complementary strands. [Modified from a drawing provided by Howard Nash.]

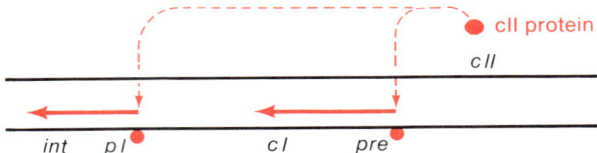

Figure 18-11   The role of the cII protein in stimulating synthesis of *cI* and *int* mRNA from *pre* and *pI*, respectively. The mRNA molecules are drawn as red arrows that indi-cate the direction of RNA synthesis. The arrows are located nearer the DNA strand that is copied. Dashed arrows point to the sites of binding of the cII protein (red dots).

Induction and Prophage Excision

A lysogen is generally quite stable and can replicate nearly indefinitely with-out release of phage. However, if a lysogenic bacterium were to become damaged, it would be to the advantage of the phage to become derepressed and initiate a lytic cycle. This does occur and the signal for derepression (**prophage induction**) is damage to the DNA. So far, all inducing agents, of which ultraviolet radiation has been studied most extensively, cause DNA damage that results in the activation of the SOS repair pathway (Chapter 10). Recall from Chapter 15 (the SOS Regulon) that induction of the SOS response occurs when the RecA protein is activated to become a protease, after which RecA cleaves the LexA repressor. The RecA protease also cleaves the λ cI repressor (also the repressors of many temperate phages), and this causes induction. Following cleavage, derepression occurs, the early pro-moters *pL* and *pR* (Figure 18-4) become available to RNA polymerase, and a lytic cycle ensues.

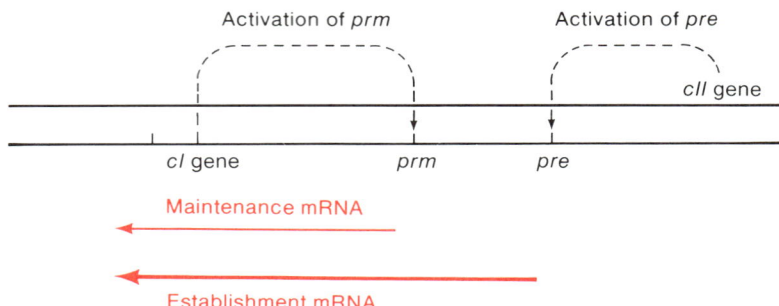

Figure 18-12   The *cI* gene of phage λ, and adjacent regions, showing the two pro-moters for synthesis of *cI*-encoded mRNA. The heavy black lines are the two DNA strands. The thin and heavy red arrows rep-resent two mRNA molecules, both of which are transcribed from the lower DNA strand. The cII protein is translated from a mRNA molecule (not shown) transcribed rightward from the upper strand; this protein activates *pre*, yielding the establishment mRNA (heavy red arrow). The cI protein activates *prm,* yielding the maintenance mRNA.

One product of transcription from *L* is a protein called **excisionase** (designated Xis), encoded by the gene *xis*. Excisionase has recently been purified but only a few of its properties are known. It forms a complex with integrase allowing the latter to recognize the prophage attachment sites *BOP′* and *POB′*; once bound to these sites, integrase can make cuts in the core sequence and re-form the *BOB′* and *POP′* sites. The physical role of the Xis protein has recently been elucidated in experiments that demonstrated a tight complex in which *POB′* is wrapped around the Xis protein; this complex is thought to contain the Int protein also. Formation of the complex is required before *POB′* and *EOP′* can interact. Thus, the Xis-Int complex reverses the integration reaction, causing excision of the prophage (Figure 18-7). Note that the reactions between all attachment sites can now be written.

$$BOB' + POP' \underset{Int,Xis}{\overset{Int}{\rightleftharpoons}} BOP' + POB'$$

The Xis-Int complex fails to catalyze the rightward reaction, so when excess excisionase is present, excision is an irreversible reaction.

The product of excision of a prophage from the *E. coli* chromosome is an intact chromosome and a circular λ molecule, which is also the arrangement present in an infected cell immediately after infection.

Two Promoters for the Synthesis of the λ cI Repressor

At the outset of a lysogenic cycle repressor molecules must be synthesized more rapidly than DNA replication increases the number of operators. If there were too many operators for the number of repressor molecules, transcription leading to phage development and cell death would occur. Thus, when the lysogenic pathway is to be followed, the repression system has an initial burst of repressor synthesis just after infection. In contrast, a lysogen contains only a single λ DNA molecule (the prophage), so less (but some) cI repressor is needed to maintain repression. Clearly, transcription of the *cI* gene must occur in a lysogen (this modifies an earlier statement that in a lysogen the prophage is in an "off" state), but transcription does not need to be very strong.

In λ high-level and low-level synthesis of the repressor is achieved by the existence of two promoters (Figure 18-12)—(1) the establishment promoter, *pre*, which functions in an infected cell and which requires activation by the cII protein, and (2) the maintenance promoter, *prm*, which functions in a lysogen and *is regulated by the repressor itself*. Thus, when lysogenization occurs (that is, when the phage has made enough cII protein to activate *pre*), there is a burst of synthesis of establishment mRNA and the repressor from it activates the *prm* promoter. As the amount of repressor increases, transcription of the *cII* gene is repressed and no more cII protein is made. The cII protein is unstable, so soon its concentration becomes too low to maintain *pre* in an active state and synthesis of *pre* mRNA terminates. Synthesis of *prm* mRNA continues throughout successive generations and translation of this mRNA maintains a concentration of the repressor sufficient to repress the prophage. As we have already stated, less repressor is needed in a lysogen; thus, for the sake of efficiency, less should be made when transcription is occurring from *prm*. Actually, in the fully "on" state the two promoters are equally active in binding RNA polymerase; however, less repressor is made from the transcript initiated at *prm* because the *pre* transcript has a strong ribosome binding site and the *prm* transcript binds only weakly to the ribosome. Thus, the amount of repressor made from the two mRNA molecules is determined by the efficiency of translation. A summary of the events in the sequence of repressor synthesis is given in Table 18-2.

We have just said that the repressor is needed to turn on transcription from *prm*. However, when the repressor concentration is very high, transcription from *prm* does not occur because the repressor also negatively regulates its own transcription. This regulatory mechanism enables λ to maintain a fairly constant repressor concentration, which is advantageous for two reasons: (1) In nature bacteria have continually varying growth rates owing to fluctuations in the supply of nutrients. Regulation ensures that the repressor concentration never diminishes so much that induction occurs spontaneously. (2) If unregulated, the repressor concentration might become so great that stimulated induction could not occur when needed.

One might ask why *prm* should be stimulated by the repressor. Since less repressor is made from *prm* than from *pre,* then if λ only had a lysogenic pathway, synthesis of *prm* mRNA could be constitutive. However, if a lytic cycle is also to be possible, then constitutive synthesis of *prm* mRNA should be avoided—such synthesis might allow premature repression to occur and this would prevent initiation of the lytic cycle.

In order to understand the mechanisms of these regulatory circuits, knowledge of the structure of the operators is required; this is presented in the following section.

Table 18-2 An outline of the sequence of events in cI repressor synthesis in the lysogenic pathway

1. Infection of cell
2. Transcription of rightward mRNA from *pR*
3. Activation of *pre*
4. Transcription and translation of the cI repressor from *pre* (high-level synthesis)
5. Activation of *prm* by the cI repressor
6. Transcription and translation of the cI repressor from *prm*
7. Turning off of *pR* by the cI protein
8. Degradation of the cII protein, so that *pre* becomes inactive
9. Continued synthesis of the cI protein from *prm* (low-level synthesis)

Structure of the Operator and Binding of the Repressor and the Cro Product

The operators *oL* and *oR* can be subdivided into six regions: *oL1*, *oL2*, *oL3*, *oR1*, *oR2*, and *oR3* (Figure 18-13). The base sequences of these regions are similar and each region is capable of binding cI repressor. The relative affinities of each region for the repressor are not the same and are

$$oL1 > oL2 > oL3 \quad \text{and} \quad oR1 > oR2 > oR3.$$

Binding to *oL1* and *oL2* and to *oR1* and *oR2* is sequential and cooperative. Binding to *oR3* and *oL3* is not cooperative.

At low concentrations of repressor, *oL1* and *oR1* are occupied. Since the promoters *pL* and *pR* are adjacent to *oL1* and *oR1*, respectively, when repressor is bound to these sites, RNA polymerase has reduced access to both promoters. The block is even more complete when *oL2* and *oR2* are also filled, which is the state of the operators in a lysogen. The promoter *prm* is between *oR2* and *oR3*, so if the repressor concentration is very high, *prm* is blocked; this is the means by which the repressor negatively regulates its own synthesis. Furthermore, *prm* is not accessible to RNA polymerase unless *oR1* is occupied by repressor. Recent work has given some insight into the mechanism of *prm* activation. X-ray-diffraction analysis of the cI–operator complex (similar to those described for the Cro-operator complex, Chapter 7) indicate that repressor bound to *oR1* may provide a protein-protein binding interaction with RNA polymerase, which would significantly strengthen the binding of the polymerase to *prm*.

Thus, if a cell infected by a λ phage is destined for lysogeny, the following events occur:

1. Transcription from *pL* and *pR* begins.
2. The cII protein is translated from the *pR* transcript.

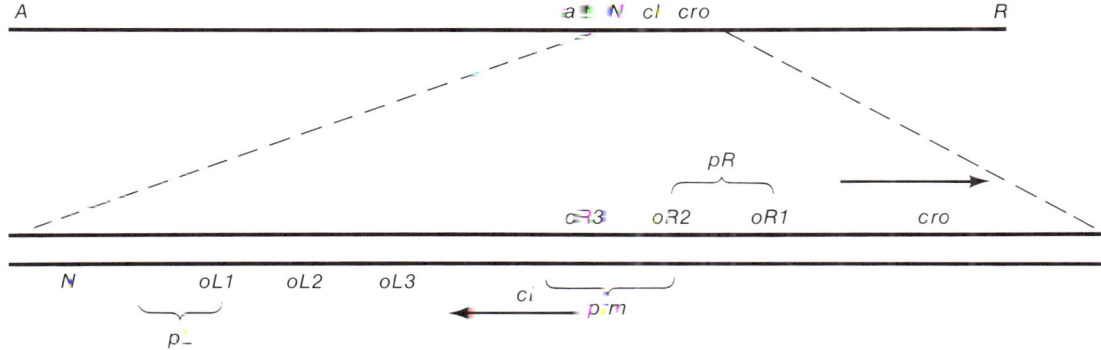

Figure 18-13 A close look at the operator-promoter region of λ. The arrows denote the direction of transcription of the *cro* and *cI* genes. See also Figure 18-4.

3. The cII protein enables RNA polymerase to transcribe from the *pre* promoter in order that repressor will be synthesized.
4. Repressor binds to *oL1* and *oR1*, turning off transcription and hence synthesis of the cII protein. The cII protein is unstable and soon there is no transcription from *pre*.
5. Occupation of *oR1* (item 4) causes *prm* to be activated, so the repressor continues to be synthesized from this promoter, albeit at a lower rate.
6. The cI repressor continues to accumulate, so *oL2* and *oR2* become occupied and repression of transcription from *pL* and *pR* becomes complete.
7. Transcription from *prm* continues unless the repressor concentration becomes so high that the protein also binds to *oR3*. Henceforth, the activity of *prm* is turned on and off to accommodate mild fluctuations in repressor concentration.

As we have described the system, there is no possibility for a lytic cycle. Entry into the lytic cycle requires the Cro protein, which is necessary for several features of the lytic cycle and to prevent synthesis of the cI repressor. The Cro protein is also a repressor that binds to the operators *oL* and *oR*. Its mode of action is based on its affinity for the subsites in the operators; *its affinity is opposite to that of the cI repressor*—namely,

$$oR3 > oR2 = oR1 \quad \text{and} \quad oL3 > oL2 > oL1.$$

The sequence is the following:

1. The *cro* gene is transcribed from *pR* (Figure 18-13).
2. After transcription of the *cro* gene, the *cII* gene, whose product is required to activate the *pre* promoter, is transcribed.
3. Once the Cro protein is synthesized, it binds to *oR3*; hence, activation of *prm* is prevented.
4. The concentration of the Cro protein increases and *oR2* and *oR1* gradually become occupied. When these sites are filled, RNA polymerase loses access to *pR*, so synthesis of the Cro protein eventually stops.

Items 3 and 4 indicate that the Cro protein represses its own synthesis as well as repressing synthesis of the cII protein.

Recall from Figures 17-19, 17-20, and 18-11 that the lysogenic cycle requires leftward transcription from *pre* and *pI*, and the lytic cycle requires extensive rightward transcription. With this in mind, we can summarize the competitive roles of the cI and Cro proteins. As shown in Figure 18-13, cI and Cro engage in a mutually exclusive competition for the operator sites and thereby control the transcription events that involve *oR* (except for the earliest synthesis of N mRNA, *oL* is primarily involved in excision, as we will soon see). Binding to *oR1* and *oR2* blocks *pR*, and binding to *oR3* inhibits transcription from *prm*. Cro and cI bind to these sites in a different order and hence constitute a bidirectional switch. When cI predominates, rightward transcription of *cro* from *pR* is off because *oR1* and *OR2* are occupied; leftward transcription of *cI* from *prm* is autoregulated by cI (on, when there is little cI and only *oR1* and *oR2* are filled: off, when there is excess cI and *oR3* is also occupied). When there is more Cro than cI, *oR3* is occupied first, so leftward transcription is always off; rightward transcription from *pR* continues unless the amount of Cro is quite high, because binding of Cro to *oR1* is fairly weak. Thus, excess cI yields stable leftward transcription and the lysogenic pathway ensues, and a relative excess of Cro produces rightward transcription and the lytic pathway. The question that remains is what determines whether cI or Cro are in excess at the point in the life cycle when the choice must be made. In the next section we will see that since the cII protein is the regulator of *cI* transcription, cII is the critical element.

Decision Between the Lytic and Lysogenic Cycles

The mechanism that determines the choice between the lytic and lysogenic cycles has not yet been fully worked out. However, it is clear that the cII protein plays a central role. This protein has three important functions—(1) it activates the *pre* promoter, (2) it activates *pI*, the *int* promoter, and (3) it delays synthesis of late mRNA (which encodes phage heads, phage tails, and the lytic system). For a lytic cycle to occur either the synthesis or the activity of cII protein needs to be reduced to the point that *pre* is not activated. If cII abundance or activity initiates the decision between lysis or lysogeny, these features must be responsive to various signals. The critical property of the cII protein is its low stability: its half life is about two minutes, owing to various proteolytic activities in the cell. Another phage protein, critical to cII, is cIII, which increases the half-life of cII, in part by binding to cII. In a way that is unclear, cIII senses the ratio of phage to bacteria, enhancing lysogeny when the ratio is high. The bacterial proteins HflA and HflB also affect the stability by some effect on the protease activities and these proteins respond in some way to the physiological state of the cell. These points of control affect cII *activity*. Another control is cII *synthesis*, and the bacterial proteins HimA and HimD regulate the translation of the mRNA encoding cII. Interestingly, HimA and HimD form the cimeric protein, integration host factor, needed in the integrase reaction. The complex decision phenomenon requires more study before it will be understood.

Excision and Its Avoidance during Lysogenization

In the lysogenic pathway, before cI repressor is made, transcription from *pL* occurs, so both the *int* and *xis* genes are transcribed. If both integrase and

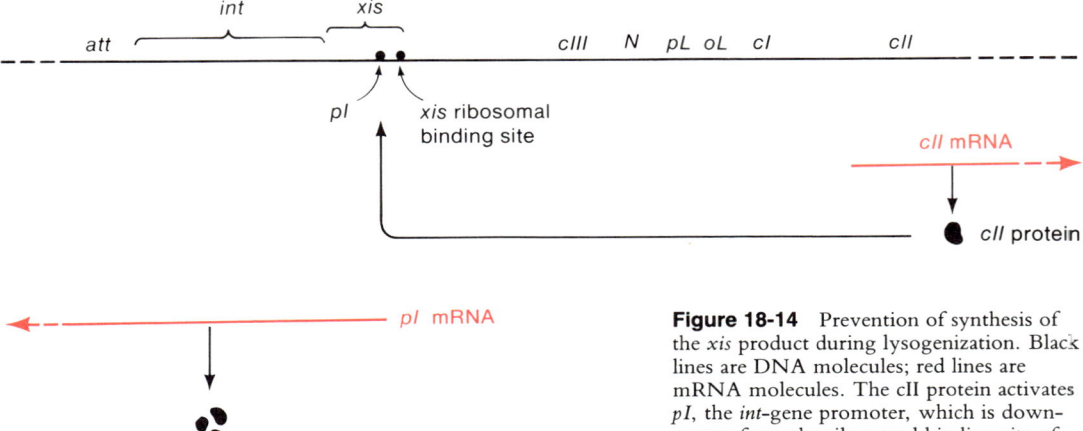

Figure 18-14 Prevention of synthesis of the *xis* product during lysogenization. Black lines are DNA molecules; red lines are mRNA molecules. The cII protein activates *pI*, the *int*-gene promoter, which is downstream from the ribosomal binding site of the Xis protein.

excisionase were present, a prophage that had just been inserted into the chromosome would immediately be excised, which clearly must be avoided. A simple mechanism prevents this excision (Figure 18-14).

As part of the progress along the lysogenic pathway, the activity of the cII protein turns on synthesis of the cI repressor from the *pre* promoter and of integrase from the *pI* promoter. The cI repressor then acts on *pL* and turns off transcription of the *int* and *xis* genes from *pL*. The promoter *pI* is located within the *xis* gene downstream from the ribosomal binding site for synthesis of the *xis* product. Thus, this mRNA molecule provides integrase but not excisionase.

OTHER MODES OF LYSOGENY

Most temperate phages form lysogens in the way described for λ—namely, a prophage is inserted at a unique site in the host chromosome. Phages have been observed for which there are several chromosomal *att* sites but this is rare. We mentioned at the beginning of the chapter that *E. coli* phage P1 is markedly different in that its prophage is not inserted into the chromosome. Following infection, P1 DNA circularizes and is repressed. In the lysogenic mode it remains as a free supercoiled DNA molecule, roughly one or two per cell. Once per bacterial life cycle the P1 DNA replicates and somehow this replication is coupled to chromosomal replication. When the bacterium divides, each daughter cell receives a P1 circle; how this orderly assortment is accomplished is unknown. The mechanism of prophage maintenance is not as foolproof in phage P1   as in temperate phages that insert their DNA into a chromosome; for example, in each round of cell division about 1 cell per 1000 fails to receive a P1 circle. It is not known whether this is due to occasional failure in replication or to imperfect segregation of plasmids into the daughter cells.

TRANSDUCING PHAGES

In the phage systems discussed so far, two rules for packaging DNA into a phage head have been encountered. These are the *headful rule* used by phage T4 (Chapter 17), in which a fixed amount of DNA is packaged, and the *defined termini rule,* in which two cuts are made in a concatemeric DNA molecule at fixed sites, as in the case of phage λ (Chapter 17). In the case of the headful mechanism nothing has been described that would prevent packaging of host DNA if a free end could be provided to initiate the process. However, this is not a common event since phage-encoded nucleases often degrade host DNA very rapidly; little or no host DNA is left by the time packaging occurs. In the case of phage λ, packaging of host DNA should also be rare because it is not likely that the host DNA would contain appropriately spaced base sequences equivalent to *cos* sites. We will see shortly that sometimes a λ phage particle contains host DNA linked to phage DNA.

Several phages are known for which host DNA can be packaged. Such phages are called **transducing phages** and a phage particle containing host DNA is called a **transducing particle.** There are two types of these phages: **generalized transducing phages,** which can produce particles containing only bacterial DNA, and **specialized transducing phages,** which occasionally produce particles containing both phage and bacterial DNA sequences. Both types of transducing particles can inject their DNA into a host bacterium and thereby transfer host DNA from one bacterium to another; this transfer process is called **transduction.** Formation of specialized, but not generalized, transducing particles is a result of lysogenization, but it is convenient to discuss both types together.

Properties of Generalized Transducing Particles

An excellent example of a generalized transducing particle is that produced when phage P1 infects *E. coli*. Phage P1 has both a lysogenic and a lytic cycle but only the lytic cycle is relevant to transduction. The life cycle of P1 is not very different from those of the lytic phages discussed in Chapter 17. The features pertinent to this discussion are: (1) the burst size (which is about 100), (2) the molecular weight of the phage DNA (9×10^6), and (3) the fact that P1 encodes a nuclease that causes degradation of *E. coli* DNA. However, this nuclease acts very slowly compared to that of T4, so when packaging begins, the host DNA consists of fragments whose molecular weights range, for the most part, from 10^7 to 10^8. Packaging of one of these fragments into a phage head occurs in about 1 percent of the particles produced. The P1 packaging system is not very fastidious, so the size of the DNA molecule does not have to be exactly the same in all particles. The result is that in the population of particles produced when infected bacteria lyse, there are generalized transducing particles, the major class of which contains a single fragment of bacterial DNA. Fragmentation of the host DNA is a random process and the transducing particles contain fragments derived from all regions of the host DNA. Thus a sufficiently large population of P1-phage progeny will contain at least one particle possessing each host gene. On the average, for any particular gene, there is roughly one transducing particle per 10^6 viable phage. Generalized transducing particles do not produce P1-phage progeny because they contain no P1 DNA; however, the bacterial DNA is injected into the host cell.

Let us now examine the events that ensue when a transducing particle infects a bacterium. Consider a transducing particle that has emerged from an infected wild-type *E. coli* and that contains the gene for leucine (*leu*) synthesis. For P1, such a particle is denoted P1 *leu$^+$*. Consider also that a P1 *leu$^+$* adsorbs to a bacterium whose genotype is *leu$^-$* and injects its DNA. The bacterium will survive because no phage genes are injected and its exonucleases may indeed degrade the injected linear DNA fragment. Another possibility is that the *leu$^+$* segment can be incorporated into the host DNA by genetic recombination, resulting in replacement of the *leu$^-$* allele by a *leu$^+$* allele. In this way, the genotype of the host cell would be converted from *leu$^-$* to *leu$^+$*. This is how transducing particles are detected (and how they were discovered). That is, phage infect a culture of X$^+$ bacteria and the resulting lysate is then used to infect an X$^-$ culture. The infected X$^-$ bacteria are plated on agar lacking X; if colonies grow, they must be X$^+$, and transduction must have occurred. This entire process is depicted in Figure 18-15.

Production of transducing particles

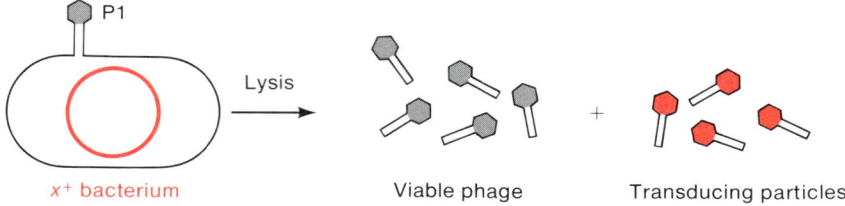

Transduction of *x$^-$* bacterium by P1*x$^+$*

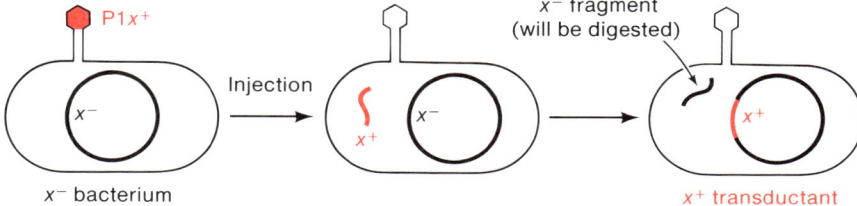

Figure 18-15   Conversion of an *x$^-$* bacterium to an *x$^+$* genotype by P1 transduction. In the upper drawing there should be about one transducing particle per 10^6 viable phage.

Properties of Specialized Transducing Particles

In the generalized transducing system just discussed, transducing particles form during a lytic cycle when host DNA fragments are packaged. Since fragmentation is random, all possible host sequences are represented in a heterogeneous population of generalized transducing particles, if the population is large enough. Transducing particles of a different type can also be produced during excision of an integrated prophage. These particles differ from generalized transducing particles in three ways: (1) the transducing particles contain both host and phage DNA linked in one continuous molecule; (2) only one or at most two regions of the host DNA, specifically those regions that flank the prophage, are found in these particles; and (3) a single transducing particle can serve as a template for production of a homogeneous population of identical transducing particles. These particles are called **specialized transducing particles.**

The mechanism by which specialized transducing particles form is shown in Figure 18-16, which depicts the formation of the galactose- and biotin-transducing forms of phage λ—namely, λ*gal* and λ*bio*.

When a λ prophage is induced, an orderly sequence of events ensues in which the prophage DNA is precisely excised from the host DNA. In the case of phage λ this is accomplished by the combined efforts of the *int* and *xis* genes acting on the left and right prophage attachment sites. At a very low frequency—namely, in about one cell per 10^6–10^7 cells—an excision error is made, two incorrect cuts are made—one within the prophage and the other cut in the bacterial DNA. The pair of abnormal cuts will not always yield a length of DNA that can fit in a λ phage head—it may be too large or too small. However, if the spacing between the cuts produces a molecule between 79 percent and 106 percent of the length of a normal λ phage-DNA

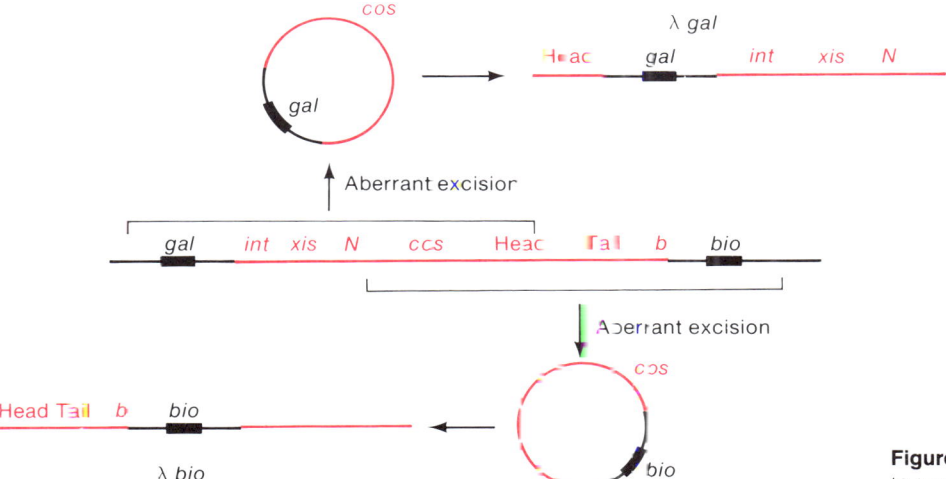

Figure 18-16 Aberrant excision leading to production of λ*gal* and λ*bio* phages.

molecule, packaging can occur. Since the prophage is located between the *E. coli gal* and *bio* genes and because the cut in the host DNA can be either to the right or the left of the prophage, transducing particles can arise carrying the *bio* genes (cut to the right) or the *gal* genes (cut to the left). Formation of the λ*gal-* and λ*bio*-transducing particles entails loss of λ genes. The λ*gal* particle lacks the tail genes, which are located at the right end of the prophage; the λ*bio* particle lacks the *int, xis, . . .* genes from the left end of the prophage. The number of missing phage genes of course depends on the position of the cuts that generated the particle and thus correlates with the amount of bacterial DNA in the particle. The missing phage genes come from the prophage ends, but because of the permutation of the gene order in the prophage and the phage particle, the deleted phage genes are always from the central region of the phage DNA (Figure 18-16). The term deletion is usually used to describe a DNA molecule in which wild-type base sequences are absent; however, in a specialized transducing particle the missing phage genes are replaced by bacterial genes. Thus, these are called **deletion-substitution particles.**

Since specialized transducing particles lack phage genes, they might be expected to be nonviable. This is indeed true of the λ*gal* type, which lack genes essential for synthesis of the phage tail and, in the case of the larger deletion-substitutions, the genes for the phage head also. Thus, these particles are incapable of producing progeny; they are defective and this is denoted by the symbol d—thus, a *gal*-transducing particle is written λd*gal* (and sometimes λd*g*). The λd*gal* particle contains the cohesive ends and all of the information for DNA replication and transcription; it therefore goes through a normal life cycle, including bacterial lysis. In fact, if the head genes are not deleted, the concatemeric branch produced by rolling circle replication is cleaved by the Ter system. However, tails are not added to the filled head and no viable (plaque-forming) particles are produced in the lysate. Note that there is no discrepancy between the formation of a λd*gal* transducing particle and its ability to reproduce itself. A λd*gal* particle fails to reproduce only because it lacks the tail genes, but such a particle arises from a normal prophage having a full set of genes.

The situation is quite different with λ*bio* particles for these usually lack only nonessential genes—*int, xis,* etc. These genes are needed for the lysogenic cycle but not for the lytic cycle so these particles are able to replicate and to form plaques. To denote this, the letter p, for plaque-forming, is added; that is, the particle is called λp*bio*.

Specialized Transduction of a Nonlysogen

There are several mechanisms by which a specialized transducing particle can transduce a mutant bacterium. The mechanisms are the same for the λ*gal* and λ*bio* particles, so we will use only λ*gal* as an example. Consider a

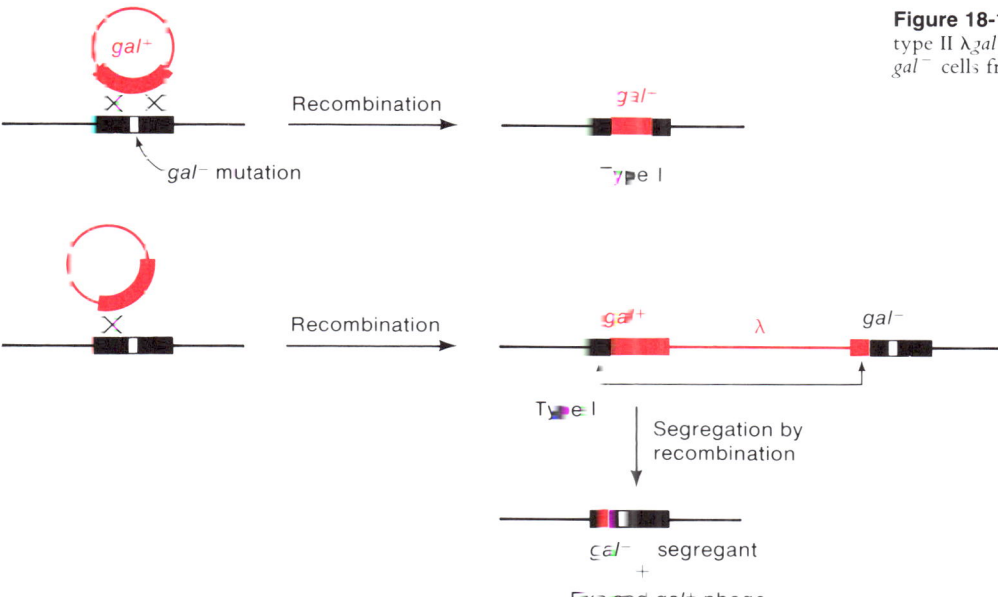

Figure 18-17 Production of type I and type II λ*gal* transductants and segregation of *gal*⁻ cells from a type II cell.

Gal⁻ cell that is infected with a lysate resulting from induction of a Gal⁺ culture lysogenic for λ. This lysate will consist overwhelmingly of the normal λ phage but will contain a tiny fraction of the λ*gal*⁺ phage. Conditions are used that lead to the establishment of immunity repression; thus, the cell is not killed by a lytic response. If these infected cells are plated on agar that distinguishes Gal⁺ from Gal⁻ colonies, by their color (e.g., purple for Gal⁺, white for Gal⁻), purple colonies are found at a very low frequency (about 0.001 percent of the infected cells). These purple colonies are of two types (Figure 18-17):

Type I consists of nonlysogenic cells all of which are Gal⁺. If the colony is dispersed and individual bacteria are allowed to form colonies, all colonies are purple. These contain stable *gal*⁺ bacteria and have arisen as a result of two genetic exchanges, as shown in the figure.

Type II cells contain a prophage. If a type II colony is dispersed and individual cells are plated, about 1 percent of the colonies are white (and hence Gal⁻) and also no longer have a prophage. Type II cells have arisen by a single exchange within the *gal* genes. These cells now contain two copies of the *gal* genes, one *gal*⁺ and one *gal*⁻. These transductants are called **heterogenotes.** The type I cells, which contain only one *gal*⁺ gene, are haploid

Plasmids

Plasmids are extrachromosomal circular DNA molecules found in most bacterial species and in some species of eukaryotes. Normally a particular plasmid is dispensable to its host cell, though many plasmids contain genes that may be essential in certain environments. For example, the R plasmids carry genes that confer resistance to numerous antibiotics, so in nature a cell containing such a plasmid can better survive. Plasmids of this type are often detected by the genes that they carry; however sometimes a plasmid is discovered incidentally as a small circular DNA molecule present in a DNA sample isolated from a cell extract.

Plasmids have many interesting biological properties, which will be described in this chapter. However, they are also extraordinarily useful tools for the molecular biologist, as was indicated in the discussion of the *lac* operon in Chapter 15, and for genetic engineering (Chapter 22).

GENERAL PROPERTIES AND TYPES OF PLASMIDS

Plasmids, like phages, are heavily dependent on the metabolic functions of the host cell for their reproduction, though each plasmid has genes and regulatory sites that distinguish one type of plasmid from another. For example, plasmids normally use most of the replication machinery of the host.

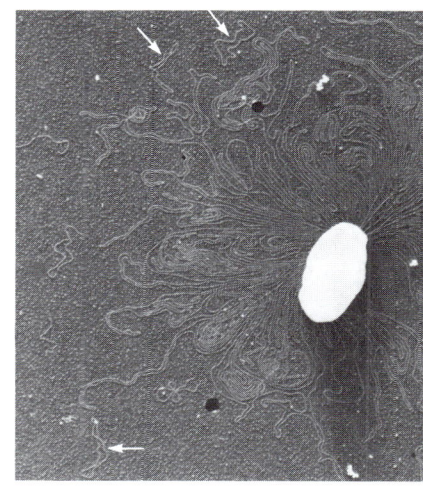

A ruptured *E. coli* cell showing three small circular plasmids (arrows). [Courtesy of David Dressler and Huntington Potter.]

However, each type of plasmid has its own genes for regulating the time of synthesis and the number of plasmid copies per cell (which may range from 1 or 2 for the **stringent** or **low-copy-number** plasmids to 10–100 for the **relaxed** or **high-copy-number** plasmids). Furthermore, the segregation of plasmid replicas into daughter bacterial cells during cell division is carefully regulated. That is, if a cell that is ready to divide contains two plasmid DNA molecules, each daughter cell receives one molecule; the failure of this process, which is indicated by the appearance of a plasmid-free cell, occurs at a frequency of only about 1 per 10^4 cells per generation. This stability is conferred by a type of plasmid-chromosome association that is not understood.

In this chapter the discussion will be confined to plasmids of *E. coli* except when otherwise noted. Many types of plasmids are found in a variety of *E. coli* strains but three main types—the F, R, and Col plasmids—have been studied. These plasmids share some properties but for the most part are quite different. The presence of an F, R, or Col plasmid in a cell is indicated by the following characteristics acquired by the cell:

1. *F, the sex plasmid.* Ability to transfer chromosomal genes (that is, genes not carried on the plasmid) and the ability to transfer F itself to a cell lacking the plasmid.
2. *R, the drug-resistance plasmid.* Resistance to one or more antibiotics and often the ability to transfer the resistance to cells lacking R.
3. *Col, the colicinogenic factor.* Ability to synthesize **colicins**—that is, proteins capable of killing closely related bacterial strains that lack the Col plasmid.

Further discussion of each of these plasmid types will be presented throughout the chapter.

With only a single exception (the killer-plasmid of yeast, which is an RNA molecule) all known plasmids are supercoiled circular DNA molecules (Figure 19-1). The molecular weights of the DNA range from about 10^6 for the smallest plasmid to slightly more than 10^8 for the largest one. Table 19-1 lists the molecular weights for several plasmids that are actively being studied. How plasmid DNA is isolated has already been described in Chapter 5.

The isolation of DNA usually entails a deproteinization step. When the DNA molecules of some *E. coli* plasmids are isolated without such a step, about half of the supercoiled DNA molecules contain three tightly bound protein molecules. This DNA-protein complex is called a **relaxation complex.** If this complex is heated or treated with alkali, proteolytic enzymes, or detergents, one of these proteins, which is a nuclease, nicks one DNA strand, thereby "relaxing" the supercoil to the nicked circular form (Figure 19-2). This nick occurs in only one strand and at a unique site. During relaxation the two smallest proteins are released but the largest protein becomes covalently linked to the 5′-P terminus of the nick. If, prior to relaxation, the supercoiled DNA is nicked by any of a number of laboratory techniques,

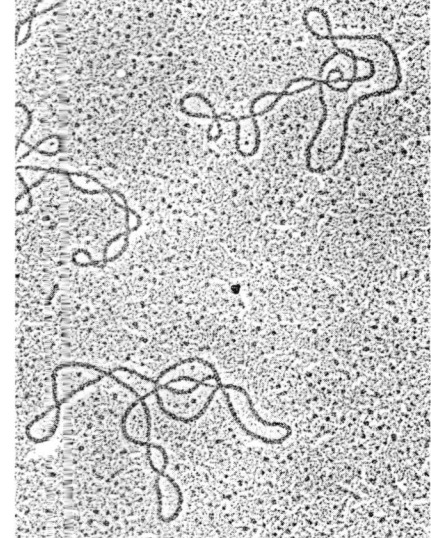

Figure 19-1 Two supercoiled plasmid DNA molecules.

Table 19-1 Several plasmids and selected properties

Plasmid	Mass, $\times 10^6$	No. copies/chromosome	Self-transmissible	Phenotypic features
Col plasmics				
ColE1	4.2	10–15	No	Colicin E1 (membrane changes)
ColE2 *(Shigella)*	5.0	10–15	No	Colicin E2 (DNase)
ColE3	5.0	10–15	No	Colicin E3 (ribosomal RNase)
Sex plasmids				
F	62	1–2	Yes	F pilus
F'*lac*	95	1–2	Yes	F pilus; *lac* operon
R plasmids				
R100	70	1–2	Yes	Cam-r Str-r Sul-r Tet-r
R64	78	(limited)	Yes	Tet-r Str-r
R6K	25	12	Yes	Amp-r Str-r
pSC101	5.8	1–2	No	Tet-r
Phage plasmid				
λdv	4.2	≈50	No	λ genes *cro, cI, O, P*
Recombinant plasmids				
pDM500	9.8	≈20	No	*Drosophila melanogaster* histone genes
pBR322	2.9	≈20	No	High-copy-number
pBR345	0.7	≈20	No	ColE1-type replication
Yeast plasmid				
2μm	4.0	≈60	No	No known genes

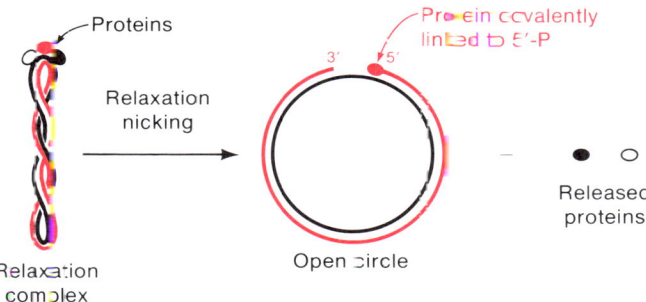

Figure 19-2 Nicking of one strand of a supercoiled DNA relaxation complex.

the relaxation nuclease is unable to make its site-specific nick, indicating that the nuclease is active only on supercoiled DNA. This nicking is thought to play a role in conjugational transfer of the plasmid to another bacterium (see next section).

TRANSFER OF PLASMID DNA

Many plasmids can transfer a replica of the plasmid from a donor cell to a recipient cell. For example, if a single *E. coli* cell containing the F plasmid is added to a culture of growing F^- cells, after 15–20 generations of cell growth, a large fraction of the cells contain F. This is a result of transfer of a replica of F from an F^+ cell to an F^- cell without loss of F by the F^+ cell. After transfer, the F replica remaining in the original cell can replicate again, and another replica can be transferred to a second F^- cell. Every recipient acquires F and therefore becomes a donor from which F can be transferred to other F^- cells. Transfer can occur once or twice per cell generation, so F quickly spreads throughout a bacterial population.

An Outline of the Transfer Process and Some Definitions

The plasmid transfer process can be divided into four stages:

1. Formation of specific donor–recipient pairs (**effective contact**).
2. Preparation for DNA transfer (**mobilization**).
3. DNA transfer.
4. Formation of a replicative functional plasmid in the recipient (**repliconation**).

Many plasmids fail to carry out all of these processes, so the following terminology has been developed:

A **conjugative plasmid** is defined as a plasmid carrying genes that determine the effective contact function.

A **mobilizable plasmid** can prepare its DNA for transfer.

A **self-transmissible plasmid,** such as F, is both conjugative and mobilizable.

A cell may contain two different plasmids—for example, F and ColE1. F is both conjugative and mobilizable (it is self-transmissible), and ColE1 is mobilizable and nonconjugative. In a cell containing both plasmids, F can provide the missing conjugative function to ColE1, and ColE1 can thereby be transferred. This process—namely, one in which a nonconjugative plasmid is transferred via the effective contact provided by a conjugative plasmid—is called **donation;** its hallmark is efficient transfer of the nonconjugative plasmid. A conjugative plasmid can also help a nonmobilizable plasmid to be transferred, though not by complementation because the mobility genes and elements are usually plasmid specific (to be discussed). Instead, the process, which occurs at low frequency, occurs via genetic recombination between the two plasmids to form a single transferable DNA molecule; it is called **conduction,** in contrast with donation. The frequency (high versus low) of helped transfer of a non–self-transmissible plasmid is a useful criterion for determining whether a plasmid is mobilizable by donation or conduction.

F, whose molecular weight is about 63 × 10^6, contains at least 19 genes needed for transfer; these are called *tra* genes. A mutation in any one of these genes eliminates self-transmissibility so presumably this is the minimal number needed for any self-transmissible plasmid. Since this number of genes requires about 15 × 10^6 molecular weight unit of DNA (assuming an average molecular weight of the gene products of 40,000), the small plasmids are not self-transmissible. The smallest known self-transmissible plasmid has a molecular weight of 17 × 10^6. F will serve as a model for the study of the transfer process.

A Detailed Look at Transfer of F

The first step in effective contact is pair formation between a donor and a recipient cell. This requires a hairlike protein appendage, called a **sex pilus,** on the donor cell (Figure 19-3). The pili on F- and R-containing cells are called **F pili** and **R pili,** respectively. Pili retract into the donor cell after pairing, so the pilus probably serves first to bring the pair into initial contact and then to draw the cells together into close contact.

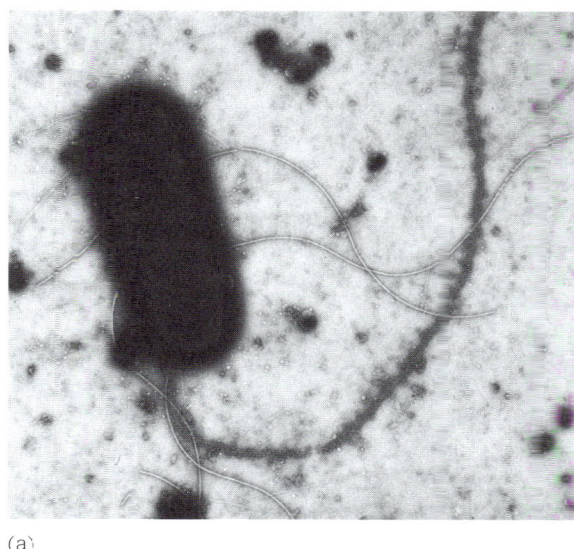

(a)

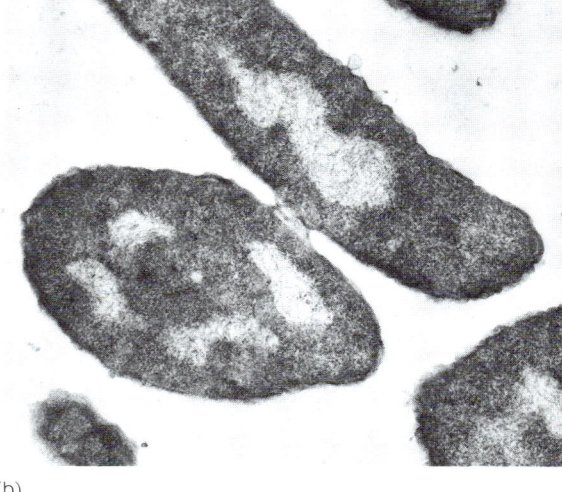

(b)

Figure 19-3 (a) An *E. coli* cell showing a single sex pilus, which is coated with the F-specific phage R17 to make the very thin pilus visible as a rough dark appendage. The five heavy light fibers are flagella. The very faint thin hairs are called fimbrae. [Courtesy of Barry Eisenstein.] (b) Electron micrograph of two *E. coli* cells during conjugation. The small cell is an *F*⁻ cell; the larger cell contains F′*lac*. [With permission, from L. Caro, *J. Mol. Biol*. (London: Academic Press, Ltd., 1966), pp. 16, 269.]

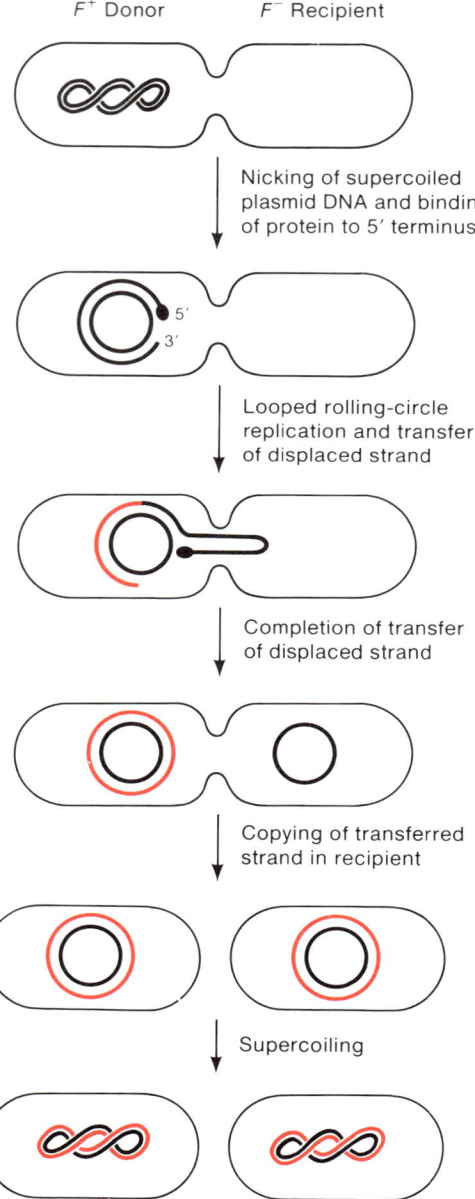

Figure 19-4 A model for transfer of F plasmid DNA from an F^+ cell by a looped rolling-circle mechanism. The displaced single strand is transferred to the F^- recipient cell, where it is converted to double-stranded DNA. Chromosomal DNA and proteins of the relaxation complex in completed DNA molecules are omitted for clarity.

Mobilization begins when a plasmid-encoded protein, which is probably the nicking protein of the relaxation complex, makes a single-strand break in a unique base sequence called the **transfer origin** or *oriT*. (Relaxation complexes have not been detected for all transmissible plasmids but presumably there is a protein serving the function just described.) This nick initiates rolling circle replication and the linear branch of the rolling circle is transferred. It is thought that the nicking protein remains bound to the 5′ terminus and that the replication mode is like the looped rolling-circle mechanism used by phage φX174 (Chapter 9). The sequence of events during transfer is shown schematically in Figure 19-4.

Note that DNA synthesis occurs both in donor and recipient cells. The synthesis in the donor, called **donor conjugal DNA synthesis,** serves to replace the single strand that is transferred. Synthesis in the recipient cell (**recipient conjugal DNA synthesis**) converts the transferred single strand to double-stranded DNA.

In the usual situation during transfer the transferred strand is simultaneously replaced by donor conjugal synthesis and converted to double-stranded DNA in the recipient. This would indicate that DNA synthesis and transfer are coupled were it not for the following observations: (1) Transfer occurs even if donor synthesis is inhibited by appropriate host mutations; (2) donor synthesis occurs even if transfer is prevented by appropriate plasmid mutations; (3) transfer occurs even though recipient conjugal synthesis is inhibited by mutations in the recipient cell. These observations raise the question of the identity of the motive force for transfer since clearly it is not DNA replication. It has been suggested (and there is preliminary evidence) that DNA helicase activity in the donor both unwinds the DNA prior to replication and provides the driving force for transfer.

PLASMID REPLICATION

A plasmid can only replicate within a host cell; one might therefore expect that all plasmids native to the same host species would have the same mode of replication. However, as with phages, which also replicate only within a host cell, there is enormous variation in both the enzymology and mechanics of plasmid DNA replication. This is explained in the following two sections.

Variations in the Use of Host Proteins in Replication Mechanisms

Plasmids rely heavily on the host replication apparatus for their replication. In fact, some plasmids seem to use host gene products exclusively. For example, the small plasmid, ColE1, can be replicated *in vitro* by adding purified

ColE1 DNA to a cell extract prepared from cells that do not contain ColE1 or any other plasmid and hence do not contain any plasmid-encoded gene product. Other plasmids require some plasmid gene products. For instance, temperature-sensitive mutants of F are known that fail to replicate at 42°C; however, since bacterial DNA is synthesized normally at this temperature, cells containing the temperature-sensitive plasmid produce F^- daughter cells after several generations of growth at 42°C.

All plasmids examined to date replicate semiconservatively and maintain circularity throughout the replication cycle. However, there are significant differences in the replication pattern from one plasmid to the next. Some of the variants are listed below:

1. *Directionality.* Both purely unidirectional and purely bidirectional replication have been observed. In addition, there are plasmids in which both modes are present. For example, the plasmid RK6 replicates first in one direction and then later in the opposite direction from the same origin.

2. *Termination.* In unidirectionally replicating plasmids termination necessarily occurs at the origin after one cycle of replication. Bidirectionally replicating molecules are of two types though, both of which have been observed. In one type, termination occurs when the growing forks reach the same region, as with phage λ (Chapter 9). Others have a fixed termination site that is sometimes reached by one growing fork before the other fork reaches it. The signal for termination is unknown.

3. *Replicating form.* In the most carefully studied plasmids, replication occurs by the so-called *butterfly* mode, first observed for animal viruses (Figure 19-5). In a partially replicated molecule the replicated portions are untwisted, as is usually the case in θ replication, but the unreplicated portion is supercoiled. When the replication cycle is completed, one of the circles must be cleaved (probably by DNA gyrase). The result after one round of replication is one nicked molecule and one supercoiled molecule. The nicked molecule is then sealed and, somewhat later, supercoiled. Whether this is a general mechanism for plasmid replication is not known.

Control of Copy Number

Some plasmids are present in cells in low-copy number—one or a few per cell—whereas others exist in large numbers—from 10 to 100 per cell. The following simple experiment indicates that the copy number is established and regulated by controlling the rate of initiation of DNA synthesis. If a cell is transformed with a single copy of a low-copy-number plasmid, the plasmid DNA replicates only once or twice before cell division. However, if a

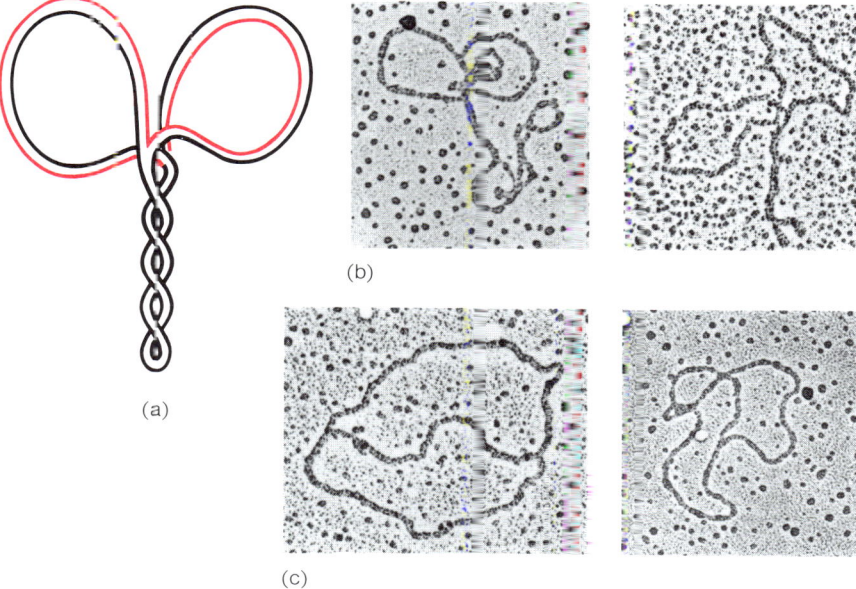

(a)

(b)

(c)

Figure 13-8 Replication of ColE1 DNA. (a) Diagram of a butterfly molecule. The newly synthesized strands are red. (b) Electron micrographs of butterfly molecules. (c) Nicked butterfly molecules, showing that a nick converts a butterfly molecule to a θ molecule. [Courtesy of Donald Helinski.]

cell is transformed with a single DNA molecule of a high-copy-number plasmid, the plasmid DNA replicates repeatedly until the proper copy number is reached.

The generally accepted explanation for the regulation of copy number is that there is a plasmid-encoded inhibitor that is a negative regulator of the initiation of replication. (We avoid the term repressor because, as we will see, in at least one case, the inhibitor does not act on the DNA.) Let us first see how this idea explains maintenance of copy number from generation to generation. In the current theory it is assumed that the activity of the inhibitor is dependent on concentration. Thus, as a cell grows (enlarges), the inhibitor concentration drops, replication is not inhibited, and the number of plasmid DNA molecules doubles. At this point there will exist twice the initial number of inhibitor genes; therefore, by protein synthesis the inhibitor concentration also doubles, causing replication to stop. A similar sequence of events would occur if there were initially only one copy of a high-copy-number plasmid—that is, replication would continue until there is sufficient inhibitor to turn off synthesis. How can this explanation account for differences in copy number? The most likely possibility is that for the high-copy-number plasmids the inhibitory activity requires a higher concentration of inhibitor than is the case for the low-copy-number plasmids. Thus, only when the number of plasmids per cell is high is the "gene dosage" high enough to cause inhibition. The following experiment supports this view by providing evidence that each plasmid controls its own copy number.

A hybrid plasmid pSC134 was constructed (by recombinant DNA techniques; Chapter 22) that consists of a complete copy of each of two plasmids, ColE1 and pSC101. The copy numbers for these plasmids are 18 and 6, respectively. The following three facts about pSC134 were obtained (Figure 19-6):

1. Plasmid pSC134 replicates from the ColE1 origin and has a copy number of 16—roughly equal to that for ColE1.
2. If pSC134 is put into a $polA^-$ cell (ColE1 cannot replicate in a $polA^-$ cell), the pSC101 origin is used and the copy number becomes 6, the value for pSC101. These two results show that the copy number correlates with the replication origin that is being used.
3. If pSC101 DNA is taken into a bacterium containing 16 copies of pSC134 (using $CaCl_2$ transformation), the pSC101 cannot replicate. This lack of replication shows that the pSC101 inhibitor is being made by pSC134.

The interpretation of these three results is the following. If there were a copy number less than 6, both pSC101 and ColE1 origins would be active and replication from both origins would increase the number. If the number were greater than 6, the pSC101 origin would be inactive because the pSC101 inhibitor concentration would be above its inhibitory level. Synthesis of pSC101 would continue, though if it were autogenously regulated, the concentration would not exceed that produced by a copy number of 6. The ColE1 origin would remain active, for it would only be shut off if the copy number were to exceed 16. In a $polA^-$ cell ColE1 cannot replicate; consequently, the copy number is totally controlled by the concentration of pSC101 inhibitor, and thus it will not exceed its normal value. This interpretation is completely consistent with the facts, and with several other types of experiments that also support the inhibitor model of copy-number regulation.

The following important point must be understood about the inhibitor model. Since all of the plasmids in a particular cell are identical, an inhibitor molecule cannot distinguish one molecule from another. Thus, when the concentration of inhibitor is low enough that all DNA molecules are not inhibited, the few that are inhibited are drawn randomly from the population. This means that if one plasmid replicates to form two daughter plasmids, an individual daughter plasmid has the same probability of replicating a second time as any other molecule has of replicating (a first time). That is, at any instant, a molecule is chosen for replication by random selection from the entire population of plasmids.

The mechanism of inhibition is fairly well understood for ColE1 (Figure 19-7). The process of initiation of ColE1 synthesis begins with transcription of a long primer RNA 651 bp upstream from the replication origin. The

Replication of normal plasmid in *polA⁺* or *polA⁻* cells:

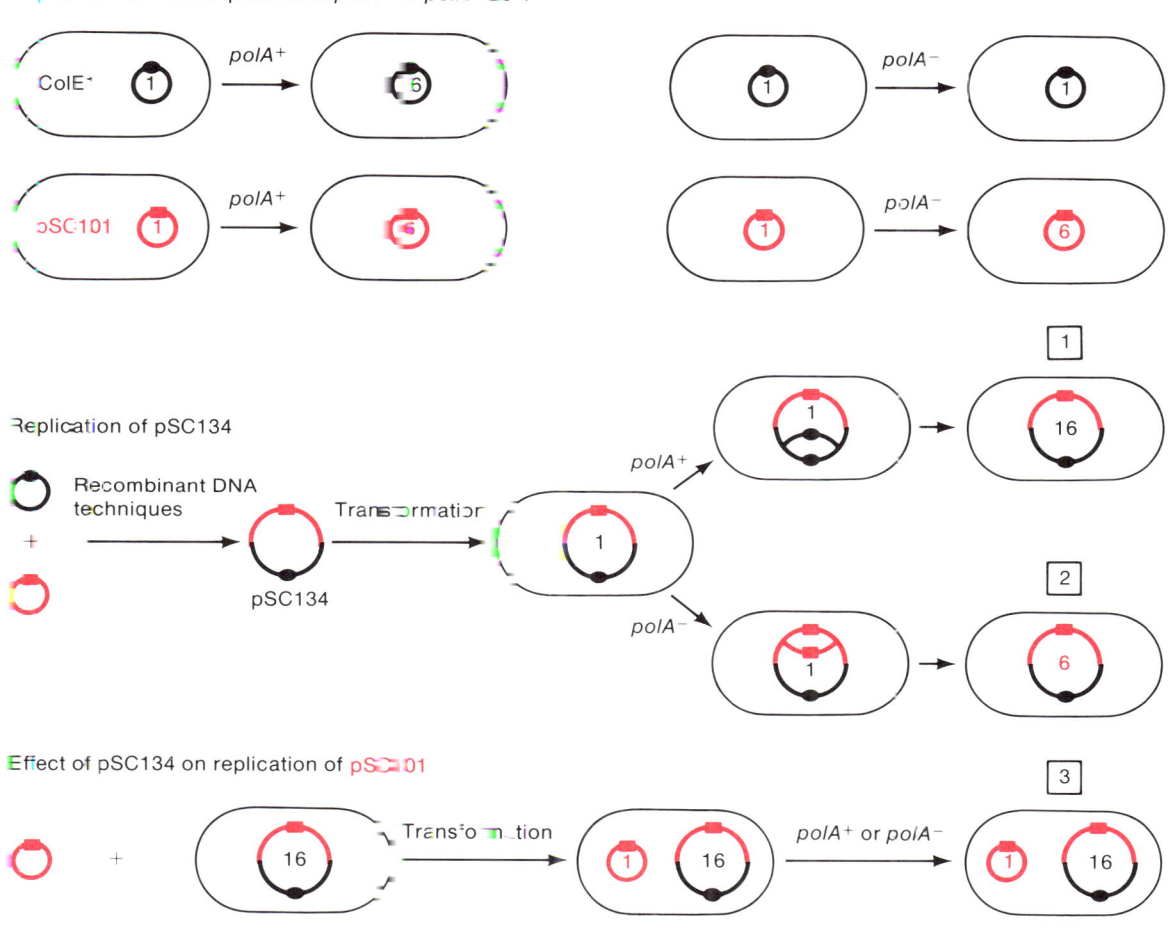

Replication of pSC134

Recombinant DNA techniques

Transformation

Effect of pSC134 on replication of pSC101

Transformation

polA⁺ or *polA⁻*

pSC101 fails to replicate

Figure 19-6 Diagram depicting the replication of ColE1, pSC101, and the hybrid plasmid pSC134, starting with one copy per cell. The black solid circle and the red solid square designate the replication origins of ColE1 and pSC101, respectively. The numbers (1, 6, and 16) in the DNA molecules indicate the number of copies in the cell. The boxed numbers (1, 2, and 3) refer to similarly numbered items in the text.

RNA molecule, called RNA II, is at first synthesized like a typical RNA—that is, only a few bases at the 3′ terminus remain paired with the DNA. Because of internal complementarity the released RNA forms a structure with several stems and loops. Successful formation of this secondary structure somehow signals the newly made RNA to remain in contact with the DNA (555 bases from the transcription initiation site). After about 100 bases form an RNA-DNA hybrid region, the enzyme RNase H (which cuts RNA only in an RNA-DNA hybrid) makes a single cut in the RNA at the repli-

Figure 19-7   Initiation of replication of ColE1.

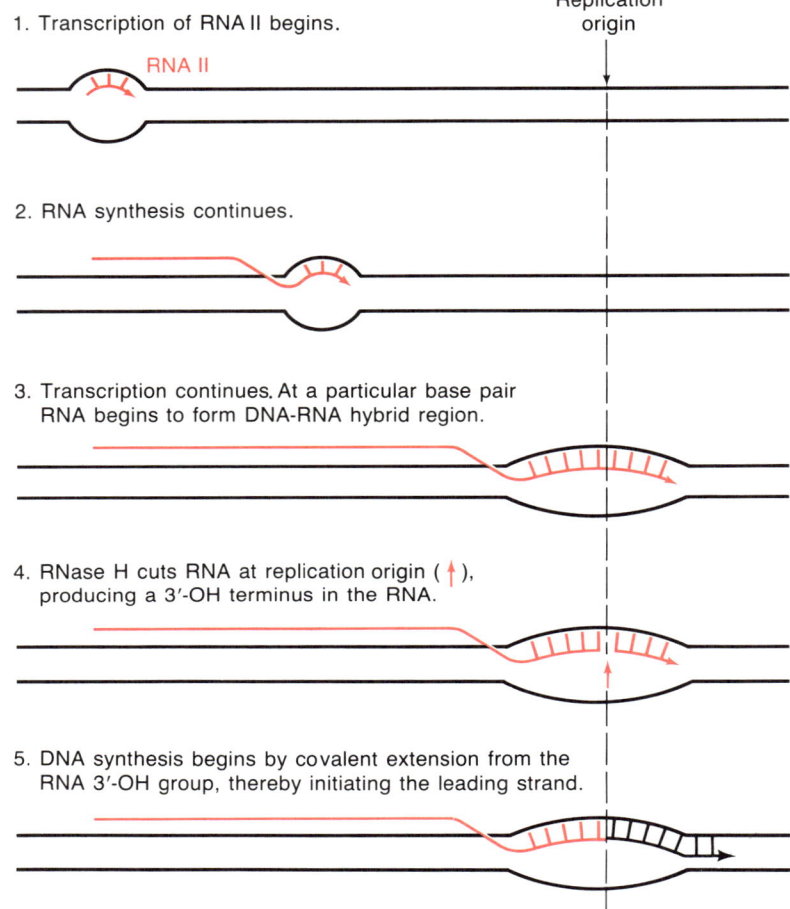

1. Transcription of RNA II begins.

2. RNA synthesis continues.

3. Transcription continues. At a particular base pair RNA begins to form DNA-RNA hybrid region.

4. RNase H cuts RNA at replication origin (↑), producing a 3′-OH terminus in the RNA.

5. DNA synthesis begins by covalent extension from the RNA 3′-OH group, thereby initiating the leading strand.

cation origin, producing a 3′-OH terminus. DNA synthesis then ensues by addition of a deoxynucleotide to the 3′-OH group.

Another RNA molecule is also transcribed in the origin region. This molecule, called RNA I, is initiated 110 bp to the right (using the right-left orientation of Figure 19-7) from the start point for transcription of RNA II. It is 108 bases long and is transcribed leftward. That is, it is copied from the DNA strand that is complementary to the DNA strand that is the template for RNA II, and thus RNA I is complementary to the 5′ portion of RNA II. Hence, RNA I can form a double-stranded structure with RNA II and thereby prevent formation of the secondary structure needed for maintenance of the RNA–DNA hybrid that is ultimately cut by RNase H. RNA I is therefore an inhibitor of initiation of ColE1 DNA replication and participates in regulation of copy number. Another regulatory element is important—a small plasmid-encoded protein called Rom (*RNA* one inhibition *modulation*). In a way that is not clear, Rom determines the rate of formation of the double-stranded (RNA I)-(RNA II) segment. If the concentration of Rom is too low, double-stranded RNA does not form, RNA II achieves the stem-and-loop conformation, RNase H cleaves RNA II, and DNA synthesis is initiated.

High-copy-number plasmids exhibit a phenomenon called **amplification.** If chloramphenicol, or any other inhibitor of protein synthesis, is added to a culture of plasmid-containing bacteria, initiation of replication of chromosomal DNA, but not plasmid DNA, is inhibited, and the number of plasmids per cell increases to 1000 or more. The inhibitor model explains this phenomenon. Bacterial DNA synthesis is inhibited because of the inability to synthesize initiation proteins (e.g., DnaA protein). However, if the plasmid utilizes only stable bacterial replication proteins (such as pol I, pol III, RNase H, RNA polymerase) or stable plasmid-encoded proteins, plasmid DNA synthesis can continue. In fact, since the copy-number regulator (for example, Rom) is a concentration-dependent inhibitor, the amount of inhibitor will be insufficient (in the absence of protein synthesis) and plasmid replication will be initiated repeatedly, without regulation. Increased availability of bacterial replication proteins, owing to lack of bacterial DNA synthesis, and metabolic instability of RNA molecules, such as RNA I, contribute to the excessive synthesis.

Incompatibility

Pairs of closely related plasmids usually cannot be stably maintained in the progeny of a single cell; such plasmids are said to be **incompatible.** The inhibitor model for initiation of plasmid replication also explains this phenomenon.

Let us first consider a cell that contains two plasmids say, F and ColE1, having different inhibitors. Replication of each type of plasmid will proceed independently of one another since the inhibitor of one type (e.g., F) does not regulate the replication of the other type (e.g., ColE1). Thus, F and ColE1 are compatible. Alternatively, one says that they belong to different **incompatibility groups.**

The situation is quite different with two plasmids A and B whose inhibitors are either identical or are similar enough that the inhibitor of A can regulate replication of B and vice versa. Let us consider a cell having one copy of A and one of B and which has enlarged sufficiently that initiation occurs (Figure 19-8). Since the two plasmid copies are selected at random for replication, the result of the first replication event is a cell having either one copy of A and two of B (1A,2B) or one of B and two of A (2A,1B). When the second replication event occurs, each cell will have four plasmids, but, depending on the plasmid that is replicated, the plasmid composition

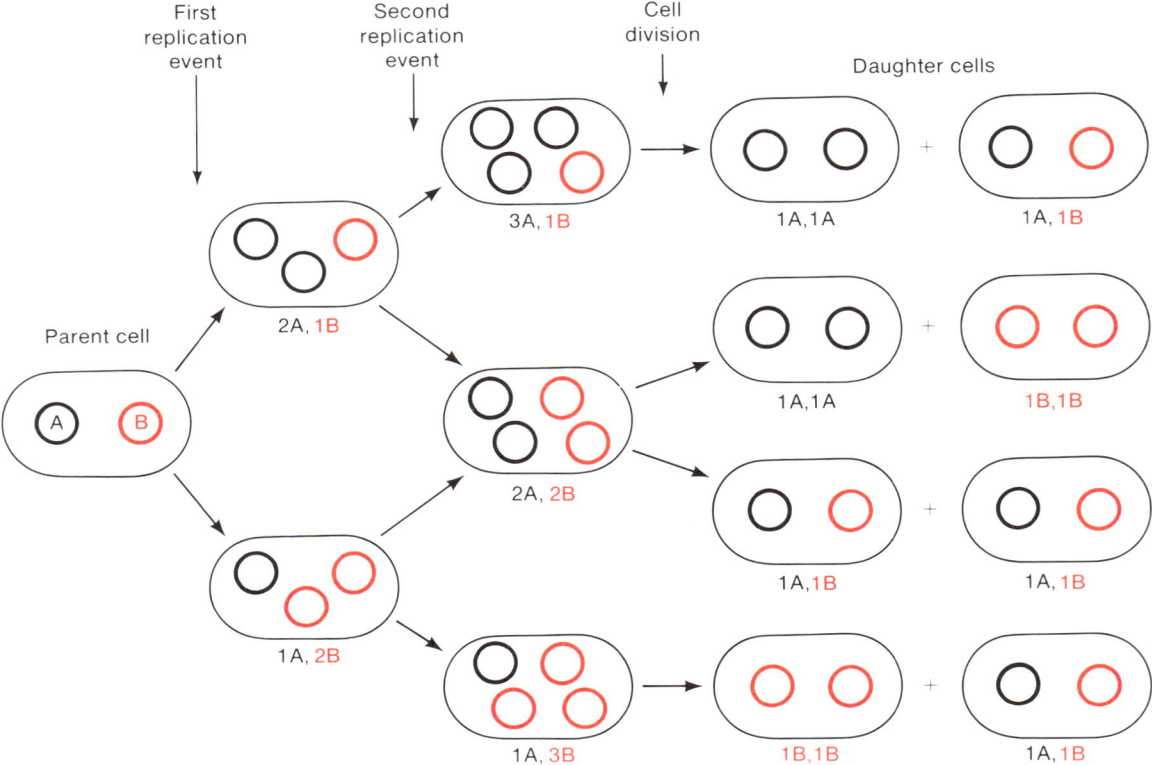

Figure 19-8   Four possible pairs of daughter cells that can arise from division of a cell containing one copy each of the incompatible plasmids A and B. See text for details.

may be (1A,3B), (2A,2B), or (3A,1B), as shown in the figure. At this point the cell, which has twice the initial number of plasmids, can divide. The plasmid composition of the two daughter cells will be one of the following:

> (1A,3B) becomes (1A,1B) + (1B,1B).
> (2A,2B) becomes (1A,1A) + (1B,1B) or (1A,1B) + (1A,1B).
> (3A,1B) becomes (1A,1A) + (1A,1B).

Note that two possible types of cells, namely, (1A,1A) and (1B,1B), contain only one of the two kinds of plasmids; daughter cells obtained from these cells will of course continue to have only one kind of plasmid. In subsequent cell divisions each cell still containing one A and one B will, with 50 percent probability, produce daughter cells lacking one of the plasmids. Thus, as a single cell initially containing two incompatible plasmids divides, the percentage of the progeny population containing only one plasmid type will increase with each generation.

Thus, incompatibility is a result of (1) two plasmids having a similar inhibitor and (2) the random selection of plasmids for DNA replication.

PROPERTIES OF PARTICULAR BACTERIAL PLASMIDS

In this section we examine the properties of the F, R, and Col plasmids and survey a few other plasmids.

The Sex Plasmid F and its Derivatives

F is a circular DNA molecule having a molecular weight of 62.5×10^6 (94.5 kilobase pairs). It has been extensively mapped genetically and physically. Of particular interest are the sequences IS2, IS3, and $\gamma\delta$ shown in Figure 19-9. These regions are called transposons and will be discussed later in this chapter and in Chapter 21.

An important property of F is its ability to integrate into the bacterial chromosome to generate an Hfr cell (Chapter 1). Integration is a reciprocal exchange much like that occurring when phage λ lysogenizes a bacterium. Integration of F into the E. coli chromosome has the following characteristics that differ from λ prophage formation: (1) The exchange site in F is not unique; whereas exchange occurs predominantly in the IS3 element located near 94.5 on the physical map, exchange also occurs in the other IS3 element and in $\gamma\delta$ as shown in Figure 19-10. (2) There are many sites in the chromosome at which integration can occur and more than 20 major sites are known. The affinity of F for each site is not the same in that the frequency of formation of a particular Hfr strain varies from one site to the next. Each of these sites is believed to be a copy of IS2, IS3, or $\gamma\delta$ in the chromosome,

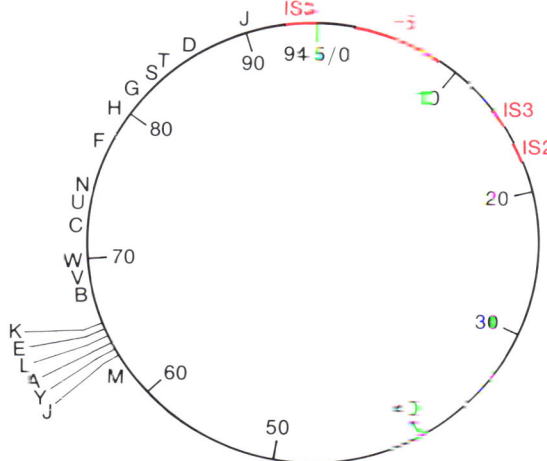

Figure 19-9 A map of the sex plasmid F. Points are given in kilobase pairs. The single capital letters refer to the midpoints of the locations of the corresponding *tra* genes. The sequences $\gamma\delta$, IS2, and IS3 are shown in red.

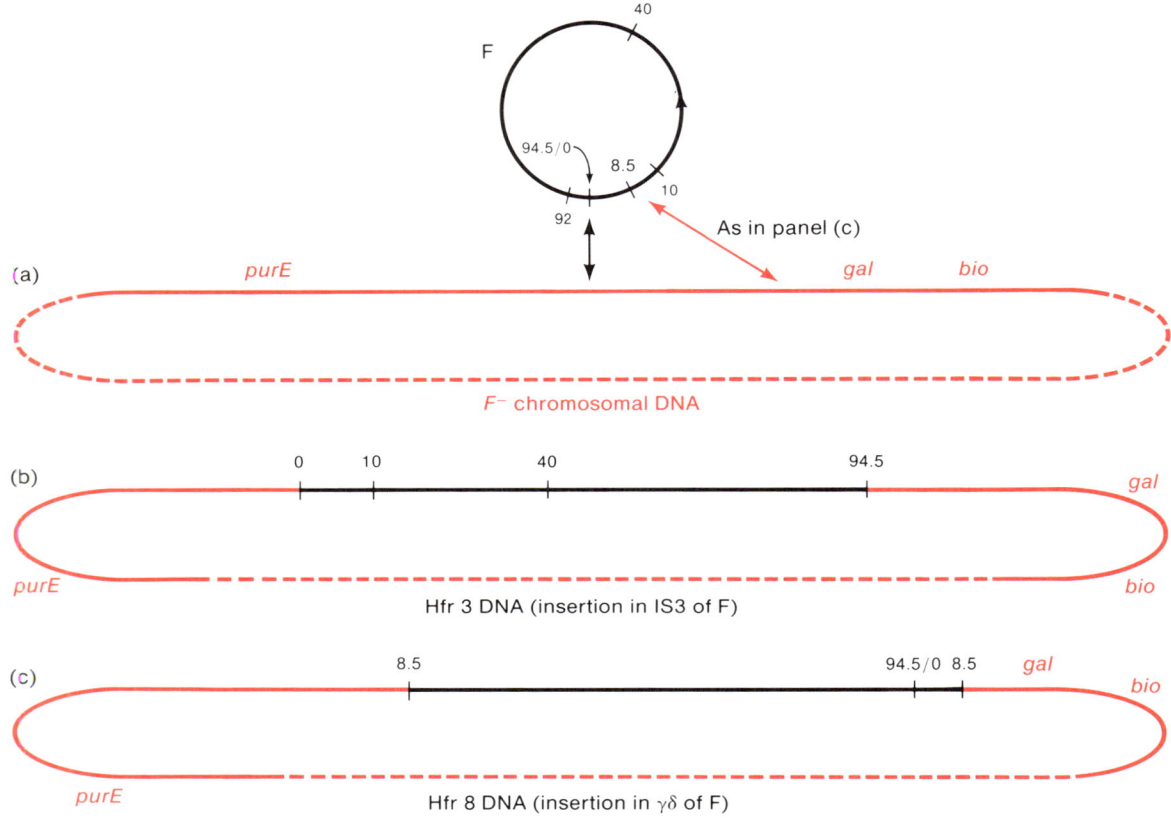

Figure 19-10 Formation of two different Hfr strains. (a) The F^- chromosome and F. The exchange in both molecules occurs at the points indicated by the black double-headed arrow. (b) The result of the exchange shown in panel (a). (c) The result of the exchange indicated in panel (a) by the double-headed red arrow.

and the affinity represents the probability of exchange with these elements. (3) F can integrate in both clockwise and counterclockwise orientations. Since transfer of F to a female cell occurs from a single origin and in a single direction, there are Hfr strains that transfer the chromosome in either direction. In fact, there are a few instances in which F can integrate in either orientation at a single site.

Excision of F also occurs, though this is quite rare. Often excision is imperfect and one cut is made at one of the two termini within the integrated F and a second cut is in the adjacent chromosomal DNA (Figure 19-11). This process is akin to the production of specialized transducing particles from λ lysogens (Chapter 18). The F plasmids containing chromosomal genes are called F′ plasmids. The terminology used to describe F′ plasmids and their genetic properties are described in Chapters 1 and 15.

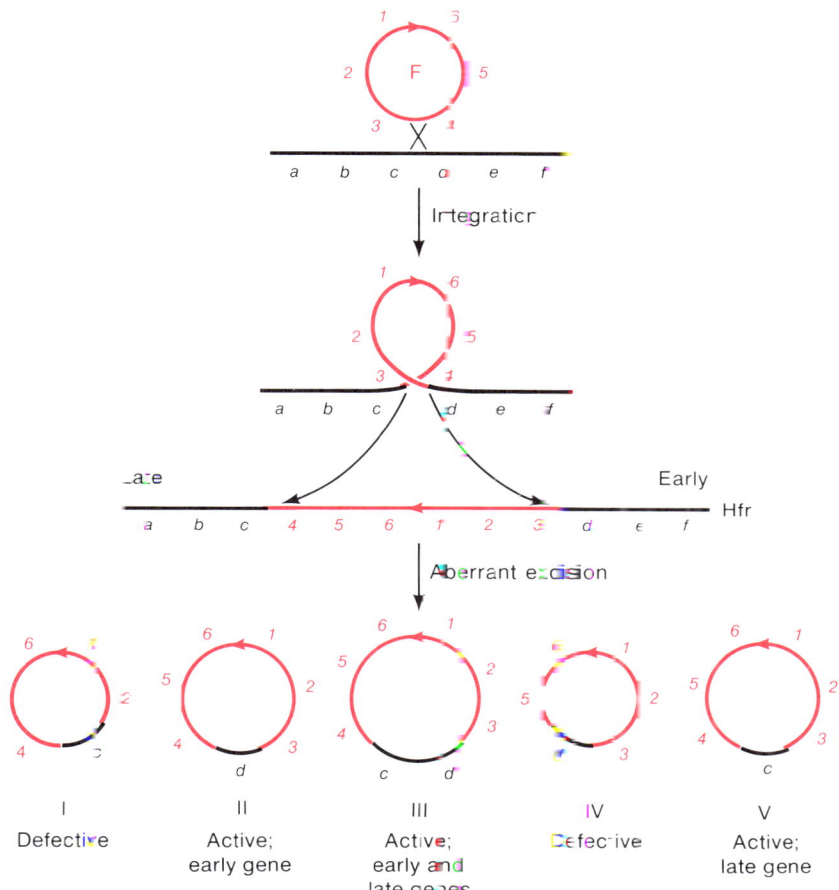

Figure 19-11 Formation of various F′ plasmids by aberrant excision from a particular Hfr strain. Plasmids I and IV have lost F genes and hence are defective. If the plasmids are replication-defective, they cannot be maintained and hence will not be detected. The usual means of detection of F′ plasmids is by the presence of genes, normally transferred late by an Hfr cell, at a time sufficiently early that the genes could not have been transferred by the Hfr cell. Thus, I and IV are normally not detected because the defects in these plasmids are defects in transfer; similarly, a type II plasmid will not be found because it only contains early genes.

F′ plasmids have been very valuable tools in molecular genetic research primarily in dominance studies, done with partial diploids, as was seen in the analysis of the *lac* operon (Chapter 15). They have also been useful in studying the mechanism of DNA transfer.

The Drug-resistance (R) Plasmids

The drug-resistance, or R, plasmids, were originally isolated from the bacterium *Shigella dysenteriae* during an outbreak of dysentery in Japan and have since been found in *E. coli* and other bacteria. Their defining characteristics are that they confer resistance on their host cell to a variety of fungal antibiotics and are usually self-transmissible. Most R plasmids consist of two contiguous segments of DNA (Figure 19-12). One of these segments is

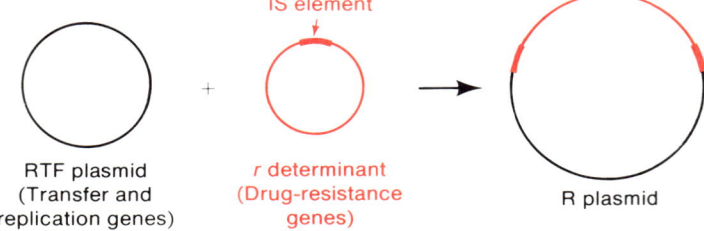

Figure 19-12   The components of an infectious R plasmid. In Chapter 21 we will describe the mechanism for formation of two IS elements in the R plasmid from the single one present in the *r* determinant.

called **RTF (resistance transfer factor);** it carries genes regulating DNA replication and copy number, the transfer genes, and sometimes the gene for tetracycline resistance *(tet),* and has a molecular weight of 11×10^6. The other segment, sometimes called the **r determinant,** is variable in size (from a few million to more than 100×10^6 molecular weight units) and carries other genes for antibiotic resistance. Resistance to the drugs penicillin (Pen), ampicillin (Amp), chloramphenicol (Cam), streptomycin (Str), kanamycin (Kan), and sulfonamide (Sul), in combinations of one or more, appears commonly. Small drug-resistance plasmids, lacking the ability to transfer but still containing the *tet* gene, are also known; one of these, pSC101, whose molecular weight is 5.8×10^6, is commonly used in genetic engineering (Chapter 22). The two-component R plasmids are reminiscent of F′ plasmids, but the drug-resistance genes are not acquired by integration of RTF followed by aberrant excision; instead, they result from acquisition of transposons, as will be seen in Chapter 21.

The Colicinogenic or Col Plasmids

Col plasmids are *E. coli* plasmids able to produce colicins, proteins that are capable of preventing growth of a bacterial strain that does not contain a Col plasmid.* There are many types of colicins, each designated by a letter (e.g., colicin B) and each having a particular mode of inhibition of sensitive cells (Table 19-2). Colicin production is detected by an assay similar to that for detecting phage. A colicin-producing cell is placed on a lawn of sensitive cells; the colicin inhibits growth of nearby bacteria, producing a clear area, known as a **lacuna,** in the turbid layer of bacteria.

*The Col plasmids are one class of a general type of plasmid called a bacteriocinogenic plasmid, which produce bacteriocins in many bacterial species. Bacteriocins, of which colicins are one example, are proteins that can bind to the cell wall of a sensitive bacterium and inhibit one or more essential processes such as replication, transcription, translation, or energy metabolism.

Table 19-2   Properties of several *E. coli* colicins

Colicin	Action of colicin
Colicin B Colicin Ib	Damages cytoplasmic membrane
Colicin E1 Colicin K	Uncouples energy-dependent processes by an unknown effect on the cell membrane
Colicin E2	Degrades DNA
Colicin E3	Cleaves 16S rRNA

Colicins are probably of two types—true colicins and defective phage particles. The latter class is inferred from studies of many purified colicins. Only a few colicins are simple proteins, others look like phage tails when examined by electron microscopy and are thought to be gene products transcribed from remnants of ancient prophages. The hypothesis is that repeated mutation has resulted in loss of the genes for replication, head production, and lysis, but genes encoding a repressor system and the tail proteins have survived intact. Presumably these phages shared with T4 the property that adsorption without DNA injection causes an inhibition of macromolecular synthesis

The Col plasmids range in size from a molecular weight of a few million, for those that are not self-transmissible, to more than 60×10^6, for the self-transmissible plasmids. The best-studied Col plasmid is ColE1, whose molecular weight is 4.4×10^6. Its complete sequence of 6646 base pairs is known. It is used extensively in recombinant DNA research (Chapter 22).

ColE1 is a mobilizable but nonconjugative plasmid and has provided the most definitive evidence that transfer requires a plasmid-encoded nuclease. For ColE1 the nuclease is encoded by the *mob* gene, and a specific base sequence called *bom* (basis of mobility), which contains the cutting site. The critical experiment is shown in Figure 19-13, in which the compatible plasmids F and ColE1 cohabit a single bacterium and F donates the conjugal functions that ColE1 lacks, thereby enabling ColE1 to be transferred. Panel (a) shows the state of ColE1 in an F^- cell. The *mob* gene is transcribed, the *mob* product nicks in the *bom* site (the actual cutting site is called *nic*), and the ColE1 supercoil is converted to a nicked circle. Transfer cannot occur because ColE1 lacks the ability to form pili and hence to form conjugate pairs. In panel (b) the cell contains both F and ColE1. F causes synthesis of the pilus and the transfer apparatus; and ColE1 is transferred. Mutants (*mob⁻*) of ColE1 that fail to be transferred when F donates the transfer apparatus have been isolated. Genetic analysis shows that the *mob⁻* mutation is recessive, which indicates that the *mob* gene encodes a protein. Furthermore, the mutant plasmid, when isolated, is not in the form of a relaxation complex.

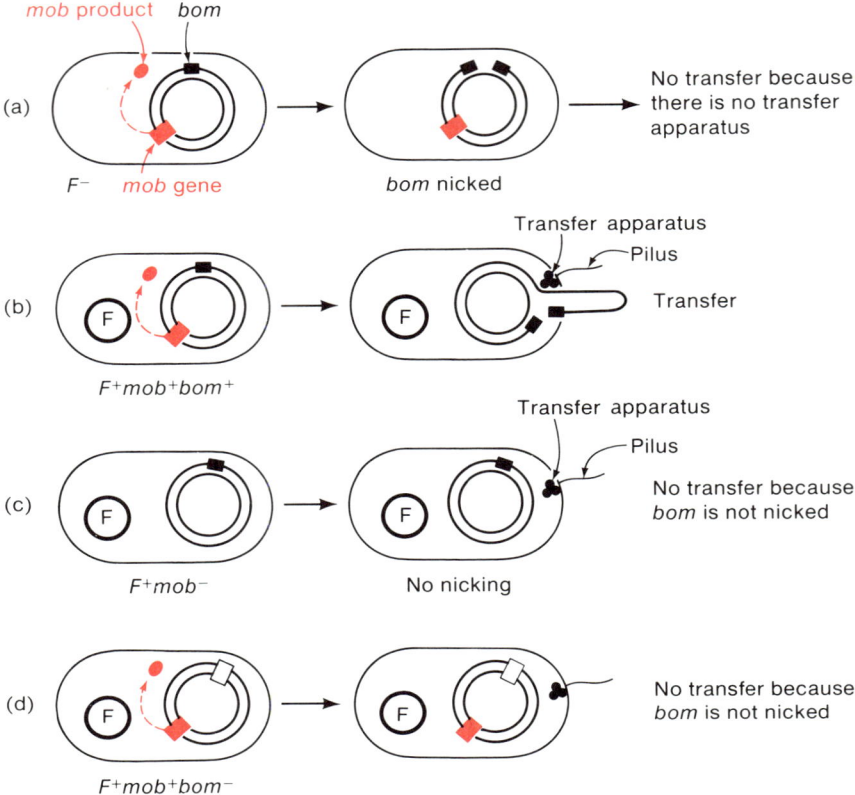

Figure 19-13   Conditions in which F can mobilize a ColE1 plasmid. The *mob* gene and the *mob* product are drawn in red. The *bom* site in the DNA is drawn as a solid box, when functional, and as an open box, when mutant or deleted. Transfer only occurs when ColE1 makes an active *mob* product that acts on a functional *bom* site and F provides the transfer apparatus.

The failure of F to help the *mob*⁻ mutant to transfer is shown in panel (c), in which the F pilus and transfer apparatus are formed but nicking of the ColE1 DNA does not occur. Another ColE1 mutant, having the Mob⁻ phenotype, but which has a *cis*-dominant mutation, is a deletion that removes the *bom* site. The *mob* protein is made but nicking fails to occur and there is no transfer, as shown in panel (d).

The *Agrobacterium* Plasmid Ti

A crown gall tumor found in many dicotyledonous plants is caused by the bacterium *Agrobacterium tumefaciens*. The tumor-causing ability resides in a plasmid called Ti. When a plant is infected, some of the bacteria enter and grow within the plant cells and lyse there, releasing their DNA in the cell. From this point on, the bacteria are no longer necessary for tumor formation. By an unknown mechanism a small fragment of the Ti plasmid, containing the genes for replication, becomes integrated into the plant cell chromosomes. The integrated fragment breaks down the hormonally regulated system that controls cell division and the cell is converted to a tumor cell. This plasmid has recently become very important in plant breeding because specific genes can be inserted into the Ti plasmid by recombinant DNA techniques, and sometimes these genes can become integrated into the plant chromosome, thereby permanently changing the genotype and phenotype of the plant. New plant varieties having desirable and economically valuable characteristics derived from unrelated species can be developed in this way. This will be discussed further in Chapter 22.

PLASMIDS IN EUKARYOTES

Plasmids have been detected in eukaryotes, though they seem to be much less common than in bacteria. Several plasmids are known in yeast, a unicellular eukaryote. One of the more intriguing ones is the **killer particle,** which is a double-stranded RNA molecule of molecular weight about 15×10^6. It is the only known plasmid that does not contain DNA. This particle contains ten genes for replication and several others for synthesis of the killer substance, a colicinlike material.

The best studied yeast plasmid is the so-called 2μm plasmid, a DNA molecule with a length of 2 μm and a molecular weight of 4×10^6. There are about 50 copies of this plasmid per cell and all are found in the yeast nucleus. As is the case for all nuclear DNA, the plasmid DNA is coated with histones. The plasmid DNA apparently does not integrate into the host DNA. The base sequence of the 2μm plasmid includes two inverted repeats, each containing 599 base pairs; these are separated by two unique sequences, containing 2346 and 2774 base pairs, as shown in Figure 19-14. There are

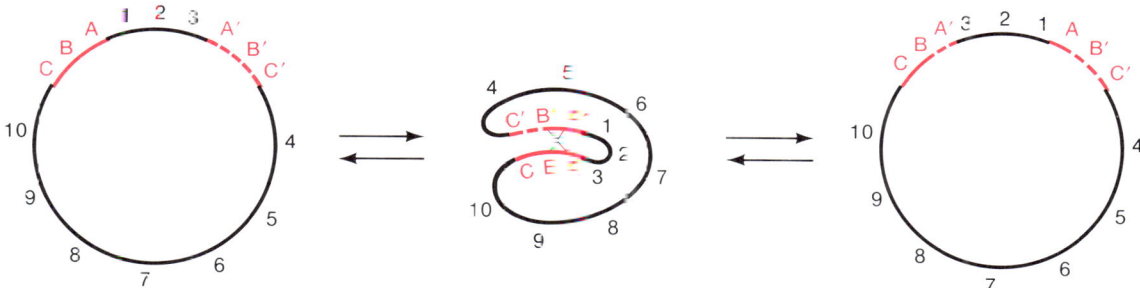

Figure 19-14 Mechanism of interchange of the two forms of the yeast 2μm plasmid. The red segments are the inverted repeats. The four segments are not drawn to scale.

two classes of the plasmid in a cell, which differ by the orientation of one of the unique sequences. These classes, which are equal in number, are interchangeable by frequent recombinational events between the inverted repeats, as shown in the figure. The interconversion is catalyzed by a plasmid gene called *flp*. The function of the interconversion is unknown and in the laboratory *flp*⁻ plasmids replicate within the host yeast cells, apparently without defect. It seems likely that in nature interconversion is of value since 40 percent of the coding capacity of the plasmid is devoted to the *flp* gene.

Replication of the 2μm plasmid has one characteristic that is quite different from other plasmids. It is a high-copy-number plasmid and presumably the copy number is regulated by an inhibition system related to the one described earlier for *E. coli* plasmids pSC101 and ColE1. As explained earlier, when this mechanism prevails, plasmids are selected at random for replication; however, for the 2μm plasmid this is not the case. Instead, each DNA molecule is replicated once, and no more, in each division cycle of the host cell. How this is regulated is unknown.

Whereas most eukaryotic plasmids, like those in prokaryotes, are circular, a few plasmids have been isolated from corn (maize), sorghum, and several fungi that are linear DNA molecules. Such molecules have the problem of termination of replication faced by any linear DNA. Examination of the base sequence of these plasmids shows unique terminal structures, such as palindromes, hairpins, and repetitive sequences, much like the telomeres of yeast and of *Tetrahymena*. Presumably, these plasmids terminate replication in ways similar to that used by chromosomes.

Genetic Recombination Between Homologous DNA Sequences

In Chapter 10 we saw that the ability to exchange DNA segments has short-term survival value; in a cell whose DNA has suffered radiation damage, undamaged segments of daughter molecules can be shuffled to yield an undamaged template for subsequent DNA synthesis. This process was called recombination repair. The ability to exchange DNA segments also has long-term value. Evolution depends on the generation of genetic variation, the first step of which is mutation. The mutations from different lineages are combined in mating and emerge recombined in the progeny. This variation is enriched by exchanging segments of homologous chromosomes, a process called **crossing over,** which can yield recombinant chromosomes. The mechanisms by which DNA exchange occurs is the subject of this chapter. The mechanisms we shall discuss have been derived from studies with microorganisms—principally, bacteria, phages, and fungi. The general principles uncovered in these studies probably apply to the higher organisms as well, though there may be differences in detail, as we have seen with many biochemical mechanisms.

The terminal portions of two synaptic chromosomes in a meiotic cell of *Lilium longiflorum*. Joined homologous regions at which recombination occurs can be seen. [Courtesy of Marta Walters.]

TYPES OF RECOMBINATION

Genetic recombination was first recognized in *Drosophila* in the early part of the 20th century. Later it was studied extensively in fungi. However, experiments to determine the molecular basis of the phenomenon had to await the discovery of recombination in bacteria and phages. Several types of recombination are known in bacteria, three of which enable bacteria to acquire DNA from other cells and to incorporate that DNA, or a fragment thereof, into the cellular DNA. One example of this process is **transformation** (Chapter 8); here, DNA released to the environment by death of a donor bacterium enters a recipient cell. A second example is **transduction** (Chapter 18), in which a phage carrying cellular DNA, picked up by growth in a donor bacterium, injects this DNA into a recipient bacterium. A third is **conjugation** (Chapter 19), in which DNA from an Hfr cell (Chapter 1) is transferred from one cell to another by cell-cell contact. In each type, the transferred DNA is not capable of continual replication, and its maintenance requires that the DNA become physically inserted into the host DNA, usually by a substitution mechanism.

In phage systems two phage DNA molecules, contained in a single cell, interact; then, DNA exchange occurs that probably resembles in many ways the exchange between chromatids during meiotic recombination in eukaryotes. However, the recombination mechanisms in prokaryotes and eukaryotes must have differences, since eukaryotic DNA is in the form of chromatin; few, if any, physical experiments have been carried out in eukaryotic systems, so nothing is known about this likely complication.

An essential feature of each type of recombination mentioned above is that the processes require, in each pair of interacting molecules, extensive regions (hundreds of base pairs) of identical or nearly identical base sequences. That is, the two interacting sequences must be **homologous.** Thus, these processes are called **homologous recombination** (Figure 20-1). Another feature of this type of recombination is the necessity for a particular host

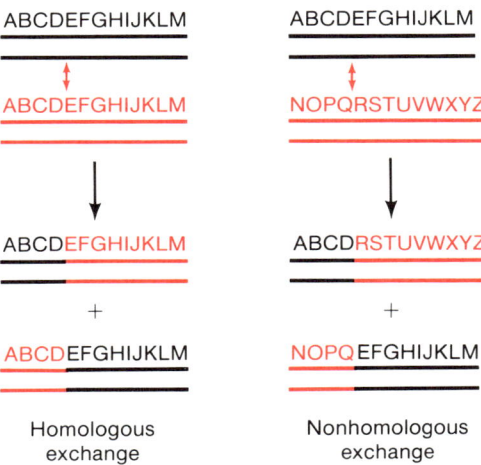

Figure 20-1   The distinction between homologous and nonhomologous recombination. The red arrows indicate the points of genetic exchange.

Table 20-1 Characteristics of four types of genetic recombination

Type	Homology required?	RecA product required?	Sequence-specific enzyme required?
Site-specific	Yes	No	Yes
Homologous	Yes	Yes	No
Transposition	No	No	Yes
Illegitimate	No	No	?

gene, which in *E. coli* is called *recA*. The *recA* product, which is discussed later in the chapter, is needed for DNA-pairing. Similar genes have been found in other bacteria, so homologous recombination is often termed *recA*-dependent.

Three types of recombination do not require either *recA*-like genes or regions of homology. For example, site-specific recombination in λ (Chapter 18) utilizes the Int enzyme and occurs between sequences having very limited homology (7 bp). Another type of recombination, in which particular segments of DNA seem to move from one part of a DNA molecule to another or from one chromosome to another in a diploid cell, has been documented. This phenomenon, called **transposition**, does not require homology and is independent of the *recA* gene. It is described in Chapter 21. A fourth type of recombination, which is an exchange between nonhomologous sequences, is called "illegitimate" recombination, a name which is simply a reflection of our total lack of knowledge of the process. Properties of these four types are summarized in Table 20-1.

In this chapter we will be concerned only with homologous (*recA*-dependent) recombination.

BREAKAGE-AND-REJOINING AND HETERODUPLEXES

Early studies of meiotic recombination in maize by light microscopy indicated that chromosomes are broken and reassembled in the process. When DNA was later identified as the genetic material and recombination between naked DNA molecules in phage crosses was discovered, it was realized that the mechanism of recombination must be quite complicated. Since recombination could occur within a gene, not even one base pair could be lost. Thus, how two enormous double-stranded DNA molecules could align and pair, and how breakage and rejoining could occur in precisely the same place in two DNA molecules became significant questions. One proposal was that DNA did not break. (This proposal did not conflict with the observation of chromosome breakage, since at the time it was not known that a eukaryotic chromosome contains a single DNA molecule.) It was suggested that during replication of one molecule, the two molecules pair in some way over a distance of only a few base pairs; then the replication apparatus switches templates and starts to copy the second molecule. The resulting base sequence in a daughter molecule consists of partial sequences of each parent. This model, known as copy choice, was rendered unlikely, at least for meiotic recombination, when it was learned that in meiosis DNA replication is completed well before any chromosome breakage and rejoining occurs.

An important experiment, a modification of which is shown in Figure 20-2, made it clear that at least for *E. coli* phage λ, recombination occurred by means of breakage of the parental DNA molecules and rejoining of fragments. In this experiment, two types of λ phage were prepared: A^+R^-, whose DNA contained a heavy isotope in both strands, and A^-R^+, whose DNA contained a light isotope in both strands. In addition, the DNA of both phage carried mutations (int^-, red^-) to eliminate all recombination except that determined by the bacterial recombination genes; this enabled one to examine the mechanism of the bacterial recombination system specifically. Mutations in the phage and in the bacteria also prevented the initiation of normal DNA synthesis in order that density changes caused by physical exchange of parental DNA were not obscured by shifts caused by replication. Bacteria were infected with both phage, and the infection was allowed to proceed until lysis occurred. Progeny phage, which contained only parental material (because replication initiation was blocked), were centrifuged to equilibrium in a CsCl solution and the density distribution of the genotypes of all phage particles was determined. Figure 20-2(a) shows

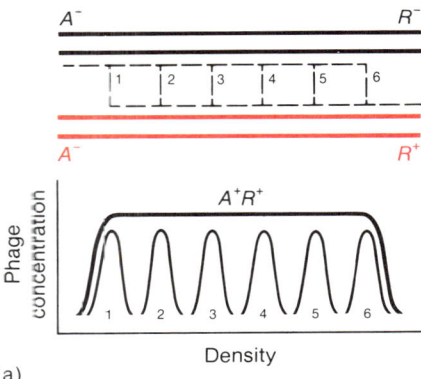

(a)

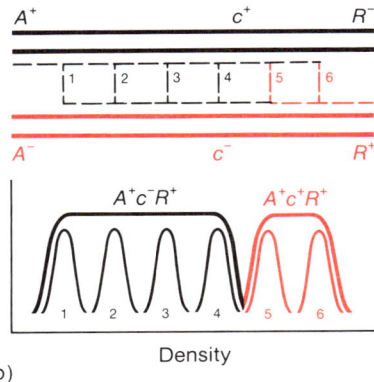

(b)

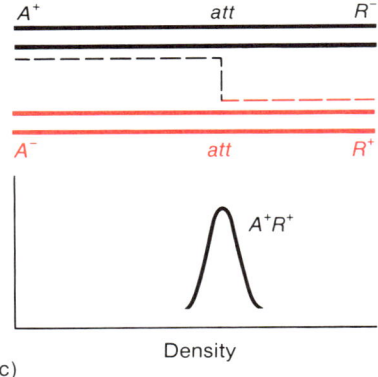

(c)

Figure 20-2 An experiment demonstrating breakage of DNA molecules and reunion to form recombinants. In the upper part of each panel, heavy lines represent high-density (black) and low-density (red) *E. coli* phage λ carrying the indicated alleles. After crossing the phage, using conditions that prevent DNA replication, as described in the text, progeny phage are centrifuged to equi-

librium in CsCl. The thin lines in the distribution curve in the lower part of each panel show the expected density distribution for each of the numbered exchanges shown in the upper part of the panel. The heavy curves are the expected and observed distributions for all recombinant progeny taken together. (a) The uniform distribution obtained with two markers. (b) When three

markers are used, the density distribution shows that A^+R^+ phage are also c^+ only if breakage and reunion occurs to the right of the c marker, as expected. (c) A cross in which only the λ *int* gene is active. A single narrow band is produced, because breakage and reunion occur only at *att*.

that recombinant phage (A^+R^+ and A^-R^-) were found to range in density from fully heavy to fully light; that is, *each recombinant particle contained material from both parents*, indicating that physical exchange of material had occurred in forming the recombinants. When a central marker (in the c gene) was included (panel (b)), the density distribution showed that the c^+ marker from the A^+ parent appeared in A^+R^+ recombinants only if breakage and reunion occurred to the right of the marker. Other experiments were later carried out using phage in which only the λ *int* gene was active (the phage was *red*$^-$, the bacterium was *recA*$^-$), again in the absence of DNA replication. Int-promoted recombination occurs only between λ attachment sites, which are not located exactly in the center of the phage DNA molecule. Thus, in contrast with the results in panel (a), the phage recombinants separated into only two distinct densities, which result from breakage of each DNA molecule at its attachment site followed by rejoining the fragments (Figure 20-2(c)).

A second important observation was made in the λ experiment, namely, that phage particles in which breakage and rejoining had occurred near the c gene were frequently heterozygotes, that is, they carried both the c^- and c^+ markers. Since a λ particle contains only a single DNA molecule, the heterozygote must have a heteroduplex region covering the c gene—that is, a region in which one strand carries the c^+ marker and the complementary strand carries the c^- marker (Figure 20-3). It should be noted that phage with a heteroduplex region were found only because DNA replication was

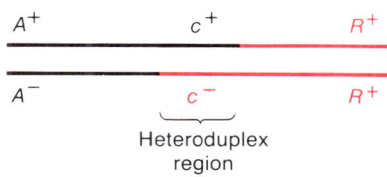

Figure 20-3 One of the c^+/c^- heterozygous phage recovered in the experiment of Figure 20-2(b).

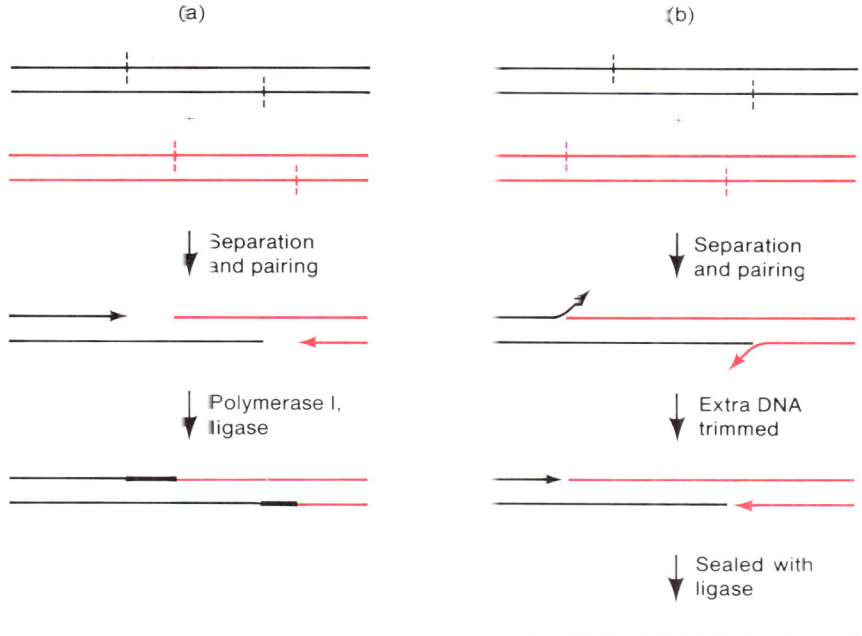

Figure 20-4 The effect of cutting of single strands at various positions. Sites of cutting are indicated by short vertical dashed lines. (a) Joining leaves gaps. The gaps are later filled (heavy black lines) by the sequential action of polymerase I and DNA ligase. (b) Joining leaves extra DNA. The arrowhead indicates a 3'-OH group. In each panel, only one recombinant molecule is shown. Note that if one recombinant has gaps, the other would have extra DNA and vice versa.

inhibited in the phage cross. Replication would have caused the two strands to segregate from one another and heterozygotes would not have been present.

A variety of other experiments in which phage–phage and other types of recombination have been examined consistently show that breakage and rejoining is an essential step in all modes of homologous recombination and that a structure with a heteroduplex region is always an intermediate (evidence for a heteroduplex region in meiotic recombination will be presented shortly). The question of alignment of molecules and matching of sites of cutting was not addressed by these experiments. However, simple mechanisms that avoid matching of sites were proposed. Figure 20-4 shows a prototype model that makes use of properties of known enzymes. This proposed mechanism is quite incomplete but does illustrate several essential features of DNA exchange.

BRANCH MIGRATION

In Chapter 4 breathing of DNA was described. Breathing refers to the phenomenon of random and transient breaking and reforming of hydrogen bonds. **Branch migration,** which refers to the displacement of one basepaired strand in a double helical region by another strand also able to basepair, is a phenomenon based on breathing. It can be understood by examining the synthetic DNA polymer shown in Figure 20-5. The W strand of this molecule is continuous. Two single strands complementary to W, I and II, whose combined length exceeds that of the W strand, are both hydrogenbonded to the W strand, as shown. The heavy black region and the red segment of the C strands have the same base sequence and are complementary to the dashed region of the W strand. When breathing temporarily opens the right terminus of strand I, one of two events can occur—strand I can snap back or strand II can increase the length of its hydrogen-bonded region, as shown in panel (b) of the figure. Breathing can occur again, resulting in restoration of the state of the molecule in panel (a) or in further change, as in panel (c). Breathing never truly stops, so the state of the molecule shifts continually. In most models of recombination, branch migration will play an important role.

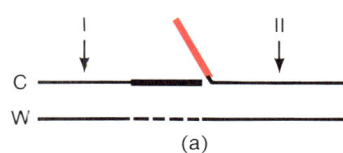 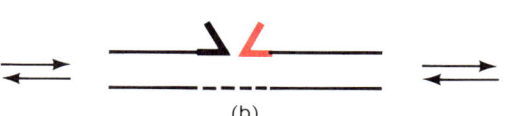 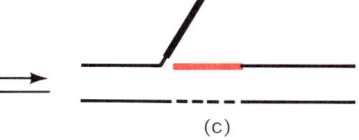

Figure 20-5   Branch migration. The heavy black and red lines are both complementary to the dashed segment of the lower strand. In panel (a) the heavy black region is hydrogen-bonded to the dashed segment. Breath-ing causes the free end of the black region to become occasionally unbonded, allowing the red region to pair with the dashed seg-ment (panel (b)). By repeated breathing more of the red region can become base-paired (panel (c)) or the molecule can revert to its state in panel (a). Many intermediate states of the type shown in panel (b) are possible.

MISMATCHED BASE PAIRS AND THEIR RESOLUTION

Formation of a heteroduplex region containing both a mutation and the corresponding wild-type site generates a mismatched base pair. Such mismatches do not persist; they are eliminated either by replication or **mismatch repair.** Replication removes a mismatched base in a straightforward way, but if replication does not resolve the mismatch, the mismatch repair system comes into play.

Mismatch repair was first demonstrated in an experiment summarized in Figure 20-6. DNA was isolated from two λ phage mutants. The molecules were denatured and the complementary strands were separately purified. Strands containing one mutant allele were renatured with complementary strands containing the other mutant allele. Bacteria with defective recombination genes were transfected with the renatured DNA. On incubation of the infected cells in growth medium progeny phage developed. Infected cells were examined for the production of wildtype phage. Replication alone would yield both singly mutant phages, in roughly equal numbers; however, if mismatch repair had occurred before replication, some wildtype phage

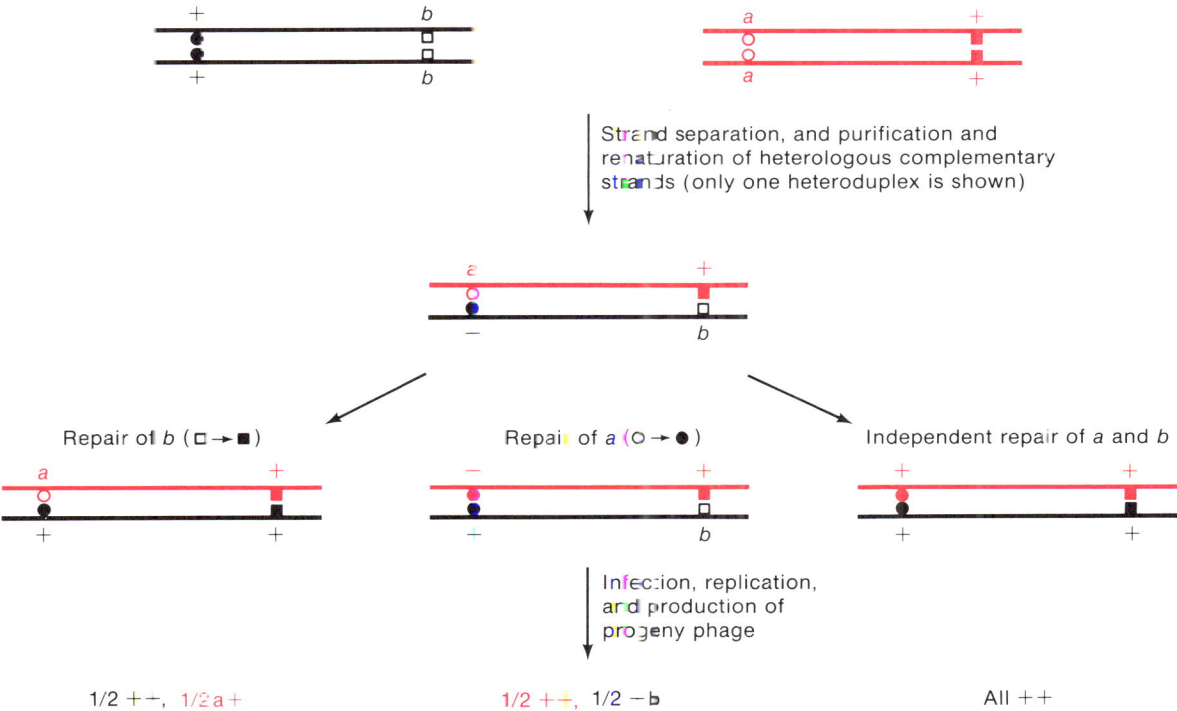

Figure 20-6 The mismatch repair experiment.

would be produced. Three types of cells yielding wildtype phage were observed: those containing wildtype and the a mutant; those with wildtype and the b mutant; and those containing only wildtype phage. How these presumably arose is shown in the figure.

Mismatch repair is accomplished by excision of a segment of one strand of a heteroduplex region. A variation of the experiment of Figure 20-6 was performed to estimate the number of base pairs removed in the process. A third marker, between the pairs of markers already used in the experiment and closely linked to one of them, was included in the experiment. Phage progeny from individual infected cells were examined to see whether repair at the third marker often accompanied repair at its neighbor—that is, whether the a^+b^+ progeny of a heteroduplex with the genotype $a^+b^-c^-/a^-b^+c^+$, in which b and c are closely linked, are also often $a^+b^+c^+$ (correction of one heteroduplex strand from b^- to b^+). It was found that the frequency of co-correction depended on the separation of the two markers and their relative positions. Examination of a series of markers indicated that when mismatch repair is initiated at a particular mismatched site, repair tends to proceed along the strand in the 5′-to-3′ direction for about 3000 base pairs. *In vitro* experiments with partially purified enzymes have indicated that a single-strand break is made on the 5′ side of a mismatch and a large tract of bases is removed from the strand being repaired. The resulting gap is then filled in by a DNA polymerase; the intact strand is used as a template. An interesting feature of these experiments was the fact that a preferred direction of correction was observed for several of the markers; that is, for some markers, wildtype was corrected to mutant, whereas for others the correction was from mutant to wildtype. No satisfactory explanation was ever provided for this phenomenon; presumably, base sequences surrounding a mismatch determine which strand is cut.

PAIRING OF DNA MOLECULES

Pairing of DNA molecules is essential to all modes of homologous recombination. An understanding of how it occurs began with studies of bacterial conjugation (Chapters 1 and 19).

In 1965 a mutant *E. coli* F^- strain was isolated that was unable to serve as a recipient in an Hfr cross. The gene defined by the mutation was *recA*, and in 1976 the RecA protein was isolated. Proteins with similar properties have been detected in other organisms, but more is known about the *E. coli* enzyme than about the others, so our discussion will be restricted to that protein.

The RecA protein has two principal biochemical activities: (1) it binds to single-stranded DNA, and (2) it is an enzyme that cleaves proteins. Its DNA-binding activity is the feature that is relevant to recombination (the other property is regulatory, as shown in Chapter 15.) When acting as a

DNA-binding protein, the RecA protein mediates nonspecific pairing of DNA molecules and homology-dependent **strand invasion.** Figure 20-7 shows several RecA-mediated DNA-DNA interactions that have been carried out with purified RecA protein and DNA molecules at room temperature. The structures shown are very stable and are held together by A·T and G·C base pairs between complementary base sequences. Study of the three interactions shown in the figure (and others that are more complex) and of pairs of DNA molecules that will not interact has shown that stable pairing is dependent on two things:

1. One molecule must be single stranded or have a single-stranded region.
2. Either one of the molecules must have a free end.

The first requirement can be (and often probably is) provided by supercoiling of the participant lacking a free end. The second requirement eliminates the possibility that a single-stranded circle could invade a double-stranded circle.

The RecA-mediated interactions shown in Figure 20-7 are the end result of a sequence of three steps: presynaptic binding of RecA protein to single-stranded DNA, synapsis, and post-synaptic strand exchange (Figure 20-8). These stages were first elucidated in a study of the RecA-mediated pairing of a double-stranded circle obtained from a phage-infected cell and the linear form of the phage (+) strand.

1. *Polymerization of RecA protein on single-stranded DNA.* If single-stranded DNA is mixed with RecA protein, a nucleoprotein filament forms in which

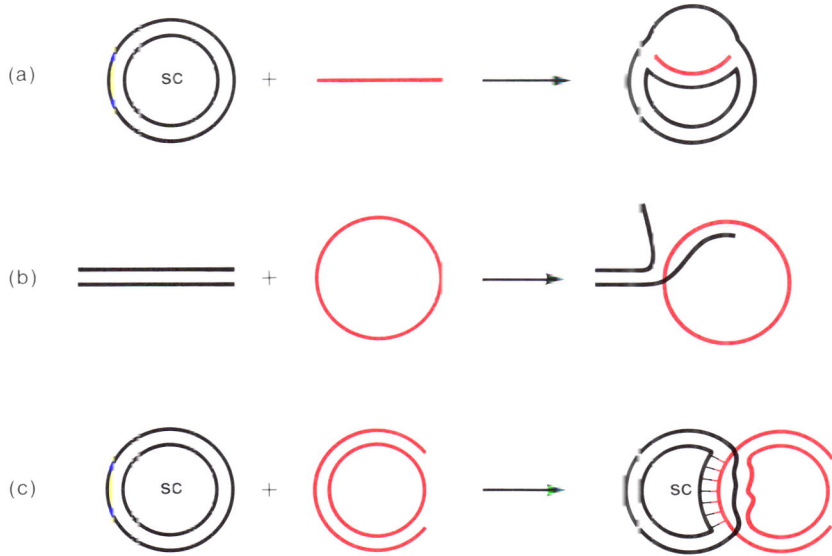

(a)

(b)

(c)

Figure 20-7 Three interactions mediated by the RecA protein; sc indicates that the circle is supercoiled.

Figure 20-8 RecA-mediated pairing that leads to assimilation of a complementary strand.

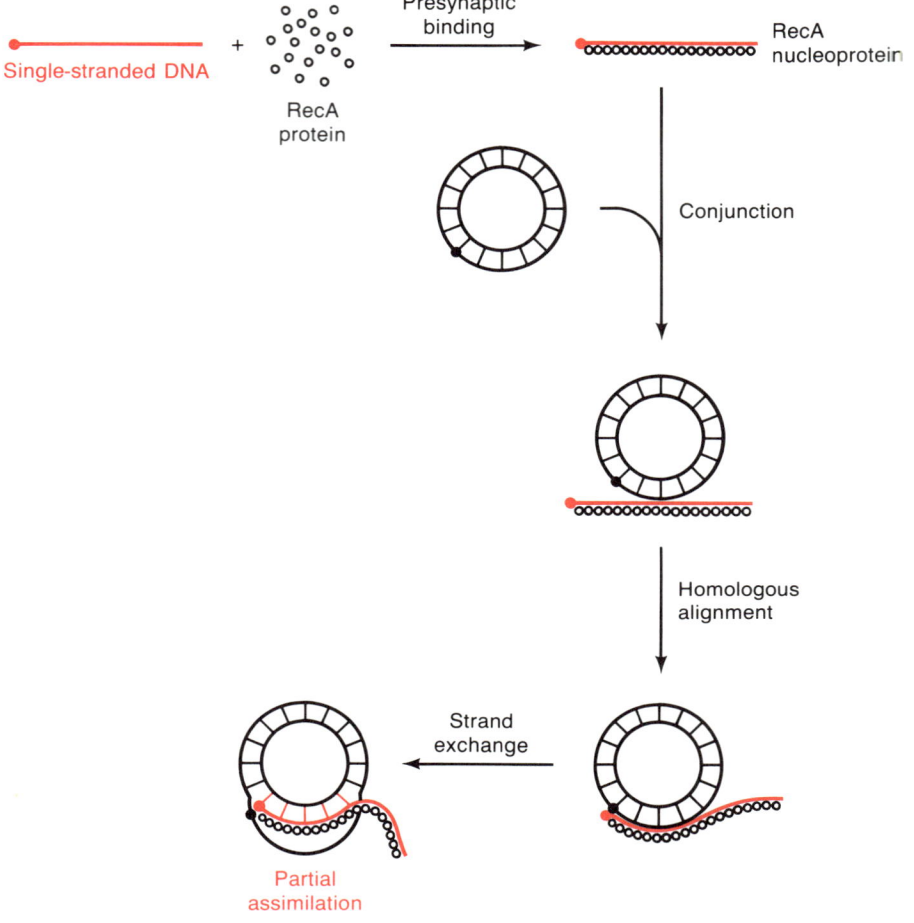

the single-stranded DNA is coated with RecA protein. The structure has a banded appearance (Figure 20-9), indicating that it has considerable regularity. Each band contains 6–8 RecA monomers and corresponds to 20 nucleotides. The DNA is extended by a factor of 1.5 compared to a single strand in double-stranded DNA.

2. *Synapsis.* In the presence of ATP the nucleoprotein forms a complex with the double-stranded circle. The initial interaction is not between homologous regions. After the initial sequence-independent interaction (termed **conjunction**), the two DNA molecules move relative to one another until homologous sequences come into contact; this is called **homologous alignment.** When homologous sequences are aligned, the two strands are not yet intertwined.

3. *Post-synaptic strand exchange*. Homologously aligned, but not intertwined, strands are bound together only weakly, and the structure is quite unstable. *In vitro* a correctly intertwined double helix can be created by *E. coli* topoisomerase I, but whether this occurs *in vivo* is unknown. However, once a homologously aligned region forms at the end of the single strand, RecA actively promotes the displacement of a strand from the double-stranded molecule (using cleavage of ATP for energy) and assimilation of the new strand. The mechanism is not yet understood, but it is clear that RecA acts as a helicase, unwinding the double-stranded DNA in advance of the forming heteroduplex. Assimilation is a polar process, moving only in the $5' \rightarrow 3'$ direction, a fact that must be built into all models of recombination.

Each of the interactions shown in Figure 20–7 (the three single-strand and double-strand interactions) can be explained by the multistep RecA-mediated process just described. However, this does not seem to have any obvious bearing on the homologous pairing of two double-stranded DNA molecules, which seems to occur in meiotic recombination, phage–phage recombination, transduction, and conjugation. Since it seems clear that both single-stranded DNA and a free end are required, it will be seen that almost all models of recombination include an early step in which one DNA strand is nicked and, in a variety of ways, unwound from the nick.

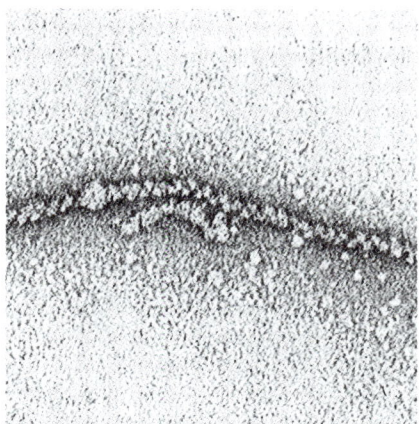

Figure 20-9 An electron micrograph of single-stranded DNA to which RecA protein is bound. These molecules were taken from a sample of presynaptic filaments that were active in homologous pairing. [From Flory et al., *Proc. Nat. Acad. Sci.*, 1984, 81:7026; courtesy of Charles Radding.]

RECOMBINATION IN BACTERIAL TRANSFORMATION

The RecA-mediated events just described, coupled with branch migration and mismatch repair, constitute almost the entire mechanism of recombination in bacterial transformation. In transformation exogenously added DNA recombines with chromosomal DNA. In transformation with *Pneumococcus* (the best-understood case) uptake of the transforming DNA into the cell is accompanied by digestion of one of the input strands; consequently, only a single strand is available for interaction with the recipient DNA. The single strand is, in an unknown way, protected from nuclease attack—possibly it is coated with SSB protein. Physical experiments in which density–labeled (heavy) DNA is used to transform a bacterium whose DNA is light show that a single strand of the donor DNA, or a portion thereof, is linearly inserted into the recipient DNA (Figure 20–10(a)). The currently accepted model for this process is shown in Figure 20–10(b). Aided by a bacterial protein, probably equivalent to the *E. coli* RecA protein, the incoming single-stranded fragment invades the DNA of the recipient, presumably from the 5' end, as described in the section on RecA-mediated pairing. By an unknown mechanism the displaced single strand is cut. Branch migration then occurs at the end of the assimilated DNA which gradually increases the fraction of the invading strand that is base-paired to the recipient strand. Trimming enzymes remove the free ends, which may be in donor or recip-

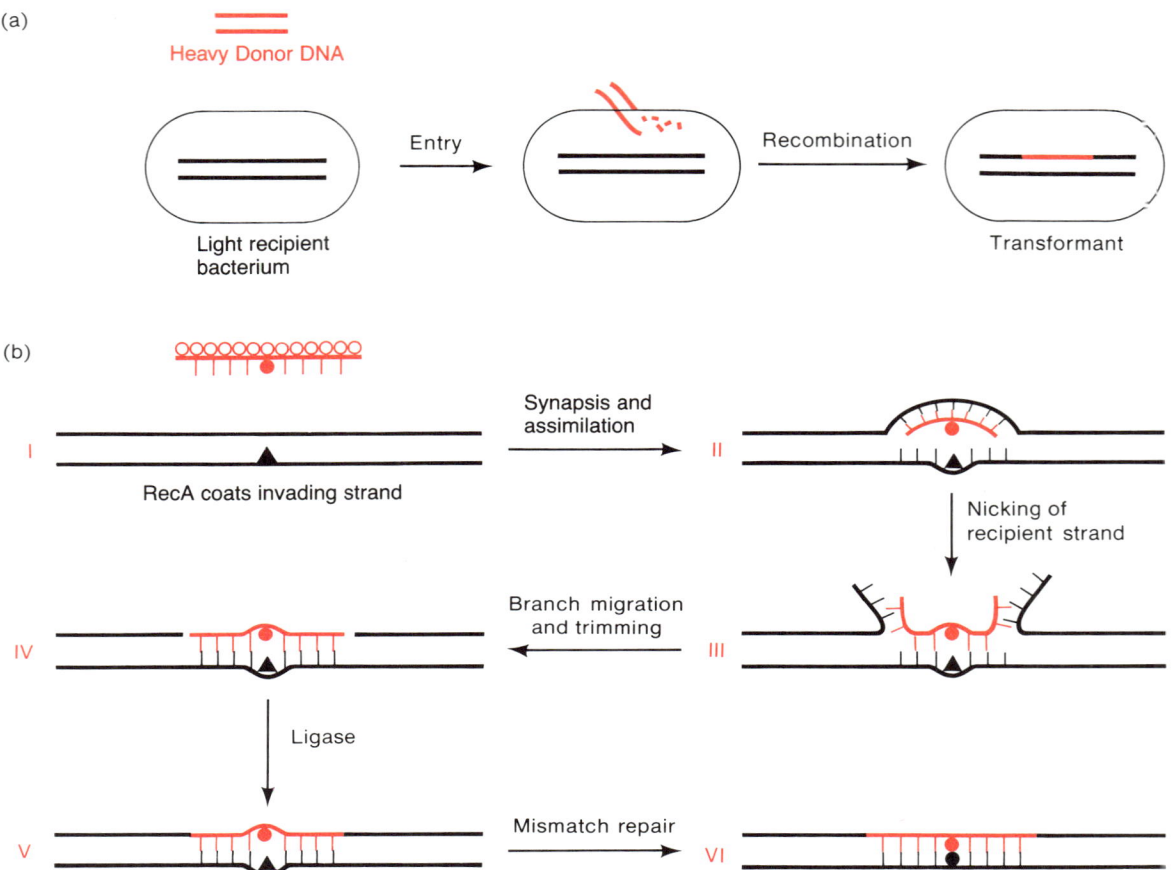

Figure 20-10 Bacterial transformation. (a) Conversion of double-stranded donor DNA to a single strand, which is then inserted. (b) The proposed mechanism for integration of single-stranded donor DNA.

ient DNA, and ultimately ligase seals the nicks. The result is a heteroduplex region containing a mismatched base pair. The outcome of the process—that is, whether the donor marker is or is not recovered in the progeny cells—depends on whether mismatch repair occurs and, if so, whether the donor or recipient base is removed. For some markers, known as high-efficiency markers, either the mismatch is always corrected to match the donor genotype or there is little or no repair. In the latter case, following cell division one cell has the donor genotype and one has the recipient genotype. Because the plating conditions used to detect recombinants are usually chosen to allow growth only of recombinants (for example, an antibiotic is added or a necessary amino acid is absent), the colony that forms consists exclusively of recombinant bacteria. For the low-efficiency markers, mismatch repair usually removes the mismatched base from the donor and the cell retains the recipient genotype. Proof that this is the correct explanation for the two types of markers comes from the recent isolation of mutants presumably defective in mismatch repair; when these mutants are transformed, all markers appear to be transformed as high-efficiency markers because one of the two daughter cells following cell division always has the donor genotype.

EXCHANGE BETWEEN HOMOLOGOUS DOUBLE-STRANDED MOLECULES

So far, we have seen mechanisms for exchange between single- and double-stranded DNA (transformation) and between specific, mostly nonhomologous, sites in double-stranded molecules (Int-promoted recombination). The most prevalent type of exchange is between homologous double-stranded molecules—for example, in bacterial conjugation, transduction, phage crosses, and meiotic crossing over—but the mechanisms are incompletely understood in these systems. In this section several genetic phenomena are described that have been important in developing theories of homologous exchange. Later this information will be used to develop two models of exchange.

When examining crossing over between distant markers (for example, in different genes), the mechanism of exchange seems straightforward and models such as those shown in Figure 20-4 are sufficient to explain the data. However, when exchange between close genetic markers is studied, in *intragenic* recombination, one sees that the mechanism of crossing over is considerably more complex than that shown in the figure. In the following unit, some of these observations are examined.

Evidence for a Heteroduplex Region in Meiotic Recombination

One profitable approach to understanding the molecular mechanism of meiotic exchange has been to study the genetics of the *Ascomycetes,* fungi in which the products of each meiotic event are easily available to the experimenter because they are packaged in a sac called an **ascus.** The sequence of events in meiosis is, in outline, the following. DNA replication occurs in a diploid premeiotic cell, yielding a cell with four double-stranded DNA copies of each chromosome. Two rounds of meiotic division occur to yield four haploid germ cells. In some fungi, those having asci with four spores, such as the yeast *Saccharomyces cerevisiae,* each of the meiotic products is present in a single spore and all four products are contained in one ascus. In *Neurospora crassa* and *Sordaria brevicollis*—two popular organisms for studying meiotic recombination—the second meiotic division is followed by a mitotic division before spores form, so eight spores are contained in a single a_ (Figure 20-11). Furthermore, the spores are arranged in a linear order deter- mined by the successive meiotic and mitotic divisions, and the genotype of each spore is the genotype of one of the eight individual polynucleotide strands present at the end of meiosis. Study of *Ascobolus immersus,* a fungus with an eight-spore unordered ascus, has also contributed significantly to our understanding. However, for our purposes, analysis of ordered asci will be easier to understand, and in this section theoretical examples will be confined to such systems.

The usual experimental approach is spore analysis: spores are individ- ually dissected from an ascus and then transferred to a nonselective growth medium on which each forms a colony, and the genotype of each colony is determined by further plating tests. Occasionally, markers are used that

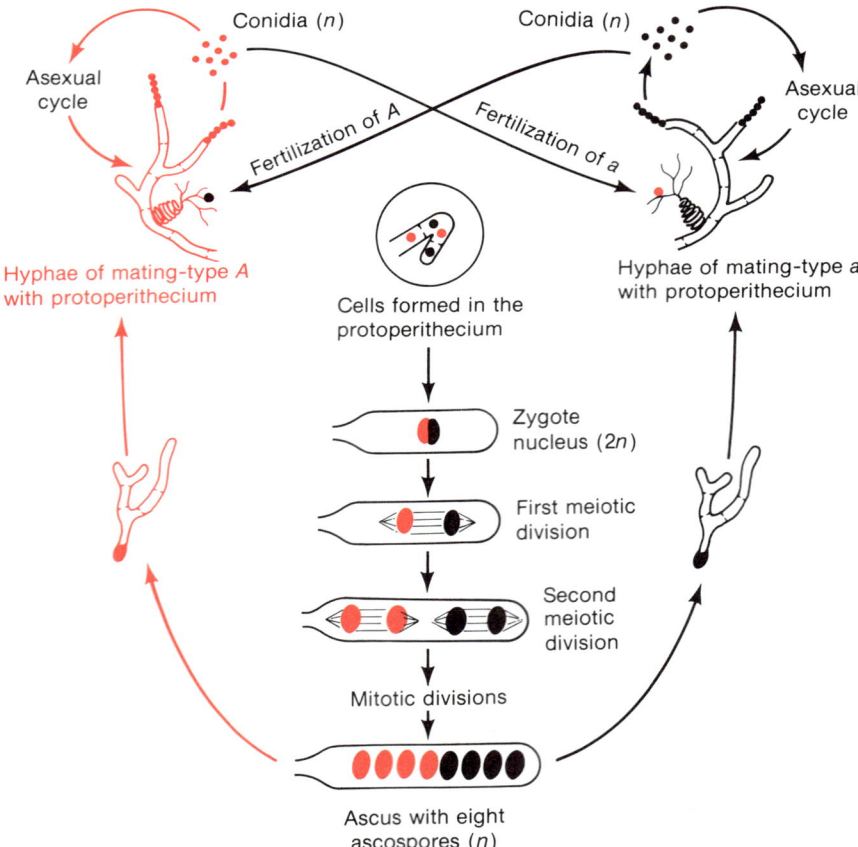

Figure 20-11   The life cycle of *Neurospora crassa,* showing how an ascus with eight ordered spores is formed. Premeiotic DNA replication occurs in the zygote before the first meiotic division.

confer visible phenotypes (color) on the colonies, eliminating the need for further testing.

Figure 20-12 reviews the composition of an ordered eight-spore ascus derived from a diploid heterozygous for a gene *A*. Three features should be noted about this and subsequent figures:

1. The centromeres (Chapter 7) do not separate following premeiotic DNA replication and hence hold the daughter molecules together. Separation occurs after the first meiotic division.

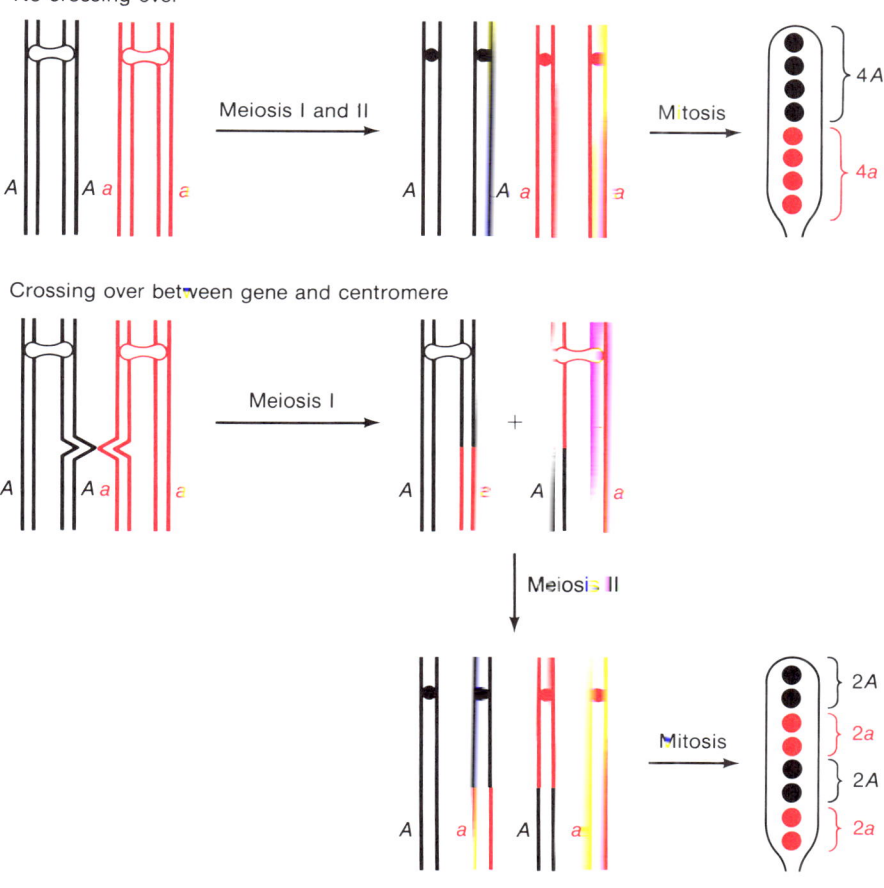

Figure 20-12 Composition of an eight-spore ordered ascus with one marker segregating, with and without crossing over between the centromere and the marked locus. The four DNA molecules present at the outset are the products of premeiotic DNA replication. The centromeres do not separate during this round of replication and hence hold the daughter molecules together; separation occurs in the second meiotic division. Crossing over occurs in the four-strand stage, before the first meiotic division.

2. Chromosome pairing occurs after DNA replication, yielding a unit consisting of four DNA molecules (this is the starting point in the figures).
3. Crossing over occurs during the 4-molecule stage, before any chromosome separation or cell division.

If crossing over between the centromere and the marker does not occur, meiosis results in the formation of a tetrad of four DNA molecules—two *A* and two *a*—which after the mitotic step yields an ascus containing four *A* spores and four *a* spores. This is the classic Mendelian **4 : 4 segregation.** If crossing over occurs between the gene and the centromere, as shown in the lower panel of the figure, segregation is still 4 : 4, though the order of the spores in the ascus differs. Note that the order of the spores is sufficient to indicate that an exchange (actually any odd number of exchanges) has occurred between the gene and the centromere.

Figure 20-13 shows the composition of an ascus when two genes *A* and *B* are considered and the initial diploid cell (the zygote in Figure 20-12) is

Figure 20-13 Composition of an eight-spore ordered ascus with two markers segregating, with and without crossing over between the markers.

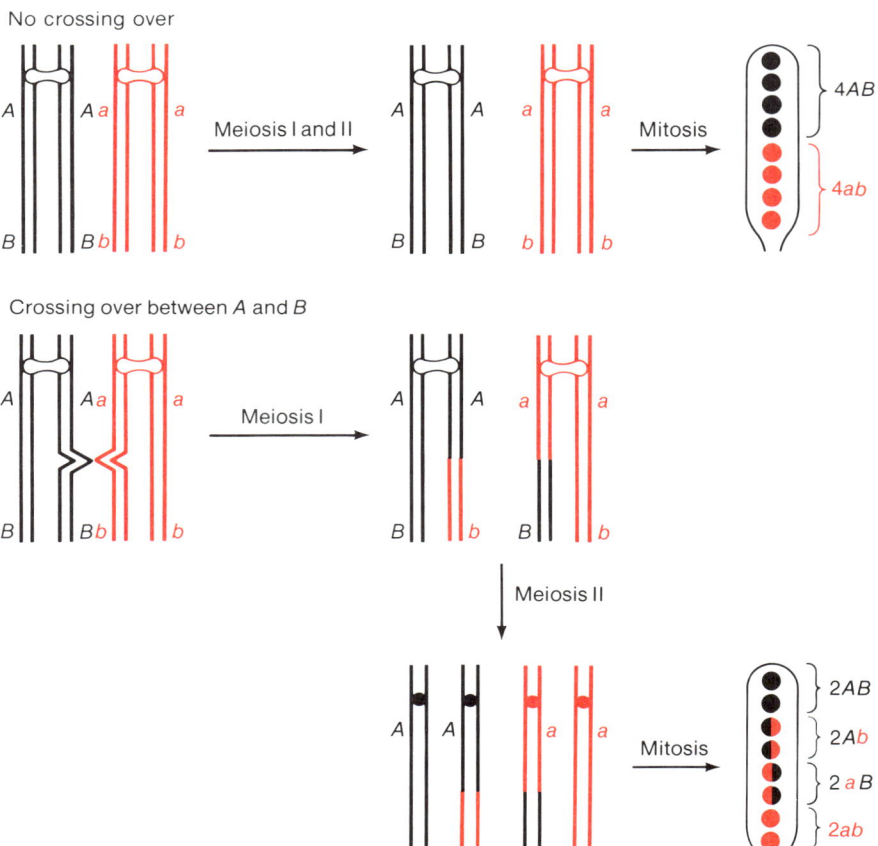

heterozygous with chromatid composition AB and ab. If an ascus forms without crossing over, four AB spores and four ab spores result—again, 4 : 4 segregation. The lower part of the figure shows the result of crossing over in the regions between the genes. Three features of the genotypes of the spores should be noted. (1) The reciprocal recombinant genotypes—Ab and aB—are both present in the same ascus. (2) Both parents are present in equal numbers, indicating that crossing over occurs after DNA replication, as we have stated. (3) Four copies of each of the alleles, A, a, B, and b, are present, though the genotypes of the spores have been recombined; that is, each gene has segregated in the Mendelian 4 : 4 ratio.

"Aberrant" asci are also observed, and those formed when three closely linked markers are segregating provide insight into the exchange process. Figure 20-14 illustrates the cross $Abd \times aBD$, in which recombination is observed to have occurred in the regions between a and d. We consider the

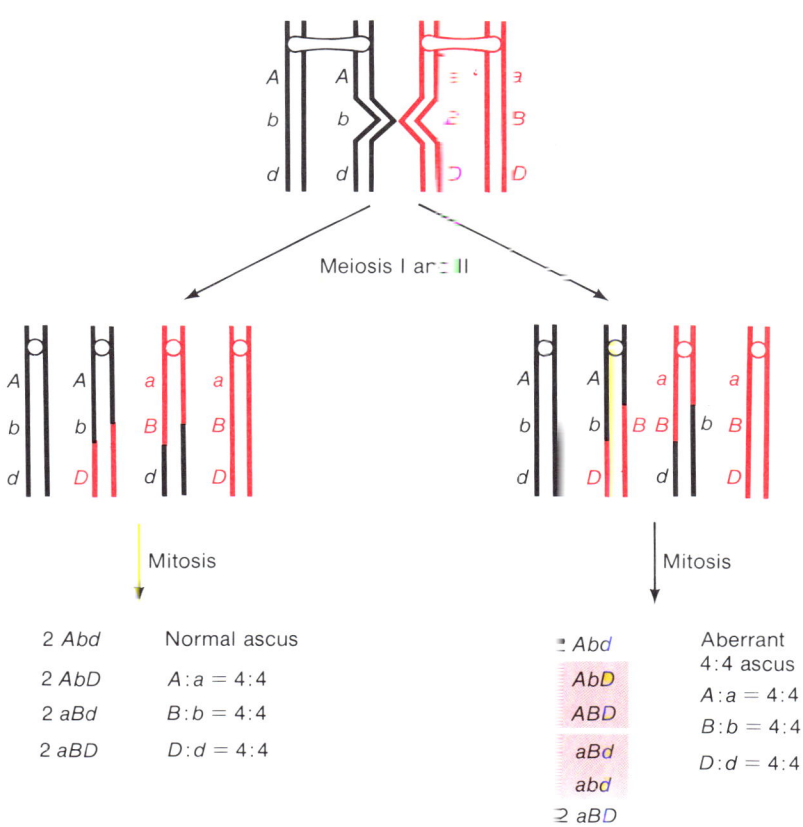

Figure 20-14 Postmeiotic segregation. Composition of one type of aberrant 4 : 4 ascus. The two pairs of spores shaded in pink have nonidentical members. The het-

eroduplex regions shown in the postmeiotic chromosomes account for the aberrant pairs; however, no mechanism for forming these regions is implied.

specific case of a crossover between *b* and *d* (though the same considerations apply to a crossover between *a* and *b*). Most of the asci in which such recombination is evident have the composition shown at the left hand side of the figure. The ratio of dominant and recessive alleles in the asci is 4 : 4, as expected, and each pair of spores is identical. However, some asci are also found that have the composition shown at the right side. Again, the allelic ratios are 4 : 4, but two pairs of spores do *not* have identical members—namely, *AbD–ABD* and *aBd–abd*. These asci are called **aberrant 4 : 4** asci.

Recall that in an eight-spore ordered ascus adjacent spores contain the daughter DNA molecules formed by replication of each meiotic product during the final postmeiotic mitosis. *Thus, if two markers of an adjacent spore pair are different, the individual polynucleotide strands of the molecule formed after the second meiotic division must also have been different.* Since the *a* and *d* alleles are the same in adjacent spore pairs and only the *b* alleles are different, a simple explanation for the difference would be the following: the molecule formed after the second meiotic division contained a heteroduplex region, as shown in the middle part of Figure 20-14, and during postmeiotic replication the heteroduplex regions separated, giving rise to daughter molecules that differ in the alleles that were present in each heteroduplex region. Indeed, the existence of aberrant 4 : 4 asci is accepted as definitive evidence for the existence of heteroduplex regions, and the aberrant 4 : 4 asci are said to result from **postmeiotic segregation.** Furthermore, because the flanking markers are in a recombinant array, it is assumed that a heteroduplex region is formed by crossing over. Note that a heteroduplex region could also be present in the molecules that gave rise to the normal 4 : 4 ascus seen in the left part of the figure, but the region would not have included the *B* and *b* alleles, and hence would not be evident.

Conversion Asci and the Role of Mismatch Repair

Asci with other than normal and aberrant 4 : 4 segregation are also observed. Figure 20-15 illustrates the phenomenon. Again, exchange has occurred in the regions between *A* and *d*. Most of the asci are normal 4 : 4 asci, but some, in which a crossover has occurred between the *b* and *d* markers, are of a different type:

1. Ascus II has two important features: (i) one pair of *b* markers is missing (the pair *ABD* replaces *AbD*), and (ii) like the aberrant 4 : 4 asci, one spore pair has dissimilar members (*aBd* and *abd*). In this type of ascus, the *B : b* ratio is 5 : 3.
2. In ascus III, one of the spore pairs (the second from the top) lacks the expected *b* allele. The ratios of the outside markers are *A : a = D : d = 4 : 4*, but the ratio of *B : b* is 6 : 2.

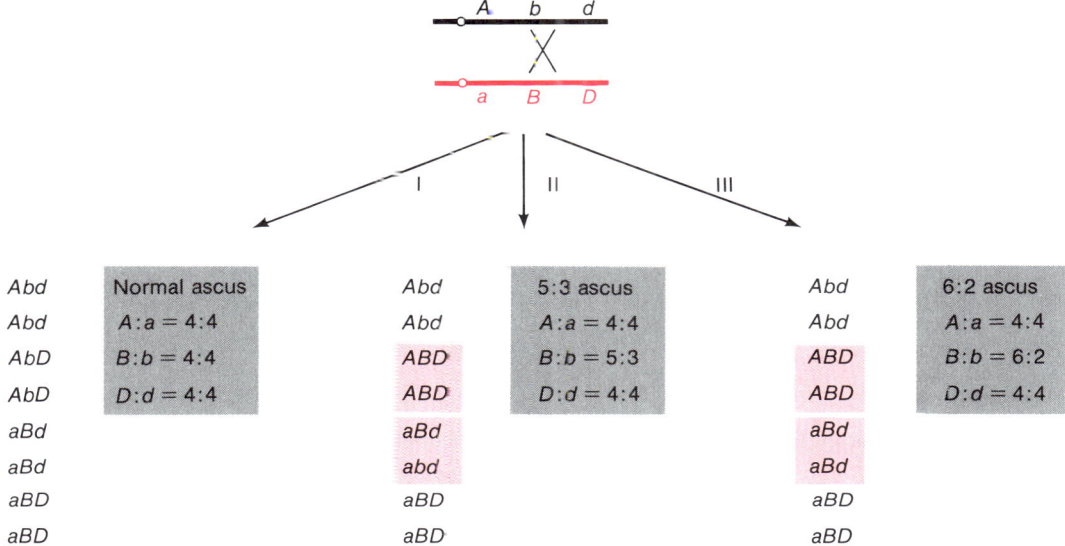

Figure 20-15 Two types of aberrant asci (II and III) in which gene conversion (b to B) has occurred. The significant spore pairs are shaded in pink. The cross and the composition of the normal asci are the same as that in Figure 20-14.

In early descriptions of these unusual asci, it was said that the *b* allele had been **converted** to *B*; thus, the phenomenon of 6 : 2 and 5 : 3 segregation of an allele is termed **gene conversion.** Asci are also found in which a *B*-to-*b* conversion has occurred (note that *B* to *b* and *b* to *B* are both called conversion).

The preceding discussion indicated that conversion asci appear among asci selected for recombination of flanking markers. However, if *all* asci from a cross are examined nonselectively, without regard to the composition of the flanking markers, conversion asci are also found. A key observation in developing a theory of crossing over is that *in roughly 50 percent of the conversion asci, the flanking markers are in a recombinant array.* This observation, which has been made in several fungal species, in *Drosophila,* and for many different markers, suggests that the processes of gene conversion, as well as postmeiotic segregation, and crossing over are associated. We will see shortly that the association is a result of sharing common steps—namely, pairing, exchange of single polynucleotide strands, and formation of a heteroduplex region.

To examine some of these common steps, alternative events that follow pairing of two chromatids and exchange of DNA in the four-chromatid stage will be considered, again with reference to the cross illustrated in Figure 20-15. Having invoked the formation of heteroduplex regions to explain postmeiotic segregation, we can use them also to explain conversion. The difference between the two phenomena is in the steps following formation of a heteroduplex region.

The heteroduplex region resulting from a crossover event is assumed to contain the markers *b* and *B,* which differ in one base pair, so a mismatched base pair is formed. Figure 20-16 illustrates three possible responses to such a mismatch:

1. In event I there is no mismatch repair. Thus, replication of each pair of mismatched strands yields one DNA molecule with the wildtype allele and one with the mutant allele. Since two molecules carry mismatches, mitosis produces two pairs of spores with nonidentical

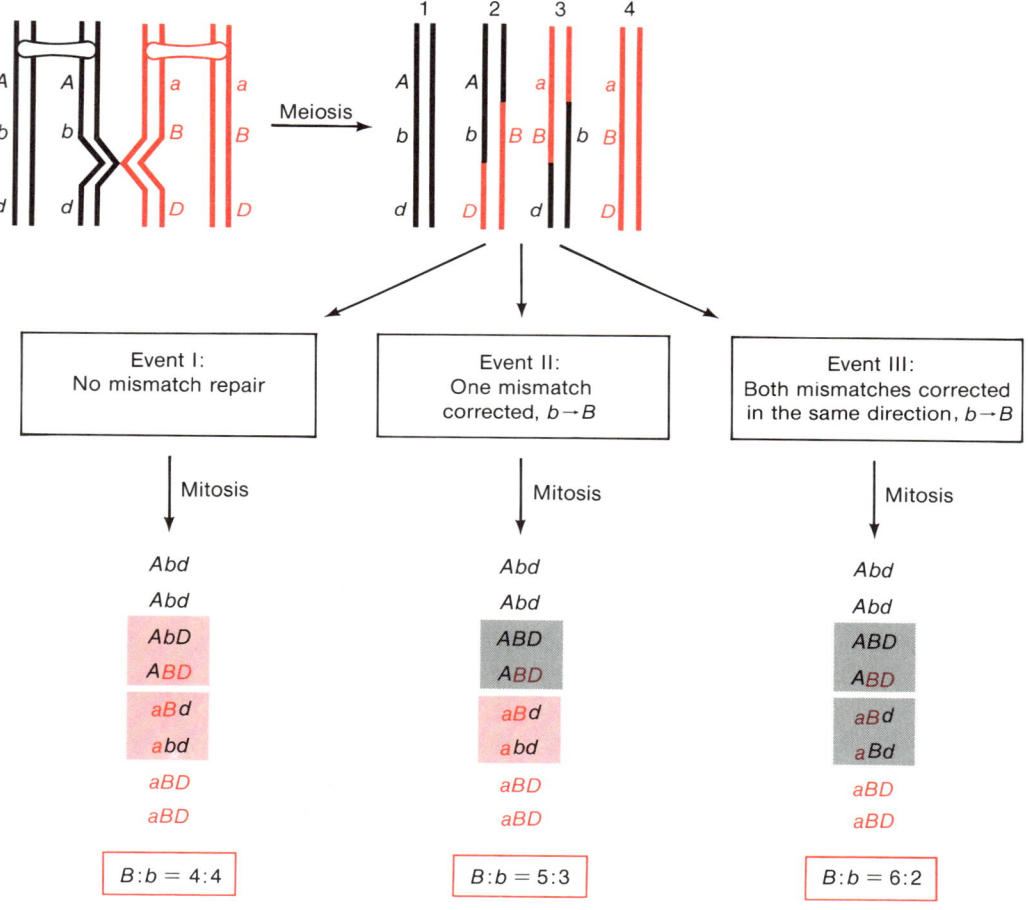

Figure 20-16   The proposed role of mismatch repair in the formation of three types of aberrant asci. Spore pairs with nonidentical members are shaded in pink; pairs that result from mismatch repair are shaded in gray.

members. The ratio $B : b$ is 4 : 4 and the ascus is of the aberrant 4 : 4 type.

2. In event II the mismatched base pair in the heteroduplex region is repaired from b to B in molecule 2 (that is, b is excised and the strand with B is used as a template), with the result that one of the pair of spores with different members is not present. The ratio $B : b$ is 5 : 3.

3. In event III both mismatches are repaired in the same direction, b to B. All spore pairs have identical members and a 6 : 2 ascus results.

The scheme shown in Figure 20-16 has not yet been demonstrated by *in vitro* experiments but is supported by a wealth of genetic data.

Although the events in Figure 20-16 show how the association of aberrant 4 : 4 and conversion asci with recombination of flanking markers may be explained, recall that in half of the aberrant asci, the flanking markers are *not* in a recombinant array. How can one account for these nonrecombinant asci? The generally accepted explanation, which is presented in the following section, is that heteroduplex regions are formed during one stage of DNA pairing and that the two cuts in the individual polynucleotide strands needed to separate the paired molecules can occur in two arrangements—one yielding recombinant asci and the other yielding nonrecombinant asci.

MODELS FOR HOMOLOGOUS RECOMBINATION

In this section two hypotheses for the mechanism of crossing over in fungi are presented. The first is the seminal model (1964) of Robin Holliday, in which formation of a heteroduplex region followed by mismatch repair was first suggested. An important consequence of the Holliday model was that it predicted a structure, now shown to exist, that is a feature of all current thinking. As additional information was obtained, Holliday's model has been modified several times. The second model presented in this section is one of the most important modifications (1975). Although details of this second model will probably have to be adjusted as more data are obtained, the general features are thought to be a good description of the events that occur in crossing over in fungi.

The Holliday Model

The Holliday model and its later modifications attempted to account both for the existence of normal asci, gene conversion, and postmeiotic segregation in the fungus *Ustilago* and for the frequent association of the abnormal asci with crossing over. Holliday's specific model in shown in Figure 20-17.

The stages of the model are the following:

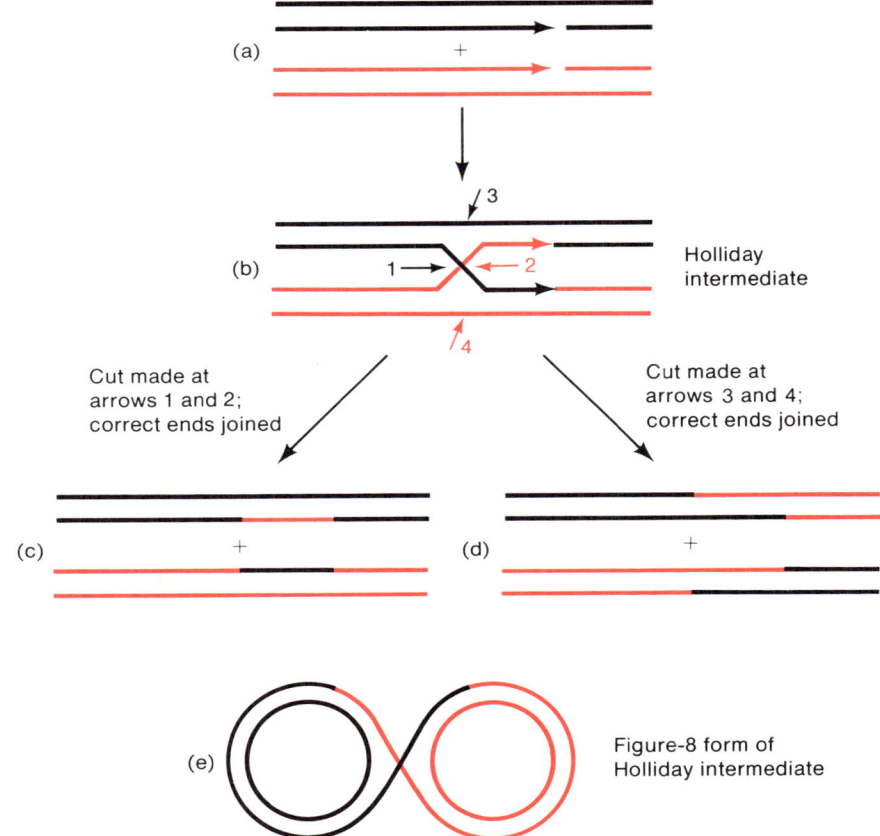

Figure 20-17 Formation of the Holliday intermediate (b) from two nicked DNA molecules (a) and the structure of two sets of recombinants (c and d). (e) is the same as (b) except that the components are circular; note that the figure-8 consists of two monomeric single-stranded DNA circles and one dimeric single-stranded DNA circle.

1. **Single-strand breakage.** Two DNA molecules come together and two single-strand breaks are made in homologous positions in each molecule by an unspecified mechanism.
2. **Unwinding.** At each break the double helix unwinds somewhat, releasing single polynucleotide strands (not shown in the figure).
3. **Invasion and assimilation.** The released strands are exchanged between the two DNA molecules by pairing with the unbroken complementary strand in the other molecule, forming a unit consisting of two DNA molecules connected by crossed single strands, as shown in the figure.
4. **Separation of the molecules.** A pair of cuts, symmetrically made, separates the two DNA molecules. The cuts can be made at either

of two different locations. If they are made at corresponding positions in the crossed strands ("East–West" cuts), a pair of nonrecombinant molecules, shown in panel (c) of the figure, results. If, on the other hand, the cuts are made at homologous positions in the uninterrupted parent strands ("North–South" cuts), the recombinant molecules shown in part (d) will result. This cutting process, which separates the two participating DNA molecules, is called **resolution.**

The Holliday model provides a molecular basis for the association between aberrant segregation and crossing over. Gene conversion and postmeiotic segregation are assumed to result from repair of mismatches, or lack thereof, on one or both chromosomes in a symmetric structure containing heteroduplex DNA, as indicated in Figure 20-16. The model also explains how 5 : 3 and 6 : 2 segregation can occur without recombination of flanking markers (Figure 20-18). As we have just seen, a Holliday junction can be resolved by pairs of cuts that either do not produce recombination of flanking markers (East-West, cuts, type (c) in the figure) or do (North–South cuts, type (d)). If both types of cuts are equally likely, half of the 5 : 3 and 6 : 2 asci will be recombinant for flanking markers, as observed.

Existence of a crossed-strand structure such as is shown in Figure 20-17(b) is an important prediction of the Holliday model. Crossed-strand regions have been observed in DNA molecules isolated from several organisms and have come to be called **Holliday junctions.** They are most easily recognized when the junction is in an "open-box" configuration (Figure 20-19). Panel (a) shows how a 180° rotation of one component of a Holliday structure with respect to the other would yield molecules with an open-box array. A molecule of this type is shown in panel (b).

Figure 20-18 Formation of recombinant and nonrecombinant aberrant 4 : 4, 5 : 3, and 6 : 2 asci from two types of crossover products.

Figure 20-19 (a) Rotation of the upper half of a Holliday structure about the axis shown (consider the bottom half to be held in place) to generate an open-box structure. (b) Electron micrograph of a crossover intermediate between a phage λ DNA molecule and a linear DNA fragment. The Holliday junction is indicated by an arrow and enlarged in the inset. The molecules have been partially denatured in order to identify corresponding regions of the two molecules. [Courtesy of Manuel Valenzuela.]

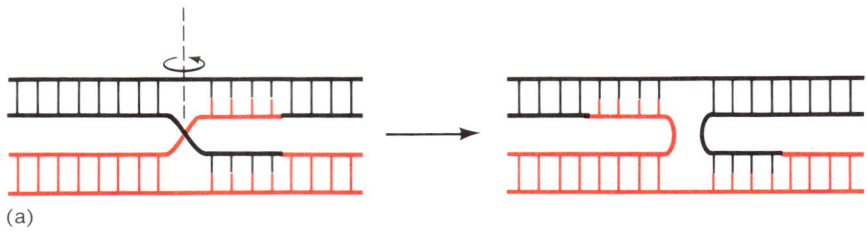

(a)

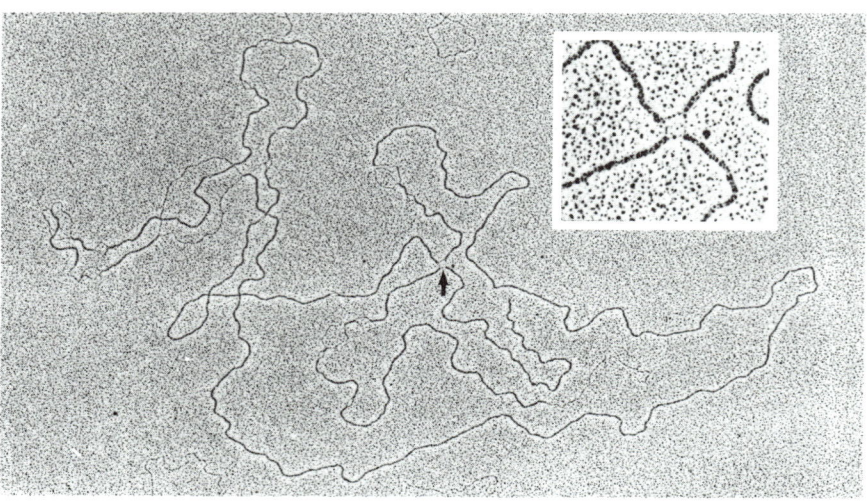

(b)

An interpretation of the region surrounding the Holliday structure. The drawing at the left shows the open-box configuration seen in (b); the drawing at the right shows a possible crossed structure.

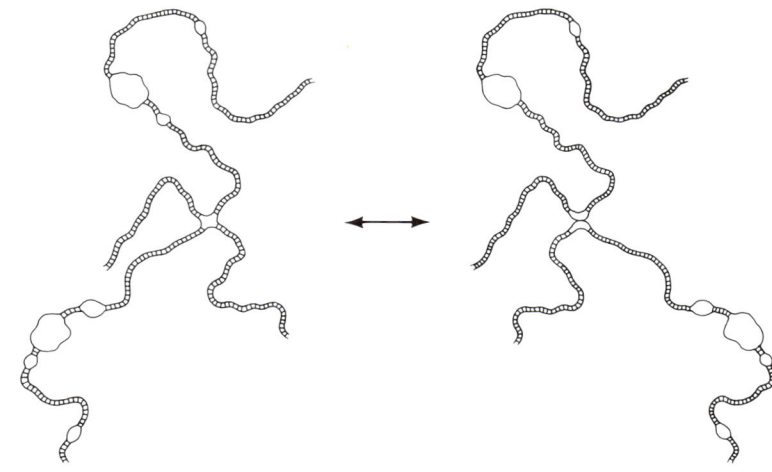

(c)

The molecules pictured are linear to match the linear DNA of eukaryotic chromosomes. However, the Holliday model may also apply to RecA-mediated recombination between circular DNA molecules in prokaryotes. Panel (e) of Figure 20-17 shows for a pair of circular molecules the configuration of the Holliday structures of panel (b). The Holliday intermediate for a pair of circles is called a figure-8. Such molecules have been observed (Figure 20-20(a)) by electron microscopy of DNA isolated from cells containing many copies of a plasmid and from RecA$^+$ cells infected with phage φX174. Proof that these figure-8 molecules do indeed have the predicted structure comes from treatment of the figure-8 with a restriction endonuclease that makes a single cut at a unique position in each circle. If the two circles of a figure-8 are joined in homologous regions, the molecule should be cleaved into an X-shape having two pairs of arms all of equal length (Figure 20-20(b,c)).

After the Holliday model was first proposed, it was modified to include branch migration (not known to exist in 1964), which could allow the Holliday junction to move and the heteroduplex regions to be elongated. This modification was necessary to account for data from four-factor crosses in which two central markers are often both converted with high efficiency (**co-conversion**). In yeast, co-conversion occurs in regions that are up to

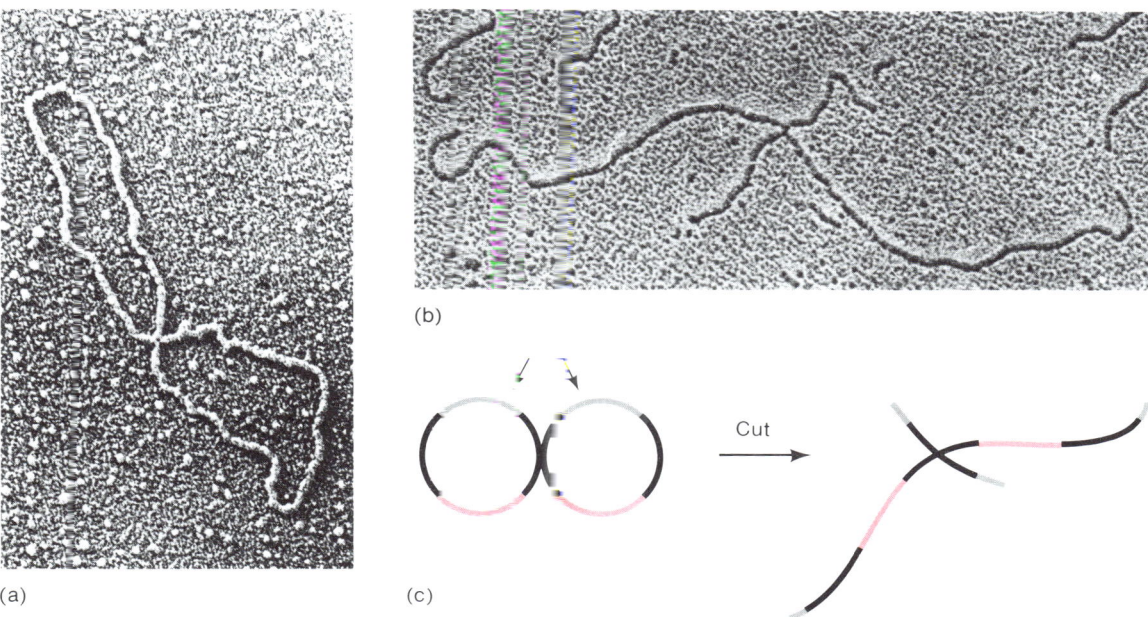

(b)

(a)

(c)

Cut

Figure 20-20 (a) Electron micrograph of a figure-8 molecule formed by recombining two φX174 monomeric circles. [With permission, from R. Warner, *J. Mol. Biol.* (1975) 91, 109. Copyright: Academic Press Inc. (London) Ltd.] (b) An X form obtained by cutting a figure-8 molecule with a restriction endonuclease that makes a single cut at a unique position. [Courtesy of Robert Warner.] (c) An interpretation of the molecules in panels (a) and (b).

1000 base pairs long, indicating that heteroduplex regions must be at least that large.

The rate of branch migration was determined by a study of the cleaved joined circles shown in Figure 20-20(c). If branch migration occurs, the position of the joint in the X will drift back and forth. The direction of branch migration and the distance that the joint drifts before reversing direction are random. In time the joint will reach a terminus, and the two joined molecules will separate. A simple statistical theory relates the number of X molecules remaining after various times with the rate of branch migration. Such measurements were made by removing aliquots of a sample of X molecule at various times and examining the molecules by electron microscopy. The fraction of molecules in X form continually decreased, which provided proof that a Holliday junction can branch-migrate and gave a measure of the rate of migration.

A notable feature of the Holliday model is its symmetry. That is, *each* product of crossing over shown in Figure 20-17 possesses a heteroduplex region. This prediction has been tested in several fungal systems. The frequent occurrence of aberrant 4 : 4 asci in *Sordaria brevicollis* indicates that heteroduplex regions can indeed form at the same site on both chromatids. Similarly, studies at the *b2* locus in *Ascobolus immersus* showed that aberrant 4 : 4, 5 : 3, and 6 : 2 segregation can occur at a single site on different copies of the chromosome. However, all data have not been supportive, as can be shown for the *W17* locus of *Ascobolus* in Figure 20-21. In this cross many recombinant asci with 5 : 3 segregation of central markers were examined. The Holliday model predicts that to form such asci the marker must be contained in a heteroduplex region and that such a region will be present in both chromatids 2 and 3. Asci with *B*-to-*b* corrections should be of two types—those in which the correction is on chromatid 2 and those in which it is on chromatid 3. However, it was observed that only one type of 5 : 3 ascus was found at this locus, from which it was concluded that at this locus heteroduplex regions are not present on both chromatids. A related phenomenon has been obtained with yeast—namely, for some sites 5 : 3 asci are frequent but aberrant 4 : 4 asci (the hallmark of heteroduplex regions) are very rare. A 5 : 3 ascus contains one nonidentical pair of spores, so at least one heteroduplex region must have been present. The Holliday model interprets the conversion pair as necessarily arising from repair of a mismatch in a heteroduplex region. However, the lack of aberrant 4 : 4 asci, which indicates that the formation of two heteroduplex regions is rare in yeast, suggests that the conversion pair has arisen without formation of a heteroduplex region. The conclusion from these observations is inescapable: if there is a single mechanism of crossing over, it must be capable of producing either a symmetric pair of heteroduplex regions or a single heteroduplex region. These observations were the impetus for developing the model described next, a model in which the initial interaction is asymmetric and subsequent steps determine whether the asymmetry will persist or a symmetric array of heteroduplex regions will be formed.

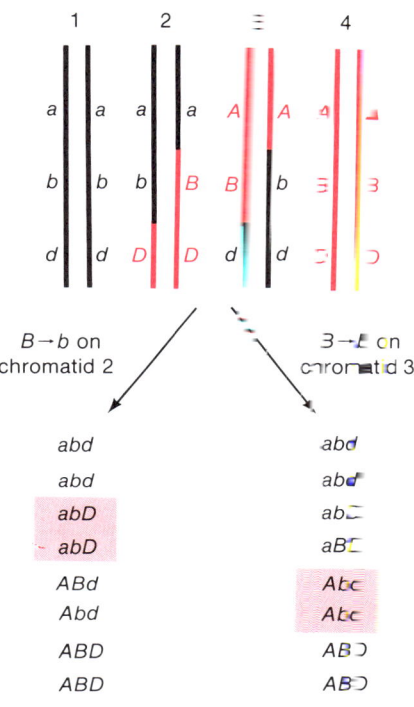

Figure 20-21 Production of two types of 5 : 3 asci from a pair of chromatids that have generated two heteroduplex regions. At the *W17* locus of *Ascobolus*, both types are not found.

The Asymmetric Strand-Transfer Model

The **asymmetric strand-transfer model** retains several features of the Holliday model—namely, joining of homologous single strands, existence of heteroduplex regions, the production of a Holliday junction, the resolution of the junction by East-West versus North-South breakage, and mismatch repair. The principal difference is that the initial pairing is asymmetric, in accord with the yeast and *Ascobolus* data. That is, in the initial interaction one molecule donates a single polynucleotide strand to a recipient molecule and a heteroduplex region is formed only in the invaded molecule. By replicative extension of the invading strand and by branch migration the heteroduplex DNA is enlarged and by a novel rearrangement (isomerization) a heteroduplex region is formed in both molecules, allowing for the reciprocal events common in other fungi. A second difference is that some conversion asci are formed without mismatch repair.

The asymmetric strand-transfer model is shown in Figure 20-22. The individual steps are the following:

1. **Nicking.** A single-strand break (nick) with a 3′-OH group and a 5′-P group is formed (in an unspecified way) on one polynucleotide strand of one of the participating double-stranded molecules. The

Figure 20-22 The asymmetric strand-transfer model. The dashed lines represent newly synthesized DNA. [Based on M. Meselson and C. Radding. *Proc. Nat. Acad. Sci., USA*, 72 (1975): 358.]

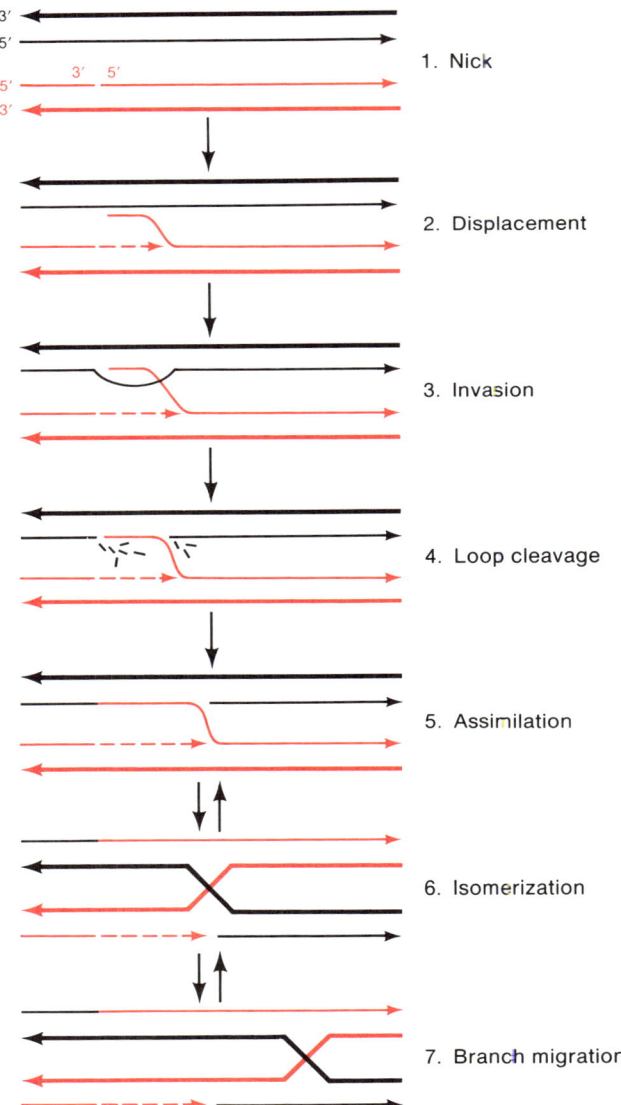

breakage is assumed to occur in only one of the molecules, in contrast with the Holliday model, in which each molecule receives a break.

2. **Displacement**. A DNA polymerase, acting at a free 3′-OH group, adds nucleotides at the nick and displaces the original polynucleotide strand. The displacement generates a single-stranded appendage that is used for complementary pairing in the next stage.

3. **Invasion.** The single-stranded appendage invades the double-stranded molecule, presumably aided by a RecA-type protein, as described in Figures 20-7 and 20-8.

4. **Loop cleavage.** In response to invasion the displaced strand is enzymatically removed, presumably by an initial break in the displaced polynucleotide strand of the invaded molecule; this is followed by either endonuclease or exonuclease action (or both).

5. **Assimilation.** The 3'-OH group of the gap formed in step 4 can be extended by a DNA polymerase (the reaction is the same as that in step 2) and joined to the invading strand by a DNA ligase. Furthermore, the gap can be enlarged to allow more of the invading strand to base-pair with the continuous strand of the invaded molecule. At this point the asymmetric intermediate becomes complete. Pairing is stabilized, but *only one heteroduplex region exists* (the one in the upper molecule, because replication has filled the gap that removing and transferring a strand in the lower molecule has caused. In the Holliday model the gap would be filled by a polynucleotide strand transferred downward from the upper molecule and two heteroduplex regions would be present.

6. **Isomerization.** A rearrangement of the strands of the participating DNA molecules occurs and a Holliday junction forms. (This rearrangement is difficult to visualize without three-dimensional molecular models; however, it is straightforward and is accomplished without steric problems or strain.) Note that the crossover strands in the Holliday structure become new backbone strands and former backbone strands become new crossed strands of the Holliday junction, putting the arms flanking the strand crossover in a recombinant configuration. At this point still only one heteroduplex region exists.

7. **Branch migration.** The Holliday junction migrates (for example, toward the right, as shown in the figure) and thereby generates a symmetric array with two heteroduplex regions. Note that because of the movement of the base-paired regions in the assimilation and branch-migration steps, when resolution occurs, the position of the cuts will typically not be at the site of either strand invasion or the initial break, in contrast with the Holliday model.

8. **Resolution.** Separation of the particular molecules occurs as in the Holliday model, again with a choice of North-South versus East-West pairs of cuts.

Examination of the products shows how the asymmetric strand-transfer model explains many of the genetically observable features of meiotic recombination in fungi that have been discussed. Figure 20-23(a) shows the alternative products of resolution of the asymmetric intermediate formed by isomerization without branch migration. With North-South cleavage, the flanking markers are nonrecombinant, and with East-West cleavage, they

(a) Asymmetric intermediate via isomerization without branch migration

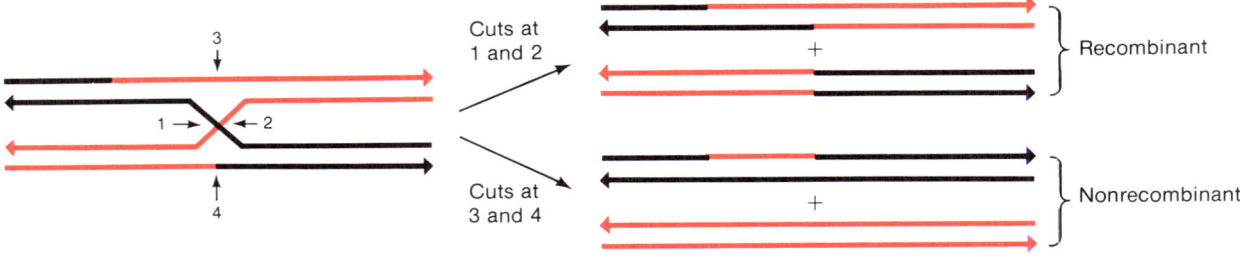

(b) Symmetric intermediate via isomerization and branch migration

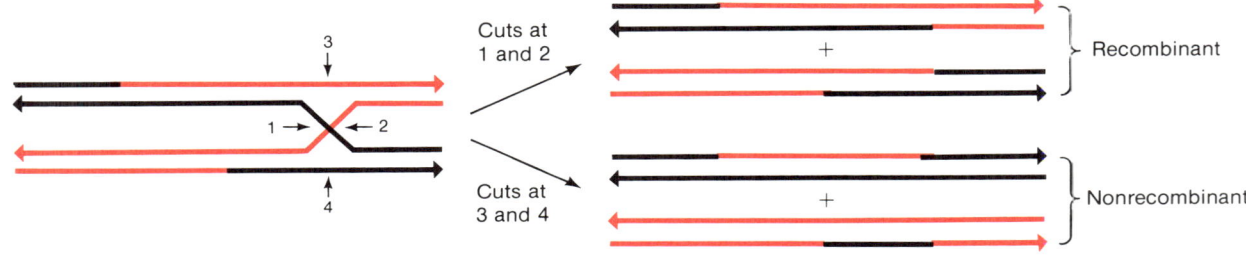

(c) Symmetric intermediate via branch migration without isomerization

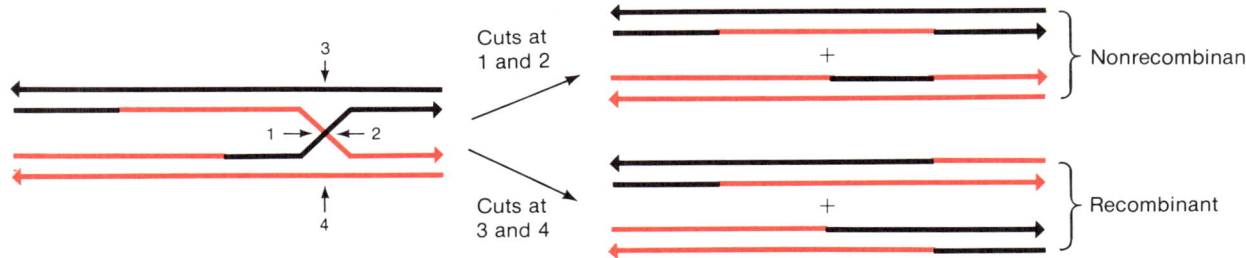

Figure 20-23 Alternative products of resolution of an (a) asymmetric intermediate formed by isomerization without branch migration, (b) a symmetric intermediate formed by branch migration of the intermediate in (a), and (c) a symmetric intermediate formed by branch migration without isomerization, according to the asymmetric strand-transfer model.

are recombinant. In both cases, the pair of molecules contains only one heteroduplex region. The recombinant array on the left is sufficient to explain the asymmetric 5 : 3 segregation in the *W17* locus of *Ascobolus* described in Figure 20-21: the single pair of nonidentical spores in Figure 20-21 comes from the upper molecule with the heteroduplex region (Figure 20-23(a)), and the converted pair of spores in Figure 20-21 would arise not as a result of repair of a mismatch, but rather of the DNA replication that occurs in steps 2–5 of Figure 20-22. Figure 20-23(b) shows the alternate pair of products of the symmetric intermediate formed by isomerization followed by branch migration. Two heteroduplex regions are present, and aberrant 4 : 4, 5 : 3, and 6 : 2 asci can arise by the mechanism of Figure 20-16. The asymmetric strand-transfer model does not exclude the possibility of forming the symmetric intermediate via branch migration without isomerization; in that case (panel (c)) recombinant and nonrecombinant products are formed instead by North-South and East-West cleavages, respectively.

How can one explain the difference between, on the one hand, the *Sordaria* asci in which a symmetric arrangement would have to be present and, on the other hand, the yeast data and the *Ascobolus* *W17* data, which requires an asymmetric intermediate? One possibility, suggested in the asymmetric strand-transfer model, is that the relative times of resolution and of branch migration determine the outcome. That is, one assumes that in yeast and at the *W17* locus, the enzymes that promote asymmetric exchange bind more tightly to their substrates and detach only late in the period of time allowed for exchange. Detachment of the enzymes is followed by rapid isomerization or branch migration, and resolution soon ensues. At other sites in *Ascobolus* and in *Sordaria* detachment is presumed to occur earlier, leaving time for the propagation of symmetric exchange before resolution occurs. Within a given species local base sequences, which could affect the activities of the enzymes and the efficiency of isomerization, branch migration, and resolution, could explain locus differences.

The asymmetric strand-transfer model describes many features of crossing over in fungi, though certain genetic observations require special modifications of the theory. As additional data accumulate in the future, modifications may need to be made to increase the understanding of meiotic crossing over. It is possible that the asymmetric strand-transfer model might prove to be incorrect and that a new explanation for crossing over would be required.

Resolution of Figure-8 Molecules

In Figure 20-17 it was shown that a Holliday intermediate formed from two linear molecules can be resolved in two different ways. If the cuts that separate the duplexes are in the crossed-strand region, the result is equivalent to a single-strand insertion (panel (c)), whereas, if the cuts are elsewhere, two molecules with overlap joints result (panel (d)). If the recombining molecules are circles instead of linear molecules, there are also two possible outcomes, but this time they are structurally distinct.

Figure 20-24 shows three ways that a figure-8 molecule can be cleaved. For clarity, we have rotated the lower loop of the figure-8 through 180° to form the open structure II. We assume first that the structure is susceptible

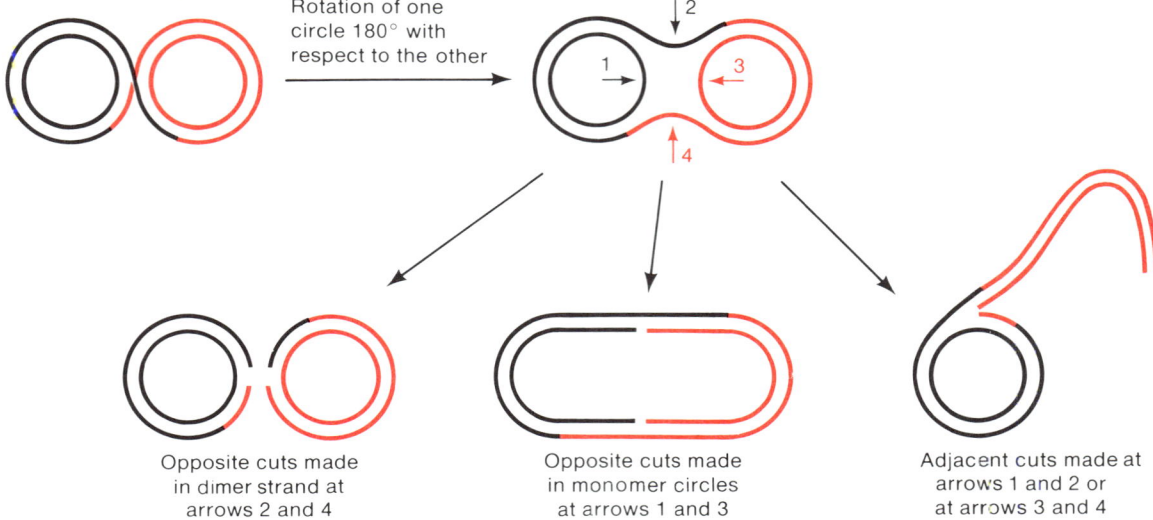

Opposite cuts made
in dimer strand at
arrows 2 and 4

Opposite cuts made
in monomer circles
at arrows 1 and 3

Adjacent cuts made at
arrows 1 and 2 or
at arrows 3 and 4

Figure 20-24 Resolution of a figure-8 molecule.

to strand breakage and that breaks occur only in single-stranded regions. This means that the breaks must be in one of the four segments indicated by the arrows. Second, we assume that the breaks always occur in pairs because one break alone cannot resolve the figure-8. There are three possible pairs: (1) opposed in the dimer strand, (2) opposed in the monomer circles, and (3) adjacent (one in a monomer circle and the other in the dimer strand). The consequences of these three pairs are two monomers, a dimer, and a rolling circle, respectively, as shown in the figure. These three forms may play very different roles in different systems. For example, if the original participants are chromosomes of either bacteria or eukaryotes, which only have monomer length, either formation of the dimer and of a rolling circle must be prevented or some further processing is required. On the other hand, in a phage system the formation of a monomer may be a dead end; for example, with phage λ, since two *cos* sites are needed for packaging (Chapter 17), only a dimer or a rolling circle (which continues to replicate) could lead to recombinant phage. At present there is no evidence whatsoever for recombination-mediated rolling circle formation in any system. However, dimerization of small plasmids frequently occurs in a RecA$^+$ bacterium.

THE RecBC PROTEIN

In *E. coli* in bacterial conjugation and in phage transduction a protein encoded by the genes *recB* and *recC* is needed in addition to the RecA protein. The RecBC protein is a multifunctional protein with an endonuclease activity and a powerful exonuclease activity against both single- and double-stranded DNA. The exonuclease activity was first detected, so the protein is also called exonuclease V. Under certain experimental conditions the nuclease activities can be inhibited and the enzyme instead exhibits the ability to unwind double-stranded DNA. ATP is hydrolyzed when unwinding occurs. The unwinding that occurs when the nuclease activity is inhibited is unusual, as shown in Figure 20-25. The enzyme binds to one end of the DNA molecule and unwinds the DNA. It moves along the strand but remains in contact with two regions of that strand, allowing the end of the strand to pass through the enzyme, thus forming the single-stranded loop shown in panel (b). As the length of the single-stranded tail increases, homologous base-pairing occurs with the other strand to form the doubly-looped structure shown in panel (c). If the nuclease activity is not inhibited, nicking occurs in the single strands several thousand nucleotides part. The RecBC protein also forms a complex with DNA polymerase I but the properties of this complex are not known. The numerous enzymatic activities of the RecBC enzyme have been incorporated into many models for homologous recombination but its precise role has not been demonstrated experimentally for any particular stage of recombination.

At least one of the activities of RecBC is detrimental to development of phage λ DNA in infected *E. coli*. The phage encodes a protein, called Gam, that binds to and inactivates the RecBC protein. Without such inactivation the late replication stage, rolling circle replication (Chapter 17) cannot be maintained.

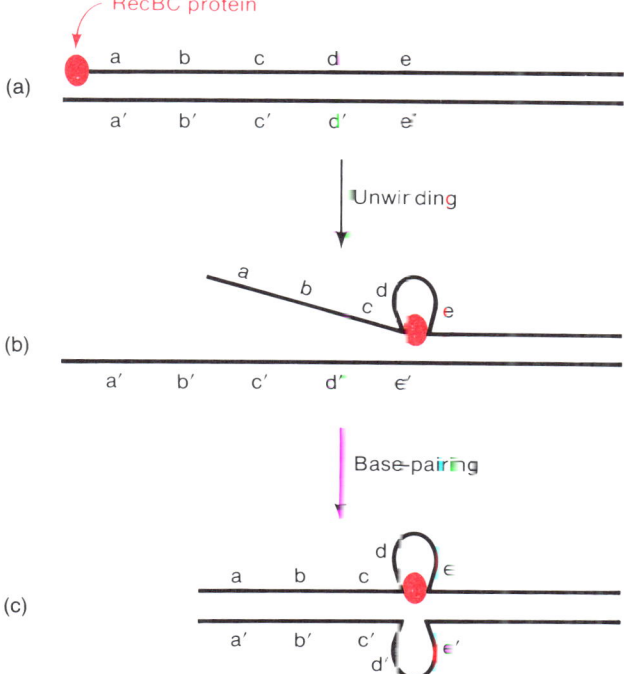

Figure 20-25   The proposed mode of unwinding of double-stranded DNA by the RecBC protein.

TRANSFORMATION IN YEAST

Transformation in yeast occurs by a mechanism quite different from transformation in *Pneumococcus*. It has been studied mainly by using an *E. coli* plasmid with a short stretch of yeast DNA as donor and a yeast cell as a recipient. The yeast DNA lacks a replication origin and the *E. coli* origin is inactive in yeast, so the appearance of a yeast gene from the plasmid in a yeast cell requires recombination. The following features of yeast transformation are significantly different from other transformation systems:

1. The entire plasmid is incorporated in the cell and the exchange occurs only between homologous yeast sequences. The position of the exchange can be anywhere in the homologous region.
2. Transformation in yeast is far less efficient than transformation in *Pneumococcus*, but it can be stimulated several thousandfold by introducing a *double*-strand break (by a restriction enzyme) in the yeast sequence. In contrast, double-strand breaks in *Pneumococcus* DNA (or the donor DNA in all known bacterial transformation systems), reduces transformation frequency greatly.
3. In yeast, in addition to enhancing transformation, a double-strand break produces a bias in the exchange site. For example, if the plasmid contains two yeast genes, *A* and *B*, and a double-strand break is made in gene *A*, the plasmid DNA is inserted into the homologous chromosomal gene *A*; if the cut is instead in *B*, insertion into chromosomal gene *B* occurs. In either case, the cells ends up with two copies of gene *A* and two copies of gene *B*. This result indicates that either the double-strand break is repaired in the course of recombination or that two nearby exchanges occur, or both.

The following experiment gave some insight into the recombination process in yeast transformation (Figure 20-26). In this experiment the donor DNA carried the A^- and B^+ alleles and the yeast cell was A^+B^-. Two cuts were made in the plasmid DNA, such that most of the A^- gene was removed. This cleavage, which produced in essence a large gap (or deletion) in gene *A*, stimulated transformation several thousand fold, like a double-strand break. Remarkably, the transformant not only contained two copies of the *B* gene but also two copies of the *A* gene. Furthermore, both copies of the *A* gene were the A^+ allele of the recipient—the deleted form was absent. This duplication indicated that the gap produced by the two double-strand breaks is repaired and that the *repair is a replicative process* in which the *A* allele of the recipient is the template. In other experiments in which the allele in the recipient was $-$ and the plasmid allele was $+$, two copies of the allele of the recipient were always found. The model shown in Figure 20-27, called the **double-strand-break-repair model,** has been proposed to explain yeast transformation. The significant features of the model are the following:

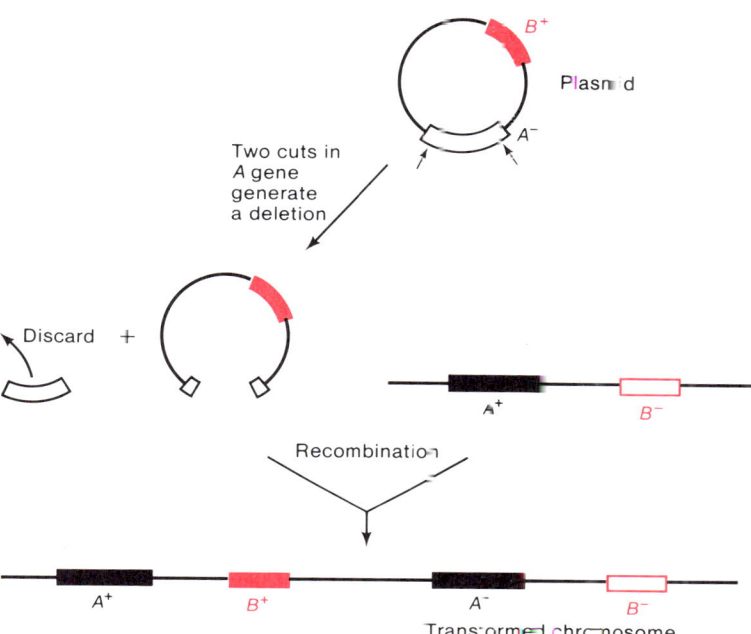

Plasmid

Two cuts in
A gene
generate
a deletion

Discard   +

A⁺   B⁻

Recombination

A⁺   B⁺   A⁻   B⁻

Transformed chromosome

Figure 20-26   The result of transformation of yeast by a plasmid carrying a gene in which a segment has been removed.

1. Double-strand breaks are repaired by two independent rounds of copying a single strand.
2. Two heteroduplex regions are formed on either side of the region in which the double-strand break is repaired.
3. Two Holliday crossed-strand intermediates form.
4. The process is completed by the resolution of both Holliday junctions; the resolution is of the same type as that proposed in the Holliday and asymmetric strand-transfer models.

In the first step of the model a double-strand break is made in what will be the invading strand. The rejoining process is initiated by attack on each 5′-P terminus by a 5′-exonuclease, removing nucleotide (a few or possibly hundreds to thousands of nucleotides) and yielding two long single-stranded extensions terminated by a 3′-OH group. In the next step one of the single strands invades the uncut strand, pairing with a complementary strand and forming a triple-stranded loop, exactly as in the strand-invasion step of the asymmetric strand-transfer model. A DNA polymerase adds nucleotides to the 3′-OH end of the invading single strand, and the loop is enlarged by continued displacement of the initially displaced strand in the invaded DNA. Once the loop is enlarged sufficiently, base pairing of the second single-stranded appendage becomes possible. Extension from the second 3′-OH terminus replaces the DNA removed by the initial exonucleolytic digestion and thereby repairs the double-strand break. Note that at this point the paired

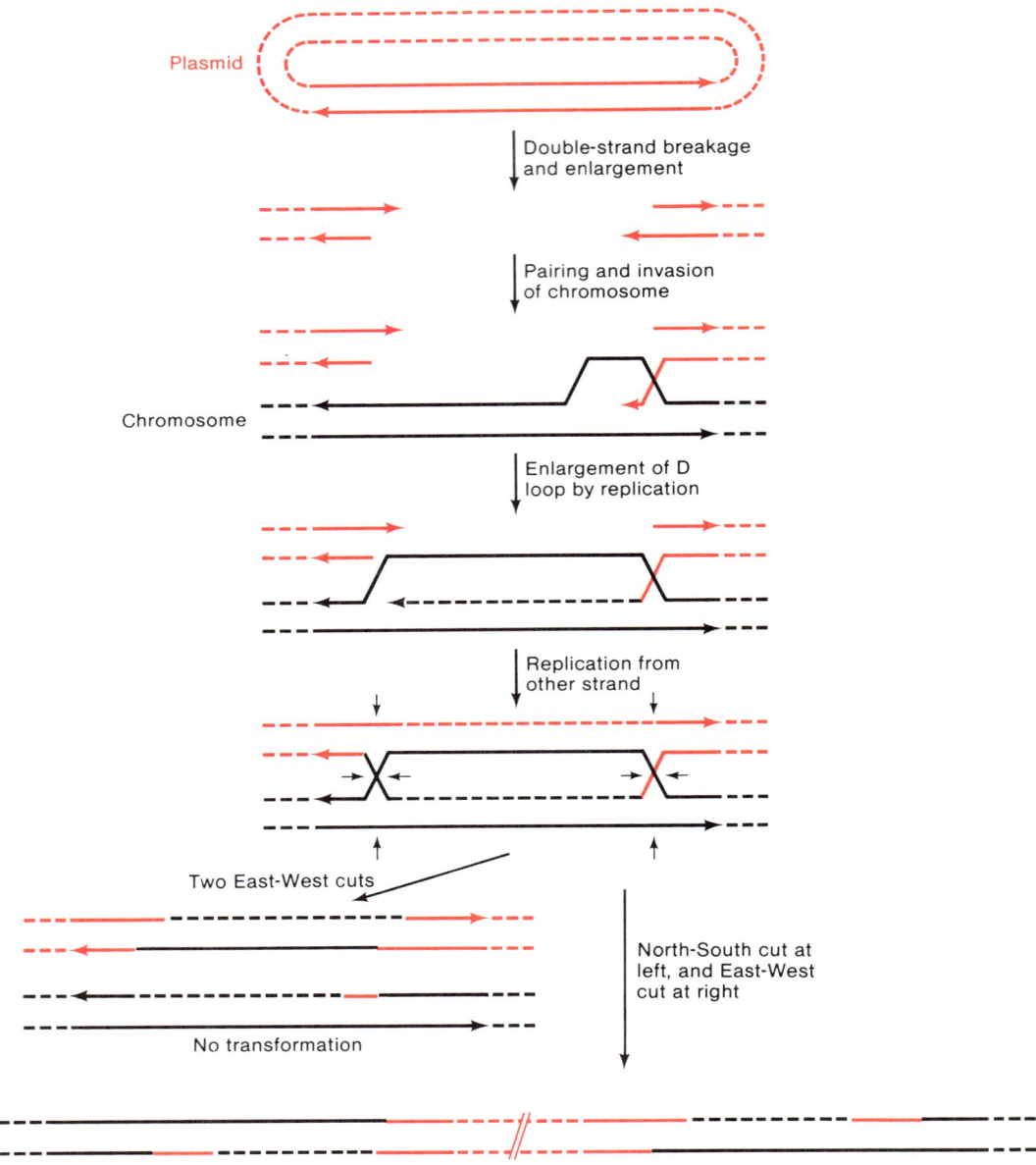

Figure 20-27 A proposed mechanism of transformation in yeast.

unit will have two Holliday junctions, each of which may be able to branch-migrate. As in the Holliday and asymmetric strand-transfer models, the Holliday junctions are ultimately resolved by pairs of single-strand breaks. As drawn in the figure, a North–South cut in the left junction and an East-West cut in the right junction will yield a transformant.

A small group of researchers has suggested that although a great deal of evidence supports the asymmetric strand transfer model as the mechanism of meiotic recombination in yeast, meiotic recombination may occur instead via the double-strand-break-repair model. The observation that raised this possibility was the isolation in yeast of a mutation, *RAD52*, that failed to repair double-strand breaks induced in yeast chromosomes by x rays. That is, irradiated *RAD52* mutants were killed by lower doses of x rays than *RAD*+ strains. Later studies showed that neither meiotic recombination nor transformation occurs in *RAD52* mutants, which suggests that x-ray repair and the recombination process probably share a common step. However, until the function of the *RAD* gene is elucidated, how repair, meiotic recombination, and transformation are related will not be known. Proponents of the double-strand break repair model have shown how it can explain such phenomena as post-meiotic segregation and gene conversion. However, whether meiotic recombination in yeast does proceed in the way suggested is at present highly controversial, and considerably more experiments are needed to test the model.

Transposable Elements

A long-standing unspoken assumption about gene organization and chromosome structure is that each gene has a definite and unvarying location in a particular chromosome. This assumption has been strengthened over the years by the ability to generate gene maps by both genetic and physical techniques. However, hints of gene rearrangements have come forth repeatedly over the past few decades. For example, gene inversion occurs at low frequency in both bacteria (Chapter 11) and in eukaryotes (observed repeatedly in *Drosophila* and yeast). In Chapter 15 we also saw an example of gene relocation in the phenomenon of phase variation in *Salmonella*. In the 1940s in a genetic analysis of the appearance of the kernels of maize Barbara McClintock discovered regulatory elements that moved from one site to another and thereby affected gene expression. Her work was followed 30 years later by the recognition that bacteria contained mobile segments of DNA that relocated at low frequency; these segments were sometimes called "jumping genes." Study of the features of the base-sequence of the bacterial elements allowed the identification of a variety of such mobile elements in eukaryotes, first in *Drosophila* and then in other eukaryotes. The frequency with which the elements move in both bacteria and eukaryotes is quite low, but nonetheless the concept of a static genome has gradually been replaced by the current view of the "genome in flux."

Because a chromosome is a continuous DNA molecule, movement of mobile elements is a genetic exchange process, a kind of recombination.

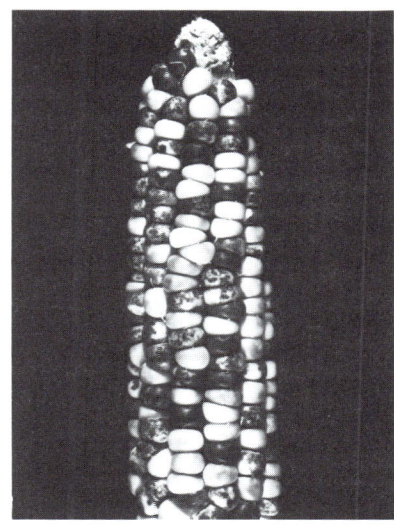

An ear of corn (maize) showing the sectoring of the kernels studied by Barbara McClintock and caused by the transposable elements Dissociation and Activator. [Courtesy of B. McClintock and the Cold Spring Harbor Laboratory Library Archives.]

However, it differs from recombination systems already encountered in this book in which exchanges occur between homologous DNA sequences, in that homology plays no part in the process. In bacteria, the evidence is clear, for homologous exchange depends on the *recA*-gene product or its equivalent, whereas, in contrast, movement of these elements occurs in *recA* ⁻ cells.

The movement discussed in this chapter is called **transposition,** and the mobile segments are called **transposable elements.**

Although transposable elements were first discovered in maize, most of the available molecular information concerns the bacterial elements, which will be the main focus of this chapter.

AN OVERVIEW OF TRANSPOSITION*

Both prokaryotes and eukaryotes contain DNA sequences that appear to move from one part of the genome to another. These sequences are called transposable elements, but more commonly in bacteria, transposons. In this overview, we summarize three features of these elements: their movement (transposition), the common characteristics of the sequences of these elements, and the biological effects of transposition.

1. *The transposition process.* Movement of transposable elements occurs by a variety of mechanisms, none of which have been unambiguously elucidated at the molecular level. In prokaryotic transposition one copy of the element remains at the original site and another copy is generated at a new site, which can be anywhere in the genome. Since one copy becomes two, transposition in prokaryotes must be a replicative process. However, relocation also requires breakage of the chromosome at the new site and linkage of the transposon to the ends that are generated, so transposition must also include recombinational steps. In most models of transposition the recombinational events generate one or two replication forks, which accounts for the duplication of the element; however, none of these models has been proved. Sequencing of many transposons and of the regions surrounding them shows that transposition does not require any homology; this observation is consistent with the fact that it occurs in *recA⁻* cells with the same frequency as in *recA* ⁺ cells. The enzymes responsible for the breakage-and-rejoining events are encoded in the transposons. In contrast, in eukaryotes movement of some transposable elements entails excision of the element and relocation at a new site; thus, for these elements transposition is unlikely to be replicative. For other elements, one copy remains at the original site and a duplicate appears at a new site, as in bacteria. For these elements the

---

*This section, which only summarizes phenomena, is for courses in which details are not included. All information is given in detail later in the chapter. Figures, as needed, are found in the main body of the chapter.

replicative steps apparently consist of production of an RNA molecule by transcription, synthesis of a double-stranded DNA copy by reverse transcriptase, and insertion of the DNA at the new site.

A significant feature of transposition in both prokaryotes and eukaryotes is the presence of a short (less than 10 bp) sequence that flanks the transposon when transposition is complete (see Figure 21-6, later in the chapter). This sequence is present in single copy at the site of entry of the transposon and prior to entry, and is duplicated in the course of transposition; it is called a target sequence. The presence of identical sequences flanking a segment of DNA is considered the hallmark of a transposable element even when transposition itself is not detected. The mechanism for duplication of the target sequence is not known, but there are many reasonable models.

In prokaryotes the number of transposition events is limited in an unknown way, so the number of copies of a transposon in a cell never exceeds about 10 to 20. In contrast, in eukaryotes control of transcription seems to be less stringent (perhaps only for those elements that transpose replicatively via an RNA intermediate) and the number of elements can be very large, up to several hundred thousand; in *Drosophila* and the higher eukaryotes transposable elements constitute the major part of the middle-repetitive DNA (Chapter 4) and amount to 5–30 percent of the genomic DNA.

2. *Features of the sequences of transposable elements.* Transposable elements range in size from a few hundred to a few thousand nucleotide pairs. The simple transposable elements consist of a major central portion that contains the transposition enzymes and short flanking inverted-repeat sequences about 15–75 bp long (Figure 21-2). The composite transposons of bacteria consist of a longer central portion flanked by two simple transposable elements (Figure 21-4); the terminal simple elements can be in direct or inverted repeat, but since these simple transposons have inverted-repeat termini, the composite elements are also terminated by inverted repeats. The central portion of bacterial transposons contain a variety of antibiotic-resistance genes, and it is by these genes that the elements are usually detected. The central region also usually contains genes and sites needed for transposition.

3. *Biological effects of transposition.* A variety of genetic effects are associated with transposition, and it is by these effects that transposable elements are usually identified. (a) When an element transposes, it may insert within a gene of the host DNA and thereby produce a mutation. In fact, transposon-induced mutagenesis is a valuable technique in microbial genetics. (b) Some of the simple bacterial transposons contain stop codons in all reading frames and Rho-dependent transcription-termination sequences. Thus, insertion of these elements between a promoter and a codon sequence or between genes of a polycistronic operon prevents transcription or translation of downstream genes. In fact, *E. coli* transposons were first detected as a class of nonrevertible polar mutations. (c) Other transposon-mediated events include

fusion of plasmid circles (replicon fusion), insertion of F plasmids into the *E. coli* chromosome to generate Hfr cells (another example of replicon fusion), genetic deletions, and genetic inversions. (d) In eukaryotes transposition is sometimes associated with switching expression of certain genes on and off, as mentioned for maize kernel color in the introduction to the chapter. (e) If a simple or composite transposon moves to a nearby site—for example, a few thousand nucleotide pairs away—the DNA in that region of the genome then consists of a large segment flanked by two transposons. This array is equivalent to a larger composite transposable element and indeed the entire segment can be transposed. Thus, in this way transposition can lead to duplication of coding segments of the genome by the mechanisms suggested in Chapter 16 (Figures 16-3 through 16-5).

TERMINOLOGY

In prokaryotes transposable elements are usually called **transposons.** Some transposons contain easily recognizable bacterial genes, particularly those for antibiotic resistance. A member of this class is designated by the abbreviation Tn followed by a number (e.g., Tn5), which distinguishes different transposons. By agreement all newly discovered bacterial transposons are designated in this way even if no recognizable bacterial gene is present. When it is necessary to refer to the genes carried on a transposon, these are indicated by standard genotypic designations, for example, Tn1(*amp-r*), in which *amp-r* indicates that the transposon carries the genetic locus for resistance to ampicillin. The transposons first discovered did not contain any known host genes, and for historical reasons, they were called **insertion sequences** or **IS elements** and designated IS1, IS2, and so on. Transposons have occasionally been designated in nonstandard ways—for example, $\gamma\delta$, the element contained in the F plasmid. In eukaryotes no consistent system for naming transposable elements is used; for instance, three transposable elements in *Drosophila* are called *copia, 497,* and *P,* and the best-studied elements in maize were named Dissociation and Activator.

A transposable element is frequently located within a particular gene and this creates a mutation in that gene, which is usually given a number. The following notation is used to designate mutation 135 in the *galT* gene produced by transposon number 4:

$$galT135::\text{Tn}4$$

Note the use of the double colon.

INSERTION SEQUENCES

The first transposable elements that were discovered in bacteria (in 1967) were the IS elements, and how they were discovered is of some interest. It is now known that they are just one type of transposon but nonetheless some of the essential characteristics of transposons and the various means of detecting transposition can be easily seen by examining this special class.

A collection of *gal*⁻ and *lac*⁻ mutations with unusual properties were found in *E. coli*. The mutations had the following features:

1. They were highly **polar mutations** in that each mapped in the first gene of the operon but proteins of the downstream genes were not synthesized. In each case the polarity resulted from the presence of a Rho-dependent termination sequence.

2. These mutations could not be reverted by base-analogue or frameshift mutagens, so the mutations could not be base substitutions or single-base additions or deletions.*

3. If a plasmid was transferred to a cell containing one of these polar mutations, similar polar mutations occasionally appeared in genes carried on the plasmid. The locations of the chromosomal and plasmid polar mutations could be quite different. For example, in the *gal* operon of a chromosome and a *lac* operon in the plasmid.

4. Physical studies of several plasmids containing such polar mutations in different genes provided the crucial evidence regarding the nature of the mutations. These experiments showed that in each case *a segment of DNA had been inserted in the operon.* It was hypothesized that these inserted segments were mobile elements that moved from one region of a DNA molecule chromosome to another.

Electron microscopic examination of the DNA of specialized transducing particles carrying the polar mutations was especially informative. Hybridization of the DNA from a transducing particle bearing the mutation with the DNA of a transducing particle lacking a polar mutation showed a loop (Figure 21-1), which indicated that the mutation is an insertion whose length equals the length of the loop. Comparison of different transducing particles containing polar insertions obtained from a particular strain showed

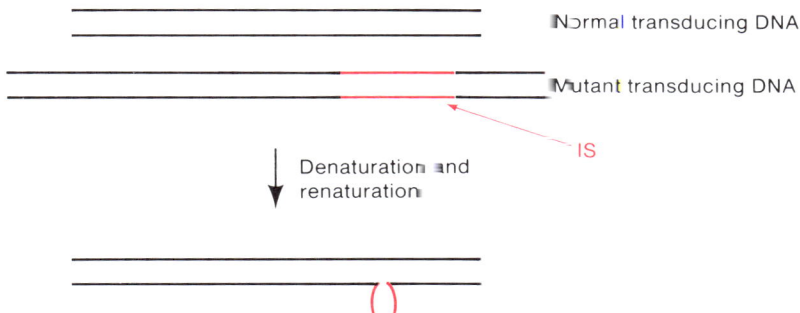

Normal transducing DNA

Mutant transducing DNA

IS

Denaturation and renaturation

Figure 21-1 A hybridization test showing that an IS element is a segment of inserted DNA. After renaturation half of the molecules have re-formed the normal and mutant transducing molecules (only the heteroduplex has been drawn).

---

*Less-polar mutations had been known since 1954, but as these were chain termination mutations, they were usually revertable by base analogues.

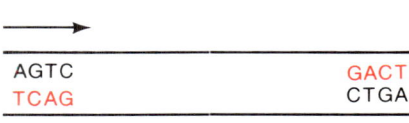

```
AGTC                    GACT
TCAG                    CTGA
```

Figure 21-2 An example of a terminal inverted repeat in a DNA molecule. The arrows indicate the inverted base sequences. Note that the sequences AGTC and CTGA are *not* in the same strand. In a so-called direct repeat, the sequence in the upper strand would be AGTC . . . AGTC.

that the loop of the DNA of one particle could hybridize with the loop of another, indicating that the *same* sequence is present in each mutated site.

Studies of a large number of transducing particles containing genes from numerous regions of the *E. coli* chromosome into which IS elements were inserted and of plasmids containing insertion sequences demonstrated the existence of several different IS elements. A few of these elements were isolated from transducing-particle DNA and their base sequences were determined. A number of significant features of these elements became apparent from the sequence analyses:

1. Those elements that produced polar mutations contained Rho-dependent transcription stop signals (except for IS1) and also chain-termination mutations in all possible reading frames. These properties were sufficient to account for the polar effects.
2. The termini of each IS element had inverted repeat sequences consisting of 16–41 base pairs, the number depending on the element (Figure 21-2).
3. At many sites of insertion (for example, in the *E. coli gal* operon, the sequences were inserted in either the left-to-right or right-to-left orientation. This is expected because of the symmetry of the inverted repeat configuration of the termini.
4. Different IS elements contained different numbers of bases. Radioactive copies of the elements were also prepared and hybridized with denatured *E. coli* DNA and with denatured F plasmid DNA to determine the number of copies of the elements in the chromosome and in F. The results are summarized in Table 21-1.
5. The elements contained at least two apparent coding sequences initiated by an AUG and terminated with an in-phase stop codon. By using *in vitro* protein-synthesizing systems with isolated IS DNA as template, proteins have been made. Some of these proteins have also been isolated from cells containing the element. Study of a large number of bacterial transposable elements indicates that each encodes at least two proteins.

Table 21-1 Properties of some *E. coli* insertion elements

Element	Number of copies and location*	Number of base pairs
IS1	5–8 in chromosome	768
IS2	5 in chromosome, 1 in F	1327
IS3	5 in chromosome, 2 in F	ca. 1400
IS4	1 or 2 in chromosome	1428
IS5	Unknown	1195
γδ	1 or more in chromosome, 1 in F	5700

*The number of copies may vary with the strain.

DETECTION OF TRANSPOSITION IN BACTERIA

The IS elements just described are a special class of bacterial transposon. In this section we examine the more general features of transposons and how these elements are observed.

In contrast with IS elements, which are normally detected by polarity effects, the more common class of bacterial transposon contains recognizable genes—for instance, genes for antibiotic resistance—that are often not present elsewhere in the genome. These genes provide the means for detecting transposition. An example is shown in Figure 21-3.

In this experiment an R plasmid having an *amp-r* transposon is transferred by conjugation to a *recA⁻* recipient bacterium that carries a transposon having the *kan-r* (kanamycin-resistance) gene. On continued growth in the presence of both antibiotics, transposition occasionally occurs and yields an R plasmid carrying both drug-resistance markers and hence both transposons. The presence of this two-transposon plasmid is detected by transferring the plasmid by conjugation to a Kan-s Amp-s bacterium and plating on agar containing both antibiotics. As a further test, the plasmid containing the two markers could be hybridized to the plasmid molecule and examined

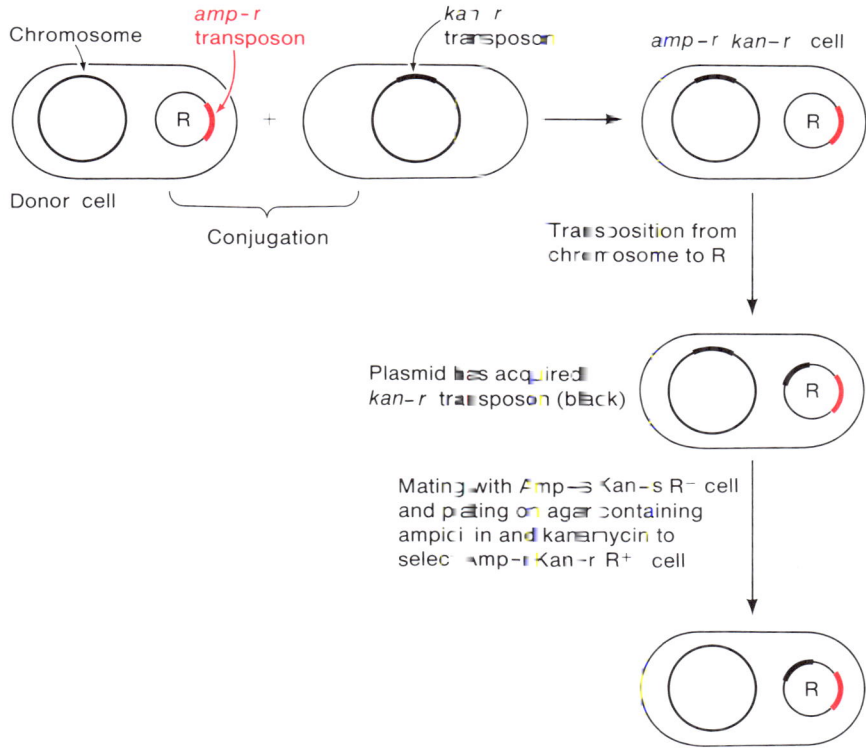

Figure 21-3 An experiment to demonstrate transposition. A donor bacterium (left) containing an *amp-r* R plasmid transfers R to a recipient cell (middle) containing a *kan-r* transposon. In subsequent generations the *kan-r* transposon is transposed to the R plasmid. The presence of the second transposon on R is demonstrated by simultaneous transfer of the *amp-r* and *kan-r* genes. The *amp-r* gene is shown as a transposon, as are most drug-resistant determinants on plasmids, but this fact is not essential to the experimental demonstration of transposition.

by electron microscopy; loops such as those in Figure 21-1 would be seen, indicating the presence of an inserted DNA sequence.

When transposition occurs, most transposons can be inserted at a very large number of positions. The most direct evidence comes from several types of experiments in which mutants formed by insertion of a particular transposon are isolated. For example, consider an Amp-s *E. coli* culture transformed by a DNA fragment carrying an *amp-r* transposon. The fragment is incapable of replication, so Amp-r cells can arise only by transposition of the *amp-r* element to the chromosome. Plating the cells on nutrient agar containing ampicillin enables one to isolate colonies derived from cells in which transposition occurred. If many hundreds of Amp-r bacterial colonies are examined and tested for both nutritional requirements and ability to utilize particular sugars as a carbon source, it is observed with most transposons that a colony can be found bearing a mutation in almost any gene that is examined. This indicates that insertion sites for transposition are scattered throughout the *E. coli* chromosome.

Although there are certainly many sites of insertion in *E. coli,* and even in several phages, it is clear that the locations are not randomly distributed; certain positions seem to be excluded and others ("hotspots") are used repeatedly. This can be seen by examining the insertion sites in very small regions whose base sequences are known. Such studies show that there is a rather broad range of insertion specificities from one transposon to the next. At one extreme is IS4, for which 20 independent insertion events have been observed in the *galT* gene, all of which are probably at the *same* nucleotide site. At the other extreme is Mu, which can insert almost anywhere. Most transposons exhibit insertion specificities between these extremes.

TYPES OF BACTERIAL TRANSPOSONS

Because of particular related features, bacterial transposons have been divided into several distinct classes. One is the IS group, which we have already discussed. Two more are described below.

1. *Composite transposons.* A variety of transposons carrying antibiotic-resistance genes are flanked by two identical or nearly identical copies of an IS element. These transposons are called composite type I transposons. The IS elements in these composite units can be in an inverted or direct repeat configuration (Figure 21-4). Since the two ends of an IS element are themselves inverted repeats, the relative orientation (inverted or direct) of the flanking IS elements of a composite transposon does not alter its terminal sequences—that is, they remain the same in the inverted repeat array. The ability to transpose is determined by the terminal IS elements. Three frequently-studied composite transposons are Tn5, Tn9, and Tn10.

Most of the composite type I transposons were first detected as genetic elements that could be transposed from one plasmid to another or to a phage.

Inverted repeat

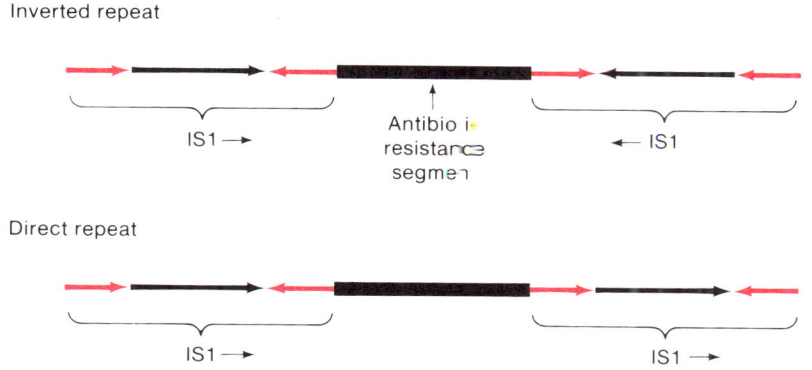

Antibiotic
resistance
segment

IS1 →   ← IS1

Direct repeat

IS1 →   IS1 →

Figure 21-4 Two composite type-I transposons flanked by IS1 in either inverted or direct repeat. The central black region contains the antibiotic-resistance genes and the genes necessary for transposition.

When the DNA of some of these plasmids was denatured and renatured, single-stranded circles containing a stem-and-loop structure were observed by electron microscopy. Such a structure is indicative of an inverted repeat, as shown in Figure 21-5.

The composite transposons have two interesting features: (1) Some of the terminal IS elements are able to transpose by themselves. (2) For many composite transposons, if the transposon resides in a small plasmid, it behaves like two different transposons. This is because within the frame of reference of the terminal IS elements the DNA they flank and the DNA that surrounds them are interchangeable. For example, consider a composite transposon having the structure IS–A–IS, where A is the central sequence. If the transposon inserts in a plasmid whose sequence we call B, the structure of the transposed plasmid will be

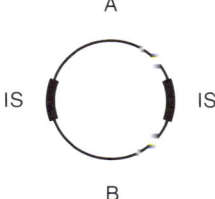

A

IS   IS

B

Because the plasmid is circular, both A and B are flanked by the IS elements and transposition of either IS–A–IS or IS–B–IS can occur.

2. *The Tn3 transposon family.* The Tn3 family of transposons consists of quite large elements (about 5000 bp). Each transposon carries three genes, one encoding β-lactamase (which confers resistance to ampicillin) and two others needed for transposition (which will be discussed later). All Tn3-like transposons contain short (38-bp) inverted repeats and none are flanked by IS-like elements.

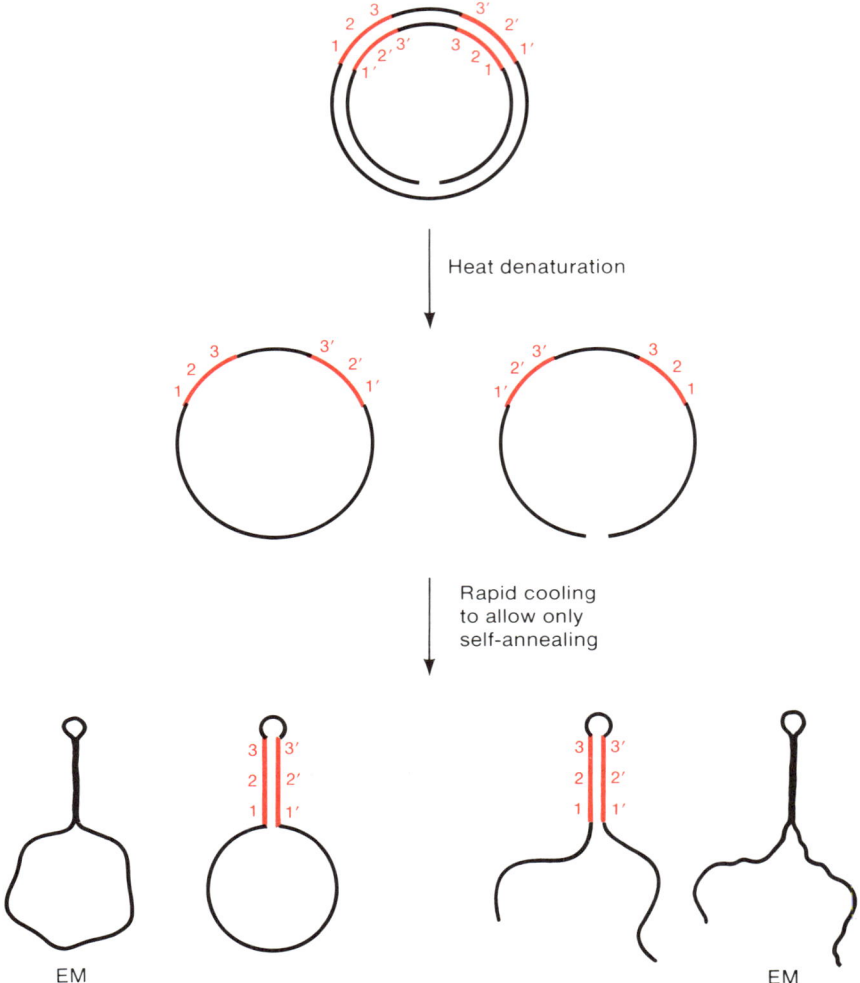

Figure 21-5 Formation of a stem-and-loop structure by denaturation and intramolecular annealing of a plasmid carrying an inverted repeat. The configurations labeled EM indi-cate the appearance of such a molecule in an electron microscope; light and heavy lines represent single- and double-stranded regions, respectively.

TRANSPOSITION

The end result of the transposition process is the insertion of a transposon between two nucleotide pairs in a recipient DNA molecule. Base sequence analysis of many transposons and their insertion sites reveals that there is no sequence homology, which is consistent with the lack of a requirement for the *E. coli recA* system. However, the analysis of sequences of the regions in which the inserted element joins the recipient DNA yielded several surprises, which we document in this section.

Duplication of a Target Sequence at an Insertion Site

A general characteristic of the insertion process is that insertion of a transposon always involves the duplication of a short nucleotide sequence (3–12 bp) in the recipient DNA molecule, called the **target sequence,** and the inserted transposon is sandwiched between the repeated nucleotides. This arrangement is shown in Figure 21-6. We repeat, emphatically, that *only one copy of the duplicated target sequence is present in the recipient DNA prior to insertion of the transposon and it is not present in the transposon itself.*

The length of the target sequence varies from one transposon to the next but *is the same for all insertions of a particular transposon.* For example, an inserted IS1 is always flanked by a 9-bp sequence, whereas another transposon is always flanked by a duplication of 5 bp. Note that *for a particular transposon, the target sequence is different for each insertion site; only the length of the duplicated sequence is constant.*

The following basic mechanism, whose details are unknown and which is shown only in outline (Figure 21-7), is believed to be responsible for the duplication of the target sequence.

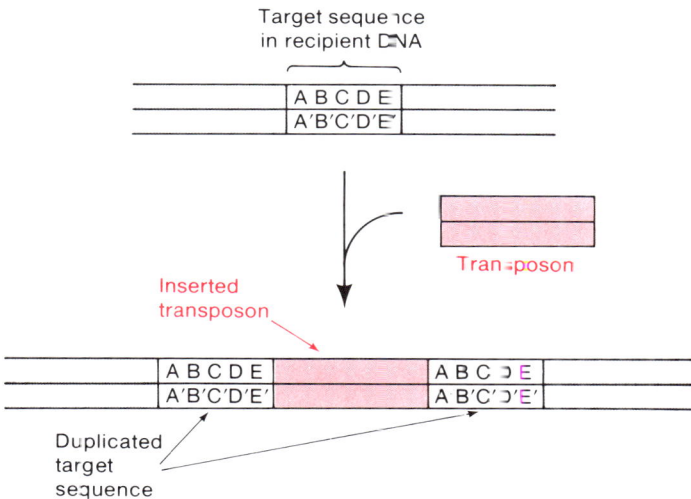

Figure 21-6 Insertion of a transposon generates a duplication of the target sequence. The transposon is never present as a free DNA segment but comes from another region of the genome, as shown in Figure 21-9.

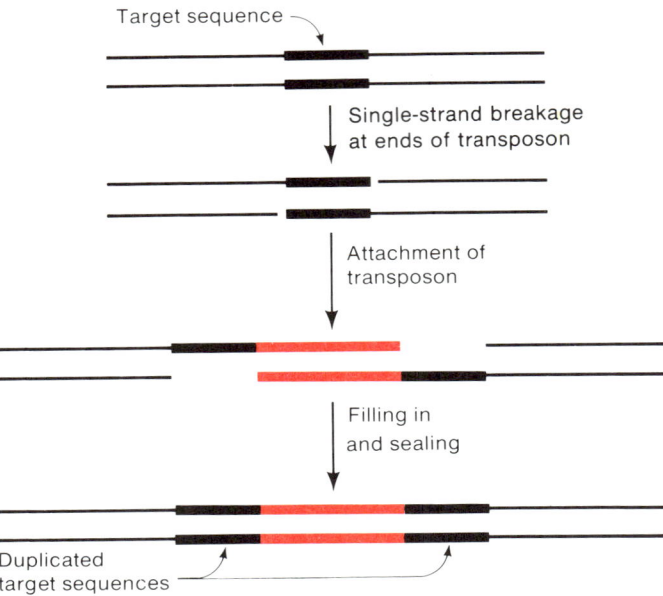

Figure 21-7   A schematic diagram indicating how target sequences might be duplicated. The transposon is never present as a free DNA segment but comes from another region of the genome, as shown in Figure 21-9.

An enzyme, possibly encoded in the transposon, acts on a target sequence in the recipient DNA by making two single-strand breaks, one in each strand, at the ends of the target sequence. The transposon DNA then is attached to the free ends generated by the nicks, such that one strand of the transposon is joined to one strand of the target sequence and the other strand of the transposon is joined to the other strand at the opposite end of the target sequence. This joining leaves two gaps, one across from each strand of the target sequence. Each gap is then filled in and the nicks are sealed; thus, the two strands of the recipient DNA are again continuous and there are two copies of the target sequence flanking the transposon. Note that no homology is utilized in joining the terminal sequence of the transposon to the recipient DNA. Later in this chapter a more detailed proposal for generating the duplicated target sequence will be described.

Since every transposon that is integrated is flanked by a duplicated target sequence and since this is true even of those transposons maintained at fixed sites (for example, those in the F plasmid), it is reasonable to inquire whether this duplication is also necessary for transposition from a donor site. This is definitely not the case since DNA sequences have been constructed (using recombinant DNA techniques) in which a transposon is not flanked by a directly repeating sequence, and these transposons are still able to transpose.

The Structure of Transposons and the Nature of the Transposon–Target-Sequence Joint

In studying numerous transposons base-sequencing analysis has been carried out on segments of DNA of known base sequence containing a particular transposon. Two important facts have emerged from these analyses:

1. A transposon has well-defined ends. This is shown by the observation that the transposon sequences adjacent to the two target sequences are identical for every insertion of a particular transposon (Figure 21-8). However, each kind of transposon has a specific sequence, so the transposition process consist of joining the ends of the transposon to the ends produced by nicking at opposite sides of the target sequence.

2. The two termini of the transposon consist of a single sequence in an inverted repeat array (one exception is known). This is also shown schematically in Figure 21-8. Note that for transposon I the sequence at the left end of the upper strand and the right end of the lower strand is 1234, reading each time in the $5' \rightarrow 3'$ direction. Depending on the transposon, the number of bases in the inverted repeat is 8 to 38.

A great deal of genetic evidence indicates that the termini of a transposon are essential for transposition. For example, transposable elements with

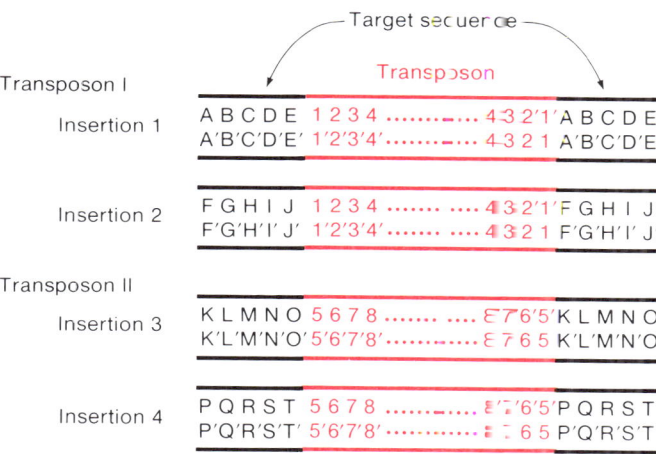

Figure 21-8 Diagram showing that two insertions by the same transposon have the same sequences at the junction of the transposon and the target sequence. That is, 1234 is at the left terminus of all insertions by transposon I. For transposon II the sequence is always 5678. Each number and each letter represent a base; the prime denotes a complementary base. The identifications are merely symbolic; 5678 designates a sequence different from 1234 and *not* one that follows 1234 in numerical order.

Figure 21-9 Transposition of a transposon from plasmid A to a plasmid B.

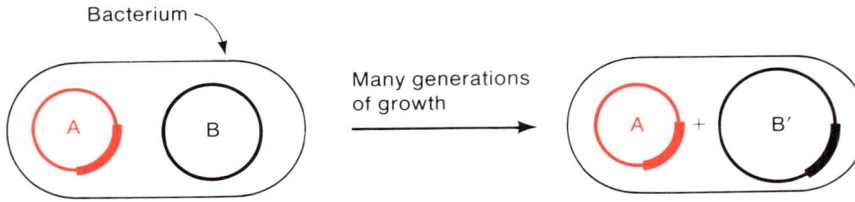

single base–pair changes, small deletions, and small insertions in the inverted repeats have been isolated and these alterations invariably prevent transposition.

Replication of a Transposon in the Course of Transposition Between Two Plasmids

Consider a cell possessing two plasmids, A and B, one of which contains a transposon. At a frequency of about 10^{-7} events per generation, the phenomenon shown in Figure 21-9 occurs. That is, after many generations of growth, cells are produced in which plasmid B (now called B′) also possesses the transposon originally in A. An important observation is that *the transposon is contained in both plasmids.* Thus the original transposon sequence is duplicated in the transposition process, indicating that *transposition is a replicative process.* (We will see shortly that this is not the case for eukaryotic elements.) A possible role for replication is to make a free DNA molecule that is a copy of the transposon. This molecule could then insert into a new site. However, free transposon DNA has never been observed and is believed not to occur.

A related phenomenon, which is important in understanding transposition, has also been observed in cells of the type just described. At a frequency also of about 10^{-7} events per cell generation, the two plasmids fuse to form a single plasmid called a **cointegrate.** (This process is sometimes also called **replicon fusion.**) This hybrid plasmid is not simply a fusion of the two plasmids, because it contains two (not just one) copies of the transposon. Both copies are in the same orientation (that is, in direct repeat) and are located precisely at the junction between the donor and recipient plasmid sequences (Figure 21-10).

Figure 21-10 Formation of a cointegrate by transposon-mediated fusion of two plasmids A and B. Two copies of the transposon are generated in the process. The arrows denoted an arbitrary direction to show that both copies of the transposon in the cointegrate are in direct repeat.

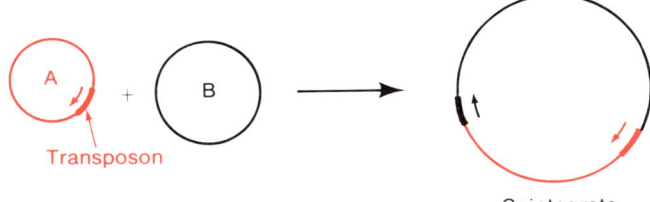

The fact that the cointegrate has two copies of the transposon, whereas only one copy was present before cointegration occurred, again indicates that the transposon has replicated.

The Cointegrate as an Intermediate in Transposition of Tn3

Certain experimental results with transposons in the Tn3 family suggest that formation of a cointegrate is an intermediate step in the transposition of these elements.

The structure of Tn3 is shown in Figure 21-11. Tn3 has a pair of terminal inverted repeats, like all transposons, and these flank three genes. The three gene products have been isolated; from their size and from the base sequence of Tn3, we know that the genes include all of the bases between the inverted repeats. The leftmost gene encodes a gigantic protein (1015 amino acids), denoted TnpA, called a **transposase;** it is responsible for formation of cointegrates. The rightmost gene encodes β-lactamase, an enzyme that inactivates ampicillin. The central gene encodes a small protein (185 amino acids) called a **resolvase** and symbolized TnpR; it has two functions— repression of the synthesis of TnpA, and catalysis of a site-specific exchange that resolves cointegrates in a second step of the transposition process. Near the boundary of the genes encoding TnpA and TnpR is a sequence of DNA called the **internal resolution site** or ***res***; it includes the base sequence at which TnpR acts. Interestingly the sequence is similar to the phage λ *att* site.

Genetic experiments have yielded the following information. All *tnpA⁻* mutants are unable either to transpose or to form cointegrates. Also, neither *tnpR⁻* mutations nor deletions of the internal resolution site interfere with the ability to form cointegrates, but they completely prevent transposition. That is, in a cell containing a plasmid with a *tnpR⁻* copy of Tn3 and a second

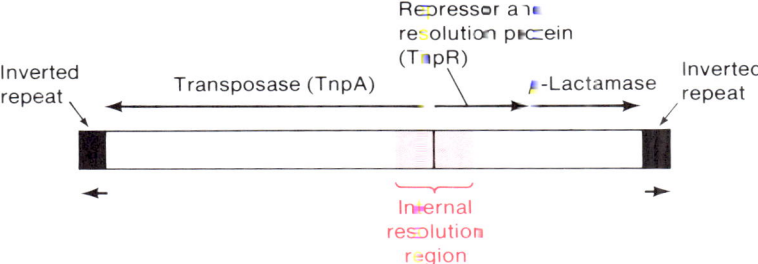

Figure 21-11 The physical map of transposon Tn3. There is a total of 4957 base pairs; each inverted repeat contains 38 base pairs. The three genes are indicated above the arrows, which show the direction of transcription. The internal resolution region *res* is discussed in the text.

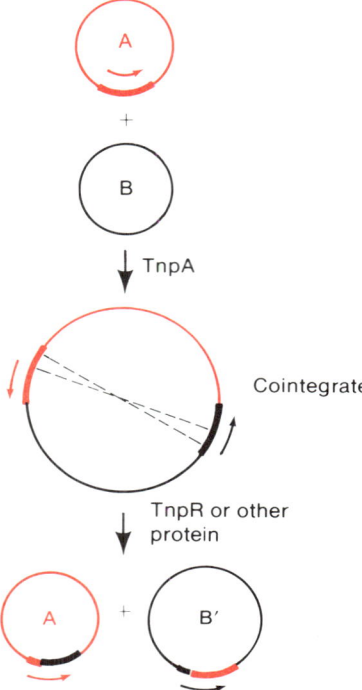

Figure 21-12 A model for transposition utilizing a cointegrate intermediate.

plasmid lacking Tn3, the process shown in Figure 21-10 can occur, but the process shown in Figure 21-9 does not occur. These observations suggest that Tn3-mediated transposition proceeds by a two-step mechanism—first, TnpA induces formation of a cointegrate and then TnpR catalyzes a site-specific exchange at the internal resolution site. A schematic diagram of this sequence of events is shown in Figure 21-12.

Many transposons outside of the Tn3 family do not carry a TnpR-like site-specific recombination system but are still able to transpose. These systems may use a homology-dependent exchange system since, once a cointegrate has formed, there are two identical sequences—namely, copies of the transposon—so any homology-dependent process could carry out the exchange shown in Figure 21-12. However, since transposition of these elements occurs in $recA^-$ cells, the homology-dependent recombinase would have to be encoded in the transposon. That a homology-dependent process could resolve the cointegrate is supported by the fact that a mutant of Tn3 lacking TnpR or the internal resolution site can carry out the transposition process of Figure 21-9 (that is, reduction of a cointegrate to individual plasmids) if a functional $recA$ gene is present in the host cell; in fact, if the $recA^+$ allele is introduced by a transducing particle into a $recA^-$ cell in which a cointegrate has already formed, resolution can occur.

A more detailed model for transposition is presented in the next section for the reader interested in current speculation. The section may be omitted in most courses.

Two Models for Transposition in Bacteria

The mechanism of transposition is not known, but current proposals abound. The facts that must be accounted for are the following: (1) the retention of the transposon by the donor at the initial site of the transposon; that is, that the process is replicative, (2) the existence of cointegrates, (3) the generation of a short repeated sequence of target DNA on each side of the newly integrated element, and (4) the requirement by some transposons for an internal resolution site or, in others, for a homology-dependent exchange.

Many models for transposition have been proposed; some are simply subtle variations of one another and others are quite different. Two features are common to all models, namely:

1. Two single strands are joined together (ligated) without base pairing to determine the position of joining.
2. A replication fork is created by the ligation event.

The reader should note these features in the models described.

The most widely favored model is shown schematically in Figure 21-13. In this model the cointegrate is an essential intermediate.

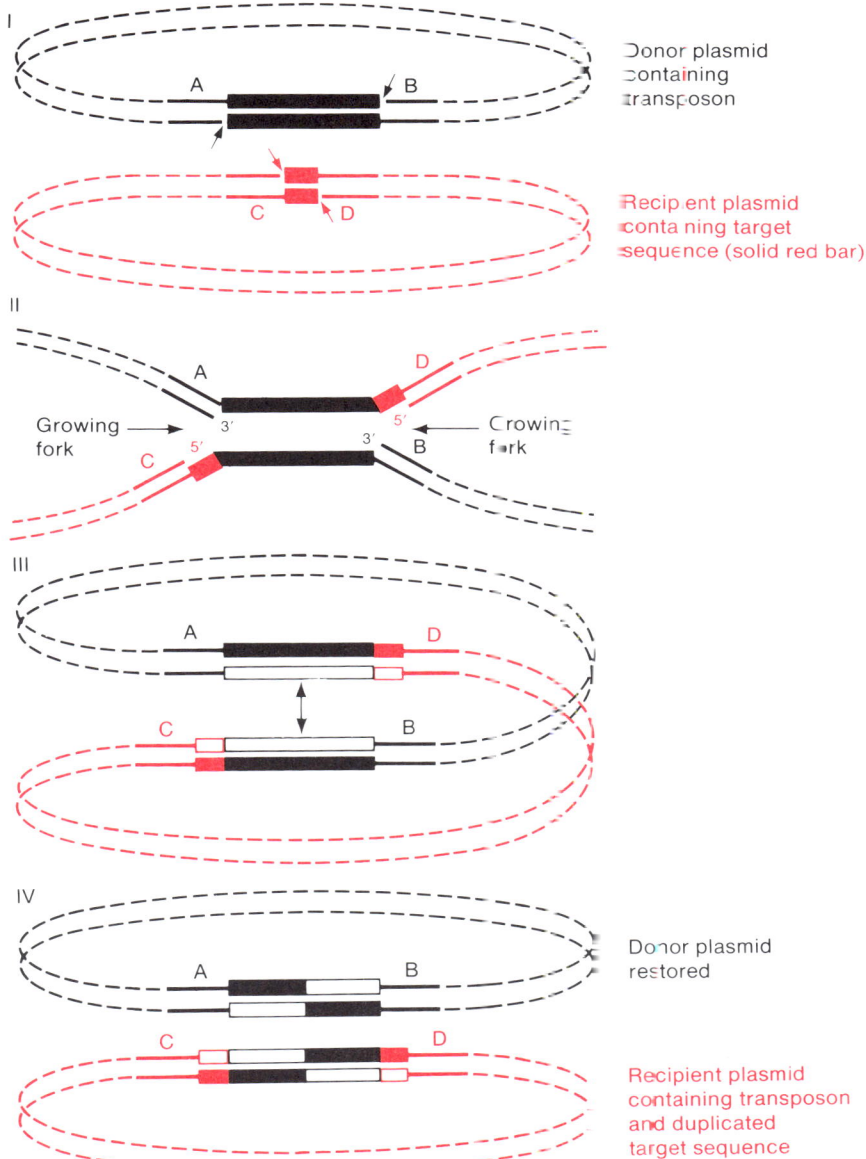

I

Donor plasmid
containing
transposon

Recipient plasmid
containing target
sequence (solid red bar)

II

Growing
fork

Growing
fork

III

IV

Donor plasmid
restored

Recipient plasmid
containing transposon
and duplicated
target sequence

Figure 21-13 A model for transposition proposed by J. Shapiro. The model is drawn for transposition from one plasmid to another, as this is a common arrangement that is studied. (I) Nicks are made at sites in the DNA molecule indicated by arrows. (II) Strand separation between nicks and joining of nonhomologous strands. Two replication forks result. (III) DNA synthesis (hollow lines) forms this structure. A site-specific strand exchange occurs between the internal resolution sites (arrows), or possibly a homology-dependent strand exchange occurs anywhere in the transposon. (IV) Plasmids separate. Capital letters denote base sequences.

We consider a donor plasmid containing a transposon and a recipient plasmid. (Neither the donor nor the recipient need to be plasmids or even circular. However, the figure is drawn that way because it is the arrangement most often studied.) In the recipient there is a short base sequence, the target sequence, at which integration of the transposon is to occur. No assumption is made concerning how the transposon and the target sequence are to be linked. The following steps, numbered as in Figure 21-13, have been proposed:

I.   A pair of single-strand breaks is made (by the transposase?) in the target sequence. These breaks are staggered; thus, cleavage of the target sequence yields two complementary single strands.

II. A pair of single-strand breaks is made in complementary strands at opposite ends of the transposon; this generates two free ends. Each 3′ free end is attached by a single strand to a protruding strand (which carries a 5′ group) of the staggered cut in the target site; this generates two replication forks.

III. Replication begins by synthesis of a strand complementary to the protruding end of the target sequence. When replication is completed, a cointegrate is formed containing two copies of the transposon.

IV. Finally double-strand exchange occurs between the two copies of the transposon. In Tn3, at least, this exchange probably takes place in the internal resolution site. The result of the exchange is to separate the cointegrate into the donor and recipient units, each of which now has a copy of the transposon. In the recipient the transposon is flanked by a direct repeat of the target site; the length of this repeat is the number of base pairs between the staggered cuts.

Certain experiments have suggested that for some transposons the cointegrate is not an obligatory intermediate but only one of two possible outcomes of transposition. These observations have led to another class of models, called **asymmetric models.** These models differ from the one just described in two major respects: (1) the two cuts at the end of the transposon are not made simultaneously but at separate stages of the insertion process, and (2) as just stated, a cointegrate is not an intermediate but one of two alternative end products of transposition (the other being a simple insertion). One asymmetric model is shown in Figure 21-14.

A great deal of research effort is currently being expended to determine the validity of these and other models.

GENETIC PHENOMENA MEDIATED BY TRANSPOSONS IN BACTERIA

Transposons mediate a variety of genetic phenomena such as gene rearrangement and plasmid-chromosome integration. These will be described in this section.

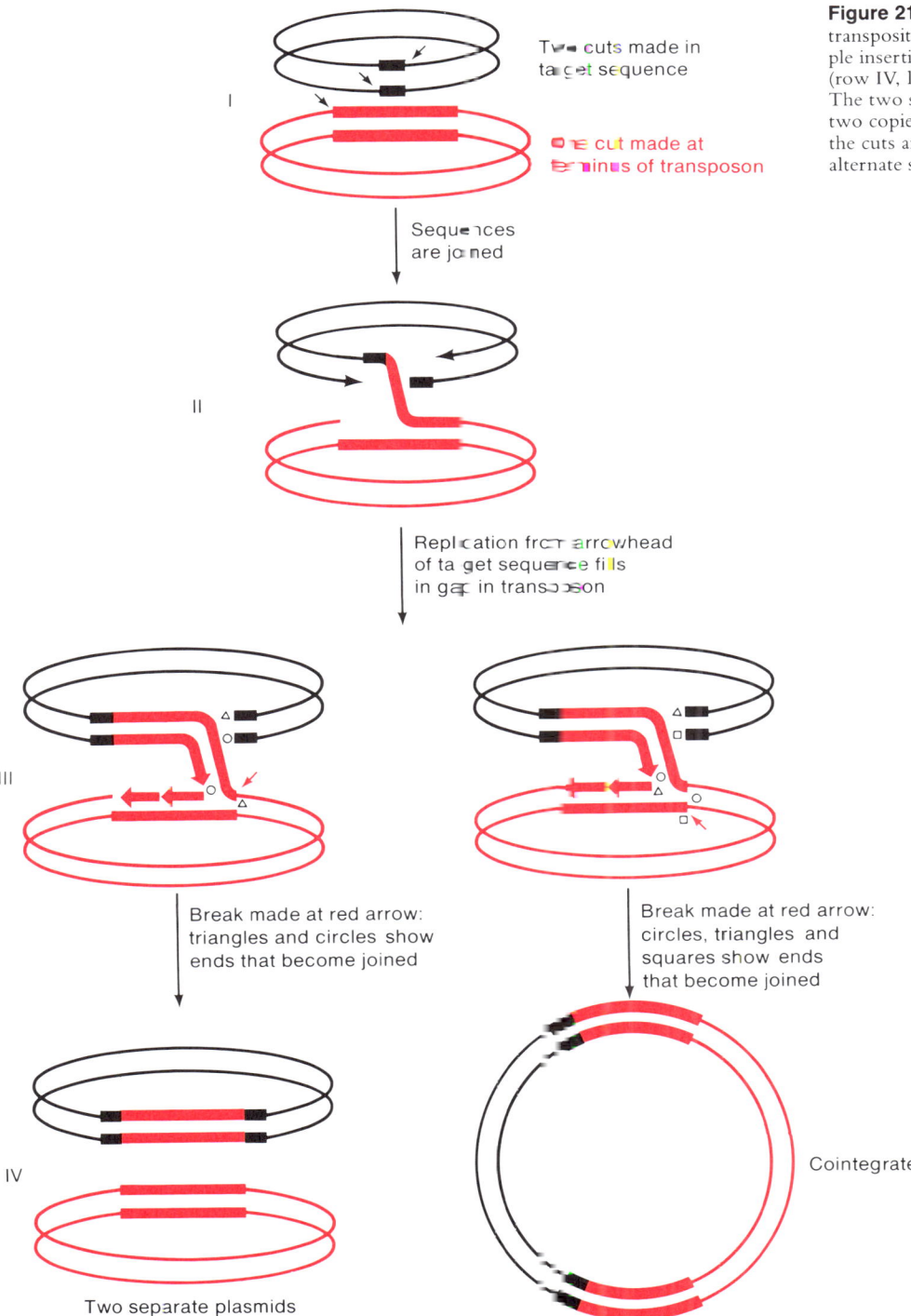

I — Two cuts made in target sequence

One cut made at terminus of transposon

Sequences are joined

II

Replication from arrowhead of target sequence fills in gap in transposon

III

Break made at red arrow: triangles and circles show ends that become joined

Break made at red arrow: circles, triangles and squares show ends that become joined

IV — Two separate plasmids

Cointegrate

Figure 21-14   An asymmetric model for transposition that can be terminated by simple insertion or formation of a cointegrate (row IV, left and right panels respectively). The two structures in row III are identical; two copies are drawn for clarity in showing the cuts and joints needed to achieve the alternate structures of row IV.

Deletions and Inversions Caused by Transposons

When a transposon is present in a chromosome or plasmid, the frequency of nearby deletions is increased about 100–1000-fold. When such a deletion occurs, it is often found that (1) the transposon is still present, (2) the deletion originates at the end of the element, and (3) the transposon is bounded by only one copy of the target sequence (that is, there is no sequence in direct repetition surrounding the transposon). These features are shown in Figure 21-15.

The frequency of inversions is also enhanced in the vicinity of transposons. In this process, beginning with a sequence

$$\ldots \text{transposon-}a\ b\ c\ d\ e\ f \ldots$$

in which a–f are genes adjacent to the transposon, an event occurs that leads to a sequence such as

$$\ldots \text{transposon-[5]-}b\ a\text{-transposon-[5]-}c\ d\ e\ f \ldots$$

in which [5] denotes an arbitrary 5-bp target sequence. Note that $a\ b$ is inverted, and that a new target sequence and the transposon are duplicated and surround the inverted sequence. The molecular mechanisms for both deletion and inversion generation are not known but can be explained by the model of Figure 21-13 and other models as well.

Role of IS Elements in Hfr Formation

Hfr cells are formed by integration of F. Several lines of evidence (principally the determination of the structure of F' plasmids) indicate that an integrated F sequence is always flanked by two copies (in direct repeat) of one of the IS elements found in the F plasmid. This and the fact that the IS elements in F are also found in the chromosome (Table 21-1) shows that Hfr formation involves transposons.

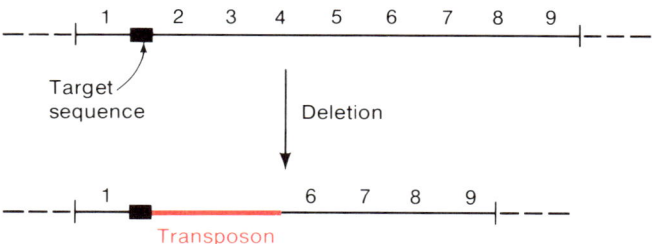

Figure 21-15 Characteristics of a deletion mediated by a transposon. Segments 2, 3, 4, and 5 are deleted and replaced by a transposon.

There are two ways by which Hfr cells might arise in a cell containing F. One is by homologous recombination between two identical IS elements, one in the chromosome and the other in F. If the recombinational event is physically reciprocal, this mechanism would yield flanking IS elements and might utilize the bacterial Rec system, a transposon gene product, or conceivably some product of a gene in F. Alternatively, an Hfr cell could arise by formation of a cointegrate mediated by an IS element in F, and by a duplication of a target sequence in the chromosome (i.e., replicon fusion). It is important to note that the consequences of these two mechanisms are the same—namely, F flanked by two copies of the IS element in direct repeat. It is likely that Hfr strains arise by both mechanisms for the reasons that follow.

There are many sites of integration of F. However at some sites, called **hot spots,** integration occurs more frequently than at other sites (called minor sites). The base sequences of several regions of the chromosome that include hot spots have been determined and in each case an IS element has been found. Furthermore, most Hfr cell lines have been $recA^+ F^+$ cells, so it seems likely that these Hfr strains were produced by $recA$-mediated homologous recombination. Integration at the minor sites occurs at a much lower frequency and often in $recA^-$ cells; the resultant Hfr strains probably arise by the cointegrate pathway. In $recA^+$ cells the homologous-recombination pathway probably predominates.

TRANSPOSABLE ELEMENTS IN EUKARYOTES

A variety of transposable elements have been detected in eukaryotes. They have become evident in two ways.

1. *Genetic analysis.* In studies of unstable effects on the expression of various genes in maize Barbara McClintock observed that the activities of certain genetic loci that affected the appearance of the kernels are disturbed by a controlling element that inhibits gene activity only when it is adjacent to the locus; significantly, changes in level of gene activity seemed to occur by movement of the element to and from its inhibiting position. Furthermore, the inhibitory activity of this element required the presence of a second element which also seemed to be mobile. In the 1980s almost 40 years after her discovery, cloning and base sequencing of the relevant genetic regions showed that the controlling elements are DNA sequences with inverted repeats, each flanked by a short sequence in direct repeat, the hallmark of a transposable element. Unstable mutations in other eukaryotes have also led to the discovery of transposable elements in other eukaryotes, using similar procedures.

2. C_0t *analysis.* Studies of eukaryotic DNA have demonstrated the presence of a fraction of repetitive sequences known as middle repetitive DNA.

Isolation of this fraction and determination of the base sequences of the various families of repetitive elements have shown that the sequences are often bounded by long inverted terminal repeats and short direct repeats, which suggests that they are transposable elements. Examination of independently cloned segments from different species of *Drosophila,* from different strains of the same organism, and from different isolates of the same strain have shown that the positions of these sequences are not always the same, suggesting that they are mobile elements. This procedure has led to the discovery of more transposable elements than genetic analysis did.

Transposition in Eukaryotes

Our understanding of transposable elements in eukaryotes is much less complete than that of the bacterial transposons. Many elements have been sequenced but, as yet, no gene products have been identified. Some of the elements transpose in many genetic backgrounds and are said to be autonomous; they probably encode a transposase. Others are unable to transpose unless an autonomous element is present. These nonautonomous elements are analogous to the nontransmissible plasmids described in Chapter 20 that can only transfer when a transmissible plasmid is also present in the cell. The nonautonomous elements probably do not encode a transposase. Enough information is available to indicate that eukaryotic transposable elements differ in several ways from bacterial transposons. The more important differences are the following:

1. Transposition occurs by means of excision and relocation. The evidence for this conclusion is that if a transposable element is adjacent to a particular gene and affects expression of the gene, transposition to a new location is accompanied by loss of the element at the original site. When excision is precise, expression of the gene is restored to normal; however, sometimes, excision is imprecise.
2. Point 1 plus several other findings suggest that transposition is not a DNA replicative process in eukaryotes.
3. Some elements have been isolated from cells as free circular DNA molecules.
4. One element has recently been shown to transpose via an RNA intermediate. Other elements seem to encode reverse transcriptase (Chapter 5), an enzyme capable of using single-stranded RNA as a template for synthesis of double-stranded DNA, though an RNA intermediate has not been demonstrated in transposition of this class.

These differences indicate that the transposition mechanism in eukaryotes is probably quite different from that in bacteria. Possibly, several mechanisms are used.

Particular Eukaryotic Transposable Elements

The yeast *Saccharomyces cerevisiae* contains two related transposable elements called Ty1 (length, 6.3 kilobase pairs; about 35 copies/genome) and Ty917 (about 6/genome). Their sequences are similar; Ty917 is smaller and may be a deletion mutation of Ty1. Each element contains terminal 338-bp sequences called δ and is usually flanked by a 5-bp sequence in direct repeat, the signature of a transposable element. Roughly 100 copies of δ are also found scattered throughout the genome. The presence of these so-called "solo" δ elements has two possible explanations: (1) They result from recombination between homologous sections of two δ's flanking a Ty element. A reciprocal exchange would remove Ty, one δ, and leave behind one intact δ. (2) δ is an IS-like element and the Ty elements are composite transposable elements. As we saw for IS elements, polar mutations are associated with insertion of Ty1 and a transcriptional start site is near the end of δ. Ty1 has very recently been shown to transpose via an RNA intermediate, which suggests a possible role of the enzyme, reverse transcriptase, in transposition.

In most *Drosophila* species 5–10 percent of the genome consists of about 40 distinct families of sequences having the features of transposable elements. A well-studied sequence is called **copia,** the name referring to the copious amount of RNA transcribed from this sequence. *Copia* contains about 5000 base pairs and is terminated by a 267-bp sequence in direct repeat; each of these terminal sequences is IS-like in having 17-bp sequences in inverted repeat; thus, copia is like the composite type I transposon shown in the lower part of Figure 21-4. In addition to *copia* there are about 30 related but not identical sequences, called *copia*-like. *Copia* and these related elements range in number from 21–60 per genome and account for about three percent of *Drosophila* DNA. Although *copia* is extensively transcribed, no gene product has yet been identified. A *copia*-like element, called 17.6 is believed to encode the enzyme reverse transcriptase, suggesting that an RNA molecule may be an intermediate in transposition of these elements.

Copia has been found in two forms—integrated and as free circular DNA. The circular molecules are of two sorts: (1) molecules with two terminal repeats—that is, the same sequence as the integrated element; (2) molecules with one terminal repeat, which presumably arose by reciprocal recombination during excision. Whether either circular form is an intermediate in transposition is unknown. Transposition of *copia* affects the activity of certain genes, presumably by movement of regulatory sites, such as promoters and other protein-binding sites, that are part of the element.

The **FB** or **foldback elements** of *Drosophila* have two unusual features. First, their terminal repeats are exceedingly long. For example, whereas each terminal repeat of *copia* is about 5 percent of the complete sequence, the inverted repeats of FB4, which consists of 4089 bp, each contain more than 1000 bp. When *Drosophila* DNA is denatured and renatured, these elements form stem-and-loop structures in which the stem contains more than half

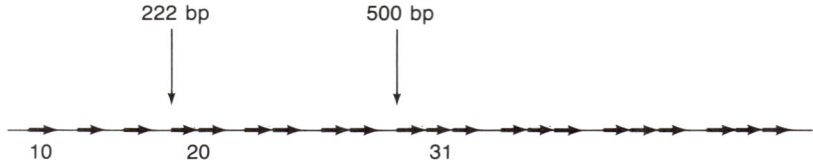

Figure 21-16 A diagram showing the sequence arrangement of a *Drosophila* fold-back element. The number of repeated copies is much greater than that shown.

of the bases (the DNA folds back, hence the name of the elements). Second, the terminal inverted repeats consist of a 10-bp sequence that is repeated in a surprising way. The sequence is repeated numerous times, separated by other unrelated sequences. Starting at nucleotide 222, a 21-bp sequence, which consists of the 10-bp sequence in tandem with a closely related sequence, then repeats many times, separated by very AT-rich sequences. At nucleotide 499 a 31-bp sequence begins, again consisting of three related copies of the 10-bp sequence in tandem; this 31-bp sequence repeats, with spacers, up to nucleotide 1018. This is shown in Figure 21-16.

Copia, FB, and other transposable elements affect the expression of some *Drosophila* genes (for example, eye color), causing changes or even gene inactivation when transposing to adjacent sites or into the gene. However, it is unlikely that the elements are essential to the organism, since *Drosophila simulans,* a species of fly closely related to *D. melanogaster,* appears to lack transposable elements. The two species occupy the same habitats in nature, and in competition experiments carried out in the laboratory, neither species seems to have any advantage over the other.

Retroposons

Retroposons are mobile elements that transpose via an RNA intermediate. Some elements considered to be retroposons have never been observed to transpose, but because they are flanked by short sequences in direct repeat and are found at numerous sites throughout the genome, they are believed to be transposable elements.

The best-understood retroposons are the **retroviruses,** animal viruses that contain RNA in the virus particle and have an unusual life cycle (Chapter 23). Following entry of the viral RNA into a host cell and early in viral development, the RNA is converted to a double-stranded DNA molecule, whose sequence is identical to that of the RNA except for the termini. The terminal regions of the DNA molecule are two directly repeated sequences, each of which can be subdivided into an IS-like element, that is, two short regions in inverted repeated flanking a central region. Overall, the base sequence of the retroviral DNA resembles that of a type I composite bacterial transposon. As part of the viral reproductive process the DNA circularizes and inserts into the chromosome. Insertion can occur at any site and the inserted DNA is flanked by a short sequence in direct repeat. The best-understood retrovirus, Rous sarcoma virus, will be discussed in Chapter 23.

Primate DNA contains about 300,000 copies of a family of base sequences consisting of about 300 bp (5 percent of the total DNA). Each unit contains a tetranucleotide that can be cleaved by the restriction enzyme AluI, so the sequences are called **Alu sequences.** Each copy is flanked by a direct repeat of a few base pairs; the resemblance of the overall array to a target sequence flanking a transposon has suggested that Alu sequences are transposable elements, though transposition has not been observed. Study of the base sequence of the Alu elements has provided some insight into their origin. Each Alu sequence is a dimeric structure composed of two related and tandemly arranged monomeric units, each about 130 bp long with an additional 31 bp in the middle of the sequence normally drawn at the right. The right (3′) end of the sequence also is a short tract of A's. Analysis of the base sequence of the monomer indicates that it is a DNA copy of a mammalian RNA molecule (7SL RNA) that plays a role in protein synthesis; homology is not perfect but extends to about 70 percent of the bases. Highly repeated sequences that are widely dispersed have been observed in a variety of animals (they are about the same length as the Alu monomer and do not have a dimer structure), and each proves to be a copy of an RNA molecule. For example, the rat ID sequences are 73 percent homologous with tRNA^{Ser}, and the bushbaby (*Galago*) Alu-like family is 68 percent homologous to the initiator tRNA^{Met}. The following hypothesis has been suggested to account for the wide dispersal and the very large number of these sequences per genome. At some time in the distant past, a gene encoding an RNA became adjacent to a sequence of A's at its 3′ end. This could have resulted from a transposition event, some other genetic exchange, or accumulated mutations. The gene was then transcribed. The oligo(A) tract was able to base-pair with the U-rich sequence found at the 3′ terminus of many of the primary transcripts of these RNA genes. The 3′ A of this short double-stranded terminus then served as a primer for synthesis of a DNA copy of the RNA by reverse transcriptase, which is probably widely distributed in eukaryotic cells, though at very low concentration (it is made by some eukaryotic transposable elements, by most integrated retroviral DNA sequences, and has been found in particles in *Drosophila*). A diffusible transposase (which could have been encoded in any transposable element in the cell) then inserted the DNA copy into the chromosome, producing the usual short direct repeat at each end of the insert. Because the promoter for genes transcribed by RNA polymerase III is internal to the RNA (Chapter 12), the inserted sequence contained the promoter and more RNA with a 3′ terminal oligo(A) could be made. The process was probably repeated at a very low rate but in the course of million of years the number of copies became very large. Since synthesis of such a huge amount of a particular tRNA would probably be deleterious to the cell, it seems essential that the sequence of the RNA would have to change. Mutations accumulated in the course of many generations, reducing the homology to 60–70 percent; cells in which such spontaneous mutations did not occur probably have not survived. One may ask how such useless DNA has persisted over such long periods of time. Such DNA has been called "selfish" in that its sole function is to self-replicate. As long as it does not harm its host, it survives. Do these sequences continue to transpose via an RNA intermediate? Presumably yes, but at a vary low rate; however, an increase via transposition of the number of copies per genome from 300,000 to 300,001 would not be detectable.

LIMITATION OF TRANSPOSITION

In bacteria transposition is a replicative process, so the number of elements seems to be capable of increasing indefinitely. This would seem to be deleterious to the element, since ultimately the host genome would be damaged beyond its ability to carry out necessary cellular functions. Transposition is very likely to be limited by some as-yet-unknown system. Evidence for such a system comes from experiments in which a bacterial transposon is placed in a cell (by plasmid transfer, phage infection, or transformation) that has no copies of the element. Transposition begins within several generations and occurs repeatedly. Soon the rate of transposition decreases and ultimately the cell acquires a stable number of copies of the transposon; the particular number is characteristic of each element. The mechanism of this inhibition of transposition is obscure.

22

Recombinant DNA and Genetic Engineering

Throughout this book we have seen how restriction enzymes have enabled one to map DNA and to isolate particular DNA fragments. We have mentioned briefly in Chapter 5 that by restriction enzyme techniques one can also join fragments and clone particular DNA segments. In this chapter we examine in greater detail the joining techniques—the recombinant DNA technology, or **genetic engineering,** as it is more commonly called. In addition, we shall describe how genetic engineering can provide a source of particular proteins in quantities hitherto considered to be nearly impossible to obtain and how the new technology is radically changing certain aspects of medicine and clinical practice.

THE JOINING OF DNA MOLECULES

The basic procedure of the recombinant DNA technique consists of two stages—(1) joining a DNA segment (which is of interest for some reason) to a DNA molecule that is able to replicate, and (2) providing a milieu that allows propagation of the joined unit (Figure 22-1). When this is done, the genes in the donor segment are said to be **cloned** and the carrier molecule is the **vector** or **cloning vehicle.** In Chapter 5 a procedure was described in which a vector and a DNA fragment of interest were cleaved with a restriction enzyme and joining was accomplished by annealing the comple-

A genetically engineered plant developing from tissue infected with a recombinant Ti plasmid. [Courtesy of P. Zambryski.]

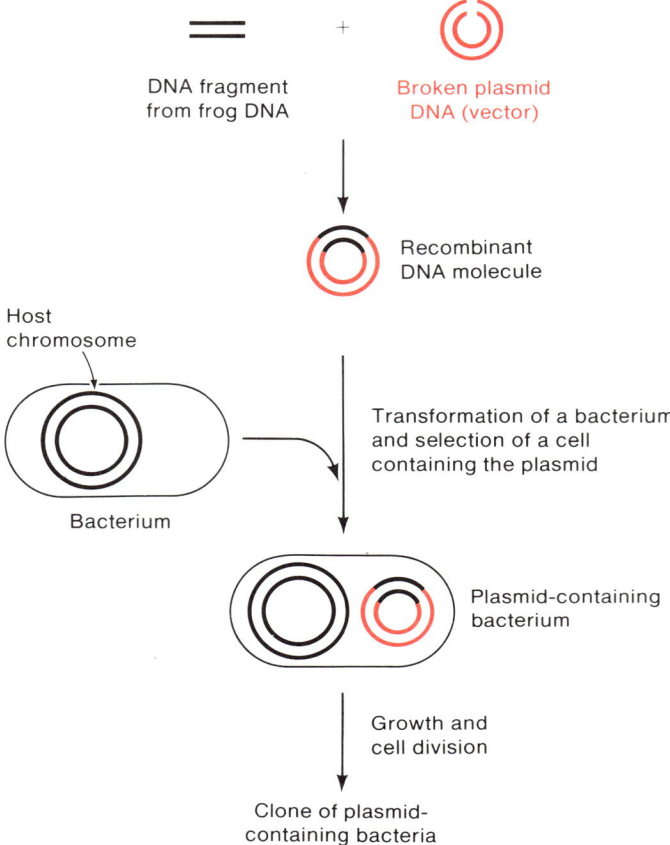

Figure 22-1 An example of cloning. A fragment of frog DNA is joined to a cleaved plasmid vector. The hybrid plasmid trans- forms a bacterium. Henceforth, frog DNA is carried to all progeny bacteria by replication of the plasmid.

mentary single-stranded ends (Figure 5-7). In this section we will examine this procedure further, as well as other procedures used for joining. Several types of vectors will also be described.

Vectors

To be useful, a vector must have three properties:

1. It must be able to replicate.
2. There must be some way to introduce vector DNA into a cell.
3. There must be some means of detecting its presence, preferably by a plating test in petri dishes.

Table 22-1   Some cloning vehicles in use at present

Designation	Genotype or characteristics
Plasmid	
pSC101	*tet-r*
ColE1	*immE1*
pBR322	*tet-r amp-r*
pUC8/9	*amp-r lac*$^+$
pHC79	*amp-r tet-r* λ*cos*
Phage	
λgt4·λB	Can lysogenize; is thermally inducible
λ–Charon	*lacZ*$^+$
λΔz1	Insertion occurs in *lacZ* gene.
M13mp7	Contains *lacP*, *lacO*, and *lacZ*; single-stranded DNA; useful for sequencing.
Virus	
SV40	Virus infects animal cells. Maximum size of added DNA is no more than 3×10^6 molecular weight units.

Abbreviations: *tet*, tetracycline; *imm*, immunity; *amp*, ampicillin; *lacP*, *lac* promoter; *lacO*, *lac* operator; *lacZ*, gene for β-galactosidase; *cos*, joined cohesive sites of phage λ.

The three most common types of vectors in use are plasmids, *E. coli* phage λ, and viruses, because the DNA of each of these vectors has all three of the aforementioned properties. That is, each has a replication origin and carries genes that are identifiable by simple plating or biochemical tests, and the DNA can be made to penetrate particular cells by using one or more variations of the CaCl$_2$ transformation technique described in Chapter 5. Only a small number of vectors are in common use; some of these vectors and their properties are listed in Table 22-1.

Joining of DNA Fragments by Addition of Homopolymers

The field of recombinant DNA research began in 1972, just before the properties of restriction enzymes were understood, with the development of a general method for joining any two DNA molecules. This method used the enzyme **terminal nucleotidyl transferase,** an unusual DNA polymerase obtained from animal tissue, which adds nucleotides (by means of triphosphate precursors) to the 3'-OH group of an extended single-stranded segment of a DNA chain. The reaction differs from that of ordinary polymerases in that *it does not need a template strand.* In order to generate the extended single strand with a 3'-OH terminus one need only treat the DNA molecule with a 5'-specific exonuclease to remove a few terminal nucleotides. In a reaction mixture consisting of exonuclease-treated DNA, dATP, and the enzyme, poly(dA) tails will form at both 3′-OH termini (Figure 22-2). If, instead, dTTP were provided, the DNA molecule would have poly(dT) tails. Two molecules can be joined if poly(dA) tails are put on one DNA molecule and poly(dT) tails on the second molecule and the poly(dA) is allowed to anneal to the poly(dT), as shown in the figure. Completion of the joined molecule is accomplished by gap-filling with DNA polymerase I and sealing with DNA ligase.

This method, which is called **homopolymer tail joining,** is useful when DNA molecules lacking complementary ends are to be joined (recall from Chapter 5 that most restriction enzymes produce complementary ends). Such molecules may be the result of digestion with restriction enzymes that yield blunt ends or may be prepared by mechanical breakage of large DNA molecules; c-DNA (complementary DNA), which is a DNA molecule prepared in the laboratory by copying an RNA template and which is extremely important in the recombinant DNA technology (to be described shortly), also has blunt ends.

Blunt-End Ligation

E. coli phage T4 encodes a DNA ligase, which is produced in large quantities in an infected cell. This enzyme is a typical ligase in that it seals nicks in

Figure 22-2 The joining of two molecules with complementary homopolymer tails.

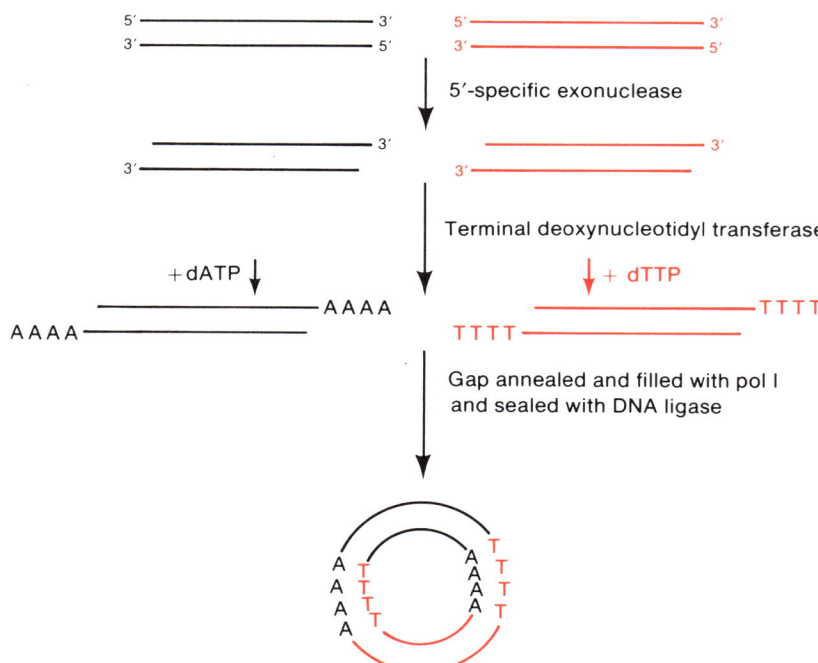

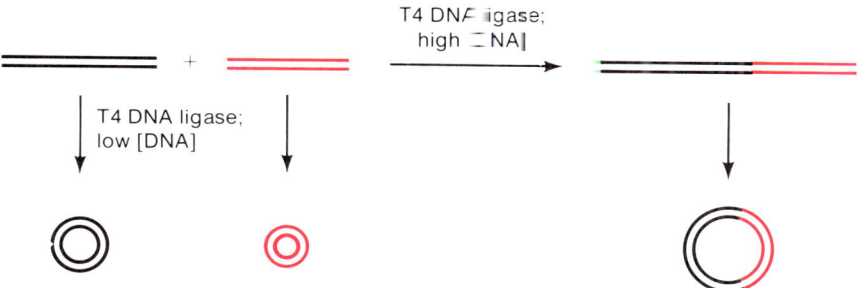

Figure 22-3 Two blunt-end joining reactions. At low concentrations of DNA intramolecular circularization is favored.

double-stranded DNA having 3'-OH and 5'-P termini, but has the additional property of joining two DNA molecules having completely base-paired ends (Figure 22-3). How this reaction occurs is unknown. Whereas blunt-end ligation is quite useful in many situations, it has a significant disadvantage compared to homopolymer joining, namely, the two ends of any fragment can be joined to form a nonrecombinant (and generally useless) circle. Such joining does not occur with homopolymer tail joining since the two ends are identical rather than complementary. This problem can, however, be avoided by use of a high DNA concentration, as shown in the figure.

Joining with Linkers

In some cases it is useful to be able to join one molecule with blunt ends to a second molecule produced by a restriction endonuclease that generates single-stranded termini. This is possible if a short DNA segment (a **linker**) containing a restriction site is coupled to both ends of the blunt-ended molecule. How this is done is shown in Figure 22-4, in which a blunt-ended molecule is inserted at the EcoRI site of a vector. This procedure is useful because at a later time the sequence contained in the blunt-ended fragment can be recovered from the vector by treatment with a restriction enzyme that cuts the site in the linker. Such recovery is not possible if joining is done with homopolymer tails.

Generation of Fragments by Hydrodynamic Shear

Consider a gene G that is to be cloned in a plasmid. This is accomplished by inserting a DNA fragment containing G into the plasmid DNA. The simplest method is to treat both the plasmid DNA and the DNA containing G with the same restriction enzyme and anneal the resulting fragments (Chapter 5). However, it may be that every known restriction enzyme that cuts the plasmid also makes a cut within gene G and thereby inactivates the gene. Clearly some means of cleavage other than the use of restriction enzymes is

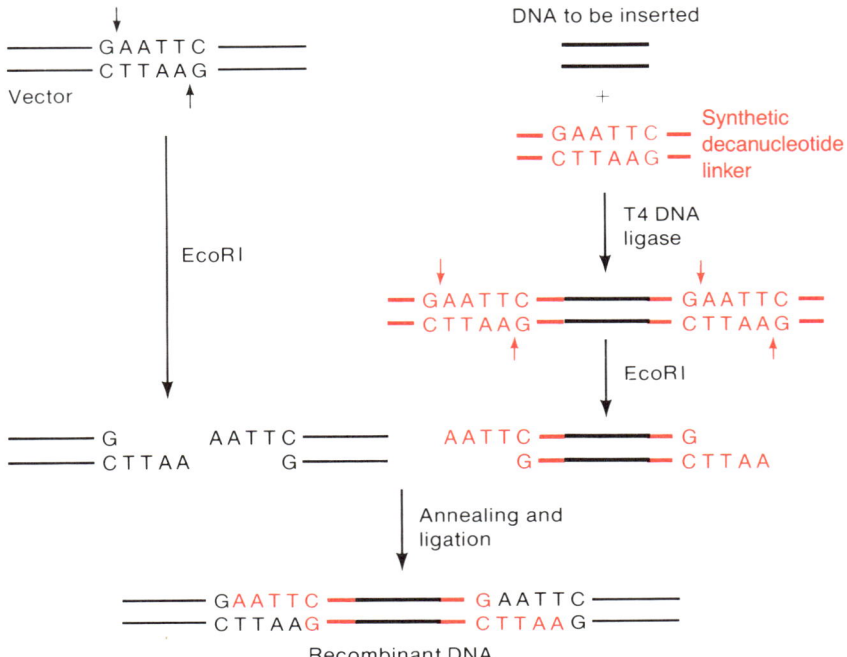

Figure 22-4   Formation of recombinant DNA through use of a linker. The short arrows indicate the sites of cutting by the EcoRI enzyme.

required; it has been found that the gene inactivation caused by site–specific cleavage can be eliminated if the donor DNA containing G is fragmented not enzymatically but mechanically by hydrodynamic shear forces.

If a DNA solution is passed very rapidly through a hypodermic needle or a small orifice or is stirred vigorously, the DNA molecules are broken into fragments in the size range 3000–30,000 bp. In some DNA molecules, gene G, which might typically consist of 1000–2000 bp, will be cleaved but, on the average, the gene will remain intact.

The fragments resulting from shearing usually have blunt ends. In order to be joined efficiently to a plasmid broken by a restriction enzyme, an appropriate linker may be added, as was shown in Figure 22-4, though blunt-end joining can be carried out.

Scrambling of Fragments During Joining

Joining of fragments does not always produce a functional DNA sequence. For example, consider a linear DNA molecule that is cleaved into four fragments, A, B, C, and D. Reassembly could yield the original molecule, whose sequence we will assume to be ABCD, but since B and C have the same pair of cohesive ends, a molecule with the sequence ACBD could form with the same probability as ABCD. Other molecules having multiple copies of one of the fragments—for example, ABED or ACCD—are also possible. Thus, only about one-fourth of the resulting sequences will be functional. Similarly, if there were five fragments A, B, C, D, and E, the molecules A(BCD)E, A(BDC)E, A(CBD)E, A(CDB)E, A(DBC)E, and A(DCB)E, and those with duplicated and triplicated fragments, are all equally likely. The probability of obtaining a functional sequence when reannealing fragments obtained from two different DNA molecules is of course lower. Thus, if plasmid DNA is cleaved into four fragments and frog DNA is cleaved into 1000 fragments, most assembled plasmids will be scrambled and non-functional. On the other hand, if a functional plasmid is selected by genetic means and if an insertion is possible, say, between B and C, the probability of having a plasmid containing any frog DNA fragment will be quite high; if a *particular* frog DNA fragment is desired, the probability will be about 1000 times lower.

INSERTION OF A PARTICULAR DNA MOLECULE INTO A VECTOR

As we have described the procedures so far, a collection of fragments obtained in various ways (for example, by restriction enzyme digestion or shearing) is allowed to anneal with a vector, yielding a large number of hybrid molecules. If a particular gene is to be cloned, the vector possessing that gene must be isolated from the class of all vectors possessing foreign DNA. Although for many genes this procedure is adequate, there are certainly many cases in which the clone of interest is either so rare or so difficult to detect that it would be preferable if, prior to hybridization, the fragment containing the gene of interest could be purified. In this section three procedures are described in which a particular DNA molecule can be cloned.

c-DNA and the Use of Reverse Transcriptase

Proof of correct insertion of a particular gene into a vector is most easily detected by observing expression of the gene. For example, the *lac* operon can be detected in a plasmid by virtue of its ability to synthesize β-galactosidase. However, knowing whether a particular eukaryotic gene has been inserted into a bacterial DNA vector can be a problem because, in general, a eukaryotic gene cannot be transcribed and translated into a functional protein when cloned in a bacterium. By the technique described in this section the insertion of some eukaryotic genes can become a quite straightforward procedure.

Let us assume that we wish to clone the rabbit gene for β-globin (a subunit of hemoglobin) in a bacterial plasmid.* The recombinant plasmid could be formed by treating both rabbit cellular DNA and the *E. coli* plasmid DNA with the same restriction enzyme, mixing the fragments, annealing, ligating, and finally infecting *E. coli* with the DNA mixture containing the hybrid plasmid. The main problem is to identify the bacterial colony containing this plasmid. If the gene to be cloned is a bacterial *gal* gene, one can use a Gal⁻ host and select a Gal⁺ colony on an appropriate growth medium. However, it is not possible for a bacterium containing the β-globin gene to synthesize β-globin because the gene contains introns (Chapter 12), and all known bacteria lack the enzymes needed to form eukaryotic mRNA molecules from a primary transcript. Other tests might be performed to find the desired cell, such as hybridization of bacterial DNA with purified β-globin mRNA, but since only about one plasmid-containing cell in 10^5 will contain the β-globin gene, this method will be very tedious. Therefore, a more convenient procedure is desirable. One useful procedure, which is applicable to certain classes of genes, is described in the remainder of this section.

Some types of animal cells, of which one example is the reticulocyte, which produces β-globin, make only one or a very small number of proteins. In these cells the specific mRNA molecules, which are already processed, constitute a large fraction of the total mRNA synthesized in the cell and, for this reason, mRNA samples can usually be obtained that consist almost exclusively of a single mRNA species—in this case, β-globin mRNA. If genes of this type—that is, those whose gene products are the major cellular proteins—are to be cloned, the purified mRNA of each type of cell can serve as a starting point for creating a recombinant plasmid containing only the gene of interest.

The animal viral enzyme **reverse transcriptase** is able to use an RNA molecule as a template and synthesize a double-stranded DNA copy (for the mechanism, see Chapter 5, Figure 5-13). A double-stranded DNA molecule prepared in this way is called **complementary DNA** or **c-DNA.** If the RNA molecule that forms the template has been processed (that is, the introns have been removed) before it is isolated, the corresponding c-DNA will contain an uninterrupted coding sequence. In Chapter 5 the procedure used showed the starting RNA with an AAA-OH terminus and it was pointed out that in general a repeating oligonucleotide would be appended to the RNA prior to addition of the enzyme. However, with most eukaryotic mRNAs this is unnecessary because of the presence of the natural 3′-poly(A) terminus. The c-DNA generally has blunt ends and is usually joined by the linker method shown in Figure 22-4.

---

*The phrase "to clone a particular gene" is common usage to mean "to form a vector containing a particular gene."

Insertion of a Particular DNA Fragment Obtained from a Large Donor Molecule by Restriction Cutting

The study of the regulation of gene expression in phages has been helped considerably by the cloning of particular phage gene in a plasmid. Cloning of a phage gene is facilitated if one can first purify the fragment containing the gene. For instance, Figure 5-3 (Chapter 5) showed a restriction map for a particular phage; suppose that one wants to clone a gene that is known to be in fragment C. One way to accomplish this is to add the complete enzymatic digest of the DNA to a cleaved vector, allow the DNA to anneal, and seek a cell containing a plasmid that has the gene of interest. Since there are five fragments, at most one-fifth of the plasmids containing phage DNA will have the gene. In some cases, it is valuable to increase this fraction. This can be done by purifying fragment C and using only the purified fragment in the annealing mixture. The most direct way to isolate a fragment is to extract it from a gel after electrophoretic separation. This can be done in the following way (Figure 22-5). The restriction digest is electrophoresed for a period of time, the electric field is turned off, and the DNA bands are located by soaking the gel in a buffer containing ethidium bromide (Figure 4-28, Chapter 4). The fluorescence of ethidium bromide is markedly enhanced when the molecule binds to DNA, so the DNA can be located by shining near-ultraviolet light on the gel and noting the positions of intense fluorescence. A small well is cut in the agar ahead of the band of interest and filled with buffer. The electric field is applied again and the motion of the fragments is followed visually. When the fragment of interest migrates into the buffer-containing well, the electric field can be turned off again and the buffer, now containing the DNA fragment, can be removed.

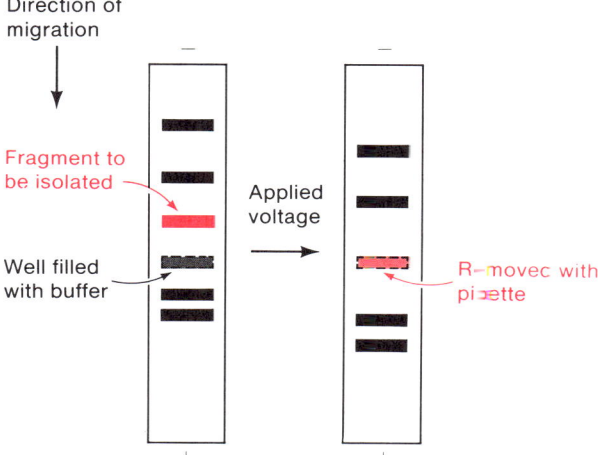

Figure 22-5 Method for isolating a particular restriction fragment.

DETECTION OF RECOMBINANT MOLECULES

The joining of DNA molecules is a straightforward process, as we have seen in the preceding sections. However, if a vector is cleaved by a restriction enzyme and annealed with an unfractionated collection of restriction fragments, many types of molecules result—examples are a self-annealed vector that has not acquired any fragments, a vector with one or more fragments, and a molecule consisting only of many joined fragments. Molecules of the third class cannot stably transform $CaCl_2$-treated cells because, in general, such molecules lack a replication origin and replication genes; thus, molecules of this class are not a problem. However, some means is needed to ensure that a plasmid (or a phage) detected after $CaCl_2$ transformation does possess an inserted DNA fragment. Furthermore, even if this can be demonstrated, there is no guarantee that a plasmid containing donor DNA has the DNA segment of interest. Thus, some test is required for detecting the desired plasmid. Similar problems exist for phage and viral vectors. In this section several procedures that provide solutions to these problems are described.

Methods Which Insure That a Plasmid Vector Will Contain Inserted DNA

The plasmid pBR322 is a widely used vector. It is small (4363 bp), and has two different antibiotic-resistance markers—namely, resistance to tetracycline (*tet-r*) and to ampicillin (*amp-r*). Thus, plasmid-containing transformants are easily detected by growth of a DNA-transformed culture on agar containing one of these antibiotics. Also, pBR322 has seven different types of restriction-enzyme cleavage sites at which foreign DNA can be inserted.

A common procedure for detecting insertion is **insertional inactivation,** which is carried out as follows (Figure 22-6). The BamHI and SalI sites are within the *tet* gene. Thus, insertion at either of these sites yields a plasmid that is *amp-r tet-s,* because the *tet* gene is inactivated. If wild-type (Amp-s Tet-s) cells are transformed with a DNA solution in which the cleaved pBR322 has been annealed with restriction fragments and the cells are plated on agar containing ampicillin, surviving colonies, which have to be Amp-r, must possess the plasmid. These colonies are then further tested for sensitivity to tetracycline. Because pBR322 carries the *tet-r* allele, an Amp-r colony will also be Tet-r unless the *tet-r* allele has been inactivated by insertion of foreign DNA. Thus, an Amp-r Tet-s cell must contain not only pBR322 DNA but donor DNA as well.

This procedure can be simplified by growing the transformed cells, prior to plating, in a medium containing cycloserine and tetracycline. Cycloserine kills growing cells, but Tet-s cells are merely inhibited, not killed, by tetracycline. Thus, in this growth medium Tet-r cells (which grow) are killed and Tet-s cells (which are inhibited) survive. Plating of cells treated in this

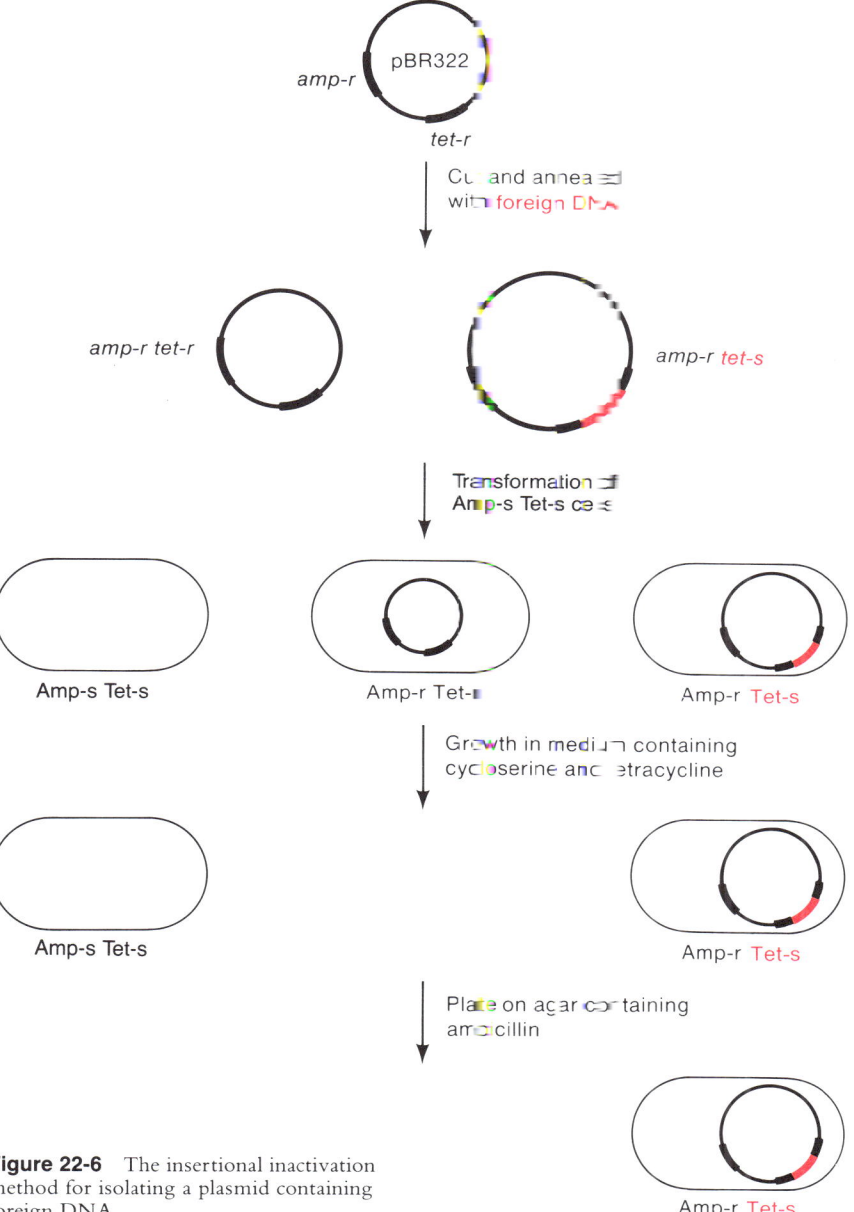

Figure 22-6   The insertional inactivation method for isolating a plasmid containing foreign DNA.

way on agar containing ampicillin yields Amp-r Tet-s colonies; these all possess pBR322 containing a donor DNA fragment.

The PstI site in pBR322 is in the *amp* gene. Thus, the insertional inactivation procedure can also be used with insertion at this site, and an Amp-s Tet-r colony is selected.

Three other procedures select against reconstituted plasmids lacking foreign DNA. One is to treat a cleaved plasmid with **alkaline phosphatase,** an enzyme that removes 5′–P termini, leaving 5′–OH groups. Following this treatment a functional plasmid cannot be reconstituted unless foreign DNA is inserted, as shown in Figure 22-7. A second procedure, homopolymer tail joining, automatically selects against re-formation of the plasmid because the two 3′–OH termini of the cleaved plasmid both have homopolymers containing the same base. A third procedure utilizes a special type of plasmid called a cosmid; this method is the following.

Cosmids (phage λ *cos*–site-carrying plasmids) are novel vectors that combine the features of a plasmid and phage λ to increase the probability of selecting a recombinant plasmid carrying foreign DNA. A typical cosmid is a circular ColE1 plasmid (Chapter 20) carrying both the gene for resistance to the drug rifampicin (*rif-r*) and the *cos* site of phage λ (Chapter 17). The cosmid has two cleavage sites for the HindIII restriction enzyme, which separate the *rif* gene from the *cos* site and the ColE1 region (Figure 22-8).

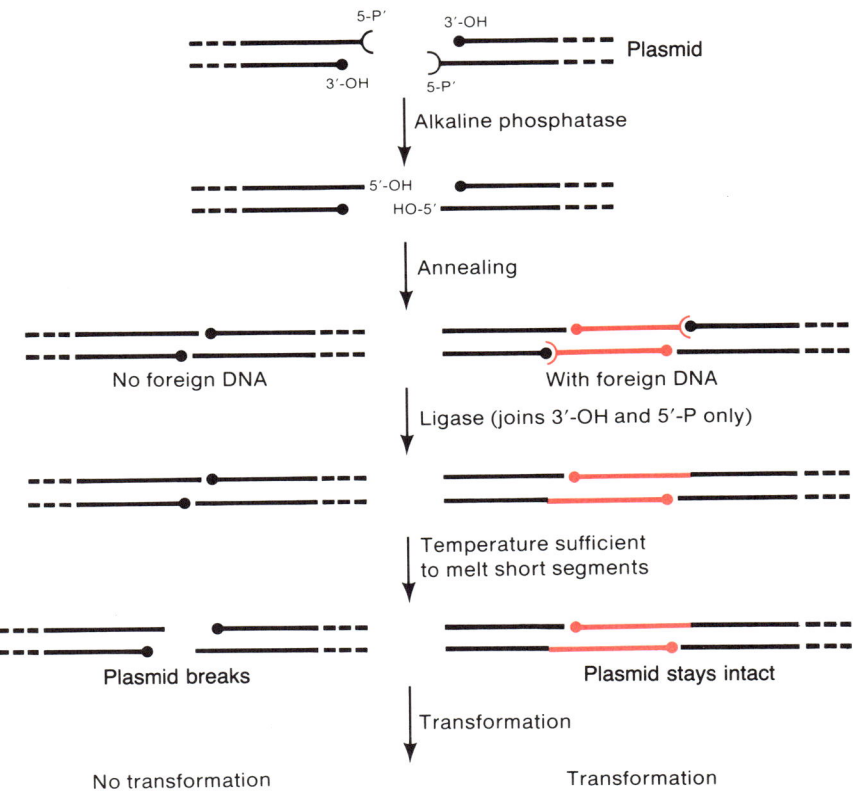

Figure 22-7 Method to prevent re-formation of a cleaved plasmid. The plasmid DNA, but not the foreign DNA, is treated with alkaline phosphatase, an enzyme that converts terminal 5′-P groups to 5′-OH groups.

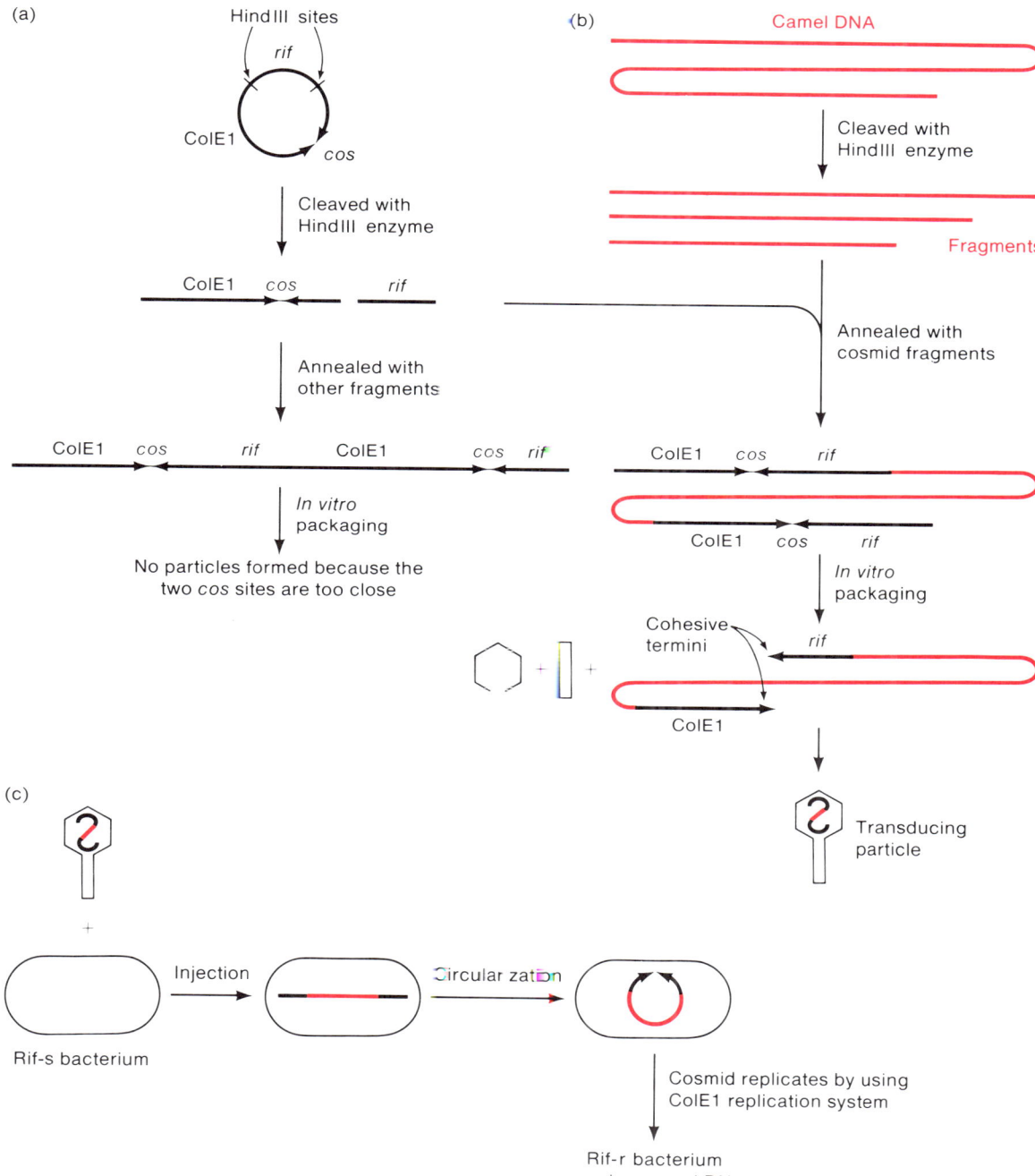

Figure 22-8 Use of a cosmid as a vector. (a) Cleavage of cosmid. (b) Formation of transducing particle. (c) Transduction.

In Chapter 17 the *cos*-site-cutting (Ter) system responsible for packaging phage λ DNA into the phage head was described. The Ter system can act on a DNA molecule only if the following two conditions are satisfied: (1) the DNA molecule must contain two *cos* sites, and (2) the *cos* sites must be separated by no less than 38,000 bp and no more than 54,000 bp. The use of cosmids as vectors is based on a technique for packaging λ DNA *in vitro* and the requirements just stated.

The cosmid shown in Figure 22-8 cannot serve as a substrate for *in vitro* packaging because the DNA has only one *cos* site. If the cosmid is treated with the HindIII restriction enzyme, the resulting linear fragments still cannot be packaged. Panel (a) of the figure shows that if the molecules are annealed with one another, among the many resulting molecules, one is a dimer (or higher multimer) containing two *cos* sites. However, the Ter system still cannot act on this molecule because the *cos* sites are separated by only a few thousand base pairs. Panel (b) shows that if a mixture of cleaved cosmid DNA and HindIII restriction fragments obtained from the DNA of another organisms (e.g., camel DNA) is allowed to reanneal, of the many molecules that arise, one is a linear dimer having two *cos* sites separated by a sufficient amount of foreign DNA that their distance is suitable for the λ packaging system. Thus, *in vitro* packaging yields a collection of transducing particles containing a linear fragment of recombinant cosmid DNA, terminated at each end by the normal cohesive ends of λ DNA. Note that packaging only occurs if the cosmid contains foreign DNA (though it would be possible to package a multimer containing many copies of the *rif* gene in tandem). Panel (c) shows the result of infecting Rif–s *E. coli* with the cosmid-carrying transducing particles. (There is no phage development because the particles contain no phage genes.) The cosmid DNA is injected and circularizes via the λ cohesive termini. Once the cosmid DNA has circularized and has been ligated, it can replicate, using the ColE1 replication system. Growth of the infected cells on agar containing rifampicin selects cells containing the *rif* gene, the ColE1 region, and the foreign DNA.

Methods Which Insure That a Phage Vector Will Contain Foreign DNA

There are two procedures in use for cloning genes in *E. coli* phage λ. Both are based on the fact that λ has several centrally located genes (namely, the *b2, int, xis, red, gam,* and *cIII* genes) that are not needed for lytic growth. These genes are in the so-called nonessential region (see Chapter 17).

The λ variant λgt4·λβ has two EcoRI restriction sites near the termini of the nonessential region (wild-type λ has five EcoRI sites but, in forming λgt4·λβ, three of these sites were removed by genetic manipulation). To use this vector the DNA is cleaved with the EcoRI enzyme and the three fragments are separated by gel electrophoresis. The terminal fragments are isolated and the central fragment is discarded (Figure 22-9). The terminal fragments can be joined via the complementary single-stranded termini but the resulting DNA molecule, whose length is 72 percent that of wild-type λ DNA, will be noninfective, because the minimum length of DNA that can be packaged in a λ phage head is 77 percent of the wild-type value. However, if foreign DNA is inserted, the resulting hybrid DNA becomes infective (that is, produces progeny phage, because the DNA can be packaged). Thus, if an *E. coli* culture is infected with DNA annealed from a mixture of the two terminal fragments and foreign DNA, any phage that is produced will contain foreign DNA.

Another λ variant contains a section of the *lac* operon that includes the *lacZ* gene, the *lacP* promoter, and the *lacO* operator. There is a restriction site in the *lac* operator, so insertion of foreign DNA into the phage at that site yields a phage with constitutive synthesis of β-galactosidase. There are several substrates of β-galactosidase that are noninducing and produce colored compounds if cleaved by β-galactosidase. If such a substance is incorporated into agar, and a mixture of *lacO⁻* and *lacC* (constitutive) cells is plated on the agar, the *lacO^c* colonies will be colored because only these cells synthesize β-galactosidase. Similarly, if *lacO⁺* and *lacO^c* phages are plated on a lawn of *lacO⁺* cells on agar containing such a substance, *lacO⁺* phage will produce colorless plaques but *lacO^c* phage (which contain inserted foreign DNA) will yield colored plaques.

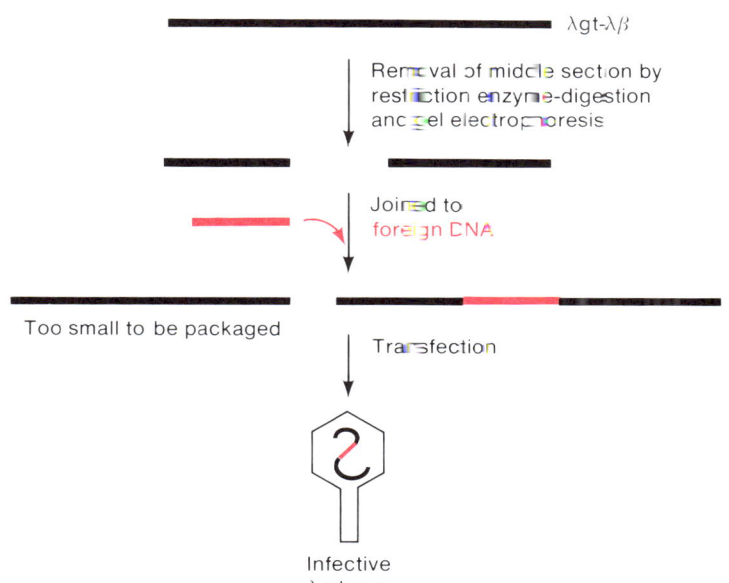

Figure 22-9 Packaging of recombinant DNA in a λ phage.

Physical Methods of Identifying a Plasmid Containing Foreign DNA

A plasmid containing a fragment of foreign DNA is larger than the original uncleaved plasmid and can be identified by this change. A particular increase in molecular weight is also a useful criterion when a specific fragment of c-DNA is to be inserted, if the c-DNA is unique, having been prepared from a purified mRNA.

A simple electrophoretic technique, described in Chapter 5 and depicted in Figure 5-5, is useful to detect an increase in molecular weight. Lysates of single colonies of plasmid-containing cells are electrophoresed (usually sixteen different colonies at a time); there is enough plasmid DNA in a single colony to be visible as a single band moving far ahead of the chromosomal DNA. The plasmid DNA moves a distance related to its molecular weight—the larger the DNA, the smaller the distance moved in a given time interval, so the larger plasmid DNA molecules, which contain donor DNA, move more slowly than those lacking donor DNA; the colonies containing donor DNA are then easily identified.

If the plasmid is thought to contain a particular segment of foreign DNA, this can be confirmed by two different hybridization procedures, if mRNA is available. If radioactive mRNA is used and it is complementary only to the desired foreign DNA, the total DNA content from a bacterium containing the plasmid is denatured and fixed to a filter, and the radioactive mRNA is added. After renaturation the filter is washed and the presence of bound radioactivity indicates that the plasmid contains integrated foreign DNA complementary to the tester mRNA (filter hybridization, Chapter 12).

When purified plasmid DNA is available, an elegant electron-microscopic procedure, called **R-looping,** can be used (Figure 22-10). Plasmid

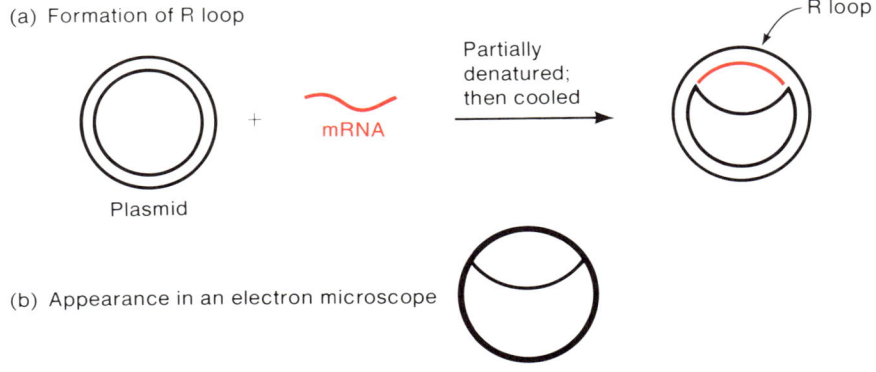

(a) Formation of R loop

Plasmid + mRNA → Partially denatured; then cooled → R loop

(b) Appearance in an electron microscope

Figure 22-10   Formation of an R loop. Since circular DNA molecules are invariably supercoiled, a single-strand break (not shown) is usually introduced prior to partial denaturation so the molecule will be more extended and more easily visualized.

DNA is partially denatured in a buffer containing the mRNA, in which a DNA-RNA hybrid is more stable than double-stranded DNA. Thus, if complementary RNA is present, this RNA will invade the helix, producing a bubble containing a single-stranded (DNA) branch and a double-stranded (DNA-RNA) branch.

Mass Screening for Plasmids Containing a Particular DNA Segment

The simplest procedure for detecting a particular foreign gene is complementation. In the first such experiment reported, a particular gene in the histidine biosynthetic pathway of yeast was detected by joining a plasmid with fragments of the yeast chromosome and transforming a particular His⁻ *E. coli* mutant with a collection of recombinant plasmid DNA molecules, some of which contained the yeast gene, and then selecting for His⁺ colonies by growth on agar lacking histidine. This method works only for genes that are able to complement and is not generally successful when eukaryotic DNA is cloned in bacteria, because usually either the eukaryotic DNA is not expressed or there is no corresponding bacterial gene. The two procedures described in this section are more generally applicable.

The **colony** or ***in situ* hybridization assay** allows detection of the presence of a particular gene (Figure 22-11). Colonies to be tested are transferred from an agar surface to a filter paper. A portion of each colony remains on the agar, which constitutes the reference plate. The paper is treated with NaOH, which lyses the cells and releases denatured DNA, and then the paper is dried, which fixes the denatured DNA to the paper. The dry paper is flooded with ^{32}P-labeled mRNA, which is complementary to the gene being sought, and is subjected to conditions that favor renaturation. The paper is then washed to remove unbound [^{32}P]mRNA; bound radioactivity remains on the paper only if [^{32}P]mRNA has hybridized. The paper is dried and placed on autoradiographic film, and blackening of the film locates the position of the colony of interest, which can then be selected from the reference plate.

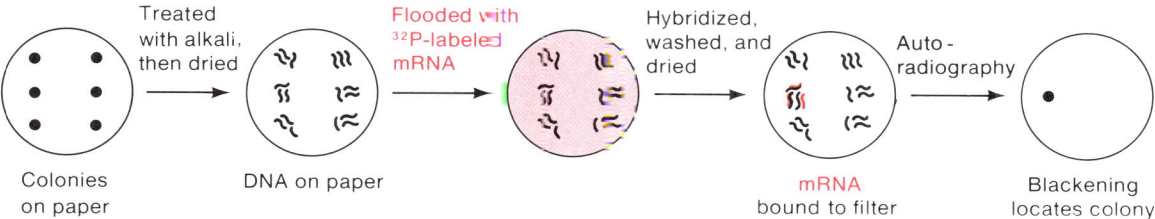

Figure 22-11   Colony hybridization. The reference plate, from which the colonies on paper were obtained, is not shown.

A similar assay is done with phage vectors. Phage are plated on a lawn of bacteria and after plaques develop, paper is placed on the agar. Some phage from each plaque stick to the paper. The paper is then treated with alkali, as above, in order to fix and denature the DNA, and the hybridization procedure just described is used to detect the plaque whose phage contain the gene of interest. The desired plaque is then removed from the reference plate.

If the protein product of the gene of interest is synthesized, two immunological techniques allow the protein-producing colony to be identified. If the protein is excreted, the **radioactive antibody test** (Figure 22-12) is used. In this test a plastic disk to which antibodies to the gene product are

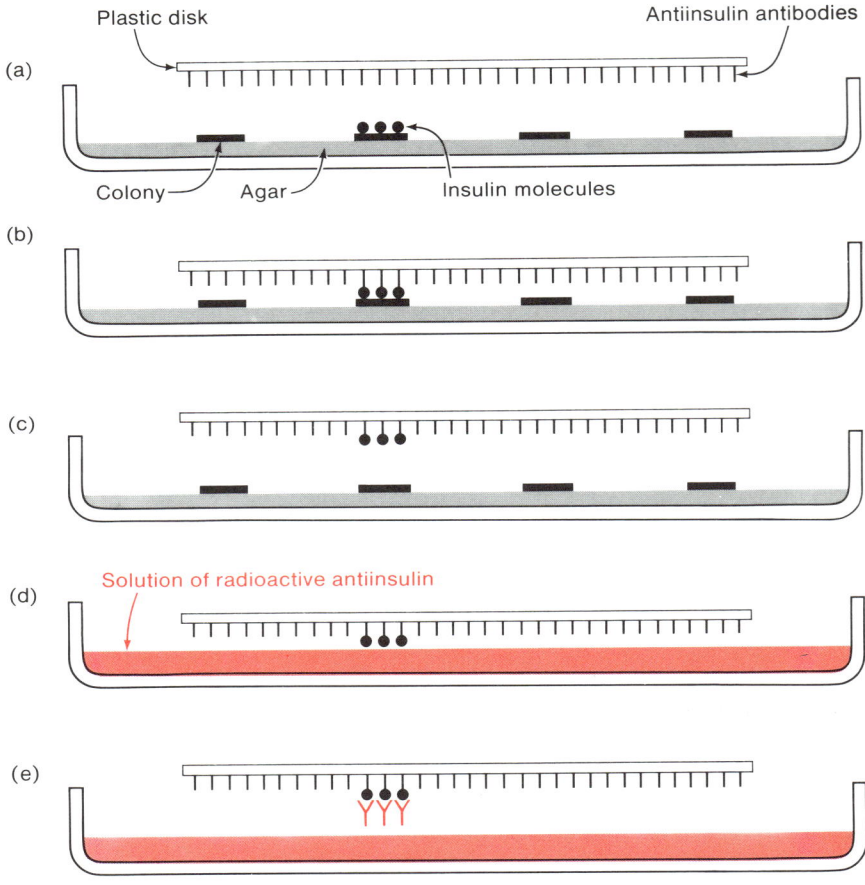

Figure 22-12 Radioactive-antibody test. (a, b) A plastic disk coated with antiinsulin antibody is touched to the surface of colonies on a petri dish. (c) Insulin molecules from an insulin-producing colony bind to antibodies. (d) The disk is then dipped into a solution of radioactive antiinsulin. (e) Radioactivity adheres to the disk at positions of insulin-producing colonies.

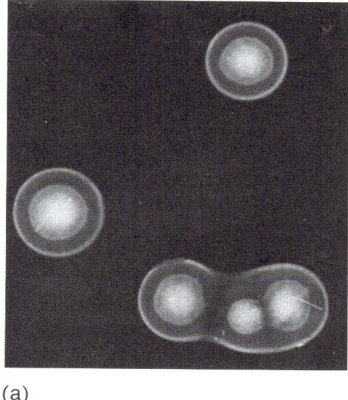
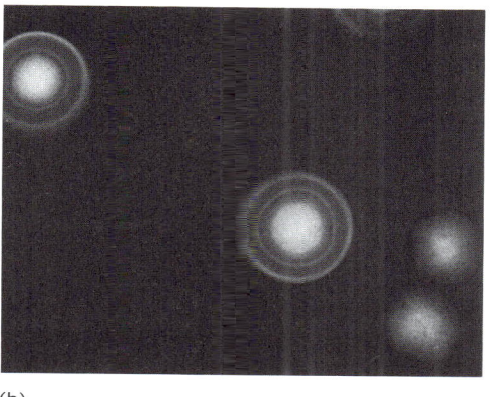

(a)   (b)

Figure 22-13   The immunoprecipitation test. Colonies making β-galactosidase are formed on agar containing antibody to the enzyme. (a) Precipitin bands form around each colony. (b) A mixture of Lac$^+$ colonies (with bands) and Lac$^-$ colonies (no bands).

bound is pressed onto the colonies. If the protein is present, it will bind to the antibody. The disk is then placed in a solution containing radioactive antibody, which sticks to the bound protein on the disk. The disk is then washed and autoradiographed. The location of the radioactivity shows which colony synthesized the gene product.

An **immunoprecipitation test** can also be used to identify a protein-producing colony by adding an antibody to the protein to the agar. If the protein is excreted, an antibody-antigen precipitate called **precipitin** forms a visible white ring around the colony producing the protein (Figure 22-13). Two slight modifications of this procedure allow one to detect a protein that is not excreted. In one modification a host cell is used that is lysogenic for λcI857, a λ mutant containing a heat-sensitive repressor that is inactivated by heating to 42°C. A copy of a reference plate containing colonies to be tested is made by transferring the colonies onto agar containing antibody. When the colonies on the antibody plate become visible, the temperature is raised to 42°C. This induces lysis of many of the cells about one hour later, thereby releasing the cell contents. The protein of interest then reacts with the antibody and forms a precipitin ring at the site of the colony. In the second modification, agar containing both antibody and the enzyme lysozyme is poured on the colonies and allowed to harden. The lysozyme lyses the cells on the surface of the colony, releasing the intracellular proteins; again, the one of interest forms a precipitin ring.

CLONING OF SINGLE-STRANDED DNA

For many purposes, for example, base-sequence analysis, it is desirable to be able to clone a particular single strand of DNA. This is possible with phage M13 as a vector. Recall from Chapter 17 that M13 is a *single-stranded* circular DNA phage with a double-stranded circular intermediate (RF). If a foreign gene is spliced into the RF and cells are transfected with the recombinant RF, the resulting phage will contain only one of the complementary strands (Figure 22-14).

Cloning with M13 makes use of a phenomenon called α-complementation. The active form of the *E. coli lacZ* gene product, β-galactosidase, is a tetramer consisting of four identical subunits of 1021 amino acids each. A small deletion (called **ΔM15**) of amino acids 11–41, which are not in the active site, prevents formation of the tetramer. For unknown reasons, the presence of certain small NH_2-terminal fragments consisting of the 92 N-terminal amino acids enables the tetramer to form. Cloning with M13 is made possible by a genetically engineered form of the phage with the following features:

1. The phage carries a HindII fragment of the *lac* operon that includes the *lac* promoter and operator, and the NH_2-terminal region of the *lacZ* gene. Recall that M13 does not kill its host, instead slowing cell multiplication, and that plaques result from the localized slow growth of the cells. Thus, when the M13 mutant carrying the HindII fragment infects a Lac$^-$ host cell carrying the ΔM15 deletion, cells within the plaque will be phenotypically Lac$^+$. The presence of Lac$^+$ cells in the plaque can be seen by adding two things to the agar: an inducer such as IPTG (Chapter 15) and a chromogenic substrate that produces a blue product when cleaved by β-galactosidase. That is, the mutant M13 forms blue plaques on this medium.

2. The HindII fragment has been engineered to contain a short sequence with cleavage sites for the restriction enzymes EcoRI, SstI, SmaI,

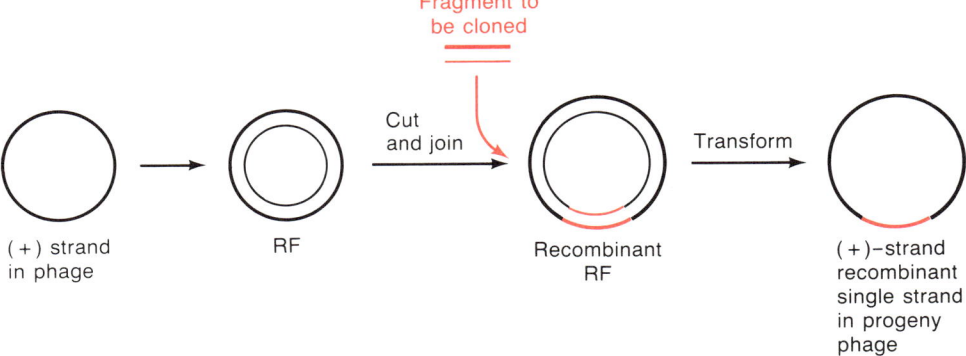

Figure 22-14 Cloning of a single strand in M13 by insertion of a double-stranded fragment into the RF.

BamHI, XbaI, SalI, PstI, and HindIII, in that order; this is called a **polycloning site** and increases the versatility of the vector by allowing the use of numerous restriction enzymes. Insertion of DNA into any of these sites destroys the α-complementing ability of the HindII fragment, and results in the M13 hybrid phage forming a white plaque.

The engineered M13 is used in the following way. RF obtained from cells with the phage is cleaved with two restriction enzymes active in the polycloning site. DNA to be cloned is also obtained by cleavage with the same enzymes. Since the restriction fragment has different single-stranded termini, it can only be inserted in one orientation. The ΔLM15 host is then transfected with ligase-treated DNA. White plaques are isolated, for only these contain phage with inserted DNA. Phage in one plaque are then propagated and DNA is isolated from phage particles. Since the phage DNA is single-stranded, it contains only one of the complementary strands of the cloned fragment. In order to be able to isolate both individual strands, another M13 variant was engineered to contain the polycloning segment in the opposite orientation. Cloning of the same restriction fragment separately in both vectors yields two different recombinant phage (Figure 22-15). Isolation of DNA from each phage yields one of the complementary strands of the cloned fragment.

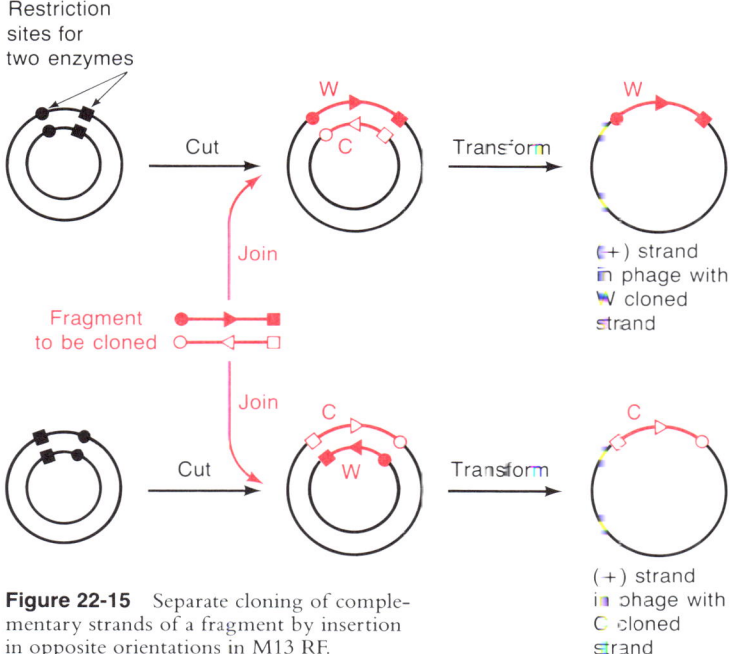

Figure 22-15 Separate cloning of complementary strands of a fragment by insertion in opposite orientations in M13 RF.

GENE LIBRARIES

Many laboratories utilize recombinant DNA techniques repeatedly as a means of isolating particular genes or DNA segments from a single organism. It is quite time-consuming to go through the complete cloning procedure each time a new DNA segment is needed. Thus, collections have been established of hybrid plasmid-containing bacteria that are of sufficient size that each segment of the cellular DNA is represented once (or, on occasion, twice) in the collection. Such collections are called **colony banks** or **gene libraries.** In this section we describe how a gene library is established and how to determine how many clones are needed in order that each sequence will be represented.

One of the first libraries was a collection of *E. coli* colonies containing the ColE1 plasmid in which yeast DNA had been cloned. To form the library, purified ColE1 DNA was cleaved with the EcoRI nuclease, which makes a single cut in this DNA molecule. The yeast DNA was not cleaved with a restriction enzyme; doing that would certainly have prevented many genes from being in the library as intact units inasmuch as some genes would be cleaved by EcoRI. Instead, yeast DNA was broken at random by hydrodynamic shear forces. In order to link the ColE1 and yeast DNA molecules, poly(dT) tails were added to the 3′-OH termini of the cleaved ColE1 DNA and poly(dA) tails were added to the yeast DNA fragments. The molecules were joined by homopolymer tail joining, and the bacteria were transformed. Cells containing the ColE1 factor are resistant to colicin E1 added to the agar, so plasmid-containing cells were readily isolated. The colonies were then transferred to blocks containing small wells filled with storage medium.

The following simple calculation enables one to determine how many colonies are required to make a complete library. Consider a DNA sample containing fragments of such a size that each fragment represents a fraction f of the genomic DNA. The probability P that a particular sequence is present in a collection of N colonies is $P = 1 - (1 - f)^N$ or

$$N = \ln (1 - P)/\ln (1 - f).$$

Let us assume that we want every sequence to be in the library with a probability of 0.99—that is, $P = 0.99$. If the donor DNA is haploid yeast (whose molecular weight is about 10^{10}) and the average molecular weight of the sheared DNA is 8×10^6, then $f = 8 \times 10^6/10^{10} = 0.0008$ and $N = [\ln (1 - 0.99)]/[\ln (1 - 0.0008)] = 5754$. Thus, if the library contains about 5800 colonies, there is a 99 percent probability that any yeast gene will be present in at least one colony. Furthermore, if a few colonies are selected at random for further study, it is exceedingly likely that their yeast DNA sequences do not overlap. For cells with a large DNA content (such as *Drosophila*) the required number of colonies is about 300,000; this can be reduced roughly n-fold by increasing the size of the constituent fragments by a factor of n.

If a clone containing a particular DNA segment is needed, it can in principle be found by any of the screening procedures described earlier in this chapter.

PRODUCTION OF PROTEINS FROM CLONED GENES

One of the goals of genetic engineering is the production of a large quantity of a particular protein that is otherwise difficult to obtain (for example, proteins for which there are normally only a few molecules per cell). In principle the method is straightforward: the gene is inserted adjacent to a promoter and tested to determine whether it is oriented such that the coding strand will be transcribed. With a high–copy-number plasmid or an actively replicating phage, synthesis of a gene product may reach a concentration of about 1 to 5 percent of the cellular protein.

In practice, the goal of producing large quantities of a desired protein is not always met, because of a variety of theoretical and technical problems. Some of these problems and how one seeks solutions are explained in the following sections. Some successful procedures will also be described.

Expression of a Cloned Gene

Expression of a cloned gene requires that the gene be both transcribed and translated. Let us see how one designs a system to be sure that this occurs.

Often a restriction-enzyme cleavage site separates the coding portion of a gene from its promoter, and unless another promoter is provided, the gene cannot be transcribed after it is inserted into a vector. In that case vectors engineered to contain a ribosome binding site and a promoter near the insertion site are used. Usually the easily regulated promoter-operator region of the *E. coli lac* operon (Chapter 15) or the *cI* repressor-*oL-pL* region of *E. coli* phage λ (Chapter 17) are used. For instance, in the first case, transcription can be initiated by the addition of an inducer of the *lac* operon. With the λ system, a temperature-sensitive repressor (called *cI857*) is used.

Coupling a restriction fragment containing a gene of interest to a functioning promoter and a ribosomal binding site does not guarantee that the gene will be correctly expressed (Figure 22-16). A problem is that both ends of a restriction fragment have the same single-stranded termini, so insertion into a vector can occur in two possible orientations. Since only one DNA strand of any gene is a coding strand, only one orientation will yield useful mRNA. Insertion in both orientations occurs with equal probability; therefore, one must select a recombinant vector that has the gene in the correct orientation. If synthesis of the desired gene product occurs, the correct orientation is detected easily. However, if there is no synthesis, methods described in earlier sections must be used.

Figure 22-16 Insertion of a cloned gene in two orientations with respect to a promoter *p* in the vector.

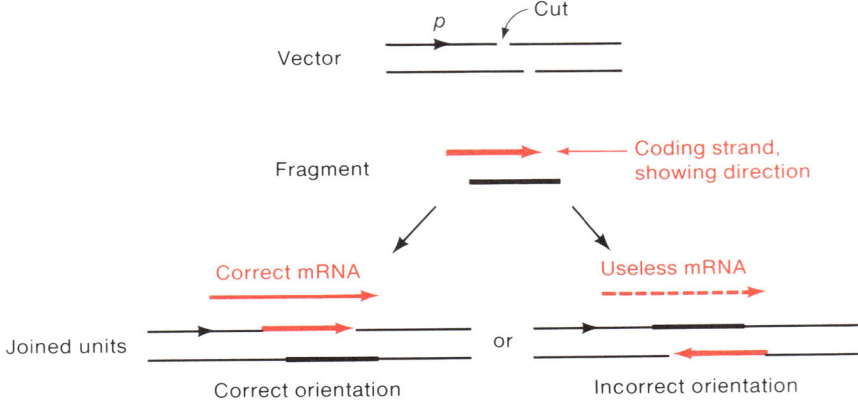

The problem of gene expression is more severe when eukaryotic genes are to be cloned. Particular problems are the following:

1. The eukaryotic promoter may not be recognized by a bacterial RNA polymerase.
2. The mRNA transcribed from eukaryotic genes lacks the Shine-Dalgarno sequence needed for binding to bacterial ribosomes.
3. The mRNA may contain introns that must be excised.
4. The gene product often must be processed.
5. Eukaryotic proteins are often recognized by bacterial proteases as foreign and are cleaved.

Problems 1 and 2 can usually be eliminated by coupling the gene to the *lac* or λ promoter, as just described. The intron problem is avoidable if c-DNA can be prepared. However, for most eukaryotic genes the needed processed mRNA cannot be isolated in pure form. The fact that many finished proteins are modified forms of the initial translation product limits the usefulness of bacteria for the production of eukaryotic proteins, though in some cases (soon to be described) the processing can be carried out *in vitro* after the protein is isolated. Bacterial proteases that destroy eukaryotic proteins can presumably be eliminated (partially, at least) by mutation. Some examples of animal proteins synthesized in *E. coli* will be described shortly.

Cloning in yeast seems to solve most of these problems. Yeast RNA polymerase is able to recognize many animal promoters and processing systems in yeast can often remove introns from animal genes and process some animal proteins. Furthermore, there is much less degradation of foreign eukaryotic proteins. Cloning in yeast has been carried out with the yeast 2μm plasmid (Chapter 21). However, with the advanced technology of genetic engineering in *E. coli* it has proved more valuable to engineer a plasmid capable of multiplying in both *E. coli* and *Saccharomyces*. Such a plasmid is called a **shuttle vector.** It has the following properties.

Shuttle Vectors

A **shuttle vector** is a vector that can replicate in different organisms. The first shuttle vector, which contained *E. coli* and yeast components, was used to clone yeast genes. If a yeast gene were cloned into an *E. coli* plasmid and then *E. coli* cells were transformed by the recombinant plasmid, in general the yeast gene would not be expressed, for the usual reasons—lack of recognition of yeast promoters in *E. coli,* incorrect processing, and other problems already mentioned in Chapters 12 and 16. When cloning of yeast genes was first attempted, these difficulties limited the techniques for detecting colonies containing yeast DNA to *in situ* hybridization. Shuttle vectors, which contain sequences from an *E. coli* plasmid and a particular region of the yeast genome, were designed to avoid this problem (Figure 22-17). Essential features of this shuttle vector are: two replication origins (one active in yeast and one in *E. coli*), two selective markers (*trp,* detectable in yeast, and *amp-r,* detectable in *E. coli*), and restriction sites next to a yeast promoter. Such a vector can be cleaved and and a yeast DNA fragment can be inserted, but, most important, the hybrid vector will transform *trp*⁻ yeast cells, producing Trp⁺ cells. The gene of interest carried on the fragment can also be detected by virtue of its expression in yeast.

A problem with these vectors is that they are not particularly stable in yeast, because they lack a centromere (the portion of a chromosome by which the chromosome is attached to the mitotic spindle, Chapter 7); thus, replicas are not efficiently segregated into daughter cells. To avoid this problem, the recombinant vector is isolated from a culture obtained from a colony of successfully transformed yeast and then used to transform *E. coli,* in which its presence can be detected by an antibiotic-resistance marker (Figure 22-17). In *E. coli* the recombinant vector can be maintained indefinitely. The sequence of steps in the use of this shuttle vector is listed below:

1. Insert a eukaryotic gene in the cleaved restriction site in the yeast segment.

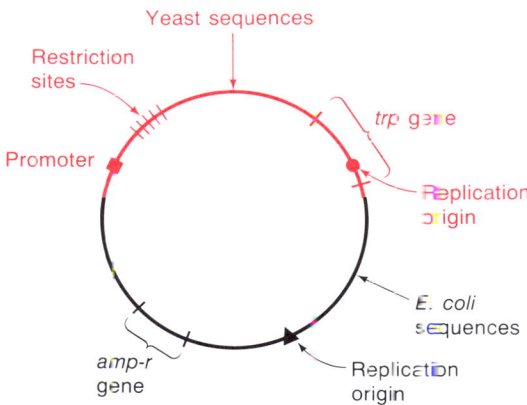

Figure 22-17 A yeast–*E. coli* shuttle vector.

2. Transform Trp$^-$ yeast and plate on a medium lacking tryptophan.
3. Select Trp$^+$ colonies.
4. Test Trp$^+$ colonies for expression of the eukaryotic gene.
5. Isolate a colony with the expressed gene.
6. Isolate the plasmid.
7. Transform Amp-s *E. coli.*
8. Select an Amp-r colony.

The Amp-r colony will contain the shuttle vector with the inserted eukaryotic gene.

Mammalian Expression Vectors

Animal viruses go through a large number of rounds of replication when infecting an animal cell. This fact has been used to develop vectors, commonly called **mammalian expression vectors,** which can produce eukaryotic proteins in large quantities. These vectors are also a type of shuttle vector in that the basic vector (without any inserted DNA) is usually grown in *E. coli,* simply because large amounts of the vector can be obtained with ease. An example of such a vector is pgD–DHFR, constructed for the expression of glycoprotein D, a protein of herpes simplex virus. This protein is of particular interest because it has antigenic sites in common with cells infected with various herpes viruses. It is hoped that inoculation of humans with this protein will protect them from subsequent herpes infections (see later discussion of synthetic vaccines.) If this is successful, large quantities of glycoprotein D would be needed, and such quantities could be provided by cultured cells transformed with a vector containing the glycoprotein-D (*gd*) gene. Plasmid pgD–DHFR is an engineered vector consisting of the following fragments:

1. A portion of the *E. coli* plasmid pBR322 containing the replication origin and the *amp-r* gene. These parts enable the plasmid to be maintained and propagated in *E. coli.*
2. The replication origin of the tumor virus SV40, which enables the plasmid to replicate in several mammalian cell lines.
3. A c-DNA insert of the mouse dihydrofolate reductase gene (*dhfr*). This is a selectable gene, whose presence enables mammalian cells to grow in medium lacking hypoxanthine, glycine, and thymidine–a medium in which mutant cells lacking the gene fail to grow.
4. The SV40 early promoter, located upstream from several restriction sites at which a fragment of DNA can be inserted. The promoter contains the TATA box and the mRNA capping site.

This plasmid has been used to synthesize glycoprotein D in the following way. Restriction maps of herpes simplex virus DNA were obtained for numerous restriction enzymes. An enzyme was selected that made cuts surrounding the *gD* gene and generated a fragment containing a stop codon, a transcriptional stop sequence, and a polyadenylation site. Herpes simplex DNA was cleaved with BamHI, electrophoresed in agarose, and the fragment containing the *gD* gene was isolated from the gel. The vector pgD-DHFR was also cleaved with BamHI and ligated with the purified herpes fragment. A mutant strain of Chinese hamster ovary cells, which lacked the *dhfr* gene, was transformed with the vector, and Dhfr$^+$ colonies (that is, colonies containing the vector) were isolated by growth in selective medium. These colonies were tested for the presence of glycoprotein D by indirect immunofluorescence; that is, cells were exposed to rabbit antibody directed against glycoprotein D and then to goat antibody directed against rabbit immunoglobulin G. The goat antibody had been covalently linked to a fluorescent molecule, so colonies making glycoprotein D were identified by their fluorescence.

Synthesis of Somatostatin Using a Synthetic Gene

In this section a successful attempt to synthesize a eukaryotic protein is described. The student should note how detailed knowledge of gene expression gained from fundamental research on bacterial regulation and protein synthesis contributed to the success.

Somatostatin is a 14-residue polypeptide hormone synthesized in the hypothalamus. It has been made in *E. coli* by a procedure that is applicable to most short polypeptides. Through chemical techniques, a double-stranded DNA molecule was synthesized that contained 51 base pairs; the base sequence of the corresponding mRNA is

AUG-(42 bases encoding somatostatin)-UGAUAG.

This mRNA has a methionine codon (which will not be used for initiation), a coding sequence for somatostatin, and two stop codons. The vector was a plasmid containing the *lac* promoter and the NH$_2$-terminal segment of the *lac* gene. The vector was cleaved in the *lacZ* segment (Figure 22-18) by a restriction enzyme that leaves blunt ends; the synthetic DNA molecule was then blunt-end-ligated to the cleaved plasmid. Addition of a *lac* inducer yielded a protein consisting of the NH$_2$-terminal segment of β-galactosidase *coupled by methionine* to somatostatin. This protein was purified and treated with cyanogen bromide, a reagent that cleaves proteins only at the carboxyl side of methionine. Thus, the methionine linker remained attached to the β-galactosidase fragment and somatostatin was released. Following the methionine-coupling by cyanogen bromide cleavage is a useful trick for separating any polypeptide from a bacterial protein to which it is fused, as long as the polypeptide itself does not contain methionine. Another more generally useful linker is (Asp)$_4$-Lys. In a sequence (Asp)$_4$-Lys-X, in which X is another amino acid, the protease enterokinase cleaves between lysine and X. Since such a sequence is not widespread, this linker is more valuable than methionine.

Figure 22-18   Synthesis of somatostatin from a chemically synthesized gene joined to the plasmid pBR322 *lac*.

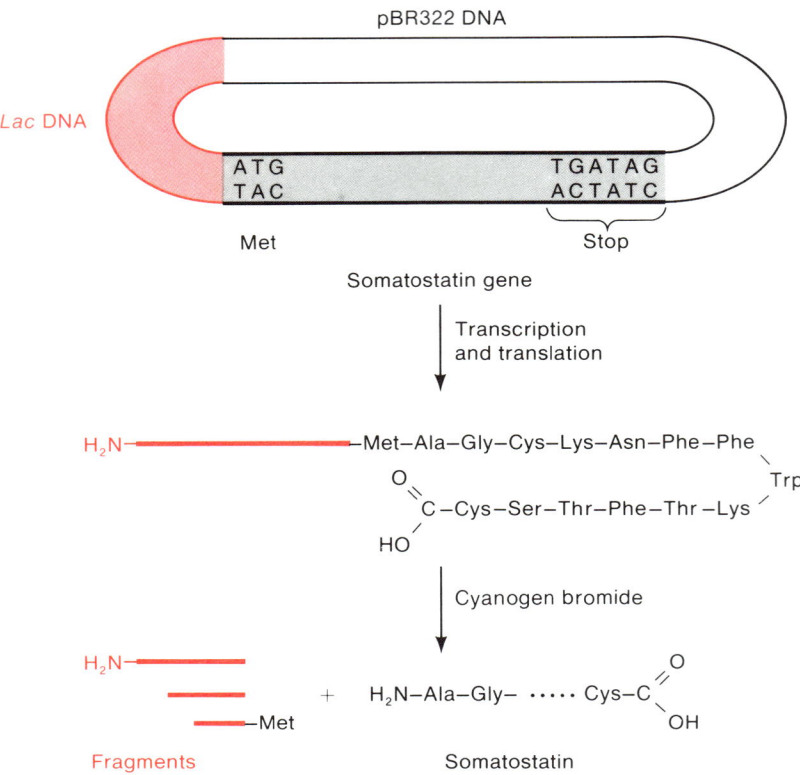

APPLICATIONS OF GENETIC ENGINEERING

So far, we have discussed the utility of the recombinant DNA technology mostly in terms of the production of useful proteins, and this is certainly of great importance. In earlier chapters it was seen that cloned genes were used as probes in hybridization experiments to detect RNA species, which is a particularly valuable technique in the study of regulation in eukaryotes. Genetic engineering also has other uses that we document in this section.

1. *Studies of regulation by subcloning.* A variant of the cloning technique, **subcloning,** is used to identify regulatory sequences in both bacteria and eukaryotes. The technique begins with cloning of a functional genetic unit, which typically contains the coding sequence, regulatory sequences, and often irrelevant adjacent DNA. The cloned sequence is then reisolated and screened for cutting sites by several restriction enzymes. The isolated sequence is trimmed to successively smaller sizes by one or more enzymes and then recloned in a suitable plasmid. The plasmid DNA is used in a transfection

experiment and the phenotype of the recipient cell is observed. A change in phenotype occurs when regulatory regions or structural genes are cut. For example, if the *E. coli lac* operon were trimmed in this way, one would identify an element adjacent to the structural genes (the *lacI* gene) as a regulatory element, because when removed, *lac* expression would become constitutive. Further trimming would yield a Lac⁻ phenotype and loss of ability to make *lac* mRNA and thereby identify the promoter.

2. *Construction of industrially important bacteria.* Bacteria with novel phenotypes can be produced by genetic engineering, sometimes by combining the features of several other bacteria. For example, several genes from different bacteria have been inserted into a single plasmid that has then been placed in a marine bacterium, yielding an organism capable of metabolizing petroleum; this organism has been used to clean up oil spills in the oceans. Furthermore, many biotechnology companies are at work designing bacteria that can synthesize industrially important chemicals. Bacteria have been designed that are able to compost waste more efficiently and to fix nitrogen (to improve the fertility of soil) and an enormous effort is currently being expended to create organisms that can convert biological waste to alcohol. An interesting bacterium is a strain of *Pseudomonas fluorescens,* which lives in association with maize and soybean roots. A lethal gene from *Bacillus thuriengis,* a bacterium pathogenic to the black cutworm has been engineered into this bacterium. The black cutworm causes extensive crop damage and is usually combatted with noxious insecticides. In preliminary studies inoculation of soil with the engineered *Ps. fluorescens* resulted in death of the cutworm. This type of genetic engineering will surely have a great impact on world economy, environmental quality, and the quality of life.

3. *Genetic engineering of plants.* Altering the genotypes of plants is an important application of recombinant DNA technology. In Chapter 19 we discussed the bacterium *Agrobacterium tumefaciens* and its plasmid Ti, which produces crown gall tumors in plants. These tumors result from integration of the plasmid DNA into the plant chromosome. It is possible by genetic engineering to introduce genes from one plant into this plasmid and then, by infecting a second plant with the bacterium, transfer the genes of the first plant to the second plant. (Actually genes are first cloned in an *E. coli* plasmid and then recloned in Ti.) Attempts are being made to perform plant breeding in this way. An example is the attempted alteration of the surface structure of the roots of grains such as wheat, by introducing certain genes from legumes (peas, beans), in order to give grains the ability of the legumes to establish root nodules of nitrogen-fixing bacteria. If successful, this would eliminate the need for the addition of nitrogenous fertilizers to grain-growing soils.

The first engineered recombinant plant of commercial value was developed in 1985. An economically important herbicide (weed killer) is glyphosate, which inhibits a particular essential enzyme in many plants. However, most herbicides cannot be applied to fields growing crops because both the crop and the weeds would be killed. (The chlorinated acids that selectively kill dicotyledonous plants but not the grasses, such as maize and the cereals, are out of favor because of their persistence in soil, toxicity to animals, and possible carcinogenicity in humans.) The target gene of glyphosate is also present in the bacterium *S. typhimurium*. A resistant form of the gene was obtained by mutagenesis and growth of *Salmonella* in the presence of glyphosate; the gene was cloned in *E. coli,* and then recloned in *Agrobacterium*. Infection of plants with purified Ti containing the glyphosate-resistance gene has yielded varieties of maize, cotton, and tobacco that are resistant to glyphosate. Thus, fields of these crops can be sprayed with glyphosate at any stage of growth of the crop.

4. *Production of drugs*. Genetic engineering is having an impact on clinical medicine. The initial focus was on developing organisms that would overproduce antibiotics, thereby reducing production costs, and this has been accomplished for several antibiotics. Of greater significance is the production of biologically active compounds that hitherto could only be isolated in miniscule amounts from enormous amounts of tissue. We have already seen the example of somatostatin. Another example is the antiviral agent α-interferon, which could not even be tested clinically before genetic engineering provided a source of pure material. This substance reduces the duration of viral infection, is effective against herpesvirus infection of the eye, reduces the incidence of attacks of multiple sclerosis, suppresses atherosclerosis in rats on high-cholesterol diets, and is being examined for its antitumor activity. In 1985, tumor necrosis factor, a powerful anticancer substance in rats was cloned, and animal studies began shortly afterward. Interleukin II, a substance that stimulates multiplication of certain cells in the immune system, has also been cloned; it is being tested on patients with AIDS and other viral diseases and shows promise in shrinking cancers in humans.

5. *Synthetic vaccines*. A major breakthrough in disease prevention has been the development of synthetic vaccines. Production of certain vaccines such as anti–hepatitis B, has been difficult because of the extreme hazards of working with large quantities of the virus. The danger would be avoided if the viral antigen could be cloned and purified in *E. coli* or yeast, because the pure antigen could be given as a vaccine. Several viral antigens have been cloned, but because of either thermal instability or poor antigenicity when pure, attempts to make vaccines in this way have been unsuccessful. However, the use of vaccinia virus (the anti–smallpox agent) as a carrier has been fruitful. The procedure makes use of the fact that viral antigens are on the surface of virus particles and that some of these antigens can be engineered into the coat of vaccinia. The use of vaccinia virus is not straightforward because the virus is huge, containing a single DNA molecule with a molec-

ular weight of 160×10^6, far too large for controlled cutting by restriction enzymes and subsequent splicing. Therefore, the viral antigen genes are first cloned in an *E. coli* plasmid. Animal cell lines that support the growth of vaccinia are then infected with both normal vaccinia DNA and the plasmid DNA containing the viral-antigen gene. Genetic recombination occurs in the infected cells and some progeny vaccinia particles are formed that contain the cloned gene and hence the foreign viral antigen in the vaccinia coat. Various procedures are used to isolate these vaccinia hybrids. By 1985 vaccinia hybrids with surface antigens of hepatitis B, influenza virus, and vesicular stomatitis virus (which kills, cattle, horses, and pigs) have been prepared and shown to be useful vaccines in animal tests. A surface antigen of *Plasmodium falciparum,* the parasite that causes malaria, has also been placed in the vaccinia coat; this may lead to an antimalaria vaccine.

6. *Gene therapy.* Retroviruses are also being tested as vectors that might be used to alter the genotype of animal cells. These viruses insert viral DNA into the chromosome of infected cells (Chapter 23). If the virus contained a cloned gene, that gene might be expressed in the cell. However, the use of retroviruses is not simple because many contain a gene that converts the recipient cell to a cancer cell (Chapter 24). Thus, retroviral vectors usually have this gene removed, which also provides the space for a cloned gene. Some preliminary experiments have been done.

Human cells deficient in the synthesis of purines have been obtained from patients with Lesch-Nyhan syndrome and grown in culture. These cells have been converted to normal cells by transformation with recombinant retroviruses. The exciting potential of this technique lies in the possibility of correcting genetic defects—for example, restoring the ability of a diabetic individual to make insulin or correcting immunological deficiencies. This technique has been termed **gene therapy.** However, it must be recognized that retroviruses are not well understood and are potentially dangerous.

Gene therapy is not yet a practical technique, because major problems exist; for example, there is no reliable way to ensure that a gene is inserted in the appropriate target cell or target tissue. In addition, some means is needed to regulate the expression of the inserted genes. At present, U.S. government regulations require that research in gene therapy be confined to infection of somatic (body) cells because of unknown hazards that might accompany conversion of germ-line cells.

Since the early 1980s an increasingly greater fraction of researchers in biology have turned their attention toward genetic engineering. This is exemplified by the formation of more than 100 biotechnology companies since that time and the creation of scientific journals and newspapers devoted exclusively to genetic engineering. With the legal consequences of developing new products, patent lawyers are even being required to take courses in biology. It is likely that this growing billion-dollar industry will some day employ many readers of this book.

Eukaryotic Viruses

Hundreds of viruses are known, each of which infects a particular species of animal or plant cells. Many of these are responsible for serious diseases but others are benign intracellular parasites

Originally, viruses were studied as a means of understanding disease, but it quickly became clear that they possessed a rich variety of interesting biochemical features. Just as a byproduct of the study of phages was the elucidation of many fundamental processes in bacteria, so virus research has provided a window into eukaryotic cells. Some phenomena common to eukaryotes—for example, gene splicing, transcriptional enhancers, and polyproteins—were first observed while studying viruses. In addition, unique biochemical processes (such as reverse transcription) were discovered.

BASIC STRUCTURE OF EUKARYOTIC VIRUSES

There are four morphological types of viruses—naked icosahedral, naked helical, enveloped icosahedral, and enveloped helical (Figure 23-1). Each consists minimally of a coat covering a nucleic acid. The complete virus particle is called a **virion.** A particular species of virus contains only a single type of nucleic acid, which may be single- or double-stranded RNA or single- or double-stranded DNA (Table 23-1). Usually the virion contains a single nucleic acid molecule, but some virions have several molecules, each with one or occasionally two genes.

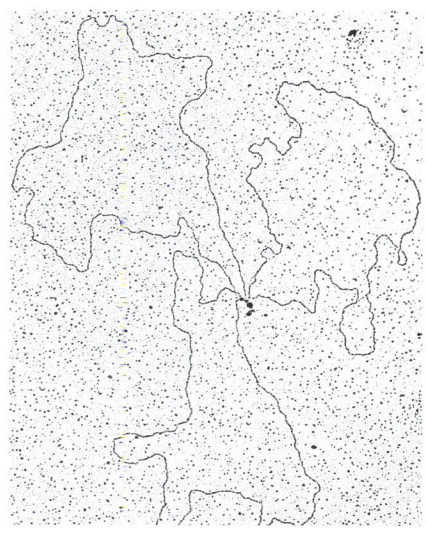

Three adenovirus DNA molecules, each bearing two terminal proteins, have circularized via dimerization of the proteins, and then, again by protein-protein interaction, the three DNA molecules have joined together in trifoliate form. [Courtesy of Ellen Daniell.]

737

Figure 23-1 Four basic structures for viruses.

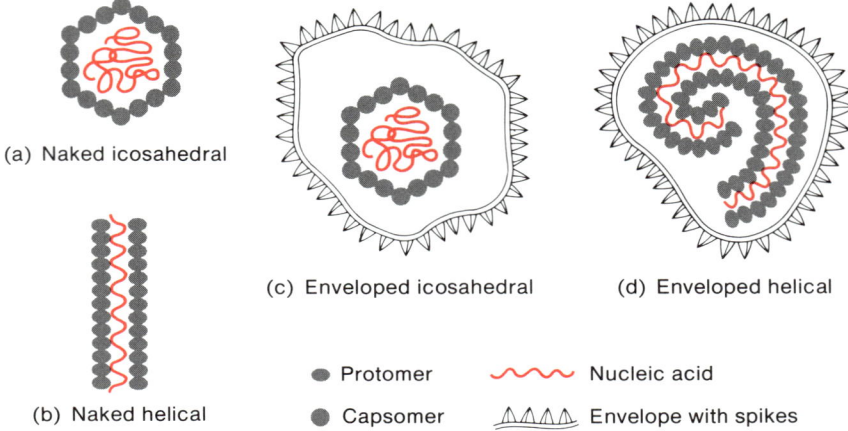

(a) Naked icosahedral

(b) Naked helical

(c) Enveloped icosahedral

(d) Enveloped helical

● Protomer 〜〜 Nucleic acid

● Capsomer ΔΔΔΔ Envelope with spikes

Table 23-1 Nucleic acids of several viruses

Type of nucleic acid	Representative virus	Molecular weight $\times 10^6$	Approximate number of genes
Single-stranded DNA	Adeno-associated	1.5	4–5
	Parvovirus	1.5	4–5
Double-stranded DNA	Polyoma, SV40	3	6–7
	Papilloma	6	9
	Adenovirus 2	23	30
	Herpes simplex	100	150
	Vaccinia	160	240
Single-stranded RNA	Satellite tobacco necrosis	0.4	1–2
	Tobacco mosaic	2	6
	Polio	2.5	7–8
	Influenza	4*	12
Double-stranded RNA	Reovirus	15*	22
	Rice dwarf	15	22

* These virions contain several RNA fragments. The total molecular weight is given. The range for the individual fragments is 0.03×10^6 to 3.4×10^6.

The coding capacity of the nucleic acid molecule of the naked viruses is limited; therefore, the coat of a naked virion consists primarily of one (plant viruses) or two (animal viruses) structural proteins. This building block is called a **protomer.**

In a naked helical virus the protomers are arranged in a helical array and surround one molecule of single-stranded RNA. An electron micrograph of the most thoroughly studied naked helical virus, tobacco mosaic virus, is shown in Figure 23-2.

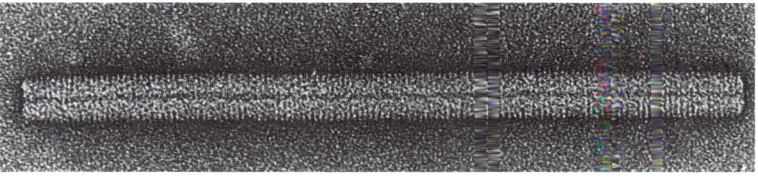

(a)   (b)

Figure 23-2   (a) Electron micrographs of tobacco mosaic virus (TMV) showing the disklike arrangement of the proteins and the hollow core. (b) Virus fragments "end-on," showing the hollow core. The magnification in (b) is about 1/3 that of (a). [Courtesy of Robley Williams.]

The naked icosahedral viruses have more complex structures than the helical viruses (Figure 23-3(a,b)). In these virions the protomers are organized aggregates called **capsomers.** The capsomers contain either five protomers—**pentons**—or six protomers—**hexons.** In icosahedral viruses infecting animals two different polypeptide chains are used to form pentons and hexons but in plant viruses both pentons and hexons use the same protein. The capsomers are organized in several ways to form an icosahedral (20-faced) capsid. The capsid encloses the nucleic acid and the capsid-nucleic acid unit is called a **nucleocapsid.** If no envelope is present, the nucleocapsid itself is the virion.

The enveloped virions do not have well-characterized structures. In these, the nucleocapsid, which may be either helical or icosahedral, is encased in a loose membranous envelope. The shape of the virion is variable because the envelope is not rigid. The envelope is derived from the cell membrane, but usually has virus-specific proteins included among proteins of the envelope. In an enveloped helical virion, the nucleocapsid is coiled within the envelope. Figure 23-3(b,c) shows two examples of enveloped viruses.

Many virions contain virus-encoded enzymes not present in the host cell. Sometimes these enzymes are nucleic acid replicases or are factors needed for adsorption of the virion to the host cell; however, in many cases, the precise role of these enzymes is unknown

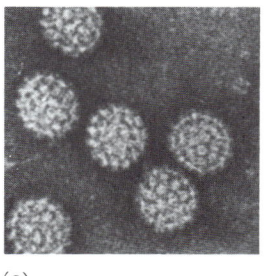
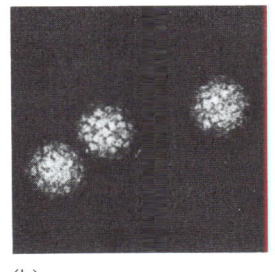
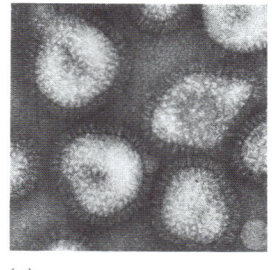
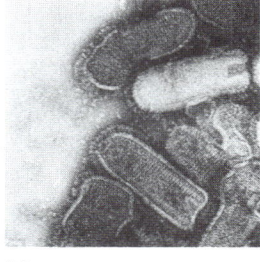

(a)   (b)   (c)   (d)

Figure 23-3   Electron micrographs of two naked icosahedral viruses—(a) human wart virus and (b) tomato bushy stunt virus—and two enveloped viruses—(c) influenza virus and (c) vesicular stomatitis virus. [Courtesy of H.W. Howatson (a) and Robley Williams (b, c, d).]

Figure 23-4 The closed hairpin structure of vaccinia virus DNA. The short vertical lines represent base pairs.

VIRAL NUCLEIC ACIDS

Viral nucleic acids have an enormous range in molecular weight, as shown in Table 23-1. Furthermore, viral DNA and RNA have many forms. Single-stranded linear DNA is found in one small family of viruses. Double-stranded DNA molecules—linear, circular-nonsupercoiled, and circular-super-coiled—are all common, just as they are in phages. However, there are also unusual forms that have not been observed in other organisms. For example, the DNA of vaccinia virus is double-stranded with closed ends (Figure 23-4), and hepatitis B virus has partially double-stranded DNA. Base sequences containing inverted repeats are an essential feature of the linear DNA molecules. For example, all linear single-stranded DNA molecules of parvoviruses as well as the double-stranded DNA of adenovirus have an inverted repetition at the termini; these repeated sequences are essential for replication.

Several forms of viral RNA are known. The most common form is single-stranded RNA; in a few viruses inverted terminal repeats form a terminal double-stranded region, which plays a role in replication, as is the case with single-stranded DNA viruses. Many mammalian retroviruses contain two identical RNA molecules that are held together near the 5′ termini by hydrogen bonds; these retroviruses also contain two host tRNA molecules, which are hydrogen-bonded to the viral RNA. Some single-stranded RNA viruses contain several nonidentical single strands, each carrying separate genetic information (influenza virus is an example); such viruses are said to possess a **segmented genome.** Alfalfa mosaic virus is an especially interesting example of a virus with a segmented genome: it has four segments of RNA and each is packaged separately into a virion (Figure 23-5). Successful infection requires that one RNA of each type enter the cell. Viruses with segmented genomes, the components of which are packaged in distinct

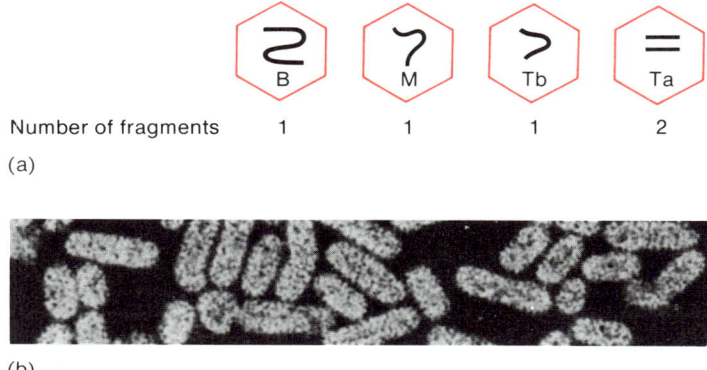

Number of fragments   1   1   1   2

(a)

(b)

Figure 23-5 (a) The four virions of alfalfa mosaic virus, a multiple component virus (covirus). (b) Electron micrograph showing that the particles having different contents of nucleic acid have the same width but different lengths. [Courtesy of R. Hull.]

virions (alfalfa mosaic virus), are called **heterocapsidic** viruses; if all fragments are present in a single virion (influenza virus), the virus is called **isocapsidic.** Plant viruses with segmented genomes are called **multiple-component viruses** or **coviruses.**

The single-stranded RNA molecules in viruses are of two types: (+) strand and (−) strand.

> A viral RNA molecule is said to be a (+) strand if it has the same polarity as viral mRNA and contains sequences that can be translated into viral protein after the RNA enters the cell.
>
> A (−)-strand RNA molecule is a noncoding strand and thus it must be copied by an RNA-dependent RNA polymerase to yield a translatable mRNA.

Viruses having each of these types of RNA necessarily have different life cycles.

Usually the 5′ end of each virion (+) strand or viral mRNA molecule has a cap similar to that found in mRNA molecules in eukaryotic cells. (Capping was in fact first discovered in viral RNA.) An exception is poliovirus RNA, which is uncapped. In (+)-strand viruses having segmented genomes (e.g., rodaviruses) each fragment is capped. There is no cap on the 5′ ends of a double-stranded viral RNA molecule, though mRNA molecules copied from these duplex molecules are capped. Neither the viral RNA nor the mRNA copied from the double-stranded RNA viruses (e.g., reovirus) has a poly(A) tail. The 3′ terminus of the RNA of the single-stranded RNA animal viruses has a poly(A) tail, which is usually shorter than the tail found in animal cellular mRNA.

Viruses having double-stranded RNA always have segmented genomes and are isocapsidic. The best-studied viruses of this type are the wound-tumor plant viruses, which have twelve segments, and reovirus, an animal virus having ten segments (discussed shortly).

Clearly there is great variation in the kinds of nucleic acids present in the RNA viruses. We will see that the various structures require different strategies for protein and mRNA production and their regulation, primarily because most eukaryotic mRNA molecules are monocistronic; this property of the mRNA also introduces different requirements for nucleic acid replication and protein processing.

THE BASIC LIFE CYCLE OF VIRULENT VIRUSES

The life cycle of all virulent viruses consists of the following stages:

1. Adsorption.
2. Entry of the nucleic acid into the cell.
3. Transcription, translation, and replication.

4. Maturation of particles.

5. Release of particles.

The life cycle typically takes 6–48 hours (in contrast with 20–60 minutes for phages), and 10^3–10^5 virions are released per cell per generation. These values are noticeably larger than the burst sizes of most phages (Table 23-2). In the next chapter tumor viruses, which have a different life cycle, will be described.

Viral life cycles are so varied that no single mechanism can account for any one of the stages just listed. Nonetheless, in the following sections some of the major characteristics of these stages and a few common mechanisms are described briefly (except for adsorption, which occurs in a great many ways and is poorly understood). The mechanisms of transcription, translation, and replication will be discussed in detail elsewhere in the chapter.

The number of mechanisms for transport of the viral nucleic acid into a cell may be as great as the number of viruses. For some viruses only the nucleic acid enters the cell; others carry their own polymerase, which is essential, so both nucleic acid and protein molecules (or possibly nucleoproteins) are transported across the cell membrane. With a few enveloped animal viruses there is some evidence that an early stage of penetration is fusion of the membranous envelope of the virus with the cell membrane; for a while the nucleocapsid is surrounded by cell membrane components. Other enveloped viruses enter pits on the cell surface. The pits pinch off, bringing the virus into the cell in a vesicle. The vesicle then fuses with an intracellular vacuole called an endosome and shortly thereafter fuses with another vesicle, a lysosome. The low pH within the lysosome causes a fusion of the viral

Table 23-2 Comparison of burst sizes and infectivity of selected phages and viruses

Phage or virus	Burst size	Fraction of particles capable of infection*
Phages		
λ	40–60	1.0
T4	200–300	1.0
T7	200	1.0
Viruses		
Adenovirus	10^4	0.1
Herpes simplex	10^4	0.1
Papilloma	10^4	0.0001
Polio	500	0.001–0.03
SV40	10^4–10^5	0.005–0.01

* For viruses, the value depends on the particular bioassay.

and lysosomal membranes, which results in ejection of the nucleocapsid into the cytoplasm. Ultimately, for both enveloped and naked viruses, proteolytic digestion of the capsid, called **uncoating,** occurs, and the nucleic acid is released.

The life cycles of virulent animal and plant viruses do not always terminate with programmed cell death and lysis, in contrast with the life cycles of most virulent bacteriophages. Upon infection by some viruses an infected cell will become usually unhealthy, grow very slowly, and perhaps ultimately die, but it is equally common for cells infected with certain viruses to grow indefinitely, releasing progeny viruses continually

Virus particles are assembled from viral nucleic acids and proteins and are produced in very large numbers. (The mode of assembly of tobacco mosaic virus has already been described in Chapter 7.) For some viruses maturation and release are coupled processes. There are three basic mechanisms for release of viruses: (1) release from vacuoles (liquid-filled sacs), (2) lysis, and (3) budding. The first two are characteristic of the naked viruses, whereas budding is the rule for the enveloped viruses. Each of these mechanisms exhibits many variations. In part the differences are imposed by the location of viral assembly—some viruses replicate and are assembled in the cytoplasm, whereas others are nuclear viruses. Cytoplasmic particles often accumulate in vacuoles located just under the cell membrane and are released by rupture of these vacuoles, but without disruption of the entire cell. The nuclear viruses generally do not cause vacuolization. Instead, the viruses form large quasicrystalline arrays in the nucleus and sometimes also in the cytoplasm. About 24–48 hours after infection the cell starts to disintegrate and the virions are released. Some viruses that multiply in the cytoplasm also lyse cells by this disintegration mode. The enveloped viruses are released by **budding** (Figure 23-6). In this mode of release virus-encoded proteins, which are synthesized in the cytoplasm, are incorporated into the cell membrane, causing a reorganization of the membrane structure. The nucleocapsid is thought to adhere closely to these viral proteins. This causes the membrane to form a protruding sphere around the nucleocapsid. The cell membrane then re-forms under the bud, releasing the enveloped nucleocapsid. Thus, the envelope is a portion of the cell membrane that contains viral proteins and no cellular proteins.

In our discussion of the phage life cycle a stage was defined in which the phage "takes over" the bacterial host, converting the bacterium exclusively to a phage-making machine. Part of this takeover is a cessation or reduction of synthesis of host DNA, RNA, and protein early in the life cycle of the phage. Events of this type are evident in the life cycle of a few viruses, but this is certainly not generally true. In fact often the virus and host coexist for some time and only in the late stages of infection are host functions impaired. In the life cycles of some viruses there are definite periods of time in which the host is severely altered, but it is not clear that this is necessary for optimal production of viruses.

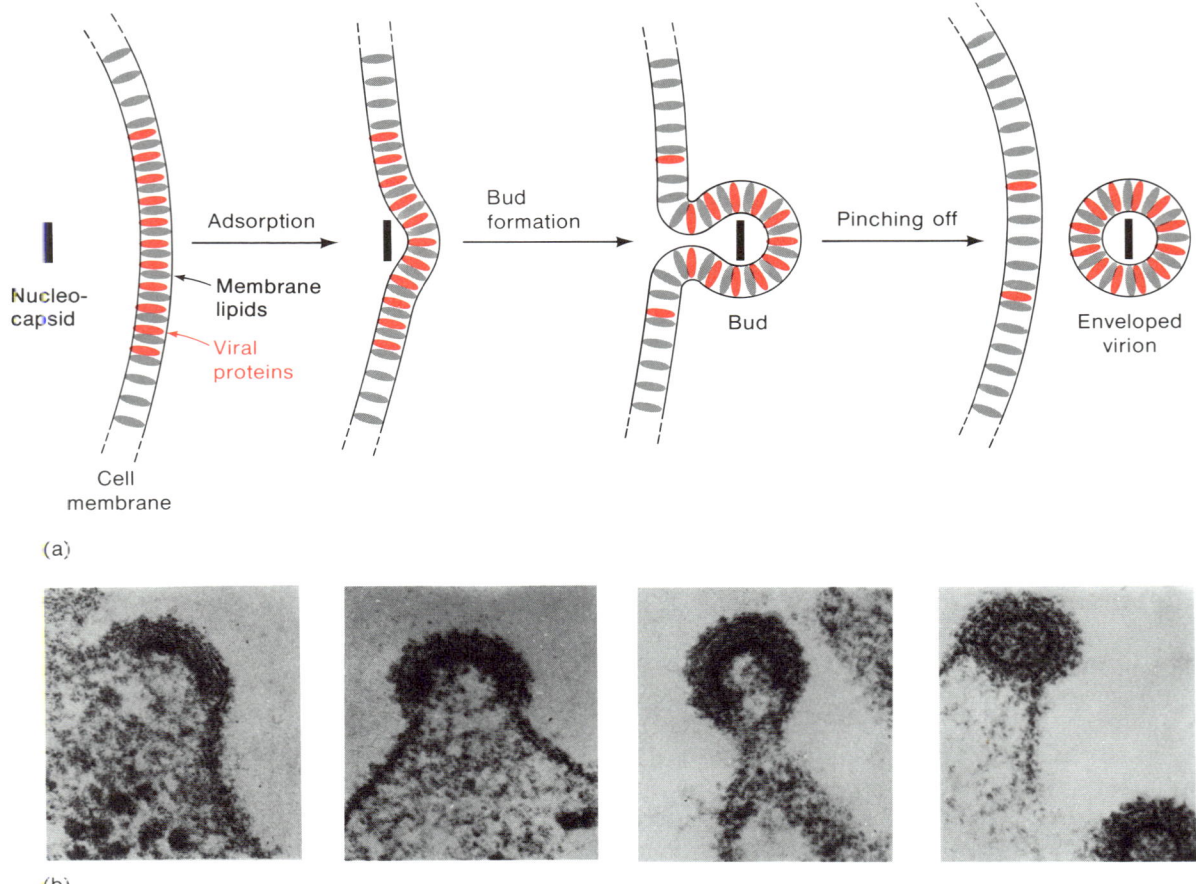

(a)

(b)

Figure 23-6 (a) Formation of an enveloped nucleocapsid by budding. In reality, the membrane is much thinner than the width of the nucleocapsid, and the lipids and viral proteins are not in an array as ordered as that shown. (b) Electron micrograph of the budding process. [Courtesy of Edward Kellenberger.]

CLASSIFICATION OF ANIMAL VIRUSES

Animal viruses will be the major topic of this chapter. So many animal viruses are used in molecular biological research and their names are so complicated that the list in Table 23-3 is provided for further reference. This table lists the major families of animal viruses, the particular viruses encountered most frequently in molecular biological research, and the derivation of their names.

Table 23-3   Viruses used in molecular biological research

Virus family	Derivation of name	Type of nucleic acid	Representative virus	Tumor-forming?
Adenovirus	*Adeno*id tissue	ds DNA	Adenovirus	Some
Herpesvirus	*Herpein,* Gr., "to creep"	ds DNA	Herpes zoster (chicken pox) Herpes simplex I (cold sores)	Some
Poxvirus	Pock-forming	ds DNA	Vaccinia, fibroma	Some
Picornavirus	*Pico-* ("little") + "RNA"	ss RNA	Polio virus	No
Myxovirus	*Myxo,* Lat., "mucus"	ss RNA	Influenza	No
Rhabdovirus	*Rhabdos,* Gr., "rod"	ss RNA	Vesicular stomatitis	No
Togavirus, or arbovirus	*Toga,* Lat., "covering" or "coat"; *arthropod-borne*	ss RNA	Sindbis, Semliki forest, Eastern equine encephalitis	No
Reovirus	*Respiratory enteric* orphan	ds RNA	Reovirus	No
Polyoma*	Lat., *Poly-* ("many") + *-oma* ("tumor")	ds DNA	Polyoma, SV40	Yes
Papilloma*	*Papilla,* Lat., "nipple" or "wart" (wart-forming)	ds DNA	Shope papilloma	Yes
Retrovirus	*Retro,* Lat., "reverse"	ss RNA	Various leukosis, and leukemia viruses, Rous sarcoma, mouse mammary tumor	Yes

Note: ss, single-stranded; ds, double-stranded.

* *Papilloma, polyoma,* and the *vacuolating* viruses together form the papovavirus family.

ANIMAL RNA VIRUSES

Several points must be considered when studying animal RNA viruses:

1. The mechanism of production of mRNA is the principal feature by which the RNA viruses can be distinguished, in the sense that the particular mechanism determines the basic pattern of replication and translation.
2. Most eukaryotic mRNA molecules are monocistronic and eukaryotic ribosomes are rarely able to translate more than one polypeptide chain from a single mRNA molecule. However, all known RNA viruses direct the synthesis of several proteins.
3. The $(-)$-strand RNA viruses need to synthesize a $(+)$ strand (an mRNA) before translation can occur, yet host enzymes that can copy an RNA molecule do not exist.
4. A single-stranded virion RNA molecule cannot serve as a template for direct synthesis of virion RNA; thus, a replication intermediate containing or consisting of a complementary strand must be made.

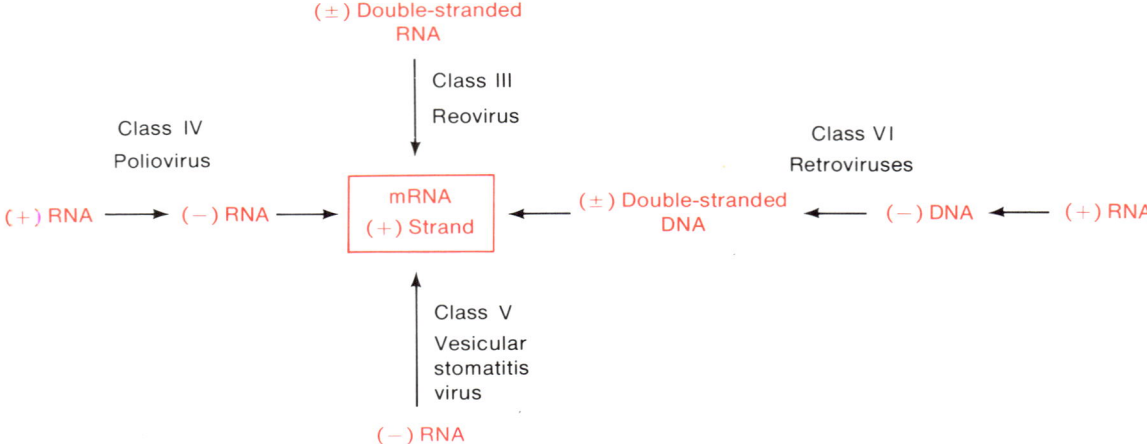

Figure 23-7 Four mechanisms for producing mRNA. The taxonomic classes corresponding to these mechanisms are indicated, as are the viruses discussed in the text. Classes I and II are DNA viruses and are discussed elsewhere.

Replication and the production of viral mRNA are interdependent processes for many RNA animal viruses and are best discussed together. There are four main mechanisms for producing mRNA, and many variations. These four mechanisms are depicted in Figure 23-7. It is convenient to select a particular virus of each type and examine the mode of replication, production of mRNA, and translation for each. The viruses that have been selected are also indicated in the figure.

Viral Polymerases

Many virions carry their own polymerases. The (−)-strand RNA viruses could not otherwise function because animal cells do not contain RNA-dependent RNA polymerases and the infecting (−) strand is by definition not translatable. A virion polymerase is also essential for the double-stranded RNA viruses because the (+) strand in the duplex is unavailable for ribosome binding and translation, and no known cellular enzyme uses double-stranded RNA as a template for RNA synthesis. There is no obvious need for a virion-bound polymerase for (+)-strand RNA viruses. However, the retroviruses, which have a DNA intermediate (because of a unique feature of their life cycles), are in need of a virion-bound enzyme because animal cells can rarely synthesize DNA from an RNA template (exceptions are those containing certain transposable elements).

Vesicular Stomatitis Virus (VSV)

Vesicular stomatitis virus, or VSV (Figure 23-3(d)), is a pathogen of cattle. It contains a single (−)-strand RNA molecule consisting of about 12,000 nucleotides and lacking both a 5′ cap and a 3′-poly(A) tail. Five proteins are encoded in this molecule. These proteins are the following:

N   The major capsid protein.
L, NS Found in the capsid and needed for RNA synthesis.
M   The envelope protein.
G   The spike protein.

The gene order is 3′-N-NS-M-G-L-5′ (Figure 23-8).

For synthesis of these five proteins VSV must form (+)-strand mRNA and the problem of the inability of animal cells to translate polycistronic mRNA must be overcome. The mechanism that evolved in VSV is to synthesize five different mRNA molecules, as shown in Figure 23-8. During

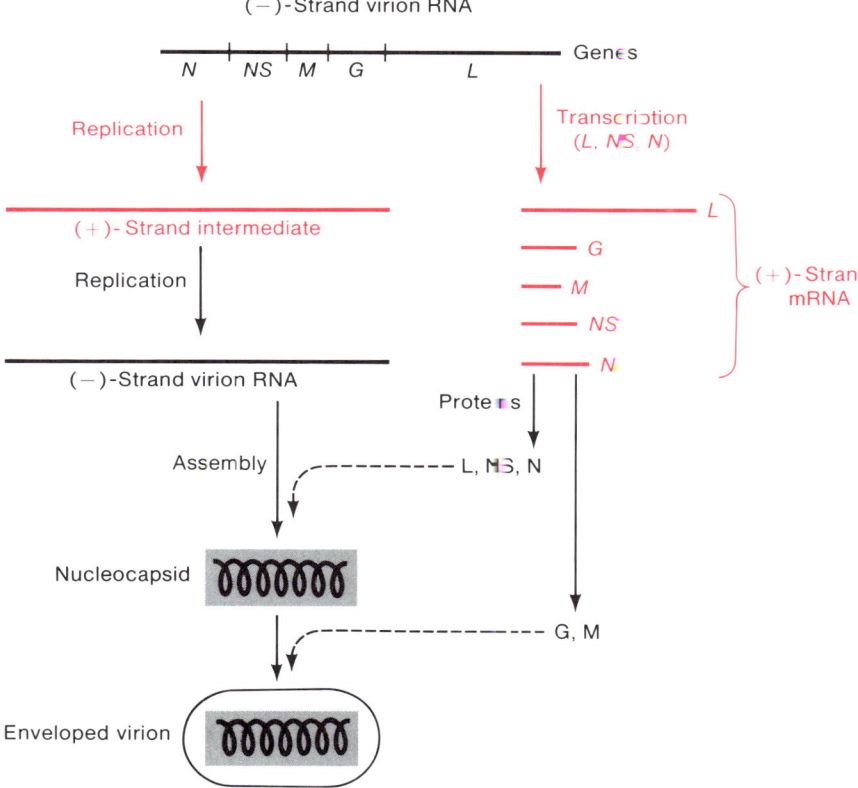

Figure 23-8 Schematic diagram for replication and transcription of vesicular stomatitis virus.

uncoating of the virus the (−)-strand virion RNA is coated with N-protein molecules, which are contained in the capsid and released during uncoating. The L and NS enzymes bind to the N-coated virion RNA and are thereby brought into the cell. These enzymes then synthesize the five mRNA molecules. Each of these mRNA molecules is capped, has a 3′-poly(A) terminus, and contains only one cistron. It is not known with certainty whether a single large (+) strand is synthesized and then cleaved into five fragments or if the five molecules are made by successive starting and stopping. Preliminary evidence suggests that distinct (+)-strand molecules are synthesized by a sequence of successive starting and stopping events.

Replication of VSV also proceeds by first synthesizing a (+) strand by copying the virion RNA. These full-length strands are also made by the L and NS proteins but they are not cleaved (serving no mRNA function); instead, they are immediately copied, again by the L and NS enzymes, to make (−)-strand virion RNA. At the time (−)-strand RNA is made, a large amount of N protein has been translated from the mRNA strands made earlier; the virion RNA and the N protein rapidly associate to form nucleocapsids.

The processes of synthesizing (+)-strand mRNA molecules, which are cleaved, and (+)-strand templates for making (−)-strand virion RNA are somehow separated. This is known because mutations exist that inhibit one process but not the other.

Poliovirus

Poliovirus (Figure 23-9), one of the best-understood viruses, contains a single (+)-strand RNA molecule containing about 7600 nucleotides. The virus replicates in the cytoplasm of sensitive cells, forming large crystalline inclusions of mature virions (similar inclusions of adenovirus are shown in Figure 23-30). The virions are released when the damaged cell lyses. The virion RNA has a 3′-poly(A) tail but the 5′ end is uncapped; instead the 5′ end is covalently linked to a small protein (called **VPg** for genome-linked viral protein), whose function is to prime replication. The RNA molecule encodes eight proteins—four coat proteins, an RNA replicase, and a protease, the terminal protein, and one protein of unknown function. No enzymes are carried in the capsid. The virion RNA has some mRNA activity but does not direct the synthesis of all of the proteins. At no time is the RNA segmented; thus some mechanism is needed to generate several proteins from one mRNA molecule.

Replication of Poliovirus RNA

Immediately after entry into the cell, the virion RNA strand is partially translated and an RNA replicase is synthesized. The replicase is processed from a larger polypeptide chain, as we will soon see. This enzyme catalyzes synthesis of (−)-strand RNA (Figure 23-10) whose sole function is appar-

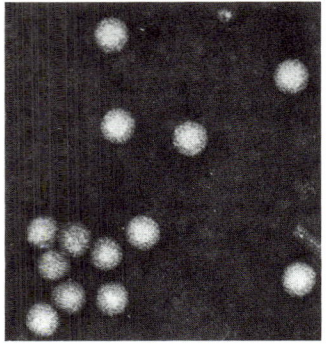

Figure 23-9 Electron micrograph of poliovirus. [Courtesy of Robley Williams.]

ently to serve as a template for further synthesis of (+)-strand RNA. Copying of the (−) strand begins at the 3′ terminus. Initiation of synthesis of a second (+) strand does not await completion of the first strand and usually five (+) strands are simultaneously copied from a single (−) strand, to yield a multibranched replication intermediate. The (+) strands that are formed serve three functions: (1) as mRNA, (2) as a template for further synthesis of (−) strands, and (3) as virion RNA for progeny viruses.

Two features of the poliovirus replication process require further comment. These features are listed and discussed below:

1. *The nascent RNA is not totally base-paired to the template.* In the molecule shown in Figure 23-10, consisting of one intact (−) strand and an incomplete (**nascent**) (−) strand, it might be expected that the nascent RNA would base-pair with the complementary segment of the (−) strand to form a double-stranded region. It is believed that the nascent RNA is coated with a single-stranded RNA-binding protein that prevents this pairing.

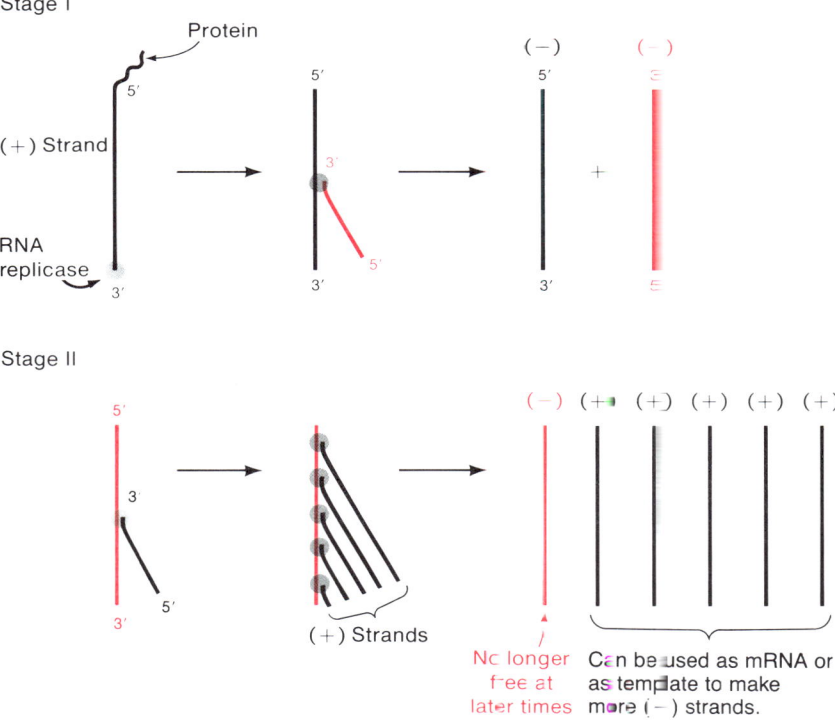

Figure 23-10 Replication of poliovirus RNA. In stage I the (−)-strand template is made. In stage II many (+) strands are made. For clarity, the terminal protein has been shown only on the initial viral RNA. In reality, it is present on the 5′ ends of all molecules shown; it is removed from progeny (+) strands destined to be mRNA; see text for discussion of the protein.

2. *The amount of poliovirus RNA increases exponentially for several hours and then increases linearly with time.* Figure 23-10 shows the replication of poliovirus RNA occurring in two stages. At the end of stage I, which is responsible for synthesis of (−)-strand RNA, there are two molecules. In stage II, (+) strands are made with the (−)-strand template synthesized in stage I. The parental (+) strand can be used as a template again to synthesize another (−) strand, as can the (+) strands made in stage II. The figure shows that in stage II several (+) strands are being synthesized simultaneously. However, at early times in the infection there is a deficiency of replicase molecules, so the first time stage II is used, only one progeny (+) strand forms. Thus, after the first part of stage-II synthesis is completed, there are four RNA molecules. A repeat of simultaneous rounds of stages I and II yields eight molecules, a third yields sixteen, and so forth. Furthermore, there is a significant time delay between each round of replication because only one minute is required to synthesize a poliovirus RNA molecule, yet the apparent doubling time is fifteen minutes. This doubling, which is the exponential process, continues for about two hours until there are about 600 molecules. During these rounds of doubling, the number of replicase molecules has gradually increased by repeated translation and processing, as has the number of capsid proteins. The multibranched replicative form then predominates, turning out (+) strands linearly with time. The newly made (+) strands are either being translated or being packaged, so (−) strands are only infrequently made. Thus, at late times, the major synthesis is production of (+) strands with linear kinetics.

In vitro the viral RNA replicase is unusual among RNA-polymerizing enzymes in that it needs a primer. When poliovirus RNA is the template, this requirement is satisfied by the 5′-terminally linked protein, VPg, which becomes part of the RNA replication-initiation complex. However, since replication begins at the 3′ terminus, the viral RNA must fold to bring VPg to the 3′ terminus. Exactly how VPg primes initiation is not known. Because of this requirement, all progeny poliovirus RNA strands, both (+) and (−) strands, carry a 5′ VPg (both types of strands bear a 5′-terminal uridine). VPg is attached by an unknown mechanism via a tyrosine to the terminal uridine. What is known is that the product of gene *N3,* a small (mol. wt. = 12,000) highly hydrophobic protein, binds to cytoplasmic membranes. There it is cleaved to form VPg, which contains only 22 amino acids, and linkage to poliovirus RNA then occurs; this series of events accounts for the fact that replication of poliovirus RNA occurs exclusively on cytoplasmic membranes.

Poliovirus mRNA molecules lack VPg: apparently, it interferes with translation. Since VPg is present on each (+) strand during its synthesis, it must be removed in the formation of mRNA; this is accomplished by the viral protease N5.

Production of Poliovirus Proteins: Polyproteins

Poliovirus mRNA encodes seven proteins. Because in eukaryotes only a single polypeptide chain can be translated from most mRNA molecules, a special mechanism is needed to produce these proteins.* An indication that there are unusual features of protein synthesis in poliovirus-infected cells came from two observations:

1. Before the number of poliovirus genes was known, it was found by pulse-labeling infected cells with [^{35}S]methionine (which labels proteins) that ten poliovirus proteins are synthesized and that the total molecular weight of these proteins exceeds the coding capacity of the mRNA. This suggested that some proteins are cleavage products of others.
2. In a pulse-chase experiment using [^{35}S]methionine some labeled proteins disappeared during the chase and the label simultaneously reappeared in the form of proteins of lower molecular weight than those labeled initially. These data clearly indicated that cleavage of poliovirus proteins occurs.

Several experiments have led to the following picture of the production of poliovirus proteins (Figure 23-11). A single giant polypeptide, called a

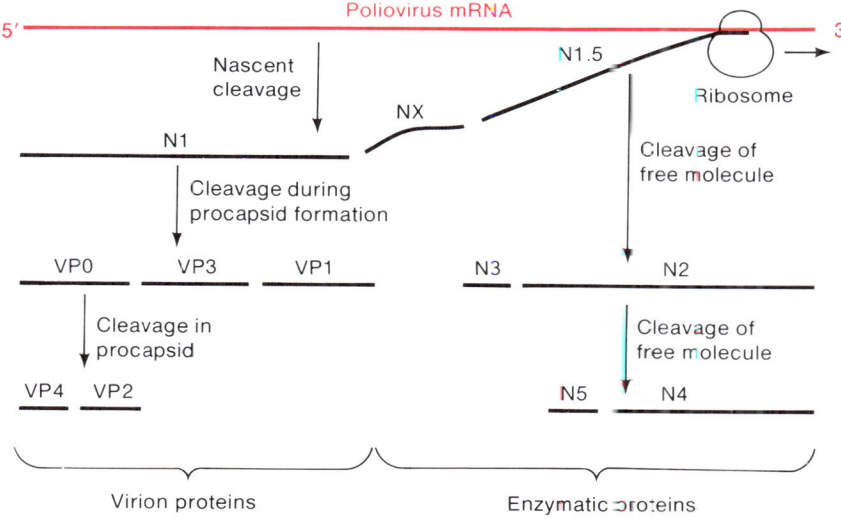

Figure 23-11 Synthesis of poliovirus proteins by successive cleavage of precursors. The labels VP and N stand for *v*irion *p*rotein and *n*onvirion protein respectively. N3 is the precursor for Vpg, N4 is the replicase, N5 is the protease, and NX has an unknown function.

*Interestingly, translation of poliovirus mRNA does not begin at the first AUG, but the seventh or eighth. The others are apparently embedded in sequences that make them inactive for ribosome binding.

polyprotein, is translated from the mRNA. Two cleavages, those surrounding protein NX, are made in the polyprotein immediately after the two peptide bonds to be cleaved have formed. Thus, proteins N1 and NX are released from the growing chain of the polyprotein before protein N1.5 is fully synthesized. These cleavages are called nascent cleavages since they are made in a growing (nascent) polypeptide chain. Once N1.5 is released, a small NH_2-terminal fragment, N3 (the precursor to VPg), is removed, converting N1.5 to N2, which is again cleaved to form N4 and N5. Protein N4 is the RNA replicase and N5 is the protease that removes VPg from (+) strands to form mRNA. NX is probably cleaved further, but its function is unknown.

Protein N1 is cleaved in a series of steps during morphogenesis of the capsid (Figure 23-12). Five N1 protein molecules aggregate to form a pentamer. Then, cleavage of each N1 occurs to form five units, each of which consists of one VP0, VP1, and VP3 protein and which is called a propenton. Twelve of these propentons aggregate to form the procapsid, an empty particle having the size and shape of the virion. A poliovirus (+) strand enters the procapsid to form a provirion. Finally, VP0 is cleaved to form VP2 and VP4, and a subtle rearrangement of VP1, VP2, VP3, and VP4 occurs to generate the intact virion.

In summary, the mechanisms for production of distinct proteins by vesicular stomatitis virus and poliovirus differ in the following ways:

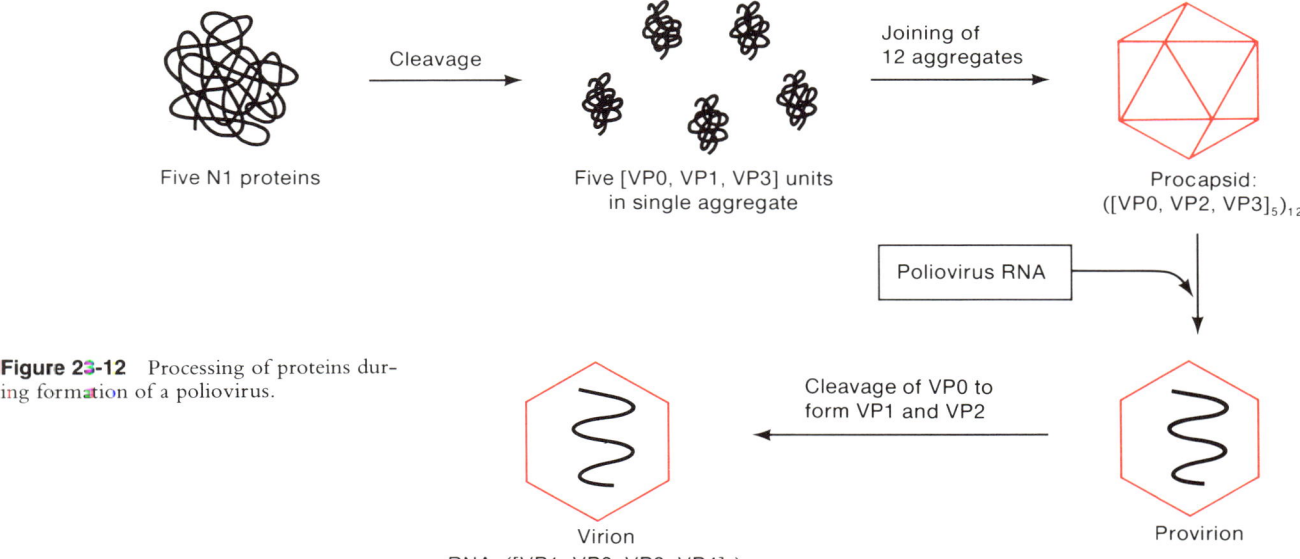

Figure 23-12 Processing of proteins during formation of a poliovirus.

1. Poliovirus uses a single mRNA molecule and cleaves the single polypeptide translated from it to form the individual viral proteins.
2. Vesicular stomatitis virus synthesizes monocistronic mRNA molecules from (−)-strand virion RNA.

Reovirus

Reovirus contains ten double-stranded RNA molecules and exemplifies a third type of viral genetic system, somewhat more complex than those just described. The most striking difference between reovirus and other viruses is that the virion RNA is not released from the virion as free RNA molecules at any time in the course of infection.

The ten virion RNA molecules, each of which encodes a single polypeptide chain, are joined as part of a complex nucleoprotein core that is contained within an icosahedral shell. The shell is removed after the virion enters a cell and an RNA replicase, which is one of the components of the core, is activated (Figure 23-13). The replicase catalyzes synthesis of (+)-strand molecules (mRNA) from each of the double-stranded segments. Capping of each mRNA molecule occurs, catalyzed by enzymes within the core. A poly(A) tail is not added, making reovirus mRNA the only known viral

Figure 23-13   The life cycle of reovirus.

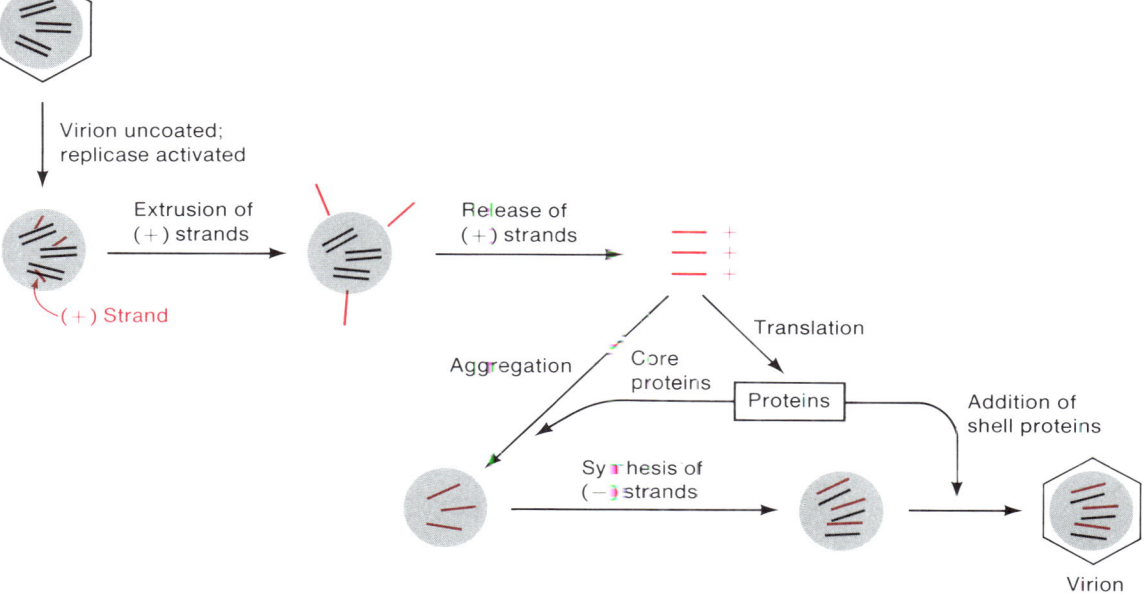

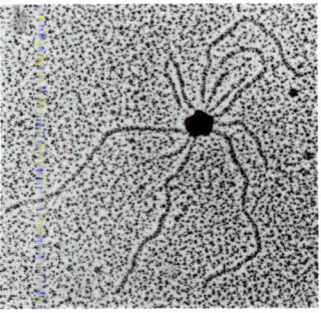

Figure 23-14 Reovirus core. The mRNA molecules appear as threads emerging from the dark core. [From N. M. Bartlett et al., *J. Virol.* (1974), 14: 324.]

mRNA that does not have a poly(A) tail. The newly synthesized (+) strands are then extruded from the core (Figure 23-14) but the double-stranded segments remain in the core. Each of the (+) strands is then translated into a single polypeptide chain. Some of the ten different polypeptides are cleaved to form the finished proteins.

Replication of reovirus requires the production of the ten double-stranded segments. This is accomplished by copying (+) strands rather than by semi-conservative replication of double-stranded RNA. Formation of a new virion requires that one of each of the ten fragments be packaged in the capsid. In reovirus the assortment is regulated and precedes replication. In a way that is not understood, one each of the ten (+) strands is joined with core proteins to form a precore and the (+) strands are copied sequentially in a definite order, which is probably determined by the core proteins. The result of this replication is an aggregate containing a complete set of double-stranded segments and the core proteins. (Note that at no time are there free (−) strands.) The finished core then rapidly forms and the icosahedral shell assembles around it.

A striking feature of the reovirus life cycle is that the (+) strand of the virion RNA is never used. The (−) strand is the template for synthesis of all (+) strands, and *progeny* (+) strands are templates for the double-stranded virion RNA.

Retroviruses

Retroviruses contain two identical single-stranded RNA molecules linked in a complex. These viruses are unusual in that the growth cycles have a stage in which the flow of information is reversed (that is, RNA→DNA) with respect to the usual mechanism (DNA→RNA). Another characteristic of the growth cycles of most retroviruses is a step in which the DNA-containing intermediate is inserted into the host chromosome. Some retroviruses cause tumors and hence have aroused special interest. In this section we will not be concerned with their tumor-producing ability and will consider only the replicative mode in which progeny virions are produced.

The best-understood retrovirus, at this time, is **Rous sarcoma virus,** or **RSV,** which infects chickens. In 1963 Howard Temin observed that the multiplication of RSV could be prevented by inhibitors of DNA synthesis and by actinomycin D, an inhibitor of DNA-dependent RNA synthesis; on the basis of these observations he proposed that the viral growth cycle includes a DNA intermediate, which he called a DNA **provirus.** This hypothesis, which included the novel notion that DNA can be made using an RNA template, was considered quite doubtful until 1970 when Temin and David Baltimore independently demonstrated the presence in several avian and murine (mouse) retroviruses of the enzyme **reverse transcriptase,** an enzyme that makes double-stranded DNA from a single-stranded RNA template.

Since then it has been shown that reverse transcriptase is not confined to retroviruses. It is also made by hepatitis B virus and cauliflower mosaic virus, both of which have an RNA intermediate in their life cycles; it may also be encoded in certain eukaryotic transposable elements (Chapter 21). The RSV enzyme is also a useful tool in genetic engineering (Chapters 5 and 22).

Reverse Transcriptase

Reverse transcriptase has three enzymatic activities:

1. It copies an RNA molecule to yield double-stranded DNA·RNA, using a primer and joining deoxynucleoside 5′-triphosphates in a 3′-5′ linkage.
2. It copies a primed single strand of DNA to form double-stranded DNA.
3. It degrades RNA in a DNA-RNA hybrid; this is called its RNase H activity.

In vitro, initiation of copying RNA requires a primer, as explained in Chapter 5. In the production of c-DNA in the laboratory the primer is supplied as an oligonucleotide. However, such primers are not available in nature, and much more complicated schemes are needed for initiation *in vivo.* How RSV primes the copying activity will be shown shortly.

The Growth Cycle of Rous Sarcoma Virus: An Outline

The basic growth cycle of a retrovirus is shown in Figure 23-15. The growth cycle is divided into two phases:

1. *Early phase.* Reverse transcriptase, which is *contained in the virion,* catalyzes the formation of a double-stranded DNA molecule from

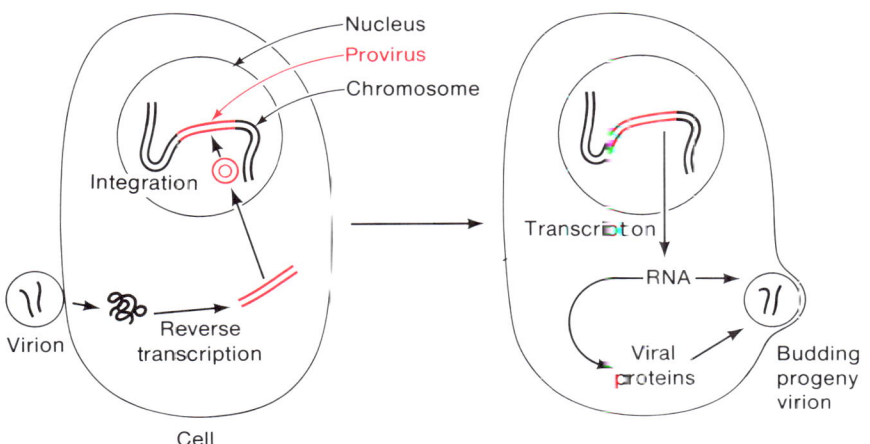

Figure 23-15   Life cycle of a retrovirus. Virion RNA enters the cell and by reverse transcription is converted to double-stranded DNA, which is circularized and then integrated into the host chromosome, becoming a provirus. Transcription then produces RNA for protein synthesis and packaging into progeny virions.

the single-stranded virion RNA. This molecule circularizes and is then inserted into the host chromosome. The gene order in the inserted DNA is the same as the gene order in the viral RNA, in contrast with phage λ (Chapter 17), so the sites of circularization and integration of the viral DNA must be the same.

2. *Late phase*. Transcription and translation occur, yielding two finished proteins and two polyproteins. The two polyproteins are then cleaved, yielding six more polypeptide chains, two of which aggregate to form reverse transcriptase. Four of these proteins form the capsid and two form the envelope. Virion RNA is synthesized by host-cell RNA polymerase II and molecules containing encapsidatable sequences are packaged. Completed virions are extruded continuously from the infected cell by budding without any obvious harm to the cell.

In the following sections several features of the biology of Rous sarcoma virus are examined further.

Properties of RSV Virion RNA

As mentioned earlier, RSV does not contain a single RNA strand but four strands hydrogen-bonded together (Figure 23-16). The larger two strands (called the 38S or monomer strands) carry all of the genetic information and are identical. The two small strands are host tRNA^{Trp} molecules. This species of tRNA is found in many avian retrovirus virions; others carry host tRNA^{Pro} instead. The location of the tRNA molecules is significant for replication, as we shall see shortly. The 3′-terminal section of each tRNA molecule is hydrogen-bonded to the 38S strand 101 bases from the 5′ terminus of the 38S strand in a region termed pbs. Other nearby regions are also hydrogen-bonded, thereby joining the two 38S strands. As drawn in the figure, the strands in the region appear to be parallel rather than anti-

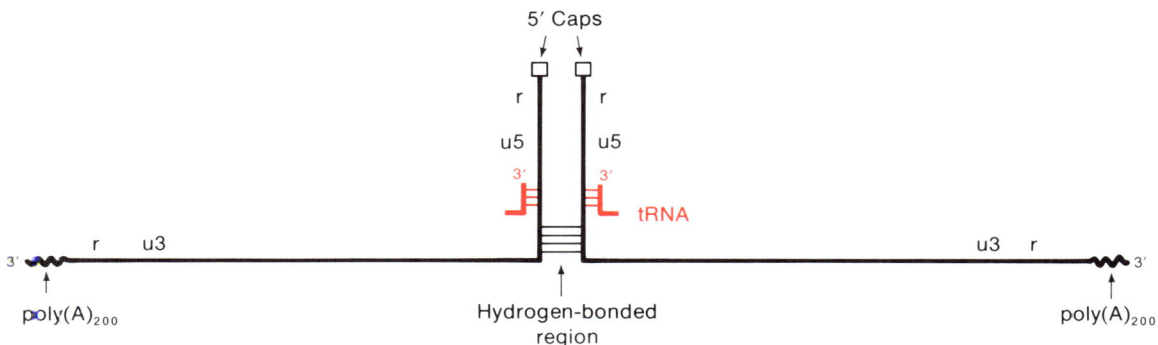

Figure 23-16 Structure of Rous sarcoma virus RNA. Hydrogen-bonding joins the two strands and the tRNAs to the strands. Detailed understanding of this unusual region is poor. The symbols, r, u3, and u5 represent base sequences described in the text; DNA copies of these sequences are denoted R, U3, and U5, respectively.

parallel. However, this is not the case; the identical regions form an anti-parallel structure, though with many unusual features that are poorly understood.

Another important feature of the Rous sarcoma virus RNA is that each 38S molecule contains two identical noncoding unique sequences of 21 bases. These sequences are designated r (Figure 23-16) and are located immediately adjacent to the cap at the 5′ termini and to the poly(A)$_{200}$ sequence at the 3′ termini of the 38S strands. Two other noncoding terminal sequences, which will be discussed shortly, are u5, located next to the r sequence at the 5′ terminus, and u3, next to the 3′-terminal sequence.

Copying of Virion RNA to Form a Double-Stranded DNA Intermediate

All intermediates in the formation of a double-stranded DNA molecule from virion RNA contain only one 38S RNA strand.

Synthesis of RSV DNA precedes transcription and translation, which is possible because the virus carries its own replicase, reverse transcriptase, in the virion. This enzyme, like many replication enzymes, needs a primer; the retroviruses have solved the priming problem in part by including a primer, namely tRNA, in the virion. However, the primer is near the 5′ end of the 38S RNA; as a consequence, only a small segment of DNA can be polymerized onto the 3′-OH group of the tRNA before the growing DNA strand reaches the 5′ terminus of the template RNA molecule.

The mechanism of synthesis of a double-stranded DNA intermediate from viral RNA is complex. An important feature is that the base sequence of the DNA copy is not identical to that of the viral RNA, as shown in Figure 23-17. The sequence U3-R-U5, which is a combination of the 5′ r-u5 segment and the 3′ u3-r segment of the RNA, is found at both ends of the double-stranded DNA molecule—*the DNA is terminally redundant.* The sequence U3-R-U5 is commonly denoted **LTR** for *long terminal repeat.*

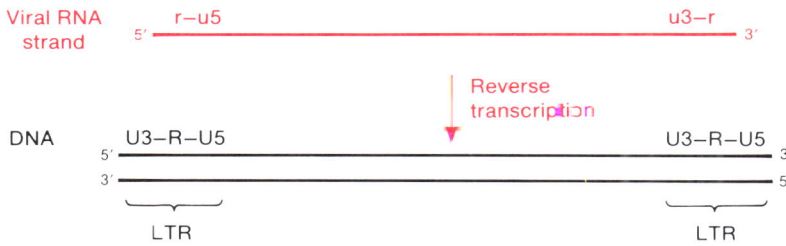

Figure 23-17 The arrangement of terminal sequences (lowercase letters) on one 38S viral RNA strand and on the DNA copy (capital letters). The sequence U3-R-U5 is called LTR, or, long terminal repeat.

The complete mechanism is shown in Figure 23-18. The steps are the following:

1. Reverse transcriptase extends one of the tRNA molecules from its 3′-OH terminus to the 5′ end of the template, resulting in a DNA copy of r–u5.
2. The RNA bases corresponding to the r–u5 sequence are removed, which renders the complementary sequence R′–(U5)′ in the newly synthesized DNA available for hydrogen-bonding. The cap and poly(A) tail are also assumed to be removed at this point.
3. The tRNA and its complementary sequence melt, separating the newly synthesized DNA from the RNA. The DNA segment then binds to the other end of the RNA by forming hydrogen bonds between r and R′.
4. Reverse transcriptase adds nucleotides to the 3′ terminus of the DNA, forming a nearly completely double-stranded molecule.
5. The sequence u3–r at the 3′ terminus of the RNA is digested away.
6. Reverse transcriptase again adds nucleotides to the 3′ terminus of the RNA, completing the duplex and forming the first LTR.
7. All RNA is digested by the RNase H activity of reverse transcriptase.
8. The first LTR is melted and the short fragment forms a (PBS)(PBS)′ duplex at the opposite end of the molecule. This pairing generates a template for formation of the second LTR.
9. Reverse transcriptase adds nucleotides to both 3′ termini, forming a complete duplex terminated at each end by an LTR.

The final molecule shown in Figure 23-18 is integrated into the chromosome and there it is transcribed. At some point in the life cycle a template must be copied to regenerate virion RNA. This apparently occurs as a transcriptional step, whose details are unknown.

Integration of Viral DNA

The mechanism of formation of the provirus is not known in any detail, but the structure of the provirus suggests a possible and somewhat surprising mechanism. Integration has been studied by cloning proviruses from many infected cells into *E. coli* phage λ. Proviral DNA was isolated and cleaved by restriction enzymes, and the base sequences in the regions surrounding the two junctions between host and viral DNA were determined. These sequences were then compared with sequences at the termini of unintegrated viral RNA. This experiment has not been carried out with RSV but conducted instead with seven other retroviruses. The results for spleen necrosis virus (SNV), which has been studied in some detail, are depicted in Figure 23-19. Panel (a) shows a schematic diagram of the unintegrated viral RNA. A long 569-base segment (the LTR) terminates each end, and five terminal bases at each end form an inverted repetition. The two vertical dotted lines indicate the limits of the base sequence present in the provirus; that is, the terminal TT and AA bases of the viral form are not integrated. Panel (b)

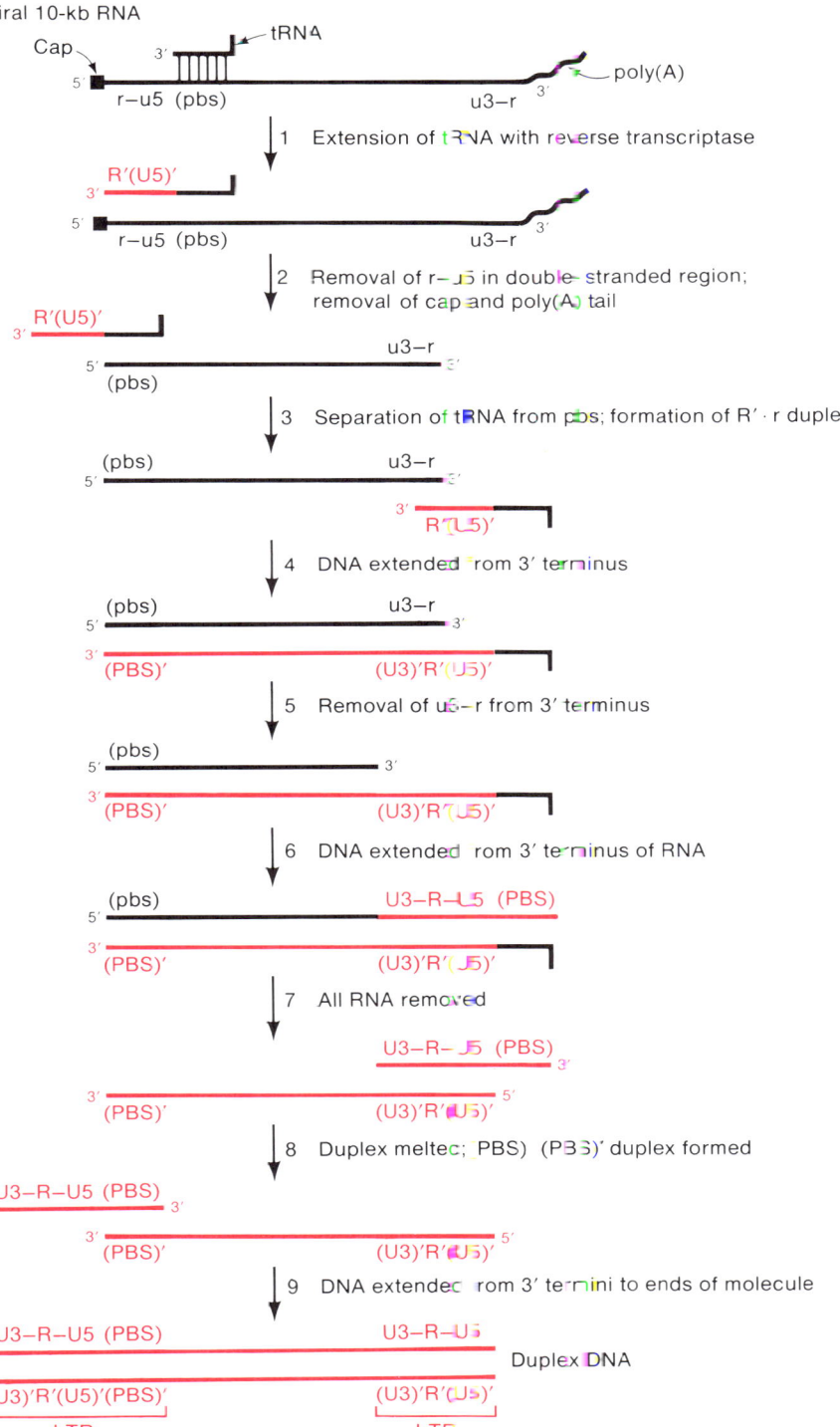

Figure 23-18 Scheme for generating a duplex DNA molecule having two terminal LTR sequence from Rous sarcoma viral RNA. The sequence pbs is complementary to the tRNA; lower- and uppercase letters and numbers refer to RNA and DNA sequences, respectively. A prime indicates a complementary sequence. RNA and DNA are black and red, respectively. Sequences are not drawn to scale.

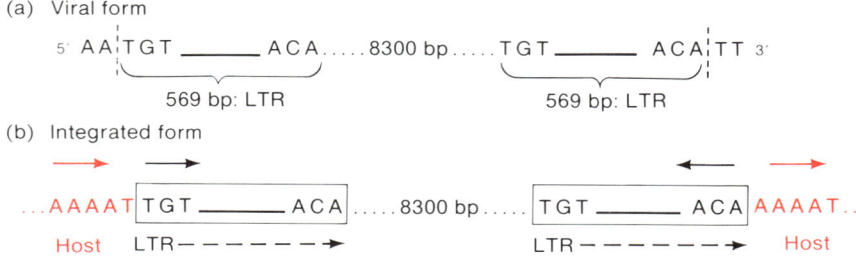

Figure 23-19   Sequences in unintegrated and integrated spleen necrosis virus DNA. Viral and host sequences are black and red, respectively. Pairs of arrows of the same kind (black, red, dashed) represent corre- sponding sequences. If the arrows point in the same direction, the sequence is a direct repeat; if they point in the opposite direc- tion, the sequences are inverted.

shows the structure of one strand of the integrated proviral DNA. Three important features are:

1. The 569-base sequence is repeated.
2. The complete inserted region, including the repeated 569 bases, is terminated by 5'-TG and CA-3'.
3. The host sequences at each joint are sequences of five bases in direct repeat.

Thus the sequence of the integrated DNA may be written as

$$\overbrace{}^{\text{LTR}} \qquad \overbrace{}^{\text{LTR}}$$
$$\text{5H—3V—563V—3V—8300V—3V—563V—3V—5H}$$

in which H and V stand for host and virus, respectively. The following pattern can be observed: (1) viral DNA terminated by two long sequences (LTR) in direct repeat, each of which contains a short viral sequence (3V) in inverted repeat; and (2) a short host sequence in direct repeat. The pattern has been observed for every species of retrovirus examined; only the length and base sequences of the short repeated sequence vary from one species to the next. Furthermore, the integrated DNA has the termini 5'-TG and CA- 3' for all species of retrovirus observed so far, which suggests that there may be a common integration mechanism for all retroviruses. When independent proviruses of a single species have been examined, it has been found that the directly repeated short host sequence differ in each isolate; thus, integration can occur at many sites in the host DNA.

The sequences of the provirus and the adjacent regions are reminiscent of the properties of transposable elements. That is, both integrated retroviral DNA and transposable elements have terminal inverted repeats and are flanked by a small direct repeat of cellular DNA (4–12 base pairs). Furthermore, the yeast element Ty1 and the *Drosophila* elements all end with the dinucleotides TG and CA. It is far too improbable that these are chance similarities, so it has been hypothesized that the mechanisms of transposition and of integration of retroviruses are similar. There is no evidence that transposition of retroviral DNA actually occurs. In the retroviruses, a chromosomal target sequence is duplicated but one does not know whether the double-stranded DNA molecule synthesized by reverse transcriptase is itself inserted or whether a copy of it is transposed from the nucleoplasm of the cell to a chromosomal site. Nonetheless, it is widely believed that retroviruses are derived from ancient transposable elements that gained the ability to replicate and be encapsidated.

Some information is available about the integration mechanism, though at present (1986) the story is incomplete. Several experiments have shown that a circle with two tandem LTRs is an intermediate in integration. This circle is formed by blunt-end ligation of the LTRs of the linear form. Since the integrated form has two terminal LTRs (though with a loss of four base pairs, as shown in Figure 23-19), the LTR-LTR junction of the circular intermediate has been designated an *att* site, analogous to that of phage λ (Chapter 18). Integration of the circle must require at least three enzymatic activities: an endonuclease to cleave both *att* and the chromosomal DNA, a ligase for joining the viral DNA to the cellular DNA, and a DNA polymerase to duplicate the short flanking direct repeats in the cellular DNA. The 3′ end of the *pol* gene (the reverse transcriptase gene) also encodes an endonuclease that acts on *att;* in analogy to λ it has been termed an integrase. The other required enzymes have not been identified. Since additional factors might also be required, a complete model for integration has not yet been formulated.

Once integrated, the proviral DNA is there to stay—it is not excised (except rarely by recombination between LTRs). The presence of the provirus does not inhibit cell division, so daughter cells inherit the integrated provirus. These daughter cells (and cells in subsequent generations) continue to produce active virions.

Transcription and Translation of RSV

Transcription begins the second phase of the RSV life cycle, as shown in Figure 23-15. There is a single transcript (designated 38S RNA) which includes all of the viral genes (Figure 23-20). Some of these transcripts remain intact and are used both as a source of viral RNA and as a mRNA for synthesis of a polyprotein from which the Gag and the Pol proteins and the integrase are cleaved. Many of the 38S molecules are cleaved to yield mRNA molecules that encode the products of the *env* and *src* genes. The Gag protein is cleaved into four smaller proteins that are contained in the viral core. The Pol protein, which is the β subunit of reverse transcriptase, is also cleaved to yield two fragments; one of these fragments, called the α chain, joins with the β subunit to form active reverse transcriptase. The other is the integrase. The *env*-gene product is the viral protein of the envelope of the virion. The Src

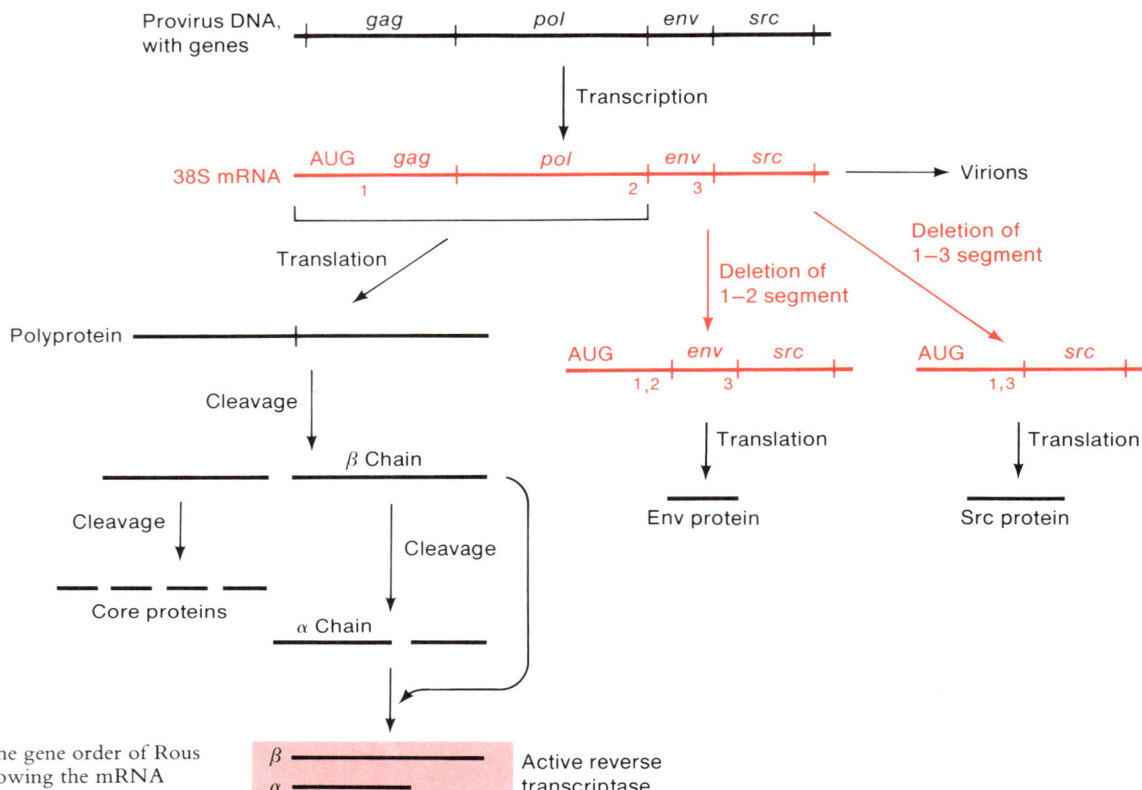

Figure 23-20 The gene order of Rous sarcoma virus, showing the mRNA molecules (red) and the proteins.

protein determines the tumor-producing response and will be discussed in Chapter 24.

It was mentioned earlier that insertion of viral DNA (actually the activity of reverse transcriptase) is important in the life cycle of the virus. One reason for this apparent requirement has become clear from an analysis of the base sequences of the LTR and the 38S mRNA molecule. The coding sequence for the *gag* gene begins to the right of the leftmost U5 sequence shown in Figure 23-17. However, the promoter for the 38S mRNA is in the U3 region. Thus, in viral RNA the promoter is disconnected from the coding region. Construction of the LTR by joining U3 to the R–U5 sequence brings the promoter to a position upstream from the coding sequence, thereby allowing synthesis of 38S mRNA. However, the situation is more complicated. It has been shown that mutants that cannot integrate can produce progeny virus; however, the number of progeny is about 0.001 that produced with wild-type virus. The critical difference is transcription. For completely obscure reasons unintegrated DNA is a poor template for transcription compared to integrated DNA. Presumably, this is a result of some conformational change accompanying integration.

ANIMAL DNA VIRUSES

One might expect that DNA viruses would lack some of the problems we have seen with the RNA viruses, since both replication and transcription of DNA viruses should follow the same patterns seen with the DNA bacteriophages. Furthermore, DNA viruses should not require special enzymes, such as reverse transcriptase: theoretically a DNA virus could consist solely of a DNA molecule that specifies one or two capsid proteins. The tiny parvoviruses do follow such a simple pattern, but we will see that the life cycles of most DNA viruses are considerably more complex. Some DNA viruses have evolved with more than 100 genes, which gives them greater versatility than those with only a few genes, and these viruses are enormously complicated. We shall examine only a few of the many types of DNA viruses in this section, selecting only those whose life cycles are reasonably well understood. These viruses are the papovaviruses, and adenovirus. For discussion of the well-studied parvoviruses, herpesviruses, poxviruses (for example, vaccinia), and papilloma viruses, the reader should consult the references at the end of the book.

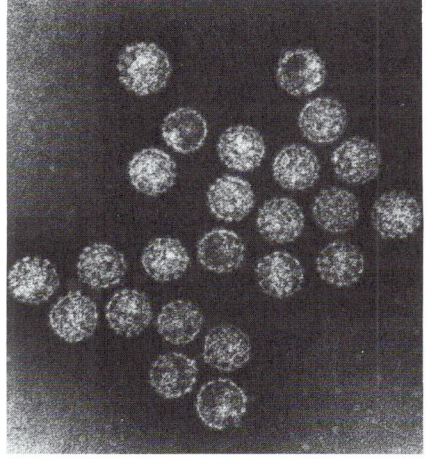

Papovaviruses: Polyoma and SV40

Polyoma virus and simian vacuolating virus 40, or SV40 (Figure 23-21), are two well-characterized papovaviruses. Both induce tumors very efficiently when infecting certain hosts; in other hosts they are virulent, resulting in a bursting of the infected cell and release of progeny viruses (lytic cycle). In this chapter we will consider the lytic cycle only.

Properties of Papovavirus Virions and DNA

The virions of all papovaviruses are naked icosahedrons containing a single circular DNA molecule; when purified, the DNA is supercoiled (see below). The capsid consists of a major capsid protein VP1 and two minor proteins VP2 and VP3. The DNA is associated with all of the host histone proteins except for histone H1. These proteins cause the DNA molecule to contract and form the beaded nucleosomal structure characteristic of eukaryotic chromatin (Figure 23-22, and see Chapter 7). These beaded structures are called **minichromosomes.** Bound histones cause a slight unwinding of the double helix (Chapter 7); since they are bound throughout the replication cycle, at the completion of a round of replication, daughter molecules are slightly underwound. This underwinding is not evident in a minichromosome, as it is taken up by the winding of the DNA around the histones; however, when the DNA is purified and the histones are removed, the DNA assumes the supercoiled configuration.

The DNA molecules of both polyoma and SV40 have been completely sequenced; each contains about 5200 base pairs.

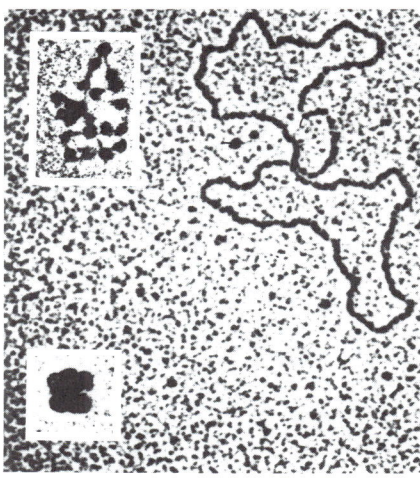

Table 23-4   Proteins encoded by SV40 virus

Protein	Timing of synthesis	Function
T antigen	Early	Initiation of DNA replication
t antigen	Early	Unknown
VP1	Late	Major capsid protein
VP2	Late	Minor capsid protein
VP3	Late	Minor capsid protein
Agnoprotein	Late	Unknown

In the sections that follow, only SV40 will be discussed. At the end of the discussion a few comments will be made about the minor differences between SV40 and polyoma.

Life Cycle of SV40

SV40 grows lytically in monkey cells. Its life cycle can be divided into early and late phases. Six virus-encoded proteins (Table 23-4) are synthesized. Following infection there is a long (12 hour) latent period in which the viral DNA is inactive; during this period the SV40 DNA gradually migrates to the nucleus of the host cell. Neither transcription nor replication can occur outside of the nucleus, presumably because the necessary host polymerases are only contained in the nucleus.

In the early phase of infection the major viral proteins synthesized are the T and t antigens (Figure 23-23).* Following synthesis of early mRNA

Figure 23-23   The time scale for various processes that occur in the life cycle of SV40.

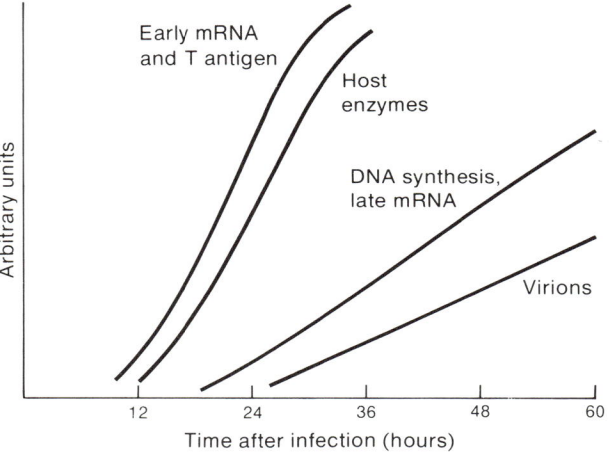

*The names T and t antigen are derived from an observation made by immunofluorescence microscopy that these proteins are located in the nuclei of cells that have been converted to *t*umor cells by viral infection.

the synthesis of several host enzymes is stimulated—DNA polymerase, thymidine kinase, dTMP kinase, and dCMP deaminase. These host enzymes are all part of the pathway for DNA synthesis; T antigen is also necessary for initiation of viral DNA synthesis, as will be seen shortly. Several hours later, host DNA synthesis is also stimulated, but it is not clear whether this is essential to the life cycle of the virus or just an inevitable concomitant. Approximately 12 hours after synthesis of viral and host replication enzymes, viral DNA synthesis begins. Shortly after the onset of viral DNA synthesis, viral late mRNA is synthesized. Viral DNA synthesis is apparently a prerequisite to late mRNA synthesis, as with *E. coli* phage T4 and the adenoviruses. The molecular basis for the requirement for DNA synthesis is unknown.

After 24 hours, late proteins and viruses aggregate and form virions. The assembly process is fairly well understood but will not be described.

After about three days the infected cells lyse, releasing 10^5 virions per cell. Lysis probably results from mechanical damage caused by very large crystalline arrays of virions rather than by a lysing enzyme.

Physical Map of SV40 DNA

Figure 23-24 shows a physical map of SV40, which includes the location of the replication origin and the primary transcripts. This map has been constructed by cleaving virion DNA with many restriction enzymes, ordering

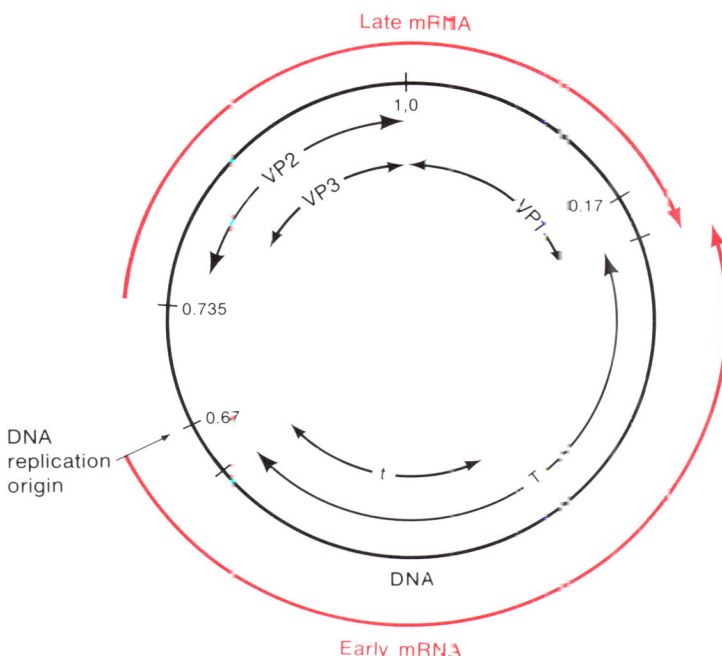

Figure 23-24 Physical map of SV40 showing the location of the five genes, *T, t, VP1, VP2,* and *VP3,* the replication origin, and the early and late mRNA molecules. The splicing patterns of some of these proteins are given in Figure 23-38.

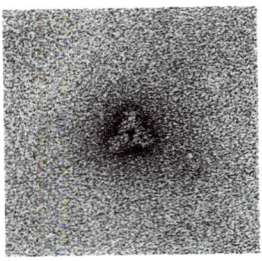

Figure 23-25　T antigen molecules of SV40 viewed by negative-contrast electron microscopy. [Courtesy of Robley Williams.]

the fragments to form a restriction map, and hybridizing fragments with labeled SV40 mRNA isolated from infected cells. The gene products corresponding to the mRNA molecules have been identified by using purified mRNA in an *in vitro* translation system and identifying the proteins synthesized by a particular mRNA molecule. The enzyme EcoRI makes a single cut in SV40 and the position of this cut has been chosen as the reference or zero point of the map.

SV40 DNA Replication

Replication of SV40 DNA is fairly well understood except for the mechanism of initiation. The replication origin is at position 0.67 of the SV40 map; here T antigen binds (Figure 23-25). T antigen is essential for the initiation of SV40 DNA synthesis; this was first demonstrated by the fact that a temperature-sensitive mutant fails to initiate DNA synthesis at a high temperature. How T antigen causes initiation is unknown. T antigen has several other functions: (1) it represses early transcription, (2) it activates late transcription, and (3) in some species of cells it causes a conversion of the cells to tumor cells (to be discussed in Chapter 24).

An interesting observation suggested that the replication origin must have an unusual structure. When supercoiled (deproteinized) virion DNA is exposed to the enzyme S1 nuclease, an enzyme that attacks only single-stranded DNA, a single cut is made at the origin. The enzyme is inactive against the nonsupercoiled form of the DNA. In Chapter 4 it was pointed out that a significant fraction of the base pairs of a supercoiled DNA molecule are broken at any moment and that these melted regions are continually changing. It would be expected that a high-A + T region would remain single-stranded most of the time if the DNA is supercoiled; from the result of S1 cleavage it could be guessed that the replication origin of SV40 would be rich in A·T pairs. This idea was confirmed when the base sequence of SV40 DNA was completely determined. Figure 23-26 shows the base sequence of the replication origin. This region does have an unusual structure, consisting of a 27-bp G + C-rich sequence made up of an inverted repetition of 13 bp (pink in the figure) surrounding a single G·C pair (arrow) and flanked by 17 A·T pairs. A T antigen tetramer binds first to the early-region side of the high-G + C palindrome; then two additional T antigen tetramers bind to fill the G + C region and then the region to the late-gene side of the A + T

Position
0.67
↓

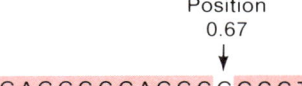

```
CAGAGGCCGAGGCGGCCTCGGCCTCTGCATAAATAAAAAAAATTA
GTCTCCGGCTCCGCCGGAGCCGGAGACGTATTTATTTTTTTAAT
```

Figure 23-26　Base sequence at the replication origin of SV40, showing an inverted repetition of 13 base pairs separated by a single G·C pair at position 0.67.

region. Replication begins at the single G-C pair on the axis of symmetry of the inverted repetition. It is of interest that this is also very near the site of initiation of early mRNA synthesis, as will soon be discussed.

Study of viral DNA replication has been made possible by a DNA fractionation procedure that separates viral DNA from host DNA. The viral DNA, which consists of both replicating and nonreplicating molecules, can then be examined directly by electron microscopy. The following important features of SV40 DNA replication have been observed:

1. Replication is bidirectional and, like *E. coli* phage λ, terminates by collision of the growing forks at a point opposite the origin.
2. Partially replicated DNA molecules contain intact parental strands, and the unreplicated portion remains supercoiled (Figure 23-27). This intermediate, in which the hybrid progeny loops appear open, is

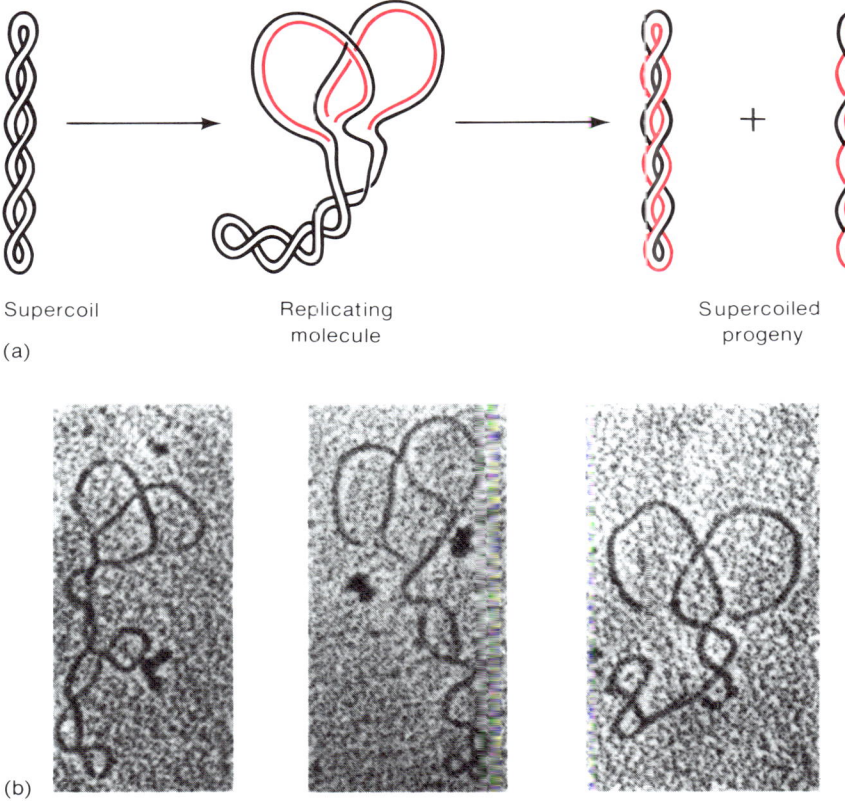

Supercoil

(a)

Replicating molecule

Supercoiled progeny

(b)

Figure 23-27 Replication of the DNA of a papovavirus. (a) A replicating molecule retains its superhelical form in the unreplicated portion. (b) Three replicating molecules. [R. Jaenisch et al. *Nature.* 233: 72–75. Reprinted by permission. Copyright, Macmillan Journals Ltd.]

sometimes termed a **butterfly molecule.** Some unwinding of the parental helix is necessary, and without some relief of the torque developed by the unwinding, the structure would become a tightly tangled mass. This torque is relieved by a topoisomerase.

3. Late in the life cycle, large amounts of long concatemeric SV40 DNA are found. These molecules could be branches of rolling circles, yet rolling circles have rarely, if ever, been isolated.* These concatemers can be used to infect monkey cells; it is found that they are cleaved into monomers and then function to produce progeny virus. The significance of the concatemers is unknown but it is possible that, like *E. coli* phage λ, they are the precursors to virion DNA.

An *in vitro* SV40-DNA replication system is also available. It has provided the following information:

1. One daughter strand is made continuously and the other discontinuously.

2. Each precursor fragment on the lagging strand is primed by a 10-base ribonucleotide chain synthesized by a primase like the *E. coli* enzyme. The base sequences of the primers vary widely, indicating that a special sequence is not recognized.

3. Removal of the primer is not (as in *E. coli*) coupled to joining of the precursor fragments but comes at a much earlier time.

4. The polymerizing enzyme is DNA polymerase α, the major host-DNA replication enzyme.

5. The immediate products of termination are a mixture of linked covalent circles (catenanes) and separated circles with a gap of a few nucleotides in one strand. The source of the gap is unknown. Gap repair and topoisomerase action on the catenane results in the formation of covalent circles. These presumably form minichromosomes, which after deproteinization appear supercoiled (Figure 23-27). If an insertion is made in the DNA such that the replication forks collide at an abnormal site, gapped circles do not form; catenanes result, which are ultimately separated.

Transcription in an SV40 Infection

Transcription of SV40 DNA is a complex process involving temporal (early-to-late) regulation and extensive mRNA processing, which selectively removes start and stop codons. The early and late mRNA molecules are transcribed from different strands of the DNA molecule; they are synthesized in opposite directions. Furthermore, they are initiated at several nearby sites in regions adjacent to the replication origin. Figure 23-28(a) shows that a single early

---

*Circles with short linear branches have on occasion been observed. However, if the length of a branch is less than the circumference of the circle, it is difficult to know whether the molecule is a rolling circle or a theta molecule that is broken in a replication fork.

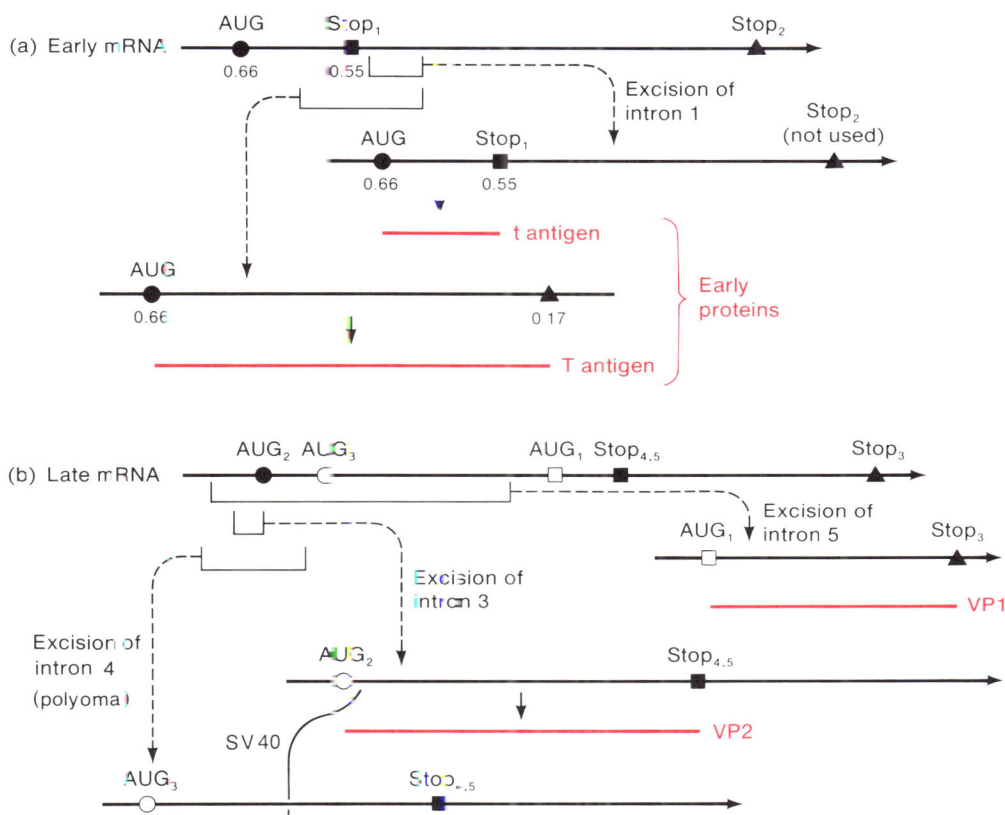

Figure 23-28 Processing of SV40 (a) early and (b) late mRNA. Horizontal brackets indicate excised introns; a dashed arrow from above leads from an unprocessed mRNA molecule to the processed molecule lacking the intron indicated by the arrow. Open and closed squares, circles, and triangles designate identical positions on each mRNA molecule. The mRNA is black; protein is red.

RNA molecule, which encodes both the T and t antigens, is synthesized from a promoter to the left of position 0.66 in the physical map. A single AUG codon is responsible for initiation of synthesis of both proteins. This RNA is processed in two different ways to yield monocistronic mRNA molecules, one encoding T antigen and the other encoding t antigen. In one mode of processing, intron 1 is excised, to yield $mRNA_1$, from which the t antigen is translated by starting at the AUG and terminating at $stop_1$. For synthesis of T antigen, intron 2 is removed; it has the same 3′ terminus as intron 1 but has a 5′ terminus nearer to the 5′ end of the mRNA. Excision of this larger intron removes the $stop_1$ codon, so that translation continues to $stop_2$, to yield the T antigen, which is much larger than t antigen. Note that the NH_2-terminal segments of t and T antigens are identical.

Three proteins—VP1, VP2, and VP3 (and a fourth called agnoprotein)—are encoded in the late RNA. The pattern of splicing determines which protein is translated, and the use of overlapping reading frames enables the virus to store a great deal of information in a rather small segment of DNA. (A more extreme example of such economy will be seen in the next section, where adenovirus is described.) Several classes of late mRNA with multiple 5′ ends are transcribed from an initiation site at position 0.735 in the map. Figure 23-28(b) shows the different modes of splicing that generate the three monocistronic mRNA molecules needed to translate the three proteins. Note that no protein is translated from unspliced RNA. VP2 and VP3 are translated in the same reading frame and both have the same carboxyl termini. Thus, VP3 is actually a fragment of VP2. In SV40 it is translated from an internal AUG of VP2 mRNA. In polyoma it is translated from a mRNA fragment from which the start codon for translation of VP2 has been removed by processing enzymes. VP1 is read in a different reading frame from VP2 and VP3. This reading frame is established by excising a very large segment of RNA that contains the AUG codons for both VP2 and VP3. The only remaining start signal establishes the reading frame for VP1.

The pattern of replication of polyoma viral DNA is the same as that for SV40 DNA, and transcription is almost identical (see point above about VP3). The major difference is that polyoma has an additional mode of processing early RNA molecules, which enables it to synthesize a protein called middle-t antigen. The function of this protein is discussed in Chapter 24.

The amount of early RNA is determined in part by an enhancer (in fact, this was the first enhancer discovered). Deletion of this 72-bp unit reduces early transcription about 100-fold. This enhancer has already been discussed in Chapter 12.

Large DNA Viruses: Adenovirus

So far, viruses having only a few gene products have been described and consequently the number of their mRNA molecules has been small. There are many large viruses and, as expected, these have very complex life cycles—sufficiently complex that in no case can one say that the life cycle is understood. In this section we will only present the highlights of one large DNA virus, adenovirus.

The adenoviruses form a large group of well–studied, naked icosahedral viruses (Figure 23-29), each containing one double-stranded DNA molecule with 35–36,000 bp. The DNA encodes at least 35 proteins of which 10 proteins are contained within the virion. Four virion proteins are associated with the DNA—one interacts with the phosphates, a second organizes the DNA into a multilobed form and binds the DNA to the 12 pentons of the capsid, a third is covalently linked to the 5′ termini of the DNA molecule,

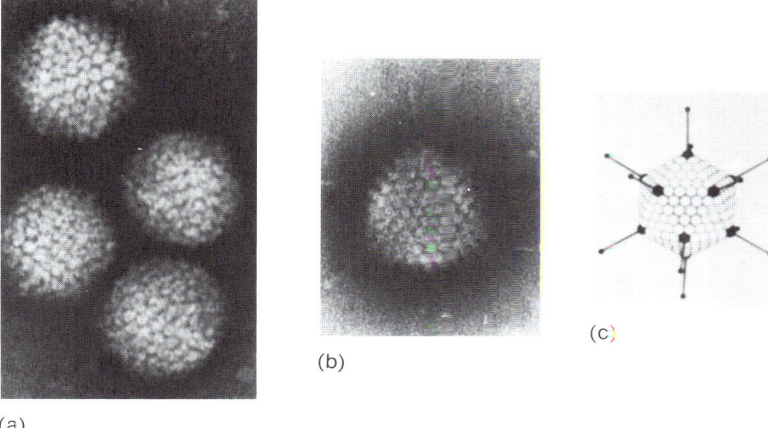

Figure 23-29 Adenovirus particles. (a) Electron micrograph of several virions, viewed by the negative-contrast technique. Each virion consists of 252 capsomers (which appear as white circles in the micrograph) arranged to form an icosahedron. 240 of the capsomers have six neighbors and are called hexons (represented as white spheres in the diagram in (c)); 12 vertex capsomers have five neighbors and are called pentons; each vertex capsomer consists of a base and a projection, known as a spike. (b) An electron micrograph of a single particle, in which the spike can be seen. (c) A plastic model of the virion showing hexons (white spheres), pentons (black spheres), and spikes. [Courtesy of M. Wurtz and Edward Kellenberger.]

and a fourth has an unknown function. The linked protein will be discussed in detail shortly. The virus forms large crystalline arrays in the nuclei of infected cells (Figure 23-30).

Much of the initial interest in adenovirus arose as a result of observations that most adenoviruses, which cause respiratory infections in humans and simians, are oncogenic (tumor-forming) in newborn rats and mice. Furthermore, rodent cells in culture are converted to transformed cells (Chapter 24) by adenovirus. Thus, the adenoviruses have been studied as models of viral carcinogenesis. Splicing of mRNA was first observed in adenovirus. With that, it became clear that these viruses are also useful for studying eukaryotic transcription and regulation of transcription.

The Termini of Adenovirus DNA

Adenovirus DNA is a linear molecule with an inverted terminal repetition that extends precisely to the end of the molecule (Figure 23-31). The size of this sequence ranges within the adenovirus family from 103 to 162 base pairs. There is considerable homology from one virus to another within the first 50 bases, which are exceedingly rich in A·T pairs. This homology, evident throughout the adenovirus family in the face of extensive sequence-divergence elsewhere in adenovirus DNA suggests that the sequence is of

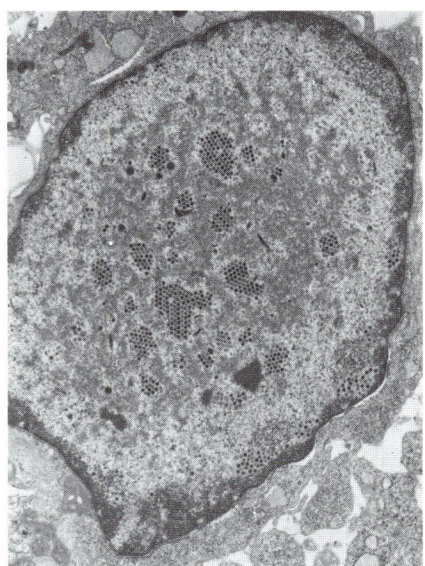

Figure 23-30 Adenovirus crystals in the nucleus of an infected L cell. [Courtesy of Alvin Winter and Keith Brown.]

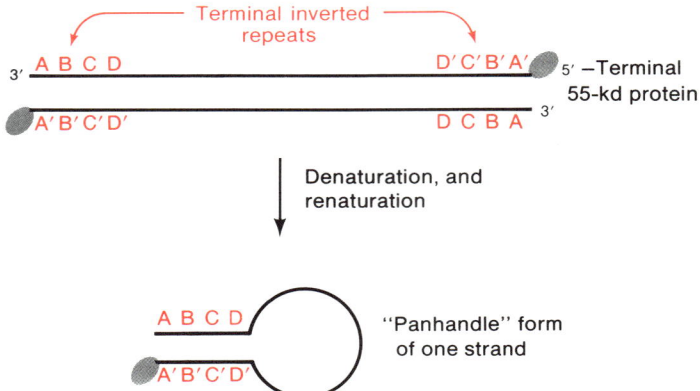

Figure 23-31   The structure of adenovirus DNA and the panhandle form that has been created in the laboratory by successive denaturation and renaturation.

importance; indeed it is probably used in the initiation of DNA replication.

Another striking feature of the terminus of adenovirus DNA was discovered when it was noted that DNA isolated from the virion spontaneously forms circles and oligomers. This is reminiscent of phage λ DNA, whose complementary single-stranded termini can produce these structures, but adenovirus DNA does not have single-stranded termini. Treatment of the DNA with a protease eliminates the ability to form these structures. The explanation for circularization and oligomer formation is that each 5' terminus of adenovirus DNA is covalently linked to a terminal protein (the 55-kd protein) and these proteins can form dimers and even higher multimers (see photograph at the opening of the chapter). The protein-DNA linkage is a phosphodiester bond between the OH group of a serine residue in the 55-kd protein and the 5'-phosphoryl group of the terminal deoxycytidine of each DNA strand. The protein plays an important role in DNA replication. (Terminal proteins have been observed in other DNA-containing organisms. One of the best-understood is in *Bacillus subtilis* phage φ29; other organisms include parvovirus H1, mouse minute virus, and hepatitis B virus.)

Replication of Adenovirus DNA

Studies of the replication of adenovirus DNA have showed the following:

1. Initiation occurs at either terminus of the DNA molecule. Since the two ends are inverted repetitions, they are indistinguishable and initiation occurs at both ends with equal frequency. Furthermore, some molecules are initiated at both ends. An RNA primer is not used for initiation; how this need is bypassed will be discussed shortly.

2. Growth of the daughter strand continues to the other end of the molecule. In the course of this growth, a parental strand is displaced by the growing daughter strand. In a molecule initiated at both ends, both parental strands are displaced. This means that replication intermediates have extensive linear single-stranded branches, a molecular state that is rather unusual in phage and bacterial systems. Such intermediates comprise the bulk of the intracellular DNA during the replication phase of adenovirus (Figure 23-33).

Figure 23-33 shows the currently accepted mode for adenovirus DNA replication. There are two pathways for synthesis. In the major pathway synthesis and displacement go to completion (via a type I intermediate—see the figure), generating a daughter duplex molecule and releasing a parental single strand. The latter then serves as a template for synthesis of a complementary strand, so two semiconservatively replicated molecules result. Complementary synthesis begins at the 3 end of the released strand and generates a type II intermediate, which is observed in electron micrographs as a double-stranded segment to which a long single strand is attached. Continued growth of this intermediate yields the complete double-stranded molecule. The minor pathway results because initiation occasionally occurs at both ends of the molecule. In such a doubly initiating molecule, when two oppositely moving replication forks meet, the parental strands can no longer be held together by base-pairing and will separate, forming two type II molecules whose replication is completed by simple extension; this event is termed "fork annihilation." The bracketed forms n Figure 23-33 will be discussed shortly.

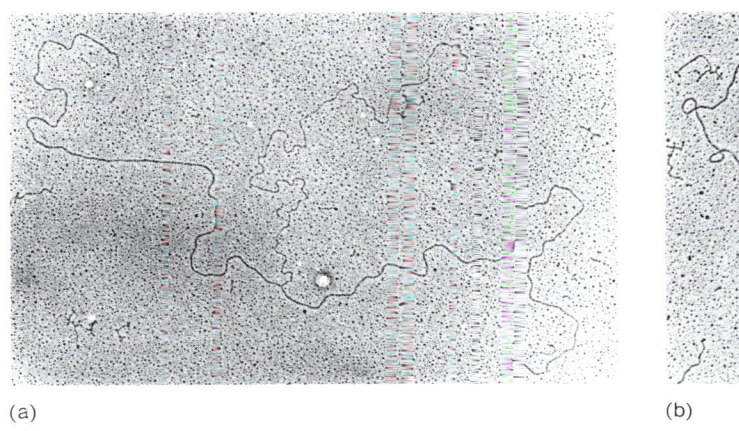

(a)   (b)

Figure 23-32 Electron micrographs of replicating adenovirus DNA (a) A type I molecule in which a single-stranded branch is being displaced from a parental duplex. (b) A type II molecule in which a fully displaced single strand is being converted to double-stranded DNA; the arrow shows the position of the growing 3′ terminus. [Courtesy of Thomas Kelley.]

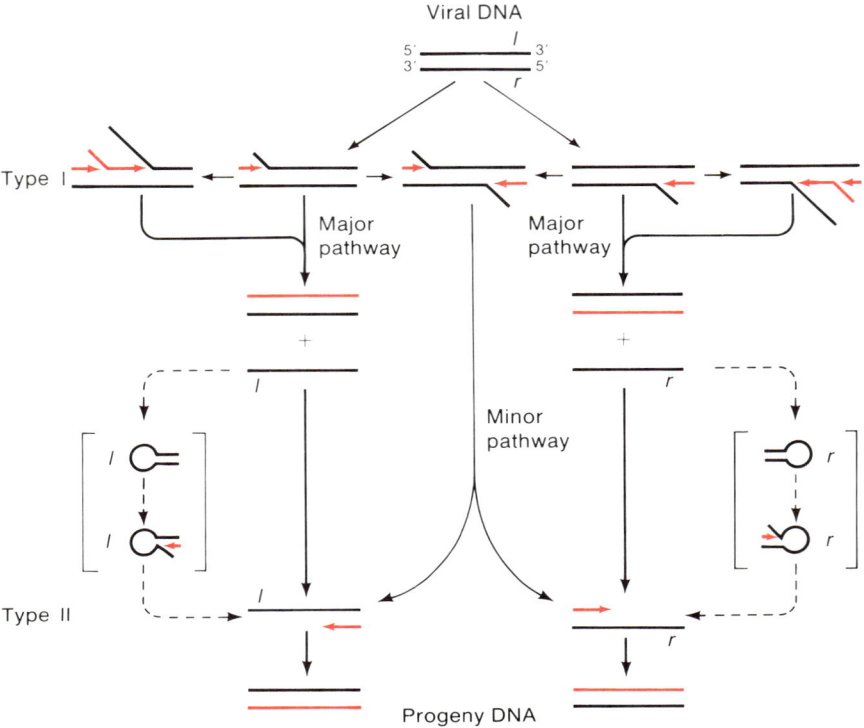

Figure 23-33   Scheme for replication of adenovirus DNA. Heavy arrows indicate the major pathway; light arrows, minor path- way; dashed arrows, hypothetical reactions. See the text for details.

The mechanism of chain elongation of adenovirus DNA differs from the other double-stranded phage and viral systems we have considered so far. The *l* and *r* strands of adenovirus DNA are replicated separately, each in the direction appropriate for DNA polymerase activity. Thus, as the model has been conceived, there is no need for discontinuous replication—that is, for precursor fragments. There is in fact considerable experimental support for continuous replication of both strands.

Initiation of replication requires special consideration for two reasons: (1) initiation at termini creates the same types of problems faced by *E. coli* phage T7 (Chapter 9), and (2) an RNA primer is not used. Furthermore, as the model has been presented, there are two classes of initiation events— those occurring at the termini of double-stranded DNA (generating type I molecules) and those at the 3′ terminus of a single strand (generating type II molecules). It has been suggested that the different kinds of initiation may be more apparent than real for, by virtue of the inverted terminal repetition, a single strand should be able to form a terminal double-stranded region, as shown in the brackets in Figure 23-33. These regions are known as "panhandle" forms. The forms will have a terminal double-stranded region identical to that found in linear double-stranded adenovirus DNA; initiation of a panhandle will therefore be identical to initiation of virion DNA. Although panhandle molecules can be created in the laboratory by successive denaturation and renaturation, they have not yet been isolated from infected cells, and it is not known if they exist at all *in vivo*.

Priming is accomplished by a protein molecule from which the terminal 55-kd protein (which is linked to nongrowing viral DNA) is derived. The 55-kd terminal protein is synthesized from a viral gene as a larger 80-kd protein. The 5′ terminus of a growing single strand is covalently linked to the 80-kd protein. This protein is used in initiation by the mechanism depicted in Figure 23-34. The primary initiation event is the formation of an ester

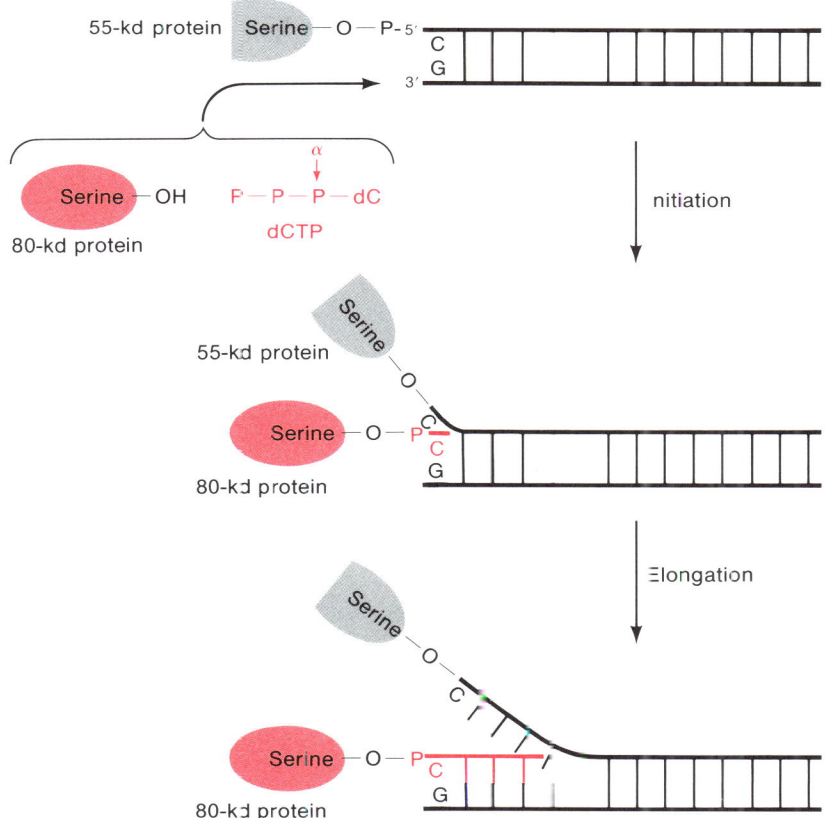

Figure 23-34 The functions of the terminal proteins in initiation and elongation of adenovirus DNA. Possibly, the incoming 80-kd protein to which dCTP is linked binds to the terminal 55-kd protein in an early recognition step.

linkage between the α-phosphoryl group (the phosphate attached to deoxyribose) of dCTP and the β-OH group of a serine residue in a 80-kd protein molecule. This linkage is concomitant with association of the 80-kd protein with the terminus of the parental DNA. Such association might be mediated via binding to a particular base sequence in the template strand or binding to the 55-kd protein on the other parental strand. The latter possibility is believed to be correct since DNA freed of the 55-kd protein is a poor template for initiation *in vitro*. Growth of the daughter chain then proceeds from the 3'-OH group of the dCMP primer. At a later time a viral protein cleaves the attached 80-kd protein, converting it to a 55-kd protein.

Transcription of Adenovirus DNA

The transcription pattern of adenovirus DNA is complex.* Not only are there early and late classes of RNA but, in contrast with the smaller viruses discussed so far, there are at least five early transcription units, each initiating at a unique promoter. Furthermore, these five units are not activated simultaneously, initiation covering a period of about three hours, and several of these units are negatively regulated by their gene products. Finally, each early RNA is spliced prior to translation. The pattern of late mRNA production is even more complex. One example of this complexity is that the synthesis of the majority of late mRNA requires that viral DNA replication begin. In contrast with the case of phage T4 (Chapter 17), it is only the onset of DNA replication, rather than replication itself, that is needed, because if replication is inhibited shortly after it starts, late mRNA continues to be made. Actually, some late mRNA is initiated continually, but, prior to the onset of DNA replication, it is terminated about 8300 bases downstream from the promoter. Late RNA is also extensively processed to form many different mRNA molecules.

The splicing patterns of the adenovirus primary transcripts provide excellent examples of maximizing the information content of eukaryotic DNA, which is always limited by its ability to translate more than one protein from a processed mRNA molecule. Most adenovirus primary transcripts contain several usable genes in one of two respects: (1) they contain overlapping genes that are sorted out by splicing *within the coding region,* or (2) they consist of serially arranged cistrons any one of which can be joined to the 5' leader by deletion of intervening cistrons. Six early and several intermediate and late primary transcripts are converted into more than 17 early mRNAs and more than 19 late mRNAs by use of various polyadenylation sites and of distinct patterns of splicing.

The early regions 1A and 1B exemplify two mechanisms for achieving protein diversity by RNA splicing. The 1A primary transcript utilizes dif-

---

*The reader should not attempt to memorize the details of this section, but should try to get an overall picture.

ferent splicing patterns to delete segments of the coding regions, thereby producing distinct proteins having common amino and carboxyl termini. The relative amounts of each protein changes during the course of infection, so variable splicing is a regulatory mechanism. The 1B region is spliced to create deletions of various start and stop codons, resulting in the synthesis of totally different proteins from different reading frames. Addition of cycloheximide, an inhibitor of protein synthesis, at various times after infection prevents some of the changes in splicing patterns. This inhibition indicates that newly synthesized proteins are positive regulators of splicing. It is not known whether these are host- or virus-encoded proteins since a protein in region 1A stimulates transcription of numerous host genes.

The late mRNAs are also formed in a complicated way. A major late promoter is activated, generating a long primary transcript extending to the end of the DNA. Five major and several minor sites are polyadenylated. A single leader, to be used in all mRNAs, is produced by splicing three segments near the 5′ end of the primary transcript. The leader is then spliced to nearly 20 different sites. The result is the production of five major families of mRNAs, each family sharing the same 3′ terminus and all molecules having the same 3′ leader (Figure 23-35). The longer mRNAs of a family carry several cistrons, but only the 5′-proximal one is translated. The timing

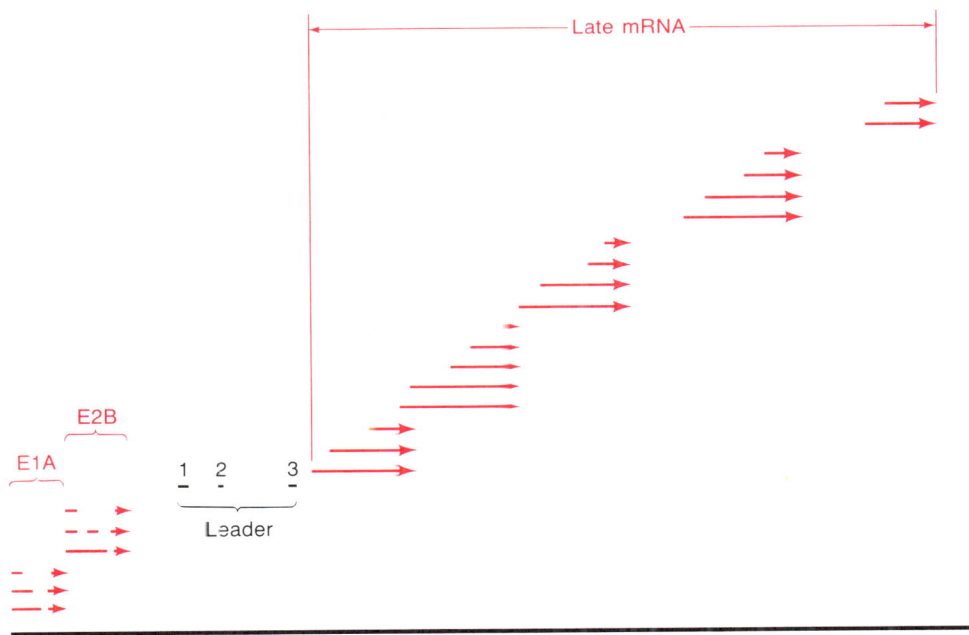

Figure 23-35 Part of the transcription map of adenovirus, showing two classes of early mRNA (E1A and E2B) and five classes of late mRNA. Fragments 1, 2, and 3 are spliced to yield the leader for all of the late mRNAs.

and frequency of the different splices determines the timing of synthesis of each late protein made and puts all proteins under the control of a single promoter. This type of translational control is common in eukaryotic systems but does not occur in prokaryotes; in the latter, the probability of translation of different cistrons in a polycistronic mRNA is determined by the strength of the ribosome binding sites and the proximity of a start codon to a preceding stop codon.

The complete base sequence of adenovirus 2 (35,937 bp) is known. This sequence shows that each intron has the GU . . . AG pattern characteristic of all eukaryotic introns. It might be expected that the sequences surrounding a family of 3′ splice sites that join to a particular 5′ splice site would show common features. However, this has not been observed; at present there is no simple scheme by which splicing patterns can be predicted from the sequence.

INFECTIOUS DNA

Under certain conditions some bacteria are able to be infected with phage DNA, which can initiate an infectious cycle. The process usually requires that the bacteria be converted to spheroplasts by exposure to lysozyme (Chapter 1) before addition of the DNA or that the DNA be added to cells suspended in a solution of $CaCl_2$ (Chapter 5).

Purified nucleic acids of many DNA viruses are also infectious and usually the cells do not require any pretreatment or special solutions to take in the DNA. (When pretreatment is needed, $Ca_3(PO_4)_2$ is found to be more efficient than $CaCl_2$; since the DNA precipitates in the presence of this salt, the treatment is sometimes called the $Ca_3(PO_4)_2$ precipitation technique.) Negative-strand viral RNA is not infectious by itself because of the lack of essential enzymes not present in the host and usually carried into the cell with the capsid—for example, a lack of replicase or reverse transcriptase. Positive-strand viral RNA is frequently infectious. Plant viral RNA is also infectious.

The ability to initiate a life cycle with purified viral nucleic acid has been quite important in experimental virology. Most animal viruses engage in genetic recombination very infrequently, so for many years it was difficult to create the kinds of multiple mutants that have been so valuable in studying phages. However, by using recombinant DNA techniques, multiple mutants of DNA viruses are obtained in a straightforward way. Two mutant molecules are cleaved by the same restriction enzyme, fragments are purified, and then by mixing the appropriate fragments, a DNA containing both mutations can be formed. Infection with purified DNA yields doubly mutant viruses.

Tumor Viruses and Oncogenes

In 1911 Peyton Rous demonstrated that sarcomas (a type of tumor) could be induced in chickens by a virus now called Rous sarcoma virus. Since then many tumor-inducing (**oncogenic**) viruses have been found in animals, though in humans they appear to be rare. Oncogenic viruses are found in almost all families of DNA viruses but among the RNA viruses only some of the retroviruses produce tumors. The tumor-forming ability of the DNA and RNA tumor viruses differ in a significant way:

1. A particular virus may cause a productive infection in cells of one species (permissive cells) and a tumor in another (nonpermissive cells), but an oncogenic retrovirus usually causes tumors in most species in which it can grow.
2. When a culture of nonpermissive cells is infected with a DNA tumor virus, the infection is abortive in most cells and only a small fraction of the population is converted to a tumor cell. On the other hand, for the tumor-causing retroviruses most infected permissive cells are tumor cells.
3. Both DNA tumor viruses and retroviruses integrate viral DNA in the host chromosome. However, the relative probability of forming a provirus is quite different. In the growth cycle of a retrovirus, integration of viral DNA is obligatory (Chapter 23), whereas, in the case of a DNA virus, multiplication of the virus does not require proviral integration, and integration occurs only occasionally.

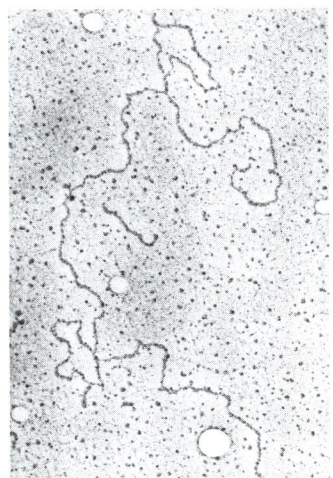

The RNA of Moloney leukemia virus, a cancer-causing mouse retrovirus. The molecule is a dimer, with each monomer joined near the ends (see Figure 23-16). The monomer has an inverted repeat, which forms a stem-and-loop; these two regions enable one to see the symmetry of the dimer. [Courtesy of K. G. Murti.]

We will have more to say about integration shortly.

The first breakthrough in the study of tumor induction by viruses was the discovery of the **transformed cell.** These are cells that differ in a variety of properties from normal cells and can often form a tumor if injected into an appropriate animal. We begin this chapter with a description of these cells.

TRANSFORMED CELLS

Animal cells, with the exception of some specific blood cells and cells of the immune system, grow in culture only on surfaces. If a piece of a typical tissue is placed on a glass or specially treated plastic surface, various types of fibroblast cells grow out from the tissue mass. Alternatively, the tissue can be treated with the protease trypsin, which separates the tissue into individual cells; in a suitable medium, the fibroblasts in the suspension adsorb to the surface and grow. (Most of the other cell types grow very poorly or not at all.) Such a culture, which is called a **primary culture,** has the property that the cells divide only for a few generations and then most of the cells die and disintegrate; this is called **crisis.** If the few remaining cells are maintained in culture for many months (during which time there is little or no division)*, ultimately some growth begins until a culture is developed that can grow. Such a culture is called an **established cell line.** This new culture differs from the original cells in the tissue in many ways, of which one is especially noticeable: they have become immortalized in the sense that they can grow indefinitely, in contrast with the primary cells, which passed through a few generations and died. These cells are clearly not normal, but nonetheless are used in most studies of cell growth.

If cells of a typical established cell line are placed in a culture vessel, growth and multiplication occur until a confluent monolayer of cells is formed; then growth stops. This cessation of growth is called **density-dependent growth,** formerly called **contact inhibition.** When growth stops, most of the cells are in an orderly array (Figure 24-1(a)). These cells can be removed from the surface by a brief treatment with trypsin. If the cells are then placed on a new surface and fresh growth medium is provided, again the cells grow to form a confluent monolayer.

If an established cell line that is nonpermissive for growth of a particular tumor virus is infected with that virus, the result of culturing such cells is quite different from that of uninfected cells. In this case, when confluency is reached, some of the cells, which have been growing in somewhat disordered arrays, continue to grow and ultimately form a jumbled, multilayered mass on top of the confluent background. If these cells are removed and dispersed by exposure to trypsin and then allowed to grow on a fresh surface, the resulting culture shows disordered, multilayered growth (Figure 24-1(b)). Cells having this property are said to be **transformed.** Transformation may be brought about in other ways also—for example, by exposure

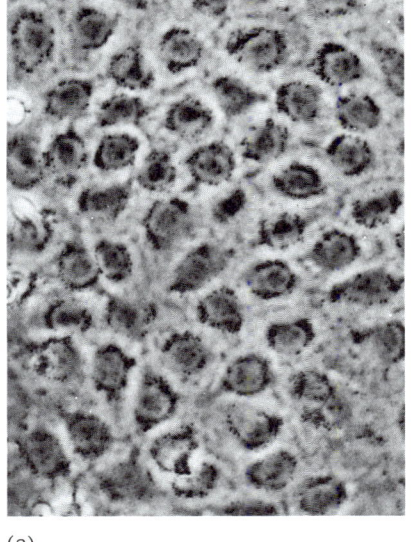 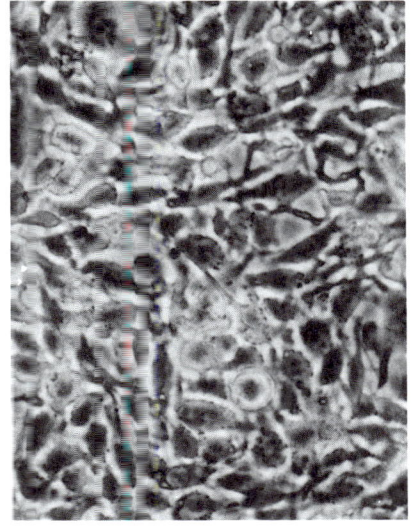

(a) (b)

Figure 24-1 Confluent layers of (a) normal mouse 3T3 cells and (b) 3T3 cells transformed by polyoma virus The transformed cells not only have a different shape, but also grow in less-organized fashion than normal cells [Courtesy of Walter Eckhart.]

to radiation, by chemical carcinogens (to be discussed later), and even by forcing cells to grow for many generations at high cell density.*

Transformed cells have several other properties that distinguish them from normal cells, namely:

1. They can grow in suspension (rather than only on a surface).
2. They require a lower concentration of blood serum in the growth medium.
3. They have an altered surface and a more fluid cell membrane.
4. The number of chromosomes always exceeds the normal diploid number; they may be triploid, tetraploid or have extra copies of individual chromosomes.

However, the most outstanding property of transformed cells is that *they often form tumors if the cells are injected into an animal of the same species* (i.e., transformed mouse cells into a mouse). A clue to the molecular basis of transformation comes from a universal feature of virus-transformed cells: each transformed cell contains integrated viral DNA.

*This difference in density can be quite small. For example, if during the early stages of establishment of a mouse cell line, the cells are maintained at a concentration of 3×10^5 cells/ml, a "normal" cell line results; if the cells are maintained at 1.2×10^6 per ml (just four times as great, but in a fairly crowded state), the established cell line consists of transformed cells.

DETECTION OF INTEGRATED VIRAL DNA

In the study of a virus-induced transformation system, methods are available for demonstrating that a transformed cell contains integrated viral DNA and also for determining the number and locations of the DNA sequences. These methods are the topic of this section.

The first indication that virus-transformed cells contain viral DNA came from hybridization experiments: ^{3}H-labeled RNA transcribed *in vitro* from purified polyoma DNA hybridized with DNA isolated from polyoma-transformed cells but not with DNA from nontransformed cells. The viral DNA could either have been integrated or in plasmid form (as we saw in Chapter 18 for the two classes of lysogenic phages), and a simple experiment distinguished these alternatives. Cellular DNA was isolated from polyoma-transformed cells and then broken at random positions into fragments with a broad range of total molecular weights, generally larger than polyoma DNA (molecular weight, 4.8×10^6). The fragments were fractionated by zonal centrifugation and each fraction was tested for the presence of polyoma DNA by denaturation and hybridization with ^{3}H-labeled polyoma RNA. If the viral DNA was a plasmid, all of the polyoma DNA would sediment as a single component with a molecular weight of 4.8×10^6. However, if the polyoma DNA was integrated and the cellular DNA was broken at random, a fragment containing polyoma DNA could be of any size; it was found that ^{3}H-labeled polyoma RNA would hybridize to fragments of any size, so polyoma DNA was integrated.

More elegant experiments done with SV40 virus demonstrated conclusively that in cells transformed with SV40 virus SV40 DNA is integrated. These experiments also enabled one to count the number of proviruses and to determine whether integration occurs at a unique position in the chromosome and in the viral DNA. The method in each of these experiments was to fragment DNA from transformed cells with a restriction enzyme, separate the fragments by gel electrophoresis, and identify the fragments containing viral DNA by Southern blotting with ^{32}P-labeled denatured viral DNA as a probe.

The first experiment (Figure 24-2) showed that the viral DNA was integrated at a single site in a transformed cell. Total DNA from a transformed cell was digested with a restriction enzyme that makes no cuts in SV40 DNA but many cuts in cellular DNA, and was then electrophoresed. The figure shows the data expected for an SV40 plasmid and for a provirus located in one or two different positions. The observed data corresponded to integration at a single site. This experiment has been repeated with DNA isolated from many different clones of transformed cells. The electrophoregrams have usually had a single band, though a few of the clones had two bands. Thus, one may conclude that SV40 DNA is usually integrated at only one site but that some transformed cells contain two SV40 sequences at different sites. When many clones transformed by a particular virus were

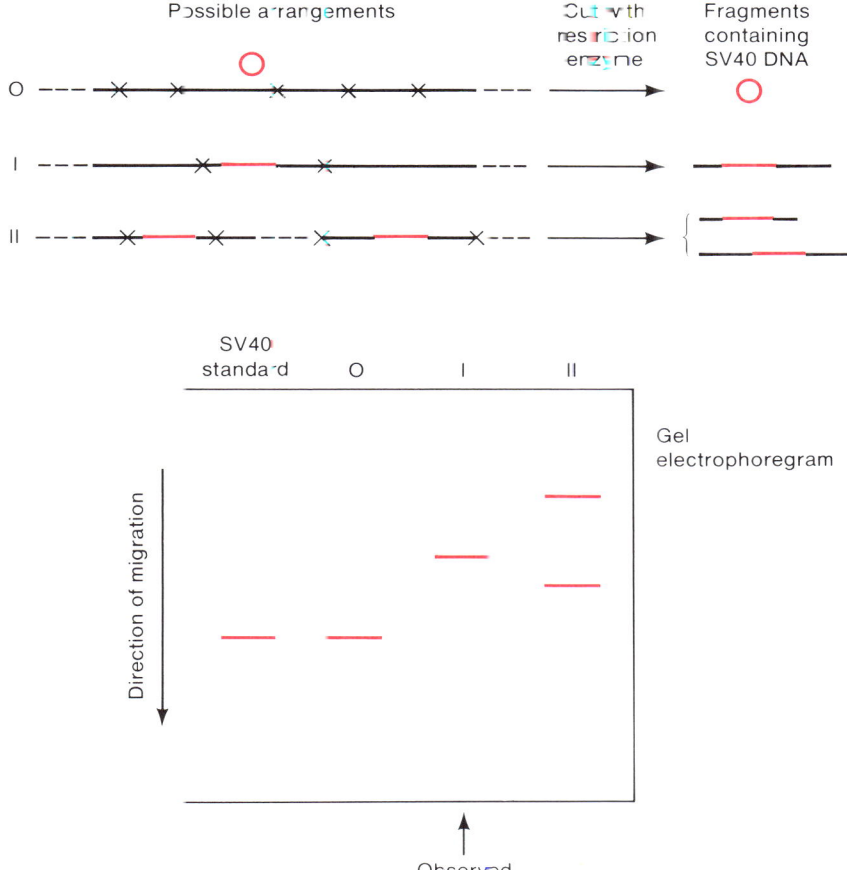

Figure 24-2   An experiment showing that SV40 DNA is integrated in a transformed cell. The black solid lines represent cellular DNA; the dashed lines represent large segments that are irrelevant to the experiment; SV40 DNA is shown in red. The x's are sites of cutting by the restriction enzyme. If the SV40 DNA were a plasmid (0), electrophoresis would show a single band migrating at the same rate as pure SV40 DNA. If a single provirus were present (I), it would be expected that an electrophoretic band would move less far (have higher molecular weight) because the DNA would be linked to cellular DNA. If there were two proviruses (II), there would be two bands, because in general the restriction sites would be spaced differently around the two proviruses. A supercoiled SV40 DNA standard is used because its migration rate is almost identical to that of linear, unit-length SV40 DNA.

examined, it was found that the band or bands of one clone were very rarely at the position for other clones, as shown in Figure 24-3. This means that there are many potential integration sites; possibly, the sites are randomly disposed.

If a restriction enzyme that makes several cuts in the provirus is used, it is possible to determine whether insertion occurs at a particular point in

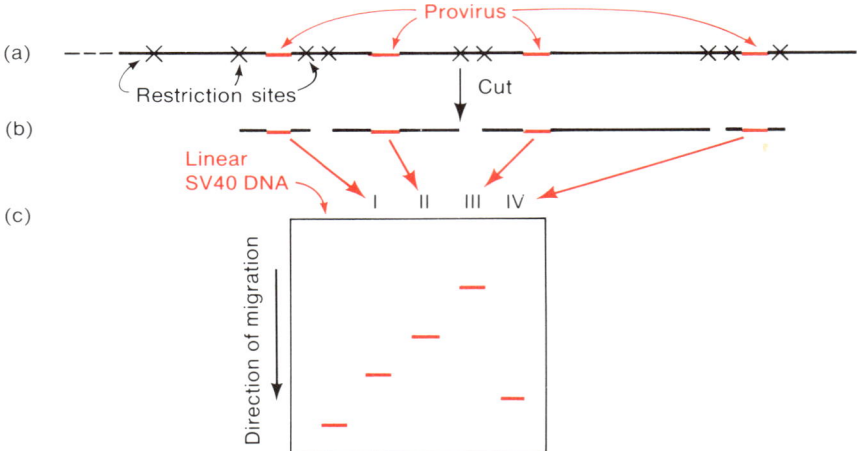

Figure 24-3   Position of bands in a gel for different provirus locations. (a) A portion of the cellular DNA showing four possible positions of a provirus in four different transformed cell lines. (b) Restriction fragments containing a provirus for the four dif- ferent cell lines. The fragments have differ- ent sizes because the proviruses are not always the same distance from the nearest restriction sites. (c) The gel. Each band (I–IV) represents a separate clone of trans- formed cells.

the viral DNA. How this is done is shown in Figure 24-4. DNA is isolated from many transformed clones and digested. Because the enzyme cuts the viral DNA, fragments smaller than the size of uncut SV40 viral DNA will always be present. These fragments are a subset of the fragments obtained by digesting circular viral DNA with the same enzyme. If the integration event occurs at a unique site, the same fragments will be found for all clones. If the site is variable and the enzyme makes n cuts, $n-1$ fragments will be seen and there will be n subsets. This is the observed result for SV40-trans- formed cells, which indicates that SV40 does not have a fixed site of integration.

It is clear that SV40 DNA can be integrated at many different sites in cellular DNA. An interesting question is whether these sites all reside on the same chromosome. The answer, which is that they do not, was obtained by applying the technique of **cell fusion.** Sendai virus has the property, which is extremely useful in the laboratory, of inducing fusion of Sendai- infected cells.* Fusion can occur between cells of both the same and of different species—for instance, mouse and human cells can fuse. Fused mouse- human cells randomly lose human chromosomes during the first few cell divisions until stable cell lines result having a complete set of mouse chro-

---

*Sendai virus is an enveloped DNA virus that does not kill its host but promotes such a weakening of the host cell membrane that two infected cells frequently merge to form one fused cell or **heterokaryon.** As a precaution against uncontrolled infection in human subjects, nonviable or inactive Sendai virus is used, for it promotes fusion also.

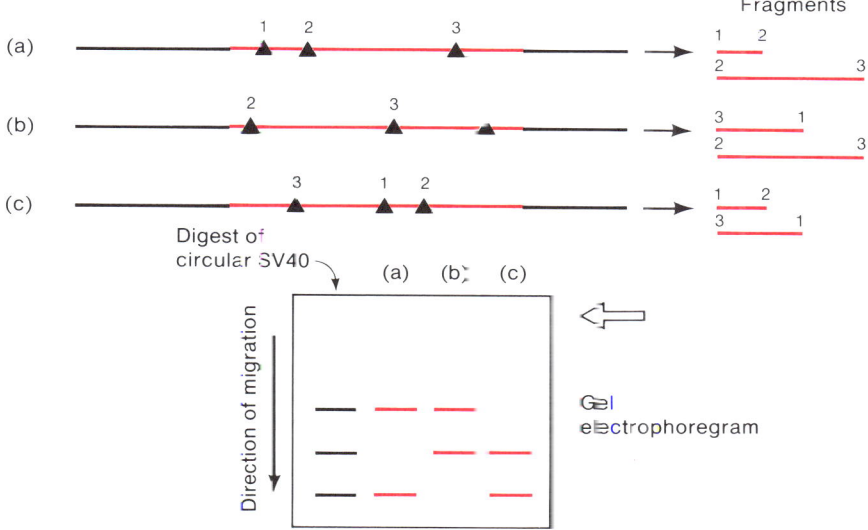

Figure 24-4 Hypothetical electrophoregrams of restriction digests of the DNA from three SV40-transformed cell lines; it is assumed that the integration site in the viral DNA can vary. If the site were constant, the patterns would be identical. The two bands containing cellular DNA are not shown. These molecules would usually be much larger than the bands shown and would be in the region indicated by the open arrow.

mosomes but only a few human chromosomes. Normal mouse cells have also been fused with SV40-transformed human cells and clones resembling mouse cells but with the transformed phenotype have been isolated. Mouse and human chromosomes, as well as individual human chromosomes, are easily distinguished by their morphology during metaphase, so transformed mouse clones can be examined to determine which human chromosomes are present. Analysis of many clones, each possessing one ore more different human chromosomes, has clearly indicated that SV40 DNA can be located in many chromosomes.

STRUCTURE OF INTEGRATED VIRAL DNA

When phage DNA integrates into a bacterial chromosome, the resulting prophage is a faithful copy of the phage DNA, though it may be permuted. The main point is that phage DNA sequences are neither lost nor gained during integration. We have seen in Chapter 23 that when a retrovirus integrates, a few bases at either end of the virion base sequence are lost and a large segment of viral DNA (the LTR) is duplicated. Thus, integration of retroviral DNA is accompanied by subtle rearrangements; however, coding sequences are not lost, which is essential since, with retroviruses, only integrated DNA can be a source of progeny virus.

The situation is quite different with the DNA-containing tumor viruses. Integration is not an essential part of their life cycles and in fact might only occur when an infection is abortive. Thus, when integration occurs, there is no selective pressure to retain all of the coding sequences, because the ability to integrate does not have survival value for the virus. In fact it has been found that integration is usually accompanied by loss of genetic information: segments of DNA may be deleted, inverted, or scrambled. The significance of these changes is unknown, and clearly integration with these viruses is a very complex process.

The most extreme case of loss of genetic information that has been observed occurs with adenovirus. Clones transformed by adenovirus always contain integrated viral DNA, but the viral sequences are never an intact adenovirus DNA molecule. Analysis of many clones shows that different clones contain different viral sequences. However, a particular small segment of adenovirus DNA is always present in an adenovirus-transformed cell. This important finding suggests that a particular gene or set of genes is responsible for the transformed phenotype.

POSSIBLE CONSEQUENCES OF INTEGRATION

There are several ways that viral integration can lead to transformation:

1. Integration can occur within a host gene, thereby inactivating the gene; if the gene is regulatory, deregulation of some cellular process can occur.
2. Integration can separate a gene from its promoter and thereby prevent expression of the gene.
3. Integration can put a host gene under the influence of a viral promoter and cause overproduction of the host-gene product.
4. A viral gene product, expressed in the provirus, can directly or indirectly cause transformation.
5. Integration can ultimately lead to chromosome breakage and rearrangement, which can alter gene expression.

It is likely that items 3–5 are important.

Tumor viruses differ from lysogenic phages in that they lack repressors and, in fact, transcription of the integrated DNA may occur continually. A phage repressor is needed in a lysogen in order to prevent activation of functions that kill the host bacterium. With retroviruses this is unnecessary since these viruses do not kill the host cell. The DNA tumor viruses can transform without a repressor for another reason. All oncogenic DNA viruses have an early and a late phase of transcription and, for those viruses having functions that are lethal to the host cell, these functions are confined to the late phase. Recall that transformation with DNA viruses occurs only in nonpermissive cells; these are cell types in which late transcription or processing of late RNA does not occur.

Failure of a virus to kill a cell is not sufficient to explain the complex cellular changes accompanying transformation. Thus, for many years there has been an active search for tumor-inducing viral genes. These genes are described in the following section.

VIRAL ONCOGENES

In the preceding section it was pointed out that viral transformation could result from the effects of viral DNA on the expression of cellular genes or from products of a tumor-inducing viral gene. Examples of both influences are known. In this section we will examine the oncogenic viral genes of retroviruses, which are part of a general class of tumor-causing genes called **oncogenes.**

Protein Kinases and Transformation by Retroviruses

In 1970 temperature-sensitive transformation mutants of Rous sarcoma viruses (RSV) were isolated. Mutant viruses grew productively equally well at both permissive and nonpermissive temperatures yet they transformed only at permissive temperatures. Moreover, if cells transformed at permissive temperatures were shifted to a nonpermissive temperature, the transformed cells reverted back to normal morphology after one or two days. Since nontransformed, transformed, and revertant cells all produced the same virus, these experiments clearly dissociated the replicative function from the transforming function of RSV. At the same time, nonconditional deletion mutants of RSV were isolated that had lost the transforming function altogether. The nontransforming mutants replicated and formed infectious particles that were physically like wild-type RSV. The gene responsible for the transforming function was called *src* (for *sarcoma-producing*). The location of the gene is shown in Figure 23-20. Comparison of the RNA of wild-type RSV and the nontransformable mutants proved the variants to be *src*⁻ deletion mutants that lacked about 1600 bases of RNA. This provided the biochemical basis for the analysis of the *src* gene. An exciting experiment confirmed the fact that *src* is alone responsible for transformation. A restriction fragment was isolated that contained only the *src* gene. This fragment was added to cultured chick cells using a variant of the $CaCl_2$ technique. Some of the cells were transformed and these were found to contain the *src* fragment incorporated in a chromosome. Similar transformation genes have been found in several other retroviruses.

Probably the most unusual feature of the *src* gene is that a single virus-encoded protein is able to cause the large number of effects associated with the transformed phenotype. The explanation is that the *src*-gene product is an enzyme capable of modifying a large number of proteins.

The product of the *src* gene is a single protein called pp60-v-src; the 60 refers to the fact that its molecular weight is 60,000, and v indicates that the gene is derived from a virus (a distinction that will soon be seen to be important). The protein is an enzyme that is able to transfer a phosphate group from ATP to the amino acid tyrosine in cellular proteins. An enzyme that phosphorylates a protein is called a **protein kinase.** Most enzymes of this sort act on the amino acid serine; the *src* enzyme is unusual in its preference for tyrosine and is called a **tyrosine-specific protein kinase.** Similar enzymes have been found to be the products of the transforming genes of other retroviruses.

Protein kinases are common enzymes in animal cells. They function mainly as regulators of the activity of other proteins. For some proteins, phosphorylation brings out enzymatic activity of the protein, whereas for other proteins it has an inactivating effect. A given protein kinase is usually able to phosphorylate a large number of proteins, so the regulatory activity of these kinases can be quite far-reaching. It is easy to imagine that the synthesis of an enzyme like the *src* protein kinase could change the growth properties of a cell in a dramatic way, for it could activate or inactivate a large number of enzymes, some of which could be involved in growth regulation. Studies with temperature-sensitive mutants of oncogenes in several retroviruses have shown that tyrosine kinase activity and transformation go hand in hand. At permissive temperatures the tyrosine kinase activity is evident, cells are transformed, and some cellular proteins contain phosphorylated tyrosine. At nonpermissive temperatures there is neither tyrosine kinase activity nor transformation, and there are no phosphorylated tyrosines.

It is obvious that to understand the significance of the kinase activity, the essential proteins that are phosphorylated must be identified. Many proteins that are phosphorylated by the *src* kinase have been isolated, but the functions of most of these proteins have not yet been identified. Some information about these proteins will be given in a later section.

Oncogenes in Cancer Cells

Recognition that a single gene such as *src* could cause transformation led to a search for oncogenes in cancer cells obtained by surgical excision from cancer patients. Recall that the *src* gene can cause transformation in two ways: by integration into a chromosome of an entire RSV DNA molecule or a fragment of DNA containing only the *src* gene. The latter result suggested that it might be possible to find an oncogene in a cancer cell by testing DNA fragments isolated from human cancers for their ability to transform a cultured normal cell.

The first experiment utilized DNA from a line of *human* bladder cancer cells. The DNA was fragmented and used to transfect a culture of *mouse* cells (NIH 3T3 cells, an established cell line used repeatedly in such studies). The result of this simple experiment was that some of the mouse cells were transformed, which indicated that an oncogene was present in the bladder cancer cells. In a second experiment the DNA fragments were separated into a large number of fractions and each was tested for transforming activity. It was found that only a single DNA fragment could induce the transformation. Therefore, this fragment contained the human bladder cancer oncogene. Another set of experiments used human lymphoma cells, and a lymphoma oncogene was isolated.

Once the human bladder oncogene was discovered, the question of where it came from was raised immediately. One theory of cancer, the mutational theory, predicts that the bladder oncogene should be a mutant form of a normal gene. Indeed this proved to be the case, for it was possible to isolate from normal bladder cells a DNA fragment whose coding sequence contains a base sequence that differs from the bladder oncogene in only one base: a guanine in the normal DNA is replaced by a thymine in the oncogene. Thus, for this particular human bladder cancer, the oncogene is a mutant form of a normal gene. Similar results have been obtained for several, but not all, oncogenes.

In the next section we will see that the *src* gene has a cellular counterpart.

The Cellular Counterpart of *src*

Southern blotting experiments, using genomic DNA of normal chick cells and radioactive *src* DNA as a probe has indicated that the *src* gene of RSV has a counterpart in normal cells. A similar search for *src*-like genes in other organisms showed that a similar base sequence is present in the DNA of other birds, as well as fishes, mammals, and humans. In fact, it appears that all vertebrates possess DNA sequences similar to *src*. To distinguish the two classes of sequences, the viral sequence has been renamed *v-src*, and the cellular sequence is called *c-src*. The protein made by *c-src* is also a tyrosine kinase, but its enzymatic activity is less than 10 percent that of the *v-src* kinase.

Two alternative possibilities have been suggested for the origin of *c-src* and *v-src*: (1) *c-src* arose in a wide variety of organisms by repeated infection of these organisms with retroviruses through evolutionary time, that is, *c-src* is a derivative of *v-src*; (2) *v-src* was somehow picked up by the virus from cellular DNA in an event that gave the virus tumor-forming ability (that is, *v-src* is a derivative of *c-src*). The fact that a *src* deletion mutant of RSV can multiply normally suggests that the second explanation may be correct, and study of the structure of *c-src* and *v-src* provided verification. Retroviral genes do not contain introns. Analysis of the base sequences of the cellular and viral *src* genes has shown that *c-src* sequences are always interrupted and *v-src* sequences are never interrupted. Since there are many examples of loss of introns and none of acquisition of introns, it has been concluded that *v-src* was acquired from a cell containing *c-src* at some time in the past. Certain experiments suggest that a DNA copy of fully processed *c-src* mRNA was an intermediate. The presence of *c-src* in a wide variety of vertebrates suggests that *c-src* is an essential gene of these animals, probably having a regulatory role in the development of normal organisms.

Other retroviral oncogenes have counterparts in cellular DNA. In each case, the cellular genes are interrupted by intervening sequences and hence have the structure of typical cellular genes. It seems to be a general phenomenon that retroviruses picked up their oncogenes from normal (and probably essential) cellular genes. Such cellular genes are called **protooncogenes.**

The bladder oncogene differs from its protooncogene by a single base. It is important to know whether this is a general phenomenon or if the sequence of the cellular counterpart could be identical to that of the oncogene. In fact, both types of pairs of oncogenes and protooncogenes have been observed. When the sequences do differ, one may conclude that the protooncogene does not by itself cause transformation and a mutation produces the cancerous form. However, when the sequences are identical, one must explain how the gene can be harmless in a normal cell, yet so dangerous in a cancer cell. A variety of hypotheses have been made, only some of which are supported to any extent by experimental results.

1. A single copy of the protooncogene is harmless, but when a cell has one or more extra copies (for example, provided by the integrated viral DNA), the increased amount of gene product puts cell growth and division out of control. This is not unreasonable, especially since quite often several copies of the viral DNA are integrated.
2. The amount of protein made by the viral oncogene is much greater than that made by the cellular gene, either because the viral gene is regulated in a different way or because the viral gene has a different location from the cellular gene.

These possibilities will be discussed in a later section of this chapter.

Carcinogens and Oncogenes

At one time it was believed that all cancers arose from *viral* genes that are contained in normal cells and that are activated by chemical carcinogens and radiation. The evidence supporting this view was that many agents that induce transformation also cause retrovirus particles to appear in animal cell cultures. The conclusion was that many cells harbor latent retroviruses and that these viruses are activated by carcinogens. Whereas it is possible that some cancers are caused in this way, after many years of trying to prove this hypothesis, strong support has never been obtained. Furthermore, a great many cancers, especially human cancers, show no evidence either for virus production or participation of viruses in any way, and in a large number of experiments in which normal cells are treated with carcinogens and transformation occurs, virus particles are not seen.

Many carcinogens are mutagens (Chapter 11), which suggests that a carcinogen might act by causing a mutation in a protooncogene, thereby converting it to an oncogene. As a test of this hypothesis, the following experiment was recently done. Rats were treated with the carcinogen nitrosomethylurea, and mammary carcinomas formed in many of the rats. Tumors were excised, and the DNA was isolated from the cancer cells and fragmented. This DNA was then added to cultures of mouse 3T3 cells (as in the original experiment that demonstrated the existence of the bladder oncogene) to test for transformation, and transformed cells resulted. Analysis of the DNA used showed that the transforming DNA contained a rat mammary oncogene. A counterpart was sought and found in normal rat tissue. Determination of the base sequences of the mammary oncogene and the cellular gene gave two remarkable results: (1) the mammary protooncogene and the bladder protooncogene had the same sequence, and (2) the mammary oncogene differed from its normal counterpart by a single base, which was at the same position as the altered base in the human bladder cancer oncogene. Thus, it could be concluded that (1) the cancers in the human bladder and in the rat mammary gland were caused by mutations of the same gene, and (2) the carcinogen that caused the mammary cancer in the rat had mutated the normal gene. Since this experiment was done, numerous examples of mutation of a gene by a carcinogen to form an oncogene have been observed.

Relation Between Animal and Viral Oncogenes

If oncogenes are altered forms (or aberrantly expressed forms) of normal genes that usually participate in growth regulation, it seems likely that the number of different oncogenes should be fairly small. Thus, we should expect that if a large number of tumor viruses and tumors are screened for oncogenes, the same ones should appear repeatedly. In the preceding section, one example of this was seen: the oncogene in human bladder carcinoma is the same as that in rat mammary carcinoma. Furthermore, recall that the two oncogenes were obtained from *human* and *rat* tumors but tested in *mouse* 3T3 cells; since these oncogenes transformed the mouse cells, the ability of an oncogene to transform cells crosses species barriers (at least in mammals), suggesting that growth regulation in all mammals may have common features. Note also the possibility, raised by the presence of *c-src* in normal tissue, that viruses and normal cells may carry the same set of oncogenes and protooncogenes.

To see whether the number of oncogenes is indeed small, oncogenes were isolated from a large number of retroviruses and from cancerous human tissue, and their counterparts were isolated from normal tissue. The base sequences of all of these genes were determined. The first evidence for commonality came from a study of the oncogene originally detected in Kirsten sarcoma virus, which causes sarcomas in rats. A similar oncogene, named **ras** (*rat sarcoma*), was also found in Harvey sarcoma virus. Except for a single base change, the *ras* sequence was the same as that reported above for human bladder carcinoma and was also found in human lung and colon carcinomas. A similar sequence but with a different base change was found in a human neuroblastoma and a fibrosarcoma. This variant was called *N-ras*. Studies of a large number of viruses and tumors (human cancers of colon, lung, pancreas, skin, breast, brain, and white cells, and a variety of rat and mouse tumors) has led to the identification of more than 20 different oncogenes, which are listed in Table 24-1. Their curious names are all abbreviations or acronyms for the virus or tissue of origin. Analysis of the activities

Table 24-1 Some oncogenes and their sources

Oncogene	Source*
abl	Mouse leukemia, human leukemia cells
B-lym	Chicken and human lymphoma cells
erbA	Chicken leukemia
erbB	Chicken leukemia
ets	Chicken leukemia
fes	Cat sarcoma
fgr	Cat sarcoma
fms	Cat sarcoma
fos	Mouse sarcoma
fps	Chicken sarcoma
Ha-ras	Rat sarcoma; human and rat carcinoma cells
Ki-ras	Rat sarcoma; human carcinoma, sarcoma, and leukemia cells
mil	Chicken sarcoma
mos	Mouse sarcoma, mouse leukemia cell
myb	Chicken leukemia and human leukemia cells
myc	Chicken leukemia; human lymphoma cells
N-ras	Human leukemia and carcinoma cells
raf	Mouse sarcoma
rel	Turkey leukemia
ros	Chicken sarcoma
sis	Monkey sarcoma
ski	Chicken sarcoma
src	Chicken sarcoma
yes	Chicken sarcoma

*Indicates a retrovirus unless indicated as a cell.

of their gene products shows that several are tyrosine kinases (like *src*), two are threonine kinases, a few are DNA-binding proteins, and some have features in common with growth factors (discussed later in the chapter).

ONCOGENES AND THE MULTISTEP INDUCTION OF CANCER

In the experiments and tests described so far, the presence of an active oncogene seems to be able to confer cancerous properties on a cell. This fact is in conflict with a large body of evidence that induction of cancer requires at least two steps, initiation and promotion. Whereas these steps are not unambiguously defined and hypotheses about their molecular bases have been changing rapidly, **initiation** is believed to involve acquisition of a mutation, and **promotion** is a later stage in which the mutation is expressed. Loss of growth control, which is essential for cancer induction, may result from promotion but probably requires additional steps. The key to understanding how one oncogene can cause transformation is to note that all tests for oncogenic potential have been made with an established line of cells obtained from a mouse, namely, NIH 3T3 cells. This cell line, which had been maintained for thousands of generations, was chosen because it had been used for many years to study viral transformation and chemical carcinogenesis. The 3T3 line became popular for these studies because it was easily transformed. A little thought should suggest to the reader that 3T3 cells probably already experienced one or more early steps, a logical conclusion since 3T3 cells are certainly abnormal in that they grow indefinitely in culture.

That multiple steps are involved in cancer induction can be examined in culture by determining whether a known oncogene could transform other cells that had not been immortalized. Unfortunately, studies of this sort have given various results, and depend on the criterion used to identify the transformed state. For example, in one study with *freshly isolated* normal skin cells (not yet immortalized) of a mouse the cells could not be transformed in culture with purified *ras* DNA, using contact inhibition and ability to grow without a surface as a criterion for transformation. If the skin cell culture was first treated either with x rays or a chemical carcinogen, and a few immortalized cell lines were isolated from these treated cells and then transfected with *ras* DNA, transformation occurred. This result implies that *ras,* or more precisely, the *ras* gene product, determines a later step in oncogenesis. However, in other experiments in which primary cells were transfected with *ras* and ability to grow in low concentrations of serum was used as a criterion for transformation, the cells were transformed. At present, all that can be said about *v-ras* is that it does trigger one or more of the steps in oncogenesis and that the phenomenon is poorly understood.*

---

*The study of oncogenes is proceeding at such a rapid and frenetic pace, because of its great importance, that theories seem to arise and fall at an unprecedented rate.

Some retroviral oncogenes are responsible for cancer in animals yet do not test as oncogenes with mouse 3T3 cells. This seems to apply to an oncogene called *myc,* found in a chicken leukemia virus. *Myc* does not transform 3T3 cells, yet induces transformation of other cell lines. However, mouse primary skin cells can be transformed by sequential treatment with both *myc* DNA and *ras* DNA. In such an experiment, *myc* substitutes for the x rays or the chemical carcinogen in the experiment described in the previous paragraph. This results indicates that there are at least two steps in oncogenesis and that the number of steps required may vary from one cell type to another. However, once again the main conclusion is that the phenomenon is not really understood very well.

It is important to note that the oncogenes that transform NIH 3T3 cells do not all have the base sequence of *ras*. Since a variety of oncogenes transform 3T3 cells, it is clear that many different events in a cell can cause conversion of this cell line to cancer cells.

ACTIVATION OF ONCOGENES

We have pointed out that counterparts to oncogenes are found in normal cells and that often the distinction between a normal gene and an oncogene is a single base pair, in which case the oncogene is a mutant form of the normal gene. Although we have no information about the biological function of the oncogene products, it is worth considering the kinds of events that might take place:

1. A single base change could cause activation by increasing or decreasing the activity of the oncogene protein. For example, if the protein were a regulatory gene whose function is to turn on synthesis of a growth protein, increased activity of the oncogene product could result in excess synthesis of the growth protein. On the other hand, the function of the oncogene product could be to turn *off* synthesis of a growth inhibitor; in this case, inactivation of the oncogene protein would eliminate its inhibitory activity, and excess growth protein would be made.

2. An oncogene protein could be a growth-stimulating molecule, whose activity is regulated by other (unknown) proteins. A change in the oncogene protein (caused by the base-pair change) could cause a loss of its ability to respond to the regulator; in this case, the growth-stimulating activity of the oncogene product could become excessive.

3. The oncogene protein might be unstable, with its activity regulated by the relative rates of synthesis and degradation of the protein. An alteration in its structure might prevent or reduce its rate of degradation, without affecting its biochemical activity. In this case, its total activity in the cell would increase.

When an oncogene does not differ from its cellular counterpart, the hypothetical mechanisms for oncogene action listed above do not necessarily apply. In studies of cell biology it has become abundantly clear that certain biological functions are not determined simply by the presence of a particular gene product, but by the concentration of the product. This raises the possibility that the conversion of a protooncogene to an active oncogene, thereby inducing the cancerous state, is occurring simply by changes in concentration. We now consider how a change in concentration might come about and also provide some evidence that this does occur in some cases.

The amount of a gene product contained in a cell is determined by its rate of synthesis (also, its rate of degradation, but this possibility will not be considered here). The rate of synthesis of most proteins is determined by the rate of production of mRNA. A variety of features of DNA structure affect the synthesis of mRNA—for example, promoter strength and regulatory regions. Clearly, the activity of a gene can be changed by altering the adjacent promoter and/or regulatory sequences, and indeed, this phenomenon has been observed: some oncogenes differ from the normal counterparts only in that base changes exist in the promoter and these changes presumably alter the rate of RNA synthesis.

Another important mechanism by which the expression of a protooncogene can be altered is by moving the gene to a new location, for example, by relocating it next to a different promoter, or by separating it from an adjacent regulatory element. This is presumably one way by which a protooncogene becomes oncogenic when located in a virus. That is, a promoter in the virus may have a significantly different strength from the promoter adjacent to a protooncogene in a normal cell. Alternatively, the virus may contain an enhancer (this would be more likely with a DNA virus than a retrovirus). Thus, when the viral DNA is inserted into the chromosome of an infected cell, the viral oncogene may be expressed at a much greater rate than the protooncogene, which is still at its normal location (Figure 24-5).

A variety of experiments have been carried out that indicate that relocating a protooncogene can convert it to an oncogene. For example, the normal mouse protooncogene, c-ras (and another called c-mos), has been attached to a viral regulatory sequence, without any alteration of the base sequence of the oncogene itself, as is the case with v-ras. This genetically engineered DNA thereby gained the ability to transform mouse 3T3 cells. In another experiment, a DNA fragment containing a protooncogene was isolated, and the terminal sequences, which presumably contained the regulatory elements that are adjacent to the gene, were removed; the protooncogene also gained the ability to transform 3T3 cells.

Chromosome changes are one of the hallmarks of cancer cells. The most common type of change is **translocation,** a chromosome alteration in which two chromosomes have apparently exchanged segments. For example, in Burkitt's lymphoma, fragments in chromosomes 8 and 14 are exchanged in 90 percent of the patients, between chromosomes 8 and 2 in 5 percent of the patients, and between chromosomes 8 and 22 in another 5

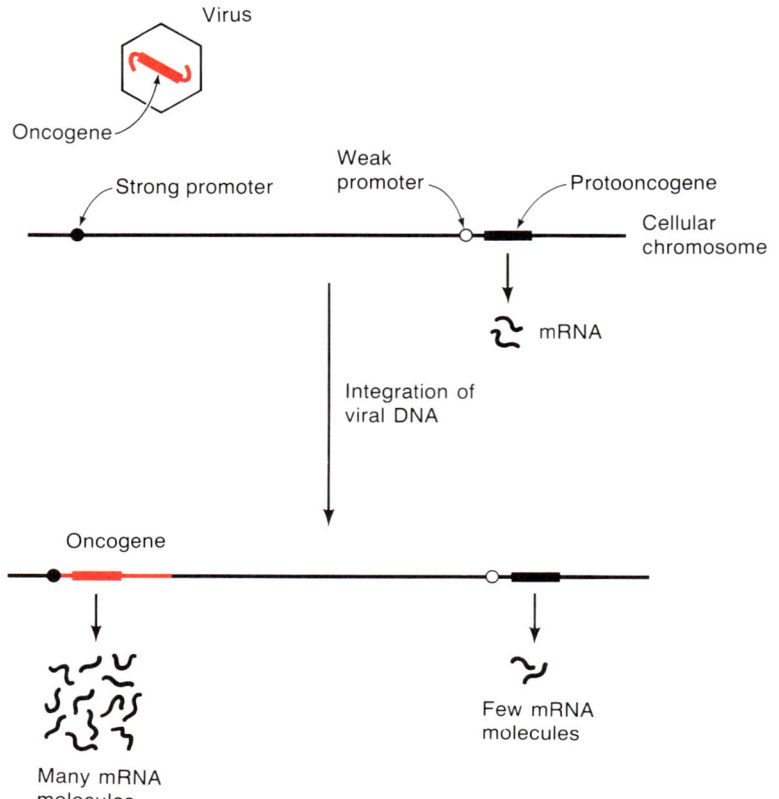

Figure 24-5 One way by which a viral oncogene whose coding sequence is the same as that of its cellular counterpart could produce excess product. The viral gene is adjacent to a stronger promoter than the cellular gene; an alternative is that the strong promoter could be in the viral DNA.

percent. Note that in each case a segment of chromosome 8 has been moved. The protooncogene *c-myc* is located on chromosome 8. In each of the translocations, *c-myc* is relocated adjacent to a gene encoding an antibody. This relocation apparently places the protooncogene under control of the genes that regulate antibody synthesis, so that *c-myc* is made in large quantities. (Whether this is significant for the etiology of Burkitt's lymphoma is unknown.) Numerous examples of translocations of protooncogenes have been observed. For example, patients with chronic myelogenous leukemia have a translocation between chromosomes 22 and 9. This causes the protooncogene *abl* (also present as the oncogene of Abelson mouse leukemia virus) to move from its normal location in chromosome 9 to chromosome 22; at this new location the rate of synthesis of *c-abl* mRNA is 8-fold greater than when the gene is in chromosome 9.

One must not be misled in thinking that simply increasing the amount of mRNA of a particular protooncogene is sufficient to induce cancer (recall that two steps are required to induce cancer). In fact, in several experiments synthesis of mRNA of a protooncogene has been amplified by genetic engineering and malignant transformation does not occur, clearly indicating a requirement for other events, which remain to be discovered.

ONCOGENE PROTEINS

Oncogenes and protooncogenes encode proteins, and certainly the activity of these proteins must be involved in transformation. In this section we examine several classes of oncogene proteins (Table 24-2), including, when possible, how they might work.

Protein Kinases

Earlier in this chapter we indicated that several oncogenes encode protein kinases, specifically those that phosphorylate tyrosine or threonine. The main approach to understanding how the kinase activity leads to transformation has been to seek proteins that contain phosphorylated tyrosines or threonines and see what these proteins do. These studies have not led to any understanding of the transformation process but do indicate why some of the changes that accompany transformation come about. One of the first such proteins identified was p36, a protein found in cells infected with Rous sarcoma virus. It is located on the inner (cytoplasmic) side of the cell membrane. In some cell types it is involved in some way with anchoring of microfilaments (rodlike cytoplasmic fibers, composed of the protein actin, that contribute to the shape of cells). Cancer cells have disorganized microfilaments, but the significance of this observation is unknown. Another protein whose tyrosines are phosphorylated is vinculin, a protein present in small patches (adhesion plaques) on the cell surface. These patches help cells adhere to surfaces and to one another and serve as attachment sites for bundles of microfilaments. Phosphorylated vinculin is less able to organize the microfilaments. Possibly, this alteration of vinculin is responsible for the

Table 24-2 Classes of oncogene proteins

Protein	Oncogene
Tyrosine kinase	abl, fes, fg , fps, ros, src, yes
Other protein kinases	erbB, fms, mil, mos, raf
Nuclear proteins	B-lym, fos, myb, myc, ski
Growth factor	sis
GTP-binding proteins	Ha-ras, K-ras, N-ras

decreased adhesion (a malignant feature) of cancer cells and the fact that cancer cells are usually rounder than normal cells. The Src kinase also phosphorylates the lipid phosphatidylinositol, a cell-membrane component. The altered lipid is then cleaved into two substances, diacylglycerol and triphosphoinositide. Diacylglycerol activates a serine protein kinase and thereby causes alteration of a very large number of proteins. The other fragment causes a release of calcium bound in the membranes. The significance of this release is also unclear, but many cancer cells have aberrant calcium metabolism. Both of these events could alter the control of cell division and other events occurring in transformation.

Growth Factors

A variety of substances affect the growth rate of particular cell types; some of these are called **growth factors.** Recently a connection has been seen between certain oncogenes and growth factors. The *v-erbB* oncogene is carried by avian erythroblastosis virus, which causes red blood cell leukemia. It is a large gene, part of which encodes a tyrosine protein kinase. However, of possibly greater interest is the other part of the protein, which is closely related to the cell surface receptor for epidermal growth factor (EGF), a stimulator of cell division for a wide variety of cell lines. The normal receptor, which is believed to be encoded in the cellular counterpart, *c-erbB,* is a large protein that spans the cell membrane. Only a portion of the receptor (the part inside the membrane and inside the cell) is encoded by *v-erbB*. It is thought that when *v-erbB* was acquired by the virus, only a part of the *c-erbB* gene was picked up. It is possible that the altered receptor produced by *v-erbB* disrupts the regulatory mechanisms in which the receptor is a crucial link. Interestingly, when EGF binds to the altered receptor, the tyrosine kinase is activated. The reader surely must have the impression that we are seeing in these observations something very significant, yet still well hidden.

The oncogene *sis,* which is carried by simian sarcoma virus, encodes a protein similar to the platelet-derived growth factor (PDGF), which is made by the protooncogene *c-sis*. Cells infected with simian sarcoma virus continually produce the altered PDGF, which binds to the receptors on the cell surface. Not all cells produce the PDGF receptor. Possibly, transformation only occurs when a cell that already makes the receptor is infected with the virus. This could cause autostimulation of cell division. Experimental confirmation of these hypotheses is not yet available.

The *ras* Proteins

An underlying theme in biology is that fundamental processes are usually carried out in much the same way for all organisms, though there are subtle differences between the prokaryotes and the eukaryotes. For example, all organisms use DNA and RNA polymerases to make DNA and RNA, glycolysis is a universal reaction for metabolizing glucose, and the mechanisms for protein synthesis differ only in detail between prokaryotes and eukaryotes. Since cell division is a property of all cells, one would expect some common features in the mechanism and regulation of this division.

Studies of cell division in bacteria have progressed at a more rapid rate than with animal cells, since in a given time period bacteria divide about 50 times more often than animal cells in culture. Also, bacteria have only a single chromosome, so the mechanism is probably simpler than the complex mitotic system of animal cells. An inherent difference between multiplication of bacteria and animal cells is that bacteria multiply continually, whereas animal cells organized in tissue do not. However, the yeasts (and other single cell eukaryotes) are in an intermediate situation: they are free-living organisms like bacteria, but they are nonetheless eukaryotes with many chromosomes. Since the prokaryotes and the eukaryotes diverged hundreds of millions of years ago, one may hope that certain systems that regulate cell division in animal cells may have counterparts in yeast. If so, one would have an experimental handle on the regulation of cell division, because yeast cells can be manipulated experimentally with much greater ease than animal cells. Recently, analogues of *src* and *ras* have been detected in yeast and in *Drosophila*. Yeast has two *ras* genes. Inactivation of both genes by mutation prevents growth; either gene in active form is sufficient for normal growth. One of the most exciting findings of the mid 1980s was the discovery that growth of a yeast that contains two defective *ras* genes is restored by uptake of human DNA containing the human *c-ras* gene. This gives credence to the hypothesis that *c-ras* is a gene that regulates growth and division in animal cells. Interestingly, only the protooncogene will correct the defect in yeast; none of the altered forms, such as those with a different single base change present in various retroviruses, can substitute for a defective yeast gene. These observations are so recent (1985) that little is yet known about the functioning of the normal yeast *ras* genes. However, a tantalizing observation is that the normal *ras* protein activates adenylate cyclase, the enzyme that makes cyclic AMP.

Transformation Proteins in DNA Viruses

Transformation proteins are also made by the DNA tumor viruses, adenovirus, polyoma, and SV40.

Cells transformed by adenovirus contain only fragments of adenovirus DNA, and a specific region of adenovirus DNA is always present in a transformed cell, as mentioned earlier. Transfection with a restriction fragment containing the specific region, which includes the genes *EIA* and *EIB,* induces transformation. It is thought that both of these genes are oncogenes. However, these gene products encode essential early proteins needed for replication and hence are fundamentally different from the nonessential oncogene proteins of retroviruses.

The polyoma gene *a* and the SV40 gene *A,* both of which encode T antigens (Chapter 23), are necessary both for replication of the viral DNA in a permissive host and for transformation of a nonpermissive host. However, a relation between DNA replication and transformation proved to be much more complex than first thought, for several reasons, one of which is that these genetic loci encode several proteins (as indicated for SV40 in Figure 23-24) whose coding sequences overlap. The SV40 early region encodes two proteins, t antigen and T antigen, and the polyoma locus encodes three— t antigen, middle-t antigen, and T antigen. Study of SV40 mutations of the *A* gene in the t and T regions has shown that in certain cell lines only T antigen is needed for transformation. (In some cell lines t antigen is also needed but the reason is unknown.) Furthermore, T antigen is a multifunctional protein having several activities located in different parts of the molecule. A defective mutant in the T region has been isolated, which prevents the viral DNA from replicating at the nonpermissive temperature, yet transformation occurs at that temperature. Thus, for SV40, transformation does not require that the viral DNA replicate. The situation with polyoma virus is more complex.

Recognition that there may be several types of oncogenes, for example, the *myc* and *ras* types, has made transformation by the DNA viruses a little more understandable. Polyoma large *T* and adenovirus *EIA* genes, seem to be like *myc;* the polyoma middle-*t* and adenovirus *EIB* genes are like *ras*. Significantly, the DNA virus genes that seem to have the same effect in transformation function do not all have the same base sequence, which again indicates that different genes can provide the same function.

Virus-induced protein kinases have been sought in cells transformed by DNA viruses, but to date, none have been found with certainty.

Bibliography

Chapter 1

Davis, B. D., et al. 1980. *Microbiology*. Harper & Row.

Frieden, E. 1972. "The chemical elements of life." *Scient. Amer.*, July, p. 52.

Hinkle, P. C., and R. E. McCarty. 1978. "How Cells Make ATP." *Scient. Amer.*, March, p. 104.

Lehninger, A. L. 1971. *Bioenergetics*. Benjamin/Cummings.

Lehninger, A. L. 1982. *Principles of Biochemistry*. Worth.

Puck, T. T. 1972. *The Mammalian Cell as a Microorganism* Holden-Day.

Chapter 2

Hayes, W. 1968. *The Genetics of Bacteria and Their Viruses* Wiley.

Snyder, L. A., D. Freifelder, and D. L. Hartl. 1985. *General Genetics*. Jones and Bartlett.

Chapter 3

Brewer, J. M., A. J. Pesce, and R. B. Ashworth. 1974. *Experimental Techniques in Biochemistry*. Prentice-Hall.

Cooper, T. G. 1977. *The Tools of Biochemistry*. Wiley.

Dickerson, R. E., and I. Geis. 1980. *The Structure and Action of Proteins*. Harper & Row.

Freifelder, D. 1982. *Physical Biochemistry*. W. H. Freeman.

Lehninger, A. L. 1982. *Principles of Biochemistry.* Worth.

Meselson, M., et al. 1957. "Equilibrium sedimentation of macromolecules in density gradients." *Proc. Nat. Acad. Sci.,* 43, 581.

Tanford, C. 1978. "The hydrophobic effect and the organization of living matter." *Science,* 200, 1012.

Wold, F. 1971. *Macromolecules: Structure and Function.* Prentice-Hall.

Younghusband, H. B., and R. B. Inman. 1974. "The electron microscopy of DNA." *Ann. Rev. Biochem.,* 43, 605.

Zwan, J. 1967. "Estimation of molecular weights by polyacrylamide gel electrophoresis." *Analyt. Biochem.,* 21, 155.

Chapter 4

Bauer, W. R. 1978. "Structure and reactions of closed duplex DNA." *Ann. Rev. Biophys. Bioeng.,* 7, 287.

Bauer, W. R., et al. 1980. "Supercoiled DNA." *Scient. Amer.,* July, p. 118.

Britten, R. J., and D. Kohn. 1968. "Repeated segments of DNA." *Scient. Amer.,* April, p. 24.

Cozzarelli, N. 1980. "DNA gyrase and supercoiling of DNA." *Science,* 207, 953.

Crick, F. H. C. 1976. "Linking numbers and nucleosomes." *Proc. Nat. Acad. Sci.,* 73, 2639.

Davidson, J. N. 1977. *The Biochemistry of the Nucleic Acids.* Academic Press.

Freifelder, D., ed. 1978. *The DNA Molecule: Structure and Properties.* W. H. Freeman.

Freifelder, D. 1982. *Physical Biochemistry.* W. H. Freeman.

Kolata, G. 1983. "Z-DNA moves toward real biology." *Science,* 222, 496.

Nordheim, A., et al. 1981. "Antibodies to left-handed Z-DNA bind to interband regions of *Drosophila* polytene chromosomes." *Nature,* 294, 417.

Ts'o, P. O. P. 1974. *Basic Principles in Nucleic Acid Chemistry.* Academic Press.

Watson, J. D., and F. H. C. Crick. 1953. "Molecular structure of nucleic acid. A structure for deoxyribose nucleic acid." *Nature,* 171, 737.

Younghusband, H. B., and R. B. Inman. 1974. "The electron microscopy of DNA." *Ann. Rev. Biochem.,* 43, 605.

Chapter 5

Dillon, J. R., A. Nasim, and E. R. Nestmann. 1985. *Recombinant DNA Methodology.* Wiley.

Glover, D. M. 1984. *Gene Cloning: The Mechanics of DNA Manipulation.* Chapman and Hall.

Hackett, P.B, J. A. Fuchs, and J. W. Messing. 1984. *An Introduction to Recombinant DNA Techniques.* Benjamin/Cummings.

Linn, S. M., and R. J. Roberts, eds. 1982. *Nucleases.* Cold Spring Harbor.

Maxam, A. M., and W. Gilbert. 1977. "A new method for sequencing DNA." *Proc. Nat. Acad. Sci.,* 74, 560.

Rodriguez, R. L., and R. C. Tait. 1983. *Recombinant DNA Techniques.* Addison-Wesley.

Sanger, F., S. Nicklen, and A. R. Coulson. 1977. "DNA sequencing with chain-terminating inhibitors." *Proc. Nat. Acad. Sci.*, 74, 5463.

Chapter 6

Anfinsen, C. G. 1973. "Principles that govern the folding of protein molecules." *Science*, 181, 223.

Baldwin, R. L. 1978. "The pathway of protein folding." *Trends Biochem. Sci.*, 3, 66.

Bernhard, S. 1968. *The Structure and Function of Enzymes*. Benjamin.

Capra, J. D., and A. B. Edmundson. 1977. "The antibody-combining site." *Scient. Amer.*, January, p. 50.

Clothia, C. 1984. "Principles that determine the structure of proteins." *Ann. Rev. Biochem.*, 53, 537.

Dickerson, R. E., and I. Geis. 1980. *The Structure and Action of Proteins*. Harper & Row.

Edelman, G. M. 1973. "Antibody structure and molecular immunology." *Science*, 180, 830.

Fersht, A. 1977. *Enzyme Structure and Mechanisms*. W. H. Freeman.

Freifelder, D. 1982. *Physical Biochemistry*. W. H. Freeman.

Haschemeyer, R. H., and A. E. V. Haschemeyer. 1973. *Proteins: A Guide to Study by Physical and Chemical Methods*. Wiley.

Kendrew, J. C. 1961. "The three-dimensional structure of a protein molecule." *Scient. Amer.*, December, p. 96.

Koshland, D. E. 1973. "Protein shape and biological control." *Scient. Amer.*, October, p. 52.

Lehninger, A. L. 1982. *Principles of Biochemistry*. Worth.

Perutz, M. 1964. "The hemoglobin molecule." *Scient. Amer.*, November, p. 64.

Perutz, M. 1978. "Hemoglobin structure and respiratory control." *Scient. Amer.*, December, p. 92.

Phillips, D. C. 1966. "The three-dimensional structure of an enzyme molecule." *Scient. Amer.*, November, p. 78.

Schachman, H. K. 1974. "Anatomy and physiology of a regulatory enzyme." *Harvey Lectures*, 68, 67.

Schultz, G. E., and R. H. Schirmer. 1979. *Principles of Protein Structure*. Springer-Verlag.

Silverton, E. W., M. A. Navia, and D. R. Davies. 1977. "Three-dimensional structure of an intact human immunoglobulin." *Proc. Nat. Acad. Sci.*, 74, 5140.

Stryer, L. 1981. *Biochemistry*. W. H. Freeman.

Chapter 7

Alberts, B., et al. 1983. *Molecular Biology of the Cell*. Garland.

Anderson, W. F., et al. 1981. "Structure of the Cro repressor from bacteriophage λ and its interaction with DNA." *Nature*, 290, 754.

Blackburn, E. H., and J. W. Szostak. 1984. "The molecular structure of centromeres and telomeres." *Ann. Rev. Biochem.*, 53, 163.

Bretscher, M. S. 1973. "Membrane structure: some general principles." *Science,* 181, 622.

Butler, P. J. G., and A. J. Klug. 1978. "The assembly of a virus." *Scient. Amer.,* November, p. 62.

Capaldi, R. A. 1974. "A dynamic model of cell membranes." *Scient. Amer.,* March, p. 26.

Eisenberg, D. 1984. "Three-dimensional structure of membrane and surface proteins." *Ann. Rev. Biochem.,* 53, 595.

Eyre, D. R. 1980. "Collagen: molecular diversity in the body's protein scaffold." *Science,* 207, 1315.

Fox, C. F. 1972. "The structure of cell membranes." *Scient. Amer.,* February, p. 30.

Frederick, C. A., et al. 1984. "Kinked DNA in crystalline complex with EcoRI endonuclease." *Nature,* 309, 327.

Gomperts, B. D. 1977. *The Plasma Membrane.* Academic Press.

Gross, J. 1961. "Collagen." *Scient. Amer.,* May, p. 120.

Kornberg, R. D. 1977. "Structure of chromatin." *Ann. Rev. Biochem.,* 46, 931.

Kornberg, R., and A. Klug. 1981. "The nucleosome." *Scient. Amer.* February.

Lodish, H. F., and J. E. Rothman. 1979. "The assembly of cell membranes." *Scient. Amer.,* January, p. 48.

Ohlendorf, D. H., et al. 1982. "The molecular basis of DNA- protein recognition inferred from the structure of cro repressor." *Nature,* 298, 718.

Ollis, D. L., et al. 1985. "Structure of large fragment of *E. coli* DNA polymerase I complexed with dTMP." *Nature,* 313, 762.

Pabo, C. O., and R. T. Sauer. 1984. "Protein-DNA recognition." *Ann. Rev. Biochem.,* 53, 293.

Ramachandran, G. N., and A. H. Reddi, eds. 1976. *Biochemistry of Collagen.* Plenum.

Singer, S. J., and G. L. Nicholson. 1972. "The fluid mosaic model of the structure of cell membranes." *Science,* 175, 720.

Steck, T., and K. Drlica. 1984. "Bacterial chromosome segregation: evidence for DNA gyrase involvement in decatenation." *Cell,* 36, 1081.

Takeda, Y., et al. 1983. "DNA-binding proteins." *Science,* 221, 1020.

Tanford, C. 1980. *The Hydrophobic Effect. Formation of Micelles and Biological Membranes.* Wiley-Interscience.

Weissman, G., and R. Claiborne, eds. 1975. *Cell Membranes.* H. P. Publishing.

Wharton, R. P., and M. Ptashne. 1985. "Changing the binding specificity of a repressor by redesigning an α helix." *Nature,* 316, 601.

Chapter 8

Avery, O. T., C. M. MacLeod, and M. McCarty. 1944. "Studies on the chemical nature of the substance inducing transformation of pneumococcal types." *J. Exp. Med.,* 79, 137.

Cairns, J., G. S. Stent, and J. D. Watson, eds. 1966. *Phage and the Origins of Molecular Biology.* Cold Spring Harbor.

Freifelder, D., ed. 1978. *The DNA Molecule*. W. H. Freeman.

Hershey, A. D., and M. Chase. 1952. "Independent function of viral protein and nucleic acid in growth of bacteriophage." *J. Gen. Physiol.*, 36, 39.

Judson, H. 1979. *The Eighth Day of Creation*. Simon and Schuster.

McCarty, M. 1985. *The Transforming Principle: Discovering that Genes Are Made of DNA*. Norton.

Mirsky, A. E. 1968. "The discovery of DNA." *Scient. Amer.* June, p. 78.

Portugal, F. H., and J. S. Cohen. 1977. *A Century of DNA: A History of the Discovery of the Structure and Function of the Genetic Substance*. MIT Press.

Watson, J. D. 1968. *The Double Helix*. Athenaeum.

Chapter 9

Abdel-Monem, M., and H. Hoffman-Berling. 1980. "DNA unwinding enzymes." *Trends Biochem. Sci.*, 5, 128.

Cairns, J. 1966. "The bacterial chromosome." *Scient. Amer.*, January, p. 36.

DePamphilis, M., and P. Wassarman. 1980. "Replication of eukaryotic chromosomes: a close-up of the replication fork." *Ann. Rev. Biochem.*, 49, 627.

Geider, K., and H. Hoffman-Berling. 1981. "Proteins controlling the helical structure of DNA." *Ann. Rev. Biochem.*, 50, 233.

Gellert, M. 1981. "DNA topoisomerases." *Ann. Rev. Biochem.*, 50, 879.

Gilbert, W., and D. Dressler. 1968. "DNA replication: the rolling circle model." *Cold Spring Harb. Symp. Quant. Biol.*, 33, 473.

Harland, R. 1983. "Initiation of DNA replication in eukaryotic chromosomes." In *DNA Makes RNA Makes Protein*. (T. Hunt et al., Eds.) Elsevier Biomedical.

Kornberg, A. 1980. *DNA Replication*; also, *1982 Supplement*. W. H. Freeman.

Lehman, I. R. 1974. "DNA ligase: structure, mechanism, and function." *Science*, 186, 790.

McHenry, C., and A. Kornberg. 1977. "DNA polymerase III holoenzyme of *E. coli*: purification and resolution into subunits." *J. Biol. Chem.* 252, 6478.

McKnight, S. L., and O. L. Miller, Jr. 1977. "Electron microscopic analysis of chromatin replication in the cell blastoderm of a *Drosophila melanogaster* embryo." *Cell*, 12, 795.

Meselson, M., and F. W. Stahl. 1957. "The replication of DNA in *E. coli*." *Proc. Nat. Acad. Sci.*, 44, 671.

Morrison, A., and N. R. Cozzarelli. 1981. "Contact between DNA gyrase and its binding site on DNA: features of symmetry and symmetry revealed by protection from nuclease." *Proc. Nat. Acad. Sci.*, 78, 1416.

Nossal, N. 1983. "Prokaryotic DNA replication systems." *Ann. Rev. Biochem.*, 52, 581.

Ogawa, T., and T. Okazaki. 1980. "Discontinuous DNA replication." *Ann. Rev. Biochem.*, 49, 421.

Prescott, D. M., and P. L. Kuempel. 1972. "Bidirectional replication of the chromosome in *E. coli*." *Proc. Nat. Acad. Sci.*, 69, 2842.

Riley, D., and H. Weintraub. 1979. "Conservative segregation of parental histones during replication in the presence of cycloheximide." *Proc. Nat. Acad. Sci.,* 76, 328.

Rowen, L., and A. Kornberg. 1978. "Primase, the *dnaG* protein of *E. coli:* an enzyme which starts DNA chains." *J. Biol. Chem.,* 253, 758.

Scheuermann, R. H., and H. Echols. 1984. "A separate editing exonuclease for DNA replication: the ϵ subunit of *E. coli* DNA polymerase III holoenzyme." *Proc. Nat. Acad. Sci.,* 81, 7747.

Schnös, M., and R. Inman. 1970. "Position of branch points in replication of lambda DNA." *J. Mol. Biol.,* 51, 61.

Scovassi, A. I., P. Pievani, and U. Bertazzoni. 1983. "Eukaryotic DNA polymerases." In *DNA Makes RNA Makes Protein.* (T. Hunt et al., eds.) Elsevier Biomedical.

Seidman, M. M., A. J. Levine, and H. Weintraub. 1980. "The asymmetric segregation of parental nucleosomes during chromosome replication." *Cell,* 18, 439.

Tomizawa, J. 1978. "Replication of colicin E1 plasmid DNA *in vitro.*" In *DNA Synthesis: Present and Future.* I. Molineux and K. Kohiyama, eds. Plenum.

Tomizawa, J., and G. Selzer. 1979. "Initiation of DNA synthesis in *E. coli.*" *Ann. Rev. Biochem.,* 48, 999.

Tye, B. K., et al. 1977. "Transient accumulation of Okazaki fragments as a result of uracil incorporation into nascent DNA." *Proc. Nat. Acad. Sci.,* 74, 154.

Valenzuela, M., et al. 1976. "Lack of a unique termination site in lambda DNA replication." *J. Mol. Biol.,* 102, 569.

Wang, J. 1985. "DNA topoisomerases." *Ann. Rev. Biochem.,* 54, 665.

Watson, J. D., and F. H. C. Crick. 1953. "Genetic implications of the structure of desoxyribonucleic acid." *Nature,* 171, 964.

Wickner, S. H. 1978. "DNA replication proteins of *E. coli.*" *Ann. Rev. Biochem.,* 49, 421.

Chapter 10

Friedberg, E. C. 1985. *DNA Repair.* W. H. Freeman.

Grossman, L., et al. 1975. "Enzymatic repair of DNA." *Ann. Rev. Biochem.,* 44, 19.

Hanawalt, P. C., et al. 1979. "DNA repair in bacteria and mammalian cells." *Ann. Rev. Biochem.,* 48, 783.

Hanawalt, P. C., E. C. Friedberg, and C. F. Fox. 1978. *DNA Repair Mechanisms.* Academic Press.

Hanawalt, P. C., and R. B. Setlow. 1975. *Molecular Mechanisms for Repair of DNA.* Plenum.

Haseltine, W. A. 1983. "Ultaviolet light repair and mutagenesis revisited." *Cell,* 33, 13.

Haynes, R. H., and B. A. Kunz. 1982. "DNA repair and mutagenesis in yeast." In *Molecular Biology of the Yeast Saccharomyces: Life Cycle and Inheritance.* J. Strathern, et al., eds. p. 371. Cold Spring Harbor.

Little, J. W., et al. 1980. "Cleavage of the *E. coli lexA* protein by the *recA* protease." *Proc. Nat. Acad. Sci.,* 77, 3225.

Little, J. W., and D. W. Mount. 1982. "The SOS regulatory system of *E. coli.*" *Cell*, 29, 11.

Livneh, Z., and I. R. Lehman. 1982. "Recombinational bypass of pyrimidine dimers by the RecA protein of *E. coli.*" *Proc. Nat. Acad. Sci.* 79, 3171.

Roberts, J. J. 1978. "The repair of DNA modified by cytotoxic, mutagenic, and carcinogenic chemicals." *Adv. Radiat. Biology,* 6, 211.

Setlow, R. B. 1966. "Cyclobutane-type pyrimidine dimers in polynucleotides." *Science,* 153, 379.

Walker, G. C. 1985. "Inducible DNA repair systems." *Ann. Rev. Biochem.,* 54, 425.

Ward, J. 1975. "Molecular mechanisms of radiation-induced damage to nucleic acids." *Adv. Radiat. Biol.,* 5, 182.

West, S. C., et al. 1982. "Postreplication repair in *E. coli:* strand-exchange reactions of gapped DNA by RecA protein." *Molec. Gen.,* 187, 209.

Chapter 11

Ames, B. W. 1979. "Identifying environmental chemicals causing mutations and cancer." *Science,* 204, 587.

Ames, B. W., et al. 1973. "Carcinogens are mutagens: a simple test system combining liver homogenates for activation and bacteria for detection." *Proc. Nat. Acad. Sci.,* 70, 2381.

Coulondre, R., et al. 1978. "Molecular basis of base-substitution hotspots in *E. coli.*" *Nature,* 274, 775.

Drake, J. W. 1970. *The Molecular Basis of Mutation.* Holden-Day.

Drake, J. W., and R. H. Baltz. 1976. "The biochemistry of mutagenesis." *Ann. Rev. Biochem.,* 45, 11.

Gusella, J. F., et al. 1983. "A polymorphic DNA marker genetically linked to Huntington's disease." *Nature,* 306, 234.

Haynes, R. H., and B. A. Kunz. 1985. "A possible role for deoxyribonucleotide pool imbalances in carcinogenesis." In *Basic and Applied Mutagenesis.* A. Muhammed and R. C. von Borstel, eds. Plenum.

Heidelberger, C. 1975. "Chemical carcinogens." *Ann. Rev. Biochem.,* 44, 79.

Kunz, B. A. 1982. "Genetic effects of deoxyribonucleotide pool imbalances." *Environ. Mutag.,* 4, 695.

Orgel, L. E. 1965. "The chemical basis of mutation." *Adv. Enzymol.,* 27, 289.

Smith, M. 1982. "Site-directed mutagenesis." *Trends Biochem. Sci.,* 7, 440.

Streisinger, G., et al. 1966. "Frameshift mutations and the genetic code." *Cold Spring Harb. Symp. Quant. Biol.,* 31, 77.

Zoller, M. J., and M. Smith. 1982. "Oligonucleotide-directed mutagenesis using M13-derived vectors: an efficient and general procedure for the production of point mutations in any fragment of DNA." *Nucl. Acids Res.,* 10, 6487.

Chapter 12

Abelson, J. 1979. "RNA processing and the intervening sequence problem." *Ann. Rev. Biochem.,* 48, 1035.

Adhya, S., and M. Gottesman. 1978. "Control of transcription termination." *Ann. Rev. Biochem.*, 47, 967.

Altman, S. 1981. "Transfer RNA processing enzymes." *Cell*, 23, 3.

Bass, B. L., and T. R. Cech. 1984. "Specific interaction between the self-splicing RNA of *Tetrahymena* and its guanosine substrate: implications for biological catalysis by RNA." *Nature*, 308, 820.

Bear, D. G., et al. 1985. *Sequence Specificity in Transcription and Translation.* (R. Calendar and L. Gold, eds.) Alan R. Liss.

Brown, D. D. 1984. "The role of stable complexes that repress and activate eukaryotic genes." *Cell*, 37, 359.

Bujard, H. 1980. "The interaction of *E. coli* RNA polymerase with promoters." *Trends Biochem. Sci.*, 5, 274.

Cech, T. R. 1983. "RNA splicing: three themes with variations." *Cell*, 34, 713.

Chambon, P. 1981. "Split genes." *Scient. Amer.*, May, p. 60.

Cozzarelli, N., et al. "Purified RNA polymerase III accurately and efficiently terminates transcription of 5S RNA genes." *Cell*, 34, 829.

de Crombrugghe, B., S. Busby, and J. Buc. 1984. "Cyclic AMP receptor protein: role in transcription activation." *Science*, 224, 831.

Dynan, W., and R. Tjian. 1983. "Transcription enhancer sequences: a novel regulatory element." In *DNA Makes RNA Makes Protein.* (T. Hunt et al., eds.) Elsevier. Biomedical.

Freifelder, D. 1982. *Physical Biochemistry.* 2nd Ed. W. H. Freeman. Chapters 7 and 9 have detailed descriptions of hybridization techniques and the Southern blotting method.

Gluzman, Y., and T. Shenk, eds. 1983. *Enhancers and Eukaryotic Gene Expression.* Cold Spring Harbor.

Khoury, G., and P. Gruss. 1983. "Enhancer elements." *Cell*, 33, 313.

Lathe, R. 1978. "RNA polymerase of *E. coli.* "*Curr. Topics Microbiol. Immunol.*, 83, 37.

Lewin, B. 1985. *Genes II.* Chapter 20. Wiley.

Losick, R., and M. Chamberlin 1976. *RNA Polymerase.* Cold Spring Harbor.

McClure, W. R. 1985. "Mechanism and control of transcription initiation in prokaryotes." *Ann. Rev. Biochem.*, 54, 171.

Miller, O. L., Jr. 1973. "The visualization of genes in action." *Scient. Amer.*, March, p. 34.

Moses, P. B., and P. Model. 1984. "A Rho-dependent transcription termination signal in bacteriophage f1." *J. Mol. Biol.*, 172, 1.

Ogden, R. C., et al. 1983. "The mechanism of tRNA splicing." In *DNA Makes RNA Makes Protein.* (T. Hunt et al., eds.). Elsevier. Biomedical.

Padgett, R. A., et al. 1985. "Splicing messenger RNA precursors: branch sites and lariat RNAs." *Trends Biochem. Sci.*, 10, 154.

Ruskin, B., and M. R. Green. 1985. "Role of the 3' splice site consensus sequence in mammalian pre-mRNA splicing." *Nature*, 317, 732.

Shatkin, A. J. 1976. "Capping of eukaryotic mRNA." *Cell*, 9, 645.

Turner, P. 1985. "Controlling role for snurps." *Nature,* 316, 105.

von Hippel, P. H., et al. 1984. "Protein-nucleic acid interaction in transcription: a molecular analysis." *Ann. Rev. Biochem.,* 53, 389.

Chapter 13

Adamiak, R. W., and P. Gornicki. "Hypermodified nucleosides of tRNA: synthesis, chemistry, and structural features of biological interest." *Prog. Nucl. Acid Res. Mol. Biol.,* 32, 27.

Altman, S., ed. 1978. *Transfer RNA.* MIT Press.

Anderson, S., et al. 1981. "Sequence and organization of the human mitochondrial genes." *Nature,* 290, 457.

Celis, J. E., and J. D. Smith. 1979. *Nonsense Mutations and tRNA Suppression.* Academic Press.

Clark, B. F. C. 1977. "Correlation of biological activities with structural features of transfer RNA." *Prog. Nucl. Acid Res. Mol. Biol.,* 20, 1.

Cold Spring Harbor Laboratory. 1966. *The Genetic Code.* Vol. 31, Symposium on Quantitative Biology.

Crick, F. H. C., et al. 1961. "General nature of the genetic code for proteins." *Nature,* 192, 1227.

Fersht, A. 1980. "Enzymatic editing mechanisms in protein synthesis and DNA replication." *Trends Biochem. Sci.,* 5, 262

Goodmann, H. M., et al. 1968. "Amber suppression: a nucleotide change in the anticodon of tyrosine transfer RNA." *Nature,* 217, 1019.

Jukes, T. 1978. "The amino acid code." *Adv. Enzym.,* 47, 375

Jukes, T. 1983. "Evolution of the amino acid code: inferences from mitochondrial codes." *J. Molec. Evol.,* 19, 219.

Khorana, H. G. 1968. "Nucleic acid synthesis in the study of the genetic code." In *Nobel Lectures: Physiology or Medicine.* Vol. 4 (1973). American Elsevier.

Kozak, M. 1984. "Point mutations close to the AUG initiation codon affect the efficiency of translation of rat preproinsulin *in vivo.*" *Nature,* 308, 241.

Nirenberg, M. 1963. "The genetic code." *Scient. Amer.,* March, p. 80.

Rich, A., and S. Kim. 1978. "The three-dimensional structure of transfer RNA." *Scient. Amer.* January. p. 52.

Schimmel, P. R., S. Putney, and R. Starzyk. 1983. "RNA and DNA sequence recognition and structure-function of aminoacyl tRNA synthetases." In *DNA Makes RNA Makes Protein.* (T. Hunt et al., eds.) Elsevier Biomedical.

Schimmel, P., and A. G. Redfield. 1980. "Transfer RNA in solution: selected topics." *Ann. Rev. Biophys. Bioeng.,* 9, 181.

Schimmel, P. R., and D. Soll. 1979. "Aminoacyl-tRNA synthetases: general features and recognition of transfer RNA's." *Ann. Rev. Biochem.,* 48, 601.

Sherman, F. 1982. "Suppression in the yeast *Saccharomyces cerevisiae.*" In *The Molecular Biology of the Yeast Saccharomyces.* J. Strathern, E. Jones, and J. Broach, eds. Cold Spring Harbor.

Soll, D., J. Abelson, and P. Schimmel, eds. 1980. *Transfer RNA.* Volumes 1 and 2. Cold Spring Harbor.

Chapter 14

Bermek, E. 1978. "Mechanisms in polypeptide chain elongation on ribosomes." *Prog. Nucl. Acid Res. Mol. Biol.*, 2, 63.

Brimacombe, R., et al. 1976. "The ribosome of *E. coli.*" *Prog. Nucl. Acid Res. Mol. Biol.*, 18, 1.

Caskey, C. T. 1980. "Peptide chain termination." *Trends Biochem. Sci.*, 5, 234.

Chambliss, G., ed. 1980. *Ribosomes: Structure, Function, and Genetics.* University Park Press.

Clark, B. 1980. "The elongation step of protein biosynthesis." *Trends Biochem. Sci.*, 5, 207.

Fersht, A. R. 1983. "Enzymatic editing mechanisms in protein synthesis and DNA replication." In *DNA Makes RNA Makes Protein.* (T. Hunt et al., eds.) Elsevier Biomedical.

Gold, L., et al. 1981. "Translational initiation in prokaryotes." *Ann. Rev. Microbiol.*, 35, 365.

Grunberg-Manago, M., and F. Gros. 1977. "Initiation mechanisms of protein synthesis." *Prog. Nucl. Acid Res. Mol. Biol.*, 20, 209.

Hunt, T. 1980. "The initiation of protein synthesis." *Trends Biochem. Sci.*, 5, 178.

Jagus, R., et al. 1981. "The regulation of initiation of mammalian protein synthesis." *Prog. Nucleic Acid Res. Mol. Biol.*, 25, 128.

Kim, S. 1978. "Three-dimensional structure of transfer RNA and its functional implications." *Adv. Enzymol.*, 46, 279.

Kozak, M. 1980. "Evolution of the scanning model for initiation of protein synthesis." *Cell*, 22, 7.

Krayefsky, A. A., and M. K. Kukhenova. 1979. "The peptidyl transferase center of ribosomes." *Prog. Nucl. Acid Res. Mol. Biol.*, 23, 2.

Kurland, C. G. 1977. "Structure and function of the bacterial ribosome." *Ann. Rev. Biochem.*, 46, 173.

Lake, J. 1980. "Ribosome structure and tRNA binding sites." In *Ribosomes: Structure, Function, and Genetics.* G. Chambliss, ed. University Park Press.

Meyer, D. 1983. "The signal hypothesis." In *DNA Makes RNA Makes Protein.* (T. Hunt et al., eds.). Elsevier Biomedical.

Nierhaus, K. N. 1982. "Structure, assembly, and function of ribosomes." *Curr. Topics Micro. Immunol.*, 97, 81.

Nomura, M., A. Tissieres, and P. Lengyel, eds. 1974. *Ribosomes.* Cold Spring Harbor.

Ojala, D., J. Montoya, and G. Attardi. 1981. "tRNA punctuation model of RNA processing in human mitochondria." *Nature*, 290, 470.

Prince, J. B., R. B. Gutell, and R. A. Garrett. 1983. "A consensus model of the *E. coli* ribosome." *Trends Bioch. Sci.*, 8, 359.

Schimmel, P. R., and D. Soll. 1979. "Aminoacyl tRNA synthetase: general features and recognition of transfer RNAs." *Ann. Rev Biochem.*, 48, 601.

Sherman, F., and J. W. Stewart. 1982. "Mutations altering initiation of translation of yeast iso-1-cytochrome *c:* contrasts between eukaryotic and prokaryotic initiation processes." In *The Molecular Biology of the Yeast Saccharomyces*. (J. Strathern et al., eds.) Cold Spring Harbor.

Shine, J., and L. Dalgarno. 1974. "The 3′ terminal sequence of *E. coli* 16S ribosomal RNA: complementarity to nonsense triplets and ribosomal binding sites." *Proc. Nat. Acad. Sci.*, 71, 1342.

Spirin, A. S. 1985. "Ribosomal translocation: factors and models." *Prog. Nucl. Acid. Res. Mol. Biol.*, 32, 75.

Steitz, J. A. 1979. "Genetic signals and nucleotide sequences in messenger RNA." In *Biological Regulation and Development*. (R. F. Goldberger, ed.) Vol. 1, p. 349. Plenum.

Steitz, J. A., and K. Jakes. 1975. "How ribosomes select initiation regions in mRNA: base pair formation between the 3′ terminus of 16S rRNA and the mRNA during initiation of protein synthesis in *E. coli*." *Proc. Nat. Acad. Sci.*, 72, 4734.

Weissbach, H., and S. Pestka, eds. 1977. *Molecular Mechanisms in Protein Synthesis*. Academic Press.

Wool, I. 1979. "The structure and function of eukaryotic ribosomes." *Ann. Rev. Biochem.*, 48, 719.

Chapter 15

Clark, B. F. C., et al., eds. 1978. *Gene Expression*. Pergamon.

Gilbert, W., and B. Müller-Hill. 1966. "Isolation of the *lac* repressor." *Proc. Nat. Acad. Sci.*, 56, 1891.

Irani, M. H., et al. 1983. "A control element within a structural gene: the *gal* operon of *E. coli. Cell*, 32, 783.

Jacob, F., and J. Monod. 1961. "Genetic regulatory mechanisms in the synthesis of proteins." *J. Mol. Biol.*, 3, 318.

Johnson, H. M., et al. 1980. "Model for regulation of the histidine operon of *Salmonella*." *Proc. Nat. Acad. Sci.*, 77, 508.

Lee, N., W. O. Gielow, and R. G. Wallace. 1981. "Mechanism of *araC* autoregulation and the domain of two overlapping promoters, *pC* and *pBAD* in the L-arabinose regulatory region of *E. coli*." *Proc. Nat. Acad. Sci.*, 78, 752.

Lindahl, L., and J. M. Zengel. 1982. "Expression of ribosomal genes in bacteria." *Adv. Genet.*, 21, 53.

Little, J., et al. 1980. "Cleavage of the *E. coli lexA* protein by the *recA* protease." *Proc. Nat. Acad. Sci.*, 77, 3225.

Majumdar, A., and S. Adhya. 1984. "Demonstration of two operator elements in *gal: in vitro* repressor binding studies." *Proc. Nat. Acad. Sci.*, 81, 6100.

Maniatis, T., and M. Ptashne. 1976. "A DNA operator." *Scient. Amer.*, January. p. 64.

Miller, J. H. 1980. "Genetic analysis of the *lac* repressor." *Cur. Topics Microbiol. Immunol.*, 90, 1.

Miller, J. H., and W. S. Reznikoff, eds. 1978. *The Operon*. Cold Spring Harbor.

Nierlich, D. P. 1978. "Regulation of bacterial growth, RNA, and protein synthesis." *Ann. Rev. Microbiol.*, 32, 393.

Nomura, M. 1984. "The control of ribosome synthesis." *Scient. Amer.* January, p. 1.

Nomura, M., R. Gourse, and G. Baughman. 1984. "Regulation of the synthesis of ribosomes and ribosomal components." *Ann. Rev. Biochem.,* 53, 75.

Oxender, D. L., G. Zurawski, and C. Yanofsky. 1979. "Attenuation in the *E. coli* tryptophan operon: role of RNA secondary structure involving the tryptophan coding region." *Proc. Nat. Acad. Sci.,* 76, 5524.

Platt, T. 1978. "Regulation of gene expression in the tryptophan operon in *E. coli.*" In *The Operon.* J. H. Miller and W. S. Reznikoff, eds. Cold Spring Harbor.

Ptashne, M., and W. Gilbert. 1970. "Genetic repression." *Scient. Amer.,* June, p. 36.

Reznikoff, W., R. Winter, and B. C. Hurley. 1974. "The location of the repressor-binding site in the *lac* operon." *Proc. Acad. Sci.,* 71, 2314.

Stark, G. R., and G. M. Wahl. 1984. "Gene amplification." *Ann. Rev. Biochem.,* 53, 447.

Ullman, A., and A. Danchin. 1980. "Role of cyclic AMP in regulatory mechanisms of bacteria." *Trends Biochem. Sci.,* 5, 95.

Wallace, R. G., et al. 1980. "The *araC* gene of *E. coli:* transcription and translation start points and complete nucleotide sequence." *Gene,* 12, 179.

Watson, M. D. 1983. "Attenuation: translational control of transcription termination." In *DNA Makes RNA Makes Protein.* (T. Hunt et al., eds.) Elsevier Biomedical.

Yanofsky, C. 1981. "Attenuation in the control of expression of bacterial operons." *Nature,* 289, 751.

Yanofsky, C., R. L. Kelley, and V. Horn. 1984. "Repression is relieved before attenuation in the *trp* operon of *E. coli* as tryptophan starvation becomes increasingly severe." *J. Bact.,* 158, 1018.

Chapter 16

Axel, R., et al. 1979. *Eukaryotic Gene Regulation.* Academic Press.

Bienz, M. 1985. "Transient and developmental activation of heat-shock genes." *Trends Biochem. Sci.,* 10, 157.

Bostock, C. 1980. "A function for satellite DNA?" *Trends Biochem. Sci.,* 5, 117.

Brent, R. 1985. "Repression of transcription in yeast." *Cell,* 42, 3.

Brown, D. 1981. "Gene expression in eukaryotes." *Science,* 211, 667.

Capasso, O., and N. Heintz. 1985. "Regulated expression of mammalian histone H4 genes *in vivo* requires a trans–acting transcription factor." *Proc. Nat. Acad. Sci.,* 82, 5622.

Chisholm, R. 1983. "Gene amplification during development." In *DNA Makes RNA Makes Protein.* (T. Hunt et al., eds.) Elsevier Biomedical.

Collins, F. S., and S. M. Weissman. 1984. "The molecular genetics of human hemoglobin." *Prog. Nucl. Acid. Res. Mol. Biol.,* 31, 317.

de Crombrugghe, B., and I. Pastan. 1983. "Structure and regulation of a collagen gene." In *DNA Makes RNA Makes Protein.* (T. Hunt et al., eds.) Elsevier Biomedical.

Dynan, W. S., and R. Tjian. 1985. "Control of eukaryotic mRNA synthesis by sequence-specific DNA-binding proteins." *Nature,* 316, 774.

Gough, N. 1981. "The rearrangement of immunoglobulin genes." *Trends Biochem. Sci.,* 6, 203.

Guarente, L. 1984. "Yeast promoters: positive and negative elements." *Cell,* 36, 799.

Hood, L. E., I. L. Weissman, and W. B. Wood. 1978. *Immunology.* Benjamin.

Jelinek, W. R., and C. W. Schmid. 1982. "Repetitive sequences in eukaryotic DNA and their expression" *Ann. Rev. Biochem.,* 51, 813.

Karlsson, S., and A. W. Nienhaus. 1985. "Developmental regulation of human globin genes." *Ann. Rev. Biochem.,* 54, 1071.

Kolata, G. 1981. "Gene regulation through chromosome structure." *Science,* 214, 775.

Kolata, G. 1984. "New cues to gene regulation." *Science,* 224, 588.

Kolata, G. 1985. "Fitting methylation into development." *Science,* 228, 1183.

Leighton, T., and W. F. Loomis, eds. 1982. *The Molecular Genetics of Development.* Academic Press.

Long, E. O., and I. B. Dawid. 1980. "Repeated genes in eukaryotes." *Ann. Rev. Biochem.,* 49, 727.

Marx, J. L. 1981. "Antibodies: getting their genes together" *Science,* 212, 1015.

Numa, S., and S. Nakanishi. 1983. "Corticotropin-β-lipotropin precursor." In *DNA Makes RNA Makes Protein.* (T. Hunt et al., eds.) Elsevier Biomedical.

Ochoa, S., and C. de Haro. 1979. "Regulation of protein synthesis in eukaryotes." *Ann. Rev. Biochem.,* 48, 549.

O'Malley, B. W., and W. T. Schroeder. 1976. "The receptors of steroid hormones." *Scient. Amer.,* February, p. 32.

Revel, M., and Y. Goner. 1978. "Posttranscriptional and translational controls of gene expression in eukaryotes." *Ann. Rev. Biochem.,* 47, 1079.

Safer, B., and W. F. Anderson. 1978. "The molecular mechanism of hemin synthesis and its regulation in reticulocytes." *CRC Rev. Biochem.* 6, 261.

Sakano, H., et al. 1979. "Sequences in the somatic recombination of immunoglobulin light chain genes." *Nature,* 280, 288.

Seidman, J. G., and P. Leder. 1978. "The arrangement and rearrangement of antibody genes." *Nature,* 276, 790.

Tata, J. R. 1976. "The expression of the vitellogenin gene." *Cell* 9, 1.

Taylor, J. H. 1984. *Methylation and Cellular Differentiation.* Springer-Verlag.

Wahli, W., et al. 1981. "Vitellogenesis and the vitellogenin gene family." *Science,* 212, 298.

Weissbrod, S., and H. Weintraub. 1979. "Isolation of a subclass of nuclear proteins responsible for conferring a DNase I-sensitive structure in globin chromatin." *Proc. Nat. Acad. Sci.,* 76, 630.

Chapter 17

Adhya, S., et al. 1981. "Regulatory circuits of bacteriophage lambda." *Prog. Nucl. Acid Res. Mol. Biol.,* 26, 103.

Arber, W., and S. Linn. 1967. "DNA modification and restriction." *Ann. Rev. Biochem.,* 38, 467.

Casjens, S. 1985. *Virus Assembly*. Jones and Bartlett.

Earnshaw, W. C., and S. R. Casjens. 1980. "DNA packaging by the double-stranded DNA bacteriophages." *Cell*, 21, 319.

Echols, H. 1980. "Bacteriophage development." In *Molecular Genetics of Development*. (T. Leighton and W. F. Loomis, eds.) Academic Press.

Hershey, A. D., ed. 1971. *The Bacteriophage Lambda*. Cold Spring Harbor.

Kaiser, A. D., et al. 1975. "DNA packaging steps in bacteriophage lambda head assembly." *J. Mol. Biol.*, 91, 175.

Lewin, B. 1977. *Gene Expression. III. Plasmids and Phages*. Wiley-Interscience.

Luria, S. E., et al. 1978. *General Virology*. Wiley.

Robussay, D., and E. P. Geiduschek. 1977. "Regulation of gene action in the development of lytic bacteriophage." *Comprehensive Virol.*, 8, 1.

Studier, F. W. 1973. "Organization and expression of bacteriophage T7 DNA." *Cold Spring Harb. Symp. Quant. Biol.*, 47, 999.

Szybalski, W. 1977. "Initiation and regulation of transcription and DNA replication in coliphage lambda." In *Regulatory Biology*. (J. C. Copeland and G. A. Marzluff, eds.) Ohio State University Press.

Zinder, N., ed. 1975. *RNA Phages*. Cold Spring Harbor.

Chapter 18

Campbell, A. 1976. "How viruses insert their DNA into the DNA of a host cell." *Scient. Amer.*, December, p. 102.

Campbell, A. 1977. "Defective bacteriophages and incomplete prophages." *Comprehensive Virol.*, 8, 259.

Craig, N., and H. A. Nash. 1983. "The mechanism of phage λ site-specific recombination: site-specific breakage of DNA by Int topoisomerase." *Cell*, 35, 795.

Devoret, R. 1979. "Bacterial test for potential carcinogens." *Scient. Amer.*, August, p. 40.

Hershey, A. D., ed. 1971. *The Bacteriophage Lambda*. Cold Spring Harbor.

Herskowitz, I., and D. Hagen. 1980. "The lysis-lysogeny decision of phage λ: explicit programming and responsiveness." *Ann. Rev. Genetics*, 14, 399.

Landy, A., and W. Ross. 1977. "Viral integration and excision: structure of the lambda *att* site." *Science*, 197, 1147.

Luria, S. E., et al. 1978. *General Virology*. Wiley.

Miller, H. I., et al. 1979. "Site-specific recombination of bacteriophage λ: the role of host gene products." *Cold Spring Harb. Symp. Quant. Biol.*, 43, 1121.

Miller, H. I., et al. 1980. "Regulation of the integration-excision reaction by bacteriophage λ." *Cold Spring Harb. Symp. Quant. Biol.*, 45, 439.

Mizuuchi, M., and K. Mizuuchi. 1980. "Integrative recombination of bacteriophage λ: extent of the DNA sequence involved in the attachment site function." *Proc. Nat. Acad. Sci.*, 77, 3220.

Nash, H. A. 1978. "Integration and excision of bacteriophage λ." *Curr. Topics Microbiol. Immun.*, 78, 171.

Nash, H. A., et al. 1980. "Strand exchange in λ integrative recombination: genetics, biochemistry, and models." *Cold Spring Harb. Symp. Quant. Biol.*, 45, 417–428

Oppenheim, A., and A. B. Oppenheim. 1978. "Regulation of the *int* gene of bacteriophage λ: activation by the *cII* and *cIII* gene products and the roles of the *pI* and *pL* promoters." *Molec. Gen. Genetics,* 165, 39.

Ptashne, M., et al. 1980. "How the λ repressor and Cro work." *Cell,* 19, 1.

Ross, W., and A. Landy. 1983. "Patterns of λ Int recognition in the regions of strand exchange." *Cell,* 33, 261.

Weisberg, R. A., et al. 1977. "Bacteriophage λ: the lysogenic pathway." *Comprehensive Virol.,* 8, 197.

Chapter 19

Broach, J. R. 1982. "The yeast plasmid 2μm circle." In *The Molecular Biology of the Yeast Saccharomyces.* J. Strathern, et al., eds. Cold Spring Harbor.

Broda, P. 1979. *Plasmids.* W. H. Freeman.

Bukhari, A. I., J. A. Shapiro, and S. L. Adhya. 1978. *DNA Insertion Elements and Plasmids.* Cold Spring Harbor.

Clark, A. J., and G. J. Warren. 1979. "Conjugal transmission of plasmids." *Ann. Rev. Genetics,* 13, 99.

Clowes, R. C. 1972. "Molecular structure of bacterial plasmids." *Bact. Rev.,* 36, 361.

Clowes, R. C. 1973. "The molecule of infectious drug resistance." *Scient. Amer.,* April, p. 18.

Deonier, R. C., and N. G. Davidson. 1976. "The sequence organization of the integrated F plasmid in two different Hfr strains of *E. coli*." *J. Mol. Biol.,* 107, 207.

Gunge, N. 1983. "Yeast DNA plasmids." *Ann. Rev. Microbiol.,* 37, 253.

Gurney, D. G., and D. Helinski. 1975. "Relaxation complexes of plasmid DNA and protein. III. Association of protein with the 5′ terminus of the broken DNA strand in the relaxed complex of plasmid ColE1." *J. Biol. Chem.,* 250, 8796.

Hayes, W. 1968. *The Genetics of Bacteria and Their Viruses.* Blackwell.

Holloway, B. W. 1979. "Plasmids that mobilize bacterial chromosomes." *Plasmid,* 2, 1.

Hooykaas, P. J. J., and R. A. Schilperoort. 1985. "The Ti plasmid of *Agrobacterium tumefaciens:* a natural genetic engineer." *Trends Biochem. Sci,* 10, 307.

Kingsman, A., and N. Willets. 1978. "The requirement for conjugal DNA synthesis in the donor strand during F′lac transfer." *J. Mol. Biol,* 122, 287.

Lewin, B. 1977. *Gene Expression. 3. Plasmids and Phage.* Wiley-Interscience.

Low, K. B., and R. D. Porter. 1978. "Modes of gene transfer and recombination in bacteria." *Ann. Rev. Genetics.,* 12, 249.

Mitsuhashi, S., ed. 1980. *R Factor: Drug Resistance Plasmid.* University Park Press.

Novick, R. P. 1980. "Plasmids." *Scient. Amer.,* December, p. 102.

Willets, N., and R. Skurray. 1980. "The conjugation system of F-like plasmids." *Ann. Rev. Genetics,* 14, 41.

Chapter 20

Alberts, B., and C. F. Fox, eds. 1980. *Mechanistic Studies of DNA Replication and Recombination*. Academic Press.

Bianchi, M. E., and C. M. Radding. 1983. "Insertions, deletions, and mismatches in heteroduplex DNA made by RecA protein." *Cell*, 35, 511.

Cunningham, R. P., et al. 1979. "Single strands induce RecA protein to unwind duplex DNA for homologous pairing." *Nature*, 281, 191.

Cunningham, R. P., et al. 1980. "Homologous pairing in genetic recombination: RecA protein makes joint molecules of gapped circular DNA and closed circular DNA." *Cell*, 20, 223.

Eisenstark, A., 1977. "Genetic recombination in bacteria." *Ann. Rev. Genetics*, 11, 369.

Holliday, R. 1964. "A mechanism for gene conversion in fungi." *Genet. Res.*, 5, 282.

Howard-Flanders, P., S. C. West, and A. Stasiak. 1984. "Role of RecA protein spiral filaments in genetic recombination." *Nature*, 309, 215.

Lacks, S. 1962. "Molecular fate of DNA in genetic transformation in *Pneumococcus*." *J. Mol. Biol.*, 5, 119.

McEntee, K., et al. 1979. "Initiation of genetic recombination catalyzed *in vitro* by the RecA protein of *E. coli*." *Proc. Nat. Acad. Sci.*, 76, 2615.

Meselson, M., and C. M. Radding. 1975. "A general model for genetic recombination." *Proc. Nat. Acad. Sci.*, 72, 358.

Potter, H., and D. Dressler, 1976. "On the mechanism of genetic recombination: electron-microscopic observation of recombination intermediates." *Proc. Nat. Acad. Sci.*, 73, 3000.

Radding, C. M. 1978. "Genetic recombination: strand transfer and mismatch repair." *Ann. Rev. Biochem.*, 47, 847.

Radding, C. M. 1982. "Homologous pairing and strand exchange in genetic recombination." *Ann. Rev. Genet.*, 16, 405.

Radding, C. M., et al. 1983. "Three phases in homologous pairing: polymerization of *recA* protein on single-stranded DNA, synapsis, and polar strand exchange." *Cold Spring Harb. Symp. Quant. Biol.*, 47, 821.

Sigal, N., and B. Alberts, 1972. "Genetic recombination: the nature of a crossed-strand exchange between two homologous DNA molecules." *J. Mol. Biol.*, 71, 789.

Smith, H. O., et al. 1981. "Genetic transformation." *Ann. Rev. Biochem.*, 50, 41.

Stahl, F. W. 1979. *Genetic Recombination*. W. H. Freeman.

Stahl, F. W. 1979. "Specific sites in generalized recombination." *Ann. Rev. Genetics*, 13, 7.

Valenzuela, M., and R. B. Inman. 1975. "Visualization of a novel junction in bacteriophage lambda DNA." *Proc. Nat. Acad. Sci.*, 72, 3024.

Weisberg, R. A., and S. Adhya. 1977. "Illegitimate recombination in bacteria and bacteriophage." *Ann. Rev. Genetics*, 11, 451.

West, S. C., et al. 1981. "Homologous pairing can occur before DNA strand separation in generalized genetic recombination." *Nature,* 290, 29.

Wu, A. M., et al. 1983. "Unwinding associated with synapsis of DNA molecules by recA protein." *Proc. Nat. Acad. Sci.,* 80, 1256.

Chapter 21

Bukhari, A. I. 1981. "Models of DNA transposition." *Trends Biochem. Sci.,* 6, 56.

Bukhari, A. I., et al. 1977. *DNA Insertion Elements and Plasmids.* Cold Spring Harbor.

Calos, M. P., and J. H. Miller. 1980. "Transposable elements." *Cell,* 20, 579.

Campbell, A. 1981. "Evolutionary significance of accessory DNA elements in bacteria." *Ann. Rev. Microbiol.,* 35, 55.

Campbell, A., et al. 1979. "Viruses and inserting elements in chromosomal evolution." In *Concepts of the Structure and Function of DNA, Chromatin, and Chromosomes.* (A. S. Dion, ed.) Symposia Specialists.

Cohen, S. N., and J. A. Shapiro. 1980. "Transposable genetic elements." *Scient. Amer.,* February, p. 40.

Engels, W. R. 1983. "The P family of transposable elements in *Drosophila.*" *Ann. Rev. Genetics,* 17, 315.

Green, M. M. 1980. "Transposable elements in *Drosophila* and other Diptera." *Ann. Rev. Genetics,* 14, 109.

Grindley, N. D. F., and R. R. Reed. 1985. "Transpositional recombination in prokaryotes." *Ann. Rev. Biochem.,* 54, 863.

Keller, E. 1981. 'McClintock's maize." *Science 31.* August.

Kleckner, N. 1977. "Translocatable elements in procaryotes." *Cell,* 11, 11.

Kleckner, N. 1981. "Transposable genetic elements." *Ann. Rev. Genetics,* 15, 341.

McClintock, B. 1965. "The control of gene action in maize." *Brookhaven Symp. Biol.,* 18, 162.

Nevers, P., and H. Saedler, 1977. "Transposable genetic elements as agents of gene instability and chromosomal rearrangements." *Nature,* 268, 109.

Potter, S. R. 1982. "DNA sequence of a foldback transposable element in *Drosophila.*" *Nature,* 297, 201.

Rogers, J. 1985. "Origins of repeated DNA." *Nature,* 317, 765

Starlinger, P. 1977. "DNA rearrangements in procaryotes." *Ann. Rev. Genetics,* 11, 103.

Starlinger, P., and H. Saedler. 1976. "IS elements in microorganisms." *Curr. Topics Microbiol. Immunol.,* 75, 111.

Ullu, E. 1983. "The human Alu family of repeated DNA sequences." In *DNA Makes RNA Makes Protein.* (T. Hunt et al., eds.) Elsevier Biomedical.

Chapter 22

Abelson, J., and E. Butz. 1980. "Recombinant DNA." *Science* 209, 1317.

Anderson, W. F., and E. G. Diacumakos. 1981. "Genetic engineering in mammalian cells." *Scient. Amer.*, July, p. 106.

Anderson, W. F. 1984. "Prospects for human gene therapy." *Science,* 226, 401.

Beers, R. F., and E. G. Bassett, eds. 1977. *Recombinant Molecules: Impact on Science and Society.* Raven.

Brown, D. D. 1973. "The isolation of genes." *Scient. Amer.,* August, p. 20.

Caskey, C. T., and R. L. White. 1983. *Recombinant DNA Applications to Human Disease.* Cold Spring Harbor.

Curtiss, R. 1976. "Genetic manipulation of microorganisms: potential benefits and hazards." *Ann. Rev. Microbiol.,* 30, 507.

Dillon, J. R., A. Nasim, and E. R. Nestmann. 1985. *Recombinant DNA Methodology.* Wiley.

Gilbert, W., and L. Villa-Komaroff. 1980. "Useful proteins from recombinant bacteria." *Scient. Amer.,* April, p. 74.

Glover, D. M. 1984. *Gene Cloning: The Mechanics of DNA Manipulation.* Chapman and Hall.

Hacket, P. B., J. A. Fuchs, and J. W. Messing. 1984. *An Introduction to Recombinant DNA Techniques.* Benjamin/Cummings.

Lobban, P., and A. D. Kaiser. 1973. "Enzymatic end-to-end joining of DNA molecules." *J. Mol Biol.,* 78, 453.

Maniatis, T., E. F. Fritsch, and J. Sambrook. 1982. *Molecular Cloning.* Cold Spring Harbor.

Maniatis, T., et al. 1978. "The isolation of structural genes from libraries of eucaryotic DNA." *Cell,* 15, 687.

Mertz, J., and R. Davis. 1972. "Cleavage of DNA: RI restriction enzyme generates cohesive ends." *Proc. Nat. Acad. Sci.,* 69, 3370.

Rodriguez, R. L., and R. C. Tait. 1983. *Recombinant DNA Techniques.* Addison-Wesley.

Schell, J., and M. van Montagu. 1983. "The Ti plasmids as natural and as practical gene vectors for plants." *Biotechnol.,* April, 175.

Scott, W. A., and R. Werner. 1977. *Molecular Cloning of Recombinant DNA.* Academic Press.

Seeberg, P. H., et al. 1978. "Synthesis of growth hormone by bacteria." *Nature,* 276, 795.

Chapter 23

Bishop, J. M. 1978. "Retroviruses." *Ann. Rev. Biochem.,* 47, 35.

Bishop, J. M. 1980. "The molecular biology of RNA tumor viruses: a physician's guide." *New England J. Med.,* 303, 675.

Botchan, M., et al. 1980. "Integration and excision of SV40 DNA from the chromosome of a transformed cell." *Cell,* 20, 143.

Broker, T. R., and L. T. Chow. 1983. "Patterns and consequences of adenoviral

RNA splicing." In *DNA Makes RNA Makes Protein.* (T. Hunt et al., eds.) Elsevier Biomedical.

Crawford, L. V. 1980. "The T antigens of simian virus 40 and polyoma viruses." *Trends Biochem. Sci.*, 5, 39.

Dos, G. C., S. K. Niyogi, and N. P. Salzman. 1985. "SV40 promoters and their regulation." *Prog. Nucl. Acid. Res. Mol. Biol.*, 32, 210.

Eckhardt, W. 1977. "Genetics of polyoma virus and simian virus 40." *Comprehensive Virol.*, 9, 1.

Eckhardt, W. 1981. "Polyoma T antigens." *Adv. Cancer Res.*, 35, 1.

Fenner, F., et al. 1974. *The Biology of Animal Viruses.* Academic Press.

Graessman, A., M. Graessman, and C. Mueller. 1981. "Regulation of SV40 gene expression." *Adv. Cancer. Res.*, 35, 111.

Kitamura, N., et al. 1981. "Primary structure, gene organization, and polypeptide expression of poliovirus RNA." *Nature,* 291, 547.

Knight, C. A. 1975. *Chemistry of Viruses.* Springer-Verlag.

Lebowitz, P., and S. Weisman. 1979. "Organization and transcription of the simian virus 40 genome." *Curr. Topics Microbiol. Immunol.*, 87, 43

Luria, S. E., et al. 1978. *General Virology.* Wiley.

Mitra, S. 1980. "DNA replication in viruses." *Ann. Rev. Genetics,* 14, 347.

Nevins, J. R. 1982. "Adenovirus gene expression: control at multiple steps of mRNA biogenesis." *Cell,* 28, 1.

Panganiban, A., and H. M. Temin. 1983. "The terminal nucleotides of retrovirus DNA are required for integration but not virus production." *Nature,* 306, 155.

Panganiban, A., and H. M. Temin. 1984. "Circles with two tandem LTRs are precursors to integrated retrovirus DNA." *Cell,* 36, 673

Shimotuno, K., et al. 1980. "Sequence of retro provirus resembles that of bacterial transposable elements." *Nature,* 285, 550.

Silverstein, S. G., et al. 1976. "The reovirus replicative cycle." *Ann. Rev. Biochem.*, 45, 375.

Temin, H. M. 1972. "RNA-directed DNA synthesis." *Scient. Amer.*, January, p. 24.

Temin, H. M. 1980. "Origin of retroviruses from cellular movable genetic elements." *Cell,* 21, 599

Varmus, H. E. 1982. "Form and function of retroviral proviruses." *Science,* 216, 812.

Williams, R. C., and H. W. Fisher. 1974. *An Electron Micrographic Atlas of Viruses.* Thomas.

Wimmer, E. 1982. "Genome-linked proteins of viruses." *Cell,* 28, 199.

Chapter 24

Bishop, J. M. 1982. "Cellular oncogenes and retroviruses." *Ann. Rev. Biochem.*, 52, 301.

Bishop, J. M. 1983. "Cancer genes come of age." *Cell,* 32, 1018.

Bos, J. L., and A. J. van der Erb. 1985. "Adenovirus region EIA: transcription modulation and transforming agent." *Trends Biochem. Sci.,* 10, 310.

Cold Spring Harbor. 1979. "Viral Oncogenesis." *Cold Spring Harb. Symp. Quant. Biol.* Vol. 44.

Hunter, T. 1985. "Oncogenes and growth control." *Trends Biochem. Sci.,* 10, 275.

Hynes, R. 1980. "Cellular location of viral transformation proteins." *Cell,* 21, 601.

Ratner, L., S. F. Josephs, and F. Wong-Staal. 1985. "Oncogenes: their role in neoplastic transformation." *Ann. Rev. Microbiol.,* 39, 419.

Tooze, J. 1981. *The Molecular Biology of Tumor Viruses.* Cold Spring Harbor.

Weinberg, R. A. 1980. "Integrated genome of avian viruses." *Ann. Rev. Biochem.,* 49, 197.

Index